D0812351

CHEMICAL PROCESS INDUSTRIES

fourth edition

CHEMICAL PROCESS INDUSTRIES

R. NORRIS SHREVE

Late Professor Emeritus of Chemical Engineering
Purdue University

JOSEPH A. BRINK, JR.

Consulting Chemical Engineer
Professor and Chairman of Chemical Engineering
Washington State University

McGRAW-HILL BOOK COMPANY

New York St. Louis San Francisco Auckland Bogotá
Düsseldorf Johannesburg London Madrid Mexico
Montreal New Delhi Panama Paris São Paulo
Singapore Sydney Tokyo Toronto

Library of Congress Cataloging in Publication Data

Shreve, Randolph Norris, 1885–1975.
Chemical process industries.

Includes bibliographies and index.
1. Chemistry, Technical. I. Brink, Joseph A., date. joint author. II. Title.
TP145.S5 1977 660.2 77-4236
ISBN 0-07-057145-7

234567890 FGRFGR 78654321098

The editors for this book were Jeremy Robinson and Esther Gelatt, and the production supervisor was Teresa F. Leaden. It was set in Bodoni Book.

Printed and bound by Fairfield Graphics.

CONTENTS

FOREWORD

A new dimension has been added to Shreve's *Chemical Process Industries,* the most successful and widely distributed single volume ever published in the fields of *chemical engineering, industrial chemistry,* and the *chemical process industries.* Joseph A. Brink, Jr., adds his name to the masthead; he brings to the book a vast experience in the chemical process industries and in chemical engineering education. Best known for the development of superlative equipment for mist elimination, Professor Brink has mastered scores of the chemical process industries in order to solve their problems with his units. To report these successes, he has frequently been asked to write articles for the technical press of the United States and other countries.

"Shreve," or now "Shreve and Brink," has greatly expanded its coverage, especially of the newer chemical process industries—by complete additions and major revisions showing current practice in this ever-changing combination of chemistry, chemical engineering, manufacturing, and sales. This fourth edition, a masterpiece of selection, reporting, and presentation, will be available to new generations of students. They will carry it with them, as have their predecessors, to each new job they do. It will inform them of practices in the new industry and provide a base of comparison with the problems of their own particular plants.

The name *chemical process industries* has been popularized for some 30 years by this durable book, revised by demand each decade to mirror faithfully the changing status of the art. Tens of thousands of chemical process industry offices will place this new edition alongside the dictionary and the telephone book, to serve as a source of dependable and ever-needed information. No one of the 100,000 students who used the first three editions has learned all that is in this book, either in school years or in professional life. Instead, each has found in *CPI* examples of mathematics and physics applications and of successful process and plant development.

Engineering education encompasses more and more of the sciences as our profession becomes more complex. However, the younger chemical engineer must not regard the computer-derived solution of a combined problem in reactions, heat, and mass transfer, for example, as an end in itself. He or she must be familiar with the successful realization of these scientific principles in industries actually made possible by this type of computation used with engineering judgment. Even the most sophisticated chemical engineering curriculum must give the student some cultural background of the process industries—how sulfuric acid is produced, how petroleum refining and the caustic soda industry have been affected by increasing demands of their former by-products in making polymers, and the like. To be well rounded, the chemical engineer needs, in addition to academic knowledge and achievement, an appreciation of the profession's role in the chemical process industries.

To me, a practicing chemical engineer in scores of countries, the earlier editions have been constant travel companions. They have provided the best and smallest practical condensations and presentations of ready information on overall chemical manufacture. Now, modernized and expanded, *CPI* presents an even better treatise—complete with economic data, bibliographies, flowcharts, important material balances, charts of controlling functions, and other aids. *As text, guidebook, and treatise, the fourth edition of* Chemical Process Industries *will serve well the chemical engineers who remake much of these same industries in the balance of this century.* They will join in thanks to the authors for their services as teachers and compilers of so much information on the practice of chemical engineering.

DONALD F. OTHMER
Distinguished Professor of Chemical Engineering
Polytechnic Institute of New York

PREFACE

The illness and subsequent death of Dr. R. Norris Shreve make it necessary for the fourth edition of *CPI* to appear without his full collaboration. The book was originally the product of his initiative, and the first three editions were molded by him. This fourth edition follows the format of previous editions.

Each chapter covers a given *chemical process industry*—such as glass, paper, rubber, or sulfuric acid—in somewhat the following order. After a brief introduction, aimed at epitomizing the industry, some attention is given to the historical background of the particular process. This is followed by a consideration of *uses and economics*, including statistical tabulations by which the industry's importance can be judged. Production trends, whether on the increase or decrease, are of more importance than mere statements that so many pounds or so many dollar's worth of a given substance are being manufactured. Trends are shown by parallel columns for different years and by statistical curves. Under *manufacture*—as this is a book on chemical engineering—*energy change, unit operations,* and *chemical conversions* are brought to the reader's attention. For some important processes, principal unit operations and unit chemical conversions are tabulated. Dividing the industries into these units helps greatly in the transference of information from one industry to another. Indeed, the many flowcharts do this in a visual manner. This breakdown clearly shows the reader that filtration and evaporation and hydrogenation and nitration are employed in many industries. *Raw materials,* their sources and their economic and chemical relationship to the manufacturing procedures, are also discussed.

This book has several potential uses. It is a college text designed to give the young chemical engineer some comprehension of the various fields into which he will enter, or with which he will be associated. It is a reference book for practicing chemical engineers, chemists, other engineers, and scientists in industry. Indeed, many professionals who are not chemical engineers have found that it helps them understand the chemical engineering aspects of company operations. The chemical process industries employ mechanical, electrical, and civil engineers and scientists—as well as chemical engineers—and these professionals can profit greatly from material in *Chemical Process Industries* that applies to their work and to their companies.

Since the third edition of *CPI* appeared in 1967, legislation and events have greatly affected the various industries. Environmental health and control restrictions have resulted in modification processes and products. The energy crises of the 1970s have altered the economics of processes and the design of plants. These changes have necessitated many revisions in the fourth edition of *CPI.* Dr. George A. Hawkins, Vice President Emeritus for Academic Affairs at Purdue University, drafted the revisions to Chapter 4, in which energy considerations are so important. Dr. Carl W. Hall, Dean of Engineering at Washington State University, prepared the first section of Chapter 25 on Food Processing, not covered in previous editions. This important industry utilizes more and more engineering input, and Dr. Shreve had planned to have it covered in the fourth edition. The contributions of these outstanding engineers are greatly acknowledged.

I wish to thank Professor George T. Austin and his wife, Helen Austin, for revising and reviewing many of the chapters. Acknowledgment is also made to the patience and help of my wife, Dorothy Shea Brink, over the years it has taken to write this book. Elizabeth Prentiss, secretary to Dr. Shreve, typed most of the manuscript and her various efforts are greatly appreciated. Acknowledgments to many others appear on the following page.

JOSEPH A. BRINK, JR.

ACKNOWLEDGMENTS

It is impossible to write a book with the wide coverage that *Chemical Process Industries* has had in its four editions without the utmost cooperation of companies in the process industries and of their employees, many of whom have been lifelong friends or students of the authors. Specific acknowledgments are given in footnotes and on the flowcharts. More than 100 organizations, and over 200 individuals, have supplied technical data, flowcharts, cuts, photographs, and advice. It is indeed a professional obligation and a personal pleasure to recognize the assistance received, particularly in this fourth edition, from the following, among many others: AMERICAN PETROLEUM INSTITUTE, *Miriam B. Yazge;* ASHLAND CHEMICAL COMPANY, *M. L. Deviney;* BOISE CASCADE, *John A. Falkowski;* CALGON CORP., *D. E. Noll, B. Q. Welder, L. J. Weaver;* COLONIAL SUGAR COMPANY, *Henry G. Gerstner;* CONSOLIDATED NATURAL GAS SERVICE CO., *Theodore E. Ross, Roy A. Sisken, Robert C. Weast;* COORS PORCELAIN COMPANY, *Michael J. Fenerty;* CORNING GLASS WORKS, *George W. McLellan, L. R. Schlotzhauer;* DENVER RESEARCH INSTITUTE, *Charles H. Prien;* DIAMOND SHAMROCK CORP., *Gary A. Klein, James B. Worthington;* DOW CHEMICAL U.S.A., *Bob C. Mayo;* E. I. DuPONT, *Robert C. Forney, W. D. Lawson, Donald W. Lyon;* ESSO RESEARCH & ENGINEERING CO., *E. J. Barrasso, C. W. Smith;* EXXON COMPANY, U.S.A., *E. W. Squires;* FREEPORT MINERALS CO., *L. F. Good;* GENERAL ELECTRIC CO., *F. P. Bundy;* GENERAL TIRE & RUBBER CO., *R. W. Watkins;* GREAT LAKES CARBON CORP., *M. P. Whittaker;* GREAT LAKES RESEARCH CORP., *L. A. Bryan, M. P. Whittaker;* HERCULES INC., *C. W. Eilo;* J. M. HUBER CORP., *P. W. Brandon;* IMS AMERICA, LTD., *William A. Lockwood;* INMONT CORP., *Harry Burrell;* ELI LILLY & CO., *Mart T. Straub, H. W. Rhodehamel;* MOBIL OIL CO., *Stanley Johann;* MONSANTO CO., *Clayton F. Callis, C. Y. Shen, James R. Fair;* MONSANTO ENVIROCHEM SYSTEMS, INC., *J. R. Donovan;* NORTON CO., *L. J. Trostel;* OLIN CORP., *Swen W. Englund;* OREGON STATE UNIVERSITY, *J. R. Shay;* PPG INDUSTRIES, *F. E. Batton, Harry Hyman, A. E. Thompson, Joe Y. Keller;* PROCTOR & GAMBLE CO., *R. E. Hall, Bruce Martin;* PURDUE UNIVERSITY, *George A. Hawkins, Theodore J. Williams, Roy L. Whistler, Henry B. Haas;* SALT INSTITUTE, *Frank O. Wood;* SEM CORP., GLIDDEN-DURKEE, *Martha L. Embree;* STANDARD OIL OF CALIFORNIA, *Thomas C. Austin;* STANFORD RESEARCH INSTITUTE, *F. Y. Chan;* STAUFFER CHEMICAL CO., *H. A. Betaque;* TEXASGULF, INC., *E. H. Conroy, J. R. West;* 3M COMPANY, *M. W. Miller, R. M. Adams;* UNION CARBIDE, CARBON PRODUCTS DIV., *G. B. Spence, H. B. Allport, J. R. Schley, L. L. Winter;* UNION CARBIDE CORP., LINDE DIV., *R. W. Hirsch, B. B. Murphy, Charles R. Baker, George M. Lukchis, Walter J. Olszewski, David Sorensen, John W. Terbot;* UNITED STATES DEPARTMENT OF AGRICULTURE, *Harold Tarkow;* WASHINGTON STATE UNIVERSITY, *George and Helen Austin, Carl W. Hall, John C. Sheppard.* The following individuals have made substantial contributions: *C. A. Burchsted, M. M. Eakins, Harold E. Marsh, Jr., H. A. McLain, R. C. Specht,* and *U. B. Yeager.*

chapter 1

CHEMICAL PROCESSING

This book is concerned with the chemical processing of raw materials into usable and profitable products. Certain of these marketable products will be consumer goods and enter directly into the economic life of the United States. Others will be intermediates or chemicals for the manufacture of consumer items. Manufacturers of chemicals utilize about 20% of the total chemical output, inasmuch as the products of one become raw materials for further fabrication in another; thus *the industry*, as a whole, *is its own best customer.*

For the most part, users of this book will be chemical engineers, chemists employed in various phases of the chemical process industries, and students in chemical engineering curricula. Abbreviating the American Institute of Chemical Engineers[1] definition, we may state that *chemical engineering* is that branch of engineering concerned with the development and application of manufacturing processes in which chemical or certain physical changes of materials are involved.

In the course of the development of chemical engineering, a tremendous impetus was given to the basic structure by suggested classifications, and later by their implementation, whereby many of the physical changes of what is involved in the engineer's work in the chemical process industries were, from about 1910 on, called *unit operations.* This classification was instituted by a group of chemical engineering professors at Massachusetts Institute of Technology headed by W. H. Walker. The name "unit operations," however, was suggested by A. D. Little, who had been an industrial partner of Walker's. For thirty or more years, the concept of unit operations was a tremendous stimulus, because it classified these physical changes. This caused filtration, heat transfer, and many other changes to be studied individually, and the findings in application to one industry or the manufacture of one chemical to be applied to many other such chemicals. Shortly after 1930 there arose a somewhat similar classification called *unit processes,* as named by P. H. Groggins. These pertain to the chemical changes, just as unit operations pertain to physical changes. Not as much work was done on the extension of unit processes as for unit operations, though Groggins, the originator of this book, and others investigated and wrote about unit processes and made a very definite classification of chemical changes, which served as a most important reference for the chemical process industries based on chemical changes or reactions.

In the last two decades, however, there have been many improvements in the various divisions of the rapidly developing chemical engineering field. Such new and improved factors involving both physical and chemical changes are presented in the fifth edition of Perry's[2] Chemical Engineers' Handbook and elsewhere. Table 1.1, listing physical changes, epitomizes the Perry classification of the physical principles of chemical engineering and distinguishes between the scientific foundations of chemical engineering and the equipment used to industrialize the processes. Analogously, the

[1]Newman, Development of Chemical Engineering Education, *Trans. AIChE,* suppl. to **34**(3a), 6 (1938), and **32,** 568 (1936).
[2]Chemical Engineers' Handbook, Perry and Chilton (eds.), 5th ed., McGraw-Hill, 1973. Hereafter referred to as "Perry."

TABLE 1.1 *Chemical Engineering Physical Principles*

Perry, section	
5	Fluid and Particle Mechanics Flows, measurements, fluid and particle dynamics
6	Transport and Storage of Fluids Pumping and storage of liquids and gases by pumps, compressors, blowers, valves, piping, storage and process vessels
7	Handling of Bulk and Packaged Solids Conveying, packaging, storage (pneumatic and fluidized conveying)
8	Size Reduction and Size Enlargement Crushing, grinding, agglomerization, granulating, compacting
9	Heat Generation and Transport Fuels, furnaces, combustion, power generation and transmission
10	Heat Transmission Heat transmission by condition, convection, and radiation
11	Heat-transfer Equipment Evaporators, heat exchangers
12	Psychrometry, Evaporative Cooling, Air Conditioning, and Refrigeration Normal and cryogenic
13	Distillation (theory and basic principles) Vapor-liquid equilibrium, continuous and batch binary and multiple distillation; extractive, azeotropic distillation, molecular distillation
14	Gas Absorption
15	Liquid Extraction Phase equilibriums, extraction systems, design methods
16	Adsorption and Ion Exchange Theory, operation, design, methods of sorption
17	Miscellaneous Separation Processes Leaching, crystallization, sublimation, gaseous diffusion, dialysis, electrodialysis
18	Liquid-Gas Systems (equipment) Gas-liquid contacting, phase dispersion, phase separation
19	Liquid-Solid Systems (equipment) Filters and centrifuges, thickeners and clarifiers, paste mixers and agitators, ion exchange
20	Gas-Solid Systems (equipment) Contacting equipment for heat and mass transfer
21	Gas-Gas, Liquid-Liquid, and Solid-Solid Systems (equipment) Sampling, screening, froth flotation, electrostatic separation

Source: Adapted from Perry where the underlying principles are given for unit operations. [Perry and Fair, History of a Handbook, *Chem. Eng.* (N.Y.), **81**(4), 129 (1974), gives an interesting account of the five editions of Perry.]

expression "unit processes," used to describe chemical changes, is being superseded by *chemical conversions*, long used by the petroleum industry. Chemical changes, or chemical conversions, are listed in Table 1.2 according to the scientific classification (chemistry), referring to Perry and other sources.

This book emphasizes these *chemical conversions* (Table 1.2), which may be defined as chemical reactions applied to industrial processing. They include (1) the basic chemistry of the particular reaction, (2) the equipment in which the reaction takes place, and (3) the management of the entire process at a sufficiently low cost to be competitive and efficient and make a profit. They have not been as scientifically and quantitatively formulated or classified as have the physical operations, because of their diversity and complexity, but they do provide a powerful approach of wide usefulness in designing new manufacturing processes.

The characteristics of chemical conversions as applied to the manufacture of chemicals may be summarized as follows:

1. Each chemical conversion points out the *unitary*, or like, aspects in a group of numerous individual reactions. These unitary aspects, apart from the basic chemical family, may be a similarity in energy change, pressure, reaction time, equilibrium, or raw materials.

2. Frequently there is *factory segregation* by related chemical conversion processes, wherein a building or section of a building is devoted to the making of many chemicals by means of a given chemical conversion such as diazotization and coupling, nitration, hydrogenation, esterification, fermentation, or alkylation.

3. There frequently is a close relationship among the types of *equipment* used for making many different products by means of a single chemical conversion. For instance, the cast-iron well-agitated reactor, provided with cooling coils, called a *nitrator*, is used for nitration conversion in the manufacture of a number of chemicals, such as nitrobenzene, nitronaphthalene, and TNT.

4. Where production is small or products variable, *equipment* may be conveniently and economically transferred from the making of one chemical to that of another based on the same chemical conversion. It is the aim of a production manager to keep all the equipment constantly in use. To do this, frequently it is necessary to make first one chemical and then another in the same general reactor—a sulfonator, for example. This *multiple use of equipment* is most easily managed with chemical-conversion segregation.

5. The chemical-conversion classification enables a chemical engineer to think from group performance to that of a new individual chemical in the related class. It is necessary chiefly to know and to remember *principles* rather than specific performances. This method of approach greatly facilitates the making of any chemical by having available past knowledge regarding the generalized data of a specific conversion. This procedure obviates memorizing individual observations.

6. Since the basis of chemical-conversion classification is a chemical one, stress is placed upon the *chemical reaction*. Because materials are usually expensive and energy relatively cheap, a slight increase in chemical yield will materially affect the profit of the manufacturing sequence.

7. *Inorganic and organic procedures* need not be set apart industrially. Because the equipment and the manufacturing problems are frequently so similar for both organic and inorganic chemicals,

TABLE 1.2 ***Principal Chemical Conversions***[3]

Acylation	Condensation	Hydrolysis and hydration (saponification, alkali fusion)
Alcoholysis	Dehydration	Ion exchange
Alkylation	Diazotization and coupling	Isomerization
Amination by reduction	Double decomposition	Neutralization
Ammonolysis	Electrolysis	Nitration
Aromatization or cyclization	Esterification (sulfation)	Oxidation (controlled)
Calcination	Fermentation	Polymerization
Carboxylation	Friedel-crafts (reactions)	Pyrolysis or cracking
Causticization	Halogenation	Reduction
Combustion (uncontrolled oxidation)	Hydroformylation (oxo)	Silicate formation
	Hydrogenation, dehydrogenation, and hydrogenolysis	Sulfonation

[3]See Perry, sec. 4, reaction kinetics, reactor design, and thermodynamics; Groggins (see Selected References for this chapter) for fullest coverage, including most of the above-named chemical conversions; McGraw-Hill Encyclopedia for excellent summaries, and Encyclopedia of Chemical Technology (ECT) for full coverage of each chemical conversion, alphabetically arranged; and the current periodical *Chemical Week*. Throughout this book, the manufacture of individual chemicals is emphasized where the *greatest tonnage* is used, classified and described under the appropriate chemical conversion (see Index).

it may be industrially important to group them together. For example, there is surprisingly little difference in the conditions and equipment used to hydrogenate nitrogen to ammonia, or carbon monoxide to methanol, except in the raw materials charged and the catalyst employed.

8. The *design* of equipment can often be simplified by the generalizations arising from a like chemical-conversion arrangement rather than by considering each reaction separately. That which experience has indicated for a number of reactions allied under a like chemical conversion is an excellent guide for a new reaction in this same grouping. In handling chemical conversions, the more that is understood about the underlying physical chemistry of the equilibriums and the reaction rates, the better will be plant conception, the higher the conversion, and the lower the costs. It is vitally important to know how fast a reaction will go and how far. Many times, as pointed out in connection with the manufacture of sulfuric acid by the contact process (Chap. 19), conditions that increase the rate decrease the equilibrium. Therefore, as shown in the design of the sulfur trioxide converter, conditions are first secured to cause a high rate of reaction, which allows the apparatus to be relatively small, and then, toward the end, are changed to favor the equilibrium.

This book, "Chemical Process Industries," has been characterized by a wide use of *flowcharts* to present the best method for outlining the various chemical processes carried out in industry. In the first two editions, these processes were broken down into a sequence of unit operations and unit processes. In this edition, this is continued, but shortened, with the abbreviation Op used for a coordinated sequence of unit operations and Ch for a chemical conversion (unit process).

Some of the flowcharts are supplemented by a succinct list of unit steps, which have been used so successfully in past editions in presenting an outline of these physical and chemical changes. This is considered to be a needed explanation for the actual flowchart. The flowcharts used in this book are only résumé charts giving the high points of changes from raw materials to finished products. However, when chemical engineers design a chemical process for any given chemical, they use a rough block type of flowchart and proceed through more complex flowcharts to the very detailed flowcharts required for actual construction. Of course, such a flowchart is uncalled for in this book.

Because of space limitations, very little has been included pertaining to the chemical industries of foreign countries. No attempt is made here to supply the names of companies in various branches of the chemical industry, although there is an occasional reference to a specific company. A full listing of such companies is obtainable from a number of trade directories and from the pages of various chemical publications, including the annual Chemical Engineering Catalog, which is also an excellent source of information about equipment.

Most chemical engineers do not have intimate contact with more than one industry. With this in mind, the text emphasizes not details, but the *broad principles of the systems* characteristic of a specific chemical process industry. These should be part of the working knowledge of even the engineer who becomes a specialist, for it is sometimes possible to translate to one field a principle that has been put into practice in another process. For the growing number of chemical engineers who enter sales, executive, or management positions, a broader acquaintance with the chemical industry in its entirety is essential. For all these, the specialist, the salesperson, and the manager, the flowcharts will present in a connected logical manner an overall viewpoint of many processes, from raw materials to salable products, which have been developed so adequately by our competitive system under the economic stimulus and sound protection of our patent laws.

All chemical engineers should be familiar with the current selling prices of the principal chemicals with which they are concerned. These can be readily obtained from such journals as *Chemical Week* and *Chemical Marketing Reporter*. Similarly, statistics with dollar values are necessary for proper valuation of a process or of a chemical. Such statistics are included, but because of compiling delays are generally earlier than 1974.

The presentation of the chemical process industries around flowcharts and energy changes should lead to a logical following-through of a connected series of steps rather than an attempt to memorize purely descriptive matter. These will emphasize the *why* rather than the *how* of industrial procedures, or *thinking creatively* rather than *memorizing.*

The texts and advertising pages of such journals as *Chemical Engineering, Chemical and Engineering News, Chemical Engineering Progress,* and *Chemical Week,* together with specialized journals such as *Modern Plastics, Sugar, Hydrocarbon Processing,* and many others, should be consulted by the chemical engineer for up-to-date information and fundamental data on equipment.

To summarize, *chemical processing may be defined as the industrial processing of chemical raw materials, leading to products of enhanced industrial value.* Generally, this involves a chemical conversion (or reaction), as in the manufacture of sulfuric acid from sulfur by oxidation and hydration, but the production of fibers from chemicals is also included, such as nylon from hexamethylenediamine and adipic acid by more complicated chemical reactions. In all these chemical changes, physical operations are intimately involved, such as heat transfer and temperature control (Table 1.1), which are necessary to secure good yields required by competitive industry as, for example, in the oxidation of sulfur dioxide to sulfur trioxide. In fewer instances, but of great importance to the chemical industry, no chemical reaction may be involved, but only physical changes. Illustrations of the latter type of chemical processing are distillations to separate and purify petroleum fractions and to obtain pure alcohol after fermentation. Another example is mass transfer of liquids or solids in suspension from place to place or to a chemical reactor.

INFORMATION SOURCES

In addition to channels of communication, the alert chemical engineer has at his or her command a number of other sources of information which can be tapped. Among them are *libraries,* large and small, public and private. A majority of the larger science-oriented industries have special libraries serving their own needs.[4] In smaller organizations or in other situations where special subject material is unavailable, engineers can secure aid in other ways, such as asking a local library for interlibrary loan service or maintaining membership in organizations offering limited services to members, such as the Chemists' Club and the Engineering Societies Library. *Mechanized information handling* is being adopted by more and more organizations in some form or other. The form of information storage and retrieval may vary from adapting a single phase of an operation, such as the recording of test results of compounds prepared, to a broad system of information retrieval. The first publicly accessible, on-line chemical dictionary became available from the National Library of Medicine in 1974.[5] It is an adjunct to the toxicology information retrieval service and provides a mechanism by which 60,000 chemicals can be searched by computer. In 1974, 70 organizations were subscribers, including industry, government and universities. In the future additional such services will probably become available.

PERIODICALS

American Chemical Society (ACS), P.O. Box 3337, Columbus, Ohio 43210.

Industrial and Engineering Chemistry (Ind. Eng. Chem.)
Industrial and Engineering Chemistry Process Design and Development (Ind. Eng. Chem. Process Des. Dev.)
Industrial and Engineering Chemistry Fundamentals (Ind. Eng. Chem. Fundam.)
Industrial and Engineering Chemistry Product Research and Development (Ind. Eng. Chem. Prod. Res. Dev.)
Journal of Chemical and Engineering Data (J. Chem. Eng. Data)
Chemical and Engineering News (Chem. Eng. News)
Chemtech

[4]Strauss, Strieby, and Brown, Scientific and Technical Libraries: Their Organization and Management, Wiley-Interscience, 1964.

[5]*Chem. Eng. News,* Jan. 28, 1974, p. 22.

American Institute of Chemical Engineers (AIChE), 345 East 47th St., New York, N.Y. 10017.
Chemical Engineering Progress (Chem. Eng. Prog.)
AIChE Journal (AIChE J.)
International Chemical Engineering (Int. Chem. Eng.)
Gulf Publishing Company, P.O. Box 2608, Houston, Tex. 77001.
Hydrocarbon Processing (Hydrocarbon Process.)
McGraw-Hill Publications, 1221 Avenue of the Americas, New York, N.Y. 10020.
Chemical Engineering [*Chem. Eng. (N.Y.)*]

SELECTED REFERENCES

Acrivos, A. (ed.): Modern Chemical Engineering: Physical Operations, vol. 1, Reinhold, 1963.
Bennett, C. O., and J. E. Myers: Momentum, Heat, and Mass Transfer, 2d ed., McGraw-Hill, 1974.
Bird, R. B., *et al.:* Transport Phenomena, Wiley, 1960.
Brown, G. G., *et al.:* Unit Operations, Wiley, 1961.
Chemical Engineering Catalog (annual).
Chemical Engineering Thesaurus, AIChE, 1963 (8,000 chemical engineering terms).
Churchill, S. W.: The Interpretation and Use of Rate Data, McGraw-Hill, 1974.
Clark, G. L., and G. G. Hawley: Encyclopedia of Chemistry, 2d ed., Reinhold, 1966.
Considine, D. M.: Chemical and Process Technology Encyclopedia, McGraw-Hill, 1974.
Cooper, A. R., and C. V. Jeffreys: Chemical Kinetics and Reactor Design, Prentice-Hall, 1973.
Cremer, H. W. (ed.): Chemical Engineering, Practice Series, Academic, 1956–1961, 12 vols.
Dean, J. A. (ed.): Lange's Handbook of Chemistry, McGraw-Hill.
Dickey, G. D.: Filtration, Reinhold, 1961.
Facts and Figures for the Chemical Process Industries, *Chem. Eng. News* (yearly).
Faith, W. L., *et al.:* Industrial Chemicals, 3d ed., Wiley, 1965.
Francis, A. W.: Liquid-Liquid Equilibriums, Wiley, 1963.
Gilbert, E. E.: Sulfonation and Related Reactions, Interscience, 1965.
Gogler, H. S.: The Elements of Chemical Kinetics and Reactor Calculations, Prentice-Hall, 1974.
Groggins, P. H. (ed.): Unit Processes in Organic Synthesis, 5th ed., McGraw-Hill, 1958.
Handbook of Chemistry and Physics, Chemical Rubber.
Henglein, F.: Chemical Technology, Pergamon, 1967.
Himmelblau, D. M.: Principles and Calculations in Chemical Engineering, 3d ed., Prentice-Hall, 1974.
Kemper, J. D.: Engineer and His Profession, Holt, 1967.
Kent, J. A. (ed.): Riegel's Industrial Chemistry, 7th ed., Reinhold, 1974.
Kingzett: Chemical Encyclopedia, 9th ed., Van Nostrand, 1967.
Kirk, R. E., and D. F. Othmer: Encyclopedia of Chemical Technology, 2d ed., Wiley-Interscience, 1963–1972, 1st ed., 1956–1962.
McGraw-Hill Encyclopedia of Science and Technology, McGraw-Hill (annual supplement), 4th ed. (in press)
Modell, M., and R. C. Reid: Thermodynamics and Its Applications, Prentice-Hall, 1974.
Perry, R. H. *et al.* (eds.): Chemical Engineers' Handbook, 5th ed., McGraw-Hill, 1973.
Randolph, J.: Mixing in the Chemical Industries, Noyes, 1967.
Ranz, W. E.: Describing Chemical Engineering Systems, McGraw-Hill, 1970.
Rudd, D.: Process Synthesis, Prentice-Hall, 1973.
Schoen, H. M. (ed.): New Chemical Engineering Separation Techniques, Wiley-Interscience, 1963.
Smith, J. M.: Chemical Engineering Kinetics, McGraw-Hill, 1970.
Smith, J. M., and H. C. Van Ness: Introduction to Chemical Engineering Thermodynamics, 3d ed., McGraw-Hill, 1975.
Stephenson, R. N.: Introduction to Chemical Process Industries, Reinhold, 1967.
Treybal, R. E.: Liquid Extraction, 2d ed., McGraw-Hill, 1963.
Whitewell, J. C. and R. K. Toner, Conservation of Mass and Energy, McGraw-Hill, 1973.

chapter 2

CHEMICAL PROCESSING AND THE WORK OF THE CHEMICAL ENGINEER

In the 1970s over 50,000 chemical engineers were working in the United States.[1] This total included private industry, government, universities, and nonprofit institutions. About 190,000 persons were working as chemists and chemical engineers. A survey of chemical engineers made for the AIChE by the Opinion Research Corporation showed that 77% were employed in private industry[2] and 23% worked in government, education, etc., or were self-employed. This survey also included the functional areas in which chemical engineers are employed in industry and showed that they typically work in more than one area:

	1963	*1973*[3]
Research	47%	22%
Development	. . .	26
Administration	38	26
Process	31	28
Design	26	31
Production	24	12
Consulting	14	23
Marketing and business	12	12
Maintenance	10	5
Purchasing	8	7
Other	13	15

The chemical engineer must be trained and expected to operate a chemical industry in all of its phases.

To carry out chemical processing commercially, involving *chemical conversions* and *physical operations* in any chemical plant, presupposes factory-scale equipment and chemical engineering experience. To keep the factory itself from corroding away, proper construction materials should have been selected by the designing chemical engineer. Efficient regulation of chemical conversions requires instruments for recording and controlling procedures. To avoid harmful impurities in raw materials, to follow the course of chemical reactions, and to secure the requisite yield and purity of

[1]Chemistry in the Economy, p. 458, ACS, 1973; see also Roethel and Counts, Realignments of the Chemical Profession Continue, *Chem. Eng. News,* Nov. 15, 1971. p. 90.

[2]Profile of Members, *Chem. Eng. Prog.*, **70**(3), 56 (1974).

[3]Note that the total percentage for 1973 adds up to 207%, since the engineers surveyed worked in more than one functional area. On the average each worked in about two areas; for example, an engineer may work in research and also in administration.

products, careful process control by periodic analyses is required, as well as modern instrumentation and automatic controls, all based on standards predetermined by research, development, and design. To transmit goods in a clean and economical manner from the manufacturer to the customer, suitable containers must be provided. To effect the safety of the workers and the plant, all procedures must be carried out in a nonhazardous manner. To secure the processes from excessive competition and to ensure an adequate return for the large sums spent on research and plant, many of the chemical steps and equipment in the factory are protected for the limited period granted by U.S. patent laws. To guarantee progress, to continue profits, and to replace obsolescent processes and equipment, much attention and money must be spent upon continuing R&D. To prevent contamination of streams and interference with the rights of neighbors, factories must avoid the discharge of toxic materials either into the air or into the streams in their locality.

Table 2.1 lists the various basic principles of the systems in the chemical process industries, supplementing and supporting the widely employed physical operations and chemical conversions presented in Chap. 1 (Tables 1.1 and 1.2). This chapter describes the supporting fundamentals that

TABLE 2.1 ***Chemical Processing and the Work of the Chemical Engineer***

1. Basic chemical data
   - *a.* Yields of the reaction; how far the reaction will go
   - *b.* Conversion of the reaction; how efficient the reaction is as shown by material balances
   - *c.* Kinetics, catalysis; speed of reaction; how fast a reaction will go
   - *d.* Thermodynamics; energy changes involved in the chemical reaction and that required as actual external heat or power
2. Batch versus continuous processing
3. Flowcharts, to represent a chemical process and to serve as a base for design and operation in pilot plant and factory
   - *a.* Unit operations or physical changes (Table 1.1)
   - *b.* Chemical conversions (Table 1.2)
   - *c.* Equipment (see item 4)
   - *d.* Material balance and energy balance, labor and utilities
   - *e.* Labor and utilities
4. Chemical process selection design and operation (involving application to actual plant of all the principles in this tabulation)
   - *a.* Process selection
   - *b.* Statistical analysis for efficient planning and operation for pilot and production plants
   - *c.* Pilot plant as a step between the chemist in the laboratory and the chemical engineering production
   - *d.* Equipment
   - *e.* Corrosion by the reactants or the products, or how long the equipment will last, materials of construction
   - *f.* Process instrumentation, automation (see item 5)
   - *g.* High vacuum
   - *h.* High pressure
   - *i.* Cryogenics
   - *j.* Factory segregation of allied or similar procedures or equipment, e.g., still house or nitration house
5. Chemical process control and instrumentation
   - *a.* Computer and other control of reactions to ensure uniform products
   - *b.* Automation: data collecting, process dynamics (see item 4*f*)
6. Chemical process economics
   - *a.* Competing processes
   - *b.* Materials: material balances
   - *c.* Energy: energy balances
   - *d.* Labor: laborsaving devices
   - *e.* Overhead: insurance, management, consultants, taxes, etc.
   - *f.* Capital (money): interest and depreciation
   - *g.* Capital: obsolescence
   - *h.* Repairs and maintenance
   - *i.* Overall costs
7. Market evaluation
   - *a.* Statistics and growth
   - *b.* Stability
   - *c.* Purity and uniformity of product
   - *d.* Physical condition of product
   - *e.* Packaging: containers for product
   - *f.* Sales and sales service
8. Plant locations
9. Safety: hazards such as fire or toxic materials
10. Construction of plant
    - *a.* Registration of engineers
11. Management for productivity and creativity
    - *a.* Training of operators
12. Research and development
13. Patents
14. Chemical systems
15. Information sources

characterize modern chemical processing. Naturally, these cannot be thoroughly presented at length, but references are given to books and journal articles. Furthermore, throughout this book, under the various chemical process industries, many examples will be used to illustrate these supporting fundamentals.

BASIC CHEMICAL DATA

Chemistry is the basic science upon which the chemical process industries rest. The function of the chemical engineer is to apply the chemistry of a particular process through the use of coordinated scientific and engineering principles. To do this efficiently, the results of the chemist must be taken from the research laboratory and welded into an economical chemical process. The chemical engineer is always concerned with the economic aspects of the chemistry of a process, hence with *yields*, *conversions*, and *rates* expressed differently but not identically with equilibriums, residence times, and reaction velocities. The operational efficiency of chemical plants is interpreted in terms of the yield and conversion. These terms may be defined as follows:

$$\text{Percentage yield} = 100 \times \frac{\text{moles of main product}}{\text{moles of main product equivalent to net disappearance of chief reactant}}$$

$$\text{Percentage conversion} = 100 \times \frac{\text{moles of main product}}{\text{moles of main product equivalent to chief reactant charged}}$$

For example, in the synthesis of ammonia at 150 atm and 500°C, the yield is frequently above 98%, whereas the conversion will be limited by the equilibrium to about 14%, which means that 86% of the charge does not react and must be recirculated. Somewhat similar figures prevail for methanol synthesis. The aim of the chemical engineer concerned with cost is to have the conversion figures approach or equal the yield as closely as possible. This is not economical with the ammonia reaction, because of the usual low equilibriums. As a corollary of the low ammonia and methanol conversions, four to five times larger conversion equipment is required than would be necessary if the conversion figures were nearer those of the yield. By changing the operating conditions, the equilibriums may be increased, hence the conversion enhanced. In the case of ammonia synthesis, this could be effected by raising the operating pressure, but would require more costly equipment. In addition, sufficient residence time in the reaction equipment must be provided to permit the reaction to approach equilibrium (Chap. 18).

KINETICS[4] The speed of a chemical reaction is sometimes too slow to be economical, and a study of catalysis becomes necessary. It was not until Haber and Bosch showed that the rate of the hydrogenation of nitrogen to furnish ammonia could be greatly increased by contact with a catalyst of iron, promoted by small percentages of K_2O and Al_2O_3, that the chemical reaction for the synthesis of ammonia became commercially economical.

MATERIAL BALANCES, ENERGY CHANGES, AND ENERGY BALANCES must be in the forethought of any chemical engineer planning to commercialize a reaction. Thermodynamic principles also provide physical and chemical data on reactants and products. Furthermore, thermodynamics[5]

[4]Smith, Chemical Engineering Kinetics, 2d ed., McGraw-Hill, 1970; Perry, sec. 4; Cooper and Jeffreys, Chemical Kinetics and Reactor Design, Prentice-Hall, 1973; Gogler, The Elements of Chemical Kinetics and Reactor Calculations: A Self-Paced Approach, Prentice-Hall, 1974.

[5]Caswell and Smith, Thermodynamics, *Ind. Eng. Chem.* **57**(12), 45 (1965); Coull and Stewart, Equilibrium Thermodynamics, Wiley, 1964; Modell and Reid, Thermodynamics and Its Application in Chemical Engineering, Prentice-Hall, 1974; Reed and Coubbins, Applied Statistical Mechanics: Thermodynamic and Transport Properties of Fluids, McGraw-Hill, 1973.

concerned with free energy gives the conditions under which a reaction is possible. Throughout this book many of the reactions given are followed by the heat of reaction ΔH. This is expressed as $-\Delta H$ when the system loses or evolves heat.

BATCH VERSUS CONTINUOUS PROCESSING

In the trend toward continuous processing as opposed to batch operations,[6] especially in large-scale processing, the instrumentation of such chemical processing involves not merely the recording of temperature, pressure, and volume, but requires both the maximum in trend controls and automatic corrections for undesirable deviations from established standards. Modern labor and timesaving computers are being employed in ever-increasing instances to control the complicated procedures of certain chemical processes, especially continuous ones. Batch processing is still extensively practiced, particularly when the production is small or where safety demands that small quantities be handled at one time, as in the case of many explosives. Furthermore, batch operations provide optimum but varying kinetics for slow reactions, versus steady state in continuous processing, and frequently are more easily controlled.

Initial processing of many chemicals is done by batch operations but, when markets enlarge, a change may be made to continuous processing. The reduction in plant cost *per unit of production* is often the major force behind the change. As the volume of production increases, the chemical engineer will calculate that point where the expense of labor, research, instrumentation, and continuous equipment will justify a continuous process, with its lower unit investment, operating cost, and more uniform quality.

FLOWCHARTS

Flowcharts show the coordinated sequence of unit chemical conversions and unit operations and as such depict fundamentals of a chemical process. They indicate the points of entrance for raw materials and any necessary energy, likewise, the points of removal of the product and the by-products. In the evaluation of the process from the initial conception to the detailed flowchart for the design and operation of the final plant, many flowcharts should be drafted. Initially, they will be the very crude block type, growing continuously more detailed and quantitative and containing more information as the process develops into the system required by modern chemical process industries. No other description of a chemical process is as concise or as revealing regarding equipment, operating details, and general reactions as that presented by skillfully drawn flowcharts, which should include not only materials but labor and utilities as well.

CHEMICAL PROCESS SELECTION, DESIGN, AND OPERATION

Intrinsic in any chemical process is adequate and flexible initial design. This can be simple or complicated, depending upon the plant and the procedures required. For complicated plants there are specialists, sometimes called process engineers, who are experts in the various modern aspects of chemical process design (Table 2.1). Practical experience is a necessity for the senior design engineer to enable him or her to foresee and to solve such plant problems as maintenance and operation of the varied types of equipment needed. Experienced consultants, either individuals or chemical engineering concerns, are available to design and erect chemical process plants.

[6]Sanders, Process Instruments, *Chem. Eng. News*, Oct. 13, 1969, p. 33; Mayfield, Backup for Batches, *Chem. Eng. (N.Y.)*, **81**(12), 79 (1974); Herring and Shields, Development: The Automated Way, *Chem. Eng. Prog.*, **61**(6), 94 (1965).

PILOT PLANTS[7] As a rule, pilot-plant development experiments are necessary to close the gap between laboratory results and the chemical process. Therefore design and pilot-plant experimentation should go forward hand in hand to save cost and time. The chemical design engineer should plan pilot-plant runs carefully by using *statistical analysis* of the procedures necessary to determine the answers to questions involving efficient industrial operation. This should be done with the least expenditure of time and effort; pilot plants are not cheap. Furthermore, particularly in new operations, the pilot plant should be constructed with equipment identical in material with that to be used by the commercial plant, in order to ascertain questions of corrosion and "to commit blunders on a small scale and make profits on a large scale."[8]

It is also much cheaper to correct errors by experimentation in the pilot plant. For certain new chemical processes, the pilot plant is kept functioning even after the main plant is in operation, to carry on studies for improving the process as suggested by the research or operating departments. It is expensive to experiment with plant-scale processing. The modern chemical engineer also keeps in mind the latest data available in order to scale up the pilot plant[9] proportionally to the large one. Finally, it may be desirable to obtain samples of a new product from the pilot plant to test its potential market.

EQUIPMENT This book contains no separate chapters on equipment. It is considered more important to emphasize equipment in conjunction with descriptions of the various processes and with flowcharts representing these processes. On the other hand, any chemical engineer should start early to become familiar with industrial equipment such as pumps, filter presses, distillation towers, nitrators, and sulfonators. The Chemical Engineering Catalog[10] includes convenient information concerning the actual equipment that can be supplied by various manufacturers.

CORROSION, MATERIALS OF CONSTRUCTION In chemical factories the successful consummation of chemical reactions and the maintenance of equipment depend not only upon the strength of the materials, but also upon their proper selection to resist corrosion and to withstand the effects of elevated temperatures and pressures. Mechanical failures are seldom experienced unless there has been previous corroding or weakening by chemical attack on the material used for the construction of either the equipment or the building. Erosion is occasionally a factor in the deterioration of equipment, particularly in the bends of continuous pipe reactors. This can be greatly reduced by avoiding small radii whenever bends occur.

In some cases corrosion cannot be prevented; it can only be minimized. The advance of chemical and metallurgical engineering has provided many corrosion-resistant materials: rubber-covered steel, resin-bonded carbon, and tantalum, to resist hydrochloric acid; stainless steel to resist the action of aqueous nitric acid and organic acids even under pressure; and nickel or nickel-clad steel to resist caustic solutions, hot or cold. Various polymeric organic materials, resins, and plastics have become important materials in the fight against corrosion. Among the *construction materials* used by chemical engineers are to be found many of the commonest products as well as some of the rarest—brick, iron, cement, and wood, on the one hand, and platinum, tantalum, and silver, on the other. Frequently, actual testing will be in order, which should not be carried out with pure laboratory chemicals but with commercial grades, not only of the chemicals to be used in the factory, but also of the actual construction material to be tested. It frequently happens that a small amount of contaminant in a

[7]Davis, Statistically Designed Pilot Plants, *Chem. Eng. Prog.*, **58**(2), 60 (1962); Clark, Economic Pros and Cons of Pilot Plants, *Chem. Eng. (N.Y.)*, **71**(8), 169 (1964); Hamilton *et al.*, Pilot Plants and Technical Considerations, *Chem. Eng. Prog.*, **58**(2), 51 (1962); Harrison and Lengemann, Make Your Pilot Plants, *Chem. Eng. (N.Y.)*, **71**(13), 129 (1964); Knapp, Effective Techniques, *Ind. Eng. Chem.*, **54**(2), 58 (1962); Gernand, Streamlined Data Gathering Systems, *Chem. Eng. Prog.*, **61**(6), 62 (1965).

[8]Baekeland, *Ind. Eng. Chem.*, **8**, 184 (1916).

[9]Johnstone and Thring, Pilot Plants, Models and Scale-up Methods in Chemical Engineering, McGraw-Hill, 1957.

[10]"CEC" is distributed annually to *practicing* chemical engineers by Van Nostrand Reinhold.

commercial raw material affects corrosion appreciably. An example of this is the attack upon aluminum by dilute nitric acid carrying a trace quantity of halogen. In recent years many data have been collected and added to the literature pertaining to corrosion.[11]

CHEMICAL PROCESS CONTROL AND INSTRUMENTATION[12]

Automatic and instrument-controlled chemical processes are here. "Packaged plants" are advertised for hydrogen, oxygen, sulfuric acid, and many other chemicals. In the United States and elsewhere data processing and computing instruments are actually taking over the running of complex chemical processing systems in the heavy chemical, rubber, and petroleum industries. The chemical engineer should not choose instruments simply to record temperatures or pressures, but as reliable tools to control and maintain desired operating conditions. In large-scale continuous operations, the function of the workers and the supervising chemical engineer is largely to maintain the plant in proper running order. Under this type of maintenance, instruments play an important part. When chemical manufacturing is on a small scale, or when it is not adaptable to continuous procedures, the *batch sequence* should be used, requiring more supervision on the part of the workers and the chemical engineer because the conditions or procedure usually differ from start to finish.

Instrumentation for indicating, recording, and controlling process variables is an almost universal outstanding characteristic of modern chemical manufacture. In many chemical plants instrument expense represents up to 15% of the total expenditure for the plant. Instrumentation has been forced into this position of eminence by the increase in continuous procedures, by the increase in the cost of labor and supervision, and by the availability of all types of instruments and computers, including:

1. *Current information* (indicating instruments), from mercury thermometers and thermocouples for temperature, ordinary scales for weighing, and pressure gages for pressure.

2. *Continuous records,* from special instruments for recording temperature, pressure, weight, viscosity, flow of fluids, percentage of carbon dioxide, and many other physical and chemical data. Recording of electric energy required frequently indicates a certain state in carrying out a chemical procedure when a change should be made, as in "beating" pulp in a Jordan (Chap. 33) or crystallizing sugar in a vacuum pan (Chap. 30).

3. *Complete automation or computer*[13] *control* of different variables of a chemical process system. Special instruments for maintaining a desired pressure, temperature, pH, or flow of material are necessary but complicated devices. George R. Marr states: "Sensitive reactions, novel equipment arrangements, or unusually complex control schemes represent systems so complex that it is essentially impossible for the human mind to conceive or calculate the process behavior during start-up or after disturbance away from steady state. The ease with which the computer can be programmed by the engineer, its speed and accuracy in solving differential equations, and the insight it provides as to the nature of process behavior are three main reasons for the success of the analog computer in the processing industries." This division of control is a specialty in itself, but is growing rapidly with the availability of optimization controllers and better application of mathematics.

[11]Perry, sec. 23; Wilson and Oates, Corrosion and the Maintenance Engineer, Hart, 1968; Fontana and Greene, Corrosion Engineering, McGraw-Hill, 1967.

[12]Perry, sec. 22; Considine, D. M. and S. D. Ross, Process Instruments and Controls Handbook, 2d ed., McGraw-Hill (1973); Murrill, Automatic Control of Processes, International, 1967; Shinskey, Process Control Systems, McGraw-Hill, 1967; Considine, Process Instruments and Controls, McGraw-Hill, 1974.

[13]Davis, Computer Process Control, *Chem. Eng. (N.Y.)*, **81**(19), 52 (1974); Nisenfeld, Cost Comparisons of Analog-control and Computer-control Systems, *Chem. Eng. (N.Y.)*, **82**(17), 104(1975); Adoptive Control Emerges for Practical Use, *Chem. Eng. News*, Jan. 13, 1975, p. 28; Baker, Direct Digital Control of Batch Processes Pays Off, *Chem. Eng. (N.Y.)*, **76**(27), 121 (1969); Williams, What's Next for Process Control Computers?, *Chem. Eng. Prog.*, **68**(4), 45, 1972; Lawrence and Buster, Computer-process Interface, *Chem. Eng. (N.Y.)*, **79**(14), 102 (1972).

Chemical control has a threefold function in factory procedures: (1) analysis of incoming raw materials; (2) analysis of reaction products during manufacturing, i.e., so-called *process control;* and (3) analysis of outgoing finished products. The chemical manufacturer should not only know the character of the raw materials he is buying, but should set up strict quality specifications to ensure the minimum or complete absence of certain undesirable impurities. For instance, the presence of arsenic in the acid employed for hydrolyzing starch would be deleterious to dextrose for food purposes, although of lesser significance if used to produce dextrin adhesives. Well over 90% of the raw materials of the chemical process industries are probably purchased on the basis of *chemical analysis.*

CHEMICAL PROCESS ECONOMICS

Engineers must always be conscious of costs and profits. In fact, they are frequently required to make estimates of new designs, expansions, or obsolescence, and in so doing, will find it impossible to disregard cost-consciousness, which must be a part of their work. They must continue to keep abreast of economic factors that may affect their products. One of the primary objectives of the engineer's endeavor should be to deliver the best product or the most efficient services at the lowest cost to his or her employer and to the consuming public.[14]

COMPETING PROCESSES Since change is an outstanding characteristic of chemical procedures, potential alteration of any process is of importance not only when the plant is first designed, but should always be kept in mind by chemical engineers. Indeed, one of the functions of an R&D division is to keep abreast of progress and to make available knowledge of improvements or even fundamental changes leading to the making of any given product in which the organization is interested. This R&D component should also keep informed regarding developments in other companies and be in a position to advise management of the relative competitive position of actual or anticipated processes or products (see R&D p. 20). Judgment based on competitive facts must be exerted in most of the important discussions of the chemical engineer. For instance, there are many technical processes for making phenol, two for sulfuric acid, three for alcohol, and many for acetic acid. Choosing the one from among many that is best for a particular location or time is a practical decision, aided by the various formulations under the principles listed in Table 2.1. These will help, but in the final analysis the engineer, with judgment based on experience, will arrive at the most successful decision. Table 2.2 lists different and competing processes for various chemical products.

MATERIAL BALANCES Yields and conversions of the chemical process form the basis for the *material balances,* which in turn are the foundation for cost determination. When obtainable, materials and quantities from standard practice are tabulated under the flowcharts.

ENERGY An engineer is concerned with the direction and control of *energy.* This energy may be expended in the moving of raw materials by ship, rail, or pipeline; it may be employed in the form of heat of steam or electricity; or it may be the energy given out in exothermic reactions, or that which is absorbed in endothermic chemical reactions. The chemical engineer works with chemical changes involving chemical reactions but, on the other hand, under modern competitive conditions, serious consideration should be given to the other types of energy expenditure connected particularly with the process involved. Chemical conversions cause changes in chemical energy and resultant effects. Unit operations include physical changes in energy or position such as heat flow, fluid flow, or separation; these frequently are an essential part of the fundamental chemical conversions. Experience shows that the chemical engineer should also consider other broad energy expenditures, such as those represented by the transportation of raw materials. All these enter into the cost. (See Chap. 4 for many details on energy.)

[14]Perry, sec. 25; Bauman, Fundamentals of Cost Engineering in the Chemical Industry, Reinhold, 1964; Bierman and Smidt, The Capital Budgeting Decision, 2d ed., Macmillan, 1966; Jelen, Cost and Optimization Engineering, McGraw-Hill, 1970; Peters and Timmerhaus, Plant Design and Economics for Chemical Engineers, 2d ed., McGraw-Hill, 1968; Holland, Watson, and Wilkinson, How to Budget and Control Manufacturing Costs, *Chem. Eng. (N.Y.),* **81**(10), 105 (1974).

TABLE 2.2 *Selected Outline of Multiple Competitive or Alternative Processes for a Given Chemical**

Chap.	*Chemical*	*Process*
38	Acetaldehyde	(1) $CH\vdots CH + H_2O \longrightarrow$
		(2) LPG $+ O_2 \longrightarrow$
		(3) $C_2H_5OH \xrightarrow{\text{air}}$
31	Acetic acid	(1) Fermentation (vinegar)
38		(2) $C_2H_2 + H_2O \longrightarrow CH_3CHO \xrightarrow{\text{air}}$
		(3) $C_2H_5OH \xrightarrow{\text{air}} CH_3CHO \xrightarrow{\text{air}}$
		(4) $Ca(Ac)_2 + H_2SO_4 \longrightarrow$
		(5) $C_xH_y + \xrightarrow{\text{air}} CH_3CHO \xrightarrow{\text{air}}$
		Natural gas $\longrightarrow C_2H_4, C_2H_5OH + CH_3CHO \longrightarrow$
		(6) Partial oxidation of H_xC_y and separation of various acids
35	Acetic anhydride	(1) Distill ethylidine diacetate $\longrightarrow$
		(2) $C_2H_2 + CH_3 \cdot COOH \longrightarrow$
		(3) $CH_3CHO + \xrightarrow[\text{Mn acetate}]{\text{air}}$
		(4) $H_2C{:}CO + CH_3COOH \longrightarrow$
38	Acetone	(1) $Ca(Ac)_2 + \Delta$ (obsolete)
31		(2) Fermentation $\longrightarrow$
34		(3) Isopropyl alcohol $\xrightarrow{-H_2O}$
		(4) Cumene $\xrightarrow{\text{air}}$
7	Acetylene	(1) $CaC_2 \xrightarrow{H_2O}$
		(2) $CH_4 \xrightarrow{\Delta}$
		(3) $C_xH_y \xrightarrow{\Delta}$
35	Acrylonitrile	(1) $CH\vdots CH \xrightarrow{HCN}$
		(2) $CH_2{:}CH \cdot CH_3 \xrightarrow{O_2} \xrightarrow{NH_3 + \frac{1}{2}O_2}$
		(3) $CH_2{:}CH \cdot CH_3 \xrightarrow{NH_3 + 1.5O_2}$
		(4) $CH_2—CH_2$ (epoxide, O bridging) $\xrightarrow{HCN} \xrightarrow{-H_2O}$
35	Adipic acid	(1) Cyclohexanol $+ \xrightarrow{HNO_3 \text{ or } O_2}$
		(2) Cyclohexane $+ \xrightarrow{HNO_3 \text{ or } O_2}$
18	Ammonia	(1) $N_2 + 3H_2 \longrightarrow$
5		(2) Coal pyrolysis $\longrightarrow$
39	Aniline	(1) Benzene $\xrightarrow{HNO_3} C_6H_5NO_2 \xrightarrow{Fe}$
		(2) Benzene $\xrightarrow{HNO_3} C_6H_5NO_2 \xrightarrow{H_2}$
		(3) Benzene $\xrightarrow{Cl} C_6H_5Cl \xrightarrow{NH_3}$
39	Anthraquinone	(1) Phthalic anhydride + chlorobenzene $\xrightarrow{NH_3}$
5		(2) Coal tar $\longrightarrow$
37	Benzene	(1) Cyclization, demethylation $\longrightarrow$
5		(2) Coal pyrolysis $\longrightarrow$
36	Butadiene	(1) Ethyl alcohol $\xrightarrow{-H_2O}$ (obsolete)
		(2) Acetaldehyde $\longrightarrow$ 1:3 butanediol $\xrightarrow{-H_2O}$
		(3) $CH\vdots CH + CHO_2O \longrightarrow \xrightarrow{H_2}$ 1.4 butanediol $\xrightarrow{-H_2O}$
		(4) $\left.\begin{matrix} C_4H_{10} \\ C_4H_8 \end{matrix}\right\} \xrightarrow[\text{600–630°C}]{18\%\ Cr_2O_3 \text{ on } Al_2O_3\text{-}H_2}$
31	1-Butanol	(1) Fermentation $\longrightarrow$
38		(2) Crotonaldehyde $\xrightarrow{H_2}$

TABLE 2.2 ***Selected Outline of Multiple Competitive or Alternative Processes for a Given Chemical**** *(Continued)*

Chap.	Chemical	Process
8	Carbon black	(1) Pyrolysis (partial combustion): $CH_4 \longrightarrow$
		(2) Pyrolysis (partial combustion): fuel oil $\longrightarrow$
13	Caustic soda	(1) Causticization, soda ash $\longrightarrow$
		(2) Electrolysis NaCl $\longrightarrow$
38	Ethyl alcohol	(1) Fermentation $\longrightarrow$
31		(2) $C_2H_4 + H_2SO_4 \xrightarrow{H_2O} C_2H_5OH + H_2SO_4$
37	Ethylene	(1) Dehydration from $C_2H_5OH \longrightarrow$
		(2) Dehydrogenation C_2H_6, $C_3H_8 \longrightarrow$
38	Ethylene glycol	(1) Ethylene oxide + $\xrightarrow{H_2O}$
		(2) Ethylene chlorohydrin + $\xrightarrow{Na_2CO_3}$
38	Ethylene oxide	(1) Ethylene $\xrightarrow[Ag]{O_2}$
		(2) Chlorohydrin $\xrightarrow{NaOH}$
28	Fatty alcohols	(1) Fatty oils $\xrightarrow{\text{Na red.}}$
		(2) Fatty oils $\xrightarrow[\text{high psi}]{H_2}$
34	Formaldehyde	(1) Methane + $\xrightarrow{O_2}$
		(2) LPG $\xrightarrow{O_2}$
		(3) Methanol $\xrightarrow{\text{air}}$
29	Glycerin	(1) Soap mother liquors $\longrightarrow$
		(2) Syn. propylene $\xrightarrow[NaOH]{Cl_2,\ H_2O}$
		(3) Hydrogenolysis of dextrose $\longrightarrow$
		(4) Fermentation (ICI,SD) $\longrightarrow$
35	Hexamethylenediamine	(1) Butadiene $\xrightarrow{Cl_2} \xrightarrow{HCN} \xrightarrow{H_2}$
		(2) Acrylonitrile $\xrightarrow{H_2}$
		(3) Adiponitrile $\xrightarrow{H_2}$
20	Hydrochloric acid	(1) Salt and sulfuric acid (niter cake) $\longrightarrow$
		(2) $H_2 + Cl_2 \longrightarrow$
35	Hydrocyanic acid	(1) $CH_4 + NH_3$ + air $\longrightarrow$
		(2) $NaCN + H_2SO_4 \longrightarrow$
20	Hydrogen peroxide	(1) Quinones + $O_2 \longrightarrow$
		(2) LPG + $O_2 \longrightarrow$
		(3) Isopropanol + $O_2 \longrightarrow$
		(4) $H_2O(H_2SO_4) \xrightarrow{\text{electrolysis}}$
36	Isoprene	(1) From isopentane $\longrightarrow$
		(2) From propylene $\longrightarrow$
		(3) From isobutylene + $H_2CO \longrightarrow$
		(4) From acetylene + acetone $\longrightarrow$
38	Methanol	(1) Wood distillation $\longrightarrow$
		(2) Syn. CO + $H_2 \longrightarrow$
		(3) LPG + $O_2 \longrightarrow$
37	Naphthalene	(1) Demethylation $\longrightarrow$
5		(2) Coal pyrolysis $\longrightarrow$
34	Phenol	(1) Distillation (tar) $\longrightarrow$
		(2) $C_6H_6 + H_2SO_4 \xrightarrow{NaOH}$
		(3) $C_6H_6 + Cl_2 \xrightarrow{NaOH}$
		(4) Cumene $\xrightarrow{\text{air}} \xrightarrow{H_2SO_4}$ + acetone
		(5) Toluene $\xrightarrow{O_2}$ benzoic $\xrightarrow{-CO_2\ +\ \text{air}}$ + HCOOH

TABLE 2.2 ***Selected Outline of Multiple Competitive or Alternative Processes for a Given Chemical**** *(Continued)*

Chap.	*Chemical*	*Process*
16	Phosphoric acid	(1) $Ca_3(PO_4)_2 + H_2SO_4 \longrightarrow$
		(2) $P + \text{air} \longrightarrow P_2O_5 \xrightarrow{H_2O} H_3PO_4$
34	Phthalic anhydride	(1) Naph. $\xrightarrow[+\text{ air}]{V_2O_5}$ catalytic (fixed bed or fluidized)
		(2) *o*-Xylene + air $\longrightarrow$
22	R.D.X. or cyclonite	(1) "Hexa" + strong HNO_3 nitration $\longrightarrow$
		(2) "Hexa" cont. NH_4NO_3; HNO_3; ac. anhy. $\longrightarrow$
13	Soda ash	(1) Le Blanc $\longrightarrow$
		(2) Solvay $\longrightarrow$
		(3) From Trona $\longrightarrow$
36	Styrene	(1) Ethylbenzene (Dow) $\longrightarrow$
		(methods for $\longrightarrow$ Ethylbenzene)
19	Sulfuric acid	(1) Chamber $\longrightarrow$
		(2) Contact $\longrightarrow$
7	Synthesis gas	(1) Natural gas, oxygen (partial combustion) $\longrightarrow$
		(2) H_xC_y, oxygen (partial combustion) $\longrightarrow$
		(3) Natural gas, steam $\xrightarrow[\text{reforming}]{H_2O}$
35	Terephthalic acid	(1) *p*-Xylene $\xrightarrow{HNO_3}$
		(2) $C_6H_6 + K_2CO_3 \xrightarrow[340°C]{CO_2 \text{ pres. Cd}}$
37	Toluene	(1) Cyclization, demethylation $\longrightarrow$
5		(2) Coal $\xrightarrow{\Delta}$

*Most chemicals can be made from alternate raw materials and by several different processes. Hence there is keen competition and great flexibility. See text of CPI. Note that this is a selected outline.

LABOR The chemical process industries have moved rapidly into laborsaving techniques, particularly by rapid extension of continuous processing, by the more recent but extremely important adoption of optimizing control with computers, and by introducing laborsaving devices and procedures. At the same time managers of chemical process industries have been able to set aside from retained earnings large sums to replace obsolete plants or for repairs and maintenance (Table 2.3).

OVERALL COST The cost of processing cannot finally be obtained until the plant is in operation, but an experienced chemical engineer can closely estimate it. Indeed, he or she can calculate quite accurately from the *material balances* the raw-material cost and estimate the labor, taking into consideration, of course, modern laborsaving devices. On the other hand, most errors in cost estimating are due to an *underestimate* of overhead and sales and sales service expenses, as well as the considerable capital required to finance the actual running, including raw materials and inventory of finished product. After the plant has been designed, bids can be obtained for the equipment, which will indicate the amount of capital necessary for construction. Likewise, capital must be supplied initially or out of retained earnings for replacement, modernization, and repairs and expansion.

MARKET EVALUATION

During the development of the chemical industry, experts have been trained, and it is necessary to call upon them for a market evaluation when a new product is under appraisal.

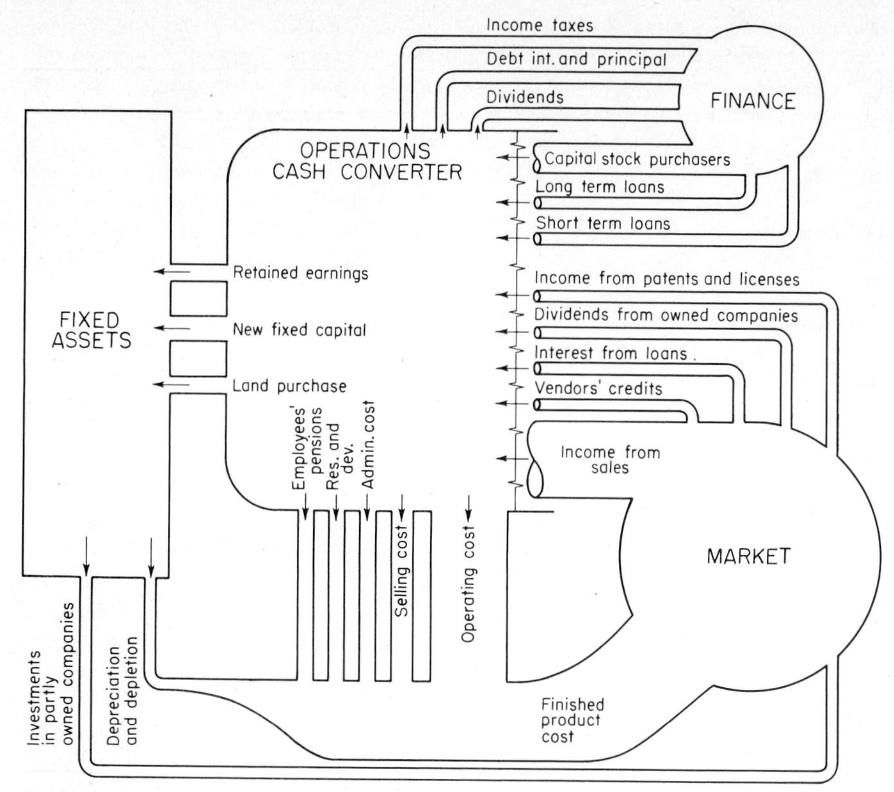

Fig. 2.1 Business cash flowchart. Economic chart similar to process flowcharts. [*Ind. Eng. Chem.*, **53**(6), 59A (1961).]

PRODUCT COUNTDOWN Once construction is under way and the fixed capital goes in, it is too late to recall questions about the new product that should have been asked earlier. The time for complete facts, clearly presented, is when the management team makes its decision to commercialize. *Statistics* to evaluate growth and stability are available for the chemical process industries from many sources and should be a guide for these aspects of market evaluation. However, judgment is necessary in using statistics, as in most engineering and business situations.

TABLE 2.3 ***Chemical Process Industries Capital Spending for New Plants (In billions of dollars)****

	1974†	*1973*	*1972*	*1971*	*1970*
Chemicals and allied products	5.31	4.46	3.45	3.44	3.44
Food and beverages	3.59	3.11	2.55	2.69	2.84
Paper	2.50	1.86	1.38	1.25	1.65
Petroleum	7.18	5.45	5.25	5.85	5.62
Rubber	1.67	1.56	1.08	0.84	0.94
Textiles	0.80	0.77	0.73	0.61	0.56

Sources: Dept. of Commerce; *Chem. Eng. News*, June 3, 1974, p. 32.
*Domestic capital spending. †Estimate.

PURITY AND UNIFORMITY OF PRODUCT The ultimate in the trend toward pure chemicals has been in the fields of nucleonics and pharmaceuticals. For example, the purity of uranium must be high, and absence of neutron absorbers such as boron be reduced to a few parts per 10 million. This has required chemical engineers to design and install many new procedures, for example, organic solvent extraction to purify uranium nitrate to the requisite degree (Chap. 21). Research in the laboratory checked by pilot-plant operations should be directed to obtain optimum conditions to ensure uniformity of product and process performance and, if possible, to reduce by-products, since sales costs of the latter may not be justified. Instrumentation can aid in the maintenance of these conditions, but it is largely the function of the superintendent of chemical engineering to see that they are carried out in actual plant operation.

The physical condition of the products has a great influence on marketability. This involves crystal structure, fineness, and color for solids and liquids. For instance, both light and dense soda ash are sold and required by different markets. Gasoline is dyed various colors to provide a distinctive trademark, but also to conceal an objectionable dark tinge which does not in any way affect its performance. Here it is cheaper to dye the product than to remove the dark color.

PACKAGING The most economical containers are refillable bulk ones, such as tank cars, boxcars, gondola cars, barges, and even tank steamers. Within this category might even be pipelines for the transportation of petroleum products, acid, brine, and many other liquid chemicals. Coal and other solids have been transported in pipelines, in suspension in liquids, usually water. Similar pipeline transportation in brine is being considered for potassium chloride. The United States has been most progressive in the development of bulk means of transportation, not only for entire trainloads of coal, oil, and gasoline, but also for molasses. Large tank cars or barges make many trips between seller and purchaser and provide very low-cost containers. Many chemicals are shipped in smaller containers, which may be either of the returnable or of the one-trip variety.

SALES AND SALES SERVICE The salesperson should be the eyes, ears, and nose to aid in economic forecasting. In many chemical companies, some of the most profitable and salable articles arise from the suggestions of salespeople. They should also be able to instruct customers as to how to use their employers' products profitably. All this is reflected in the ever-increasing number of chemical engineers entering this important area.

PLANT LOCATIONS

Prudent and proper location for a chemical plant or a branch plant is determined largely by raw materials, transportation, and markets. Yet many other factors enter into the choice, such as power, water, availability of efficient labor, cost of land, and waste disposal. There is a very strong tendency on the part of chemical concerns to leave congested cities and to move either to smaller towns or to the country. Indeed, legal restrictions occasionally force such moves.

SAFETY: HAZARDS SUCH AS FIRE OR TOXIC MATERIALS

Nothing is so destructive to a plant as fire. Precautions to prevent fire or to fight fire must be taken into consideration in the design of any chemical process. Likewise, employees must be protected against toxic chemicals. Safety measures not only keep an employee regularly on the job, and equipment working, but actually save money by reducing premiums paid by employers for liability and fire insurance. Too frequently, familiarity with chemicals breed carelessness; hence well-run plants have safety devices and continuing programs for alerting against danger those actually working with a given process. Adequate safety and fire protection measures require expert guidance.

In December, 1970, the Occupational Safety and Health Act of 1970, which has become widely known as OSHA, became law. The law (also known as Public Law 91-596 or the Williams-Steiger Act) became effective April 28, 1971. OSHA was enacted to provide improved safety and healthful working conditions throughout industry and covers about 57 million workers in over 4 million establishments. Under OSHA, the Secretary of Labor has the responsibility to formulate standards for all areas of industry including the chemical process industries. Since the act became effective, the Secretary of Labor has promulgated consensus standards which had been previously adopted by the American National Standards Institute (ANSI) and the National Fire Protection Association (NFPA).[15]

OSHA is a federal law, and it is strictly enforced. The employer must comply or suffer penalties up to $10,000 and 6 months imprisonment for repeated and willful violations or may even be required to close down the plant until regulations are met. Under OSHA there are three types of standards:

1. Initial standards (already issued).
2. Emergency temporary standards.
3. Permanent standards.

Emergency standards are issued to protect employees from serious danger from toxic or harmful substances or from new hazards. Permanent standards can be set only after formal proceedings where interested persons have a chance to submit their views and objections.

The first OSHA citation was made on May 14, 1971, just 2 weeks after the law became effective, against a chemical company with a mercury cell chlor-alkali plant in West Virginia.[16] The company involved traced the problem to minute leaks in lines containing mercury and quickly solved the problem.

In general the chemical industry has had an excellent safety record and prior to OSHA stressed safety without the law. With the new law even greater efforts are required.[17] By February 1974, 600 carcinogens (chemicals that cause cancer in humans or animals) were on the toxic substance list of the National Institute for Occupational Safety and Health, OSHA's research arm.[18] By OSHA's estimate, potentially about 2,000 workers are exposed to 14 chemicals which OSHA considers cancer hazards. There is considerable difference of opinion about just how dangerous many of these materials are, and it may take years before formal proceedings are completed and permanent standards set.

CONSTRUCTION OF PLANT

For small and large companies construction engineering organizations are available that will build a plant and will also participate in its design. On the other hand, some large chemical companies have their own construction departments and erect their own plants. The chief advantage of the latter procedure is that the workers who are actually going to operate the equipment can be more intimately connected with the construction and thus familiarize themselves with the plant. Three-dimensional scale models have produced important savings in construction costs by revealing flaws not evident from the usual engineering drawings. In building plants, under the laws of most states, at least the top engineers must be "registered." Such registration guarantees that the responsible engineers have been trained and examined to guarantee technical competency and personal responsibility. Young engineers should be encouraged to become registered engineers as soon as possible.

[15]*Fed. Regist.*, **36**(105), pt. II, May 29, 1971; *Fed. Regist.*, **37**(202), pt. II, Oct. 18, 1972.

[16]Brink and Kennedy, Mercury Pollution Control, Proceedings of the 1972 Clean Air Conference, Melbourne, Australia, May 15–18, 1972; Citation for Excess Mercury, *Chem. Eng. News*, June 7, 1971.

[17]Synder, Implementing a Good Safety Program, *Chem. Eng. (N.Y.)*, **81**(11), 112 (1974).

[18]Final Rules Set for Exposure to Carcinogens, *Chem. Eng. News*, Feb. 11, 1974, p. 12.

MANAGEMENT FOR PRODUCTIVITY AND CREATIVITY

Accompanying the dramatic rise in productivity, the chemical process industries have become so complex, because of technological change, that many plants are developing at least their plant *superintendents* and plant *managers* from among their chemical engineers. Indeed, certain companies arrange for promising young engineers to spend time studying management at various institutions. The plant manager's first responsibility is to so run a plant (trained personnel and efficient machines) that it will turn out market-acceptable goods at a profit. The manager is indeed a *creator of profits*. A leader in the pharmaceutical industry, Eli Lilly, said to his management group, "There is no excellent performance without high morale. When an organization loses its capacity to evoke high individual performance, its great days are over. Morale is like liberty. It takes constant work to preserve and deserve it."

TRAINING FOR PLANT PROCEDURES The best place to do this is at the pilot plant, where the supervisors and other selected personnel can take part in procuring pilot-plant data. Such experience provides an excellent foundation for the skilled operators now required. Operation questions to be asked are: "Can it be eliminated? Can it be combined? Can it be changed? Can it be simplified?"

LABOR The conduct of chemical plants requires, as a rule, skilled labor, with a limited requirement for ordinary back-breaking work. Most of the help needed is for workers who can repair, maintain, and control the various pieces of equipment and instruments necessary to carry out *chemical conversions* and *physical operations*. As each year passes, the chemical industry, by virtue of a wider use of instruments and a greater complexity of equipment, requires more and more highly skilled labor.

RESEARCH AND DEVELOPMENT[19]

It is certainly true that adequate and skilled research with patent protection guarantees future profits. In the chemical process industries some of the outstanding characteristics are changing procedures, new raw materials, new products, and new markets. Research creates or utilizes these changes. Without this forward-looking investigation or research, a company would be left behind in the competitive progress of its industry. The progress of industry opens up new markets for even the most fundamental established products. The results and benefits of research may be tabulated as follows:

1. New and improved processes.
2. Lower costs and lower prices of products.
3. New services and new products never before known.
4. Change of rarities to commercial supplies of practical usefulness.
5. Adequate supply of chemicals previously obtained only as by-products.
6. Freedom for American users from foreign monopoly control.
7. Stabilization of business and industrial employment.
8. Products of greater purity.
9. Products of superior service, e.g., light-fast dyes.

The number of people employed in R&D is shown in Table 2.4, and the cost is shown in Table 2.5. For chemicals and allied products the cost per employee was about $50,000 in 1976.

[19]Hunter and Hoff, Planning Experiments to Increase Research Efficiency, *Ind. Eng. Chem.*, **59**(3), 43 (1967); Corrigan and Beavers, Research and Development, *Chem. Eng.* (*N.Y.*), **75**(1), 56 (1968); Federal R&D Spending on the Rise Again, *Chem. Eng. News*, Feb. 18, 1974, p. 13.

TABLE 2.4 ***R&D Employment (In thousands of employees)****

	1973	*1972*	*1971*	*1970*	*1969*	*1967*	*1965*	*1963*
Chemicals and allied products	41.6	40.9	42.8	40.2	40.3	38.7	40.0	38.3
Industrial chemicals	19.5	19.7	22.4	21.9	22.6	22.7	25.7	22.9
Drugs	12.5	11.8	11.6	11.4	10.1	9.3	7.7	6.9
Other chemicals	9.6	9.5	8.8	6.9	7.6	6.7	6.6	8.5
Petroleum	8.2	8.3	9.2	9.9	10.0	10.4	9.7	8.9
Rubber	5.8	5.8	5.9	6.8	6.3	5.8	5.8	5.8
Paper	4.9	4.9	5.0	4.9	4.8	2.8	2.4	2.5
Textiles	1.8	1.8	1.8	2.9	2.6	1.9	1.2	1.0
All industry	359.9	349.9	366.8	384.1	387.1	367.2	343.6	327.3

Sources: National Science Foundation; *Chem. Eng. News*, June 3, 1974, p. 50. *Full-time equivalents.

PATENTS

In order to encourage new discoveries for the benefit of the nation, the Constitution provides for patents. These are limited monopolies, extending over 17 years, and are given in exchange for something new and beneficial. However, since it takes an average of about 7 years from the time an invention is patented until it is commercialized, the monopoly exists practically for only about 10 years. Patents are necessary in our competitive system of free enterprise in order that research funds can be generously spent for improvements on old processes and for new and useful discoveries. This system enables a concern to reimburse itself for the large expenditure for R&D. The U.S. patent system also discourages secret processes by guaranteeing a limited monopoly only where there is adequate disclosure.

Dean A. A. Potter, as Executive Director of the National Patent Planning Commission, wrote the following about patents:[20]

> A clear understanding is essential of the difference between an invention, a patent, and a marketable product. Invention is the act of finding something that is new. A patent is a grant of exclusive right to the inventor to his invention for a limited period of time. An invention is not a product, and the patent by itself does not produce a product. To produce a marketable product a new idea in the form of an invention must be developed and embodied in a form suitable for manufacture, and appropriate tools must be available so that the product can be manufactured at a cost acceptable to the public. The patent serves to protect the inventor and those who develop, manufacture, and sell the product from the uncontrolled competition of parties who have not shared the burden of invention and its commercialization.

TABLE 2.5 ***R&D Expenses by Industry (In millions of dollars)***

	1971	*1972*	*1975*
Chemicals and allied products	1,635	1,713	n.a.*
Industrial chemicals	858	870	1,025
Petroleum	488	458	525
Electrical equipment	2,232	2,442	3,000
Aerospace	1,012	945	1,150
All industries	10,643	11,347	13,950

Sources: National Science Foundation; *Chem. Eng. News*, June 3, 1974, p. 50. *Not available.

[20]Potter, Engineer and the American Patent System, *Mech. Eng.*, **66**, 15–20 (1944).

Fig. 2.2 Du Pont experimental station located in suburban Wilmington, Del. In 1975, Du Pont spent $336 million on R&D, and approximately 5,000 scientists and engineers were engaged in these activities. (*E. I. Du Pont de Nemours & Co.*)

Inventions, by bringing new products to the commercial world, benefit the public at large and for all time. The inventor receives only a *limited* reward; however, he or she *creates*[21] something that did not exist before. In the chemical industries, the American patent system is accountable for much of the recent growth because it encourages research upon which growth is founded. The obtaining of a patent is a procedure requiring skilled and experienced guidance. The patent must be for something *new*, the nature of which must be fully *disclosed*, and the essentials of the invention must be properly covered by the claims.

CHEMICAL SYSTEMS

PROCESS SYSTEM ENGINEERING Chemical systems are the modern way of viewing any one of the multitudinous chemical processes in the chemical process industries. Because of the increasing complexity of the chemical process industries and in order to draw in the modern use of computers, other instruments, and mathematics as applied to CPI, it is appropriate to summarize general fundamentals, with the emphasis upon chemical systems and their supporting computers and other instruments. Undoubtedly, the greatest single factor promoting the growth of process systems engineering[22] was in the introduction of computers for process-industry calculations. Finally, Dr. Theodore J. Williams of Purdue University points out in his "Systems Engineering for the Process Industries"

[21]Willson, Don't Overlook Patents, *Chem. Eng. (N.Y.)*, **71**(3), 79 (1964) (retrieval key to a vast storehouse of technical information); Gould, R. (ed.), Patents for Chemical Inventions, *Adv. Chem. Ser.*, no. 46 (1964) (excellent, many references); Hurd, The Flow of Patents, *Chemtech* **1,** 210 (1971); Invention and the Patent System, Technology, Economic Growth, and the Variability of Private Investment, U.S. Govt. Printing Office, 1964; Kent, Current Patent Office Procedures, *Chemtech*, **2,** 599 (1972).

[22]Prados, Systems Engineering: Its Present Status in the Process Industries, *Chem. Eng. Prog.* **58**(5), 37 (1962), and seyen following articles: Peterson, Chemical Reaction Analyses, Prentice-Hall, 1965. (See footnotes under preceding section on Chemical Process Control and Instrumentation for further such references.) Williams, About the Future, *Chem. Eng.* (*N.Y.*), **68**(9), 87 (1961); *ibid.*, Process Control, *Ind. Eng. Chem.*, **57**(12), 33 (1965); Coughanour and Koppel, Process Systems, McGraw-Hill, 1965.

Fig. 2.3 Research center of the western division of Dow Chemical located at Walnut Creek, Calif. Dow had 235 research employees at this location in 1976. Worldwide R&D spending by Dow in 1976 was planned at $190 million. (*Dow Chemical Co.*)

that "more and more, engineers are realizing that they can no longer think of a process plant as a collection of individually designed operations and processes. It is becoming increasingly evident that each separate unit of a plant influences all others in both obvious and subtle ways."

SELECTED REFERENCES

Adams, J. A., and D. F. Rogers: Computer-Aided Heat Transfer Analysis, McGraw-Hill, 1973.
American Society for Testing Materials (ASTM): Standards, Symposiums, Charts, Indexes, Manuals, 1965. (List of publications.)
Backhurst, J. R., and T. H. Parker: Process Plant Design, Elsevier, 1973.
Bauman, H. C.: Fundamentals of Cost Engineering in the Chemical Industry, Reinhold, 1964.
Berenson, C. (ed.): Administration of the Chemical Enterprise, Wiley-Interscience, 1963.
Clauser, H. R., *et al.* (eds.): Encyclopedia of Engineering Materials and Processes, Reinhold, 1963.
Conover, J. A.: Automatic Control for Chemical Processes, AIChE, 1968.
Considine, D. M.: Encyclopedia of Instrumentation and Control, McGraw-Hill.
Coughanour, D. R., and L. B. Koppel: Process Systems Analysis and Control, McGraw-Hill, 1965.
DeGarmo, E. P.: Materials and Processes in Manufacturing, 2d ed., Macmillan, 1962.
Evans, L.: Selecting Engineering Materials for Chemical Plants, Wiley, 1974.
Ewing, G. W.: Instrumental Methods of Chemical Analysis, 4th ed., McGraw-Hill, 1975.
Fawcett, H. H., and W. S. Wood: Safety and Accident Prevention: Chemical Operation, Wiley-Interscience, 1965.
Francis, A. W.: Liquid-Liquid Extraction, Wiley, 1964.
Hamilton, A., and H. L. Hardy: Industrial Toxicology, Publishing Sciences Group, 1974.
Handbook of Chemistry and Physics, Chemical Rubber.
Holman, J. P.: Heat Transfer, 3d ed., McGraw-Hill, 1972.
Holland, F. A., Watson, F. A., and J. K. Wilkinson: Introduction to Process Economics, Wiley, 1974.
Jurah, J. M., *et al.* (eds): Quality Control Handbook, McGraw-Hill, 1974.
Kuester, J. L., and J. H. Mize: Optimization Techniques with Fortran, McGraw-Hill, 1973.
Laitinen. H. A., and W. E. Harris: Chemical Analysis, McGraw-Hill, 1975.
Landau, R.: Chemical Plant, Reinhold, 1966.
Levenspiel, O.: Chemical Reaction Engineering, Wiley, 1962.
Lion, K. S.: Elements of Electrical and Electronic Instrumentation, McGraw-Hill, 1975.
Luyben, W. L.: Process Modeling, Simulation, and Control for Chemical Engineers, McGraw-Hill, 1973.
Mead, W. J. (ed.): Encyclopedia of Chemical Process Equipment, Reinhold, 1964.
Mellon, M. G.: Chemical Publications: Their Nature and Use, 4th ed., McGraw-Hill, 1965.
Nagiev, M. F.: The Theory of Recycle Processes in Chemical Engineering, Macmillan-Pergamon, 1964.
Park, W. R.: Cost Engineering Analysis, Wiley, 1973.

Patents for Chemical Inventions, no. 46, ACS, 1964.
Ray, W. H.: Process Optimization, Wiley, 1973.
Rhodes, T., and G. Carroll: Industrial Instruments for Measurements and Control, McGraw-Hill, 1972.
Sherwood, T. K.: Process Design, M.I.T., 1963.
Sittig, M.: Organic Chemical Process Encyclopedia, Noyes, 1967 (patents).
———: Chemical Guide to the U.S., Noyes, 1967.
Snell, F. D., and C. T. Snell: Dictionary of Commercial Chemicals, 3d ed., Van Nostrand, 1962.
Tomeshov, N. D.: Corrosion, Macmillan, 1966.
Treybal, R. E.: Mass Transfer Operations, McGraw-Hill, 1968.
Uhl, V. W., *et al.*, Mixing Theory and Practice, Academic, 1967.
Williams, T. J.: Systems Engineering for the Process Industries, McGraw-Hill, 1961.
Woods, D. R.: Financial Decision Making in the Process Industry, Prentice-Hall, 1975.

chapter 3

WATER CONDITIONING AND ENVIRONMENTAL PROTECTION

Water conditioning and wastewater treatment have long been essential functions of municipalities. However, the importance of suitably preparing water for the chemical industry is now fully recognized. Industrial wastewaters present a complex and challenging problem to the chemical engineer. Besides moral and community considerations, laws prohibiting and limiting the pollution of streams and air require these problems to be considered a necessary operating expense. Although the solution is specific with each industry (indeed, almost with each plant or factory), a few general principles may be observed: increasing reuse of wastewaters, control of pollution and, if feasible, recovery of by-products at their source to lessen the expense of treatment and lagooning of wastes to keep pollution at a minimum level or to effect a saving in neutralization costs.

As is well known, the quality and the quantity of available water are very important in choosing the location of a chemical plant. Both the surface water and the groundwater should be considered. The latter is usually more suitable for cooling purposes, because of its uniformly low summer and winter temperature, but such water is generally harder, may cause scale, hence may interfere with heat transfer. The impurities contained in water[1] vary greatly from one section of the country to another. Hard waters are those containing objectionable amounts of dissolved salts of calcium and magnesium. These are usually present as bicarbonates, chlorides, or sulfates. These salts give insoluble precipitates with soap, and calcium sulfate, carbonate, and silicate form clogging scales with low thermal conductivity in boilers. Magnesium silicate, as well as calcium carbonate, may reduce heat transfer in process heat exchangers.

Although, on an average day, 4,300 billion gal of water falls on the United States (30 in./year average), of which 70% is recycled to the atmosphere by evaporation or transpiration of plants, the increasing growth of population and of industry makes for local shortages. The problems[2] of quantity, quality, reuse, and pollution are complex and usually require expert study to decide between alternative sources of water and optimum treatment to minimize total cost of use. The decision generally depends on the use, whether for power generation, heating, cooling, or actual incorporation into a product or its manufacturing process.

Hardness is usually expressed in terms of the dissolved calcium and magnesium salts calculated as calcium carbonate equivalent. Water hardness may be divided into two classes: *carbonate* and *noncarbonate*, also frequently known as *temporary* and *permanent*. Temporary hardness can usually

[1]Nordell, Water Treatment for Industrial and Other Uses, 2d ed., Reinhold, 1961; Chanlett, Environmental Protection, McGraw-Hill, 1973.

[2]McGauhey, Engineering Management of Water Resources, McGraw-Hill, 1968; American Water Works Association, Water Quality and Treatment, 3d ed., McGraw-Hill, 1971; Fair, Geyer, and Okun, Water and Wastewater Engineering, vol. 1, Wiley, 1966; Fair, Geyer, and Okun, Water Purification and Wastewater Treatment and Disposal, Wiley, 1968.

be greatly reduced by heating; permanent hardness requires the use of chemical agents. Carbonate, or temporary hardness, is caused by bicarbonates of calcium and magnesium; noncarbonate, or permanent hardness, is due to the sulfates and chlorides of calcium and magnesium. In addition to hardness, varying amounts of sodium salts, silica, alumina, iron, or manganese may also be present. The total dissolved solids may range from a few parts per million in snow water to several thousand parts per million in water from mineral springs. The common units used in expressing water analyses are parts per million (ppm) and grains per gallon (gr/gal). One grain per gallon is equivalent to 17.1 ppm. Other water impurities that may be present are suspended insoluble matter (classed usually as turbidity), organic matter, color, and dissolved gases. Such gases are carbon dioxide (largely as bicarbonate), oxygen, nitrogen and, in sulfur waters, hydrogen sulfide.

REUSE

Forty percent of the United States population consumes water that has been used at least once for industrial and domestic purposes.[3] Industry may withdraw large quantities of water from the original source and use it in varied ways, but actually consumes relatively little,[4] so that most of it is returned to the source for reuse by others. Reuse is growing, and it is estimated that this alone might counteract the increase in the industrial-water requirement predicted for 1980. In many areas, treated sewage effluent is of a better quality than the natural water supply and, with improved waste treatment, industrial reuse becomes more and more feasible. Although roughly 80% of all industrial water is used for cooling, there has been a gradual increase in air cooling as a substitute for more expensive water. However, when water is cheap, it is most efficiently cooled in a tower, with countercurrent air flow. Industrial reuse reduces chemical, thermal, and resultant biological stream pollution and brings the handling of water more under the direct control of the plant manager.

WATER CONDITIONING

Water conditioning must be adapted to the particular use for which the water is designed, and problems should be referred to the experts in this field. The use of elevated pressures (2,500 psi and higher for steam generation) requires the employment of extremely carefully purified boiler feedwater. And each industry has its special water-conditioning requirement; e.g., laundries require zero hardness to prevent the precipitation of calcium and magnesium soap on clothes. Calcium, magnesium, and iron salts cause undesirable precipitates with dyes in the textile industries and with the dyes in paper manufacture.

HISTORICAL In this country an abundance of fairly soft water has long been available from surface supplies of the industrial Northeast. Cities such as New York and Boston obtain comparatively soft water from rural watersheds over igneous rocks. However, as the Middle West and the West were developed, it became necessary to use the harder water prevailing in many areas, particularly those rich in limestone. This hard water needs to be softened for many uses. Furthermore, as the advantages of really soft water are recognized, more and more fairly soft waters are being *completely softened* for laundries, homes, textile mills, and certain chemical processes. Thomas Clark of England, in 1841, patented the lime process for the removal of carbonate, or temporary, hardness, followed by Porter, who developed the use of soda ash to remove the noncarbonate, or permanent, hardness of water. In 1906, Robert Gans, a German chemist, applied zeolites to commercial use for water-softening purposes. Only since the 1930s has softening been extended to municipal supplies to any appreciable extent. Figure 3.1 shows in a general way the variations in hardness of waters.

[3]Marshall, Today's Wastes: Tomorrow's Drinking Water?, *Chem. Eng. (N.Y.)*, **69**(16), 107 (1962); Needed: An Extra 250 Billion GPD Water by 1975, *Chem. Eng. News*, Mar. 24, 1958, p. 50; Water Reuse, *Chem. Eng. News*, Mar. 21, 1966, p. 91.
[4]Clarke, Industrial Re-use of Water, *Ind. Eng. Chem.*, **54**(2), 18 (1962).

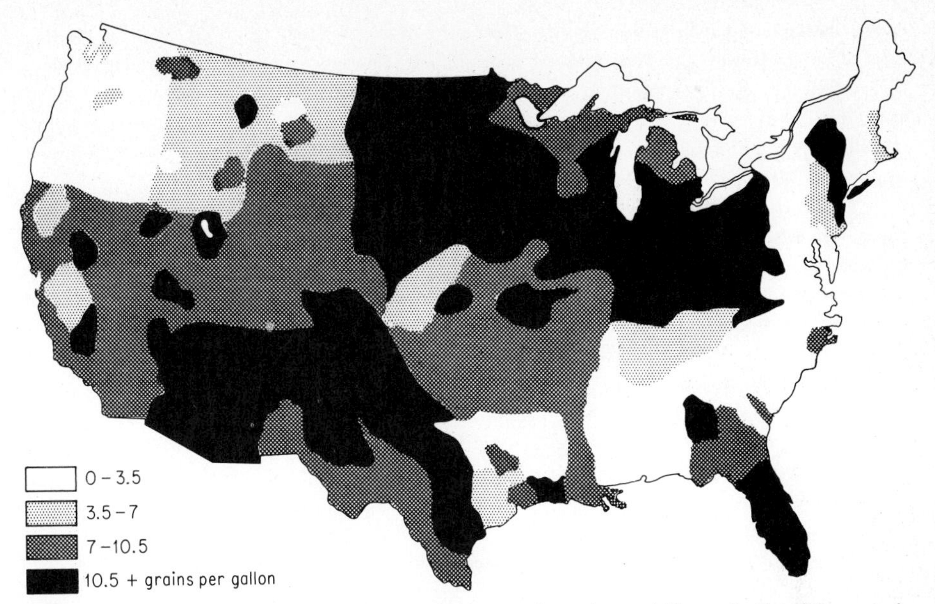

Fig. 3.1 A hard-water map of the United States shows that few parts have soft water (white areas). Most of the country has water of varying hardness (gray and black areas).

METHODS OF CONDITIONING WATER

The purification and softening of water may be accomplished by different methods, depending on the prospective use. *Softening* is the term applied to processes which remove or reduce the hardness of water. *Purification*, as distinguished from softening, usually refers to the removal of organic matter and microorganisms from water. *Clarification* may be very important and may be combined with cold-water softening by precipitation.

ION EXCHANGE In 1852, Way discovered that the removal of ammonia from aqueous liquids passing through certain soils was really an exchange with the calcium of a particular type of silicate occurring in the soils. The real stimulus for organic exchange resins came when Adams and Holmes[5] published their paper on purely synthetic organic exchange resins and described anion-exchange resins. Now ion exchange has become a valuable chemical conversion process. Its utilization on a large industrial scale is widespread, including the commercial production of demineralized water having low electrical conductivity. Ion exchange is really "a chemical reaction in which mobile hydrated ions of a solid are exchanged, equivalent for equivalent, for ions of like charge in solution. The solid has an open fishnetlike structure, and the mobile ions electrically neutralize charged, or potentially charged, groups attached to the solid matrix, called the ion exchanger. Cation exchange occurs when the fixed charged groups (functional groups) of the exchanger are negative; anion exchange occurs when the immobilized functional groups are positive."[6]

The first products used industrially for ion exchange were naturally occurring zeolites such as greensand, which have a very low exchange capacity per cubic foot. The next improvement was the introduction of organic ion exchangers for which natural products such as coal, lignite, and peat were sulfonated. As shown by Table 3.1, however, most high-capacity ion-exchange resins are based

[5]*J. Soc. Chem. Ind., London*, **54**, 1–6T (1935).

[6]McGraw-Hill Encyclopedia, vol. 7, p. 241, 1966; Guccione, Synthesis of Ion-exchange Resins, *Chem. Eng. (N.Y.)*, **70**(8), 138 (1963) (flowchart); Perry, sec. 16, Adsorption.

upon polystyrene-divinylbenzene (SDVB). Over 80% of ion-exchange resins are used for the treatment of water. However, the nonwater uses are of great importance and are growing. The literature covering all the versatile applications of ion exchange is extensive.[7] They can be chosen to remove and purify, for example, uranium, yttrium, or streptomycin (Chap. 40), from dilute aqueous solutions, or to remove impurities from aqueous solutions, as in sugar sirup to improve crystallization, or in food technology, or for catalysts, or simply to dry nonpolar solvents.

Sodium-cation-exchange process is the most widely employed method for softening water.[8] During the softening process, calcium and magnesium ions are removed from hard water by cation exchange for sodium ions. When the exchange resin becomes almost all changed to calcium and magnesium compounds, it is regenerated to restore the sodium resin with salt solution and in the pH range between 6 and 8. Regeneration is with common salt, at a chemical efficiency which varies between the limits of 0.275 and 0.5 lb of salt per 1,000 gr of hardness removed as compared with the stoichiometric efficiency of 0.17 lb of salt. The sodium or hydrogen-cycle cation exchangers for water treatment are usually of the SDVB sulfonated synthetic-resin type (Table 3.1). This is exceptionally stable at high temperatures and pH and is resistant to oxidizing conditions. The exchange capacity is up to 40,000 gr of $CaCO_3$ per cubic foot of ion exchanger with a hydrogen cycle and up to about 35,000 gr of $CaCO_3$ per cubic foot with a sodium cycle. The usual practical operating capacities are not so high.

The symbol R represents the cation-exchanger radical in the following reactions for softening:

$$\left.\begin{matrix}Ca\\Mg\end{matrix}\right\}\left\{\begin{matrix}(HCO_3)_2\\SO_4\\Cl_2\end{matrix}\right. + 2NaR \longrightarrow \left.\begin{matrix}Ca\\Mg\end{matrix}\right\} R_2 + Na_2\left\{\begin{matrix}(HCO_3)_2\\SO_4\\Cl_2\end{matrix}\right.$$

Calcium and/or magnesium	Bicarbonate, sulfate, and/or chloride	Sodium cation exchanger	Calcium and/or magnesium cation exchanger	Sodium	Bicarbonate, sulfate and/or chloride
(soluble)		(insoluble)	(insoluble)	(soluble)	

When the ability of the cation-exchanger bed to produce completely softened water is exhausted, the softener unit is temporarily cut out of service, backwashed to cleanse and hydraulically regrade the bed, regenerated with a solution of common salt (sodium chloride), which removes the calcium and magnesium in the form of their soluble chlorides and simultaneously restores the cation exchanger to its sodium state, rinsed free of these soluble by-products and the excess salt, and then returned to service ready to soften another equal volume of hard water. The regeneration reactions may be indicated as follows, using salt (or H_2SO_4 on a hydrogen cycle):

$$\left.\begin{matrix}Ca\\Mg\end{matrix}\right\} R_2 + 2NaCl \longrightarrow 2NaR + \left.\begin{matrix}Ca\\Mg\end{matrix}\right\} Cl_2$$

Calcium and/or magnesium exchanger	Sodium chloride	Sodium cation exchanger	Calcium and/or magnesium chloride
(insoluble)	(soluble)	(insoluble)	(soluble)

[7]Kunin and McGarvey, Research Keynotes Advances in Ion Exchange, *Ind. Eng. Chem.*, **54**(7), 49 (1962); Michalson, Ion Exchange, *Chem. Eng.* (*N.Y.*), **70**(6), 163–182 (1963); Kunin, Ion Exchange in Chemical Synthesis, *Ind. Eng. Chem.*, **56**(1), 35 (1964).

[8]Newman, Water Demineralization Benefits from Continuous Ion Exchange Process *Chem. Eng. (N.Y.)*, **74**(28), 72 (1967); Godfrey and Larkin, Ion Exchange Tames Radioactive Waste Solutions, *Chem. Eng. (N.Y.)*, **77**(16), 56 (1970); Iammartino, New Ion-Exchange Options, *Chem. Eng. (N.Y.)*, **80**(1), 60 (1973); Nordell, *op. cit.*, chap. 15, Sodium Cation Exchange (Zeolite) Water Softening Process.

TABLE 3.1 ***Water Treatment Takes Bulk of Ion-exchange Resin***

Uses of styrene-divinylbenzene resins	*Consumption (in thousands of cubic feet per year)*
Cation types:	
Industrial water treating	175.0
Household water treatment	375.0
Nonwater uses:	
Epoxidation	8.0
Uranium extraction	4.8
Nuclear grade	1.0
Pharmaceutical grade	9.0
Sugar treating	6.0
All others	26.2
Anion types:	55.0
Resins other than styrene-divinylbenzene	20.0
Total	680.0

Source: Chem. Eng. News, Sept. 3, 1962 (estimate).

The equipment for the process, shown in Fig. 3.2, is a large, closed, cylindrical tank in which the zeolite is supported on graded gravel. The water to be softened may flow down through the tank. Auxiliary apparatus includes both brine- and salt-storage tanks. Washing and regeneration may be carried out automatically as well as manually. These softeners are installed in water lines and operated under whatever water pressure is necessary. As the exchanger bed also exerts a filtering action, any sediment from the water or from the salt must be washed off by an *efficient backwash.* This step suspends and hydraulically regrades the resin bed. The water from ion-exchange treatment is usually practically zero in hardness. In cases where very hard bicarbonate water is encountered, it is often desirable to treat the water first by the lime process, followed by cation exchange. The lime process actually reduces dissolved solids by *precipitating* calcium carbonate from the water, whereas the cation resin *exchanges* calcium and magnesium ions for sodium ions. The great advantage of

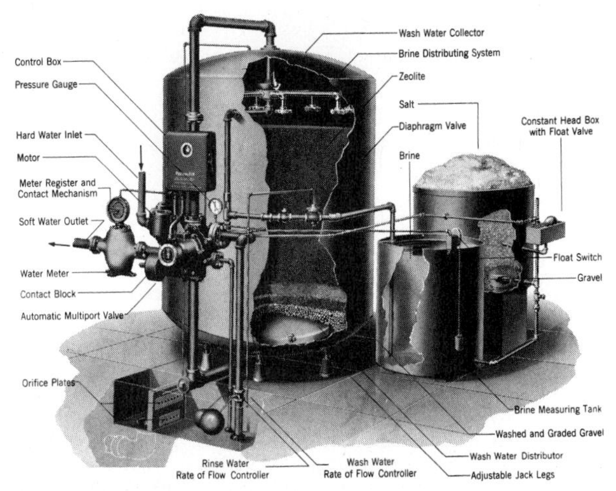

Fig. 3.2 Automatic sodium-cation-exchanger water softener. (*Permutit Co.*)

these softeners is their convenience and the fact that they furnish water of zero hardness *without attention* or *adjustment* until regeneration is required even though the raw water varies in hardness from one day to the next.

The hydrogen-cation-exchange process closely resembles the sodium cation procedure, except that the exchange resins contain an exchangeable *hydrogen ion* and can be employed to remove all cations. The symbol R represents the complex radical for the hydrogen-cation exchanger in the following reactions *for exchanges with bicarbonates:*

$$\left.\begin{matrix}\text{Ca}\\ \text{Mg}\\ \text{Na}_2\end{matrix}\right\}(\text{HCO}_3)_2 + 2\text{HR} \longrightarrow \left.\begin{matrix}\text{Ca}\\ \text{Mg}\\ \text{Na}_2\end{matrix}\right\}\text{R}_2 + 2\text{H}_2\text{O} + 2\text{CO}_2$$

Calcium, magnesium, and/or sodium Bicarbonate	Hydrogen cation exchanger	Calcium, magnesium, and/or sodium cation exchanger	Water	Carbon dioxide
(soluble)	(insoluble)	(insoluble)		(soluble)

The reactions *with sulfates and chlorides,* using the symbol R to represent the organic radical of the exchanger, may be indicated as follows:

$$\left.\begin{matrix}\text{Ca}\\ \text{Mg}\\ \text{Na}_2\end{matrix}\right\}\left\{\begin{matrix}\text{SO}_4\\ \text{Cl}_2\end{matrix}\right. + 2\text{HR} \longrightarrow \left.\begin{matrix}\text{Ca}\\ \text{Mg}\\ \text{Na}_2\end{matrix}\right\}\text{R}_2 + \text{H}_2\left\{\begin{matrix}\text{SO}_4\\ \text{Cl}_2\end{matrix}\right.$$

Calcium, magnesium, and/or sodium	Sulfate and/or chloride	Hydrogen cation exchanger	Calcium, magnesium, and/or sodium cation exchanger	Sulfuric and/or hydrochloric acid
(soluble)		(insoluble)	(insoluble)	(soluble)

Regeneration with sulfuric acid is the most widely used and most economical method of regeneration. The reactions, in condensed form, may be indicated as follows:

$$\left.\begin{matrix}\text{Ca}\\ \text{Mg}\\ \text{Na}_2\end{matrix}\right\}\text{R}_2 + \text{H}_2\text{SO}_4 \longrightarrow 2\text{HR} + \left.\begin{matrix}\text{Ca}\\ \text{Mg}\\ \text{Na}_2\end{matrix}\right\}\text{SO}_4$$

Calcium, magnesium, and/or sodium cation exchanger	Sulfuric acid	Hydrogen cation exchanger	Calcium, magnesium, and/or sodium sulfate
(insoluble)	(soluble)	(insoluble)	(soluble)

Acidic water is not desirable for most purposes, and therefore the effluent from the hydrogen cation-exchange treatment is either neutralized or, if demineralization is required, passed through an anion-exchange material, as shown in Fig. 3.3.

Anion exchangers are of two resin types, highly basic or weakly basic. Both types will remove strongly ionized acids such as sulfuric, hydrochloric, or nitric, but only highly basic anion exchangers will also remove weakly ionized acids such as silicic and carbonic. For the anion exchange of a strongly ionized acid, where R_4N represents the complex anion-exchanger radical, the reactions follow (some of the R's may be hydrogen):

$$\text{H}_2\text{SO}_4 + 2\text{R}_4\text{NOH} \longrightarrow (\text{R}_4\text{N})_2\text{SO}_4 + 2\text{H}_2\text{O}$$

Regeneration:

$$(\text{R}_4\text{N})_2\text{SO}_4 + 2\text{NaOH} \longrightarrow 2\text{R}_4\text{NOH} + \text{Na}_2\text{SO}_4$$

Highly basic anion exchangers are regenerated with caustic soda, and weakly basic anion exchangers may be regenerated with caustic soda, soda ash, or sometimes ammonium hydroxide.

DEMINERALIZATION

Demineralization systems are very widely employed, not only for conditioning water for high-pressure boilers, but also for conditioning various process and rinse waters. The ion-exchange systems chosen vary according to (1) the volumes and compositions of the raw waters, (2) the effluent-quality requirements for different uses, and (3) the comparative capital and operating costs. Briefly, if silica removal is not required, the system may consist of a hydrogen-cation-exchanger unit and a weakly basic anion-exchanger unit, usually followed by a degasifier to remove, by aeration, most of the carbon dioxide formed from the bicarbonates in the first step. When silica removal is required, the system may consist of a hydrogen-cation-exchanger unit and a strongly basic anion-exchanger unit, usually with a degasifier between the units, as shown in Fig. 3.3, to remove carbon dioxide ahead of the strongly basic anion-exchanger unit. For uses where the very highest quality of effluents is required, this may be followed by secondary "polishing" unit(s) consisting of either (1) a hydrogen-cation-exchanger unit and a strongly basic anion-exchanger unit or (2) a single unit containing an intimately mixed bed of a hydrogen cation exchanger and a strongly basic anion exchanger.

The only other process for removing all the ions in water is *distillation*. Both distilled water and deionized water should be handled in special pipes to keep the soft water from dissolving small amounts of the metal and thus becoming contaminated. Block tin has been employed for this purpose for many years but has the disadvantage of extreme softness. Aluminum and molded polyvinyl chloride pipes are now satisfactory substitutes for conducting "pure" water. Polyethylene, polypropylene, and polycarbonates are also sometimes used.

PRECIPITATION PROCESSES

LIME-SODA PROCESS The use of slaked lime and of soda ash to remove hardness in water has long been important. Modern application has been divided into the cold-lime process and the hot-lime–soda process. The calcium ions in hard water are removed as $CaCO_3$, and the magnesium ions as $Mg(OH)_2$. Typical equations for these reactions are

For carbonate hardness:

$$Ca(HCO_3)_2 + Ca(OH)_2 \longrightarrow \underline{2CaCO_3} + 2H_2O$$

$$Mg(HCO_3)_2 + Ca(OH)_2 \longrightarrow MgCO_3 + \underline{CaCO_3} + 2H_2O$$

Then, since $MgCO_3$ is fairly soluble,

$$MgCO_3 + Ca(OH)_2 \longrightarrow \underline{Mg(OH)_2} + \underline{CaCO_3}$$

Fig. 3.3 Demineralization equipment in two steps. (*Permutit Co.*)

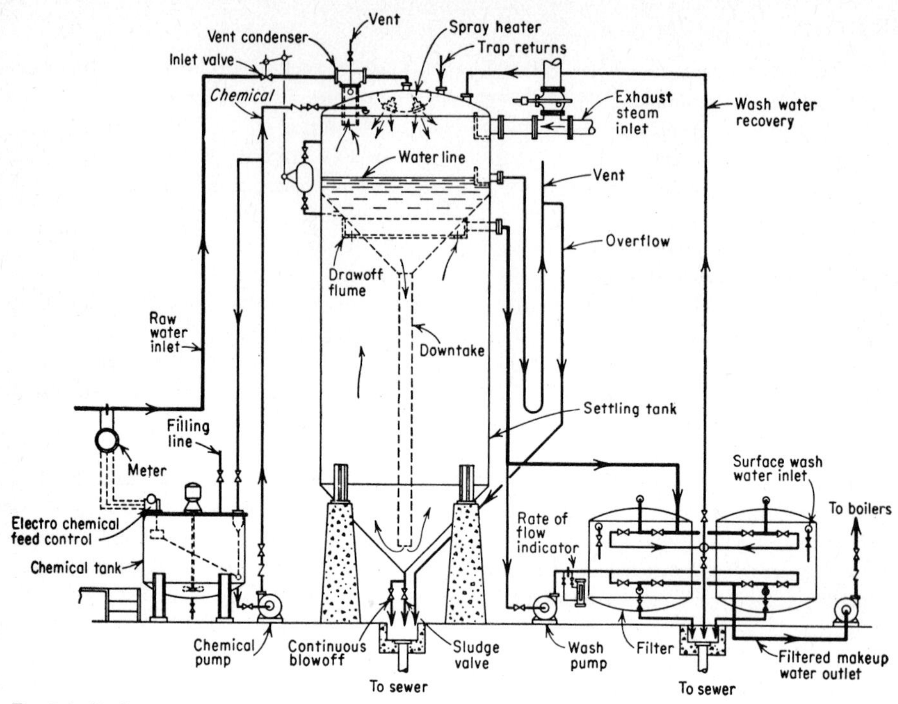

Fig. 3.4 Hot-lime–soda water softener. (*Permutit Co.*)

For noncarbonate soluble calcium and magnesium salts:

$$MgCl_2 + Ca(OH)_2 \longrightarrow \underline{Mg(OH)_2} + CaCl_2$$

$$CaCl_2 + Na_2CO_3 \longrightarrow \underline{CaCO_3} + 2NaCl$$

$$CaSO_4 + Na_2CO_3 \longrightarrow \underline{CaCO_3} + Na_2SO_4$$

$$MgSO_4 + Na_2CO_3 + Ca(OH)_2 \longrightarrow \underline{Mg(OH)_2} + \underline{CaCO_3} + Na_2SO_4$$

From these reactions it is apparent that, for carbonate hardness, each unit of calcium bicarbonate requires 1 mol of lime, whereas for each unit of magnesium bicarbonate, 2 mol of lime are needed. For noncarbonate hardness, likewise, magnesium salts require more reagent (1 mol each of soda ash and lime) and calcium salts require only 1 mol of soda ash.

The following material quantities have been figures for removing 100 ppm of hardness from 1 million gal of water:

Calcium bicarbonate hardness (expressed as $CaCO_3$), 521 lb lime.
Magnesium bicarbonate hardness (expressed as $CaCO_3$), 1,040 lb lime.
Calcium noncarbonate hardness (expressed as $CaCO_3$), 900 lb soda ash.
Magnesium noncarbonate hardness (expressed as $CaCO_3$), 900 lb soda ash, 520 lb lime.

The *cold-lime* process is employed chiefly for partial softening and ordinarily uses only cheaper lime for its reagent reactions. It can reduce calcium hardness to 35 ppm if proper opportunity is given for precipitation. This cold-lime process is particularly applicable to the partial softening of municipal water (Figs. 3.5 and 3.6), to the conditioning of cooling water where calcium bicarbonate hardness may be the scale former, and to the processing of certain paper-mill waters where calcium

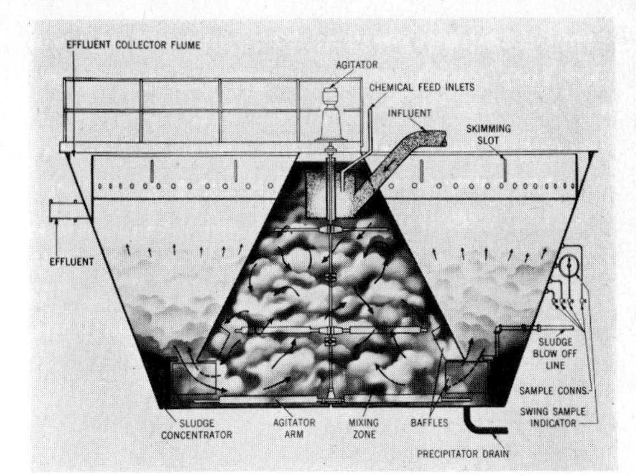

Fig. 3.5 Drawing of precipitator, showing formation of precipitate and upward filtration through sludge blanket. (*Permutit Co.*)

bicarbonate is troublesome. Magnesium carbonate hardness can be removed in any desired economical amount, but, if a low residual is wanted, an excess of hydroxyl ions will be needed to depress the solubility of the magnesium hydroxide. Usually, to aid in the process, a coagulant, aluminum sulfate or ferric sulfate, is added to minimize the carryover of suspended particles and to a lesser degree to reduce the afterdeposits from the supersaturated effluent.

A successful method of reducing *supersaturation* in the cold-lime–soda process is contacting previously precipitated sludge (Fig. 3.5). When this material is exposed to the raw water and chemicals, the like surfaces, or "seeds," accelerate the precipitation. The result is a more rapid and more complete reaction with larger and more easily settled particles in the newly formed precipitate. The equipment developed for this contact, by Infilco, Inc., is called the *Accelerator.* The Permutit Spaulding Precipitator[9] consists of two compartments, one for mixing and agitating the raw water with the softening chemicals and with the previously formed sludge, and the other for settling and filtering the softened water as it passes upward through the suspended blanket of sludge. Machines of these types reduce sedimentation time from 4 h to less than 1 h and usually effect savings in chemicals employed.

The *hot-lime–soda process* is employed almost entirely for conditioning boiler feedwater. Since it operates near the boiling point of the water, the reactions proceed faster, coagulation and the precipitation are facilitated, and much of the dissolved gas such as carbon dioxide and air is driven out. The hot-lime–soda treatment for softening may consist of the following coordinated sequences, as illustrated by Fig. 3.4:

Analysis of the raw water (Ch).

Heating of the raw water by exhaust steam (Op and Ch).

Mixing and proportioning of the lime and soda ash in conformance with the raw-water analysis (Op).

Pumping of the lime-soda mixture into the raw water (Op).

Reacting of the lime soda, facilitated by mixing with or without previous heating (Ch and Op).

Coagulation or release of the "supersaturation" by various methods, such as slow agitation or contact with "seeds" (Ch and Op).

Settling or removal of the precipitate with or without final filtration (Op).

Pumping away of the softened water (Op).

Periodic washing away of the sludge from the cone tank bottom (and from the clarifying filters) (Op).

[9]Nordell, *op. cit.*, pp. 495–501.

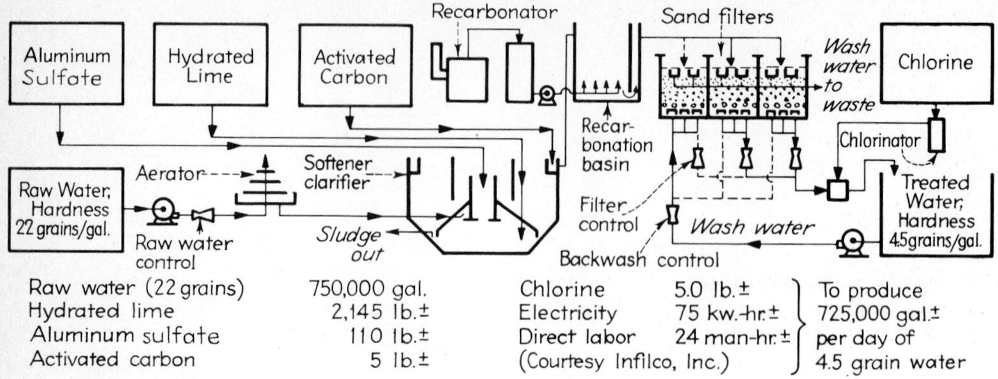

Fig. 3.6 Municipal water-treatment flowchart. (*Infilco, Inc.*)

If the chemical reactions involved could be carried to completion, the hardness of the water would be reduced to the theoretical solubilities of the calcium carbonate and magnesium hydroxide. This would leave a hardness of approximately 20 to 25 ppm in pure water, and a little less at the pH 10 to 11 involved in the lime-soda process. The newest development is the use of resinous ion exchangers for more complete softening of the effluent of a hot-lime process. This is particularly necessary when feeding high-pressure boilers.

PHOSPHATE CONDITIONING Various phosphates are employed, usually in conjunction with one of the previously described procedures. This process is used in internal conditioning of boiler water, on the one hand, and conditioning of cooling and process water, on the other. Orthophosphates, such as trisodium phosphate, and complex phosphates, such as sodium hexametaphosphate, are both used in steam boilers to precipitate whatever small amounts of calcium ion reach the boiler water through the pretreatment system or by leakage through the condenser. Sodium hexametaphosphate is advantageous where the boiler water naturally tends to become too alkaline, because it reduces this excess alkalinity by reverting to an acid orthophosphate in the boiler. Threshold treatment of cooling and process water with a few parts per million of sodium hexametaphosphate depends on other properties of this complex phosphate. When added to water which would normally deposit calcium carbonate scale when made more alkaline or when heated, it inhibits precipitation. Sodium hexametaphosphate is also widely used for minimizing the corrosion and pickup of iron by water in circulating cooling systems, in plant-water distribution systems, and in municipal systems.[10]

SILICA REMOVAL Silica is not removed by hydrogen cation exchange or sodium zeolite exchange and usually is only partially removed in cold- or hot-lime–soda processes. It may be a very objectionable impurity, since it can form a tenacious scale. Silica may be removed from feedwater by the use of dolomitic lime or activated magnesia in the softener. If preliminary coagulation and settling are carried out, the use of a ferric coagulate will remove some silica. These are especially suitable where the silica concentration of makeup water is high. Such methods do not entirely remove the dissolved silica, but they do lower its concentration to a point where adequate blowdown will eliminate the danger of scale in a boiler.

CORROSION Deaeration of water is often necessary to condition water properly for industrial purposes, although this is unnecessary for municipal waters. *Dissolved oxygen* hastens corrosion by a number of reactions,[11] depending on conditions. The following is a typical presentation of an

[10]Illig, Glassy Phosphate in Water Treatment, *J. Am. Water Works Assoc.*, **49**, 805–816 (1957); see Chap. 16, Phosphorus Industries.

[11]For corrosion of metals in contact with liquids, see Nordell, *op. cit.*, pp. 100ff.

important phase of iron water corrosion accelerated by oxygen under alkaline or neutral conditions. Iron in contact with water exerts a certain solution pressure and sets up the anodic half reaction:

$$Fe(s) \longrightarrow Fe^{2+}(aq) + 2e$$

This ceases after a certain potential is reached. However, oxygen can react with water to give OH ions at the cathode:

$$O_2(g) + 2H_2O(l) + 4e \longrightarrow 4OH^-(aq)$$

The Fe^{2+} and the OH^- ions react, and the electrons are neutralized by the flow of current between the adjacent anode and cathode:

$$Fe^{2+}(aq) + 2OH^-(aq) \longrightarrow Fe(OH)_2(s)$$

The initial reactions then proceed further. This electrochemical corrosion can be summarized:

$$2Fe(s) + O_2(g) + 2H_2O(l) \longrightarrow 2Fe(OH)_2(s)$$

Naturally, air and water can change ferrous hydroxide to ferric hydroxide. Anything that stops the foregoing sequences will stop the corrosion. This may be accomplished by removal of the dissolved oxygen, by electrode polarization, by organic inhibitors, or by protective salts. Such protective salts are chromates, silicates, phosphates, and alkalies, which probably act as anodic inhibitors by forming a film over the anodic or active areas and thus interrupt the electrochemical sequence. Water ordinarily saturated with air at 50°F contains about 8 ml of oxygen per liter. Oxygen is removed by spraying or by cascading the water down over a series of trays contained in a tank. During the downward flow, the water is scrubbed by steam rising upward. An open feedwater heater of the spray type will usually lower the dissolved oxygen content to below 0.3 ml/l. Scrubbing devices will remove even this small amount, or it can be chemically combined, using an oxygen scavenger like sodium sulfite or hydrazine:

$$O_2 + 2Na_2SO_3 \longrightarrow 2Na_2SO_4$$

$$O_2 + N_2H_4 \cdot H_2O \longrightarrow 3H_2O + N_2$$

Such complete deoxygenation is desirable to minimize corrosion in the modern high-temperature, high-pressure boiler.

WATER PURIFICATION This usually signifies the removal of organic material and harmful microorganisms from municipal supplies. Coagulation and filtration through sand or hard coal and oxidation by aeration are usually sufficient to remove organic matter. This treatment also removes some of the microorganisms. As a further decrease is usually considered necessary in order to produce safe or potable water, treatment with chlorine is indicated. Large quantities of this compound are consumed in this manner to protect the health of the nation. Chloramine and chlorine dioxide are also employed. The former is made by feeding ammonia into the chlorinated water:

$$2NH_3 + Cl_2 \longrightarrow NH_2Cl + NH_4Cl$$

This produces a better-tasting water in certain instances.

MUNICIPAL WATER CONDITIONING Treatment of municipal water supplies is usually necessary to produce potable and safe water.[12] Prior to widespread municipal water treatment, epidemics, particularly of typhoid fever, were caused by contaminated water. The requisites of a safe municipal water supply are freedom from pathogenic microorganisms and from suspended solids. It is also desirable, but not necessary, that the water be soft and not have an objectionable taste or odor.

Figure 3.6 shows a flowchart for a municipal water-treating plant in which both purification

[12]Chanlett, *op. cit.*, chap. 3.

and softening are carried out. The raw water is aerated to remove iron, odor, and taste, partly softened with lime, and the precipitate coagulated and filtered. Chlorine may be added to destroy pathogenic microorganisms, and activated carbon may be employed to remove odors and to improve flavor. More than 1,300 municipalities have adopted softening as part of their water treatment, but the public as yet does not realize the economic advantages of this process. The savings in soap alone are often more than sufficient to pay for the cost, though these savings have been greatly reduced by the replacement of soap by synthetic detergents for washing purposes. The many other advantages of soft water are therefore essentially free to the user.

INDUSTRIAL WATER CONDITIONING The necessary quality of water for industrial purposes depends upon the special use to which it is to be put. In most cases, hard water cannot be utilized without treatment. Many process industries employ several kinds of water, each of which serves a particular demand and is previously treated specially for this purpose. One of the most important industrial applications of water is for boiler feed.[13] The quality of the water required for various operations in the process industries varies widely. Untreated water can be utilized satisfactorily for many purposes, especially if it is not too hard. Many processes, however, require very soft water. The textile industry must have softened water to ensure even dyeing. Other industries require, or find it advantageous to use, deionized or distilled water. The demineralization process has reduced the cost of pure water and has been installed in many process industries. Ensuring the proper quality of water for an industrial operation may cost more than providing the needed quantity.

DESALTING

Desalting, or desalinization, is commonly applied to any process used to effect (1) partial or (2) complete demineralization of highly saline waters such as sea water (35,000 ppm of dissolved salts) or brackish waters. Concerning process 1, this applies to lowering the saline content to a degree which renders the water suitable for drinking purposes (preferably 500 ppm salines or less) and other general uses. Concerning process 2, this applies mainly to furnishing water suitable for use in high-pressure boilers and for certain other industrial uses. The ion-exchange demineralization processes described earlier in this chapter are not applicable to the desalting of highly saline waters; so other processes must be used.

The federal office of Saline Water program was initiated to study various desalting processes,[14] and its work has been expanded to the building of demonstration plants to appraise the practical performance of five processes: (1) the long-tube-vertical multiple-effect distillation[15] process, (2) the multistage flash-evaporation[16] process, (3) the freezing[17] process, (4) the electrodialysis[18] process, and (5) the forced-circulation vapor-compression process. In processes 1, 2, and 5 saline waters are purified by evaporation; the vapor is condensed in such a manner to recover and reuse as much of its heat content as possible (Fig. 3.7), and the concentrated brine is discharged to waste. Such processes have been employed for decades on ships and elsewhere. Process 3 is carried out by freezing highly

[13]Nordell, *op. cit.*, chaps. 8 and 17.

[14]Boehm, The Economic Realities of Water Desalting, *Fortune*, Jan. 1962, p. 97; More Stress Put on Basics in Saline Water Technology, *Chem. Eng. News*, June 10, 1963, p. 46 (good diagrams); Desalination Industry Airs Complaints, *Chem. Eng. News*, Nov. 10, 1969, p. 44.

[15]Guccione, Old Methods for Fresh-water Needs, *Chem. Eng.* **69**(24), 102 (1962) (process flowsheet); Sinek and Young, Heat Transfer in Falling Film Long Tube Vertical Evaporators, *Chem. Eng. Prog.*, **58,** 74 (1962); Dykstra, Falling Film Process, *Chem. Eng. Prog.*, **61,** 80 (1965).

[16]Guccione, Water Desalting in Multi-stage Flash Distillation, *Chem. Eng. (N.Y.)*, **69**(25), 122 (1962) (process flowsheet); Improved Process to Convert Saline Water, *Chem. Eng. News*, June 22, 1964, p. 50 (flowchart); Saline Conversion Plant at San Diego Being Moved to Guantanamo, Cuba to Supply 750,000 Gals. per Day, *ibid.*, Feb. 24, 1964, p. 59.

[17]Labine, Making Fresh Water from Salt Water, *Chem. Eng. (N.Y.)*, **67**(12), 152 (1960) (process flowsheet); Freeze-desalting Route Set for Federal Program, *Chem. Eng. (N.Y.)*, **69**(25), 78 (1962).

[18]Chopey, Water Desalting Plant to Prove out Electrodialysis, *Chem. Eng. (N.Y.)*, **69**(11), 104 (1962) (process flowchart).

Fig. 3.7 Sea-water-conversion plant. View of a 1 million gal/day facility at Freeport, Tex., shows 7 of 12 large effects, key units in the production of fresh water from the Gulf of Mexico. Built by the Chicago Bridge & Iron Co., following the design of W. L. Badger & Associates, for the Dept. of the Interior, Office of Saline Water, the plant is the federal government's first sea-water-conversion demonstration facility. The plant's conversion process is long-tube-vertical multiple-effect distillation. The first-effect temperature is 250°F, and the final-effect temperature is 110°F. The cost of water produced by the Freeport plant is less than $1 per 1,000 gal. (*Chicago Bridge & Iron Co.*)

saline waters so as to form a slurry of ice crystals and brine, from which the ice crystals are separated, rinsed, and melted. Process 4, involving no phase change, but ion-exchange membranes in an electric field, depends on the fact that, when a direct electric current is passed through saline water in a series of closely spaced, alternately placed, cation-exchanger and anion-exchanger membranes, cations pass through the cation-exchanger membranes and anions through the anion-exchanger membranes, resulting in a salinity decrease in one space and a salinity increase in the next space, and so on throughout the stack. The increased-salinity water may be run to waste, and the decreased-salinity water may be recirculated through the stack or passed through a series of stacks. This process does not produce completely demineralized water, but reduces the salinity of brackish water so as to make it suitable for drinking and general uses.

Development work has also been going on for a number of years on the use of reverse osmosis for desalination systems.[19] It is expected that this approach will find uses in the purification of brackish waters, since analysis of data on U.S. municipal water supplies indicates there are about 1,150 communities which have water supplies containing greater than 1000 ppm of total dissolved solids. A 100,000-gal/day plant is in operation at Plains, Texas.[20]

ENVIRONMENTAL PROTECTION

The protection of the environment is requiring more attention from chemical engineers every year. During the last decade environmental factors affected almost every sector of the chemical process industries as well as industry and business in general. In order to bring all of its existing facilities up to the pollution control standards in effect as of Jan. 1, 1974, it has been estimated that U.S.

[19]Mattson and Tomsic, Improved Water Quality, *Chem. Eng. Prog.*, **65**(11), 62 (1969); Leitner, Reverse Osmosis for Water Recovery and Reuse, *Chem. Eng. Prog.*, **69**(6), 83 (1973).

[20]Chanlett, *op. cit.*, p. 98.

businesses need to spend $24.67 billion over the coming years.[21] Details of this estimate are given in Table 3.2. In 1973 industry spent a total of $1.61 billion on pollution control R&D, and expenditures of $2.23 billion are planned for 1977. Additional large sums of money are required for the operation of existing pollution control processes and equipment.

The current large expenditures for pollution control in the United States reflect mainly the intervention of the federal government with new strict laws. These laws are enforced by the Environmental Protection Agency (EPA). Federal water pollution laws were enacted in 1948, 1956, 1961, 1965, 1966, and 1970, but they were primarily involved in aiding states in establishing limitations for interstate waters.[22] The nation's initial program covering all navigable waterways was passed in 1972 and is known as the *Water Pollution Control Act of 1972.* This law established effluent guidelines for both private industry and municipal sewage treatment plants on a national level. In addition, the law set the ambitious goal of prohibiting the discharge of any type of pollutant into U.S. waters by 1985. Whether or not this goal should or will be realized has become the object of considerable debate.[23] A focal point of the 1972 law is the *National Pollutant Discharge Elimination System* (NPDES), which regulates point sources of water pollution. The law states that a NPDES permit is required for the discharge of any type of pollutant. A permit does not condone pollution but rather controls the type and amount of effluent from each source. Target dates for reducing the amount of discharge in accordance with a goal of zero discharge are arranged with the recipient of the permit, who is legally bound to obey the conditions under possible threat of a stiff penalty.

Federal legislation on air pollution control in the United States followed a pattern somewhat similar to that enacted for water. It was 1955 before the first federal law was enacted as the *1955 Air Pollution Act.* This provided for research on air pollution effects by the Public Health Service, some technical assistance to the states, and the training of personnel in the area of air pollution control. In 1963 the *Clean Air Act* provided federal financial assistance but left the control problem to the states. The *1965 Amended Clean Air Act* established the first federal emission standards for cars. The *1967 Air Quality Act* covered the first requirement for ambient air standards, but these were to be set by the states. Finally, in 1970, the 1967 act was amended into strong legislation, which has become known as the *Clean Air Act.* This act provided for:

1. Additional research efforts and funds.
2. Additional state and regional grant programs.
3. National ambient air quality standards to be set by the federal government.
4. Complete designation of air quality control regions.
5. Implementation plans to meet the standards to be set in 3 above.
6. Standards of performance for new stationary sources to be set by the federal government.
7. National emission standards for hazardous air pollutants.

The 1970 act required that industry monitor air pollutants, maintain emission records, and make them available to federal officials. It also covered automobile emission standards, the development of low-emission vehicles, and aircraft emission standards. On December 23, 1971 the first new source performance standards were set by the EPA.[24] These standards covered five types of sources: fossil fuel-fired steam generators (250 million Btu/h input or greater), incinerators, nitric acid plants, sulfuric acid plants, and portland cement plants. Periodically additional such standards

[21] U.S. Industry Needs $24.67 Billion to Bring Its Existing Facilities into Compliance, *J. Air Pollut. Control Assoc.*, **24**(7), 682 (1974).

[22] Ziemba, A Look at the National Permit Program, *Ind. Wastes (Chicago)*, **22**(3), 44 (1974).

[23] Controversy Envelops Environment Standards, *Chem. Eng. News*, Mar. 4, 1974, p. 15; EPA Sees No Economic Blocks to Clean Water, *Chem. Eng. News*, Feb. 4, 1974, p. 16; Showdown Shapes Up over Pollution Laws, *Chem. Eng. News*, Jan. 28, 1974, p. 17; EPA Searches Its Soul, *Chem. Eng. (N.Y.)*, **81**(12), 48 (1974).

[24] *Fed. Regist.*, Dec. 23, 1971, p. 24876.

TABLE 3.2 *Capital Investment for Air and Water Pollution Control*

Industry	1974 Needs*	1973† Total	1973† Air	1973† Water	1977§ Total	1977§ Air	1977§ Water	Percent of capital spending, % 1973	Percent of capital spending, % 1977
Iron and steel	2.18	206	136	70	488	289	199	11.7	20.1
Nonferrous metals	1.48	301	256	45	213	178	35	18.0	7.9
Electrical machinery	0.70	105	48	57	111	54	57	3.7	3.1
Machinery	0.88	145	80	65	251	161	90	4.2	5.1
Autos, trucks, and parts	0.95	233	128	105	244	166	78	11.2	10.0
Aerospace	0.09	23	14	9	19	11	8	10.2	2.7
Other transportation equipment	0.09	9	7	2	20	12	8	4.3	6.1
Fabricated metals	0.09	136	53	83	171	84	87	7.2	7.3
Instruments	0.22	31	20	11	96	48	48	2.9	5.4
Stone, clay, and glass	0.67	134	89	45	163	102	61	9.0	8.8
Other durables	1.51	131	89	42	176	108	68	6.5	8.5
Total durables	9.52	1,454	920	534	1,952	1,213	739	7.6	7.8
Chemicals	1.45	455	183	272	609	286	323	10.2	9.8
Paper	2.06	424	275	149	373	236	137	22.8	15.8
Rubber	0.13	97	70	27	72	34	38	6.2	4.2
Petroleum	1.90	692	441	251	852	525	327	12.7	11.2
Food and beverages	1.16	194	102	92	269	122	147	6.3	6.9
Textiles	0.27	27	10	17	76	41	35	3.5	8.7
Other nondurables	0.09	48	40	8	64	52	12	3.1	4.8
Total nondurables	7.06	1,937	1,121	816	2,315	1,296	1,019	10.3	9.6
All manufacturing	16.58	3,391	2,041	1,350	4,267	2,509	1,758	8.9	8.7
Mining	0.47	207	159	48	165	77	88	7.6	4.3
Railroads	0.51	44	22	22	93	59	34	2.2	3.0
Airlines	0.73	224	224	0	121	121	0	9.3	16.4
Other transportation	0.11	34	29	5	66	57	9	2.0	3.5
Communications	0.14	256	128	128	506	253	253	2.0	3.0
Electric utilities	5.27	1,212	845	867	3,841	2,241	1,600	7.6	13.2
Gas utilities	0.14	41	22	19	92	56	36	1.5	2.3
Commercial¶	0.72	278	235	43	168	96	72	1.3	0.7
All business	24.67	5,687	3,705	1,982	9,319	5,469	3,850	5.7	7.0

Source: Seventh Annual Survey of Pollution Control Expenditures, McGraw-Hill, 1974; see *J. Air Poll. Control Assoc.*, **24**(7), 682 (1974) for additional data. *Total cost of bringing industries' existing facilities up to pollution control standards in effect as of Jan. 1, 1974. In billions of dollars. †All 1973 figures are actual expenditures. In millions of dollars. §All 1977 figures are planned expenditures. In millions of dollars. ¶Commercial estimate based on large chain, mail order, and department stores, insurance companies, banks, and other commercial businesses.

are set for other sources. In general the standards are strict and are based on the best pollution control technology available at the time they are set.

INDUSTRIAL AND SEWAGE WASTEWATER TREATMENT

MUNICIPAL WASTE WATERS Efficient *sewage disposal* is important to the health of any community. The easy method of disposal in the past was by dilution; the waste was dumped into an available body of water such as a river or lake, where the already present oxygen would in time destroy the organic sewage. However, dilution is no longer tolerated even in places where it is satisfactory.[25] There is simply not enough water with which to dilute the large amount of wastes from

[25]Othmer, The Water Pollution Control Act: Reaching toward Zero Discharge, *Mech. Eng.*, **96**(9), 32 (1973).

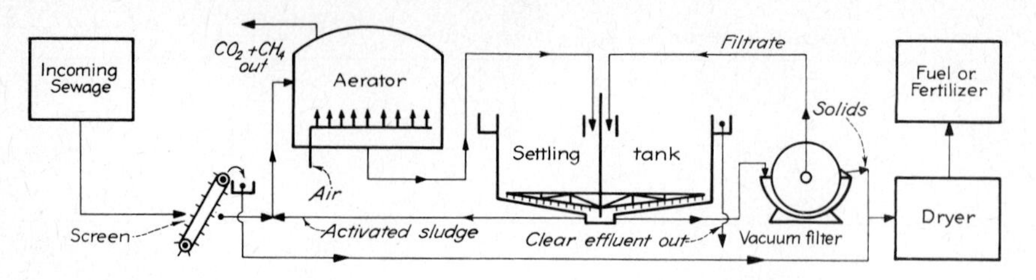

Average technical data: Air required, 0.35 to 1.94 cu. ft. per gal. sewage; detention in aeration tank, 4-6 hrs.; detention in settling tank, 1½-2 hr.; sludge return to aerator, 10-50 percent of total

Fig. 3.8 Activated-sludge sewage disposal.

our increasing population and industry. Pollution controls soon will also restrict the dumping of sewage into the sea or the barging out to sea of sewage solids, which become much more concentrated in sewage treatment plants. The goal of future sewage treatment plants will be *zero discharge* of pollutants.

The impurity in a particular sewage or, in other words, the amount of treatment required, is usually measured on one of two bases: (1) the amount of suspended solids, which needs no explanation, and (2) the biochemical oxygen demand (BOD), which measures the amount of impurity by the amount of oxygen required to oxidize it. Present methods of sewage treatment are usually divided into *primary* or physical treatment, *secondary* or biochemical treatment, and *tertiary* treatment. Primary treatment in a simple primary plant is designed to remove from the sewage 30 to 60% of the suspended solids and BOD. The effluent is usually chlorinated to destroy bacteria and viruses. In the primary plant the influent is screened to allow the passage of solids not larger than 1 to 2 in. Primary treatment then removes the "settleable" solids. Fine particles may be agglomerated to increase their size and permit settling, so that *coagulation* of fine particles makes larger ones by *flocculation*, and these are then removed by *sedimentation*. The total process is a *clarification* operation. (See Fig. 3.8.) Coagulants are added to assist in these operations. The chemical additives include inorganic salts, usually of iron or aluminum (ferric sulfate or aluminum sulfate with lime) which, under alkaline conditions, form hydrated colloidal flocs of their hydroxides. Polyelectrolytes (organic polymers) are also used.

Primary treatment alone is now considered inadequate, since it leaves much of the fines as well as all material in solution. With secondary treatment, dissolved organic materials are oxidized to reduce the BOD by 85 to 90%. This imitates nature's conversion by animal microorganisms which feed on them. Gases are formed, and a much smaller mass of solids remains. Biochemical oxidation may be accelerated in the secondary treatment by a trickling filter or an activated-sludge system, as shown in Fig. 3.8. *Activated sludge* provides one of the most effective methods for removing both suspended and dissolved substances from sewage. Activated sludge contains aerobic microorganisms which digest raw sewage. Some activated sludge from the previous run is introduced into the raw sewage, and air is blown in, not in excess but only in the amount needed. Disposal of the solids removed by any of these processes depends upon local conditions. In some cases they are buried, burned, or sold as fertilizer after filtering and drying. The liquids remaining after the removal of the solids are usually chlorinated to destroy harmful microorganisms and then discharged into nearby streams.

In recent years, oxygen-producing companies have developed processes using oxygen rather than air to speed up the aerobic treatment of sewage.[26] Union Carbide with its Unox System, Air

[26]Fallwell, Oxygen Finding a Big Outlet in Treatment of Wastewater, *Chem. Eng. News*, July 15, 1974, p. 7; Weber, Oxygenation System for Accelerated Sewage Treatment, *Mech. Eng.*, **96**(3), 2B (1974).

Fig. 3.9 UNOX system at Morganton, N.C. The system is supplied oxygen by a 26 tons/day pressure swing adsorption (PSA) oxygen plant. The PSA system is a recent advance in air separation technology and is uniquely suited for smaller wastewater treatment plants. (*Union Carbide Corp.*)

Products and Chemicals with its Oases System, FMC with its Maron System, and Airco are active in this field. More than 100 oxygen wastewater treatment systems were being designed, were under construction, or were in operation throughout the United States in 1974. High-purity oxygen enjoys a basic economic advantage over air because of energy considerations. The volume of oxygen dissolved per unit of electric power consumed is at least four times that of oxygen in air, but the amount of energy needed to separate oxygen from air is relatively low. Oxygenation systems have other cost advantages, e.g., with regard to treatment capacity. With oxygen systems, the size of the plant may be reduced from one-half to one-third the size of one operating on air. Also, there is lower rate of sludge generation, and the new systems can adjust quickly to sudden overloads of waste, a frequent problem in secondary treatment plants. Another side benefit is that covered aeration tanks are used, as shown in Fig. 3.9, so that odors are virtually eliminated. A comparison of air versus oxygen systems for secondary treatment is given in Table 3.3.

The world's largest pure oxygen wastewater treatment systems are scheduled for start-up in late 1978 and will be installed at the Detroit Metro Water Department's River Rouge plant.[27] Air Products Oases System will be the oxygen source for the 600-million-gal/day addition which will double the capacity of the existing plant. The cost will be about $40 million, of which 20 to 25% represents the cost of the oxygen system. The design includes a 450-ton/day cryogenic plant that can produce 98% oxygen at an estimated power cost of 12 to 13 kW/(ton)(day). Seventy-five percent of the cost of the plant is said to be committed by the federal government. The plant will be a two-train system, which will consume 400 tons/day of oxygen in 200 × 400 ft, 30-ft-deep aeration tanks and includes a surface aeration system as an energy conservation device in the unusually deep tanks. Conventional tanks are usually only 10 to 15 ft deep. Deep tanks have certain economical advantages.[28]

Tertiary sewage treatment involves further processing steps after secondary treatment, usually for the removal of pollutants with no BOD. After secondary treatment, water still contains phosphorus, nitrogen, and carbon compounds in solution, which can serve as nutrients for the overabundant growth of algae and other aquatic plants. The enrichment of waters with nutrients is

[27] *Chem. Week*, Oct. 16, 1974, p. 39.
[28] Oldshue, The Case for Deep Aeration Tanks, *Chem. Eng. Prog.*, **66**(11), 73 (1970).

TABLE 3.3 *Sewage Treatment with Oxygen vs. Air*

	Unox oxygenation system	*Conventional air aeration systems*
Mixed liquor dissolved oxygen level, mg/l	4–8	1–2
Aeration detention time (raw flow only), h	1–3	3–8
MLSS* concentration, mg/l	4,500–8,000	1,000–3,000
MLVSS† concentration, mg/l	3,500–6,000	900–2,600
Volumetric organic loading, lb BOD/(day)(1,000 ft^3)	150–250	30–60
Food/biomass ratio, lb BOD/lb MLVSS	0.4–1.0	0.2–0.6
Recycle sludge ratio, lb recycle/lb feed	0.2–0.5	0.3–1.0
Recycle sludge concentration, mg/l	15,000–35,000	5,000–15,000
Sludge volume index (Mohlman)	30–70	100–150
Sludge production, lb VSS/lb BOD removed	0.4–0.55	0.5–0.75

Source: Linde Division, Union Carbide Corp. *Mixed liquor suspended solids. †Mixed liquor volatile suspended solids.

referred to as *eutrophication* (derived from the Greek words *eu*, meaning "well," and *trophein*, "to nourish"). Since phosphorus, a major ingredient required for growth, is used in detergents (see Polyphosphates, Chap. 16), excess growth of algae in natural waters receiving domestic waste discharges are generally blamed on the high phosphorus content of the water.[29] Since complete removal of phosphorus from domestic waste cannot be accomplished by only limiting the use of phosphorus compounds in detergents, considerable work has been done in developing processes for removing phosphates from municipal wastewater. The most common types of chemical treatment are precipitation with lime and/or metallic hydroxides such as aluminum. It is about 90 to 95% effective and the cost is less than 5 cents/1,000 gal.[30] Under some conditions, microorganisms can absorb quantities of phosphates greatly in excess of their usual needs, a so-called luxury uptake. This uptake is reversible, as decay of the cells can release the phosphate.[31] A treatment plant, such as the one used in the District of Columbia,[32] has a process which removes the sludge from the final clarifier rapidly and feeds it into a phosphorus stripper operated in the fashion of an anaerobic digester. About two-thirds of the phosphate in the sludge is released in the digester and can be removed by chemical means. About 90% removal of phosphorus can be accomplished. The total cost of phosphate removal has been estimated to be from 3.80 to 3.02 cents/1,000 gal for plants with sizes varying from 1 to 1,000 million gal/day, respectively.

In 1974, more than 200 municipal sewage plants were using chemical additives to remove phosphates from sewage, a sharp jump from 10 such plants operating in 1972.[33] A 1974 survey by the EPA of 2,300 waste treatment plants indicated that 202 use chemicals for phosphate removal, 60 use chemicals for suspended solids removal, and 4 use chemicals for heavy-metal control. The survey showed that 134 plants add iron salts, 72 use alum, 81 treat with lime, and 115 add polymers. Many plants use more than one additive.

INDUSTRIAL WASTEWATERS The disposal of wastewater is of widespread national concern. In the chemical field alone there is a large volume of literature.[34] The problem of adequately handling industrial wastewaters is more complex and much more difficult than that involving sewage. Increas-

[29]Griffith *et al.*, Environmental Phosphorus Handbook, Wiley-Interscience, New York, 1973.

[30]*Environ. Sci. Technol.*, **2,** 182 (1968).

[31]Menar and Jenkins, *Environ. Sci. Technol.*, **4,** 1115 (1970).

[32]Levin *et al.*, *J. Water Pollut. Control Fed.* **44,** 1940 (1972).

[33]Phosphates Don't Have to Come Out of the Wash, *Chem. Week*, Aug. 21, 1974, p. 37.

[34]Waste Water 100% Recycled, *Chem. Process.* (*Chicago*), **35**(1), 13 (1972); Water Pollution Law Draws Flood of Complaints, *Chem. Week*, Oct. 16, 1974, p. 37; Two New Water Monitors Available, *Chem. Eng. News*, Oct. 21, 1974, p. 33; Firings, Carbon Adsorption Knocks Down BOD to 62 lb./day Limit, *Chem. Process.* (*Chicago*), **37**(1), 7 (1974); Ostrowski and Kunel, Nearly 33 Million gpd of Water Returned to River Cleaner than EPA Standards, *Chem. Process.* (*Chicago*), **37**(9), 9 (1974).

ingly stringent federal, state, and local regulations have been enacted prohibiting or limiting the pollution of streams, lakes, and rivers. Economic and technical studies are necessary to determine the least expensive way to comply with legal requirements and to reduce expenses or to show a profit through the recovery of salable materials. Other factors, such as reduction in real estate values, danger to inhabitants, and destruction of wildlife, are also involved. The great variety of chemical wastes produced in the nation's factories forces specific treatment in many instances. A few general practices are in use in many fields. One is that of storing, or lagooning, wastes. This may serve many different purposes. In factories having both acid and basic wastes, it reduces the cost of neutralization. In plants having wastewater containing large amounts of organic material (e.g., some paper mills) this results in a decrease in suspended matter and a reduction in the BOD. The use of flocculating agents (alum, $FeSO_4$) to remove suspended solids, and aeration to reduce the BOD, are also common to many industries.

A general problem in all industries is the disposal of wastes obtained as a result of water-softening treatment. Lime sludges may be lagooned and settled, or they may be dewatered and calcined for reuse. This sludge also finds some application in absorbing oil from other wastes. Brine used in regenerating ion-exchange plants is best stored and then added to streams by controlled dilution at high water. When the industry uses raw materials of complicated organic nature, an activated-sludge process may be used to treat the wastes. This process can be adapted to wastes from canneries, meat-packing plants, milk-processing plants, rendering plants and others.[35]

Tannery wastes may be treated by flocculation and sedimentation or filtration. Brewery wastes are subjected to trickling filters to reduce the BOD and remove most of the suspended solids. Paper mills have a serious problem, especially in treating sulfite wastes (Chap. 33). The processing of wastes from large chemical plants is exceedingly complex because of the variety of chemicals produced. The Dow Chemical Co. at Midland, Mich., for example, manufactures 400 chemicals in 500 processing plants and laboratories, resulting in a total of 200 million gal/day of wastewater. Equalization of acid and basic wastes and other general practices are followed, but many of the wastes are given treatment *at the source*, with an eye to recovery of valuable materials and by-products. Kodak[36] ensures limited pollution of the Genesee River by the use of clariflocculation basins, sludge filtration, and cake disposal and is employing ion exchange to regenerate phosphoric acid used as the electrolyte in anodizing aluminum sheet. Increasing emphasis in industrial-waste treatment is being placed on the recovery of useful materials. Fermentation wastes (slop), after evaporation and drying, are being sold as animal food. The use of ion exchangers promises the recovery of chromium and other metals from plating procedures. Ferrous sulfate (q.v.) is being obtained from pickling operations to a limited extent.

AIR POLLUTION[37]

Air pollution, or atmospheric contamination, is an acute problem throughout the country. Among the many causes of air pollution are industrial operations, transportation vehicles, and incineration of rubbish and waste by individuals. Many contaminants may be eliminated at the point of production, for instance, at the exhaust pipe of an automobile, before smog is formed. For this, catalytic *afterburners* have been developed.[38] Catalytic exhaust systems were installed on 1975 model cars in the

[35]Dloughy and Dahlstrom, Food and Fermentation Waste Disposal, *Chem. Eng. Prog.*, **65**(1), 52 (1969); Dahlstrom, Lash and Boyd, *Chem. Eng. Prog.*, **66**(11), 42 (1970); Barker and Schwarz, Engineering Processes for Waste Control, *Chem. Eng. Prog.*, **65**(11), 58 (1969).

[36]Waste-treating Methods Show Their Mettle, *Chem. Week*, Sept. 25, 1974, p. 41.

[37]A number of books have been written on air pollution. See Selected References at the end of this chapter.

[38]Iammartino, Detroit's Catalyst Choices, *Chem. Eng. (N.Y.)*, **80**(27), 24 (1973); White, Progress amidst Problems, *Chem. Eng. (N.Y.)*, **81**(22), 7 (1974).

United States to meet the tough emission guidelines of the EPA for hydrocarbons and carbon monoxide. Stringent limits on nitrogen oxide emissions are scheduled for 1977 model year cars. The use of catalytic exhaust systems requires the use of lead-free gasoline, so that the catalyst will not be poisoned by the lead. By the summer of 1974 the production and distribution of lead-free gasoline kept up with the production of new cars with catalytic systems. The suppliers of catalytic converters and the types of systems are given in Table 3.4. As shown in this table, considerable effort is being made by chemical companies to solve the air pollution problem.

Air pollutants leaving chemical processes and other industrial installations may be gases, mists (liquid particles less than 10 μm in diameter), spray particles (liquid particles larger than 10 μm), dusts, fumes, or combinations of these. Electrostatic precipitators are widely used for dust collection, along with baghouses, cyclones, and scrubbers.[39] Mist eliminators have been developed and applied widely in the chemical process industries for the collection of liquid particles.[40]

Of the various gaseous pollutants from industrial processes, sulfur dioxide has probably received the most attention. Sulfur dioxide has been discharged into the atmosphere in large quantities from electric power plants which burn coal and oil containing sulfur. Other sources have included ore-roasting processes for the production of lead, copper, and zinc, and sulfuric acid plants (see Chap. 19). Of these various sources, based on the amount of sulfur dioxide discharged, the electric power industry problem has been studied the most extensively. One approach has been the desulfurization of fuel prior to its use in a boiler.[41] This approach has been used for years by the petroleum industry to produce low-sulfur oil; however, it has not been developed to date for the large-scale desulfurization of coal. A number of stack gas sulfur dioxide removal processes have been worked out for large boilers.[42] In 1973 units were operating in 42 commercial applications, using 13 versions of 7 basic processes. These units ranged in size from less than 5 to as large as 250 MW. However, most of these were in Japan, and progress toward sulfur dioxide control in the United States has been slower. The chemical processes under development in the United States include scrubbing with a magnesium oxide slurry, limestone slurry scrubbing, and catalytic conversion of sulfur dioxide to sulfur trioxide.[43] The cost for most of these systems will be high. For example, for a 1000-MW power plant, the capital cost, in 1974 dollars, would be in the range of $30 to $65 million depending on the process used.

The chemical industry faces many other air pollution problems such as odor control.[44] Problems of this type are often solved by the use of activated-carbon beds, incineration, or catalytic combustion.

[39]Celenza, Designing Air Pollution Control Systems, *Chem. Eng. Prog.*, **66**(11), 31 (1970); Cyclone Dust Collector Erected Quickly in Field, *Chem. Process. (Chicago)*, **37**(1), 17 (1974); Environmental Forecast: Increased Precipitation, *Chem. Week*, Aug. 18, 1971, p. 77; Culhane, Production Baghouses, *Chem. Eng. Prog.*, **64**(11), 65 (1968); Owen, Selecting In-plant Dust-control Systems, *Chem. Eng. (N.Y.)*, **81**(21), 120 (1974); see Perry, pp. 20-74 to 20-121, for a very complete coverage of dust collection equipment.

[40]Brink, Mist Eliminators for Sulfuric Acid Plants, in Sulfur and SO_2 Developments, pp. 36–40, *Chemical Engineering Progress* Technical Manual, AIChE, 1971; Brink, Removal of Phosphoric Acid Mist, in Gas Purification Processes for Air Pollution Control, G. Nonhebel (ed). 2d ed., Newnes-Butterworth, London, pp. 549–563, 1972; Brink and Dougald, Particulate Removal from Exhaust Gases, *Pulp Pap.*, **47**(1), 51 (1973); Brink and Dougald, Air Pollution Control in Vinyl Plastic Manufacturing, *Pollu. Eng.*, **5**(11), 31 (1973); see Perry, pp. 18-82 to 18-93, for a complete coverage of mist collection equipment.

[41]Meridith and Lewis, Desulfurization and the Petroleum Industry, **64**(9), 57 (1968); Cortelyou, Mallatt, and Meridith, A New Look at Desulfurization, *Chem. Eng. Prog.*, **64**(1), 53 (1968); O'Hara, Jentz, Rippee and Mills, Producing Clean Boiler Fuel from Coal, *Chem. Eng. Prog.*, **70**(6), 70 (1974); Putnam and Manderson, Iron Pyrites from High Sulfur Coals, *Chem. Eng. Prog.* **64**(9), 60 (1968).

[42]Beychoic, Coping with SO_2, *Chem. Eng. (N.Y.)*, **81**(22), 79 (1974).

[43]Miller, Cat-ox Process at Illinois Power, *Chem. Eng. Prog.*, **70**(6), 49 (1974); Epstein, *et al.*, Limestone Wet Scrubbing of SO_2, *Chem. Eng. Prog.*, **70**(6), 53 (1974); Strom and Downs, A Systematic Approach to Limestone Scrubbing, *Chem. Eng. Prog.*, **70**(6), 55 (1974); Koehler, Alkaline Scrubbing Removes Sulfur Dioxide, *Chem. Eng. Prog.*, **70**(6), 63 (1974); LaMantia, *et al.*, Dual Alkali Process for Sulfur Dioxide Removal, *Chem. Eng. Prog.*, **70**(6), 66 (1974).

[44]Herr, Odor Destruction: A Case History, *Chem. Eng. Prog.*, **70**(5), 65 (1974); Lovett and Cunniff, Air Pollution Control by Activated Carbon, *Chem. Eng. Prog.*, **70**(5), 43 (1974); Beltran, Engineering/Economic Aspects of Odor Control, *Chem. Eng. Prog.*, **70**(5), 57 (1974).

TABLE 3.4 ***Catalytic Converters for the Big Three U.S. Automakers***

Catalyst supplier	*Volume*	*Model year*	*Noble metals*	*Substrate supplier*
General Motors Corp.				
Houdry Div., Air Products and Chemicals Inc.	8–10 million lb/year	1975–1977	Platinum and palladium (5:2)	Rhone-Progil
American Cyanamid Co., Japan Catalytic International, Inc.	8–10 million lb/year	1975–1977	Platinum and palladium (5:2)	American Cyanamid Co.
Engelhard Industries Div., Engelhard Minerals & Chemicals Corp.	8–10 million lb/year	1975–1977	Platinum and palladium (5:2)	Rhone-Progil
Davison Chemical Div., W. R. Grace & Co.	12–15 million lb/year	1975–1977	Platinum and palladium (5:2)	In house*
Ford Motor Co.				
Engelhard Industries Div., Engelhard Minerals & Chemicals Corp.†	60% of needs‡	1975–1977	Platinum and palladium (2:1)	Corning Glass Works and American Lava Corp.
Catalyst Systems Div., Matthey Bishop, Inc.	40% 20%	1975 1976–1977	Platinum§	Corning Glass Works and American Lava Corp.
Chrysler Corp.				
Automotive Products Div., Universal Oil Products Co.¶	100% of needs**	1975–1979	Platinum and palladium (5:2)	Corning Glass Works

Source: Chem. Eng. (N.Y.), **80**(27), 24 (1973). *Davison will in part use American Cyanamid Co. Technology. †Engelhard will also provide monolith catalysts to Japan's Nissan Motor Co. and Toyota Motor Co., and to Sweden's A. B. Volvo. ‡Total capacity for the plants is about 4 million units/year. §Matthey Bishop says a second catalytic ingredient is used, but declines to identify it. ¶UOP will also provide pellet catalysts for Japan's Nissan Motor Co., Toyota Mitsubishi Motors Corp., and Daihatsu Motor Co. **Plant capacity is about 8 million units/year.

SELECTED REFERENCES

Alfred, J. V. T.: Environmental Side Effects of Rising Industrial Output, Lexington Books, 1970.
American Institute of Chemical Engineers: Latest International Data, AIChE, 1974.
Baum, B., and C. H. Parker: Solid Waste Disposal Recycle and Pyrolysis, vol. 2, Ann Arbor Science, 1974.
Bennett, G. F.: Water—1973, AIChE, 1974.
Brown, C., *et al.*: Decision-Making in Water Resource Allocation, Lexington Books, 1973.
Camp, T. R.: Water and Its Impurities, Reinhold, 1963.
Campbell, M. D., and J. H. Lehr: Water Well Technology, McGraw-Hill, 1973.
Chanlett, E. T.: Environmental Protection, McGraw-Hill, 1973.
Ciaccio, L. L. (ed.): Water and Water Pollution Handbook, vol. 4, Dekker, 1972.
Cooper, H. B. H., and A. T. Rossano, Jr.: Source Testing for Air Pollution Control, McGraw-Hill, 1971.
Dickinson, D. (ed.): Practical Waste Treatment and Disposal, Wiley, 1974.
Eckenfelder, W. W., and D. J. O'Connor: Biological Waste Treatment, Pergamon, 1962.
Gurnhan, C. F. (ed.): Industrial Wastewater Control, Academic, 1965, 2 vols.
Hall, W. A., and J. A. Dracup: Water Resources Systems Control, McGraw-Hill, 1970.
Helfferich, F.: Ion Exchange, McGraw-Hill, 1962.
Hersh, C. K.: Molecular Sieves, Reinhold, 1961.
Hirshleifer, J., *et al.*: Water Supply: Economics, Technology, and Policy, University of Chicago Press, 1960.
Hopkins, E. S., *et al.*: Water Purification Control, 4th ed., Williams & Wilkins, 1966.
Jacobs, M. B.: Chemical Analysis of Air Pollutants, Wiley-Interscience, 1960.
Jackson, F. R.: Recycling and Reclaiming of Municipal Solid Wastes, Noyes, 1975.
James, L. D., and R. R. Lee: Economics of Water Resources Planning, McGraw-Hill, 1970.
Jones, H. R.: Wastewater Cleanup Equipment, Noyes, 1974.
Linsley, R. K., and J. B. Franzini: Water Resources Engineering, 2d ed., McGraw-Hill, 1972.
Maier, F. J.: Manual of Water Fluoridation Practice, McGraw-Hill, 1963.
Manual on Industrial Water and Industrial Waste Water, 3d ed., ASTM, 1963.
Martin, L. F.: Industrial Water Purification, Noyes, 1974.
McCoy, J. W.: The Chemical Treatment of Cooling Water, Chemical Publishing, 1974.

McGauhey, P. H.: Engineering Management of Water Quality, McGraw-Hill, 1968.
McKee, J. E., and H. W. Wolf (eds.): Water Quality Criteria, Department of General Services, Sacramento, Calif., 1963.
Metcalf & Eddy, Inc.: Wastewater Engineering, McGraw-Hill, 1972.
Nemerow, N.L.: Theories and Practices of Industrial Waste Treatment, Addison-Wesley, 1963.
Nemerow, N. L.: Scientific Stream Pollution Analysis, McGraw-Hill, 1974.
Perkins, H. C.: Air Pollution, McGraw-Hill, 1974.
Post, R. G., and R. L. Seale (eds.): Water Production, Nuclear Energy, University of Arizona Press, 1966.
Rich, L. G.: Environmental Systems Engineering, McGraw-Hill, 1973.
Rossano, A. T., Jr.: Air Pollution Control, McGraw-Hill, 1969.
Rubel, F. N.: Incineration of Solid Wastes, Noyes, 1974.
Saline Water Conversion, *Adv. Chem. Ser.*, no. 38, (1962).
Seneca, J. J., and M. K. Taussig: Environmental Economics, Prentice-Hall, 1974.
Speece, R. E., and J. F. Malina, Jr. (eds.): Applications of Commercial Oxygen to Water and Wastewater Systems, University of Texas Press, 1973.
Spiegler, K. S. (ed.): Principles of Desalination, University of California Press, Berkeley, 1966.
Stern, A. C.: Air Pollution, Academic, 1962.
Thurman, A., and R. K. Miller: Secrets of Noise Control, Wiley, 1974.
Twort, A. C.: Water Supply, Elsevier, 1964.
Walton, W. C.: Groundwater Resource Evaluation, McGraw-Hill, 1970.
Wolff, G. R. (ed.): Environmental Information Sources Handbook, Simon & Schuster, 1974.
Wulfinghoff, M.: Wastes—Solids, Liquids and Gases, Chemical Publishing, 1974.

chapter 4

ENERGY, FUELS, AIR CONDITIONING, AND REFRIGERATION

As a rule, a chemical engineer is not enough of a specialist to be capable of adequately designing plants for power production, refrigeration, or air conditioning. Since the chemical process industries consume more than a third of the energy used by *all* the manufacturing industries, chemical engineers should be familiar with the broad technical aspects of the production of power, cold, and heat. They should also be prepared to work with power, refrigeration, and air conditioning engineers for the proper coordination of the production of these essential tools and their use in chemical processes in order to attain the cheapest manufacturing costs. Frequently, the cost of power, particularly if it is to be used electrochemically, is the deciding factor in choosing the location of a factory. Process industries under the direction of chemical engineers are in most instances outstanding consumers of steam for evaporation, heating, and drying. Consequently, these industries need large quantities of steam, usually in the form of low-pressure or exhaust steam from turbines. Occasionally, however, certain exothermic reactions, as in the contact sulfuric acid process, can be employed to generate steam for use. If only electricity is desired from a steam power plant, naturally the turbines are run condensing; if, on the other hand, both steam and power are needed, as in the chemical process industries, it is economical to lead the high-pressure steam directly from the boilers through non-condensing turbines, obtaining exhaust steam from these prime movers to supply the heat necessary for drying, evaporation, and endothermic chemical reactions throughout the plant.

ENERGY

The increased consumption of energy per capita and the search for new sources for the future have become two of the foremost objectives of science, engineering, and government throughout the world.

It is estimated that, in 1850, coal, oil, and gas supplied 5% of the world's energy, and the muscles of humans and animals, 94%. Today, coal, oil, gas, and nuclear sources account for approximately 94%, water power about 1%, and the muscles of humans and animals the remaining 5%. Of the total amount of coal, oil, and gas that has been burned for the benefit of humans up to this time, less than 10% was burned in all the years prior to 1900, and 90% in this century. Humans have become energy-hungry beings.

One of the nation's leading analysts of resources, Admiral Hyman G. Rickover, recently stated:

> Our country with only 5 percent of the world's population uses one-third of the world's total energy input; this proportion would be even greater except that we use energy more efficiently than other countries. Each American has at his disposal each year energy equivalent to that obtainable from 8 tons of coal. This is six times the world's per capita energy consumption. . . . With high energy consumption goes a high standard of living. Thus, the enormous energy from fuels feeds machines which make each of us master of an army of mechanical slaves. Man's muscle power is rated at $\frac{1}{20}$ horsepower continuously. Machines furnish every American industrial worker with energy equivalent to that of 244 men, while the equivalent of at least 2,000 men push each automobile along the road, and every family is supplied with 33 faithful household helpers. Each locomotive engineer controls energy equivalent to that of 100,000 men; each jet pilot [that] of 1 million men. Truly the humblest American enjoys the services of more slaves than were once owned by the richest nobles, and lives better than most ancient kings. In retrospect, and despite wars, revolutions and disasters, the 100 years gone by may well seem like the 'Golden Age.'

In all systems wherein energy conversions take place, temperature is usually an important variable. Table 4.1 indicates the approximate continuous temperature associated with various energy conversion processes.

Not only are fossil fuels, coal, oil, and gas, being used extensively, but nuclear energy is being used; also, solar devices, fuel cells, and fusion are being actively developed, and many new "unconventional sources" are being studied in order to furnish the needed energy sources for the future.

PROJECTED ENERGY DEMANDS

Table 4.2 indicates historical and projected U.S. energy demands on the basis of three assumed growth rates. This and other tables to follow, as well as other information, are based on a 4-year study made by the National Petroleum Council entitled "U.S. Energy Outlook." This study drew on resources of over 1,000 representatives of academia, government, energy industries, and financial institutions.

The initial appraisal made by the National Petroleum Council indicated that U.S. energy consumption would probably increase at an average rate of 4.2% per year during the years 1971 to 1985. It was predicted that the United States would face an increasingly tighter energy supply and higher costs during the same period. Potential variations in future energy demands were established based on sets of assumptions different from those used to fix the initial appraisal value. The most significant

TABLE 4.1 ***High-temperature Sources***

Source	*Temp., K, continuous*	*Remark on temperature*
Bunsen burner	1400	
Blowtorch	1900	
Controlled nuclear reaction	3000	Limited by construction materials
Oxyacetylene flame	3380	Hottest low-cost chemical flame available
Combustion of aluminum powder in pure oxygen	3800–4400	Theoretical maximum; pressure 1 and 10 atm, respectively
Solar furnace	4000	Maximum estimated
Combustion of carbon subnitride	5000–6000	Maximum estimated
Electrical induction furnace	To 5000	Limited by crucible construction material
Plasma arc*	2000–50,000	Depends on arc type and current

Source: Ind. Eng. Chem., **55**(1), 18 (1963). *From Plasma, "Fourth State of Matter."

TABLE 4.2 ***Projections of U.S. Total Energy Demand for Three Different Assumed Growth Rates***

Condition	Average annual growth rate in percent gain			Total energy demand, quadrillion Btu/year*		Total equivalent in: Oil, billion bbl/year		Gas, trillion ft^3/year		Coal, billion tons/year	
	1970–1981	1981–1985	1971–1985	1980	1985	1980	1985	1980	1985	1980	1985
High growth rate	4.5	4.3	4.4	103.5	130.0	17.8	22.4	103.5	130.0	4.31	5.41
Intermediate rate	4.2	4.0	4.2	102.6	124.9	17.6	21.5	102.6	124.9	4.27	5.20
Low growth rate	3.5	3.3	3.4	95.7	112.5	16.5	19.4	95.7	112.5	3.98	4.68

Sources: Nichols, Balancing Requirements for World Oil and Energy, *Chem. Eng. Prog.*, **70**(10), 36 (1974); U.S. Energy Outlook, A Report of the National Petroleum Council's Committee on U.S. Energy Outlook, December 1972. *One quadrillion Btu (10^{15} Btu) units is equivalent to the amount of energy associated with 172 million bbl of oil, 1 trillion ft^3 of natural gas, or 41.6 million tons of coal.

long-range determinants of energy demand were: economic activity, gross national product, cost of energy, and population and environmental controls.

When considering the energy requirements of the United States it is necessary also to consider the energy requirements of the other non-Communist countries. The projection is shown in Table 4.3.

Table 4.4 shows the currently predictable energy demand estimates for the United States and other non-Communist countries for the years 1980 and 1985 in 10^{15} Btu units. It will be necessary to reconsider this prediction today in view of the present state of the arts and the restrictions now existing in the United States on the use of high-sulfur coal, the sharply increased price of imported oil, and the increase in nuclear generating facilities.

FOSSIL FUELS

Fossil fuels can be divided into three classes: solid, liquid, and gaseous. To these as energy sources should be added water and nuclear power. The actual and comparative costs of different energy supplies vary in different parts of the country. Solar and wind sources are as yet too trivial to report statistically. Worldwide, coal is the important and growing fuel used for power purposes, but there

TABLE 4.3 ***Non-Communist Countries' Total Energy and Oil Consumption Demands*** *(Exclusive of the United States)*

Year	Average total energy demand (oil equivalent), billion bbl/year	Average oil demand, billion bbl/year	Oil demand as percent of total energy demand	Average per capita consumption, barrels per capita per year: Total energy	Oil
1970	15.7	9.1	58	6.9	4.0
1975	21.2	13.7	65	8.3	5.4
1980	29.0	18.8	65	10.3	6.7
1985	39.6	25.2	64	12.5	7.9

Sources: U.S. Energy Outlook, *op. cit.*; Nichols, *op. cit.*

TABLE 4.4 ***Projections of Non-Communist Countries' Total Energy Demands***

	Total energy demand, Quadrillion Btu/year			*U.S. percentage of total non-Communist countries' energy demand*
Year	*United States*	*Other non-Communist*	*Total*	
1980	102.6	168.6	271.2	38
1985	124.9	230.2	355.1	35

Sources: Based on the U.S. intermediate growth rate shown in Table 4.2.

is a trend toward the use of a cleaner fuel, such as fuel oil or natural gas, and toward the development of better methods of coal combustion, which result in less contamination of the atmosphere. This is especially true in large towns and cities. Liquid fuels are derived mainly from petroleum and follow coal in importance as a source of heat for power generation. Petroleum products also furnish almost all the energy for the numerous internal-combustion engines in this country. Fuel gas is now mostly natural gas (Chap. 6).

SOLID FUELS Coal is the most important of the solid fuels, with an annual consumption of approximately 0.339 billion tons. Tables 4.5 and 4.6 show that the United States has 149.6 billion tons of known commercially available reserves of coal and will use only 0.339 billion tons/year unless high-sulfur coal can be burned by electrical utilities or the coal can be treated before or after burning to reduce atmospheric pollution by sulfur. To replace the annual shortage of about 4.7 billion bbl of oil in 1985 will require the conversion of over a billion tons of coal per year, assuming that a commerical liquefaction or gasification process is available. The popular method divides coal into the following classes: anthracite, bituminous, subbituminous, and lignite, with further subclassifications into groups. Anthracite was a valuable domestic fuel because of its clean burning characteristics but is now largely exhausted. The principal uses of bituminous coal are combustion for energy and *carbonization* (Chap. 5) for coke, tar, coal chemicals, and coke-oven gas. Table 5.1 lists in extensive outline the various fields in which coal is an important raw material and the areas in which exploratory development is being undertaken.

Pulverized coal has been used to an increasingly large extent during recent years in suspension firing for power-plant installations, because of the high thermal efficiency with which it can be burned, the low cost of operation and maintenance, and its great flexibility. All these factors more than balance the increased cost of preparing the fuel. In burning pulverized coal the fly ash leaves the boiler and is carried along with the waste gases. It is removed from the flue gases by an electric precipitator or other device. It has been used in bricks, building blocks, and concrete. Large tonnages of fly ash are sintered to produce a lightweight aggregate.

Coke is a fine fuel, but it is expensive at present for industrial use except in blast furnace operation, where it is a chemical raw material as well as a fuel. Coke production still parallels pig-iron production, though the amount of coke used per ton of iron continues to decrease. Other solid fuels,

TABLE 4.5 ***Underground Coal Reserves in the United States and 1970 Production*** *(Minable by underground mining methods)*

Economically available reserves, billion tons	*Recoverable reserves, billion tons*	*1970 production, billion tons*	*Life of recoverable reserves at 1970 production rate, year*
209.2	104.6	0.339	309

Sources: U.S. Energy Outlook, *op. cit.;* Nichols, *op. cit.*

TABLE 4.6 ***Surface Coal Reserves in the United States and 1970 Production*** *(Minable by surface mining methods)*

Recoverable reserves, billion tons	*1970 production, billion tons*	*Life of reserves at the 1970 production rate, years*
45.0	0.264	170

Sources: U.S. Energy Outlook, *op. cit.*: Nichols, *op. cit.*

such as coke "breeze," wood, sawdust, bagasse, and tanbark, are used where they are available cheaply or where they are produced as by-products.

LIQUID FUELS *Fuel oil* is the only important commercial liquid fuel used for power purposes. It is the portion of the crude oil that cannot be economically converted by the refiner to higher-priced products such as gasoline. It consists of a mixture of the liquid residues from the cracking processes and fractions of a suitable boiling point obtained from the distillation of crude oil. Fuel oil is classified according to its properties, such as flashpoint, pour point, percentage of water and sediment, carbon residue, ash, distillation temperature, and viscosity. All these are determined by tests that have been standardized by the ASTM. The flashpoint is relatively unimportant for determining the behavior of the fuel in the burner. Oil-burning equipment usually shows higher thermal efficiency (75%) than coal-burning boilers, and labor costs are usually less. However, the latent heat loss of steam produced by combustion of the hydrogen of the fuel is about two times greater than such losses from bituminous coal. Other liquid fuels include tar, tar oil, kerosine, benzol, and alcohol, which are consumed to a much smaller extent than fuel oil. Gasoline is consumed mainly in internal-combustion engines.

Based upon available assumptions, it was estimated that the total non-Communist world oil consumption would range between 351 and 392 billion bbl of crude oil during the 1971–1985 period, of which the United States would require 94 to 115 billion bbl, accounting for 17 to 29% of the total.

In view of the 1972 conclusion that the United States must increase its reliance on imports of petroleum to fulfill its total energy needs to 1985, the breakdown of estimated non-Communist sources projected through 1985 as shown in Table 4.7 is very important.

Table 4.8 indicates the balance of production and consumption within the United States and its Western Hemisphere neighbors.

TABLE 4.7 ***Estimated Developable U.S. and Non-Communist Foreign Oil*** *(In billions of barrels per year)*

Area	*1970 (actual)*	*1975*	*1980*	*1985*
United States (based on lower growth rate)	4.12	3.58	4.27	4.49
Canada	0.58	0.84	1.35	1.72
Latin America	1.93	2.12	2.56	2.85
Western Europe	0	0.55	1.09	1.46
North Africa	1.64	1.90	2.19	2.56
West Africa	0.91	1.39	1.83	2.37
Middle East	6.21	10.95	14.78	18.43
Far East/Oceania	0.73	1.10	1.46	2.01
Total non-Communist world supply	16.31	22.41	29.53	35.88
Total non-Communist world demand	14.60	20.28	26.55	32.09

Sources: U.S. Energy Outlook, *op. cit.*; Nichols, *op. cit.*

TABLE 4.8 ***Estimated Western Hemisphere Oil Production and Consumption (1960–1985)***
(In billions of barrels per year)

	1960	*1965*	*1970*	*1975*	*1980*	*1985*
Oil consumption (excluding exports)						
United States	3.58	4.20	5.37	6.68	8.14	9.42
Canada	0.33	0.40	0.55	0.69	0.84	1.04
Latin America	0.62	0.77	1.02	1.42	1.86	2.46
Total	4.53	5.37	6.94	8.79	10.84	12.92
Oil production						
United States	2.92	3.29	4.12	3.58	4.23	4.31
Canada	0.18	0.33	0.55	0.80	1.10	1.35
Latin America	1.39	1.72	1.93	2.12	2.45	2.56
Total	4.49	5.33	6.60	6.50	7.77	8.21
Oil production minus consumption	−0.04	−0.04	−0.34	−2.29	− 3.07	− 4.71

Sources: U.S. Energy Outlook, *op. cit.;* Nichols, *op. cit.*

Tables 4.7 and 4.8 show that the United States now has a great and growing dependence on the Middle East for oil to supplement its expected production of energy from gas, coal, and nuclear sources, as projected in 1972. But the folly of U.S. dependence on imported oil reflected in these tables was learned by bitter experience during the 1973–1974 embargo and the subsequent exorbitant increase in cost with its impact upon our economy and the unsolved problem of imbalance of payments with the Middle East countries. So we must go beyond the now obsolete viewpoint of 1972 and concentrate on developing self-sufficiency in the United States. Fortunately, the U.S. energy resources potential is large and could sustain higher rates of production of all fuels.

The long-term oil potential for the United States is shown in Table 4.9.

The shale oil potential is not included in the preceding estimates of ultimate oil resources. Since so great a portion of the oil shale lies on public lands, reasonable access to these lands must be allowed if the resources are to be explored and developed. The maximum production of 750,000 bbl/day that can be developed by 1985 will require less than 6 billion of the 129 billion bbl high-quality oil shale deposits. With existing technology and in the economy of today this development would require an estimated capital investment of several billion of dollars in plants.

Tar sands[1] contain 0 to 18% heavy bitumen and sand, silt, and water. The oil content of such sands is lower than that of oil shale, but large deposits containing up to 14% bitumen have been identified. Athabasca (Canada) sands alone have been estimated to contain 2.85×10^{11} bbl of recoverable oil. This is enough to supply the total energy requirements of the United States for 23 years at the 1970 consumption rate, so deposits are not trivial. Mining and removing the tar in an environmentally acceptable way under arctic conditions is proving to be difficult.

Lignite is a fuel intermediate in composition between peat and coal. Large, accessible deposits occur in South Dakota, and there are significant deposits in seven other states. Estimated reserves are 448,083 million short tons,[2] representing 59% of the present bituminous coal reserve and 104% of the subbituminous reserve. This material is not widely used at present, although a commercial barbecue briquette process[3] is based on it, but it is potentially of great value. Some lignites contain significant amounts of uranium.

[1]Hottel and Howard, New Energy Technology, MIT, 1971.
[2]Minerals Yearbook 1972, Dept. of the Interior, 1974.
[3]*Chem. Eng.* (*N.Y.*), **70**(15), 108 (1963).

TABLE 4.9 ***Oil-in-Place Resources in the United States*** *(In billions of barrels)*

Region	*Ultimate discoverable oil in place*	*Oil in place discovered to 1/1/71*
Lower 48 states, onshore	561.8	384.9
Offshore and South Alaska	128.6	16.3
Alaska, north slope, onshore, and offshore	120.0	24.0
Total	810.4	425.2

Sources: U.S. Energy Outlook, *op. cit.;* Nichols, *op. cit.*

Peat (wet, partially decomposed organic matter) deposits are worked in other parts of the world, but are unlikely to become a significant power source in the United States because the energy content per unit weight is not great, deposits are mainly in isolated areas, and the drying problem is severe.

GASEOUS FUELS Gas is burned as a source of heat in domestic installations and also in industry, especially where it is obtained as a by-product. Blast-furnace gas resulting from the smelting of iron is an outstanding example of a by-product gas employed for heating the blast stoves, with the remainder burned under boilers or for heating coke ovens. Fuel gases are discussed in Chap. 6, where a tabulation is given of Btu values and other properties.

The long-term gas potential for the United States is shown in Table 4.10.

The reserves of U.S. oil and gas are not inexhaustible. The cost involved in finding, developing, and supplying the quantity of oil needed through 1985 will probably increase sharply over the intervening years. It is well recognized that there is not an endless supply of so-called low-cost oil. Increased crude oil-producing capacity will be more costly, since much of the producing capacity will come from offshore and arctic regions involving higher costs of exploration, production, transportation, and meeting stringent environmental standards.

COMBUSTION Most modern industrial plants burn coal either on mechanically operated grates and stokers or in the pulverized form. These present-day procedures allow the ratio of air to fuel to be properly controlled, thus ensuring efficient combustion and reducing heat losses through stack and ash. When fuel oil is burned, it is frequently necessary to provide heaters to lower the viscosity of the oil sufficiently for proper burner operation. The flue-gas analysis is valuable in controlling combustion, since the proportions of CO_2, CO, and O_2 in the flue gas will indicate incomplete combustion or excess air.

TABLE 4.10 ***Recoverable Gas Supply in the United States*** *(In trillions of cubic feet)*

Region	*Ultimate discoverable gas*	*Gas discovered to 1/1/74*
Lower 48 states, onshore	963.1	413.1
Lower 48 states, offshore	260.1	45.9
Alaska	277.4	5.1
Total	1500.6	464.1

Sources: U.S. Energy Outlook, *op. cit.;* Nichols, *op. cit.*

POWER GENERATION

Near the end of the seventeenth century the first successful attempt to generate steam under pressure in an enclosed vessel was made. Since then the use of steam has increased to such an extent that today steam furnishes the major portion of the total power developed in this country. The most recent developments have been toward improvements in boiler construction, with the aim of producing higher-pressure steam, in central power stations. Such high pressures increase overall efficiency in the production of electric power. The limiting factor is the failure of materials at the high temperatures and pressures involved. Earlier, $4\frac{1}{2}$ to 5 lb of the best coals was required per kilowatthour, whereas currently, some power plants produce a net kilowatthour from 10 oz of coal. Figure 4.1 is the elevation of a large utility water-tube boiler.

There are two main types of boilers, the fire tube and the water tube (Figs. 4.1 and 4.2). The *fire-tube boiler* is usually of small or medium capacity and is designed for the generation of steam at moderate pressure. In this type of boiler the fire passes through the tubes. Fire-tube boilers have a

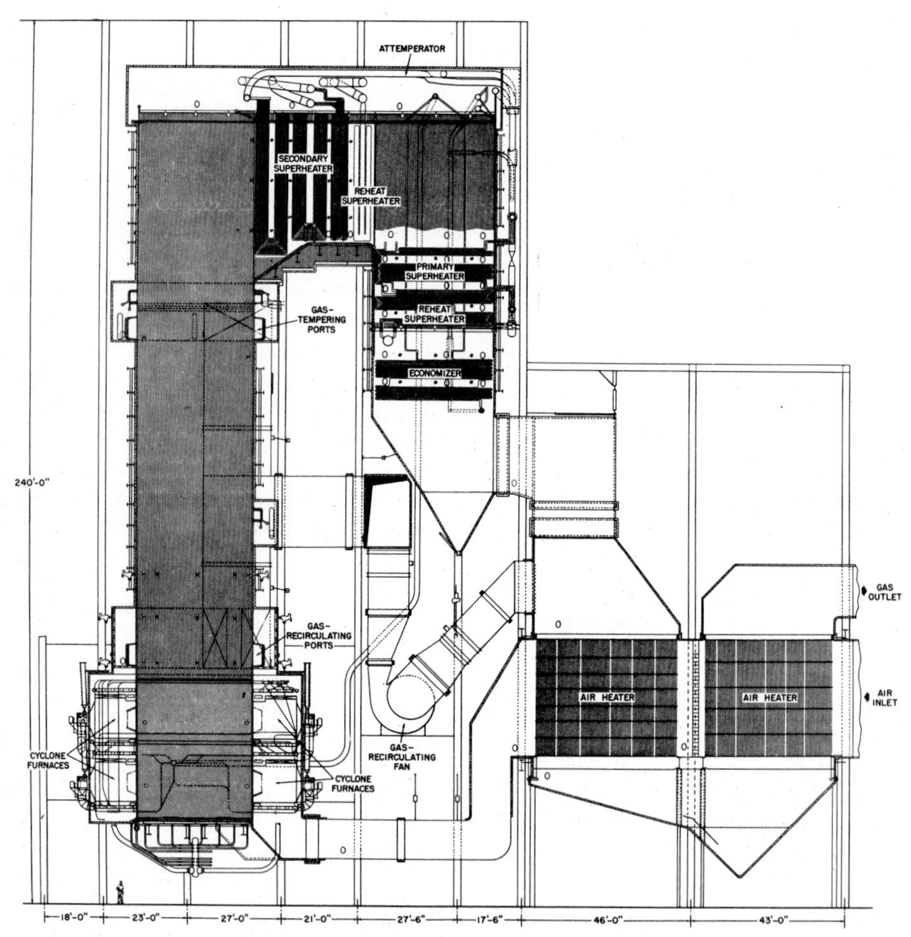

Fig. 4.1 Elevation of a large utility water-tube boiler. Capacity 4.3 million lb/h; outlet pressure 2,620 psi; temperature 1005°F; height 240 ft. (*Babcock & Wilcox Co.*)

Fig. 4.2 Portable water-tube boiler. (*Babcock & Wilcox Co.*)

low first cost and a relatively large reservoir of hot water. This is of particular advantage in small chemical plants, where there may be a sudden demand on the steam plant. Locomotive boilers are also of this type.

Water-tube boilers are used almost exclusively in stationary installations where service demands a large amount of evaporation at pressures above 150 psi. The water is in the tubes and can be converted to steam more quickly than with the fire-tube boiler ("quicker steaming"). High efficiencies are obtained by this type. Boiler feedwater should be conditioned before introduction into the boilers (Chap. 3). Poor boiler water results in foaming, caustic embrittlement, corrosion, and scale formation, with a consequent loss of steam made and in efficiency. The higher the pressure, the more important the use of properly conditioned water.

Figure 4.2 represents a portable, shop-assembled, package modern water-tube boiler used for generating steam at 250 psi, with a capacity of up to 100,000 lb of saturated steam per hour.

ELECTRIC POWER FROM STEAM Steam is often regarded as being made solely for the generation of power. However, in the process industries, particularly in those pertaining to the making of chemicals, so much heat is obtained by condensation of steam that the dual use of steam for power and for heat is of paramount importance. An economical engineer endeavors to coordinate and balance the two. In these chemical processes, the object should be to expand the steam through the steam engine or turbine and then employ the exhaust or lower-pressure steam for heating purposes. In this procedure the prime movers are run noncondensing. For turbines, where dry steam is necessary to prevent blade erosion, the entering steam is superheated and condensation prevented. The turbine is a particularly flexible prime mover for chemical plants, since it can be designed not only to furnish low-pressure exhaust steam but, by bleeding off, to supply steam at higher presssures.

HEAT BALANCE The economics of the dual use of heat and power concerns the desirability of coordinating the generation of steam for power and for process heat so that the power needed is obtained as a by-product of the process steam demand. The first step in balancing these two energy demands should be a careful, accurate survey of the heat and power requirements of the various processes. Since superheat usually retards the rate of heat transfer in process operations, the condition of the steam at the turbine exhaust or at the bleeding point should be such that it will have sufficient superheat to overcome transmission losses but yet have little or no superheat when it reaches the point of use. This will result in full utilization of the latent heat of steam for heating purposes.

The fundamental importance of this principle of coordination of the steam for power and the steam for process heating can be seen by inspection of Table 4.11, which gives the Btu converted to power by expansion of steam at various pressures and superheat to saturated steam at 15 psig, as contrasted with the Btu supplied by the condensation of this exhaust steam. As process steam must be furnished for heat-transfer operations, it may be seen from Table 4.11 that the power available by expansion is probably the most economical source of power for a plant. Consequently, the ideal plant expands all its steam through turbines for power and then leads the exhaust steam into the factory for the various heat-transfer operations. When there is no contamination of this process steam, the water from the condensation should be brought back for the boiler feed makeup water. All chemical industries should be studied for this dual use of steam. Those pertaining to the manufacture of sugar (Chap. 30), caustic soda (Chap. 13), and salt (Chap. 12), and to fermentation (Chap. 31), have applied this dual-energy balance very profitably. To balance the power required, obtained from the expansion of steam, with the process demands for steam exhausted from the steam engine or turbine, so that there is no excess of either, should be the aim of the engineers in charge of any plant. This ideal can frequently be realized in the chemical process industries.

HEAT TRANSMISSION OTHER THAN BY STEAM Where indirect heating is necessary, some method other than steam should be used for temperatures above about 400 or 450°F, where the pressure of steam becomes too high for economical design of the plant. Since steam owes its importance as a heating medium to its convenience and cleanliness, but particularly to its large heat of condensation, the use of superheated steam is not indicated as a heating medium except in rare cases. Hence, where the conditions lie outside the range of steam, the engineer turns progressively, as the temperature rises, to such other means as

Direct firing, above 300°F: low cost and convenient but a fire hazard.
Indirect firing, above 300°F: low operating cost but elaborate setting and a fire hazard.
Direct gas heating, above 300°F: moderate cost and excellent control but a fire hazard.
Hot oil, 300 to 600°F: good control but high first cost and oil carbonization.
Dowtherm, 400 to 750°F: good control and moderate operating cost but higher first cost.
Mercury vapor, 600 to 1200°F: good control and moderate operating cost but highest first cost.
Mixed salts, 250 to 900°F: good control and good heat transfer at high temperatures.
Electricity, above 300°F: most accurate control but operating cost usually high.

Direct or indirect heating with coal or gas is frequently surprisingly efficient when the furnace is well designed; however, the open flame may be a fire hazard. Under conditions where oil is not carbonized, this heat-transfer medium has been so widely employed that furnaces for heating the oil and equipment for the heat transfer are both available on the market as standardized and tested designs. Dowtherm (diphenyl 26.5%–diphenyl oxide 73.5%, eutectic mixture) is stable at higher temperatures than oil and has the added advantage that it can be employed as a vapor, where its latent heat of condensation can be used as well as its sensible heat. Mercury is very successful in controlling the heat of reaction in certain reactors. For years, mixtures of inorganic salts have been accepted as heat-transfer media, but the modern large-scale demand for such salts to *remove* heat from petroleum cracking processes such as the Houdry catalytic one was necessary to justify a careful study of the properties of the mixed salt consisting of approximately 40% $NaNO_2$, 7% $NaNO_3$, and 53% KNO_3. Strange to say, tests indicated no danger even when a stream of hot petroleum was injected into the molten nitrate-nitrite bath. Electricity in contact, immersion, and radiant heaters are most convenient, accurate, safe, and efficient, though generally costly, heating media.

NUCLEAR ENERGY Table 4.12 indicates that U.S. uranium resources minable at reasonable cost are adequate to meet requirements through 1985 if development is not hindered. After 17 years of commerical nuclear power use in the United States and more than 30 years of military nuclear

TABLE 4.11 ***Comparison of Energy from the Isentropic Expansion of Steam to an Exhaust Pressure of 29.7 psia (15 psig) for Power and Heat Transfer Processes***

Pressure							*End of expansion*				
Psia	*Psig*	*Saturation temperature, °F*	*Degrees of superheat*	*Vapor temperature, °F*	*Enthalpy, BTU/lb*	*Entropy, BTU/(lb)(°F)*	*Enthalpy, BTU/lb*	*Moisture content, %*	*Temp., °F*	*Work of expansion, BTU/lb*	*Heat available for processes, BTU/lb*
400	385.3	444.70	500	944.70	1494	1.742	1195	....	313	299	977
400	385.3	444.70	400	844.70	1442	1.704	1164	....	251	278	946
350	335.3	431.82	350	781.82	1408	1.692	1156	0.6	....	252	938
300	285.3	417.43	500	917.43	1484	1.766	1215	....	352	269	997
300	285.3	417.43	300	717.43	1378	1.682	1150	1.4	....	228	932
250	235.3	401.04	250	651.04	1345	1.674	1145	2.0	....	200	927
200	185.3	381.86	400	781.86	1415	1.758	1207	....	335	208	989
200	185.3	381.86	200	581.86	1312	1.667	1140	2.5	....	172	922
150	135.3	358.48	150	508.48	1278	1.664	1140	2.5	....	138	922
100	85.3	327.86	200	527.86	1294	1.722	1180	....	280	114	962
100	85.3	327.86	100	427.86	1243	1.669	1141	2.5	....	102	923
50	35.3	281.03	200	481.03	1274	1.779	1223	....	370	51	1005
50	35.3	281.03	90	371.03	1220	1.718	1178	....	276	42	960

Source: Data computed from Keenan, Keyes, Hill, and Moore, Steam Tables—Thermodynamic Properties of Water Including Vapor, Liquid, and Solid, Wiley, 1969. Superheated steam is used to prevent erosion of turbine blades by excess moisture in the steam.

TABLE 4.12 *Resources of Uranium U_3O_8 in the United States as Estimated by the Atomic Energy Commission Based on the Cost of Production as of January 1, 1972*

Cost of production, $/lb	*Reasonably assured reserves, tons*	*Estimated additional potential reserves, tons*	*Total*
8 or less	273,000	460,000	733,000
10 or less	423,000	650,000	1,073,000
15 or less	625,000	1,000,000	1,625,000

Sources: U.S. Energy Outlook, *op. cit.*; Nichols, *op. cit.*

power use, there has not been a single injury or monetary loss to a member of the public. No other industry can boast such a fine safety record.

The fast breeder reactor appears to be an enormous source of energy for the future. In the conventional reactor the U_3O_8 is burned up and no useful by-product recovery is made. In contrast, the breeder reactor generates plutonium.

The French constructed a 250,000-kW fast breeder plant in 1973, and the British Atomic Energy Authority built a breeder reactor at Dounreay in northern Scotland. These early reactors were followed by others in Russia, Belgium, and the Netherlands. Progress made in foreign countries has relegated the United States to a low status in the world drive to develop technology for the second generation of nuclear power.

HYDROELECTRIC POWER Many chemical industries require large amounts of low-priced electric power for their operation. These companies usually locate near a hydroelectric plant where cheap power is available, especially to the large, steady, day-and-night consumer. Such industries are the electrochemical ones covered by Chap. 14 on Electrolytic Industries and Chap. 15 on Electrothermal Industries. These hydroelectric plants are situated where a head of water is available either from a waterfall or from a dam. This water is used to drive a water wheel of some sort, to which is attached either a direct- or alternating-current generator. The initial cost of a hydroelectric plant is much greater than that of a corresponding steam power plant, but the running cost is much lower.

OTHER POSSIBLE SOURCES OF ENERGY

FUEL CELLS[4] These devices for the "direct" generation of electric energy have very intriguing possibilities. In a fuel cell a fuel such as hydrogen, natural gas, or propane can be converted directly without moving parts into twice the amount of electricity that would result from the burning of a corresponding amount of fuel indirectly through boilers, turbines, and generators to produce electricity. Efficiencies would be 40 to 80%, versus 25 to 40%.

The fuel cell differs from the storage battery in that usually its gaseous or liquid fuel and its oxidizer are *led* in from outside, whereas the storage battery *stores* its solid fuel and oxidizer on plates, where they are consumed. A fuel cell operates electrochemically or, more literally, "chemicoelectrically," as shown in the cell diagramed in Fig. 4.3. This cell is actually a reactor, wherein hydrogen as the feed stream or fuel is conducted into the empty space paralleling the porous, electric conducting anode. This anode can be made of porous carbon with a metal catalyst like platinum, which *chemically changes the hydrogen atoms to positively charged hydrogen ions and electrons.* The electrons leave the anode, perform work, and enter the cathode (Fig. 4.3). Meanwhile, the positively

[4]Sweeney and Heath, Fuel Cell: Its Promise and Problems, API, Houston, Tex., May 9, 1961: Schultz, Fuel Cell Developments, Society of Petroleum Engineers of AIME, Dallas, Tex., Oct. 8, 1961; Institute of Gas Technology, Chicago (38 refs.); Moos *et al.*, Fuel Cells, *Ind. Eng. Chem.*, **54**(1), 65 (1962) (good review and summary); Fuel Cells: Fact and Fiction, *Chem. Eng. (N.Y.)*, **81**(11), 62 (1974).

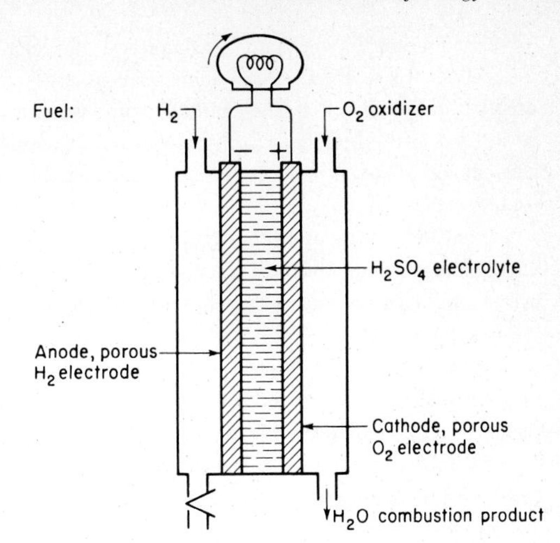

Fig. 4.3 A hydrogen-oxygen fuel cell in which the current is carried internally via mobile hydrogen ions. (*Sweeney and Heath.*)

charged hydrogen ions migrate through the electrolyte (for instance, 50% KOH or H_2SO_4), attracted by the oxygen from the cathode. To complete the reaction, the oxygen pulls in the electrons also, and water is generated and discharged from the cell. The cell is so arranged that the electrons must move up the anode, leave the cell through a wire, and enter the cell again at the cathode. While the electrons are outside the cell, they form the electric current and do work.

$$H_2 \xrightarrow{\text{anode}} 2H^+ + 2e$$

$$O_2 \xrightarrow{\text{cathode}} 2O$$

$$2H^+ + O \text{ (or } \tfrac{1}{2}O_2) + 2e \longrightarrow H_2O$$

Cell overall:

$$H_2(g) + \tfrac{1}{2}O_2(g) \xrightarrow[\text{2 electrons}]{25°\text{C, 1 atm}} H_2O(l) \qquad \Delta F = -56.7 \text{ kcal}$$

$$H_2(g) + \tfrac{1}{2}O_2(g) \xrightarrow{25°\text{C, 1 atm}} H_2O(l) \qquad \Delta H = -68.4 \text{ kcal}$$

$$\text{Theoretical efficiency} = \frac{56.7}{68.4} \times 100 = 82.9\%$$

The decrease in free energy indicates that the hydrogen and oxygen want to react: the fuel cell provides a mechanism for this. Hydrogen-oxygen fuel cells have been operated at about 25°C, but a single cell produces only about 1 V of direct current. Therefore many cells are needed in series to produce a useful current.

Natural gas is available in many places in the United States and elsewhere at about one-sixth the cost for its energy in comparison with electricity supplied from power companies. Thus it is much cheaper to consume natural gas or propane than to use hydrogen or a mixture of hydrogen and carbon monoxide obtained by reacting natural gas with carbon-producing synthesis gas. It is also much more economical to use air rather than oxygen as the oxidizer. Many other chemicals are being studied for use in fuel-cell operation. The questions to be resolved are relative costs and the actual design of an efficient cell. The fuel cell is one of the most intriguing devices for the production of useful energy now being investigated, particularly for space applications. Their use in the production of partially oxidized products, as well as electricity, is being investigated. An example is the oxidation of methanol for formic acid.

Examples of fuel cells under commercial development are shown in Fig. 4.4 and Fig. 4.5. The fuel cell shown in Fig. 4.4 can achieve about 40% efficiency in generating electricity. When it is used for an apartment or office building with heat recovery, over 80% of the fuel's heating value can be utilized in the building. Conventional power plants utilize only about 30% of the energy. Fuel cells should make an extremely important contribution toward solving energy and environmental problems of the future.

FUSION Ultimately the future of humankind may depend upon the fusion process. In an energy-hungry world that will eventually run out of fossil fuels, and one beset by organized concern over harmful by-products of fission reactors, fusion offers one of the few well-founded hopes for low-risk production of almost unlimited amounts of energy.

A nuclear fusion reactor is a device in which controlled, self-sustaining nuclear fusion reactions may be carried out in order to produce useful power. The reaction is carried out in a very hot but tenuous fuel gas mixture of hydrogen, helium, or lithium isotopes.

GEOTHERMAL ENERGY SOURCES Whether geothermal energy will materialize on a large-scale basis is an open question. Only in Italy, Mexico, New Zealand, and the United States are power stations in operation. All of these operate at a very modest generating rate of a few megawatts. Donald E. White estimated that the world's ultimate geothermal capacity down to a depth of 10 km is roughly 1.11×10^{14}/KWh. This is indeed a negligible amount in comparison with coal, oil, gas, and nuclear fuels.

PIPELINE HYDROGEN Hydrogen may become an important secondary energy form if problems of generating, storage, and transportation can be solved. Like electricity, hydrogen requires some other source of energy for its production. However, once produced, hydrogen holds the promise of significant technical and economic advantages over other alternatives for various applications as

Fig. 4.4 United Technologies Powercel 18, 40-kW experimental fuel cell generator, operating on natural gas and air, achieves 40% overall efficiency at part load. Unit is designed for on-site installations. (*United Technologies Power Systems Div.*)

Fig. 4.5 Model of United Technologies 26-MW dispersed generator under development for electric utility application, designed to generate electricity at 40% efficiency. It will use naphtha fuel. (*United Technologies Power Systems Div.*)

the United States becomes increasingly short of conventional fossil fuels and more dependent on other energy sources such as nuclear power.

SOLID WASTE ENERGY Increases in energy prices, escalating hauling costs, and decreasing availability of landfill sites make the incineration of solid waste and heat-recovery union an attractive alternative to conventional solid-waste disposal methods. Table 4.13 shows the approximate heating value and density of various solid wastes.

WIND ENERGY Few engineers feel that windmills and other wind machines are capable of producing large amounts of energy for the future. The large amount of material required to construct such machines, their low output per unit of investment, and their unreliability make installations unpromising.

WATER AND TIDAL POWER Water and tidal power resources would probably yield a few tenths of 10^{18} Btu. In general, these power sources would be of regional interest but are not an option for large-scale energy supply.

SOLAR ENERGY The problem with using solar power for large-scale operations is low power density. The solar input above the atmosphere averaged over day and night in all zones of the globe amounts to 340 W/m^2. Of this, 47% actually reaches the earth. An estimate of the global average for utilizing solar power is 20 W/m^2. The average home receives on its roof about 500 times more energy from the sun than the electricity it uses. The difficulty lies in utilizing and storing this energy day and night during the entire year. Actually, temperatures of thousands of degrees Fahrenheit have been obtained from sunlight-concentrating mirrors. Even now, solar stoves in warm countries could save precious fossil fuel or heat water on the roofs (as some Japanese do). It is to be hoped that scientific studies on the photosynthesis of plants through chlorophyll or through *Chlorella* algae

TABLE 4.13 *Approximate Heating Value and Density of Some Representative Wastes*

	Btu per pound (as fired)	*Pounds per yard (as discarded)*
Bitumen waste	16,570	1,500
Brown paper	7,250	135
Cardboard	6,810	180
Cork	11,340	320
Corrugated paper (loose)	7,040	100
Hardboard	8,170	900
Latex	10,000	1,200
Meat scraps	7,600	400
Nylon	13,620	200
Paraffin wax	18,620	1,400
Plastic-coated paper	7,340	135
Polyethylene film	19,780	20
Polypropylene	19,860	100
Polystyrene	17,700	175
Polyurethane foam	17,580	55
Resin-bonded fiber glass	19,500	990
Rubber	14,610	1,200
Shoe leather	7,240	540
Tar paper	11,500	450
Textile nonsynthetic waste	8,000	280
Textile synthetic waste	15,000	240
Vegetable food waste	1,800	375
Wood	9,000	300

Source: Chem. Eng. (*N.Y.*), **81**(18), 38 (1974).

(3% efficiency) will lead to practical solar energy utilization. Because both conventional and solar systems must be installed to obtain year-around reliability, investment tends to be excessive per unit of utilizable power.

AIR CONDITIONING

The use of air conditioning in industrial plants has become more and more common in the past few years. Control of the temperature, humidity, and cleanliness of the air is very important in many chemical processes, particularly in artificial-fiber and paper manufacture. Textile fibers are quite sensitive to changing conditions of the air. The comfort of workers is also important to efficient industrial organizations. This latter fact has led to the use of air conditioning in plants and offices where it is not essential for product quality.

REFRIGERATION

Refrigeration is the process of producing cold, particularly referring to cooling below atmospheric temperature. It is a vital factor in many chemical processes where cold or the removal of heat is necessary for optimum reaction control. Examples are the manufacture of azo dyes, the separation of an easily frozen product from liquid isomers or impurities, and the food and beverage industries. Further examples are the catalytic manufacture of ethyl chloride from liquid ethylene and anhydrous hydrogen chloride under pressure and at $-5°C$, the production of "cold" rubber by polymerization

TABLE 4.14 ***Properties of Refrigerating Agents***

Refrigerant	*Boiling point at 760 mm, F*	*Critical temperature, F*	*Critical pressure, atm*	*Latent heat at 760 mm, Btu/lb*
Ammonia (NH_3)	−28	270.5	111.5	589
Carbon dioxide (CO_2)	−108.8*	88.0	73.0	126†
Sulfur dioxide (SO_2)	+14	315.0	77.7	167
Methyl chloride (CH_3Cl)	−11	289.6	65.8	184
Ethyl chloride (C_2H_5Cl)	+55	369.0	52.0	168
Freon-12 (CCl_2F_2)	−21	232.7	39.6	71.9
Propane (C_3H_8)	−44.4	206.2	42.0	159

*Sublimes. †Latent heat at −20°F and 220.6 psia.

at 41°F or lower, and the freezing of mercury at −100°F into complex molds which are coated by repeated dipping in a ceramic slurry, the mercury then being allowed to melt and run out. Refrigeration operations involve a change in phase in a body so that it will be capable of abstracting heat, exemplified by the vaporization of liquid ammonia and the melting of ice. Mechanical refrigeration can be divided into two general types: the *compression system* and the *absorption system.* Both systems cause the refrigerant to absorb heat at the low temperature by vaporization and to give up this heat at the higher temperature by condensation. The absorption system is used mainly in household units, but it finds economical industrial application where exhaust steam is available.

An ammonia refrigerating plant is a typical illustration of the vapor-compression system and is depicted in Fig. 4.6. Table 4.14 gives properties of usual refrigerating agents.

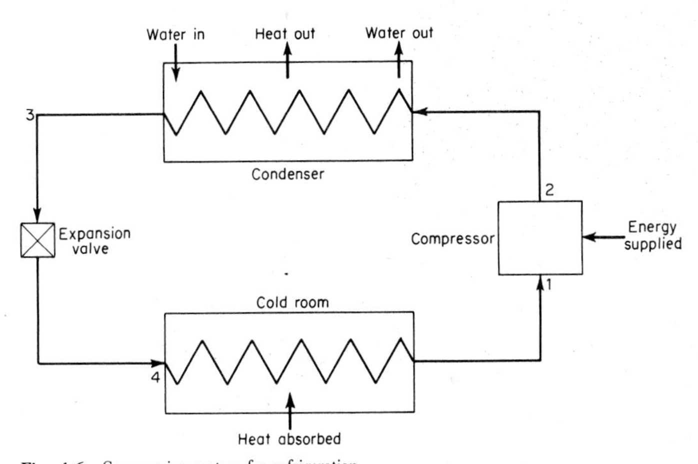

Fig. 4.6 Compression system for refrigeration.

SELECTED REFERENCES

Berman, E. R.: Geothermal Energy 1975, Noyes, 1975.
Daniels, F.: Direct Use of Sun's Energy, Yale University Press, 1965.
Frances, D.: Fuels and Fuel Technology, Pergamon, 1965, 2 vols.

Industrial Fuel Markets 1973–1974, Cahners Publishing, 1973.
Inglis, K. A.: Energy: From Surplus to Scarcity?, Halsted, 1974.
Jackson, F. R.: Energy from Solid Waste, Noyes, 1974.
Laub, J. M.: Air Conditioning and Heating Practice, Holt, 1963.
Macrakis, M. S. (ed.): Energy: Demand, Conservation, and Institutional Problems, M.I.T., 1974.
Miliaras, E. S.: Power Plants with Air-Cooled Condensing Systems, M.I.T., 1974.
Mitchell, W., Jr. (ed.): Fuel Cells, Academic, 1963.
Palmer, H. B., and J. M. Beer (eds.): Combustion Technology, Academic, 1974.
Penner, S. S., and L. Icerman: Energy, vol. I, Demands, Resources, Impact, Technology, and Policy, Addison-Wesley, 1974.
Rickles, R. N.: Energy in the City Environment, Noyes, 1973.
Rimberg, D.: Utilization of Waste Heat from Power Plants, Noyes, 1974.
Szilard, J. A.: Descaling Agents and Methods, Noyes, 1972.
Tooley, P.: Fuels, Explosives and Dyestuffs, J. Murray, London, 1971.
Unconventional Power Sources, Office of Technical Services, 1962.
U.S. Energy Outlook: New Energy Forms, National Petroleum Council, 1973.
William, C. R.: Energy: From Nature to Man, McGraw-Hill, 1974.
Williams, J. R.: Solar Energy: Technology and Applications, Ann Arbor Publishers, 1974.
Young, G. J., and H. R. Linden (eds.): Fuel Cell Systems, ACS, 1965.
Ziel, R., and M. Eagleson: The Twilight of World Steam, Grosset & Dunlap, 1973.

chapter 5

COAL CHEMICALS

Chemicals from coal were initially and mostly obtained by destructive distillation of coal, furnishing chiefly aromatics. In recent years substantial production of aromatics, particularly benzene, toluene, xylene, naphthalene, and methylnaphthalenes, has been obtained by processing petrochemicals (Chap. 38). With the advancing application of chemical conversion of coal, many more chemicals can be made from coal whenever it is economical to do so. However, coal chemicals, except for metallurgical coke, are now in a very competitive field. Table 5.1 summarizes chemicals produced from coal as presented in this chapter and elsewhere in this book (see also Fig. 5.2 for many coal chemicals). Coal is not only the country's fundamental fuel, but shares with petrochemicals the furnishing of the basic raw materials for many essential industries from dyes, medicines, pesticides, and elastomers to modern plastics. "Coal also forms the world's largest reserve of concentrated organic raw materials, and it serves not only as a chemical supplier but as a cheap source of heat and power needed for processing."[1] Although gas from coal and aromatics from coal have had production curtailed by petrochemical and natural-gas competition, much new research and development is being conducted by the Bureau of Mines, Office of Coal Research, and by private industry. In Chap. 6, on Fuel Gases, reference is made to the intense efforts to manufacture a fuel gas of high Btu to compete even now for local and peak demands for such gas. These efforts may also supply a substitute for natural gas when its supply diminishes in the decades ahead. Other research is directed to making new or old chemicals from coal, lignite, and shale through improved technology and to continuing investigations with the aim of eventually supplying competitively from coal, motor fuel, and other organic chemicals.[2]

THE DESTRUCTIVE DISTILLATION[3] OF COAL

When coal is thermally pyrolyzed or distilled by heating without contact with air, it is converted into a variety of solid, liquid, and gaseous products. The nature and amounts of each product depend upon the temperature used in the pyrolysis and the variety of coal. In ordinary practice, coke-oven temperatures are maintained above 1650°F but may range anywhere from 950 to 1800°F. The principal product by weight is coke. If a plant uses temperatures from 850 to 1300°F, the process

[1]Rose, H. J., private communication; OCR Calls for Petrochemicals from Coal, *Chem. Eng. News*, Jan. 6, 1969, p. 40.

[2]Coal Gasification: Just Shy of Commercial, *Chem. Eng. News*, May 19, 1969, p. 11; Perry, Coal Conversion Technology, *Chem. Eng.* (*N.Y.*), **81**(15), 88 (1974); Eyes Stay on Coal Conversion, *Chem. Eng.* (*N.Y.*), **81**(15), 58 (1974); Burke, They're Making a Solid Effort to Get Clean Coal Liquids, *Chem. Week.*, Sept. 11, 1974, p. 38.

[3]Lowry, Chemistry of Coal Utilization, Suppl. Vol., chap. 11, pp. 461–494, Wiley, 1963 (121 refs.); High Temperature Carbonization, vols. 1 and 2, Wiley, 1945 (hundreds of early references); ECT, 2d ed., vol. 4, p. 400, 1964; Sass, Garrett's Coal Pyrolysis Process, *Chem. Eng. Prog.*, **70**(11), 72 (1974).

TABLE 5.1 *Chemicals from Coal*

Chemical conversion process	Products and procedures (numbers refer to chapters)
a. Carbonization, pyrolysis of coal, lignite, and carboniferous shales (destructive distillation)	Coal-tar aromatics, benzene and homologs, phenol and homologs, naphthalene, anthracene, phenanthrene, etc. (5) High-temperature coke (5), low-temperature coke (5) Carbon for pigments (8), carbon for electrodes (5,8), structural material (8), activated carbon (8)
b. Reduction and refining of ores	Iron, ferro alloys, etc., aluminum (14), magnesium (14)
c. Gasification (cf. *g* and *j*) Blue and producer gas are very minor	Coal gas (6), blue water gas (6), producer gas (6), peak gas (6), synthesis gas (7), C^*_2 liquid and dry ice (7)
d. Combustion for comfort heating and power generation	Electric power utilities (4), comfort heating, retail deliveries (4), fly ash (4), sintered ashes for filter and concrete (4)
e. Combustion for process heating	Heat for manufacture of lime (10), cement (10), ceramics (9), steel and rolling mills
f. Reduction, chemical (cf. *k*)	Sodium sulfite (11,12), sodium sulfide (12), barium sulfide (20), phosphorus (16)
g. Hydrogenation and hydrogenolysis, catalytic	Carbide process for aromatics (5), hydrogenation of coal Bergius process (5), Fischer-Tropsch liquid fuels (5,6), catalytic methanation of synthesis and pyrolysis gases (5,6)
h. Demethylation	Benzene from toulene or xylene (37), naphthalene from methylnaphthalenes (37)
i. Hydrolysis, alkaline	Mixed aromatics (5)
j. Oxidation, partial (controlled) and complete	Synthesis gas (7) for NH_3, CH_3OH, etc., hydrogen (7), coal acids (5), carbon monoxide (7)
k. Electrothermal	Graphite (8) and electrodes (8), abrasives: silicon carbide (15), calcium carbide (15), cyanamide (18), carbon disulfide (15,38)
l. Sulfur recovery	H_2S from gas (6,7) Pyrite from coal
m. Sulfonation	Ion exchange, water softeners (3)
n. Solvent extraction of coal	Ashless coal, montan wax (28), humic acids, coumarone resins (34)

Sources: Lowry, Chemistry of Coal Utilization, Wiley, vols. 1 and 2, 1945, Suppl. Vol., 1963; Future for Fossils, *Chem. Week*, Oct. 9, 1974, p. 46; Perry, *op. cit.* (several flowcharts); Iammartino, *Chem. Eng.* (*N.Y.*), **81**(21), 68 (1974); McMath *et al.*, Coal Processing: A Pyrolysis Reaction for Coal Gasification, *Chem. Eng. Prog.*, **70**(6), 72 (1974); Chemical Communication Shows Promise for Coal, *Chem. Eng. News*, Sept. 2, 1974, p. 16; UOP Set to Scale Up Oil-from-Coal Process, *Chem. Eng. News*, Sept. 23, 1974, p. 7.

is termed *low-temperature* carbonization; with temperatures above 1650°F it is designated *high-temperature* carbonization. In low-temperature carbonization the quantity of gaseous products is small and that of the liquid products is relatively large, whereas in high-temperature carbonization the yield of gaseous products is larger than the yield of liquid products, the production of tar being relatively low. The liquid products are water, tar, and crude light oil; the gaseous products are hydrogen, methane, ethylene, carbon monoxide, carbon dioxide, hydrogen sulfide, ammonia, and nitrogen. The products other than coke are collectively known as coal chemicals (*coproducts*, or *by-products*).

The destructive distillation of coal, or its carbonization, is really a striking example of *chemical conversion*, or the *unit process of pyrolysis*. This chapter outlines the equipment needed to carry

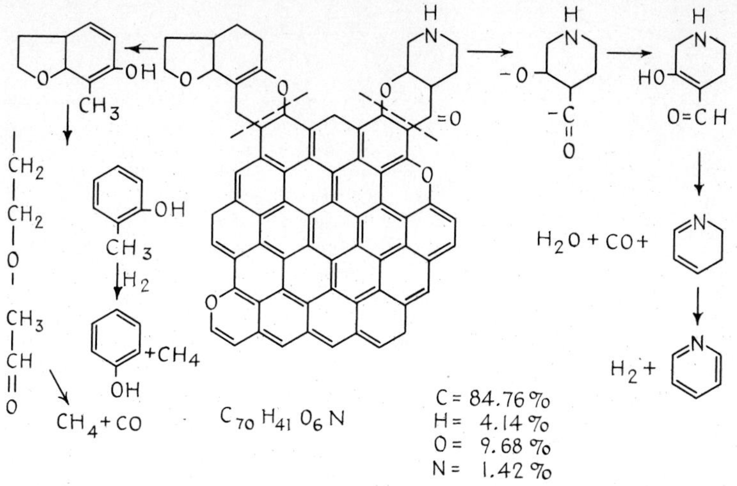

Fig. 5.1 Example of coal pyrolysis. (*After Fuchs and Sandhoff.*)

out, on a commercial scale, the basic chemical changes that take place. The chemical theory of the pyrolysis of coal[4] indicates the following step-by-step decomposition:

1. As the temperature is raised, the aliphatic "carbon to carbon bonds are the first to break."
2. "Carbon to hydrogen linkages are severed next as the temperature of 600°C (1100°F) is approached and exceeded."
3. "The decompositions during carbonization are essentially reactions effecting the elimination of heterocycle complexes and progressive aromatization."
4. "The average molecular weights of the volatile intermediate products constantly decrease as the temperature of carbonization rises. This decrease is marked by the evolution of water, carbon monoxide, hydrogen, methane, and other hydrocarbons."
5. "Final decompositions are at a maximum between 600 and 800°C," (1110 and 1470°F). Fuchs and Sandhoff give several examples of coal pyrolysis, of which Fig. 5.1 is typical.

Hill and Lyon[5] suggest that "coal consists of large heterocyclic nuclei-monomers, with alkyl side chains held together by three dimensional C—C groups, and includes functional oxygen groups."

HISTORICAL It is known that coke was an article of commerce among the Chinese over 2,000 years ago, and in the Middle Ages it was used in the arts for domestic purposes. Nevertheless, it was not until 1620 that the production of coke in an oven was first recorded. Up until the middle of the nineteenth century, coal tar and coal-tar products were regarded as wastes. Synthesis of the first coal-tar color, by Sir William Perkin in 1856, caused a great demand for crude coal tar, and it became a commercial product of increasing value. Perkin, with his discovery of the brilliant violet dye mauve (Chap. 39), while attempting the synthesis of quinine through oxidation of crude aniline in England, laid the foundation of the world's coal-dye industry. In 1792 the first successful experiment involving the production of gas from coal was carried out by William Murdock, who made it possible to light the streets of London with gas in 1812. The first battery of Semet-Solvay ovens was erected in Syracuse, N.Y., in 1893.

[4]Fuchs and Sandhoff, Theory of Coal Pyrolysis, *Ind. Eng. Chem.*, **34**, 567 (1942); Lowry, *op. cit.*, Suppl. Vol., pp. 379–384.

[5]Hill and Lyon, A New Chemical Structure for Coal, *Ind. Eng. Chem.*, **54**(6), 37 (1962); Lowry, *op. cit.*, Suppl. Vol., pp. 233–270, 290–295.

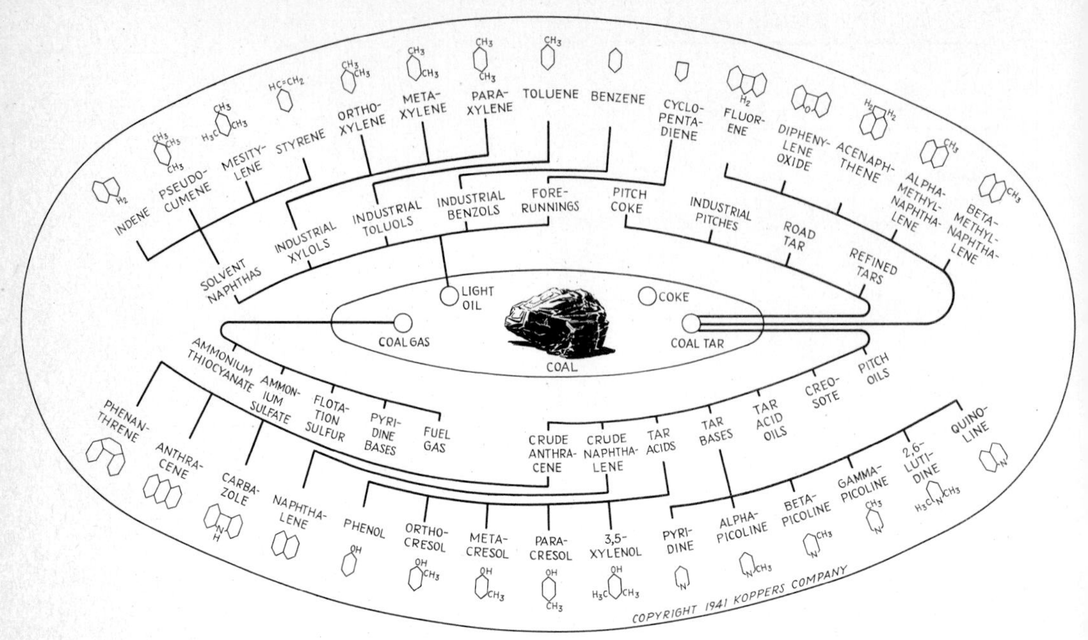

Fig. 5.2 How many grandchildren has a lump of coal? (*Koppers Co., Inc.*)

USES AND ECONOMICS *Coke* is the product of largest tonnage from the distillation of coal. The demand for coke depends on the demand for steel, so the amount of coal-tar production reflects the demand for steel. About 98% of coal-tar production is from coproduct ovens.[6] See Table 5.2 for data on crude aromatics from coal tar and petroleum. Until recent years aromatic crudes and the pure compounds derived therefrom were based on coal tar. Now that the petroleum industry supplies so many aromatics and natural gas through its manifold pipelines, supplanting other fuel gases, the chief incentive in distilling coal is to supply coke for steel. The *liquid products,* comprising coal tar and ammonia liquor, are not so large in volume as the solid products of coal distillation, but are of importance to chemical-recovery ovens. A considerable volume of coal tar is still used as fuel in open-hearth furnaces and for roofing and roads. Aromatics from petroleum and coal tar are made into dyes, intermediates, medicinals, flavors, perfumes, resins, rubber chemicals, and thousands of other useful products that are almost indispensable in our present-day civilization.

COKING OF COAL

The two main types of coking procedures were the beehive and the coproduct. Beehive coking is the old, primitive method. In coproduct ovens the carefully blended coal charge is heated on both sides so that heat travels toward the center and thus produces shorter and more solid pieces of coke than are made in the beehive oven. No burning takes place within the oven, the heat being supplied completely from the flues on the sides. About 40% of the oven gas, after being stripped of its coproducts, is returned and burned for the underfiring of the battery of ovens, and some is used for fuel gas locally.

[6]Formerly called by-product ovens, a term still deeply entrenched in commercial usage. *Chemical-recovery* ovens and *slot-type* ovens are also terms in usage.

TABLE 5.2 ***U.S. Production of Tar and Tar Crudes***

Product	Production			
	1953	*1964*	*1970*	*1972*
Crude light oil, 1,000 gal	304,091	248,669	244,107	214,201
Intermediate light oil, 1,000 gal	1,062	5,392	5,066	3,704
Light-oil distillates				
Benzene, specification and industrial grades, total, 1,000 gal	272,744	730,238	1,133,520	1,252,442
Tar distillers, 1,000 gal	32,108			
Coke-oven operators, 1,000 gal	177,593	118,944	93,492	79,849
Petroleum operators, 1,000 gal	63,043	611,294	1,040,028	1,172,593
Toluene, all grades, total, 1,000 gal	156,248	495,040	829,607	915,872
Tar distillers, 1,000 gal	4,677			
Coke-oven operators, 1,000 gal	36,036	25,521	17,041	14,571
Petroleum operators, 1,000 gal	115,535	469,519	812,566	901,301
Xylene, total, 1,000 gal	113,474	343,198	537,637	739,332
Coke-oven operators, 1,000 gal	9,928	7,119	4,501	3,351
Petroleum operators, 1,000 gal	102,886	336,079	533,136	735,981
Solvent naphtha, total, 1,000 gal	15,661		4,539	
Tar distillers, 1,000 gal	9,376		832	
Coke-oven operators, 1,000 gal	6,285	4,484	3,707	2,815
Other light-oil distillates				
Total, 1,000 gal	14,657			
Tar distillers, 1,000 gal	8,560			
Coke-oven operators, 1,000 gal	6,097	9,101		6,905
Pyridine crude bases, dry basis, 1,000 gal	551	464		
Naphthalene, crude, 1,000 lb	275,799	425,690	428,086	410,075
Crude tar-acid oils, 1,000 gal	27,757		18,989	9,731
Creosote oil, dead oil, 1,000 gal	145,300	113,272	128,933	139,308
All other distillate products, 1,000 gal	18,469		90,038	
Tar, road, 1,000 gal	109,832	55,696	53,005	29,807
Tar, miscellaneous uses, 1,000 gal	36,415	19,968		14,395
Pitch of tar				
Soft and medium, 1,000 tons	1,174	886	859	582
Hard, 1,000 tons	710	991	899	786
Pitch of tar coke and pitch emulsion, 1,000 tons	38			

Source: Synthetic Organic Chemicals, 1953, 1964, 1970, and 1972 U.S. Tariff Commission. See annual reports for further details.

Beehive coking The beehive oven consists of a beehive-shaped brick chamber provided with a charging hole at the top of the dome and a discharging hole in the circumference of the lower part of the wall. The coal is introduced through the hole in the dome and spread over the floor. The heat retained in the oven is sufficient to start the distillation. The gases given off from the coal mix with the air entering at the top of the discharge door and burn; the heat of combustion is sufficient for pyrolysis and distillation.

Coproduct coking The coproduct coke oven is a narrow chamber, usually about 38 to 40 ft long, 13 ft high, and tapering in width from 17 to 18 in. at one end and to 15 or 16 in. at the other. The ovens hold from 16 to 24 tons of coal. These ovens are used for carbonizing coal only in large amounts and are built in batteries of 10 to 100 ovens. The general arrangements for the operation of a coproduct coke oven with its various accessories, followed by the initial treatment of its coproducts, are depicted[7] in Fig. 5.3.

The coproduct coke oven is one of the most elaborate and costly of masonry structures and is erected with the closest attention to engineering details, so that it can withstand the severe strains

[7]Lowry, *op. cit.*, Suppl. Vol., chap. 11, pp. 461–493 (121 refs.).

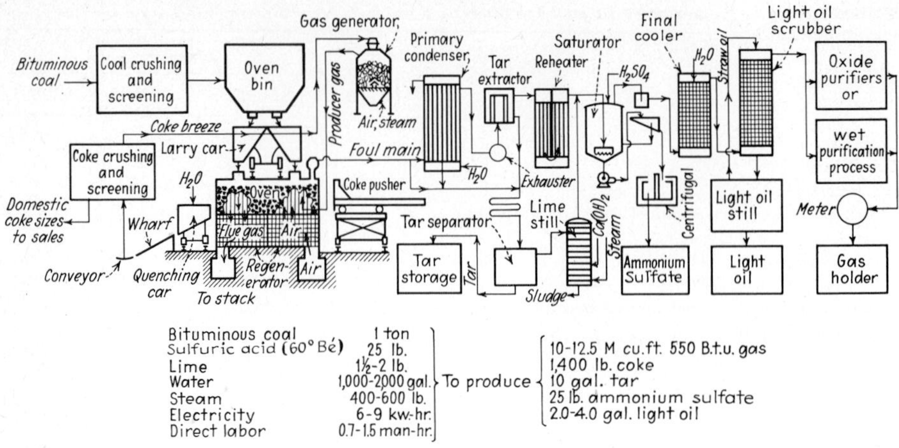

Fig. 5.3 Coproduct coke-oven procedures.

incurred in its use and remain gastight, even after the great expansion during heating up. The oven block is built of refractory brick, with heating flues between the coking ovens, as shown in Fig. 5.4.

The *individual* coproduct coke oven operates intermittently, but each oven is started and stopped at different times, so that the operation of the entire block continuously produces gas of good average composition. A charge of finely crushed coal is dropped from a larry car through charging holes (usually four, though the generalized flowchart of Fig. 5.3 shows only two) in the top and into the oven, where the walls are approximately at 2000°F. The surface of the coal in the oven is leveled,

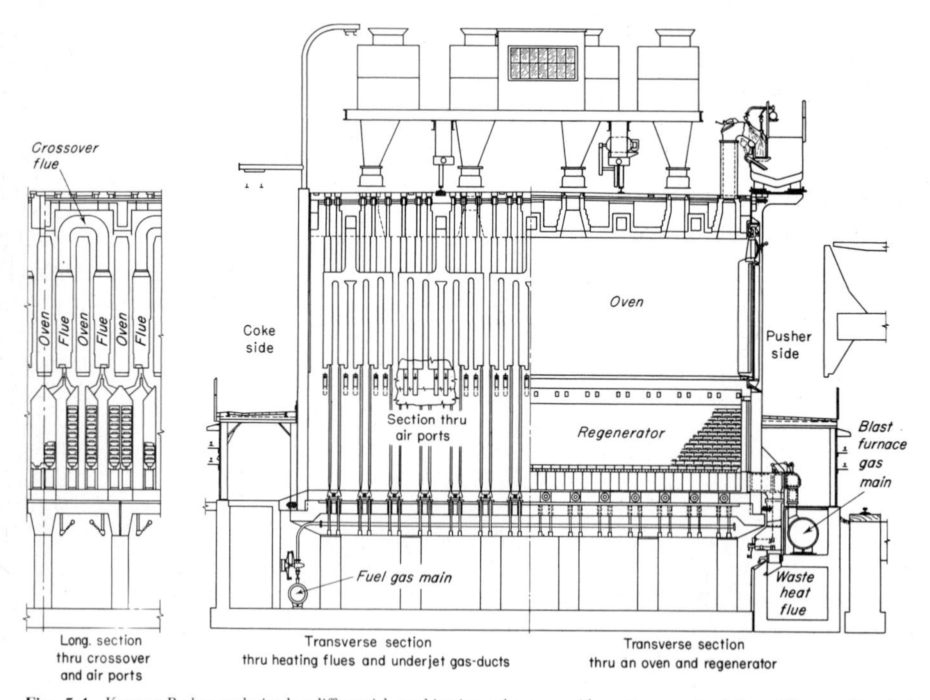

Fig. 5.4 Koppers-Becker underjet low-differential combination coke oven with waste-gas recirculation. (*Koppers Co., Inc.*)

and the charging holes are then covered. Heating is carried out, and the charge left in the oven until it is completely coked and the evolution of volatile matter has ceased. The average temperature at the center of the charge at the end of the heating period is usually about 1800°F, and the average flue temperature is about 2350°F. Temperatures vary with the conditions of operation, the coking time, the width of the oven, the kind of coal, its moisture content, and its fineness of division. Coproduct ovens are operated to make the best quality furnace coke for steel mills, and no longer to make a large volume of high-grade coal tar or gas. At the end of the coking time (approximately 17 h) the doors at the ends of the oven are opened and the entire red-hot mass is pushed out in less than a minute by a ram electrically driven through the oven from end to end. The coke falls into a quenching car which holds the charge from one oven.

The gas from the destructive distillation of the coal, together with entrained liquid particles, passes upward through a cast-iron gooseneck into a horizontal steel trough which is connected to all the ovens in a series. This trough is known as the *collecting main*, and is sometimes called a hydraulic main. As the gas leaves the ovens, it is sprayed with weak ammonia water. This condenses some of the tar and ammonia from the gas into liquid. The liquids move through the main along with the gases until a settling tank is reached, where separation is effected according to density. Some of the

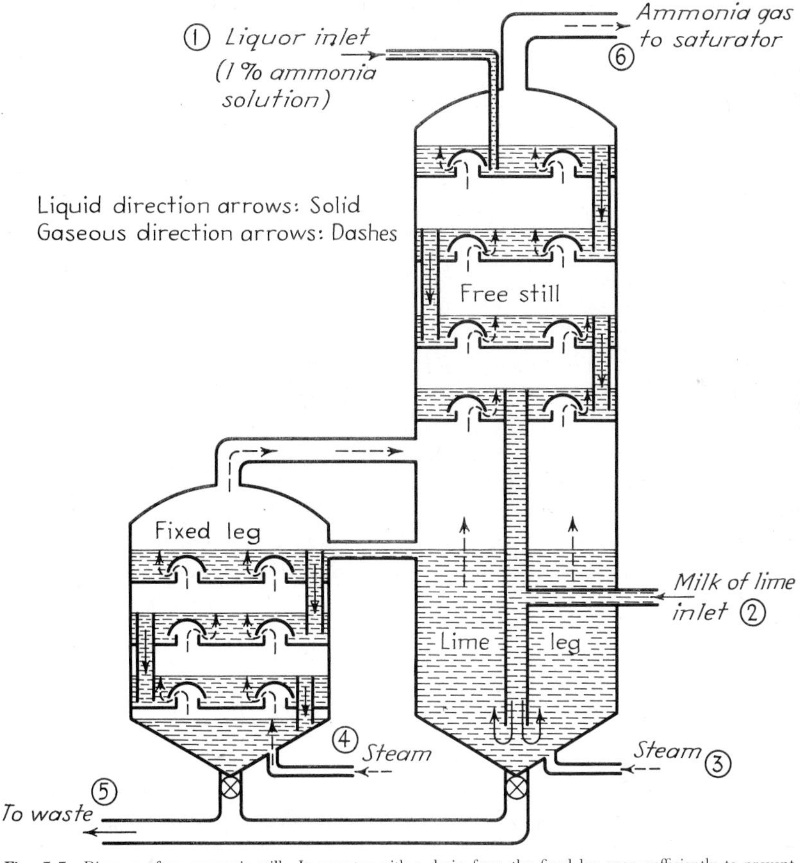

Fig. 5.5 Diagram of an ammonia still. It operates with a drain from the fixed leg open sufficiently to prevent accumulation of liquor in this section. The drain from the lime leg is closed except for cleanout.

ammonia liquor is pumped back into the pipes to help condensation; the rest goes to the ammonia still, which releases the ammonia for subsequent chemical combination in the saturator. All the tar flows to storage tanks for tar distillers or for fuel.[8]

Figure 5.3 can be divided into different steps, representing the flow of material through the various pieces of equipment, wherein the proper physical operation or chemical conversion takes place. Figure 5.3 may be thus broken down into the following sequences:

Coal is transferred, crushed, and screened (Op).
Coal is charged to a hot, empty oven (Op).
Coal is chemically transformed to coke and volatiles by pyrolysis (Ch).
Hot coke is pushed out of the oven, quenched, and transported (Op).
Condensable products of distillation are liquefied and collected in the hydraulic main (Op).
Foul gas is cooled, and tar extracted (Op).
Ammonia is removed from gas as ammonium sulfate (Ch).
Gas is cooled and subjected to benzol and toluol removal by absorption in straw oil (Op).
Hydrogen sulfide is removed (Ch).
Purified gas is metered and transferred to consumers (Op).

The tar separated from the collecting main and the tar extractor or electrostatic precipitators is settled from ammonia liquor and, together with light oil, subjected to the sequences represented by the flowchart (Fig. 5.6).

Koppers-Becker ovens are the most widely used vertical-flue ovens in the United States. The elevation is shown in Fig. 5.4. These ovens are all of the regenerative type and have individual heat regenerators, generally from side to side, underneath the oven. The reversal of the air and waste gas flow is made from side to side every half hour. The gas burns upward, crosses over the top of the oven, comes downward on the other side, and exits through the regenerative oven underneath. This travel is completely reversed periodically.

RECOVERY OF COAL CHEMICALS[9] The gaseous mixture leaving the oven is made up of permanent gases which form the final purified coke-oven coal gas for fuel, accompanied by condensable water vapor, tar, and light oils, with solid particles of coal dust, heavy hydrocarbons, and complex carbon compounds. The gas is passed from the foul main (Fig. 5.3) into the primary condenser and cooler at a temperature of about 165°F. Here the gases are cooled to 85°F by means of water. The gas is conducted to an exhauster, which serves to compress it. During the compression the temperature of the gas rises to as high as 120°F. The gas is passed to a final tar extractor, where the tar is thrown out by impingement of gas jets against metal surfaces. In newer plants the tar extractor may be replaced by electrostatic precipitators. On leaving the tar extractor, the gas carries three-fourths of the ammonia and 95% of the light oil originally present when it left the oven.

The gas is led to a *saturator* (Fig. 5.3), which contains a solution of 5 to 10%[10] sulfuric acid, where the ammonia is absorbed with the formation of solid and crystalline ammonium sulfate. The saturator is a lead-lined, closed vessel, where the gas is fed in by a serrated distributor underneath the surface of the acid liquid. The acid concentration is maintained by the addition of 60°Bé sulfuric acid, and the temperature is kept at 140°F by the reheater and the heat of neutralization. The crystallized ammonium sulfate is removed from the bottom of the saturator by a compressed-air injector, or a centrifugal pump, and drained on a table, from which the mother liquor is run back into the saturator. The salt is dried in a centrifuge and bagged, usually in 100-lb sacks (Chap. 18).

[8] Lowry, *op. cit.*, Suppl. Vol., pp. 462–471, for other types of ovens.

[9] Details for this recovery of the many coal chemicals are to be found in Lowry. *op. cit.*, Suppl. Vol., pp. 618–628, for tar processing, pp. 629–674, for light-oil processing; ECT, 2d ed., vol. 2, p. 299, 1963.

[10] At the start of a charge, the acidity is often as high as 25%, but ammonium bisulfate is liable to be formed when free sulfuric acid exceeds 10 to 11%.

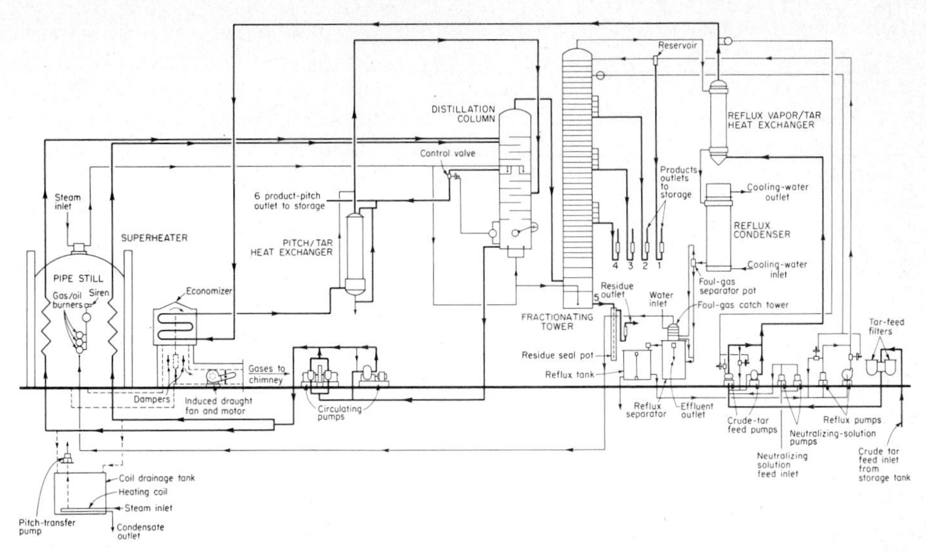

Fig. 5.6 Flowchart of coal-tar continuous distillation (without tar dehydration). The pipe still is of the radiant type. The crude-tar feed inlet is at the lower right. This is filtered and pumped through the reflux vapor-tar heat exchanger, economizer, pitch-tar heat exchanger into the top of the lower third of the distillation column and out at the bottom, to the circulating pumps and into the pipe still (where the crude tar joins 4 to 5 volumes of the circulation pitch and on finally to near the top of the distillation column). The vapors, steam-distilled and superheated, pass overhead from the top of the tray-type distillation column, enter the bottom of the bubble-cap fractionating column, and are separated into four fractions: 1,2,3,4, with a residue, 5, leaving at the bottom of this fractionating column. The product pitch, 6, cascades from the top of the distillation column down through the superheated steaming section to establish the desired pitch hardness and to strip this pitch of the higher-boiling volatile oils. It is then withdrawn from above the middle of the distillation column and conducted through the pitch-tar heat exchanger to storage. The products are:

Side stream 1: Light oil, bp to 170°C.
Side stream 2: Carbolic oil, bp 170–205°C.
Side stream 3: Naphthalene oil, bp 205–240°C.
Side stream 4: Creosote or wash oil, bp 240–280°C.
Residue 5: The residue or anthracene, bp 270–340°C.
Pitch 6: Residuum or pitch, bp 325–400°C.

There are standby steam pumps in the event of failure of the motor-driven pumps. The Wilton principle of recirculation of hot pitch at 350°C, mixed with the crude tar at 120°C, rises the temperature of the pitch-tar leaving the pipe still to only 350°C and avoids thermal cracking and vaporization in the pipe still. (*Chemical Engineering, Wiltons Ltd., London.* See special pamphlets from this company for details.)

The gas leaving the saturator at about 140°F is taken to final coolers or condensers, where it is scrubbed with water until its temperature is 75°F. During this cooling some naphthalene separates and is carried along with the wastewater and recovered. The gas is passed into a *light-oil* or *benzol scrubber* (Fig. 5.3), over which is circulated, at about 75°F as the absorbent medium, a heavy fraction of petroleum called *straw oil,* or sometimes a coal-tar oil. The straw oil is sprayed into the top of the absorption tower while the gas flows up through the tower. Most scrubbers use spiral metal packing, although wooden grids were formerly used. The straw oil is allowed to absorb about 2 to 3% of its weight of light oil, with a removal efficiency of about 95% of the light-oil vapor in the gas.

The rich straw oil, after being warmed in heat exchangers by vapors from the light-oil still and then by hot debenzolized oil flowing out of the still, is passed to the stripping column, where the straw oil, flowing downward, is brought into direct contact with live steam. The vapors of the light oil and steam pass off and upward from the still through the heat exchanger previously mentioned and into a condenser and water separator. The lean, or stripped, straw oil is returned through the heat exchanger to the scrubbers. The gas, after being stripped of its ammonia and light oil, has the

TABLE 5.3 *Approximate Yields per Ton of Coal Carbonized (Depending on Coal and Conditions Used)*

	High-temperature, lb	Low-temperature, lb
Furnace coke	1,430	
Coke breeze	93	
Semicoke (12% volatiles)	....	1,440
Tar	78	150
Ammonium sulfate	20	18
Light oil (removed from gas by oil scrubbing; see Table 5.4)	20	16
Gas	3,500	250

Source: Ind. Eng. Chem, **48,** 352 (1956). See also Fig. 5.3. Note that 4 lb of technical ammonium sulfate is equivalent to 1 lb of ammonia.

sulfur removed in purifying boxes, which contain iron oxide on wood shavings, or by a solution of ethanolamine (Girbotol) in scrubbing towers, the best present-day practice. These procedures are fully described in Chaps. 6 and 7.

U.S. Steel has built a coke-oven plant at Clairton, Pa.,[11] which employs new processing, furnishing a clean gas for compression and hydrogen recovery and avoiding coproduction of ammonium sulfate. The key to the new process is a recycled solution of ammonium phosphate that absorbs NH_3 from the cooled (compressed?) coke-oven gas. In a typical absorption cycle, lean 40% phosphate solution (the molar ratio of NH_3 to H_3PO_4 is less than 1.5) soaks up all but traces of NH_3. The rich ammonium phosphate solution is reboiled in a distillation tower in which NH_3 vapor and a lean phosphate solution are separated for reuse.

LOW-TEMPERATURE CARBONIZATION[12] In this century a large amount of experimental work has been carried out on the carbonization of coal at temperatures ranging from 750 to 1100°F, with the main object of obtaining maximum yields of liquid products and producing semicokes containing from 8 to 20% volatile matter. Here again the characteristics and yields of the various products depend upon the coal, the temperature, and the treatment. Tables 5.3 and 5.5 show the difference in gas content and in yield products for high- and low-temperature systems. The Disco plant at McDonald, Pa., is the only plant in this country to utilize a low-temperature carbonization process[13] and has a daily capacity to convert high-volatile coal into 800 tons of Disco char, a domestic fuel. The tar is sold and refined to produce tar-acid oil, tar acids, "creosote,"[14] and fuel pitch. The gas, after liquid-product removal, is used for firing. The maximum temperature used in this process is 1050°F. Low-temperature carbonization has been important in several European countries, especially England, for many decades, but has never been popular in the United States.

DISTILLATION OF COAL TAR[15]

Coal tar is a mixture of many chemical compounds, mostly aromatic, which vary widely in composition (Figs. 5.2, 5.6, and 5.9). It is a coproduct of the destructive distillation or pyrolysis of coal. Most of the tar in this country is produced by steel companies as a coproduct from blast-furnace coke

[11]Cool Gain for Ammonia Recovery, *Chem. Week*, Feb. 27, 1965, p. 39; U.S. Pat. 3,024,090.
[12]Lowry, *op. cit.*, Suppl. Vol., chap. 10, pp. 395–460 (138 refs.).
[13]Lowry, *op. cit.*, Suppl. Vol., p. 439.
[14]By definition of the American Wood Preservers Association; however, creosote is a distillate of high-temperature coke-oven tar.
[15]Cf. Bituminous Materials in Road Construction, pp. 26ff, Dept of Scientific and Industrial Research, HMSO, London.

production. The quality and quantity of tar from this operation will vary, depending on the rate of production of the ovens and the nature of the coal used. The specific gravity will vary from 1.15 to 1.2, and the quantity of tar will vary from 8 to 12 gal of tar per ton of coal. A typical light-tar composition is shown in Fig. 5.9. The end product of the distillation (Fig. 5.6) of coal tar is pitch, usually more than 60% of the crude tar. The objects of the distillation are to produce a salable end product, with a separation of the valuable products into useful cuts. In a modern refinery equipped with a fractionating column (Figs. 5.6 and 5.8) for the primary distillation, these products must usually be sharply fractionated, with a minimum of overlapping. The process also should be thermally economical, and the furnace (Figs. 5.6 and 5.7) so designed and constructed that repairs and fuel usage are at a minimum.

METHODS OF DISTILLATION There have been many improvements in coal-tar distillation over the years. These can be divided into three general classifications:

1. The 3,000- to 10,000-gal batch still, which has been much improved and used for special end products such as pipe enamel (Fig. 5.7).
2. The continuous still, with a single distillation column, using side streams (Fig. 5.6).
3. The continuous unit, using multiple columns with reboilers (Fig. 5.8).

PRODUCTS OF DISTILLATION Modern practice, as exemplified by the pipe still and fractionating columns, is producing such clean cut fractions that often little further purification is necessary. However, in Fig. 5.6 and below are given the fractions obtained in an ordinary continuous distillation, which will vary with the coal and with conditions.

a. Light oils usually comprise the cut up to 390°F. They are first crudely fractionated and agitated at a low temperature with concentrated sulfuric acid, neutralized with caustic soda, and redistilled, furnishing benzene, toluene, and homologs (Table 5.4).

b. Middle oils, or *creosote oils,* generally are the fraction from 390 to 480 or 520°F, which contains naphthalene, phenol, and cresols. The naphthalene settles out upon cooling, is separated

Fig. 5.7 Tar-distillation batch tank still, heated by gas or oil burning in large return bend flue with about 70% thermal efficiency. Two stirrers. (*Reilly Tar & Chemical Co.*)

Fig. 5.8 Fractionating distilling column (147 ft high) at Clairton, Pa., producing high-quality benzene, toluene, and xylene. (*U.S. Steel Corp.*)

by centrifuging, and is purified by sublimation. After the naphthalene is removed, phenol and other tar acids[16] are obtained by extraction with a 10% caustic soda solution and neutralization, or "springing," by carbon dioxide. These are fractionally distilled.

c. Heavy oil may represent the fraction from 480 to 570°F, or it may be split between the middle oil and the anthracene oil.

d. Anthracene oil is usually the fraction from 520 or 570°F up to 660 or 750°F. It is washed with various solvents to remove phenanthrene and carbazole; the remaining solid is anthracene.

MISCELLANEOUS USES OF COAL TAR In 1976 coal tar used as fuel amounted to about 15 to 20% of the total consumption of tar for the year. Coal tar is also utilized for roads and roofs. For these purposes the tar is distilled up to the point where thermal decomposition starts. This "base tar" is then oiled back with creosote oil to ensure satisfactory rapid drying. Somewhat similar tars are used to impregnate felt and paper for waterproofing materials.

FRACTIONATION AND PURIFICATION OF COAL TAR CHEMICALS Largely because of the present competition among aromatic chemicals from petroleum (Chap. 38), interest in aromatics

[16] "Tar acids," in the language of the tar distiller, represent phenol and its homologs, which are soluble in caustic soda.

TABLE 5.4 *Typical Composition of Light Oil from Gas*

	Gallons per ton of coal
Benzene	1.85
Toluene	0.45
Xylene and light solvent naphtha	0.30
Acid washing loss (mostly unsaturateds)	0.16
Heavy hydrocarbons and naphthalene	0.24
Wash oil	0.20
Total crude light oil	3.20
Pure motor fraction	2.50

Source: Lowry, Chemistry of Coal Utilization, vol. 2, p. 940; more detailed tabulation, Suppl. Vol., pp. 633–642.

from coal tar has temporarily decreased (see Table 5.2 for main products). Also, synthetic processes from acetaldehyde and ammonia are supplying the increased demands for pyridine, of which coal tar at one time was the sole supplier. This is also true of phenol (Chap. 34). In Europe,[17] which has much coal and little petroleum, there is continued interest in coal chemicals. Table 5.6 lists some solid chemicals that *could* be obtained from coal tar. The product of largest potential is *phenanthrene* (the second most abundant in coal tar), of which Franck estimates that 250,000 tons can be recovered in the Western world yearly as soon as profitable uses can be found. This is based on a total recovery of 10 million tons of crude tar and with a 50% yield. *Anthracene,* or rather anthraquinone (Chap. 39), is the basis of many vat dyes, but here again it is cheaper to synthesize this compound (Chap. 39) from phthalic anhydride and benzene or preferably chlorobenzene.

COAL TO CHEMICALS

This section considers the rest of the processes listed in Table 5.1, except those designated for discussion in other chapters. In all cases further details are available in the references, especially in the monumental three volumes edited by Lowry.

Solvent extraction of coals and lignites has been tried at temperatures below and above 300°C and with and without mild hydrogenation. Although various resins and waxes result, the processes have met only minor commercial acceptance.[18] *Alkaline* hydrolysis[19] has likewise been investigated, with meager results. *Partial oxidation* yields the tremendously important *synthesis gas* (Chap. 7).

TABLE 5.5 *Variation of Gas from Low- and High-temperature Coal Carbonization*

Gas	*Coking temperature, 500 C, %*	*Coking temperature, 1000 C, %*
CO_2	9.0	2.5
C_nH_m	8.0	3.5
CO	5.5	8.0
H_2	10.0	50.0
CH_4 and homologs	65.0	34.0
N_2	2.5	2.0

[17]Franck, The Challenge in Coal Tar Chemicals, *Ind. Eng. Chem.*, **55**(5), 38 (1963). Many coal chemicals described.
[18]Lowry, *op. cit.*, Suppl. Vol., p. 1089.
[19]Lowry, *op. cit.*, Suppl. Vol., p. 1090.

TABLE 5.6 *Some Important Constituents of European Coal Tar*

Compound	*Percent*	*Compound*	*Percent*
Naphthalene	10	Diphenyl	0.4
Phenanthrene	5	Indole	0.2
Fluoranthene	3.3	2-Phenylnaphthalene	0.3
Pyrene	2.1	Isoquinoline	0.2
Fluorene	2.0	Quinaldine	0.2
Chrysene	2.0	Acridine	0.6
Anthracene	1.8	Phenanthridine	0.2
Carbazole	1.5	7,8-Benzoquinoline	0.2
2-Methylnaphthalene	1.4	Thianaphthene	0.3
Dibenzofuran	1.0	Diphenylene sulfide	0.3
1-Methylnaphthalene	1.0		

Source: Ind. Eng. Chem., **55**(5), 38 (1963).

Dow[20] has experimented with caustic oxidation (oxygen), obtaining high-molecular-weight polyfunctional aromatic coal acids, which have found limited use in thermosetting resins and water-soluble films. *Sulfur*[21] *recovery* from coal is still small and variable, but in foreign lands that lack the U.S. sulfur raw materials (H_2S from gases and sulfur from salt domes), pyrite has been recovered from coal and is used to the extent of about 10,000 tons yearly in England and Germany. *Sulfonation*[22] has been employed to a limited extent to manufacture ion-exchange material for water softening.

HYDROGENOLYSIS (HYDROGENATION-PYROLYSIS) Many development investigations[23]have been carried out on direct and catalytic hydrogenation of coal both in the United States and abroad. Most of these experiments are really hydrogenolyses or hydrogenations (methanation) of the pyrolysis products of coal. They were designed to yield a high-Btu gas to compete with natural gas (see peak gas, Chap. 7) or to make motor fuel in petroleum-poor countries. The results gave such a gas, but at a high cost with much of the coal left as residual carbon. The motor-fuel objectives were largely for wartime demands. The present attack on coal to secure other and hopefully cheaper coal chemicals tends to follow *catalytic* hydrogenation and other processing, often grouped together as "coal refining" and combined with liquid separation, coking, and hydrocracking in the presence of hydrogen, without aiming for the uneconomical total hydrogenation of the carbon.[24]

COAL RESEARCH The Office of Coal Research, created to conduct research on mining, preparation, and utilization of coal, including chemicals, is financing industry in many efforts to upgrade coal and coal chemicals.

The energy crisis of 1973–1974, the increasing U.S. demand for fuel, and the fourfold rise in oil cost greatly improve the outlook for coal as a feedstock in place of oil.[25] It appears that it will be only a question of time until coal replaces a significant amount of oil as feedstock in the United States.

[20]Montgomery, Coal Acids, *Chem. Eng. News*, Sept. 28, 1959, p. 96; Perry, *op. cit.;* Burke, *op. cit.*

[21]Burice, *op. cit.;* Institute for Gas Technology, Clean Fuels from Coal, Symposium, Sept. 10–14, 1973; Project Independence: An Economic Evaluation, MIT Energy Laboratory Policy Study Group, Mar. 15, 1973.

[22]Broderick and Bogard, *U.S. Bur. Mines Rep. Invest.*, 3559, 1951; Lowry, *op. cit.*, Suppl. Vol., p. 1091.

[23]Farber, Pipeline Coke-Oven Charging, *Chem. Eng.* (*N.Y.*), **80**(29), 36 (1973); Perry, *op. cit.*

[24]South Africa Banks on Coal for Chemical Buildup, *Chem. Week*, Sept. 25, 1974, p. 33. AE&CI (formerly called African Explosives and Chemical Industries), South Africa's leading chemical firm set aside $465 million for coal-based production units in 1973–1974.

[25]Chemicals from Coal: Best Bet in Energy Crisis, *Chem. Week*, June 12, 1974, p. 11.

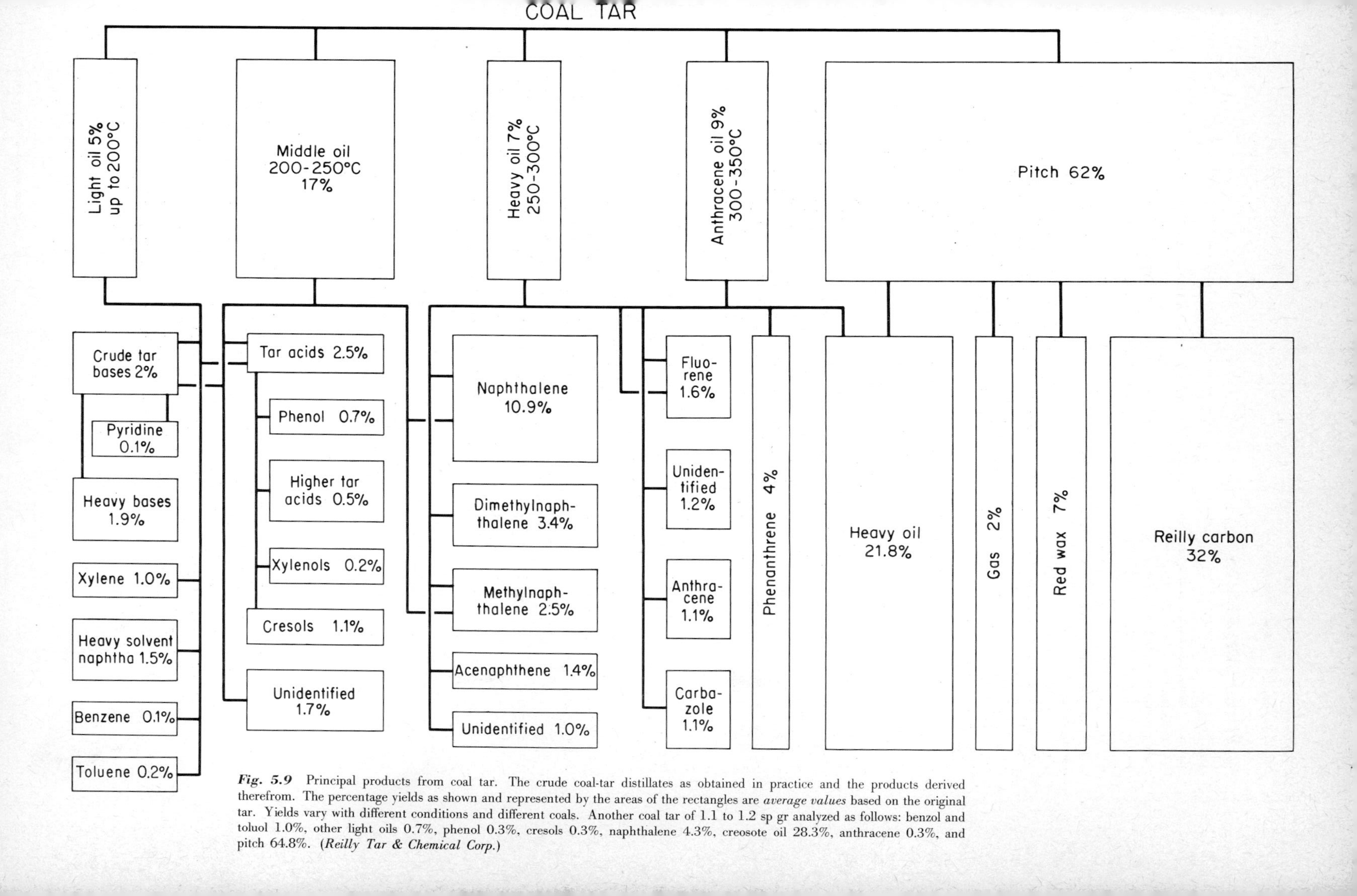

Fig. 5.9 Principal products from coal tar. The crude coal-tar distillates as obtained in practice and the products derived therefrom. The percentage yields as shown and represented by the areas of the rectangles are *average values* based on the original tar. Yields vary with different conditions and different coals. Another coal tar of 1.1 to 1.2 sp gr analyzed as follows: benzol and toluol 1.0%, other light oils 0.7%, phenol 0.3%, cresols 0.3%, naphthalene 4.3%, creosote oil 28.3%, anthracene 0.3%, and pitch 64.8%. (*Reilly Tar & Chemical Corp.*)

SELECTED REFERENCES

Abramovitch, R. A. (ed.): Pyridine and Its Derivatives, Wiley, 1974.
Claxton, G.: Benzoles: Production and Use, National Benzole and Applied Products Association, London, 1961.
Coal Processing Technology: AIChE, 1974.
Coal's New Horizons, *Proceedings of a Conference at Southern Research Institute*, Birmingham, Ala., 1961.
Francis, W.: Coal: Its Formation and Composition, 2d ed., E. Arnold, London, 1961.
Given, P. H. (ed.): Coal Science, ACS, 1966.
Goldman, G. K.: Liquid Fuels from Coal, Noyes, 1972.
Gould, R. F. (ed.): Literature of Chemical Technology, chap. 24, Coal Carbonization, ACS Monograph, 1967.
Hoiburg, A. J. (ed.): Coal Tars and Pitches, Wiley, 1967.
Lowry, H. H. (ed.): Chemistry of Coal Utilization, Suppl. Vol., Wiley, 1963.
Review of Benzole Technology, 1962 and 1963, National Benzole and Allied Products Association, London, 1963.
Review of Coal Tar Technology, Coal Tar Research Association: Gomersal Leeds, England (semiannual review).
U.S. Bureau of Mines: Various publications on coal and coal hydrogenation. (Current lists available on request.)

chapter 6

FUEL GASES

During the 1950s a deep and far-reaching change took place in the fuel-gas industries, involving the domination of these widespread markets by natural gas. This was made economical through the countrywide installation of gas pipelines, reaching from large gas fields to most homes and factories in the United States. Local peak demands in winter are satisfied by the use of natural gas stored in neighboring underground depleted production wells, by the use of liquefied natural gas (LNG), by the use of liquid petroleum gas (LPG), or by peak production. LPG also meets needs in areas pipelines cannot reach. The convenience, cleanliness, and reasonable price of natural gas have been a boon to Americans. This change has competitively restricted coke-oven gas to those coproduct areas where coke is made from coal for the steel and foundry industries (Chap. 5). Also, this competition has reduced water gas mostly to use in the making of peak gas and synthesis gas (Chap. 7). Producer gas and retort coal gas have almost disappeared in the United States.

The need for a gas-from-coal process for peak shaving was eliminated with the general acceptance of LNG in the late 1960s as a safe method for storing natural gas for peak shaving where underground storage was not available. A propane-air system continues to be used during periods of high demand when only relatively small volumes of additional gas are required. Tables 6.1 and 6.2 summarize the composition and heating values of fuel gases. The choice in any case involves the composition, heating value, and cost of producing and distributing the gas. The Btu of a gas is a summation of the heats of combustion of its constituents and can be very accurately calculated from these values. Table 6.3 lists the individual heats of combustion and explains the variation in the heats of combustion of the gases given in Table 6.1.

HISTORICAL The first recorded use of combustible gas was made by the Chinese about A.D. 900, when natural gas was piped through bamboo tubes and used for lighting. The first production of coal gas took place about 1665, in England, and its first utilization was for lighting purposes in 1792. Similar endeavors in this country began in Philadelphia about 1796. It was not long after this that gas companies began to be organized and the manufacture of gas was put on a businesslike basis. The discoveries of water gas, or *blue gas,* in 1780, and of producer gas were essential steps in the development of this industry. The tremendous exploitation of our natural-gas fields gave the final impetus to the gas industry as we know it today.

This exploitation, however, has finally resulted in local natural-gas shortages, which began to appear in 1968. In that and following years, the volume of natural gas produced from wells was greater than the volume of new gas discovered. Because of this reduction in proven natural-gas reserves, exploration for gas on land and offshore increased. Gas has been found in Alaska, but it could be 1980 before a gas line is completed to transport it to the lower 48 states. Some LNG is being exported from and imported into the United States in specially built ships over distances of more than 4,000 mi. The exports, contracted for before our shortages occurred, will not increase,

TABLE 6.1 *Percent Composition and Heating Values of Various Fuel Gases**

Fuel gas	*Natural gas (mid-continent)*	*Natural gas (Pennsylvania)*	*Coke-oven gas*	*Blue water gas*	*Carbureted water gas*	*Bituminous producer gas*
Carbon monoxide	....	....	6.3	42.8	33.4	27.0
Carbon dioxide	0.8	....	1.8	3.0	3.9	4.5
Hydrogen	....	....	53.0	49.9	34.6	14.0
Nitrogen	3.2	1.1	3.4	3.3	7.9	50.9
Oxygen	....	....	0.2	0.5	0.9	0.6
Methane	96.0	67.6	31.6	0.5	10.4	3.0
Ethane	....	31.3				
Illuminants	....	....	3.7	....	8.9	
Gross Btu/ft^3	967	1,232	588	308	536	150

*Extensive tables, Tables 9-15, 9-16, 9-18, and 9-19, are given by Perry where more properties are tabulated.

but the imports will increase considerably. A few plants for gasifying petroleum products are operating, but their number will be limited because of oil shortages and high oil prices. Several coal-to-gas plants are being planned. These plants (oil and coal) will make pipeline-quality gas with about 1000 Btu/ft.^{3}

New processes for the production of clean gas from coal with about 150 Btu/ft^3 to be used in combined-cycle power plants to produce electricity are being planned. A combined-cycle power plant uses both gas and steam turbines. Methods for gasifying oil shale, solid municipal waste, sewage, animal waste, waste wood products, and material that can be grown for the purpose, such as trees and algae, are also being investigated. The United States has just started to use manufactured gas to supplement its still very large reserves of natural gas. The new gasification processes[1] being considered will gasify more of the organic material in coal than the old processes.

NATURAL GAS

Not only is natural gas (Table 6.2) the predominating fuel gas, but it has become a most important *chemical raw material* for various syntheses. These will be presented in Chap. 38, Petrochemicals, as well as elsewhere in this book, so fundamentally important has natural gas become (Chaps. 7, 8, 18). For example, in 1972 the chemical process and allied industries used 845.6 trillion Btu, and the carbon black and lamp industry used 59.2 trillion Btu as natural gas (approximately 1,000 Btu/ft^3). Most large chemical companies have established plants in Texas or Louisiana near or right on the gas fields, to supply the chemical processes with the cheapest possible natural gas.

USES AND ECONOMICS The proved recoverable reserve[2] of natural gas was at a maximum at the end of 1967, when it was 293 trillion ft^3. At the end of 1972, it was down to 266 trillion ft^3, including gas in Alaska. The amount of natural gas produced in 1967 was 18.4 trillion ft^3, and in 1972 it was 22.5 trillion ft^3, indicating an increase in demand. At the end of 1972, the U.S. *potential* natural-gas supply was 1,146 trillion ft^3 justifying the increase in exploration efforts. However, much of the potential supply[3] lies in high-cost areas such as Alaska, at drilling depths to 30,000 ft and water depths to 1,500 ft. Of the 17,109 trillion Btu sold to customers in 1972, residential sales were 5148 trillion Btu, commercial sales were 2280 trillion Btu, and industrial sales amounted to

[1]Perry, The Gasification of Coal, *Sci. Am.*, March 1974, pp. 19–25.
[2]1972 Gas Facts, American Gas Association.
[3]*Oil Gas J.*, Dec. 31, 1973, p. 72.

TABLE 6.2 ***Natural Gas Distributed in Various Cities in the United States***
(Components of gas, percent by volume)

City	*Methane*	*Ethane*	*Propane*	*Butanes*	*Pentanes*	CO_2	N_2	*Btu/ft³*
Birmingham, Ala.	93.14	2.50	0.67	0.32	0.12	1.06	2.14	1024
Columbus, Ohio	93.54	3.58	0.66	0.22	0.06	0.85	1.11	1028
Dallas, Tex.	86.30	7.25	2.78	0.48	0.07	0.63	2.47	1093
Denver, Colo.	81.11	6.01	2.10	0.57	0.17	0.42	9.19	1011
New Orleans, La.	93.75	3.16	1.36	0.65	0.66	0.42	0.00	1072
San Francisco, Calif.	88.69	7.01	1.93	0.28	0.03	0.62	1.43	1086

Source: Gas Engineers' Handbook, American Gas Association, 1962. *Gross or higher heating value at 30 in. Hg, 60°F, dry. To convert to a saturated basis, deduct 1.73%, i.e., 17.3 from 1,000, 19 from 1,100. Hexanes and higher hydrocarbons analyzed from 0.00 to 0.05. See also Perry, table 9-15.

8797 trillion Btu. The average price for the residential customer in 1972 was nearly $1.19 per million Btu, for the commercial customer it was nearly $0.91 per million Btu, and for the industrial customer it was $0.45 per million Btu.

The average price of natural gas at the wellhead in 1972 was $0.186 per 1,000 ft³. However, the cost of some new-found gas is $0.50 and higher at the wellhead. Also, the cost of supplemental gas from oil and coal is expected to be in the range of $1.00 to $3.00 per 1,000 ft³. By 1985, domestic production from wells is still expected to provide most of our gas supply, as indicated in Table 6.4. The price of gas will, however, increase.

Various by-products from raw natural gas are of industrial significance, such as methane, ethane, propane, butane, LPG, and natural gasoline. For helium, see Chap. 7. These are mostly separated from the crude gas as liquids by oil absorption, followed by fractionation under pressure; some plants still use compression and adsorption. They are marketed in steel cylinders, tank trucks, tank cars, and special sea-going vessels throughout the United States. The separation and utilization of natural gasoline, or *casing-head* gasoline, as it is often called, will be presented under petroleum refining in Chap. 37.

DISTRIBUTION After necessary purification, which will be described, natural gas is compressed to about 1000 psig and sent through transmission mains which network the country. It is recompressed at periodic distances by gas compressors driven by various prime movers, frequently gas engines or turbines. As it nears the consumer, the distribution pressure is reduced to as low as a

TABLE 6.3 ***Heats of Combustion***

	Btu per cubic foot at 60°F and 1 atm		*Btu per pound*	
Substance	*Gross (high)*	*Net (low)*	*Gross (high)*	*Net (low)*
Hydrogen	325	275	61,100	51,623
Carbon monoxide	321	322	4,347	4,347
Methane	1013	913	23,879	21,520
Ethane	1786	1641	22,304	20,432
Propane	2592	2385	21,661	19,944
Butane	3370	3113	21,308	19,630
Carbon	....	....	14,093	14,093

Note: The net, or low, heating value designates the heat of combustion with the water vapor uncondensed. If the products are cooled to the initial temperature and the water is condensed, its latent heat of condensation is liberated and the larger, or gross, heating value is obtained. Cf. Perry, tables 9-15 and 9-18. All tables not in full agreement.

TABLE 6.4 ***Possible Sources of Gas Supply in 1985***

Source	*Billion cubic feet per day*
Domestic production	60.0
Imports from Canada	5.5
Imports of LNG	6.5
Coal gasification	3.5
Oil gasification	3.0
Total gas supply	78.5

Source: Outlook for Energy in the United States to 1985; Chase Manhattan Bank, Energy Economics Division, June 1972, p. 43.

few inches of water pressure at the gas appliance. This natural gas is employed as fuel or raw material in cities throughout the country, some cities being 2,000 mi or more from the producing wells. The natural-gas distribution system in this country now involves 959,000 mi of transmission and distribution pipelines. Ships and tank trucks transport LNG to distant markets and those not reached by pipelines. The whole natural-gas industry in this country has a plant investment of $59 billion, much of which is in pipelines.

PEAK DEMANDS[4] It is not unusual for the winter natural-gas load for a company to be several times that of its summer load, so that it is desirable to be able to store gas near densely populated areas to be used during periods of high demand. The best procedure, widely practiced, is to store natural gas underground in depleted oil or gas reservoirs, at pressures that existed when the well was first drilled (as high as 4,000 psig), or in aquifiers where the displaced water acts as a seal. This stored gas is withdrawn when needed. Natural gas is also liquefied, at −260°F, to obtain a 600 to 1 volume reduction, and stored, usually in above-ground double-walled insulated steel containers. During periods of high demand it is vaporized and distributed. To a limited extent, it is possible to vaporize stored propane, mix it with air, and add the mixture to natural gas as another method of peak shaving. When mixing gases, the combustion properties of the final composition must always be considered.

NATURAL-GAS PURIFICATION[5] In addition to the industrially valuable propane and butane, raw natural gas contains undesirable water and hydrogen sulfide which must be removed before it can be placed in transmission lines. Four important methods are employed for the dehydration of gas: compression, treatment with drying substances, adsorption, and refrigeration. A plant for water removal by *compression* consists of a gas compressor, followed by a cooling system to remove the water vapor by condensation. The treatment of gas with *drying* substances has found widespread usage in this country. The agents employed for this purpose are activated alumina and bauxite, silica gel, sulfuric acid, glycerin, diethylene glycol,[6] and concentrated solutions of calcium chloride or sodium thiocyanate. Plants of this type usually require a packed tower for countercurrent treatment of the gas with the reagent, together with a regenerator for the dehydrating agent. Gas may also be dehydrated by passing it over *refrigerated* coils. In general, this method is more costly than the others but, where exhaust steam is available to operate the refrigeration cycle, refrigeration charges can be reduced. If the water present in most fuel gas is not removed, unduly high corrosion will occur in the transmission lines, and trouble may also result from the formation of hydrates, which

[4]Peak Shaving Report, *Pipeline Gas J.*, Dec. 1973, pp. 25–30.
[5]Wolke and Webber, Natural Gas Processing Developments, *AIChE Petrochem. Pet. Refiner*, **57**(34), 50 (1961) (excellent).
[6]Polderman, Glycols as Hydrate Point Depressants in Natural Gas Systems, University of Oklahoma and Union Carbide Chemical Co., 1958.

TABLE 6.5 ***Sulfur-removal Processes for Gases*** *(Only the Girbotol or ethanolamine process is much used in the United States)**

Process or reagent	*Reaction*	*Regeneration*
Girbotol*	$2RNH_2 + H_2S \rightleftharpoons (RNH_3)_2S$	Steam stripping
Phosphate (Shell)	$K_3PO_4 + H_2S \rightleftharpoons KHS + K_2HPO_4$	Steam stripping
Iron oxide	$Fe_2O_3 \cdot xH_2O + 3H_2S \longrightarrow Fe_2S_3 + (x + 3)H_2O$	Oxidation with air
	$2Fe_2S_3 + 3O_2 + xH_2O \longrightarrow 2Fe_2O_3 \cdot xH_2O + 6S$	
Thylox†	$Na_4As_2S_5O_2 + H_2S \longrightarrow Na_4As_2S_6O + H_2O$	Oxidation with air
Seaboard†	$Na_2CO_3 + H_2S \rightleftharpoons NaHCO_3 + NaHS$	Air blowing
Caustic soda	$2NaOH + H_2S \longrightarrow Na_2S + 2H_2O$	None
Lime	$Ca(OH)_2 + H_2S \longrightarrow CaS + 2H_2O$	None
Alkazid	$RCHNH_2COONa + H_2S \rightleftharpoons RCHNH_2COOH + NaHS$	Steam stripping

*Mono-, di-, or triethanolamine. †Operating data: ECT, 2d ed., vol. 7, pp. 83ff., 100ff., 1965; CPI 2, pp. 109–112; Lowry, The Chemistry of Coal Utilization, Suppl.. Vol., pp. 1017–1022, Wiley, 1963.

can cause line stoppages. The freezing of valves and regulators in cold weather may also cause difficulties.

State standards require the *removal* of H_2S and other sulfides from gas, not only because of the corrosion problem, but because the oxides of sulfur formed during combustion would be a nuisance to gas consumers. Also, the sulfur compounds, particularly H_2S and sulfur have commercial value and are employed as a source of sulfur or SO_2 for sulfuric acid plants. In raw natural gas, the amount of H_2S present may range from 0 to 100 gr/100 ft^3 and much higher in some fields. Refinery gas from sulfur crudes will also carry high amounts of sulfur. Table 6.5 summarizes the various processes that have been employed commercially to remove H_2S from gases. However, the *Girbotol* procedure (Girdler Corp.) is now used in most cases, though some of the others[7] listed in the table are still economical in foreign lands. A flowchart is presented for this procedure[8] in Fig. 6.1. See also Fig. 19.2. Although the early available triethanolamine process was the first to be used, it has been largely replaced. Monoethanolamine has a greater capacity per unit volume of solution than diethanolamine for absorption of acid gas, and the initial plant investment and operating cost are lower. It is rather difficult, however, to make general statements on the selection of amines for gas purification. Diethanolamine as a 10 to 20% aqueous solution is commonly employed for H_2S removal from refinery gases. For desulfurization of natural gas, a 10 to 30% aqueous solution of monoethanolamine is normally used. Ethanolamines differ in their selectivity for absorption of H_2S and CO_2, and this property, as well as oxygen and sulfur compounds present in a gas stream, frequently determine the choice between monoethanolamine and diethanolamine. For example, in treating refinery gas, diethanolamine is selected when carbon oxysulfide is known to be present, since this amine does not react so readily as monoethanolamine with this impurity to form nonregenerable compounds. If simultaneous *dehydration* and desulfurization is desired, natural gas may be scrubbed with a combination solution consisting of amine, water, and diethylene glycol. Solution compositions for this

[7]For details of iron oxide, Seaboard, Thylox, and other processes, see CP1 2 and ECT, 2d ed., vol. 7, pp. 833ff., 1965.

[8]Reed, Improved Design, Operating Techniques for Girbotol Absorption Processes, *Pet. Process.*, **2**(12), 907 (1947); Updegraff and Reed, Twenty-Five Years of Progress in Gas Purification, *Pet. Eng.* (Los Angeles), September 1954; Hydrogen Sulfide, Girdler Corp., Louisville, Ky. (pamphlet); Reed and Updegraff, Removal of H_2S from Industrial Gases, *Ind. Eng. Chem.*, **42,** 2269 (1950) (34 refs.); Cronan, Natural Gas Processing Plant, *Chem. Eng. (N.Y.)*, **66**(8), 148 (1959) (includes quantities of raw materials and products); Brennan, Amine Treatment of Sour Gas, *Chem. Eng.* (*N.Y.*), **69**(22), 94 (1962) (at Person Field, Tex.).

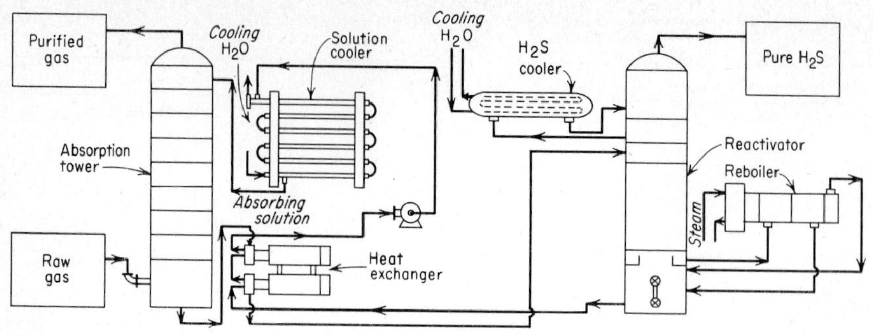

Fig. 6.1 Gas purification by the Girbotol aliphatic amine absorption process.

purpose are from 10 to 36% monoethanolamine, 45 to 85% diethylene glycol, and the remainder water. The general reaction of amines with H_2S, which is reversible on heating, may be expressed as

$$2RNH_2 + H_2S \rightleftharpoons (RNH_3)_2S$$

The apparatus consists of an absorption tower followed by a stripping tower and various pumps, condensers, and heat exchangers. The fouled liquid from the absorber passes through the heat exchangers to the stripper, where it meets rising steam and is heated to about 250°F and has its content of H_2S and other gases removed.

Natural gas with a high nitrogen content can be upgraded by a cryogenic[9] process which drys feed gas at 2,000 psig, cooled to −200°F, and expanded to 500 psig. The nitrogen gas is separated from the LNG in a column. The natural gas is vaporized, and both this and the separated nitrogen gas leave the system via heat exchangers against incoming gas.

COKE-OVEN GAS

Coke-oven[10] gas is now only produced as a coproduct. The processes for the distillation of coal are discussed in Chap. 5. The following is merely an outline of the treatment of gas obtained from chemical-recovery coke ovens. The coproducts in the hydraulic main first go via the foul main through a primary condenser and a baffled tar extractor, as shown in Fig. 5.3, where the coke-oven gas[11] is separated from the ammonia liquor and the coal tar. The gas has its ammonia removed as ammonium sulfate by bubbling through sulfuric acid. Coal-tar products such as benzene, toluene, and some naphthalene are scrubbed by straw oil in a packed light-oil tower or scrubber. The uses of coke-oven gas are largely for fuel gas in the many steel-plant operations with which almost all U.S. coke-oven plants are associated. Some of the gas may be used for underfiring the coke ovens themselves. In a few localities coke-oven gas is purified and sold to the local utility company for distribution to the community.

WATER GAS (BLUE GAS)

Water gas is often called *blue gas* because of the color of the flame when it is burned. Water gas has met the same fate as coal gas, except that it is used (or rather its reaction) for the first step in

[9]Jameson and Mann, Nitrogen Removal: Its Costs and Profits, *Pet. Refiner*, April 1961, p. 121.
[10]Retort coal gas is still made and used in Europe, but not in the United States.
[11]Operating results on coal gas are given in Lowry, The Chemistry of Coal Utilization, Suppl. Vol., p. 892, Wiley, 1963.

making *synthesis gas* (Chap. 7), with the substitution by a petroleum hydrocarbon (Fig. 7.5) for the coal that was generally employed for city gas. The production of water gas is carried out by the reaction[12] of steam on incandescent coke or coal at temperatures of about 1000°C (1832°F) and higher, where the rate and equilibrium are favorable, according to the principal equation:

$$C(amorph) + H_2O(g) \longrightarrow CO(g) + H_2(g) \qquad \Delta H_{1832°F} = +53{,}850 \text{ Btu}$$

Another reaction also occurs, apparently at a temperature several hundred degrees lower:

$$C(amorph) + 2H_2O(g) \longrightarrow CO_2(g) + 2H_2(g) \qquad \Delta H_{1832°F} = +39{,}350 \text{ Btu}$$

These reactions are endothermic and therefore tend to cool the coke bed rather rapidly, thus necessitating alternate "run" and "blow" periods. During the run period, the foregoing blue-gas reactions take place, and salable, or *make*, gas results; during the blow period, air is introduced and ordinary combustion ensues, thus reheating the coke to incandescence and supplying the Btu's required by the endothermic useful gas-making reactions plus the various heat losses of the system. The reactions are

$$C(amorph) + O_2(g) \longrightarrow CO_2(g) \qquad \Delta H_{1832°F} = -173{,}930 \text{ Btu}$$

$$CO_2(g) + C(amorph) \longrightarrow 2CO(g) \qquad \Delta H_{1832°F} = +68{,}400 \text{ Btu}$$

The manufacture of blue water gas or *synthesis* gas, from solid carbonaceous fuels treated simultaneously with water and oxygen is possible because of reductions in cost of tonnage oxygen.[13] Blue water gas, is usually identified with synthetic fuels, or the synthesis of other chemicals such as alcohols and oxygenated compounds. A mixture of CO and H_2 is also used to produce hydrogen for ammonia (Chap. 18) and other hydrogenations.

To enhance the heating value of blue water gas, oil is atomized into the hot blue gas. A plant producing this *carbureted blue gas* has, in addition to an ordinary generator, a carburetor and a superheater, as shown in Fig. 6.2. These are large refractory-lined steel shells with brick checkerwork, at least in the superheater, to provide intensive heat-transfer surfaces. The combustion gases formed during the blow period serve to heat the arches, side walls, and checkerwork. Then, during the run, oil is sprayed into the carburetor, is cracked or pyrolyzed, and is passed on, mixed with the water gas. In most cases it is necessary to have tar traps to remove the heavy water-gas tar formed during the vaporization of the oil. One installation producing 520 to 540 Btu/ft^3 of gas, using 23 to 25 lb of coal and 2.7 to 3.1 gal of heavy oil per 1,000 ft^3, is detailed in Fig. 6.2 and Table 6.6. This is one answer to the demand for manufactured gas for peak consumption wherein the use of bituminous coal and heavy oils is efficient. The carburetor is *not* filled with checkerwork, but is provided with a brick orifice only; thus trouble is avoided when heavy petroleum oils are used, which fill up and block checkerwork with the coke formed and reduce heat transfer. The plant in Fig. 6.2 is called an automatic *reverse-flow* water-gas machine, because the usual flow of gas through the carburetor and superheater is reversed by elevating the carburetor somewhat and changing the flues. This results in better pyrolysis of the carburetor oil as it is sprayed in countercurrent to the hot blue gas and permits the use of heavier oils; this also results in better heat balances and fewer repairs necessitated by carbon blocks or stoppages. Heavy oils can now be burned directly in many industrial applications. This has caused the price of heavy oil to increase, at the same time reducing the need for manufactured gas of this type.

[12]Parker, A Thermal Study of the Process of Manufacture of Water Gas, *J. Soc. Chem. Ind., London*, **46,** 721 (1927); Scott, Mechanism of the Stream Reaction, *Ind. Eng. Chem.*, **33,** 1279 (1941); for rate and equilibrium results, see Lowry, *op. cit.*, vol. 1, pp. 1673–1749, and Suppl. Vol., pp. 892ff.

[13]Powell, Symposium on Production of Synthesis Gas, *Ind. Eng. Chem.*, **40,** 558 (1948); Martin, Synthesis Gas from Coal, *Chem. Ind. (N.Y.)*, **66,** 365 (1950); ECT, 2d ed., vol. 8, pp. 778ff., 772ff., 1965.

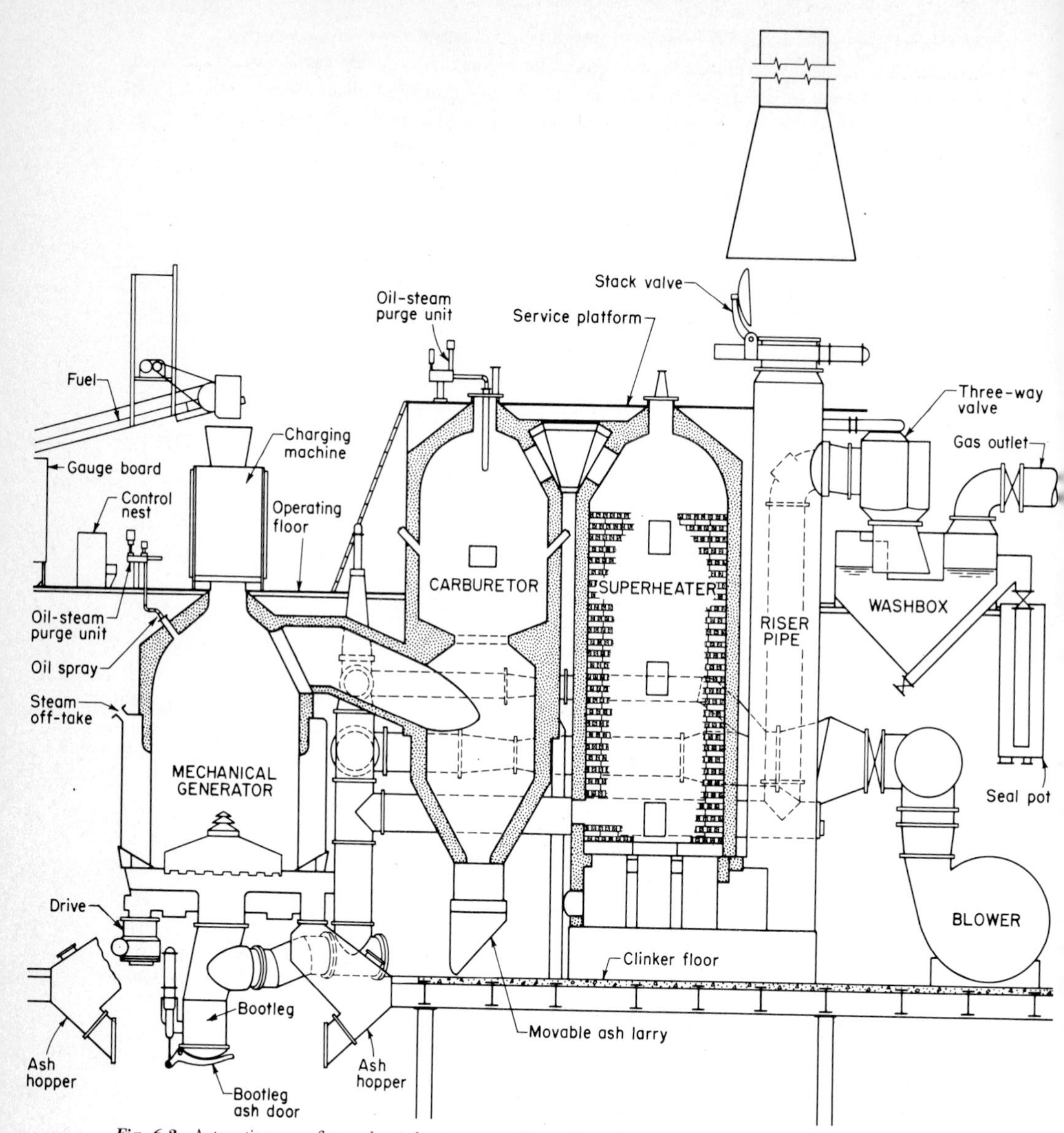

Fig. 6.2 Automatic reverse-flow carbureted water-gas machine. (*Semet-Solvay Engineering Div.*)

PRODUCER GAS[14]

Producer gas is made by passing air and steam through a thick bed of hot fuel. The primary purpose of the steam (25 to 30% of the weight of the coke) is to use up as much as possible the exothermic

[14]Lowry, *op. cit.*, vol. 1, pp. 1586–1672, and Suppl. Vol., pp. 892ff., 951–965; *ECT*, 2d ed., vol. 8, p. 765 (cuts), 1965.

TABLE 6.6 ***Breakdown of 3.5-min Cycle on Reverse-flow Carbureted Water-gas Machine in Fig. 6.2*** *(Fuel is coproduct coke)*

Blow (operation involving 30%, or 63 s)

Primary air is admitted at the base of the mechanical generator and passes up through the fuel bed for the purpose of bringing the coke up to gas-making temperature. Secondary air is admitted into the large connection between the generator and carbureter to complete combustion of blast gases. Resulting products of combustion move up the side walls of the carbureter, over and down the superheater, and up and out the riser pipe stack, imparting much of their heat to the refractory brickwork along the way.

Blow run (operation involving 9%, or 19 s)

This follows immediately after the blow, while the fire temperature is at its peak and the blast gases contain the highest percentage of CO. Blow run is accomplished by closing the secondary air valve and the stack valve and allowing the producer gas to pass through the machine into the washbox out to the relief holder.

Up run (operation involving 32%, or 67 s)

Steam is admitted at the base of the generator and passes up through the red-hot coke, forming blue gas. Oil is sprayed downward into the carbureter countercurrent to the upward flow of hot blue gas. Oil gas is produced by the pyrolysis of the oil in an atmosphere of blue gas and from the radiant heat. The blue and oil gases mix and pass on to the superheater, where the pyrolysis of the gasified oil is completed and the gases fixed (or made permanent).

Back run (operation involving 24%, or 51 s)

Steam is admitted into the top of the riser pipe, passing up the superheater, where it is superheated, down through the carbureter, reacting with any carbon which might have been deposited during the oil-admission period, then reacting with the carbon in the generator fuel, finally passing out the bottom of the generator through the cast-iron back-run pipe through the three-way valve into the washbox, to the relief holder. Fuel is automatically charged during this portion of the cycle after the back-run, or reformed, oil has been shut off.

Final up run (operation involving 3%, or 6 s)

This puts a blanket of steam between the blue gas in the base of the generator and the air that follows. The carbureter, superheater, and riser pipe are already filled with back-run steam.

Blow purge (operation involving 2%, or 4 s)

This purges the machine of blue gas and steam and produces some CO, all of which is swept through the machine, through the washbox, into the relief holder. This is accomplished by opening the generator air valve prior to opening the stock valve, which releases the products of combustion to the atmosphere.

Source: Semet-Solvay Engineering Div., Allied Chemical Corp.

energy from the reaction between carbon and oxygen to supply the endothermic reaction between the carbon and steam. The reactions[15] may be written as follows:

$$C(amorph) + O_2(g) \longrightarrow CO_2(g) \qquad \Delta H_{60°F} = -174{,}600 \text{ Btu} \qquad (1)$$

$$CO_2(g) + C(amorph) \longrightarrow 2CO(g) \qquad \Delta H_{60°F} = +70{,}200 \text{ Btu} \qquad (2)$$

$$C(amorph) + H_2O(l) \longrightarrow CO(g) + H_2(g) \qquad \Delta H_{60°F} = +70{,}900 \text{ Btu} \qquad (3)$$

$$C(amorph) + 2H_2O(l) \longrightarrow CO_2(g) + 2H_2(g) \qquad \Delta H_{60°F} = +71{,}600 \text{ Btu} \qquad (4)$$

$$CO(g) + H_2O(l) \longrightarrow CO_2(g) + H_2(g) \qquad \Delta H_{60°F} = +700 \text{ Btu} \qquad (5)$$

The initial reaction in the producer is the formation of CO_2 and N_2 (1). As the gases progress up the bed, the initial CO_2 is reduced to CO (2), and the water vapor (made from liquid water in the jacket) is partly decomposed to give H_2, CO, and CO_2. The allowable temperature of the fuel bed, which depends on the fusion points of the fuel ash, usually ranges from 1800 to 2800°F, and the minimum

[15]Equilibrium and rate conditions for these reactions are given by Fulweiler, *op. cit.*, p. 737 and by Lowry, *op. cit.*

bed height ranges from 2 to 6 ft for most of the commonly used fuels. Producer gas is a low-Btu fuel gas, once used for all types of industrial heating purposes near where the gas is generated, a prime example being its application to heat coke ovens and other furnaces. Its use in the United States is almost obsolete. Producer-gas manufacture with low-cost-tonnage oxygen is also showing new possibilities. Here the Btu per cubic foot is doubled or tripled.

LIQUEFIED PETROLEUM GASES

Liquid propane and butane gas are employed as standby and peak-load supplements in municipal and industrial[16] systems using natural and manufactured gas, and as a complete gas supply in some communities and industries. LPG is used for flame weeding, tobacco curing, grain drying, and in motor vehicles and tractors. The petrochemical industry is the second largest consumer (Chap. 38). Since butane (3,260 Btu/ft^3) does not vaporize below 31°F, it is preferred in the South, whereas propane (2,520 Btu/ft^3) vaporizes down to -44°F and is used in the North. Sometimes a mixture of these two is employed. LPG comes from the ground as a constituent of wet natural gas or crude oil or as a by-product from refining. For example, a natural gasoline plant treats raw "wet" natural gas through absorption by "washing" with gas oil and fractionating out the usable fractions.

SYNTHETIC NATURAL GAS

Gas made from organic material to match the composition and combustion properties of natural gas is called synthetic natural gas (SNG) or pipeline-quality gas. There is now a need for this type of manufactured gas, which can be added to or is interchangeable with natural gas, because the demand for natural gas is greater than the amount that can be produced from wells. The raw materials most readily available for making this gas are oil and coal, with coal having the advantage because our total requirements can be supplied by domestic production. Gasification processes using coal and oil have been the subject of much research by the Institute of Gas Technology in Chicago, the American Gas Association, the Bureau of Mines, the Office of Coal Research, many energy-related industries, and private research institutions. The Institute of Gas Technology's HYGAS plant in Chicago and Consolidation Coal's CO_2 acceptor plant in Rapid City, S.D., are examples of large-scale coal-to-gas pilot plants in operations. The Bituminous Coal Research BI-GAS plant at Homer City, Pa., the Union Carbide/Battelle agglomerating ash process near Columbus, Ohio, and the Bureau of Mines synthane plant at Bruceton, Pa., are examples of large-scale coal-to-gas pilot plants under development. The processes being developed today have not advanced to a point where large-production-scale plants can be built.

Although the above processes have features that are different, they accomplish the same objective as indicated in the very simplified coal gasification flowsheet in Fig. 6.3. Gas with about 150 Btu/ft^3 can be produced by using coal, air, and steam in the gasification reaction. Some of the coal is burned with air to H_2O and CO_2 to provide the heat required by the gasification step. To make the gas product a clean-burning fuel, similar to producer gas in Table 6.1, the sulfur (H_2S) is removed. Low-Btu gas of about 300 Btu/ft^3 is made in the same manner, except that oxygen is used rather than air, or some other method is used for supplying the heat of reaction. For example, an inert solid can be heated in a furnace, using coal and air, and the hot inert material then transferred to the gasifier. The principal difference between the 150- and 300-Btu/ft^3 gas is that the 150-Btu/ft^3 gas contains about 50% nitrogen that entered the system as part of the air.

[16]Cronan, Natural Gas Processing Plant, *Chem. Eng. (N.Y.)*, **61**(8), 148 (1959) (complete flowchart for LPG natural gasoline and sulfur).

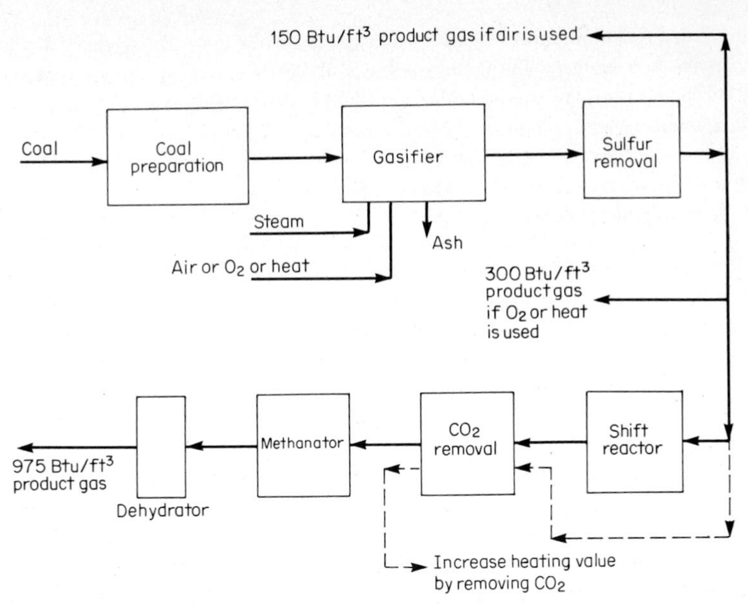

Fig. 6.3 Simplified coal gasification flow sheet. (*Consolidated Natural Gas Service Co.*)

To make a gas with 900 to 1000 Btu/ft^3, a low-Btu gas is first made by gasifying coal with steam and oxygen or heat and removing the sulfur. This gas is processed further in a water-gas-shift reactor to change the ratio of hydrogen to carbon monoxide to 3:1 by the following reaction:

$$CO + H_2O \longrightarrow H_2 + CO_2$$

The carbon dioxide formed in this step is then removed, and the hydrogen and carbon monoxide are converted to methane using a catalyst. The operating pressure in the gasifier, depending on the process, can be atmospheric to over 1000 lb, and temperatures can vary from about 1500 to about 3000°F. The higher pressure and lower temperature result in a larger amount of methane being formed in the gasifier.

AVAILABLE COAL GASIFICATION PROCESSES Two processes available today for making gas from coal are the Koppers-Totzek and the Lurgi processes. There have been 20 Koppers-Totzek plants built from 1949 to 1972, using mostly pulverized lignite and coal (this process can use all types of coal) with oxygen and steam as raw materials. The gas produced in this slagging gasifier is primarily carbon monoxide and hydrogen which can be shifted and methanated to produce pipeline-quality gas. None of these Koppers-Totzek plants is located in the United States.

The Lurgi pressurized (up to 400 psi) coal gasification process has a moving-bed system that cannot use strongly caking coals, although modifications to make caking coals acceptable are being investigated. The first Lurgi plant was built in Europe in the 1930s, and several more plants were put into operation before and during World War II to make synthesis gases for the German war effort. Additional plants were built in England, India, South Africa, and Korea during the 1950s and 1960s. To date, a total of 14 plants has been built, none in the United States. All these plants are designed to make either synthesis gas or low-Btu gas. SNG can be made with additional processing.

The Lurgi process requires that the coal be properly sized and then fed into the gasifier through a pressurized lock hopper. Steam and oxygen are fed into the reactor under the ash grate. As the coal gravitates downward and is heated, devolatilization commences and, from a temperature of

1150 to 1400°F onward, devolatilization is accompanied by gasification of the resulting char. Ash is removed through the ash lock hopper. The crude gas leaves the gasifier at temperatures between 700 and 1100°F, depending upon the type of coal. It contains carbonization products such as tar, oil, naphtha, phenols, ammonia, etc., and traces of coal and ash dust. This crude gas is passed through a scrubber where it is washed by circulating gas liquor and then cooled to a temperature at which the gas is saturated with steam. As higher-boiling tar fractions are condensed, the wash water contains tar to which the coal and ash dust is bonded. The steam-saturated gas is passed to a waste heat boiler in which waste heat at a temperature of 320 to 360°F is recovered. The gas liquor condensed in this boiler is pumped to the scrubber, and surplus gas liquor is routed to a tar–gas-liquor separator. The mixture of tar and dust is returned to the gasifier for cracking and gasification. The gas leaving the gasifier is mainly CO_2, CO, CH_4, H_2, and H_2O. The resulting H_2/CO ratio of the gas is not suitable for subsequent methane synthesis, so a shift conversion reaction ($CO + H_2O \longrightarrow CO_2 + H_2$) is necessary.

Impurities are removed from the gas by the Rectisol process. It employs a physical gas absorption system using organic solvents, preferably methanol, at low temperatures between approximately +30 and −80°F. The Rectisol unit has three sections: one for the removal of gas naphtha, unsaturated hydrocarbons, and other high-boiling gas impurities, one for the removal of CO_2, H_2S, and COS, and one to remove the last portion of CO_2 and to dehydrate the gas following the methane synthesis. The final product gas can have a heating value of 970 Btu/ft^3 and is almost all methane.

Because this process for the gasification of coal is available today, at least six plants with a capacity of 250 million ft^3 of gas per day are in various stages of planning. The one nearest completion is being built in New Mexico by the El Paso Natural Gas Company and is expected to start initial gas production soon.

OIL GASIFICATION As already indicated, there is now a need for SNG, so that some companies have built or are planning to build SNG plants using known oil-based processes. The appeal of oil gasification, especially when naphtha is used, is that such a plant requires a smaller investment and a shorter time to construct than a coal-to-gas plant. Generally, the degree of difficulty in gasifying oil increases as the weight of the oil used increases. When making gas from oil most of the cost of the product gas is due to the cost of oil. Therefore, when the Middle East imposed an oil embargo on the United States late in 1973 and the cost of crude oil more than doubled, the cost of gas made from oil became high. Because of the high price and limited supply of oil, of the 47 oil-to-gas plants being considered as of February 1974, only 3 are in operation, 5 are in the testing or start-up stage, 13 have been temporarily suspended, and 7 have been canceled.

An example of a naphtha gasification process is one being offered by The Gas Council of Great Britain International Consultancy Service. It is the catalytic rich gas (CRG) double methanation process. In this process route the CRG reactor is followed by two separate stages of methanation. The purified naphtha feedstock, after admixture with steam to give a steam/naphtha ratio of 2 lb/1 lb, is preheated to 450°C (840°F) and gasified in a CRG reactor. The rich gas leaving this reactor at 505°C (940°F) is cooled in a waste heat boiler to 300°C (570°F) before entering the first methanator in which part of the hydrogen reacts with the carbon oxides to cause a temperature rise of 74°C (133°F). The gas is then not only cooled in a second waste heat boiler, but part of the undecomposed steam is also rejected to allow further methane formation to take place in the second methanator; the temperature rise in this reactor is 40°C (72°F). The catalyst used for the methanation stages is identical with the CRG catalyst. It has a very low threshold temperature and, since it is protected from sulfur poisoning, has a very long life. The final stages of the process then comprise carbon dioxide removal and drying. There is a progressive reduction in the hydrogen and an increase in the methane content. After the removal of carbon dioxide the gas contains 98% methane with a calorific value of just under 1000 Btu/ft^3.

SELECTED REFERENCES

Abell, C.: Butane-Propane Power Manual, Jenkins, 1952.
American Gas Association: Current data available, including Gas Engineers' Handbook, Industrial Press.
Coal Processing Technology, AIChE, 1974.
D'Ouville, E. L. and M. L. Kalinowski (eds.): Literature of the Combustion of Petroleum, ACS, 1958.
Gumz, W.: Gas Producers and Blast Furnaces: Theory and Methods of Calculation, Wiley, 1950.
Katz, D. L., *et al.:* Handbook of Natural Gas Engineering, McGraw-Hill, 1959.
Kohl, A. L., and F. C. Riesenfeld: Gas Purification, McGraw-Hill, 1960.
Lewis, B., and G. von Elbe: Combustion, Flames and Explosions of Gases, 2d ed., Academic, 1961.
Liston, A. M., *et al.:* Dynamic Natural Gas Industry, University of Oklahoma, 1963.
LNG Information Book, American Gas Association, 1973.
Lom, W. L.: Liquefied Natural Gas, Halsted, 1974.
Massey, L. G.: Coal Gasification, ACS, 1974.
McDermott, J.: Liquid Fuels From Oil Shale and Tar Sands, Noyes, 1972.
McDermott, J.: Liquefied Natural Gas Technology, Noyes, 1973.
Neuner, E. J.: Natural Gas Industry, University of Oklahoma, 1960.
Shnidman, L.: Gaseous Fuels, 2d ed., American Gas Association, 1954.
Sliepcevich, F. M., and H. T. Hashemi: LNG Technology, *Chem. Eng. Prog.*, **63**(6), 51 (1967).
Symposium Papers: Clean Fuels from Coal, Institute of Gas Technology, 1973.
Williams, A. F., and W. L. Lom: Liquefied Natural Gas, Halsted, 1974.

chapter 7

INDUSTRIAL GASES

Industrial gases have performed varied and essential functions in our economy. Some are raw materials for the manufacture of other chemicals. This is particularly true of oxygen, nitrogen, and hydrogen. Nitrogen preserves the flavor of packaged foods by reducing chemical action leading to rancidity of canned fats. Some gases are essential medicaments, like oxygen and helium. However, many of these gases, their liquids, and their solids have a common application in creating cold largely by absorbing heat, upon evaporation, by performing work, or by melting. In past decades, the outstanding examples of this have been liquid carbon dioxide and Dry Ice. On the other hand, with the modern expansion of industry, a new division of engineering has arisen called *cryogenics.*[1] This widely embracing term pertains to the production and use of extreme cold at the range of temperatures below −150°F. The term "cryogenics" has been applied very extensively during the last decade and is exemplified by the use of liquid hydrogen, oxygen, and fluorine in missiles for military and space projects. New cryogenic techniques have been worked out, reducing the cost of liquefaction and pertaining to equipment, storage, shipment, and use of cold liquids and gases. The economic advantage of using cryogenic liquids is apparent when it is realized that 1 ft^3 of liquid oxygen, for example, is equivalent to 860 ft^3 of gaseous oxygen. Each cylinder of gaseous oxygen contains only 17 lb of oxygen, or 220 ft^3 (at 60°F, 1 atm), but at a pressure of 2,200 lb in a steel tank weighing 150 lb. Cryogenic engineers store and ship these liquids in tanks or large tank cars[2] built on the two-walled vacuum bottle (Dewar) principle. The value of cryogenic[3] gases and liquids is about $1 billion per year. These cryogenic, or supercold, temperatures cause fundamental changes in properties of materials. Cryogenics is being applied to rocket propulsion, infrared photoptics, and electronic data processing, with newer applications in magnetics and high-vacuum pumping. The major application of cryogenics to the chemical field is in the manufacture of nitrogen for ammonia, and in metallurgy, where the use of oxygen hastens (by 25% or more) the production of steel in open-hearth furnaces, converters, and even in blast furnaces for pig iron. Cryogenics,[4] in manufacturing low-temperature liquids, has long applied the following *fundamental principles*, and currently, most extensively to air and its constituents:

1. Vapor compression with liquefaction if below the critical temperature of the gas in question.
2. Interchange of heat in heat exchangers such as double-pipe units; refrigeration.
3. Adiabatic expansion of the compressed air to cool by the Joule-Thomson effect, due to energy being consumed in overcoming the attraction between molecules. This must be carried out below the inversion temperature of the respective gas (see Helium). Precooling below the ambient temperature is necessary only for helium isotopes, hydrogen isotopes, and neon.

[1]From two Greek words meaning "the making of cold."

[2]New Tank Car Hauls Liquid H_2 from Florida to California, *Chem. Eng. (N.Y.)*, **68**(18), 66 (1961). At −423°F, the loss is only 0.3% per day.

[3]Cryogenics: A Decade to $1 Billion, *Chem. Eng. News*, Aug. 26, 1968, p. 21; Specialty Gases: Large and Growing, *Chem. Eng. News*, Oct. 25, 1971, p. 14; Fallwell, Oxygen Finding a Big Outlet in Treatment of Wastewater, *Chem. Eng. News*, July 15, 1974, p. 7. See Chap. 3 for the use of oxygen in sewage treatment.

[4]See Perry, p. 11-10, 11-48 to 11-52, 12-49 to 12-53.

4. Cooling of compressed gases by allowing them to perform work under adiabatic conditions (Claude modification). One kilogram of air at 200 atm and 10°C when simply expanding to 5 atm produces a cooling effect of 10 kcal (Joule-Thomson effect), whereas, if the same quantity of air expands to 5 atm in a cylinder *doing work,* the cooling effect is 23 kcal, assuming that the working engine has an efficiency of 70%. This latter effects a cooling of 135°C.

5. Cooling of the liquids by evaporation.

6. Separation, for example, of liquid oxygen from the more volatile nitrogen in a rectifying column. Nitrogen boils at −195.8°C, and oxygen at −183°C at 1 atm. The difference of 12.8° in boiling point permits fractional separation of the two liquids.

CARBON DIOXIDE

Carbon dioxide in liquid and solid forms has been known for over a century. Although Thilorier produced solid carbon dioxide in 1835 from the liquid material, it was not until 1924 that the solid product gained industrial importance through its first and still most important use for refrigeration. The production of carbon dioxide in 1974 was about 1,350,000 tons total for the gaseous, liquid, and solid forms.

USES[5] By far the largest use of the solid form is for refrigerating ice cream, meat, and other foods. An added advantage is that a carbon dioxide atmosphere reduces meat and food bacteria spoilage. The solid form is also important as a source of carbon dioxide for inert atmospheres and occasionally for carbonated beverages. There are many other specialty uses, such as chilling aluminum rivets and shrink-fitting machine parts. Liquid carbon dioxide is also used for safer blasting of coal, the liquid being used to fill the drill holes and the blasting secured by a controlled heater inserted in the liquid, resulting in larger lumps of coal. The largest outlet of liquid carbon dioxide is for carbonated beverages. It is also important as a fire-extinguishing material. Gaseous carbon dioxide has many applications in the chemical industry, such as in the making of salicylic acid (Chap. 40) and as a raw material for soda ash (Chap. 13).

Carbon dioxide has advantages over ordinary acids in neutralizing alkalies, because it is easily shipped in solid form, is noncorrosive in nature, and is light in weight. Chemically, it is equivalent to more than twice its shipping weight in sulfuric acid and about five times its weight in hydrochloric acid. With respect to food refrigeration, solid carbon dioxide is primarily a transport refrigerant. Its advantages cannot be attributed to any single factor but result from its dryness, its relatively high specific gravity, its excellent refrigerating effect, its low temperature, and the insulating and desiccating action of the gas evolved. Generally, about 1,000 lb of solid carbon dioxide will refrigerate an average car for a transcontinental rail trip without recharging. A similar load of water ice would require 3,700 lb initially plus several supplemental chargings along the way. Liquid nitrogen has also entered this refrigerating field.

MANUFACTURE OF PURE CO_2[6] Although there are many sources of CO_2, the following three are the most important for commercial production:

1. Flue gases which result from burning carbonaceous material (fuel oil, gas, coke) and contain about 10 to 18% CO_2.

2. By-product from the fermentation industries, resulting from dextrose breakdown into alcohol and carbon dioxide, the gas containing about 99% CO_2.

3. By-product of lime-kiln operations, where carbonates are calcined to oxides; the gases analyze from 10 to 40% CO_2.

[5]Yaws, Li, and Kuo, Carbon Oxides, *Chem. Eng. (N.Y.),* **81**(20), 115 (1974).

[6]Perry, p. 16-36 and p. 16-42 for adsorption of carbon dioxide, pp. 3-162 to 3-164 for thermodynamic properties; ECT, vol. 3, p. 125, and vol. 4, p. 353, 1964; Montrone and Long, Choosing Materials for CO_2 Absorption Systems, *Chem. Eng. (N.Y.),* **78**(2), 94 (1971).

TABLE 7.1 *Comparison of Physical Properties of Solid CO_2 and Water Ice*

Property	*Solid CO_2*	*Water ice*
Specific gravity	1.56	0.90
Sublimation point or melting point	−109.6°F	32°F
Critical temperature	87.8°F	689°F
Critical pressure	1,071 psia	2,860.6 psia at 32°F
Latent heat of fusion, lb	82.0 Btu	144 Btu
Latent heat of vaporization, lb	158.6 Btu	1,074.29 Btu at 2°F
Weight of gas, ft^3	0.117 lb	0.000304 lb
Weight solid, ft^3	90 lb	57 lb
Latent heat of sublimation, lb	248 Btu	
Refrigerating effect, lb	275 Btu	144 Btu

An absorption system is used for concentrating the CO_2 gas obtained from sources 1 and 3 to over 99%. In all cases the almost pure CO_2 must be given various chemical treatments for the removal of minor impurities which contaminate the gas. One of the reversible reactions long used for the concentration of CO_2 is

$$Na_2CO_3 + CO_2 + H_2O \rightleftharpoons 2NaHCO_3$$

This reaction is forced to the right by increasing the partial pressure of the CO_2 and by reducing the temperature. It is forced to the left by heating the sodium bicarbonate solution. The absorption efficiency of 10 to 18% CO_2 is not very good. Other usually more economical reversibly absorbing solutions[7] are hot, concentrated potassium carbonate solution and monoethanolamine (Girbotol process, Chap. 6). The CO_2 available in these cases comes from the combustion of fuel oil in a boiler plant generating the required steam; it happens that the steam is the critical, expensive, and controlling item. To carry out this manufacture, the principal sequences detailed in the following procedure are employed:

Oil, natural gas, or coke burned, giving heat for 200 psig steam and furnishing 10 to 15% CO_2 at 650°F (Ch and Op).

Flue gas cooled, purified, and washed by passing through two water scrubbers (Op).

CO_2 removed by countercurrent selective absorption into an aqueous solution of ethanolamines (Ch).

CO_2 and steam leave the top of the reactivator, passing through a CO_2 cooler to condense the steam, which returns to the tower as reflux (Op).

CO_2 at about 30 psig purified in a permanganate scrubber from traces of H_2S and amines and dried (Ch).

CO_2 compressed, cooled, and liquefied (Op).

For "liquid" draw off from CO_2 receiver.

For Dry Ice:

Liquid CO_2 subjected to reduction to atmosphere pressure, with consequent partial solidification (Op). The evaporated revert gas returns through the precooler.

Recirculation, with recompression and recooling of CO_2 (Op).

CO_2 "snow" compressed to cake (Op).

Dry Ice cakes generally sawed to 10-in. cubes of approximately 50-lb weight (Op).

A typical flowchart, for continuously producing either liquid or solid and embracing the sequences of these unit operations and chemical conversions, is shown in Fig. 7.1. In the merchandising of liquid CO_2, the energy (hence expense) involved in handling the cylinders, full and

[7]See this chapter under Hydrogen Purification.

empty, is so great that this one fact has required the installation of many relatively small plants economically located in consuming centers. However, bulk cryogenic shipment of relatively low-pressure liquid *cold* CO_2 is now being practiced to reduce container cost and weight. If heat exchangers are properly designed, it is necessary to buy little outside energy for solid CO_2 and none for liquid CO_2 (Fig. 7.1). The live steam generated by the boiler is sufficient to power the turbines for pumping and compression, and the exhaust steam from turbine-driven compressors boils off the CO_2 from the amine solution in the reboiler. The condensate from the reboiler is returned to the water boiler. *These* CO_2 *plants are splendid examples of well-balanced energy and chemical requirements.*

Another source of CO_2 is the fermentation industry, as described in Chap. 31. If yeast is used, alcohol and CO_2 are produced, and certain other microorganisms generate solvents and a gaseous mixture of H_2 and CO_2. The yield of CO_2 varies with the mode of fermentation. From starchy material, such as corn, is obtained from 1 bu, for example, $2\frac{1}{2}$ gal of 190-proof ethyl alcohol and 17 lb of CO_2. The recovery and purification of CO_2 from fermentation differ from the absorption system in that the temperature seldom exceeds 105°F, so that no special cooling is necessary and the CO_2 content of the gas usually starts above 99.5%. When the fermentors are sealed for the recovery of the gases, a purer and higher yield of CO_2 per gallon of mash is obtained, and the yield of alcohol is increased by at least 1% by alcohol recovery from the CO_2 scrubbers.

A typical CO_2 recovery from fermentation is illustrated in the flowchart in Fig. 7.2. Here purification consists of oxidation of the organic impurities and dehydration by means of chemicals in liquid form. The gas from the fermentors is passed through three scrubbers containing stoneware spiral packing and on to the gasometer. The first scrubber contains a weak alcohol solution which acts as a preliminary purifier and removes most of the alcohol carried by the gas. The next two scrubbers, in which the scrubbing medium is deaerated water, remove almost all the water-soluble impurities. The scrubbing liquid is pumped either to the stills or to the fermentors for alcohol recovery.

From the gasometer the gas is conducted to a scrubber containing $K_2Cr_2O_7$ solution, which oxidizes the aldehydes and alcohols in the gas, and is then cooled. In the second scrubber, which contains sulfuric acid, oxidation is completed and the gas is dehydrated. The CO_2 leaving the acid

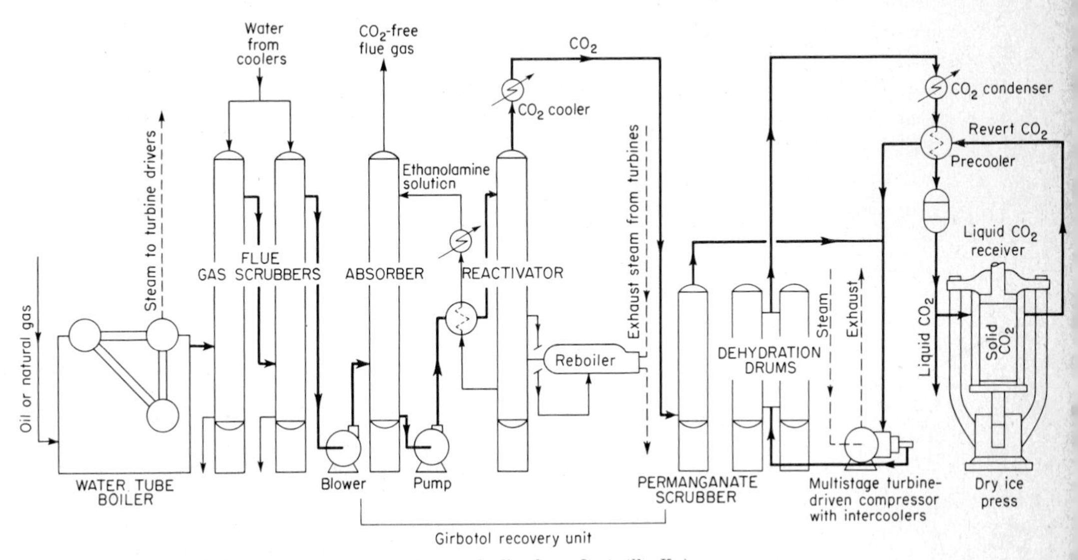

Fig. 7.1 Flowchart for CO_2 from fuel oil or natural gas. (*Girdler Corp., Louisville, Ky.*)

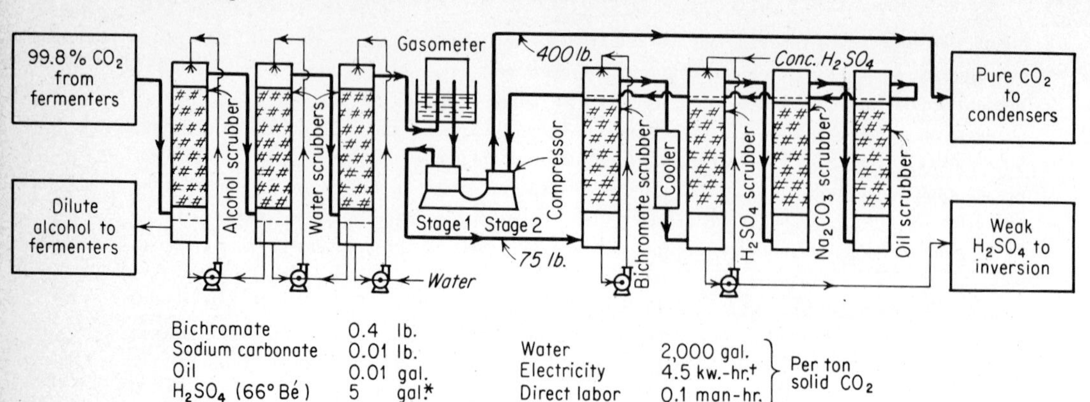

Bichromate	0.4 lb.			
Sodium carbonate	0.01 lb.	Water	2,000 gal.	Per ton solid CO_2
Oil	0.01 gal.	Electricity	4.5 kw.-hr.†	
H_2SO_4 (66° Bé)	5 gal.*	Direct labor	0.1 man-hr.	

*Weak H_2SO_4 all recovered for use in molasses fermentation
†Not including solid CO_2 compressing system

Fig. 7.2 Flowchart for fermentation CO_2 purification.

scrubber contains some entrained acid which is removed in a tower filled with coke and over which a Na_2CO_3 solution is circulated. When the acid is neutralized, CO_2 is released. Before going to the compressor, the gas passes through a scrubber containing a small amount of glycerin, which absorbs the oxidized products and delivers an odorless gas to the compressor. The sulfuric acid, after being used for deodorization and drying, is pumped to the distillery, where it serves for pH control.

HYDROGEN

USE AND ECONOMICS Hydrogen has long been an important *gaseous* raw material for the chemical and petroleum industries. The annual production of hydrogen is in excess of 2.5 trillion ft^3. Table 7.2 gives the hydrogen requirements for making some typical chemicals. In the production of hydrogen (exclusive of that for NH_3 and CH_3OH) a few percent is of high purity (99.5%). High-purity hydrogen is the so-called *merchant hydrogen* supplied in cylinders or trucks. It is frequently manufactured by electrolysis of alkalized water or from steam hydrocarbon reforming. Somewhat over one-third of the total production is used for ammonia. Other important chemical uses involve hydrogenating edible oils, electric machinery and electronics, missiles, and heat-treating ovens for metals and start-up of catalytic crackers. New uses for hydrogen are expected in the future.[8]

HYDROGEN MANUFACTURE Hydrogen is derived almost exclusively from carbonaceous materials, primarily hydrocarbons, and/or water. These materials are decomposed by the application of energy, which may be either electric, chemical, or thermal. Examples include electrolysis of water, steam reforming[9] of hydrocarbons, and thermal dissociation of natural gas (Chap 8). Hydrogen is also produced by partial oxidation of hydrocarbons and by such less important methods[10] as the

[8]Nilsen, A Solar/Hydrogen Economy?, *Chem. Eng. (N.Y.)*, **81**(25), 44, (1974); Hydrogen: Likely Fuel of the Future, *Chem. Eng. News*, June, 1972, p. 14; Johnson and Litwak, Outlook for H_2 and CO, *Chem. Eng. Prog.* **65**(3), 21 (1969); Hydrogen Economy Concept Gains Credence, *Chem. Eng. News*, Apr. 1, 1974, p. 15; On-site Plants Swap Hydrogen for Waste Oil, *Chem. Week*, July 11, 1973, p. 49.

[9]Hydrogen Consumption Climbs for New Heights, *Chem. Eng. (N.Y.)*, **68**(10), 62 (1961) (flowchart for natural gas); Hydrogen Use and Output Moving Up Fast, *Chem. Eng. News*, Nov. 27, 1967, p. 20.

[10]Hydrogen, Girdler Corp., Louisville, Ky., 1962 (pamphlet); Hydrogen, CW Report, *Chem. Week*, May 19, 1962, p. 120. Both these references give further details on all the processes. A flowchart is given in the Girdler pamphlet for the palladium diffusion process, p. 29. See Ammonia, Chap. 18.

TABLE 7.2 ***Hydrogen Required for Typical Products***

Raw material except hydrogen	*Product*	*Hydrogen required, cu ft, 60°F**	*Raw material except hydrogen*	*Product*	*Hydrogen required, cu ft, 60°F**
Phenol	Cyclohexanol	25,000	Olein	Stearin	2,600
Nitrogen	Ammonia	84,000	Diisobutylene	Isooctane	6,700
Naphthalene	Tetralin	12,000	Carbon monoxide	Methanol	230

Source: Stengel and Shreve, Economic Aspects of Hydrogenation, *Ind. Eng. Chem.*, **32,** 1212 (1940). *All hydrogen requirements are per ton of product, except synthetic methanol (per gallon) and isooctane (per barrel).

steam-iron process, water gas and producer gas processes, and separation from coke-oven[11] gas and refinery off-gas streams. Diffusion through a palladium-silver alloy furnishes very high-purity hydrogen.

ELECTROLYTIC METHOD[12] The electrolytic process produces high-purity hydrogen and consists in passing direct current through an aqueous solution of alkali, decomposing the water according to the following equation:

$$2H_2O(l) \longrightarrow 2H_2(g) + O_2(g) \qquad \Delta H = +136 \text{ kcal}$$

The theoretical decomposition voltage for this electrolysis is 1.23 V at room temperature; however, because of the overvoltage of hydrogen on the electrodes and also cell resistance itself, voltages of 2.0 to 2.25 V are usually required. A typical commerical cell electrolyzes a 15% NaOH solution, uses an iron cathode and a nickel-plated-iron anode, has an asbestos diaphragm separating the electrode compartments, and operates at temperatures from 60 to 70°C. The nickel plating of the anode reduces the oxygen overvoltage. Most types of cells produce about 7.0 ft^3 (9.408 ft^3, theoretically) of hydrogen and half as much oxygen per kilowatt hour. The gas is about 99.7% pure and is suitable even for hydrogenating edible oils. The cells are of two types: the bipolar, or filter-press, type, where each plate is an individual cell, and the unipolar, or tank, type, usually containing two anode compartments with a cathode compartment between them. The unipolar cells may be open or closed tanks. In most installations the oxygen produced is wasted unless it can be used locally. Hydrogen is also obtained from other electrolytic processes, for example, electrolysis of salt brine (cf. Chaps. 13 and 14).

STEAM-HYDROCARBON REFORMING PROCESS This process consists of catalytically reacting a mixture of steam and hydrocarbons at an elevated temperature to form a mixture of H_2 and oxides of carbon. The following basic reactions occur:

$$C_nH_m + nH_2O \rightleftharpoons nCO + \left(\frac{m}{2} + n\right) H_2$$

$$CO + H_2O \rightleftharpoons CO_2 + H_2$$

Although the equations are shown for the general case of any hydrocarbon feed, only light hydrocarbons have been successfully used in commercial practice. Natural gas is most common, and propane and butane (LPG) are also frequently used. With the use of a specially prepared catalyst, naphtha is also a suitable feedstock.

[11]New Nitrogen Plant, Geneva Works, Hydrogen Extracted from 252 Coke Ovens, *U.S. Steel Quart.*, **11**(3), 1 (1957); U.S. Steel to Increase Chemicals Capacity (Hydrogen, etc.), *Chem. Eng. News*, Oct. 5, 1964, p. 25.

[12]Pressure Lowers Electrolytic H_2 Cost, *Chem. Eng.* (*N.Y.*), **67**(5), 60 (1960) (cell is operated at 30 atm); Solid Electrolytes Offer Route to Hydrogen, *Chem. Eng. News*, Aug. 27, 1973, p. 15.

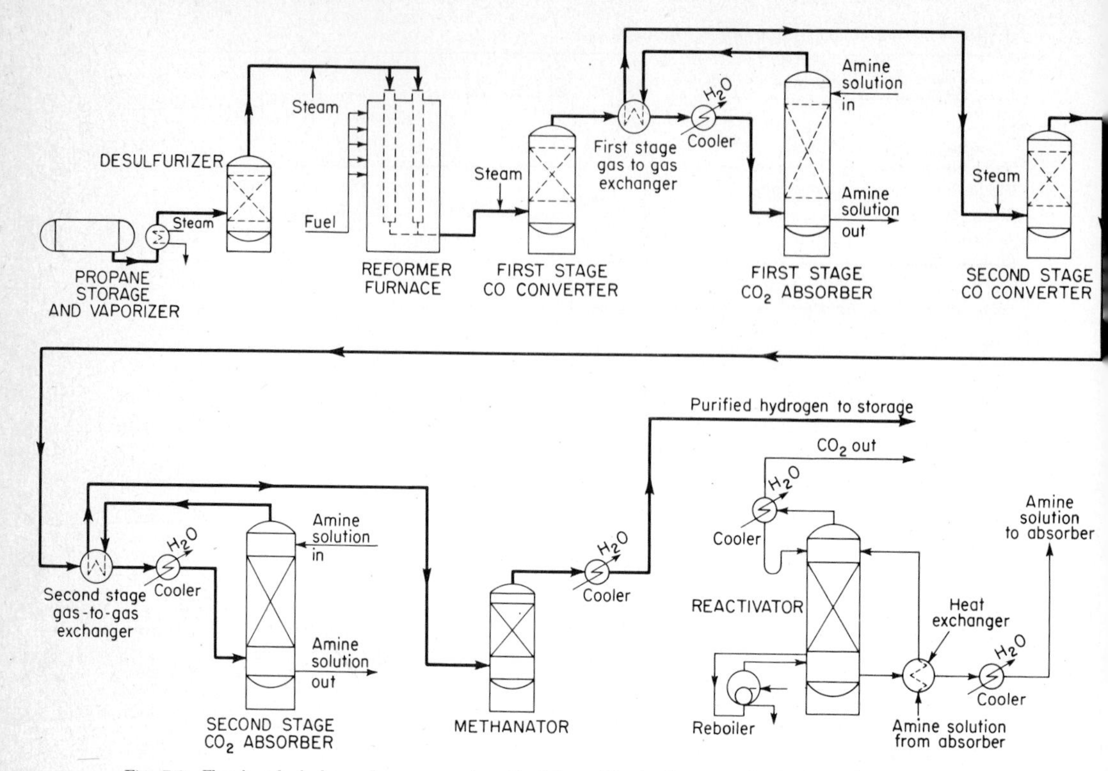

Fig. 7.3 Flowchart for hydrogen from propane (or with slight modification from natural gas). See Table 7.3. (*Girdler Corp., Louisville, Ky.*)

The first reaction is the reforming reaction. It is highly endothermic, and the moles of product exceed the moles of reactant so the reaction goes to completion at high temperature and low pressure. Excess steam is normally used. Although the basic purpose is to prevent carbon formation, it also helps force the reaction to completion.

The second reaction is the water-gas-shift reaction. It is mildly exothermic and is favored by low temperature but unaffected by pressure. Excess steam also forces this reaction to completion and is so used. A catalyst is usually employed. Both these reactions occur together in the steam-reforming furnace at temperatures of 1400 to 1800°F. The composition of the product stream depends upon the process conditions, including temperature, pressure, and excess steam, which determine equilibrium, and the velocity through the catalyst bed, which determines the approach to equilibrium. A typical product contains approximately 75% H_2, 8% CO, and 15% CO_2; the remainder consists of nitrogen and unconverted methane.

For the production of additional H_2, the reformer is followed by a separate stage of water-gas-shift conversion. Additional steam is added, and the temperature is reduced to 600 to 700°F to obtain more favorable equilibrium conditions. A single stage converts 80 to 95% of the residual CO to CO_2 and H_2. Because the reaction is exothermic, the reactor temperature rises which enhances the reaction rate but has an adverse effect on the equilibrium. When high concentrations of CO exist in the feed, the shift conversion is usually conducted in two or more stages, with interstage cooling to prevent an excessive temperature rise. The first stage may operate at higher temperatures to obtain high reaction rates, and the second at lower temperatures to obtain good conversion.

TABLE 7.3 ***Product Analysis—Utility Requirements for Hydrogen by Steam Hydrocarbon Process*** *(The purification includes two stages of carbon monoxide conversion and carbon dioxide removal and one stage of methanation)*

Product hydrogen analysis	*Low-pressure reforming, vol %*	*Pressure reforming, vol %*
Carbon monoxide	0.001	0.001
Carbon dioxide	0.001	0.001
Methane	0.400	1.200
Hydrogen (minimum)	99.598	98.798
Raw Material and Utility Requirements per 1,000 ft³ of Hydrogen		
Process material:		
Propane, sulfur-free	2.75 gal	2.80 gal
Natural gas, 1000 Btu/std ft³	250 ft³	255 ft³
Fuel: natural gas, propane, or oil	215,000 Btu	230,000 Btu
Steam, 70 psig:		
Propane for process	280 lb	150 lb
Natural gas for process	240 lb	135 lb
Cooling water, 30°F rise:		
Propane for process	1,350 gal	850 gal
Natural gas for process	1,200 gal	800 gal
Power, exclusive of lighting	0.5 kWh	0.6 kWh
Estimated cost of replacement catalysts and chemicals	1.0¢	0.75¢

This process using propane is depicted in Fig. 7.3, and the material requirements are listed in Table 7.3. See also synthesis gas. This process[13] can be broken down into the following coordinated sequences:

Propane is vaporized by steam (Op).

Propane vapor is desulfurized by contact with activated carbon to prevent deactivation of the catalyst (Op).

Propane vapor mixed with steam is reformed (Fig. 7.4) over a nickel catalyst at about 1500°F in alloy tubes in a combustion furnace (to furnish the heat for this endothermic reaction) (Ch).

The gases H_2, CO, and some CO_2 are cooled to about 700°F, and the partial pressure of water increased by addition of steam or condensate, and passed over an iron oxide catalyst in the first-stage CO converter, where 90 to 95% of the CO is converted to CO_2 with more H_2 (Ch). (This is the so-called shift,[14] or water-gas-conversion, reaction.)

These hot gases are first cooled by heat exchange with the gases leaving the first-stage CO_2 absorber before entering the second-stage CO converter, and finally by water to about 100°F (Op).

The cooled gases are scrubbed with a monoethanolamine solution in the first-stage Girbotol absorber to remove essentially all the CO_2 (Ch). Following this, the gases heated by exchange with the gases leaving the first-stage CO converter are passed through the second-stage CO converter (Ch), also followed by second-stage CO_2 absorber (Ch).

The gases low in CO_2 and CO are heated to about 600°F by exchange with gases from the second-stage CO converter and passed to the methanator over a nickel catalyst to convert essentially all the carbon oxides to methane by reaction with H_2 (Ch).

The product H_2 gas from the methanator is cooled to 100°F (with water), leaving it pure except for saturation with water vapor. The lean amine solution from the Girbotol reactivator is first pumped

[13]Further details and other flowcharts as well as technical pamphlets on hydrogen should also be consulted. These are available from Girdler Corp., Louisville, Ky., Chemical Construction Co., New York, N.Y., and M. W. Kellogg Co., Houston, Texas.

[14]Moe, Design of Water-gas Shift Reactors, *Chem. Eng. Prog.*, **58**(3), 33 (1962). Rate equations for producing hydrogen.

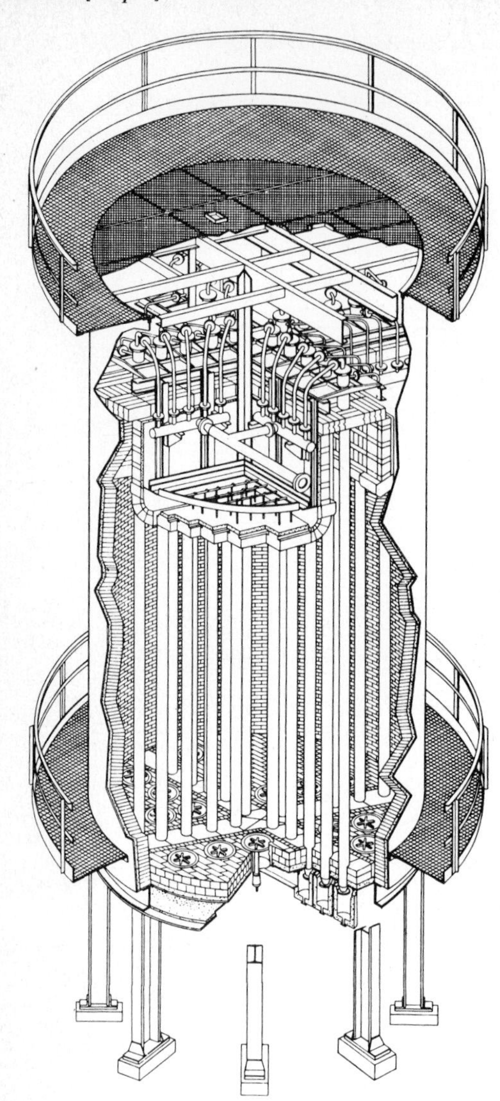

Fig. 7.4 Domestic representation of furnace for catalytic reforming of hydrocarbons with steam to H_2 and CO. (*Chemical Construction Corp.*)

through the second-stage CO_2 absorber to reduce the CO_2 content of the H_2 gas to a minimum and then pumped to the first-stage CO_2 absorber to remove the bulk of the CO_2 in the H_2 gas. The rich solution is returned to the reactivator through heat exchangers, and the CO_2 is stripped out by steam (Ch).

The CO_2, about 35 lb/1,000 ft^3 H_2, can be recovered for sale or use (Op).

CATALYSTS In many reactions connected with hydrogen manufacture, various catalysts[15] are necessary, such as nickel for steam-hydrocarbon reforming, iron oxide for CO conversion or the shift reaction, nickel for methanation, and nickel for ammonia cracking or dissociation.

[15]Girdler, Catalysts, Chemical Products Division, Chemetron Corp., Louisville, Ky., 1960 (pamphlet); Catalyst Cuts Hydrogen Costs, *Chem. Eng. News*, Feb. 11, 1963, p. 46; Habermehl, Improved Catalysts Reduce Costs, *Chem. Eng. Prog.*, **61**(1), 57 (1965).

PARTIAL OXIDATION PROCESSES These rank next to steam-hydrocarbon processes in the amount of hydrogen made. They can use natural gas, refinery gas, or other hydrocarbon gas mixtures as feedstocks, but their chief advantage is that they can also accept liquid hydrocarbon feedstocks such as gas oil, diesel oil, and even heavy fuel oil. There are three commercial versions of the process. It was originally developed by Texaco,[16] and variations were introduced later by both Shell and Montecatini,[17] which differ primarily in the design and operation of the partial oxidation burner. All employ noncatalytic partial combustion of the hydrocarbon feed with oxygen in the presence of steam in a combustion chamber at flame temperatures between 1300 and 1500°C. When methane is the principal component of the feedstock, the reactions involved are:

$$CH_4 + 2O_2 \longrightarrow CO_2 + 2H_2O$$

$$CH_4 + CO_2 \longrightarrow 2CO + 2H_2$$

$$CH_4 + H_2O \longrightarrow CO + 3H_2$$

The first reaction is highly exothermic and produces enough heat to sustain the other two reactions, which are endothermic. The net reaction is as follows:

$$CH_4 + \tfrac{1}{2}O_2 \longrightarrow CO + 2H_2$$

This reaction is exothermic, and so the overall process is a net producer of heat; for efficient operation, heat recovery (using waste heat boilers) is important.

The product gas has a composition which depends upon the carbon/hydrogen ratio in the feed and the amount of steam added. Pressure does not have a significant effect on composition, and the process is usually conducted at 20 to 40 atm, permitting the use of more compact equipment and reducing compression costs. The following composition is typical for a fuel oil feedstock:

	Mole percent
H_2	48.0
CO	46.1
CO_2	4.3
CH_4	0.4
N_2, etc.	0.3
H_2S	0.9
	100.0

This gas has a much higher carbon oxides/hydrogen ratio than steam reformer gas.

In the commercial application of the Texaco process, preheated oil is mixed, under pressure, with oxygen and preheated steam and fed to the partial oxidation burner. About 260 ft^3 of 95 to 99% oxygen is used per 1000 ft^3 of H_2 + CO produced. The product gas is cooled by a direct-contact water quench followed by a water scrubber and a filter, all of which serve to remove unreacted carbon from the product gas. The carbon is removed from the quench water, for example, by filtering, for disposal or reuse, or it may be discarded as a water slurry.

The remaining process steps for conversion of the partial oxidation product gas to hydrogen are the same as for the steam-hydrocarbon reforming process: water-gas-shift conversion, CO_2 removal via monoethanolamine scrubbing, and methanation.

COAL GASIFICATION PROCESSES Coal as a source of hydrogen will assume increasing importance in the future as reserves of gaseous and liquid hydrocarbon feedstocks decrease. Although

[16]Eastman, Synthetic Gas by Partial Oxidation, *Ind. Eng. Chem.*, **48,** 1118 (1956); Girdler, Hydrogen Pamphlet, p. 12; Labine, How the Air Force Liquefied Hydrogen, *Chem. Eng. (N.Y.)*, **67**(2), 86 (1960) (process flowchart of Texaco partial oxidation of crude oil with shift reaction, including catalytic change of ortho- to parahydrogen before liquefaction).

[17]ECT, vol. 11, p. 352, 1966; Hydrogen Gasification, *Adv. Pet. Chem. Refin.*, **10,** 167 (1965).

coal was used in early water-gas and producer-gas plants for the manufacture of a H_2-CO mixture, these plants cannot compete with more modern processes. Although many coal gasification processes are in various stages of development, only the three commercialized versions will be described. These are the Lurgi process, the Koppers-Totzek process, and the Winkler process.[18] Each reacts coal with steam and oxygen to produce a water-gas composition. Differences lie mainly in the reactor vessels.

The Lurgi process takes place in a moving-bed gasifier at pressures up to 30 atm. Dried, graded coal is introduced into the top of the gasifier through a lock hopper and is spread onto the bed below via a distributor. The bed is supported by a revolving grate through which steam and oxygen are fed and through which the ash drops to a lock hopper below for periodic removal. Devolatilization occurs initially, followed by gasification as the temperature reaches 1150 to 1600°F. Complete oxidation of a portion of the coal in the lower bed supplies heat for the process. The crude gas leaving the chamber contains oils, tars, and other devolatilization products, plus dust and ash, and is cleaned by oil scrubbing.

The Koppers-Totzek process gasifies finely divided coal entrained in a stream of oxygen and steam at nearly atmospheric pressure in a horizontal, cylindrical gasifier, which is refractory-lined and steam-jacketed for cooling. The entrained mixture is introduced through either two or four opposing burners. Gasification takes place at temperatures up to 3500°F, and the ash at these temperatures is converted to molten slag, half of which drops out into a quench tank, the rest leaving with the crude product gas at 2750°F. At this temperature, no oils, tars, or other devolatilization products can exist and thus are not contained in the product. The gasifier is usually equipped with water sprays at the outlet to cool the gas partly and to solidify the slag.

The Winkler process gasifies crushed coal in a fluidized bed with steam and oxygen at atmospheric pressure. Bed temperatures reach 1500 to 1850°F, which is high enough to react all tars and condensable hydrocarbons; up to 3% methane may exist. Most of the ash leaves the gasifier in the product gas. Secondary steam and oxygen are added above the fluid bed to convert unreacted carbon leaving the bed. A radiant boiler is included in the upper section of the gasifier to cool the gas to 350 to 400°F before leaving, to prevent sintering of fly ash on the walls of the exit duct.

The Lurgi process is the only one which operates under pressure and provides process operating economy when the product is processed under pressure. However, unlike the others, it requires sized coal, and it cannot handle coking coal. The Koppers-Totzek process uses all the coal and can handle any type and rank of coal. The Winkler process has difficulty handling coking coals unless they are suitably pretreated.

Typical product gas compositions are as follows:

	Composition (dry basis), mol %		
	Lurgi	*Koppers-Totzek*	*Winkler*
H_2	38.0	36.7	41.8
CO	20.2	55.8	33.3
CO_2	28.6	6.2	20.5
CH_4	11.4	0.0	3.0
C_2H_6	1.0	0.0	0.0
H_2S or COS	0.5	0.3	0.4
N_2	0.3	1.0	1.0

Conversion of these gases to high-purity hydrogen requires the water-gas-shift conversion and purification processes previously described under steam-hydrocarbon reforming.

[18]Perry, Coal Conversion Technology, *Chem. Eng. (N.Y.)*, **81**(15), 88 (1974).

CRACKED AMMONIA A mixture of nitrogen, 1 volume, and hydrogen, 3 volumes, may be prepared from the cracking or dissociation of ammonia. The process consists in vaporizing the liquid ammonia from cylinders, heating it to 1600°F, passing it over an active catalyst, and then cooling it in heat exchangers, where the incoming gas may be vaporized. Brandt states, "A single 150 lb cylinder of anhydrous ammonia will produce 6,750 cu ft of cracked ammonia. This is equivalent to the contents of some 33 hydrogen cylinders."

SYNTHESIS GAS

By use of the preceding described processes properly adjusted, it is possible to make various synthesis gas mixtures, of which the following are most commonly needed:

	Volume ratios		
Synthesis gas for:	H_2	CO	N_2
Ammonia	3	0	1
Methanol	2	1	0
Fischer-Tropsch	2	1	0
Oxo	1	1	0

These may be exemplified for synthesis gas for ammonia (Chap. 18) by one of the following process sequences or variants:

Alternative 1:

a. Steam hydrocarbon process, including secondary reforming for addition of nitrogen.
b. Carbon monoxide conversion.
c. Bulk carbon dioxide removal by water scrubbing, hot potassium carbonate, or Girbotol.
d. Final carbon dioxide removal by Girbotol, caustic scrubbing, or ammonia wash.
e. Final carbon monoxide removal by methanation or copper-liquor scrubbing.

Alternative 2:

a. Partial oxidation process.
b. Carbon monoxide conversion.
c. Bulk carbon dioxide removal by water scrubbing, hot potassium carbonate, or Girbotol.
d. Final carbon dioxide removal by caustic or ammonia wash.
e. Final purification by low-temperature separation and liquid-nitrogen wash.

Specifically, the *synthesis gas* for ammonia production is made by reforming natural gas as follows: Natural gas is passed over zinc oxide at 675 to 750°F to remove all traces of sulfur compounds. The desulfurized gas is mixed with approximately 3.5 volumes of steam and passed to the reforming furnace, which contains a supported nickel catalyst. The catalyst is kept at 1300 to 1500°F by external heat. The natural gas is reformed by the steam endothermically to form carbon monoxide, and hydrogen, lesser amounts of carbon dioxide, unconverted methane (about 7%), oxygen, and nitrogen.

$$CH_4(g) + H_2O(g) \longrightarrow CO(g) + 3H_2(g) \qquad \Delta H_{1500^\circ F} = +54.2 \text{ kcal}$$

This product gas is mixed with enough air to complete the conversion and to add the nitrogen required for ammonia synthesis before entering a second reforming unit. The product gas is passed into a shift converter (see steam-hydrocarbon process), and these gases are compressed in three stages usually to approximately 250 psi for scrubbing with water or an ethanolamine solution to remove carbon dioxide. It is further compressed in two stages to 1,800 psi and scrubbed with a solution of copper ammonium formate to take out essentially all the remaining carbon monoxide and carbon dioxide. The gas is then washed with a 5% caustic solution, leaving a product with the theoretical H_2/N_2 ratio of 3:1 required for ammonia synthesis.

HYDROGEN PURIFICATION

CARBON MONOXIDE REMOVAL High concentrations of carbon monoxide are commonly reduced by conversion to hydrogen via the water-gas-shift reaction; see Fig. 7.3 and the accompanying text. Low concentrations can be converted via catalytic methanation. Removal by scrubbing in solutions of complex copper ammonium salts has been used commercially.[19]

CARBON DIOXIDE AND HYDROGEN SULFIDE REMOVAL Many processes[20] for the removal of acid gases have been employed commercially.

1. *Monoethanolamine (MEA) or Girbotol process.* This process has been illustrated in flowcharts (Fig. 7.1 and Fig. 7.3). A water solution of MEA is reacted with CO_2 in an absorber vessel under pressure at room temperature. The MEA solution is then heated and fed to a reactivation column where the MEA-CO_2 complex is dissociated by stripping with steam at 200 to 250°F and near atmospheric pressure. The CO_2 and steam leave the top of the reactivation column, and the regenerated solution leaves the bottom, is cooled, and is pumped back to the absorber. The chemical reactions involved are:

$$2NH_2CH_2CH_2OH + CO_2 + H_2O \rightleftharpoons (HOCH_2CH_2NH_3)_2CO_3$$

$$NH_2CH_2CH_2OH + CO_2 + H_2O \rightleftharpoons HOCH_2CH_2NH_3HCO_3$$

Contrasted with physical absorption types of CO_2 removal processes, the MEA process is capable of reducing the CO_2 concentration in the process gas to less than 0.01% by volume.

A major problem associated with the use of MEA solutions is their corrosive effect on process equipment. Corrosion is most severe at elevated temperatures and where the acid gas concentration in solution is the highest. Control of corrosion is achieved by the use of stainless steel at potential trouble spots, by limiting the concentration of MEA in aqueous solution in order to limit the CO_2 in solution, by excluding oxygen from the system, and by the removal of oxidation and degradation products using side-stream distillation. A recent innovation is the "Amine Guard"[21] which uses a corrosion inhibitor and has completely eliminated corrosion as an operating problem even with increased CO_2 loadings.

2. *Hot potassium carbonate process*[22] This process, developed originally by the Bureau of Mines, is particularly useful for removing large quantities of CO_2. Although it can remove CO_2 down to 0.1% by volume in the gas being purified, it is generally more economical at purity levels of 1% or greater. By absorbing CO_2 under pressure in hot solution close to its boiling point and regenerating it at the same temperature but at near-atmospheric pressure, steam consumption is reduced and heat exchangers are eliminated. Process improvements have been obtained through the

[19]Kohl and Riesenfeld, Gas Purification, chap. 14, McGraw-Hill, 1960; Girdler, Hydrogen, p. 16; for sulfide removal, cf. Table 6.4.

[20]Kohl and Riesenfeld, Gas Purification, *Chem. Eng.* (*N.Y.*), **66**(12), 127 (1959).

[21]Butwell, Hawkes, and Mago, Corrosion Control in CO_2 Removal Systems, *Chem. Eng. Prog.*, **69**(2), 57 (1973).

[22]Girdler, Hydrogen, p. 19 (flowchart); Trail, Reynolds, and Alexander, Carbon Dioxide Removal by Hot Potassium Carbonate and Amine Scrubbing, *AIChE Symp. Ser.*, **34**, 57, 81 (1961); Catalyzed Scrubbing Improves Gas Cleanup (for Synthesis Gas), *Chem. Eng. (N.Y.)*, **69**(24), 60 (1962) (flowchart).

use of catalytic additions and promoters to the solution, such as in the Giammarco-Vetrocoke[23] and the Catacarb processes.[24]

3. *Physical solvent processes.*[25] These are processes in which CO_2 is removed by physical solution in a solvent, which is often proprietary. Examples of this class of process are the Rectisol process using cold (approximately $-60°C$) methanol, the fluor solvent process using a nonaqueous organic solvent such as propylene carbonate, the Sulfinol process using an organic solvent, sulfolane (tetrahydrothiophene dioxide), the Selexol process using the dimethyl ether of polyethylene glycol, and the Purisol process using *N*-methyl-2-pyrrolidone. Aqueous ammonia solutions have been used for removal of CO_2 and H_2S, particularly in ammonia synthesis plants where ammonia is available. Water has also been used, but CO_2 solubility is poor, requiring high-pressure operation, and hydrogen losses, because of solubility, are high.

ADSORPTIVE PURIFICATION Fixed-bed adsorption can remove such impurities as CO_2, H_2O, CH_4, C_2H_6, CO, Ar, and N_2, among others. One type of process is the thermal-swing process in which the impurity is adsorbed at a low temperature and desorbed thermally by raising the temperature and passing a nonadsorbable purge gas through the bed to aid desorption and carry the desorbed gas from the bed. For continuous operation two beds are necessary; while one bed is on-stream, the second bed is being regenerated.

A second type of process is the pressure-swing adsorption process[26] in which the impurity is adsorbed under pressure and desorbed at the same temperature but at a lower pressure. A purge gas may be used. For continuous operation, at least two beds are required. Its principal advantage is that it can operate on a shorter cycle than the thermal-swing process, thereby reducing vessel sizes and adsorbent requirements. It is capable of purifying a typical crude hydrogen stream to a high-purity hydrogen product containing 1 to 2 ppm total impurities.

CRYOGENIC LIQUID PURIFICATION An impure hydrogen stream can be partially purified by simply cooling to the appropriate cryogenic temperature,[27] where the impurity will condense and can be separated as a liquid stream. This is often used for bulk removal of light hydrocarbons from hydrogen in a refinery off-gas. Product purity obtainable depends upon the vapor pressure of the impurity, and high purities are not readily attainable in most cases.

For additional purification, Linde[28] removes low-boiling contaminants in a hydrogen gas stream at 300 psia and $-180°C$ by washing successively with liquid methane (to remove nitrogen and carbon monoxide) and liquid propane (to remove methane) to give 99.99% hydrogen. Final purification employs activated carbon, silica gel, or molecular sieves. Low temperatures[29] for washing with liquid nitrogen or for fractionation are also used to remove impurities.

The parahydrogen content of bulk liquid hydrogen should be greater than 95%. The terms ortho and para refer to the direction of nuclear spin in the hydrogen molecule. Hydrogen normally contains 25% para at room temperature but slowly converts to nearly pure para at liquid temperature. During liquefaction, hydrogen should be catalytically[30] converted from the ortho to the para form to

[23] *Chem. Eng. (N.Y.)*, **67**(19), 166 (1960) (flowchart).

[24] Eickmeyer, Catalytic Removal of CO_2 (Hot Carbonate), *Chem. Eng. Prog.*, **58**(4), 89 (1962).

[25] Kohl and Riesenfeld, Gas Purification, chaps. 4, 6, and 14, McGraw-Hill 1960; Herbert, *VDI Z.*, **98**(28), 1647 (1956); Kohl and Buckingham, Fluor Solvent, *Hydrocarbon Process. Pet. Refiner*, **39**(5), 193 (1960); Sulfinol, *Chem. Eng. (N.Y.)*, **70**(19), 78 (1963); Physical Solvent Stars in Gas Treatment/Purification (Selexol), *Chem. Eng. (N.Y.)*, **66**(13), 54 (1970) (flowchart).

[26] Stewart and Heck, Pressure Swing Adsorption, *Chem. Eng. Prog.*, **65**(9), 78 (1969).

[27] Haslain, Which Cycle for H_2 Recovery?, *Hydrocarbon Process.*, **51**(3), 101 (1972).

[28] Baker and Paul, Purification of Liquefaction-grade Hydrogen, *Chem. Eng. Prog.*, **59**(8), 61 (1963); Production and Distribution of Liquid Hydrogen, *Adv. Pet. Chem. Refin.* **10,** (1965).

[29] Girdler, Hydrogen, p. 21.

[30] Schmauch and Singleton, Technical Aspects of Ortho-Para Hydrogen Conversion, *Ind. Eng. Chem.*, **56**(5), 20 (1964); Schmauch *et al.*, Activity Data on Improved Para-Ortho Conversion Catalysts, *Chem. Eng. Prog.*, **59**(8), 55 (1963); Lipman, Cheung, and Roberts, Continuous Conversion Hydrogen Liquefaction, *Chem. Eng. Prog.*, **59**(8), 49 (1963).

prevent subsequent exothermic conversion and evaporation of the product in storage. The heat of conversion (168 cal/g) is great enough compared with the heat of vaporization (106 cal/g) to cause considerable loss.

OXYGEN AND NITROGEN

USES AND ECONOMICS In 1973, the production of oxygen in the United States amounted to 385 billion ft^3 or 31.87 billion lb, second only to that of sulfuric acid. This represents a 9.1% increase over the preceding year, which is slightly less than the 11.6% average growth rate over the preceding decade.

Oxygen is produced almost exclusively by the liquefaction and rectification of air in highly efficient, well-insulated, compact plants. Supply of raw material is not a problem and, because transportation of the product adds considerably to the cost, plants are usually located close to the point of consumption.[31] An air separation plant may be located on the customer's property and, as such, is known as an on-site plant. It may be owned and operated off-site by the designer and builder of the plant and the product delivered by a short pipeline. At some air separation complexes, several customers are serviced by a single pipeline many miles in length. In such cases, oxygen assumes many of the characteristics of a utility.

Air separation plants vary in size from small units producing less than 25 tons/day to giant plants producing more than 1000 tons/day.[32] Single-train plants rated at 2000 tons/day of oxygen are now being designed and built in this country. The bulk of the oxygen is produced at high purity (99.5%). Low purity usually refers to the 95 to 99% purity range. For high-purity oxygen, the impurity is argon except for traces of rare gases, hydrocarbons, and carbon dioxide.

A major use of oxygen is by the steel industry[33] for open-hearth or basic oxygen furnaces in the production of steel. Other uses by the steel industry include scale removal from billets by an oxyacetylene flame, and oxygen lances for cutting out imperfections. The chemical industry is a large consumer of oxygen, using it in such applications as acetylene and ethylene oxide production and in ammonia and methanol production via partial oxidation of hydrocarbons.

Other uses include metalworking applications, enhancement of combustion processes in nonferrous metallurgical processes, medical purposes in hospitals, and aviators' breathing oxygen. In an aerospace application, the first stage of the Saturn V rocket required 3.3 million lb of liquid oxygen for fuel oxidation. Consumption for this purpose by NASA peaked in the mid-1960s at about 10 billion ft^3/year and has since declined with the completion of the Apollo program.

Environmental and energy problems are producing new areas of application for oxygen. Municipalities as well as industry are now aerating wastewater streams with oxygen instead of air in the activated sludge process for secondary treatment.[34] Oxygen is also used for the disposal and conversion of refuse to usable by-products, and for the conversion of coal to other more suitable fuel forms. It is foreseen that these applications will require oxygen in unprecedented amounts that may even surpass usage by the steel industry.

Nitrogen ranked seventh among all chemicals in 1973 with the production of 16.38 billion lb, an increase of 16.5% over the preceding year. A 10-year average growth rate of 16.1% annually makes it the fastest growing among the top 15 chemicals produced in the United States. Nitrogen has two major uses. One is for blanketing purposes for exclusion of oxygen and moisture. For this,

[31]ECT, 2d ed., vol. 14, p. 390.

[32]Guccione, New Look at Giant Air Separation Plants, *Chem. Eng. (N.Y.)*, **70**(19), 150 (1963) (describes 800 and 600 tons/day oxygen plants); Liquefying Oxygen at Low Pressure, *ibid.*, **72**(4), 144 (1965) (flowchart).

[33]Price, New Technology of Iron and Steel, *Chem. Eng. (N.Y.)*, **71**(19), 179 (1964); Browning, New Processes Focus Interest on Oxygen, *Chem. Eng. (N.Y.)*, **75**(5), 88 (1968).

[34]Direct Oxygenation of Wastewater, *Chem. Eng. (N.Y.)*, **78**(27), 66 (1971); see Chap. 3.

Fig. 7.5 180 tons/day cryogenic oxygen plant at Detroit, Mich. The oxygen plant supplies a UNOX system to improve the secondary activated sludge wastewater treatment process. (*Union Carbide Corp.*)

it must be dry and have an extremely low oxygen content (less than 10 ppm). The other use is to obtain extremely low temperatures, down to −345°F. The largest consumption of nitrogen is in the manufacture of ammonia. Production statistics usually do not include this usage, because the nitrogen usually originates from feedstock and air rather than nitrogen. Other usage by the chemical industry is in applications to exclude oxygen or moisture, such as blanketing of polymerization processes, or for applications as a diluent, for example, in the control of reaction rates.

It is extensively used by the steel industry for blanketing and bright-annealing. The food processing industry uses nitrogen blanketing to prevent food spoilage and as a refrigerant in the processing and refrigerated transport of frozen foods. Other uses for liquid nitrogen include low-temperature metal treatment, shrink-fitting of parts, deflashing of molded plastic and rubber parts, and in cryobiology for storage of biological materials such as whole blood and bull semen and as a refrigerant in cryosurgical procedures.

MANUFACTURE Oxygen is produced principally by the liquefaction and rectification of air.[35] Production by the electrolytic dissociation of water is of little significance. Air is a mixture of many substances, of which eight are found in unvarying concentration; seven of these are of commercial importance. Table 7.5 lists the major constituents of air and their properties.

The production of high-purity gaseous oxygen from air[36] via the cryogenic process is illustrated by the flowsheet in Fig. 7.6 and in Table 7.4. Filtered air is compressed to approximately 75 psig in a centrifugal compressor and aftercooled. After separating out any liquid water, the air enters the reversing heat exchanger and is cooled to nearly its dew point in countercurrent heat exchange with the outgoing gaseous products. As the air cools, moisture is first condensed and frozen on the walls of the heat exchanger passage. At lower temperatures carbon dioxide freezes and is also deposited on the heat exchanger passage walls. The air emerging from the reversing heat exchanger is completely dry and has had over 99% of the carbon dioxide removed. Gas-phase, fixed-bed adsorption is used to remove the remaining carbon dioxide and, more importantly, any hydrocarbons entering with the air which would be hazardous in the presence of liquid oxygen. The cleaned air

[35]Latimer, Distillation of Air, *Chem. Eng. Prog.*, **63**(2), 35 (1967) (extensive treatment of air separation technology); Advances in Large Scale Oxygen Production, *Adv. Pet. Chem. Refin.*, vol. 9, chap. 1, Interscience, 1964.

[36]McGraw-Hill Encyclopedia, vol. 7, p. 535 (1966) (most excellent diagrams).

TABLE 7.4 ***Typical Low-pressure Air Separation Plant*** *(See Fig. 7.6)*

PRODUCT

500-tons/day of contained oxygen, 95% pure, at 5 psig

PLANT REQUIREMENTS

Electric power:	6900 kW
Cooling water:	2000 gal/min

INVESTMENT Approximately $5,600,000

APPROXIMATE PRODUCTION COST

	Annual	*Per ton*
Direct cost:		
Utilities:		
Power: @ mills/kWh	$ 763,000	$ 4.31
Cooling water: @ 5¢/1000 gal	51,000	0.29
Direct labor		
Two workers per shift @ $4.50/h	78,800	0.44
Supervision @ 15%	11,800	0.07
Plant maintenance (1.5% of investment)	84,000	0.47
Payroll overhead (18.5% of payroll)	31,300	0.18
Operating supplies	12,000	0.07
Total direct cost	$1,031,900	5.83
Indirect cost:		
Maintenance, supplies, and labor	108,800	0.61
Fixed cost:		
Taxes and insurance (2% of investment)	113,400	0.64
Depreciation at 6.66%	377,600	2.13
Total operating cost	$1,631,700	$ 9.21
Return on investment (12%)	672,000	3.79
Total production cost	$2,303,700	$13.00

Source: Linde Division, Union Carbide Corp., mid-1974 costs.

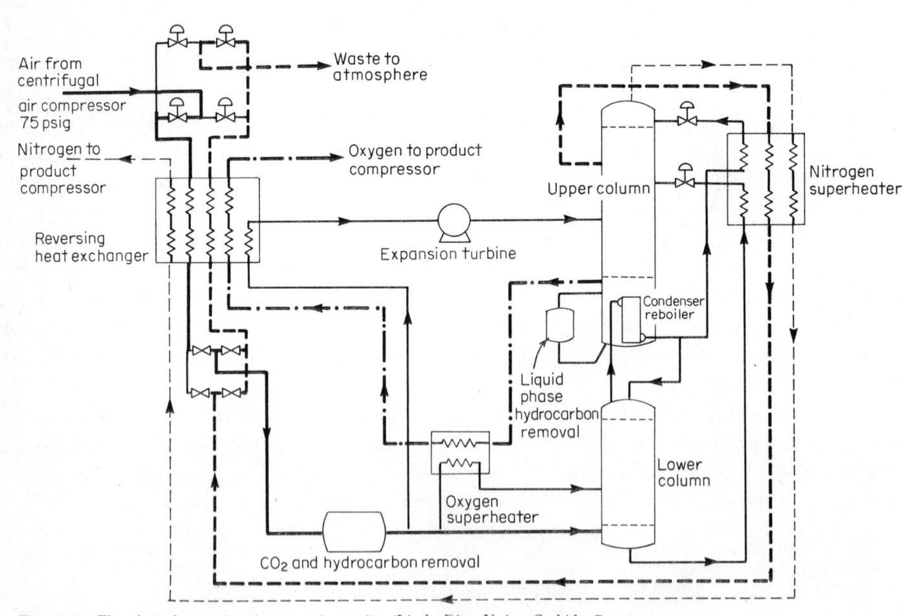

Fig. 7.6 Flowchart for on-site air separation unit. (Linde Div., Union Carbide Corp.)

TABLE 7.5 ***Properties of Air and Constant Constituents***

Gas	*Volume, %*	*Triple point,°K*	*Boiling point, °K*	*Critical temp., °K*	*Critical pressure, atm*
Nitrogen	78.084	63.156	77.395	126.1	33.49
Oxygen	20.946	54.363	90.19	154.7	50.1
Argon	0.934	83.78	87.27	150.8	48.3
Hydrogen	0.00005	13.96	20.39	33.19	12.98
Neon	0.001818	24.55	27.07	44.4	26.9
Helium	0.0005239	*	4.216	5.20	2.26
Krypton	0.0001139	115.95	119.80	209.4	54.3
Xenon	0.0000086	161.3	165.05	289.8	57.64
Carbon dioxide†	0.02–0.04	216.6	194.68‡	304.2	72.85

Source: Scott, Cryogenic Engineering, Van Nostrand, 1962, pp. 268–321; Cook, Argon, Helium and the Rare Gases, Interscience, 1961. *Has no triple point. †Variable. ‡Sublimation temperature.

is then fed to the bottom tray of the lower column of the double-column rectifier. This is shown in more detail in Fig. 7.7.

The double-column rectifier consists of two tray-type distillation columns which are thermally connected at the center through a heat exchanger which serves as a condenser for the lower column and a reboiler for the upper. Because nitrogen is more volatile than oxygen, it will ascend each column and oxygen will descend. Thus, on the reboiler side of the upper column, there is a pool of high-purity, boiling, liquid oxygen, while on the condenser side of the lower column, nearly pure nitrogen is being condensed. Because the normal boiling point of oxygen is 12.8°C higher than that of nitrogen, the pressure in the lower column must be high enough to raise the condensing temperature of nitrogen sufficiently to provide a positive temperature-driving force in the main condenser. The condensed nitrogen is split as it leaves the main condenser; one portion returns to the lower column as reflux, and the other is diverted to the upper column, through the nitrogen superheater, also for use as reflux. An oxygen-rich (35%) liquid stream leaves the bottom of the lower column and, after being subcooled in the nitrogen superheater, serves as the main feed stream for the upper column. The two liquid streams entering the upper column are first subcooled to reduce flashing when throttled to the lower pressure of the upper column. The oxygen product is removed as saturated vapor from the main condenser, and a high-purity nitrogen product is removed as saturated vapor from the top of the upper column. The remaining gas is removed as a low-purity waste nitrogen stream several trays from the top of the upper column.

Carbon dioxide and light hydrocarbons tend to accumulate in the liquid oxygen in the main condenser. These constituents are removed by recirculating the main condenser liquid through a silica gel adsorption trap to prevent the buildup of carbon dioxide and hazardous hydrocarbons.

The two nitrogen and the oxygen streams are superheated to approximately 100°K in their respective superheaters and are delivered to the reversing heat exchangers for warm-up to the ambient temperature in heat exchange with the incoming air. The high-purity nitrogen and high-purity oxygen are warmed in separate nonreversing passages, while the waste nitrogen flows through passages which periodically reverse with the air passages. The waste nitrogen thus flows past the solid carbon dioxide and frozen moisture previously deposited from the air and causes the deposits to sublime into the waste nitrogen and to be carried out of the heat exchanger. Periodic reversal of the air and waste nitrogen passages ensures that the heat exchanger is maintained in a clean, operable condition.

Refrigeration to overcome heat influx to the process is provided by expanding a portion of the air stream through a centrifugal expansion turbine. Ideally, this is an isentropic process which reduces the enthalpy of the air being expanded and rejects the energy from the process via the turbine shaft. Usable work can be recovered by coupling an electric generator, for example, to

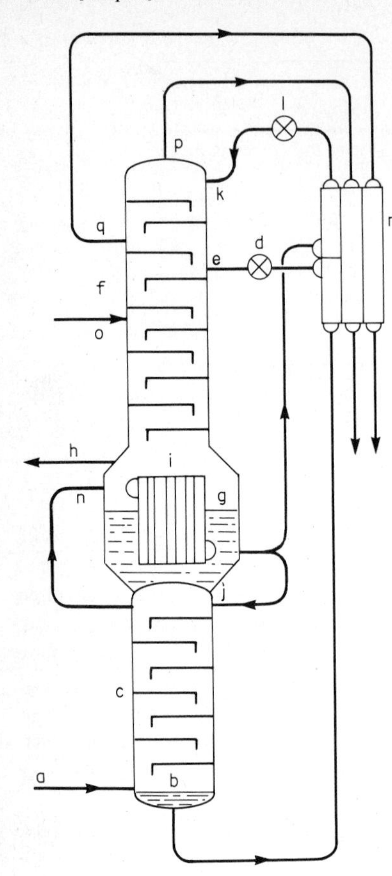

Fig. 7.7 Diagram of Linde double column and main condenser. This rectifier separates liquid air into the more volatile nitrogen (bp 77.4°K) and the less volatile oxygen (bp 90.2°K). (*a*) cold-air feed at 70 to 80 psig; (*b*) kettle liquid rich in oxygen (about 35% O_2); (*c*) lower or high-pressure column 70 to 80 psig; (*d*) throttling valve for kettle liquid; (*e*) liquid feed to upper column; (*f*) upper or low-pressure column, 3 to 12 psig; (*g*) accumulated oxygen liquid; (*h*) exit for product oxygen, 99.5% pure; (*i*) plate-and-fin main condenser-reboiler; (*j*) liquid nitrogen reflux to lower column; (*k*) liquid nitrogen reflux to upper column; (*l*) throttling valve for nitrogen reflux; (*m*) plate-and-fin heat exchanger, liquid-nitrogen reflux and kettle liquid to product nitrogen and waste nitrogen; (*n*) cold nitrogen vapor to main condenser; (*o*) cold exhaust air from expansion turbine: (*p*) cold product nitrogen gas; (*q*) cold waste nitrogen. (*Linde Div., Union Carbide Corp.*)

the turbine shaft. Clean, cold air for the turbine is withdrawn downstream of the absorber and reheated in a separate passage of the reversing heat exchanger to the proper temperature for introduction to the turbine. The reheating also serves to control the temperature pattern to help maintain clean operation of the reversing heat exchanger. Air exhausting from the turbine is fed to the proper tray of the upper column.

Heat exchangers are usually of the very compact brazed aluminum plate and fin construction. Distillation trays may be either perforated or bubble cap. Materials of construction are aluminum, stainless steel, copper, and copper alloys. All low-temperature equipment is assembled, with piping, in one or more cold boxes heavily insulated with nonflammable materials.

Some oxygen is produced by a noncryogenic, pressure-swing adsorption process[37] using a molecular sieve as the adsorbent. This produces oxygen of $90\%^+$ purity in the capacity range of 1 to 80 tons/day.

RARE GASES OF THE ATMOSPHERE Oxygen and nitrogen are the primary components of the atmosphere, but air also contains nearly 1% argon and lesser amounts of neon, krypton, and xenon (Table 7.5). Argon, neon, krypton, and xenon are all produced commercially as by-products from large cryogenic air separation plants. The distillation of liquid air is normally performed in

[37]Davis, Oxygen Separated from Air Using Molecular Sieve, *Chem. Eng.* (*N.Y.*), **79**(23), 88 (1972).

the double-column arrangement illustrated in Fig. 7.7. The rare gases are produced in side columns operated in conjunction with the standard double-column plant.

A combination of a double column and side rectification columns is used to produce argon. Since argon boils at a temperature just below oxygen, its concentration level builds up in the upper column at a point above the oxygen product level. The argon-rich vapor is withdrawn from the upper column at a point about one-third of the distance between the oxygen product and the nitrogen waste and is fed to a side argon column. The feed vapor is primarily an argon-oxygen mixture (with some nitrogen) and because of the close boiling points of the components, the argon column is operated at a relatively high reflux ratio. The liquid reflux from the argon column is returned to the upper column at the same point as the vapor withdrawal. The crude argon product is withdrawn from the top of the argon column. The side argon column is driven by expanded kettle liquid obtained from the bottom of the lower column. The vaporized kettle vapor is fed to the upper column at the same point as the low-pressure turbine air feed. The argon column condenser usually includes a small liquid drain to the upper column. The purpose of this liquid drain is to avoid hydrocarbon buildup in the argon condenser liquid reservoir, since this condenser is not protected by hydrocarbon adsorption traps, as is the case with the main plant condenser. All process streams handled by the argon column are returned to the main plant columns, with the exception of the crude argon product. The crude argon is further processed in a section of the plant usually referred to as the argon refinery. The crude argon product contains oxygen and nitrogen impurities. The oxygen impurity is removed by the addition of hydrogen and subsequent catalytic combustion and gas drying to recover the water. The nitrogen impurity is removed by another distillation step that produces the refined argon product at a purity of 99.999%.

The primary application of argon is in metallurgical processes. One major use is as a shielding gas in the welding of metals such as aluminum and stainless steel and in the refining of exotic metals such as zirconium, titanium, and many alloys. Another significant use of argon is as a filler gas for incandescent light bulbs. A new process for the production of stainless steel is the "argon-oxygen decarburizing" process, which requires large quantities of argon.

Since neon boils at a considerably lower temperature than nitrogen, it normally collects in the dome of the main condenser as a noncondensable gas. It can be recovered by the addition of a side column. Neon has well-known applications as a filler gas for display lights. Much larger quantities are used in high-energy research. It is also finding increasing uses in instrumentation and as a safe low-temperature cryogen for specialty applications. A new use is evolving in deep-sea diving, where mixtures of neon and helium have shown many advantages.

Since krypton and xenon have high boiling points relative to oxygen, they normally accumulate in the liquid oxygen sump of the upper column of the main plant. The primary application of krypton is as a light bulb filler gas. Its thermal properties are more favorable than those of argon and lead to more effective light bulbs. Xenon is finding use in a revolutionary x-ray system. Additionally, both krypton and xenon are used in instrumentation and research applications.

HELIUM

This industry began with the discovery of helium in the Hugoton field in Kansas, about 1900, but it is now also available from certain fields located in Oklahoma and Texas. The first plants were constructed to supply helium for the lighter-than-air airships of the Allies in World War I to replace the highly flammable hydrogen. Helium has 92.5% of the lifting power of hydrogen. Another less widely known use has been in a mixture with oxygen to provide a synthetic atmosphere for deep-sea divers and tunnel workers. The advantage is that helium is much less soluble in the body fluids than is nitrogen, which is present in air; this substitution prevents "bends," to a large extent. In aerospace applications, helium is employed to purge and pressurize spacecraft. This was the largest market for helium in the mid-1960s.[38] Since then, there has been a rise in the relative importance of other markets, such as protective atmospheres in the fabrication of titanium, zironium, and other metals, the growing of transistor crystals, and as a shielding gas for welding. Other uses include leak detection and cryogenic applications. The annual consumption increased from 630 million ft^3 in 1962 to 950 million ft^3 in 1966 and then declined steadily to 447 million ft^3 in 1971. Consumption in 1974 was estimated at about 500 million ft^3. The Bureau of Mines, in a government conservation program, is storing crude helium in underground reservoirs and, by 1974, 35 to 40 billion ft^3 was stockpiled and the high costs have resulted in a controversial picture for helium.[39]

Operating data have been published for the Keyes plant for helium gas, with an illustrated flowchart.[40] The Keyes plant removes the 2% helium from the natural gas in the pipeline that crosses the site, with an extraction efficiency of 92 to 95%. The pipeline gas enters at 450 to 650 psi and is first scrubbed, to take out water and condensable hydrocarbons, and then passed through a gas cleaner, which removes pipeline dust. From the cleaner, the gas goes to absorption towers, where CO_2 is removed by a solution of MEA and diethylene glycol, and finally passes through a bauxite dryer. The purified gas enters, for the crude-helium-separation step, large boxlike units, 40 ft high, with a cross section of 10 by 10 ft, operating in parallel. After entering one of the units, the gas is first chilled to $-250°F$ by heat exchange with exiting crude helium and depleted natural gas. The chilled stream is expanded into a separator-rectifier column, where, with the aid of a side stream of cold low-pressure nitrogen in a coil, the natural gas is liquefied and separated. The rectified, or crude-helium, gas (75% helium, 25% nitrogen) passes through a heat exchanger counter to incoming gas.

The depleted natural gas as a liquid passes from the bottom of the separator-rectifier through an expansion valve and also helps chill incoming feed gas, after which it leaves the unit and is compressed and returned to the gas pipeline. To provide cold and to be usable in a cold side stream, high-pressure nitrogen is chilled by expansion and split into two parts, one of which is further cooled to about $-290°F$ by expansion through two centrifugal turbo expanders located outside the separation unit box, and this cold stream then further chills the unexpanded portion of the original nitrogen stream.

To purify the crude-helium gas, the trace of hydrogen is first removed in a reactor with a small amount of air, where it is oxidized to water over a platinum catalyst, following which the gas is dried and compressed by two stages to 2,750 psig. The final purification consists of first chilling the gas in an exchanger with nitrogen liquid to liquefy most of the nitrogen, and then removing the last traces of nitrogen and hydrogen by passing the gas through adsorption beds of activated coconut charcoal. The finished 99.995% pure product is shipped in small cylinders at a pressure of 3,700 psi.

Plants[41] are in operation to supply liquid helium (bp $-452.1°F$ at 1 atm compared with $-423°F$ for hydrogen). The first plant at Amarillo, Tex., produced 100 l/h. The Linde Co. Division of Union Carbide assembled its plant inside a Dewar (i.e., thermos bottle) type of vessel 20 ft high and 9 ft in diameter. The difficulty is that *compressed helium does not exert the Joule-Thomson cooling effect upon expanding until* $-448°F$ *is reached.* Helium gas is compressed to 270 psig at 80°F, where it enters the Dewar cold box. Here it is cooled to $-305°F$ in the first of three fin-type

[38]Chopey, What's Next for Helium?, *Chem. Eng.* (*N.Y.*), **81**(12), 40 (1974); Helium Use Could Drop 25% this Year, *Chem. Eng. News*, July 3, 1967, p. 18.

[39]Chopey, *op cit.;* Helium Conservation Drains Federal Funds, *Chem. Eng. News*, Dec. 15, 1969, p. 46; Industrial Gases, *Chem. Eng. News*, Nov. 11, 1968, p. 18.

[40]Labine, New Plant (Keyes, Okla.) Counters Helium Shortage, *Chem. Eng.* (*N.Y.*), **67**(15), 96 (1960); Linde Co., Permeation Recovers Helium from Natural Gas, *Chem. Eng. News*, Apr. 29, 1963 p. 48 (new process).

[41]Chopey, Liquid-helium Plant Tackles Gas That's Hardest of All to Liquefy, *Chem. Eng.* (*N.Y.*), **69**(20), 76–78 (1962) (pictured flowchart); Liquid Helium Plant, *Chem. Eng.* (*N.Y.*), **70**(15), 86 (1963); Liquid Helium Comes on Strong, *Chem. Week*, June 6, 1964, p. 57; Haul Helium Faster, *Chem. Week*, Aug. 20, 1966, p. 57.

aluminum heat exchangers against recycle helium and cold (−316°F) gaseous and liquid nitrogen. Next trace impurities are removed in a gel trap, and the gas stream is further cooled to −415°F in a heat exchanger against recycle helium. The effluent is split into two streams, so that the large stream can cool by expanding from 266 to 4 psig in an expander engine, thereby cooling itself to −436°F, and thus providing a heat-exchange medium for further chilling the smaller fraction. Finally, the combined stream is chilled to −448°F by the recycle stream. Here the Joule-Thomson effect is positive and, by expanding from 265 to 4.5 psig, cools itself in a separating vessel to liquefy about 15% of the helium at −451.2°F. The helium gas is recycled.

The liquid-helium plant of the Kansas Refined Helium Co. at Otis, Kan. has an annual capacity of 6.1 million l of liquid helium,[42] equivalent to 180 million ft^3. This plant supplies the Air Reduction Co. (Airco) with liquid helium for distribution nationally. Airco has 10,000-gal trailer trucks as well as special cryogenic containers for delivering this helium.

ACETYLENE

USES AND ECONOMICS[43] Acetylene is employed with oxygen to give a high welding temperature and in the manufacture of an ever-increasing number of industrial chemicals such as vinyl chloride, acrylonitrile, polyvinylpyrrolidone, trichloroethylene, and acetic acid. Although some of these compounds are also derived from other sources such as ethylene, only one chemical is made mostly from acetylene: chloroprene and its polymer neoprene. Studies of acetylene reactions at high pressures (the Reppe high-pressure technique) are very significant in that vinylation, ethynylation, and polymerization reactions have opened up a new field of chemistry by introducing many new compounds.

MANUFACTURE Until recently all acetylene was made by the reaction of calcium carbide with water:

$$CaC_2(c) + 2H_2O(l) \longrightarrow Ca(OH)_2(c) + C_2H_2(g) \qquad \Delta H = -30 \text{ kcal}$$

There are two principal methods for generating acetylene from calcium carbide. The batch carbide-to-water, or wet, method takes place in a cylindrical water shell surmounted by a housing with hopper and feed facilities. The carbide is fed to the water at a measured rate until exhausted. The calcium hydroxide is discharged in the form of a lime slurry containing about 90% water. For large-scale industrial applications "dry generation," a continuous process featuring automatic feed, is popular. Here 1 lb of water per pound of carbide and the heat of the reaction (166 Btu/ft^3 acetylene) are largely dissipated by water vaporization, leaving the by-product lime in a dry, fairly easily handled state. Part of this can be recycled to the carbide furnaces. Continuous agitation is necessary to prevent overheating, since the temperature should be kept below 300°F and the pressure lower than 15 psig.

The newest methods of manufacturing acetylene are through the *pyrolysis*, or *cracking, of natural gas* or liquid hydrocarbon feeds. The processes of most interest include partial oxidation, using oxygen, thermal cracking, and an electric arc to supply both the high temperature and the energy. An electric-arc procedure was used commercially at Huls, Germany. The free energy of acetylene decreases at higher temperatures. At 1600°K (1327°C) and higher, acetylene is more

[42]*Chem. Week*, Aug. 20, 1966, p. 57.

[43]Acetylene, *Chem. Week*, Mar. 26, 1960, p. 45; *Chem. Eng. News*, July 22, 1963, p. 54 (excellent use and production data); Miller, Acetylene, vol. 1, Academic, 1965; Hardie, Acetylene Manufacture Uses, Oxford, 1965; Sittig, Acetylene, 1965, Noyes, 1965; Lorber *et al.*, Acetylene Recovered from Ethylene Feedstock, *Chem. Eng.* (*N.Y.*), **78**(15), 33 (1971); Baur, Acetylene from Crude Oil, *Chem. Eng.* (*N.Y.*), **76**(3), 82 (1969) (process flowsheet); Acetylene from Coal Soon, *Chem. Eng.* (*N.Y.*), **76**(6), 76 (1969).

stable than other hydrocarbons[44] but decomposes into its elements. Hence conversion, or splitting, time must be incredibly short (milliseconds). The amount of energy needed is very large and in the region of the favorable free energy.

$$2CH_4(g) \longrightarrow C_2H_2(g) + 3H_2(g) \qquad \Delta H_{1500°C} = +96.7 \text{ kcal}$$

$$CH_4(g) \longrightarrow C + 2H_2(g) \qquad \Delta H = +20.3 \text{ kcal}$$

However, the decomposition of CH_4 into its elements starts at 850°K (578°C), hence compete with the degradation to acetylene. To lessen this degradation after raising the CH_4 (or other hydrocarbon) to a high temperature of about 2700°F for milliseconds, the reaction mass must be water-quenched almost instantaneously. Many flowcharts with technical data are depicted in ECT (*loc. cit.*) and also in the 1963 Petrochemical Handbook.[45]

The partial combustion of natural gas is probably the most used by experienced chemical manufacturers (Carbide and Carbon, Tennessee Eastman, Monsanto, American Cyanamid, and Rohm & Haas). This is presented as a flowchart in Fig. 7.8 and embodies the chemical conversions just given. The process can be broken down into the following coordinated sequences:

Oxygen (90 to 98%) and natural gas are preheated separately to about 1200°F, using fuel gas (Op and Ch).

The two hot gases are conducted to and mixed in a burner or converter in a molar ratio of 0.60:100 for oxygen-methane (Op).

The furnace or burner for this partial combustion consists of three parts: a mixing chamber, a flame or chemical-conversion zone, followed by a quench chamber with quench oil or water sprays (Op, Ch). The chemical conversion is an almost instantaneous partial (two-thirds) combustion of the methane.

Overall reaction of the methane (combustion and splitting) is 90 to 95% (Ch), whereas the oxygen is 100% converted. Residence time is 0.001 to 0.01 s. The acetylene and gases are cooled rapidly by quench oil or water sprays (Op) to 100°F and have the following typical composition, in percent:

Acetylene	8.5	Methane	4
Hydrogen	57	Higher acetylenes	0.5
Carbon monoxide	25.3	Inerts	1.0
Carbon dioxide	3.7	Total	100

The soot is removed in a carbon filter (Op).

The clean gases are compressed to 165 psig (Op).

Acetylene is removed in a column (packed) by a selective solvent, e.g., dimethylformamide. Carbon dioxide is flashed and stripped overhead out of the rich solvent in a column (packed), where the acetylene is fractionated out, giving a 99% + product with a 30 to 36% yield from the carbon in the natural gas.

Higher acetylenes and water are stripped out under reduced pressure, and the solvent is reused (Op).

[44]ECT, vol. 1, pp. 171–211, 1963 (tables and other thermodynamic and technical data); Leroux and Mathieu, Kinetics of the Pyrolysis of Methane to Acetylene, *Chem. Eng. Prog.* **57**(11), 54 (1961); Sittig, *op. cit.*, pp. 78–100 (many flowcharts, diagrams, and references); cf. Acetylene Flame Technology, *Chem. Eng. Prog.* **61**(8), 49–67 (1965) (four articles); Kampter *et al.*, Acetylene from Naphtha Pyrolysis, *Chem. Eng.* (*N.Y.*), **73**(5), 80 and 93 (1966) (flowchart and operating costs of high-temperature cracking).

[45]*Hydrocarbon Process. Pet. Refiner,* **42**(11), 129–240 (1963); Faith, Keyes, and Clark, Industrial Chemicals, 3d ed., pp. 28–36, Wiley, 1965; Lobo, Acetylene Costs Today, *Chem. Eng. Prog.* **57**(11), 35 (1961); Howard *et al.*, Acetylene Production by Partial Combustion, *Chem. Eng. Prog.*, **57**(11), 50 (1961).

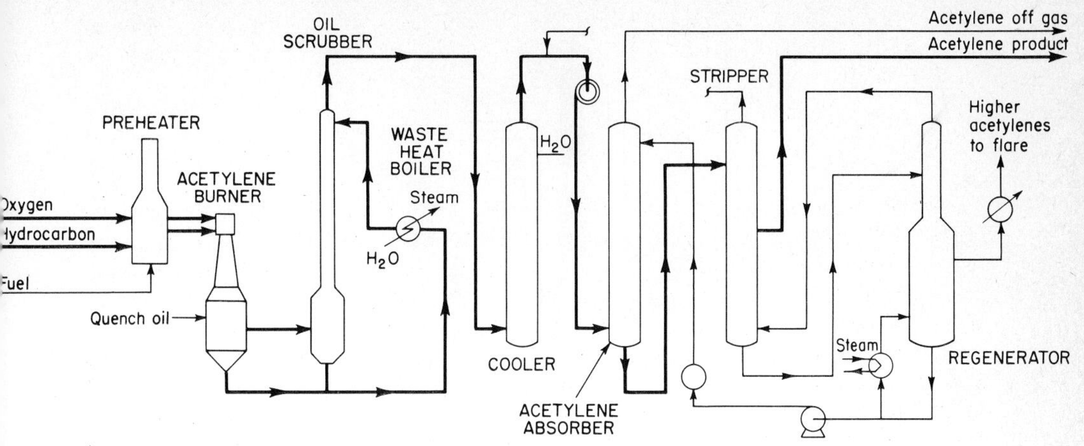

Fig. 7.8 Flowchart for acetylene by partial oxidation from hydrocarbon feedstock. (*Chemical Construction Co.*)

Another method for pyrolyzing hydrocarbons to acetylene is that of Du Pont[46] at Montague, Mich., using an *electric arc*, for which are claimed high yields of acetylene and lower laydown of carbon as well as of other by-products. The Du Pont burner is outlined in Fig. 7.9, keyed to three important patented innovations, which probably are the basis for its efficiency. The magnetic coil creates a 7,000-rpm electromagnetic field that causes a stretch-out of the arc in the direction of gas flow. The quench is rapid, being caused by a hydrocarbon-quench inlet, thereby cooling the gases to about 2000°F in a fraction of a second, and the water-quench spray to less than 600°F. The capital costs are said to be low, and the heat recovery high. The overall hydrocarbon yields are claimed to be as high as 77.7% (based on carbon balances), with increased acetylene concentration in the product gas of 21 to 22% on a volume basis.[47]

SULFUR DIOXIDE[48]

Sulfur dioxide may be produced by the burning of sulfur or by the roasting of metal sulfides in special equipment. It may be obtained also by recovery from the waste gases of other reactions. Its production and the subsequent compression and cooling to form liquid sulfur dioxide, which boils at −10°C, are shown in Fig. 7.10. With very careful control of the amount of air entering the combustion chamber, sulfur dioxide can be produced up to 18% by volume at a temperature of 1200°C. As the gases from the combustion chamber pass through the heat exchanger, they heat the water for the boilers. The cooled gases, containing from 16 to 18% sulfur dioxide, are pumped into the absorbers through acidproof pumps. The strength of the solution from the absorbers is dependent upon the temperature and the strength of the gases entering, but the concentration usually runs about 1.3%, with the temperature close to 30°C. A very small amount of sulfur dioxide is lost in the exhaust from the second absorber—about 0.02%. The temperature of the vapors coming from the steaming tower depends upon its design, but usually runs about 70°C. The vapors are cooled and passed through a drying tower in which 98% sulfuric acid is used. Other drying agents may be employed, and some plants by special procedure eliminate the use of this sulfuric acid dryer altogether.

[46]New Burner Opens Door to Arc, *Chem. Week*, Jan. 18, 1964, p. 64; Patents: Brit. 938,823, Can. 573,701, U.S. 3,073,769.

[47]Since 2 volumes of methane is required to yield 1 volume of acetylene plus 3 volumes of hydrogen by-product, the volume percent of acetylene in a theoretically perfect methane conversion is only 25%. *Chem. Week*, Jan. 18, 1964.

[48]Yaws, Li and Kuo, Sulfur Oxides, *Chem. Eng.* (*N.Y.*), **81**(14), 85 (1974); Potter and Craig, Commercial Experience with an SO_2 Recovery Process, *Chem. Eng. Prog.*, **68**(8), 53 (1972); Profit in Stack Gas, *Chem. Week*, July 20, 1968, p. 53.

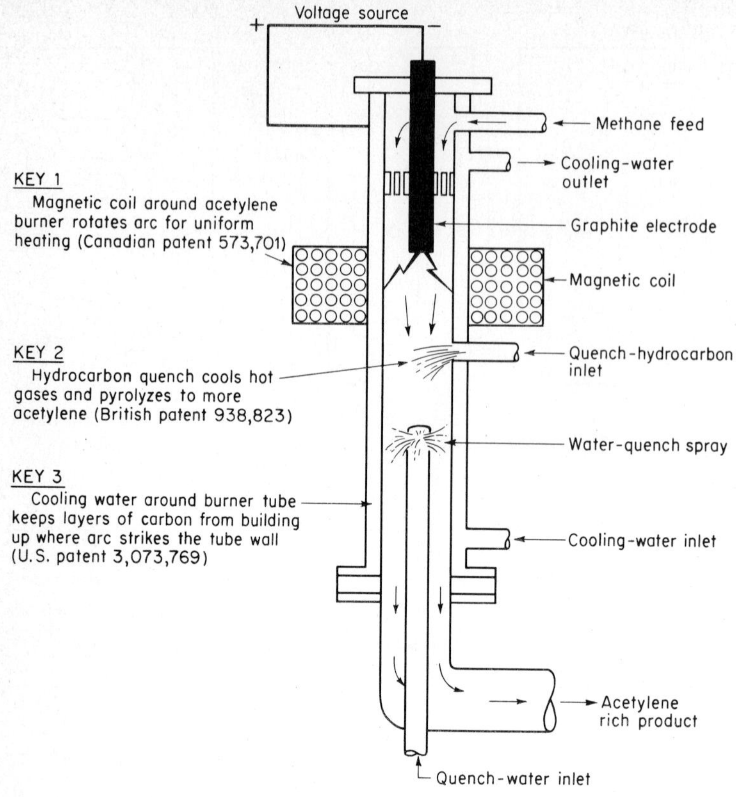

Fig. 7.9 Du Pont's arc burner for pyrolysis to acetylene. (*Chem. Week, Jan. 18, 1964.*)

The sulfur dioxide is liquefied by compressing to about 7 atm and cooling. It is stored or put into cylinders.

Sulfur dioxide is shipped as a liquid under 2 or 3 atm pressure. It is obtainable in steel cylinders of from 50- to 100-lb capacity, in 1-ton tanks, or in 15-ton tank cars. Its uses are numerous. A quite pure commercial grade, containing not more than 0.05% moisture, is suitable for most applications. A very pure grade, however, containing less than 50 ppm of moisture, is supplied for refrigeration. Sulfur dioxide also serves as raw material for the production of sulfuric acid. It finds application as a bleaching agent in the textile and food industries. Following the use of chlorine in waterworks and in textile mills, sulfur dioxide is an effective antichlor for removing excess chlorine. It is an effective disinfectant and is employed as such for wooden kegs and barrels and brewery apparatus and for the prevention of mold in the drying of fruits. Sulfur dioxide efficiently controls fermentation in the making of wine. It is used in the sulfite process for paper pulp, as a liquid solvent in petroleum refining, and as a raw material in many plants, e.g., in place of purchased sulfites, bisulfites, or hydrosulfites.

CARBON MONOXIDE

Carbon monoxide is one of the chief constituents of synthesis gas, as described in this chapter. It is obtained in pure form through cryogenic procedures, with hydrogen as a coproduct. It is an important

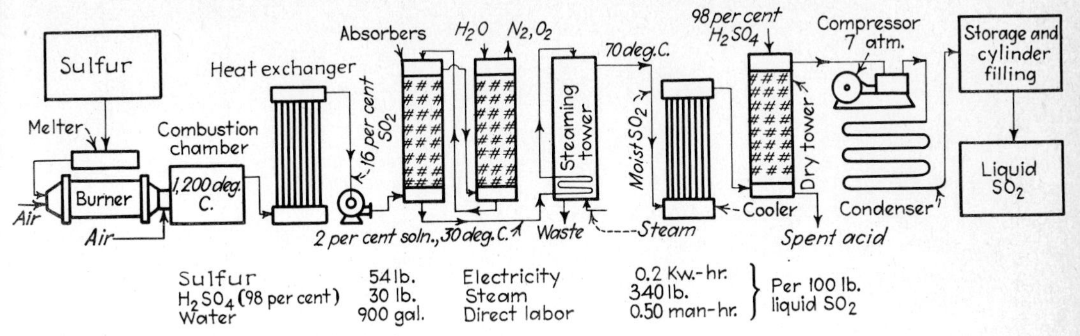

Fig. 7.10 Flowchart for liquid sulfur dioxide.

raw material in the production of methanol and other alcohols and of hydrocarbons. It is a powerful poison. It is used for making diisocyanate and ethyl acrylate by the following reactions:

$$2CO + 2Cl_2 \longrightarrow \underset{\text{Phosgene}}{2COCl_2} \xrightarrow{\text{toluene-2,4-diamine}} \underset{\text{Diisocyanate}}{CH_3C_6H_3(NCO)_2} + 4HCl$$

$$CO + CH{:}CH + C_2H_5OH \xrightarrow[\text{HCl}]{\text{Ni(CO)}_4} \underset{\text{Ethyl acrylate}}{CH_2{=}CH\overset{O}{\overset{\|}{C}}OC_2H_5}$$

NITROUS OXIDE

Nitrous oxide is generally prepared by heating very pure ammonium nitrate to 200°C in aluminum retorts.

$$NH_4NO_3(c) \longrightarrow N_2O(g) + 2H_2O(g) \qquad \Delta H = -8.8 \text{ kcal}$$

The purification consists in treatment with caustic to remove nitric acid and with dichromate to remove nitric oxide. It is shipped in steel cylinders as a liquid at a pressure of 100 atm. It is used as an anesthetic, usually mixed with oxygen.

SELECTED REFERENCES

Adler, L. B., and H. B. Shuey: Acetylene Handling, AIChE, 1963.
Cryotech 1973 Proceedings: The Production and Use of Industrial Gases, IPC Science and Technology Press (London), 1974.
Din, F. (ed.): Thermodynamic Functions of Gases, Butterworth, London, 1956–1961, 3 vols.
Gould, R. F. (ed.): Literature of Chemical Technology, chap. 3, Industrial Gases, ACS Monograph, 1967.
McClintock, M.: Cryogenics, Reinhold, 1964.
Miller, S. A.: Acetylene: Its Properties, Manufacture and Uses, Academic, 1967, 3 vols.
Scott, R. B., *et al.* (eds.): Technical Uses of Liquid Hydrogen, MacMillan, 1964.
Sittig, M.: Cryogenics Research and Applications, Van Nostrand, 1963.
Timmerhaus, K. D. (ed.): Advances in Cryogenic Engineering, vol. 9, Plenum Press, 1964.

chapter 8

INDUSTRIAL CARBON

The element carbon exists in three allotropic modifications, amorphous carbon, graphite, and diamond, which are employed industrially.[1] In general, carbon is chemically inert and is infusible at atmospheric pressure. Graphite and diamond resist oxidation even at high temperatures. Some industrial applications depend upon the chemical inertness of carbon. On the other hand, amorphous carbon can be activated, in which case it has a great capacity for selective adsorption from either the gaseous or liquid state. For nonmetallic substances, carbon and graphite possess very satisfactory electrical and heat conductivity. The heat conductivity of graphitized, coke-base carbon is superior to that of most metals. Both fabricated-amorphous and graphitic-carbon products have very low coefficients of thermal expansion and good thermal conductivity, giving them high resistance to thermal shock.

Carbon is an extremely versatile and useful element. Its applications grow each year. In 1928, Mantell[2] wrote, "The possibility of the expansion of carbon industries is very great, for they are intimately bound up with our highly complex life of the present day." This expansion has been realized.[3] The usefulness of carbon is apparent only when we single out the many phases of our everyday living where this adaptable element is essential: in the black pigment of the ink of our books, magazines, and newspapers; in carbon paper; in pencils; as the black color in many paints, automobile finishes, and shoe blacking; as a strengthening and toughening constituent of rubber tires, tubes, and other rubber goods; and as an *essential element in the construction of a very large amount of electrical and nuclear equipment,* from the household vacuum cleaner to the largest dynamos and nuclear reactors. The carbon arc is used in the production of visible and ultraviolet radiation in an increasingly large number of industrial processes dependent on photochemical reactions.

Nonfabricated industrial carbon is represented by lampblack, carbon black, activated carbon, graphite, and industrial diamonds. The first three are examples of *amorphous carbon. Lampblack* is soot formed by the imcomplete burning of carbonaceous solids or liquids. It is gradually being replaced in some uses by *carbon black,* the most important of these amorphous forms and also the product of incomplete combustion. *Activated carbon* is amorphous carbon that has been treated with steam and heat until it has a very great affinity for adsorbing many materials. *Graphite* is a soft, crystalline modification of carbon that differs greatly in properties from amorphous carbon and from diamond. *Industrial diamonds,* both natural and synthetic, are used for drill points, special tools, glass cutters, wire-drawing dies, diamond saws, and many other applications, where this hardest of all substances is essential.

[1]Carbon as coal is considered in Chaps. 4 to 6.

[2]Mantell, Industrial Carbon, Van Nostrand, 1st ed., 1928, and 2d ed., 1946.

[3]Riley, New World of Carbon and Graphite, *Mater. Des. Eng.*, **56**(9), 113 (1962); Riley, Graphite, in Ceramics for Advanced Technologies, Hove and Riley (eds.), pp. 14–75, Wiley, 1965; Cahn and Harris, Newer Forms of Carbon and Their Uses, *Nature*, **221**, 132–141 (1969).

LAMPBLACK

Lampblack,[4] or soot, is an old product, made for many years by the restricted combustion of resins, oils, or other hydrocarbons. It has gradually been replaced in the pigment trade, particularly in the United States, by carbon black, which has superior tinting strength and coloring qualities, but in the manufacture of carbon brushes for electrical equipment and lighting carbons, lampblack is still a very important constituent. Its color is bluish-gray black, rather than the deep black of carbon black; this steely black is desirable for some metal polishes and for pencils. Annual production is estimated to be 20 to 30 million lb.

In the United States either tar oils or petroleum oils are burned with restricted air to form soot, or lampblack; a suitable furnace is shown in Fig. 8.1. The soot is collected in large chambers from which the *raw* lampblack is removed, mixed with tar, molded into bricks, or *pugs*, and calcined up to about 1000°C to destroy bulkiness. The calcined pugs are ground to a fine powder. Some furnaces remove empyreumatic impurities by calcination in the gas stream of the furnaces where formed. Lampblack process development in recent years has evolved along lines similar to that of the oil furnace carbon black process (to be discussed later).

Carbon brushes for use in electric machinery are made by mixing lampblack with pitch to form a plastic mass. Petroleum coke or graphite may be added to this mix to impart special properties. Plates or blocks from which the brushes are later machined are formed by extrusion or high-pressure molding, and these *green* plates are then baked at a high temperature for several days to drive off volatile matter. Some grades are heated in electric furnaces at temperatures as high as 3000°C (5432°F) to convert the amorphous carbon into the soft, crystalline graphite which reduces the friction, increases the life, and generally improves the quality of the product (cf. Manufactured Graphite).

When carbons for the production of special arc lights are desired, a mixture of petroleum coke and thermally decomposed carbon is extruded in the form of a tube. This tube is baked at 1450°C, and a *core* of selected material is forced into its center and calcined again. The type of radiation emitted by the arc is largely dependent on the core material employed, which is frequently a mixture of lampblack flour, rare-earth oxides, and fluorides, with coal tar as a binder. Many millions of such lighting carbons are sold each month in peacetime for movie projectors. In war years, a great quantity of such lighting carbons was used in searchlights.

CARBON BLACK

The first factory for the making of carbon black[5] in this country was built at New Cumberland, W.Va., in 1872. The black was produced by cooling a burning gas flame against soapstone slabs and scraping off the carbon produced. A short time later, in 1883, the roller process was patented and, in 1892, McNutt perfected the channel process. Early production was moderate, 25 million lb/year. In 1904 the reinforcing effect carbon black imparts to rubber was reported by S. C. Mote in England. In the summer of 1912, the B. F. Goodrich Co., convinced of the value of carbon black in the rubber industry, sampled a carload of the material and later that year placed an order for 1 million lb/year. The advent of this new market stimulated the industry. Expansion was rapid, and natural gas was the initial raw material. The first thermal black process was patented in 1916, and production began in 1922; the gas furnace process made its commercial debut in 1928. In November 1943, the first oil

[4]ECT, 1st ed., vol. 3, pp. 80–84, 1949, and 2d ed., vol. 4, pp. 254–256, 1964; Mantell, Carbon and Graphite Handbook, chap. 5, Interscience, 1968.

[5]Until this date, lampblack was the only commercially available form of "carbon black," which is the most commonly accepted generic term for the entire family of colloidal carbons. Carbon black and lampblack are similar chemically, but they differ greatly in particle size and shape. ECT, 2d ed., vol. 4, pp. 243–282, 1964, and Suppl. Vol., pp. 91–108, 1971.

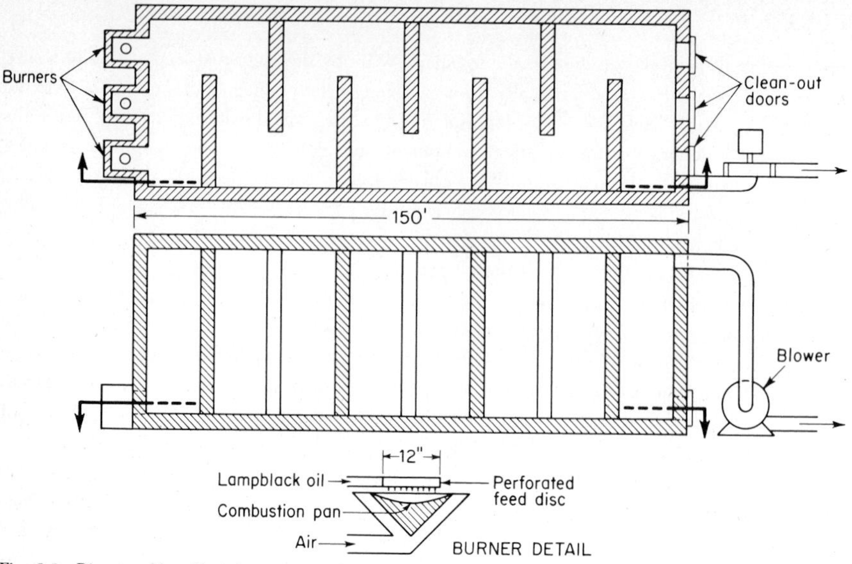

Fig. 8.1 Diagram of lampblack house.

furnace black plant went on stream in Texas. In 1974 the Charles Goodyear Award Medal of the Rubber Division of the American Chemical Society was presented to J. C. Krejci in recognition of his pioneering work on the oil furnace process. The selling price of carbon black has been greatly reduced over the years, in keeping with the expansion of the industry, dropping from $2.50 per pound in 1872 to an average of about 6 cents per pound in 1967. However, prices increased again to about 7 cents/lb by 1973 and rose to 11.5 cents per pound in 1974, reflecting both higher petroleum production costs and, more particularly, the U.S. energy shortage generated by the Middle East war of October 1973. The growth of the U.S. carbon black market during the early 1970s is shown in Fig. 8.2.

Carbon black is a very finely divided, essentially nonporous type of carbonaceous material which is produced in a precisely controlled pyrolytic petrochemical process. The best rubber-grade blacks are spherical carbon particles (ranging from about 10 to 75 nm in diameter) fused together in grapelike clusters. Each particle is composed of several thousand microcrystallite bundles which are stacked together in a random (turbostratic) order.[6] Each bundle consists of three to five micrographitic planes or platelets, elliptical-shaped and with axis lengths of about 1.5 to 1 nm, which are stacked in a not quite parallel manner. The 0.348- to 0.356-nm interlayer spacing is slightly greater than that of graphite (0.3354 nm).

About 50 to 80% of the carbon atoms in carbon black are thought to be present in the microcrystallite bundles, while the remainder is disorganized carbon. At the carbon surface, oxygen, sulfur, and sometimes other elements interact with the hydrogen positions of the polynuclear aromatic platelets, forming complexes which are generally analogous to classes of compounds of organic chemistry (e.g., carboxylic acid, phenolics, quinones, etc.). The oxygen content of most rubber-grade oil furnace blacks is usually below 1%, but is higher (3 to 8%) for ink-grade types.

USES AND ECONOMICS About 94% of the domestic carbon black consumption is in rubber products, especially tires (60% of total), heels, and mechanical goods, where it is used to improve

[6]Deviney, Surface Chemistry of Carbon Black and Its Implications in Rubber Chemistry, *Adv. Colloid Interface Sci.*, **2**, 237–259 (1969).

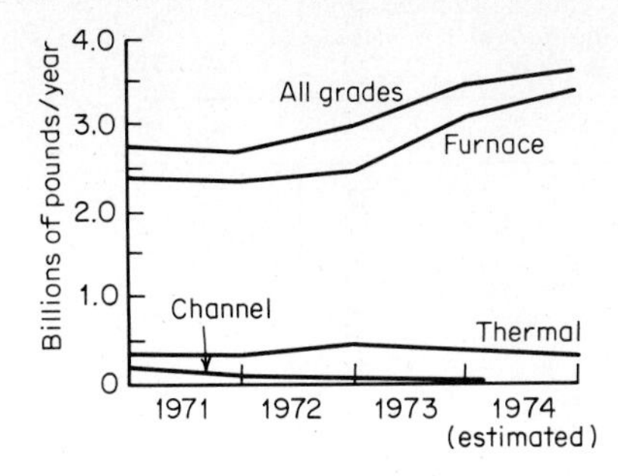

Fig. 8.2 Growth of U.S. carbon black market, 1970–1974. (*Ashland Chemical Co.*)

wearing qualities by imparting toughness. Carbon black is significantly responsible for the adoption of synthetic rubber [especially styrene-butadiene (SBR) copolymer] for automobile tires during peacetime. High-abrasion furnace blacks, when combined with "cold-polymerized" SBR, produce tire treads that consistently give better wear than controls compounded with natural rubber and channel blacks. The rubber compound in modern tires contains about 35% carbon black, which accounts for the rubber industry's role as the leading consumer of carbon black. Most printing inks are also based on carbon black. In this case the blacks must be uncompressed, grind easily, and absorb oil readily. The pigments in black plastics, paints, lacquers, and enamels are usually carbon black. Color-grade blacks used for inks and paints (representing 1.5 and 0.6%, respectively, of the total black usage) were once produced almost exclusively by the channel process but have been largely replaced by furnace blacks, particularly chemically and air-oxidized types. Even typewriter ribbons and carbon papers employ this pigment.

MANUFACTURE The four basic carbon black manufacturing processes are either of the partial combustion type (the channel, oil furnace, or gas furnace process) or of the cracking type (the thermal process). By the early 1970s, the oil furnace process had become dominant, and most (about 90%) of the world's carbon black is manufactured by this process, which uses aromatic fuel oils and residues as feedstock (cf. Table 8.1). The ever-increasing cost of natural gas led to the shutting down of almost all channel black plants (only one very small U.S. channel black plant, and three overseas, were still in operation at the end of 1973).

For historical interest, the *channel black process* will be briefly described. These blacks are produced when underventilated natural gas flames impinge upon 8- to 10-in. "channel irons," which are slowly reciprocated over scrapers to remove soot deposits. The type of black produced was controlled by burner-tip design, burner-to-channel distance, and air supply (degree of partial combustion). Yields were always poor, varying from 5% for some rubber grades to only 1% for high-color ink grades. Particle diameters ranged from 10 to 30 nm (compared with 18 to 75 nm for oil furnace blacks), and the oxygen content was greater than, and the sulfur percentage less than, that of corresponding furnace blacks.

In addition to the low yield and natural gas cost factors, the rapid demise of the channel process was caused by high steel and capital investment requirements, severe pollution problems, and somewhat poorer performance in synthetic rubber (compared with furnace blacks).

In the *thermal black process,* natural gas is cracked to carbon black and hydrogen at 1100 to 1650°C in a refractory-lined furnace in a two-cycle (heating and "making" or decomposition) operation. The reaction is

$$CH_4 \longrightarrow C(amorph) + 2H_2(g) \qquad \Delta H = +20.3 \text{ kcal}$$

The large thermal blacks (140 to 500 nm) are essentially spherical, contain little chemisorbed surface oxygen, and are mainly used in special rubber applications. Yields are about 40 to 50%, and

TABLE 8.1 *Salient Statistics for Carbon Black Production from Natural Gas and Liquid Hydrocarbons in the United States, 1973, 1972, and 1971*

	1973	*1972*	*1971*
Number of producers	8	8	9
Number of plants	34	34	37
Capacity of operating plants, lb/day	11,573,181	11,411,559	10,902,330
Quantity produced, 1000 lb			
By states:			
Louisiana	1,207,708	1,077,977	1,078,732
Texas	1,511,127	1,425,874	1,326,153
Other states	781,106	697,258	612,250
Total	3,499,941	3,201,109	3,017,135
By processes:			
Channel	14,222	22,378	46,354
Furnace	3,485,719	3,178,731	2,970,781
Quantity sold, 1,000 lb			
Domestic:			
To rubber companies	3,114,565	2,953,779	2,678,151
To ink companies	84,364	82,532	75,201
To paint companies	21,667	21,408	18,693
For miscellaneous purposes	92,998	88,989	81,482
Total	3,313,594	3,146,708	2,853,527
Exports:	192,665	111,328	163,246
Total sales	3,506,259	3,258,036	3,016,773
Losses	1,052	1,406	421
Stocks held by producers, December 31:			
Channel	2,350	7,677	9,743
Furnace	227,975	230,018	286,285
Total stocks	230,325	237,695	296,028
Value, at plants, of carbon black production:			
Total, thousands of dollars		248,361	232,049
Average per pound, cents	8.12	7.76	7.69
Quantity of natural gas used, million ft^3	49,682	53,939	63,699
Average value of gas per 1000 ft^3, cents	24.19	19.54	17.50
Quantity of liquid hydrocarbons used, 1,000 gal	623,236	590,753	547,704
Average yield per gallon, lb	5.22	4.96	4.92
Average value of liquid hydrocarbons per gallon, cents	9.03	8.13	7.96

Source: 1973 Mineral Industry Survey, Dept. of the Interior, 1975.

thermal blacks accounted for slightly less than 10% of the total U.S. carbon black production in 1973; future prospects are uncertain because of projected natural-gas costs.

In the *oil furnace process,* a highly aromatic feedstock oil (usually a refinery catalytic cracker residue, although some coal tar–derived material is also used) is converted to the desired grade of carbon black by partial combustion or pyrolysis at 1400 to 1650°C in a refractory (mainly alumina)-lined steel reactor. Two carbon black properties of vital importance in rubber applications—particle size and "structure" (the degree of particle fusion or reticulation into three-dimensional chain networks)—are closely controlled through nozzle design, reaction chamber geometry, temperature, residence time, and intensity of gaseous turbulence.

A schematic of the oil furnace process and a reactor is shown in Fig. 8.3. Basically, the heavy aromatic feedstock oil is atomized and sprayed through a specially designed nozzle into a highly

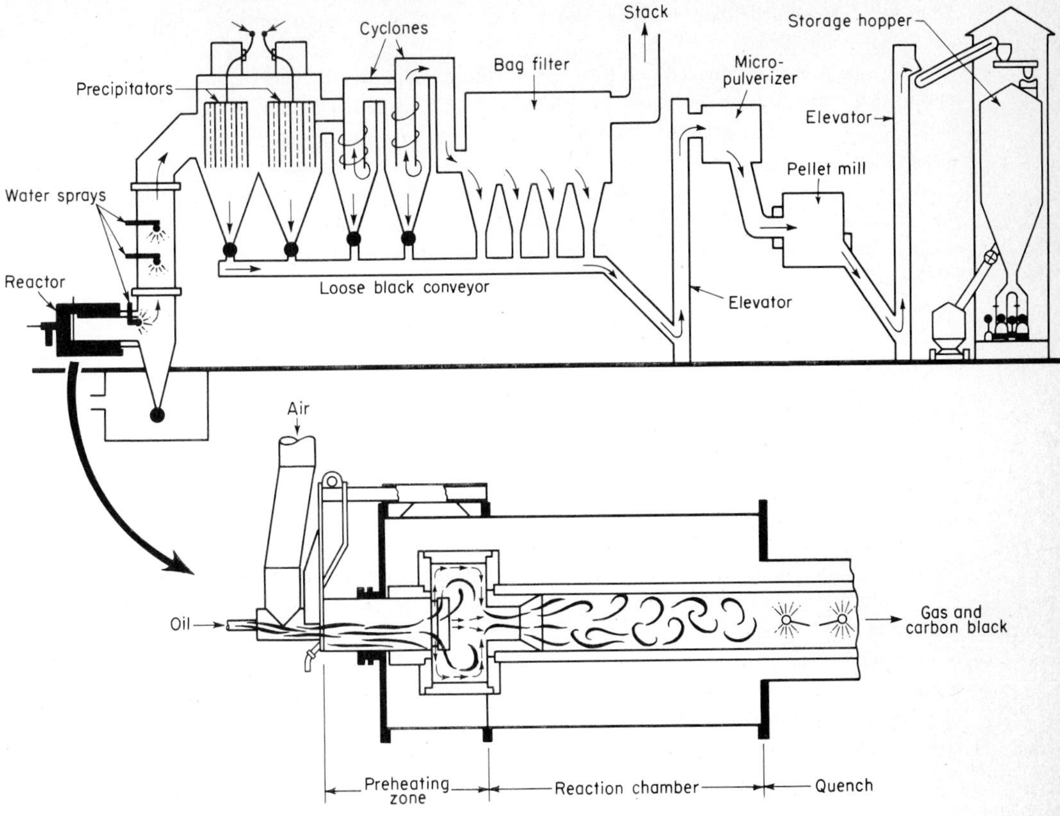

Fig. 8.3 Flowchart, carbon-black furnace process. (*Chem. Week, June 16, 1962, p. 79.*)

turbulent combustion gas stream formed by preburning natural gas or oil with about a 50% excess of air. Insufficient air remains for feedstock combustion, and the oil spray is pyrolyzed in a few microseconds into carbon black with yields (based on carbon) varying from 65 to 70% for large carcass–grade types down to about 35% for small tread–grade types, which impart maximum abrasion resistance to SBR and other synthetic elastomers.

The black "smoke" from the reaction chamber enters the tunnel where, 1 to 3 m downstream, it is water-quenched to about 200°C. Tunnel and combustion chamber dimensions vary, but frequently are from 2 to 6 m for length and from 0.1 to 1 m for internal diameter. Further downstream, precipitators, cyclones, and to a greater extent bag filters (insulated to remain above 100°C) are used to separate the product black from water vapor and the combustion off-gases. With modern fiberglass bag filters, carbon black recovery efficiency is essentially 100%. Following pulverization and grit removal steps, the loose, fluffy black is normally pelletized for convenient bulk handling and to reduce the tendency to dust. In the commonly used wet pelletizing process, the black is agitated and mixed with water in a trough containing a rotating shaft with radially projecting pins, which form the pellets. The wet pellets, usually in the 25 to 60-mesh range, are dried in a gas-fired rotating-drum dryer. In the dry pelletizing process, pellets are formed by gentle agitation and rolling (10 rpm) of the black in horizontal rotating drums (about 3 m in diameter and 6 to 12 m long) for periods of

12 to 36 h. An oil pelletizing process is also used, to a minor extent, for specialty applications (e.g., certain ink-grade blacks).

Furnace blacks are divided into tread grade (18- to about 38-nm diameter) and carcass grade (44- to about 75-nm diameter) types. Specific grades are now designated according to a three-digit numbering system (ASTM D 2516) in which the first digit refers to the particle diameter range, as measured with an electron microscope. For example, the old SAF (super abrasive furnace) black is now N-110, the first digit designating a 11- to 19-nm particle diameter, while the former FEF-LS (fast extrusion furnace–low structure) is now N-539, the 5 designating a particle diameter of 40 to 48 nm. The *gas furnace process,* similar to the oil furnace process but based on natural gas as feedstock, has been almost completely replaced by the latter, which is capable of reproducing all gas furnace–grade blacks.

Acetylene black, a special type of thermal black used in dry cell batteries and for imparting electrical and thermal conductivity to plastics and rubber, is produced by acetylene decomposition (in the absence of air) at about 800°C in a continuous exothermic process. This material (of about 42-nm particle diameter) has a higher carbon content and structure level than furnace and channel blacks.

ACTIVATED CARBON[7]

Activated carbon first came into prominence through its use as an adsorbent in gas masks in World War I. However, the knowledge that carbon produced by the decomposition of wood can remove coloring matter from solutions dates back to the fifteenth century. The first commercial application of this property, however, was not made until 1794, when charcoal filters were used in a British sugar refinery. About 1812, bone char was discovered by Figuer. Activated carbons can be divided into two main classes, those used for adsorption of gases and vapors, for which a granular material is generally employed, and those used in purification of liquid, for which a powdered material is desired.

USES There is no particular activated carbon that is effective for all purposes (Table 8.2). As a decolorant, activated carbon, with its very great surface area and pore volume, is hundreds of times more efficient than charcoal and at least 40 times more than bone black. It is estimated that 5 lb of activated carbon has an active area of 1 mi^2 or more. The amount of material adsorbed by activated carbon is surprisingly large, amounting frequently to from a quarter to an equal weight of such vapors as gasoline, benzene, and carbon tetrachloride. These substances can be *recovered and reused.* Adsorption[8] is a physical phenomenon, depending largely upon surface area and pore volume. The pore structure limits the size of molecules which can be adsorbed, and the surface areas developed limit the amount of material which can be adsorbed, assuming suitable molecular size.

The major use[9] of activated carbon is in solution purification, such as the cleanup of cane, beet, and corn sugar solutions, and for the removal of tastes and odors from water supplies, vegetable and animal fats and oils, alcoholic beverages, chemicals, and pharmaceuticals. The recovery of streptomycin represents a typical application of the continuous treatment of liquids.

The vapor-adsorbent type of activated carbon was first used in military gas masks because of its ability to adsorb certain poisonous gases, and it is now widely employed in both military and

[7]ECT, vol. 4, 2d ed., pp. 149–158, 1964.

[8]For an excellent discussion of adsorption, see Perry, sec. 16, Adsorption and Ion Exchange.

[9]Schontz and Parry, The Activated Carbon Industry, *Ind. Eng. Chem.*, **54**(12), 24 (1962); Fornwalt and Hutchins, Purifying Liquids with Activated Carbon, *Chem. Eng.* (*N.Y.*), **79**(8), 179 (1966); Browning, New Water-cleanup Roles for Powdered Activated Carbon, *Chem. Eng.* (*N.Y.*), **79**(4), 36 (1972); Erskine and Schuliger, Activated Carbon Processes for Liquids, *Chem. Eng. Prog.*, **67**(11), 41 (1971); Barnebey, Activated Charcoal in the Petrochemical Industry, *Chem. Eng. Prog.* **67**(11), 45 (1971).

TABLE 8.2 *Applications of Activated Carbon*

Adsorbing gases or vapors (gas-adsorbent carbon)

1. Adsorbent in military and industrial gas masks and other devices.
2. Recovery of gasoline from natural gas.
3. Recovery of benzol from manufactured gas.
4. Recovery of solvents vaporized in industrial processes such as manufacture of rayon, rubber products, artificial leather, transparent wrappings, film, smokeless powder, and plastics, and in rotogravure printing, dry cleaning of fabrics, degreasing of metals, solvent extraction, fermentation, etc.
5. Removing impurities from gases such as hydrogen, nitrogen, helium, acetylene, ammonia, carbon dioxide, and carbon monoxide.
6. Removing organic sulfur compounds, H_2S, and other impurities from manufactured and synthesis gas (usually impregnated with either Fe or Cu salts).
7. Removing odors from air in air conditioning, stench abatement, etc.
8. Absorbing radio-active emanations from nucleonic reactors for the time sufficient so that the decay is completed while still trapped in the activated carbon bed.

Decolorizing and purifying liquids (decolorizing carbon)

1. Refining of cane sugar, beet sugar, glucose, and other sirups.
2. Refining oils, fats, and waxes such as cottonseed oil, coconut oil.
3. Removing impurities from food products such as gelatin, vinegar, cocoa butter, pectin, fruit juices, and alcoholic beverages.
4. Removing impurities from pharmaceutical and other chemical products, including acids.
5. Water purification—removal of taste, odor, and color.
6. Removing impurities from used oils, dry-cleaning solvents, electroplating solutions, sirups, etc.
7. Removal of metals from solution—silver, gold, etc.

Catalyst and catalyst support (gas-adsorbent carbon)

1. Support for $HgCl_2$ catalyst for manufacture of vinyl chloride.
2. Support for zinc acetate catalyst in the manufacture of vinyl acetate.
3. Manufacture of phosgene.
4. Carrier for hydrogenation catalysts, etc.

Medicine

1. Internal medicine for adsorption of gases, toxins, and poisons.
2. Administering adsorbent medicinals.
3. External adsorbent for odors from ulcers and wounds.

Source: Carbon Products Div., Union Carbide.

industrial gas masks. Activated carbon is used in air conditioning systems to control odors in large restaurants, auditoriums, and airport concourses. An important field of application is the industrial recovery and control of vapors (Fig. 8.4). The recovery of such vapors amounts to billions of pounds per year, with a recovered value of several hundred million dollars. Activated carbon is able to adsorb practically any organic solvent at about 100°F and release it when heated to 250°F or higher for organic solvents. Activated carbon can now be made in an extruded form, which in vapor adsorbing presents only about half the air resistance of the older granulated heterogeneous powder. The pressure drop through carbon, whether pelleted or granular, depends primarily on the average particle size. Pelleted material ensures more uniform packing, hence more even distribution of air flow.

MANUFACTURE Many carbonaceous materials, such as petroleum coke, sawdust, lignite, coal, peat, wood, charcoal, nutshells, and fruit pits, may be used for the manufacture of activated carbon, but the properties of the finished material are governed not only by the raw material but by the method of activation used. Decolorizing activated carbons are usually employed as powders. Thus the raw materials for this type are either structureless or have a weak structure. Sawdust and

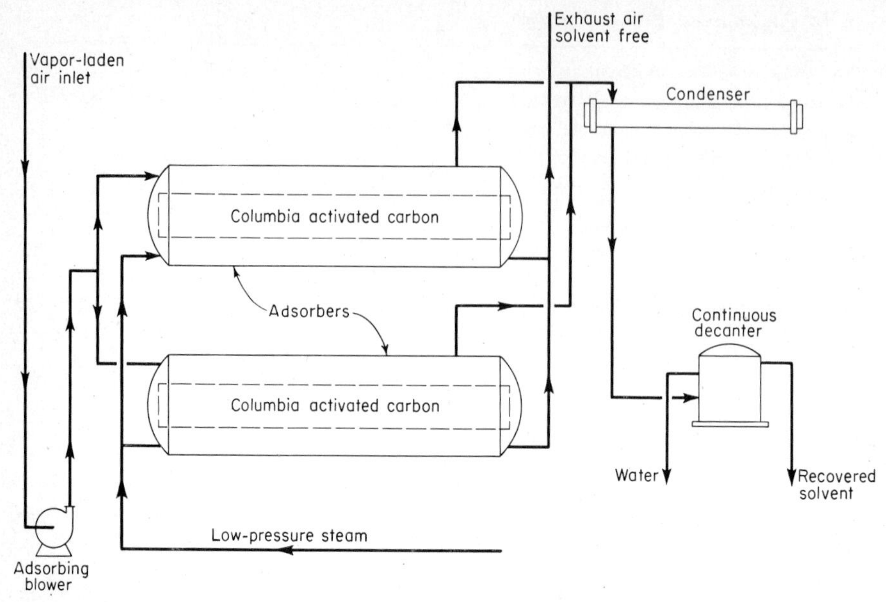

Fig. 8.4 Flowchart of one type of solvent recovery plant employing Columbia activated carbon. A typical plant of this type in recovering a hydrocarbon vapor at a rate of 1,100 lb/h from 13,000 ft^3/min of vapor-laden air has the following approximate average utility requirements per pound of recovered solvent: steam 3.5 lb, water 7.5 gal, electric power 0.05 kWh. (*Carbon Products Div., Union Carbide Corp.*)

lignite yield carbons of this kind. Vapor-adsorbent carbons are used in the form of hard granules and are generally produced from coconut shells, fruit pits, and briquetted coal and charcoal.

Activation is a physical change wherein the surface of the carbon is tremendously increased by the removal of hydrocarbons. Several methods are available for this activation. The most widely employed are treatment of the carbonaceous material with oxidizing gases such as air, steam, or carbon dioxide, and the carbonization of the raw material in the presence of chemical agents such as zinc chloride or phosphoric acid.

Gaseous-oxidation activation employs material that has been carbonized at a temperature high enough to remove most of the volatile constituents but not high enough to crack the evolved gases. The carbonized material is subjected to the action of the oxidizing gas, usually steam or carbon dioxide, in a furnace or retort at 1475 to 1800°F. Conditions are controlled to permit removal of substantially all the adsorbed hydrocarbons and some of the carbon, so as to increase the surface area. The use of chemical impregnating agents causes the carbonization to proceed under conditions that prevent the deposition of hydrocarbons on the carbon surface. The raw material, sawdust or peat, is mixed with the chemical agent, dried, and calcined at temperatures up to 1560°F. When the carbonization has been completed, the residual impregnating agent is removed by leaching with water.

Revivification. After activated carbon has become saturated with a vapor or an adsorbed color, either the vapor can be steamed out, condensed, and recovered as shown in Fig. 8.4, or the coloration can be destroyed and the carbon made ready for reuse. The oldest example of this process uses the decolorizing carbon long known as *bone char*, or *bone black*.[10] This consists of about 10% carbon deposited on a skeleton of tricalcium phosphate and is made by the carbonization in closed retorts at 1380 to 1740°F of fat-free bones.

NATURAL GRAPHITE[11]

Graphite, as it occurs naturally, has been known to humans for many centuries. Probably its first use was for decorative purposes in prehistoric times. By the Middle Ages it was being employed for writing and drawing purposes. The name *graphite* was given to this substance by the mineralogist Werner in 1879. Natural graphite occurs throughout the world in deposits of widely varying purity and crystallite size and perfection. For import purposes, natural graphites are classified as *crystalline* and *amorphous*. The latter is not truly amorphous but has an imperfect lamellar microcrystalline structure.

Foundry facings constitute the largest single use of natural graphite; this application, steelmaking, refractories, and crucibles account for approximately 70% of U.S. consumption. Lubricants, pencils, brake linings, and batteries account for approximately 20%. The U.S. production of natural graphite has been small in recent years. Imports in 1972 were 64,000 tons from a world production of 406,000 tons.

MANUFACTURED GRAPHITE AND CARBON

Natural graphite was the only kind available, except in laboratory quantities, until 1896, when Edward G. Acheson invented the first successful process for the commercial production of artificial graphite, as an outgrowth of his work on silicon carbide. He discovered that, when most forms of amorphous carbon are placed together with certain catalysts such as silica or alumina in an electric furnace and subjected to a temperature of approximately 3000°C, they are converted into the allotrope graphite. It is now known that the industrial process converts amorphous carbon directly to graphite in an electric furnace. In the carbon industry, the term "carbon" is normally used for carbonaceous materials which have been heated to 1000 to 1300°C; the term "graphite" is used for materials heated to greater than 2500°C.

USES[12] Since they combine chemical resistance with a variety of physical properties desirable in chemical-plant structural materials, carbon and graphite are among the most widely used nonmetallics in this field. They are both highly resistant to thermal shock, and graphite has unusually high thermal conductivity (Fig. 8.5). This, combined with excellent machinability, makes graphite and impervious graphite the preferred choice for many items of chemical equipment. Impervious graphite (or impervious carbon) is made by impregnating the somewhat porous base graphite (or carbon, if impervious carbon is to be produced) with a synthetic resin. The impregnation employs a vacuum-pressure cycle for complete resin penetration. The resin generally used is a phenolic type, which is polymerized by curing at elevated temperatures in steam-heated autoclaves.

The chemical resistance of impervious graphite and impervious carbon materials produced in this manner is practically identical with that of the original stocks. The graphitic materials are recommended for use at material body temperatures up to 338°F (170°C). Graphite is recommended for use with practically all mineral acids, salt solutions, alkalies, and organic compounds at temperatures up to boiling. The only exceptions are when strong oxidizing conditions exist, such as with sulfuric acid above 96% concentration, nitric acid, strong chromic acid, bromine, fluorine, and iodine. Cements for assembling graphite use resins similar to those which produce the impervious material. Permanent weldlike joints are formed with corrosion resistance and strength equivalent to or stronger than that of the basic material.

Carbon and graphite are most commonly used in the form of brick and large blocks or plates for chemical-process applications. Single-piece shapes are made routinely in diameters up to 67 in. and in lengths up to 80 in. Figure 8.6 shows typical large-block construction with carbon and

[11]ECT, vol. 4, 2d ed., pp. 312–335, 1964.
[12]Schley, Impervious Graphite for Process Equipment, *Chem. Eng.* (*N.Y.*), **81**(4), 144, and **81**(6), 102 (1974).

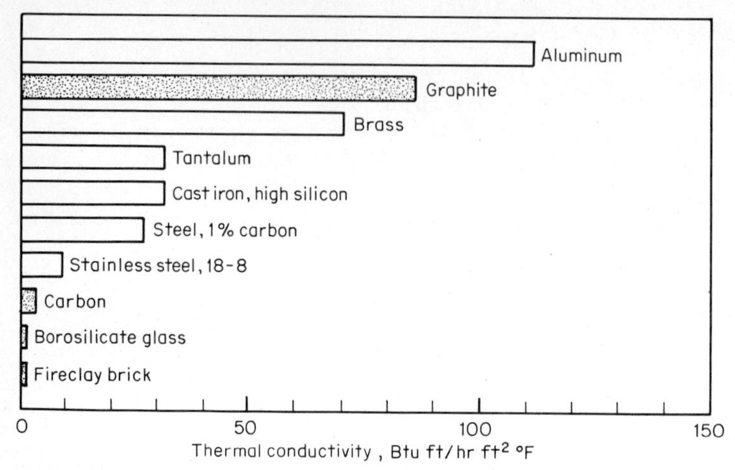

Fig. 8.5 Thermal conductivity of common materials of construction.

graphite. The quonset-hut type of structure is a graphite chamber for the combustion of phosphorus with oxygen to form P_2O_5, which is then hydrated in the silolike carbon structure alongside to form phosphoric acid. Similar construction is used for linings of sulfuric acid concentrators and for electrostatic precipitators, where dimensional stability is important. Tank linings of carbon brick are encountered in metal pickling, cleaning, and finishing work, particularly where exposure to hydrofluoric acid is involved, and in the digestion of phosphate rock with sulfuric acid in the manufacture of phosphate fertilizers.

Fig. 8.6 The right-hand structure is built completely of electric-furnace graphite, whereas the tower at the left is assembled of amorphous carbon blocks. The unit is for the production of phosphoric acid. (*Carbon Products Div., Union Carbide Corp.*)

Mold-grade graphite is the most important of the carbon and graphite structural materials to the chemical industry. Graphite's easy-machining qualities, together with its availability in many sizes and shapes, permits construction of a complete range of process equipment, including heat exchangers, pumps, valves, pipes and fittings, towers, and absorbers. Graphite's high thermal conductivity (Fig. 8.5) is particularly important in heat-exchange applications. Most common of the standard constructions are the conventional shell-and-tube heat exchangers, shown in Fig. 8.7, which range in size to units having as much as 15,000 ft^2 of transfer surface. Operating pressures range up to 100 psi, with steam pressures of 50 psig. Other uses of impervious graphite include plate- and block-type heat exchangers, cascade coolers, and centrifugal pumps.

Recently, new flexible and compressible types of graphite containing no binders, resins, or other additives have been developed, which have the chemical inertness and lubricity typical of pure graphite. These materials can be made in sheets and tapes, as well as in laminates and foam. Uses in chemical-process equipment include high-temperature, corrosion-resistant packing and gaskets, and high-temperature thermal insulation, where the unique anisotropic properties can be used to advantage.

Electrothermic applications such as the production of steel in the electric open-arc furnace, as well as the production of ferroalloys and phosphorus in submerged arc furnaces, account for the largest consumption of preformed carbon and graphite electrodes. *Electrolytic* applications include anodes for fused-salt electrolysis to produce aluminum, magnesium, and sodium metals, as well as electrolysis of aqueous brines to produce chlorine, caustic soda, and chlorates. Flowcharts for these applications are given in other chapters. Carbon and graphite's strength at elevated temperatures, their immunity to thermal shock, their resistance to the corrosive effects of molten slags, and the failure of most molten metals to wet them account for their metallurgical uses in the crucible and bosh zones of blast furnaces, cupola linings, runout troughs, molds, dies, and crucibles. *Electrical* and *mechanical* uses are in brushes, resistors, contacts, electron-tube parts, seal rings, and welding and projector carbons.

Pyrolytic graphite has exceptionally low porosity and crystallites which are highly aligned with a substrate surface. Uses include crucibles, fixtures for processing metals and alloys such as germanium and gallium arsenide, rocket nozzles, and coatings for nuclear fuel particles.

Carbon fibers have stiffness-to-weight and strength-to-weight ratios exceeding those of common structural materials. Fiber prices have decreased from $550 per pound in 1965 to less than $50

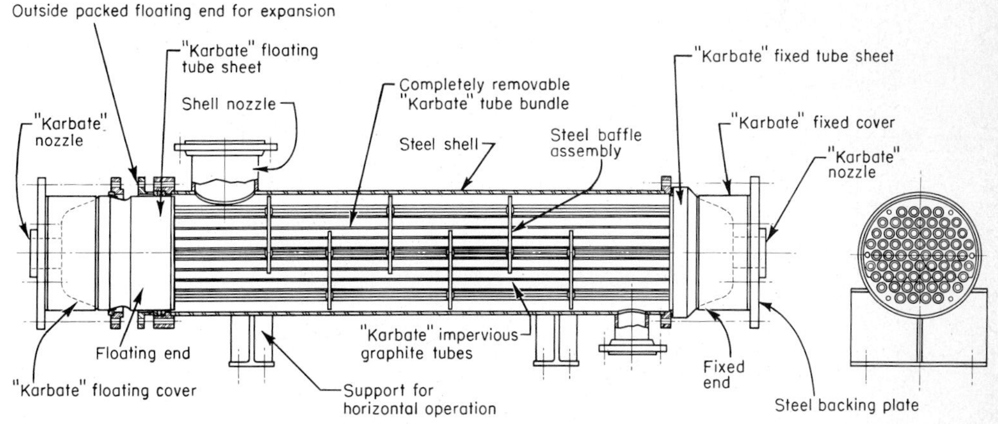

Fig. 8.7 Sectional view, impervious graphite shell and tube exchanger, single-tube pass. (*Carbon Products Div., Union Carbide Corp.*)

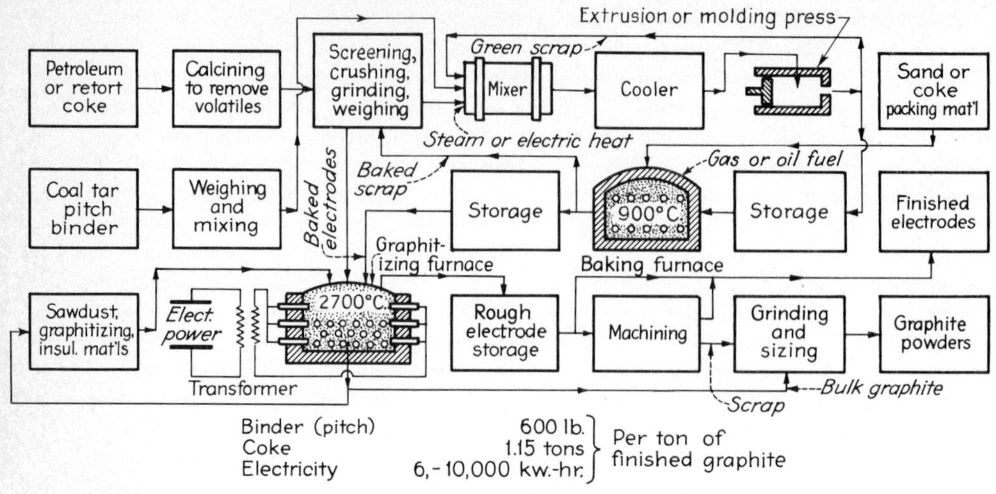

Fig. 8.8 Flowchart for the manufacture of artificial-graphite electrodes and powders.

per pound in 1974. Synthetic resin composites are used in aerospace structures, where weight savings are important, and in high-performance sporting goods. All-carbon composites, made by pyrolyzing the resin matrix, are used in aerospace applications and are being investigated as friction materials in brakes. Projected fiber costs of less than $5 per pound should stimulate large-volume applications in the construction and transportation fields.

Graphite of extremely high purity is used in many *nuclear reactors* as a moderator and reflector. The moderator holds the nuclear fuel in position and "moderates" or slows down the fast-moving neutrons to speeds at which they can be captured, thus enabling the chain reaction to continue. The reflector surrounds the moderator and scatters escaping neutrons back into the core. In some reactors, graphite is also used as a matrix in which to disperse and contain the nuclear fuel.

MANUFACTURE[13] *Manufactured graphite* is made electrically from retort or petroleum coke, in line with the flowchart depicted in Fig. 8.8. The reaction for this allotropic change is essentially

$$C(amorph) \longrightarrow C(graphite) \qquad \Delta H = -2.5 \text{ kcal}$$

The *unit changes* involved in commercializing these reactions, as charted in Fig. 8.8, may be summarized for electrodes and other graphite articles:

Coke (petroleum or retort) is selected and shipped to the graphite plant (Op).

Carbon material is calcined (1250°C) to volatilize impurities (Op and Ch).

Raw materials are carefully analyzed.

Raw materials are ground, screened, weighed, mixed with binder and formed into green electrodes, and arranged in the furnace (Op).

Green electrodes are baked at 900°C to carbonize the binder and furnish amorphous electrodes (Ch).

Amorphous electrodes are graphitized in the electric furnace at high temperature (2700°C) (Ch).

Graphite is shipped to industrial demands, and the scrap is powdered (Op).

In manufacturing graphite the furnace used consists essentially of a core of the coke being graphitized, surrounded by a heavy layer of sand, coke, and sawdust as insulation. The floor and

[13]ECT, 2d ed., vol. 4, pp. 158–202, 1964; Mantell, Carbon and Graphite Handbook, chaps. 15 and 19, Interscience, 1968.

ends of the furnace are built mainly of concrete, with cooling coils at the ends to reduce the temperature of the electrodes in contact with air, to prevent them from burning. The side walls, built of loose blocks and plates, are torn down after each run. The average charge is from 50,000 to 200,000 lb of material. The resistance of the charge makes possible a temperature of up to 3000°C. As the coke graphitizes, the voltage, because of lowered resistance, drops from 200 to 40 V. The high temperature employed volatilizes most of the metallic impurities. A complete cycle of the furnace consists of loading, 1 day; power on, 4 to 5 days; cooling, about 14 days; unloading, 2 days. Thus about 22 days are required from charge to charge. After cooling, the furnace is torn down and the graphitized carbon taken out. The insulating sand, coke, and any silicon carbide contamination are used over again. Recently, direct current has been employed[14] also, and at least one company conducts the graphitizing in a double-row countercurrent graphitizing furnace. The opposite row, after graphitizing, imparts its heat to the incoming carbon to be graphitized. To operate such a furnace, all blocks to be graphitized must be of about the same size.

In the manufacture of amorphous carbon electrodes,[15] the petroleum coke or anthracite coal used is crushed and hot-mixed with a pitch binder (Fig. 8.9). The resulting soft mass is either extruded or molded to the finished electrode shape. These green electrodes are then baked at about 950°C in either a gas-fired furnace or an electric furnace. After cooling and removal from the furnace, they are turned down and threaded on a lathe in the same manner as that used for graphite electrodes. A similar process is involved in making brushes for motors; the carbon is formed into small plates, which are then cut to the necessary shape for making the brushes. Graphite electrodes have less electric resistance and consequently can conduct more current per cross section, but they cost more than the usual amorphous carbon electrodes.

Pyrolytic graphite[16] is produced by the thermal cracking of hydrocarbons such as methane, propane, and acetylene at reduced pressures or diluted by an inert carrier gas. Structural shapes up to about 1 cm in thickness are formed by passing the gas over a heated mandrel of the desired shape. Small particles, e.g., nuclear fuel particles, are coated in a heated, fluidized bed in which the hydrocarbon–inert carrier gas mixture is used to fluidize the bed. The graphite structure, hence the properties, can be varied widely by changing the operating conditions.

Carbon fibers[17] are made by the pyrolysis of organic precursor fibers, usually rayon, polyacrylonitrile (PAN), or pitch. Fibers carbonized at 1300 to 1600°C have a modulus typically of 230 GN/m^2 (33 million psi) and a tensile strength of 2.1 to 2.8 GN/m^2 (300 to 400,000 psi). Graphitization to 2800°C increases the modulus by a factor of 2 or 3; the strength may either increase or decrease, depending on the fiber type.

INDUSTRIAL DIAMONDS

Because diamond is the hardest of all known substances, it has great industrial importance in cutting, shaping, and polishing hard substances. Diamond is used extensively for grinding tungsten carbide tooling, glass, and nonferrous metals, and for sawing and drilling concrete and stone. Tungsten-filament wire for electric lights and millions of miles of other kinds of wire are drawn through diamond dies. The United States is the largest consumer of industrial diamonds in the world. The world production of industrial diamonds is of the order of 90% of the total diamond production in weight.

[14]Olson and Price, A D-C Graphitizing Furnace, *Elec. Eng.* (*N.Y.*), **82,** 260–263 (1963); Dow Changes to DC for Graphitizing, *Chem. Eng. News,* Jan. 14, 1963, p. 46.

[15]Detailed flowcharts for both carbon and graphite electrodes are given by Mantell, Electrochemical Engineering, pp. 380 and 633, McGraw-Hill, 1960.

[16]Smith and Leeds, Pyrolytic Graphite, in Modern Materials: Advances in Development and Applications, Gonser, (ed.), vol. 7, pp. 139–221, Academic, 1970; Bokros, Deposition, Structure, and Properties of Pyrolytic Carbon, in Chemistry and Physics of Carbon, Walker (ed.), vol. 5, pp. 1–118, Dekker, 1969.

[17]ECT, 2d ed., Suppl. Vol., pp. 109–120, 1971; a flowchart for a PAN process is given in Iammartino, Graphite Fibers Forge Ahead, *Chem. Eng.* (*N.Y.*), **79**(3), 30 (1972).

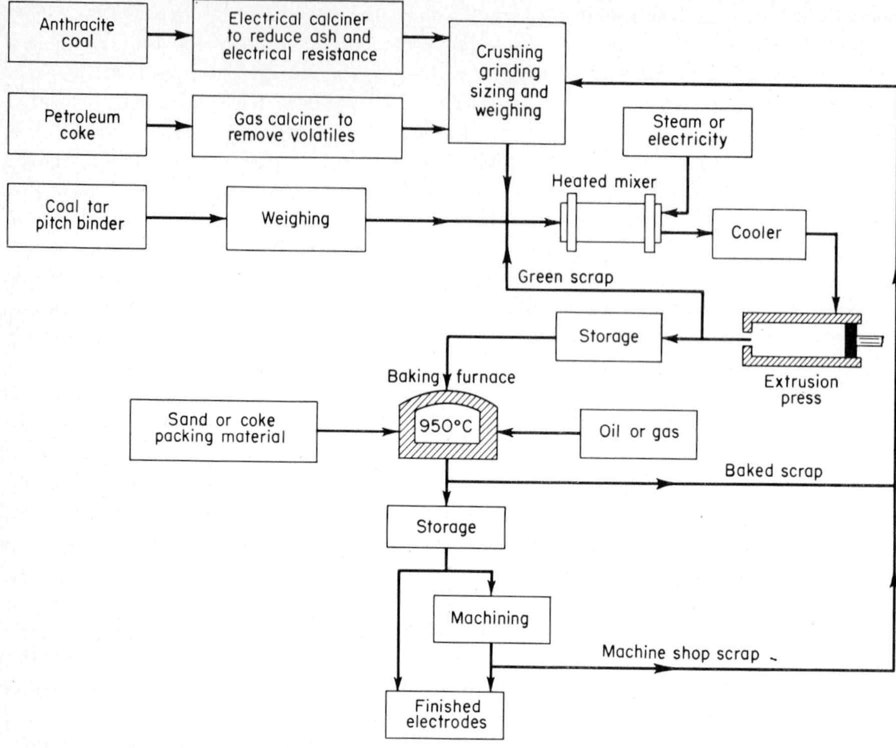

Fig. 8.9 Flowchart for manufacture of amorphous-carbon electrodes.

According to U.S. government statistics[18] in 1973 the United States imported about 19 million carats (a carat is 200 mg) of industrial diamond valued at about $66 million. About 16 million carats was exported or reexported, valued at about $59 million. The U.S. production of industrial diamond was estimated to be over 22 million carats, including synthesized powder and dust, some reclaimed material, and that released from the U.S. stockpile. Diamond consumption in the United States in 1972 (estimate) was about 16 million carats of crushed bort and dust (including synthesized powder) and about 2.3 million carats of stones.

Three natural types are employed: (1) *crystalline* and cleavable diamonds, more or less off-grade and off-color; (2) *bort,* translucent to opaque, gray or dark brown, with a radiated or confused crystalline structure; and (3) *carbonado,* frequently known as *black diamond* or *carbon,* which occurs in an opaque, tough, crystalline aggregate without cleavage. Bort is the most widely used of industrial diamond stones, and most of it comes from Africa. Synthetic diamond has become the standard of performance for grinding wheels, saws, and drills, and is a major factor in the economy of the industry.

An extensive use of industrial diamonds is in the grinding of shaped carbide or ceramic tips for turning and boring tools. Bonded-diamond wheels for grinding hard, abrasive materials such as cemented carbide tools, ceramics, and glass are made from synthesized diamond powder or crushed bort, held by a matrix of resin, vitreous material, or sintered metal. Tools set with industrial diamond are a practical necessity in truing and dressing worn abrasive wheels in all industries where precision grinding is performed. Diamond lathe tools, because of the hardness and high heat conductivity of

[18]Clarke, Bureau of Mines, Division of Non-Metallic Minerals.

diamond, allow greatly increased machining speeds in turning nonferrous metal parts to close tolerances and fine finishes. Since the continuous drawing of wire through reduction dies causes much wear, dies for accurate wire drawing are made of diamonds. Such dies are generally made from fine-quality single-crystal stones with an appropriate hole drilled through them. Recently, successful wire-drawing dies have been made from diamond compacts which consist of fine diamond powder grains thoroughly bonded together. The holes are started by using sharp diamond points, or by a fine laser beam, and are finished by lapping with fine diamond dust. When worn, dies must be redressed to the next larger wire size. Diamond dies are generally used only for drawing wire smaller than 2.5 mm in diameter. Diamond bits for drilling through hard rock formations for oil, gas, or water, for boring blast-charge holes in mines and quarries, and for drilling holes in and taking cores from concrete structures are set with small whole diamonds in a sintered metal bond. Here the diamonds do the actual work, the steel heads simply serving to hold them in the cutting position. Diamond drill crowns allow actual cores of rock to be taken from great depths and examined to yield information on what minerals or deposits are present. The increasing trend toward automation is causing the domestic consumption of industrial diamonds to be the fastest growing in the abrasives industry.

In mounting diamond crystals in tools they should be oriented in the crystal direction which provides the greatest abrasion resistance.[19] The most favorable orientation can be done "by eye" if the shape of the crystal is such that the crystal directions can be recognized. Otherwise x-ray diffraction must be used. Proper orientation of the diamonds can improve the performance of a tool manyfold over that of one having a random diamonds orientation.

In 1957 General Electric Co. commercially introduced synthetic diamond grit for resin-bonded grinding wheels based on a process discovered by scientists in its research laboratory in 1954.[20] The same basic process is used currently in factories in South Africa, Ireland, Sweden, Japan, and Russia.

The process requires pressures and temperatures in the region of thermodynamic stability for diamond (versus graphite, see Fig. 8.10) and a molten catalyst-solvent metal consisting of a group VIII metal or alloy. Special ultra high-pressure apparatus is used, the moving members of which are forced together by large hydraulic presses. One of the 1,000-ton hydraulic presses used at the General Electric Research and Development Center is shown in Fig. 8.11. It is a batch process requiring a period of a number of minutes. Different types and sizes of diamond particles, or crystals, require different conditions of pressure, temperature, catalyst-solvent, and reaction time. The crude diamonds are chemically cleaned and sized by sieving. Many different types are now currently available, ranging from the finest polishing powders up to 25/35-mesh blocky crystals which yield excellent service in saws and drills for ceramic, stone, and concrete cutting.

Fine diamond polishing powders can be made directly from certain graphites by shock-compression processes.[21] A shock wave of the order of 300- to 1200-kbar intensity and 10^{-6}s duration is passed through the graphite. In the Du Pont process the graphite particles are dispersed in a metal matrix, which helps to transmit the pressure wave and to thermally quench the extremely hot carbon components of the system immediately after compression. The thermal quench helps prevent graphitization of the diamond formed during the compression. The metal is removed by acids, and the extraneous graphite by selective oxidation processes.

Techniques have been developed for bonding diamond powder into strong, relatively tough compacts. These, when sharpened and mounted properly, serve as effective cutting tips for tools

[19]Custers and Weavind, The Manufacture of Diamond Tools in Theory and in Practice, 1956; X-ray Orientation of Diamonds for Diamond Tools, Diamond Research Laboratory, Johannesberg, South Africa, 1958.

[20]Bundy, Hall, Strong, and Wentorf, Jr., *Nature,* **176,** 51 (1955).

[21]DeCarli and Jamieson, *Science,* **133,** 182 (1961); Mears and Bowman, paper presented at Industrial Diamond Association of America, Technical Session, Mar. 7, 1966, Boca Raton, Fla.; U.S. Pat. 3,401,019, E. I. Du Pont de Nemours & Co., Cowan, Dunnington, and Holtzman (1966); Trueb, *J. Appl. Phys;* **39,** 4707 (1968).

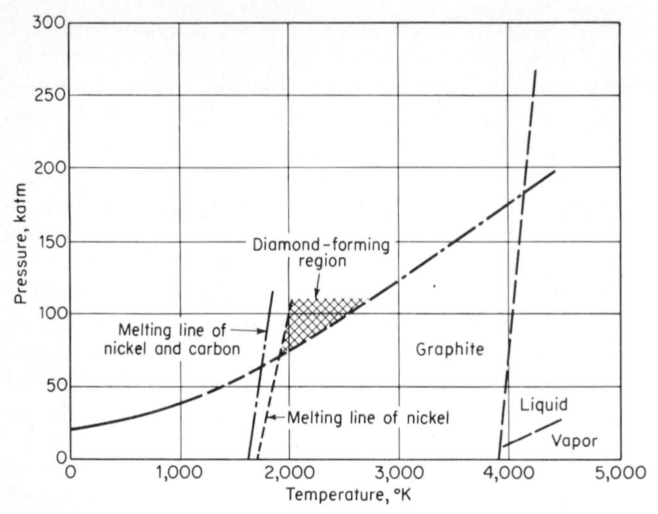

Fig. 8.10 Diamond-forming region for nickel-carbon system (katm = 1,000 atm = 1,000 kbars). (*ECT, vol. 4, p. 294, 1964.*)

8.11 One-thousand-ton hydraulic press used to provide the squeezing force on the belt superpressure apparatus for synthctic diamonds. (*General Electric Co.*)

to machine very hard workpieces. These diamond compacts also serve well as cores for small wire-drawing dies.

The new diamondlike abrasive material, cubic boron nitride,[22] is now synthesized, marketed, and used commercially. It is made from graphitic boron nitride at pressures and temperatures similar to those used in diamond synthesis, but with different catalyst-solvent materials. Compacts of cubic boron nitride are also made. Cubic boron nitride abrasive grain is particularly effective in grinding hard ferrous metals with which diamond tends to react. In general, diamond abrasive works best on hard nonmetallic materials, while cubic boron nitride abrasive does best on hard metallic materials.

Gem-quality diamonds can be grown in the laboratory,[23] but the process is slow and expensive and is not considered economically feasible in the usual market. However, these diamonds have the highest thermal conductivity of any known material, more than matching that of type II natural diamond crystals,[24] and they may become useful as heat sinks for high-power-density microelectric solid-state devices and for high-intensity laser beam windows.

SELECTED REFERENCES

Austin, O. K.: Commercial Manufacture of Carbon Black in Reinforcement of Elastomers, G. Kraus (ed.), chap. 5, Wiley-Interscience, 1965.
Cacciotti, J. J.: Graphite and Its Properties, General Electric Co., Flight Propulsion Laboratory, Cincinnati, Ohio. Report R58AGT591, issued July 1958, revised June, 1960. (232 ref.)
Gould, R. F. (ed.): Literature of Chemical Technology, Carbon Black, chap. 36, ACS Monograph, 1967.
Harker, H., and H. Marsh, Carbon, Elsevier, 1967.
Hassler, J. W.: Purification with Activated Carbon, Chemical Publishing, 1974.
MacInnes, D.: Synthetic Gem and Allied Crystal Manufacture, Noyes, 1973.
Mantell, C. L.: Carbon and Graphite Handbook, Interscience, 1968.
Morton, M. (ed.): Rubber Technology, 2d ed., Van Nostrand Reinhold, 1973.
Nightingale, R. E. (ed.): Nuclear Graphite, Academic, 1962.
Ubbelohde, A. R., and F. A. Lewis: Graphite and Its Crystal Compounds, Oxford, 1960.
Walker, P. H., and P. A. Throuer (eds.): Chemistry and Physics of Carbon, A Series of Advances, vol. 2, Dekker, 1973.

[22]Wentorf, *J. Chem. Phys.*, **26,** 956 (1957).
[23]Wentorf, *J. Chem. Phys.*, **75,** 1833 (1971).
[24]Strong and Chrenko, *J. Chem. Phys.*, **75,** 1838 (1971).

chapter 9

CERAMIC INDUSTRIES

The traditional ceramic industries, sometimes referred to as the *clay products* or *silicate industries*. have as their finished materials a variety of products that are essentially silicates.[1] In recent years new products have been developed as a result of the demand for materials that withstand higher temperatures, resist greater pressures, have superior mechanical properties, possess special electrical characteristics, or can protect against corrosive chemicals. The following types of products are considered in this chapter:

1. *Whitewares.* China, earthenware, pottery, porcelain, stoneware, and vitreous ware.
2. *Structural clay products.* Building brick, face brick, terra cotta, sewer pipe, and drain tile.
3. *Refractories.* Firebricks; silica, chromite, magnesite, magnesite-chromite brick: silicon carbide and zirconia refractories; aluminum silicate and alumina products.
4. *Specialized ceramic products.*
5. *Enamels and enameled metal.*

The glass, cement, and artificial-abrasives industries are dealt with in separate chapters, since the emphasis of this book is on manufacturing procedures peculiar to the major chemical process industries.[2]

HISTORICAL Pottery making is one of the most ancient of human industries. Burnt clayware has been found dating from about 15,000 B.C. and was well developed in Egypt 10 centuries later. Museums contain, as a record of culture, clay products created independently by various races. Demands for superior materials have produced a broader spectrum of systems; greater emphasis has been placed on the cross-fertilization of *silicate chemistry* with metallurgy and solid-state physics, coupled with many computer-controlled processes and advancing automation, which characterize modern methods of fabrication. Recently, new processes have been under development for brickmaking from inorganic wastes such as fly ash from power plants, foundry sand, mine tailings, furnace slag, and a large variety of other materials that lie in mammoth piles throughout the country.[3] Other new developments are appearing in the patent literature.[4]

[1]Kingery, Introduction to Ceramics, Wiley, 1960.

[2]ECT, vol. 4, p. 759, 1964: "Ceramics (plural noun) comprise all engineering materials or products that are chemically inorganic, except metals and metal alloys, and are usually rendered serviceable through high temperature processing. The word *ceramic* may be used as a singular noun but is more often used as an adjective meaning inorganic, non-metallic. A primary difference between ceramics and metals is the nature of their chemical bonding. The difference is responsible for characteristic properties and behavior of ceramics. Composites such as ceramics-metal combinations and others involving organic materials are often included as ceramic materials."

[3]New Processes Turn Wastes into Bricks, *Chem. Eng. News*, Sept. 13, 1971, p. 49.

[4]Angstadt and Bell, Production of Alumina and Portland Cement from Clay and Limestone, U.S. Pat. 3,642,437, Feb. 15, 1972; Burk *et al.*, Alumina Extraction from Alumino-silicate Ores and Potassium Sulfate Ores, U.S. Pat. 3,652,208, Mar. 28, 1972; Huska *et al.*, Method and Apparatus for Converting Aluminum Nitrate Solution to Alpha Alumina, U.S. Pat. 3,647,373, Mar. 7, 1972.

TABLE 9.1 ***Salient Clay and Clay Products Statistics for the United States**** *(In thousands of short tons and thousands of dollars)*

	1968	*1969*	*1970*	*1971*	*1972*
Domestic clays sold or used by producers	57,348	58,694	54,853	56,666	59,456
Value	$246,938	$264,415	$267,912	$274,431	$303,022
Exports	1,519	1,574	2,076	1,973	1,847
Value	$44,134	$45,767	$66,116	$65,329	$66,216
Imports for consumption	97	82	87	64	67
Value	$1,951	$1,750	$1,802	$1,501	$1,309
Clay refractories, shipments (value)	$229,660	$257,507	$256,384	$236,563	$274,679
Clay construction products, shipments (value)	$590,776	$608,982	$554,431	$641,567	$772,236

Source: Minerals Year Book 1972, Dept. of the Interior, 1974. *Excludes Puerto Rico.

USES AND ECONOMICS Inasmuch as sales data on whitewares are not usually available, and because many ceramic products are used as components of finished materials classified in other categories, it is difficult to determine their sales volume. However, an extensive effort at studying this industry is made each year by the Bureau of Mines. The economic statistics are summarized in Table 9.1. Clays were produced in 47 states and Puerto Rico in 1972. The states leading in clay output in 1972 were Georgia, 6.2 million tons, Texas, 5.2 million tons, and Ohio, 4.1 million tons. Georgia also leads in total value of clay output with $132.3 million. Seventy-seven percent of all clay produced is consumed in the manufacture of heavy-clay construction products. Of these products, building brick, sewer pipe, and drain tile account for 38% of the total clay usage, portland cement accounts for 21%, and lightweight aggregate accounts for 18%. As shown in Table 9.1, the value of construction products was over $722 million in 1972. Refractories use large quantities of clays, and usage is increasing primarily because of use of plastic, ramming, and castable mixes. Clays are used as fillers in many products such as paper, rubber, plastics, paint, and fertilizers. Fuller's earth is the principal clay used in adsorbent applications. Other important uses include drilling mud, floor and wall tile, pelletizing iron ore, and pottery.

BASIC RAW MATERIALS[5]

The three main raw materials used in making classic, or "triaxial," ceramic products are clay, feldspar, and sand. Clays are more-or-less impure hydrated aluminum silicates that have resulted from the weathering of igneous rocks in which feldspar was a noteworthy original mineral. The reaction may be expressed:

$$\underset{\text{Potash feldspar}}{K_2O \cdot Al_2O_3 \cdot 6SiO_2} + CO_2 + 2H_2O \longrightarrow K_2CO_3 + \underset{\text{Kaolinite}}{Al_2O_3 \cdot 2SiO_2 \cdot 2H_2O} + \underset{\text{Silica}}{4SiO_2}$$

There are a number of mineral species called *clay minerals,* but the most important are kaolinite ($Al_2O_3 \cdot 2SiO_2 \cdot 2H_2O$), montmorillonite ($(Mg, Ca)O \cdot Al_2O_3 \cdot 5SiO_2 \cdot nH_2O$), and illite ($K_2O$, MgO, Al_2O_3, SiO_2, H_2O, all in variable amounts). From a ceramic viewpoint clays are plastic and moldable when sufficiently finely pulverized and wet, rigid when dry, and vitreous when fired at a suitably high temperature. Upon these properties depend the manufacturing procedures.

Accompanying clay minerals in the clays of commerce are varying amounts of feldspar, quartz, and other impurities such as oxides of iron. In nearly all the clays used in the ceramic industry, the

[5]ECT, vol. 4, p. 762, 1964, presents a detailed discussion of raw materials. Nepheline syenite, $(Na,K)(Al, Si)_2O_4$, a quartz-free igneous rock, is used extensively in whitewares; it is a more active flux than feldspar; Kingery, *op. cit.*, tables 2.1 and 2.2; Agnello *et al.*, Kaolin, *Ind. Eng. Chem.*, **52,** 370 (1960); Conrad, Ceramic Formulas, Macmillan, 1973; Mercade, Purification of Clay by Selective Flocculation, U.S. Pat. 3,701,417, Oct. 31, 1972.

basic clay mineral is kaolinite, although bentonite[6] clays based on montmorillonite are used to some extent where very high plasticity is desired. This property of plasticity, or workability, is influenced most by the physical condition of the clay and varies greatly among different types of clays. Clays are chosen for the particular properties desired and are frequently blended to give the most favorable result. Clays vary so much in their physical properties and in the impurities present that it is frequently necessary to upgrade them by the *beneficiation* procedure. Figure 9.1 shows the steps necessary for such a procedure, wherein sand and mica are removed. Almost all the steps in this flowchart apply to physical changes, or *unit operations*, such as size separation by screening or selective settling, filtration, and drying. However, the colloidal properties are controlled by appropriate additives such as sodium silicate and alum. Beneficiation processes also include froth flotation. Chemical purification is used for high-purity materials such as alumina and titania.

There are three common types of *feldspar*, potash ($K_2O \cdot Al_2O_3 \cdot 6SiO_2$), soda ($Na_2O \cdot Al_2O_3 \cdot 6SiO_2$), and lime ($CaO \cdot Al_2O_3 \cdot 6SiO_2$), all of which are used in ceramic products to some extent. Feldspar is of great importance as a fluxing constituent in ceramic formulas. It may exist in the clay as mined, or it may be added as needed. Half of the feldspar used in the United States comes from North Carolina. The third main ceramic constituent is sand or *flint*. Its essential properties for the ceramic industries are summarized, along with the similar characteristics of clay and feldspar, in Table 9.2. For light-colored ceramic products, sand with a low iron content should be chosen.

In addition to the three principal raw materials, a wide variety of other minerals, salts, and oxides is used as fluxing agents and special refractory ingredients. Some of the more *common fluxing agents* which lower temperatures are

Borax ($Na_2B_4O_7 \cdot 10H_2O$)
Boric acid (H_3BO_3)
Soda ash (Na_2CO_3)
Sodium nitrate ($NaNO_3$)
Pearl ash (K_2CO_3)
Nepheline syenite
Calcined bones
Apatite [$Ca_5(F,Cl,OH)(PO_4)_3$]
Fluorspar (CaF_2)
Cryolite (Na_3AlF_6)
Iron oxides
Antimony oxides
Lead oxides
Lithium minerals
Barium minerals

Some of the more common *special refractory ingredients* are

Alumina (Al_2O_3)
Olivine [$(FeO,MgO)_2SiO_2$]
Chromite ($FeO \cdot Cr_2O_3$)
Aluminum silicates ($Al_2O_3 \cdot SiO_2$)
(kyanite, sillimanite, andalusite)
Dumortierite ($8Al_2O_3 \cdot B_2O_3 \cdot 6SiO_2 \cdot H_2O$)
Magnesite ($MgCO_3$)
Lime (CaO) and limestone ($CaCO_3$)
Zirconia (ZrO_2)
Titania (TiO_2)
Hydrous magnesium silicates,
e.g., talc ($3MgO \cdot 4SiO_2 \cdot H_2O$)
Carborundum (SiC)
Mullite ($3Al_2O_3 \cdot 2SiO_2$)
Dolomite [$CaMg(CO_3)_2$]
Thoria (ThO_2)

Many other raw materials are used in various combinations; at least 450 have been classified.[7]

CHEMICAL CONVERSIONS, INCLUDING BASIC CERAMIC CHEMISTRY

All ceramic products are made by combining various amounts of the foregoing raw materials, shaping, and heating to firing temperatures. These temperatures may be as low as 700°C for some overglazes

[6]See Perry, pp. 21-65 to 21-69, for flotation.
[7]Raw Materials, *Ceram. Age*, **79**(10), 21 (1963).

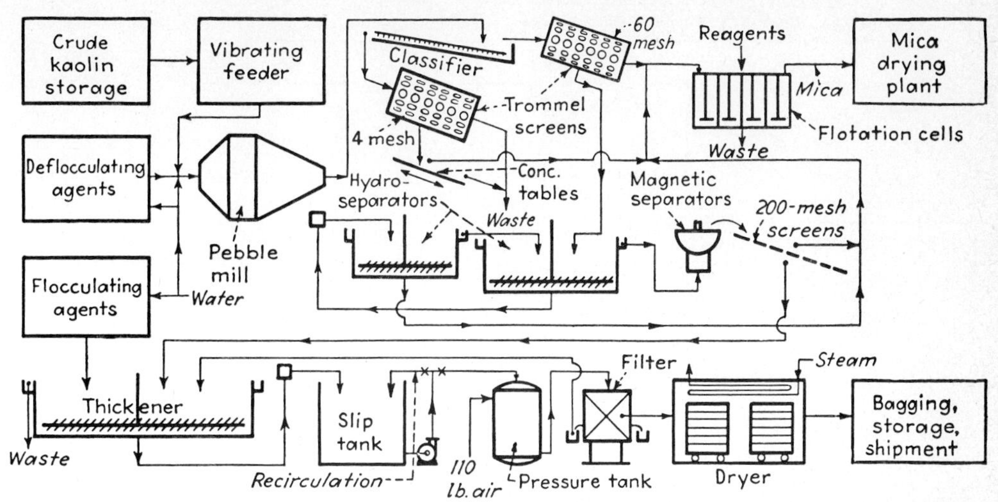

Note: Quantities cannot be given since clay recovery varies from 8 to 18 percent, depending on crude clay used. Plant shown here designed for 30 tons per day output regardless of crude clay variations

Fig. 9.1 China-clay beneficiation. (*Harris Clay Co.*)

or as high as 2000°C for many vitrifications. Such temperatures cause a number of reactions, which are the chemical bases for the *chemical conversions:*

1. Dehydration, or "chemical water smoking," at 150 to 650°C.
2. Calcination, e.g., of $CaCO_3$ at 600 to 900°C.
3. Oxidation of ferrous iron and organic matter at 350 to 900°C.
4. Silicate formation at 900°C and higher.

Some of the initial chemical changes are relatively simple, such as calcination of $CaCO_3$ and the dehydrations and decompositions of kaolinite. Other reactions, such as silicate formations, are quite complex and change with the temperature and constituent ratios as depicted by Figs. 9.2 and 9.6. Phase-rule[8] studies as exemplified by these figures have been of *revolutionary importance* in interpreting empirical observations in the ceramic industries and in making predictions for improvements. For instance, the data in Fig. 9.2 on the $Al_2O_3 \cdot SiO_2$ system have led to the important development of processes for mullite refractories. This diagram shows that any percentage of liquefaction can be obtained, dependent on a definite temperature, except at the monovariant points. Thus, if the progressive melting is kept from going too far by controlling the rise in temperature, sufficient solid skeletal material will remain to hold the hot mass together. This $Al_2O_3 \cdot SiO_2$ diagram shows that mullite is the only stable compound of alumina and silica at high temperatures.

Ceramic products are all more or less refractory, i.e., resistant to heat, and the degree of refractoriness of a given product is determined by the relative quantities of refractory oxides and fluxing oxides. The principal refractory oxides are SiO_2, Al_2O_3, CaO, and MgO, with ZrO_2, TiO_2, Cr_2O_3, and BeO used less commonly. The principal fluxing oxides are Na_2O, K_2O, B_2O_3, and SnO_2, with fluorides also used as fluxes in certain compositions.

The common ingredient of all ceramic products is clay (kaolinite, usually), and therefore the chemical reactions which occur on heating clay are quite important. The first effect of the heat is

[8]Kingery, *op. cit.*, contains detailed discussion, tables of physical properties, and other phase diagrams; Levin *et al.*, Phase Diagrams for Ceramists, American Ceramic Society, 1950.

TABLE 9.2 ***Basic Raw Materials for Ceramics***

	Kaolinite	*Feldspar*	*Sand or flint*
Formula	$Al_2O_3 \cdot 2SiO_2 \cdot 2H_2O$	$K_2O \cdot Al_2O_3 \cdot 6SiO_2$	SiO_2
Plasticity	Plastic	Nonplastic	Nonplastic
Fusibility	Refractory*	Easily fusible binder	Refractory*
Melting point	3245°F, 1785°C	2100°F, 1150°C	3110°F, 1710°C
Shrinkage on burning	Much shrinkage	Fuses	No shrinkage

*Infusible at highest temperature of coal fire (1400°C).

to drive off the water of hydration; this occurs at about 600 to 650°C and absorbs much heat, leaving an amorphous mixture of alumina and silica, as shown by x-ray studies.

$$Al_2O_3 \cdot 2SiO_2 \cdot 2H_2O \longrightarrow Al_2O_3 + 2SiO_2 + 2H_2O$$

In fact, a large proportion of the alumina can be extracted with hydrochloric acid at this stage. As heating is continued, the amorphous alumina changes quite sharply at 940°C to a crystalline form of alumina, γ-alumina, with the evolution of considerable heat. At a slightly higher temperature, beginning at about 1000°C, the alumina and silica combine to form mullite ($3Al_2O_3 \cdot 2SiO_2$). At a still higher temperature, the remaining silica is converted into crystalline cristobalite. Therefore the fundamental overall reaction in the heating of clay is as follows

$$\underset{\text{Kaolinite}}{3(Al_2O_3 \cdot 2SiO_2 \cdot 2H_2O)} \longrightarrow \underset{\text{Mullite}}{3Al_2O_3 \cdot 2SiO_2} + \underset{\text{Cristobalite}}{4SiO_2} + 6H_2O$$

The equilibrium state of Al_2O_3-SiO_2 mixtures as a function of temperature is summarized in the phase-equilibrium diagram of this system shown in Fig. 9.2. The presence of fluxes tends to lower the temperature of formation of mullite and of attainment of the equilibrium conditions given in the equilibrium diagram.

An actual ceramic body contains many more ingredients than clay itself. Hence the chemical reactions are more involved, and there will be other chemical species besides mullite and cristobalite present in the final product. For example, various silicates and aluminates of calcium, magnesium, and possibly alkali metals may be present. However, the alkali portion of feldspar and most of the fluxing agents become part of the glassy, or vitreous, phase of the ceramic body. All ceramic bodies

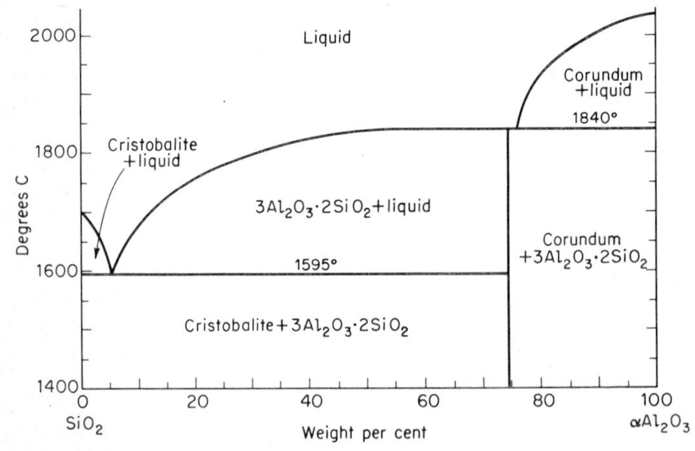

Fig. 9.2 Phase diagram of the system alpha Al_2O_3-SiO_2. Mullite has the formula $3Al_2O_3 \cdot 2SiO_2$.

undergo a certain amount of vitrification, or glass formation, during heating, and the degree of vitrification depends upon the relative amounts of refractory and fluxing oxides in the composition, the temperature, and the time of heating. The vitreous phase imparts desirable properties to some ceramic bodies, e.g., by acting as a bond and imparting translucency in chinaware. Even in refractories some vitrification is desirable to act as a bond, but extensive vitrification destroys the refractory property. Thus it is seen that any ceramic body is composed of a vitreous matrix plus crystals, of which mullite and cristobalite are two of the most important.

The *degree of vitrification*, or the progressive reduction in porosity, provides the basis for a useful classification of ceramic products as follows:

1. *Whitewares.* Varying amounts of fluxes, heat at moderately high temperatures, varying vitrification.
2. *Heavy-clay products.* Abundant fluxes, heat at low temperatures, little vitrification.
3. *Refractories.* Few fluxes, heat at high temperatures, little vitrification.
4. *Enamels.* Very abundant fluxes, heat at moderate temperatures, complete vitrification.
5. *Glass.* Moderate fluxes, heat at high temperatures, complete vitrification (Chap. 11).

WHITEWARES

Whiteware is a generic term for ceramic products which are usually white and of fine texture. These are based on selected grades of clay bonded together with varying amounts of fluxes and heated to a moderately high temperature in a kiln (1200 to 1500°C). Because of the different amounts and kinds of fluxes, there is a corresponding variation in the degree of vitrification among whitewares, from earthenware to vitrified china. These may be broadly defined as follows:

Earthenware, sometimes called *semivitreous* dinnerware, is porous and nontranslucent with a soft glaze.

Chinaware is a vitrified translucent ware with a medium glaze which resists abrasion to a degree; it is used for nontechnical purposes.

Porcelain is a vitrified translucent ware with a hard glaze which resists abrasion to the maximum degree. It includes chemical, insulating, and dental porcelain.

Stoneware, one of the oldest of ceramic wares, was in use long before porcelain was developed; in fact, it may be regarded as a crude porcelain, the raw material being of a poorer grade and not so carefully fabricated.

Sanitary ware, formerly made from clay, was usually porous; hence a vitreous composition is presently used. Prefired and sized vitreous grog is sometimes included with the triaxial composition.

Whiteware tiles, available in a number of special types, are generally classified as *floor tiles,* resistant to abrasion and impervious to stain penetration, glazed or unglazed, and as *wall tiles* (in a variety of colors and textures), which also have a hard, permanent surface.

MANUFACTURE OF PORCELAIN To represent a typical manufacturing procedure in this group, porcelain as defined above is chosen. There are three lines of production: *wet-process porcelain,* used for production of fine-grained, highly glazed insulators for high-voltage service; *dry-process porcelain,* employed for rapid production of more open-textured, low-voltage pieces; and *cast porcelain,* necessary for the making of pieces too large or too intricate for the other two methods. These three processes are based on the same raw materials, the differences in manufacturing being largely in the drying and forming (shaping) steps. The wet process is illustrated in Fig. 9.3. This flowchart may be broken down into the following sequences:

Raw materials of proper proportions and properties to furnish porcelain of the desired quality are weighed from overhead hoppers into the weighing car (Op).

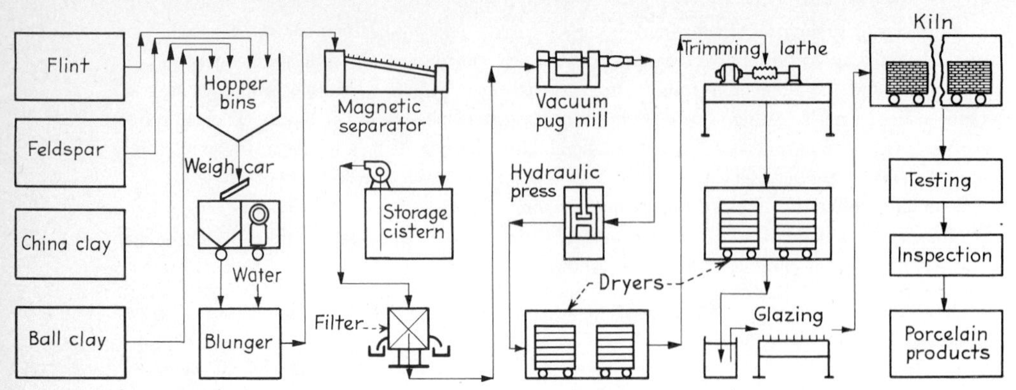

Fig. 9.3 Porcelain manufacture by the wet process of Westinghouse.

Feldspar, clays, and flint are mixed with water in the blunger (clay-water mixer) and then passed over a magnetic separator, screened, and stored (Op).

Most of the water is removed (and wasted) in the filter press (Op). All the air is taken out in the pug mill, assisted by a vacuum and by slicing knives. This results in denser and stronger porcelain (Op).

The prepared clay is formed into blanks in a hydraulic press or by hot-pressing in suitable molds (Op).

The blanks are preliminarily dried, trimmed, and finally completely dried, all under carefully controlled conditions (Op).

A high surface luster is secured by glazing with selected materials (Op).

The vitrification of the body and the glaze is carried out in tunnel kilns, with exact controls of temperature and movement (Ch).

The porcelain articles are protected by being placed in saggers[9] fitted one on top of the other in cars. This represents a one-fire process wherein body and glaze are fired simultaneously. The porcelain pieces are rigidly tested electrically and inspected before storage for sale (Op).

Much *tableware* is manufactured by more complicated procedures than illustrated by the somewhat related porcelain product. Some objects are shaped by being thrown on the potter's wheel where skilled hands work the revolving plastic clay into the desired form; some objects are *cast* from the clay slip in molds of absorbent plaster of paris. After drying, they are removed and further processed. Complex shapes such as artware and laboratoryware can be produced by this method (Fig. 9.4*b*). The mass production of simple round objects, like cups, saucers, and plates, is carried out economically by *jiggering*, where the plastic clay is pressed into or on a single revolving mold, the potter often being aided in shaping the outer surface and in removing excess clay by a lever, which is lowered over the mold, shaped in the profile of the object desired (Fig. 9.5*b*). Mechanized jiggers have been developed; however, high-energy mixing and rapid infrared drying are necessary in successful automation. After drying, whiteware may be fired in three ways; the common method is to fire before glazing to a sufficiently high temperature to mature the body.

Glazing is important in whitewares and particularly so for tableware. A glaze is a thin coating of glass melted onto the surface of more-or-less porous ceramic ware. It contains ingredients of two distinct types in different proportions: refractory materials such as feldspar, silica, and china clay,

[9]Saggers are made by mixing coarse granules obtained from grinding old saggers with new clay and water in a pug mill and discharging the doughlike mass from an extrusion machine in loaf form. These loaves are placed in a molding machine, where box shape is imparted, and then fired. They may be used several times.

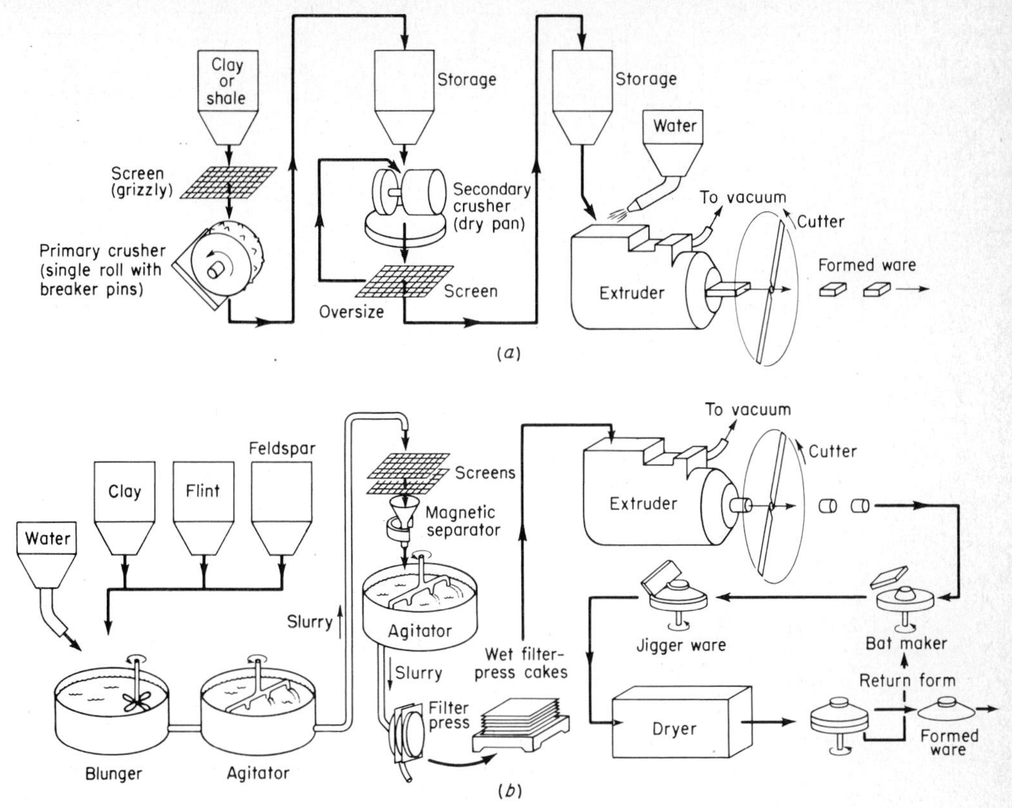

Fig. 9.4 Simplified flowcharts of forming processes. (*a*) Extrusion of structural clay bricks; (*b*) jiggering of dinner plates. (*ECT, 1964.*)

and fluxes such as soda, potash, fluorspar, and borax. Different combinations of these materials and the different temperatures at which they are fixed give a wide range of texture and quality. Nepheline syenite permits firing at a lower temperature. The glaze must be bonded to the ware, and its coefficient of expansion must be sufficiently close to that of the ware to avoid defects such as "crazing" and "shivering." The glaze may be put on by dipping, spraying, pouring, or brushing. Decoration of such ware may be "underglaze" or "overglaze." *Glost firing* is the technical term for the firing of the glaze. Earthenware should be glazed between 1050 and 1100°C; stoneware between 1250 and 1300°C.

STRUCTURAL-CLAY PRODUCTS

Low-cost but very durable products, such as building brick, face brick, terra cotta, sewer pipe, and drain tile, are frequently manufactured from the cheapest of common clays with or without glazing. The clays used generally carry sufficient impurities to provide the needed fluxes for binding. When the clay is glazed, as in sewer pipe or drain tile, this may be done by throwing salt ("salt glaze") upon the kiln fire. The volatilized salt reacts to form the fusible coating or glaze.

MANUFACTURE OF BUILDING BRICK The raw materials are clays from three groups: (1) red burning clay, (2) white burning clay, and (3) buff burning clay, usually a refractory. The requirements for face-brick clay are freedom from warping, absence of soluble salts, sufficient hardness

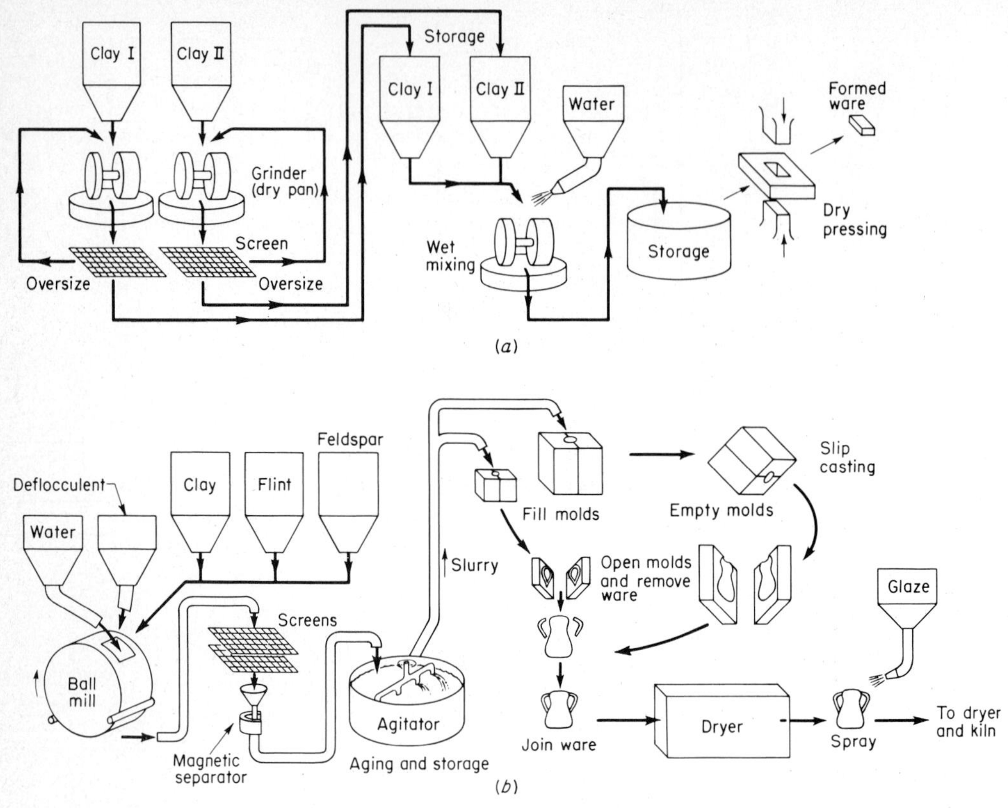

Fig. 9.5 Simplified flowcharts of forming processes. (*a*) Dry pressing of refractories; (*b*) slip-casting artware. (*ECT, vol. 4, p. 781, 1964.*)

when burned at a moderate temperature, and general uniformity in color upon burning. Requirements for brick are much less stringent; red burning clay is usually used. Bricks are manufactured by one of three processes:[10] the soft-mud, the stiff-mud, or the dry-press. In the now predominant *stiff-mud* procedure, the clay is just wet enough (12 to 15%) to stick together when worked. This clay is forced out through an extruder (Fig. 9.5*a*). *De-airing* increases the workability, plasticity, and strength of undried brick by reducing voids. Bricks may be re-pressed to make face brick; re-pressing ensures a more uniform shape and overcomes the internal stresses set up by the extruder. Bricks are *dried* in various ways: outdoors, in sheds, or in tunnel dryers. After drying, the bricks are *fired* in kilns of such type as described later in the chapter to a temperature from 875°C to somewhat above 1000°C. Combining the dryer and the kiln is a recent trend. The stiff-mud process is employed for the manufacture of practically every clay product, including all types of brick, sewer pipe, drain tile, hollow tile, fireproofing, and terra cotta. The clay in some cases can be worked directly from a bank into the stiff-mud machine, but a more desirable product will result if the clay is ground and tempered before use. The type of clay locally available often determines the ceramic product which can be made economically. Structural-clay product fabrication has become highly mechanized; unmodernized plants cannot remain competitive.

[10]See CPI 2 for soft-mud and dry-press procedures.

REFRACTORIES

Refractories, termed *acid, basic,* and *neutral,* and also superrefractories embrace those materials used to withstand the effect of thermal, chemical, and physical effects met with in furnace procedures. Refractories are sold in the form of firebrick; silica, magnesite, chromite, and magnesite-chromite brick; silicon carbide and zirconia refractories; aluminum silicate and alumina products. The fluxes required to bind together the particles of the refractories are kept at a minimum to reduce vitrification. The possibility of shaping articles made from bodies without clay and even with no natural plasticity led to the manufacture of single-component ceramics with superior qualities, e.g., pure oxide refractories. These are monocrystalline and self-bonded as compared with the conventional vitreous-bonded refractories.

PROPERTIES OF REFRACTORIES In making the refractory best suited for a definite operation it is necessary to consider the materials, the working temperature of the furnace where the refractory is needed, the rate of temperature change, the load applied during heats, and the chemical reactions encountered. Generally, several types of refractories are required for the construction of any one furnace, because usually no single refractory can withstand all the different conditions that prevail in the various parts of furnaces.

Chemical properties. The usual classification of commercial refractories divides them into acid, basic, and neutral groups, although in many cases a sharp distinction cannot be made. Silica bricks are decidedly acid, and magnesite bricks are strongly basic; however, fire-clay bricks are generally placed in the neutral group, though they may belong to either of these classes, depending upon the relative silica-alumina content. It is usually inadvisable to employ an acid brick in contact with an alkaline product, or vice versa. Neither chemical reactions nor physical properties are the only criteria of acceptable behavior; both should be considered. Chemical action may be due to contact with slags, fuel ashes, and furnace gases, as well as with products such as glass or steel.

Porosity is directly related to many other physical properties of brick, including resistance to chemical attack. The higher the porosity of the brick, the more easily it is penetrated by molten fluxes and gases. For a given class of brick, those with the lowest porosity have the greatest strength, thermal conductivity, and heat capacity.

Fusion points are found by the use of pyrometric cones of predetermined softening points. Most commercial refractories soften gradually over a wide range and do not have sharp melting points because they are composed of several different minerals, both amorphous and crystalline. The fusing points of these pyrometric cones are available in the literature.[11] Typical fusion points of refractories, both for pure substances and for technical products, are given in Table 9.3.

Spalling. A fracturing, or a flaking off, of a refractory brick, or block, due to uneven heat stresses or compression caused by heat is known as *spalling.*[12] Refractories usually expand when heated. Bricks that undergo the greatest expansion at the least uniform rate are the most susceptible to spalling when subjected to rapid heating and cooling.

Strength. Cold strength usually has only a slight bearing on strength at high temperatures. Resistance to abrasion or erosion is also important for many furnace constructions, such as co-product coke-oven walls and linings of the discharge end of rotary cement kilns.

Resistance to temperature changes. Bricks with the lowest thermal expansion and coarsest texture are the most resistant to rapid thermal change; also, less strain develops. Bricks that have been used for a long time are usually altered properties, and are often melted to glassy slags on the outside surface or even more or less corroded away.

[11]ASTM Designation C 24–42. Many other specifications for refractories and refractory materials are given in the ASTM standards.

[12]Norton, Refractories, 3d ed., chap. 15, McGraw-Hill, 1949, is devoted entirely to spalling and has many literature references.

TABLE 9.3 ***Fusion Temperatures of Refractories***

Material	*Temperature, °C*	*Temperature, °F*
Fire-clay brick	1600–1750	2912–3182
Kaolinite ($Al_2O_3 \cdot 2SiO_2 \cdot 2H_2O$)	1785	3245
Silica brick	1700	3090
Silica (SiO_2)	1710	3110
Bauxite brick	1732–1850	3150–3362
High-alumina clay brick	1802–1880	3276–3416
Mullite ($3Al_2O_3 \cdot 2SiO_2$)	1810	3290
Sillimanite (Al_2SiO_5)	1816	3300
Forsterite ($2MgO \cdot SiO_2$)	1890	3434
Chromite ($FeO \cdot Cr_2O_3$)	1770	3218
Chrome brick	1950–2200	3542–3992
Alumina (Al_2O_3)	2050	3720
Spinel ($MgO \cdot Al_2O_3$)	2135	3875
Silicon carbide (SiC)	2700	4890
Magnesite brick	2200	3992
Zirconia brick	2200–2700	3992–4892
Boron nitride*	2720	4930*

Source: Norton, Refractories, 3d ed., pp. 356–361, McGraw-Hill, 1949. For other data on refractory materials, see Perry, p. 23–73
Chem. Eng.* (*N.Y.*), **70(22), 110 (1963).

Thermal conductivity. The densest and least porous bricks have the highest thermal conductivity because of the absence of air in voids. Though heat conductivity is wanted in some furnace constructions, as in muffle walls, it is not so desirable as some other properties of refractories, such as resistance to firing conditions. Insulation is desired in special refractories.

Heat capacity. Furnace heat capacity depends upon the thermal conductivity, the specific heat, and the specific gravity of the refractory. The low quantity of heat absorbed by lightweight brick works as an advantage when furnaces are operated intermittently, because the working temperature of the furnace can be obtained in less time with less fuel. Conversely, dense, heavy fire-clay brick is best for regenerator checkerwork, as in coke ovens, glass furnaces, and stoves for blast furnaces.

MANUFACTURE OF REFRACTORIES Included are these *physical operations* and *chemical conversions:* grinding and screening, mixing, pressing or molding and re-pressing, drying, and burning or vitrification. Usually, the most important single property to produce in manufacture is high bulk density, which affects many of the other important properties, such as strength, volume stability, slag and spalling resistance, as well as heat capacity. On the other hand, for insulating refractories, a porous structure is required, which means low density.

Grinding. Obviously, one of the most important factors is the size of the particles in the batch. It is known that a mixture in which the proportion of coarse and fine particles is about 55:45, with only few intermediate particles, gives the densest mixtures. Careful screening, separation, and recycling are necessary for close control. This works very well on highly crystalline materials but is difficult to obtain in mixes of high plasticity.

Mixing. The real function of mixing is the distribution of the plastic material so as to *coat* thoroughly the nonplastic constituents. This serves the purpose of providing a lubricant during the molding operation and permits the bonding of the mass with a minimum number of voids.

Molding. The great demand for refractory bricks of greater density, strength, volume, and uniformity has resulted in the adoption of the dry-press method of molding with mechanically operated presses (Fig. 9.4*a*). The dry-press method is particularly suited for batches that consist primarily of nonplastic materials. In order to use high-pressure forming, it is necessary to de-air the bricks during pressing to avoid laminations and cracking when the pressure is released. When pressure is applied, the gas is absorbed by the clay or condensed. The use of a vacuum is applied through vents in the mold box. Large special shapes are not easily adapted to machine molding.

Drying is used to remove the moisture added before molding to develop plasticity. It should be noted that the elimination of water leaves voids and causes high shrinkage and internal strains. In some cases drying is omitted entirely, and the small amount necessary is accomplished during the heating stage of the firing cycle.

Burning may be carried out in typical round, downdraft kilns or continuous-tunnel kilns. Two important things take place during burning: the development of a permanent bond by *partial vitrification* of the mix, and the development of stable mineral forms for future service, as shown in the phase diagrams Figs. 9.2 and 9.6. The changes that take place are removal of the water of hydration, followed by calcination of carbonates and oxidation of ferrous iron. During these changes the volume may shrink as much as 30%, and severe strains are set up in the refractory. This shrinkage may be eliminated by *prestabilization* of the materials used.

VARIETIES OF REFRACTORIES About 95% of the refractories manufactured are nonbasic, with silica (acid) and fire-clay (neutral) brick predominant. Although a refractory is usually thought of in terms of its ability to withstand temperature, it is really only in exceptional cases that heat is the sole agent that effects the final destruction usually caused by chemical action at the operating temperature.

Fire-clay brick. Fire clays are the most widely used of all available refractory materials, since they are well suited for a variety of applications. Fire clays range in chemical composition from those with a large excess of free silica to those with a high alumina content. The steel industries are the largest consumers of refractories for the linings of blast furnaces, stoves, open hearths, and other furnaces. Other industries having use for them are foundries, lime kilns, pottery kilns, cupolas, brass and copper furnaces, continuous ceramic and metallurgical kilns, boilers, gas-generating sets, and glass furnaces.

Silica brick contains about 95 to 96% SiO_2 and about 2% lime added during grinding to furnish the bond. Silica bricks undergo permanent expansion, which occurs during firing and is caused by an *allotropic transformation* that takes place in the crystalline silica. When reheated, silica bricks again expand about 1.5%, but the effect is reversible and the bricks return to size when cooled. Silica bricks are manufactured in many standard sizes by power pressing. They have a very homogeneous texture, are free from air pockets and molding defects, and possess low porosity. These are highly desirable properties for resistance to slag penetration. The physical strength of silica bricks when heated is much higher than that of those made from clay. Consequently, they are suitable for arches in large furnaces. Furnaces using these must, however, be heated and cooled gradually, to lessen spalling and cracking. Open-hearth furnaces have silica bricks in their main arch, side walls, port arches, and bulkheads, but the newest installations have superduty silica bricks and basic bricks. Superduty silica bricks have higher refractoriness and lower permeability to gases than conventional silica bricks. Because of their high thermal conductivity, silica bricks have been utilized in coproduct coke ovens and gas retorts.

High-alumina refractories are used increasingly to meet the demand for materials that can withstand severe conditions for which the older fire-clay and silica bricks are not suitable. High-alumina bricks are made from clays rich in bauxite and diaspore. The refractoriness and temperature of incipient vitrification increase with the alumina content (see the phase diagram in Fig. 9.2). Another valuable property of high-alumina bricks is that they are practically inert to carbon monoxide and are not disintegrated by natural-gas atmospheres up to 1000°C. Bricks with a high percentage of alumina are classed among the superrefractories, and those of almost-pure alumina (+97%) may be considered among the recently developed special refractories, termed *pure oxides* (q.v.). High-alumina bricks are employed in the cement industry, in paper-mill refractories, and in modern boiler settings. They are also used in the lining of glass furnaces, oil-fired furnaces, and high-pressure oil stills, in the roofs of lead-softening furnaces, and in regenerator checkers of blast furnaces.

Basic refractories. The important basic bricks are made from magnesia, chromite, and forsterite. To achieve the required strength and other physical properties, basic bricks are usually power-pressed and are either chemically bonded or hard-burned. The disadvantages of lack of bond and volume stability in *unburned* basic or other bricks have been overcome by three improvements in manufacturing: (1) interfitting of grains has been developed to a maximum by using only *selected particle sizes* combined in the proper proportions to fill all the voids; (2) forming pressure has been increased to 10,000 psi, and de-airing equipment used to reduce air voids between the grains; and (3) use of a refractory chemical bond.

Magnesia refractories are made from domestic magnesites, or magnesia extracted from brines (Chap. 10). Magnesia bricks do not stand much of a load at elevated temperatures, but this difficulty has been overcome by blending with chrome ores.[13] Many blends are possible, ranging from predominantly magnesia to predominantly chrome. In the nomenclature the *predominant* blend constituent is given first. There is a large price variation, because of the composition variation; these bricks are among the most expensive. Chemically bonded magnesite-chrome bricks are frequently supported with mild steel to hold the brickwork and minimize spalling loss. These refractories are used in open-hearth and electric-furnace walls, in the burning zones of cement kilns, and in the roofs of various nonferrous reverberatory furnaces. Hard-burned chrome-magnesite bricks have many important physical properties because of their special composition, particle sizes, high forming pressure, and high firing temperature. They are of particular interest in basic open-hearth furnaces.

Forsterite ($2MgO \cdot SiO_2$) is employed both as a bond and as a base for high-temperature refractories. Where forsterite forms the base, refractories are generally made from alumina. In the manufacture of forsterite refractories, dead-burned magnesite is usually added to convert some accessory minerals to forsterite, which is the most stable silicate at high temperatures, as shown by Fig. 9.6. For example, enstatite or clinoenstatite, occurring in the rock olivine, as mined, is converted to forsterite:

$$\underset{\text{Enstatite}}{MgO \cdot SiO_2} + \underset{\text{Magnesia}}{MgO} \longrightarrow \underset{\text{Forsterite}}{2MgO \cdot SiO_2}$$

Such superrefractories have the advantages of a high melting point, no transformation during heating, and unsurpassed volume stability at high temperatures. No calcining is necessary in their preparation. The most important use of forsterite is in glass-tank superstructures and checkers, since its high chemical resistance to the fluxes employed and good strength at high temperatures allow enhanced tank output. These refractories also find many other industrial applications, e.g., in open-hearth end walls and copper-refining furnaces.

Insulating brick is of two types: for backing refractory bricks and for use in place of regular refractory bricks. Most bricks used for backing are made from naturally porous diatomaceous earth, and those of the second type, usually called lightweight refractories, are similar in composition to heavy bricks and owe their insulating value to the method of manufacture. For instance, waste cork is ground and sized; then it is mixed with fire clay, molded, and burned. In the kiln the cork burns out, leaving a highly porous, light brick. These lightweight refractories may be used safely for temperatures of 2500 to 2900°F, whereas diatomaceous-earth brick are not suitable above 2000°F under ordinary conditions.

[13]Havinghorst and Swift, Manufacture of Basic Refractories, *Chem. Eng.* (*N.Y.*), **72**(17), 98 (1965) (quantitative flowchart).

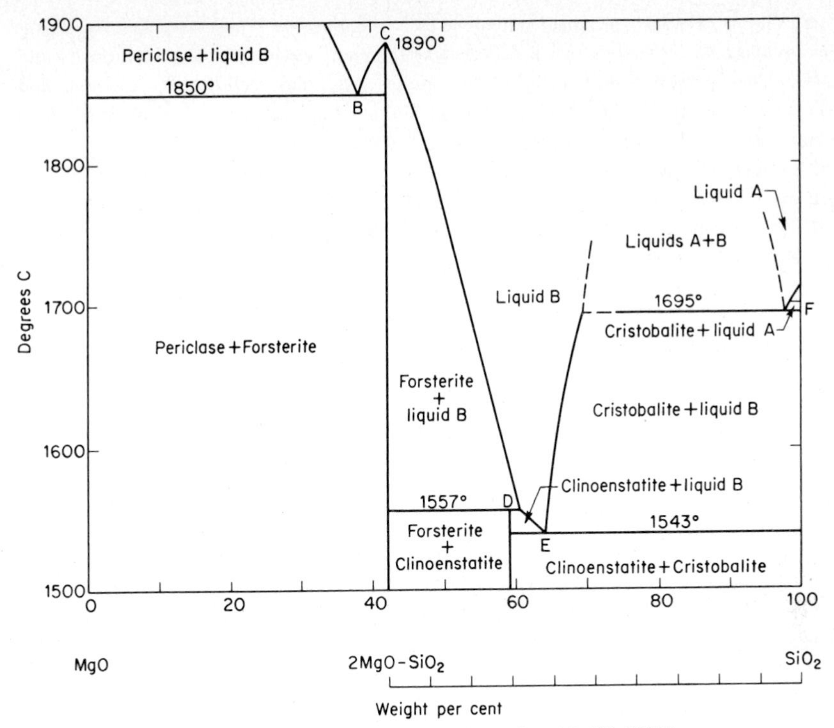

Fig. 9.6 Phase diagram of the system $M_gO \cdot SiO_2$. (*J. Am. Ceram. Soc.*, **16,** 493, 1933.)

Silicon carbide. Superrefractories are noted for their chemical resistance and ability to withstand sudden temperature changes. In their manufacture the crude material from the silicon carbide furnace (Chap. 15) is ground, and less than 10% ceramic bond is added. The latter may be clay or finely divided silicon carbide itself. These bricks are extremely refractory and possess high thermal conductivity, low expansion, and high resistance to abrasion and spalling. They are strong mechanically and withstand loads in furnaces to temperatures up to 1650°C. Such refractories are used chiefly in muffles because of their thermal conductivity. Their ability to absorb and release heat rapidly and their resistance to spalling under repeated temperature changes make them desirable for recuperators.

Refractories from crystalline alumina or aluminum silicates. Research has revealed that mullite and corundum have high slag resistance and remain in the crystalline state at temperatures of 1600°C and higher (see Fig. 9.2 for a phase diagram). High-temperature kilns now furnish alumina bricks that closely approach pure corundum in properties and mullite bricks made from calcined Indian kyanite, with the old clay bond replaced by a mullite bond consisting of interlocking crystals. Such refractories are used where severe slagging occurs.

Electrocast, or Corhart, refractories. Many kinds of bonded refractories are made from electrically *fused* mullite. They are manufactured by introducing a mixture of diaspore clays of high-alumina (to furnish $3Al_2O_3 \cdot 2SiO_2$ ratio) content into the top of an electric furnace. If necessary, these are adjusted to furnish the stable mullite ratio of $3Al_2O \cdot 2SiO_2$. Molten aluminum silicate at 3400°F is tapped from the furnace at intervals and run into molds built from sand slabs. The molds containing these blocks are annealed from 6 to 10 days before the blocks are usable. The refractory obtained from this process has a vitreous, nonporous body which shows a linear coefficient of expansion of about one-half that of good firebrick. The blocks cannot be cut or shaped, but may be ground on Alundum wheels; however, skill in casting has progressed rapidly, and now many sizes and intricate shapes are available. This electrocast has *only* 0.5% voids, in contrast to the usual 17 to 29% of fire-clay blocks. Cast refractories are employed in glass furnaces, in linings of hot zones of rotary kilns, in modern boiler furnaces exposed to severe duty, and in metallurgical equipment such as forging furnaces. These refractories have the advantages of long life and minimum wear, against which must be balanced their greater initial cost. A newer and superior variety of fused cast refractories, with from 83 to 95% alumina, is available. Here the interstitial glass content is greatly reduced, along with voids induced during the usual casting process.[14]

Pure oxide refractories.[15] The refractories industry is constantly faced with increased demands for products which will withstand higher temperatures and more severe operating conditions. To meet these requirements, a group of special pure oxide refractories has been developed. Their superior qualities are based mostly on freedom from fluxes. Refractory oxides of interest in order of increasing cost per unit volume include alumina, magnesia, zirconia, beryllia, and thoria. All have been developed commercially for light refractory products. The first three have certain properties in common: (1) they are of high purity (a minimum of 97% alumina, magnesia, or stabilized zirconia), and (2) they are principally composed of electrically fused grain. Beryllia is not used commercially for heavy wear because of its high cost and volatilization above 3000°F in the presence of water vapor. Thoria has a number of disadvantages, particularly since its radioactivity places it under the control of the Atomic Energy Commission.

Of these pure oxide[16] refractories the material that has the widest application is *sinter alumina.* It is used successfully at temperatures up to about 3400°F. *Magnesia* is a basic refractory and is easily reduced at high temperatures. Its applications are limited to oxidizing atmospheres at temperatures not much over 4000°F. Since pure *zirconia* undergoes a *crystalline change from monoclinic to tetragonal* form at about 1800°F, accompanied by a drastic volume change on inversion, stabilization of the crystal structure to the cubic form which undergoes no inversion, is necessary. This is accomplished by adding certain metallic oxides (particularly CaO and MgO). Processing temperatures in the range of 4600 to 4700°F are now commercial with available fused, stabilized zirconia. The only present large-scale use is as kiln furniture in the firing of barium titanate resistors, but future substantial usage seems probable.

SPECIALIZED CERAMIC PRODUCTS[17]

CERAMIC COMPOSITES Structures of metallic *honeycombs,* or webbings, impregnated with a ceramic phase, derive strength from high-alloy metals and good thermal properties from ceramic foams. The temperature limits of such bonded materials are exceedingly high; they are employed for aerospace hardware such as heat shields, rocket nozzles, and ram-jet chambers. *Cermets* compose one of a group of composite materials consisting of an intimate mixture of ceramic and metallic components, usually in the form of powder. These are compacted and sintered in order to obtain

[14]Alper, Refractory Products Manufactured by an Electric Furnace, Symposium on Electrochemistry and Electrometallurgy, Grenoble, France, Oct. 18, 1963 (extensive illustrations, diagrams, bibliography, and uses, including fused chrome-magnesia refractory).

[15]Hauth, Ceramic Oxides, *Chem. Eng. (N.Y.)*, **70**(25), 185 (1963); Kingery, Oxides for High Temperature Applications, Proceedings of an International Symposium, Stanford Research Institute, McGraw-Hill, 1960; Columbium, Tantalum, etc., *Chem. Week,* May 9, 1961, p. 87.

[16]Alumina is a versatile chemical and is used in its various forms and purities as follows: (1) as raw material for metallic aluminum; (2) as "impure fused alumina" when bauxite is fused; (3) as "pure fused alumina" when pure dried alumina is fused; (4) as single-crystal Al_2O_3, or sapphire (q.v.); (5) as sinter alumina when grains are united by incipient surface melting or sintering.

[17]Lee, Ceramics, pp. 50–151, Reinhold, 1961; Hibberd, Jr., Composites: Materials of the Future, *Chem. Eng. (N.Y.)*, **70**(23), 203 (1963); McGraw-Hill Encyclopedia, vol. 2, p. 655, 1966; New Fiber Forms, *Chem. Week,* June 9, 1962, p. 61.

certain physical properties not found solely in either of the components; for example, they are used in linings for brakes and clutches because of the greater weights and higher speeds involved and also nonlubricating bearings in the temperature range from 370 to 815°C.

Ceramic fibers. This is a new textile form for resistance to thermal shock conditions and usefulness in hypothermal environments. A longer staple fiber is available, which can be made into roving, yarn, cloth, braid, wicking, rope, and hollow sleeving. These fibers are also used for thermal insulation, for zone curtains in annealing furnaces, and for high-temperature—low-voltage wiring circuits of space and aircraft equipment. Ceramic-fiber-reinforced metals, metal-impregnated ceramics, ceramic-reinforced plastic, and metal-ceramic laminates have been developed to withstand high temperatures.

FERROELECTRIC AND FERROMAGNETIC CERAMICS The most common ceramic type in this class is barium titanate ($BaTiO_3$). Titania and its compounds exhibit unusual properties useful in electrical applications, the most important of which involves high capacity at various frequencies. The shortage of mica during World War II gave impetus to the development of *synthetic capacitators.* Procedures used in fabricating titania and titanate bodies are ceramic in character. Ferromagnetic ceramic materials have been responsible for important advances in the design of electronic equipment; they are used in television sets, computers, magnetic switches, wideband transformers, recorders, and memory devices.

PYROCERAM AND OTHER NEW DEVELOPMENTS The hybrid products Pyroceram and Fotoform (Chap. 11) are manufactured first as glass and then converted to a ceramic with a predominantly crystalline body. *Single crystals,* such as sapphires, quartz, and diamonds, are treated in other chapters. New compositions for nuclear uses such as fuel elements are treated in Chap. 21.

ENAMELS AND ENAMELED METAL

Porcelain, or vitreous enamel, is a ceramic mixture containing a large proportion of fluxes, applied cold and fused to the metal at moderate red heat. *Complete vitrification* takes place. The application of enamel to gold, silver, and copper is one that dates back to the ancients. Long valued as a material of great beauty in the field of decorative art, it has come into general commercial use because it provides a product of great durability and wide application; it is easy to clean and resists corrosion. Current uses are in plumbing fixtures, cooking utensils, industrial equipment, and glass-enameled steel for chemical use. To offset losses in the household-appliances industry, in favor of baked-on organic coatings, new markets are under development, which include the use of porcelain in electroluminescent lighting and in the automobile industry. In the latter there is a multimillion-dollar market for frits in manufacturing tailpipes and mufflers; about 1 lb of frit is used for each muffler and $\frac{3}{4}$ lb for each tailpipe.

RAW MATERIALS For enamels, these must not only have high purity but also fineness, suitable mineral composition, proper grain shape, and other physical characteristics, depending on the specific enamel. The raw materials used in the enamel industry may be divided into six different groups: refractories, fluxes, opacifiers, colors, floating agents, and electrolytes. *Refractories* include such materials as quartz, feldspar, and clay, which contribute to the acidic part of the melt and give body to the glass. *Fluxes* include such products as borax, soda ash, cryolite, and fluorspar, which are basic in character and *react* with the acidic refractories to form the glass. They tend to lower the fusion temperature of enamels. *Opacifiers* are compounds added to the glass to give it the white opaque appearance so characteristic of vitreous enamels. They are of two principal types: insoluble opacifiers (titanium dioxide, tin oxide, and zirconium oxide) and devitrification opacifiers (cryolite and fluorspar). In 1945 titanium dioxide-opacified enamels were commercially developed, and they have gained general acceptance in the industry because of high opacity and good acid resistance. Their chief advantage is that they may be applied as thinner coatings than the best previous opacifiers and therefore are more resistant to chipping, as well as being smoother and of greater reflectance. Devitrification opacifiers also act as fluxes, rendering the enamel more fusible. *Color* materials may be oxides, elements, salts, or frits, and may act either as refractories or as fluxes. *Floating agents* such as clay and gums are chosen to suspend the enamel in water. A pure plastic clay is required. To peptize the clay and properly suspend the enamel, *electrolytes* are added, such as borax, soda ash, magnesium sulfate, and magnesium carbonate.

MANUFACTURE OF THE FRIT The preparation of the enamel glass, or frit, is similar to the first stages of the manufacture of ordinary glass. The raw materials are mixed in the proper proportions and charged into a melting furnace maintained near 2500°F, from 1 to 3 h. After the batch has been uniformly melted, it is allowed to pour from the furnace into a quenching tank of cold water, shattering the melt into millions of friable pieces. This is called *frit.* Enamel is normally made in a wet process by grinding the ingredients, principally a mixture of frit and clay, the latter as a suspension aid, in a ball mill, and then passing through a 200-mesh screen.

PREPARATION OF METAL PARTS The success of enameling depends on the nature and uniformity of the metal base to which the enamel is fused and the obtaining of a parallelism between the coefficients of expansion of the enamel and the metal. In the cast-iron enameling industry castings are frequently made in the same factory in which they are enameled. The sheet-metal enameler usually purchases sheets to meet a definite specification. Before the liquid enamel (suspension in water) is applied to the metal, the surface must be thoroughly cleaned of all foreign matter, so that the enamel coating will adhere well to the metal and also not be affected itself. Sheet metal is cleaned by pickling in 8% sulfuric acid at 140°F after the iron has been annealed.

APPLICATION OF THE ENAMEL Sheet-iron coats are generally applied by dipping or slushing, since the ware is usually coated on all sides. Slushing differs from draining in that the enamel slip is thicker and must be shaken from the ware. A third method of application is spraying. Any of these three processes may be employed wet or dry. Cast-iron treatment is very similar to that for sheet iron. The enamel is air-dried, and colors are brushed and stenciled on if wanted. The enamel is usually applied in two coats for premium ware.

FIRING All enamels must be fired on the ware to melt them into a smooth, continuous, glassy layer. The requirements for successful firing of a good enamel are (1) proper firing temperature, 1400 to 1500°F; (2) time, 1 to 15 min; (3) proper support of the ware; (4) uniform heating and cooling of the ware; and (5) an atmosphere free from dust. Enamel coats average about 0.0065 in. as compared with 0.026 in. a few years ago, but increasing numbers of products are being fabricated with a single coat, less subject to chipping, because of the use of better opacifiers. The special enamel- or glass-lined equipment so extensively used in chemical plants is tested by high-frequency waves to exclude defects which only this method will detect but which in the course of use would offer an avenue for acid penetration.

KILNS[18]

The vitrification of ceramic products and the prior *chemical conversions dehydration, oxidation,* and *calcination* are carried out in kilns that may be operated in a periodic or continuous manner. Muffle kilns are also employed. All the newer installations are continuous-tunnel kilns, which have many advantages over batch kilns, such as lower labor costs, greater thermal efficiency, shorter processing-time cycle, and better operating control. Gas, coal, and oil are the most economical fuels, hence are more commonly used for firing; electricity is used in some instances.

CONTINUOUS KILNS The most important kilns are the continuous car *tunnel kilns* used for the firing of brick, tile, porcelain, tableware, and refractories. There are two general types of such

[18]Norton, *op. cit.*, pp. 242, 244, 252, tabulates fuel consumption for kilns; Kingery, *op. cit.*, pp. 60–61, gives illustrations and explanations of various types of kilns.

kilns: the direct-fired type, where combustion gases burn directly among the wares, and the indirect (muffle) type, where the products of combustion are not allowed to contact the wares. The wares are loaded directly onto open cars or enclosed in saggers which keep them clean. The cars pass through the tunnel counterflow to the combustion gases from the high-fire zone. The continuous *chamber kiln* consists of a series of connected chambers. The heat from one chamber is passed to another countercurrent to the wares. Because the chambers are burned in succession, the operation is continuous. There is always one chamber cooling, another being fired, and another being heated by the waste heat of the two other chambers. This type of kiln is used to burn brick and tile.

PERIODIC KILNS *Downdraft kilns,* which are round or rectangular in shape, are used in burning face brick, sewer pipe, stoneware, tile, and common brick. In them the heat is raised from room temperature to the finishing temperature for each burning operation. The kiln is "set" (filled with wares to be burned), the heating is started, and the temperature is raised at a definite rate until the firing temperature is reached. The downdraft kiln is so named because the products of combustion go down in passing over the wares set in the kiln. The *updraft kiln* has been most commonly used in burning potteryware but is rapidly being replaced by tunnel kilns. Common bricks are burned in *scove kilns,* which are really variations on the updraft kiln. The kiln itself is built from green brick, and the outside walls are daubed, or "scoved," with clay.

SELECTED REFERENCES

Andrews, A. I.: Porcelain Enamels, 2d ed., Garrard, 1961.
Bradley, W. F. (ed.): Proceeding of the 12th National Conference on Clays and Clay Minerals, Macmillan-Pergamon, 1964.
Budnikov, P. O.: Technology of Ceramics and Refractories, Prentice-Hall, 1964.
Burke, J. E. (ed.): Progress in Ceramic Science, Macmillan-Pergamon, 1961–1964, 3 vols.
Conrad, J. W.: Ceramic Formulas, Macmillan, 1973.
Eitel, W.: Silicate Science, vol. 1, Silicate Structures, Academic, 1964.
Gould, R. F. (ed.): Literature of Chemical Technology, chap. 5, Ceramics, and chap. 6, Refractory Ceramics for Aerospace, American Ceramic Society, 1964.
Hove, J., and W. C. Riley: Ceramics for Advanced Technologies, Wiley-Interscience, 1965.
Huminik, J., Fr.: High-temperature Inorganic Coatings, Reinhold, 1963.
Kriegel, W. W., and H. Palmour, III (eds.): Mechanical Properties of Engineering Ceramics, Interscience, 1961.
Lee, P. W.: Ceramics, Reinhold, 1961.
Salmang, H.: Ceramics, Butterworth, London, 1961.
Singer, F. S.: Industrial Ceramics, Chemical Publishing, 1965.
Van Vlack, L. H.: Physical Ceramics for Engineers, Addison-Wesley, 1964.

chapter 10

PORTLAND CEMENTS, CALCIUM, AND MAGNESIUM COMPOUNDS

The industrial uses of limestone and cements have provided important undertakings for chemists and engineers since the early years when lime mortars and natural cements were introduced. In modern times one need only mention reinforced-concrete walls and girders, tunnels, dams, and roads to realize the dependence of present-day civilization upon these products. The convenience, cheapness, adaptability, strength, and durability of cement products have been a foundation of these applications. The production of portland cement in the United States is now over 81 million tons with a value of over $1.6 billion (Table 10.1).

PORTLAND CEMENTS

In spite of the modern concrete roads and buildings everywhere around us, it is difficult to realize the tremendous growth of the cement industry during the past century. Humans had early discovered certain natural rocks which, through simple calcination, gave a product that hardened on the addition of water. Yet the real advance did not take place until physicochemical studies and chemical engineering laid the basis for the modern efficient plants working under closely controlled conditions with a variety of raw materials.

HISTORICAL Cement dates back to antiquity, and one can only speculate as to its discovery.[1] A cement was used by the Egyptians in constructing the Pyramids. The Greeks and Romans used volcanic tuff mixed with lime for cement, and a number of these structures are still standing. In 1824 an Englishman, Joseph Aspdin, patented an artificial cement made by the calcination of an argillaceous limestone. He called this *portland* because concrete made from it resembled a famous building stone obtained from the Isle of Portland near England. This was the start of the portland cement industry as we know it today. The hard clinkers resulting from burning a mixture of clay and limestone or similar materials is known by the term portland cement to distinguish it from natural or pozzolan and other cements.

USES AND ECONOMICS Before 1900 concrete was relatively little used in this country because the manufacture of portland cement was an expensive process. Thanks to the invention of laborsaving machinery, cement is now low in cost and is applied everywhere in the construction of homes, public buildings, roads, industrial plants, dams, bridges, and many other structures. Table 10.1 indicates the large volume of this industry.

[1]Bogue, The Chemistry of Portland Cement, 2d ed., chap. 1, Reinhold, 1955; Bogue and Gilkey, Cement, ECT, vol. 4, pp. 684–710, 1964 (excellent summary); Skalny and Daugherty, Everything You Always Wanted to Know About Portland Cement, *Chemtech*, **2**(1), 38 (1972).

TABLE 10.1 ***Portland Cement Shipments by Types, 1972*** *(In thousands of short tons and thousands of dollars)*

	Quantity	*Value*	*Average value per ton*
General use and moderate heat (Types I and II)	75,452	1,512,214	20.04
High-early-strength (Type III)	2,827	61,508	21.76
Sulfate-resisting (Type V)	581	11,672	20.09
Oil-well	671	14,626	21.80
White	459	20,795	45.31
Portland-slag and portland pozzolan	438	8,412	19.21
Expansive	177	5,213	29.45
Miscellaneous*	827	19,341	23.39
Total or average†	81,432	1,653,779	20.31

Source: Minerals Yearbook 1972, Dept. of the Interior, 1974. *Includes Type IV. †Data may not add to totals shown because of independent rounding.

CEMENT MANUFACTURE

There are various types of cements:

1. *Portland cement* has been defined as[2] "the product obtained by pulverizing clinker consisting essentially of hydraulic calcium silicates, to which no additions have been made subsequent to calcination other than water and/or untreated calcium sulfate, except that additions not to exceed 1.0% of other materials may be interground with the clinker at the option of the manufacturer. . . ." Five types (Table 10.3) of portland cement are recognized in the United States, which contain varying amounts of the clinker compounds listed in Table 10.2.

Type I. *Regular* portland cements are the usual products for general concrete construction. There are other types of this cement, such as white, which contains less ferric oxide, oil-well cement, quick-setting cement, and others for special uses.

Type II. *Moderate-heat-of-hardening* and *sulfate-resisting* portland cements are for use where moderate heat of hydration is required or for general concrete construction exposed to moderate sulfate action. The heat evolved from these cements should not exceed 70 and 80 cal/g after 7 and 28 days, respectively.

Type III. *High-early-strength* (HES) cements are made from raw materials with a lime-to-silica ratio higher than that of Type I cement and are ground finer than Type I cements. They contain a higher proportion of tricalcium silicate (C_3S) than regular portland cements. This, with the finer grinding, causes quicker hardening and a faster evolution of heat. Roads constructed from HES cement can be put into service sooner than roads constructed from regular cement.

TABLE 10.2 ***Clinker Compounds***

Formula	*Name*	*Abbreviation*
$2CaO \cdot SiO_2$	Dicalcium silicate	C_2S
$3CaO \cdot SiO_2$	Tricalcium silicate	C_3S
$3CaO \cdot Al_2O_3$	Tricalcium aluminate	C_3A
$4CaO \cdot Al_2O_3 \cdot Fe_2O_3$	Tetracalcium aluminoferrite	C_4AF
MgO	Magnesium oxide in free state	MgO

Note: See Table 10.6 for amounts in cement.

[2]ASTM Specifications C 150–60 and C 175–61.

TABLE 10.3 *Chemical Specifications for Portland Cements*

Constituent	*Regular, Type I*	*Moderate-heat-of-hardening, Type II*	*High-early-strength, Type III*	*Low-heat-of-hydration, Type IV*	*Sulfate-resisting, Type V*
Silicon dioxide (SiO_2), min %	....	21.0			
Aluminum oxide (Al_2O_3), max %	....	6.0	....	....	*
Ferric oxide (Fe_2O_3), max %	....	6.0	....	6.5	*
Magnesium oxide (MgO), max %	5.0	5.0	5.0	5.0	4.0
Sulfur trioxide (SO_3), max %:					
When $3CaO \cdot Al_2O_3$ is 8% or less	2.5	2.5	3.0	2.3	2.3
When $3CaO \cdot Al_2O_3$ is more than 8%	3.0	....	4.0		
Loss on ignition, max %	3.0	3.0	3.0	2.3	3.0
Insoluble residue, max %	0.75	0.75	0.75	0.75	0.75
Tricalcium silicate ($3CaO \cdot SiO_2$), max %	....	....	....	35	
Dicalcium silicate ($2CaO \cdot SiO_2$), min %	....	....	....	40	
Tricalcium aluminate ($3CaO \cdot Al_2O_3$), %	....	8	15	7	5

Source: ASTM Designation C 150–61. *The tricalcium aluminate shall not exceed 5%, and the tetracalcium aluminoferrite ($4CaO \cdot Al_2O_3 \cdot Fe_2O_3$) plus twice the amount of tricalcium aluminate shall not exceed 20%.

Type IV. *Low-heat* portland cements contain a lower percentage of C_3S and tricalcium aluminate (C_3A), thus lowering the heat evolution. Consequently, the percentage of tetracalcium aluminoferrite (C_4AF) is increased because of the addition of Fe_2O_3 to reduce the amount of C_3A (Table 10.5). Actually, the heat evolved should not exceed 60 and 70 cal/g after 7 and 28 days, respectively, and is 15 to 35% less than the heat of hydration of regular or HES cements.

Type V. *Sulfate-resisting* portland cements are those which, by their composition or processing, resist sulfates better than the other four types. Type V is used when high sulfate resistance is required. These cements are lower in C_3A than regular cements. In consequence of this, the C_4AF content is higher.

Air entrainment. The use of air-entraining agents (minute quantities of resinous materials, tallows, and greases) is important. These agents increase the resistance of the hardened concrete to scaling from alternate freezing and thawing and the use of de-icers (such as $CaCl_2$). Federal specifications permit the addition of an air-entraining material to each of the three types which are then designated IA, IIA, and IIIA.

Masonry cements are commonly finely ground mixtures of portland cement, limestone, and air-entraining agents. Some (also based on portland cement) contain hydrated limes, diatomaceous earth, and small additions of calcium stearate, petroleum, or high-colloidal clays.

2. *Pozzolan cement.*[3] Since the beginning of the Christian era the Italians have successfully employed pozzolan cement, made by grinding 2 to 4 parts of a pozzolan with 1 part of hydrated lime. A pozzolan is a material which is not cementitious in itself but which becomes so upon admixture with lime. Natural pozzolans are volcanic tuffs; an artificial one of prominence is fly ash.

3. *High-alumina cement,* essentially a calcium aluminate cement, is manufactured by fusing a mixture of limestone and bauxite, the latter usually containing iron oxide, silica, magnesia, and other impurities. It is characterized by a very rapid rate of development of strength and superior resistance to sea water and sulfate-bearing water.

4. *Special,* or *corrosion-resisting, cements* and *mortars* are used in large quantities for the fabrication of corrosionproof linings for chemical equipment such as brick-lined reactors, storage

[3]ASTM Designation C 340–58; Lea, Chemistry of Cement and Concrete, 2d ed., St Martins, 1956 (rev. ed. of Lea and Desch).

TABLE 10.4 *Raw Materials for Portland Cement in the United States (In thousands of short tons)*

	1970	*1971*	*1972*
Cement rock	22,824	23,074	23,799
Limestone	83,230	85,857	90,003
Marl	1,669	1,741	2,080
Clay and shale	11,833	11,808	12,158
Blast-furnace slag	853	713	759
Gypsum	3,491	3,750	4,094
Sand and sandstone	2,193	2,226	2,774
Iron materials	777	693	839
Miscellaneous	341	479	414
Total	127,211	130,341	136,920

Source: Minerals Yearbook 1972, Dept. of the Interior, 1974.

tanks, absorption towers, fume ducts and stacks, pickling tanks, floors, sumps, trenches, and acid digesters, to name but a few. Many kinds of *special cements* (see this heading) are described in the literature, but furan, phenolic, sulfur, and silicate cements are the most important.

5. *Controlled cement* to prevent shrinking and cracking upon setting is being introduced by Permanente Cement and Medusa. By combining 10 to 20% calcium sulfoaluminate (from bauxite, gypsum, and limestone) with portland cement, the tendency of concrete to crack can be virtually eliminated.

RAW MATERIALS Portland cement is made by mixing and calcining calcareous and argillaceous materials in the proper ratio. Table 10.4 summarizes the raw materials consumed. Formerly a large amount of cement was burned from argillaceous limestone, known as *cement rock*, found in the Lehigh district of Pennsylvania and New Jersey. This material was first used as a natural cement and, when found deficient in lime, was corrected by adding a small proportion of limestone. In addition to natural materials, some plants use artificial products such as blast-furnace slag and precipitated calcium carbonate obtained as a by-product in the alkali and synthetic ammonium sulfate industry. Sand, waste bauxite, and iron ore are sometimes consumed in small amounts to adjust the composition of the mix. Gypsum (4 to 5%) is added to regulate the setting time of the cement.

UNIT OPERATIONS, CHEMICAL CONVERSIONS, ENERGY REQUIREMENTS Essentially, the unit operations prepare the raw materials in the necessary proportions and in the proper physical state of fineness and intimate contact so that chemical conversions can take place at the calcining temperature in the kiln to form, by double decomposition or neutralization, the compounds listed in Tables 10.2, 10.5, and 10.6. K_2O and Na_2O are present in small amounts and form compounds with CaO, Al_2O_3, SiO_2, and SO_3, for example, $KC_{23}S_{12}$, which has properties similar to C_2S. Alkali oxides do not alter the properties of cement, because of the similarity of the properties of the compounds formed in them and those of similar compounds that do not contain them. Other reactions take place, such as dehydration of clay and decarbonization or calcination of limestone, both being endothermic and having values of 401 and 713 Btu/lb, respectively. The clinker formation is exothermic, with a probable value of 200 Btu/lb of clinker.[4] Probably the net heat requirements are in the neighborhood of 900 Btu/lb of clinker. The coal consumption, however, indicates an expenditure of 3000 or 4000 Btu/lb of clinker.[5] This heat is evolved in the kiln in carrying out the following reactions, as tabulated by Lea and Desch:[6]

[4]Lea, *op. cit.*, chap. 7, pp. 112–150.
[5]Cf. Lacey and Woods, Heat and Material Balances for a Rotary Cement Kiln, *Ind. Eng. Chem.*, **27**, 379 (1935).
[6]Lea and Desch, *op. cit.*, 1st ed., p. 111, and 2d ed., pp. 117, 126.

Temperature, °C	Reaction	Heat change
100	Evaporation of free water	Endothermic
500 and above	Evolution of combined water from clay	Endothermic
900 and above	Crystallization of amorphous dehydration products of clay	Exothermic
900 and above	Evolution of carbon dioxide from calcium carbonate	Endothermic
900–1200	Main reaction between lime and clay	Exothermic
1250–1280	Commencement of liquid formation	Endothermic
1280 and above	Further formation of liquid and completion of formation of cement (compounds)	Probably endothermic on balance

It should be noted that most of the reactions in the kiln proceed in the solid phase but, toward the end, the important fusion occurs. Lea and Desch assume that no appreciable formation of C_3S occurs below 1250°C, which is given as the temperature at which liquid formation begins. It is this final reaction that yields C_3S, the chief strength-producing constituent of cement, and reduces the free lime to a sufficiently small amount. This final fluid phase amounts to 20 to 30% of the reactions. All these reactions are involved in the "burning of cement."[7]

MANUFACTURING PROCEDURES: THE WET AND DRY PROCESSES In both processes *closed-circuit grinding* is preferred to *open-circuit grinding* in preparing the raw materials because in the former the fines are passed on and the coarse material returned, whereas in the latter the raw material is ground continuously until its mean fineness has reached the desired value (see Fig. 10.2 for some grinding hookups). However, Perry[8] presents more details of actual grinding circuits with power requirements. The *wet process,* though the original one, is being displaced by the *dry process,* especially for new plants, because of the saving in heat, accurate control, and mixing of the raw mixture it affords. At the end of 1973, dry process plants accounted for 58% of the total cement production capacity.[9] These are illustrated in the generalized flowchart in Fig. 10.1. In the wet process the solid material, after dry crushing, is reduced to a fine state of division in wet tube or ball mills and passes as a slurry through bowl classifiers or screens. The slurry is pumped to correcting tanks, where rotating arms make the mixture homogeneous and allow the final adjustment in composition to be made. In some plants, this slurry is filtered in a continuous rotary filter and fed into the kiln. The *dry process* is especially applicable to natural cement rock and to mixtures of limestone and clay, shale, or slate. In this process the materials may be roughly crushed, passed through gyratory and hammer mills, dried, sized, and more finely ground in tube mills, followed by air separators. Proportioning equipment is included in the sequences. Before entering the kiln a *thorough* mixing and blending by air or otherwise takes place (Fig. 10.1). This dry, powdered material is fed directly to rotary kilns, where the previously mentioned chemical reactions take place. Heat is provided by burning oil, gas, or pulverized coal, using preheated air from cooling the clinker. The tendency in recent years has been to lengthen the rotary kiln in order to increase its thermal efficiency. Dry-process kilns may be as short as 150 ft, but in the wet process, 300- to 600-ft kilns are not uncommon. The internal diameter is usually from 8 to 20 ft. The kilns are rotated at from $\frac{1}{2}$ to 2 rpm depending on size. The kilns are slightly inclined, so that materials fed in at the upper end travel slowly to the lower firing end, taking from 1 to 3 h. In order to obtain greater heat economy, part of the water is removed from wet process slurry. Some of the methods used employ slurry filters and Dorr thickeners. Efficient air pollution control equipment such as baghouses or

[7]Martin, Chemical Engineering and Thermodynamics Applied to the Cement Rotary Kiln, Technical Press, London, 1932. This book is filled with data and calculations pertaining to cement. Chapter 9 considers the heat absorbed in making clinker. Chapter 15 presents the concept of entropy in burning cement as "high-grade heat," or heat at the right thermal pressure or temperature.

[8]Perry, Cement, Lime, and Gypsum, pp. 8-50 to 8-52. The entire section 8 should be carefully studied in this connection.

[9]Iammartino, Cement's Changing Scene, *Chem. Eng.* (*N.Y.*), **81**(13), 103 (1974).

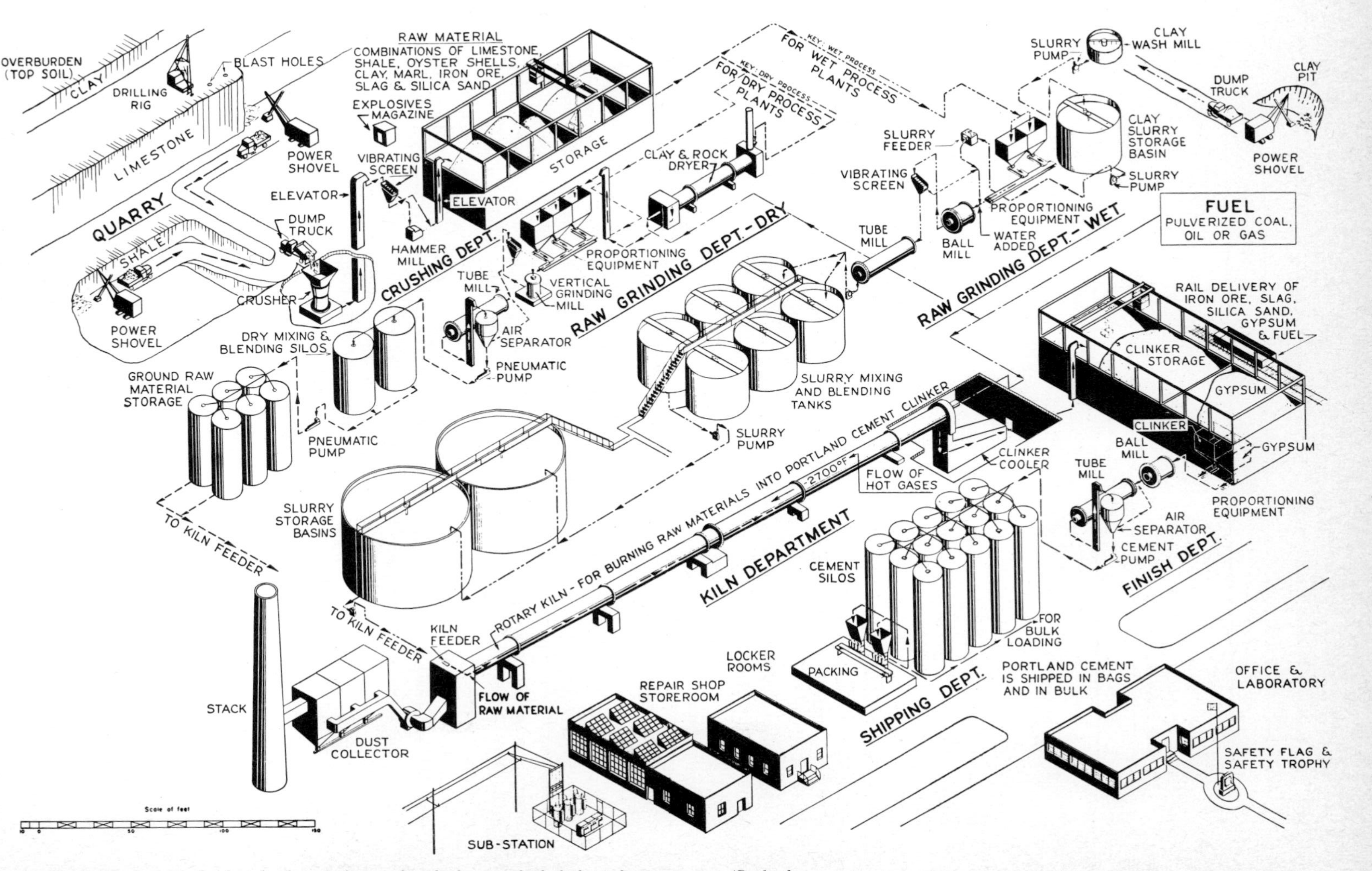

Fig. 10.1 Isometric flowchart for the manufacture of portland cement by both dry and wet processes. (*Portland Cement Association.*)

Per 376-lb bbl of finished cement	Wet process	Dry process
Fuel, Btu	1,270,000	1,120,000
Electrical energy, kWh	22.6	24.1
Water, gal	70	30
Direct labor, work-hours	0.17	0.17
Raw materials for both processes per 376 lb cement, lb		
Limestone		498
Shale		124
Gypsum		16
Total		638

electrostatic precipitators are now required for kilns.[10] Waste-heat boilers are sometimes used to conserve heat and are particularly economical for dry process cement, since the waste gases from the kiln are hotter than those from the wet process and may reach 800°C. Because the lining of the kiln has to withstand severe abrasions and chemical attack at the high temperatures in the clinkering zone, the choice of a refractory lining is difficult. For this reason high-alumina and high-magnesia bricks are widely used. Computers[11] are now used to improve kiln control. The final product formed consists of hard, granular masses from $\frac{1}{8}$ to $\frac{3}{4}$ in. in size, called *clinker.* The clinker is discharged from the rotating kiln into the air-quenching coolers, which quickly bring its temperature down to approximately 100 to 200°C. These coolers simultaneously preheat the combustion air. Pulverizing, followed by fine grinding in the tube ball mills and automatic packaging, completes the process. During the fine grinding, retarders, such as gypsum, plaster, or calcium lignosulfonate, and air-entraining, dispersing, and waterproofing agents are added. The clinker is ground dry by various hookups, as illustrated in Fig. 10.2 and by Perry.

COMPOUNDS IN CEMENTS Portland cements contain a mixture of compounds (previously listed) present in amounts partly dependent on the degree of attainment of equilibrium conditions during burning. Tables 10.5 and 10.6 give analyses of various types of portland cement and the average composition of about 102 regular cements. From these analyses it is seen that portland cement composition approaches a rough approximation to the system $CaO\text{-}SiO_2$ and successively closer approximations to the systems $CaO\text{-}SiO_2\text{-}Al_2O_3$, $CaO\text{-}SiO_2\text{-}Al_2O_3\text{-}Fe_2O_3$, and $CaO\text{-}SiO_2\text{-}Al_2O_3\text{-}Fe_2O_3\text{-}MgO$. A complete understanding of portland cement would require knowledge of the phase-equilibrium relations of the high-lime portions of all the two-, three-, four-, and five-component systems involved. Of these, all 12 of the principal two- and three-component systems, and the parts of the four-component systems, $CaO\text{-}SiO_2\text{-}Al_2O_3\text{-}Fe_2O_3$ and $CaO\text{-}SiO_2\text{-}Al_2O_3\text{-}MgO$, in which portland cement compositions are located, are known.[12] In 1946 part of the five-component system, $CaO\text{-}MgO\text{-}Al_2O_3\text{-}Fe_2O_2\text{-}SiO_2$, was studied by Swayze;[13] such studies are being continued. Modern cement technology owes much to Rankin and Wright[14] of the Geophysical Laboratory. It can be observed for the regular cement compositions presented in Tables 10.5 and 10.6 that a change in CaO percentage from 61.17 to 66.92 alters the C_3S percentage from 35.3 to 70.6 and that of C_2S from 0 to 33.2.

SETTING AND HARDENING OF CEMENT Although many theories have been proposed to explain the setting and hardening of cement, it is generally agreed that *hydration* and *hydrolysis*[15]

[10]Culhane, Production Baghouses, *Chem. Eng. Prog.*, **64**(1), 65 (1968); Immartino, *op. cit.*

[11]*Chem. Eng.* (*N.Y.*), **69**(22), 62 (1962); Guccione, New Developments in Cement Making, *Chem. Eng.* (*N.Y.*), **71**(24), 112 (1964).

[12]Bogue, *op. cit.*, chaps. 11 to 20, deals with the phase equilibria of clinker components.

[13]Swayze, *Am. J. Sci.*, **244**(1), 63 (1946).

[14]The Ternary System: $CaO\text{-}Al_2O_3\text{-}SiO_2$, *Am. J. Sci.*, **39,** 1–79 (1915).

[15]For hydration studies see Lea and Desch, *op. cit.*, pp. 205–208, to 1956; for later studies, see Steinour, Progress in the Chemistry of Cement, 1887–1960, pp. 7–10, Portland Cement Association, Chicago. *J. PCA Res. Devel. Lab.*, **3**(2), 2–11 (1961) (35 refs.); Bogue, *op. cit.*, chaps. 21 to 30 (detailed study of chemistry of cement utilization).

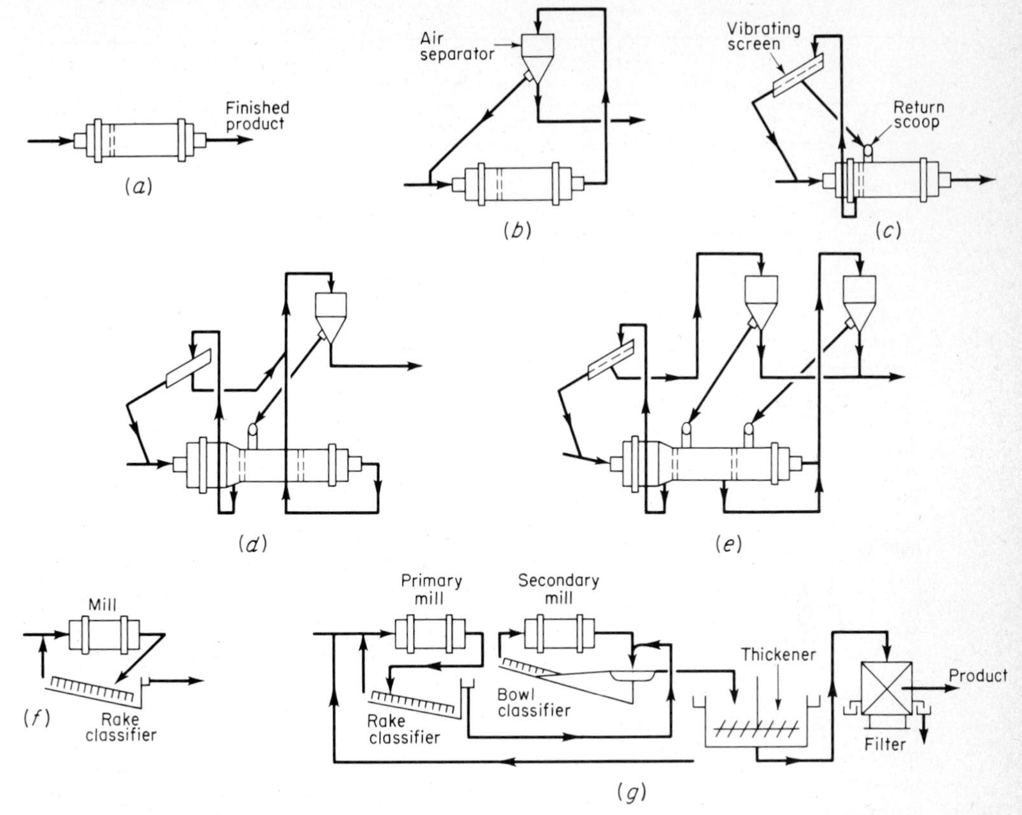

Fig. 10.2 Dry- and wet-grinding hookups.

are involved. The hydration products have very low solubility in water. If this were not true, concrete in contact with water would be rapidly attacked. In recent years much attention has been given to the heat evolved during the hydration of cement. The various compounds contribute to the heat of hardening (basis, equal weights, i.e., gram for gram) after 28 days, as follows:[16]

$C_3A > C_3S > C_4AF > C_2S$

Table 10.7 shows why low-heat-of hardening cements are made low in C_3A and C_3S but high in C_2S. This is accomplished (1) by adding more Fe_2O_3, which takes the Al_2O_3 out of circulation as C_4AF, thereby diminishing the amount of C_3A, and (2) by decreasing the CaO/SiO_2 ratio. Notice these facts in the analyses in Table 10.5. This low-heat-of-setting cement was used in the construction of Hoover Dam to avoid cracking the structure from heat stresses during setting and cooling. As an additional safeguard, the structure was cooled during setting by circulating cold water through 300 mi of lightweight 1-in. pipe placed in the concrete mass.[17] Tables 10.8 and 10.9 present further facts regarding the functions of the different compounds in the setting and hardening of cement. To

[16]Woods, Steinour, and Strake, Effect of Composition of Portland Cement on Heat Evolved during Hardening, *Ind. Eng. Chem.*, **24,** 1207 (1932); cf. Bogue and Lerch, Hydration of Portland Cement Compounds, paper 27, Portland Cement Association Fellowship at National Bureau of Standards, Washington, 1934; Lea and Desch, *op. cit.*, p. 245. Some investigators (Verback and Foster) obtained a reverse order for C_3S and C_4AF at early ages.

[17]Robertson, Boulder Dam, *Ind. Eng. Chem.*, **27,** 242 (1935).

TABLE 10.5 ***Analyses of Portland Cements*** *(In percentage)*

	CaO	SiO_2	Al_2O_3	Fe_2O_3	*MgO*	*Alkali oxides*	SO_3
			Regular Cement (Average of 102)				
Minimum	61.17	18.58	3.86	1.53	0.60	0.66	0.82
Maximum	66.92	23.26	7.44	6.18	5.24	2.9	2.26
Average	63.85	21.08	5.79	2.86	2.47	1.4	1.73
			High-early-strength (Average of 8): High C_3S				
Minimum	62.7	18.0	4.1	1.7	....	....	2.2
Maximum	67.5	22.9	7.5	4.2	....	....	2.7
Average	64.6	19.9	6.0	2.6	....	....	2.3
			Low-heat-of-hardening (Average of 5): Lower C_3S and C_3A, Higher C_2S and C_4AF				
Minimum	59.3	21.9	3.3	1.9	....	....	1.6
Maximum	61.5	26.4	5.4	5.7	....	....	1.9
Average	60.2	23.8	4.9	4.9	....	....	1.7

hold up the "flash set" caused by C_3A, the gypsum added as a retarder causes the formation of $C_3A \cdot 3CaSO_4 \cdot 31H_2O$.

Concrete[18] based on cement is broadening in application and increasing in quantity each year. Mention need be made only of low-heat-of-hardening, quick or retarded hardening, low and high density, prestressed concrete beams for greater strength and to save steel. The Portland Cement Association, which has offices in many cities, will furnish extensive data on this subject.

SPECIAL CEMENTS For many corrosive conditions portland cement is unsuitable or not economical. Hence many special cements have been developed, of which many types are industrially important. The types of special cements based on organic plastics now generally known as *adhesives* are presented in Chap. 25, along with glues. Since 1900, *sulfur cements* have been available commercially as simple mixtures of fillers, and since 1930 as homogeneous plasticized filled sulfur ingots possessing low coefficients of expansion. Sulfur cements are resistant to nonoxidizing acids and salts but should not be used in the presence of alkalies, oils, greases, or solvents. The crystalline change in sulfur structure at 200°F limits their use. Thiokol-plasticized, silica-filled sulfur cements have been accepted as a standard material for joining bricks, tile, and cast-iron pipe.

Silica-filled chemically setting *silicate cements* withstand a higher temperature (1000°F) than the preceding cements. Usually 2 parts by weight of finely divided silica powder is used to 1 part of

TABLE 10.6 ***Percentage Composition of 102 Regular Portland Cements in Terms of Compounds*** *(Calculated from Table 10.5)*

	C_3S	C_2S	C_3A	C_4AF	$CaSO_4$	*MgO*
Minimum	35.3	0	0	4.7	1.4	0.60
Maximum	70.6	33.2	15.5	18.4	3.8	5.24
Average	51.7	21.4	10.5	8.7	2.9	2.47

Note: These compound compositions are computed from the oxide analysis as follows: (1) Put all SO_3 in $CaSO_4$; (2) put all Fe_2O_3 in C_4AF; (3) put remaining Al_2O_3 in C_3A; (4) divide remaining CaO and SiO_2 stoichiometrically between C_2S and C_3S; (5) assume MgO is in free state and neglect Na_2O and K_2O. This calculation treats any free CaO as though it were combined.

[18]Expansive Cement Opens Era of New Concretes, *Chem. Eng. News*, Aug. 10, 1964, p. 38, describes use of calcium sulfoaluminates in its expanding form to furnish a concrete with an expansion of 0.1% of initial curing.

TABLE 10.7 ***Heat of Hydration*** *(In calories per gram)*

	Days				
Compound	*3*	*7*	*28*	*90*	*180*
C_4AF	29	43	48	47	73
C_3A	170	188	202	188	218
C_2S	19.5	18.1	43.6	55.2	52.6
C_3S	98.3	110	114.2	122.4	120.6

Source: Ind. Eng. Chem., **24,** 1207 (1932).

sodium silicate (35 to 40°Bé). Silicate cements are resistant to all concentrations of inorganic acids except hydrofluoric. Silicate cements are not suitable at pH values above 7 or in the presence of crystal-forming systems. Two typical applications are the joining of bricks in chromic acid reaction tanks and in alum tanks.

LIME

HISTORICAL The manufacture of lime and its application can be traced back to the Roman, Greek, and Egyptian civilizations, but the first definite written information concerning lime was handed down from the Romans. In his book "De Architectura," Marcus Pollio, a celebrated engineer and architect who lived during the reign of Augustus (27 B.C. to A.D. 14), deals quite thoroughly with the use of lime for mortar involved in the construction of harbor works, pavements, and buildings. In colonial America the crude burning of limestone was one of the initial manufacturing processes engaged in by the settlers; they used "dugout" kilns built of ordinary brick or masonry in the side of a hill, with a coal or wood fire at the bottom and a firing time of 72 h. These kilns can still be seen in many of the older sections of the country. It was not until recent years that, under the influence of chemical engineering research, the manufacture of lime developed into a large industry under exact technical control, resulting in uniform products at lower cost.

USES AND ECONOMICS Lime itself may be used for medicinal purposes, insecticides, plant and animal food, gas absorption, precipitation, dehydration, and causticizing. It is employed as a

TABLE 10.8 ***Strength Contribution of Various Compounds in Portland Cement*** *(The relative strengths are the apparent relative contributions of equal weights of the compounds listed.)*

1 day	$C_3A > C_3S > C_4AF > C_2S$
3 days	$C_3A > C_3S > C_4AF > C_2S$
7 days	$C_3A > C_3S > C_4AF > C_2S$
28 days	$C_3A > C_3S > C_4AF = C_2S$
3 months	$C_2S > C_3S = C_3A = C_4AF$
1 year	$C_2S > C_3S > C_3A = C_4AF$
2 years	$C_2S > C_3S > C_4AF > C_3A$

Note: The calcium silicate hydrate of cement paste is probably a tubermorite-type gel and is the principal source of strength in the hardened cement. Steinour, Chemistry of Cement, *PCA Res. Dept. Bull.* 130; *J. PCA Res. Dev. Lab.*, **3**(2), 2–11 (1961).

TABLE 10.9 ***Function of Compounds***

Compound	*Function*
C_3A	Causes set but needs retardation (by gypsum)
C_3S	Responsible for early strength (at 7 or 8 days)
C_2S and C_3S	Responsible for final strength (at 1 year)
Fe_2O_3, Al_2O_3, Mg, and alkalies	Lower clinkering temperature

reagent in the sulfite process for papermaking, dehairing hides, the manufacture of high-grade steel and cement, water softening, recovery of by-product ammonia, and the manufacture of soap, rubber, varnish, refractories, and sand-lime brick. Lime is indispensable for use with mortar and plaster and serves as a basic raw material in the production of calcium salts and for improving the quality of certain soils. Either directly or indirectly, limestone and lime are employed in more industries than any other natural substance. Lime production was 20.3 million tons in 1972 with a value of $341.1 million. The estimated production for 1976 is 22 million tons.[19] Lime is sold as a high-calcium quicklime containing not less than 90% CaO and from 0 to 5% magnesia; small percentages of calcium carbonate, silica, alumina, and ferric oxide are present as impurities. The suitability of lime for any particular use depends on its composition and physical properties, all of which can be controlled by the selection of the limestone and the details of the manufacturing process. Much lime must be finely ground before use.[20]

Depending on composition, there are several distinct types of limes. Hydraulic limes are obtained from the burning of limestone containing clay, and the nature of the product obtained after contact with water varies from putty to set cement. High-calcium content limes harden only with the absorption of carbon dioxide from the air, which is a slow process; hydraulic limes also harden slowly, but they can be used under water. For chemical purposes high-calcium is required, except for the sulfite paper process, where magnesian lime works better. Although in many sections of the country high-calcium lime is preferred by the building industry for the manufacture of mortar lime plaster, there are places where limestone containing magnesium is burned or where even dolomitic stone is calcined. Typical compositions of such stones are from 35 to 45% CaO and from 10 to 25% MgO. These products, called *magnesian* limes, or dolimes,[21] are favored by some plasterers, who claim they work better under the trowel. In the metallurgical field, "refractory lime," as dead-burned dolomite or as raw dolomite, is employed as a refractory patching material in open-hearth furnaces; it is applied between heats to repair scored and washed spots on the bottom of the furnace. *Hydrated lime* is finding increased favor in the chemical building trades over the less stable quicklime, despite its increased weight. Quicklime is almost invariably slaked or hydrated before use. Because of the better slaking and the opportunity to remove impurities, factory hydrate is purer and more uniform than slaked lime prepared on the job.

LIME MANUFACTURE

Lime has always been a cheap commodity because limestone deposits are readily available in so many sections of the United States. It is produced from limestone near centers of consumption so that freight costs are low.

RAW MATERIALS The carbonates of calcium and magnesium are obtained from deposits of limestone, marble, chalk, dolomite, or oyster shells. For chemical usage, a rather pure limestone

[19]Lime Makers See Big Growth in Pollution-related Outlets, *Chem. Eng. News*, Sept. 2, 1974, p. 7.
[20]Perry, pp. 8-50 to 8-52.
[21]To prevent "popping out," the MgO must be completely hydrated, e.g., by steam hydration at 160°C and under such pressure as 60 psi.

is preferred as a starting material because of the high-calcium lime that results. Quarries are chosen which furnish a rock that contains as impurities low percentages of silica, clay, or iron. Such impurities are important because the lime may react with the silica and alumina to give calcium silicates or calcium aluminosilicates which possess not undersirable hydraulic properties. The lumps sometimes found in "overburned" or dead-burned lime result from changes in the calcium oxide itself, as well as from certain impurities acted upon by excess heat, recognized as masses of relatively inert, semivitrified material On the other hand, it often happens that rather pure limestone is calcined insufficiently, and lumps of calcium carbonate are left in the lime. This lime is called "underburned" lime.

ENERGY CHANGES, PHYSICAL OPERATIONS, CHEMICAL CONVERSIONS[22] Energy is required for blasting out limestone, for transporting and sizing it, and for the burning or calcining process.[23] The reactions involved are for

Calcining:

$$CaCO_3(c) \rightleftharpoons CaO(c) + CO_2(g) \qquad \Delta H_{1200-1300°C} = +4.25 \text{ million Btu/ton of lime produced}$$

Hydrating:

$$CaO(c) + H_2O(l) \longrightarrow Ca(OH)_2(c) \qquad \Delta H = -15.9 \text{ kcal}$$

During calcining the volume contracts, and during hydrating it swells. For calcination the average fuel ratios, using bituminous coal, are 3.23 lb of lime from 1 lb of coal in shaft kilns and 3.37 lb in rotary kilns. As is shown above, the calcination reaction is reversible. Below 650°C the equilibrium decomposition pressure of CO_2 is quite small. Between 650 and 900°C the decomposition pressure increases rapidly and reaches 1 atm at about 900°C.[24] In most operating kilns the partial pressure of CO_2 in the gases in direct contact with the outside of the lumps is less than 1 atm; therefore initial decomposition may take place at temperatures somewhat less than 900°C. The decomposition temperature at the center of the lump is probably well above 900°C, since there the partial pressure of the CO_2 not only is equal to or near the total pressure, but also must be high enough to cause the gas to move out of the lump, where it can pass into the gas stream. The total *heat* required for *calcining* per ton of lime produced may be divided into two parts: sensible to heat to raise the rock to decomposition temperature, and latent heat of dissociation. Theoretical heat requirements per ton of lime produced, if the rock is heated only to a calcining temperature of 900°C, are approximately 1.3 million Btu for sensible heat and 2.6 million Btu for latent heat. Actual calcining operations, because of practical considerations, e.g., lump size and time, require that the rock be heated to between 1200 and 1300°C, thereby increasing sensible-heat requirements by about 350,000 Btu. Then practical heat requirements are approximately 4.25 million Btu/ton of lime produced in a vertical kiln. About 40% is sensible heat; the rest is latent heat of decomposition.

The sequence of steps connected with manufacturing in the kiln shown in Figs. 10.3 to 10.5 are

Blasting down of limestone from a quarry face or occasionally from underground veins (Op).
Transportation from the quarry to mills, generally by an industrial railroad (Op).
Crushing and sizing of the stone in jaw and gyratory crushers[25] (Op).
Screening to remove various sizes (e.g., a 4- to 8-in. stone implies that all pieces passing a 4-in. screen or retained on an 8-in. screen have been separated out) (Op).
Carting of large stones to top of vertical kilns (Op).

[22]Perry, pp. 8-50 to 8-52.
[23]Roberts, UCM's Vertical Lime Kiln, *Chem. Eng. Prog.*, **59**(10), 88 (1963); Kohanowski, The Grate-Kiln System, *Chem. Eng. Prog.*, **60**(1), 80 (1964).
[24]Cf. ECT, 1st ed., vol. 8, p. 353, 1952.
[25]Perry, p. 8-16.

Conveying of small rocks to a rotary kiln (Op).

Conveying of fines to a pulverizer to make powdered limestone for agricultural and other demands (Op).

Burning of limestone according to size, in vertical kilns to give lump lime (Ch), or in horizontal rotary kilns to furnish fine lime (Ch).

Packaging of the finished lime in barrels (180 or 280 lb) or sheet-iron drums (Op) or conveying it to a hydrator (Op).

Hydration of the lime (Ch).

Packaging of slaked lime in 50-lb paper bags.

MANUFACTURING PROCEDURES For the burning of lump limestones, upright kilns are usually employed. However, past difficulties with uneven heating and the blocking of burning gas in the vertical kiln have been largely eliminated by recent chemical engineering changes, as exemplified in Figs. 10.3 to 10.5 with the improvements described. The horizontal revolving kiln, although costing considerably more in capital investment and in fuel, also has been improved to reduce these additional charges per ton of goods produced. Both types of lime kilns have steel shells lined with refractory brick. The following chemical processes have been or are employed in the burning of calcium carbonate to lime:

1. *Vertical shaft kilns with outside heating hearths.* These were the more primitive form and are now obsolete.

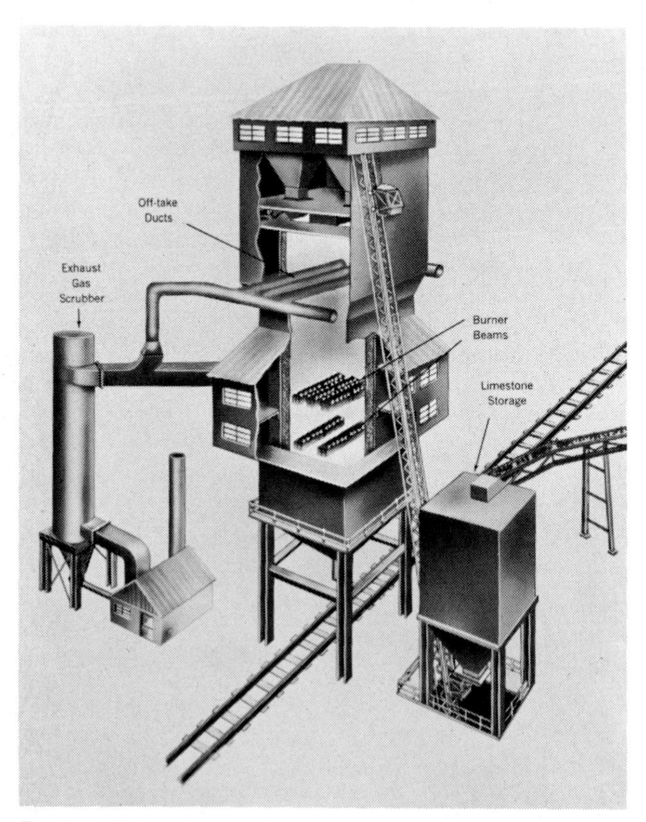

Fig. 10.3 High-capacity vertical lime kiln. (*Union Carbide Metals Co.*)

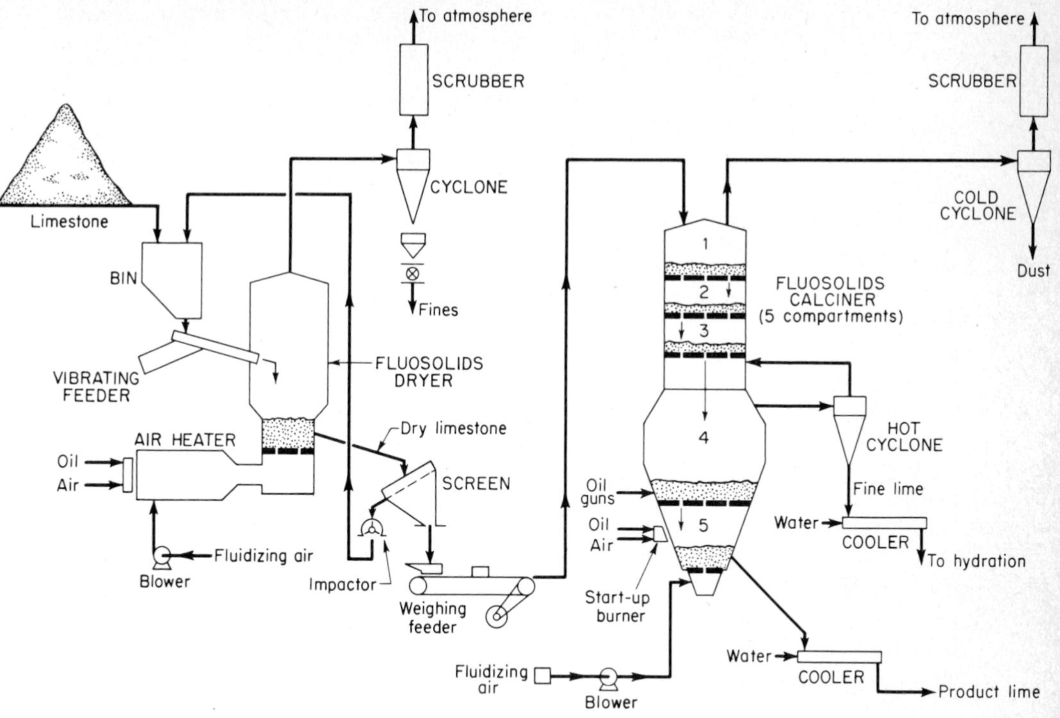

Fig. 10.4 Dorrco FluoSolids system for producing lime from pulverized limestone or calcium carbonate sludge. This is a five-compartment reactor as labeled. (*Dorr-Oliver, Inc.*)

2. *Vertical kilns burning a mixed feed* made up of limestone and coke. Here the fuel should be well integrated with the limestone. Such kilns are used in the Solvay soda ash process (Chap. 13) to provide the high CO_2 gas, as well as lime, needed for this manufacture.

3. *Union Carbide Metals*[26] *vertical kilns* with gas firing using special multiport water-cooled burners to ensure uniform distribution of the fuel with even calcination to lime. These usually recirculate part of the gas and are fully instrumented. See Fig. 10.3 for illustration of a kiln which has a capacity of from 300 to 500% of that of past similar-sized vertical kilns and requires lower capital investment than rotaries or the usual number of vertical kilns producing an equal tonnage of lime; production costs are claimed to be 20% less than those of rotary kilns of comparable production. The unique design embraces burner pipes for gas and air, mounted inside water-cooled H beams, in the calcining zone in staggered tiers, with three beams on the top tier and two on the bottom. This kiln recirculates much of the exhaust gas to ensure uniform heating and 98% decarbonation at not over about 1600°F. This not only leads to more efficient use of fuel, but furnishes a more active lime (less "overburnt").

4. *Dorrco FluoSolids system* for lime in a vertical kiln with gas-calcining of fluidizing dried carbonate as illustrated in Fig. 10.4.[27] This is an extension of the fluidized-bed reactors used so widely in the petroleum industry and, for example, in the oxidation of naphthalene to phthalic anhydride (Chap. 34).

[26]Unique Burner Design Boosts Capacity of Lime Kiln, *Chem. Eng. (N.Y.)*, **69**(18), 74 (1962); Carbide Metals Cut Lime Costs, *Chem. Eng. News*, Aug. 6, 1962, p. 28, and Aug. 11, 1962, p. 57.

[27]Perry, p. 20-51, for diagrams; Trauffer, Florida's New 200 TPD Lime Plant, *Pit Quarry*, May 1962; Krause, Softening Plant Regains Lime Sludge, *Water Works Eng.*, April 1957; cf. Pfizer, New Lime Kiln, *Chem. Eng. News*, Aug. 24, 1964, p. 40; Havinghorst, Fluid Bed Calcination, *Chem. Eng. (N.Y.)*, **71**(22), 104 (1964).

Fig. 10.5 Azbe natural-gas-fired kiln with full control over calcining-zone temperature. U.S. Pats. 2,742,276 and 3,033,545. Kiln may have calcining and preheating zones 38 ft high and may be 10 to 12 ft in diameter. It produces up to 250 tons/day of high-quality lime. This is made possible by the prevention of overburning and by regulated and sequential introduction of fuel, with recirculation of air as shown by the annotations on the figure. All fuel enters at the base of the active calcining zone, and all air from the base of the lime-cooling section. A substantial amount of hot air is withdrawn from the lime-cooling section to bypass and reduce the temperature of the hot-calcining zone. The lime-draw system operates at 5- to 20-min intervals, automatically controlled. (*V. J. Azbe, Azbe Corp., Clayton, Mo.*) ▶

5. Vertical kilns of the Azbe[28] design, illustrated by Fig. 10.5, are fired with multiple injection and control of natural gas with recirculation of air.

6. *Rotary horizontal kilns,* commonly 150 ft long, but occasionally extended to as much as 400 ft (11 ft in diameter) to ensure better overall fuel consumption. The modern horizontal kiln has various devices to properly size and preheat the entering limestone. A rotary countercurrent cooler for the product lime is used to preheat air for the burning. These kilns operate on fines or small lumps of limestone or marble or moist precipitated calcium carbonate sludge, any of which would block the efficient and uniform burning of the fuel in a vertical kiln. An 11- by 275-ft rotary kiln is described[29] burning lime carbonate slurry from causticizing plant as underflow from clarifiers or thickeners. This has 35% solids and is put through two Bird centrifuges, reducing moisture to about 35%. The slurry is introduced into the kiln by screw conveyors at 400°F and the front end temperature is 1760 to 1800°F. This lime kiln uses 8.5 million Btu per dry ton of reclaimed lime from moist carbonate slurry. Other devices used to lessen the heat requirements of a rotary are those of Dow,[30] the traveling grate-kiln system of Allis-Chalmers Mfg. Co., requiring 6 million Btu of natural gas per ton of lime, and Azbe's[31] combination of a short high-temperature rotary kiln discharging into a modified vertical kiln consisting of heating (air-gas), soaking, and cooling zones.

GYPSUM

Gypsum is a mineral that occurs in large deposits throughout the world.[32] It is hydrated calcium sulfate, with the formula $CaSO_4 \cdot 2H_2O$. When heated slightly, the following reaction occurs:

$$CaSO_4 \cdot 2H_2O(c) \longrightarrow CaSO_2 \cdot \tfrac{1}{2}H_2O(c) + 1\tfrac{1}{2}H_2O(g) \qquad \Delta H_{25°C} = +16.5 \text{ kcal}$$

If the heating is at a higher temperature, gypsum loses all its water and becomes anhydrous calcium sulfate, or *anhydrite.* Calcined gypsum can be made into wall plaster[33] by the addition of a filler material such as asbestos, wood pulp, or sand. Without additions, it is plaster of paris and is used for making casts and for plaster; it is hydraulic and hardens under water, but is also slightly soluble in water, hence must not be used in moist exposures.

CALCINATION OF GYPSUM The usual method of calcination of gypsum consists in grinding the mineral and placing it in large calciners holding 10 to 25 tons. The temperature is raised to about 120 to 150°C, with constant agitation to maintain a uniform temperature. The material in the kettle,

[28] Azbe, Vertical Gas-fired Kilns Improved, *Rock Prod.,* February 1962, p. 85, and Fundamental Mechanics of Calcination and Hydration of Lime, *Am. Soc. Test. Mater.,* Mar. 9, 1939.

[29] One Kiln Does Work of Three, *Chem. Process.* (Chicago), Oct. 8, 1962, p. 36; St. Marys Kraft Corp., St. Marys, GA.; Lime-mud Reburning, *Chem. Eng.* (*N.Y.*) **71**(19), 102 (1964); Ellwood and Denatos, Process Furnaces, *Chem. Eng.* (*N.Y.*), **73**(8), 151 (1966).

[30] Dow Chemical Co., Ludington, Mich., *Chem. Process.* (Chicago), January 1961, p. 68; for horizontal platform kiln, cf. *Chem. Eng.* (*N.Y.*). **71**(12), 89 (1964).

[31] Azbe, Short Insulated Rotary Kiln, *Rock Prod.,* July 1964, and Planning for Super Rotary Kiln, National Lime Association, May 11, 1964.

[32] Synthetic gypsum is now being made by the Allied Chemical Corp. from by-product wet-process phosphoric acid sludge.

[33] ECT, 1st ed., vol. 3, p. 443; Havinghorst, Gypsum Manufacture, *Chem. Eng.* (*N.Y.*), **72**(1), 52–54 (1965) (process flowchart).

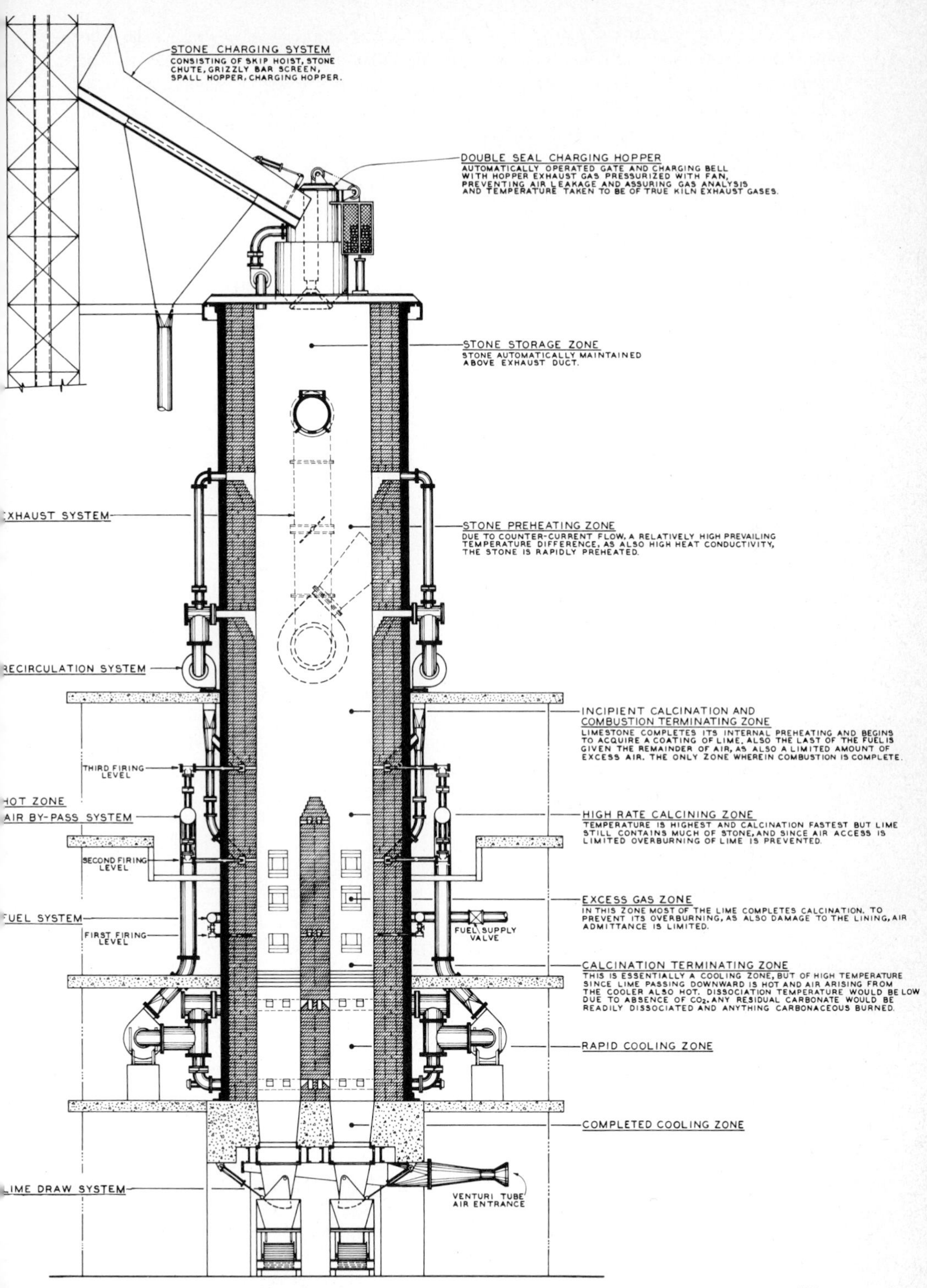

STONE CHARGING SYSTEM
CONSISTING OF SKIP HOIST, STONE CHUTE, GRIZZLY BAR SCREEN, SPALL HOPPER, CHARGING HOPPER.
DOUBLE SEAL CHARGING HOPPER
AUTOMATICALLY OPERATED GATE AND CHARGING BELL WITH HOPPER EXHAUST GAS PRESSURIZED WITH FAN, PREVENTING AIR LEAKAGE AND ASSURING GAS ANALYSIS AND TEMPERATURE TAKEN TO BE OF TRUE KILN EXHAUST GASES.
STONE STORAGE ZONE
STONE AUTOMATICALLY MAINTAINED ABOVE EXHAUST DUCT.
XHAUST SYSTEM
STONE PREHEATING ZONE
DUE TO COUNTER-CURRENT FLOW, A RELATIVELY HIGH PREVAILING TEMPERATURE DIFFERENCE, AS ALSO HIGH HEAT CONDUCTIVITY, THE STONE IS RAPIDLY PREHEATED.
RECIRCULATION SYSTEM
INCIPIENT CALCINATION AND COMBUSTION TERMINATING ZONE
LIMESTONE COMPLETES ITS INTERNAL PREHEATING AND BEGINS TO ACQUIRE A COATING OF LIME, ALSO THE LAST OF THE FUEL IS GIVEN THE REMAINDER OF AIR, AS ALSO A LIMITED AMOUNT OF EXCESS AIR. THE ONLY ZONE WHEREIN COMBUSTION IS COMPLETE.
THIRD FIRING LEVEL
HOT ZONE
AIR BY-PASS SYSTEM
HIGH RATE CALCINING ZONE
TEMPERATURE IS HIGHEST AND CALCINATION FASTEST BUT LIME STILL CONTAINS MUCH OF STONE, AND SINCE AIR ACCESS IS LIMITED OVERBURNING OF LIME IS PREVENTED.
SECOND FIRING LEVEL
EXCESS GAS ZONE
IN THIS ZONE MOST OF THE LIME COMPLETES CALCINATION. TO PREVENT ITS OVERBURNING, AS ALSO DAMAGE TO THE LINING, AIR ADMITTANCE IS LIMITED.
UEL SYSTEM
FIRST FIRING LEVEL
FUEL SUPPLY VALVE
CALCINATION TERMINATING ZONE
THIS IS ESSENTIALLY A COOLING ZONE, BUT OF HIGH TEMPERATURE SINCE LIME PASSING DOWNWARD IS HOT AND AIR ARISING FROM THE COOLER ALSO HOT. DISSOCIATION TEMPERATURE WOULD BE LOW DUE TO ABSENCE OF CO_2. ANY RESIDUAL CARBONATE WOULD BE READILY DISSOCIATED AND ANYTHING CARBONACEOUS BURNED.
RAPID COOLING ZONE
COMPLETED COOLING ZONE
LIME DRAW SYSTEM
VENTURI TUBE AIR ENTRANCE

known to the public as *plaster of paris* and to the manufacturer as *first-settle plaster*, may be withdrawn and sold at this point, or it may be heated further to 190°C to give a material known as *second-settle plaster*. First-settle plaster is approximately the half hydrate $CaSO_4 \cdot \frac{1}{2}H_2O$, and the second form is anhydrous. Practically all gypsum plaster sold is in the form of first-settle plaster mixed with sand or wood pulp. The second form is used in the manufacture of plasterboard and other gypsum products. Gypsum may also be calcined in rotary kilns similar to those used for limestone.

HARDENING OF PLASTER This hardening of plaster is essentially a hydration chemical conversion, as represented by the equation

$$CaSO_4 \cdot \tfrac{1}{2}H_2O(c) + 1\tfrac{1}{2}H_2O(l) \longrightarrow CaSO_4 \cdot 2H_2O(c) \qquad \Delta H = -0.7 \text{ kcal}$$

This equation is the reverse of that for the dehydration of gypsum. The plaster sets and hardens because the liquid water reacts to form a solid, crystalline hydrate. With reference to Fig. 10.6, it is apparent that hydration with liquid water takes place at temperatures below about 99°C and that the gypsum must be heated above 99°C for practical dehydration. Commercial plaster usually contains some glue in the water used or a material such as hair or tankage from the stockyards to retard the setting time in order to give the plasterer opportunity to apply the material.

MISCELLANEOUS CALCIUM COMPOUNDS

CALCIUM CARBONATE is a very widely used industrial chemical, in both its pure and its impure state. As marble chips, it is sold in many sizes as a filler for artificial stone, for the neutralization of acids, and for chicken grit. Marble dust is employed in abrasives and in soaps. Crude, pulverized limestone is used in agriculture to "sweeten" soils in large tonnage. Some pulverized and levigated limestone, to replace imported chalk and whiting, is manufactured quite carefully from very pure raw material and is finding acceptance.[34] *Whiting* is pure, finely divided calcium carbonate prepared by wet grinding and levigating natural chalk. Whiting mixed with 18% boiled linseed oil furnishes *putty*, which sets by oxidation and by formation of the calcium salt. Much whiting also is consumed in the ceramic industry. Precipitated, or artificial, whiting arises through precipitation, e.g., from reacting a boiling solution of calcium chloride with a boiling solution of sodium carbonate or passing carbon dioxide into a milk-of-lime suspension. Most of the latter form is used in the paint, rubber, pharmaceutical, and paper industries.

CALCIUM SULFIDE is made by reducing calcium sulfate with coke. Its main use is as a depilatory in the tanning industry and in cosmetics. In a finely divided form it is employed in luminous paints. Polysulfides, such as CaS_2 and CaS_5, made by heating sulfur and calcium hydroxide, are consumed as fungicides.

HALIDE SALTS Calcium chloride is obtained commercially as a by-product of chemical manufacture and from natural brines which contain more or less magnesium chloride. For this reason, and since large tonnages are available, it is a very cheap chemical. Its main applications are in solutions used to lay dust on highways (because it is deliquescent and remains moist) and in low-temperature refrigeration. Calcium bromide and iodide have properties similar to those of the chloride. They are prepared by the action of the halogen acids on calcium oxide or calcium carbonate. They are sold as hexahydrates for use in medicine and photography. Calcium fluoride occurs naturally as a fluorspar (Chap. 20).

CALCIUM ARSENATE is produced by the reaction of calcium chloride, calcium hydroxide, and sodium arsenate or lime and arsenic acid:

$$2CaCl_2(aq) + Ca(OH)_2(c) + 2Na_2HAsO_4(aq) \longrightarrow Ca_3(AsO_4)_2(c) + 4NaCl(aq) + 2H_2O(l)$$

$$\Delta H = -6.64 \text{ kcal}$$

[34]Rose, Condensed Chemical Dictionary, 6th ed., Reinhold, 1961. This book is invaluable for presenting outline specifications, including the spelling of chemical names.

Some free lime is usually present. Calcium arsenate is used extensively as an insecticide and as a fungicide. It is especially useful on cotton plants to poison boll weevils (Chap. 26).

CALCIUM ORGANIC COMPOUNDS Calcium acetate and lactate are prepared by the reaction of calcium carbonate or hydroxide with acetic or lactic acid. The *acetate* was formerly pyrolyzed in large amounts to produce acetone, but now it is employed largely in the dyeing of textiles. The *lactate* is sold for use in medicines and in foods as a source of calcium; it is an intermediate in the purification of fermentation lactic acid. *Calcium soaps* such as stearate, palmitate, and the abietate are made by the action of the sodium salts of the acids on a soluble calcium salt such as the chloride. These soaps are insoluble in water, but are soluble in hydrocarbons. Many of them form jellylike masses, which are constituents of greases. These soaps are used mainly as waterproofing agents.

MAGNESIUM OXYCHLORIDE CEMENT

This cement, discovered by the French chemist Sorel and sometimes called *Sorel's cement,* is produced by the exothermic action of a 20% solution of magnesium chloride on a blend of magnesia obtained by calcining magnesite and magnesia obtained from brines:

$$3MgO + MgCl_2 + 11H_2O \longrightarrow 3MgO \cdot MgCl_2 \cdot 11H_2O$$

The resulting crystalline oxychloride ($3MgO \cdot MgCl_2 \cdot 11H_2O$) contributes the cementing action to the commercial cements. The product is hard and strong, but is attacked by water; which leaches out the magnesium chloride. Its main applications are as a flooring cement with an inert filler and a coloring pigment, and as a base for such interior floorings as tile and terrazzo. It is strongly corrosive to iron pipes in contact with it. Sand and wood pulp may be added as fillers. The expense of these cements restricts their use to special purposes. They do not reflect sound. They can be made spark-proof and as such have been widely employed in ordnance plants. The magnesia used may contain small amounts of calcium oxide, calcium hydroxide, or calcium silicates, which during the setting process increase the volume changes, thus decreasing strength and durability. To eliminate this

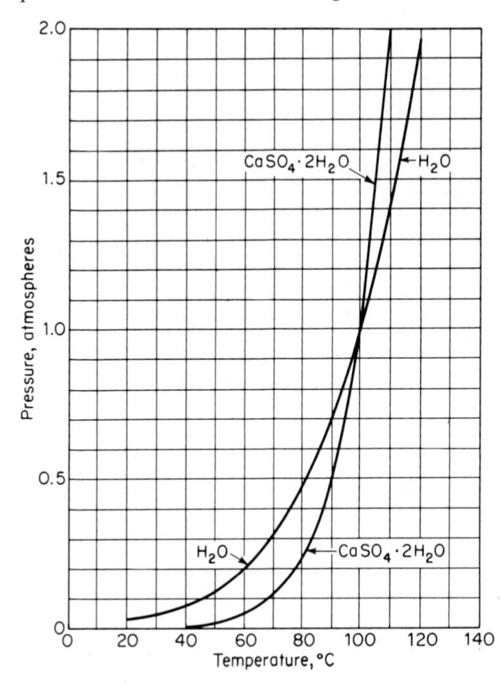

Fig. 10.6 Dehydration pressure of gypsum and vapor pressure of water. (*Calculated from Bur. Mines Tech. Pap. 625, 1941.*)

time effect, hydrated magnesium sulfate ($MgSO_4 \cdot 7H_2O$) or 10% finely divided metallic copper is added to the mixture. The use of copper powder not only prevents excessive expansion, but greatly increases water resistance, adhesion, and dry and wet strength over that of ordinary magnesium oxychloride cement. Such a product even adheres in thin layers to concrete and serves to seal cracks therein.

MAGNESIUM COMPOUNDS[35]

Magnesium is one of the most widely distributed elements, occupying 1.9% of the earth's crust. It occurs usually in the chloride, silicate, hydrated oxide, sulfate, or carbonate, in either a complex or in simple salts. Magnesium metal first became available commercially shortly before 1914, when the Germans initiated production, using magnesium chloride from the Stassfurt deposits as the raw material. Table 10.10 gives recent production figures for magnesium and its compounds.

RAW MATERIALS AND USES Important domestic sources of magnesium salts are sea water, certain salt wells, bitterns from sea brine, salines, dolomite, and magnesite ($MgCO_3$). Magnesium compounds are used extensively for refractories and insulating compounds, as well as in the manufacture of rubber, printing inks, pharmaceuticals, and toilet goods.

Magnesium oxide is finding new important uses in air pollution control systems for the removal of sulfur dioxide from stack gases.[36] New uses for magnesium may mean major growth for this metal in the future.[37]

MANUFACTURE The manufacture of magnesium compounds from salines has long been successful in Germany. As a result of thorough physical and chemical study, the International Minerals and Chemical Corp. is making magnesium chloride from langbeinite,[38] crystallizing out carnallite ($KCl \cdot MgCl_2 \cdot 6H_2O$). This double salt is decomposed to furnish magnesium chloride.

The production of magnesium compounds by separation from aqueous solutions may be divided into four processes:

1. Manufacture from sea water without evaporation, using sea water and lime as the principal raw materials. This is done by Dow Chemical Co. at Freeport and Velasco, Tex.,[39] and by Merck & Co. at South San Francisco (Fig. 10.7). The Dow Chemical Co. of Freeport and Velasco, Tex., manufactures magnesium[40] hydrate, which is dissolved in 10% hydrochloric acid to furnish a solution of magnesium chloride. This is concentrated in direct-fired evaporators, followed by shelf dryers, producing 76% magnesium chloride ready to be delivered to electrolytic cells to make metallic magnesium (Chap. 14).

2. Manufacture from bitterns or mother liquors from the solar evaporation of sea water for salt. FMC[41] employs this source, with its plant at Newark, Calif., for manufacture of various grades of magnesia.

3. Manufacture from dolomite and sea water, with factories operating at Moss Landing, Calif., by Kaiser Chemical Division,[42] as depicted by the flowchart in Fig. 10.8. At Pascagoula, Fla., H. K. Porter Co.[43] obtains magnesium salts from similar sources to manufacture refractory brick (periclase).

[35]Comstock, Magnesium and Magnesium Compounds, *U.S. Bur. Mines Inf. Circ.* 8201 (1963) (an excellent presentation, 128 pp., 129 refs., including raw materials, compounds, uses, and tabulations of all producers, domestic and foreign); Tallmadge *et al.*, Minerals from Sea Salt, *Ind. Eng. Chem.*, **56**(7), 44 (1964) (magnesium recovery); Magnesium Cast for New Roles, *Chem. Week*, Jan. 17, 1973, p. 24; Magnesium Booster: Budding Auto Market, *Chem. Week*, May 17, 1969, p. 36.

[36]Shah, MgO Absorbs Stack Gas SO_2, *Chem. Eng. (N.Y.)*, **79**(14), 80 (1974).

[37]Magnesium Casts for New Roles, *Chem. Week.*, Jan. 17, 1973, p. 24.

[38]Comstock, *op. cit.*, 43–47 (flowcharts and details).

[39]Comstock, *op. cit.*, pp. 41–42 (flowchart).

[40]Process Changes Boost Magnesium Capacity, *Chem. Eng. News*, Aug. 14, 1967, p. 54.

[41]See Comstock, *op. cit.*, p. 45, for flowchart.

[42]Havinghorst and Swift, Dolomite Calcining, *Chem. Eng.* (*N.Y.*), **72**(15), 150 (1965) (flowchart); Magnesia Extraction from Seawater, *Chem. Eng.* (*N.Y.*), **72**(16), 84 (1965).

[43]*Chem. Eng. News*, Sept. 28, 1959, p. 28.

TABLE 10.10 ***Magnesium and Magnesium Compounds Statistics, 1972*** *(In short tons and thousands of dollars)*

	Quantity	*Value*
Used in the United States		
Caustic-calcined and specified (USP and technical) magnesias	128,260	15,856
Refractory magnesia	696,102	60,331
Magnesium hydroxide (100%)	66,671	2,454
Magnesium chloride	559,709	36,202
Precipitated magnesium chloride	5,074	1,476
Dead-burned dolomite	1,125,000	21,097
Magnesium	99,455	74,094
U.S. production		
Primary magnesium	120,823	
Secondary magnesium	15,662	
World production		
Primary magnesium	225,995	
Magnesite (crude)	9,975,138	

Source: Minerals Yearbook 1972, Dept. of the Interior, 1974.

4. Manufacture from deep-well brines. Production is being increased by Dow Chemical Co. with Michigan brines, analyzing 20.7% $CaCl_2$, 3.9% $MgCl_2$, 5.73% NaCl. The small amount of bromine is freed by chlorine and, following its removal, the $Mg(OH)_2$ is precipitated by pure slaked dolime (calcined dolomite). The $Mg(OH)_2$ produced is settled, filtered, and washed to provide a slurry containing 45% $Mg(OH)_2$ of high purity. This is calcined at high temperatures to produce periclase, a sintered MgO nodule used in making refractory brick.

A typical analysis of sea water is given in Table 10.11. The production of magnesium compounds from sea water is made possible by the almost complete insolubility of magnesium hydroxide in water. Success in obtaining magnesium compounds by such a process depends upon

1. Means to soften the sea water cheaply, generally with lime or calcined dolomite.
2. Preparation of a purified lime or calcined dolomite slurry of proper characteristics.
3. Economical removal of the precipitated hydroxide from the large volume of water.
4. Inexpensive purification of the hydrous precipitates.
5. Development of means to filter the slimes.

The reactions are

$$MgCl_2(aq) + Ca(OH)_2(c) \longrightarrow Mg(OH)_2(c) + CaCl_2(aq) \qquad \Delta H = +2.260 \text{ kcal}$$

$$MgSO_4(aq) + Ca(OH)_2(c) + 2H_2O(l) \longrightarrow Mg(OH)_2(c) + CaSO_4 \cdot 2H_2O(c) \qquad \Delta H = -3.170 \text{ kcal}$$

Figure 10.8 shows a flowchart of magnesium products from sea water. Figure 10.7 presents a flowchart for producing such fine chemicals and pharmaceuticals as milk of magnesia and several basic magnesium carbonates such as $3MgCO_3 \cdot Mg(OH)_2 \cdot 4H_2O$ for tooth powders and antacid remedies, for coating table salt to render it noncaking, and for paint fillers. Certain of these basic magnesium compounds are also employed with rubber accelerators. Where calcined dolomite is used instead of calcium carbonate, only about one-half of the magnesia must come from the magnesium salts in the sea water.[44] Consequently, the size of the plant is much smaller and the cost of production

[44]Sea water has normally 2.2 g/l of equivalent MgO, actually present as $MgCl_2$ and $MgSO_4$. Hence, 1 ton of MgO theoretically requires 111,000 gal of sea water. Based on a 70% yield, about 158,000 gal would be pumped if all the magnesia came from the sea water, or about half this amount if dolomite were used.

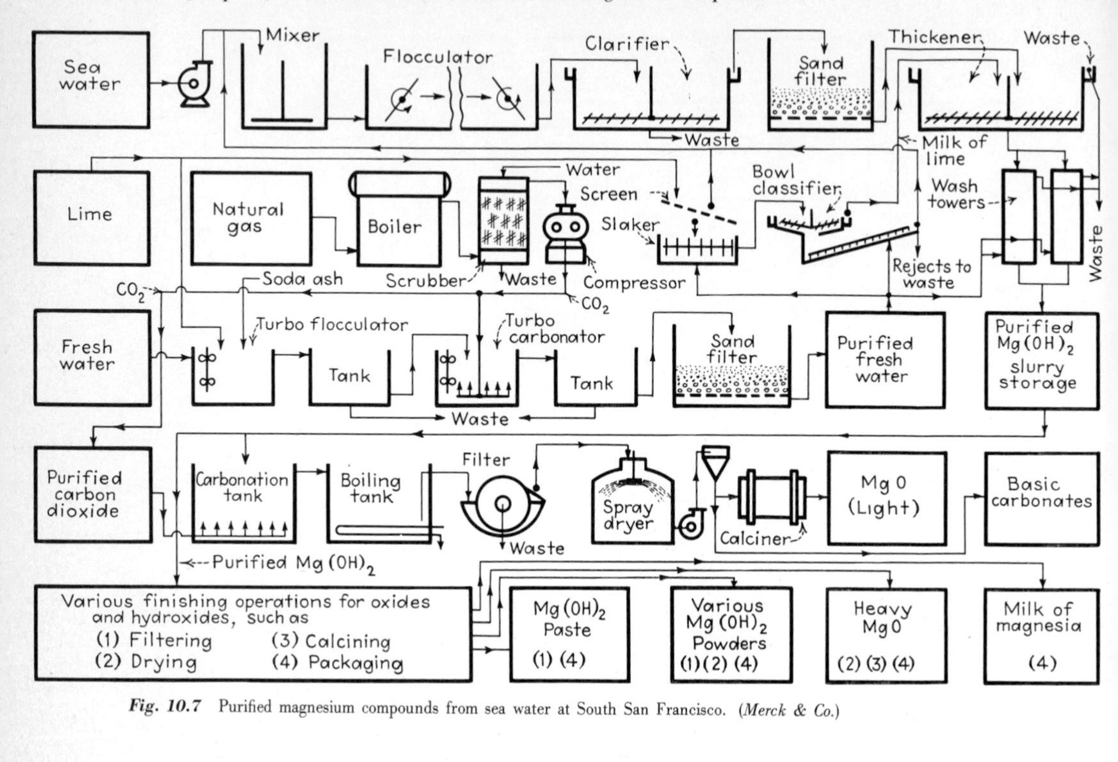

Fig. 10.7 Purified magnesium compounds from sea water at South San Francisco. (*Merck & Co.*)

probably lower. The $Mg(OH)_2$ may be calcined at about 1300°F to active, chemical MgO, or at about 2700 to 3000°F to periclase MgO. This $Mg(OH)_2$ is quite different from the slow-settling $Mg(OH)_2$ precipitated by a soluble alkali or by milk of lime.

The reactions, as illustrated in the flowchart in Figs. 10.7 and 10.8, are principally the following

Calcination:

$$2CaMg(CO_3)_2(c) \longrightarrow 2CaO(c) + 2MgO(c) + 4CO_2(g) \qquad \Delta H = +145.9 \text{ kcal}$$

Slacking:

$$2CaO(c) + 2MgO(c) + 4H_2O(l) \longrightarrow 2Ca(OH)_2(c) + 2Mg(OH)_2(c) \qquad \Delta H = -40.3 \text{ kcal}$$

Precipitation:

$$2Ca(OH)_2(c) + 2Mg(OH)_2(c) + MgCl_2(aq) + MgSO_4(aq) + 2H_2O(l) \longrightarrow 4Mg(OH)_2(c) + CaCl_2(aq) + CaSO_4 \cdot 2H_2O(s) \qquad \Delta H = -5.4 \text{ kcal}$$

Calcination:

$$4Mg(OH)_2(c) \longrightarrow 4MgO(c) + 4H_2O(g) \qquad \Delta H = +59.3 \text{ kcal}$$

Hydrochlorination:

$$Mg(OH)_2(c) + 2HCl(aq) \longrightarrow MgCl_2(aq) + 2H_2O(l) \qquad \Delta H = +11.4 \text{ kcal}$$

Only about 7% of the slaked calcined dolomite is needed for softening the sea water, the rest precipitating crystalline $Mg(OH)_2$ which is settled, filtered, and washed. This hydroxide is converted to other products.

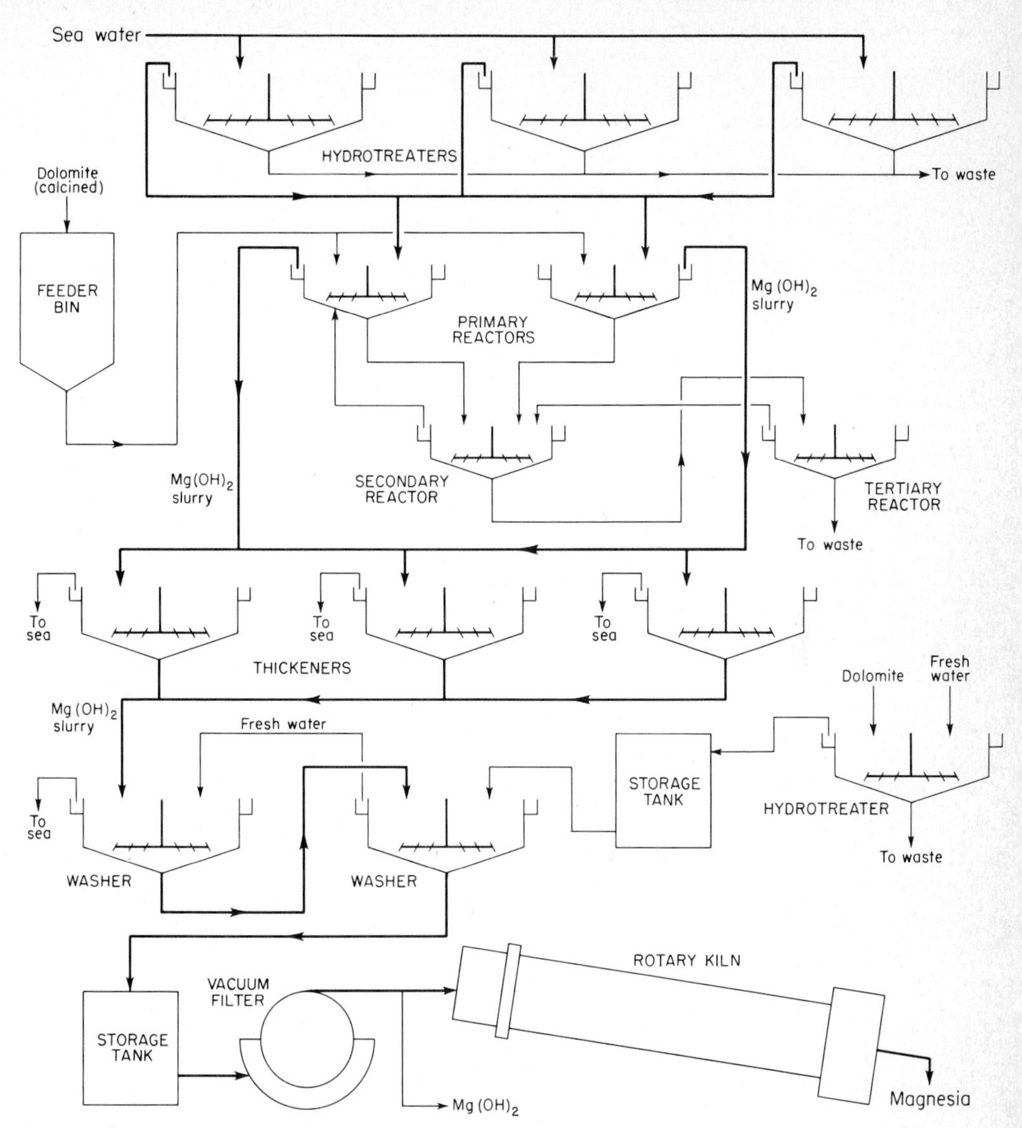

Fig. 10.8 Flowchart for $Mg(OH)_2$ from sea water and dolomite. The sea water is screened and pumped to a battery of three 125-ft-diam concrete hydrotreaters. Here enough calcined dolomite is added (not shown in the figure) to pretreat or soften the sea water by precipitating carbonates, which are removed by rakes. The softened sea water flows to two primary reactors 75 ft in diam into which is fed the main addition of calcined dolomite to precipitate the $Mg(OH)_2$, which in a fine suspension overflows into three parallel-flow 250-ft-diam thickeners, the overflow being returned to the sea. (As shown in Fig. 10.8, the underflow of the two primary reactors is fed first into a 60-ft-diam secondary reactor and then into the tertiary reactor, from which the three-times-reacted residue, now largely silicious material and unreacted dolomite, is wasted.) The thickened $Mg(OH)_2$ slurry from the three 250-ft thickeners is united and washed with calcined-dolomite-softened fresh water in two 250-ft thickeners in series. The $Mg(OH)_2$ underflow is conducted through storage to a 14- by 18-ft nylon-cloth rotary filter operating at about 22 in. vacuum and delivering about 50% $Mg(OH)_2$. This is removed from the rotary filter by a taut wire and conveyed either to sale or to one of three rotary kilns operating up to 3300°F to give different grades of MgO. (*Kaiser Chemical Div., Moss Landing, Calif.*)

TABLE 10.11 *Composition of Sea Water (In grams per liter of sp gr 1.024)*

NaCl	27.319	$Ca(HCO_3)_2$	0.178
$MgCl_2$	4.176	K_2SO_4	0.869
$MgSO_4$	1.668	B_2O_3	0.029
$MgBr_2$	0.076	SiO_2	0.008
$CaSO_4$	1.268	Iron and alumina (R_2O_3)	0.022

MAGNESIUM CARBONATES These vary from dense $MgCO_3$ used in magnesite bricks to the very low-density $4MgCO_3 \cdot Mg(OH)_2 \cdot 5H_2O$ and $3MgCO_3 \cdot Mg(OH)_2 \cdot 3H_2O$ once employed for insulation.[45] There are also other basic carbonates on the market with variations in adsorptive index and apparent density. Many of these are employed as fillers in inks, paints, and varnishes (see Fig. 10.7 for manufacture).

OXIDES AND HYDROXIDE OF MAGNESIUM On heating magnesium carbonate or hydroxide, magnesium oxide (MgO) is formed. This oxide has many uses, e.g., in the vulcanization of rubber, as a material for making other magnesium compounds, as an insulating material, as a refractory material, and as an abrasive. The magnesium hydroxide is principally made directly from sea water, as Figs. 10.7 and 10.8 show. After purification this is the well-known milk of magnesia used in medicine. *Magnesium peroxide* is available from the reaction of magnesium sulfate and barium peroxide. It is employed as an antiseptic and a bleaching agent.

MAGNESIUM SULFATE is prepared by the action of sulfuric acid on magnesium carbonate or hydroxide. It is sold in many forms, one of which is the hydrate $MgSO_4 \cdot 7H_2O$, long known as *Epsom salts.* The less pure material is used extensively as sizing and as a fireproofing agent.

MAGNESIUM CHLORIDE is made from hydrochloric acid and magnesium hydroxide as described above. The compound itself much resembles calcium chloride and has many of the same uses. In addition, it finds application in ceramics, in the sizing of paper, and in the manufacture of oxychloride cement. Its principal use in the making of metallic magnesium was described earlier in this chapter and in Chap. 14.

MAGNESIUM SILICATES A consideration of magnesium silicates[46] includes two widely used naturally occurring compounds, asbestos and talc. Asbestos is a magnesium silicate mixed with varying quantities of silicates of calcium and iron. It is a fibrous, noncombustible mineral and is used in the manufacture of many fireproof and insulating materials. Talc is a rather pure magnesium silicate in the form of $3MgO \cdot 4SiO_2 \cdot H_2O$, found naturally in soapstone. It is employed as a filler in paper and plastics and in many cosmetic and toilet preparations. Magnesium silicate[47] is also made synthetically from Na_2SiO_3 magnesium hydroxide.

SELECTED REFERENCES

Akroyd, T. N. W.: Concrete: Properties and Manufacture, Pergamon, 1962.
Boynton, R. S.: Chemistry and Technology of Lime and Limestone, Wiley, 1966.
Bresler, B.: Reinforced Concrete Engineering, Wiley, 1974.
Gould, R. F. (ed.): Literature of Chemical Technology, chap. 8, Cement, Lime, Plaster and Gypsum, ACS Monograph, 1966.
LaLonde, W. S., Jr., and M. J. Janes: Concrete Engineering Handbook, McGraw-Hill, 1961.
Larson, T. D.: Portland Cement and Asphalt Concretes, McGraw-Hill, 1963.
Martin, L. F.: Cement and Mortar Additives, Noyes, 1972.
McMillian, F. R., and L. H. Tuthill: Concrete Primer, 3d ed., American Concrete Institute, 1973.
Stowell, F. P.: Limestone as a Raw Material in Industry, Oxford, 1963.
Taylor, H. F. W. (ed.): Chemistry of Cements, Academic, 1964, 2 vols.
Waddell, J. J.: Concrete Construction Handbook, 2d ed., McGraw-Hill, 1974.
Witt, J. C.: Portland Cement Technology, 2d ed., Chemical Publishing, 1965.

[45]For "85% magnesia" see CPI 2, p. 226.

[46]Cf. phase diagram of the system $MgO \cdot SiO_2$ in chap. 9; Whittaker, Chrysotile Fibers, *Chem. Eng. News*, Sept. 30, 1963, p. 34; Badollet, ECT, vol. 2, pp. 734–747, 1963.

[47]Meinholdt *et al.*, Magnesium Silicate, *Chem. Process.* (Chicago), June 1960.

chapter 11

GLASS INDUSTRIES

Glass has many uses because of its transparency, high resistance to chemical attack, effectiveness as an electrical insulator, and ability to contain a vacuum. Glass is a brittle material and characteristically exhibits compressive strength much greater than its tensile strength. Strengthening techniques, most of which involve prestressing to introduce surface compression, have been developed to the point where glass can be employed in more arduous environments than previously. Approximately 800 different glass compositions are produced, some with particular emphasis on one property and some with attention to a balanced set of properties.

HISTORICAL[1] As in the case of many other commonplace materials of our modern civilization, the discovery of glass is very uncertain. One of the earliest references to this material was made by Pliny, who related the familiar story of how ancient Phoenician merchants discovered it while cooking a meal in a vessel placed accidentally upon a mass of trona at the seashore. The union of the sand and alkali caught the men's attention and led to subsequent efforts at imitation. As early as 6000 or 5000 B.C. the Egyptians were making sham jewels of glass which were often of fine workmanship and marked beauty. Window glass is mentioned as early as A.D. 290. The hand-blown window-glass cylinder was invented by a twelfth-century monk. During medieval times Venice enjoyed a monopoly as the center of the glass industry. It was not until the fifteenth century, however, that the use of window glass became general. No glass was made in either Germany or England until the sixteenth century. Plate glass appeared as a rolled product in France in 1688.

Glassworks in the United States were founded in 1608 at Jamestown, Va., and in 1639 at Salem, Mass. For more than three centuries thereafter, the processes were practically all manual and rule of thumb. From the chemical point of view, the only major improvements during this period were confined to purifying the batch materials and increasing the fuel economy. To be sure, some relations were established between the chemical composition of glasses and their optical and other physical properties but, all in all, the industry prior to 1900 was an art, with closely guarded secret formulas and empirical processes of manufacture based primarily upon experience.

In 1914 the Fourcault process for drawing a sheet of glass continuously was developed in Belgium. During the next 50 years engineers and scientists produced modifications of the flat-sheet drawing process aimed at reducing the optical distortion characteristic of sheet (window) glass and the cost of producing ground and polished plate glass. These efforts led to the most recent step forward in flat-glass production technology. Based on concepts patented in the United States in 1902 and 1905, a research group in England perfected the float glass process. In just over 10 years float glass has all but eliminated plate glass produced by other means and has invaded the market for window glass in a major way. Scientists and engineers entered the field in increasing numbers, and new products appeared as a result of intensive research. Automatic machines were invented to speed up production of bottles, light bulbs, etc. As a result, today the glass industry is a highly specialized

[1]Shand, Glass Engineering Handbook, 2d ed., McGraw-Hill, 1958; for an excellent general presentation, see Glass, *Chem. Eng. News*, Nov. 16, 1964, p. 81; Stookey, Glass Chemistry as I Saw It, *Chemtech*, **1**(8), 458 (1971).

TABLE 11.1 ***Value of Glass and Glassware Shipped***
(In millions of dollars)

	1972	*1963*
Flat glass, total	1,241	737
Sheet window glass	157	141
Float and plate glass	362	176
Other flat glass	163	100
Laminated glass	528	315
Glass containers	2,056	982

Source: Dept. of Commerce, 1973 Census of Manufacturers, 1975.

field employing all the tools of modern science and engineering in the production, control, and development of its many products.

USES AND ECONOMICS Glass and glassware production by types are shown in Table 11.1. The uses and applications of glass are very numerous, but some conception of the versatility of this material can be gained from a discussion of the various types, as presented in the rest of this chapter. Overall, glassmaking in the United States is about a $5-billion industry. Automobile glass represents almost half of the flat glass produced annually. The architectural trend is toward more glass in commercial buildings and in particular colored glass.[2]

MANUFACTURE

Glass may be defined, *physically*, as a rigid, undercooled liquid having no definite melting point and a sufficiently high viscosity (greater than 10^{13} P) to prevent crystallization; and *chemically*, as the union of the nonvolatile inorganic oxides resulting from the decomposition and fusing of alkali and alkaline earth compounds, sand, and other glass constituents, ending in a product with *random* atomic structure. Glass is a *completely vitrified* product, or at least such a product with a relatively small amount of nonvitreous material in suspension.

COMPOSITION In spite of thousands of new formulations for glass during the past 30 years, it is worthy of note that lime, silica, and soda still form over 90% of all the glass of the world, just as they did 2,000 years ago. It should not be inferred that there have been no important changes in composition during this period. Rather, there have been minor changes in major ingredients and major changes in minor ingredients. The major ingredients are sand, lime, and soda ash, and any other raw materials may be considered minor ingredients, even though the effects produced may be of major importance. The most important factors in making glass are viscosity[3] of molten oxides and the relation between this viscosity and composition. Table 11.2 shows the chemical composition of various glasses.

In general, commercial glasses fall into numerous classes:

1. *Fused silica, or vitreous silica.* Glass made by high-temperature pyrolysis of silicon tetrachloride. It is very resistant thermally and chemically. The high-silica glass Vycor approaches fused silica in some of its properties.
2. *Alkali silicates.* Soluble glasses used only as solutions.
3. *Soda-lime glass.* The soda-lime-silica glass of wide applications, used for windows, transparent fixtures, and all manner of containers.

[2]Market for Colored Glass Accelerates, *Chem. Eng. News*, Feb. 2, 1970, p. 14.
[3]MacAvoy, The Changing Pattern in Glass Manufacture, *Ind. Eng. Chem.*, **54**(7), 17 (1962).

TABLE 11.2 ***Chemical Composition of Typical Glasses*** *(In percent)*

No.	SiO_2	B_2O_3	Al_2O_3	Fe_2O_3	As_2O_3	*CaO*	*MgO*	Na_2O	K_2O	*PbO*	SO_3
1	67.8	....	4.4	....	....	4.0	2.3	13.7	2.3	....	1.0
2	69.4	....	3.5	1.1	....	7.2	....	17.3			
3	70.5	....	1.9	0.4	....	13.0	....	12.0	1.9		
4	71.5	....	1.5	....	....	13.0	....	14.0			
5	72.88	....	0.78		....	12.68	0.22	12.69			
6	72.9	....	0.7		....	7.9	2.8	15.0			
7	72.68	....	0.50	0.07	....	12.95	....	13.17	....	....	0.44
8	70–74	....	±2	0.09	....	....	10–13	....	13–16		
9	73.6	....	1.0	....	....	5.2	3.6	16.0	0.6		
10	73.88	16.48	2.24		0.73	....	....	6.67	Trace		
11	74.2	0.4	....	....	0.2	4.3	3.2	17.7			
12	67.2	....	....	....	0.5	0.9	....	9.5	7.1	14.8	
13	69.04	0.25	....	....	....	12.07	....	5.95	11.75		
14	64.7	10.6	4.2	....	....	0.6	....	7.8	0.3		
15	80.5	12.9	2.2	....	....	....	....	3.8	0.4		
16	96.3	2.9	0.4	....	....	....	....	<0.2	<0.2		
17	70.3		7.15	0.47	....	4.93	....	12.75	1.97		

Sources: Data from Sharp, Chemical Composition of Commercial Glasses, *Ind. Eng. Chem.*, **25,** 755 (1933), Blau, Chemical Trends, *Ind. Eng. Chem.*, **32,** 1429 (1940), and Shand, Glass Engineering Handbook, 2d ed., McGraw-Hill, 1958; cf. Table 11.3. 1, Egyptian from Thebes, 1500 B.C. (Blau); 2, Pompeian window (Blau); 3, German window, 1849, hand-blown (Blau); 4, representative window and bottle glass of 19th century (Sharp); 5, machine cylinder glass (Sharp); 6, Fourcault process sheet with 0.7% BaO (Sharp); 7, polished plate with 0.18% Sb_2O_3 (Sharp); 8, soda-lime containers (Shand); 9, an electric lamp bulb (Shand); 10, Jena, incandescent-gas chimney (Sharp); 11, tableware, lime crystal (Sharp); 12, tableware, lead crystal (Sharp); 13, spectacle with 0.9% Sb_2O_3 (Sharp); 14, Jena with 10.9% ZnO, 1911 laboratory (Sharp); 15, laboratory Pyrex 7740 (Shand); 16, 96% silica No. 790 (Shand); 17, silica glass (fused silica) (Shand).

4. *Lead glass.* The product obtained from lead oxide, silica, and alkali for decorative and optical effects.

5. *Borosilicate glass.* Boric oxide and silica glasses for optical and scientific work and for utensils.

6. *Glass ceramic.* For cook-serve-freeze utensils.

7. *Alumino-silica glasses.* Twenty percent or more alumina for higher temperatures.

8. *Fiber glass.* For textiles and reinforcing.

9. *Special glasses.* Colored glass; opal glass; translucent glass; safety: laminated and tempered glass; photosensitive glass; and specialties for chemical and industrial use.

Fused silica, sometimes referred to erroneously as *quartz glass,* is the end member of the silicate glasses. It is characterized by low expansion and a high softening point, which impart high thermal resistance and permit it to be used beyond the temperature ranges of other glasses. This glass is also extraordinarily transparent to ultraviolet radiation. Compare fused and high-silica glass under *special glasses.*

Alkali silicates are the only two-component glasses of commercial importance. They are water-soluble. Sand and soda ash are simply melted together, and the products designated sodium silicates,[4] having a range of composition from $Na_2O \cdot SiO_2$ to $Na_2O \cdot 4SiO_2$. A knowledge of the equilibrium relations[5] in these two-component systems has aided the glass technologist in understanding the behavior of more complicated systems. Silicate of soda solution, also known as *water glass,* is widely consumed as an adhesive for paper in the manufacture of corrugated-paper boxes.

[4]See Chap. 12 for fuller descriptions.

[5]See Morey, The Properties of Glass, 2d ed., Reinhold, 1954.

Other uses include fireproofing and egg preservation. The higher-alkaline varieties enter into laundering as detergents and as soap builders.

Soda-lime glass represents by far the largest tonnage of glass made today and serves for the manufacture of containers of all kinds, flat glass, automobile (window, plate, float, wire, and figured), tumblers, and tableware. There has been a general improvement in the physical quality of all flat glass, such as increased flatness and freedom from waves and strains, but the chemical composition has not varied greatly. The composition as a rule lies between the following limits (Table 11.2): (1) SiO_2, 70 to 74%; (2) CaO, 10 to 13%; (3) Na_2O, 13 to 16%. Products of these ratios do not melt at too high a temperature, and are sufficiently viscous that they do not devitrify and yet are not too viscous to be workable at reasonable temperatures. The great improvement has been in the substitution of instrument-controlled mechanical devices for the hand operator. Similarly, in container glass, the progress has been largely of a mechanical nature. However, the influence of the liquor trade has created a tendency among manufacturers to make glassware particularly high in alumina and lime and low in alkali. This type of glass melts with more difficulty but is more chemically resistant. The color of container glass is much better than formerly because of the improved selection and purification of raw materials and the use of selenium as a decolorizer.

Applications of phase-rule[6] studies have explained many of the earlier empirical observations of the glassmaker, have led to some improvements, (such as more exactness in the manufacture of soda-lime glass), and have laid the basis for new glass formulations. The phase diagrams for many systems are known and have been published, and the system Na_2O-CaO-SiO_2 has been particularly detailed.

Lead glasses are of very great importance in optical work because of their high index of refraction and high dispersion. Lead contents as high as 92% (density 8.0, refractive index 2.2) have been made. The brilliance of good "cut glass" is due to its lead-bearing composition (Table 11.2). Large quantities are used also for the construction of electric light bulbs, neon-sign tubing, and radiotrons, because of the high electric resistance of the glass. Such glass is also suitable for shielding from nucleonic rays.

Borosilicates[7] usually contain about 13 to 28% B_2O_3 and 80 to 87% silica and have low expansion coefficients, superior resistance to shock, excellent chemical stability, and high electrical resistance. Wide and ever-increasing uses for these glasses have revolutionized the ideas about the properties of glass. Among the diversified applications of these glasses are baking dishes, laboratory glassware, pipelines, high-tension insulators, and washers. Probably the most famous casting of borosilicate glass is the 200-in. disk in the giant telescope at Mt. Palomar.

Special glasses include (1) colored glass, (2) opal or translucent glass, (3) fused-silica and high-silica glass, (4) glass-ceramic, (5) safety: laminated or tempered glass, (6) fotoform, (7) optical glass, and (8) fiber glass. The manufacture of many of these glasses is presented later in this chapter.

Though for many centuries *colored glass* was used merely for decoration, today colored transparent glasses are essential for both technical and scientific purposes, and are produced in many hundreds of colors. *Colored* glass may be one of three types: (*a*) Color is produced by the *absorption* of certain light frequencies by agents in solution in the glass. The coloring agents of this group are the oxides of the transition elements, especially the first group, Ti, V, Cr, Mn, Fe, Co, Ni, and Cu. This class can be subdivided into those in which the color is due to chemical structural environment and those in which the color is caused by differences in state of oxidation. As an example of the former, NiO dissolved in sodium-lead glass yields a brown color, but in a potash glass it produces heliotrope. In the latter, chromium oxides produce colors ranging from green to orange, depending on the proportion of the basic oxide, Cr_2O_3, to the acidic oxide and the composition of glass, i.e.,

[6]Lewis, Squires, and Broughton, Industrial Chemistry of Colloidal and Amorphous Materials, pp. 285–304, Macmillan, 1942; Morey, Phase Equilibrium Relationships Determining Glass Compositions, *Ind. Eng. Chem.*, **25**, 742 (1933); Sharp, Chemical Composition of Commercial Glasses, *Ind. Eng. Chem.*, **25**, 755 (1933).

[7]Frequently sold under the name Pyrex.

TABLE 11.3 ***Comparative Properties of Glasses****

	Common lime	Pyrex borosilicate	Vycor, 96% silica	Fused silica	Pyroceram
Softening point, °F	1285	1508	2732	2600	2282
Annealing point, °F	750	1027	1710	1965	
Strain point, °F	887	950	1575	1810	
Maximum temperature for use (for limited periods), °F	. . . .	932–1022	1830–1990		
Specific gravity	2.47	2.23	2.18	2.07	2.60
Coefficient of linear thermal expansion, $°F^{-1} \times 10^{-7}$	92	18	4.4	5.4	32

*Cf. Perry, 4th ed., pp. 23–58, and 5th ed., pp. 23–60; Stookey, Glass, ECT, 1st ed., 2d suppl., p. 448, 1960; cf. Table 11.2.

whether it is basic or acidic. (*b*) Color is produced by *colloidal particles* precipitated within an originally colorless glass by heat treatment. The classic example is the precipitation of colloidal gold, producing gold-ruby glass. (*c*) Color is produced by *microscopic* or *larger particles* which may be colored themselves, such as selenium reds (SeO_2) used in traffic lights, lantern globes, etc., or the particles may be colorless, producing opals.

Opal, or translucent, glasses are clear when molten but become opalescent as the glass is worked into form, because of the separation and suspension of minute particles of various type, size, and density in the medium, which disperse the light passing through.

Opal glass[8] is often produced by growing nonmetallic crystals from nucleated silver particles developed from an original clear glass containing silver. It is employed for architectural effects, e.g., in window louvers, for transmission of specified wavelengths, and for tableware.

Safety, or laminated, glass may be defined as a composite structure consisting of two layers of glass with an interleaf of plasticized polyvinyl butyral resin. When the glass is broken, the fragments are held in place by the interlayer. Another type is a tempered, or case-hardened, safety glass, which, with a blow sufficiently hard to break it, disintegrates into many small pieces without the usual sharp cutting edges.

Fiber glass[9] (cf. Chap. 35, synthetic fibers), although not a new product, owes its enhanced usefulness to its extreme fineness (often about 0.0005 in., but down to 0.0002 in.). It can be spun into yarn, gathered into a mat, and made into insulation, tape, air filters, and a great variety of other products, such as pipe, with plastic bond.

RAW MATERIALS In order to produce these various glasses, large tonnages of glass sand are used in the United States each year. To flux this silica requires soda ash, salt cake, and limestone or lime. In addition, there is heavy consumption of lead oxide, pearl ash (potassium carbonate), saltpeter, borax, boric acid, arsenic trioxide, feldspar, and fluorspar, together with a great variety of metallic oxides, carbonates, and the other salts required for colored glass. In finishing operations, such diverse products as abrasives and hydrofluoric acid are consumed.

Sand[10] for glass manufacture should be almost pure quartz. A glass-sand deposit has, in many cases, determined the location of a glass factory. Its iron content should not exceed 0.45% for tableware or 0.015% for optical glass, as iron affects the color of most glass adversely.

Soda (Na_2O) is principally supplied by dense soda ash (Na_2CO_3). Other sources are sodium bicarbonate, salt cake, and sodium nitrate. The latter is useful in oxidizing iron and in accelerating

[8]Shand, *op. cit.*, pp. 313–316.

[9]Shand, *op. cit.*, chaps. 20 to 24; Loewenstein, The Manufacturing Technology of Continuous Glass Fibers, American Elsevier, 1973.

[10]Havinghorst, Beneficiation by Separating Glass Sand from Clay, *Chem. Eng.* (*N.Y.*), **70**(12), 158 (1963) (excellent flowcharts).

the melting. The important sources of *lime* (CaO) are limestone and burnt lime from dolomite ($CaCO_3 \cdot MgCO_3$), the latter introducing MgO into the batch.

Feldspars have the general formula $R_2O \cdot Al_2O_3 \cdot 6SiO_2$, where R_2O represents Na_2O or K_2O or a mixture of these two. They have many advantages over most other materials as a source of Al_2O_3, because they are cheap, pure, and fusible and are composed entirely of glass-forming oxides. Al_2O_3 itself is used only when cost is a secondary item. Feldspars also supply Na_2O or K_2O and SiO_2. The alumina content serves to lower the melting point of the glass and to retard devitrification.

Borax, as a minor ingredient, supplies glass with both Na_2O and boric oxide. Though seldom employed in window or plate glass, borax is now in common use in certain types of container glass. There is also a high-index borate glass which has a lower dispersion value and a higher refractive index than any glass previously known and is valuable as an optical glass. Besides its high fluxing power, borax not only lowers the expansion coefficient but also increases chemical durability. Boric acid is used in batches where only a small amount of alkali is wanted. Its price is about twice that of borax.

Salt cake, long accepted as a minor ingredient of glass, and also other sulfates such as ammonium and barium sulfate, are encountered frequently in all types of glass. Salt cake is said to remove the troublesome scum from tank furnaces. Carbon should be used with sulfates to reduce them to sulfites. *Arsenic trioxide* may be added to facilitate the removal of bubbles. *Nitrates* of either sodium or potassium serve to oxidize iron and make it less noticeable in the finished glass. Potassium *nitrate* or carbonate is employed in many better grades of table, decorative, and optical glass.

Cullet is crushed glass from imperfect articles, trim, and other waste glass. It facilitates melting and utilizes waste material. It may be as low as 10% of the charge or as high as 80%.

Refractory blocks for the glass industry have been developed especially because of the severe conditions encountered. *Sintered* zircon, alumina, mullite, mullite-alumina and *electrocast* zirconia-alumina-silica, alumina, and chrome-alumina are typical of those for glass tanks. See electrocast refractories, Chap. 9. The latest practice in regenerators utilizes basic refractories because of the alkali dust and vapors. Furnace operating temperatures are limited mainly by silica-brick crowns, which are economical to use in the industry.

CHEMICAL REACTIONS The chemical reactions involved may be summarized:

$$Na_2CO_3 + aSiO_2 \longrightarrow Na_2O \cdot aSiO_2 + CO2 \qquad (1)$$

$$CaCO_3 + bSiO_2 \longrightarrow CaO \cdot bSiO_2 + CO_2 \qquad (2)$$

$$Na_2SO_4 + cSiO_2 + C \longrightarrow Na_2O \cdot cSiO_2 + SO_2 + CO \qquad (3)$$

The last reaction may take place as in equations (4) or (5), and (6):

$$Na_2SO_4 + C \longrightarrow Na_2SO_3 + CO \qquad (4)$$

$$2Na_2SO_4 + C \longrightarrow 2Na_2SO_3 + CO_2 \qquad (5)$$

$$Na_2SO_3 + cSiO_2 \longrightarrow Na_2O \cdot cSiO_2 + SO_2 \qquad (6)$$

It should be noted that the ratios Na_2O/SiO_2 and CaO/SiO_2 are not molar ratios. The ratio may be of the type $Na_2O/1.8SiO_2$, for example. In an ordinary window glass the molar ratios are *approximately* 2 mol Na_2O, 1 mol CaO, and 5 mol SiO_2. Other glasses vary widely (Table 11.2). Chemical compounds do not exist in glass, which is essentially an amorphous, multicomponent, solid mixture, or an undercooled liquid.

Typical manufacturing sequences can be broken down into the following *unit operations* and *chemical conversions* (cf. Fig. 11.1):

Transportation of raw materials to the plant (Op).
Sizing of some raw materials (Op).

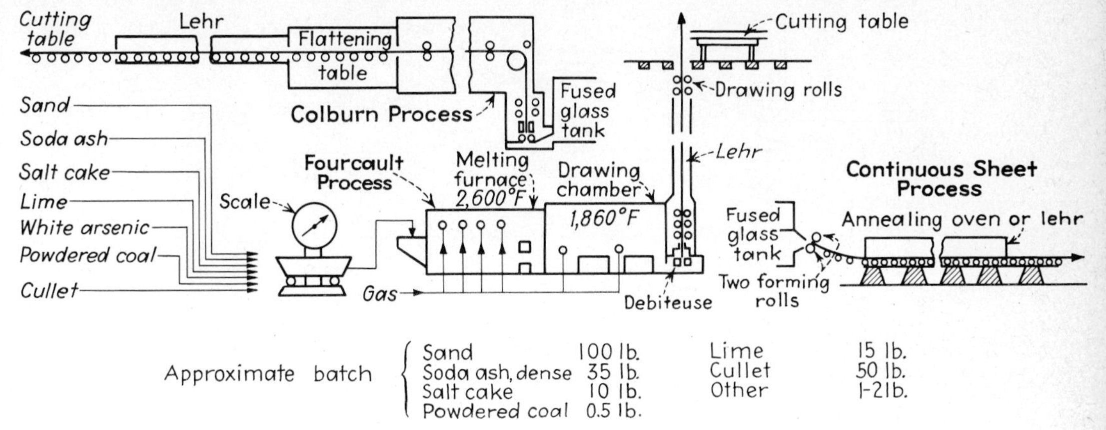

Fig. 11.1 Flowchart for flat-glass manufacture.

Storage of raw materials (Op).
Conveying, weighing, and mixing raw materials, and feeding them into the glass furnace (Op).
Reactions in the furnace to form glass (Ch).
Burning the fuel to the secure temperature needed for glass formation (Ch).
Saving of heat by regeneration or recuperation (Op).
Shaping of glass products (Op).
Annealing of glass products (Op).
Finishing of glass products (Op).

To carry out these steps, modern glass factories are characterized by the use of materials-handling machinery supplying automatic and continuous manufacturing equipment, in contrast to the "shovel and wheelbarrow" methods of older factories. In spite of the modernization of plants, however, the manual charging of small furnaces is still carried on, though a dusty atmosphere is created. The trend, however, is toward mechanical batch transporting and mixing systems so completely enclosed that practically no dust is emitted at any stage of the handling of glass or raw materials.

METHODS OF MANUFACTURE

The manufacturing procedures may be divided (cf. Fig. 11.1) into four major phases: (1) melting, (2) shaping or forming, (3) annealing, and (4) finishing.

MELTING Glass furnaces may be classified as either pot or tank furnaces (cf. Fig. 11.2, of tank furnace). *Pot furnaces,* with an approximate capacity of 2 tons or less, are used advantageously for the small production of special glasses or where it is essential to protect the melting batch from the products of combustion. They are employed principally in the manufacture of optical glass, art glass, and plate glass by the casting process. The pots are really crucibles made of selected clay or platinum. It is very difficult to melt glass in these vessels without contaminating the product or partly melting the container itself, except when platinum is used. In a *tank furnace* (Fig. 11.2), batch materials are charged into one end of a large "tank" built of refractory blocks, some of which measure 125 by 30 by 5 ft and have a capacity of 1,500 tons of molten glass. The glass forms a pool in the hearth of the furnace, across which the flames play alternately from one side and the other. The "fined"[11] glass is worked out of the opposite end of the tank, the operation being continuous. In

[11] "Fining" is allowing *molten* glass sufficient time for the bubbles to rise and leave or dissolve in the glass.

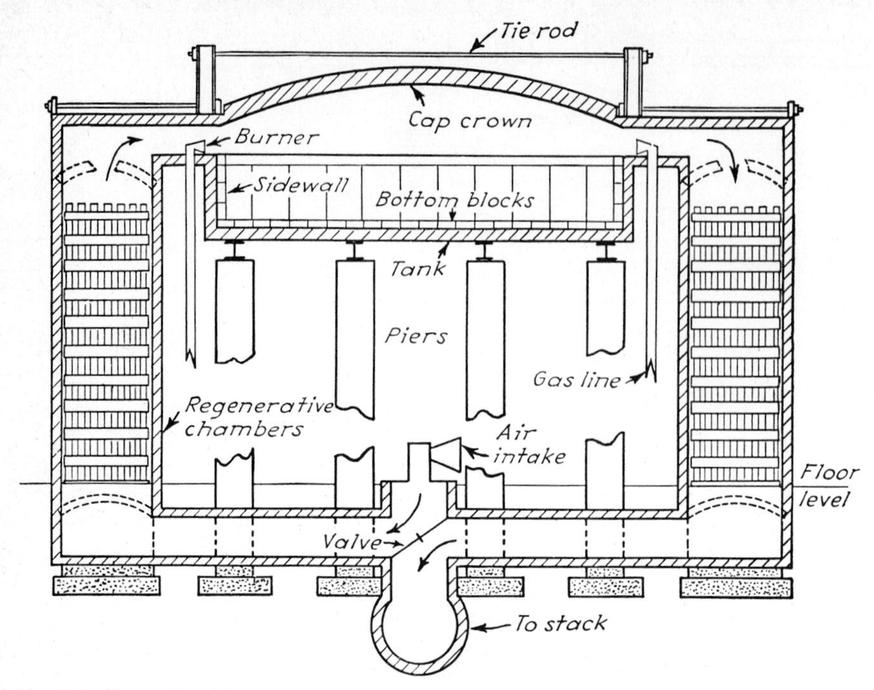

Fig. 11.2 Cross section of glass tank furnace, showing regenerative chambers.

this type of furnace, as in the pot furnace the walls gradually corrode under the action of the hot glass. The quality of the glass and the life of the tank are dependent upon the quality of the construction blocks. For this reason, much attention has been given to glass furnace refractories.[12] Small tank furnaces are called day tanks and supply a day's demand of 1 to 10 tons of molten glass. They are heated either electrothermally or by gas.

The foregoing types are *regenerative* furnaces and operate in two cycles with two sets of checkerwork chambers. The flame gases, after giving up their heat in passing across the furnace containing the molten glass, go downward through one set of chambers stacked with open brickwork or checkerwork, as shown in Fig. 11.2. A great deal of the sensible-heat content of the gases is removed thereby, the checkerwork reaching temperatures ranging from 2800°F near the furnace to 1200°F on the exit side. Simultaneously, air is preheated by being passed up through the other previously heated regenerative chamber and is mixed with the fuel gas[13] burned, the resulting flame being of a higher temperature than would have been possible if the air had not been preheated. At regular intervals of 20 to 30 min, the *flow* of the air-fuel mixture, or the cycle, is reversed, and it enters the furnace from the opposite side through the previously heated checkerwork, passing through the original checkerwork, now considerably cooled. Much heat is saved by this regenerative principle, and a higher temperature is reached.

The temperature of a furnace just starting production can be raised only certain increments each day, depending upon the ability of the refractory used to stand the expansion. Once the regenerative furnace has been heated, a temperature of at least 2200°F is maintained at all times. Melting cost is $6 or more per ton of glass. Most of the heat is lost by radiation from the furnace,

[12]See electrocast refractories in Chap. 9.

[13]If the fuel is producer gas, it is preheated also, but not if a higher Btu type is employed, such as natural gas or oil.

and a much smaller amount is actually expended in the melting. Unless the walls are allowed to cool somewhat by radiation, however, their temperature would become so high that the molten glass would dissolve or corrode them much faster. To reduce the action of the molten glass, water cooling pipes are frequently placed in the furnace walls.

SHAPING OR FORMING Glass may be shaped by either machine or hand molding. The outstanding factor to be considered in *machine molding* is that the design of the glass machine should be such that the article is completed in a very few seconds. During this relatively short time the glass changes from a viscous liquid to a clear solid. It can therefore be readily appreciated that the design problems to be solved, such as flow of heat, stability of metals, and clearance of bearings, are very complicated. The success of such machines is an outstanding tribute to the glass engineer. In the following discussion the most common types of machine-shaped glass, i.e., window glass, plate glass, float glass, bottles, light bulbs, and tubing, are described.

Window glass. For many years window glass was made by an extremely arduous hand process that involved gathering a gob of glass on the end of a blowpipe and blowing it into a cylinder. The ends of the latter were cut off, and the hollow cylinder split, heated in an oven, and flattened. This tedious manual process has now been entirely supplanted by continuous processes or their modifications, the Fourcault process and the Colburn process, as outlined in the flowcharts in Figs. 11.1, 11.3, and 11.6 and the accompanying text discussing the flow process.

In the *Fourcault* process a drawing chamber is filled with glass from the melting tank. The glass is drawn vertically from the kiln through a so-called *débiteuse* by means of a drawing machine. The débiteuse consists of a refractory boat with a slot in the center through which the glass flows continuously upward when the boat is partly submerged. A metal bait lowered into the glass through the slot at the same time the débiteuse is lowered starts the drawing as the glass starts flowing. The glass is continuously drawn upward in ribbon form as fast as it flows up through the slot, and its surface is chilled by adjacent water coils. The ribbon, still traveling vertically and supported by means of asbestos-covered steel rollers, passes through a 25-ft-long annealing chimney, or lehr. On emerging from the lehr, it is cut into sheets of desired size and sent on for grading and cutting. This is shown diagrammatically in Figs. 11.1 and 11.3.

PPG Industries operates a modified Fourcault process which produces *Pennvernon* glass. Sheets of glass 120 in. wide and thicknesses up to $\frac{7}{32}$ in. are made by varying the drawing rate from 38 in./min for single-strength[14] to 12 in. for $\frac{7}{32}$ in. This process substitutes for the floating débiteuse a submerged draw bar for controlling and directing the sheet. After being drawn vertically a distance of 26 ft, most of which is in an annealing lehr, the glass is cut. For thicknesses above single and double strength, a second annealing is given in a 120-ft standard horizontal lehr.

During 1917 the Libbey-Owens-Ford Glass Co. began drawing sheet glass by the *Colburn* process. In this process, drawing is started vertically from the furnace, as in the Fourcault process, but after traveling for about 3 ft the glass is heated and bent over a horizontal roller and carried forward by grip bars attached to traveling belts. The sheet moves over the flattening table through the horizontal annealing lehr, with its 200 power-driven rolls, onto the cutting table.

Plate glass.[15] Between 1922 and 1924 the Ford Motor Co. and PPG Industries independently developed a continuous automatic process for rough-rolled glass in a continuous ribbon (cf. continuous-sheet process in Fig. 11.1). The glass is melted in large continuous furnaces holding 1,000 or more tons. The raw materials are fed into one end of the furnace, and the melted glass, at a temperature as high as 2900°F, passes through the refining zone and out the opposite end in an unbroken flow. From the wide refractory outlet, the molten glass passes between two water-cooled forming rolls, which give it a plastic-ribbon configuration. The ribbon of glass is drawn down over

[14]Single-strength window glass has a thickness of 0.087 to 0.100 in.; double-strength glass measures 0.118 to 0.132 in.
[15]Pilkington, Manufacture of Flat Glass at Pilkington's, *Engineering*, **144**, 144, 308 (1962).

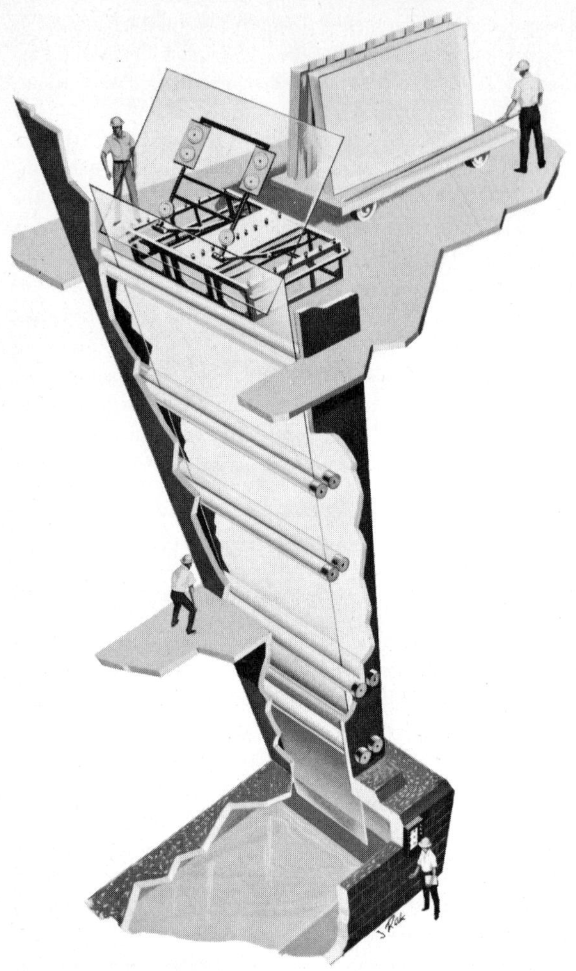

Fig. 11.3 Pictorial flowchart of modified Fourcault process for flat glass. Here the glass is continuously drawn vertically by rolls through the lehr to an automatic cutter. (*American-Saint Gobain Corp.*)

a series of smaller water-cooled rolls running at a slightly higher surface speed than the forming rolls. The stretching effect of the different speeds and the shrinking of the glass as it cools flatten the ribbon as it enters the annealing lehr. After annealing, the ribbon may be cut into sheets for grinding and polishing or, as in the newest processes, it may progress automatically for several hundred feet, undergoing annealing, grinding, polishing, and inspection, before it passes through cutting machines and is reduced to salable plates (Figs. 11.4 and 11.5). Continuous processes have very large capacities and are particularly adapted to the production of glass with a uniform thickness and a composition suited to the requirements of the automobile industry. Short runs of plate or special glass cannot be produced economically by these machines but are handled in pots and mechanically cast by machines made especially for this type of operation.

Twin grinding for plate glass is an important technical and cost improvement. The ribbon emerging from the rolls of the tank furnace after passing through the 600-ft lehr is ground simul-

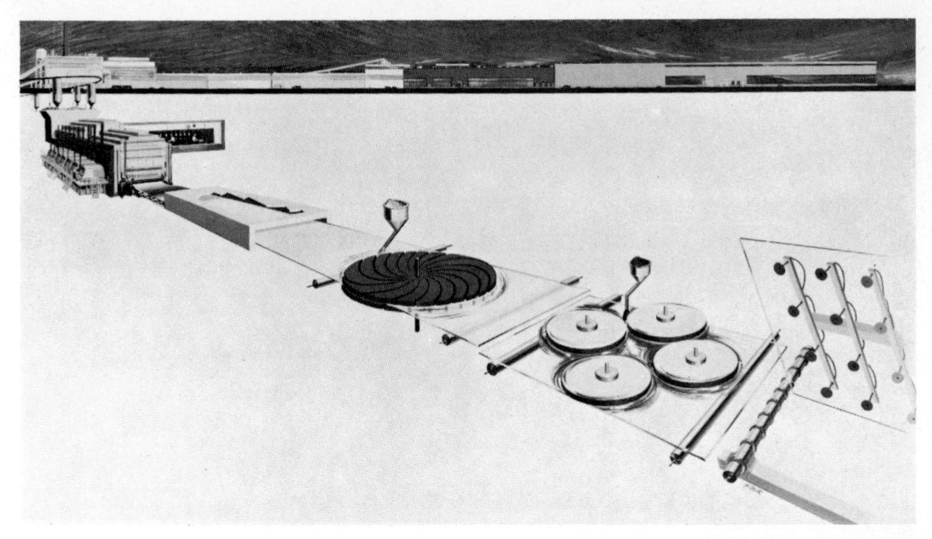

Fig. 11.4 Pictorial flowchart of continuous twin (top and below) grinding followed by cutting and polishing for plate glass. The twin grinding disks are of iron and 11 ft in diameter and use an abrasive slurry of silica sand. (*American-Saint Gobain Corp.*)

taneously above and below (Figs. 11.4 and 11.5). In several factories the twin grinding is followed by twin polishing. The ribbon is then cut to size and inspected, resulting in a more uniform plate.

Float glass,[16] developed by Pilkington Brothers in England, is a fundamental improvement in the manufacture of high-quality sheet glass. A ribbon of glass formed by rolling is floated on a bath of molten metal in a controlled atmosphere. At the entrance to the float chamber, the temperature is high enough that the glass is still soft enough to flow. The bottom surface is smoothed by contact with the metal, and the top surface flows under the action of gravity until it also is smooth. This technique of holding a glass surface at a high temperature while irregularities are smoothed out by flow is called *fire-polishing*. The surface quality produced is so close to that obtained by mechanical grinding and polishing that float glass is replacing polished plate glass in many applications. Thickness range in the float process is 0.1 to 1.0 in. (Figs. 11.6 and 11.7).

Wired and patterned glass. In patterned glass manufacture, the molten glass flows over the lip of the furnace and passes between metal rolls on which a pattern has been engraved or machined. The rolls form the glass and imprint the pattern in a single operation. Such glass diffuses the light and ensures a certain amount of privacy, which recommends it for use in rooms, doors, and shower enclosures. Such glass can be reinforced with wire during the initial forming for special safety needs, e.g., for windows near fire escapes. Both operations are pictured in Fig. 11.8.

Blown glass. Glass blowing, one of the most ancient arts, until the last century depended solely upon human lungs for power to form and shape molten glass. Modern demands for blown glass, however, have required the development of more rapid and cheaper methods of production.[17] The machine making of bottles is only a casting operation which uses air pressure to create a hollow. Several types of machines produce *parisons,* partly formed bottles or bottle blanks. One is the suction-feed type used, with certain variations, in bulb and tumbler production. Another is the gob-feed type, which has been applied to the manufacture of all types of ware made by pressing, blowing, or a combination of "press and blow."

[16]Pilkington, Development of Float Glass, *Glass Ind.*, February 1963, p. 80; *Am. Glass Rev.*, February 1963, p. 13.
[17]Shand, *op. cit.*, chap. 11, p. 197.

Fig. 11.5 Photograph of twin grinding of a continuous 120-in.-wide ribbon of plate glass. Only the top grinders are visible. The sheet of glass can be dimly seen in the left background. (*American-Saint Gobain Corp.*)

In the suction-feed type, glass contained in a shallow, circular revolving tank is drawn up into molds by suction. The mold then swings away from the surface of the glass, opens, and drops away, leaving the parison sustained by the neck. The bottle mold next rises into position around the parison, and a blast of compressed air causes the glass to flow into the mold. The latter remains around the bottle until another gathering operation has been performed; it then drops the bottle and rises to close around a fresh parison. The operations are completely automatic, and speeds of 60 units per minute are not uncommon.

The gob feeder represents one of the most important developments in automatic glassworking. In this operation the molten glass flows from the furnace through a trough, at the lower end of which

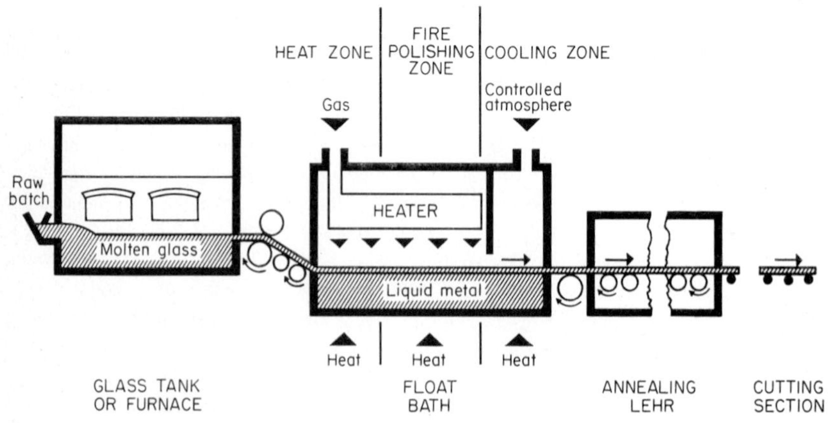

Fig. 11.6 Flowchart of a continuous process for the manufacture of float glass. (*PPG Industries.*)

Fig. 11.7 Photograph of the Pilkington *float-bath* section in which the 100-in.-wide glass ribbon is formed at the Pilkington Bros. float-glass process plant at St. Helens, Lancashire, England. The operating temperatures in the melting section are about from 1500 to 1200°C. [*Chem. Eng.*, ***71****(3), 38 (1964. PPG Industries.*]

is an orifice. The glass drops through the orifice and is cut into a gob of the exactly desired size by mechanical shears. It is delivered through a funnel into the parison mold, which starts the formation of the bottle in an inverted position, as shown in Fig. 11.9. A neck pin rises into place, and another plunger drops from the top, whereupon compressed air in the "settle blow" forces the glass into the finished form of the neck. The mold is closed on top (bottom of the bottle), the neck pin is retracted, and air is injected in the "counterblow" through the newly formed neck to form the inner cavity. The parison mold opens, and the parison is inverted as it passes to the next station, so that the partly formed bottle is then upright. The blow mold closes around the parison, which is reheated for a brief interval. Air is then injected for the final blow, simultaneously shaping the inner and outer surfaces of the bottle. The blow mold swings away, and the bottle moves on to the lehr.

Automatic bottle-blowing machines usually consist of two circular tables, known as the *parison-mold* table and the *blow* table. As the glass moves around the periphery of the table, the various operations described above take place. Table movement is controlled by compressed air which operates reciprocating pistons, and the various operations occurring on the table are coordinated with table movement by a motor-timing mechanism. This latter device constitutes one of the most vital and expensive parts of the equipment.

Light bulbs. The blowing of a thin bulb differs from bottle manufacture in that the shape and size of the bulb are determined initially by the air blast itself and not by the mold. The molten glass flows through an annular opening in the furnace and down between two water-cooled rollers, one of which has circular depressions that cause swellings on a glass ribbon coinciding with circular holes on a horizontal chain conveyor onto which the ribbon moves next. The glass sags through these holes by its own weight. Below each hole is a rotating mold. Air nozzles drop onto the surface of the ribbon, one above each of the glass swellings or conveyor holes. As the ribbon moves along, these nozzles eject a puff of air which forms a preliminary blob in the ribbon. The spinning mold now rises, and a second puff of air, under considerably less pressure than the first, shapes the blob into the mold and forms the bulb. The mold opens, and a small hammer knocks the bulb loose from the ribbon.

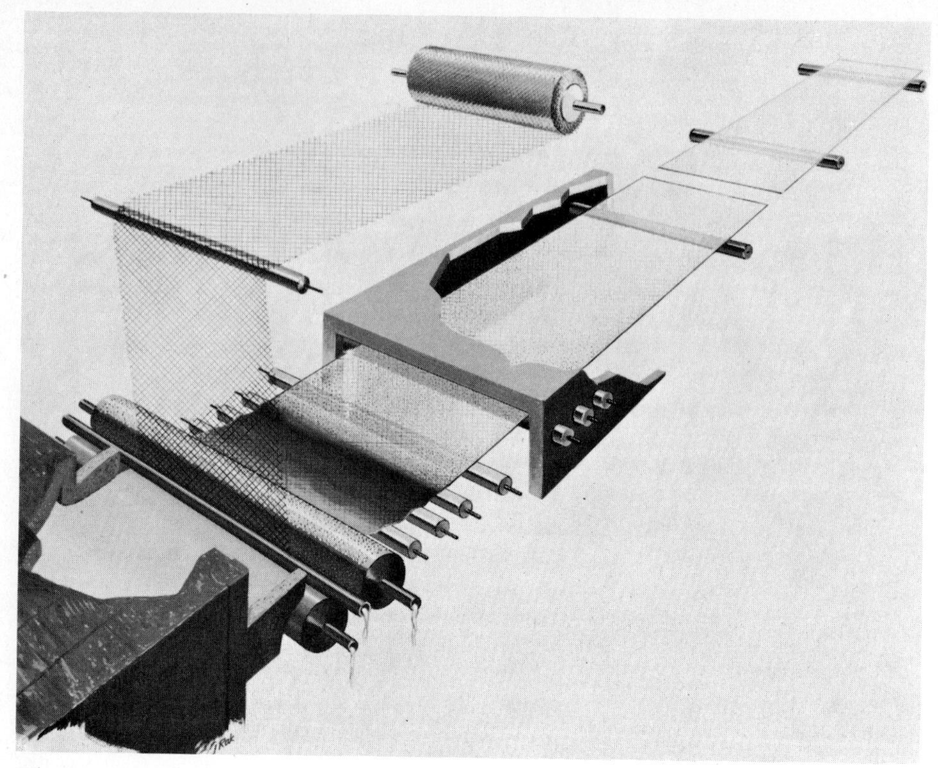

Fig. 11.8 Pictorial flowchart of a 5-ft ribbon of wired and patterned glass. (*American-Saint Gobain Corp.*)

The bulbs drop onto an asbestos belt which carries them to the lehr rack, where they slip neck down between two parallel vertical strips which support them as they are annealed. The total time for the entire series of operations, including annealing, is about 8 min. Machine speeds as high as 2,000 bulbs per minute have been attained.

Television tubes are now made as large as 27 in. across and consist of three principal parts, the face phosphorescent screen, on which the picture is produced, the envelope, and the electron gun. The phosphor is applied to the face screen of the envelope either by settling or dusting. Manufacture of the glass envelope was difficult until centrifugal casting was invented, which uses a revolving mold to produce a much more uniform wall thickness. The glass parts are sealed together, using a gas flame or gas and electricity. For *colored* television tubes, the phosphor is applied to the inner surface of the screen. A perforated mask is mounted behind the screen to direct the electron beams properly. The high temperature involved in sealing cannot be employed here, since this would cause deterioration of the phosphor.

Glass tubing.[18] In the Vello process molten glass flows into a drawing compartment, from which it drops vertically through an annular space surrounding a rotating rod or blowpipe in which air pressure is maintained, to produce tubing of the desired diameter and wall thickness.

Glass tower plates and bubble caps, prisms, and most other optical glass, most kitchenware, insulators, certain colored glasses, architectural glass, and many similar items are *hand-molded*. The process consists essentially in drawing a quantity of glass, known as the *gather*, from the pot or tank

[18]Phillips, Glass: Its Industrial Applications, Reinhold, 1960.

and carrying it to the mold. Here the exact quantity of glass required is cut off with a pair of shears, and the ram of the mold driven home by hand or by hydraulic pressure. Certain glass forming is carried out by semiautomatic methods which involve a combination of the machine- and hand-molding processes previously described. Volumetric flasks and cylindrical Pyrex sections for towers are fabricated in this manner.

ANNEALING To reduce strain, it is necessary to *anneal all glass* objects, whether they are formed by machine- or hand-molding methods. Briefly, annealing involves two operations: (1) holding a mass of glass above a certain critical temperature long enough to reduce internal strain by plastic flow to less than a predetermined maximum and (2) cooling the mass to room temperature slowly enough to hold the strain below this maximum. The lehr, or annealing oven, is nothing more than a carefully designed heated chamber in which the rate of cooling can be controlled so as to meet the

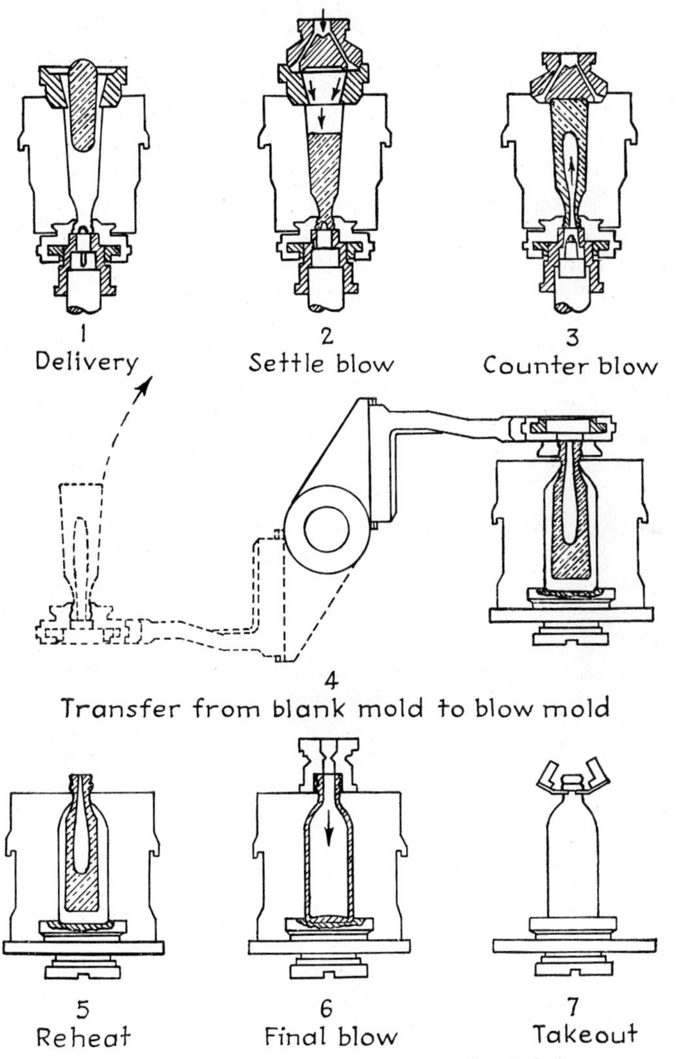

Fig. 11.9 Automatic bottle manufacture by an I-S machine. (*Glass Packer.*)

foregoing requirements. The establishment of a quantitative relationship[19] between stress and birefringence caused by the stress has enabled glass technologists to design glass to meet certain conditions of mechanical and thermal stress. With the foregoing data as a basis, engineers have produced continuous-annealing equipment, with automatic temperature regulation and controlled circulation, which permits better annealing at a lower fuel cost and with less loss of product.

FINISHING All types of annealed glass must undergo certain finishing operations which, though relatively simple, are very important. These include cleaning, grinding, polishing, cutting, sandblasting, enameling, grading, and gaging. Although all these are not required for every glass object, one or more is almost always necessary.

MANUFACTURE OF SPECIAL GLASSES[20]

R&D are at the heart of the new and improved types and properties of glass. This section illustrates some of the new glass products that have resulted.

FUSED SILICA GLASS, or vitreous silica, was made by fusing pure silica, but such products were usually blebby and difficult to produce in transparent form. It is now manufactured by Corning[21] by vapor-phase high-temperature pyrolysis of silicon tetrachloride. This type of process lends itself naturally to controls which permit chemically pure SiO_2. The raw silica produced in this manner is in the form of plates, or boules. The high temperature of the reaction tends to drive off undesired contaminants, giving fused-silica impurities in the order of 1 part in 100 million. This fused-silica glass has remarkable properties (Table 11.3), and has the lowest ultrasonic absorption of any material. Because of its low thermal expansion it is used for telescope mirrors, e.g., in a 62-in. mirror for a telescope at the U.S. Naval Observatory.

HIGH-SILICA GLASS This product, known as Vycor,[22] constitutes an important advance toward the production of a glass approaching fused silica in composition and properties. This has been accomplished with the avoidance of former limitations on melting and forming. The finished articles contain approximately 96% silica and 3% boric oxide, and the rest is alumina and alkali. Borosilicate-glass compositions of about 75% silica content are used in the earlier stages of the process, in which the glasses are melted and molded. After cooling, the articles are subjected to heat treatment and annealing, which induce the glass to separate into two distinct physical phases. One of these phases is so high in boric and alkaline oxides that it is readily soluble in hot acid solutions, whereas the other is rich in silica and therefore insoluble in these solutions. The glass article is immersed in a 10% hydrochloric acid (98°C) bath for sufficient time to permit the soluble phase to be virtually all leached out. It is washed thoroughly to remove traces of the soluble phase, as well as impurities, and subjected to another heat treatment which serves to dehydrate the body and to convert the cellular structure to a nonporous vitreous glass. In the course of these processes the glassware undergoes a shrinkage in linear dimensions amounting to 14% of its original size. Table 11.3 compares its properties with those of other glasses. This method of glass manufacture furnishes a product that can be heated to a cherry red and then plunged into ice water without any ill effects. Also, this glass has high chemical durability and is extremely stable to all acids except hydrofluoric which, however, attacks this glass considerably more slowly than others. Its shrinkage is also proportionately even, so that the original shape is preserved.

[19]Shand, *op. cit.*, pp. 103–109.

[20]Stookey, Glass, ECT, 2d suppl. pp. 435–454, 1960; MacAvoy, Changing Pattern in Glass Manufacture, *Ind. Eng. Chem.*, **54**(7), 17, (1962).

[21]Stookey, *op. cit.*, p. 445, and properties listed in table, p. 448.

[22]Corning Glass Works, U.S. Pat. 2,106,744 (1938), and 2,221,709 (1940); Phillips, *op. cit.*

GLASS-CERAMIC[23] or Pyroceram, is a material melted and formed as glass, and then converted largely to a crystalline ceramic by processes of *controlled devitrification.* Ceramics are usually bodies with crystalline and high-melting particles bound together either in a vitreous matrix or by fusion of the particles at their grain boundaries. The vitreous matrix results from chemical reactions between small amounts of fluxes included in the raw materials and the crystalline constituents. These glass ceramics, after first being machine-fabricated in the glassy state, are followed by catalytic nucleation around a minor but important ingredient such as TiO_2 through heat treatment. "Crystals are then grown on the nuclei by raising the glass to a temperature within the devitrifying range of the particular composition involved." Such crystals are much smaller and more uniform than those in conventional ceramics. Properties of ceramics produced from glass correspond more closely to those of ceramics than to those of unconverted glass, the glass-ceramic possessing superior rigidity and other mechanical and thermal properties. The bodies are usually opaque, glossy, either white or colored, and nonporous; some compositions have outstanding electrical properties. Nonporous, fine-grained, crystalline microstructure glass-ceramics have *higher* flexural strength than conventional ceramics of similar chemical composition, ranging up to 30,000 psi and above. The low-thermal-expansion formulations are *practically immune* to breakage by thermal shock. These products can be fabricated with *close dimensional* tolerances into a wide variety of shapes and sizes, employing conventional glass-forming methods. They are being used for radomes, for guided missiles, and in various electronic devices and, under the Corning trademark Pyroceram, in cook-serve-freeze utensils.

SAFETY GLASS Safety glasses[24] may be grouped into two general classes: laminated safety glass and heat-strengthened (or tempered) or case-hardened safety glass. Wired glass may also be considered safety glass (Fig. 11.8).

Laminated safety glass, which is by far the most widely used in this country, consists of two sheets of thin plate glass, each of which is about 0.125 in. thick, with a sheet of nonbrittle plastic material between. The plastic and glass are washed, and an adhesive is applied to the glass (if the plastic used requires it, such not always being the case). The glass and plastic sheet are pressed together under moderate heat to seal the edges. The glass is subjected to high temperatures and hydraulic pressures in an autoclave, in order to bring the entire interlayer into intimate contact, after which the edges of the sandwich may be sealed with a water-resistant compound.

The glass used in the manufacture of laminated safety glass has the same physical properties as ordinary glass, so that the safety features depend solely upon the ability of the plastic interlayer to hold fragments caused by accidental breaking of the glass itself. The first plastic used commercially was cellulose nitrate, which was replaced by cellulose acetate. Now practically all laminated safety glass uses polyvinyl butyral resin. This vinyl plastic is more elastic than cellulose acetate, since it stretches under relatively low stresses up to its elastic limit, after which considerable additional stress is necessary to make it fail. It remains clear and colorless under all conditions of use, is not affected by sunlight, and does not need adhesives or water-resistant compounds in manufacturing.

Tempered, or strengthened,[25] *glass* is very strong and tough. It is used for doors and windows of automobiles and for pipe. It possesses high internal stresses and, if the surface is broken, shatters into many pieces. Its manufacture involves controlled heat annealing whereby the nonuniform stresses in glass are replaced by controlled, uniform, low-level stresses. Such glass is really very high in compression and very weak in tension. *Physical tempering* is an outgrowth of the study of

[23]Shand, *op. cit.*, 364–365; ECT, 2d suppl., p. 453, 1960 (many details); Stookey, Glass Ceramics, *Chem. Eng. News*, June 19, 1961, p. 116; *Ind. Eng. Chem.*, **51**(7), 805 (1959); Forming, Color Boost Glass-Ceramics Uses, *Chem. Eng.*, Oct. 23, 1967, p. 65; Stookey, Glass Chemistry as I Saw It, *Chemtech*, **1**(8), 458 (1971).

[24]Randolph, Evolution of Safety Glass, *Mod. Plast.*, **18**(10), 31 (1941); cf. Weidlein, History and Development of Laminated Safety Glass, *Ind. Eng. Chem.*, **31,** 563 (1939).

[25]Olcott, Corning Glass Co., Chemical Strengthening of Glass, *Science*, **140,** 1189 (1963).

annealing and is less drastic than the quenching process long used for making Prince Rupert drops. The already-formed glass vessel or sheet to be strengthened by tempering or annealing is heated to some temperature, e.g., 800°F, just below its softening point, and then quenched in air, molten salt, or oil. During this tempering, a sandwich effect results when the exterior, or skin, of the glass cools rapidly and becomes hard and the interior cools more slowly and continuously and contracts after the exterior has become rigid. Thus the interior pulls on the *outside* surface, *compressing* it, whereas the *interior* develops compensating *tension* and provides a threefold increase in strength. Chemcor[26] is *chemically strengthened* glass that may have three to five times the strength of even physically tempered products. It duplicates the stresses just described by physical means (controlled quenching). This has been done by ion exchange with the outer layer of glass by immersing sodium glass in a molten-lithium salt bath, resulting in a lithium glass on the surface and a sodium glass in the interior. This alkali substitution results in a product with the surface under compression, since the lithium glass has a lower coefficient of expansion and hence shrinks less on cooling than the interior sodium glass. Such tempered glass can be bent and twisted. It does not fracture as easily as ordinary glass, and tests indicate that tableware sold as Centura, manufactured[27] by chemical tempering, is lighter in weight and three times stronger than ordinary tableware.

FOTOFORM[28] glass is a photosensitive glass that is essentially a lithium silicate modified by potassium oxide and aluminum oxide and contains traces of cerium and silver compounds as photosensitive ingredients. Under exposure to ultraviolet light, nuclei are formed by silver sensitized by the cerium, around which an image of lithium metasilicate results upon development by heat treatment to near 600°C. The acid-soluble lithium metasilicate can be removed by 10% hydrofluoric acid. If light exposure is made through a negative (photographically made from a drawing), the final result is a glass reproduction with great accuracy and intricate detail. For example, sheet-glass electrical-circuit boards can be made cheaply and accurately. The process has been termed *chemical machining* of glass.

PHOTOCHROMIC SILICATE GLASSES[29] supplement the just presented Fotoform glass but possess the following unusual properties: *optical darkening* in light from ultraviolet through the visible spectrum; *optical bleaching*, or fading in the dark; and *thermal bleaching* at higher temperatures. These photochromic properties are truly *reversible* and not subject to *fatigue*. Indeed, specimens of this photochromic glass have been exposed to thousands of cycles without any deterioration in performance. The scientific explanation of this photochromic process is the manufacture of a glass in which submicroscopic silver halide particles exist, which react differently from ordinary photographic silver halides when exposed to light. Such tiny particles are about 50 Å in diameter and have a concentration of about 10^{15} per cubic centimeter, and are embedded in rigid, impervious, and chemically inert glass, which ensures that the photolytic "color centers" cannot diffuse away and grow into stable, larger silver particles or react chemically to produce an irreversible decomposition of the silver halide, such as takes place in the photographic process when larger, opaque silver particles are formed. These two processes might be diagramed as follows:

Irreversible photographic process:

$$n\text{AgCl} \xrightarrow{\text{light}} n\text{Ag}^0 + n\text{Cl}^0$$

$$\searrow n(\text{Ag}^0) \text{ latent image silver particle}$$

[26]Olcott, *op. cit.*, p. 1191; Chemically Corseted Glass Bounces More Breaks Less, *Chem. Eng. (N.Y.)*, **69**(23), 110 (1962); Chemical Tempering Strengthens Glass, *Ind. Eng. Chem.*, **54**(11), 9 (1962); Corning Chemical Engineering Achievement, *Chem. Eng.* (*N.Y.*), **70**(23), 233 (1963), and **71**(1), 36 (1964).

[27]Corning Develops Strong Glass, *Chem. Eng. News*, Sept. 24, 1962, p. 23.

[28]Stookey, ECT, 2d suppl. p. 438, 1960.

[29]Armestead and Stookey (Corning Glass Works), Photochromic Silicate Glass Is Sensitized by Silver Halides, *Science*, **144**, 150 (1964).

Reversible photochromic process:

$$AgCl \underset{\text{light 2 or heat}}{\overset{\text{light 1}}{\rightleftharpoons}} Ag^0 + Cl^0$$ or light 1 outdoors, day
light 2 outdoors, night

$$Ag^+ + Cu^+ \underset{\text{light or heat}}{\overset{\text{light}}{\rightleftharpoons}} Ag^0 + Cu^{2+}$$

The primary photolytic reaction is the freeing of an electron from Cl^- to be caught by an Ag^+ ion, resulting in $Ag^0 + Cl^0$. The application of these photochromic glasses is expected to be in sunglasses, windows, and instruments, and processes wherein dynamic control of sunlight is desired.

OPTICAL GLASS Optical glass[30] includes only those glasses with high homogeneity and special composition, which have definite predetermined optical characteristics of such accuracy as to permit their use in scientific instruments. Spectacle glass and ordinary mirror glass are not included. Optical glass should fill certain rigid requirements: (1) Its composition should be such as to ensure the required optical properties. (2) The batch should produce glass of sufficiently low viscosity. (3) The glass should not devitrify, even upon long annealing. (4) It should produce as colorless a product as possible without the use of decolorizing agents. (5) It should be free from bubbles and striae. (6) Its properties with respect to grinding and polishing should be advantageous. (7) It should be capable of withstanding the action of the atmosphere and of maintaining its surface after long usage under all climatic conditions. In melting optical glass, reusable platinum pots are used, resulting in little or no corrosion or contamination and sometimes yielding up to 90% first-class-quality optical glass. Small platinum-lined tank furnaces are also being employed by Corning Glass Works for the continuous melting of optical glass. Essentially, the process consists of melting the batch in a T-shaped furnace heated by electrodes immersed in the liquid glass. After being fined and stirred, the glass is ready for forming optical and ophthalmic ware.

SELECTED REFERENCES

Advances in Glass Technology, Plenum, 1962–1963, 2 vols.
Doremus, R. H.: Glass Science, Wiley-Interscience, 1973.
Loewenstein, K. L.: The Manufacturing Technology of Continuous Glass Fibers, American Elsevier, 1973.
Mazelev, L. Y.: Borate Glasses, Plenum, 1960.
McMillan, P. W.: Glass-Ceramics, Academic, 1964.
Phillips, C. J.: Glass: Its Industrial Applications, Reinhold, 1960.
Rawson, H.: Inorganic Glass-Forming Systems, Academic, 1967.
Reser, M. K. *et al.*: Symposium on Nucleation and Crystallization in Glasses and Melts, American Ceramic Society, 1962.
Volf, M. B.: Technical Glasses, Plenum, 1962.
Wilhelm, E.: Silicate Science, Academic, 1965, 5 vols.

[30]Morey, *op. cit.;* Shand, *op. cit.*, pp. 8, 58, 62.

chapter 12

SALT AND MISCELLANEOUS SODIUM COMPOUNDS

Many sodium salts are of definite industrial necessity. Most of them are derived directly or indirectly from ordinary salt so far as their sodium content is concerned. In one sense, the sodium may be viewed simply as a carrier for the more active anion to which the compound owes its industrial importance. For instance, in sodium sulfide, it is the sulfide part that is the more active. Similarly, this is the case with sodium thiosulfate and sodium silicate. The corresponding potassium salt could be used in most cases; however, sodium salts can be manufactured more cheaply and in sufficient purity to meet industrial demands. In addition to the sodium salts discussed in this chapter, sodium salts are described elsewhere: sodium chlorate and perchlorate in Chap. 14; sodium hypochlorite in Chap. 13, which includes various sodium alkali compounds; sodium phosphates in Chap. 16; sodium nitrate in Chap. 18; and sodium bichromate and sodium borate in Chap. 20.

The U.S. production of sodium salts is presented in Table 12.1.

SODIUM CHLORIDE, OR COMMON SALT

HISTORICAL The salt industry is as old as mankind. Salt has long been an essential part of the human diet. It has served as an object of worship and as a medium of exchange, lumps of salt being used in Tibet and Mongolia for money. Its distribution was employed as a political weapon by ancient governments, and in Oriental countries high taxes were placed on salt. Salt is a vital basic commodity for life but is also a source of many of the chemicals which are now the mainstay of our complex industrial civilization.

USES AND ECONOMICS Sodium chloride is the basic raw material of a great many chemical compounds, such as sodium hydroxide, sodium carbonate, sodium sulfate, hydrochloric acid, sodium phosphates, and sodium chlorate and chlorite, and it is the source of many other compounds through its derivatives.[1] Practically all the chlorine produced in the world is manufactured by the electrolysis of sodium chloride. Salt is used in the regeneration of sodium zeolite water softeners and has many applications in the manufacture of organic chemicals.

The production of chlorine and sodium hydroxide accounts for 45% of salt usage in the United States. Soda ash manufacturing requires 13%; however, this will decline in the coming years as natural soda ash from Trona increases in production and more old ammonia—soda ash plants are closed down, primarily because they pollute streams. Highway use accounts for 20%, and salt for food uses and related activities is about 11%. All other chemical requirements come to about 2% of the total use.

[1]Saltmakers Scent New Success, *Chem. Week*, May 24, 1969, p. 59.

TABLE 12.1 ***U.S. Production of Sodium Salts*** *(In thousands of short tons)*

	1952	*1968*	*1972*
Sodium chloride	19,500	40,400	45,022
Sodium sulfate, anhydrous	208	724	800
Glauber's salt and crude salt cake	840	757	526
Sodium sulfite	43	240	223
Sodium hydrosulfide	20	35	30
Sodium hydrosulfite	25	40	47
Sodium sulfide	24	116	143
Sodium thiosulfate	36	30	24
Sodium silicate	519	633	663

Sources: Minerals Yearbook 1972, Dept. of the Interior, 1974; Chemical Profiles, *Chemical Marketing Reporter;* Current Industrial Reports, Dept. of Commerce.

The world production of salt in 1976 was over 170 million tons. U.S. production represented 28% of world production and U.S. consumption 30.5% of world production. Reserves are large in the principal salt-producing countries, but the grade is unknown.[2] The oceans offer an additional almost inexhaustible supply of salt.[3] After many years of relatively stable prices a surge in prices started early in 1974. Current prices can be found in a number of publications including the *Chemical Marketing Reporter.*

MANUFACTURE Salt occurs throughout the United States. In 1972, Louisiana was the salt-producing state, with 30% of the total output; Texas was second, with 21%. New York and Ohio each produced about 13%, Michigan about 10%, and California, Kansas, and West Virginia each 3%.[4]

Salt is obtained in three different ways, namely, solar evaporation of sea water on the Pacific Coast or from western salt-lake brines, from mining of rock salt, and from well brines.[5] The purity of the salt obtained from the evaporation of salt water is usually about 98 to 99%. Mined salt varies very widely in composition, depending on the locality. Some rock salt, however, runs as high as 99.5% pure. The solution obtained from wells is usually about 98% pure and depends to a great extent on the purity of the water forced down into the well to dissolve the salt from the rock bed. For many purposes, the salt obtained from mines and by direct evaporation of salt solutions is pure enough for use; however, a large portion must be purified to remove such materials as calcium and magnesium chlorides.

Solar evaporation is used extensively in dry climates, where the rate of evaporation depends on the humidity of the air, wind velocity, and amount of solar energy absorbed. San Francisco, southern California, and the Great Salt Lake area are the primary producing areas in the United States. Solar evaporation represents only a small part of the total U.S. production but a large part of the world production.

The mining of rock salt uses methods similar to coal mining. As the salt is removed, large rooms are formed, supported by pillars of salt.

Salt brine is obtained by pumping water into a salt deposit and bringing the salt to the surface for further purification.[6] In shallow wells a double-pipe system of pumping water down one pipe and recovered brine up the other has been used for years. However, in newer wells a series of wells is drilled to the base of the salt cavity, water is injected under pressure into one well, and the brine

[2]Lefond, Handbook of World Salt Resources, Plenum, 1969.
[3]Statistical Abstracts of the United States, Dept. of Commerce, Bureau of the Census (yearly).
[4]Minerals Yearbook 1972, Dept. of the Interior, Bureau of Mines, vol. 1, 1974.
[5]Morton Decides to Mine Brine, *Chem. Eng. News*, Jan. 16, 1967, p. 17.
[6]Kaufmann (ed.), Sodium Chloride; ACS Monograph No. 145, Reinhold 1968.

is removed from a nearby well when the two are interconnected by dissolving the salt between the two wells. This is known as hydraulic fracturing and has reduced the time required to achieve maximum production.

The vacuum pan system, as shown in Fig. 12.1, employing multiple-effect evaporators, is the most common method of producing salt from brine. The type of evaporator may vary, but modern units are generally forced-circulation units constructed of Monel. Evaporation processes are also under development.[7]

SODIUM SULFATE (Salt Cake and Glauber's Salt)

The production of salt cake, or crude sodium sulfate, from natural sources has been growing, whereas the amount of by-product salt cake obtained from rayon, Cellophane, hydrochloric acid, and other sources has been declining. The brines of Searles Lake at Trona, Calif., as well as deposits in Texas and Utah, account for a major share of U.S. natural production.[8] About 20% of the total U.S. consumption is imported. Canada supplies 44% of the imports; a like amount comes from Belgium and Luxembourg and lesser amounts from the Netherlands, West Germany, and Sweden. Searles Lake is estimated to contain 400 million tons and the Great Salt Lake contains about 30 million tons.

USES AND ECONOMICS About 70% of the sodium sulfate consumed in this country is used in the manufacture of kraft pulp. Salt cake, after reduction to sodium sulfide or hydrolysis to caustic, functions as an aid in digesting pulpwood and dissolves the lignin. About 18% goes into the compounding of household detergents, and the remainder into a variety of uses such as glass, stock feeds, dyes, textiles, and medicine.

MANUFACTURE Natural brines account for more than 52% of the sodium sulfate production from 7 producers. The most important by-product source recently has been rayon manufacture (Chap. 35), with 12 producers.

The equations for the production of salt cake from sulfuric acid and salt by the Mannheim process are:

$$NaCl + H_2SO_4 \longrightarrow NaHSO_4 + HCl$$

$$NaHSO_4 + NaCl \longrightarrow Na_2SO_4 + HCl$$

The procedure for commercializing these reactions is described under Hydrochloric Acid in Chap. 20. When the temperature has reached the proper level in the furnace, the finely ground salt and other raw materials are charged. The furnace runs continuously, batch after batch, until it is shut down for periodical cleaning and repair. Most products from the Mannheim furnace go into the manufacture of other materials in the same plant where the furnace is operated.

Another method of making high-grade sodium sulfate, which originated in Europe, is the Hargreaves process. The equation for the reaction is:

$$4NaCl + 2SO_2 + 2H_2O + O_2 \longrightarrow 2Na_2SO_4 + 4HCl$$

Three plants use the Mannheim and Hargreaves processes.

MANUFACTURE OF GLAUBER'S SALT Glauber's salt is made by dissolving salt cake in mother liquor, removing the impurities, clarifying, and crystallizing. The solution is then treated with a paste of chloride of lime, followed by milk of lime in sufficient quantities to neutralize the solution. The precipitated impurities of iron, magnesium, and calcium are allowed to settle, and the clear solution is run to crystallizers by side outlets. The precipitated mud is washed with water, and the water is

[7]Evaporation Process Solves Vacuum-Related Problems, *Chem. Eng. (N.Y.)*, **81**(19), 72 (1974).

[8]Sulfates Are Won from Salt Lake, *Chem. Eng.* (*N.Y.*), **78**(1), 46 (1971); Weisman, Sodium Sulfate from Brine, *Chem. Eng. Prog.*, **60**(11), 47 (1964).

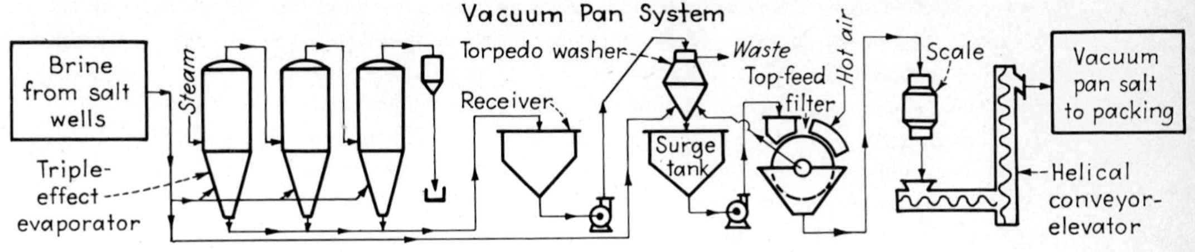

Fig. 12.1 Salt from brine.

used as makeup for the process. When the solution cools to room temperature, the pan is drained and the crystals are collected and centrifuged.

Pure anhydrous sodium sulfate is increasingly in demand by the kraft paper industry. It is made by dehydrating Glauber's salt in a brick-lined rotary furnace, by crystallizing out from a hot, concentrated solution, or by chilling and dehydrating. Figure 12.2 shows the steps in the production of anhydrous sodium sulfate from naturally occurring Glauber's salt.

SODIUM BISULFATE, OR NITER CAKE

Sodium bisulfate is commonly called *niter cake* because it was obtained by the obsolete process of reacting sodium nitrate, or niter, with sulfuric acid:

$$NaNO_3 + H_2SO_4 \longrightarrow NaHSO_4 + HNO_3$$

It may also be formed when salt is moderately heated with sulfuric acid:

$$NaCl + H_2SO_4 \longrightarrow NaHSO_4 + HCl$$

It is an easily handled dry material which reacts like sulfuric acid. Major uses are in the manufacture of acid-type toilet bowl cleaners and for industrial cleaning and metal pickling. Minor uses are in dye baths, carbonizing wool, and various chemical processes.

Total demand is about 75,000 tons/year. There are three major producers, and one producer supplies about 80% of the market.

SODIUM BISULFITE

USES AND ECONOMICS Sodium bisulfite finds industrial use either in solution or as a solid. The solid is of the anhydrous form, and the pure reagent has the formula $NaHSO_3$. The commercial

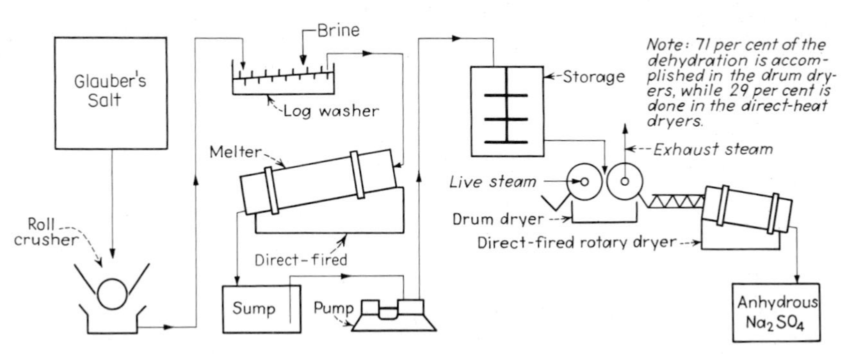

Fig. 12.2 Flowchart for naturally occurring sodium sulfate by the three-stage procedure.

product usually consists almost entirely of $Na_2S_2O_5$ (sodium pyrosulfite), or sodium metabisulfite, which is the dehydrated derivative of two molecules of sodium bisulfite. The solutions may be easily shipped, stored, and handled in 316 stainless steel. Kynar and polyvinyl chloride (PVC) are also suitable construction materials with appropriate temperature limitations.

Major uses are as chemical intermediate, in pharmaceuticals, and in food preservatives. Minor uses are as an antichlor for pulp, paper, and textiles, and in water treatment and pollution control. It is also used in the tanning industry as a reducing agent for chrome solutions and in the photographic and organic chemical industries.

About 45,000 tons ($Na_2S_2O_5$ equivalent) are produced per year by four major producers, one of which provides 40% of the product. In the textile field it is a bleaching agent and is a raw material for the manufacture of hydrosulfite solutions. It is also consumed in the paper, photographic, and organic chemical industries.

MANUFACTURE Sodium bisulfite is produced by passing 7 to 8% sulfur dioxide through mother liquors from previous processes containing in solution small amounts of sodium bisulfite and in suspension a considerable amount of soda ash. The reaction taking place is:

$$2NaHSO_3 + \underline{2Na_2CO_3} + 2H_2O + 4SO_2 \longrightarrow \underline{6NaHSO_3} + 2CO_2$$

The product is obtained as a suspension, which is removed from the solution by centrifuging. The dried product is really the sodium metabisulfite $Na_2S_2O_5$.

SODIUM SULFITE

USES AND ECONOMICS Sodium sulfite is a compound that is very easily oxidized. For this reason, it is employed in many cases where a gentle reducing agent is desired, e.g., to bleach wool and silk, as an antichlor after the bleaching of yarns, textiles, and paper, as a preservative for foodstuffs, and to prevent raw sugar solutions from coloring upon evaporation. It is very widely used in the preparation of photographic developers to prevent the oxidation of hydroquinone and other agents. It has a small application in the field of medicine as an antiseptic and as an antizymotic for internal use. About 80% of the total merchant market is in the paper industry. While merchant capacity is in excess of 250,000 tons/year, the paper mills have twice this capacity for captive use. The demand for boiler feedwater treatment is about 5%. It is used to remove oxygen from the water and thus helps to prevent corrosion and scale formation. There are seven major producers, one of which provides 65% of the product.

MANUFACTURE The most important commercial procedure for preparing this compound involves passing sulfur dioxide into a solution of soda ash until the product has an acid reaction. At this point the solution consists chiefly of sodium bisulfite. This may be converted into sodium sulfite by adding more soda ash to the solution and boiling until all the carbon dioxide has evolved. After the solution has settled, it is concentrated, and crystals of $Na_2SO_3 \cdot 7H_2O$ settle out upon cooling. Another commercial source of this compound in the crude anhydrous form is from the preparation of phenol by the fusion of sodium benzenesulfonate with sodium hydroxide.[9]

SODIUM HYDROSULFITE

USES AND ECONOMICS One of the most important chemicals in the dyeing and printing industries is sodium hydrosulfite. It is a very powerful reducing agent and has a specific action on many dyes, particularly vat dyes, reducing them to the soluble form. This reducing agent is employed for stripping certain dyes from fabrics and for bleaching straws and soaps. Formerly, one

[9]Groggins, Unit Processes in Organic Synthesis, 5th ed., p. 797, McGraw-Hill, 1958.

of the main difficulties with this compound was a lack of stability; however, this has been corrected, and it may be shipped and stored. Dye-house workers call it simply "hydro."

The pulp and paper industry is starting to switch from zinc to sodium hydrosulfite bleaching to meet EPA regulations for zinc discharge from their plants. The demand in the clay industry, where "hydro" is used to remove the red iron color from clay used in manufacturing coated papers and china is expected to grow. There are currently seven producers with nine plants.

MANUFACTURE There are two methods of manufacture, which require zinc dust. In the first process the zinc dust is allowed to reduce sodium bisulfite at room temperature:

$$2NaHSO_3 + H_2SO_3 + Zn \longrightarrow ZnSO_3 + Na_2S_2O_4 + 2H_2O$$

The products of the reaction are treated with milk of lime to neutralize any free acid, and this reduces the solubility of the $ZnSO_3$, which is filtered off. Sodium chloride is added to salt out the $Na_2S_2O_4 \cdot 2H_2O$, which is dehydrated with alcohol and dried. The crystals are stable only in the dry state. The second process consists in treating an aqueous suspension of zinc dust in formaldehyde with sulfur dioxide at 80°C. A double-decomposition reaction is then carried out with soda ash to form the sodium salt. A pure product is obtained by evaporation under vacuum. Another method is to reduce sodium sulfite solution with sodium amalgam [formed in the production of chlorine using a mercury cell (Chap. 13) at a controlled pH in the range 5 to 7].

Sodium hydrosulfite can also be produced from sodium formate formed as a by-product in the manufacture of pentaerythritol. The sodium formate in an alkaline alcoholic solution is reacted with sulfur dioxide:

$$HCOONa + 2SO_2 + NaOH \longrightarrow Na_2S_2O_4 + CO_2 + H_2O$$

Rising costs of zinc metal and adverse pollution regulations will shift manufacture away from the zinc process. Over the long term sodium hydrosulfite will be manufactured by the amalgam or formate process.

SODIUM SULFIDE

USES AND ECONOMICS Sodium sulfide is an inorganic chemical that has attained an important position in the organic chemical industry. It is consumed as a reducing agent in the manufacture of amino compounds, and it enters into the preparation of many dyes. It is also employed extensively in the leather industry as a depilatory. Sodium polysulfide is one of the necessary reactants for making Thiokol synthetic rubber. Other industries where its use is important are the rayon, metallurgical, photographic, and engraving fields.

MANUFACTURE This chemical may be obtained as 30% crystals of sodium sulfide or as 62% flakes. Solutions of it may be shipped in steel. One disadvantageous property of sodium sulfide is its deliquescence. It crystallizes with nine molecules of water, $Na_2S \cdot 9H_2O$. The oldest method of production is the reduction of sodium sulfate with powdered coal in a reverberatory furnace. Another similar process is the reduction of barite in the same manner, leaching, and double decomposition with soda ash. For the reactions involving the reduction of salt cake, very drastic conditions are needed. The reaction must be carried out above 900°C. The reactions are:

$$Na_2SO_4 + 4C \longrightarrow Na_2S + 4CO$$

$$Na_2SO_4 + 4CO \longrightarrow Na_2S + 4CO_2$$

Specially designed small reverberatory furnaces seem to give the least difficulty, because they avoid overheating the charge. The reaction takes place in the liquid phase and is very rapid. Rotary kilns of the type used for calcining limestone have been applied with some success to this reaction.

The reduction of barium sulfate is very similar. Newer plants use nickel or nickel-clad equipment to reduce iron contamination. The substance issuing from the furnace is known as *black ash.*

Another method of production involves the saturation of a caustic soda solution with hydrogen sulfide and the addition of another molecule of sodium hydroxide according to the reactions:

$$H_2S + NaOH \longrightarrow NaHS + H_2O$$

$$NaHS + NaOH \longrightarrow Na_2S + H_2O$$

Heavy-metal sulfides are filtered from the solution of sodium hydrosulfide. The solution is concentrated and then reacted with sodium hydroxide to give a high-purity product. Stainless steel is used as a construction material.

SODIUM HYDROSULFIDE

USES AND ECONOMICS Sodium hydrosulfide (NaHS) finds use in dyestuff processing, the manufacture of rayon and organic chemicals, paper pulping, and the metallurgical industry. There are six merchant producers who operate eight plants. In some instances the same equipment is used to make sodium sulfide. Total demand is about 30,000 tons per year.

MANUFACTURE This chemical may be obtained as a 44 to 46% liquid or as a 72 to 75% flaked material. It is made by treating sodium hydroxide solution with hydrogen sulfide or is obtained as a so-called by-product in the manufacture of carbon disulfide, where a mixture of carbon disulfide and hydrogen sulfide gases is treated with sodium hydroxide to produce sodium hydrosulfide and carbon disulfide. Mercaptans may also be formed with the hydrocarbons, so this product cannot be used in all applications.

SODIUM THIOSULFATE

USES AND ECONOMICS Sodium thiosulfate crystallizes in large, transparent, extremely soluble prisms with five molecules of water. It is a mild reducing agent, like sodium sulfite. It is employed as an antichlor following the bleaching of cellulose products and as a source of sulfur dioxide in the bleaching of wool, oil, and ivory. In photography, which accounts for 90% of current use, it is used to dissolve unaltered silver halogen compounds from negatives and prints, where it is commonly called "hypo." It is a preservative against fermentation in dyeing industries and serves in the preparation of mordants. Other minor uses of sodium thiosulfate are in the reduction of indigo, in the preparation of cinnabar, in the preparation of silvering solutions, and in medicine.

The use of sodium is slowly declining, especially in the photography business, where gains are being made by ammonium thiosulfate. In 1976 three plants produced sodium thiosulfate.

MANUFACTURE Of the several methods for the production of sodium thiosulfate, the most important is from sodium sulfite and free sulfur:

$$Na_2SO_3 + S \longrightarrow Na_2S_2O_3$$

The resulting liquor is concentrated and crystallized. The crystals formed ($Na_2S_2O_3 \cdot 5H_2O$) are immediately packed in airtight containers to prevent efflorescence. A second method of preparation is from sodium sulfide. Sulfur dioxide is passed into a solution of sodium sulfide and sodium carbonate of low concentration (not more than 10% of each):

$$Na_2CO_3 + 2Na_2S + 4SO_2 \longrightarrow 3Na_2S_2O_3 + CO_2$$

The sodium thiosulfate is obtained by evaporation and crystallization.

SODIUM NITRITE

USES AND ECONOMICS From the standpoint of the dye industry, sodium nitrite is a very important chemical. Here it is employed for the diazotization of amines in making azo dyes. It is used in meat processing as a preservative to prevent botulism, and when mixed with sodium nitrate it is employed in metal treatment. It is also used as a bleach for fibers, in photography, and in medicine. Formerly, it was prepared by the reaction

$$NaNO_3 + Pb \longrightarrow NaNO_2 + PbO$$

Sodium nitrite is now manufactured by passing the oxidation product of ammonia into a soda ash solution:

$$Na_2CO_3 + 2NO + \tfrac{1}{2}O_2 \longrightarrow 2NaNO_2 + CO_2$$

Any sodium nitrate also formed may be separated by crystallization.

SODIUM SILICATES[10]

USES AND ECONOMICS Sodium silicates are unique in that the ratio of their constituent parts, Na_2O and SiO_2, may be varied to obtain desired properties. At the present time there are more than 40 varieties of commercial sodium silicates, each with a specific use. Silicates that have a molar ratio of $1Na_2O/1.6SiO_2$ to $1Na_2O/4SiO_2$ are called colloidal silicates. Sodium metasilicate (Na_2SiO_3) has a ratio of 1 mol of Na_2O to 1 mol of SiO_2. A compound with a higher content of Na_2O, with $1\frac{1}{2}$ mol of Na_2O to 1 mol of SiO_2, is called sodium sesquisilicate ($Na_4SiO_4 \cdot Na_2SiO_3 \cdot 11H_2O$). Another compound still higher in Na_2O content is sodium orthosilicate (Na_4SiO_4), which contains 2 mol of Na_2O to 1 mol of SiO_2.

A major use is in the manufacture of silica-based catalysts and silica gels. Other significant uses are in the production of soaps and detergents, pigments, and adhesives, metal cleaning, and water and paper treatment.[11] A large percentage of lower colloidal silicates (water glass) is sold as 32 to 47% solutions employed as adhesives for many kinds of materials, especially paperboard used in corrugated containers. These solutions are also consumed alone or in mixtures with many other materials, e.g., as adhesives for plywoods, wallboard, flooring and metal foils. Sodium metasilicate exists in the hydrated form and is sold as a solid. It is more alkaline than soda ash and is used in metal cleaning and as a strongly alkaline detergent. Sodium sesquisilicate and sodium orthosilicate are employed in the same fields as metasilicates but where a more strongly alkaline product is needed. There are seven producing plants.

MANUFACTURE These silicates are made by fusing sodium carbonate and silica (sand) in a furnace resembling that used for the manufacture of glass, as shown in Fig. 12.3. The reaction takes place at about 1400°C:

$$Na_2CO_3 + nSiO_2 \longrightarrow Na_2O \cdot nSiO_2 + CO_2$$

In the most common commercial silicates n equals 2.0 or 3.2. Intermediate compositions are obtained by mixing, and more alkaline ratios by adding caustic soda or by initially fusing the sand with caustic soda. The product upon cooling forms a clear, light-bluish-green glass. The color is due to impurities of less than 1%, usually iron. If the material is to be sold as a solution, the product is ground and dissolved in water or by steam under pressure when the ratio of silica to alkali is above 2. This procedure is shown in Fig. 12.3 but, when liquids are made directly, the melt flows from the

[10]Vail, Soluble Silicates, ACS Monograph 116, Reinhold, 1952, 2 vols.
[11]Ready to Clean Up in Detergents, *Chem. Week*, May 12, 1971, p. 27.

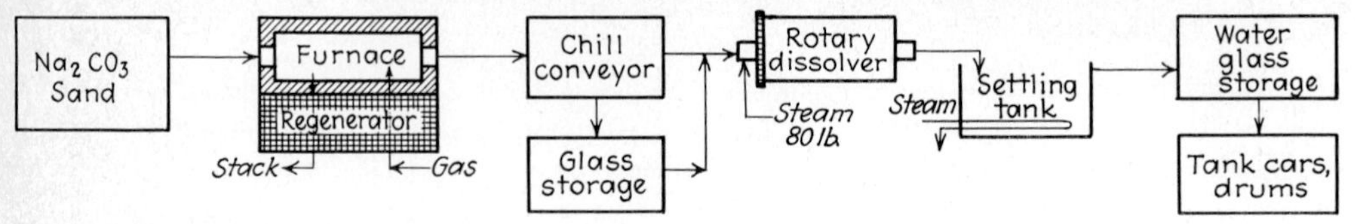

Fig. 12.3 Flowchart for the manufacture of sodium silicate.

furnace and without chilling passes into an open rotary dissolver where it is hydrated in steam and water not under pressure. The resulting solution from either method is usually sold as concentrated as possible. In the case of $1Na_2O/3 \cdot 2SiO_2$, 43°Bé, or 40% solids, is the upper limit, whereas in the case of $1Na_2O/2SiO_2$, 60° Bé, or 54% solids, is the upper limit. Sodium hydroxide can be substituted for soda ash in the production of silicates. In another process[12] sodium hydroxide, sand, water, and recycle sand slurry are reacted under moderate temperature and pressure to form sodium silicate.

SODIUM PEROXIDE

USES AND ECONOMICS Sodium peroxide is a pale-yellow, hygroscopic powder. It absorbs moisture from the air and forms a snow-white hydrate, $Na_2O_2 \cdot 8H_2O$. When added to water, this compound forms sodium hydroxide and oxygen. Its principal use is as a powerful oxidizing agent, and its next most important use is in the bleaching of wool, silk, and fine cotton articles and in chemical synthesis. Sodium peroxide reacts with carbon monoxide to form sodium carbonate, and with carbon dioxide to form carbonate with the liberation of oxygen:

$$Na_2O_2 + CO \longrightarrow Na_2CO_3$$

$$Na_2O_2 + CO_2 \longrightarrow Na_2CO_3 + \tfrac{1}{2}O_2$$

These reactions have led to its use in the regeneration of air in enclosed spaces.

MANUFACTURE Sodium peroxide is manufactured by burning sodium in an excess of air or oxygen.[13] The product has a purity of about 96%. Care must be taken in handling, since when it is placed on paper or wood, the heat of hydration of the material is sufficient to cause ignition. Fused sodium peroxide reacts with platinum, iron, copper, tin, and brass, but not with nickel. Sodium peroxide reacts violently with ether, glycerin, and glacial acetic acid.

SODIUM PERBORATE

Sodium perborate is a mild oxidizing agent used in the medical and dental fields. It is recommended as a mouthwash because of its oxidizing and cleansing effects. Other uses are as an oxidizing and bleaching agent for cosmetics, soaps, and textiles. It is made by mixing solutions of borax, sodium peroxide, and hydrogen peroxide and heating the mixture slightly and allowing it to crystallize.

$$Na_2B_4O_7 + Na_2O_2 + 3H_2O_2 \longrightarrow 4NaBO_3 + 3H_2O$$

SODIUM AMIDE

Sodium amide is another chemical that has found many special uses in organic chemistry. It is a vigorous dehydrating agent and for this reason is used in the synthesis of indigo and in the prepara-

[12] New Liquid-Process Sodium Silicate, *Chem. Eng.* (*N.Y.*), **69**(3), 76 (1962).
[13] Mellor, Comprehensive Treatise on Inorganic and Theoretical Chemistry, vol. II, suppl. II, The Alkali Metals, Wiley, 1961.

tion of pure hydrazine. It is also an intermediate in the preparation of sodium cyanide and finds application as an aminating agent. It is prepared by passing ammonia into metallic sodium at 200 to 300°C:

$$NH_3 + Na \longrightarrow NaNH_2 + \tfrac{1}{2}H_2$$

This compound should be handled with care, since it decomposes explosively in the presence of water. Also, it should not be kept, because disastrous explosions have resulted from handling material that had been stored. It should be consumed as made.

SODIUM CYANIDE AND FERROCYANIDE

Sodium cyanide is not only important in the inorganic and organic chemical fields but also has many metallurgical applications. It is used in treating gold ore, in the case-hardening of steel, in electroplating, in organic reactions, in the preparation of hydrocyanic acid, and in making adiponitrile. Sodium cyanide can be made from sodium amide heated with carbon at 800°C, the carbon as charcoal being thrown into the molten sodium amide. It is first converted to sodium cyanamide:

$$2NaNH_2 + C \xrightarrow{600°C} Na_2NCN + 2H_2$$

$$Na_2NCN + C \xrightarrow{800°C} 2NaCN$$

Another method of manufacture is to melt sodium chloride and calcium cyanamide together in an electric furnace. A commercial method consists of neutralizing hydrocyanic acid with caustic soda:

$$HCONH_2 \longrightarrow HCN + H_2O$$

$$HCN + NaOH \longrightarrow NaCN + H_2O$$

Sodium ferrocyanide is manufactured from a crude sodium cyanide made from calcium cyanamide by fusion with sodium chloride in an electric furnace. The ferrocyanide reaction is carried out in a hot aqueous solution with ferrous sulfate. Soluble sodium and calcium ferrocyanides are formed, and calcium sulfate is precipitated. The calcium ferrocyanide is changed to the sodium salt by the addition of soda ash, precipitating calcium carbonate. The slurry is filtered and washed, and the decahydrate, $Na_4Fe(CN)_6 \cdot 10H_2O$, crystallizes out on cooling.

SELECTED REFERENCES

Borchert, H., and R. O. Muir: Salt Deposits, Van Nostrand, 1964.
Braunstein, J., *et al.*, (ed.): Advances in Molten Salt Chemistry, Plenum, 1973.
Goldberg, E. D. (ed.): The Sea, vol. 5, Marine Chemistry, Wiley-Interscience, 1974.
Kaufman, D. W.: Sodium Chloride: The Production and Properties of Salt and Brine, ACS Monograph 145, Reinhold, 1960.
Kirk-Othmer (ed.): Encyclopedia of Chemical Technology, 2d ed., vol. 18, Wiley-Interscience, 1969.
Lefond, S. J., Handbook of World Salt Resources, Plenum, 1969.
Sconce, J. S.: Chlorine, ACS Monograph 154, Reinhold, 1972.
Tallmadge, J. A., *et al.*: Minerals from Sea Salt, *Ind. Eng. Chem.*, **56,** 44 (1964).

chapter **13**

CHLOR-ALKALI INDUSTRIES: SODA ASH, CAUSTIC SODA, CHLORINE

The manufacture of soda ash, caustic soda, and chlorine is one of the most important heavy chemical industries.[1] These chemicals rank close to sulfuric acid and ammonia in magnitude of dollar value of use. The applications are so diverse that hardly a consumer product is sold that is not dependent at some stage of its manufacture on chlorine and alkalies. The three products are sold almost entirely to industry for the production of soap and detergents, fibers and plastics, glass, petrochemicals, pulp and paper, fertilizers, explosives, solvents, and other chemicals.

HISTORICAL The present process for the manufacture of soda ash is the Solvay process. Before this method was developed, the LeBlanc (1773) process was in universal use. It was based on roasting salt cake with carbon and limestone in a rotary furnace and subsequent leaching of the product with water. The crude product of the reaction was called *black ash.* It was leached cold; whereupon some hydrolysis of sulfides took place. These were changed to carbonate by treatment with the carbon dioxide—containing gases from the black-ash furnace. The resulting sodium carbonate solution was concentrated to obtain crystalline sodium carbonate,[2] which was then dried or calcined. No LeBlanc plant was ever built in the United States, and none is now being operated anywhere in the world. In 1861 Ernest Solvay began developing the ammonia-soda process. At first this method had great difficulty in competing with the older and well-established LeBlanc process, but in a few years the Solvay process reduced the price of soda ash almost one-third. After a struggle, during which the producers of LeBlanc soda sold at a loss for many years, the ammonia-soda process completely displaced the LeBlanc, by 1915.

Solutions of caustic soda have been made ever since soda was used, from about the middle of the eighteenth century. The development of a marketable solid product dates back to the LeBlanc sodamakers (1853), who manufactured it by batchwise lime causticization:

$$Na_2CO_3 + Ca(OH)_2 \longrightarrow 2NaOH + \underline{CaCO_3}$$

depending upon the fact that calcium carbonate is almost insoluble in water.

The electrolytic production of caustic soda was known in the eighteenth century, but it was not until 1890 that caustic was actually produced in this way for industrial consumption. Until shortly before World War I the amount of caustic soda sold as a coproduct of chlorine from the electrolytic

[1]ECT, vol. 1, p. 668, 1963; Hou, Manufacture of Soda, 2d ed., Reinhold, 1942.

[2]A flowchart of this obsolete process is given in Badger and Baker, Inorganic Chemical Technology, 2d ed., p. 137, McGraw-Hill, 1941.

TABLE 13.1 ***United States Production of Soda Ash*** *(In thousands of tons and thousands of dollars)*

Year	*Ammonia-soda process*	*Natural sodium carbonate*	
		Production	*Value*
1948	4,575	289	6,623
1965	4,900	1,400	
1968	4,596	2,043	42,104
1971	4,275	2,865	60,774
1972	4,305	3,128	71,689
1974	3,480	4,048	

Sources: Minerals Yearbook 1972, Dept. of the Interior, 1974; Bureau of Census CIR, Series M28A; Natural Product Leads Surge in Soda Ash, *Chem. Eng. News*, Nov. 24, 1975, p. 10.

TABLE 13.2 ***Estimated Distribution of Soda Ash in the United States*** *(In thousands of tons)*

Consuming industry	*1974*
Glass	3,840
Soap and detergents	377
Other chemicals	1,657
Pulp and paper	452
Water treatment	226
Nonferrous metals	n.a.*
Miscellaneous†	527
Total	7,530

Source: Chem. Eng. News, Feb. 24, 1975, p. 11. *Not available. †Includes exports.

process was almost negligible compared with that made from soda ash by lime causticization. In 1940, however, electrolytic caustic began to exceed lime-soda caustic, and by 1962 the latter had almost disappeared.

The first patent connected with an industrial use of chlorine is dated 1799 (a quarter century after its discovery) and was for bleaching. This demand led to the development of the Weldon and Deacon processes, based on hydrochloric acid, during the last half of the nineteenth century. In the Weldon process, oxidation was conducted with manganese dioxide, and the Deacon process used air with a copper chloride catalyst.

Development of high-capacity, direct-current electrical generating equipment toward the end of the nineteenth century made the causticization process obsolete and, by the middle of the twentieth century, more than 99% of the world's chlorine was produced by the electrolytic process (Table 13.3).

USES AND ECONOMICS *Soda ash* is a lightweight solid, moderately soluble in water, usually containing 99.3% Na_2CO_3. It is sold on the basis of its sodium oxide content, which is generally 58%. Figure 13.1 and Table 13.2 summarize the more important uses. The production of soda ash from natural trona deposits now exceeds that from ammonia-soda. The synthetic process has been plagued by high costs and pollution problems and appears likely to disappear in the United States.

Pure caustic soda is a brittle, white solid which readily absorbs moisture and carbon dioxide from the air. It is sold on the basis of its Na_2O content and usually contains about 76% Na_2O or 98% NaOH. The term caustic soda is widely used because it is corrosive to the skin. Table 13.4 presents its diversified distribution. The traditional uses in the fields of soap, textiles, and petroleum refining, although still substantial, show percentage decreases, and there is increasing use in the field of chemicals, as seen in Table 13.4.

TABLE 13.3 ***United States Production of Caustic Soda and Chlorine*** *(In thousands of tons)*

Year	*Caustic soda (100% NaOH)*	*Chlorine*
1960	4,972	4,637
1965	6,842	6,517
1970	10,141	9,782
1972	10,217	9,819
1973	10,679	10,303
1978 (est. demand)	14,600	14,000

Source: Statistical Abstract of the United States.

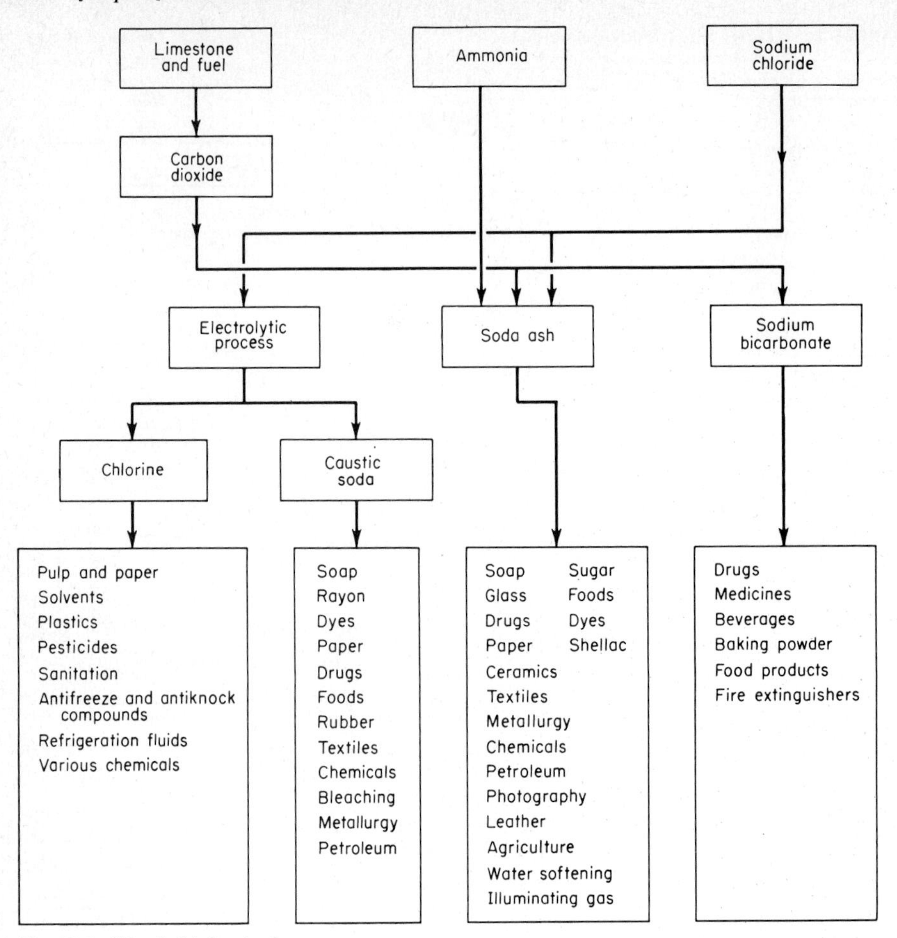

Fig. 13.1 Chlor-alkali industries chart.

TABLE 13.4 ***Distribution of Caustic Soda Use in the United States*** *(In thousands of tons)*

Consuming industry	*1950**	*1960**	*1974*
Chemical processing	850	2,150	5,221
Pulp and paper	185	480	1,816
Rayon and cellophane	525	555	454
Aluminum processing	n.a.*	n.a.*	568
Petroleum, textiles, soaps, foods, and other merchant uses	485	605	1,362
Miscellaneous, including exports	340	1,012	1,021
Total	2,510	4,972	11,350

Source: ECT, vol. 1, 1963, p. 753; 1974 values are estimated. *Not available.

TABLE 13.5 ***United States Chlorine Statistics***
(In thousands of tons)

Year	*Total capacity*	*Total production*
1900	11	1.4
1920	220	61
1940	690	605
1950	2,150	2,084
1960	4,900	4,637
1970	10,330	9,782
1972	10,439	9,819
1974*	10,950	10,990
1975*	12,739	11,644

Source: Statistical Abstract of the United States; *Chem. Eng.* (*N.Y.*), **81**(4), 80 (1974). *Estimated.

Chlorine, originally used almost entirely for bleaching, has increased in importance extremely fast, largely for the synthesis of organic chemicals, in many of which it does not appear in the end product but enters into the intermediate steps. Table 13.5 shows the production of chlorine in the United States.

Table 13.6 reviews the consumption of chlorine. Metallurgical uses include not only the beneficiation of ores and fluxing, but also the actual extraction of copper, lead, zinc, nickel, gold, and platinum. The ultimate end products are as varied as different items of clothing, jewelry, paints, foods, paper, tires, and toys.

MANUFACTURE OF SODA ASH

Soda ash is manufactured by the Solvay, or ammonia-soda, process, although there has been a rapid increase in the amount of "natural soda" on the market since World War II. This has developed principally from the opening up of mines for the exploitation of the sodium sesquicarbonate (trona) deposits in the Green River Basin in southwestern Wyoming. There has also been some increase in the carbonation of natural alkali brines at both Searles and Owens Lakes in California. The production of natural soda ash in the United States equaled that from the ammonia-soda process in 1973 and has forged rapidly ahead. The disposal of calcium chloride and brines has posed a severe problem for Solvay plants, many of which have become obsolescent. In countries not possessing large reserves of natural soda ash, ammonia-soda plants will continue to be built, but in the United States natural soda appears to have won the battle.

TABLE 13.6 ***Distribution of Chlorine Consumption in the United States*** *(In thousands of tons)*

Category	*1940*	*1950*	*1960*	*1974**
Organic chemicals (including solvents)	274	1,287	2,828	5,211
Plastics, resins	20	128	497	2,106
Pulp and paper	170	315	740	1,663
Inorganic chemicals	46	208	385	1,221
Water supply and sewage treatment	50	90	155	777
Other	35	40	30	122
Total consumption	595	2,068	4,635	11,100

Source: ECT, vol. 1, 1963, p. 705. *Estimated.

RAW MATERIALS The raw materials for the Solvay process are salt, limestone, and coke or gas. Ammonia is a cyclic reagent in the process—it enters into the reactions and is recovered, and only a small amount is lost. The most suitable limestone is hard and strong, has low concentrations of impurities (silica is the least desirable), and is graded to reasonably uniform coarse size (4 to 8 in.). Metallurgical coke burns the limestone and also furnishes additional CO_2. Common salt enters the reactions as a natural or artificial saturated brine, generally from rock salt.

REACTIONS AND ENERGY CHANGES The Solvay process is carried out by the following principal reactions:

$$CaCO_3(c) \longrightarrow CaO(c) + CO_2(g) \qquad \Delta H = +43.4 \text{ kcal} \quad (1)$$

$$C(amorph) + O_2(g) \longrightarrow CO_2(g) \qquad \Delta H = -96.5 \text{ kcal} \quad (2)$$

$$CaO(s) + H_2O(l) \longrightarrow Ca(OH)_2(c) \qquad \Delta H = -15.9 \text{ kcal} \quad (3)$$

$$NH_3(g) + H_2O(l) \longrightarrow NH_4OH(aq) \qquad \Delta H = -8.4 \text{ kcal} \quad (4)$$

$$2NH_4OH(aq) + CO_2(g) \longrightarrow (NH_4)_2CO_3(aq) + H_2O(l) \qquad \Delta H = -22.1 \text{ kcal} \quad (5)$$

$$(NH_4)_2CO_3 + CO_2 + H_2O \longrightarrow 2NH_4HCO_3 \quad (6)$$

$$NH_4HCO_3 + NaCl \longrightarrow NH_4Cl + NaHCO_3 \quad (7)$$

$$2NaHCO_3(c) \longrightarrow Na_2CO_3(c) + CO_2(g) + H_2O(g) \qquad \Delta H = +30.7 \text{ kcal} \quad (8)$$

$$2NH_4Cl(aq) + Ca(OH)_2(c) \longrightarrow 2NH_3(g) + CaCl_2(aq) + 2H_2O(l) \qquad \Delta H = +10.7 \text{ kcal} \quad (9)$$

The overall equation for the entire process can be written

$$CaCO_3 + 2NaCl \longrightarrow Na_2CO_3 + CaCl_2$$

This process cannot take place directly, but is achieved in a number of steps represented by equations (1) to (9).

The reactions are industrialized by the following sequences of *unit operations* and *chemical conversions*, as exemplified in the flowchart in Fig. 13.2.

Preparation and purification of the salt brine
Mining and solution of salt or pumping of brine (Op).
Purification of brine (Ch and Op). (Not shown in Fig. 13.2.)
Ammoniation of brine
Solution of NH_3 in the purified brine (Op and Ch).
Solution of weak CO_2 in the purified brine (Op and Ch).
Carbonation of ammoniated brine
Formation of ammonium carbonate (Ch).
Formation of ammonium bicarbonate (Ch).
Formation and controlled precipitation of sodium bicarbonate (Ch).
Filtration and washing of the sodium bicarbonate (Op).
Burning of limestone with coke to form CO_2 and CaO (Ch).
Compressing and cooling of lime kiln or lean CO_2 and piping to carbonator (Op).
Calcination of sodium bicarbonate
Conveying and charging of moist sodium bicarbonate to calciner (Op).
Calcination of sodium bicarbonate (Ch and Op).
Cooling, screening, and bagging of soda ash (for sale) (Op).
Cooling, purifying, compressing, and piping of rich CO_2 gas to carbonator (Op).
Recovery of ammonia
Slaking of quicklime (Ch).

Treatment of sodium bicarbonate mother liquors with steam and $Ca(OH)_2$ in strong NH_3 liquor still to recover NH_3 (Ch and Op).

Detailed manufacturing procedures for the ammonia-soda process are given in ECT.

MANUFACTURE OF SODIUM BICARBONATE

Sodium bicarbonate, or baking soda, is not made by refining the crude sodium bicarbonate obtained from the filters in the Solvay process. There are several reasons for this. (1) There is great difficulty in completely drying this carbonate. (2) The value of the ammonia in this crude bicarbonate would be lost. (3) Even a small amount of ammonia gives the bicarbonate an odor that renders it unfit for many uses. (4) The bicarbonate made in this way contains a great many other impurities.

To make sodium bicarbonate, a saturated solution of soda ash is first prepared and then run into the top of a column similar to the carbonating tower used in soda-ash manufacture. Carbon dioxide is sent in from compressors at the bottom of the tower, which is maintained at about 40°C. The suspension of bicarbonate formed is removed from the bottom of the tower, filtered, and washed on a rotary-drum filter. After centrifugation, the material is dried on a continuous-belt conveyor at 70°C. Bicarbonate from this process is about 99.9% pure. Sodium bicarbonate finds large usage in the manufacture of baking powder, carbonated water, and leather goods, and in fire extinguishers. Annually, about 120,000 tons is manufactured in the United States.

MISCELLANEOUS ALKALIES

Various alkalies of different strengths are consumed commercially, according to the contained amounts of NaOH, Na_2CO_3, or $NaHCO_3$. Some of these are mechanical mixtures, such as *causticized ash* (soda ash with 10 to 50% caustic) for bottle washing and metal cleaning, and *modified sodas*

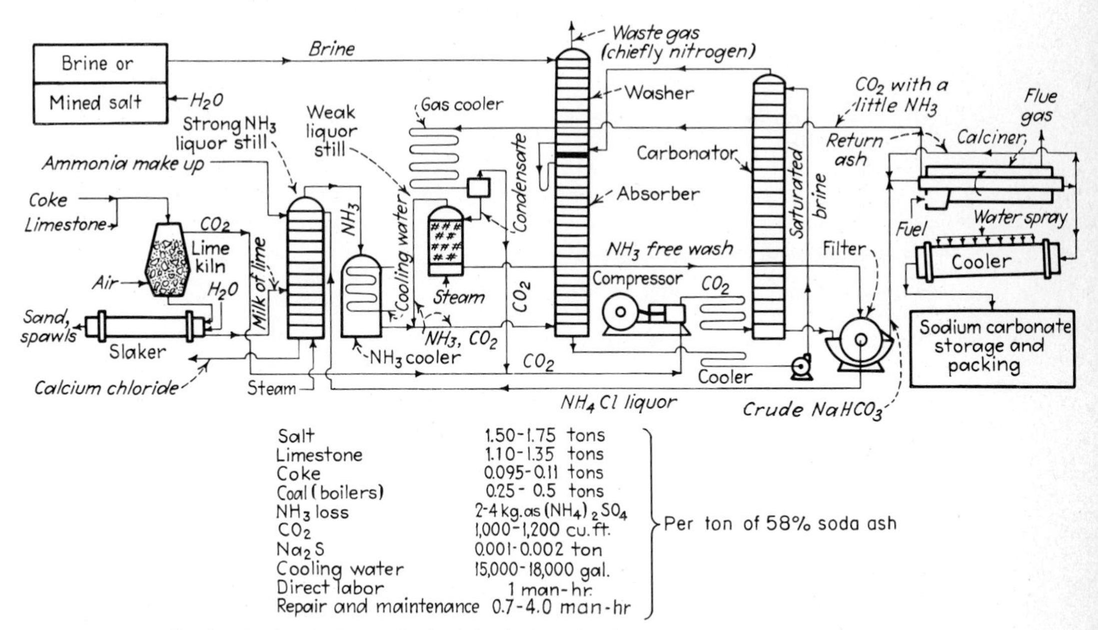

Fig. 13.2 Flowchart for the manufacture of soda ash by the ammonia-soda process.

(soda ash with 25 to 75% $NaHCO_3$) for mild-alkali demands, as in the tanning industry. *Sodium sesquicarbonate*, or the natural mineral trona, has the composition $Na_2CO_3 \cdot NaHCO_3 \cdot 2H_2O$ and is very stable. It finds use in wool scouring and in laundering. *Sal soda* ($Na_2CO_3 \cdot 10H_2O$) is also known as *washing soda* or *soda crystals.*

MANUFACTURE OF CHLORINE AND CAUSTIC SODA

Chlorine and caustic soda are produced almost entirely by electrolytic methods from fused chlorides or aqueous solutions of alkali metal chlorides. In the electrolysis of brines, chlorine is produced at the anode, and hydrogen, together with sodium or potassium hydroxide, at the cathode. Since the anode and cathode products must be kept separate, many ingenious cell designs have been invented and commercialized. However, all these designs have been either variations on a type of diaphragm or on a mercury intermediate electrode.

DIAPHRAGM CELLS[3] Diaphragm cells produced approximately three-fourths of U.S. caustic. In the operation of a typical diaphragm cell, sodium or potassium chloride, nearly saturated and at approximately 60 to 70°C, is fed into the anolyte, which flows through the diaphragm into the catholyte, where caustic is formed. Flow is continuous through the diaphragm to the catholyte caused by a differential head.

REACTIONS AND ENERGY CHANGES *Decomposition voltage and voltage efficiency.* The energy consumed in the electrolysis of the brine is the product of the current flowing and the potential of the cell. The *theoretical*, or minimum, *voltage* required for the process may be derived from the Gibbs-Helmholtz equation, which expresses the relation between the electric energy and the heat of reaction of a system.[4]

$$E = \frac{-J\,\Delta H}{nF} + \frac{T\,dE}{dT}$$

where E = theoretical decomposition voltage
ΔH = enthalpy change of reaction, cal
J = electrical equivalent of heat, or J/cal (4,182)
T = absolute temperature
F = Faraday constant, coulombs/g-equiv (96,500)
n = number of equivalents involved

The heat of reaction for the electrolysis of salt may be found from the heats of formation of the components of the overall reaction or from the negative of these values, the change in the heat contents of the system. The overall reaction is

$$NaCl(aq) + H_2O(l) \longrightarrow NaOH(aq) + \tfrac{1}{2}H_2(g) + \tfrac{1}{2}Cl_2(g)$$

This may be broken down to the following reactions for formation:

$$Na(s) + \tfrac{1}{2}Cl_2(g) \longrightarrow NaCl(aq) \qquad \Delta H = -97.1 \text{ kcal}$$

$$H_2(g) + \tfrac{1}{2}O_2(g) \longrightarrow H_2O(l) \qquad \Delta H = -68.4 \text{ kcal}$$

$$Na(s) + \tfrac{1}{2}O_2(g) + \tfrac{1}{2}H_2(g) \longrightarrow NaOH(aq) \qquad \Delta H = -112.1 \text{ kcal}$$

The net ΔH for the overall reaction results from

$$+97.1 + 68.4 - 112.1 = +53.4 \text{ kcal}$$

[3]Sconce (ed.), Chlorine, chap. 5, ACS Monograph 154, 1962; Sommers, The Chlor-Alkali Industry, *Chem. Eng. Prog.*, **61**(3), 94 (1965); Outlook for Chlorine-Caustic Production, *Chem. Eng. Prog.*, **70**(3), 59 (1974).

[4]Perry, 5th ed., p. 4–70.

When this value of ΔH is substituted in the Gibbs-Helmholtz equation and the change in voltage with temperature is neglected, the value of E is found to be 2.31 V. The omission of $T\,dE/dT$ involves an error of less than 10% for most cells.

The ratio of this theoretical voltage to that actually used is the *voltage efficiency* of the cell. Voltage efficiencies range from 60 to 75%. According to Faraday's law, 96,500 C of electricity passing through a cell produce 1 g-equiv of chemical reaction at each electrode. Because of unavoidable side reactions, cells usually require more than this amount. The ratio of the theoretical to the actual current consumed is defined as the *current efficiency*. Current efficiencies run 95 to 97% and, unless otherwise specified, are understood to be *cathode* current efficiencies. The current divided by the area in square inches upon which the current in amperes acts is known as the *current density*. A high value is desirable in cases where the product formed is subject to decomposition. The product of voltage efficiency and current efficiency is the *energy efficiency* of the cell. Another consideration is the *decomposition efficiency*, which is the ratio of equivalents produced in the cell to equivalents charged. In the usual commercial cell this decomposition efficiency is about 60 to 65%. When the cell is operated to attain a higher decomposition efficiency, the natural flow of brine through the cell decreases with the migration of OH^- ions back to anode and formation of hypochlorite, which means loss of caustic and chlorine. Also, when OH^- ions reach the anode, the following reaction occurs:

$$2OH^- \longrightarrow H_2O + \tfrac{1}{2}O_2 + 2e$$

The oxygen formed reacts with the graphite of the anodes, causing decreased anode life. In cells with metal anodes, the oxygen does not react.

TYPES OF CELLS Diaphragm cells contain a diaphragm to separate the anode from the cathode. This allows ions to pass through by electrical migration but reduces the diffusion of products. Diaphragms permit the construction of compact cells of lowered resistance, because the electrodes can be placed close together. The diaphragms become clogged with use, as indicated by higher voltage and higher hydrostatic pressure on the brine feed. They must be replaced regularly. The diaphragm permits a *flow of brine* from anode to cathode and thus greatly lessens or prevents side reactions (e.g., sodium hypochlorite formation). Cells with metal cathodes (titanium coated with rare earth oxides, platinum or noble metals, or oxides) rarely develop clogged diaphragms and operate for 12 to 24 months without requiring diaphragm replacement. It is expected that Du Pont's Nafion (a perfluorosulfonic acid polymer), and similar materials used as membranes to replace asbestos, will increase service life. Permionic membranes passing NaOH while retaining NaCl presently increase the purity of diaphragm cell caustic, eliminating a purification step to remove chlorine.[5]

Mercury cells produce purer NaOH, but a small loss of mercury to the environment presents extreme problems. Originally it was believed that this loss was unimportant, but marine life concentrates the mercury biologically, resulting in fish containing lethal amounts of methyl mercury. Ingestion of contaminated fish led to development of the so-called Minimata disease, resulting in deaths and affliction. Japan outlawed mercury cell use after 1975, and the construction of mercury cells in the United States came to an abrupt halt. Careful process control,[6] combined with treatment of water and air effluent may make it possible for mercury cell plants to meet environmental standards and survive, but most companies are hesitant to erect new units. Greatly improved diaphragm cell technology, particularly the replacement of graphite anodes with dimensionally stable titanium anodes coated with a catalytic coating, has swung interest to large (200 kA) diaphragm cells. The use of plastic has simplified cell construction, minimized and simplified maintenance, and allowed outdoor cell rooms in some cases, thus reducing costs.

[5]Dahl, Chlor-Alkali Cell Features New Ion-Exchange Membrane, *Chem. Eng.* (*N.Y.*), **84**(17), 60 (1975).
[6]Cell Systems Keep Mercury from Atmosphere, *Chem. Eng. News*, Feb. 14, 1972, p. 14.

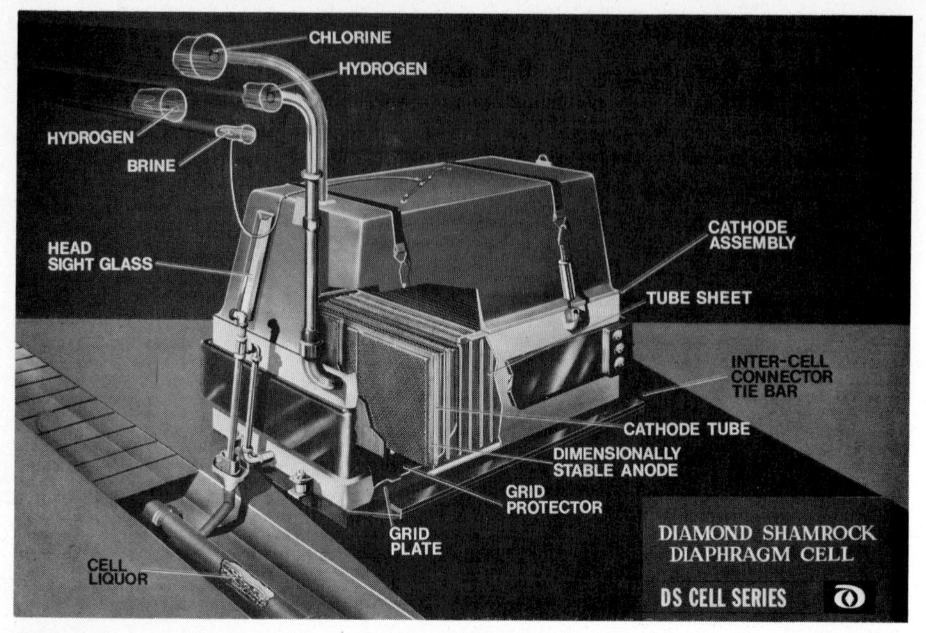

Fig. 13.3 Diamond Shamrock diaphragm cell. The DS cell series is made in three sizes. The largest is the DS-85 which handles a current up to 150 kA and produces 5.05 short tons of chlorine per day. This cell is 112 in. long, 60 in. wide, and 81 in. high. It has dimensionally stable anodes. (*Diamond Shamrock Corp.*)

Diaphragm cell liquor requires concentration (from about 12% NaOH containing 16% NaCl) and chloride removal. Increasing fuel costs have made quadruple-effect evaporators economically desirable,[7] and ammonia extraction can reduce the NaCl level to 200 to 300 ppm—rayon-grade specifications. Cells enjoying particular favor at this time include those produced by Diamond-Shamrock, PPG-DeNora, Hooker, Dow, Olin, Solvay, Hoechst-Uhde, Kureha, and Toyo Soda. The differences in total production cost among the several proprietary cells, monopolar and bipolar systems, and mercury and diaphragm cells are small. Large diaphragm cells seem to be the current trend.

The *Dow cell* (Dow Chemical Co.) has a bipolar design which usually results in the erection of blocks of 50 cells, where the butting frames are pressed together to form a tight block operating as a single unit. Each such unit is connected electrically to the next cell unit within the frames constituting the block. Chlorine and hydrogen gases are each collected in inverted troughs (concrete headers on top of the cells) and caustic at the side through a trap.

UNIT OPERATIONS AND CHEMICAL CONVERSIONS The main steps in a typical diaphragm-cell flowchart, as depicted in Fig. 13.5 may be represented by the following *unit operations* and *chemical* conversions:

Brine purification. To make a purer caustic soda and to lessen clogging of the cell diaphragm with a consequent voltage increase, purification of the NaCl solution of calcium, iron, and magnesium compounds is practiced, using soda ash with some caustic soda. Sometimes sulfates are removed with $BaCl_2$ or the hot brine is treated with hydroxyl and carbonate ions.[8] The clear brine is neutral-

[7]New Chlorine Plants Watch Their Watts, *Chem. Week.*, Nov. 27, 1974, p. 45.

[8]Shearon *et al.*, Modern Production of Chlorine and Caustic Soda, *Ind. Eng. Chem.*, **40**, 2002 (1948); Hightower, Chlorine-caustic Soda, *Chem. Eng.*, **55**(12), 112 (1948) (pictured flowchart); Integrated Alkali Industry, *Chem. Eng.* (*N.Y.*), **53**(2), 172 (1946) (pictured flowchart).

Fig. 13.4 Typical Diamond-Shamrock cell installation. (*Diamond Shamrock Corp.*)

ized with hydrochloric acid (Ch). This brine is stored, heated, and fed to the cells through a float feed system designed to maintain a constant level in the anode compartments (Op).

Brine electrolysis. Typical cells employed were described in the preceding section. Each cell usually requires 3.0 to 4.5 V; consequently, a number of them are put in series to increase the voltage of a given group (Ch).

Evaporation and salt separation. The decomposition efficiency of the cells being in the range of only 50%, about half of the NaCl charged is in the weak caustic and is recovered by reason of its low solubility in caustic soda solutions after concentration. Hence the weak, about 10 or 12%, NaOH solution is evaporated to about 50% NaOH in a double- or triple-effect evaporator with salt separators and then passes through a settler and washing filter. The salt so recovered is again made into charging brine. In the evaporators, nickel tubes are generally used to lessen iron contamination. Since the caustic soda liquor from the evaporator contains 50% NaOH, it will dissolve only about 1% NaCl and other sodium salts after cooling. This liquor may be sold, after thorough settling, as liquid caustic soda in tank cars or drums.

Final evaporation. Either the cooled and settled 50% caustic or a specially purified caustic may be concentrated in a single-effect final or high evaporator to 70 or 75% NaOH, using steam at 75 to 100 psig. This very strong caustic must be handled in steam-jacketed pipes to prevent solidification. It is run to the finishing pots. Another method of dehydrating 50% caustic utilizes the precipitation of NaOH monohydrate. This monohydrate contains less water than the original solution. Precipitation is accomplished by the addition of ammonia to the 50% solution, and this also purifies the caustic. If the 50% caustic is treated with anhydrous ammonia,[9] particularly in a countercurrent system, free-flowing anhydrous crystals separate from the resulting aqueous ammonia. Naturally, this procedure should be carried out in pressure vessels (Ch).

Finishing of caustic in pots. Although 50% caustic was at one time finished in special close-grained, cast-iron, direct-fired pots, the heat efficiency is so low that good practice now handles only 70 or 75% NaOH in this fashion. The final temperature is 500 to 600°C and boils off all but about

[9]Muskat, Method for Producing Anhydrous Caustic, U.S. Pat. 2,196,593 (1940).

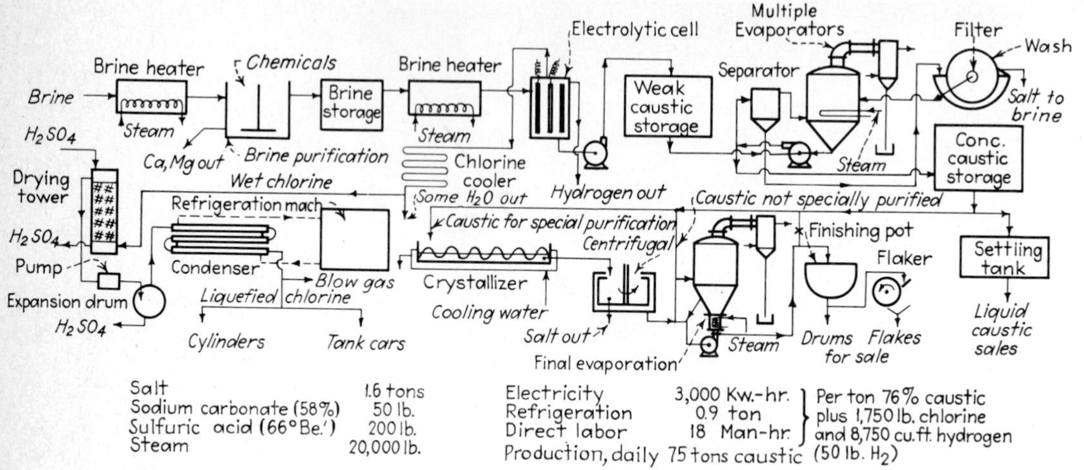

Fig. 13.5 Flowchart for diaphragm caustic soda and chlorine cell.

1% or less of the water. These pots are now being replaced by Dowtherm heated evaporators for caustic evaporation above 50%. The hot anhydrous caustic is treated with sulfur to precipitate iron and settled. The product is pumped out by lowering into the molten caustic a centrifugal pump that discharges the caustic into thin steel drums holding about 700 lb, or into the flaking machine (Op).

Special purification of caustic. Some of the troublesome impurities in 50% caustic are colloidal iron, NaCl, and $NaClO_3$. The iron is often removed by treating the caustic with 1% by weight of 300-mesh calcium carbonate and filtering the resulting mixture through a Vallez filter on a calcium carbonate precoat. The chloride and chlorate may be removed by allowing the 50% caustic to drop through a column of 50% aqueous ammonia solution. This treatment produces caustic almost as free of chlorides and chlorates as that made by the mercury process. To reduce the salt content of the caustic necessary for certain uses, it is cooled to 20°C in equipment such as that outlined in Fig. 13.5. Another crystallization method used industrially, however, involves the actual separation of the compound $NaOH \cdot 3\frac{1}{2}H_2O$ or $NaOH \cdot 2H_2O$, leaving the NaCl in the mother liquor. Another procedure reduces the salt content of the caustic soda solution by formation of the slightly soluble complex salt $NaCl \cdot Na_2SO_4 \cdot NaOH$.[10]

A standard process for the continuous extraction of NaCl and $NaClO_3$ in 50% caustic solution is countercurrent extraction in vertical columns with 70 to 95% ammonia.[11] It should be noted that mercury cells furnish caustic soda free from salt. These purification or manufacturing methods give *high-grade* caustic with less than 1% impurities (anhydrous basis). The *standard grade* contains $2\frac{1}{2}$ to 3% impurities (anhydrous basis) (Ch).

Chlorine drying. The hot chlorine evolved from the anode carries much water vapor. It is first cooled to condense most of this vapor and then dried in a sulfuric acid scrubber or tower as shown in Fig. 13.5. Up to the sulfuric acid tower, the wet chlorine should be handled in polyester, polyvinyl chloride, or a similar resistant material; after drying, iron or steel can be employed (Op and Ch).

Chlorine compression and liquefaction.[12] The dried chlorine is compressed to 35 psi, or sometimes as much as 80 psi. For lower pressures the usual type of compressor employed is the liquid-piston rotary blower[13] constructed of iron, with concentrated sulfuric acid as the sealing liquid. For

[10]Hubel and Sweetland, Removal of Salt from Caustic Soda Produced by Diaphragm Cells, *Can. Chem. Met.*, **17**, 52 (1933).
[11]Twichaus and Ehlers, Caustic Purification by Liquid-Liquid Extraction, *Chem. Ind.* (*N.Y.*), **63**, 230 (1948).
[12]Eichenhofer and Fedoroff, Chlorine Liquefaction, *Chem. Eng.* (*N.Y.*), **58**(12), 142 (1951).
[13]Perry, Fig. 6-47. Centrifugal Compressors in Chlorine Service, *Chem. Eng. Prog.*, **70**(3), 69 (1974).

larger capacities and higher pressures, centrifugal and nonlubricated reciprocating compressors are chosen. Carbon piston rings or carbon cylinder liners are usually used in the reciprocating compressors, other parts being cast iron. The heat of compression is removed and the gas condensed. The liquid chlorine is stored in small cylinders, ton cylinders, pipeline, or the 55-ton tank cars that are shipped to large consumers. Barges carrying 600 or 1,100 tons are also employed. There is always some residue, or "blow gas," made up of an equilibrium mixture of chlorine and air. The blow gas is used to make chlorine derivatives, either organic or inorganic, especially bleaching powder (Op).

Hydrogen disposal. The hydrogen is frequently made into other compounds such as hydrochloric acid or ammonia or is employed for the hydrogenation of organic compounds (Ch).

MERCURY CELLS[14] The method of producing chlorine and caustic soda in an electrolytic cell with a mercury cathode was discovered simultaneously by H. Y. Castner, an American, and K. Kellner, an Austrian, in 1892. The Mathieson Alkali Works (now Olin Corp.) operated this cell, or modifications of it, from 1894 to recent years at Niagara Falls. In the mercury cell, continuously fed brine is partly decomposed in one compartment (called the electrolyzer) between a graphite anode and a moving mercury cathode, forming chlorine gas at the anode and sodium amalgam at the cathode. The reactions are:

Anodic half-reaction:	$2Cl^-(aq) \longrightarrow Cl_2(aq) + 2e^-$
	$Cl_2(aq) \longrightarrow Cl_2(g)$
Cathodic half-reaction:	$2Na^+(aq) + Hg + 2e^- \longrightarrow 2Na(Hg)$
Overall cell reaction:	$2NaCl(aq) + Hg \longrightarrow Cl_2(g) + 2Na(Hg)$

The average electrolysis temperature is about 65°C, and the average NaCl concentration is 290 g/l. The sodium amalgam flows continuously to a second compartment (called the amalgam decomposer, secondary cell, or denuder) where it becomes the anode to a short-circuited iron or graphite cathode in an electrolyte of NaOH solution. Purified water is fed to the cell countercurrent to the sodium amalgam; hydrogen gas is formed, and the NaOH increases to 40 or 50%. (The eutectic composition of 73% NaOH can be obtained if the temperature is kept above 100°C.) The reactions are:[15]

Anodic half-reaction:	$2Na(Hg) \longrightarrow 2Na^+ + Hg + 2e^-$
Cathodic half-reaction:	$2H_2O + 2e^- \longrightarrow 2OH^- + H_2(g)$
Overall decomposition reaction:	$2Na(Hg) + 2H_2O \longrightarrow 2NaOH(aq) + H_2(g) + Hg$

Most modern cells, such as the DeNora cell (Figs. 13.6 and 13.7), are constructed with a flat-bottomed steel trough over which the mercury cathode flows uniformly. Parts of the trough not in contact with mercury are provided with a corrosion-resistive coating. The steel bottom of the cell, which in operation is electrically connected with the cathode bus bars, is covered with a sheet of mercury. The anodes are horizontal graphite plates (L in Fig. 13.6) that hang on insulated rods extending through the cover (M). The anodes are parallel and close to the mercury-brine interface and are perforated or grooved to facilitate release of chlorine in the cover. The vertical amalgam decomposer, or denuder, shown in Fig. 13.7, is characteristic of the DeNora[16] cell. This figure indicates the brine, chlorine, and hydrogen flows. No evaporation is necessary. The weak brine or anolyte is dechlorinated and resaturated with solid salt, purified, and returned to the cell.

The choice between the diaphragm and the mercury cell depends on whether solid salt or strong brine is cheaper and the purity of the caustic desired. The mercury cell produces superior-quality

[14]MacMullin, 1962, chap. 6 in Sconce (ed.), ACS Monograph 154; ECT, 2d ed., vol. 1, 1963, p. 686; MacMullin, *Chem. Eng. Prog.*, **46,** 440 (1950).

[15]The Mercury Cathode Chlor-Alkali Process, Hampel, Electrochemical Encyclopedia, Reinhold, 1964.

[16]DeNora and Gallone, The Mercury Cathode Chlor-Alkali Process, and Hampel, Electrochemical Encyclopedia, Reinhold, 1964.

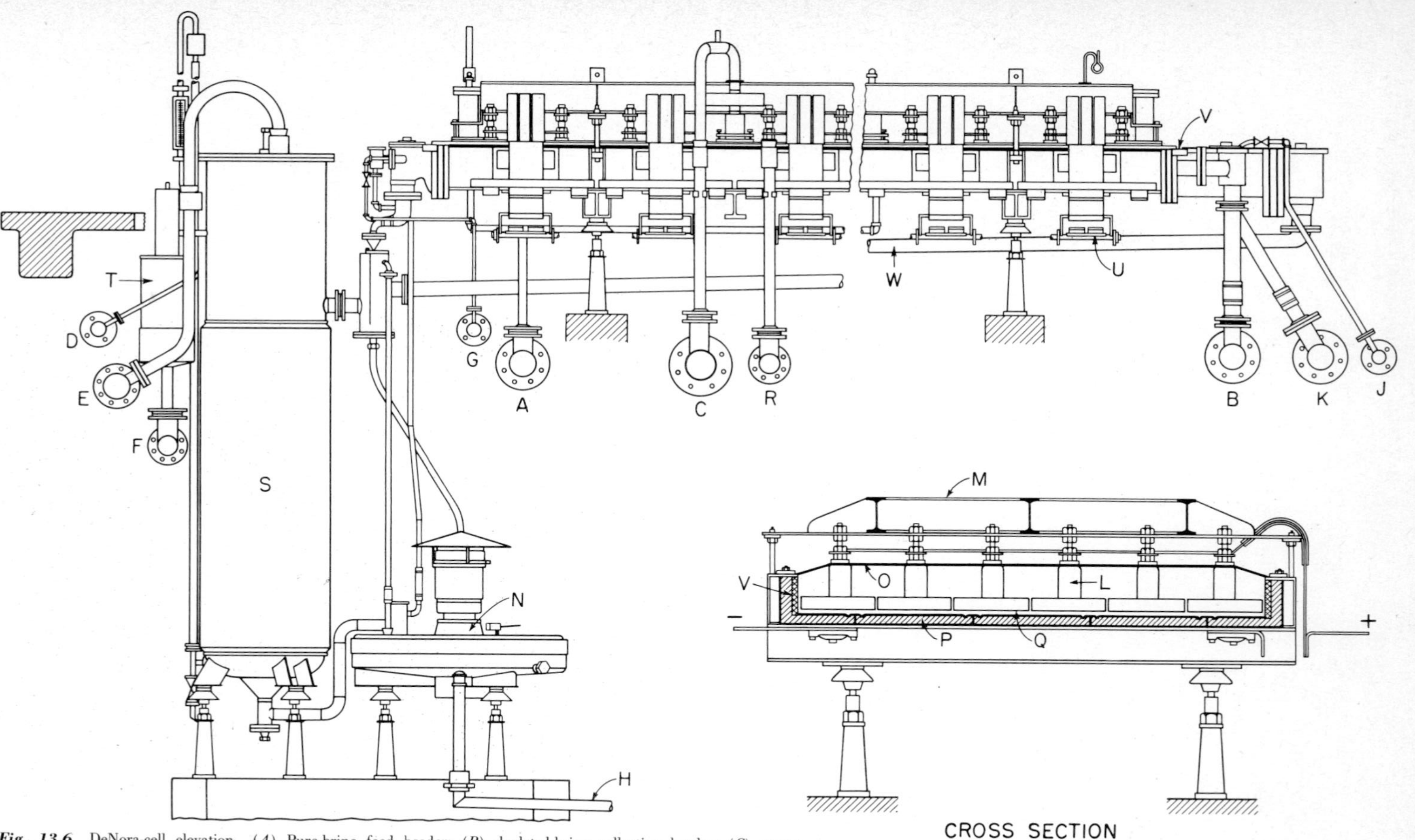

Fig. 13.6 DeNora-cell elevation. (*A*) Pure-brine feed header; (*B*) depleted-brine collection header; (*C*) strong chlorine header; (*D*) pure-water feed header to decomposer; (*E*) hydrogen-collection header; (*F*) caustic-collection header; (*G*) pure-water feed header to inlet end; (*H*) to mercury trap and sewer; (*L*) anode post-and-plate assembly; (*M*) anode support structure; (*N*) mercury pump and sump; (*O*) rubber cover; (*P*) concrete grout; (*G*) corrosion-resistant membrane; (*R*) dilute-chlorine header; (*S*) decomposer; (*T*) current breaker; (*U*) cell-shorting switch; (*V*) stone side facing; (*W*) amalgam return; (*Y*) quick-flushing device. (*DeNora.*)

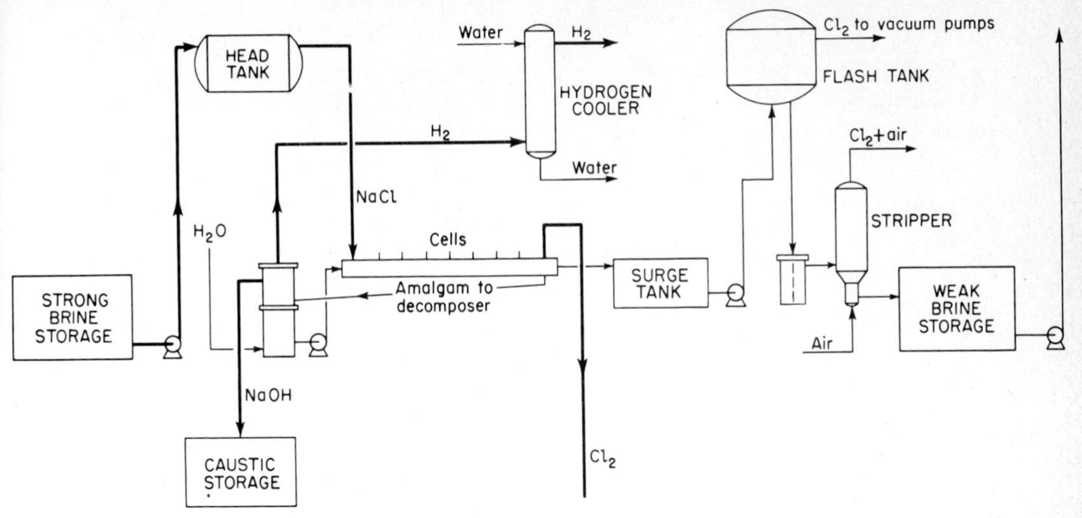

Fig. 13.7 DeNora-cell operating arrangement. Compare Fig. 13.5 for diaphragm cell. (*DeNova.*)

50% NaOH directly from the cell assembly through the decomposition of the mercury-sodium amalgam. The diaphragm cell produces a weak caustic liquor containing about 10% caustic and 15% unconverted salt. On the other hand, the mercury cell requires about 15% more energy per ton of product than the diaphragm cell. The predominance of the diaphragm cell in this country is largely due to the availability of natural brines. In other countries, such as Germany and Italy, however, the mercury process is responsible for much of the total chlorine production.

Other processes for making chlorine include the Downs cell (for making sodium, Chap. 14), hydrogen chloride decomposition (Chap. 14), the Solvay nitrosyl chloride process, and the Allied Chemical process at Hopewell, Va., where common salt is treated with nitric acid to form sodium nitrate and chlorine with nitrosyl chloride (containing 4 to 10% nitrogen tetroxide) as a by-product. The nitrosyl chloride vapor placed in contact with oxygen to produce nitrogen tetroxide and chlorine: $2NOCl + O_2 \longrightarrow N_2O_4 + Cl_2$. After liquefying and distilling the chlorine out, the nitrogen tetroxide is absorbed in water to make nitric and nitrous acids which are recycled: $N_2O_4 + H_2O \longrightarrow HNO_3 + HNO_2$. The advantage of this process is that it produces chlorine but no caustic soda. The limited demand for sodium nitrate regulates the amount of chlorine that can be made in this way. The overall reactions may be simplified as follows:

$$3NaCl + 4HNO_3 \longrightarrow 3NaNO_3 + Cl_2 + NOCl + 2H_2O$$

$$2NOCl + 3HNO_2 + 3O_2 + H_2O \longrightarrow 5HNO_3 + Cl_2$$

Du Pont has started up a 600 ton/day plant to recover chlorine from hydrogen chloride by the Kel-Chlor[17] process. This process accepts either gaseous or aqueous hydrogen chloride and converts it to chlorine, using a four-step operation based on the Deacon process but involving nitrosyl sulfuric acid ($HNSO_5$) as a catalyst. The development of oxychlorination technology has been the major factor in reducing the oversupply of by-product hydrogen chloride from organic chlorination processes.

[17]Hydrogen Chloride to Chlorine via the Kel-Chlor Process, *Chem. Eng.* (*N.Y.*), **69**(4), 57 (1974); The Kel-Chlor Process, *Ind. Eng. Chem.*, **61**(7), 23 (1969).

BLEACHING POWDER

A decreasing tonnage of chlorine goes into the production of bleaching powder, chiefly because of its instability and the large proportion of inert material. The reaction by which bleaching powder is made is

$$Ca(OH)_2 + Cl_2 \longrightarrow Ca\begin{matrix} / O-Cl \\ \backslash Cl \end{matrix} \cdot H_2O$$

This reaction is carried out below 50°C in a countercurrent fashion by passing chlorine through a rotating steel cylinder with inner lifting blades which shower the solid in the path of the gas. When allowed to stand in the air, the bleaching powder absorbs carbon dioxide. (Other inorganic acids will also liberate the HOCl.)

$$2CaCl(OCl) + CO_2 + H_2O \longrightarrow CaCl_2 + CaCO_3 + 2HClO$$

$$2HClO \longrightarrow 2HCl + O_2$$

Simply on standing, however, the following decomposition takes place:

$$2CaCl(OCl) \longrightarrow 2CaCl_2 + O_2$$

When dissolved in water, the reaction gives ionized calcium chloride and hypochlorite:

$$2CaCl(OCl) \longrightarrow Ca^{2+} + 2Cl^- + Ca^{2+} + 2OCl^-$$

The OCl^- ion decomposes, liberating oxygen. In general, bleaching powder is considered an oxidizing agent. Its activity, however, is measured in what is termed *available chlorine*, which is, by definition, the same weight as that of gaseous or liquid chlorine that would exert the same action as the chlorine compound in question. In the case of bleaching powder ($CaOCl_2$), the available chlorine is the same as the percentage of chlorine, but in case of calcium hypochlorite ($Ca(OCl)_2$), the available chlorine is twice the percentage (49.6) of chlorine in $Ca(OCl)_2$, or 99.2%. This is another way of saying that 1 mol of chlorine is equivalent in oxidizing power to 1 mol of HOCl, or to the ion OCl^-. Bleaching powder by this convention contains about 35% or less of available chlorine when freshly manufactured. The available-chlorine concept may be further explained by the reactions

For calcium hypochlorite:

$$Ca(OCl)_2 + 2HCl \longrightarrow CaCl_2 + 2HOCl$$

or

$$Ca(OCl)_2 \longrightarrow Ca^{2+} + 2OCl^-$$

One mole of bleaching powder will furnish only half this amount of OCl^- ions.

For chlorine:

$$Cl_2 + H_2O \longrightarrow HOCl + HCl$$

Calcium hypochlorite[18] itself may be made in several ways. One method that has been used is the chlorination of calcium hydroxide as in the manufacture of bleaching powder, followed by the separation of the $Ca(OCl)_2$ through salting out from solution with NaCl. It is also manufactured by the formation under refrigeration of the salt $[Ca(OCl)_2 \cdot NaOCl \cdot NaCl \cdot 12H_2O]$,[18] which is prepared by the chlorination of a mixture of sodium and calcium hydroxides. This is reacted with a

[18]MacMullin and Taylor, U.S. Pat. 1,787,048 (1930); Avery and Evans, Modified MgO Process Yields $MgCl_2$ and Calcium Hypochlorite, *Chem. Metall. Eng.*, **52**(4), 94 (1945) (pictured flowchart, p. 130).

chlorinated lime slurry, filtered to remove salt, and dried, resulting finally in a stable product containing 65 to 70% $Ca(OCl)_2$. The final reaction is

$$2[Ca(OCl)_2 \cdot NaOCl \cdot NaCl \cdot 12H_2O] + CaCl_2 + Ca(OCl)_2 \longrightarrow 4Ca(OCl)_2 \cdot 2H_2O + 4NaCl + 16H_2O$$

The great advantage of $Ca(OCl)_2$ is that it does not decompose on standing, as does bleaching powder. It is also twice as strong as ordinary bleaching powder, and is not hygroscopic.

SODIUM HYPOCHLORITE

Sodium hypochlorite is employed as a disinfectant and deodorant in dairies, creameries, water supplies, sewage disposal, and for household purposes. It is also used as a bleach in laundries. During World War I, it was employed in the treatment of wounds as a stablized isotonic solution. As a bleaching agent, it is very useful for cotton, linen, jute, artificial silk, paper pulp, and oranges. Indeed, much of the chlorine bought for bleaching cellulose products is converted to sodium hypochlorite before use. The most common method for making it is the treatment of sodium hydroxide solution with gaseous chlorine.

$$Cl_2 + 2NaOH \longrightarrow NaCl + H_2O + NaOCl$$

The other once widely used method was the electrolysis of a concentrated salt solution whereby the same product was made. These electrolytic cells do not have a diaphragm and are operated at high current density in a nearly neutral solution. The cells are designed to function at a low temperature and to bring the cathode caustic soda solution in contact with the chlorine given off at the anode.

SODIUM CHLORITE

Sodium chlorite ($NaClO_2$) was introduced in 1940 by the Mathieson Chemical Corp. (now Olin Corp.). The 80% commercial material has about 125% available chlorine. It is manufactured from chlorine through calcium chlorate to chlorine dioxide, ending with the reaction

$$4NaOH + Ca(OH)_2 + C + 4ClO_2 \longrightarrow 4NaClO_2 + CaCO_3 + 3H_2O$$

After filtering off the calcium carbonate, the solution of $NaClO_2$ is evaporated and drum-dried. $NaClO_2$ is a powerful but stable oxidizing agent. It is capable of bleaching much of the coloration in cellulosic materials without tendering the cellulose. Hence it finds use in the pulp and textile industries, particularly in the final whitening of kraft paper. Besides being employed as an oxidizer, $NaClO_2$ is also the source of another chlorine compound, chlorine dioxide, through the reaction

$$NaClO_2 + \tfrac{1}{2}Cl_2 \longrightarrow NaCl + ClO_2$$

Chlorine dioxide has $2\frac{1}{2}$ times the oxidizing power of chlorine and is important in water purification, for odor control, and for pulp bleaching.

SELECTED REFERENCES

Buehr, W.: Caustic Soda Production Technique, Noyes, 1962.
Calciner (Steam Tube) 98 Ft. Long, *Chem. Eng. (N.Y.)*, **74**(11), 131 (1967).
Gould, R. F. (ed.): Literature of Chemical Technology, chap. 1, Chlor-alkali Industry, ACS Monograph, 1967.
Guccione: Soda Ash (Trona), *Chem. Eng. (N.Y.)*, **74**(14), 60 (1967).
Hampel: Electrochemical Encyclopedia, Reinhold, 1964.
Hou, Te Pang: Manufacture of Soda, ACS, 1942.
Jones, H. R.: Mercury Pollution Control, Noyes, 1971.
Sconce, J. S. (ed.): Chlorine: Its Manufacture, Properties and Uses, ACS Monograph 154, Reinhold, 1962.
Szilard, J. A.: Bleaching Agents and Techniques, Noyes, 1973.
Trona-based Soda Ash Capacity Grows, *Chem. Eng. News*, May 1, 1967, p. 46.

chapter 14

ELECTROLYTIC INDUSTRIES

Electric energy is extensively consumed by the chemical process industries, not only to furnish power through electric motors, but to give rise to elevated temperatures and directly to cause chemical change. Energy in the form of electricity causes chemical reactions to take place in the electrolytic industries described in this chapter.[1] The heat produced thereby is the basis for the high temperature required in the electrothermal industries, which are discussed in Chap. 15. Most of the electrolytic processes have been developed since World War I; few of them are older. The materials manufactured with the aid of electricity vary from chemicals that are also produced by other methods, such as caustic soda, hydrogen, and magnesium, to chemicals that at present cannot be made economically in any other way, such as aluminum and calcium carbide (Table 14.1). The cost of electric power is usually the deciding factor in the electrochemical industries. Thus these industries have tended to become established in regions of cheap electric power based on falling water, e.g., at Niagara Falls, and in Norway. The establishment of new areas of abundant and inexpensive electricity in the West and Northwest pointed the way to the growth of electrochemical industries in those regions. But cheap electric power has tended to vanish as public demand has escalated. This has caused some companies to look beyond the borders of the United States for new plant sites.[2] Table 14.1 indicates the number of kilowatthours needed to produce 1 lb of the various materials listed.

Landis writes, "The electrochemical process has completely *revolutionized* the production of certain primary products and at such *lowered cost* as to permit the development of new secondary industries utilizing these cheaper raw materials." Examples of such lowered costs are for chlorine, sodium, and magnesium, as well as for aluminum. Direct current is used in the electrochemical industries. Since 1938 the single-anode mercury-arc rectifier has been used largely in the electrochemical industries to convert alternating current to direct current. The balance has been divided between motor-generator sets, synchronous converters, and multianode rectifiers. A more recent development has been acyclic generators, a means of generating large blocks of direct-current power without sliding brushes or the necessity for rectification. The cost of installation has been estimated to be 25% less than for conventional generating and rectifying equipment.[3] An important development in the aluminum industry is the use of direct-current generators actuated by gas engines.

ALUMINUM

Aluminum is potentially the most abundant metal in the world. It makes up 8.05% of the solid portion of the earth's crust. Every country possesses large supplies of aluminum-containing materials, but as yet processes for obtaining metallic aluminum from most of these compounds are not economical.

[1]See Perry, pp. 4-68 to 4-75, for the Thermodynamics of Electrochemical Cells; Newman, Engineering Design of Electrochemical Systems, *Ind. Eng. Chem.*, **60**(4), 12 (1968) (85 refs.).

[2]Toth and Lippman, The Quest for Aluminum, *Mech. Eng.*, September 1973, p. 24.

[3]Acyclic Unit Generates Direct Current Directly, *Chem. Eng.*, **68**(24), 58 (1961).

TABLE 14.1 ***Chemicals and Metals Made Electrochemically***

Material	*Process*	*Kilowatt-hours per pound*	*Voltage per cell or furnace*	*Yearly consumption, kWh*
Alumina, fused	Electrothermal fusion	1–1.5	100–110	Large
Aluminum	Electrolytic reduction of alumina to aluminum	10–12	5.5–7	Very large
Ammonia, synthetic	Electrolytic hydrogen; pressure hydrogenation	6.5	....	Small
Cadmium	Electrolytic precipitation of zinc-lead residues	0.8	2.6	Small
Calcium	Electrothermal reduction	22–24	....	Small
Calcium carbide	Electrothermal reduction	1.3–1.4	....	Very large
Calcium cyanamid	Electrothermal reduction	1.3	....	Large
Caustic soda, chlorine	Electrolysis of brine	1.34	3.4–4.3	Very large
Copper	Copper electrowinning	1–1.5	1.9–2.4	Large
	Electrolytic refining	0.09–0.16	0.18–0.4	Very large
Ferrochromium, 70%	Electrothermal smelting	2–3	90–120	Large
Ferromanganese, 80%	Electrothermal reduction	1.5–3	90–115	Very small
Ferromolybdenum, 50%	Electrothermal smelting	3–4	50–150	Small
Ferrosilicon, 50%	Electrothermal smelting	2–3.5	75–150	Very large
Ferrotungsten, 70%	Electrothermal smelting and refining	1.5–2	90–120	Small
Ferrovanadium	Electrothermal smelting	2–3.5	150–250	Small
Gold	Electrolytic refining	0.15	1.3–1.6	Very small
Graphite	Electrothermal change	1.5–2.0	80–200	Large
Hydrogen and oxygen	Electrolysis of alkalized water	140*	....	Small
Hypochlorite	Electrolysis of salt solution with mixing	....	....	Medium
Iron	Reduction in arc furnace	1–1.25	....	None
Iron castings	Electric melting			
	Duplex system	0.07	....	Very small
	Continuous	0.225		
Iron, sponge	Electrothermal low-temperature reduction	0.2	....	Very small
Lead	Electrolytic refining	0.09	0.35–0.6	Small
Magnesium, metallic	Electrolysis of magnesium chloride	8–13	6–7	Very large
Phosphoric acid	Electrothermal reduction to phosphorus; oxidation	2.7	....	Large
Phosphorus	Electrothermal reduction	4–5.5	....	Large
Potassium hydroxide	Electrolysis of potassium chloride solution	1–1.2	....	Small
Quartz, fused	Electrothermal fusion	5–8	....	Small
Silicon	Electrothermal reduction	6	....	Small
Silicon carbide	Electrothermal reduction	3.2–3.85	75–230	Large
Sodium, metallic	Electrolysis of fused sodium chloride	7.1–7.3	....	Large
Sodium chlorate	Electrolysis of sodium chloride solution (mixing)	2.5	2.8–3.5	Small
Steel castings	Electric melting cold charge	0.25–0.38	....	Large
Steel ingot	Electric melting or refining of molten charge	0.1	....	Large
Zinc	Electrolytic precipitation	1.4–1.56	3.5–3.7	Very large

Source: Based on House Doc. 103, 73d Congr. *Kilowatthours per thousand cubic feet.

TABLE 14.2 ***Aluminum Production for Selected Years*** *(United States)*

Year	Production, thousands of short tons	Year	Production, thousands of short tons
1913	24	1960	2,014
1920	69	1966	2,968
1930	115	1971	3,925
1938	143	1973	4,500
1943	920	1980 est.	7,000
1952	937	1990 est.	13,900

Sources: Statistical Committee, The Aluminum Association; U.S. Business Outlook, Long Term, McGraw-Hill, July 26, 1973, p. 8. *Note:* The figures given refer to domestic primary production only and do not include domestic secondary recovery (about 19% of domestic consumption) and imports (about 12% of consumption).

HISTORICAL Aluminum metal was first obtained in pure form in 1825 by Oersted, who heated aluminum chloride with a potassium-mercury amalgam. In 1854, Henri Sainte-Claire Deville produced aluminum from sodium-aluminum chloride by heating with metallic sodium. This process was operated for about 35 years, and the metal sold for $100 a pound. By 1886 the price had been reduced to $8 a pound. In 1886, Charles Hall produced the first aluminum by the present-day large-scale process, electrolysis of alumina dissolved in a fused bath of cryolite. The same year, Paul Héroult was granted a French patent for a process similar to that of Hall. By 1893, the production of aluminum had increased so rapidly by Hall's method that the price had fallen to $2 per pound. The industry grew steadily, based soundly on new and expanding markets created largely by its own study of the properties of aluminum and of the avenues for economical consumption of this new metal. With this growth in manufacture came a decrease in cost, which the Aluminum Company of America largely passed on to its customers, even reducing the price first to 17 cents and then to 15 cents per pound (1943–1947 average). The price increased to 23.5 cents per pound by 1964 and to 25 cents per pound in 1973, reflecting the inflation during this period.[4]

ECONOMICS AND USES The consumption of aluminum is increasing very rapidly (Table 14.2). Table 14.3 gives the main avenues of distribution. The unusual combination of lightness and strength makes aluminum applicable for many uses that other metals cannot fill. Weight for weight, aluminum has twice the conductivity of copper and also has high ductility at elevated temperatures. Aluminum is commonly alloyed with other metals, such as copper, magnesium, zinc, silicon, chromium, and manganese, and thus its usefulness has increased. Metallic aluminum or aluminum alloys, particularly with magnesium, are employed in structures for aircraft, automobiles, trucks, and railway cars, for electrical conductors, and for cast and forged structural parts. Properly used,[5] aluminum resists corrosion very well. Its strength and ductility increase at subzero temperatures, which is the opposite of what happens to iron and steel.

MANUFACTURE Metallic aluminum[6] is produced by the electrolytic reduction of pure alumina in a bath of fused cryolite. For pure alumina see Chap. 20. It is not possible to reduce alumina with carbon because, at the temperature required for this reduction, the aluminum comes off as a vapor

[4]*Am. Met. Mark.*, Nov. 27, 1973, p. 1; Bureau of Mines, reported in *Forbes*, Jan. 1, 1974, p. 152.

[5]Kuli *et al.*, Aluminum and Its Alloys in the CPI, *Chem. Eng. Prog.*, **60**, 45ff., 49ff., 55ff. (1964); Wyma, Aluminum Alloys, *Ind. Eng. Chem.*, **57**(8), 85 (1965) (new alloys); Aluminum for Process Equipment, *Chem. Eng. (N.Y.)*, **73**(18), 116 (1966) and **73**(20), 142 (1966); Aluminum Anodizing Process Offers Economy, *Chem. Eng. News*, July 23, 1974, p. 12.

[6]For other noncompetitive processes that have been investigated, see ECT, 2d ed., vol. 1, pp. 944ff.; Edwards *et al.*, The Aluminum Industry, vol. 1, pp. 170–281, McGraw-Hill; Go Directly to Aluminum, *Chem. Week*, Sept. 16, 1967, p. 89; Chlorination Process Upgrades Low-grade Ores, *Chem. Eng. News*, Nov. 29, 1971, p. 31; New Processes Promise Lower-cost Aluminum, *Chem. Eng. News*, Feb. 26, 1973, p. 11; Alcoa's New Electrolytic Aluminum Process Will Get Its Commercial Test, *Chem. Week*, June 6, 1973, p. 55.

and is carried away with the carbon monoxide. This metallic vapor cannot be condensed by cooling, because the reaction reverses at lower temperatures and thus converts the aluminum back to alumina. The *enthalpy change* involved in the reaction

$$Al_2O_3 + 1\tfrac{1}{2}C \longrightarrow 2Al + 1\tfrac{1}{2}CO_2 \qquad \Delta H = +261.9 \text{ kcal at } 1000°C$$

is equivalent to 2.56 kWh of energy per pound of aluminum produced. In actual practice, however, some energy is used to bring reactants up to temperature and is lost in the sensible heat of the products. Some carbon monoxide is formed in the reaction, thus increasing the positive ΔH, which amounts *practically* to from 6 to 9 kWh/lb. Consequently, this metal cannot be made economically unless low-priced electric energy is available. The carbon for the reaction comes from the anode, which requires from 0.5 to 0.6 lb of carbon per pound of metal. Although the theoretical amount required by the equation is $\frac{1}{3}$ lb, the carbon dioxide evolved contains from 10 to 50% carbon monoxide.

The steps involved in the production of the metallic aluminum are (Figs. 14.1 to 14.3):

The cell lining is installed or replaced (Op).

Carbon anodes are manufactured and used in the cell (Op).

The cryolite bath is prepared (Ch) and the composition controlled (Op).

Alumina is dissolved in the molten cryolite bath (Op).

The solution of alumina in molten cryolite is electrolyzed to form metallic aluminum (Ch). It serves as the cathode.

The carbon electrode is oxidized by the oxygen liberated (Ch).

The molten aluminum is tapped from the cells, alloyed (if desired), cast into ingots, and cooled (Op).

The electrolytic cells are huge, bathtublike, steel containers; inside each is a cathode compartment lined with either a rammed mix of pitch and anthracite coal or coke baked in place by the passage of electric current, or prebaked cathode blocks cemented together. This cathode compartment, or cavity, may be from 12 to 20 in. in depth and up to 10 ft wide and 30 ft long, depending on the type of cell and the ampere load for which it is designed. The cavity-lining thickness varies from 6 to 10 in. on the sides and 14 to 18 in. on the bottom. Thermal insulation consisting of firebrick, asbestos block, or other similar materials is placed between the cavity lining and the steel shell. Large steel bars, serving as cathode current collectors, are embedded in the bottom portion of the cavity lining and extend out through openings in the shell to connect with the cathode bus. Cell linings normally last from 2 to 4 years. When failure occurs, usually from penetrations of metal

TABLE 14.3 ***U.S. End-Use Distribution of Aluminum Shipments, 1966–1971*** *(In thousands of short tons)*

End use	*1966*	*1971*	*Percent change*
Building and construction	992	1,384	40
Transportation	976	895	−8
Containers and packaging	370	757	105
Electrical	650	714	10
Consumer durables	455	484	6
Machinery and equipment	327	321	−2
Exports	292	284	−3
Other	454	364	−20
Total	4,516	5,203	15

Source: The Aluminum Association. *Note:* The total shipments shown exceed the production as shown in Table 14.2. The differences are due to secondary recovery and imports.

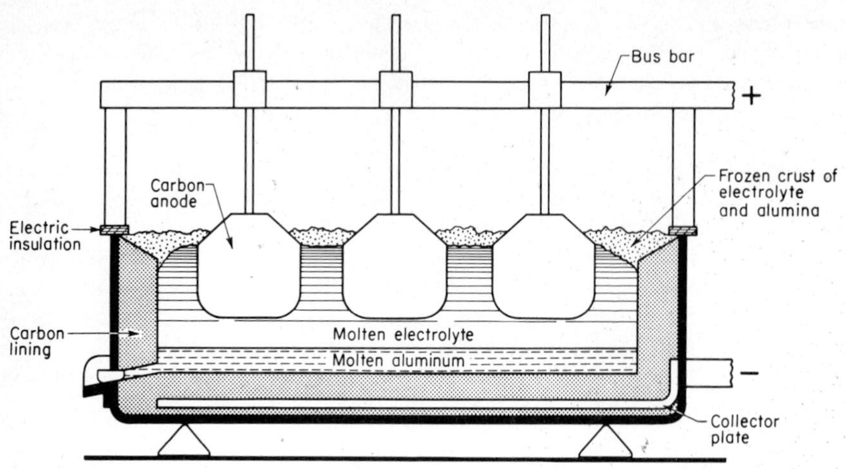

Fig. 14.1 Cross-section diagram of an aluminum reduction pot. (*Aluminum Co. of America.*)

through to the cathode collectors which it dissolves, or penetration of metal out to the steel shell where it leaks out around the collectors, the entire lining, insulation, and collector ensemble is replaced. Cell relining is an appreciable part of production expense, including not only the cost of labor, collectors, lining, and insulating materials, but also the loss of electrolyte materials absorbed by the spent lining (many producers now reclaim at least some of these electrolyte materials from the spent linings).

Two types of cells are used in the Hall-Héroult process, those with multiple prebaked anodes (Figs. 14.1 and 14.2), and those with a self-baking, or Soderberg, anode. In both types the anodes are suspended from a superstructure extending over the cell cavity and are connected to a movable anode bus so that their vertical position may be adjusted. The prebaked anode blocks are manufactured from a mixture of low-ash calcined petroleum coke and pitch or tar, formed in hydraulic presses, and baked at up to 1100°C.

Soderberg[7] anode cells have a single large anode which subtends most of the cell cavity. The anode is housed in an open steel casing with vertical sides through which it is fed down into the electrolyte. When cells are first placed in operation they are brought up to operating temperature by electric-resistance heating; the anodes are placed in contact with a layer of coke particles on the bottom of the cell cavity, and current is passed through the short-circuited cell until the desired temperature is reached. Electrolyte materials are added to the cell cavity around the anodes; as these materials gradually become molten, the anodes are raised to place the cell in operation. Normally, the anode-cathode distance is about 2 in. The molten electrolyte consists principally of cryolite ($3NaF \cdot AlF_3$) plus some excess AlF_3, 6 to 10% by weight CaF_2 (fluorspar), and 2 to 6% Al_2O_3. The cryolite is imported from Greenland, though most is produced synthetically. AlF_3 is also produced synthetically from hydrogen fluoride and aluminum hydroxide.

The control of electrolyte composition is an important operation in the aluminum production process. As the melting point of pure cryolite is 1009°C, the electrolyte contains fluorspar and some excess AlF_3 which, along with the dissolved alumina, reduces the liquidus temperature sufficiently to permit the cells to be operated in the 940 to 980°C range. The excess AlF_3 also improves cell efficiency. The NaF/AlF_3 weight ratio in cryolite is 1.50; the excess AlF_3 in the electrolyte is

[7]ECT, 2d ed., vol. 1, p. 943. In the United States both types of cells are used to approximately the same extent, but in foreign countries the Soderberg type prevails.

adjusted to yield an NaF/AlF_3 ratio in the 1.10 to 1.40 range. In the first few weeks after a newly lined cell is placed in operation, the electrolyte is rapidly absorbed into the lining and insulation, with a marked preferential absorption of a high-sodium-containing portion, tending to reduce the NaF/AlF_3 ratio below that desired. Compensation for this is made by adding an alkaline material such as soda ash:

$$3Na_2CO_3 + 4AlF_3 \longrightarrow 2(3NaF \cdot AlF_3) + Al_2O_3 + 3CO_2$$

After the first few weeks of operation of the cells, the electrolyte tends to become depleted of AlF_3 through volatilization of AlF_3-rich compounds and through reaction with residual caustic soda in the alumina and hydrolysis from moisture in the air or in added materials:

$$3Na_2O + 4AlF_3 \longrightarrow 2(3NaF \cdot AlF_3) + Al_2O_3$$

$$3H_2O + 2AlF_3 \longrightarrow Al_2O_3 + 6HF$$

The volatilized fluorides and gaseous hydrogen fluoride are collected, with other gases evolved from the cells, by gas-collecting hoods or manifolds and are passed through large ducts to central gas-treatment and recovery facilities. The loss of these materials from the electrolyte requires periodic additions of AlF_3 to maintain the desired composition. The small percentage of lime normally occurring as an impurity in the alumina is sufficient to maintain the desired concentration of fluorspar through the reaction

$$3CaO + 2AlF_3 \longrightarrow 3CaF_2 + Al_2O_3$$

During operation of the cell, a frozen crust forms over the surface of the molten bath. Alumina is added on top of this crust, where it is preheated and has its mechanically absorbed moisture driven off. The crust is broken periodically, and the alumina stirred into the bath in order to maintain a 2 to 6% concentration. The theoretical requirement is 1.89 lb alumina per pound of aluminum; in practice, the actual figure is approximately 1.91. When the alumina in the bath is depleted, the so-called *anode effect* occurs, wherein a thin film of carbon tetrafluoride forms on the anode so that the bath no longer wets the anode surface, causing an abrupt rise in cell voltage, which is indicated

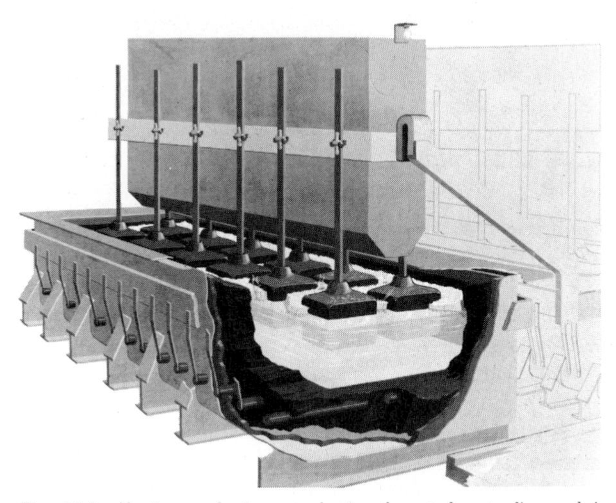

Fig. 14.2 Aluminum reduction pot, showing the actual parts diagramed in Fig. 14.1 and with the alumina hopper between the carbon anodes. (*Aluminum Co. of America.*)

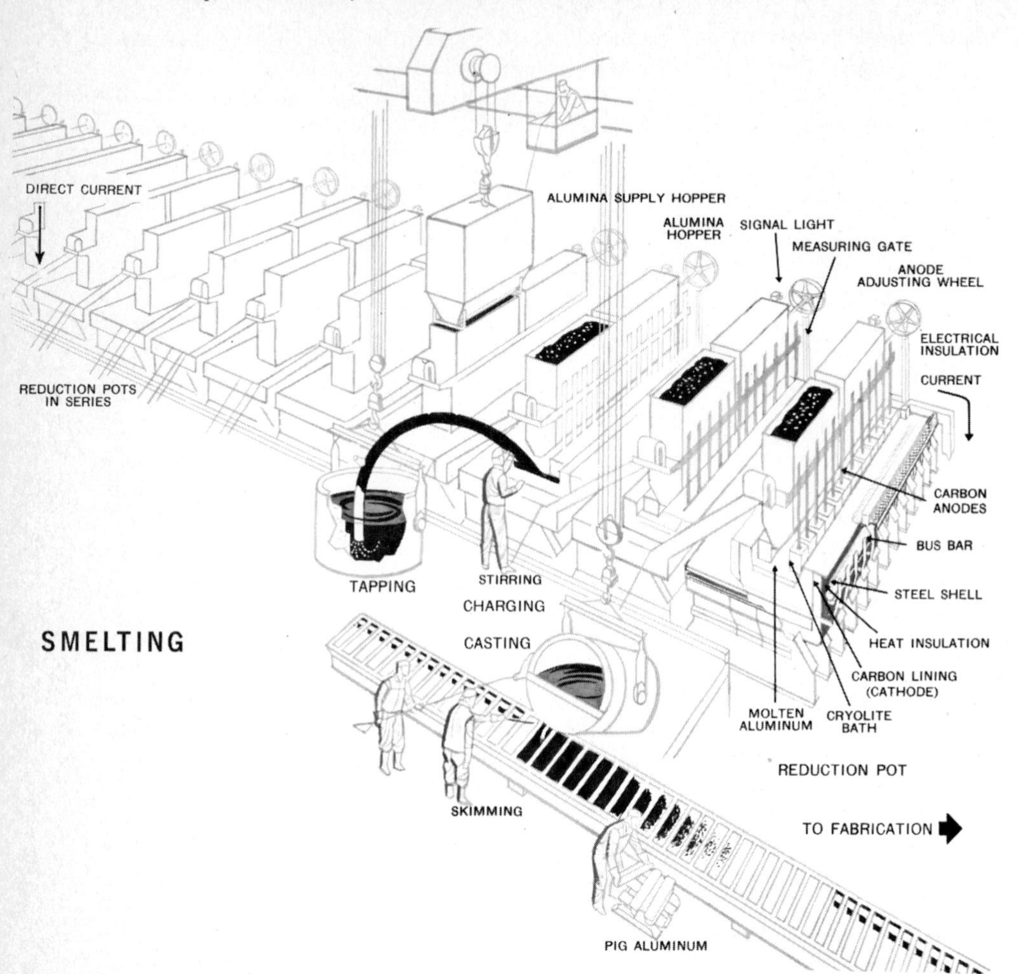

Fig. 14.3 Pictorial flowchart of large electrolytic furnaces, called reduction pots, together with the alumina supply equipment and the siphoning or tapping arrangement for removing the molten aluminum at intervals. The molten metal is transferred to holding furnaces for the removal of impurities or for alloying. It is then cast into ingots of various sizes for further fabrication. (*Aluminum Co. of America.*)

by a signal lamp or bell shunted across the cell and inoperative on normal cell voltage. When this happens, alumina is stirred into the cell, even though the time does not coincide with the routine periodic alumina addition, and the electrolysis process and cell voltage return to normal. The actual mechanism by which alumina is dissolved in the molten bath and electrolytically decomposed is still a matter of dispute; however, the end results are the release of oxygen at the anode and the deposit of metallic aluminum at the cathode. The oxygen combines with the carbon anode to yield CO and CO_2, with CO_2 predominating.

MAGNESIUM

Magnesium is a very light, silvery-white metal which has achieved widespread industrial use. Magnesium is the eighth element in order of occurrence in the world. The raw materials are widely distributed throughout the globe, particularly since sea water is a practical and very important source

Fig. 14.4 Reduction furnaces, showing electrical connections to anodes and overhead fume-collecting system. (*Aluminum Co. of America.*)

of magnesium compounds. Hence there need never be any shortage of the economical raw materials for magnesium metal and most of its salts. Magnesium salts are presented in Chap. 10.

HISTORICAL Although metallic magnesium was first isolated by Bussy in 1829, its existence was discovered by Davy in 1808. As late as 1918, almost all the magnesium produced was used for flashlight powder and in pyrotechnics. The United States output in 1918 was only 284,118 lb. Intensive research brought forth many new uses for magnesium. By 1930 it was being fabricated into complicated castings, sheets, and forgings, and a method of welding it was also perfected. Extremely lightweight, strong magnesium alloys[8] of great importance to the aircraft industry were developed, and met all demands during World War II and after. Consequently, magnesium production was expanded tremendously. The most common magnesium alloys are AZ91B: Al 6.5%, Mn 0.2%, Zn 3.0%, and the rest Mg, for pressure dye casting; and AZ31: Mn 1.5% and the rest Mg, for wrought products.

USES AND ECONOMICS The largest single use of magnesium, the world's *lightest* structural metal, is in aluminum alloying, as in space vehicles and automotive equipment. Important demands have been for use in anodes for protection against corrosion of other metals. Probably the second-largest nonstructural use for magnesium is as a reducing agent in the production of metals such as titanium, zirconium, uranium, and beryllium. During World War II, magnesium production averaged over 100,000 tons yearly; afterward it dropped to about 10,000 tons annually, but in 1964 the primary production for sale was 79,000 tons. By 1973 the primary production increased to 122,400 tons/year. Most of it is produced at Dow Chemical Co.'s Freeport, Tex., plant which has a capacity of 120,000 tons/year. As usual in the development of a material from relative rarity to a tonnage article, the price of magnesium dropped from $1.81 per pound in 1918 to 35.25 cents per pound in 1964 and down to 28.5 cents per pound by 1970.

MANUFACTURE The cheapest method of making magnesium is by the electrolytic process. During World War II magnesium was made by two other processes, the silicothermic, or ferrosilicon,

[8]Comstock, Magnesium and Magnesium Compounds, *U.S. Bur. Mines Inf. Circ.* 8201, 1963; Hanawalt, Magnesium Industry, World Magnesium Conference, Montreal, October 1963.

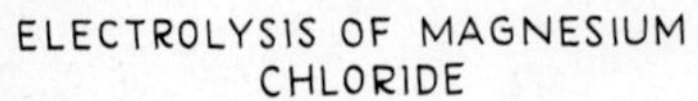

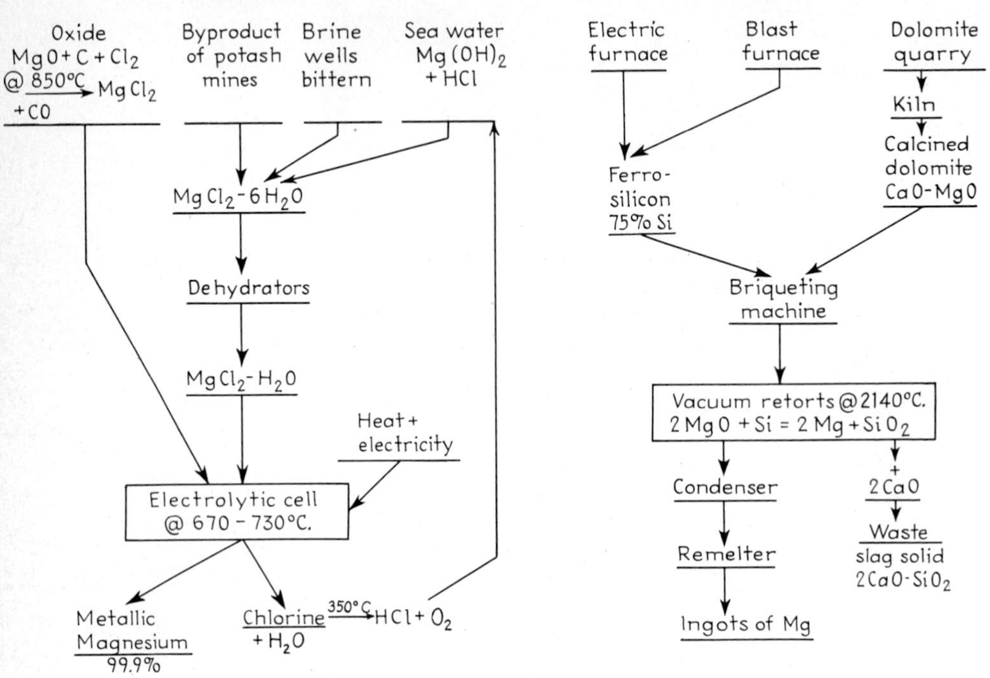

Fig. 14.5 Outline of processes for the production of metallic magnesium used during World War II.

process and the carbon reduction process. The carbon reduction process[9] never operated satisfactorily, and its use has long been discontinued. The silicothermic process is still used in areas where power is expensive and only a small magnesium capacity is required (Fig. 14.5).

ELECTROLYSIS OF MAGNESIUM CHLORIDE The magnesium chloride[10] needed is obtained (1) from salines, (2) from brine wells, and (3) from the reaction of magnesium hydroxide (from sea water or dolomite) with hydrochloric acid (Figs. 14.5 and 14.6). The pioneer producer, Dow Chemical Co., at Freeport and Velasco, Tex., makes magnesium by electrolyzing magnesium chloride from sea water, using oyster shells for the lime needed. The oyster shells, which are almost pure calcium carbonate, are burned to lime, slaked, and mixed with the sea water, thus precipitating magnesium hydroxide (Fig. 14.6). This magnesium hydroxide is filtered off and treated with hydrochloric acid prepared at high temperature from water and the chlorine evolved from the cells. This forms a magnesium chloride solution which is evaporated to solid magnesium chloride in direct-fired evaporators, followed by shelf drying. After dehydrating, the magnesium chloride[11] is fed to a Dow electrolytic cell, where it is decomposed into the metal and chlorine gas. These cells are large rectangular ceramic-covered steel pots, 5 ft wide, 11 ft long, and 6 ft deep, with a capacity of about 10 tons molten magnesium chloride and salts. The internal parts of the cell act as the cathode, and 22 graphite anodes are suspended vertically at the top of the cell. Sodium chloride is added to the

[9]For description and references, see CPI 1, pp. 310–311.

[10]Shigley, Minerals from the Sea, *J. Met.*, January 1951 (flowchart, references, and pictures); Davis, Magnesium: Maybe This Time, *Chem. Eng. (N.Y.)*, **80**(19), 84 (1973); Magnesium Cast for New Roles, *Chem. Week*, Jan. 17, 1973, p. 24.

[11]Teaming Up to Tap Brines, *Chem. Week*, Feb. 26, 1966, p. 15; Magnesium Makes Its Move, *Chem. Eng. (N.Y.)*, **76**(18), 60 (1969); Anderson, World's Nations Scramble for Sea's Riches, *Chem. Eng. News*, Mar. 4, 1974, p. 18.

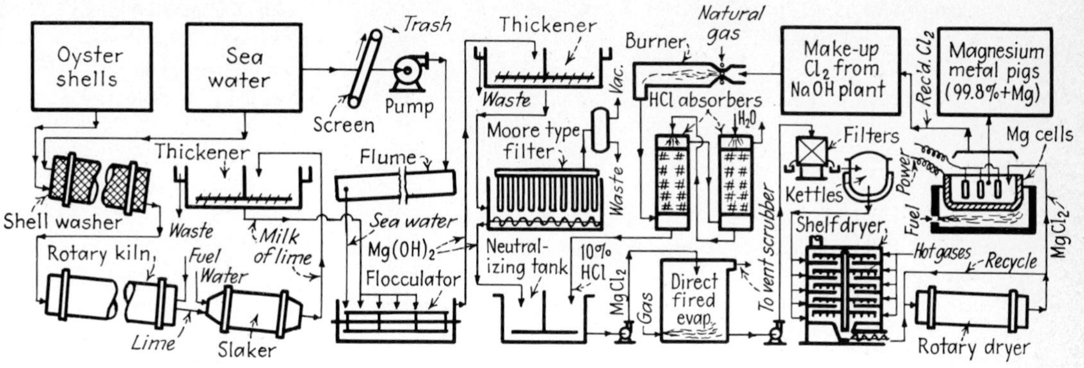

Fig. 14.6 Magnesium metal from sea water—Dow process.

bath to lower the melting point and also to increase the conductivity. The salts are kept molten by the electric current used to extract the magnesium plus external heat supplied by gas-fired outside furnaces. The usual operating temperature is 710°C, which is sufficient to melt the magnesium (mp 651°C). Each cell operates at about 6 V and 80,000 to 100,000 A, with a current efficiency of above 80%. The power requirements are 8 kWh/lb of magnesium produced. The molten magnesium is liberated at the cathode and rises to the bath surface, where troughs lead to the metal wells in front of the cell. The 99.9% pure magnesium metal is dipped out several times during the day, each dipperful containing enough metal to fill a 42-lb self-pelletizing mold.

SILICOTHERMIC, OR FERROSILICON, PROCESS[12] During World War II the U.S. government financed construction of six silicothermic plants (Fig. 14.5) with a total *rated* capacity of 70,000 tons/year, which were eventually dismantled. However, a new silicothermal plant constructed by Alamet is operating at Selma, Ala. The essential reaction of this process is:

$$2(MgO \cdot CaO) + \tfrac{1}{6}FeSi_6 \longrightarrow 2Mg + (CaO)_2SiO_2 + \tfrac{1}{6}Fe$$

The process consists of mixing ground, dead-burned dolomite with ground, 70 to 80% ferrosilicon and 1% fluorspar (eutectic) and pelleting. The pellets are charged into the furnace. High vacuum and heat (2140°F) are applied. The calcium oxide present in the burnt dolomite forms infusible dicalcium silicate that is easily removed from the retort at the end of the run. The chrome-alloy retort is equipped with a condenser tube with a removable lining. The reaction is run at very high vacuum (100 to 150 μ Hg), and the liberated magnesium is collected on the lining of the condenser. At the end of the run the furnace is partly cooled, and the magnesium is removed from the condenser, by a procedure based on the difference in contraction of magnesium and steel, and taken to the remelt furnace.

SODIUM[13]

Sodium is a silvery white, very active metal. It reacts violently with water and is usually preserved by storage under kerosine or, preferably, in containers under a nitrogen blanket. About 150 million lb of sodium enters annually into the manufacture of tetraethyl- and tetramethyllead, and the rest is

[12]Schrier, Silicothermic Magnesium Comes Back, *Chem. Eng.* (*N.Y.*), **59**(4), 148 (1952) (pictured flowchart, p. 212); Comstock, *op. cit.*, p. 52, cf. Artry, Can. Pat. 621,718 (June 6, 1961) for improvements: Magnesium, *Chem. Week*, June 17, 1961, p. 7; Barnes, Magnesium, *Chem. Eng. (N.Y.)*, **68**(11), 70 (1961).

[13]Sodium Builds on Solid Base, *Chem. Week*, Aug. 19, 1967, p. 79; Foust (ed.), Sodium-Nak Engineering Handbook, Gorden and Breach, 1972; Silverman, Sodium-Analysis, Pergamon, 1971.

employed for miscellaneous uses, e.g., Na_2O_2, hydride descaling, sodium cyanide, sodium alkyl salts, and for making other metals, such as potassium. The total production was in excess of 300 million lb in 1976.

MANUFACTURE The most important method of preparation of sodium is by electrolysis of fused sodium chloride. The reaction involved in this process is:

$$2NaCl \longrightarrow 2Na + Cl_2$$

The cell for this electrolysis consists of a closed, rectangular, refractory-lined steel box (Fig. 14.7). The anode is made of carbon, and the cathode of iron.[14] The anode and cathode are arranged in separate compartments to facilitate the recovery of the sodium and the chlorine. Sodium chloride has a high melting point (804°C), but calcium chloride is added to lower it, and the cell is operated at 600°C. The electrolyte is a eutectic of 33.2% sodium chloride and 66.8% calcium chloride. The lower temperature increases the life of the refractory lining of the cell, makes it easier to collect the chlorine, and prevents the sodium from forming a difficultly recoverable fog. A sodium-calcium mixture deposits at the cathode, but the solubility of calcium in sodium decreases with decreasing temperatures so that the heavier calcium crystals, which form as the mixture is cooled, settle back into the bath. The crude sodium is filtered at 105 to 110°C, giving a sodium 99.9%. This is frequently run molten into a nitrogen-filled tank car, allowed to solidify, and shipped.

CHLORATES AND PERCHLORATES[15]

Sodium chlorate is manufactured in large quantities (about 180,000 tons in 1976). About half of the production is for bleaching pulp, with solid rocket propellants (through ammonium perchlorate), herbicides, defoliants, and potassium perchlorate for explosives taking up most of the remainder.

MANUFACTURE *Sodium chlorate* is produced by the electrolysis of saturated acidulated brine mixed with sodium dichromate (about 2 g/l) to reduce the corrosive action of the hypochlorous acid liberated by the hydrochloric acid present. Figure 14.8 shows the essential steps in the production of this chemical. The brine solution is made from soft water or condensate from the evaporator and rock salt and purified of calcium and magnesium. The rectangular steel cell is filled with either the brine solution or a recovered salt solution, made from recovered salt-containing chlorate, dissolved in condensate from the evaporator. The temperature of the cell is maintained at 40°C by cooling water. The products of the electrolysis are actually sodium hydroxide at the cathode and chlorine at the anode, but since there is *no diaphragm* in the cell, mixing occurs and sodium hypochlorite is formed, which is then oxidized to chlorate. The overall reaction is:

$$NaCl(aq) + 3H_2O(l) \longrightarrow NaClO_3(aq) + 3H_2(g) \qquad \Delta H = +223.9 \text{ kcal}$$

The finished cell liquor is pumped to tanks, where it is heated with live steam to 90°C to destroy any hypochlorite present. The liquor is analyzed to determine the chromate content, and the required amount of barium chloride is introduced to precipitate almost all the chromate present. The graphite mud from the electrodes and the barium chromate settle to the bottom of the tank, and the clear liquor is pumped through a sand filter to evaporator storage tanks. The liquor in the storage tank is neutralized with soda ash and evaporated in a double-effect evaporator until it contains approximately

[14]Downs, Electrolytic Process and Cell, U.S. Pat. 1,501,756 (1924). Downs received the Schoellkopf medal in 1934 for the development of this cell. Zabel, Metallic Sodium, *Chem. Ind.* (*N.Y.*), **65,** 714 (1949).

[15]Schumacher, Perchlorates, ACS Monograph, Reinhold, 1960; ECT, vol. 5, p. 50, 1964 (details); Hampel, Electrochemical Encyclopedia, p. 170, Reinhold, 1964 (details); Hampel, Electrochemical Encyclopedia, p. 170, Reinhold, 1964; Reynolds and Clapper, The Manufacture of Perchlorates, *Chem. Eng. Prog.*, **57**(11), 138 (1961); Remirez, Producing Captive Sodium Chlorate in Integrated Pulp Mills, *Chem. Eng. (N.Y.)*, **74**(17), 136 (1967) (flowsheet and cell room photograph); Sodium Chlorate: Bright Pulp Industry Future, *Chem. Eng. (N.Y.)*, **74**(20), 74 (1967).

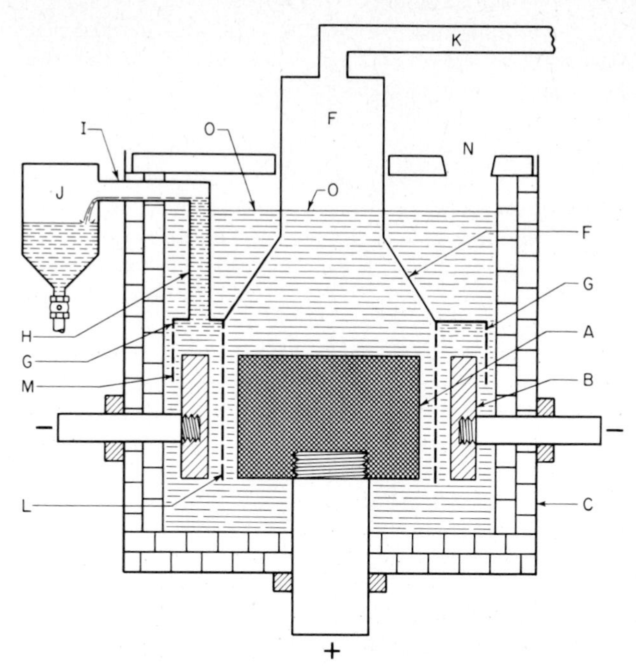

Fig. 14.7 Down's sodium cell. (*A*) Graphite anode; (*B*) iron cathode; (*F*) dome for collecting the chlorine; (*K*) pipe for conducting the chlorine away; (*G*) annular sodium collector; (*H*) and (*I*) pipes to conduct the sodium to the vessel (*J*); (*L*) and (*M*) metal screens supported by (*F*) and serving to separate the cell products; (*C*) shell of the cell made of steel but lined with refractory bricks; (*N*) charge port for salt, and (*O*) bath level.

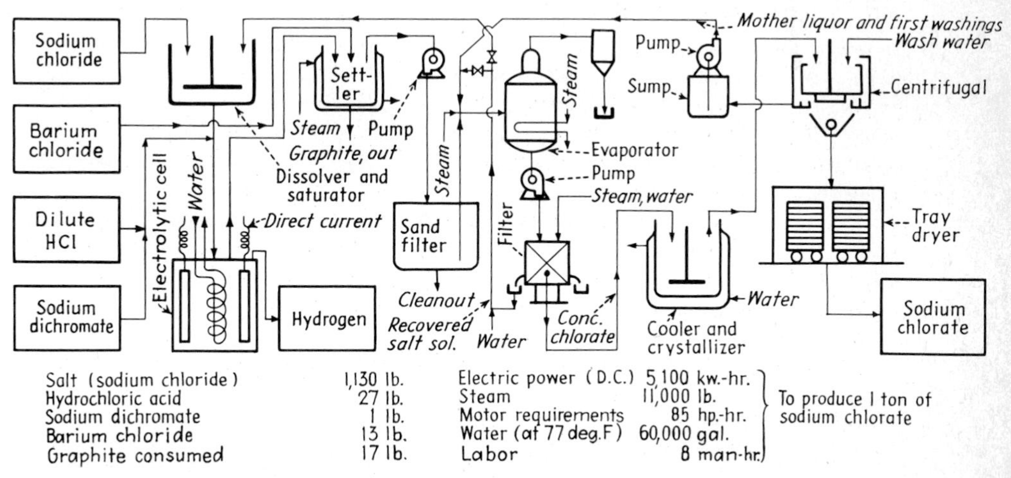

Fig. 14.8 Flowchart for the manufacture of sodium chlorate.

750 g/l sodium chlorate. After evaporation, the liquor is allowed to settle to remove the sodium chloride which constitutes the *recovered* salt and chlorate, to be used over again. The settled liquid is filtered and cooled. This cooling requires 3 to 5 days, and the crystals of sodium chlorate are spun in a centrifuge and dried.

Potassium perchlorate is made by converting sodium chlorate into sodium perchlorate in steel electrolytic cells which have platinum anodes and operate at 5.5 to 6.0 V, 2,500 A, and a temperature of 65°C. Filtered potassium chloride solution is added to the sodium perchlorate, precipitating potassium perchlorate cyrstals, which are centrifuged, washed, and dried. The mother liquor now contains sodium chloride, which can be used as cell feed for sodium chlorate manufacture.

OTHER PRODUCTS

Elsewhere in this book a few electrolytic processes are presented, as in the special chapter (Chap. 13) pertaining to caustic soda and chlorine. Electrolytic hydrogen and oxygen are included with the other procedures for hydrogen in Chap. 7 on Industrial Gases. Mention is made of electrolytic fluorine in Chap. 20. In Germany HCl is electrolyzed to chlorine and hydrogen at 150°F.[16]

ORGANIC COMPOUNDS

Electroorganic chemical preparations are not of great commercial importance at the present time, although the literature contains much information on organic reactions that can be carried out with the aid of electricity.[17] One of the few industrially important organic electrolytic reactions was the production of mannitol and sorbitol from glucose, but this electrolytic reduction has been superseded by a cheaper catalytic hydrogenation. However, Monsanto[18] has announced an electrolytic process giving adiponitrile from acrylonitrile.

PRIMARY AND SECONDARY CELLS[19]

Chemical reactions are utilized to convert chemical energy to electric energy, the converse of the procedures presented in the earlier parts of this chapter. The necessary chemicals, including the metals, cost about $140 million for the $700 million total sales of all cells. The chemical reaction can be caused to take place in a unit specially designed for the purpose of obtaining electric energy. These units are commonly called *cells*, or *batteries*. Primary cells produce electricity by means of a chemical reaction that is not easily reversible, and thus the chemicals must be replaced after the reaction has taken place. Secondary cells depend upon chemical reactions that are reversible by electric energy and thus do not require replacement of the chemical components. Obviously, then, the cost of electric energy from primary batteries is very high, especially if they are used for the production of large quantities of energy.

The most common *primary cell* today is the *dry cell*. This consists of a zinc container which serves as one electrode and a carbon rod which is the other electrode. The electrolyte is a water solution of ammonium chloride and zinc chloride adsorbed on flour or starch and manganese dioxide.

[16]Chlorine: New Process, *Chem. Eng. (N.Y.)*, **72**(3), 62 (1965).

[17]Eberson and Weinberg, Electroorganic Synthesis, *Chem. Eng. News*, Jan. 25, 1971, p. 40; Baizer (ed.), Organic Electrochemistry, Dekker, 1973; Weissberger and Weinberg (eds.), Technique of Electroorganic Synthesis, vol. 5, Wiley, 1974; Fry, Synthetic Organic Electrochemistry, Harper, 1972.

[18]Adiponitrile, Monsanto, *Chem. Eng. (N.Y.)*, **70**(22), 69 (1963); New Routes to Adiponitrile, *Chem. Eng. (N.Y.)*, **72**(20), 88 (1965). Less expensive than adipic acid, *cf.* Chap. 35. *Chem. Eng.* (*N.Y.*). **72**(23), 238 (1965) (flowchart).

[19]See Perry, pp. 4-68 to 4-75, on the thermodynamics of cells; Packing More Punch into Electric Batteries, *Chem. Eng. (N.Y.)*, **74**(2), 90 (1967).

The manganese dioxide is added as a depolarizer. Graphite is usually introduced into this mixture to increase the conductivity. The whole cell is insulated by a cardboard container.

The most important secondary cell is the *lead storage battery.* The reactions that take place on discharge between two electrodes dipping into partly diluted sulfuric acid are:

At positive lead peroxide plate (cathode):

$$PbO_2(s) + 4H^+(aq) + 2e \longrightarrow Pb^{2+}(aq) + 2H_2O(l)$$
$$Pb^{2+}(aq) + SO_4^{2-}(aq) \longrightarrow PbSO_4(s)$$
$$\overline{PbO_2(s) + 4H^+(aq) + SO_4^{2-}(aq) + 2e \longrightarrow PbSO_4(s) + 2H_2O(l)}$$

At negative sponge lead plate (anode):

$$Pb(s) \longrightarrow Pb^{2+}(aq) + 2e$$
$$Pb^{2+}(aq) + SO_4^{2-}(aq) \longrightarrow PbSO_4(s)$$
$$\overline{Pb(s) + SO_4^{2-}(aq) \longrightarrow PbSO_4(s) + 2e}$$

The overall reaction is:

$$Pb(s) + PbO_2(s) + 4H^+(aq) + 2SO_4^{2-}(aq) \longrightarrow 2PbSO_4(s) + 2H_2O$$

Electric energy is generated when these reactions take place.[20] To restore the activity or to recharge the battery, electric energy is applied and the reactions are reversed. The battery consists of lead plates in the form of a grid filled, when charged, with lead peroxide at the positive plate and sponge lead at the negative plate. The plates are insulated from each other by means of separators, which formerly were made of cedar but at present are made of glass. The battery is filled with dilute sulfuric acid. An ordinary battery consists of 13 or 15 plates per cell and has several cells in series. It delivers 2 V per cell.

Another widely used secondary cell is the *Edison battery.* The positive plate consists of nickel peroxide and flaked metallic nickel; the negative plate is finely divided iron. The electrolyte consists of a mixture of potassium hydroxide and lithium hydroxide. This battery finds widest use in heavy-duty industrial and railway applications.

The *nickel-cadmium*[21] secondary cell resembles the Edison nickel-iron cell, but it has excellent storage life, losing only about 10% of capacity in the first 48 h and then only about 3% or less per month. This cell employs nickel hydroxide and cadmium–cadmium hydroxide as the electrochemically active materials of the positive and negative electrodes, respectively. The electrolyte is an aqueous solution of potassium hydroxide. The overall cell reactions may be approximately represented as

$$\text{Charged } 2Ni(OH)_3 + Cd \longrightarrow 2Ni(OH)_2 + Cd(OH)_2 \text{ discharged}$$

Nominal discharge voltage is about $1\frac{1}{4}$ V per cell. These cells are rugged, low-gassing, efficient at low temperatures and, though more expensive, are employed for important stand-by and emergency service, portable service, and for cordless appliances such as shavers and toothbrushes. The Telstar satellite has a communication system powered by 19 nickel-cadmium batteries.

SELECTED REFERENCES

Bockris, J., *et al.:* An Introduction to Electrochemical Science, Springer-Verlag, 1974.
Care of Aluminum: Aluminum Association, 1973.

[20] To account for the electron current (from anode to cathode in the wire) the anode must be labeled negative with respect to the cathode.

[21] International Nickel Co., private communications; Booster for Battery Chemicals, *Chem. Week,* Sept. 29, 1962, p. 111.

Delahay, P., and C. Tobias (eds.): Advances in Electrochemistry and Electrochemical Engineering, Wiley-Interscience, 1961–1965, 5 vols.
Emley, E. F.: Magnesium Technology, Pergamon, 1966, 2 vols.
Gemant, A.: Electrochemistry of Ions in Hydrocarbons, Wiley-Interscience, 1962.
Gerard, G., and P. T. Stroup (eds.): Extractive Metallurgy of Aluminum, Wiley-Interscience, 1963, 2 vols.
Glasstone, S.: Fundamentals of Electrochemistry and Electrodeposition, Franklin Publishing, 1960.
Gould, R. F. (ed.): Literature of Chemical Technology, chap. 2, Industrial Electrochemistry of Non-metals, ACS Monograph, 1967.
Hampel, C. A. (ed.): Encyclopedia of Electrochemistry, Reinhold, 1964.
Hawkins, E. G. E.: Organic Peroxides: Their Formation and Reaction, Van Nostrand, 1961.
Kortum, G. F. A.: Treatise on Electrochemistry, 2d rev. ed., Elsevier, 1965.
Lowenheim, F. A.: Modern Electroplating, 3d ed., Wiley-Interscience, 1974.
Mantell, C. L.: Electrochemical Engineering, 4th ed., McGraw-Hill, 1960.
Martin, L. F.: Storage Batteries and Rechargeable Cell Technology, Noyes, 1974.
Martin, L. F.: Dry Cell Batteries, Noyes, 1973.
Muller, R. H. (ed.): Optical Techniques in Electrochemistry, Wiley, 1973.
Newman, J. S.: Electrochemical Systems, Prentice-Hall, 1973.
Pourbaix, M.: Lectures on Electrochemical Corrosion, Plenum, 1973.
Roberts, C. S.: Magnesium and Its Alloys, Wiley, 1960.
Sawyer, D. T., and J. L. Roberts: Experimental Electrochemistry for Chemists, Wiley, 1974.
Schumacher, J. C. (ed.): Perchlorates: Their Properties, Manufacture and Uses, ACS Monograph 146, Reinhold, 1960.
Van Horn, K. R. (ed.): Aluminum, American Society of Metals, 1967, 3 vols.

chapter 15

ELECTROTHERMAL INDUSTRIES

Many chemical products made at high temperatures demand the use of an electric furnace. Electric furnaces are capable of producing temperatures as high as 4100°C. This may be contrasted with the highest commercial combustion-furnace temperatures of about 1700°C. The effects of high temperature are twofold: The *speed* of the reaction is *increased*, and *new conditions* of *equilibrium* are established. These new equilibrium conditions have resulted in the production of compounds unknown before the electric furnace. Silicon and calcium carbides are examples of new products thus formed. The electric furnace affords more exact control and more concentration of heat with less thermal loss than is possible with other types of furnaces. This favorable situation is caused by the lack of flue gases and by the high temperature gradient between the source of heat and the heated mass. The electric furnace is much cleaner and more convenient to operate than the combustion furnace. It is operated by alternating current of high amperage, usually with moderate voltage, whereas the electrolytic industries require direct current.

The three chief types of electric furnaces[1] are arc, induction, and resistance. The heat in the arc furnace is produced by an electric arc between two or more electrodes, which are usually graphite or carbon, between the electrodes and the furnace charge or between two or more electrodes which are usually graphite or carbon and may or may not be consumed in the operation. The furnace itself is generally a cylindrical shell lined with a refractory material. Its use is not limited to those industries for which it is a necessity; some companies have rolling-mill operations for common-quality steels where electric-arc furnaces are the sole source of ingots.

The induction furnace may be applied only for conducting substances such as metals, where the electric energy is converted into heat by induced currents set up in the charge. The furnace can be considered as a transformer, with the secondary consisting of the metallic charge and the primary of heavy copper coils connected to the power source. Induction furnaces operate at frequencies from 60 to 500,000 Hz, but those used in commercial-scale electrothermal processes do not usually use frequencies above 6,000 Hz. The heating effect is obtained with lower field strengths as the frequency is increased. The charge should be placed around an iron core in the low-frequency furnace, but this core is unnecessary for high-frequency furnaces. When the charged material furnishes the electric resistance required for the necessary heat, the furnace is *direct-heated* resistance; when high-resistance material is added to the charge for the purpose of creating heat, the furnace is *indirect-heated.* In the electrochemical industry, the arc and the resistance furnaces are most used.

[1]Mantell, Electrochemical Engineering, 4th ed., McGraw-Hill, 1960.

ARTIFICIAL ABRASIVES

HISTORICAL Before 1891 all abrasives used were natural products such as diamonds, corundum, emery, garnet, quartz, kieselguhr, and rouge. In that year E. G. Acheson produced the first man-made abrasive in a homemade electric-arc furnace while attempting to harden clay. Acheson found these new hard purple crystals to be silicon carbide. Discovering that these crystals were hard enough to cut glass, he sold them, under the tradename Carborundum, to gem polishers for $880 per pound.[2] Another make of silicon carbide is sold under the tradename Crystolon. Fused aluminum oxide, the most extensively used abrasive, is manufactured in an electric thermal furnace. Boron nitride, or Borazon, one of the hardest substances made synthetically, is an electrothermal product.

USES AND ECONOMICS The discovery and production of artificial abrasives marked the beginning of the evolution of modern grinding tools, which are of paramount importance in the modern precise fabrication of multitudinous metal parts for automobiles, airplanes, rifles, cannon, and other manufactured items of present-day industrial endeavor. To reduce wear, harder and harder alloys are being developed by the metallurgist, many of which can be finally shaped economically only by these hard artificial abrasives. Silicon carbide and alumina are also used as a refractory material, both in the form of brick and as loose material for ramming in place. In the United States and Canada, the production of silicon carbide exceeds 100,000 short tons, and that of fused aluminum is about 200,000 tons. Impermeable silicon carbide[3] withstands 1000°C and resists corrosion, erosion, and coerosion.

SILICON CARBIDE The raw materials for the production of silicon carbide[4] are sand and carbon. Carbon is obtained from anthracite, coke, pitch, or petroleum cokes, and sand contains 98 to 99.5% silica. The equations usually given for the reactions involved are

$$SiO_2(c) + 2C(amorph) \longrightarrow Si(c) + 2CO(g) \qquad \Delta H = +144.8 \text{ kcal}$$

$$Si(c) + C(amorph) \longrightarrow SiC(c) \qquad \Delta H = -30.5 \text{ kcal}$$

The total reaction obtained by combining these equations is

$$SiO_2(c) + 3C(amorph) \longrightarrow SiC(c) + 2CO(g) \qquad \Delta H = +114.3 \text{ kcal}$$

Figure 15.1 is a flowchart for the production of silicon carbide. Sand and carbon are mixed in an approximate molar ratio of 1:3 and charged into the furnace. Sawdust, if added, increases the porosity of the charge and permits the circulation of vapors and the escape of the carbon monoxide produced. The charge is built up in the furnace around a heating core of granular carbon. This core is in the center of the 30- to 50-ft-long furnace and connects the electrodes. The walls of the furnace are loose firebrick supported by iron castings and are taken away at the end of a charge to facilitate product removal. Excessive heat loss does not occur because the outside unreacted charge serves as an insulator. A typical initial current between the electrodes is 6,000 A at 230 V, and the final current is 20,000 A at 75 V. This change in voltage is due to the decrease in the resistance of the charge as the reaction progresses. The temperature at the core is 2200°C. The temperature should not become too high, or the silicon carbide will decompose with the volatilization of silicon and the formation of graphite. Indeed, artificial graphite was so discovered. The energy efficiency is about 50%, and the chemical conversion is from 70 to 80%.

The time of the reaction is about 60 h, 36 h of heating and 24 h of cooling. After cooling, the silicon carbide crystals are removed, with a yield of about 6 to 8 tons per furnace. The larger pieces of crystals are broken, washed, and cleaned by chemical treatment with sulfuric acid and caustic soda.

[2]Cooper, Modern Abrasives, *J. Chem. Educ.*, **19**, 122 (1942); Upper, The Manufacture of Abrasives, *J. Chem. Educ.*, **26**, 676 (1949).

[3]Roe and Kwasniewski, Resist Wear and Acids with Silicon Carbide, *Chem. Eng. Prog.*, **60**(12), 97 (1964).

[4]Mantell, *op. cit.*, p. 509; ECT, 2d ed., vol. 4, p. 114.

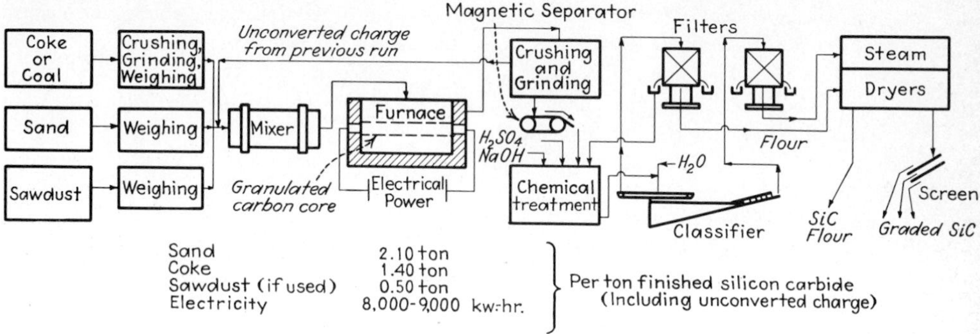

Fig. 15.1 Flowchart for the manufacture of silicon carbide.

The crystals are classified and screened, the finished product ranging from 6-mesh to fine powder. The outer unreacted part of the charge is combined with the next charge for the furnace. Part of the core used in the furnace can be made of coke suitable for graphite manufacture. After the run is completed, the graphite can be separated from the silicon carbide and converted to desired shapes.

FUSED ALUMINUM OXIDE The raw material for fused aluminum oxide abrasives such as those sold under the tradenames Alundum and Aloxite may be impure bauxite, often called *red bauxite.* The impurities, mainly iron and silicon oxides, have a great effect on the structure and properties and must be carefully controlled. Pure Bayer alumina is also fused for particular demands, e.g., where a grinding-wheel temperature must be kept low. In any case, the aluminous material should be calcined before charging into the furnace.

The vertical arc-resistance furnace consists of a circular steel shell about 7 ft high resting on a crucible carbon base about 5 ft in diameter. The shell has a slight taper at the top for easy removal from the pig. The outside surface of the furnace is water-cooled so that the unfused alumina around the walls furnishes a refractory lining for the furnace. The round carbon electrodes are lowered into the furnace, and arcs are drawn between them and the fused furnace charge. The fusion is usually started by forming a trench in the top of the starting batch and packing it with coarse coke. The

Fig. 15.2 Electric-resistance furnace in which Carborundum-brand silicon carbide is produced. (*The Carborundum Co.*)

electrodes are lowered to make contact with opposite ends of the trench. As soon as the alumina begins to fuse, it carries the current, and the starting coke is rapidly consumed. A typical charge to the furnace is calcined bauxite (89%), coke (2%), and scrap iron (9%). The charge is added as fast as it fuses between the arcs, the molten alumina carries the current, and much of the iron and silica is reduced to form a heavy alloy and sinks to the bottom. As the level of fused alumina rises, the electrodes are raised and more charge is added, until finally the furnace is full. Then the current is shut off and the entire mass cooled under controlled conditions to obtain the texture desired. Although this product has the hardness 9 on the Moh scale for corundum, it is blebbular in structure and not uniform, and therefore bearings cannot be made from it. The cooled ingots are broken up by roll crushers, washed with chemical solutions, and sieved. The product is fabricated into abrasive wheels, papers, and powders or into refractory shapes by sintering.

ARTIFICIAL CORUNDUM, OR SAPPHIRE[5] For the hard bearings necessary for watches and modern instruments, *artificial corundum*, or white sapphire, is made by crystallizing through fusion *pure* alumina in a hydrogen-oxygen upside-down flame at 3500°F by the Verneuil process. The crystal boules are cut and polished as desired. Many gems are also made by this process, such as star and transparent rubies and sapphires of varied color.

BORON CARBIDE[6] This is a very hard abrasive. It first made its appearance in 1934 under the name *Norbide*. The reaction for its production is:

$$2B_2O_3 + 7C \longrightarrow B_4C + 6CO$$

The boric oxide is caused to react with coke in a carbon resistance furnace at 2600°C. The product is about 99% B_4C. It finds specialized use as a powdered abrasive and in molded shapes such as nozzles for sandblasting.

BORON NITRIDE *Borazon* is a cubic form of boron nitride formed at 1 million psi and 3000°F to change the ordinary hexagonal boron nitride to this cubic form. Its hardness is comparable with that of diamond, and it has the advantage of resisting oxidation much better. It is made by General Electric Co.

CALCIUM CARBIDE

HISTORY The first production of calcium carbide was an accident. In 1892, T. L. Willson was attempting to prepare metallic calcium from lime and tar in an electric furnace at Spray, N.C. The product obtained, obviously not calcium, was thrown into a nearby stream, and Willson was amazed to note that it liberated great quantities of combustible gas. The first factory for the production of calcium carbide was built at Niagara Falls in 1896.[7]

USES AND ECONOMICS Calcium carbide is utilized in two principal procedures: manufacturing cyanamide by combining it with nitrogen, and preparing acetylene by reacting it with water. Cyanamide is made by heating calcium carbide in an atmosphere of nitrogen (Chap. 18). A substantial proportion of the total calcium carbide produced is converted to cyanamide. Calcium carbide was used in large quantities for the manufacture of acetylene, but hydrocarbon-based acetylene has been taking over in the United States. The U.S. production of calcium carbide fell from 625,338 short tons in 1971 to 493,339 short tons in 1972. During the 1960s production was in excess of 1 million short tons/year.

MANUFACTURE Calcium Carbide[8] is prepared from quicklime and carbon at 2000 to 2200°C.

$$CaO(c) + 3C(amorph) \longrightarrow CaC_2(l) + CO(g) \qquad \Delta H = +103 \text{ kcal}$$

[5]Flame Grown Gem Stones Enjoy Broadened Use in Optics and Jewelry, *Chem. Eng. (N.Y.)*, **68**(26), 26 (1961).
[6]Boron Carbide Process Temperature, *Chem. Eng. News*, Feb. 5, 1968, p. 262.
[7]Carbide and Acetylene, *Chem. Eng. (N.Y.)*, **57**(6), 129 (1950); Mantell, *op. cit.*, pp. 491ff.
[8]ECT, 2nd ed., vol. 4, p. 100.

The source of carbon is usually coke, anthracite, or petroleum coke. Coke is the most widely used. It should be compact and have a low ash content, a low ignition point, and high electrical resistivity, so that the bulk of the furnace charge will be highly resistant to the flow of energy. Thus the energy is concentrated, resulting in a more rapid and complete reaction. Phosphorus should be absent, since it forms a phosphide which is converted to poisonous phosphine (PH_3) when the carbide is made into acetylene. Quicklime is produced by burning limestone containing at least 97% calcium carbonate. Impurities such as magnesia, silica, and iron hamper production and give a less pure carbide.

The carbide furnace is not a true arc-resistance furnace, but has been developed from the familiar arc furnace. Ingot furnaces, similar to those producing fused aluminum oxide, have been replaced in the carbide industry by continuous or intermittent tapping furnaces producing molten carbide. The furnace consists of a steel shell with the side walls lined with ordinary firebrick and the bottom covered with carbon blocks or anthracite to withstand the extremely hot, alkaline conditions. Most of the larger furnaces use three-phase electric current and have suspended in the shell three vertical electrodes. Improvements include the "closed" furnace, where almost all the carbon monoxide from the reaction is collected and utilized, and Söderberg continuous self-baking electrodes, which permit larger-capacity furnaces than the old prebaked electrodes. The capacity range of the furnaces is generally between 5,000 to 18,000 kWh or higher, and a three-phase tapping furnace of 25,000 kWh produces about 200 tons of commercial product (usually 85% carbide) per day. The approximate consumption of materials per ton of carbide is 1,900 lb lime, 1,300 lb coke, 35 lb electrode paste, and 3,000 kWh energy (Fig. 18.1). The lime and coke are charged continuously with intermittent or continuous tapping of the liquid product directly into cast-iron chill pots of about 5 tons capacity each. The carbide is cooled, crushed, and sized and then packed in 10- to 220-lb steel drums or up to 5-ton containers for shipping.

MISCELLANEOUS ELECTROTHERMAL PRODUCTS[9]

FUSED SILICA This very useful construction material for the chemical industries is heated to fabrication temperature in an electric furnace.

SYNTHETIC QUARTZ CRYSTALS[10] These crystals have been grown successfully by a hydrothermal process at Bell Telephone Laboratories and Western Electric Co. by keeping a supersaturated solution of silica in 1 *N* sodium hydroxide in contact with seed crystal quartz plates at 30,000 psi for 3 weeks at a temperature of about 750°F. They are grown in special autoclaves, with a two-zone electrical heating system having a higher temperature for the dissolving zone, to bring into solution small chips of quartz, and then allowing growth on prepared quartz seed plates at about 100°F lower temperature. A satisfactory product results at less cost than mined quartz. Such quartz crystals are very important, indeed necessary, for use in radar, sonar, radio, and television transmitters and telephone communications. A single telephone wire can carry hundreds of messages at one time, thanks to the unscrambling powers of quartz-crystal filters.

ARTIFICIAL GRAPHITE AND ELECTRODES are described in Chap. 8, devoted to carbon, and phosphorus is presented in Chap. 16, on Phosphorus Industries. See also Table 14.1 for a summary of the various products made by electrochemical processes.

SELECTED REFERENCES

Berezhnoi, A. S.: Silicon and Its Binary Systems, Plenum, 1960.
Campbell, I. E. (ed.): High Temperature Technology, Wiley, 1956.
Gould, R. F. (ed.): Literature of Chemical Technology, chap. 7, Abrasives, ACS Monograph, 1966.
Mantell, C. L.: Electrochemical Engineering, 4th ed., McGraw-Hill, 1960.
Schwarzkopf, P., *et al.*: Cemented Carbides, Macmillan, 1960.

[9]For carbon disulfide, see Chap. 38, and CPI 2 for old electrochemical process.
[10]Higgins, 30,000 psi for Three Weeks, *Ind. Eng. Chem.*, **54**(1), 16 (1962) (details given).

chapter 16

PHOSPHORUS INDUSTRIES

The use of artificial fertilizers, phosphoric acid, and phosphate salts and derivatives has increased greatly, chiefly because of aggressive and intelligent consumption promotion on the part of various manufacturers and federal agencies. However, before full consumption of these products could be achieved, more efficient and less expensive methods of production had to be developed. During recent decades, the various phosphate industries have made rapid strides in cutting the costs both of production and distribution and have thus enabled phosphorus, phosphoric acid, and its salts to be employed in wider fields and newer derivatives to be introduced. Supplementing the development of more efficient phosphorus industries have been the epoch-making pure chemical studies of phosphorus, in its old and in its new compounds. These phosphates are not simple inorganic chemicals, as was assumed several decades ago, and their study has become a unique and complicated branch of chemistry that may some day be compared with the carbon (organic) or silicon branches of today. The properties of phosphorus chemicals are unique because of the important role of phosphorus in many biochemical processes, the ability of polyphosphates to complex or sequester many metal cations, and versatility in forming various types of organic and inorganic polymers.

HISTORICAL The use of phosphatic materials as fertilizers was practiced unknowingly long before the isolation and discovery of phosphorus by the German alchemist Brand in 1669. As early as 200 B.C., the Carthaginians recommended and employed bird dung to increase the yields from their fields. The Incas of Peru prized guano and bird dung on their islands so highly that it was made a capital offense to kill birds. Then too, we are familiar with the use of fish and bones by American Indians in their crude agricultural methods. Bones and guano continued to be the chief sources of phosphorus and phosphoric acid until after the middle of the nineteenth century, but these supplies were and still are limited. In 1842, a British patent was issued to John B. Lawes for the treatment of bone ash with sulfuric acid. This patent marked the beginning of a large acid phosphate industry which became the basis of our domestic fertilizer industry. Soon afterward various grades of phosphate ores were discovered in England. These were first finely ground and applied directly to the soil. It was soon recognized, however, that treatment of these phosphate minerals with sulfuric acid increased the availability and efficiency of the phosphate for agricultural purposes. At present, acidulation with nitric acid or strong phosphoric acid gives enhanced fertilizer values.

USES AND ECONOMICS See Table 16.1. The most important use of phosphate rock is in fertilizers. See Chap. 26 on Agrichemical Industries. Table 16.2 is a compilation of phosphate-rock treatment processes. Tricalcium phosphate in raw and/or steamed and degreased bones and in basic slag is also used after grinding as a direct phosphate fertilizer. A small percentage of the former is sometimes treated with sulfuric acid for superphosphate or as a source material for phosphate chemicals. Large tonnages of phosphate rock are converted to phosphorus or phosphoric acid and their derivatives.

TABLE 16.1 ***Phosphate Rock Production and Shipments in the United States*** *(In thousands of tons)*

	Florida-North Carolina		Tennessee		Western States		Total	
Fiscal year	*Production*	*Shipments*	*Production*	*Shipments*	*Production*	*Shipments*	*Production*	*Shipments*
1959–1960	13,600	13,340	2,000	2,020	3,280	3,290	18,890	18,650
1960–1961	14,830	14,130	2,170	2,150	3,490	3,230	20,500	19,520
1961–1962	15,760	14,920	2,850	2,940	2,900	3,110	21,510	20,970
1962–1963	15,340	15,590	2,680	2,690	3,230	3,110	21,250	21,390
1963–1964	17,600	17,250	2,730	2,780	3,470	3,770	23,800	23,410
1964–1965	20,450	19,630	2,840	2,860	4,050	4,000	27,340	26,490
1965–1966	25,420	25,270	3,080	3,050	5,970	5,280	34,470	33,610
1966–1967	30,280	28,840	2,950	2,970	5,420	4,920	38,660	36,730
1967–1968	34,060	29,760	3,120	3,030	4,850	4,950	42,030	37,740
1968–1969	31,900	29,270	3,190	3,100	4,960	4,640	40,040	37,020
1969–1970	29,810	29,660	3,180	3,220	4,300	4,570	37,290	37,460
1970–1971	31,510	31,570	2,940	2,920	4,350	4,420	38,800	38,900
1971–1972	33,170	34,860	2,350	2,440	4,400	4,460	39,920	41,760
1972–1973	34,550	36,440	2,440	2,560	4,880	4,730	41,870	43,740
1973–1974*	35,200	36,900	2,500	2,600	5,300	5,300	43,000	44,800
1978–1979*	52,100	49,300	2,500	2,500	6,500	6,500	61,150	58,300

Sources: Bureau of Mines; Phosphate Fertilizer Shortage is Worldwide, *Chem. Eng. News,* June 24, 1974, p. 10.
*Estimate by Phosphate Rock Export Association.

TABLE 16.2 ***Phosphate-rock Processing, Products, and Byproducts***

Process	*Raw materials and reagents*	*Main products and derivatives*	*By-products*
Acidulation	Phosphate rock, sulfuric acid, nitric acid, phosphoric acid, hydrochloric acid, ammonia, potassium chloride	Superphosphate, phosphoric acid (wet process), triple superphosphate, monoammonium phosphate, diammonium phosphate, monopotassium phosphate	Fluorine compounds, vanadium, uranium (limited)
Electric-furnace reduction	Phosphate rock, siliceous flux, coke (for reduction), electrical energy, condensing water	Phosphorus, phosphoric acid, triple superphosphate, various Na, K, NH_4, Ca salts; phosphorus pentoxide and halides	Fluorine compounds, carbon monoxide, slag (for RR ballast aggregate, fillers, etc.), ferrophosphorus, vanadium*
Calcium metaphosphate	Phosphate rock, phosphorus, air or oxygen, fuel	Calcium metaphosphate	Fluorine compounds
Calcination or defluorination	Phosphate rock silica, water or steam, fuel	Defluorinated phosphate	Fluorine compounds

*Vanadium is present in appreciable quantities only in the western phosphates.

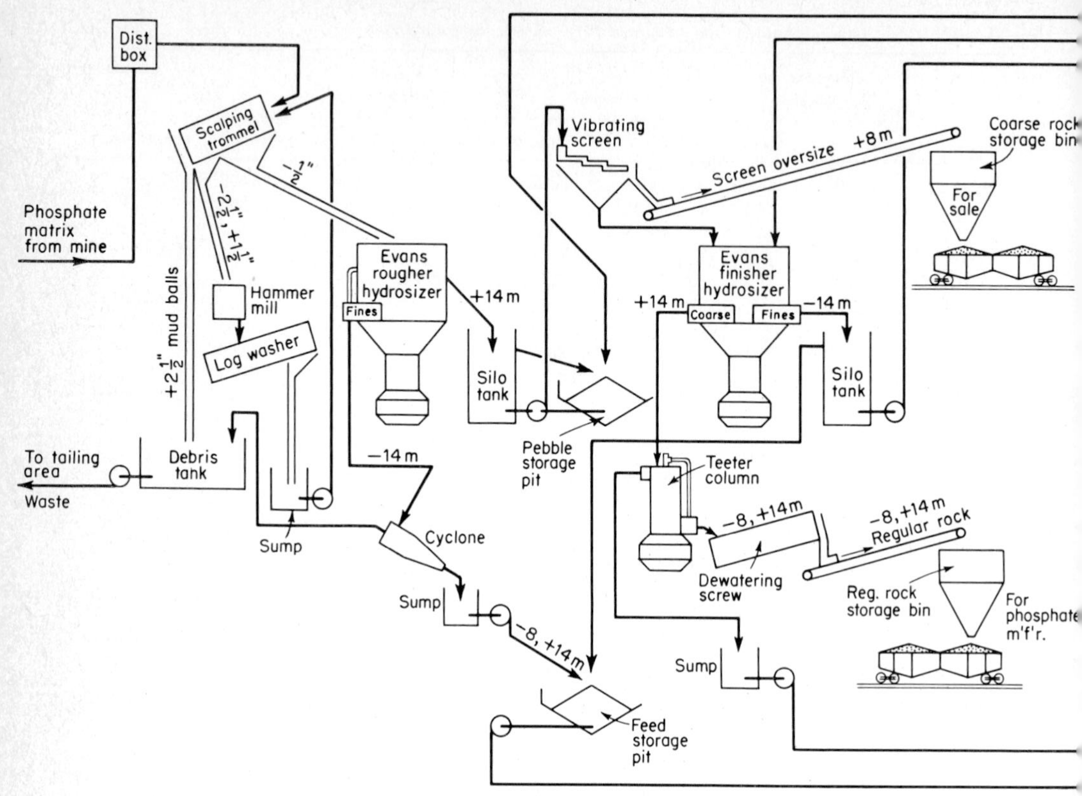

Fig. 16.1 Flowchart for the beneficiation of Florida phosphate ore.

PHOSPHATE ROCK AND SUPERPHOSPHATES[1]

Phosphate rock, when very finely pulverized, has limited use as a fertilizer itself, chiefly because of the relatively slow availability of the P_2O_5. Its main consumption, however, is as a raw material for the manufacture of phosphoric acid, superphosphate, phosphorus, and phosphorus compounds. Florida has been the principal producing area for phosphate rock, with Tennessee ranking second until recently. Now Idaho ranks second. In Florida, both hard and pebble rock phosphate are mined. Hard rock phosphate occurs as nodules and boulders in irregular pockets but with only limited exploitation. The more extensive and cheaply mined pebble deposits occur with an average overburden of 20 ft, and their phosphatic value is too low for economical processing. The pebble deposits themselves, called the *matrix*, are from 10 to 30 ft thick. This matrix is composed of clay slimes, silica sand, and phosphate pebble. Pebble sizes range from $1\frac{1}{2}$ to 2 in. (in small amounts) down to 400-mesh. The overburden is removed generally by a drag line, which dumps into a previously mined-out cut. Drag-line operations remove the matrix and drop it into an excavated area. Hydraulic guns break down the mud in the matrix and wash it into the pump suction, where it is transported through pipes by large sand pumps to the beneficiation plant (Fig. 16.1).

[1]Slack (ed.), Phosphoric Acid, Dekker, 1968; ECT, 2d ed., vol. 15, p. 232, 1968; Crerar, H_3PO_4 Route Cuts Costs, *Chem. Eng. (N.Y.)*, **80**(10), 62 (1973).

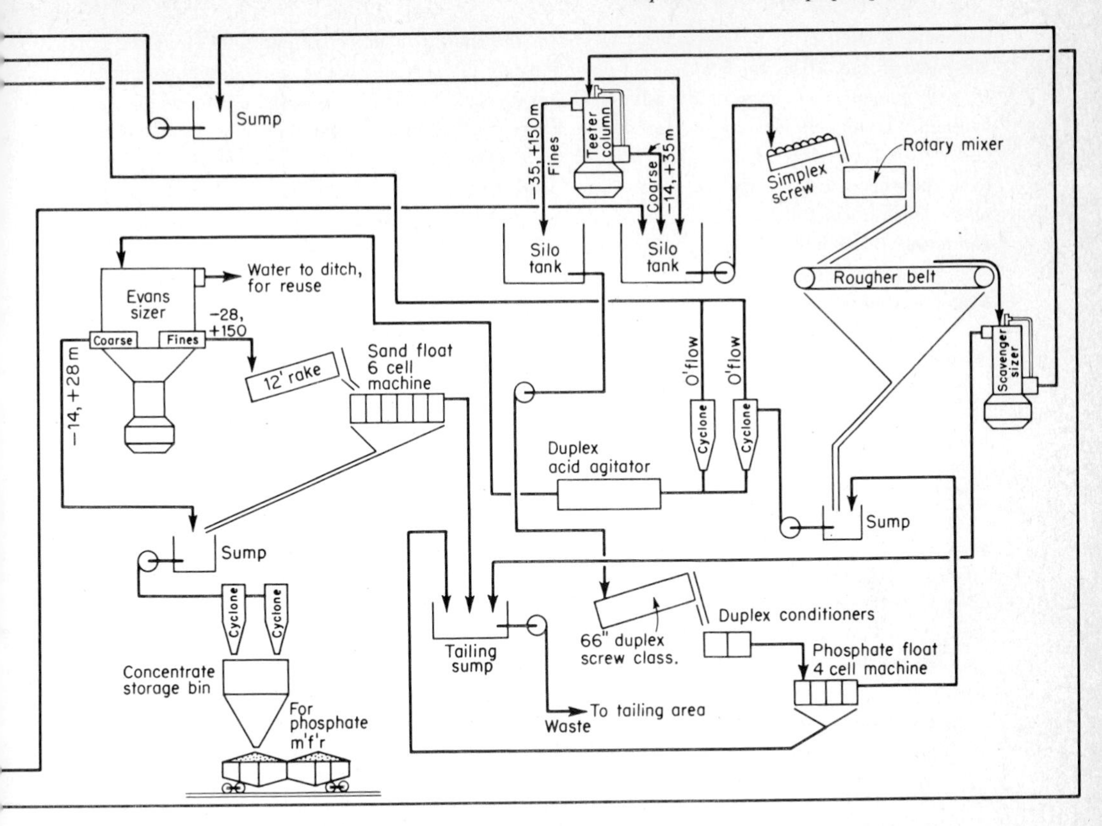

In Tennessee four types of phosphate rock are found, nodular, blue, white, and brown. Only the latter is mined at present, using open-pit, dry methods. The overburden varies greatly and has an average depth of from 6 to 8 ft. Although originally deposited in horizontal strata, in the large western fields layers of phosphate rock have been severely folded, faulted, and elevated by crustal deformations and resemble fissure veins. The rock may contain approximately 75% bone phosphate of lime[2] but, because of admixture with wall material, it usually averages nearer 70% bpl. Because of its rather soft structure, the rock has a moisture content of 4 to 6%. It is generally mined by underground methods. Western phosphate rock contains anywhere from 0.1 to 0.4 lb uranium per ton of rock. The complicated and costly recovery process now in use extracts uranium from phosphate rock via wet-process phosphoric acid, and the uranium goes into solution upon treatment with sulfuric acid. After filtration, the bulk of the uranium is found in the acid filtrate. Some is recovered, presumably by an ion-exchange technique. Fluorine and vanadium are also valuable by-products from phosphate rock.

In the Florida pebble district, initially only coarse phosphate rock was recovered, which had a high bpl, whereas the fines, with a much lower bpl, representing about equal tonnage, were wasted. In the 1920s, experimental work was started to develop a froth-flotation process which would increase the bpl of the fines or raw matrix to at least 66 or 68% from about 40% or less. This procedure is

[2]This is usually abbreviated bpl and actually means tricalcium phosphate, the chief inorganic constituent of bones.

frequently spoken of as *beneficiation*, and one of the successful processes currently employed is illustrated in Fig. 16.1. Such upgrading operations are of far-reaching and increasing importance, as easily mined or better-grade deposits of phosphate rock, and other minerals, are becoming exhausted. These operations not only produce a higher grade of product, but allow larger amounts to be recovered, beneficiated, and used, even in the case of such a low-priced product as phosphate rock. Beneficiation was largely abandoned in Tennessee when it was found that the deslimed phosphatic sand on which it worked best could be used directly in electric furnaces to produce elemental phosphorus. The bulk of the high-grade lump rock and sands in Tennessee has been exhausted, and operations would be abandoned if not for cheap electricity for the phosphorus furnaces and economical mechanical mining operations.

Charge	*Long tons/year*
Florida phosphate pebble mine matrix: specifications, ⅓ phosphate rock, ⅓ sand, ⅓ clay (approx.)	6 million
Products (in storage bins)	(2 million)
Coarse rock, 25%	500,000
Regular rock, 60%	1.2 million
Concentrate, 15%	300,000
Waste	
Tailings area (sand and clay)	4 million
Flow to ditch (water only), recycled and reused	6.5 billion gal

Labor and utilities	*Amounts*
Labor	2 hr
Power	25 kW
Water	3,250 gal
Flotation reagents	*Per long ton* lb
Amine	0.2
Caustic soda	0.9
Fuel oil	4.6
Tall oil (fatty acid)	6.0

The matrix from Florida phosphate pebble deposits is beneficiated through the general sequences illustrated in Fig. 16.1, involving various water-separation methods, such as passage through a trommel-scalping screen, a hammer mill, Evans hydrosizers with teeter columns, and additional vibrating sizing, dewatering, and flotation devices, resulting in the following sequential grading separations and concentrations:

1. *Coarse rock,* +8-mesh (storage bin), for sale (or high-grade use).
2. *Regular rock,* −8-, +14-mesh (storage bin), for phosphate manufacture.

Fines through another Evans hydrosizer and flotation to

3. *Concentrate* (storage bin), for phosphate manufacture.

The waste, with only a small percentage of phosphate ore, is run to a tailings area (from two sources) or to a ditch (for excess water). The regular rock and the concentrate are taken to the phosphate plant for acidulation or other chemical treatment, generally to increase the solubility or availability of the phosphorus compound. Domestic phosphate rocks are essentially fluorapatite admixed with various proportions of other compounds of calcium, fluorine, iron, aluminum, and silicon.

The formula of fluorapatite is $CaF_2 \cdot 3Ca_3(PO_4)_2$, equivalent to $Ca_{10}F_2(PO_4)_6$. This compound is *extremely* insoluble. The various means for making the P_2O_5 content more soluble, not necessarily in water, but in plant juices (as measured by "citrate solubility"), are manufacture of various superphosphates, the defluorination of fluorapatite by calcination at incipient fusion temperatures of 2550 to 2750°F with silica or phosphoric acid.

The uses to which phosphate compounds are put in the fertilizer field are largely dependent upon the solubilities or the availability to the plants. These products may be classified:

1. *Water-insoluble products.* Rock phosphate or fluorapatite $[CaF_2 \cdot 3Ca_3(PO_4)_2]$. This can be solubilized by the sulfuric, phosphoric, or nitric acid of the superphosphate process or by the *slow* dissolving action of plant juices.

2. *Citrate-soluble products.* Dicalcium phosphate ($CaHPO_4$), commonly called *precipitated phosphate,* or *precipitated bone;* basic slag, calcium and potassium metaphosphates, defluorinated

phosphates (calcined phosphates). Also, part of the tricalcium phosphate of bone meal is citrate-soluble. These products are soluble in ammonium citrate solution and are considered to be available for plant food.

3. *Water-soluble products.* Monocalcium phosphate $[CaH_4(PO_4)_2 \cdot H_2O]$ is the principal member of this class and is the chief ingredient of superphosphate. Various sodium phosphates, monammonium and diammonium phosphates, potassium phosphates, and some organic phosphates are also water-soluble.

SUPERPHOSPHATES[3]

The acidulation of phosphate rock to produce superphosphate has been the most important method of making phosphate available for fertilizer purposes for nearly a century. See Chap. 26 on agrichemicals. The reactions have long been given as

$$Ca_3(PO_4)_2 + 2H_2SO_4 + 4H_2O \longrightarrow \underset{\text{Monocalcium phosphate}}{CaH_4(PO_4)_2} + \underset{\text{Gypsum}}{2(CaSO_4 \cdot 2H_2O)}$$

$$CaF_2 + H_2SO_4 + 2H_2O \longrightarrow CaSO_4 \cdot 2H_2O + 2HF\uparrow$$

$$4HF + SiO_2 \longrightarrow SiF_4\uparrow + 2H_2O$$

In water:

$$3SiF_4 + 2H_2O \longrightarrow SiO_2 + 2H_2SiF_6$$

The following is a more probable expression of the main reaction:

$$\underset{\text{Phosphate rock or fluorapatite}}{CaF_2 \cdot 3Ca_3(PO_4)_2} + \underset{\text{53–57° Bé}}{7H_2SO_4} + 3H_2O \longrightarrow \underset{\text{Monocalcium phosphate}}{3CaH_4(PO_4)_2 \cdot H_2O} + 2HF\uparrow + \underset{\text{Anhydrite}}{7CaSO_4}$$

The hydrofluoric acid reacts as shown above, forming fluosilicic acid, but with incomplete removal of fluorine. An excess of sulfuric acid is consumed by such impurities in the phosphate rock as $CaCO_3$, Fe_2O_3, Al_2O_3, and CaF_2. The product increases in weight over the 70 to 75° bpl phosphate rock used, as much as 70%, resulting in a superphosphate with 16 to 20% available P_2O_5. The manufacture of superphosphate involves four steps: (1) preparation of phosphate rock, (2) mixing with acid, (3) curing and drying of the original slurry by completion of the reactions, and (4) excavation, milling, and bagging of the finished product. Although newer plants use continuous processes, some plants still conduct these operations stepwise. All plants first pulverize the rock. With modern pulverizing and air-separation equipment, most rock is ground to an average fineness of 70 to 80% through a 200-mesh screen, with the following benefits: (1) the reaction rate is faster; (2) more efficient use is made of the sulfuric acid and consequently less acid is needed; and (3) a higher grade of product in better condition is obtained.[4]

Manufacture of normal superphosphate by a continuous-den process. This is depicted by Fig. 16.2, where ground phosphate rock (90% minus 100) is fed by a weigh feeder into a double-conical mixer (TVA), where it is thoroughly mixed with metered quantities of sulfuric acid. The sulfuric acid is diluted with water in the cone to a concentration of 51°Bé; the heat of dilution serves to heat the sulfuric acid to proper reaction temperature, and excess heat is dissipated by evaporation of extra water added. The rate of water addition and acid concentration may be varied to control product moisture. The acid and water are fed into the cone mixer tangentially to provide the necessary

[3]Waggaman, Phosphoric Acid, Phosphates and Phosphatic Fertilizers, Hafner, 1969; Rushton, Wet Process Superphosphoric Acid by Vacuum Concentration, *Chem. Eng. Prog.*, **64**(5), 68 (1968); Houghtaling and Rushton, Sludge Recirculation Process for Phosphoric Acid Concentration, *Chem. Eng. Prog.*, **64**(5), 64 (1968); Kealy, Phosphatic Granulating Liquid, *Chem. Eng. Prog.*, **64**(5), 71 (1968).

[4]CPI 2, p. 345, illustrates and describes the long-used den process, together with ammoniation.

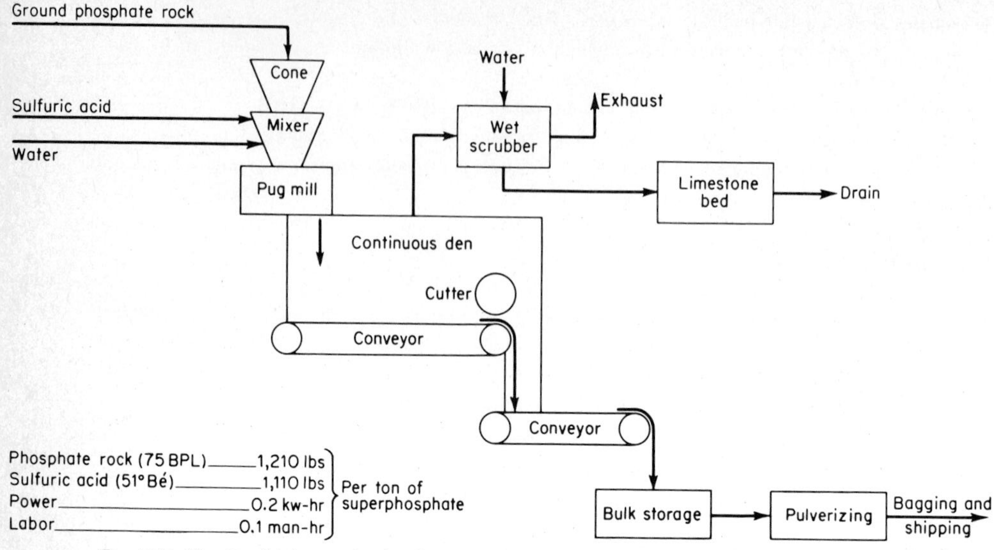

Fig. 16.2 Flowchart for the manufacture of superphosphate by the continuous-den process. (*Austin Co.*)

mixing with the phosphate rock. The fresh superphosphate is discharged from the cone mixer into a pug mill, where additional mixing takes place and the reaction starts. From the pug mill the superphosphate drops onto the den conveyor, which has a very low travel speed to allow about 1 h for solidifying before reaching the cutter. The cutter slices the solid mass of crude product so that it may be conveyed to pile storage for "curing," or completion of the chemical reaction, which takes 10 to 20 days to reach a P_2O_5 availability acceptable for plant food. The continuous den is enclosed so that fumes do not escape into the working area. These fumes are scrubbed with water sprays to remove acid and fluoride before being exhausted to the atmosphere. The scrubber water is discharged to a limestone bed to neutralize the acid.

Nitric and mixed acid acidulation of phosphate rock.[5] Europe probably first used nitric and mixed acid acidulation of phosphate rock. The substitution of nitric for sulfuric acid is desirable, since nitrogen has an essential value as plant food and can be resold at its purchase price. Also, this saves sulfur. Simple acidulation of phosphate rock with nitric acid produces a hygroscopic superphosphate, since it contains calcium nitrate. The TVA and others have studied and recommended commercial processes. In one, the phosphate rock is *extracted* by mixed nitric and sulfuric acids, followed by ammoniation, drying, and the addition of potassium chloride (optional). Another features mixed nitric and phosphoric *acidulation,* followed by the conventional steps, and others use nitric acid alone for acidulation. These processes, as well as conditioning against moisture absorption as practiced for ammonium nitrate, have led to an extension of this acidulation with nitric acid. Nitrophosphate is also gaining in Europe. Phosphate rock is decomposed with nitric acid plus a small amount of phosphoric acid. The resulting slurry is ammoniated and carbonated and, if desired, combined with potassium salts and spray-dried to yield a uniform pelletized product.

TRIPLE SUPERPHOSPHATE This material is a much more concentrated fertilizer than ordinary superphosphate, containing from 44 to 51% of available P_2O_5, or nearly three times the

[5]Davis, Meline and Graham, TVA Mixed Acid Nitric Phosphate Process, *Chem. Eng. Prog.*, **64**(5), 75 (1968); Jorquera, Nitric vs. Sulfuric Acidulation of Phosphatic Rock, *Chem. Eng. Prog.*, **64**(5), 83 (1968).

amount in regular superphosphate. Triple superphosphate is made by the action of phosphoric acid on phosphate rock, and thus no diluent calcium sulfate is formed.

$$CaF_2 \cdot 3Ca_3(PO_4)_2 + 14H_3PO_4 \longrightarrow 10Ca(H_2PO_4)_2 + 2HF\uparrow$$

The TVA continuous granular triple superphosphate production process is illustrated in Fig. 16.3. Here the ground phosphate rock (75% minus 200-mesh) and 62% phosphoric acid are metered continuously to the granulator, where reaction and granulation take place. Fines from the product screen are recycled to the granulator, and the moisture and temperature required for proper granulation are maintained by addition of water and/or steam. The granulator is a cylindrical vessel rotating about a horizontal axis and has an overflow dam at the discharge end. The phosphoric acid is fed uniformly under the bed of material through a perforated pipe. When wet-process phosphoric acid is used, it is also necessary to provide an acid preheater. The granules overflow the dam into a rotary cooler, where they are cooled and dried slightly by a countercurrent flow of air. The exhaust gases from the cooler pass through a cyclone, where dust is collected and returned to the granulator as recycle. The cooled product is screened, the coarse material being milled and returned, along with the fines, to the granulator. The product is then conveyed to bulk storage, where the material is cured 1 to 2 weeks, during which a further reaction of acid and rock occurs, which increases the availability of P_2O_5 as plant food. The exhaust gases from the granulator and cooler are scrubbed with water to remove silicofluorides. The cost per unit of P_2O_5 in this concentrate as compared with ordinary superphosphate is higher, because of greater capital investment and additional labor and processing. However, this is offset to a great extent by the ability to use a lower-grade, cheaper phosphate rock to make the phosphoric acid which is reacted with higher-grade rock. There are also substantial savings on handling, bagging, shipping, and distributing. The production of *concentrated* superphosphate has grown in short tons in terms of 100% available phosphoric acid (APA). *Normal* superphosphate production has dropped slightly.

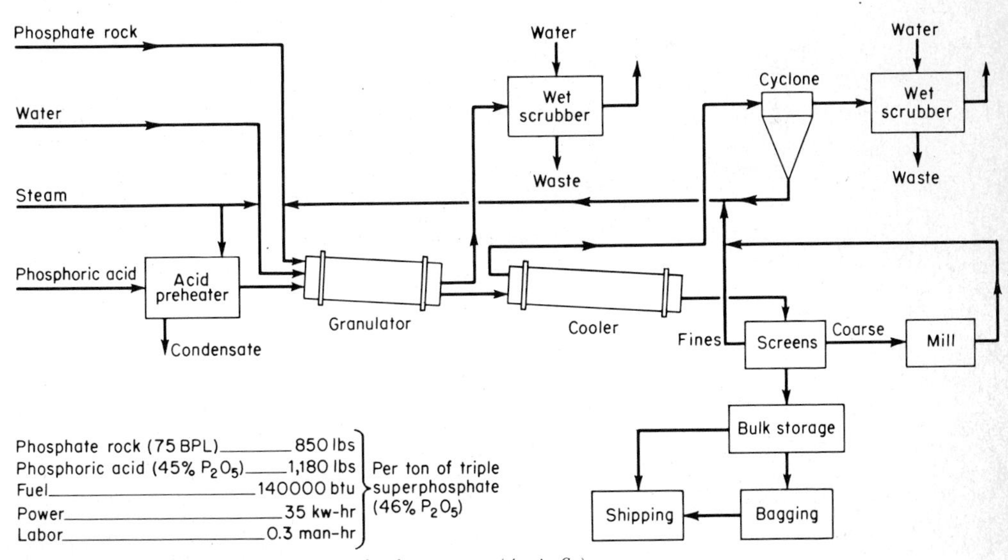

Fig. 16.3 Flowchart illustrating a triple superphosphate process. (*Austin Co.*)

CALCIUM METAPHOSPHATE[6] In 1937, the TVA developed from phosphate rock a concentrated fertilizer, $Ca(PO_3)_2$, by the following reaction:

$$CaF_2 \cdot 3Ca_3(PO_4)_2 + 6P_2O_5 + 2HPO_3 \longrightarrow 10Ca(PO_3)_2 + 2HF\uparrow$$

In this process the P_2O_5 contacts the lump rock in a vertical shaft. About 45 million lb of phosphorus equivalent has been manufactured per year. This calcium metaphosphate may be regarded as a dehydrated triple superphosphate made directly from phosphate rock. The calcium metaphosphate is quite insoluble, and it must hydrolyze to become effective:

$$x[Ca(PO_3)_2] + xH_2O \longrightarrow xCa(H_2PO_4)_2$$

DEFLUORINATED ROCK This product is mostly consumed in cattle feed supplements. It is produced by heating, as shown in the following reaction, with silica added if needed:

$$CaF_2 \cdot 3Ca_3(PO_4)_2 + H_2O + SiO_2 \longrightarrow 3Ca_3(PO_4)_2 + CaSio_3 + 2HF\uparrow$$

MANUFACTURE OF WET-PROCESS PHOSPHORIC ACID FOR FERTILIZER AND SALTS

In recent years and at present, rapid expansion in the manufacture of wet-process phosphoric acid has resulted from the increased demand for high-analysis fertilizer, triple superphosphate, and ammonium and dicalcium phosphates.

Much earlier, most of the orthophosphoric acid produced was prepared by the action of dilute sulfuric acid, 50°Bé, on ground phosphate rock or bones. This method was supplanted by the Dorr[7] strong-acid process, which produced a strong and economical acid. The equipment should be lined with lead, stainless steel, or acidproof brick, and sufficient time provided in the various agitators for the reaction to go to completion. The temperature in the digester should be kept low enough to ensure the precipitation of gypsum ($CaSO_4 \cdot 2H_2O$), and not anhydrite. If the latter is formed, it subsequently hydrates and causes plugging of pipes. Acid made by this process is used almost entirely in fertilizer production, where impurities are unimportant, or after some purification for various sodium phosphates. Pure acid is obtained from elemental phosphorus by the electric-furnace process.

Procedures for wet-process phosphoric acid have been described,[8] and Fig. 16.4 depicts a technical flowchart. The essential reactions are:

$$CaF_2 \cdot 3Ca_3(PO_4)_2 + 10H_2SO_4 + 20H_2O \longrightarrow 10CaSO_4 \cdot 2H_2O + 2HF\uparrow + 6H_3PO_4$$

or, more simply expressed,

$$Ca_3(PO_4)_2(s) + 3H_2SO_4(l) + 6H_2O(l) \longrightarrow 2H_3PO_4(l) + 3CaSO_4 \cdot 2H_2O(s) \qquad \Delta H = -76.6 \text{ kcal}$$

Under *phosphorus,* this wet process is broken down into its commercial sequences of unit operations and chemical conversions (unit processes), in comparison with similar steps for manufacturing this acid from phosphorus. Often wet-process acid is concentrated, but care must be taken to delay fouling of the tubes and reduce downtime for cleaning, as, for example, by the use of forced circulation.[9]

[6]Yates *et al.*, Improved Fertilizer Plant Design, *Chem. Eng.* (*N.Y.*), **58**(6), 135 (1951).

[7]See CPI 2, p. 353, for flowchart, and Dorr-Oliver Inc., Stamford, Conn.

[8]Wilde, Wet Phosphoric Acid, *Chem. Eng. Prog.*, **58**(4), 92 (1962); Phosphoric Acid's Vigor Optimizing Efforts, *Chem. Eng.* (*N.Y.*), **69**(11), 72 (1962); Banford, IMC's New Plant Shows Off Latest H_3PO_4 Know-how, *Chem. Eng.* (*N.Y.*), **70**(11), 100 (1963); Slack and Mason, Liquid Fertilizers from Wet-process Phosphoric Acid: Suspension of Impurities, *J. Agric. Food Chem.*, **9,** 343 (1961); Gilbert and Moreno, Dissolution of Phosphate Rock by Mixtures of Sulfuric and Phosphoric Acids, *Ind. Eng. Chem., Process Desi. Dev.*, **4,** 368 (1965); Chelminski, *et al.*, Phosphoric Acid, *Chem. Eng. Prog.*, **62**(5), 108 (1966).

[9]Meinhold, Tube Fouling Slashed in H_3PO_4 Evaporators, *Chem. Process.* (Chicago), Dec. 17, 1962, p. 28; Weisman, Submerged Combustion Equipment, *Ind. Eng. Chem.*, **53,** 708 (1961).

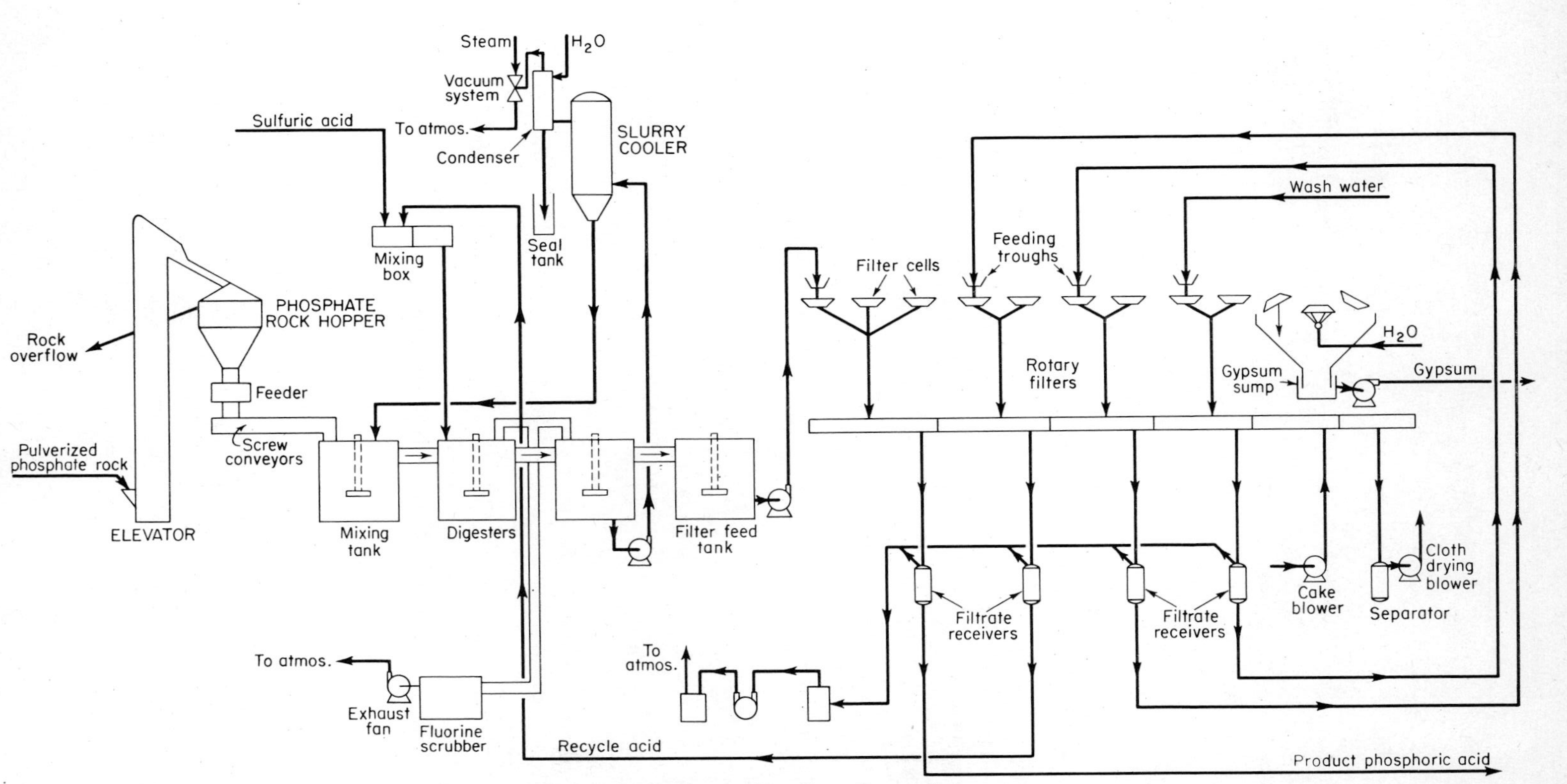

Fig. 16.4 Wet-process phosphoric acid manufacture, using Bird-Prayon tilting-pan washing filters. The overall efficiency is 94 to 98%, producing an acid of 30 to 32% P_2O_5 concentration. (*Chemical and Industrial Corp.*) See Fig. 16.5.

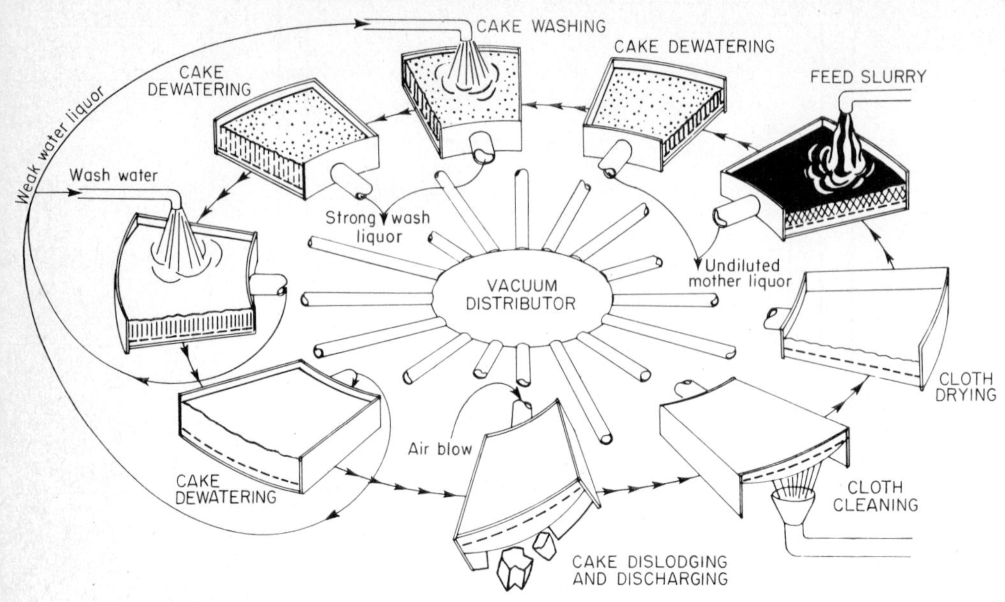

Fig. 16.5 The Bird-Prayon tilting-pan filter for phosphoric acid manufacture, picturing the simple automatic cycle through which each individual pan or cell operates continuously. The feed continuously enters the pans, which are connected to the vacuum source. The circular frame supporting the pans rotates so that each pan is moved successively under the desired number of washes. After the final wash liquor has completely drained off, the vacuum is released and the pan is inverted a full 180°. The cake drops off, its removal being ensured by a reverse blast of air through the filter medium, which is then scoured fresh and clean by a high-pressure shower while the pan is still inverted. The filter-medium and drainage area are then purged by vacuum, and the pan returned to the feed position. (*Bird Machine Co.*)

Hydrochloric[10] *acid route to phosphoric acid; Haifa process.* This process uses (1) hydrochloric acid to acidulate (in slight excess to prevent formation of monocalcium phosphate); (2) an organic solvent (a C_4 or C_5 primary alcohol) to extract the phosphoric acid; (3) water to strip out the phosphoric acid (with a small amount of solvent and hydrochloric acid); and (4) concentration to remove the small amounts of solvent and hydrochloric acid and to yield a high-grade product, 59, 85, or 75% orthophosphoric acid. This process was developed in Israel and has been applied in Japan and the United States. Calcination of the phosphate rock and/or ion-exchange purification upgrades this acid.

Granular high-analysis fertilizer. This is basically a commercial adaptation of the TVA process (Fig. 16.6), designed for the manufacture of diammonium phosphate fertilizer, either 21-54-0, using furnace phosphoric acid, or 18-46-0, with wet-process phosphoric acid, or for any of many grades of granular fertilizer (cf. Chap. 26). The last-mentioned requires the addition of facilities for metering solid raw materials and sulfuric acid to the ammoniator. Vapor or liquid anhydrous ammonia and phosphoric acid (40 to 45% P_2O_5) are metered continuously to an agitated atmospheric tank (preneutralizer) in proportions to maintain a ratio of 1.3 to 1.5 mol of ammonia per mole of phosphoric acid. This ratio is the optimum for maintaining fluidity of the slurry with a minimum quantity of water and a reasonable ammonia loss in the exit vapor from the preneutralizer. In the preneutralizer the heat of reaction elevates the temperature of the mass, evaporating approximately 200 lb of water per ton of product. The slurry formed in the preneutralizer flows into a TVA-type ammoniator-granulator at about 250°F, where it is distributed evenly over the bed of solid material

[10]Hydrochloric Based Route to Pure Phosphoric Acid, *Chem. Eng. (N.Y.)*, **69**(26), 34 (1962) (flowchart); Phosphate Markets on Road to Recovery, *Chem. Week*, Jan. 7, 1976, p. 16.

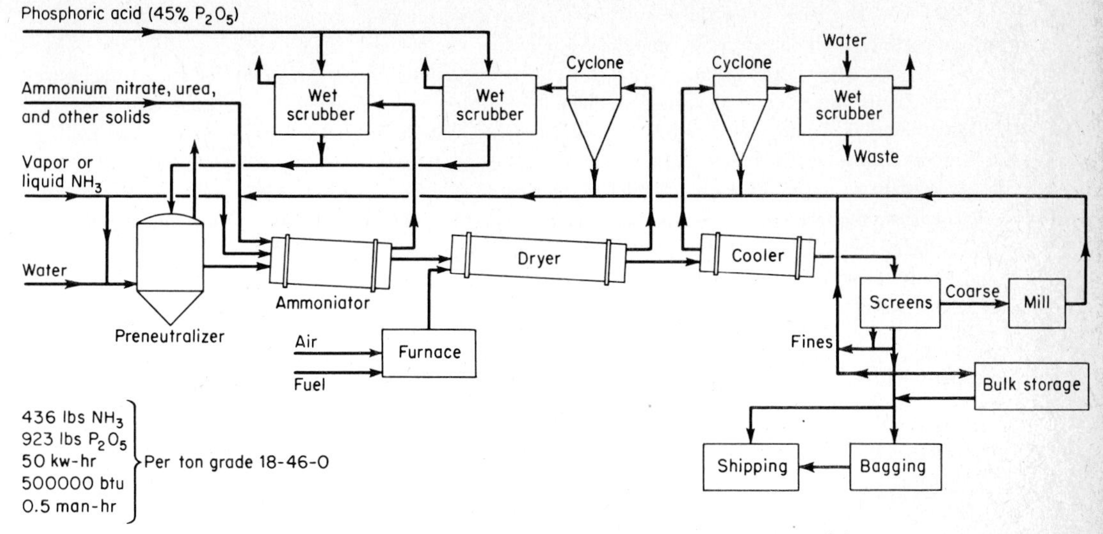

Fig. 16.6 Flowchart for the manufacture of diammonium phosphate or high-analysis fertilizer. (*Austin Co.*)

and, for most grades, reacts with additional ammonia fed through a distributor pipe below the surface of the bed to complete the reaction with a molar ratio of 2.0 (diammonium phosphate).[11] Dried recycle material from product screening is used to control the moisture, and additional solid raw materials are metered into the ammoniator, depending on the grade of fertilizer in production. The ammoniator combines the functions of chemical reaction, mixing, and formation of the proper size and shape of granular particles. The moist granules leaving the ammoniator fall into an oil- or gas-fired cocurrent rotary dryer, where the moisture content of the material is reduced to about 1%. The dry product is then cooled by a countercurrent flow of air in a rotary cooler and screened, the coarse material being milled and returned to the ammoniator-granulator. In some plants the cooler is eliminated, and hot screening is used. The fines and, if necessary, some of the product are recycled to the ammoniator to control granule formation and size. The product is conveyed to bulk storage or bagged. The exhaust gases from the dryer and cooler pass through cyclones or a wet scrubber for dust recovery. Exhaust gases from the ammoniator are scrubbed with the incoming phosphoric acid, and in some plants the phosphoric acid stream is split, a portion being used to scrub the vapor from the rotary dryer to minimize ammonia loss.

PHOSPHORUS AND PURE PHOSPHORUS DERIVATIVES

This element was first produced on a small commercial scale by treating calcined bone with sulfuric acid, filtering off the phosphoric acid, and evaporating it to sp gr 1.45. This was mixed with charcoal or coke, again heated, and the water evaporated off and calcined at white heat in retorts. The phosphorus was thus distilled off, collected under water, and purified by redistillation. The production of phosphorus today still depends on volatilization of the element from its compounds under reducing conditions. During the past decades, the method has changed chiefly in details and size of production. Elementary phosphorus is manufactured on a large scale as a heavy chemical and shipped in tank

[11]For other flowcharts, see Greek *et al.*, St. Gobain Processes Phosphate Fertilizers, *Ind. Eng. Chem.*, **52,** 639 (1960); Chopey, DAP: New Plant Ushers in Process Refinements, *Chem. Eng.* (*N.Y.*), **69**(6), 148 (1962); cf. Striplin, Superphosphoric Acid Paves Way for Fertilizer Shift, *Chem. Eng.* (*N.Y.*), **68**(10), 160 (1961).

cars from the point of initial manufacture, where raw materials are cheap, to distant plants for conversion to phosphoric acid, phosphates, and other compounds.

USES AND ECONOMICS With the commercialization of cheaper methods for producing phosphorus on a large scale, widening fields have been developed for it and its compounds. The consumption of phosphorus derivatives for use other than as fertilizers may be divided into four groups: (1) water treatment and detergents, (2) food and medicine, (3) phosphate esters, and (4) miscellaneous uses. Excluding fertilizers, the main outlet for phosphorus derivatives is *in water treatment and in soap and detergent* manufacture as various sodium phosphates. These salts, because of their ability to precipitate or sequester lime and magnesia, to emulsify or disperse solids in the detergent solution, and to augment the inherent detergent properties of soap and synthetic surface-active agents, are much used as *soap builders or detergent synergists.*

MANUFACTURE OF PHOSPHORUS AND PURIFIED DERIVATIVES

REACTIONS Phosphorus is produced by the electric-furnace method (Fig. 16.7). The following reaction is considered to take place, the raw materials being phosphate rock, silica, and coke:

$$CaF_2 \cdot 3Ca_3(PO_4)_2 + 9SiO_2 + 15C \longrightarrow CaF_2 + 6P + 15CO$$

or, more simply expressed,

$$2Ca_3(PO_4)_2 + 6SiO_2 + 10C \longrightarrow 6CaSiO_3 + P_4 + 10CO \qquad \Delta H = -730 \text{ kcal}$$

The silica is an essential raw material which serves as an acid and a flux. About 20% of the fluorine present in the phosphate rock is converted to SiF_4 and volatilized. In the presence of water vapor this reacts to give SiO_2 and H_2SiF_6:

$$3SiF_4 + 2H_2O \longrightarrow 2H_2SiF_6 + SiO_2$$

The fluorine is not recovered by manufacturers of phosphorus, but the CO is employed as a fuel in preparing the furnace charge (agglomeration). The slag tapped from the furnace is sold as ballast, or aggregate or fill. Ferrophosphorus is tapped as necessary, its quantity being dependent on the

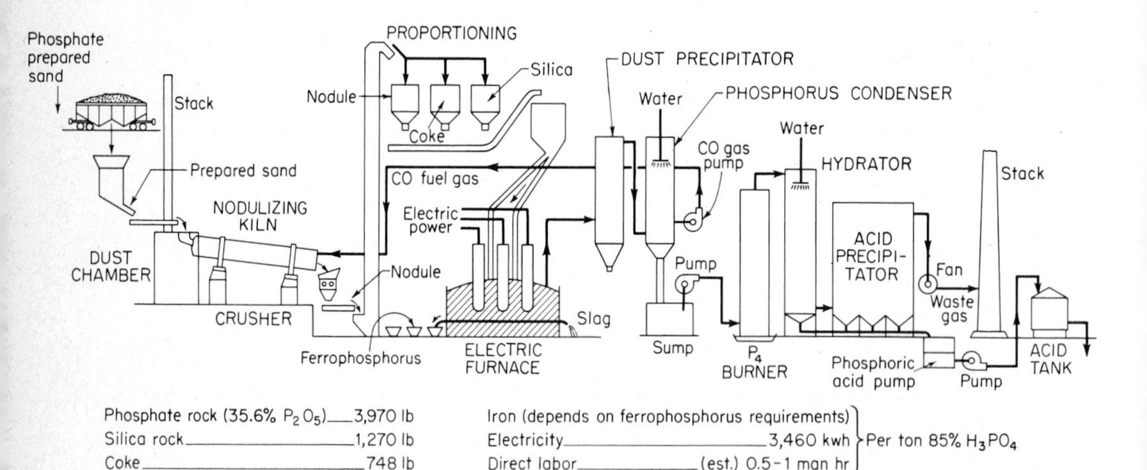

Fig. 16.7 Phosphorus production flowchart. In most cases, the hot phosphorus is pumped into an insulated tank car and shipped to the oxidizing plant where it is made into phosphoric acid and various phosphates. Note: Brink Mist Eliminators have replaced the electrostatic acid mist precipitators in most of the phosphoric acid plants in the United States.

amount of iron originally in the burden or added to it. The phosphorus is employed usually as an intermediate product, being shipped to consumption centers and there burned or oxidized to P_2O_5, which is dissolved in water to form acids or other compounds:

Phosphorus pentoxide:	$4P + 5O_2 \longrightarrow 2P_2O_5$	$\Delta H = -720$ kcal
Orthophosphoric acid:	$P_2O_5 + 3H_2O \longrightarrow 2H_3PO_4$	$\Delta H = -45$ kcal

The foregoing reactions for ordinary or orthophosphoric acid are commercialized in the following sequences of *physical operations* and *chemical conversions*, as exemplified by Figs. 16.4 and 16.5 for wet-process acid in comparison with Fig. 16.7 for phosphorus and furnace acid made therefrom.

Wet-process phosphorus and acid

Phosphate rock is finely ground and prepulped in the mixing tank with cooled recycled H_3PO_4 from the slurry cooler (Op and Ch)
Overflow from the mix tank is reacted with H_2SO_4 and recycle H_3PO_4 in an agitated digester system, forming gypsum crystals and H_3PO_4 (Ch)
Bulk of slurry is recycled through slurry cooler (Op)
Rest of slurry is conducted through the filter feed tank to the cells of the rotary Bird-Prayon tilting-pan filtration unit (Op)
In the progressive and tilting pans of this filter, the H_3PO_4 is separated from the gypsum, using three-stage countercurrent washing (Op)
Gypsum is automatically dumped (Op)
Off-gases are scrubbed to remove fluorine before venting (Op)
H_3PO_4 can be used directly (30–32% P_2O_5) or concentrated (Op)

Electric-furnace phosphorus and acid

Phosphate rock ground and sized (Op)
Rock and sand mixed with coke, sintered, and introduced into electric furnace (Op)
Mix heated and reduced at an elevated temperature (Ch)
Slag and ferrophosphorus run off separately (Op)
Phosphorus vapor and CO drawn off, phosphorus condensed (Op)
Phosphorus transported in tank cars to consuming centers (Op)
Phosphorus melted and sprayed into stainless-steel oxidation tower (Op)
Phosphorus oxidized to P_2O_5 (Ch)
P_2O_5 cooled and hydrated in stainless-steel hydrator or tower against water or dilute H_3PO_4 (Ch)
H_3PO_4 mist-precipitated in electrostatic precipitator or a Brink Mist Eliminator (Op)
H_3PO_4 filtered and purified (Op, Ch)

The electric-furnace[12] process was first employed commercially in 1920. This process permits the use of lower-grade rock than the wet-process phosphoric acid process, since the slag carries off impurities. Indeed, lower grades are frequently preferred because of the better CaO/SiO_2 balance for slag formation. The principal requirement is cheap electricity.

The phosphate rock must be charged in lump form or as +8-mesh (Fig. 16.1). Fine material tends to block the exit of the phosphorus vapors and to cause bridging and uneven descent of the furnace charge, resulting in puffs and the carrying over of excessive quantities of dust. Phosphate lumps may be prepared in the following ways: (1) pelletizing by tumbling or extrusion, (2) agglomeration by nodulizing at high temperatures, (3) sintering a mixture of phosphate fines and coke, and (4) briquetting, with the addition of a suitable binder. After agglomeration, coke breeze and siliceous flux (gravel) are added, and the materials are charged to the electric furnace. Iron slugs are added to the charge if more ferrophosphorus is desired. A flowchart with the quantities required is shown

[12]Curtis, The Manufacture of Phosphoric Acid by the Electric Furnace Method, *Trans. AIChE*, **31**, 278 (1935); Curtis *et al.*, Process Development at TVA Phosphoric Acid Plant, *Chem. Metall. Eng.*, **45**, 193 (1938); Mantell, Electrochemical Engineering, 4th ed., pp. 523–532, McGraw-Hill, 1960 (data and diagram); Highett and Striplin, Elemental Phosphorus in Fertilizer Manufacture, *Chem. Eng. Prog.*, **63**(5), 85 (1967); Bryant, Holloway and Silber, Phosphorus Plant Design, *Ind. Eng. Chem.*, **62**(4), 8 (1970).

in Fig. 16.7. The bottom of the furnace is composed of carbon blocks, and this lining extends up the wall to a point well above the slag pool. From this point, a high-grade firebrick lining is used. A domelike steel top with a cast refractory lining caps the furnace. Openings for the electrodes and for introducing raw materials are included here. The electrodes are threaded so as to facilitate replacement as the carbon is consumed. The gases and phosphorus vapor are removed at one end of the furnace. The calcium-rich slag from the furnace is usually tapped periodically and crushed for use in the manufacture of glass, for the liming of soil, and as a roadbed ballast. The ferrophosphorus is tapped separately, or it runs out with the slag when it is separated and sold as a phosphorus additive for steel. In this process 80% of the fluorine stays with the slag. The small portion that leaves with the gas is absorbed in the water used in condensing the phosphorus.

PHOSPHORIC ACID FROM PHOSPHORUS

An increasing quantity of pure, strong phosphoric acid is manufactured from elemental phosphorus by oxidation and hydration. See Fig. 16.7 and preceding reactions and sequences for manufacture. The oxidation tower, or chamber, is constructed of acid-resistant brick or stainless steel.[13] Phosphoric acid runs down the walls and absorbs about 75% of the P_2O_5 and also heat. This acid is *cooled,* some drawn off, and some recirculated. The remaining 25%, as a mist is passed to a Cottrell or Brink mist eliminator[14] for collection. Some modern plants produce superphosphoric acid, 76% P_2O_5, equivalent to 105% orthophosphoric acid, with which higher-percentage plant foods can be made, e.g., ammonium polyphosphate, 54% superphosphate, and 10-34-0 liquid fertilizer. Other plants produce 75 to 85% orthophosphoric acid, depending on the process for which it is to be used.

Phosphoric acid of high P_2O_5 content consists of mixtures of various phosphates with a certain distribution of chain lengths. Individual molecular species are difficult to prepare. Pyrophosphoric acid ($H_4P_2O_7$) can be obtained only through a slow crystallization process, and above its melting point it rapidly reverts to a distribution of various chain phosphates. If P_2O_5 is carefully dissolved in cool water, most of the phosphate will be in the form of tetrametaphosphate rings.

SODIUM PHOSPHATES

The various sodium phosphates[15] represent the largest tonnage of chemicals based on pure phosphoric acid obtained mostly from elemental phosphorus. Phosphates are phosphorus compounds in which the anions have each atom of phosphorus surrounded by four oxygen atoms placed at the corners of a tetrahedron. Chains, rings, and branched polymers result from the sharing of oxygen atoms by tetrahedra. *Orthophosphates* are based on the simple PO_4 tetrahedron as a monomeric unit and include monosodium phosphate (MSP) (NaH_2PO_4); disodium phosphate (DSP) (Na_2HPO_4), and trisodium phosphate (TSP) ($Na_3PO_4 \cdot \frac{1}{4}NaOH \cdot 12H_2O$). The first two sodium salts are made from phosphoric acid and soda ash reacted in the proper molecular proportions; the solution is purified if necessary, evaporated, dried, and milled. TSP is also made from phosphoric and soda ash, but caustic soda is necessary to substitute the third hydrogen of the phosphoric acid. These salts are employed in water treatment, baking powder (MSP), fireproofing, detergents, cleaners, and pho-

[13] Highett and Striplin, *loc. cit.;* TVA Boost for Superphosphoric Acids Poise for Takeoff, *Ind. Eng. Chem.*, **68**(23), 112 (1961).

[14] Brink, New Fiber Mist Eliminator, *Chem. Eng. (N.Y.)*, **66,** 183 (1959); Brink, chap. 15B in Gas Purification Processes for Air Pollution Control, Nonhebel (ed.), Newnes-Butterworths, London, 1972; see Perry, pp. 18-82 to 18-93, for mist collection equipment.

[15] ECT, 2d ed., vol. 15, pp. 263–273, 1968.

tography (TSP). *Condensed or molecularly dehydrated* phosphoric acids and salts have a H_2O/P_2O_5 molar ratio of less than 3 and greater than 1 and have the single chain unit P-O-P. The best known are the *polyphosphates:* pyrophosphates ($M_4P_2O_7$) and tripolyphosphates ($M_5P_3O_{10}$). When any of the condensed phosphoric acids are dissolved in water, hydrolysis to orthophosphoric acid takes place. Their salts are represented by the widely used sodium tripolyphosphate and tetrasodium and tetrapotassium pyrophosphates.

To produce sodium tripolyphosphate (Fig. 16.8), a definite temperature control is necessary. When MSPs and DSPs in correct proportions, or equivalent mixtures of other phosphates, are heated for a substantial time between 300 and 500°C and slowly cooled, the product is practically all in the form of the tripolyphosphate.

Figure 16.8 depicts the following coordinated sequences commercializing the reaction:

$$NaH_2PO_4 + 2Na_2HPO_4 \longrightarrow Na_5P_3O_{10} + 2H_2O$$

Soda ash and ±75% phosphoric acid are reacted in the mix tank (Ch).

The orthophosphates are dried either in a rotary or spray dryer (Op).

The sodium tripolyphosphate is molecularly dehydrated in a gas-fired calciner (Ch).

The tripolyphosphate is annealed, chilled, and stabilized in a continuous rotary tempering unit (Ch).

The product is milled, stored, and bagged (Op).

Certain equipment modifications have been used, such as addition of an adjustment mixer following the reactor mix tank and a spray tower for drying of the orthophosphate, together with a long, continuous rotary to carry out the dehydrating (calcining), annealing, stabilizing, and cooling in one unit.

PYROPHOSPHATES Tetrasodium pyrophosphate (TSPP) ($Na_4P_2O_7$), is used as a water softener and as a soap and detergent builder. It is manufactured by reacting phosphoric acid and soda ash to yield a DSP solution, which may be dried to give anhydrous Na_2HPO_4 or crystallized to give $Na_2HPO_4 \cdot 2H_2O$ or $Na_2HPO_4 \cdot 7H_2O$. These compounds are calcined at a high temperature in

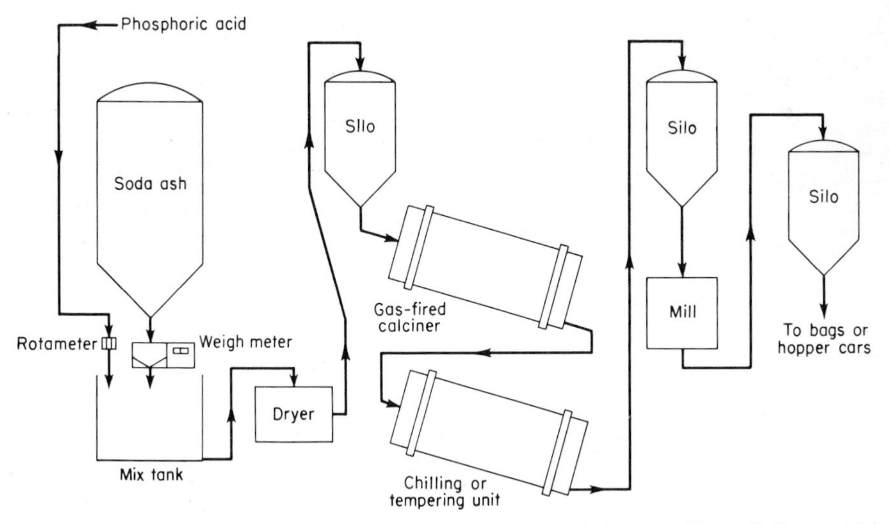

Fig. 16.8 Diagram of a plant for the manufacture of crystalline sodium polyphosphate. (*Thomas L. Hurst and John R. Van Wazer.*)

an oil- or gas-fired rotary kiln to yield TSPP in a plant such as that shown in Fig. 16.8. The reactions may be written:

$$2Na_2HPO_4 \longrightarrow Na_4P_2O_7 + H_2O$$

$$2Na_2HPO_4 \cdot 2H_2O \longrightarrow Na_4P_2O_7 + 5H_2O$$

A nonhygroscopic sodium acid pyrophosphate is used extensively as a chemical leavening agent in making doughnuts, cakes, and packaged biscuit doughs. It is manufactured by partially dehydrating monosodium acid orthophosphate at a temperature between 25 and 250°C over the course of 6 to 12 h.

$$2NaH_2PO_4 \longrightarrow Na_2H_2P_2O_7 + H_2O$$

PHARMACEUTIC AND FOOD GRADES (purer) of dicalcium phosphate dihydrate for use as a polishing agent in dentifrices, ammonium phosphate for use as yeast nutrient, calcium pyrophosphates, and monocalcium phosphate are manufactured from furnace-grade phosphoric acid.

Monocalcium phosphate $[Ca(H_2PO_4)_2 \cdot H_2O]$ is manufactured by the crystallizing, after evaporation and some cooling, of a hot solution of lime and strong furnace phosphoric acid. The crystals are centrifuged, and the highly acidic mother liquor returned for reuse. This acid salt is also made by spray-drying a slurry of the reaction product of lime and phosphoric acid. This product is used for baking powder.

BAKING POWDERS

The baking-powder industry is an important but indirect consumer of phosphate rock. The use of leavening agents to produce aeration and lightness in breads and cakes has been known since the time of the Egyptians and was handed down by the Greeks and Romans. Leavened and unleavened bread are both mentioned in the Bible. Some form of yeast or ferment acting on the carbohydrates in flour, giving CO_2 and an alcohol, was the first leavening agent used. Later, baking soda (sodium bicarbonate) was widely employed but, because it often imparted an unpleasant taste or even a yellowish color due to the alkalinity of the Na_2CO_3 formed, the search for better reagents continued. Baking powders consist of a dry mixture of sodium bicarbonate with one or more chemicals capable of completely decomposing it. The principal "baking acids" used are monocalcium phosphate monohydrate, anhydrous monocalcium phosphate, sodium acid pyrophosphate, sodium aluminum sulfate, tartaric acid, and acid tartrates. Monocalcium phosphates are consumed more than all the others, an estimate being more than 80 million lb annually in the United States. A filler or drying agent, such as starch or flour, is usually added to the active ingredients to give a better distribution throughout the dough and to act as a diluent or to prevent the reaction until water and heat are applied. The following equations represent the actions of different baking powders:

$$Na_2Al_2(SO_4)_4 + 6NaHCO_3 \longrightarrow 6CO_2 + 4Na_2SO_4 + 2Al(OH)_3$$

$$3CaH_4(PO_4)_2H_2O + 8NaHCO_3 \longrightarrow 8CO_2 + Ca_3(PO_4)_2 + 4Na_2HPO_4 + 11H_2O$$

$$KH_2PO_4 + NaHCO_3 \longrightarrow CO_2 + KNaHPO_4 + H_2O$$

$$NaH_2PO_4 + NaHCO_3 \longrightarrow CO_2 + Na_2HPO_4 + H_2O$$

$$Na_2H_2P_2O_7 + 2NaHCO_3 \longrightarrow 2CO_2 + 2Na_2HPO_4 + H_2O$$

$$KHC_4H_4O_6 + NaHCO_3 \longrightarrow KNaC_4H_4O_6 + CO_2 + H_2O$$

Baking powders must yield not less than 12% available CO_2, and most powders contain from 26 to 29% $NaHCO_3$ and enough of the acid ingredients to decompose the bicarbonate and yield from 14 to 15% CO_2. The rest, 20 to 40%, consists of corn starch or flour.

PHOSPHORUS FIRE-RETARDANT CHEMICALS

In recent years, there has been an obvious increase in the use of fire-retardants both in fire-proofing various textiles and for combatting forest fires. Phosphorus compounds are most suitable for these purposes.[16] The water resistant fire retardants commonly used for cotton textiles are a combination of trisaziridinyl phosphorus oxide (commonly referred to in the literature as APO) and tetrakishydroxymethyl phosphonium chloride (THPC). APO is prepared by reacting phosphoryl trichloride and ethyleneimine:

$$POCl_3 + \underset{NH}{CH_2\text{—}CH_2} \xrightarrow{\text{base}} \left[\underset{NH}{CH_2\text{—}CH_2}\right]_3 PO$$

THPC is readily made from phosphine and formaldehyde:

$$4CH_2O + PH_3 + HCl \longrightarrow (HOCH_2)_4PCl$$

APO is a very reactive molecule and presumably reacts with both the hydroxyl groups of cellulose and THPC to form a polymeric material. For combatting forest and brush fires, mixtures based on $(NH_4)_2HPO_4$ or $(NH_4)_2SO_4$, thickening agents, coloring matter, and corrosion inhibitors are most commonly used.[17] It is believed that the phosphorus compounds act as catalysts to produce non-combustible gas and char. Furthermore, phosphorus compounds which can yield phosphoric acid from thermal degradation are effective in suppressing glow reactions.

SELECTED REFERENCES

Allcock, H. R.: Phosphorus-Nitrogen Compounds, Academic, 1972.
Gefter, Y.: Organophosphorus Monomers and Polymers, Pergamon, 1962.
Johnson, J. C.: Antioxidants 1975, Synthesis and Applications, Noyes, 1975.
Martin, L. F.: New Fire Extinguishing Compounds, Noyes, 1972.
O'Brien, R. D.: Toxic Phosphorus Esters: Chemistry, Metabolism, and Biological Effects, Academic, 1960.
Sauchelli, V. (ed.): Chemistry and Technology of Fertilizers, ACS Monograph 148, Reinhold, 1960.
Van Wazer, J. R.: Phosphorus and Its Compounds, Interscience, 1958.

[16]Lyons, The Chemistry and Uses of Fire Retardants, Wiley-Interscience, New York, 1970.

[17]Lowden, *et al.*, Chemicals for Forest Fire Fighting, National Fire Protection Association, Boston, Mass. 1963; User Guide, Phoscheck Fire Retardant, Monsanto Co., St. Louis, Mo., 1968.

chapter 17

POTASSIUM INDUSTRIES

Potassium salts must be present in the soil in order to have normal plant growth, hence they are essential components of the fertilizers used in food production throughout the world (see Chap. 26). The yardstick by which potassium compounds are measured is the content of potash as K_2O. Various potassium salts are also basically important as raw materials for many commodities such as soaps, detergents, glass, dyes, gunpowder, and pyrotechnics. Establishment of this industry in the United States was difficult. Potash was first imported from Germany in 1869, and for over 45 years that country furnished the United States and the rest of the world with potash. After the German salts became available, there was little interest in domestic production of potash until World War I prevented its importation. At this time the scarcity of potash became a matter of national concern, and prices rose to $500 per ton of 50% muriate. Production rose from 1,000 tons of potash (K_2O) in 1915 to 45,700 tons in 1919. When the end of the war came, imports were resumed and domestic production fell.

Since that time, the discovery of vast new deposits has caused an increase in the amount and quality of potash salts produced. The American deposits are mined mechanically and brought to surface refineries. Solution of underground salines is being exploited, promising a lower-cost procedure.[1] In 1941 and 1942 the United States produced its entire requirement of potash for the first time, and the domestic industry continued to grow. In 1962 the International Minerals and Chemical Corp. struck rich potash deposits in Saskatchewan, Canada, after a 5-year drilling program.[2] Other companies also developed Canadian deposits into major sources of potash supply.[3] Potash from Canada was exported to the United States in large quantities until domestic production was cut.[4] Since 1971 imports from Canada have made up more than 50% of the total U.S. potash supply.[5]

RAW MATERIALS The largest domestic production of potassium salts has come from deep Permian sedimentary deposits of *sylvinite* [a natural mixture of sylvite (KCl) and halite (NaCl)] and langbeinite ($K_2SO_4 \cdot 2MgSO_4$) near Carlsbad, N.Mex. The sylvinite is mined and treated to yield high-grade potassium chloride, and langbeinite is processed to make potassium sulfate. Another domestic source of potassium salts is Searles Lake at Trona, Calif., which is a deposit of solid sodium salts permeated by a saturated complex brine. This brine is processed to separate high-grade potassium chloride and borax, together with numerous other saline products (Figs. 17.1 to 17.4). Deposits

[1]Piombino, Potash, *Chem. Week.*, Sept. 14, 1963, p. 73 (American production and potentials); Potash Minerals, *Chem. Eng. Prog.*, **60**(10), 19 (1964).

[2]Canadian Potash Heads for Market, *Chem. Week*, June 18, 1962, p. 31.

[3]Potash Process Peak, *Chem. Week*, May 16, 1964, p. 25; Lining Up for the Push in Potash, *Chem. Week*, May 23, 1964, p. 22; Lighting Potash Fire, *Chem. Week*, Aug. 29, 1964, p. 31; Pushing Potash Pipeline, *Chem. Week*, Dec. 5, 1964, p. 71; Frenzied Push for Potash, *Chem. Week*, Aug. 14, 1965, p. 21.

[4]Potash Problems Multiply, *Chem. Week*, Nov. 11, 1967, p. 61; Potash Problems, *Chem. Week*, Apr. 13, 1968, p. 47.

[5]The Lid's Off Potash Production, *Chem. Week*, Nov. 14, 1973, p. 29; Potash Prices Will Climb Higher, *Chem. Week*, Oct. 2, 1974, p. 19; Saskatchewan Is Moving to Take Over Privately Owned Potash Industry, *Chem. Week*, Oct. 30, 1974, p. 9.

at Moab, Utah, are being developed as a domestic source by Texas Gulf Sulphur. Bonneville, Ltd., evaporates the Salduro Marsh brines by solar heat at the western edge of the Salt Lake Basin at Wendover, Utah, crystallizing out the sodium and potassium chloride. The two salts are repulped and separated by froth flotation to yield potassium chloride. A startling new potash source is the subterranean deposits in Saskatchewan, Canada, which are being developed both by mining potash salts and by solution mining (Fig. 17.6). Subterranean deposits of potash salts occur in Europe,

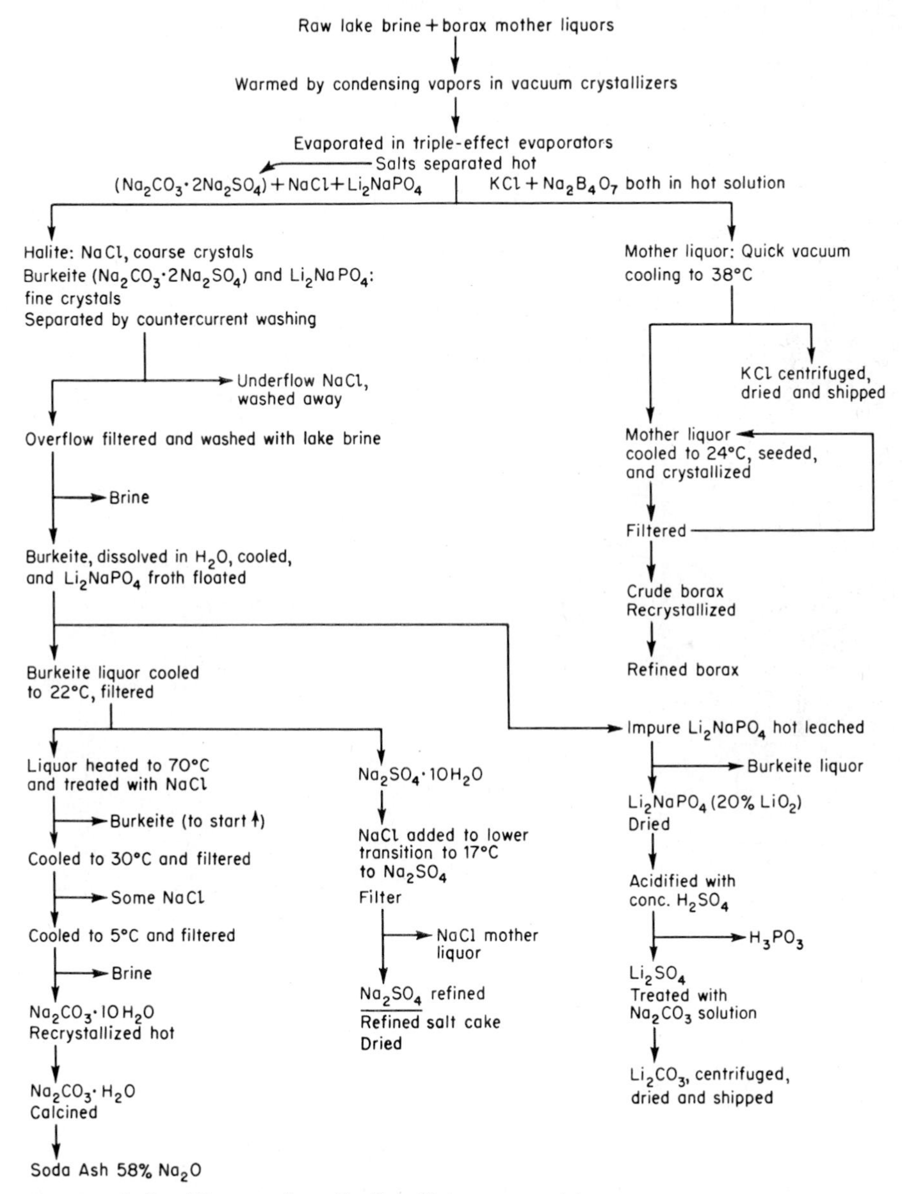

Fig. 17.1 Outline of Trona procedures. (See Chap. 20, boron compounds.)

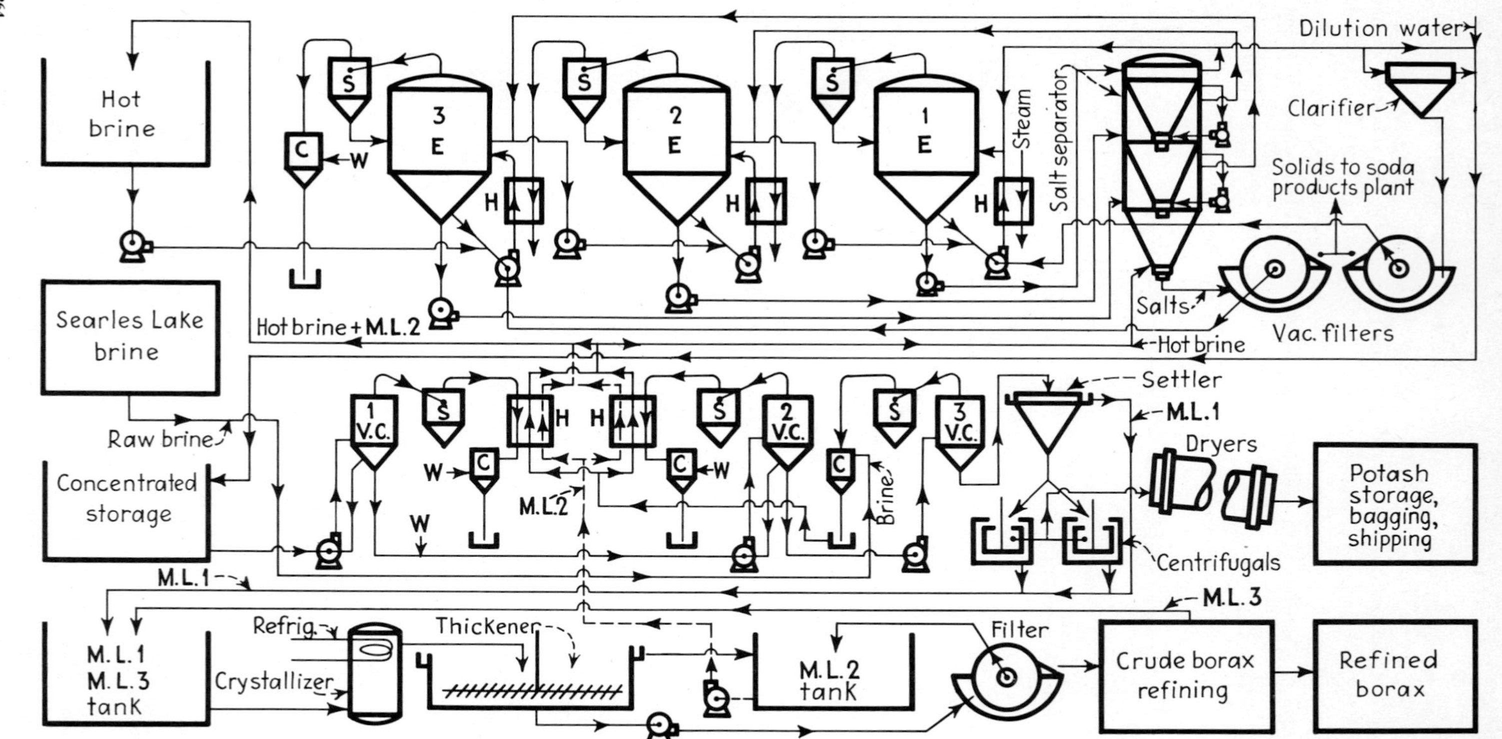

Key: **C**, Barometric condenser; **E**, Evaporator; **H**, Heater or heat exchanger; **M.L.**; Mother liquor; **S**, Separator; **V.C.**, Vacuum crystallizer; **W**, Cooling or dilution water

Fig. 17.2 Potassium chloride and borax by the Trona procedure. (See Fig. 17.1 for supplementary details.)

Fig. 17.3 Trona: overall view of the plant with the new large evaporators in the back. (*American Potash & Chemical Corp.*)

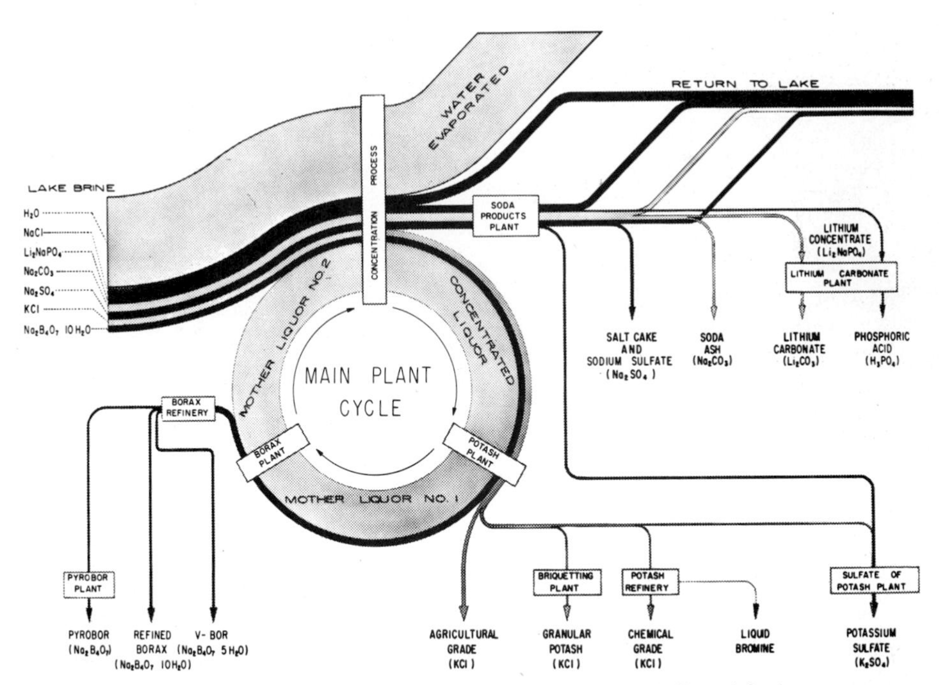

Fig. 17.4 Trona procedures: flow of brines and salts to products. (*American Potash & Chemical Corp.*)

notably in the U.S.S.R., Poland, Germany, France (Alsace), and Spain. Other sources contribute in a minor way to the annual potash production: alcohol fermentation, using molasses as a raw material, and by-products from natural salt brine. The brine of the Dead Sea also is successfully exploited as a source of potassium chloride and other saline products.

POTASSIUM CHLORIDE

In this country about 90% of the potassium chloride is produced as fertilizer grade of about 97% purity. In the fertilizer trade it is referred to as *muriate of potash.* The chemical grade, or 99.9% potassium chloride, is the basis of the manufacture of most potassium salts.

MANUFACTURE BY THE TRONA PROCESS From Searles Lake, Calif., the shipping of potash was begun in 1916 and, including numerous other products, has continued since that time on an ever-expanding scale. The lake is composed of four layers. The upper layer of crystalline salt is from 70 to 90 ft deep; the second layer, about 12 to 15 ft of mud; the third layer, about 25 ft of salt; and the bottom layer is mud interspersed with minor salt seams. In processing, the brine is pumped from the interstices in the salt body from the first and third layers. The brines have the following approximate constant composition:

Expressed as:	*Upper deposit, percent*	*Lower deposit, percent*
KCl	4.85	3.00
NaCl	16.25	16.25
Na_2SO_4	7.20	6.75
Na_2CO_3	4.65	6.35
$Na_2B_4O_7$	1.50	1.77
Na_3PO_4	0.155	
NaBr	0.109	
Miscellaneous	0.116	0.35
Total salts (approx.)	34.83	34.60
H_2O	65.17	65.40
Specific gravity	1.303	1.305
pH (approx.)	9.45	9.60

As determination of the composition of the various salts from the German potash deposits at Stassfurt by van't Hoff from 1895 to 1910 helped greatly in the development of the German potash industry, likewise the phase-rule study[6] of the much more complex systems existing at Searles Lake by Morse, Teeple, Burke, Mumford, Gale, and many others was another striking investigation carried to successful application in the Trona process. In the first-layer brine of Searles Lake the main system is $Na\text{-}K\text{-}H\text{-}SO_4\text{-}CO_3\text{-}B_2O_4\text{-}Cl\text{-}F\text{-}H_2O$. The references cited, particularly the one by Gale, demonstrate with appropriate solubility charts this important and commercial application of the *phase rule,* wherein the conditions are presented for the desired fractional crystallization.

[6]Teeple, Industrial Development of Searles Lake Brines, Reinhold, 1929; Robertson, The Trona Enterprise, *Ind. Eng. Chem.*, **21,** 520 (1929); Expansion of the Trona Enterprise, *Ind. Eng. Chem.*, **34,** 133 (1942); Gale, Chemistry of the Trona Process from the Standpoint of the Phase Rule, *Ind. Eng. Chem.*, **30,** 867 (1938). The triangular phase diagrams of the more common constituents of Searles Lake brine at 20 and 100°C are presented and described in Kobe, Inorganic Chemical Processes, pp. 71–74ff., Macmillan, 1948. These are for the brines saturated with NaCl at all points and for burkeite, Na_2SO_4, glaserite, KCl, and Na_2CO_3.

TABLE 17.1 ***Salient Statistics on Potassium Salts*** *(In thousands of short tons and thousands of dollars)*

Item	*1968*	*1969*	*1970*	*1971*	*1972*	*1980**
United States:						
Production of potassium salts, marketable	4,769	4,918	4,853	4,543	4,738	7,800
Approximate K_2O equivalent	2,722	2,804	2,729	2,587	2,659	4,390
Value	75,664	73,572	98,123	100,527	106,680	259,000
Sales of potassium salts by producers	5,091	5,340	4,703	4,578	4,653	7,680
Approximate K_2O equivalent	2,913	3,069	2,669	2,592	2,618	4,320
Value at plant	81,620	78,062	92,373	102,099	104,680	259,000
Average value per ton	16.03	14.62	19.64	22.30	22.50	33.70
Exports of potassium salts†	1,303	1,233	966	1,033	1,353	2,230
Approximate K_2O equivalent†	735	700	544	564	764	1,260
Value†	38,353	33,061	28,473	35,323	45,858	111,400
Imports for consumption of potassium salts†	3,644	3,926	4,403	4,672	4,979	8,220
Approximate K_2O equivalent†	2,166	2,332	2,605	2,766	2,961	4,900
Value†	71,910	60,703	94,734	111,844	119,666	290,800
Apparent consumption of potassium salts‡	7,432	8,033	8,140	8,217	8,279	13,660
Approximate K_2O equivalent	4,344	4,701	4,730	4,794	4,815	7,940
World: Production, marketable:						
Approximate K_2O equivalent	17,867	19,198	20,013	21,817	22,465	37,100

Source: Minerals Yearbook 1972, Dept. of the Interior, 1974. *Estimated. †Excludes potassium chemicals and mixed fertilizers. ‡Measured by sold or used plus imports minus exports.

The phase-rule chemists and the engineers at Trona have been exceedingly clever in obtaining *pure* salts from this complex brine and in building a profitable industry in a competitive field and under the handicap of being in the desert 200 mi from the nearest city. The town of Trona, Calif., houses more than 2,000 inhabitants dependent on this one industry. Figure 17.1 outlines the steps, based on the phase-rule study, necessary to commercialize this brine. Figure 17.2 gives more details of the division of the procedures leading to potassium chloride, borax, and soda products. There is no profitable market for the large tonnages of common salt obtained, and it is washed back into the lake.

In general, this successful process is founded upon many years of intensive research wherein exact conditions were worked out and then applied in the plant. In barest outline this involves the concentration of potassium chloride and borax in *hot* brine and the simultaneous separation of salt and burkeite, a new mineral with the composition $Na_2CO_3 \cdot 2Na_2SO_4$. By virtue of the "lazy" crystallization of borax, potassium chloride can be obtained by rapid cooling of the concentrated brine in vacuum coolers and crystallizers.[7] After centrifugation, the potash mother liquor is refrigerated and furnishes borax.

The somewhat detailed[8] flowchart in Fig. 17.2 can be resolved into the following sequences:

Concentration and soda-products separation. Raw brine is mixed with end liquors from the borax crystallizing house and pumped into the third effect of triple-effect evaporators (Op).

The brine is hot-concentrated, and salted out in the three effects counter to the steam flow (Op and Ch) (Figs. 17.3 and 17.4).

[7]Perry, pp. 17-8 to 17-19; Turrentine, Potash in North America, pp. 105–107, Reinhold, 1943 (the first vacuum cooler-crystallizer is pictured and described). Turrentine's invention has had a profound influence upon the cheap production of pure potassium chloride both in the United States and abroad. For evaporation in general, see Standiford, Evaporation, *Chem. Eng.* (*N.Y.*), **70**(25), 157–176 (1963) (22 refs).

[8]Mumford, Potassium Chloride from the Brine of Searles Lake, *Ind. Eng. Chem.*, **30**, 872 (1938); Kirkpatrick, A Potash Industry: At Last, *Chem. Metall. Eng.*, **45**, 488 (1938); Searles Lake Chemicals, *Chem. Metall. Eng.*, **52**(10), 134 (1945) (pictured flowchart).

The suspended salts, NaCl and $Na_2CO_3 \cdot 2Na_2SO_4$, are removed from the liquors of each effect by continuously circulating the hot liquor through cone settlers called *salt separators,* or *salt traps.* The underflow from the first-effect cone containing the salts passes through an orifice into the second-effect cone, receiving a countercurrent wash with clarified liquor from the second-effect cone. The combined salts from the first and second cones are given a countercurrent wash with liquor from the third cone as they pass through an orifice into this third cone. The combined salts of the first, second, and third cones receive a countercurrent wash with raw brine as they leave the third-effect cone (Op). All these are hot washes. The combined underflow is filtered, and the filtrate returned to the evaporators (Op).

The cake (salt: NaCl and $Na_2CO_3 \cdot 2Na_2SO_4$) is sent to the soda-products plant (Op) (middle lower part of Fig. 17.1).

The final hot, concentrated liquor is withdrawn from the overflow of the first-effect cone into an auxiliary settler called a *clarifier* (Op). The overflow from the clarifier is pumped to storage at the potash plant (Op).

The underflow from the clarifier is filtered and treated in the same manner as the previous underflow[9] (Op).

Separation of potassium chloride. The hot, concentrated liquor leaving the clarifiers is saturated with potassium chloride and borax. The potassium chloride is obtained by cooling quickly to 100°F and crystallizing in three-stage vacuum cooler-crystallizers (Op and Ch).

Enough water is added to replace that evaporated so that the sodium chloride remains in solution (Op).

The suspension of solid potassium chloride in the mother liquor is passed to a cone settler, where the thickened sludge is obtained as the underflow (Op).

The potassium chloride is dried in rotary driers, yielding 97% potassium chloride (Op). This salt is conveyed to storage, to the bagging plant, or to a recrystallizing procedure (Op).

Separation of borax.[10] The overflow, combined with the filtrate, is pumped to the borax plant for removal of borax (Op).

The potash mother liquor is cooled in vacuum crystallizers as shown in Fig. 17.2 (Op). The water is lost by evaporation and is returned to the boiling (but cooling) solution to prevent the concentration of this solution with consequent crystallization of potassium chloride with the crude borax.

The borax crystallizes out as a crude sodium tetraborate pentahydrate (Ch).

The crude borax is filtered off and washed (Op).

The filtrate is returned to the start of the evaporator cycle (Op).

When necessary, the crude borax is refined by recrystallization (Op).

This salt is centrifuged, dried, and packaged for market (Op).

The preceding description deals with the processing of the upper brine layer. In 1948 a new plant was finished at Trona for the lower brine layer. The process used involves carbonation of this brine with flue gas from the boiler plant.[11] The sodium bicarbonate separated by this reaction is calcined and converted to dense soda ash. Crude borax is crystallized from the carbonated end liquor by cooling under vacuum, and the filtrate is returned to the lake. The daily production of American Potash and Chemical Co. is approximately 750 tons of muriate of potash, 400 tons of borax and boric acid, 650 tons of salt cake, and 400 tons of soda ash. This large production was not achieved without solving many more problems than those of the phase-rule diagrams. Evaporation

[9]Mumford, *op. cit.*, pp. 876, 877; Hightower, The Trona Process, *Chem. Eng. (N.Y.)*, **58**(8), 104 (1951) (excellent flowcharts).

[10]See Chap. 20 for boron compounds for the *solvent extraction* of boric acid from Searles Lake brine and from borax mother liquors.

[11]Hightower, New Carbonation Technique: More Natural Soda Ash, *Chem. Eng. (N.Y.)*, **58**(5), 162 (1951), and **56**(4), 102 (1949).

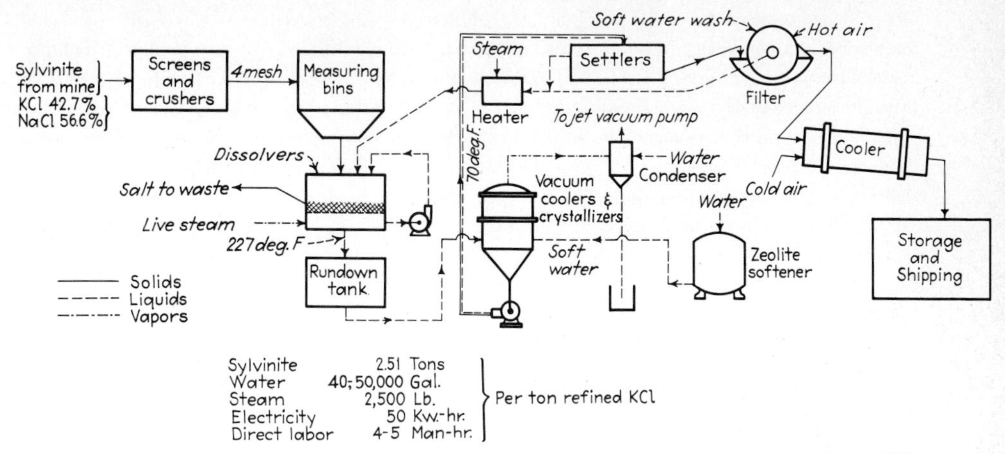

Fig. 17.5 Potassium chloride (muriate) from sylvinite, with continuous flow dissolving for hot-solution refining. (*U.S. Potash Co.*)

is at the rate of several million gallons of water per day. To effect the requisite heat transfer when salts crystallize out at the same time was a major chemical engineering problem. This problem was solved by removing the miles of piping from inside the evaporators and doing the heating under mild hydrostatic pressure with minimum evaporation in outside heaters arranged with spares around and beneath each evaporator.[12] The superheated solution is flashed into evaporators. Heat transfer is also facilitated by vacuum cooling through vaporization instead of using cooling liquids in coils, which would become fouled with encrusted solids.

Soda and lithium products[13] are a development of recent years resulting from an extension of the phase studies and sound engineering. Figure 17.1 outlines the production of soda ash, anhydrous sodium sulfate, and lithium carbonate. Figure 17.4 diagrams the flow of brines and salts at Trona.

MANUFACTURE FROM SYLVINITE In 1925 potash was discovered at Carlsbad, N.Mex.[14] After much prospecting and drilling, a shaft to mine the mineral was sunk to about 1,000 ft, and in 1932 a refinery was constructed.

The process employed by the United States Borax and Chemical Corp. depends primarily upon the fact that sodium chloride is less soluble in a hot than in a cold saturated solution of potassium chloride. Thus, when a saturated solution of the mixed salts in water is cooled from its boiling point, potassium chloride separates out, contaminated with only the sodium chloride that is entrained. The cold mother liquor is heated to 110°C by the use of heat exchangers employing exhaust steam, as shown in Fig. 17.5. The hot mother liquor is passed through a series of steam-heated turbomixer dissolvers countercurrent to a flow of ore crushed to −4-mesh size, which is moved from dissolver to dissolver by mechanical elevating equipment. The potassium chloride goes into solution, together with a small amount of sodium chloride. In this step colloidal and semicolloidal clays present in the ore are suspended in the potash-bearing solution. The enriched mother liquor passes through

[12]Manning, Capital + Vision + Research = An American Potash Industry, *Chem. Metall. Eng.*, **36,** 268 (1929).

[13]Reburn and Gale, The System Lithium Oxide–Boric Oxide–Water, *J. Phys. Chem.*, **57,** 19 (1955); Robertson, Expansion of the Trona Enterprise, *Ind. Eng. Chem.*, **34,** 133; Gale, Lithium from Searles Lake, *Chem. Ind.* (*N.Y.*), **57,** 442 (1945); Salt Lake Projects, *Chem. Eng. News*, Apr. 22, 1968, p. 11; Potash, *Chem. Week*, Nov. 19, 1969, p. 36.

[14]Smith, Potash in the Permian Salt Basin, *Ind. Eng. Chem.*, **30,** 854 (1938). For the separation of two salts with a common ion, like NaCl-KCl, see the phase diagrams figs. 3-1 and 3-2 of Kobe, Inorganic Process Industries, Macmillan, 1948; Ullmann, Enzyklopaedie der technischen Chemie, vol. 9, p. 187, 1957 (phase diagrams and flow diagrams); Gaska *et al.*, Ammonia as a Solvent (to prepare pure KCl from NaCl, KCl), *Chem. Eng. Prog.*, **61**(1), 139 (1965) (flowchart).

thickener equipment, where this insoluble mud settles out. The underflow mud is washed by countercurrent decantation in a tray thickener, and the not-clear, saturated overflow solution is pumped to vacuum coolers and crystallizers. Tailings from this process, largely sodium chloride, after passing through Bird centrifuges where adhering potash-bearing brine is removed, are carried out of the plant to waste storage. Vacuum in the coolers and crystallizers is maintained by means of steam ejectors. In these vessels the solution is cooled to 27°C, and the potassium chloride comes out of solution and, in suspension, is pumped to settling tanks where a large part of the liquor is decanted through launders to be used again. The thickened crystal mass is filtered, washed, and dried on Oliver filters. The dried cake is crushed, screened, and conveyed either to a warehouse or to cars.

The United States Borax and Chemical Corp.'s granular plant for producing 50% potassium chloride, used almost entirely by the fertilizer industry, operates as follows. Crushed and sized ore suspended in a brine saturated with both sodium and potassium chloride is carried to a bank of Wilfley tables, where sodium chloride and potassium chloride are separated by their difference in gravity. A product carrying 50 to 51% K_2O is debrined in drag classifiers and passed through gas-fired rotary dryers, from which it goes to storage and shipping. A middling product is further tabled after debrining, and the tailing is debrined and carried to a salt-storage pile.

The process of the Potash Company of America separates potassium chloride from sodium chloride chiefly by a metallurgical concentration method using a *soap-flotation* process and represents the first adaptation to water-soluble ores of the flotation principle so long familiar in the concentration of insoluble ores. The sylvinite ore, having been coarsely crushed underground, is fed to an intermediate crusher, which in turn discharges to fine crushers. The ore is then wet-ground by ball mills to 100-mesh. The mill product is treated in two series of flotation cells to float off an NaCl concentrate and depress a KCl concentrate. The NaCl crystals are washed and separated from the solution by means of a thickener and a filter and sent to a waste while the KCl concentrate, together with KCl crystals recovered from various circulating solutions, is separated by a classifier into fine and coarse fractions. Recovery of fines is accomplished by a thickener and centrifuges.

The *flotation* process is used also by International Minerals and Chemical Corp. in the refining of sylvinite to produce potassium chloride. This company also processes potassium sulfate from the mineral langbeinite ($K_2SO_4 \cdot 2MgSO_4$) by reacting it with potassium chloride to yield the sulfate and the by-product magnesium chloride. The latter is of particular interest as a raw material for the production of metallic magnesium, potassium chloride, and sulfate of potash-magnesia.

Potash from Saskatchewan, Canada. This is probably the largest new source of potash, and it is estimated that it will satisfy the world potash demand for more than 600 years. It is, however, deep down—3,500 or more feet—and layers of water-soaked deposits and quicksand must be penetrated. This was accomplished by the International Minerals and Chemical Corp.[15] It proved to be America's toughest mining venture and involved solidifying 300 ft of the prospective 18-ft-diameter shaft area by refrigeration with lithium chloride brine at 50°F. About $1\frac{1}{2}$ years was required to freeze through the 200 ft of quicksand, after which the shaft was dug and then lined with cast-iron circular segments at a cost of $2 million. The installation of hoisting machinery followed.

The refinery for processing the raw salts from the mine has been operating[16] since 1963. The ore is sylvinite (xNaCl·KCl) and carnallite ($KCl \cdot MgCl_2 \cdot 6H_2O$) (Fig. 17.6). The International Minerals and Chemical Corp. process as depicted grinds and sizes the ore. This ore is treated with a brine which leaches out the carnallite, deslimes the ore, and then separates the potassium salts into fines and coarse material. After this separation, the fines are deslimed by a hydroseparator, treated

[15]Elliott, Canada's 100 Billion Dollar Potash Pile, *Rotarian*, May, 1964; Potash Industry, *Chem. Eng. News*, Sept. 23, 1968, p. 20; The Lid's Off Potash Production, *Chem. Week*, Nov. 14, 1973, p. 29.

[16]Cross, World's Largest Supply Feeds New Potash Plant, *Chem. Eng. (N.Y.)*, **69**(23), 176 (1962) (flowchart, picture, description); Saskatchewan, *Chem. Eng. (N.Y.)*, **71**(22), 84 (1964); cf. *Chem. Eng. (N.Y.)*, **72**(3), 62 (1965); Gaska *et al.*, *Chem. Eng. Prog.*, **61**(1), 1965.

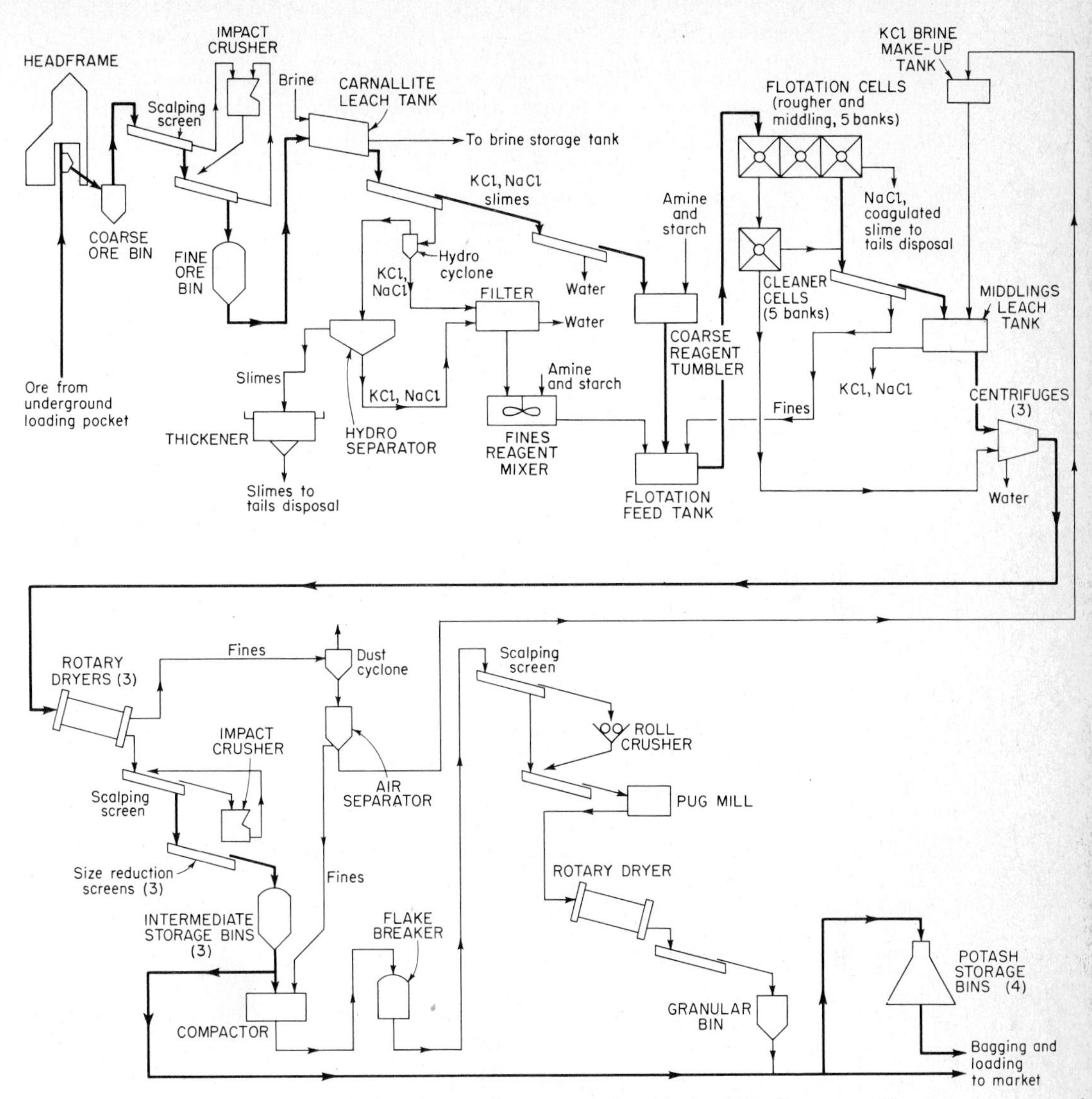

Fig. 17.6 Potash mine and refinery. Banks of flotation cells separate potash particles from NaCl. [*International Minerals & Chemical Co. Esterhazy, Saskatchewan,* Canada; *Chem. Eng.* (*N.Y.*), **68**(23), 176, 1962.]

with an "amine" and starch, and combined with the coarse salts, which are similarly treated. Both the fines and the coarse salts are conducted to flotation cells to separate the potassium salts from the sodium chloride by flotation. When the ore is treated with the brine, the carnallite is decomposed, dissolving the magnesium chloride, and the potassium chloride begins to crystallize. Fresh water added continuously prevents the brine from becoming saturated with magnesium chloride. Although some potassium chloride dissolves, the main brine effluent carries mostly sodium chloride and potassium chloride.

FERTILIZER USE Over 90% of potash salts mined enter the fertilizer field, mostly as mixed fertilizers. These are presented in Chap. 26 on agrichemicals.

VARIOUS POTASSIUM SALTS

POTASSIUM SULFATE Prior to 1939, the German potash industry was the chief source of potassium sulfate for American chemical and fertilizer industries, although considerable tonnages were being produced in this country by the interaction of potassium chloride and sulfuric acid as a side product of salt-cake manufacture. With the termination of European imports, the production of the salt was undertaken on a larger scale by the American Potash and Chemical Corp. through the interaction of burkeite ($Na_2CO_3 \cdot 2Na_2SO_4$) with potassium chloride,[17] followed in turn by the successful recovery of this salt from langbeinite by the International Minerals and Chemical Corp., as mentioned above. In agricultural use potassium sulfate is preferred for the tobacco crop of the Southeast and the citrus crop of southern California.

POTASSIUM HYDROXIDE AND CARBONATE These two chemicals are very closely related. The production of potassium hydroxide by electrolysis of potassium chloride solutions has led to the production of potassium carbonate from potassium hydroxide. The abundant supplies of refined potassium chloride now cheaply available, combined with low-cost electric energy, indicate that this will continue to be the principal method of production in the future. About 40 million lb of caustic potash is consumed per year in the soap, detergent, and dye fields. The carbonate is used mainly for hard glass, soaps, and chemicals (over 36 million lb/year) and is the preferred source of potash in the glass industry (about 30 million lb/year).

POTASSIUM NITRATE This salt is made by the double-decomposition reaction of $NaNO_3$ and KCl:

$$NaNO_3(aq) + KCl(s) \longrightarrow NaCl(s) + KNO_3(aq) \qquad \Delta H = +3.16 \text{ kcal}$$

A strong, hot solution of $NaNO_3$ is made, and solid KCl is dumped into the kettle. Upon heating, the KCl crystals change to NaCl crystals, and the hot potassium nitrate solution is run through the NaCl crystals at the bottom of the kettle. A little water is added to prevent further deposition of NaCl as the solution is cooled. A good yield of potassium nitrate results. The Southwest Potash Co.[18] makes potassium nitrate at its fertilizer plant at Vicksburg, Miss., reacting KCl from the company's Carlsbad plant with nitric acid to furnish 63,000 tons/year potassium nitrate and 22,000 tons/year chlorine. This plant cost about $8 million.[19] The overall reaction is:

$$2KCl + 2HNO_3 + \tfrac{1}{2}O_2 \longrightarrow 2KNO_3 + Cl_2 + H_2O$$

The step-by-step reactions are:

$$\text{Step 1:} \quad 3KCl + 4HNO_3 \xrightarrow{100^\circ C} 3KNO_3 + NOCl + Cl_2 + 2H_2O$$

$$\text{Step 2:} \quad NOCl + 2HNO_3 \longrightarrow 3NO_2 + \tfrac{1}{2}Cl_2 + H_2O$$

$$\text{Step 3:} \quad 2NO_2 + H_2 + \tfrac{1}{2}O_2 \longrightarrow 2HNO_3$$

The key process features are (1) a shift in the composition of the nitric acid–water azeotrope from about 70% HNO_3 to over 80% HNO_3, in the presence of KNO_3 at high concentration, and (2) the capability of oxidizing nitrosyl chloride to Cl_2 and NO_2 with nitric acid in the 70–80% concentration range, at temperatures and pressures that are practical from an operating viewpoint.

[17]Robertson, Expansion of Trona Enterprise, *Ind. Eng. Chem.*, **34**, 136 (1942).

[18]Nitrate Potash, *Chem. Eng. (N.Y.)*, **71**(1), 44 (1964); *Chem. Eng. News*, Feb. 25, 1963, p. 34; Smith and Jacobs, Production of Chlorine and Alkali Metal Nitrates, U.S. Pat. 2,963,345 (1960); cf. Haug and Albright, Reactions of Nitric Acid with Alkali Metal Chlorides in Fused Salt Solution, *Ind. Eng. Chem. Process Des. Dev.*, **4**, 241 (1965).

[19]Spealman, New Route to Chlorine and Saltpeter, *Chem. Eng. (N.Y.)*, **72**(23), 198 (1965) (flowchart and details); Hopewell $NaNO_3$, ECT, vol. 9, p. 460, 1952, 2d ed., vol. 1, p. 703, 1962; Mehring *et al.*, Potassium Nitrate, *Ind. Eng. Chem.*, **21**, 379 (1929); Synthetic Saltpeter Scores, *Chem. Week*, July 24, 1965, p. 35.

The company states that its objective is to price its fertilizer-grade potassium nitrate competitively with potassium sulfate for plants, such as tobacco, that are affected adversely by chloride in the soil. Potassium nitrate is frequently referred to as being essential in the manufacture of the slow-burning black powder used in fuses and in the ignition of smokeless powder. Other uses are in the manufacture of pyrotechnics and glass.

POTASSIUM ACID TARTRATE A domestic tartrate industry has been developed by processing pomace, which is discarded from wine fermentors and contains from 1 to 4% potassium bitartrate. Still the greater part of the supply is imported crude argols from wine vats. The pure salt is obtained from the argols by leaching and crystallizing. The main uses are as the acid ingredient of baking powder, for the preparation of tartaric acid, and in medicine.

SELECTED REFERENCES

American Potash Association. Publications.
Noyes, R.: Potash and Potassium Fertilizers, Noyes, 1966.
Sauchelli, V. (ed.): Chemistry and Technology of Fertilizers, ACS Monograph 148, Reinhold, 1960.

chapter 18

NITROGEN INDUSTRIES

Humans have increased their supply of food by feeding nitrogen compounds back into the soil. Chemists and chemical engineers have found ways of making nitrogen derivatives out of air in an economical way. The first successful process (1904), the arc process, required much cheap electric energy. However, the present solution to the nitrogen-fixation problem was obtained by reacting nitrogen with low-cost hydrogen to make ammonia under conditions requiring less power and low conversion expense. This made the arc process obsolete. Ammonia has now become one of our heavy chemicals, produced in enormous tonnage around the globe, and at such low prices as to dominate the world supply of nitrogen fertilizers and most nitrogen compounds. This is probably *one of the most important chemical engineering achievements in history.*

HISTORICAL Priestley observed that an electric spark passed through air confined over water caused a lowering of gas volume and acidification of the water. A few years later, in 1780, Cavendish repeated Priestley's experiments and found that nitrites and nitrates were formed when the experiment was carried out over an alkaline solution. The direct ammonia method was long in developing, but it now holds the paramount position. It was the research of Haber, Nernst, and their coworkers that laid the real foundation for the present exceedingly important synthetic ammonia industry, and in the years 1904–1908 equilibrium data for the ammonia-nitrogen-hydrogen system were established with fair accuracy over considerable ranges of temperatures and pressures. The development of a practical synthetic ammonia process was carried out through the chemical engineering efforts of Haber and Bosch and their coworkers, and the growth of this industry has continued. Curtis[1] describes various processes not now economical, such as the arc, cyanide, and Serpek aluminum nitride procedures. Cyanamide- and the ammonia-manufacturing procedures are currently those of technical importance for supplying the world with fixed nitrogen. The only plant in North America producing calcium cyanamide is at Niagara Falls, Canada, and is owned and operated by the American Cyanamid Co. It has an annual capacity of about 235,000 tons of calcium cyanamide.

USES AND ECONOMICS The chemical nitrogen industries include not only nitrogen fixed by humans, but also by-product ammonia from coke ovens and such natural nitrogen deposits as Chile saltpeter, both of which are subjected to manufacturing processes. Table 18.1 summarizes the statistics for the United States, and Table 26.5 illustrates the world nitrogen fertilizer situation by country, per capita and per acre. Ammonia is used in heat-treating and paper pulping, nitric acid for nitro compounds, high explosives, and propellants. Hydrogen cyanide, made largely from ammonia, is used mainly for acrylates and methacrylates, although a little is used for fumigation. Acrylonitrile uses substantial amounts of nitrogen compounds. Hydroxylamine and hydrazine also are smaller-volume nitrogen consumers. Amines of fatty acids enjoy wide use as surface-active agents and in ore flotation. World consumption of fertilizer remains far below that desirable for

[1]Curtis, Fixed Nitrogen, Reinhold, 1932; ECT, 1st ed., vol. 1, 771ff., 1947, and 2d ed., vol. 2, pp. 258–298, 1963.

TABLE 18.1 ***Ammonia and Ammonia Compound Production in the United States*** *(In thousands of short tons of contained nitrogen)*

Compound	*1968*	*1969*	*1970*	*1971*	*1972*	*1973**	*1974**	*1975**
Ammonia (synthetic)	9,968	10,502	11,369	11,538	12,511	12,737	13,094	13,588
Ammonia compounds (coking plants)								
Ammonia liquor	14	12	12	12	11	12	12	12
Ammonium sulfate	142	143	126	114	128	130	130	130
Ammonium phosphates	6	7	9	9	9	9	9	9
Total	10,130	10,664	11,516	11,673	12,650	12,888	13,245	13,739

Source: Minerals Yearbook, 1972, Dept. of the Interior. 1974. *Notes:* The 13,094,000 tons of (nitrogen-contained) ammonia for 1974 is equivalent to 15,900,000 tons of ammonia (100% ammonia basis). A 1000 tons/day ammonia plant produces 328,500 tons/year at 90% of capacity. The 1974 production of ammonia corresponds to the equivalent of 48.4 plants each with a 1000 tons/day capacity. *Estimated value.

maximum crop production. Only 13 countries[2] use over 200 kg/hectare of combined $N + P_2O_5 + K_2O$ fertilizer, and many use far below 25 kg/hectare. Presumably all countries may ultimately approach the higher use figure.

CYANAMIDE

The major use of cyanamide was at one time agricultural. Its employment as a chemical raw material has become increasingly important, however, the largest chemical use being for the preparation of the dimer dicyandiamide. This dimer is polymerized to produce the trimer melamine, which has applications in plastic resins. Other commercial derivatives are produced by way of a crude calcium cyanide[3] (48 to 50% expressed as NaCN) from a high temperature melt of cyanamide with excess carbon and NaCl. This is used directly for cyanidation of ores and to manufacture ferrocyanides. The use of cyanamide fertilizers is enjoying a minor resurgence, because these chemicals act as pesticides to snails and larvae and also enrich the soil with nitrogen. Cyanamide compounds are used to fireproof shingles and cotton fabrics.

REACTIONS AND ENERGY CHANGES[4] The essential reactions for the production of calcium cyanamide are:

$$CaCO_3(s) \longrightarrow CaO(s) + CO_2(g) \qquad \Delta H = +43.5 \text{ kcal} \qquad (1)$$

$$CaO(s) + 3C(amorph) \longrightarrow CaC_2(s) + CO(g) \qquad \Delta H = +103.0 \text{ kcal} \qquad (2)$$

$$CaC_2(s) + N_2(g) \longrightarrow CaCN_2(s) + C(amorph) \qquad \Delta H = -68.0 \text{ kcal} \qquad (3)$$

Various catalysts or fluxes are used to increase the rate of reaction or cause it to proceed at lower temperatures. The American Cyanamid Co. uses calcium fluoride, and reaction-rate studies have shown that it reduces the temperature of optimum reactivity and increases the velocity of the reaction by 4.5 times at 1000°C. Reaction (3) takes place at 900 to 1000°C. Reaction (2) is carried out in two 20,000- and two 10,000-kW furnaces at about 2000 to 2200°C. Under normal conditions reaction (2) yields up to 90% calcium carbide. Considerable overall energy is needed, principally to secure the high temperature for reaction (3) to start when it is self-sustaining and for making the calcium carbide in reaction (2). The source of carbon is usually coke. Coal is required to burn the limestone and to dry the raw materials (Fig. 18.1).

[2]Fertilizer Use throughout the World, *Chemtech.*, **2,** 570 (1972).

[3]Kastens and McBurney, Calcium Cyanamide, *Ind. Eng. Chem.*, **43,** 1020 (1951) (excellent article); Cyanamid's Cyanamide Gets a Second Chance, *Chem. Week*, Mar. 30, 1968, p. 57.

[4]Kastens and McBurney, *loc. cit.* (excellent article with many pictures; should be freely consulted). See Chap. 15 for more details on calcium carbide manufacture.

TABLE 18.2 ***Major Nitrogen Compounds Produced in the United States***
(In thousands of short tons)

Compound	*1972*	*1973**	*1974**	*1975**
Acrylonitrite	557	677	725	700
Ammonium nitrate (original solution)	6,881	6,952	7,400	7,650
Ammonium sulfate	2,471	2,500	2,650	2,800
Ammonium phosphates	6,499	6,400	7,100	7,400
Nitric acid	7,981	7,439	8,150	8,300
Urea, primary solution	3,510	3,596	3,700	3,800

Source: Minerals Yearbook, 1972, Dept. of the Interior, 1974; Bureau of the Census; Tariff Commission. *Estimated value.

The following *physical operations* and *chemical conversions* are needed to commercialize[5] the reactions on which Fig. 18.1 is based:

Limestone, coal, and coke are pulverized, separately (Op).
Limestone is calcined to quicklime (Ch).
Coke is pulverized, dried, and mixed with quicklime (Op).
Carbide is formed in an electric furnace at nearly 2000 to 2200°C and run out molten (Ch).
Carbide is cooled, crushed, and finely ground (Op).
Air is liquefied by compressing, cooling, and expansion (Op).
Nitrogen is separated from oxygen by liquid rectification (Op).
Calcium carbide is nitrified over the course of 40 h with 99.9% nitrogen at about 1000°C (Ch).
Calcium cyanamide is pulverized and treated with a small amount of water to hydrate residual CaO and CaC_2. It may be oiled to reduce dust (Op and Ch).

SYNTHETIC AMMONIA[6]

USES AND ECONOMICS Ammonia is one of the truly fundamental raw materials utilized in modern civilization. The total inorganic chemical tonnage in this country exceeds 12 million tons per year. The largest consumer of anhydrous ammonia is the fertilizer industry, which utilizes it in the production of calcium and sodium nitrate, ammonium sulfate, nitrate and phosphate, ammoniated superphosphates, urea, and aqueous ammonia. Applied by special machines, anhydrous ammonia has become a very important fertilizer for *direct* application to soils. In combination with ammonium nitrate, an increasing amount is used each year (for the world picture see Table 26.5). Ammonia is the starting point of nearly all military explosives. Scarcely an industry is untouched, since ammonia is required for the making of soda ash, nitric acid, nylon, plastics, lacquers, dyes, rubber, and other products.[7] The product is handled and shipped in two forms, ammonia liquor and anhydrous ammonia. Commercial grades of the liquor usually contain 28% ammonia. Shipment of anhydrous in this country is largely done in special tank cars and in cylinders (Table 18.1).

REACTIONS AND EQUILIBRIUMS For the reaction

$$\tfrac{1}{2}N_2(g) + \tfrac{3}{2}H_2(g) \rightleftharpoons NH_3(g)$$

an equilibrium constant can be expressed as

$$K_p = \frac{p_{NH_3}}{p_{N_2}{}^{1/2} \times p_{H_2}{}^{3/2}}$$

[5]*Chem. Metall. Eng.*, **47,** 253 (1940) (pictured flowchart).

[6]See Chap. 7, cryogenics, hydrogen, and nitrogen, and Chap. 26, mixed fertilizers; Comings, High Pressure Technology, chap. 12, McGraw-Hill, 1956 (excellent, with many references, tables, and calculations); Frear and Baber, Ammonia, ECT, 2d ed., vol. 2, pp. 258–298, 1963; Ammonia and Synthesis Gas, Noyes, 1964; company NH_3 pamphlets, e.g., M. W. Kellogg Co. and Lummus Co.

[7]Minerals Yearbook, 1972, p. 895, Dept. of the Interior, 1974.

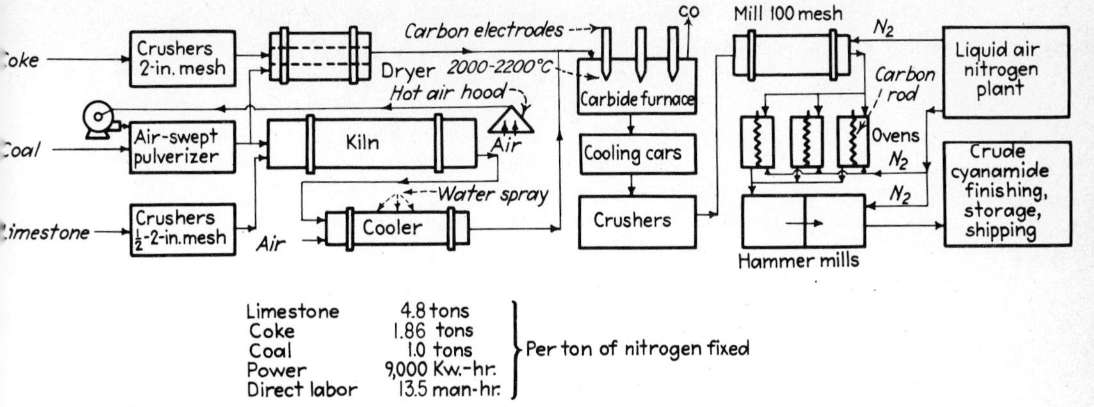

Fig. 18.1 Flowchart for the manufacture of calcium carbide and calcium cyanamide.

Since the volume of ammonia obtained is less than the combined volume of nitrogen and hydrogen, a pressure increase, according to the *principle of Le Chatelier,* gives a higher percentage of ammonia at equilibrium. The conversion percentage increases severalfold as the pressure increases from 100 to 1,000 atm, but the percentage of ammonia in equilibrium with the reacting gases decreases continually with a temperature rise up to 1100°C and reaches a minimum at this temperature. When operating below 1100°C (Fig. 18.4), the lower the temperature for a given pressure at which an ammonia converter can be run, the larger the possible percentage of ammonia in the ammonia-nitrogen-hydrogen mixture, but the longer the time required to attain equilibrium. Accordingly, temperatures in converters are kept as low as is consistent with a sufficiently rapid reaction rate to be economical with the catalyst used and with the investment required.

RATE AND CATALYSIS OF THE REACTION To be economical, the rate of this reaction must be increased, because hydrogen and nitrogen alone react very slowly. The basis of the commercial synthesis of ammonia rests upon an efficient *catalyst* to speed up the reaction rate to an economical degree. Such a catalyst[8] has been found in iron, whose reaction rate is promoted by the addition of oxides of aluminum (3%) and potassium (1%). These promoters prevent sintering.

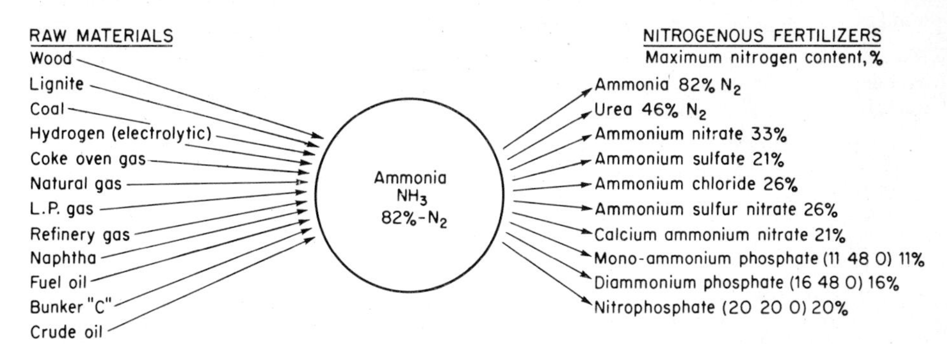

Fig. 18.2 Raw materials used throughout the world to manufacture ammonia and nitrogenous fertilizers produced from ammonia. In the United States 95% of the hydrogen required is based on natural gas or other hydrocarbons. [*Chem. Eng. Prog.*, **49**(11), 63 (1963).]

[8]Catalyst Handbook, Springer Verlag, 1970.

Many reactions are known which occur as a result of one reacting constituent combining with the catalyst to form an intermediate compound capable of reacting with a second reacting constituent to form the product of the catalytic reaction and to regenerate the catalyst. Thus it is postulated that a nitride adduct is possible with the active iron atoms on the surface of the ammonia catalyst. Iron seems to be by far the most satisfactory *catalyst* material, though it seems to lose its activity rapidly if heated to temperatures above 520°C. The catalyst[9] is promoted by metallic oxides, the activity being increased most by adding both an amphoteric oxide of metal such as aluminum, zirconium, or silicon and an alkaline oxide such as potassium oxide. P. H. Emmett summarizes: The percentage ammonia is a gaseous mixture of pure 3:1 hydrogen-nitrogen gas passed over such a doubly promoted catalyst at 100 atm pressure, 5,000 space velocity, and at 450 C, is 13 to 14 for the doubly promoted catalyst in contrast to 8 or 9 for the singly promoted one and to 3 to 5% for the pure iron catalyst. Modern catalysts are made from magnetite containing K_2O, CaO, MgO, Al_2O_3, SiO_2, and traces of TiO_2, ZrO_2, and V_2O_5. The promoters make the reduced catalyst more porous. Catalysts more active at a lower temperature would be most valuable. These catalysts are ruined by contact with many substances such as copper, phosphorus, arsenic, and sulfur that affect the electronic structure of the iron; carbon monoxide greatly reduces their activity. Hence much money must be spent in purifying the hydrogen and nitrogen for ammonia synthesis.

A detailed mechanism for the catalysis leading to ammonia is as follows, and is itself only speculation, but to a major degree meets most observed facts.[10] The controlling aspect is nitrogen activation.

$$N_2 + 2Fe = 2Fe - N_{ads}$$

$$H_2 + 2Fe = 2Fe - H_{ads}$$

$$N_{ads} + H_{ads} = NH_{ads}$$

$$NH_{ads} + H_{ads} = NH_{2ads}$$

$$NH_{2ads} + H_{ads} = NH_{3ads}$$

$$NH_{3ads} = NH_{3desorb}$$

Space velocity is the number of cubic feet of gases, corrected to standard conditions (0°C and 760 mm), that pass over 1 ft^3 of catalyst space per hour. The space velocity used in commercial operations differs considerably in different processes and in different plants using the same process. Although the percentage of ammonia in the gas stream issuing from a converter goes down as the space velocity goes up, the amount of ammonia per cubic foot of catalyst space per hour increases. Too high a space velocity, however, disturbs the thermal balance of a converter, involves increased cost of ammonia removal because of the smaller percentage present in the exit gases, and makes necessary the recirculation of large volumes of gas. Most commercial units use a space velocity between 10,000 and 20,000.

MANUFACTURING PROCEDURES Numerous variations in pressure, temperature, catalyst, and equipment from the original Haber procedure have been put into practice in plants throughout the world. Table 18.3 summarizes these and includes the procedure for obtaining the hydrogen. There is as much difference in the economical sources of hydrogen[11] as in any other variable. The principal hydrogen-manufacturing process in use in 1976 for synthetic-ammonia production in the United

[9]Nielsen, Investigation on Promoted Iron Catalysts for the Synthesis of Ammonia, Gjillerups, Copenhagen, 1956; Alexander *et al.*, Hydrogenation: NH_3 Catalysts, *Ind. Eng. Chem.*, **53**(9), 767 (1961).

[10]Alvin Stiles; cf. Nielsen, *op. cit.*; Happel, Fundamental Catalysis Research in Japan, *Chem. Eng. News*, May 30, 1966, p. 84; Subscript "ads" signifies "adduct" or "adsorbed state."

[11]For reactions and details on hydrogen manufacture, ECT, 2d ed., vol. 1, pp. 274ff., 1963; this book, Figs. 7.4 and 7.5; Synthetic Ammonia Produced from Natural Gas, *Chem. Eng.* (*N.Y.*), **52**(12), 92 (1945) (pictured flowchart, pp. 134–137); Cope, Ammonia, *Chem. Ind.*, (*N.Y.*), **64**, 920 (1949); Van Krevelen, A Few Aspects of the Development and Technology of the Nitrogen Industries, *Chem. Weekblad*, **44**, 437 (1948).

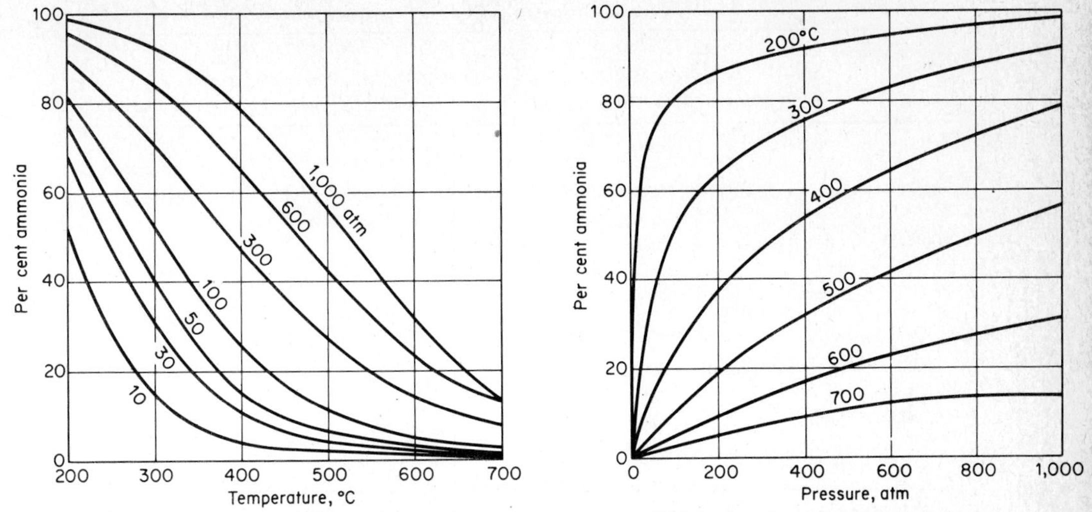

Fig. 18.3 Percentage of ammonia at equilibrium from an initial mixture of 3:1 H_2/N_2 gas at various temperatures and pressures. (*Comings, High Pressure Technology, pp. 410 ff.* Compare Frear and Baber, ECT, 2d ed., vol. 2, p. 260, 1963.)

States was the steam-hydrocarbon (natural-gas) process. Practically all the other[12] systems grew out of the pioneering work done in Germany.

AMERICAN SYSTEMS The system for ammonia synthesis developed by the Nitrogen Engineering Corp. is illustrated by the flowchart in Fig. 18.4; the steps are listed in Table 18.4. All ammonia syntheses are based on the reaction

$$\tfrac{1}{2}N_2(g) + \tfrac{3}{2}H_2(g) \longrightarrow NH_3(g) \qquad \Delta H_{18^\circ C} = -11.0 \text{ kcal}, \Delta H_{659^\circ C} = -13.3 \text{ kcal}$$

This reaction is highly exothermic, and consequently the design of the converter[13] (Fig. 18.5) must be such as to control the temperature at the point desired for the conversion deemed economical for the particular conditions chosen by the chemical engineers (cf. Fig. 18.3).
The American system as shown in the flowchart in Fig. 18.4 and in Tables 18.3 and 18.4 may be broken down into the following coordinated sequences:

Ammonia synthesis gas is prepared by high-pressure catalytic reforming of hydrocarbon feed gas (e.g., natural gas) in the primary reformer with superheated steam, and in the secondary reformer with air to furnish nitrogen. The amount of air is adjusted eventually to supply a H_2/N_2 molar ratio of 3:1 (Ch).

CO is converted by the catalytic iron oxide shift conversion with H_2O (from the reformer plus steam) to CO_2 (Ch).

CO_2 is removed by a reactivated hot K_2CO_3 solution[14] wash (with reboiler regeneration yielding CO_2) (Ch).

This is followed, as Fig. 18.4 shows, by methanation (Ch).

Figure 18.4 indicates where heat is economized in various heat exchangers (Op).

The 3:1 hydrogen-nitrogen mixture, freed of its CO, is raised to the full compression of 150 to 200+ atm and mixed with the recompressed, recirculated gases. The gases are then passed to the

[12]ECT, 2d ed., vol. 2, pp. 266ff., 1963 (flowcharts); *Hydrocarbon Process.*, **52**(11), 103 (1973).

[13]Synthesis Gas, *Chem. Eng.* (*N.Y.*), **73**(1), 24 (1966); A New High-Capacity Ammonia Converter, *Chem. Eng. Prog.*, **68**(1), 62 (1972); Ammonia Converters Grow Again, *Chem. Eng.* (*N.Y.*), **78**(24), 90 (1971).

[14]Or by an amine, MEA.

TABLE 18.3 *Synthetic Ammonia Systems Arranged in Order of Increased Conversion Pressure*

Conversion pressure, atm	*Designation*	*Temp., °C*	*Catalyst*	*Conversion, %*	*Recirculation purge inerts*	*Hydrogen sources*
120–160	Mont Cenis	400–425	Iron cyanide	9–20	Yes	Electrolytic or by-product H_2
150	American* (large)	500	Doubly promoted iron	14	Yes	Natural gas
200–300	Haber-Bosch	550	Promoted iron	8	Yes	Water gas, producer gas
200–300	Fauser-Montecatini	500	Promoted iron	12–22	Yes	Electrolytic cells and waste nitric gas
+300	American* (small)	500	Doubly promoted iron	20	Yes	Natural gas, hydrocarbon
±600	Casale	500	Promoted iron	15–25	Yes	Various
±900	Claude	500–650	Promoted iron	40–85†	No†	Coke-oven gas
900	Du Pont‡	500	Promoted iron	40–85	Yes	Natural gas

* Larger American plants producing 600 to 1,000 tons/day and more use *centrifugal compressors* alone and consequently have lowered conversion pressure to about 150 atm. Higher pressures are uneconomical for new large plants. Centrifugal compressors have diminished the variance between processes; they work up to pressures of 240–290, which sets the operating pressure. Catalysts are rapidly destroyed above 530°C. Processes today include those developed by Chemico, Kellogg, Topsoe, Udhe, Lummus, ICI, and Braun. Conversions of 9 to 30% per pass are possible with modern catalysts. † Two converters in series. Recirculation now practiced. ‡ Refers to the Du Pont plant at Belle, W. Va., built in 1927. The new Du Pont plants at Victoria, Tex., and Belle convert at a low pressure of 150 atm.

water secondary heat exchanger and ammonia refrigerator and separator for removal of residual ammonia (Op).

In the ammonia converter (Fig. 18.5) the gases are raised in a countercurrent heat exchanger to the reaction temperature and caused to react in the presence of the catalyst, after which they are cooled and most of the ammonia liquefied (Ch).

A portion of the gas is purged to prevent undue accumulation of diluents, such as methane or argon, and the rest is recompressed for recirculation (Op).

The purge gas is conducted to the primary reformer (Op).

American large lower-pressure conversion system for 600+ tons/day plants. A revolution[15] *has occurred in the ammonia business* that has been brought about through the application of vertically split, barrel-type centrifugal compressors for handling the synthesis gas at pressures up to 4,200 psi (280 atm) or greater. These compressors are compatible with Figs. 18.4 and 18.5 and with Tables 18.3 to 18.5. The enhancement of the reforming pressure of synthesis gas above 400 psig can shave as much as 25% from an ammonia plant's compression cost. By raising the temperature of the primary reforming effluent and heating the combustion air prior to injection into the endothermic secondary reformer, part of the reforming is shifted from the high-cost primary to the relatively low-cost secondary reformer. These changes are connected with and precede the main part of the system, the synthesis-gas compressor for the actual ammonia synthesis. Engineers of modern large American plants have substituted more economical centrifugal compressors for reciprocating units. The advantages are a great reduction in the capital required for a large ammonia plant and in the cost of ammonia. In smaller plants, reciprocating compressors are used for pressure desired.

[15]Weyermuller, Centrifugal Compressors Take Over in Large New Ammonia Plants, *Chem. Process.* (Chicago), December 1964, pp. 25–56; Noyes, Ammonia and Synthesis Gas, Noyes, 1964.

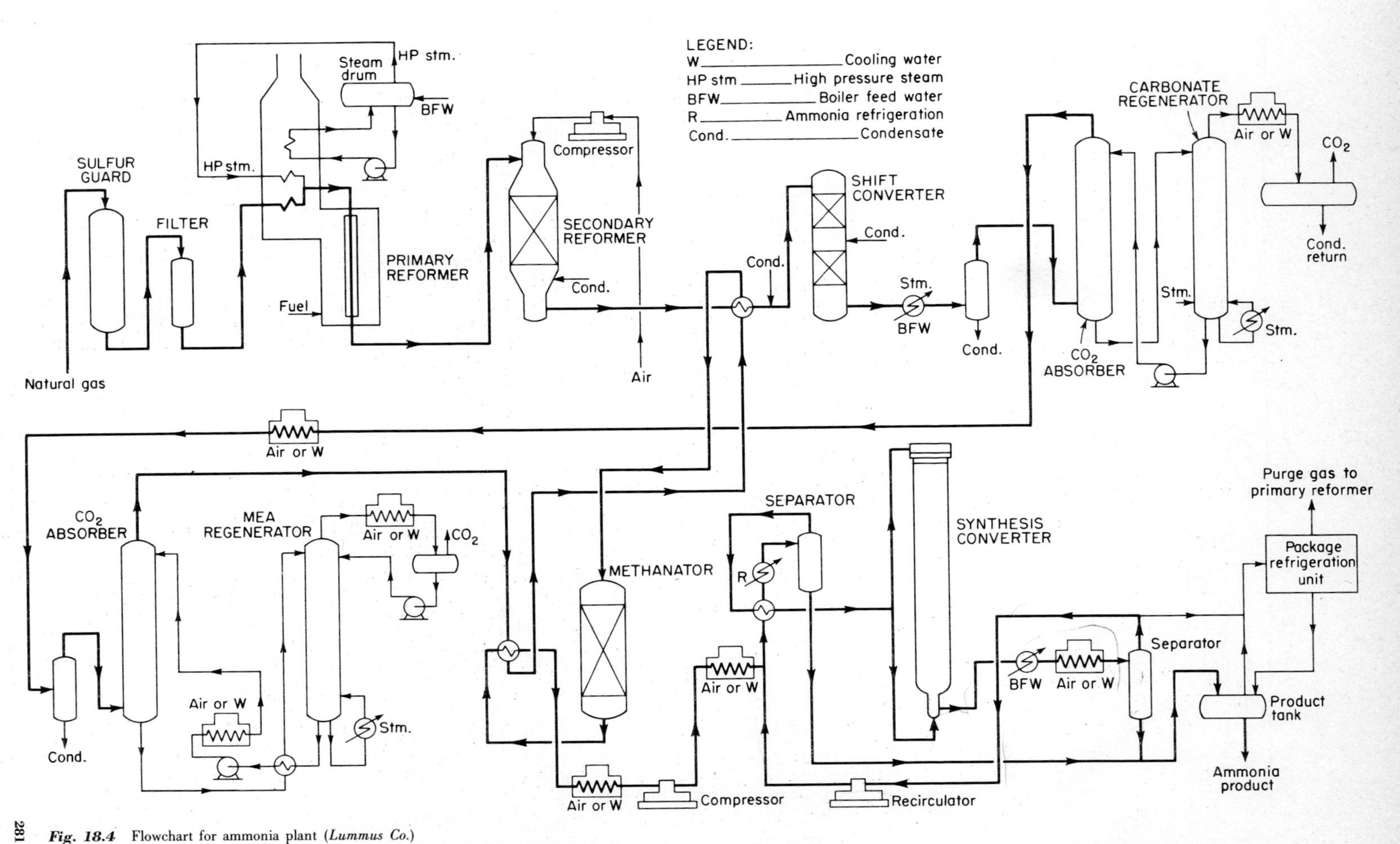

Fig. 18.4 Flowchart for ammonia plant (*Lummus Co.*)

TABLE 18.4 *Steps and Alternatives in Ammonia Synthesis (Fig. 18.4)*

A. Manufacture of raw synthesis gas ($N_2/H_2 = 1:3$)

1. *Purification* of hydrocarbon from sulfur compounds by passing through adsorptive bed (to remove catalyst poisons).
2. *Reforming* conversion of hydrocarbons,* mostly natural gas, in a primary reformer by steam and in a secondary reformer by air to supply nitrogen.

$$CH_4(g) + H_2O(g)^{\dagger} \xrightarrow[\text{400 psig, 1450°F}]{\text{Ni catalyst}} CO(g) + 3H_2(g) \qquad \Delta H = +54.3 \text{ kcal}$$

$$CH_4(g) + \text{air}(g) \xrightarrow[\text{1730°F}]{\text{Ni catalyst}} CO(g) + 2H_2(g) + qs\ N_2(g);$$

Combustion of unconverted methane with oxygen (from air) furnishes heat for the secondary endothermic reforming reaction. Other methods (minor) of reforming or other raw materials are

a. Texaco-Eastman *partial oxidation:*‡

$$CH_4(g) + \tfrac{1}{2}O_2(g) \longrightarrow CO(g) + 2H_2(g) \qquad \Delta H = -8.53 \text{ kcal}$$

b. *Coke-oven gas*

c. *Refinery* hydrogen-rich off-gas.

d. Acetylene-production off-gas.

B. *Shift* conversion of CO to CO_2 with steam, using an iron oxide–chrome catalyst.

$$CO(g) + H_2O(g) \xrightarrow[\text{400°C}]{FeO + Cr_2O_3} CO_2(g) + H_2(g) \qquad \Delta H = -9.2 \text{ kcal}$$

C. Purification by *removal* of CO_2, using monoethanolamine (MEA) or hot potassium carbonate or sulfinol in two stages.

D. Removal of residual amounts§ of CO_2 and CO to less than 10 ppm total by *methanation* over a nickel catalyst (sometimes followed by MEA or by a nitrogen wash or an ammoniated copper carbonate wash).

$$CO + 3H_2 \xrightarrow[\text{315°C}]{\text{Ni}} CH_4 + H_2O \qquad \text{(exothermic)}$$

$$CO_2 + 4H_2 \longrightarrow CH_4 + 2H_2O$$

E. Ammonia *synthesis* at various pressure and temperatures, with consequent varying conversion percentages (Table 18.3).¶

F. Cooling, condensing, and *separation* of NH_3 formed.

G. Recirculation, with *repressuring* of unreacted nitrogen and hydrogen after addition of fresh nitrogen and hydrogen.

H. *Purging* to primary reformer, to reduce inerts. Recovery of purge gas.**

Note: qs stands for quantity sufficient. *About 90% of the U.S. production of NH_3 is economically based on natural gas. Sometimes the initial well pressure of natural gas is utilized as such. Enhancement of pressure here saves equipment costs. †High-pressure superheated steam (850 psig) from reformers. ‡Hydrocarbons other than CH_4 up to bunker C fuel oil and even solid fuels can be substituted for natural gas. §Complete removal of impurities is economical, since it reduces purging and the cost of recompression. The current trend is toward large plants of 1000+ tons of NH_3 per day at the low pressure (3500 lb) reachable by economical centrifugal compressors, with a consequent lower conversion percentage and enhanced recirculation. See Chap. 7 for a fuller account of the preparation of synthesis gas not only for NH_3 but also for CH_3OH, etc. **Ammonia Plants Seek Routes to Better Gas Mileage, *Chem. Week*, Feb. 19, 1975, p. 29.

The shortage of cheap natural gas has directed attention toward other feedstocks.[16] Worldwide, 65% of feedstocks are natural gas, but naphtha, heavy hydrocarbons, and coal are becoming increasingly attractive sources. Computer control of ammonia plants increases yields by reducing process variations.[17]

AMMONIUM SULFATE has long been an important nitrogen carrier in fertilizers, though its production is decreasing. Twenty-four percent of its 2.5 million ton/year production comes from by-product ammonia scrubbed from coke-oven gas with sulfuric acid. A process is in operation at Sindri, India, where the principal raw materials for ammonium sulfate are gypsum and coke:[18]

$$(NH_4)_2CO_3(aq) + CaSO_4 \cdot 2H_2O(s) \longrightarrow CaCO_3(s) + 2H_2O + (NH_4)_2SO_4(aq)$$

Many foreign plants employ anhydrite as a raw material.

[16]Alternate Ammonia Feedstocks, *Chem. Eng. Prog.*, **70**(10), 21 (1974).

[17]Computer Control of Ammonia Plants, *Chem. Eng. Prog.*, **70**(2), 50 (1974).

[18]Ammonium Sulfate from Gypsum, *Chem. Eng. (N.Y.)*, **59**(6), 242 (1952) (pictured flowchart); Ammonium Sulfate from SO_2, *Chem. Eng. News*, Aug. 29, 1960, p. 421.

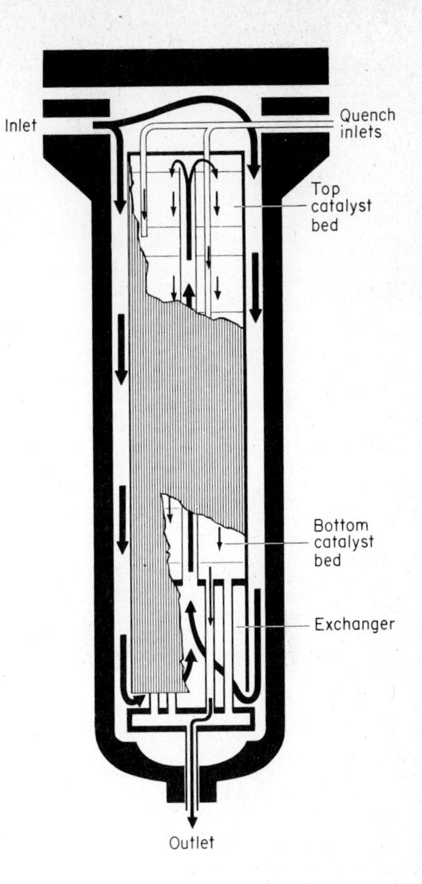

Fig. 18.5 Ammonia-synthesis converter. The converter consists of a high-pressure shell containing a catalyst section and a heat exchanger. The catalyst section is a cylindrical shell which fits inside the pressure shell, leaving an annular space between the two. The catalyst section contains several beds supported on screened grids. To maintain the catalyst at optimum temperature for maximum yield, cold-feed-gas quench is injected before each catalyst bed. The top bed contains the smallest quantity of catalyst. Since the temperature gradient is flatter in succeeding beds, bed sizes are graduated, with the largest bed at the bottom. Beneath the catalyst section is the heat exchanger. This preheats fresh inlet gas against hot reacted gas from the last catalyst bed. The top quench point permits the introduction of feed gas without preheating and provides temperature control to the first catalyst bed.

The feed gas enters at the top of the converter and flows downward between the pressure shell and the wall of the catalyst section. The gas cools the shell and is heated. The gas then enters the exchanger at the bottom of the converter and, by circulating around the exchanger tubes, is further preheated against hot effluent. Some feed gas is introduced directly to the top of the first bed, where it meets the preheated feed. The combined stream at a temperature of 700 to 800°F is then introduced into the top catalyst bed. The gas passes downward through the catalyst, with a rapid temperature rise as the ammonia reaction proceeds, and then through the catalyst supporting grid into a space between the first and second beds. Here the temperature is reduced and the ammonia content diluted by injection of cold feed gas. This permits control throughout the catalyst beds to obtain optimum temperatures for maximum yields. In a like manner, the gas flows downward through the lower catalyst beds. (*M. W. Kellogg Co.*)

TABLE 18.5 ***Economics of Ammonia Manufacture, 1000 short tons/day capacity*** *(Varying natural gas cost)*

Investment	$50,000,000		$50,000,000	
Natural gas price	$0.50/MM Btu		$1.00/MM Btu	
Power cost	$0.01/kWh		$0.01/kWh	
	Unit rate	*$/day*	*Unit rate*	*$/day*
Natural gas feedstock	850 MM Btu/h	10,200	850 MM Btu/h*	20,400
Utilities:				
Fuel	440 MM Btu/h	5,280	440 MM Btu/h	10,560
Power	640 kW	155	640 kW	155
Cooling water makeup	1,290 gal/min	465	1,290 gal/min	465
Boiler feed makeup	390 gal/min	560	390 gal/min	560
Steam	Self-sufficient		Self-sufficient	
Other charges (catalyst, labor, overhead)		1,660		1,660
Indirect return on investment		59,285		59,285
Total cost per day		$77,605		$93,085
Total cost per short ton		$77.60		$93.09

Source: Chem. Eng. Prog., **70**(10), 31 (1974). *MM Btu/h = million Btu/h

TABLE 18.6 ***Comparison of Urea Processes***

	Utility requirements per short ton of urea		
Process	*Steam, lb*	*Electricity, kWh*	*Cooling water, gal*
Conventional	2,400–3,200	132–170	16,000–29,000
SNAM	1,900	105	15,000
Stamicarbon*	2,100†	125	14,500
Weatherly	2,800‡	15‡	23,000‡
Mitsui Toatsu	1,600§	130§	16,000§
Process D	1,700	155	
Mavrovic Conventional	1,280	115	14,400

Sources: Adapted from *Chemtech.*, **1**, 37 (1971); *Chem. Eng.* (*N.Y.*), **80**(12), 72 (1973). *More recent plants are 5–10% below this. †300 lb of excess steam available at 50 psi. ‡Steam-driven compressor. §Electrically driven compressor.

AMMONIUM NITRATE is the most important nitrogen fertilizer because of its high nitrogen content (33%), the simplicity and cheapness of its manufacture, and its quick-acting nitrate nitrogen and slower-acting ammoniacal nitrogen. It is a vital ingredient in so-called "safety"-type explosives and is blended with TNT to form amatol. A minor, but important, application is in the manufacture of nitrous oxide, an anesthetic in extensive use. This gas is formed safely by heating very pure 99.5% ammonium nitrate under controlled conditions at temperatures from 200 to 260°C, furnishing a commercial product:

$$NH_4NO_3 \longrightarrow N_2O + 2H_2O$$

Where ammonium nitrate detonates when used as a constituent in commercial explosives, the following second reaction takes place with great rapidity and violence:

$$NH_4NO_3 \longrightarrow 2N_2 + 4H_2O + O_2$$

This second reaction[19] generally occurs only through initiation by a powerful priming explosive or by

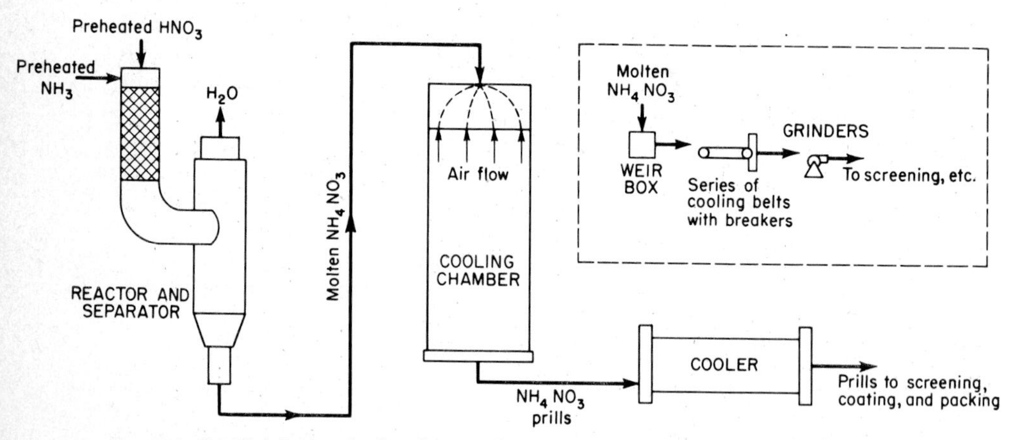

Fig. 18.6 Simplified flowchart for Stengel process for ammonium nitrate manufacture. (*Commercial Solvents Corp.* and *L. A. Stengel.*)

[19]Sykes *et al.*, Ammonium Nitrate Explosion Hazards, *Chem. Eng. Prog.*, **59**(1), 66 (1963); Feicck and Hainer, *J. Am. Chem. Soc.*, **76,** 5860 (1954).

heating under confinement in the presence of oxidizable organic matter. Ammonium nitrate is simply made by neutralizing nitric acid, oxidized from ammonia, with ammonia:[20]

$$NH_3(g) + HNO_3(aq) \longrightarrow NH_4NO_3(aq) \qquad \Delta H = -20.6 \text{ kcal}$$

The old batch process and various continuous processes have been described.[21] The Commercial Solvents Stengel process[22] saves heat and lowers equipment costs by reacting preheated nitric acid and ammonia in a continuous reactor unit where steam is separated from the ammonium nitrate melt,[23] which is caused to flow down a *short* prilling tower or which can be flaked. The ammonium nitrate has a moisture content of about 0.2%. If desired, the prills can be coated with an antihygroscopic compound (Fig. 18.6).

UREA[24]

The synthesis of urea by Wöhler in 1828, by heating ammonium cyanate, had a profound influence upon chemistry and upon civilization. It was the first time a substance produced by life had been prepared in the laboratory, thus opening up the entire field of synthetic organic chemistry. For 100 years, this epoch-making compound was relatively unimportant industrially and was made by the acid hydrolysis of calcium cyanamide. In 1933 Du Pont began production at Belle, W.Va., and remained the sole domestic producer until 1950. There are now many producers using various processes. The most important uses of urea are in high-nitrogen (46%) solid fertilizers, in protein supplement feed for ruminants,[25] and in plastics in combination with formaldehyde and furfural. Urea also finds extensive applications in adhesives, coatings, textile antishrink compounds, and ion-exchange resins. It is an intermediate in the manufacture of ammonium sulfamate, sulfamic acid, and phthalocyanine pigments. Annual production is over 3.5 million tons in the U.S.A.

The commercial processes in current use are based on two reactions:

$$CO_2 + 2NH_3 \rightleftharpoons \underset{\text{Ammonium carbamate}}{NH_4CO_2NH_2} \qquad \Delta H = -67{,}000 \text{ Btu/lb-mol}$$

$$NH_4CO_2NH_2 \rightleftharpoons NH_2CONH_2 + H_2O \qquad \Delta H = +18{,}000 \text{ Btu/lb-mol}$$

The first reaction is easily carried to completion, but the second usually has a conversion of only 40 to 70%. Since both reactions are reversible, the equilibrium depends on the temperature, pressure, and concentration of the various components. The conversion ratio increases with rising temperature; however, urea is formed only in the liquid (solution) phase, making it necessary to maintain this phase with heat and under pressure. Because the pressure increases rapidly with rising temperature, reaction temperatures over 210°C are rarely exceeded in commercial practice. This temperature corresponds to a conversion ratio of about 0.55. Because the combined reactions are highly exothermic, cooling is generally necessary.

The industrial processes for urea[26] are of three types: (1) once-through, (2) partial recycle, and (3) total-recycle processes. Figure 18.7 and Table 18.5 represent the complete aqueous total recycle

[20]*Hydrocarbon Process.*, **52**(11), 104 (1973); New Process Solves Nitrate Corrosion, *Chem. Eng.* (*N.Y.*), **74**(14), 108 (1967).

[21]ECT, vol. 2, p. 324, 1963.

[22]U.S. Pats. 2,568,901 and 2,934,412, Gulf Publishing, Commercial and Industrial Corp.; 1963 Petrochemical Handbook, p. 150, 1963 (flowchart).

[23]Short-cut and Prilled Fertilizer, *Chem. Eng. News*, Aug. 27, 1956, p. 4192 (flow diagram); At Reynolds: A New Ammonium Nitrate Unit, *Ind. Eng. Chem.*, **46,** 622 (1954).

[24]Urea Technology, A Critical Review, *Chemtech.*, **1,** 32 (1974); Urea Route Saves Utilities, *Chem. Eng.* (*N.Y.*), **80**(12), 72 (1973); Improved Urea Process is Developed, *Chem. Eng. Prog.*, **70**(2), 69 (1974); Computer Takes Control of Japanese Urea Process, *Chem. Eng. News*, Feb. 13, 1967, p. 26.

[25]Need for Feed Quickens Urea Process Battle, *Chem. Week*, July 17, 1974, p. 28.

[26]*Hydrocarbon Proces.*, **52**(11), 187 (1973).

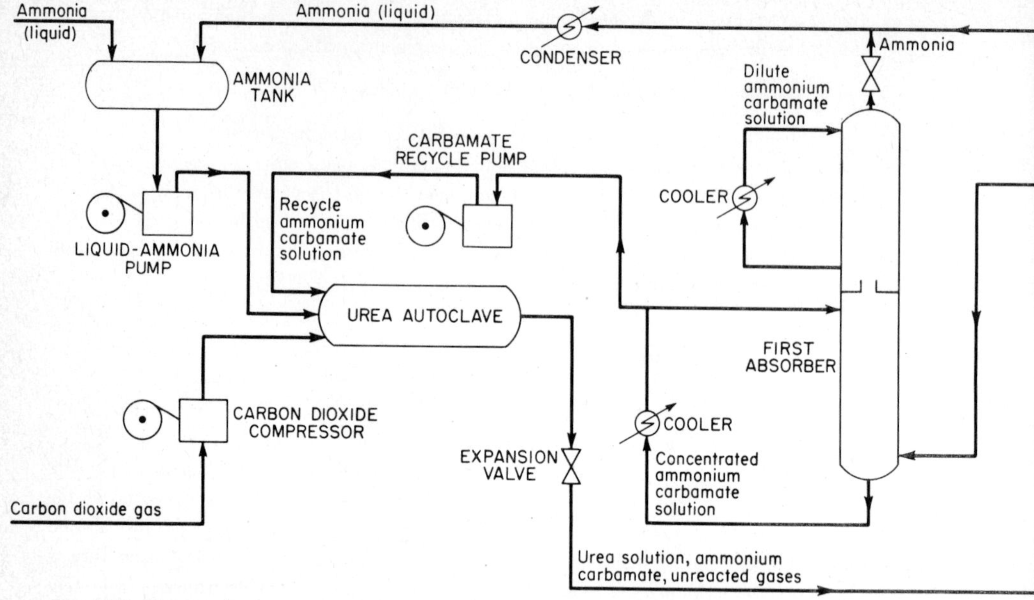

Fig. 18.7 Flowchart for the manufacture of urea, using complete recycle. (*Chemical Construction Co.* and *Chemical Engineering*.)

	Per ton prilled urea
NH_3, liquid	1,180 lb
CO_2, gaseous	1,520 lb
Conversion, carbonate to urea	40–60%
Electricity	165 kWH
Steam (150 psig, sat.)	4,000 lb
Cooling water (°F rise)	20,000 gal

process of the Chemical Construction Co. at 3,200 psi and 360 to 380°F, which breaks down into the following coordinated sequences:

Liquid NH_3 in molar excess of 3.5:1 with gaseous CO_2 is fed into the stainless-steel autoclave under a pressure of 3,200 psi and at 366 to 380°F, together with recycle carbamate in just enough water to keep it in solution and to dissolve all unreacted CO_2 as carbamate. Conversion to urea is about 60% (Op and Ch).

The expansion valve lowers the pressure of the effluent of urea and carbamate to about 300 psig (Op) en route to the tube side of the first decomposer, where the major part of the unconverted carbamate is decomposed by steam heat and the resultant gaseous NH_3 and CO_2 are separated from the urea solution, from which also some dissolved NH_3 and CO_2 are driven overhead (Ch).

This effluent (urea solution) has its pressure further reduced to 15 psig before it enters and passes through the tubes of the second-stage decomposer (Op) to drive off all NH_3 and CO_2 left. The second decomposer is heated by the vapors from the first decomposer, with final degassing by live steam (Ch).

The resulting 78% urea solution can be used as such, or it can be concentrated in a vacuum evaporator heated by steam to 99.7% purity (Op).

This concentrated urea solution is pumped through spray heads in the prilling tower (not shown), bagged, and sold (Op).

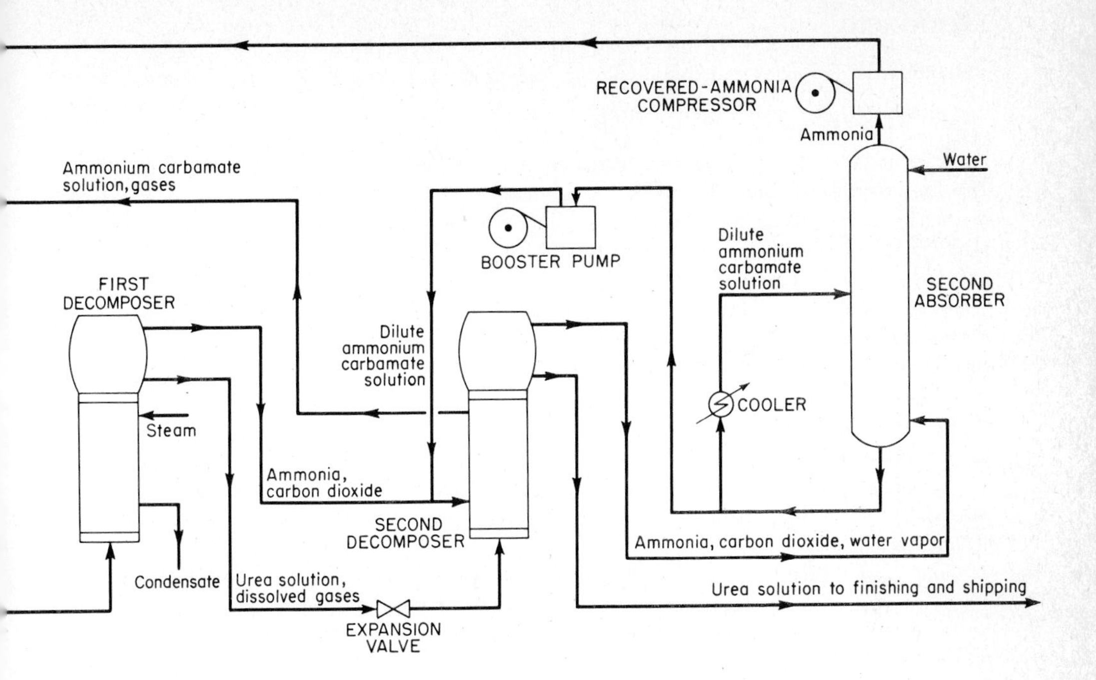

Each decomposer yields a gas stream of NH_3 and CO_2 (Ch).

The hot gases NH_3 and $CO_2 + H_2O$ from the first decomposer are partly absorbed in a dilute carbamate solution from the second absorber in the shell of the second-stage decomposer, thus providing economical exothermic heat for this second decomposition and purification of the urea stream (Ch and Op).

The gases NH_3, CO_2, and H_2O expelled from the urea solution in the tubes in the second decomposer enter the second absorber, where carbamate is formed and pumped to the shell of the second-stage decomposer (Op and Ch). A small amount of process water is added in the second absorber to maintain the carbamate in solution.

The excess NH_3 gas leaving the top of the second absorber joins the NH_3 from the top of the first absorber and is compressed, liquefied, and returned to the NH_3 storage.

The CO_2 reacts in the second absorber with NH_3, forming with H_2O a dilute carbamate solution which is pumped from the shell of the second absorber via the second decomposer to the first absorber where it is concentrated, the NH_3 going to storage and the carbamate, as recycle, to the autoclave.

A minimum amount of water is used, just sufficient to recycle all unreacted CO_2 as carbamate. With a minimum amount of water entering the autoclave, the urea yield is high. *Heat is also saved in the second decomposer.*

THERMO-UREA PLANTS Designs for large-scale plants substitute a decomposing and *hot-stripping* tower in place of the decomposers in Fig. 18.7. Here carbamate is decomposed and hot gaseous ammonia and carbon dioxide not reacted to urea are adiabatically compressed and returned to the urea reactor (autoclave).[27] Especially for large plants, savings in equipment and labor result.

[27]Cook and Mavrovic, Thermo-urea Process Eliminates Liquid Recycle, U.S. Pat. 3,200,148; *Chem. Eng. News*, Sept. 6, 1965, p. 120; Stripper Replaces Decomposer for Urea; *Chem. Eng. News*, Oct. 18, 1965, p. 46; Urea Follows Ammonia to Huge Plants, *Chem. Eng. News*, Aug. 30, 1965, p. 30; Big Urea-process Changes, *Chem. Eng.* (*N.Y.*), **72**(20), 92 (1965); Taggiasco, Simplified Process, *Chem. Eng.* (*N.Y.*), **73**(16), 52 (1966).

NITRIC ACID

HISTORICAL For many years nitric acid was made from Chile saltpeter according to the reaction

$$NaNO_3 + H_2SO_4 \longrightarrow NaHSO_4 + HNO_3$$

The current process involves commercialization of the oxidation reaction of ammonia with air. This has become economical because of the decreasing price of ammonia and the great savings in comparison with the old sodium nitrate procedure. Not only is capital investment for the original plant less, but maintenance expenses and repairs are greatly reduced.

USES AND ECONOMICS Nitric acid has direct applications as an oxidizing acid in the parting of gold and silver, in the pickling of brass, and in photoengraving. Its chief use, however, is in making nitrates in both the inorganic and organic fields, as well as nitro derivatives in all branches of organic chemistry. Inorganic nitrates important commercially are those of ammonia, sodium, copper, and silver, the first being the largest outlet for nitric acid. A growing and potentially very important use of nitric acid is as a replacement for sulfuric acid in the acidulation of phosphate rock (Chap. 26). Many nitrate and nitro compounds are used directly, especially in the explosives industry, as is true of ammonium nitrate, nitroglycerin, and nitrocellulose (actually glyceryl trinitrate and cellulose polynitrate). The nitro aromatic compounds, ammonium picrate and TNT, as well as tetryl, are most important explosives. Quite frequently, however, the nitro group is used as an entering chemical step into either aliphatic or aromatic hydrocarbons in the preparation of more useful derivatives. This is exemplified by nitrobenzene in the manufacture of aniline (cf. Chap. 39 on Intermediates and on Dyes) and by nitroparaffins in leading to amines and amino alcohols in the paraffin series (Chap. 38).

Commercial grades[28] of nitric acid are 36°Bé, or 1.330 sp gr, 52.3% HNO_3; 40°Bé, or 1.381 sp gr, 61.4% HNO_3; and 42 Bé, or 1.408 sp gr. 67.2% HNO_3.

RAW MATERIALS The essential raw materials for the modern manufacture of nitric acid are anhydrous ammonia, air, water, and platinum-rhodium gauze as a catalyst. Because of its low molecular weight, the ammonia can be shipped economically from large primary nitrogen-fixation plants to various oxidation plants at consuming centers. Furthermore, anhydrous ammonia can be shipped in steel cars, in comparison with aqueous nitric acid, which requires stainless-steel tank cars weighing much more.

REACTIONS AND ENERGY CHANGES The essential reactions for the production of nitric acid by the oxidation of ammonia may be represented as follows:

$$4NH_3(g) + 5O_2(g) \longrightarrow 4NO(g) + 6H_2O(g) \qquad \Delta H_{298°C} = -216.6 \text{ kcal} \qquad (1)$$

$$2NO(g) + O_2(g) \longrightarrow 2NO_2(g) \qquad \Delta H_{298°C} = -27.1 \text{ kcal} \qquad (2)$$

$$3NO_2(g) + H_2O(l) \longrightarrow 2HNO_3(aq) + NO(g) \qquad \Delta H_{298°C} = -32.2 \text{ kcal} \qquad (3)$$

$$NH_3(g) + O_2(g) \longrightarrow \tfrac{1}{2}N_2O(g) + \tfrac{3}{2}H_2O(g) \qquad \Delta H = -65.9 \text{ kcal} \qquad (4)$$

Several side reactions reduce somewhat the yield of reaction (1):

$$4NH_3(g) + 3O_2(g) \longrightarrow 2N_2(g) + 6H_2O(g) \qquad \Delta H_{298°C} = -302.7 \text{ kcal} \qquad (5)$$

$$4NH_3(g) + 6NO(g) \longrightarrow 5N_2(g) + 6H_2O(g) \qquad \Delta H_{298°C} = -431.9 \text{ kcal} \qquad (6)$$

$$2NO_2(g) \longrightarrow N_2O_4 \qquad \Delta H = -13.9 \text{ kcal} \qquad (7)$$

Reaction (1) is essentially a very rapid catalytic one, carried out by passing about 10% ammonia by volume, mixed with preheated air, through the multilayered, silk-fine platinum—10% rhodium

[28]Perry, p. 3–76, table 3-68. The data given are for 60°F (15.5°C).

gauze at a temperature of approximately 920°C or lower (Fig. 18.8); here, once ignited, the stream of ammonia continues to burn. The yield is 94 to 95%. The equilibrium constant for reactions (1) and (2) is 9.94×10^{14} for NO_2 production at 627°C. This reaction is usually carried out at a pressure of 100 psi. In Europe the process pressure is frequently 40 lb and the temperature is about 815°C, with consequent smaller platinum losses and a lower production rate.

Much has been published on the commercialization of these reactions.[29] As can be seen from reaction (1), there is only a small increase in volume, so that the principle of Le Chatelier does not affect the equilibrium very substantially.[30] However, the increase in pressure, by compressing the reactants, permits greater space velocity to be maintained, resulting in a plant saving until such a pressure is reached that the cost of the increased thickness of the stainless steel required more than counterbalances the saving in volume of equipment per pound produced. Pressure oxidation also furnishes an acid containing 60 to 70% HNO_3 in comparison with the 50 to 55% HNO_3 obtained from atmospheric oxidation. The speed of ammonia oxidation is extraordinarily high, giving an excellent conversion in the short contact time of 3×10^{-4} s at 750°C with a fine platinum—10% rhodium gauze catalyst. Hence, in the industrial procedure, it has been found economical to mix initially with the ammonia all the air needed for reactions (1) and (2). The oxidation of the NO to NO_2 is the slowest reaction, but the equilibrium is more favorable at lower temperatures. Hence this reaction is carried out in absorbers of considerable capacity, which are provided *with cooling* in all the upper trays. Because of a decrease in volume, this reaction is favored under pressure, according to the principle of Le Chatelier. Although these factors raise the cost of the equipment for carrying out the oxidation of nitric oxide, they increase the conversion. It is necessary in the design of a plant to know how long this reaction will take, in order to calculate the volume necessary for the equipment.

$$3NO_2 + H_2O \rightleftharpoons 2HNO_3 + NO \qquad (3)$$

Equation (3) is really an absorption phenomenon.[31] This reaction, in the opinion of Taylor, Chilton, and Handforth,[32] is the controlling one in making nitric acid, and its rate was increased by employing an absorption tower under pressure and cooling and with countercurrent graded strengths of acid for the absorption. Warm air is introduced into a short raschig-ring-packed section between the tower and acid trap. This provides for the reoxidation of the NO formed and also desorbs (bleaches) the dissolved nitrous oxides which would color the acid.

MANUFACTURING PROCEDURES Commercialization of these oxidation reactions leading from NH_3 to HNO_3 is exemplified by the flowchart in Fig. 18.8, which depicts the present-day procedure under 100 lb pressure. The energy changes involved in these reactions are given with the reactions. Much of the power required for compressing the air to a pressure of 100 lb can be regained by expanding the waste gas, mostly nitrogen, from the top of the absorption tower through a turbine. Naturally, the drive cylinder of this power-recovery compressor should be larger in diameter than the air cylinder. Actually, per ton of 100% HNO_3, the saving thus effected is 200 KWh.

[29]Catalytic Oxidation of Ammonia on Platinum, *Chem. Eng. Sci.*, **21**, 19 (1966); Three New Nitric Acid Processes, *Chem. Eng. Prog.*, **68**(4), 67 (1972); Magnetic Separators Improve Nitric Acid Yields, *Chem. Eng. Prog.*, **70**(3), 81 (1974); Concentrated Nitric Acid Made at Lower Pressure, *Chem. Eng. (N.Y.)* **79**(29), 50 (1972); cf. Krase, chap. 14 in Curtis, *op. cit.*, pp. 366–408ff, for very full description of all factors pertaining to the basic physical chemistry, together with copious references to the original literature.

[30]ECT, vol. 9, p. 333; Sorgenti and Sachsel, Nitric Acid Manufacturing Theory and Practice, *Ind. Eng. Chem.*, **52**, 101 (1960); Strelzoff, Commercial HNO_3 Processes, *Chem. Eng. (N.Y.)*, May, 1956, p. 171; Douglas, Nitric Acid, Oxford, 1961; Weatherly, Nitric Acid, *Chem. Eng. (N.Y.)*, **71**(7), 38 (1964); cf. Krase, chap. 14 in Curtis, *op. cit.*, pp. 366–408ff., for very full description of all factors pertaining to the basic physical chemistry, together with copious references to the original literature.

[31]Perry, table 14-23. See the entire section 14 on gas absorption.

[32]Manufacture of Nitric Acid by the Oxidation of Ammonia, *Ind. Eng. Chem.*, **23**, 860 (1931); Lower Pressure Is the Usual European Practice, *Chem. Eng. News*, Apr. 20, 1964, p. 68.

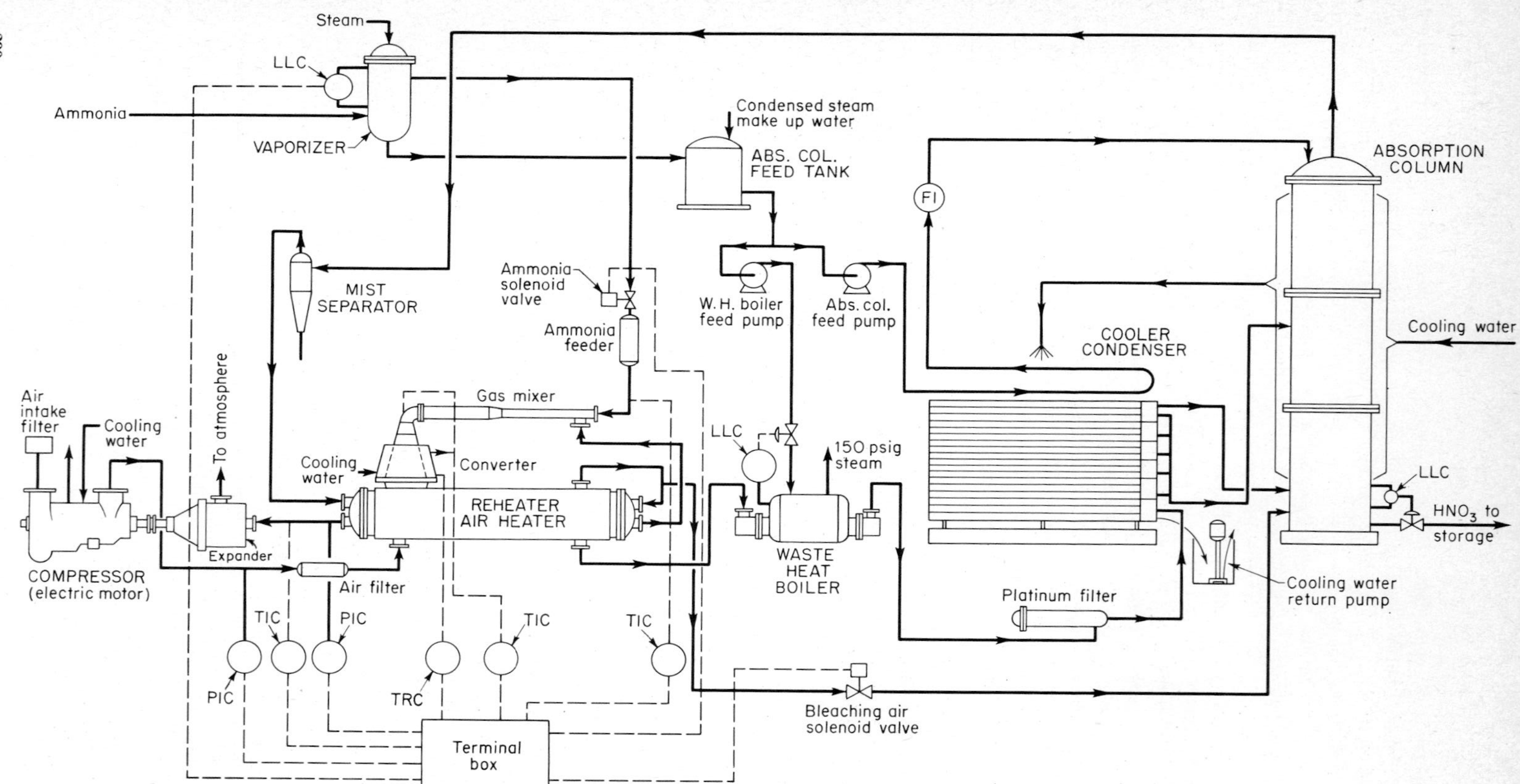

Fig. 18.8 Flowchart for the manufacture of 60% nitric acid from ammonia: Du Pont high-pressure (100-psi) self-sustaining 250 tons/day unit. The "expander" turbine operates on "waste heats" of tail gas (essentially N_2). For platinum filter, the very efficient Brink filter of Monsanto is now recommended. LLC, Liquid-level controller; LG, liquid gauge; FI, feed intake; LIC, liquid indicating controller; PIC, pressure indicating controller; TIC, temperature recording controller; WH, waste heat; abs. col., absorption column. (*E. A. Rass, Chemical & Industrial Corp.*)

	Per ton 100% nitric acid		
Ammonia	574 lb	Overall efficiency	95%
Platinum-rhodium catalyst		Labor, work-hours/day	72
(total platinum catalyst 500 troy oz)	0.0049 troy oz		
Electricity	9 kWh		
Water cooling	51,400 gal		
Abs. tower feedwater			
(pure condensate)	800 lb		
Steam made and used			
(200 psig, 500°F)	1,750 lb		

Figure 18.8 can be separated into the following coordinated sequences:

Anhydrous NH_3 is evaporated continuously and uniformly in an evaporator, using steam to supply the necessary heat of evaporation (Op). Air for reactions shown in Fig. 18.8 is compressed in a power-recovery compressor and in a steam-driven compressor to 105 psig and passed through heat exchangers and an air filter (Op).

NH_3 gas is oxidized with air to NO at a pressure of 100 lb in a converter by passing through a platinum–10% rhodium gauze at 920°C (Ch).

Then NO with the excess air necessary for the succeeding oxidizing steps is cooled in a waste-heat boiler and a water cooler and conducted to the bottom of the absorption tower (Op).

Successive oxidations and hydrations of the NO are carried out with continuous water cooling in a stainless-steel[33] absorption tower (Ch and Op).

The acid (61 to 65% HNO_3) is drawn off through an acid trap (Op).

The waste gas from the top of the absorption tower is heated in an exchanger counter to reaction gases and expanded through a compressor (expander) for part of the air, before being exhausted to the atmosphere (Op).

The "Maggie" is an extractive distillation improvement invented and first installed by Hercules Powder Co. at Parlin, N.J., for concentrating weak HNO_3. It involves $Mg(NO_3)_2$[34] instead of sulfuric acid[35] as a water binding agent to break the 68% HNO_3 azeotrope. A typical procedure takes 60% HNO_3 from the oxidation of NH_3 and feeds it into the stripping zone of the tray column. A 72% solution of $Mg(NO_3)_2$ at 140°C is added at or above the point of introduction of the weak nitric acid. Vapors arise from the stripping zone with about 87% HNO_3, which is rectified to a condensed overhead takeoff of 99.5+% HNO_3 for part of the HNO_3. The balance of the condensed acid is refluxed to the rectifying zone. The bottoms product from the stripping zone contains about 68% $Mg(NO_3)_2$ and 0.1% free HNO_3. A portion of this goes to a reboiler to provide the necessary steam for fractional distillation of the HNO_3, and the remainder to a vacuum concentrator to furnish a 72% solution which is recycled to the dehydrating zone. Stainless steel is employed for the plant, and the recoveries are 99+%. Plants producing acid up to 100% HNO_3 without use of extractive distillation have become common.[36]

SODIUM NITRATE AND POTASSIUM NITRATE

Large tonnages of sodium nitrate have been made both in America at Hopewell, Va., and in Europe by reaction of sodium chloride or soda ash from ammonia-soda process with nitric acid from ammonia

[33]Nitric Acid, *Chem. Eng.* (*N.Y.*), **57**(11), 726 (1950).

[34]Bruce and Bechtel (Hercules), ECT, 2d suppl., p. 487, 1960 (flowchart); U.S. Pat. 2,716,631; *Chem. Eng. News*, June 9, 1958, p. 50 (flowchart, photograph, and description).

[35]Spent mixed acid from nitrations can best be reclaimed and reconcentrated by the old sulfuric acid procedure.

[36]Three New Nitric Acid Processes, **68**(4), 67 (1972); Concentrated Nitric Acid Made at Lower Pressure, *Chem. Eng.* (*N.Y.*), **79**(29), 50 (1972).

oxidation. This high-grade product is sought after for both industrial and fertilizer uses. See Chap. 13, chlorine, for reactions, including by-product chlorine. The Chilean nitrate industry, until the rise of the synthetic-ammonia industry, dominated the world's nitrogen supply. These deposits occur in a desert at an elevation of from 4,000 to 9,000 ft; they extend 400 mi north and south, and have a width of from 5 to 40 mi. The *caliche* varies in thickness from 8 in. to 14 ft, with an overburden of from 1 to 4 ft. Enormous deposits are still available. The Guggenheim process as perfected consists of improved leaching and crystallization operations on a large scale. The leaching is countercurrent at 40°C, and crystallization occurs down to 5°C. The heat required for the 40°C level comes from exchangers connected to the ammonia condensers and from the exhaust gases and cooling water from diesel engines which supply the plant with electric power. The 5°C level is attained (1) to 15°C in heat interchangers counter to nitrate mother liquor and (2) to 5°C by ammonia refrigeration. This results in increased production of 25 to 30 tons of nitrate from a ton of fuel for the Guggenheim process in comparison with 6 to 7 tons by the Shanks process. The Chilean nitrate industry produces as *by-products* the world's main supply of iodine and also potassium nitrate (Chap. 17).

SELECTED REFERENCES

Astle, M.J.; Industrial Organic Nitrogen Compounds, ACS Monograph 150, Reinhold, 1961.
Comings, E. W.: High Pressure Technology, McGraw-Hill, 1956.
Lawrence, A. A.: Nitrogen Oxide Emission Control, Noyes, 1972.
Pratt, C. J., and R. Noyes: Nitrogen Fertilizer Chemical Processes, Noyes, 1965.
Safety in Air and Ammonia Plants, AIChE. Many volumes issued since 1960. (Vol. 16 issued in 1974).
Sauchelli, V. (ed.): Fertilizer Nitrogen, ACS Monograph 160, Reinhold, 1964.
Sittig, M.: Nitrogen in Industry, Van Nostrand, 1965.
Slack, A. V., and G. R. James: Ammonia, pt. 1, Dekker, 1973.
Slack, A. V., and G. R. James: Ammonia, pt. 2, Dekker, 1974.

chapter 19

SULFUR AND SULFURIC ACID

One of the most important basic raw materials in the chemical process industries is sulfur. It exists in nature both in the free state and combined in ores such as pyrite (FeS_2). It is also an important constituent of petroleum and natural gas (as H_2S). The most important application of sulfur is in the manufacture of sulfuric[1] acid.

HISTORICAL Sulfur has a history as old as any other chemical and has developed from the mystic yellow of the alchemist to one of the most useful substances in modern civilization. It was burned in early pagan rites to drive away evil spirits, and even then the fumes were used as a bleach for cloth and straw. For many years, a French company held a monopoly on sulfur by controlling the world supply from Sicily. Partly for this reason and partly because of the abundance of pyrite, sulfur itself was little used in the United States prior to 1914. Though sulfur was discovered in the U.S. gulf region in 1869, it was difficult to mine because of overlying beds of quicksand. Before 1914, the United States made most of its sulfuric acid from imported and domestic pyrite and from by-product sulfur dioxide from copper and zinc smelters. The mining of sulfur in Texas and Louisiana by the Frasch process was increased, starting about 1914, to such an extent as to provide for most domestic needs and to enter world markets. In recent years a major source of elemental sulfur has been that recovered from H_2S, a by-product in the desulfurization of sour natural gas and sour crude oil. Canada, France, and the United States are the largest producers of recovered sulfur. In 1973, Western world production of sulfur in all forms amounted to 33.75 million long tons, of which 27.6% was Frasch-produced, 41.0% recovered, and 30.3% obtained from nonelemental sources such as pyrites and smelter gases.

USES AND ECONOMICS[2] The production of elemental sulfur in the United States in 1973 was 10.0 million tons, supplied by the Frasch hot-water process (76%) and sour natural and refinery gases (24%). The consumption of sulfur in the United States in 1973 amounted to 10.4 million long tons, of which 88% was converted to sulfuric acid, 53.3% was used for fertilizers, and 34.7% for industrial acid. Industrial acid uses include the manufacture of inorganic and organic chemicals, pigments, explosives, petroleum products, rayon, and steel pickling. The remaining 12% includes such uses as the manufacture of wood pulp, carbon disulfide, insecticides and fungicides, bleaching agents, and dyestuffs, and rubber vulcanization. Various industrial and academic research groups are developing new uses for sulfur. Among some of the new uses being studied are (*a*) as an additive for asphalt, (*b*) sulfur concretes and mortars, (*c*) plant and soil treatment, (*d*) sulfur-alkali metal batteries, and (*e*) foamed sulfur insulation.[3]

[1]Duecker and West, Manufacture of Sulfuric Acid, ACS Monograph 144, Reinhold, 1959 (most important for both sulfur and sulfuric acid; many references); Sulfur and SO_2 Developments, *Chemical Engineering Progress* Technical Manual, 1971.

[2]Handler, *Eng. Min. J.*, **175**(3), 144 (1974); *Chem. Week*, Mar. 20, 1974, p. 23.

[3]Davis, *Chem. Eng. (N.Y.)*, **79**(17), 30 (1972); *Oil Week*, June 17, 1973, p. 27.

MINING AND MANUFACTURE OF SULFUR

FRASCH PROCESS A major but decreasing percentage of all the elemental sulfur in the world has been obtained from the sulfur-bearing porous limestones in the saltdome cap rocks of Texas, Louisiana, and Mexico by the Frasch process.[4] As early as the late 1890s, Herman Frasch devised his ingenious method of melting sulfur underground or under the sea and pumping it up to the surface. Ordinary oil-well equipment is used to bore holes to the bottom of the sulfur-bearing strata, a distance underground of from 500 to 2,500 ft. A nest of three concentric pipes, varying in size from 8 to $1\frac{1}{4}$ in., is slipped down the well casing. The outside pipe of this nest, 8 in. in diameter, passes through the sulfur-bearing stratum and rests on the upper portion of the barren anhydrite, as shown in Fig. 19.1. A 4-in. pipe passes through the 8-in. one so that an annular space exists between the two, extends nearly to the bottom of the sulfur-bearing rock, and rests on a collar that seals the annular space between the 8- and the 4-in. pipes. An air pipe, $1\frac{1}{4}$ in. in diameter, inside the others, reaches to a depth slightly above the collar mentioned. The 8-in. pipe is perforated at two different levels, one above and the other below the annular collar. The upper set of perforations permits the escape of hot water, and molten sulfur enters the system through the lower perforations.

For operation of the well, hot water at about 160°C is passed down the annular space between the 8- and the 4-in. pipes. It discharges through the perforations into the porous formation near the foot of the well. The sulfur-bearing rock around the well through which this water circulates is raised to a temperature above the melting point of sulfur, 119°C. Molten sulfur, being heavier than water, sinks and forms a pool around the base of the well, where it enters through the lower perforations and rises in the 4-in. pipe as depicted in Fig. 19.1. The height to which the sulfur is forced by the pressure of the hot water is about halfway to the surface. Compressed air forced down the $1\frac{1}{4}$-in. pipe aerates and lightens the liquid sulfur so that it will rise to the surface. The compressed air volume is regulated so that the production rate is equalized with the sulfur melting rate in order not to deplete the sulfur pool and cause the well to produce water. Water must be withdrawn from the formation at approximately the same rate as it is injected to prevent a buildup in pressure to the point where further injection would be impossible. Bleed wells for extracting water from the formation usually are located on the deeper flanks of the dome to withdraw the heavier cold water which accumulates there. Water-heating capacities of existing power plants range from 1 to 10 million gal/day.

On the surface, the liquid sulfur moves through steam-heated lines to a separator where the air is removed. The sulfur can be either solidified in large storage vats or kept liquid in steam-heated storage tanks. About 95% of all U.S. sulfur is shipped as liquid in insulated tank trucks, tank cars, and heated barges or ships.

SULFUR FROM FUEL GASES Hydrogen sulfide increasingly is removed[5] in the purification of sour natural and coke-oven gas and from petroleum refineries by dissolving it in potassium carbonate solution or ethanolamine, followed by heating (Chap. 6). The hydrogen sulfide thus produced is burned in especially constructed furnaces to give sulfur dioxide for sulfuric acid. However, an ever-increasing tonnage is converted to *recovered sulfur* by various modifications of the original Clause process, for which the reactions are:

$$H_2S(g) + \tfrac{3}{2}O_2(g) \longrightarrow SO_2(g) + H_2O(g) \qquad \Delta H = -123.9 \text{ kcal} \qquad (1)$$

$$SO_2(g) + 2H_2S(g) \xrightarrow{Fe_2O_3} 3S(l) + 2H_2O(l) \qquad \Delta H = -34.2 \text{ kcal} \qquad (2)$$

This is illustrated in Fig. 19.2 (flowchart).

[4]Pictured flowcharts with data are given in *Chem. Eng.* (*N.Y.*), **48**(3), 104 (1941), and **65**(23), 136 (1958); Haynes, The Stone That Burns, Van Nostrand, New York, 1959; *Oil Gas J.*, May 4, 1970, p. 125.

[5]Brennan, Amine Treating of Sour Gas: Good Riddance to H_2S, *Chem. Eng.* (*N.Y.*), **69**(22), 94 (1962); Heppenstall and Lowrison, The Manufacture of Sulphuric Acid, *Trans. Inst. Chem. Eng.*, **31,** 389 (1953); Sulfur from H_2S, *Chem. Eng.* (*N.Y.*), **59**(10), 210 (1952) (pictured flowchart); Duecker and West, *op. cit.*, chaps. 4, 5, and 9; cf. this chapter, sulfur pollution.

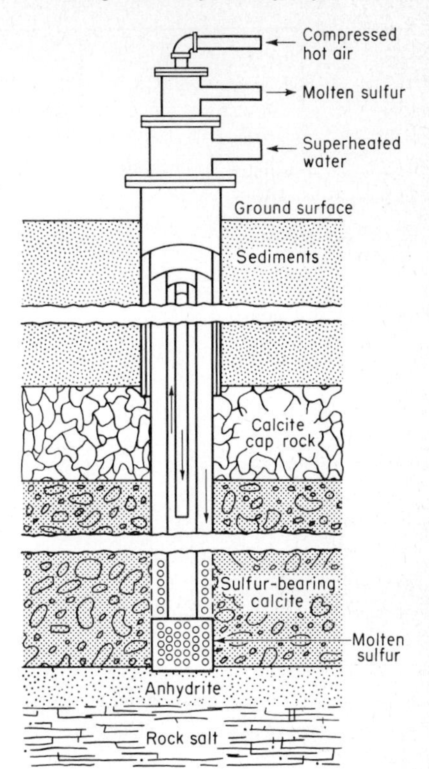

Fig. 19.1 Schematic diagram of the Frasch process for mining sulfur. (*McGraw-Hill Encyclopedia of Science and Technology.*)

Air pollution regulations now require new plants to attain over 98% conversion, and the Province of Alberta, Canada, is asking for 99.5% conversion.[6] To meet the new emission regulations, a number of processes has been developed for cleaning the residual sulfur values from the tail gases of recovery plants.[7] Other existing or potential sources of sulfur derived from fuels include coke-oven gases and synthetic crude oil from tar sands or shale oil. A 1,000-bbl/day shale oil plant is in operation in Brazil.[8] Very little sulfur is presently recovered from coal, but efforts to reduce sulfur dioxide emissions by developing clean fuels from coal could produce significant amounts of sulfur from coal in the future. Coal cleaning technology can remove about half of the pyrite in coal, but the organic sulfur can be removed only by gasification, liquefaction, and hydrogenation processes. Several such processes are in various stages of development.[9] Where sulfur-containing fuels have to be burned, sulfur oxides must be removed by stack gas cleaning methods, or by new combustion techniques designed to remove sulfur during burning.[10]

SULFUR FROM SULFIDE ORES Sulfur values derived from iron pyrites amounted to 5.01 million tons for the Western world in 1973, while nonferrous metal sulfide smelter gases contributed 4.93 million tons of sulfur equivalent. The smelting of nonferrous ores and pyrites converts sulfur to sulfur dioxide which is usually recovered for conversion to sulfuric acid, or occasionally as liquid

[6]Rowland, *Oil Week*, Oct. 23, 1974, p. 9.

[7]*Oil Gas J.*, **70**(26), 85 (1972); Chalmers, *Hydrocarbon Process.*, **53**(4), 75 (1974); Davis, *Chem. Eng.* (*N.Y.*), **79**(11), 66 (1972); Barry, *Hydrocarbon Process.*, **51**(4), 102 (1972); Ludberg, Removal of Hydrogen Sulfide from Coke Oven Gas by the Strelford Process, paper presented at the 64th Annual Meeting of the Air Pollution Control Association, Atlantic City, N.J., June 27, 1971.

[8]Pattison, *Chem. Eng.* (*N.Y.*), **74**(24), 66 (1967); *Oil Gas J.*, **70**(37), 105 (1972).

[9]Chopey, *Chem. Eng.* (*N.Y.*), **79**(16), 86 (1972); *Environ. Sci. Technol.*, **8**(6), 510 (1974).

[10]Slack, Sulfur Dioxide Removal from Waste Gases, Noyes, 1971; *Oil Gas J.*, **70**(47), 83 (1972).

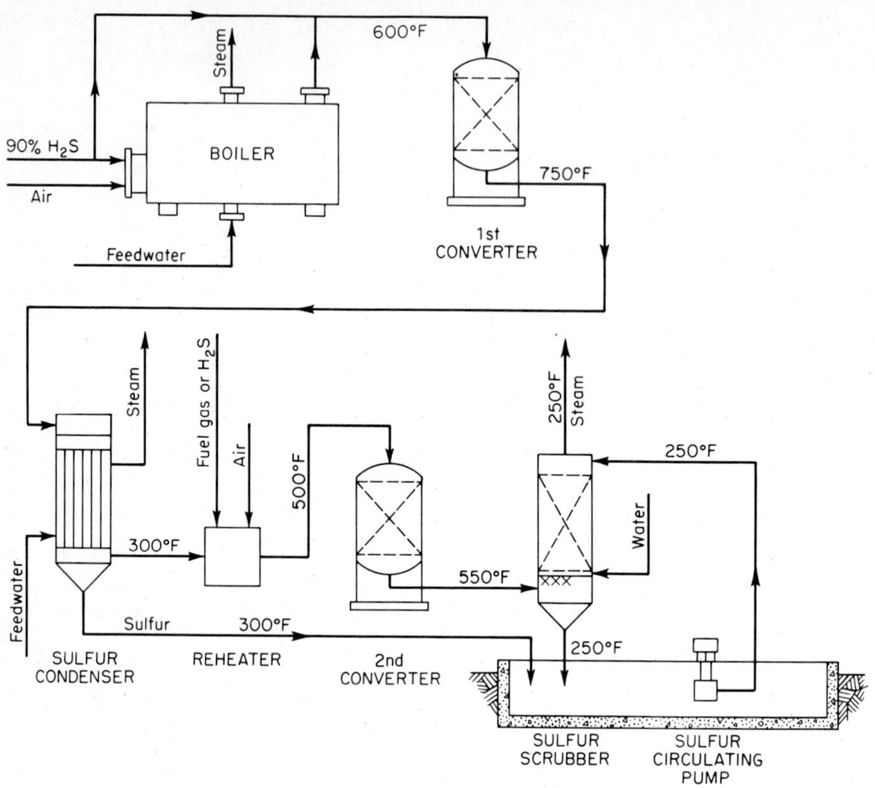

Fig. 19.2 Claus process for sulfur recovery from H_2S. (*Chemetron Corp., Girdler Construction Div.*)

Approximate utilities per long ton sulfur:		Products:	
Electric power	75 kWh	Sulfur 2,200 lb	92–98% yield
Boiler feedwater	780 gal	150 psig steam	4,400 lb
Fuel gas	70,000 Btu	35 psig steam	700 lb
Labor	1 worker per shift		

sulfur dioxide. Conventional copper smelters have limited sulfur recovery capability because certain portions of the off-gases are too low in sulfur dioxide concentration to make sulfuric acid economically. Electric furnaces and flash smelters produce more concentrated sulfur dioxide suitable for the recovery of elemental sulfur or sulfuric acid.[11]

Several commercial processes for recovering elemental sulfur from pyrite ores include the Outokumpu flash-smelter process, the Orkla process, and the Noranda process. Only the Outokumpu process is still being used commercially.

SULFURIC ACID

It seems hard to believe that a very active chemical such as sulfuric acid is at the same time one of the most widely used and most important technical products. It is of such paramount significance that the remark has frequently been made that the per capita use of sulfuric acid is an index of the

[11]Semrau, *J. Met.*, **23**(3), 41 (1971); White, *Eng. Min. J.*, **172**(7), 61 (1971).

technical development of a nation. Sulfuric acid is the agent for sulfate formation and for sulfonation, but is used more frequently because it is a rather strong and economically priced inorganic acid. It enters into countless industries, though infrequently appearing in the finished material. It is employed in the manufacture of fertilizers, leather and tin plate, in the purification of petroleum, and in the dyeing of fabrics.

HISTORICAL[12] The origin of the first sulfuric acid is unknown, but it was mentioned as far back as the tenth century. Its preparation, by burning saltpeter with sulfur, was first described by Valentinus in the fifteenth century. In 1746, Roebuck of Birmingham, England, introduced the lead chamber. Early in the nineteenth century, continuous methods replaced batch operations. Another notable improvement in the process was the invention of the Gay-Lussac tower for the recovery of nitrogen oxides in 1827, which, however, was not applied until the invention of the Glover tower in 1859. The Glover tower furnished a means of denitrating the nitrous acid from the Gay-Lussac tower without dilution, complementing the Gay-Lussac tower. Since that time both towers have been standard equipment in all lead-chamber processes.

The *contact process* was first discovered in 1831 by Phillips, an Englishman, whose patent included the essential features of the modern contact process, namely, the passing of a mixture of sulfur dioxide over a catalyst, followed by absorption of the sulfur trioxide in 98.5 to 99% sulfuric acid. Phillips' invention was not a commercial success for more than 40 years, probably because of (1) the lack of demand for fuming acids, (2) inadequate knowledge of catalytic gas reactions, and (3) the slow progress of chemical technology. Development of the dye industry resulted in a rising demand for fuming acids for the manufacture of alizarin and other organic coloring matter. In 1889 it was demonstrated that an *excess* of oxygen in the gaseous mixture for the contact process was advantageous. The contact process has been improved in all details and is now one of industry's low-cost, almost wholly automatic continuous processes (Fig. 19.4). All the new sulfuric acid plants use the contact process. The few remaining small-capacity plants employing the chamber process made 4% of the sulfuric acid produced in 1973.

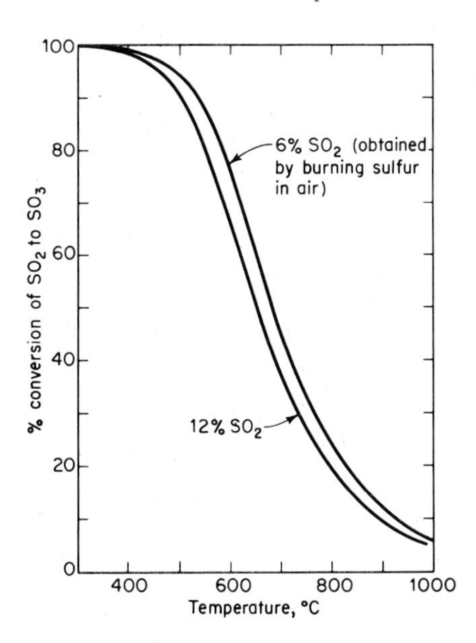

Fig. 19.3 Equilibrium-temperature relation for SO_2 conversion to SO_3. (*Monsanto Co.*)

[12]Duecker and West, *op. cit.*; Fairlie, Manufacture of Sulfuric Acid, Reinhold, 1936; both are standard reference books with bibliographies and sketches.

TABLE 19.1 *Hydrates of Sulfuric Acid*

	Formula	*Melting point, °C*	*Specific gravity*
Sulfuric acid, fuming	$H_2S_2O_7$	35	$1.9^{20°}$
100%	H_2SO_4	10.37	$1.834^{18°/4}$
Monohydrate	$H_2SO_4 \cdot H_2O$	8.48	$1.842^{15°/4}$
Dihydrate	$H_2SO_4 \cdot 2H_2O$	−38.57	$1.650^{0°/4}$

Source: Rubin and Giaque, The Heat Capacities and Entropies of Sulfuric Acid and Its Mono- and Dihydrates from 15 to 300 K, *J. Am. Chem. Soc.*, **74**, 800 (1952).

PROPERTIES OF SULFURIC ACID H_2SO_4 is a strong dibasic acid. In addition, it is also an oxidizing and dehydrating agent, particularly toward organic compounds. Its dehydrating action is important in absorbing water formed in such chemical conversions as nitration, esterification, and sulfonation, thus ensuring high yields. Solutions of the acid may be concentrated economically to about 93% by weight of H_2SO_4. Stronger acids may be made by dissolving SO_3 in 98 to 99% acid. H_2SO_4 forms many hydrates that have fairly definite melting points, as shown in Table 19.1. The irregularities in the relation between the strength of the acid and the corresponding specific gravity and freezing point are due to these hydrates.

H_2SO_4 is widely sold in the form of various solutions in water or of SO_3 in H_2SO_4. The latter, called *oleums*, are marked on the basis of the percentage of SO_3 present; 20% oleum means that, in 100 lb, there are 20 lb of SO_3 and 80 lb of H_2SO_4. This 20% oleum, if diluted with water to make H_2SO_4, would furnish 104.5 lb. For convenience, now grown into an established custom, ordinary solutions of H_2SO_4 and water, up to 93%, are sold according to their specific gravity, or their Baumé degree. Table 19.2 illustrates the acids of commerce. The usual temperature to which specific gravity, or Baumé (Bé), is referred is 60°F for H_2SO_4. The specific gravity of H_2SO_4 increases gradually to 1.844 at 60°F for 97% acid, after which it decreases to 1.835 at 60°F for 100% acid. Consequently, in this upper range, i.e., above 95%, the strengths should be determined by other means, such as electrical conductivity, titration, or temperature rise. For some of the medium-range oleums, however, specific gravity is helpful.

The normal strengths of commercial oleums fall into three categories: 10 to 35% free SO_3, 40% free SO_3, and 60 or 65% SO_3. The 20 to 35% grades can readily be produced in a single absorption tower, whereas 40% oleum generally requires two absorption towers in series. Oleum containing 60 to 65% free SO_3 must be made by distilling SO_3 gas out of 20 to 35% oleum, con-

TABLE 19.2 *Commercial Strengths of Sulfuric Acid (60°F)*

For oleums, % means-free SO_3	*Degrees Bé (60°F, or 15.6°C)*	*Specific gravity (60°F, or 15.6°C)*	*Sulfuric acid, %*
Battery acid	29.0	1.250	33.33
Chamber acid, fertilizer acid, 50 acid	50	1.526	62.18
Glover or tower acid, 60 acid	60	1.706	77.67
Oil of vitriol (OV), concentrated acid, 66 acid	66	1.835	93.19
98 acid	....	1.841	98.0
100% H_2SO_4	....	1.835	100.0
20% oleum, 104.5 acid	....	1.915	104.50
40% oleum, 109 acid	....	1.983	109.0
66% oleum	....	1.992	114.6

Source: Duecker and West, Manufacture of Sulfuric Acid, pp. 2 and 411, ACS Monograph 144, Reinhold, 1959.

TABLE 19.3 ***Sulfuric Acid Use*** *(In thousands of short tons, 100% H_2SO_4 basis)*

End use	*1971*	*1972*	*1973*
Phosphatic fertilizers	13,700	15,400	16,300
Ammonium sulfate	1,775	1,825	1,950
Total fertilizers	15,475	17,225	18,250
Aluminum sulfate	525	550	500
Hydrofluoric acid	950	1,000	1,025
Other chemicals	3,425	3,800	4,000
Inorganic pigments	1,100	1,125	1,150
Petroleum products	950	1,000	1,050
Rayon	625	650	600
Iron and steel pickling	250	250	250
Nonferrous metal processing	1,100	1,250	1,400
Other uses	4,700	4,550	4,375
Gross total consumption	29,100	31,400	32,600

Source: Texas Gulf, Inc.

densing the 100% SO_3, and then mixing it with additional 20 to 35% oleum. The freezing point of 35% oleum is about 80°F, and of 40% oleum about 94°F; consequently, small amounts of nitric acid are sometimes added to these grades to inhibit freezing during winter shipment.

USES AND ECONOMICS OF SULFURIC ACID Fertilizer manufacture, as shown in Table 19.3, is the greatest single use of sulfuric acid. About a dozen different grades[13] are supplied, each with a particular use. Grades of 53 to 56°Bé or stronger are employed in normal superphosphate manufacture, hence the chamber process can supply this acid. The 60°Bé grade is used for sulfates of ammonia, copper (bluestone), aluminum (alum), magnesium (Epsom salts), zinc, iron (copperas), etc.; mineral acids; organic acids, such as citric, oxalic, acetic, and tartaric; pickling iron and steel before galvanizing and tinning; refining and producing of heavy metals; electroplating; and preparing sugar, starch, and sirup. The 66 to 66.2°Bé grades are utilized in the purification of petroleum products, preparation of titanium dioxide, alkylation of isobutane, manufacture of many nitrogen chemicals, synthesis of phenol, recovery of fatty acids in soap manufacture, and manufacture of phosphoric acid and triple superphosphate. Oleums are needed for petroleum, nitrocellulose, nitroglycerin, TNT, and dye manufacture, and for fortifying weaker acids. There are many other uses, as can be attested to by the fact that few chemical products are manufactured without the use of sulfuric acid.

MANUFACTURE There have long been two main procedures for the making of sulfuric acid, the chamber process and the contact process. The contact process produces 96% or more of the acid made in the United States. Both processes are based on sulfur dioxide, both are catalytic, both use air as the source of oxygen for making the sulfuric trioxide, and both have been operated continuously on a large scale for many years. The chamber process produces directly a weaker acid of from 50 to 60°Bé. In the United States existing chamber plants usually continue to operate until worn out. The contact process produces 98 to 100% acids and various oleums. However, much 66°Bé, and even 60°Bé, acid is obtained by dilution.

MANUFACTURE BY THE CONTACT PROCESS

Until 1900, no contact plant had been built in the United States, although this process had become very important in Europe because of the need for oleums and high-strength acid for sulfonation,

[13]Cf. Skeen, Sulfuric Acid, *Chem. Eng.* (*N.Y.*), **57**(4), 337 (1950).

particularly in the dye industry. A substantial number of contact plants were built in the period 1900–1925, using platinum as a catalyst. In the middle 1920s, vanadium came into use as a catalyst and has gradually completely replaced platinum. All new facilities built in recent years use vanadium catalysts.

The contact process has been gradually modified to use double absorption (also called double catalysis), which increases yields and reduces stack emissions of unconverted SO_2. Recently, U.S. government regulations have specified allowable emissions of SO_2 from acid plants and require that all new plants either use the double-absorption process or else be fitted with stack gas scrubbing systems to achieve comparable emission levels. For sulfur-burning plants, allowable emissions are equivalent to 99.7% conversion of SO_2, and for plants using smelter gases to about 99.0 to 99.5% conversion. Conversions using the single-absorption contact process typically were about 97 to 98%.

Sulfur-burning plants are the simplest and cheapest, since special purification of burner gases is not required; there is no need to heat the sulfur dioxide gases, and only cooling is required, so that all the evolved heat may be recovered in the form of relatively high-pressure steam. When other raw materials such as sulfide ores and spent or sludge acids are used, extensive gas purification is required, and the heat evolved in the catalytic conversion. However, the heat evolved in roasting the ore or in burning the spent or sludge acids usually is recovered in the form of steam.

The flowchart in Fig. 19.4 can be divided into the following sequences:

Transportation of sulfur to the plant (Op).

Melting of sulfur, if not shipped molten (Op), and often filtering of the molten sulfur to remove traces of ash.

Pumping and atomizing of melted sulfur (Op).

Drying of combustion air (Ch).

Burning of sulfur (Ch).

Recovery of heat from and cooling of hot SO_2 gas (Op).

Purification of SO_2 gas by hot filtration (Op).

Oxidation of SO_2 to SO_3 in converters (Ch).

Temperature control with heat transfer, to secure high yields of SO_3 (Op).

Absorption of SO_3 in strong acid, 98.5 to 99% (Ch).

Cooling of acid from absorbers (Op).

Pumping of acid over absorption towers (Op).

The excess heat of combustion of sulfur is utilized in a waste-heat boiler to generate steam for melting the sulfur and for power purposes around the plant. Usually, steam is one of the products of the plant, since it is not all used for melting sulfur and for driving plant equipment. A typical heat and material balance for a sulfur-burning contact sulfuric acid plant is detailed by Duecker and West.[14] The steam generated in larger sulfur-burning plants (Fig. 19.4) normally materially exceeds 1 ton per ton of 100% acid produced. However, when the sulfur dioxide gas from the roasting of sulfide ore has to be cooled for purification before treatment, as shown in Fig. 19.5, somewhat less steam is produced.

REACTIONS The reactions are

$$S(g) + O_2(g) \longrightarrow SO_2(g) \qquad \Delta H = -70.9 \text{ kcal}$$

$$SO_2(g) + \tfrac{1}{2}O_2(g) \rightleftharpoons SO_3(g) \qquad \Delta H = -23.4 \text{ kcal}$$

The oxidation of SO_2 in the converters of a contact plant is a very good example of the many industrial applications of the principles of physical chemistry. As indicated above, the conversion from

[14]Duecker and West *op. cit.*, p. 214.

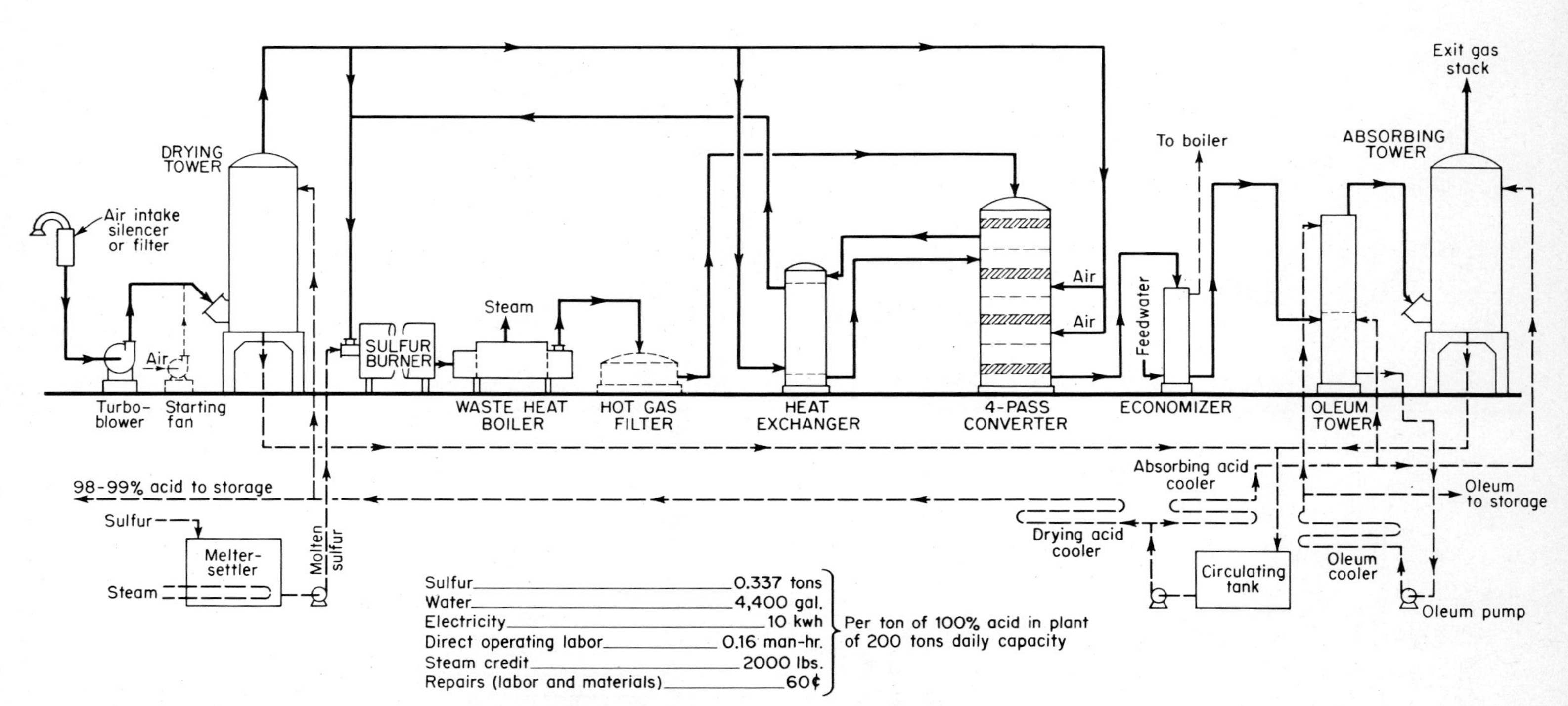

Fig. 19.4 Typical flowchart for a sulfur-burning contact sulfuric acid plant. (*Monsanto Co.*)

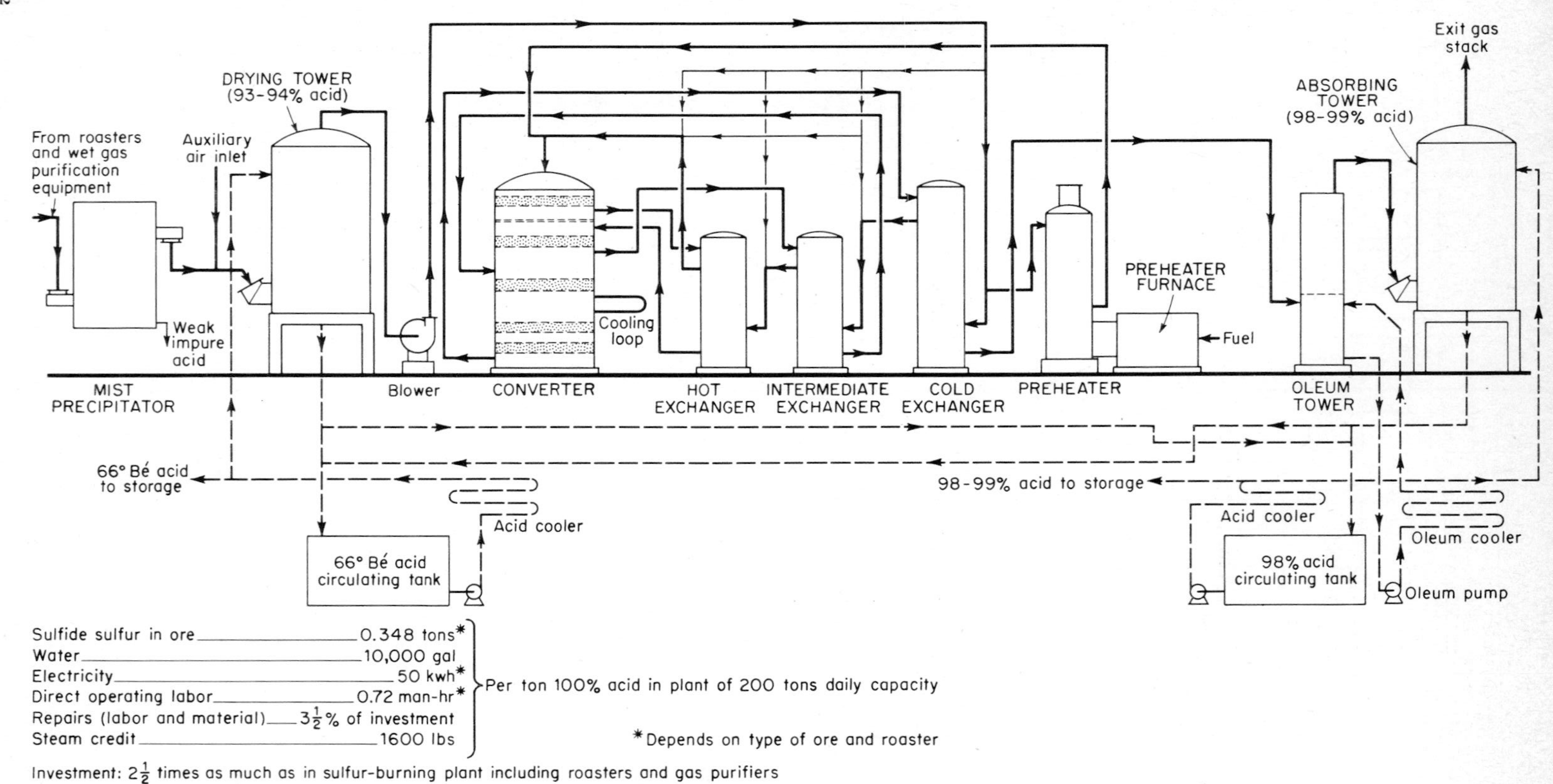

Fig. 19.5 Typical flowchart for an ore-roasting contact sulfuric acid plant. (*Monsanto Co.*)

TABLE 19.4 ***Equilibrium Constants for Sulfur Dioxide Oxidation***

Temperature, °C	K_p	*Temperature, °C*	K_p
400	397	800	0.915
500	48.1	900	0.384
600	9.53	1000	0.1845
700	2.63	1100	0.0980

Source: Z. Elektrochem., **11,** 373 (1905).

SO_2 to SO_3 is an exothermic reversible reaction. The equilibrium constant[15] for this reaction, calculated from partial pressures according to Guldberg and Waage's law of mass action, may be expressed as

$$K_p = \frac{p_{SO_3}}{p_{SO_2} \times p_{O_2}^{1/2}}$$

Values for K_p have been experimentally determined, based on p in atmospheres, as presented in Table 19.4, and are constant for any given temperature. These experimental values are in satisfactory agreement with the values of K_p calculated from thermodynamic data.

Figure 19.3 presents the equilibrium conversion of SO_2 to SO_3 as a function of temperature for two feeds of different initial SO_2 concentration. These equilibrium conversions were calculated from the experimental values of K_p (Table 19.4), assuming that all gases are ideal and that the total pressure is 1 atm. If x_F is the initial volume fraction of SO_2 in the feed gas, the partial pressures for each of the gases in the SO_2-to-SO_3 converter may be related to the equilibrium conversion y as follows:

$$p_{SO_2} = \frac{2x_F(1-y)}{2-x_F y}$$

$$p_{SO_3} = \frac{2x_F y}{2-x_F y}$$

$$p_{O_2} = \frac{0.42 - x_F(2+y)}{2-x_F y}$$

From these expressions, the definition of K_p, and the numerical values of K_p (Table 19.4), the equilibrium conversion may be calculated for various initial feed concentrations as a function of temperature. Note that the expression for oxygen is correct only for sulfur burner gas derived from air, and that the factor of 0.42 must be changed if the initial gas stream contains more or less oxygen than 21% by volume.

From the equilibrium-conversion data in Fig. 19.3, it is apparent that conversion of the SO_2 decreases with an increase in temperature. For that reason, it is desirable to carry out the reaction at as low a temperature as practicable. At 400°C, where from Fig. 19.3 the *equilibrium* condition is seen to be very favorable, being almost 100%, the *rate* of attainment of this equilibrium is slow. The rate at 500°C is 10 to 100 times faster than at 400°C; at 550°C it is still faster. Since the reverse reaction, $SO_3 \longrightarrow SO_2 + \frac{1}{2}O_2$, does not become appreciable until 550°C, it seems advisable to run the reaction initially at this temperature in order to obtain maximum conversion with a minimum of catalyst.[16] Here there is a conflict between favorable conversion equilibriums at lower

[15]*Ibid.*, pp. 134ff.

[16]Lewis and Ries, Influence of Reaction Rate on Operating Conditions in Contact Sulfuric Acid Manufacture, *Ind. Eng. Chem.*, **17,** 593 (1925), and **19,** 830 (1927).

temperatures and favorable rates at higher temperatures. The actual procedure in a contact plant takes advantage of both rate and equilibrium considerations by first entering the gases over part of the catalyst at about 410 to 430°C and allowing the temperature to increase adiabatically as the reaction proceeds. The reaction rate increases as the temperature rises but then begins to slow down as equilibrium is approached. The reaction essentially stops when about 60 to 70% of the SO_2 has been converted, at a temperature in the vicinity of 600°C. Then the gas, before it passes over the remainder of the catalyst, is cooled in a heat exchanger, in a waste-heat boiler, or by other means, until the temperature of the gases passing over the last portion of the catalyst is not over 430°C. The yields using this procedure are 97 to 98%, and the overall reaction rate is very rapid. Figures 19.4 to 19.7 illustrate how these conditions are applied in practice and how the heat of reaction is used.

By rewriting the expression for K_p in terms of mole fractions and total pressures, for the equation $2SO_2 + O_2 \rightleftharpoons 2SO_3$ we have

$$K_p = \frac{N \times n_{SO_3}{}^2}{n_{SO_2}{}^2 \times n_{O_2} \times P}$$

where n = number of moles of each component
N = total moles
P = total pressure.

By rearranging, we obtain

$$n_{SO_3}{}^2 = \frac{n_{SO_2}{}^2 \times n_{O_2} \times K_p \times P}{N}$$

From this expression it may be seen that an increase in either SO_2 or oxygen increases the conversion to SO_3, thus illustrating the law of mass action. It should be borne in mind that, in the burner gas, if the concentration of oxygen increases, that of the SO_2 decreases, and vice versa. An increase in pressure, according to the law of Le Chatelier, also increases the conversion to SO_3, but the effect is so small and costly that it is not taken advantage of industrially. The fact that the number of moles of SO_3 formed at equilibrium is inversely proportional to N, the total number of moles, shows that, if the mixture of gases going through the converter were diluted with an inert gas, such as nitrogen, the conversion to SO_3 would be decreased.

The conversion of SO_2 to SO_3 is a reversible reaction. If a portion of the SO_3 is removed, more SO_2 will be converted in order to balance the equilibrium. This fact is being used in modern sulfuric acid plants in order to increase the overall conversion efficiency and decrease the quantity of SO_2 being released to the atmosphere (Fig. 19.6). The gases leaving the converter (Fig. 19.7) after having passed through two or three layers of catalyst are cooled and passed through an intermediate absorber tower where the SO_3 is removed. The gases leaving this tower are then reheated and flow through the remaining layers of catalyst in the converter. The gases are then cooled and passed through the final absorber tower before being permitted to flow into the atmosphere. In this manner more than 99.5% of the total SO_2 is converted to SO_3 and subsequently into product sulfuric acid.

CATALYSTS[17] It should be remembered that, in all catalytic reactions, the only function of the catalyst is to *increase the rate* of the reactions. A typical such catalyst consists of diatomaceous earth impregnated with upward of 7% V_2O_5. Sometimes two grades are charged into the converter,

[17]Patent and literature references to catalysts, together with abstracts therefrom, and testimony in the various lawsuits are summarized by Fairlie, *op. cit.*, chap. 17; however, for a more up-to-date catalyst review, see Duecker and West, *op. cit.*, chap. 13, pp. 159–196, with 231 refs., pp. 196–202.

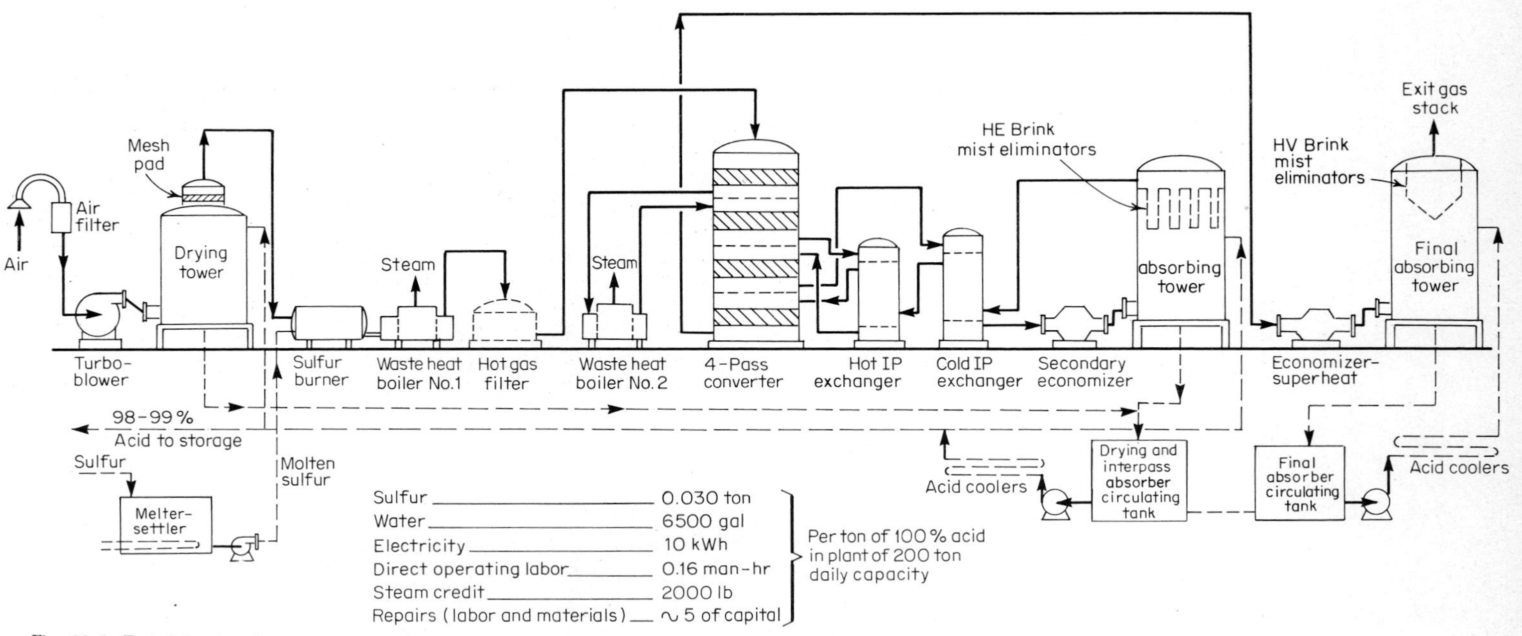

Fig. 19.6 Typical flowchart for a sulfur-burning double-absorption contact sulfuric acid plant. (*Monsanto Co.*)

a less active but harder type being used in the first pass of the converter, and a more active but softer type (increase of 30%) in passes subsequent to the first. These catalysts are of long life, up to 20 years, and are not subject to poisoning, except from fluorine, which damages the siliceous carrier. Conversions are high, up to 99.8% in double-absorber-type plants. Clean SO_2 must of course be supplied. Monsanto, American Cyanamid, Stauffer Chemical Co., and other vanadium catalysts are extensively used.

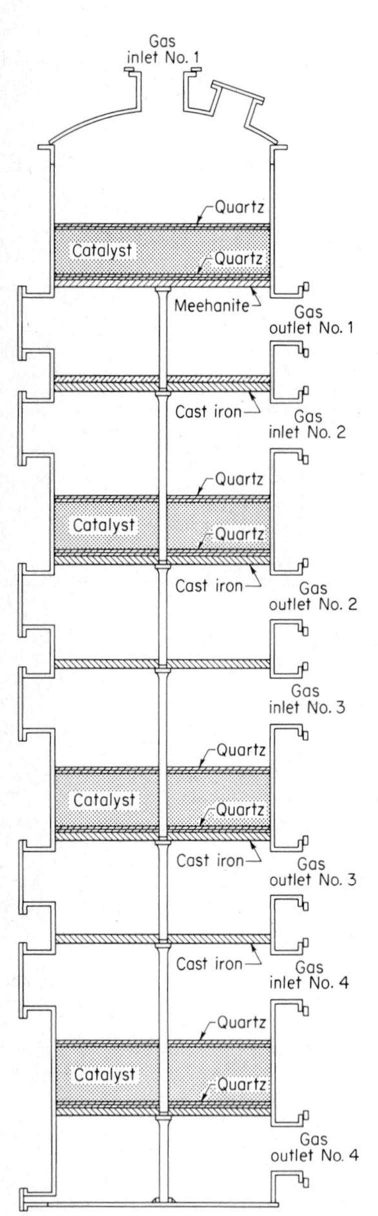

Fig. 19.7 Monsanto four-pass converter. In plants in which cold SO_2 gas must be heated by means of heat in the SO_3 gas, each of the four gas outlets except No. 3 is connected to a tubular heat exchanger. Each of the gas inlets except No. 4 is for the return connection from the exchanger. Gas outlet No. 3 and inlet No. 4 are connected to a small flue cooler from which the heat is discarded because it is insignificant in amount and not needed for preheating SO_2 gas (Fig. 19.5). In sulfur-burning plants, many variations are used for SO_3 gas cooling. Any of the gas outlets may be connected to a heat exchanger, waste-heat boiler, steam superheater, or other cooling equipment. Also, any one of them may be used for the injection of cold, dry air (Fig. 19.4). (*Monsanto Co.*)

CONTACT-PROCESS EQUIPMENT

Duecker and West[18] detail the many variations in the equipment employed and include materials of construction. All recent plants have the "outdoor" type of construction, which reduces initial capital investment. These plants are insulated but completely exposed to the elements, often having only the control room enclosed.

BURNERS The contact and the chamber processes use the same raw materials, i.e., sulfur principally and some sulfide ores. Because of its superior purity and lower transportation costs, sulfur is frequently brought in and stored in the molten condition. It is then pumped from a storage tank like any liquid and sprayed into a furnace (Fig. 19.4). Generally, about 9% SO_2 is furnished to the plant when sulfur is burned, though this percentage can be increased.

TREATMENT OF THE BURNER GAS SO_2 burner gas for the contact process may contain, in addition to dust, carbon dioxide, nitrogen, and oxygen, such impurities as chlorine, arsenic, and fluorine. The last two impurities are present only when materials other than elemental sulfur are burned. To prevent corrosion from the burner gases, it is customary to dry the air for burning the sulfur and oxidizing the SO_2 in the converter as shown in Fig. 19.4, and to dry the roaster gases before oxidizing the SO_2 as depicted in Fig. 19.5. Such drying is done in towers, usually with 93 to 98% sulfuric acid. The sulfur burner gas has much of its heat removed in waste-heat boilers for the generation of steam. If sulfide ore is roasted, efficient dust collectors, cooling and scrubbing towers, and electrical acid mist precipitators may be added.

HEAT EXCHANGERS AND COOLERS Before the gases can be taken to the first stage of the converter, they are adjusted to the minimum temperature at which the catalyst rapidly increases the speed of the reaction, usually 410 to 440°C. The gases must be cooled between catalyst stages to achieve high conversion. For this purpose, cold air may be introduced, or boilers, steam efficiency superheaters, or tubular heat exchangers may be employed. Heat exchangers usually consist of large, vertical cylinders containing many small tubes. The SO_3 gas usually passes through the tubes, and the SO_2 gas surrounds the tubes.

CONVERTERS The chemical conversion of SO_2 to SO_3 is designed to maximize this conversion by taking into consideration that

1. Equilibrium is an inverse function of temperature and a direct function of the O_2/SO_2 ratio.
2. Rate of reaction is a direct function of temperature.
3. Composition and ratio of amount of catalyst to amount of SO_3 formed affect the rate of conversion or the kinetics of the reaction.
4. Removal of SO_3 causes more SO_2 to be converted.

The commercialization of these basic conditions makes possible high overall conversion of up to 98% or more by the use of a multipass converter wherein, at an entering temperature of 410 to 430°C, the major part of the conversion (about 75%) is obtained; the exit temperature of the first catalyst bed is up to 600°C or more; depending largely on the concentration of SO_2 in the gas. Table 19.5 gives the temperatures and percent conversions for each stage of a four-pass converter.[19] The successive lowering of the temperature between stages ensures an overall higher conversion. By removing the SO_3 before the gas enters the last stage of the converter, overall conversions of over 99.5% can be obtained. Figures 19.4 to 19.8 show multipass converters and waste-heat boilers. In the multipass converters shown, about 20% of the total catalyst is frequently in the first stage, where 70 to 75% of the conversion may take place.

[18]Duecker and West, chaps. 8 and 9 for sulfur burners, with 51 refs. to literature, and chap. 14 for converters.

[19]Slin'ko and Beskov, Calculation of Contact Equipment with Adiabatic Catalyst Layers for Oxidation of Sulfur Dioxide, *Int. Chem. Eng.*, **2**(3), 388 (1962).

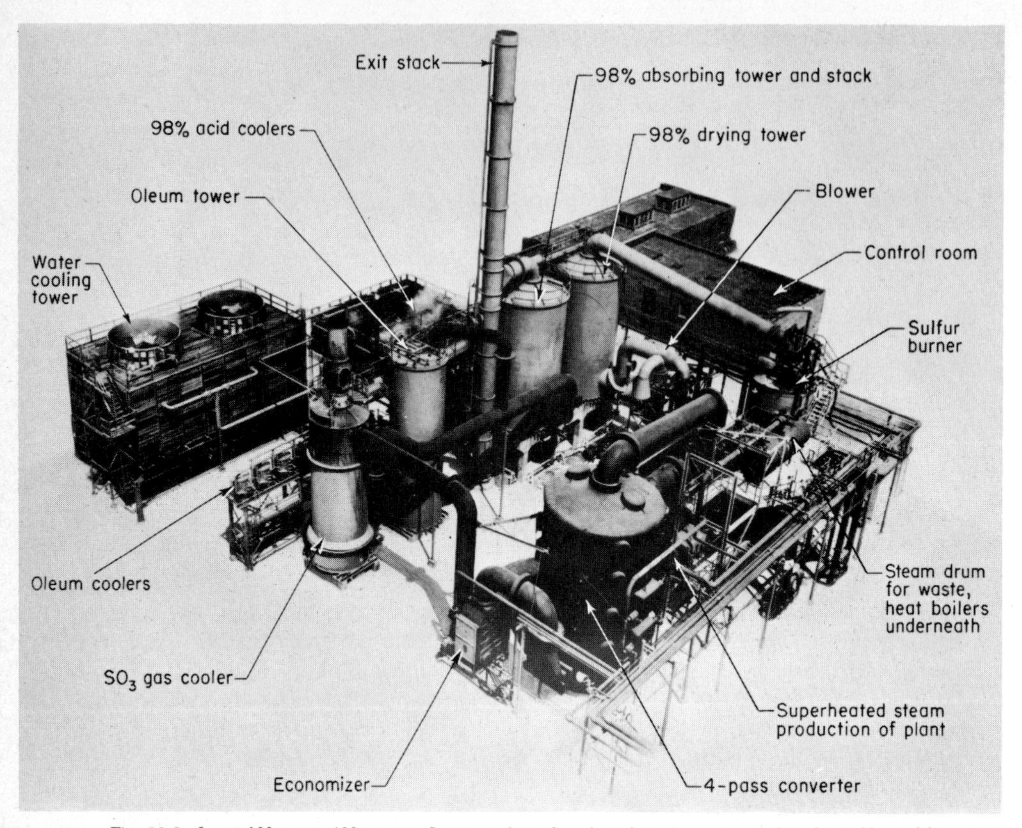

Fig. 19.8 Leonard-Monsanto 400-ton type S contact plant. Superheated steam goes to a turbine-driven blower delivering air to plants, and any surplus into works mains. The turbine is of the back-pressure type, its exhaust steam going into another works main. (*Monsanto Co.*)

In the converter in Fig. 19.7 the hot conversion gases are progressively cooled between stages (Table 19.5) and, after leaving the fourth stage of the converter, enter economizers or heat exchangers before the SO_3 enters the oleum tower or the 98 to 99% absorbing tower. The converter depicted is provided with trays and manholes for access to the catalyst control. Although the converter is the "heart" of a contact sulfuric plant, there are many variables. All these should be optimized to secure the maximum industrial results. Studies have been made and described[20] regarding the use of computers for this most important managerial manufacturing aspect.

SULFUR TRIOXIDE ABSORBERS It has long been known that a concentration of sulfuric acid between 98.5 and 99% is the most efficient agent for the absorption of SO_3, probably because acid of this strength has a lower vapor pressure than that at any other concentration. Hence acid of this strength is used for the final absorber before the waste gas is vented to the atmosphere. Water cannot be used because an acid mist is formed by the direct contact with the sulfur trioxide, which is almost impossible to absorb. Since the absorbing acid continuously becomes more concentrated, it is necessary to provide some means for diluting that part of the acid discharged from the final absorber which is to be recirculated. Also, the absorbing acid must be cooled. The recirculating acid

[20]Homme and Othmer, Sulfuric Acid–optimized Conditions in Contact Manufacture, *Ind. Eng. Chem.*, **53**, 979 (1961); Shannon, Digital Computer, *Chem. Eng.* (*N.Y.*), **72**(20), 84 (1965); Shannon, Computer Simulation, *Chem. Eng. Prog.*, **62**(4), 49 (1966).

TABLE 19.5 ***Temperatures and Conversion in Each Stage of a Monsanto Converter** (Using relatively rich SO_2 gas from sulfur)*

Location	Temperature °C	Temperature °F	Equivalent conversion, %
Gas entering first pass	410.0	770	
Gas leaving first pass	601.8	1115	
Rise in temperature	191.8	345	74.0
Gas entering second pass	438.0	820	
Gas leaving second pass	485.3	906	
Rise in temperature	47.3	86	18.4
Gas entering third pass	432	810	
Gas leaving third pass	443	830	
Rise in temperature	11	20	4.3
Gas entering fourth pass	427.0	800	
Gas leaving fourth pass	430.3	806	
Rise in temperature	3.3	6	1.3
Total rise	253.4	457	98.0

Source: Based on Duecker and West, Manufacture of Sulfuric Acid, p. 252, Monograph 144, Reinhold, 1959.

is diluted by adding spent or dilute sulfuric acid or water in the amount required, withdrawing from the system any excess acid for sale.

The 20% oleum is made in an oleum absorber,[21] as shown in Figs. 19.4 and 19.5, by passing cooled 98.3 to 98.5% acid over this tower. If an oleum of 60 to 65% is wanted, it may be prepared by distilling 30 or 35% oleum in steel boilers. The trioxide driven off is usually condensed and then added to 20% or stronger oleum to obtain the desired 60 to 65% concentration.

To remove moisture from air or SO_2 gas with 93 to 98% acid, or to absorb sulfur trioxide in 98.5 to 99% acid, packed steel towers lined with acidproof masonry are used. For oleum, the steel tower need not be lined. The packing generally consists of chemical stoneware, though at times in the distant past lumps of quartz or coke were used. The gas enters the bottom and leaves by a flue at the top. A 98.5 to 99% absorbing tower 23 ft 3 in. in diameter inside the lining with a packed volume of 7,000 ft^3 and an acid flow of approximately 2,500 gal/min can readily absorb 1,000 tons of SO_3 per 24 h with an apparent gas velocity of 122 linear ft/min measured at standard conditions of 0°C and 760 mm Hg.

PLANT COST It is difficult to give the approximate general cost of a plant, because of variations in design and in prices in different places and because of continuing inflation. Yet it may be said that the double-absorption sulfur-burning contact plant illustrated in Fig. 19.6, producing 1,000 tons/day, 100% acid basis, cost roughly $6 million in 1976. To this figure must be added whatever is required for storage, extension of utilities, and necessary offsite facilities. A metallurgical-type plant burning sulfide ores such as that shown in Fig. 19.5 would cost about three times the figure given for a sulfur-burning plant, since it would include ore roaster and extensive gas cooling and purification facilities. As shown by the utilities and labor figures under each flowchart, the operating cost of a metallurgical or ore-roasting plant would be much higher.

SULFUR TRIOXIDE[22] Each year sulfur trioxide is being used in increasing amounts, largely for sulfonation, especially in the manufacture of detergents (Chap. 29). The difficulty in the past was the instability of the sulfur trioxide. However, under the trade name *Sulfans,* stabilized forms of sulfur trioxide are commercially available, in which crystallization or conversion to a polymer is

[21]Duecker and West, *op. cit.*, pp. 226, 255–258.
[22]Enter, Trioxide Intermediates, *Chem. Week,* Apr. 18, 1965, p. 31.

inhibited by 0.25% sulfur, tellurium, carbon tetrachloride, and phosphorus oxytrichloride.[23] This product is manufactured by distillation from strong oleums, as just described.

THE CHAMBER PROCESS is obsolete. Discussions of its technology and complex chemistry can be found in the literature.[24]

RECOVERY OF WASTE SULFURIC ACID

The recovery and reuse of sulfuric acid from inorganic and organic wastes are frequently not economical, nevertheless, they are usually necessary to avoid the pollution of streams. The more concentrated and cleaner the spent acid, the easier it is to recover the values. If sufficiently clean, these spent acids can be sent to a contact plant to be fortified, though this does not remove impurities.

Some sulfuric acid is still used in the steel industry for pickling. However, because of the need to eliminate the discharge of spent liquors into streams, and because of the difficulty in treating such liquors for the recovery of acid values and to eliminate stream pollution, sulfuric acid is being replaced with hydrochloric acid for pickling. Spent liquor from the latter can be treated to recover acid values and to avoid stream pollution.

A residual liquor analogous to steel-mill pickle liquor is obtained from titanium pigment plants which use sulfuric acid to produce titanium dioxide from ilmenite. These spent liquors contain large amounts of ferrous sulfate and free sulfuric acid. Although ferrous sulfate recovery systems were operated at two titanium pigment plants,[25] they proved uneconomical and have been shut down. At these two plants the wastes are now barged to the ocean and dumped. Because of increasing pressure to avoid the contamination of streams with ferrous sulfate, certain pickle liquors are concentrated and the ferrous sulfate recovered even though this is not profitable. Titanium dioxide pigment is made by the chloride route to avoid the disposal problems associated with the use of sulfuric acid.

WASTE SPENT ACID About 2 million tons of spent acids is reused each year: (1) spent alkylation acid catalyst is black, but still relatively strong, and not too badly contaminated (about 90% H_2SO_4, 5% water, and 5% hydrocarbons); (2) nitration spent acid is diluted and only slightly contaminated; and (3) spent sludge acids result from petroleum refining. The latter are usually dirty, low in acidity, heavily contaminated, and many contain up to 75% H_2SO_4 and up to 20% or more hydrocarbons, the balance being water. Occasionally, they can be added in small percentages to spent alkylation acids or reduced to sulfur dioxide by heat with coke as a by-product, but the process is expensive.

SULFUR POLLUTION

The reduction or elimination of pollution by sulfur and sulfur compounds has been extensively studied with the hope of eventual recovery for reuse. Fuel desulfurization usually yields sulfur via hydrogen sulfide. Sulfur dioxide from nonferrous metal smelting or fuel combustion is most readily recovered as sulfuric acid, and less often as liquid sulfur dioxide, sulfur, or sulfate salts. Sulfuric acid has long been made from the richer gases at smelters with access to markets. Likewise, large tonnages of sulfur are recovered from gaseous and light liquid petroleum fractions. Amounts in

[23]McGraw-Hill Encyclopedia, vol. 13, p. 258, 1960.

[24]Duecker and West, *op. cit.*, pp. 104–118. Other interpretations with original journal references are given. Details of the chamber process are given in CPI 2 and in the Selected References.

[25]See CPI 1, pp. 357–358, for description and flowcharts; this edition, Chap. 24 (titanium dioxide flowchart); and *Chem. Week*, June 27, 1973, p. 45 for waste disposal aspects.

millions of short tons of sulfur dioxide show 25.7 emitted from fuel combustion, compared with 2.5 recovered and 7.7 emitted from industrial processes, mainly smelting, compared with 0.9 recovered.[26] Most of the sources are too small, too dilute, or too isolated for economical recovery. Nevertheless, new air pollution regulations are forcing smelters to recover up to 90% of input sulfur and are sharply limiting emissions of sulfur dioxide from fuel combustion. These regulations have brought about new facilities for desulfurizing fuel oils, revived interest in low-sulfur liquid and gaseous fuels from coal, and many demonstration plants for removing sulfur dioxide from power-plant waste gases.[27]

Emissions from new sulfuric acid plants are limited to 4 lb of sulfur dioxide and 0.15 lb of acid mist per ton of acid made. These are met either by building plants of the double-absorption type, previously described, or by adding a tail-gas cleanup step.[28]

Recovery of sulfur by the Claus process is never 100%. Allowable emissions of sulfur dioxide from these plants vary from place to place, and are often stricter for larger plants. Consequently, many new plants are built with one more converter than is shown in Figure 19.2, and various tail-gas cleanup units are being installed on both old and new plants.[29]

CONCENTRATION

The air-blown concentrator is exemplified by the Chemico *drum-type concentrator* shown in Fig. 19.9. The burner supplies hot gases at about 1250°F from the combustion of oil or fuel gas. These hot combustion gases are blown countercurrent to the sulfuric acid in two compartments in the concentrating drum and remove water as they bubble up through the acid. The off-gases at 440 to 475°F from the first compartment of the drum pass to the second compartment, along with a portion of the hot gases from the combustion furnace. They leave at 340 to 360°F to enter a gas-cooling drum, where they are cooled to 230 to 260°F in raising the dilute acid to its boiling point. Since some sulfuric acid is entrained as a mist, the hot gases are passed through a Pease Anthony venturi scrubber with a cyclone separator and are washed with feed acid for the removal of the acid mist before discharge to the atmosphere. This reduces the acid mist to perhaps 1 mg/ft^3 and involves considerably less capital investment than an electrostatic mist precipitator. See also Fig. 19.10, depicting a Brink glass-fiber mist eliminator for mist elimination in sulfuric acid concentrators and for stack gases of contact sulfuric acid plants.[30] This procedure gives an acid with a final concentration of 93% or slightly higher. The hot gases, aided by the oxidizing action of the hot sulfuric acid, also burn out many impurities that may be found in a spent acid being concentrated. Hence such air-blown concentrators are extensively employed in the concentration of spent nitrating acids from munition works. If spent acid from petroleum purification is being handled, the flow of acid from the rear to the front compartment passes through an intermediate storage tank, where a skimmer removes some of the nonvolatile carbonaceous impurities. The front and rear concentrating

[26]Galstaun *et al.*, Cost to Desulfurize Fuel Oil, Rohrman *et al.*, Sources of SO_2 Pollution, Grekel *et al.*, Package Plants for Sulfur Recovery, *Chem. Eng. Prog.*, **61**(9), 49 (1965). Many references, tables, analyses, and costs: Robinson & Robbins, Gaseous Sulfur Pollutants from Urban and Natural Sources, *J. Air Pollut. Control Assoc.*, **20,** 233 (1970); Mills and Perry, Fossil Fuel Yields Power Plus Pollution, *Chemtech*, **3,** 53 (1973).

[27]Semrau, Control of Sulfur Oxide Emissions from Primary Copper, Lead, and Zinc Smelters, *J. Air Pollut. Control Assoc.*, **21,** 185 (1971); Schuit and Gates, Chemistry and Engineering of Catalytic Hydrodesulfurization, *AIChE J.*, **19,** 417 (1973); Conn, Low Btu Gas for Power Plants, *Chem. Eng. Prog.*, **69**(12), 56 (1973); Slack, Removing Sulfur Dioxide from Stack Gases, Engineering Science & Technology, **7,** 110 (1973).

[28]Davis, Sulfur Dioxide Absorbed from Tail Gas with Sodium Sulfite, *Chem. Eng.* (*N.Y.*), **78**(27), 43 (1971).

[29]Hyne, Methods for Desulfurization of Effluent Gas Streams, *Oil Gas J.*, Aug. 28, 1972, p. 64.

[30]Brink, New Fiber Mist Eliminator, *Chem. Eng.* (*N.Y.*), **66**(11), 1959; *Chem. Process.* (Chicago), **25,** February 1962; Brink *et al.*, Fiber Mist Eliminators for Higher Velocities, *Chem. Eng. Prog.*, **60**(11), 68 (1964); *ibid.*, Mist Removal from Compressed Gases, *Chem. Eng. Prog.*, **62**(4), 60 (1966).

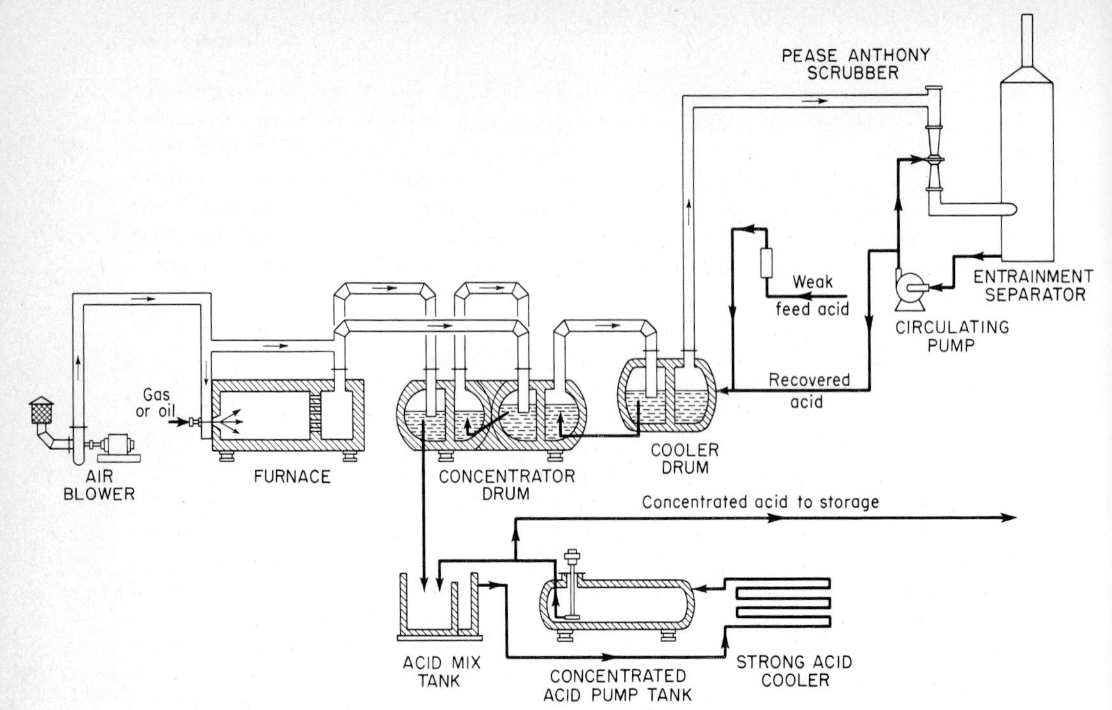

Fig. 19.9 Chemico sulfuric acid concentrator, drum type. (*Chemical Construction Co.*)

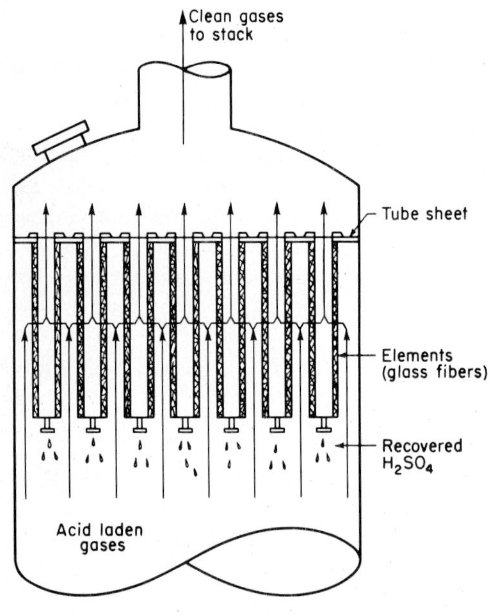

Fig. 19.10 Brink glass-fiber mist eliminator for gases containing sulfuric acid mist. (*Monsanto Co.*)

compartments of the steel drum are lined with lead and acidproof masonry. In this type of concentrator, the boiling point of the acid is reduced as much as 110°F for 93% acid by the effect of the hot air in reducing the partial pressure of the water vapor above the acid.[31]

Simonson-Mantius vacuum concentrators[32] operate at 27 to 29.6 in. Hg vacuum and have long been used to concentrate sulfuric acid up to a maximum strength of 93%. Several types[33] fit different conditions and various concentrations. They have an advantage over hot-air-blown concentrators because the vapors are condensed and noxious gases are not permitted to flow into the atmosphere.

SELECTED REFERENCES

Duecker, W. W., and J. R. West: Manufacture of Sulfuric Acid, ACS Monograph 144, Reinhold, 1959.
Fasullo, O.: Sulfuric Acid: Use and Handling, McGraw-Hill, 1964.
Manufacturing Chemists Association: Atmospheric Emissions from Sulfuric Acid Manufacturing Processes, Rept. 999-AP-13, 1965.
Meyer, B. (ed.): Elemental Sulfur, Wiley-Interscience, 1965.
Miller, D. J. and T. K. Wiewiorowski (eds.): Sulfur Research Trends, ACS, 1972.
Monsanto Enviro-chem Sulfuric Acid Plants, Monsanto Co., 1975.
Pryor, W. A.: Mechanisms of Sulfur Reactions, McGraw-Hill, 1962.
Sittig, M.: Catalyst Manufacture, Recovery, and Use, Noyes, 1972.
Sittig, M.: Sulfuric Acid Manufacture and Effluent Control, Noyes, 1971.
Slack, A. V.: Sulfur Dioxide Removal from Waste Gases, Noyes, 1961.
Stecher, P. G.: Hydrogen Sulfide Removal Processes, Noyes, 1972.
Sulfur and SO_2 Developments: AIChE, 1971.
Waeser, B. (ed.): Die Schwefelsaurefabrikation, Vieweg, Braunschweig, Germany, 1961.

[31]Duecker and West, *op. cit.*, fig. 19.8, p. 340 and, for diagrams of the Pease-Anthony venturi scrubber, pp. 91 and 92.

[32]Burke and Mantius, Concentration of Sulfuric Acid under Vacuum, *Chem. Eng. Prog.*, **43**, 237 (1947) (excellent tables and diagrams).

[33]Duecker and West, *op. cit.*, pp. 329–337. Other types described and illustrated; material and heat data on p. 335.

chapter 20

HYDROCHLORIC ACID AND MISCELLANEOUS INORGANIC CHEMICALS

Although most of these inorganic chemicals have been in use for a long time, new uses and processes continue to acquire commercial importance. During the past decade hydrochloric acid has been substituted for sulfuric acid in a recently developed steel-surface cleaning process.[1]

HYDROCHLORIC OR MURIATIC ACID

Hydrochloric acid, although not manufactured in such large quantities as sulfuric acid, is an important heavy chemical. Manufacturing techniques have changed and improved in recent years, and new procedures are now employed, such as the burning of chlorine in hydrogen.

Hydrogen chloride (HCl) is a gas at ordinary temperature and pressure. Aqueous solutions of it are known as *hydrochloric acid* or, if the HCl in solution is of the commercial grade, as *muriatic acid.* The common acids of commerce are 18°Bé (1.142 sp gr) or 27.9% HCl, 20°Bé (1.160 sp gr) or 31.5% HCl, and 22°Bé (1.179 sp gr) or 35.2% HCl,[2] which sell for $25, $27, and $32 per ton, respectively. Anhydrous HCl is available in steel cylinders at a very considerable increase in cost, because of the cylinder expense involved.

HISTORICAL Hydrogen chloride was discovered in the fifteenth century by Basilius Valentinius. Commercial production of hydrochloric acid began in England when legislation was passed prohibiting the indiscriminate discharge of hydrogen chloride into the atmosphere. This legislation forced manufacturers, using the Leblanc process for soda ash, to absorb the waste hydrogen chloride in water. As more uses for hydrochloric acid were discovered, plants were built solely for its production.

USES AND ECONOMICS The largest users of hydrochloric acid are the metal, chemical, food, and petroleum industries.[3] Industry experts estimate that the metal industries consume about 47% of the acid sold. A breakdown of the remaining uses is: chemical and pharmaceutical manufacturing and processing, 33%; food processing, 7%; oil well acidizing, 6%; and miscellaneous uses, 7%.

The major use of hydrochloric acid is in steel pickling (surface treatment to remove mill scale). Prior to 1963 almost all steel was pickled with sulfuric acid. Hydrochloric acid has taken over this market because it reacts faster than sulfuric with mill scale, less base metal is attacked by it, the

[1]*Chem. Mark. Rep.*, June 2, 1975.
[2]Perry, pp. 1-28 and 3-75.
[3]Chemical Economics Handbook, Stanford Research Institute, 738, 5030A–C.

TABLE 20.1 ***Production of Hydrochloric Acid*** *(In thousands of short tons of 100% acid)*

Source	*1950*	*1960*	*1965*	*1970*	*1974**
Salt and sulfuric	165	91	138	125	100
Chlorine and hydrogen	111	148	99	95	104
By-product and other	342	731	1,133	1,793	2,200
Total	618	970	1,370	2,013	2,404

Sources: Manual of Current Indicators, Chemical Economics Handbook, Stanford Research Institue; *Chem. Eng. News,* June 2, 1975, p. 33. *Preliminary figures.

pickled steel has a better surface for subsequent coating or plating operations, and much smaller quantities of waste pickle liquor are produced.

MANUFACTURE Hydrochloric acid is obtained from four major sources: as a by-product in the chlorination of both aromatic and aliphatic hydrocarbons, from reacting salt and sulfuric acid, from the combustion of hydrogen and chlorine, and from Hargreaves-type operations ($4NaCl + 2SO_2 + O_2 + 2H_2O \longrightarrow 2Na_2SO_4 + 4HCl$). As can be seen from Table 20.1, by-product operations furnish 90% of the acid. The old salt–sulfuric acid method and the newer combustion method each supply about half of the remainder. The Hargreaves process is used by only one company which owns probably the largest single plant and produces about 60,000 tons/year.

Reactions and energy requirements. The basic steps in the production of by-product acid include the removal of any unchlorinated hydrocarbon, followed by the absorption of the hydrogen chloride in water. A typical chlorination, for illustration, is:

$$\underset{\text{Benzene}}{C_6H_6} + Cl_2 \longrightarrow \underset{\text{Chlorobenzene}}{C_6H_5Cl} + HCl$$

Since the chlorination of aliphatic and aromatic hydrocarbons evolves large amounts of heat, special equipment is necessary for control of the temperature of reaction.[4]

The reactions of the salt–sulfuric acid process are endothermic:

$$NaCl + H_2SO_4 \longrightarrow HCl + NaHSO_4$$

$$NaCl + NaHSO_4 \longrightarrow HCl + Na_2SO_4$$

Summation:

$$2NaCl(s) + H_2SO_4(l) \longrightarrow 2HCl(g) + Na_2SO_4(s) \qquad \Delta H_{25°C} = +15.7 \text{ kcal}$$

The first reaction goes to completion at relatively low temperatures, whereas the second approaches completion only at elevated temperatures.[5] The reactions are forced to the right by the escape of the hydrogen chloride from the reaction mass. The reaction between hydrogen and chlorine[6] is highly exothermic and goes spontaneously to completion as soon as it is initiated. Absorption of the hydrogen chloride made by any process liberates about 700 Btu/lb of hydrogen chloride absorbed. This heat must be taken away in the absorber, or the efficiency will be low. Figure 20.1 illustrates the design of a hydrogen chloride absorber made from Karbate carbon.

Figure 20.2 illustrates the essential steps in the production of hydrochloric acid using the salt process and the synthetic process. The by-product process differs from these methods in the variable

[4]Oldershaw *et al.*, Absorption and Purification of Hydrogen Chloride from Chlorination of Hydrocarbons, *Chem. Eng. Prog.*, **43,** 371 (1947).

[5]Badger and Baker, Inorganic Chemical Technology, 2d ed., p. 111, references, p. 133, McGraw-Hill, 1941 (particularly good in presenting diagrams of furnaces and conditions for adsorption).

[6]Maude, Synthetic Hydrogen Chloride, *Chem. Eng. Prog.*, **44,** 179 (1948); Brumbaugh *et al.*, Synthesis and Recovery of Hydrogen Chloride Gas, *Ind. Eng. Chem.*, **41,** 2165 (1949).

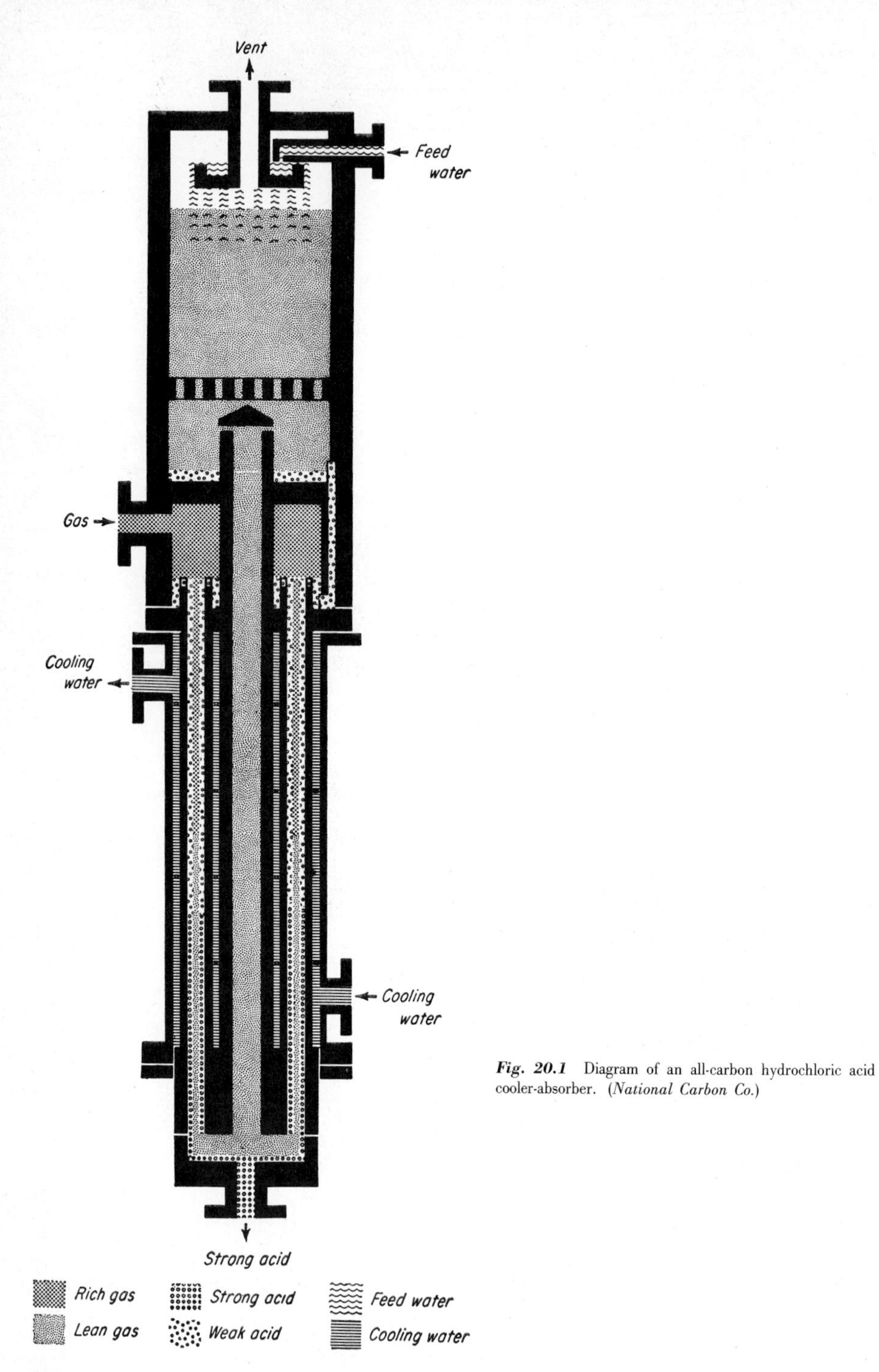

Fig. 20.1 Diagram of an all-carbon hydrochloric acid cooler-absorber. (*National Carbon Co.*)

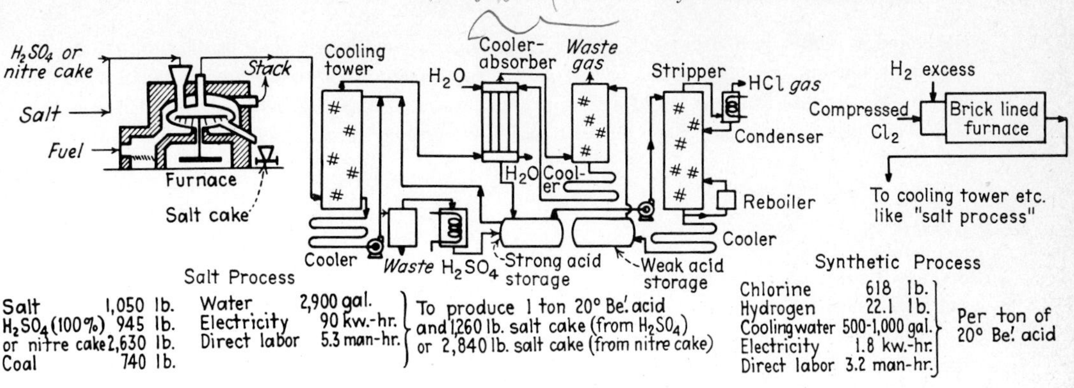

Fig. 20.2 Flowchart for hydrochloric acid manufacture.

conditions employed for the production and separation of the hydrogen chloride. The *salt process* may be divided into the following *physical operations* and *chemical conversions:*

Sulfuric acid and salt are roasted in a furnace to form hydrogen chloride and sodium sulfate (salt cake) (Ch).

The hot hydrogen chloride, contaminated with droplets of sulfuric acid and particles of salt cake, is cooled by passing it through coolers, cooled externally by water (Op).

The cooled gas is then passed upward through a coke tower to remove suspended foreign materials (Op). (This equipment is omitted in some plants.)

Purified hydrogen chloride from the top of the coke tower is absorbed in water in a tantalum or Karbate absorber (Op).

Finished hydrochloric acid is withdrawn from the bottom of the absorber, and any undissolved gas passing out the top of the absorber is scrubbed out with water in a packed tower (Op).

The most important of the salt furnaces in operation is the Mannheim furnace. This consists of a cast-iron muffle, composed of a dish-shaped top and bottom bolted together and equipped with plows to agitate the reaction mixture. The rotary furnace is growing in usage. Here sulfuric acid and salt are continuously mixed and heated.

The *synthetic process* generates hydrogen chloride by burning chlorine in a few percent excess of hydrogen. The purity of the ensuing acid is dependent upon the purity of the hydrogen and chlorine. However, since both of these gases are available in a very pure state as by-products of the electrolytic process for caustic soda, this synthetic method produces the purest hydrogen chloride of all the processes. As an example of continuing improvement of processes for well-established chemicals, a patent has been granted on a design for a burner nozzle for this process.[7] The cooling and absorption are very similar to that employed in the salt process. More than 250,000 tons/year of *anhydrous hydrogen chloride* is needed for making methyl chloride, ethyl chloride, vinyl chloride, and other such compounds. This is generally manufactured[8] by burning chlorine in excess of hydrogen, absorbing the hydrogen chloride in water. The aqueous solution is stripped of hydrogen chloride under slight pressure, giving strong gaseous hydrogen chloride which is dehydrated to 99.5% hydrogen chloride by cooling to 10°F.

Hydrochloric acid is extremely corrosive to most common metals, and great care must be taken to choose the proper materials for plant construction. This acid is usually shipped by rail or barge

[7]Hildyard and Breckon, U.S. Pat. 3,097,073 (1963).
[8]Anhydrous HCl, *Chem. Eng.* (*N.Y.*), **58**(10), 208 (1951) (pictured flowchart).

and stored in rubber-lined steel tanks. Although absorbers were formerly made of silica, in recent years tantalum and structural carbon have largely replaced fused silica. Indeed, the installation of Karbate structural-carbon heat interchangers and condensers has characterized recent hydrochloric acid plants because of higher rates of heat transfer and greater strength and resistance to corrosive attack.[9]

BROMINE

Bromine is a member of the halogen family and is a heavy, dark-red liquid. It is much less common and more expensive than chlorine.

HISTORICAL In 1826, 15 years after the discovery of iodine, bromine was discovered by Balard, a French chemist, who obtained it from the mother liquor left after separating salt from sea water by evaporation. Its manufacture from the Stassfurt deposits of carnallite was begun in 1865, from the $MgBr_2$ in the liquors left after working up $MgCl_2$ and KCl. Today Ethyl Dow operates the largest plant at Freeport, Tex. (Fig. 20.3), built during World War II.

USES AND ECONOMICS About 60% of the domestic bromine output is used in the manufacture of ethylene dibromide for antiknock fluids.[10] This bromine compound prevents lead oxide deposits in engines which would otherwise result from tetraethyllead (TEL). The appearance of compact cars and jet-fuel airplanes has interrupted TEL and ethylene dibromide growth. The Clean Air Act of 1970 mandated air quality standards that caused the EPA to make a commitment to eliminate 60 to 65% of lead antiknock compounds from gasoline by 1979. This will further reduce the amount of bromine used in the gasoline industry.[11]

Other bromine uses, however, have increased; most of this bromine is converted into organic or inorganic compounds. Although the demand for bromine in liquid form was limited mostly to laboratory reagents and small speciality uses, there is current interest in its use in swimming pools. Less than 1 ppm of bromine sterilizes the water without eye irritation or objectionable odor. Among the major inorganic compounds are alkali bromides, which are widely employed in photography and as comparatively safe sedatives in medicine. Hydrobromic acid resembles hydrochloric acid but is a more effective solvent for ore minerals, i.e., it has a higher boiling point and stronger reducing action. Methyl bromide (used as an insecticide, rodenticide, and methylating agent) and bromoindigo dyes are important organic compounds. Both methyl bromide and ethylene dibromide are finding growing use as soil and seed fumigants. Ethyl bromide is employed mainly as an ethylating agent in the synthesis of pharmaceuticals (Chap. 40). Bromine is also employed in the preparation of dyes, disinfectants, flameproofing agents, and fire-extinguishing compounds. Flame retardants have been the biggest gainers among the newer uses. As flame-retardant standards become more strict, bromine products will probably increase their share of the market. Prior to World War I, Germany supplied approximately three-quarters of the total world production (1,250 tons). By 1929 the total U.S. production was 3,207 tons plus $8\frac{1}{2}$ tons imported. Since 1938 no imports have been received. In 1952, 78,101 tons (value $30 million) was produced in this country, and by 1974 approximately 220,000 tons. The price of bromine has moved up recently to 25 to 30 cents per pound in bulk and 62 cents per pound in drums. The major part of it is shipped as ethylene dibromide.

MANUFACTURE In the United States the chief raw material is sea water, in which bromine occurs in concentrations of 60 to 70 ppm. It is also manufactured from natural brines, where its concentration may be as high as 1,300 ppm. In Germany it is produced from waste liquors resulting

[9]Lippman, Solving the Heat Exchange Problem in Cooling Hot HCl, *Chem. Metall. Eng.*, **52**(3), 112 (1945), and New HCl Rectifier Can Make CP Acid, Dry Gas, Solutions, *Chem. Eng.* (*N.Y.*), **56**(11), 110 (1949); Naidel, Hydrogen Chloride Production by Combustion in Graphite Vessels, *Chem. Eng. Prog.*, **69**(2), 53 (1973); Hulswitt, Adiabatic and Falling Film Absorption of Hydrogen Chloride, *Chem. Eng. Prog.*, **69**(2), 50 (1973).

[10]*Chem. Week*, Nov. 27, 1974, p. 26.

[11]*Chem. Eng. News*, Feb. 11, 1974, p. 11.

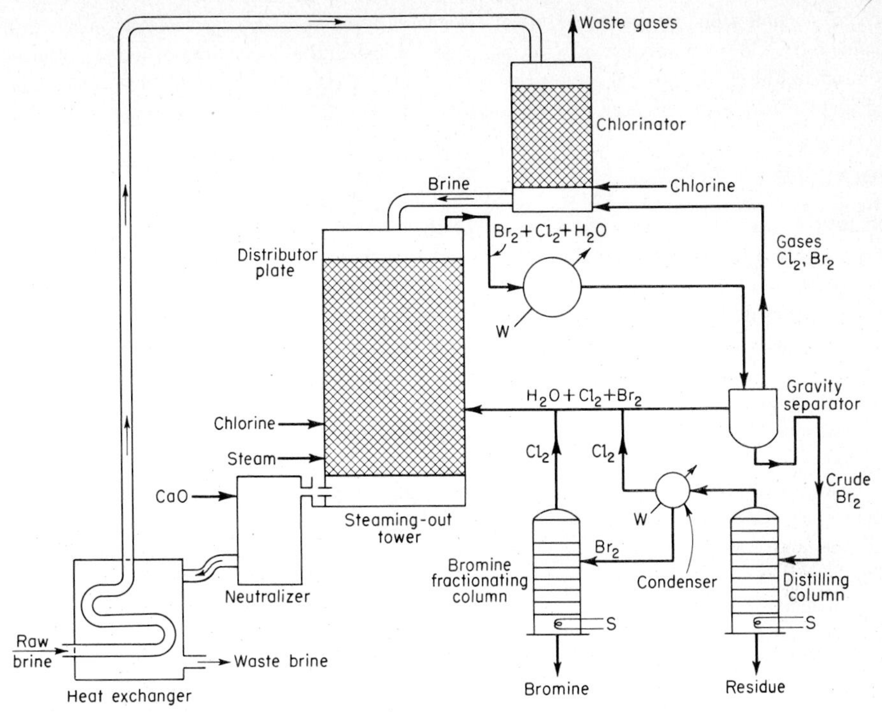

Fig. 20.3 Steaming-out process for bromine manufacture.

from the extraction of potash salts from the Stassfurt carnallite deposits, formerly the world's chief source of bromine. Various other countries have the following sources. Italy extracts bromine from one of its largest inland lakes; the U.S.S.R., from the water of Saksky Lake in Crimea; France, from deposits in Alsace; and Israel from the Dead Sea, the richest source known.

Bromine from sea water.[12] A process was worked out for the removal of a small quantity of bromine by air-blowing it out of chlorinated sea water according to the equation

$$2NaBr + Cl_2 \longrightarrow 2NaCl + Br_2\uparrow$$

Manufacture from salt brines.[13] In ocean water, where the concentration of bromine is relatively dilute, air has proved to be the most economical blowing-out agent. However, in the treatment of relatively rich bromine sources such as brines, *steaming out* the bromine vapor is the more satisfactory (Fig. 20.3). The original steaming-out process was developed by the Germans for processing the Stassfurt deposits, and with modifications is still used there, as well as in Israel and in this country. This process involves preheating the brine to 90°C in a heat exchanger and passing it down a chlorinator tower. After partial chlorination, the brine flows into a steaming-out tower, where steam is injected at the bottom and the remaining chlorine is introduced. The halogen-containing vapor is condensed and gravity-separated. The top water-halogen layer is returned to the steaming-out tower, and the crude halogen (predominantly bromine) bottom layer is separated and purified. *Crude bromine* from any of the foregoing processes can be purified by redistillation or by passing the vapors over iron filings which hold back the chlorine impurity. Alkali bromides account for an important proportion of the bromine produced. They cannot be made by the action of caustic soda on bromine,

[12]See CPI 2 for description; *Chem. Metall. Eng.*, **46,** 771 (1939); ECT, 2d ed., vol. 3, pp. 750–766, 1963.
[13]For flowchart, see ECT, 2d ed., vol. 3, p. 760, 1964.

since hypobromites and bromates are produced also. The approved method consists in allowing bromine to trickle into a mixture of water and iron borings until a heavy sirup of ferrous bromide has formed. This is siphoned off, treated with the desired alkali carbonate, filtered to remove the precipitate of iron hydroxides, evaporated, and recrystallized.

IODINE

HISTORICAL AND RAW MATERIALS It was in 1811 that Courtois, a saltpeter manufacturer of Paris, obtained a beautiful violet iodine vapor from the mother liquors left from the recrystallization of certain salts. Iodine was later found to exist almost universally in nature. It is present as iodates (0.05 to 0.15%) in the Chilean nitrate deposits. It occurs in sea water, from which certain seaweeds extract and concentrate it within their cells. Iodine has been made from these weeds by the kelp-burning process,[14] which has been practiced for years in Scotland, Norway, Normandy, and Japan. The present sources of the element, as far as the United States is concerned, however, are the nitrate fields of Chile, brine wells in Michigan, and the seaweed process of Japan, involving a total consumption of about 3,500 tons.

USES AND ECONOMICS In the United States all but 11% of the iodine used is converted directly to chemical compounds.[15] The single largest use, 39% of the total, is for organic compounds; 29% is used for potassium iodide, 2% for sodium iodide, and 19% for other inorganic compounds. Iodine is used as a catalyst in the chlorination of organic compounds and in analytical chemistry for determination of the so-called *iodine numbers* of oils. Iodine for medicinal, photographic, and pharmaceutical purposes is usually in the form of alkali iodides, prepared through the agency of ferrous iodide. In addition to the above, the element is also employed in the manufacture of certain dyes and as a germicide. Simple iodine derivatives of hydrocarbons, such as iodoform, have an antiseptic action. Organic compounds containing iodine have been used as rubber emulsifiers, chemical antioxidants, and dyes and pigments. For years, Chilean-produced iodine dominated the market and controlled the price, which was set at $4.65 per pound. Upon development of the iodine from oil-well brines in the southwestern United States, the price of iodine was reduced to approximately $2.20 per pound. This source has now decreased to one plant. Japan has recently become the world's largest iodine producer. In 1971 Japan produced more than three times the amount produced by Chile. Over four-fifths of the Japanese output was exported to the United States. The price in 1975 was $2.59 per pound for crude iodine and $4.00 per pound for resublimed material.

MANUFACTURE The major sources for the production of industrial iodine are Chilean nitrate mother liquor, brines in Midland, Mich., and natural gas brines in Japan. Five of the eighteen iodine manufacturing plants in Japan have been built since 1970. Ecological problems may seriously affect their long-term output as ground control and waste disposal regulations become more strict, but at least one of the companies has plants that satisfy environmental requirements.

FLUORINE AND FLUOROCHEMICALS

Fluorine, a pale greenish-yellow gas of the halogen family, is the most chemically active nonmetal element. It occurs in the combined form and is second only to chlorine in abundance among the halogens. The chief fluorine-containing minerals of commercial significance are fluorspar, fluorapatite (see superphosphates), and cryolite.

HISTORICAL Fluorine was discovered by Scheele in 1771 but was not isolated until 1886, by H. Moissan, after more than 75 years of intensive effort by many experimenters. The Freon refrigerants developed in 1930 fostered the commercial development of anhydrous hydrofluoric acid

[14] Dyson, Chemistry and Chemotherapy of Iodine and Its Derivatives, *Chem. Age*, **22**, 362 (1930).
[15] Minerals Yearbook 1972, vol. 1, pp. 1359–1362, Dept of the Interior, 1974.

and stimulated the growth of this new industry, which came into its own at the beginning of World War II,[16] largely because of the demands for UF_6 for isotope separation and for Freons.

USES AND ECONOMICS Elemental fluorine is costly and has comparatively limited usage. The element is employed in making sulfur hexafluoride (SF_6) for high-voltage insulation and for uranium hexafluoride. Fluorine is used directly or combined with higher metals (cobalt, silver, cerium, etc.) and halogens (chlorine and bromine) for organic fluorinations and the growing production of fluorocarbons. The largest production of fluorine compounds is that of hydrofluoric acid (anhydrous and aqueous), used in making "alkylate" for gasoline manufacture and Freon for refrigerants and aerosol bombs. It is also employed in the preparation of inorganic fluorides, elemental fluorine, and many organic fluorine- and non-fluorine-containing compounds. Aqueous hydrogen fluoride is used in the glass, metal, and petroleum industries and in the manufacture of many inorganic and acid fluorides. Three of the most unusual plastics known (Chap. 34) are Teflon, a polymerization product of tetrafluorethylene, and Kel F and Fluorothene, products formed by the polymerization of chlorofluorethylenes.

The production of hydrofluoric acid has increased rapidly to keep pace with these varied and expanding uses. In 1939, 7,421 tons were produced, in 1963, 131,050 tons,[17] and in 1974, 374,000 tons. Hydrofluoric acid sells for 37 cents in the anhydrous form and 41 cents as a 70% solution. About 45% of the hydrogen fluoride produced is used to prepare fluorocarbons, and one-third of the total goes to the aluminum industry, where synthetic cryolite, sodium aluminum fluoride, is a major constituent of the electrolyte. It is also consumed in the melting and refining of secondary aluminum. Other uses of hydrofluoric acid are found in the metals and petroleum industries.

MANUFACTURE *Fluorine gas* is generated by the electrolysis of KHF_2 under varying conditions of temperature and electrolyte composition. These variables often require cells constructed of different materials.[18] The process mentioned here is the one employed by the Pennsylvania Salt Co. This company operates commercial cells of 1,500 to 2,000 A per cell at 105 to 110°C, capable of producing 60 to 80 lb of fluorine per cell per day.

Both *aqueous* and *anhydrous hydrofluoric acid*[19] are prepared in heated kilns by the following endothermic reaction:

$$CaF_2 \text{ (fluorspar)} + H_2SO_4 \longrightarrow CaSO_4 + 2HF$$

The hot, gaseous hydrogen fluoride is either absorbed in water or liquefied; refrigeration is employed to obtain the anhydrous product needed for fluorocarbon manufacture and other uses. Although hydrofluoric acid is corrosive, concentrations of 60% and above can be handled in steel at lower temperatures; lead, carbon, and special alloys are also used in the process equipment.

FLUOROCARBONS These are compounds of carbon, fluorine, and chlorine with little or no hydrogen. Fluorocarbons containing two or more fluorines on a carbon atom are characterized by extreme chemical inertness and stability. Their volatility and density are greater than those of the corresponding hydrocarbons. More than 625 million lb of fluorocarbons was produced in 1971, of which dichlorodifluoromethane and trichlorofluoromethane Freons were the largest-volume products. About half of this quantity is used in the aerosol propellant market,[20] where fluorocarbons are solvents and pressure generators.

In 1974 scientists at the University of California at Irvine, and at the University of Michigan, hypothesized that chlorofluorocarbons used as aerosol propellants would eventually find their way to

[16]Finger, Recent Advances in Fluorine Chemistry, *J. Chem. Educ.* **28,** 49 (1951); ECT, vol. 6, pp. 656–771, 1965; Krespan, Organic Fluorine Chemistry, *Science,* **150**(3692), 13 (1965).

[17]*Chem. Eng. News.,* Jan. 6, 1975, p. 8.

[18]Symposium, Fluorine Chemistry, *Ind. Eng. Chem.,* **39,** 244 (1947); *Chem. Eng. Prog.,* **63**(6), 88 (1967).

[19]*Chem. Eng.* (*N.Y.*), **72**(2), 92 (1965).

[20]*Chem. Eng. News,* May 25, 1964, p. 26, Apr. 16, 1975, p. 32, May 19, 1975, p. 8, and May 26, 1975, p. 6; *Chem. Week,* June 11, 1975, p. 14.

the upper atmosphere. There they would react with the ozone layer that shields the earth from ultraviolet rays. A reduction in the ozone layer could allow more ultraviolet rays to reach the earth, causing a change in global weather patterns and an increase in skin cancer. This hypothesis has resulted in bills being introduced in Congress and in several state legislatures to prohibit the manufacture, sale, and use of aerosols containing chlorofluorocarbons. In 1975 Oregon passed such a bill. Study committees have been formed to determine the actual effect of these compounds on the ozone layer, but the "scare" caused a sharp decrease in the number of aerosol containers filled in 1976 as compared with 1974, and the decrease is continuing.

The other major use for refrigerants takes 44% of the production, and the remaining 6% goes into over 100 different important applications, including plastics, films, elastomers, lubricants, textile-treating agents, solvents, and fire extinguisher products. Fluorocarbons are made from chlorinated hydrocarbons by reacting them with anhydrous hydrogen fluoride, using an antimony pentachloride catalyst. For example,[21] the two Freons previously mentioned as being produced in the largest quantity are made as follows:

$$CCl_4 + HF \xrightarrow{SbCl_5} CCl_3F + HCl$$

$$CCl_4 + 2HF \xrightarrow{SbCl_5} CCl_2F_2 + 2HCl$$

Difluoromonochloromethane is made by substituting chloroform for the carbon tetrachloride:

$$CHCl_3 + 2HF \xrightarrow{SbCl_5} CHClF_2 + 2HCl$$

The fluorocarbon process shown in Fig. 20.4 may be divided into the following coordinated sequences:

Anhydrous hydrogen fluoride and carbon tetrachloride (or chloroform) are bubbled through molten antimony pentachloride catalyst. These reactions are slightly endothermic and take place in a steam-jacketed atmosphere-pressure reactor at 150 to 200°F (Ch).

The gaseous mixture of fluorocarbon and unreacted chlorocarbon is distilled to separate and recycle the chlorocarbon to the reaction (Op).

Waste hydrogen chloride is removed by absorbing in water (Op).

The last traces of hydrogen chloride and chlorine are removed in a caustic scrubbing tower (Op and Ch).

The product is dried by contacting with sulfuric acid (Op).

The fluorocarbon mixture is distilled in two columns, the first to obtain the more volatile dichlorodifluoromethane and the second to obtain monofluorotrichloromethane. Bottoms from the second column are recycled to the reactor (Op).

FLUOROSILICATES OR SILICOFLUORIDES An important potential source of fluorine compounds is phosphate rock which, in processing, yields recoverable fluorine by-products. In 1962 Stauffer Chemical announced plans for the production of hydrogen fluoride from this source.[22] The wet process for superphosphates (Chap. 16) evolves a toxic gaseous mixture of fluorine compounds, predominantly silicon tetrafluoride, hydrofluoric acid, and fluorosilicic acid (H_2SiF_6). This mixture is passed through water-absorption towers, yielding fluorosilicic acid and a silica precipitate. The H_2SiF_6 liquor is concentrated to commercial strengths (generally 30 to 35%) by recycling or distillation. An important use for this acid is in the fluoridation of municipal water supplies and also in the brewing industry as a disinfectant for copper and brass vessels. It is also employed as a preservative,

[21]U.S. Tariff Commission Reports, Synthetic Organic Chemicals, 1973; Fluorine-Chemicals Complex Getting a Big Boost, *Chem. Eng.* (*N.Y.*), **72**(2), 92 (1965) (flowchart).

[22]See Chap. 16, Phosphorus Industries.

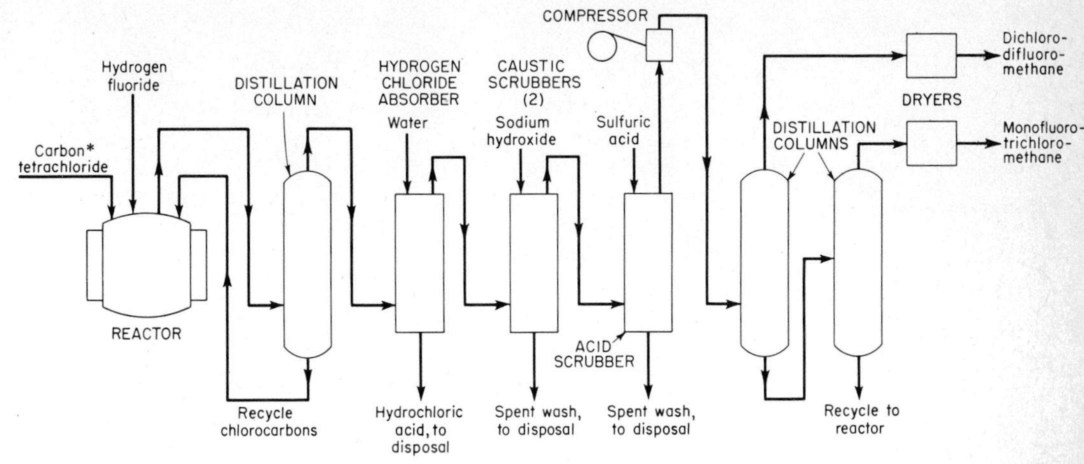

Fig. 20.4 Fluorocarbon production. [*Chem. Eng.* (*N.Y.*), ***72***(*2*), *93* (*1965*).]

in electroplating, as a concrete hardener, and in the manufacture of silicofluoride salts. The most common salt, *sodium silicofluoride* (Na_2SiF_6), is prepared by the reaction of sodium chloride or soda ash on the acid. It is consumed to the extent of 54,000 tons/year as an insecticide, laundry sour, fluxing and opacifying agent, and protective agent in the casting of light metals. Other salts such as ammonium, magnesium, zinc, copper, and barium are also prepared by neutralization and find many uses in industry.

ALUMINA

USES AND ECONOMICS Large quantities of alumina are produced yearly for the manufacture of metallic aluminum (Chap. 14). In 1972, 87% of the raw material, *bauxite*, was obtained from foreign sources.[23] Jamaica, Haiti, the Dominican Republic, Surinam, Guyana, Guinea, and Australia are the countries from which the United States imports this bauxite. Total consumption was 15.4 million long tons, about 94% of this going to alumina production; smaller uses include abrasives, chemical manufacture, refractories, and ceramic fibers.

MANUFACTURE Figure 20.5 shows the essential procedure in the production of alumina, which may be divided into the following steps:

Bauxite, a mineral containing about 55% aluminum oxide and less than 7% silica, is crushed and wet-ground to 100-mesh (Op).

The finely divided bauxite is dissolved under pressure and heat in Bayer digesters with concentrated spent caustic soda solution from a previous cycle and sufficient lime and soda ash (Ch). Sodium aluminate is formed, and the dissolved silica is precipitated as sodium aluminum silicate.

The undissolved residue (red mud) is separated from the alumina solution by filtration and washing and sent to recovery (Op). Thickeners and Kelly or Oliver filters are used.

The filtered solution of sodium aluminate is hydrolyzed to precipitated aluminum hydroxide by cooling (Ch and Op).

The precipitate is filtered from the liquor and washed (Op).

The aluminum hydroxide is calcined by heating to 1800°F in a rotary kiln (Ch).

[23]Minerals Yearbook 1972, vol. 1, pp. 189–192, Dept of the Interior, 1974.

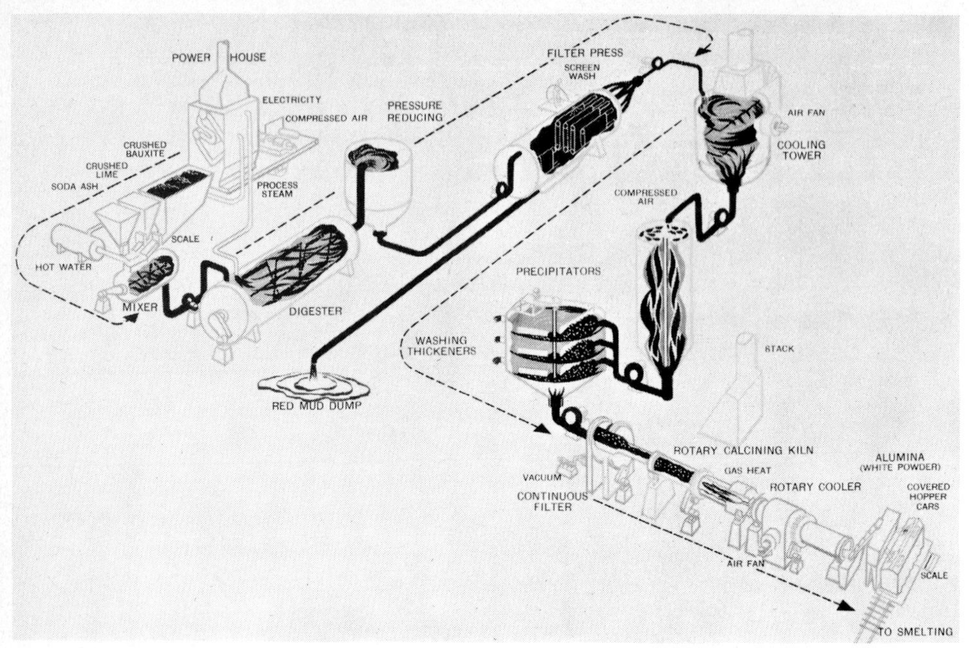

Fig. 20.5 Refining of alumina. (*Aluminum Co. of America.*)

The alumina is cooled and shipped to reduction plants (Op).
The dilute caustic soda filtered from the aluminum hydroxide is concentrated for reuse (Op).
The red mud may be reworked for recovery of additional amounts of alumina (Ch and Op).

Several new processes for producing alumina based on ores other than bauxite have been announced. A Mexican process (called the U.G. process) uses alunite, a hydrous sulfate of aluminum and potassium. It is claimed to be capable of producing 99% pure alumina from alunite containing only 10 to 15% alumina, compared with bauxite which assays 50% alumina. The alunite is crushed, dehydroxylated by heating to 1380°F, ground, and treated with aqueous ammonia. Filtration removes the alumina hydrate, and potassium and aluminum sulfates are recovered from the filtrate and sold as fertilizer. The alumina hydrate is treated with sulfur dioxide gas, and the resulting aluminum sulfate converted to alumina by heating in a kiln.[24]

The French Pechiney-Ugine Kuhlmann process treats clays and shales with concentrated sulfuric acid. Hydrochloric acid is added during the crystallization step to form aluminum chloride which crystallizes readily. Much raw material must be handled, because the clays and shales have a lower alumina content than bauxite.[25]

Other new methods involve the treatment of clays with nitric acid (Bureau of Mines) and the continuous electrolysis of aluminum chloride (Alcoa).

ALUMINUM SULFATE AND ALUMS

The manufacture of alums entails just one step additional to the aluminum sulfate process. This, along with the fact that the uses of aluminum sulfate and alums are similar and the compounds are

[24] Low-Grade Alunite Yields Alumina and Fertilizers Too, *Chem. Eng.* (*N.Y.*), **78**(9), 83 (1971).

[25] Alumina Producers Look to Alternate Raw Materials, *Chem. Eng.* (*N.Y.*), **81**(9), 98 (1974); *Chem. Week*, Mar. 6, 1974, p. 40.

largely interchangeable, justifies their concurrent discussion. The term *alum* has been very loosely applied. A true alum is a double sulfate of aluminum or chromium and a monovalent metal (or a radical, such as ammonium). Aluminum sulfate is very important industrially and, although it is not a double sulfate, it is often called either alum or papermakers' alum. Alum has been known since ancient times. The writings of the Egyptians mention its use as a mordant for madder and in certain medical preparations. The Romans employed it to fireproof their siege machines and probably prepared it from alunite ($K_2Al_6(OH)_{12}(SO_4)_4$), which is plentiful in Italy.

USES AND ECONOMICS Alums are used in water treatment and sometimes in dyeing. They have been replaced to a large extent in these applications by aluminum sulfate, which has a greater alumina equivalent per unit weight. Pharmaceutically, aluminum sulfate is employed in dilute solution as a mild astringent and antiseptic for the skin. The most important single application of it is in clarifying water, more than half of the total amount manufactured being so consumed. Sodium aluminate, which is basic, is sometimes used with aluminum sulfate, which is acid, to produce the aluminum hydroxide floc:

$$6NaAlO_2 + Al_2(SO_4)_3 + 12H_2O \longrightarrow 8Al(OH)_3 + 3Na_2SO_4$$

Second in importance is the application of aluminum sulfate in the sizing of paper. It reacts with sodium resinate to give the insoluble aluminum resinate. For the sizing of paper, aluminum sulfate must be free from ferric iron, or the paper will be discolored. The ferrous ions do no harm, since they form a soluble, practically colorless resinate, which, however, represents a loss of resinate. A small amount of aluminum sulfate is consumed by the dye industry as a mordant. Soda alum, or aluminum sulfate, is used in some baking powders. In 1974, 1,160,000 tons of commercial aluminum sulfate were produced.[26]

MANUFACTURE Practically all alums and aluminum sulfate are now made from bauxite by reaction with 60°Bé sulfuric acid. However, potash alum was first prepared from alunite by the ancients. Other possible sources of alums are shales and other alumina-bearing materials such as clay. Figure 20.6 illustrates the manufacture of aluminum sulfate.[27] The bauxite is ground until 80% passes 200-mesh; next it is conveyed to storage bins. The reaction occurs in lead-lined steel tanks, where the reactants are thoroughly mixed and heated with the aid of agitators and live steam. These reactors are operated in series. Into the last reactor barium sulfide is added in the form of black ash to reduce ferric sulfate to the ferrous state and to precipitate the iron. The mixture from the reactors is sent through a series of thickeners, operated countercurrently, which remove undissolved matter and thoroughly wash the waste so that, when discarded, it will contain practically no alum. The clarified aluminum sulfate solution is concentrated in an open, steam-coil-heated evaporator from 35 to 59 or 62°Bé. The concentrated liquor is poured into flat pans, where it is cooled and completely solidified. The solid cake is broken and ground to size for shipping. Another slightly modified procedure uses, instead of reactors and thickeners, combined reaction and settling tanks. The Dalecarlia rapid-sand filter plant, which supplies water to Washington, D.C., makes its own filter alum. In this case concentration of liquor would be an unnecessary expense; therefore the aluminum sulfate is made and used in a water solution. To make the various true alums, it is necessary only to add the sulfate of the monovalent metal to the dilute aluminum sulfate solution in the proper amount. Concentration of the mixed solution, followed by cooling, yields alum crystals.

[26] *Chem. Eng. News*, June 2, 1975, p. 33.

[27] Cf. Alum, *Chem. Eng.* (*N.Y.*), **57**(12), 172 (1950) (flowchart of American Cyanamid plant at Mobile, Ala.); Sheldon, Complete Alum Plant (process flowchart); *Chem. Eng.* (*N.Y.*), **68**(6), 132 (1961); *Chem. Eng. Prog.*, **70**(1), 55, (1974) (alum flowchart of Allied Chemical Corp.).

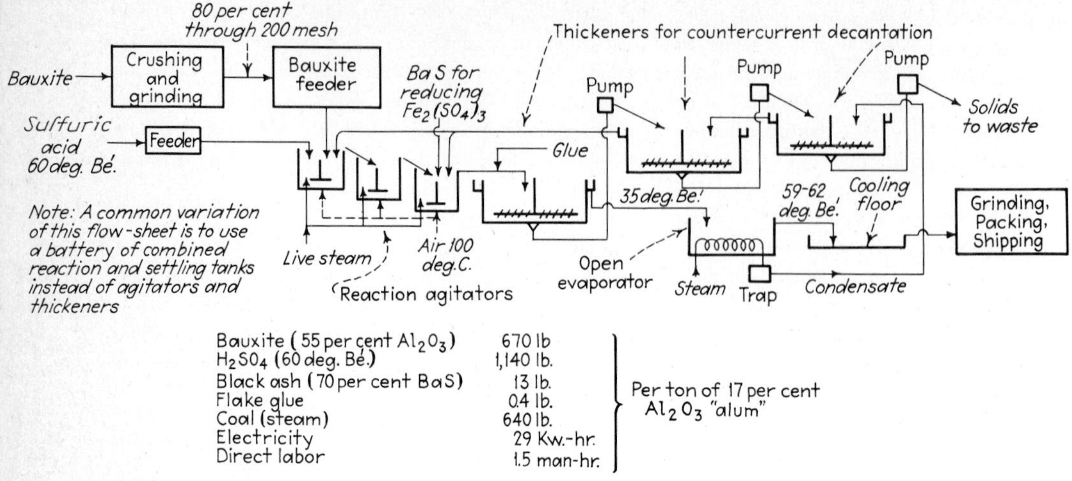

Fig. 20.6 Manufacture of aluminum sulfate by the Dorr procedure.

ALUMINUM CHLORIDE

Aluminum chloride is a white solid when pure. On heating it sublimes and, in the presence of moisture, anhydrous aluminum chloride partly decomposes with the evolution of hydrogen chloride. This salt was first prepared in 1825 by Oersted, who passed chlorine over a mixture of alumina and carbon and condensed the vapors of the aluminum chloride formed. Essentially, the same process is used today in the commercial preparation of aluminum chloride. The price in 1913 was $1.50 per pound, and the present price is 18 cents per pound in carload lots.

USES Since modern methods of manufacture have reduced the price of aluminum chloride, it has found application in the petroleum industries and various phases of organic chemistry technology. Aluminum chloride is a catalyst in the alkylation of paraffins and aromatic hydrocarbons by olefins and also in the formation of complex ketones, aldehydes, and carboxylic acid derivatives. In 1973, 36,800 tons of anhydrous aluminum chloride were manufactured. Anhydrous aluminum chloride is manufactured primarily by the reaction of chlorine vapor on molten aluminum.[28] Chlorine is fed in below the surface of the aluminum, and the product sublimes and is collected by condensing. These air-cooled condensers are thin-walled, vertical steel cylinders with conical bottoms. Aluminum chloride crystals form on the condenser walls and are periodically removed, crushed, screened, and packaged in steel containers.

COPPER SALTS

Copper sulfate[29] is the most important compound of copper, and more than 30,000 tons were produced in 1973. Commonly known as *blue vitriol,* it is prepared by the action of sulfuric acid on cupric oxide or sulfide ores. Its poisonous nature is utilized in the fungicide Bordeaux mixture, which is formed upon the mixing of copper sulfate solution with milk of lime. Copper sulfate is added to water reservoirs occasionally to kill algae. It is employed in electroplating and finds minor applica-

[28]ECT, 2d ed., vol. 2, p. 17, 1963.
[29]ECT, 2d. ed., vol. 6, p. 276–278; a flowchart is given in *Chem. Eng.* (*N.Y.*), **57**(11), 119 (1950).

tions as a mordant, germicide, and agent in engraving. Certain copper compounds are added to antifouling paints used on ship bottoms.

MOLYBDENUM COMPOUNDS

Although the largest use for molybdenum lies in metallurgy, chemical applications make up a rapidly growing segment of diversified uses. Approximately 15%[30] of the total molybdenum production of over 112 million lb/year goes into compounds. Molybdenum disulfide is dispersed in greases and oils for lubrication; in volatile carriers it is used to form dry coatings of lubricant. Sodium molybdate is an especially effective corrosion inhibitor on aluminum surfaces and is dissolved in cooling solutions to protect aluminum motor blocks in automobiles. Molybdenum salts used as catalysts include cobalt molybdate for hydrogen treatment of petroleum stocks for desulfurization, and phosphomolybdates to promote oxidation. Compounds used for dyes are sodium, potassium, and ammonium molybdates. With basic dyes, phosphomolybdic acid is employed. The pigment known as *molybdenum orange* is a mixed crystal of lead chromate and lead molybdate. Sodium molybdate, or molybdic oxide, is added to fertilizers as a beneficial trace element. Zinc and calcium molybdate serve as inhibitory pigments in protective coatings and paint for metals subjected to a corrosive atmosphere. Compounds used to produce better adherence of enamels are molybdenum trioxide and ammonium, sodium, calcium, barium, and lead molybdates. The mineral raw material is molybdenite, which, by roasting, furnishes a technical grade of molybdenum trioxide of 80 to 90% purity.

BARIUM SALTS

The most common naturally occurring barium compounds are the mineral carbonate, or witherite, which is fairly abundant in England, and the sulfate, or barite, which is common in certain sections of the United States. The most important source of barite is Arkansas.

USES The applications of barium compounds are varied and important. Each year over 900,000 tons of barite are consumed. *Barium carbonate* is sometimes employed as a neutralizing agent for sulfuric acid and, because both barium carbonate and barium sulfate are insoluble, no contaminating barium ions are introduced. The foregoing application is found in the synthetic dyestuffs industry. The glass industry uses 45% of the barium carbonate produced, and bricks and clay, 25%.[31] Witherite is used chiefly to prepare other compounds. *Barium sulfate* is a useful white pigment (Chap. 24), particularly in the precipitated form, blanc fixe. It is used as a filler for paper, rubber, linoleum, and oilcloth. Because of its opacity to x-rays, barium sulfate, in a purified form, is important in contour photographs of the digestive tract. The paint industry is the largest single consumer of barium compounds. Barium sulfide and zinc sulfate solutions are mixed to give a precipitate of barium sulfate and zinc sulfide, which is given a heat treatment to yield the cheap but good pigment *lithopone*, as described in Chap. 24. Barium chlorate and nitrate are used in pyrotechnics to impart a green flame. Barium chloride is applied where a soluble barium compound is needed. Barium saccharate is used by a large beet-sugar company in recovering sugar from discarded molasses (Chap. 30).

MANUFACTURE The preparation of soluble barium salts is simple where witherite is available. The only steps necessary are treatment with the proper acid, filtration to remove insoluble impurities, and crystallization of the salt. Since there is little witherite in the United States, barium salts are

[30] *Ind. Eng. Chem.*, **53,** 29A (1961); Minerals Yearbook 1972, vol. 1, p. 797, Dept of the Interior, 1974.
[31] *Chem. Mark. Rep.*, Mar. 3, 1975, p. 29.

prepared from barite. The high-temperature reduction of barium sulfate with coke yields the water-soluble barium sulfide, which is subsequently leached out. The treatment of barium sulfide with the proper chemical yields the desired barium salt. Purification of the product is complicated by the impurities introduced in the coke. Pure barium carbonate and barium sulfate are made by precipitation from solutions of water-soluble barium salts. Much barite is ground,[32] acid-washed, lixiviated, and dried to produce a cheap pigment or paper or rubber filler, or changed to blanc fixe.

STRONTIUM SALTS

Uses of strontium salts are small in tonnage but important; they include red-flame pyrotechnic compositions, such as truck signal flares and railroad "fusees," tracer bullets, military signal flares, and ceramic permanent magnets. Low-grade strontium deposits are available in this country, but are not currently in use; high-grade celestite is imported from Mexico and the United Kingdom. This ore (strontium sulfate) is finely ground and converted to the carbonate by boiling with 10% sodium carbonate solution, giving almost a quantitative yield:

$$SrSO_4 + Na_2CO_3 \longrightarrow SrCO_3 + Na_2SO_4$$

From reaction of the strontium carbonate with appropriate acids, the various salts result.[33]

LITHIUM SALTS

During World War II lithium hydride was reacted with sea water to produce a convenient lightweight source of hydrogen for the inflation of air-sea rescue equipment. Interest in lithium compounds grew from 1954 to 1960, when the Atomic Energy Commission (AEC) bought vast quantities of lithium hydroxide, presumably for use in the production of hydrogen bombs.[34] Termination of these contracts left a substantial excess, which in turn has fostered new commercial uses. The largest known lithium ore reserve in the world is in the King's Mountain district of North Carolina. Currently a large part of the free world's output of lithium oxide comes from this source as spodumene ($Li_2O \cdot Al_2O_3 \cdot 4SiO_2$). Crude dilithium sodium phosphate is a coproduct in the manufacture of potash, borate, and soda chemicals from the brines of Searles Lake, Calif. (Fig. 17.1). This compound is converted to lithium carbonate, which is an economical source of the oxide.

USES Lithium carbonate, the most widely used of the compounds, is employed in the production of lithium metal and frits and enamels. Together with lithium flouride, it serves as an additive for cryolite in the electrolytic pot-line production of primary aluminum. Lithium-base greases, often the stearate, are a large outlet. These lubricants are efficient over the extremely wide temperature range -60 to 320°F. Lithium hydroxide is a component of the electrolyte in alkaline storage batteries and is employed in the removal of carbon dioxide in submarines and space capsules. Lithium bromide is used for air conditioning and dehumidification. The hypochlorite is a dry bleach used in commercial and home laundries. Lithium chloride is in demand for low-temperature batteries and for aluminum brazing. Other lithium-compound uses include catalysts, glass manufacture, and of course nuclear energy. A newer use for lithium carbonate is as a drug to prevent manic depression. The Food and Drug Administration (FDA) approved its use for treating mind disorders in 1970.[35]

[32]ECT, 2d ed., vol. 3, pp. 80–98, 1964; White, The Plant That Barite Built, *Chem. Eng.*, (*N.Y.*), **56**(4), 90 (1949) (pictured flowchart pp. 128–131); Farr, U.S. Pat. 1,752,244 (1930); Shreve *et al.*, U.S. Pat. 2,030,659 (1936).

[33]Strontium Chemicals, *Chem. Eng.* (*N.Y.*), **53**(1), 152 (1946) (pictured flowchart); Strontium, New Force in Magnets, *Chem. Week*, July 6, 1968, p. 27.

[34]*Baron's*, June 8, 1964, p. 5; New Plant Pulls Lithium from Nevada Brines, *Chem. Eng. News*, Aug. 8, 1966, p. 38. The lithium hydroxide for sale by AEC is devoid of lithium-6 isotope.

[35]*Chem. Week*, Sept. 23, 1967, p. 26; and Mar. 8, 1972, p. 41.

MANUFACTURE[36] Since spodumene is by far the most important ore, the manufacture of lithium carbonate from it is presented. Spodumene ore (beneficiated to 3 to 5% Li_2O) is converted from the alpha form to the beta form by heating to over 1000°C. The latter is treated as follows:

$$Li_2O \cdot Al_2O_3 \cdot 4SiO_2 + H_2SO_4 \xrightarrow{250\text{–}300°C} H_2O \cdot Al_2O_3 \cdot 4SiO_2 + Li_2SO_4$$

The water-soluble lithium sulfate is leached out and reacted with sodium carbonate to yield lithium carbonate. Various salts are derived from the carbonate as follows:

$$Li_2CO_3 + Ca(OH)_2 + 2H_2O \longrightarrow CaCO_3 + 2LiOH \cdot H_2O \xrightarrow{\Delta} 2LiOH$$

$$Li_2CO_3 + 2HCl \longrightarrow H_2O + CO_2 + 2LiCl(\mathit{soln}) \xrightarrow{\Delta} 2LiCl(\mathit{anhyd})$$

$$2LiCl(\mathit{anhyd}) \xrightarrow{\text{electrolysis}} 2Li + Cl_2\uparrow \xrightarrow{H_2} 2LiH \xrightarrow{2NH_3} 2LiNH_2 + 2H_2\uparrow$$

BORON COMPOUNDS

The important naturally occurring ores of boron have been colemanite ($Ca_2B_6O_{11} \cdot 5H_2O$), tincal ($Na_2B_4O_7 \cdot 10H_2O$), and boracite ($2Mg_3B_8O_{15} \cdot MgCl_2$). However, boron-containing brines and kernite, or rasorite ($Na_2B_4O_7 \cdot 4H_2O$), are the present sources in the United States. Most of the output is from Kern and San Bernardino Counties, Calif.

USES The most important compound for industrial use is borax ($Na_2B_4O_7 \cdot 10H_2O$). It is widely used as a flux, in the manufacture of Pyrex glass, in the making of glazes and enamels, in the preparation of agricultural chemicals, and soaps, and in other miscellaneous processes. Boric acid (H_3BO_3) is a weak acid that is applied medicinally as a mild antiseptic and also finds some consumption in the manufacture of glazes and enamels for pottery. Boron carbide (B_4C) is the hardest artificial abrasive known.

MANUFACTURE Most of the world's *borax* comes from Searles Lake and various other deposits in southern California. Figures 17.1 and 17.2 outline this recovery based on operations of the Kerr-McGee Chemical Corp. (formerly American Potash and Chemical Corp.), which in 1963 received the Kirkpatrick award[37] in chemical engineering achievement. This process is based on the now well-known use of an organic solvent[38] to extract inorganic chemicals. The plant end liquors, still containing borates are combined with Searles Lake brines (1.7% sodium borates). From this mixture the boric acid is extracted with kerosine, carrying a patented chelating agent not yet disclosed but probably an aliphatic medium-chain-length polyol, in a mixer-settler system. In a second mixer-settler system, dilute sulfuric acid strips the borates from the chelate. The aqueous phase with boric acid, potassium sulfate, and sodium sulfate is purified by carbon treatment and evaporated in two evaporator-crystallizers; from the first, pure boric acid is separated, and from the other, a mixture of sodium and potassium sulfates. This facility can handle up to 400 gal/min. *Kernite*[39] mixed with borax is mined by the U.S. Borax and Chemical Corp. at Boron, Calif., and refined there and at Wilmington, Calif.

[36]Hader *et al.*, Lithium and Its Compounds, *Ind. Eng. Chem.*, **43**, 2636 (1951) (excellent article with flowchart); *Chem. Eng.* (*N.Y.*), **63**(3), 288 (1956).

[37]Havinghorst, AP&CC's New Process Separates Borates from Ore by Extraction, *Chem. Eng.* (*N.Y.*), **70**(23), 228 (1963) (flowchart); Chelating Agent Used to Extract Boric Acid, *Chem. Eng. News*, Oct. 7, 1963, p. 44; Solvent Extraction Gains, *Chem. Eng. News*, Apr. 6, 1964, p. 48.

[38]Shreve, Organic and other Non-aqueous Solvents as Processing Media for Making Inorganic Chemicals, *Ind. Eng. Chem.*, **49,** 836 (1957); It Knocked Boric Acid Costs, *Chem. Eng.* (*N.Y.*), **73**(8), 55 (1966).

[39]Bixler and Sawyer, Boron Chemicals from Searles Lake Brines, *Ind. Eng. Chem.*, **49**(3), 323 (1957) (mentions all companies and deposits).

RARE-EARTH COMPOUNDS

Chemical elements with atomic number 58 to 71 are so nearly identical in properties that they are separated only with difficulty and are called as a group *rare earths.*[40] The name is a misnomer, since they are neither rare nor earths. Compounds of lanthanum, yttrium, and scandium are often included in this group, as are the actinide rare earths (elements 89 and above). Cerium and thorium are the most important commercially. Although rare earths occur in certain minerals native to the Carolinas, these minerals have not been able to compete with the richer monazite sands found in Brazil and in Travancore-Cochin, India. Preliminary separation is accomplished by the use of jigging, in this case the valuable material is denser than the worthless fraction. The concentrates are leached with hot sulfuric acid of strength greater than 66°Bé, and the rare-earth metal compounds are precipitated by dilution of the sulfuric acid solution. The usual practice is to isolate thorium and cerium hydroxide, the latter containing all the other rare earths except thorium unless careful and tedious purifications are carried out.

The Molybdenum Corp. of America has developed a solvent extraction process for the production of the oxides of lanthanum, praesodymium, neodymium, yttrium, and europium in tonnage quantities. The ore, mined in California, is concentrated by hot-froth flotation and then subjected to solvent extraction.[41]

The glass industry is the largest consumer of rare-earth compounds, mostly as oxides for rapid polishing of plate glass, as well as precision optical equipment and eyeglasses. Lesser quantities of compounds are used for coloring, decoloring, and opacifying glass. Newer uses are in ultraviolet-absorbing glass, in certain high-lead glass with increased radiation stability, and in color television tubes. Fluorides and oxides are consumed in the production of carbon electrodes for arc lighting of high intensity and good color balance. Rare-earth compounds, particularly silicides, find use in the production of alloy metals. Thorium is of interest in nuclear reactions.

SODIUM DICHROMATE

The starting material for the manufacture of sodium dichromate and other chromium compounds is chromite, a chromium iron oxide containing approximately 50% Cr_2O_3, the balance being principally FeO, Al_2O_3, SiO_2, and MgO. There are no high-grade chromite deposits in the United States, and most of the ore used in the chemical industry is imported from the U.S.S.R. In war years it was proposed that domestic ore[42] should be used. Much sodium dichromate is consumed as the starting material for making, by glucose reduction, the solutions of chromium salts employed in chrome leather tanning (Chap. 25) and in chrome mordant dyeing of wool cloth. Certain pigments, such as yellow lead chromate, are manufactured basically from sodium dichromate, as are also green chromium oxides for ceramic pigments. Over half of the chromium enters the metal field as stainless steel and other high-chromium alloys and for chromium plating of other metals, the balance being about equally divided between chrome refractories and chrome chemicals.

The ore is ground to 200-mesh, mixed with ground limestone and soda ash, and roasted at approximately 2200°F in a strongly oxidizing atmosphere. The sintered mass is crushed and leached with hot water to separate the soluble sodium chromate. The solution is treated with enough sulfuric acid to convert the chromate to dichromate, with the resulting formation of sodium sulfate. Most of the sodium sulfate crystallizes in the anhydrous state from the boiling-hot solution during acidification, and the remainder drops out in the evaporators on concentrating the dichromate solution. From the

[40] McGraw-Hill Encyclopedia, vol. 11, p. 342, 1966.
[41] *Chem. Eng.* (*N.Y.*), **74**(1), 32 (1967) and **74**(24), 122 (1967); *Chem. Week.* Dec. 2, 1967, p. 51.
[42] Producing Chromate Salts from Domestic Ores, *Chem. Metall. Eng.*, **47**, 688 (1940) (flowchart).

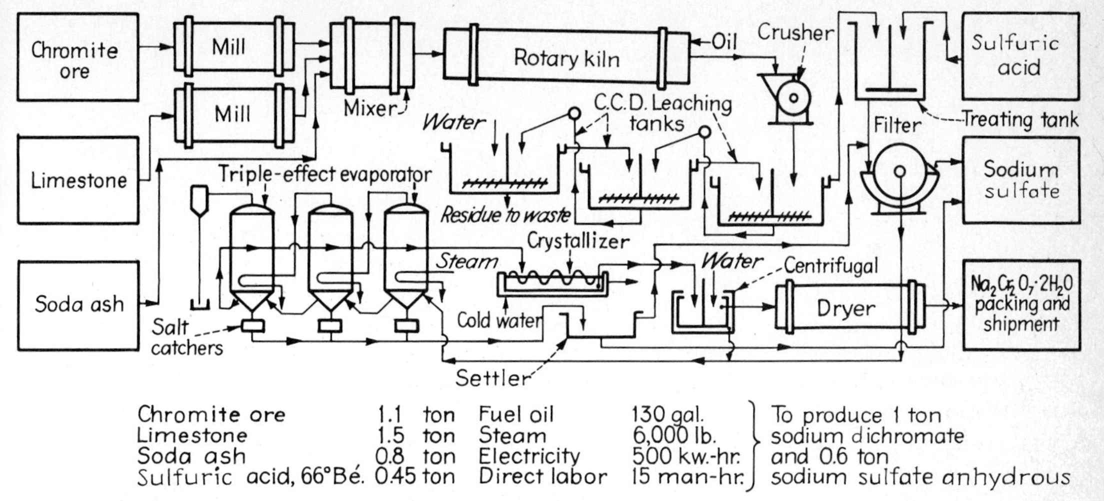

Fig. 20.7 Flowchart for manufacture of sodium dichromate from chromite ore.

evaporator the hot, saturated dichromate solution is fed to the crystallizer, and then to the centrifuge and dryer (Fig. 20.7).

HYDROGEN PEROXIDE

Hydrogen peroxide is the most widely used peroxide compound. Originally, it was produced by the reaction of barium peroxide and sulfuric acid for use as an antiseptic, but this process and use have been superseded.

USES Hydrogen peroxide applications include commercial bleaching-dye oxidation, the manufacture of organic and peroxide chemicals, and power generation. Bleaching outlets consume more than one-half of the hydrogen peroxide produced. These include bleaching of wood, textile-mill bleaching of practically all wood and cellulosic fibers, as well as of major quantities of synthetics, and paper- and pulp-mill bleaching of groundwood and chemical pulps. The advantage of hydrogen peroxide in bleaching is that it leaves no residue and generally results in an excellent product whiteness, with little or no *deterioration* of the organic matter that is bleached. Organic applications include manufacture of epoxides and glycols from unsaturated petroleum hydrocarbons, terpenes, and natural fatty oils. The resultant products are valuable plasticizers, stabilizers, diluents, and solvents for vinyl plastics and protective-coating formulations. The 90 to 100% concentrations represent a self-contained energy source which leaves no residue or corrosive gas, ideal for specialized propulsion units for aircraft, missiles, torpedos, and submarines. Hydrogen peroxide is used in liquid-fuel rockets as an oxidizing agent and also as a monopropellant. A new use is as a source of oxygen in municipal and industrial wastewater treatment systems.[43]

MANUFACTURE Hydrogen peroxide is manufactured by electrolytic and two organic oxidation processes.[44] In the electrolytic methods the active oxygen is produced by the anodic oxidation of sulfate radicals to form peroxydisulfate intermediates. Sulfuric acid electrolyte has low current efficiency (70 to 75%), but the use of ammonium sulfate causes crystallization problems, so a mixture

[43]*Du Pont Innovation,* **5**(1), 5 (1973).
[44]McGraw-Hill Encyclopedia, vol. 10, p. 28, 1966.

is used to obtain a current efficiency of 80% or higher and yet not block the cell with crystal formation. The electrolyte is fed into a typical cell held at 35°C or below with platinum metal for the anode:

$$2H_2SO_4 \rightleftharpoons H_2S_2O_8 + H_2$$

The cell product, i.e., the persalt, or peracid, and sulfuric acid in water are subjected to hydrolyses at 60 to 100°C to yield hydrogen peroxide, as shown by the following simplified reactions:

$$H_2S_2O_8 + H_2O \rightleftharpoons H_2SO_4 + H_2SO_5$$

$$H_2SO_5 + H_2O \rightleftharpoons H_2SO_4 + H_2O_2$$

One organic oxidation process uses an anthraquinone dissolved in organic solvent.[45] This quinone is catalytically hydrogenated to the hydroquinone; subsequent aeration produces hydrogen peroxide and regenerates the quinone. The reactions can be illustrated with the easily oxidizable 2-ethylanthraquinone:

O, O, Et $\xrightarrow{+H_2}$ OH, OH, Et $\xrightarrow{+O_2}$ O, O, Et $+ H_2O_2$

The hydrogen peroxide is water-extracted and concentrated, and the quinone is recycled for reconversion to the hydroquinone. A second organic process uses liquid isopropyl alcohol, which is oxidized at moderate temperatures and pressures to hydrogen peroxide and an acetone coproduct:

$$(CH_3)_2CHOH + O_2 \longrightarrow (CH_3)_2CO + H_2O_2$$

After distillation of the acetone and unreacted alcohol, the residual hydrogen peroxide is concentrated.

SELECTED REFERENCES

Chambers, R. D.: Fluorine in Organic Chemistry, Wiley-Interscience, 1973.
Eyring, L. (ed.): Chemistry, Technology, Rare Earths, Pergamon, 1964, 2 vols.
Gould, R. F. (ed.): Literature of Chemical Technology, chap. 38, Aerosols, ACS Monograph, 1967.
Hanson, L. P.: Organoboron Technology, Noyes, 1973.
Hydrochloric Acid Pickle Liquor Regeneration, *Chem. Eng.* (*N.Y.*), **74**(18), 32 (1966).
Jamrock, W. D.: Rare Metal Extraction by Chemical Engineering Techniques, Macmillan-Pergamon, 1963.
Jolles, Z. E. (ed.): Bromine and Its Compounds, Academic, 1966.
Kohn, J. A., *et al.*: Boron: Synthesis, Structure, and Properties, Plenum, 1960.
Parr, N. L.: Zone Refining and Allied Techniques, St. Martin's, 1960.
Pavlath, A. E., and A. L. Leffler: Aromatic Fluorine Compounds, ACS Monograph 155, Reinhold, 1962.
Rare-earth Source: Phosphate Rock, *Chem. Eng. News,* Oct. 24, 1966, p. 52.
Rare-earth Industry Expanding, *Chem. Eng. News,* June 12, 1967, p. 42.
Rare-earth Technology, Blind River, *Chem. Eng.* (*N.Y.*), **74**(1), 32 (1967).
Rickles, R.: Future Inorganic Growth 1970–1980, Noyes, 1967.
Rudge, A. J.: Manufacture and Use of Fluorine and Compounds, Oxford, 1962.
Saunders, B. C.: Phosphorus and Fluorine, Cambridge, London, 1957.
Sittig, M.: Fluorinated Hydrocarbon Polymers, Noyes, 1967.
Spent Pickling Liquor Recover, *Chem. Eng. News,* Nov. 14, 1966, p. 58.
Stacey, M., *et al.* (eds.): Advances in Fluorine Chemistry, Butterworth, London, 1960–1963, 3 vols.
Steinberg, H.: Organoboron Chemistry, vol. 1, Boron-Oxygen and Boron-Sulfur Compounds, Wiley-Interscience, 1964.
Wartik, T.: Borax to Boranes, ACS, 1962.
Weiss, S.: Hydrofluoric Acid Manufacture, Noyes, 1972.

[45] *Hydrocarbon Process. Pet. Refiner.*, **40**(11), 255 (1961).

chapter 21

NUCLEAR INDUSTRIES

In the early years of this century, Wilhelm Ostwald, a famed physical chemist, was lecturing at Harvard, elaborating on his book *Naturphilosophie*. Ostwald emphasized that *energy*, and not matter, was the fundamental basis of everything. Then, and for decades afterward, conservation of mass, as well as conservation of energy, were taught in colleges. Becquerel discovered radioactivity in 1898, and Richards of Harvard, in the early 1900s, determined the atomic weights of lead from different ores with the then unexplained differences much beyond experimental error. (These differences are now explained by lead isotopes.)

Reactions involving the nucleus of the atom showed a startled scientific world how to transform matter into energy. This work was started and developed in various physical and chemical laboratories in Europe and then in the United States.[1] Fission of the atom was demonstrated by Fermi, followed by Meitner and Frisch. Fission into two nearly equal fragments is accompanied by the emission of neutrons, gamma rays, and a tremendous amount of energy, and the simultaneous disappearance of a corresponding amount of mass. Climaxing these discoveries was the first nuclear *chain reaction* under the direction of Fermi at the University of Chicago on Dec. 2, 1942, using a nuclear reactor, or *pile*, of graphite with a critical mass of a few tons of extreme-purity uranium and uranium oxide. The uranium dioxide used and that converted to metal were manufactured by Mallinckrodt Chemical Works, employing ether purification of the nitrate. Westinghouse converted part of the uranium dioxide to the metal. This discovery and control of a chain of self-sustaining fission reactions based on uranium led to nuclear explosions of hitherto unheard of violence. Research work was conducted secretly before and during World War II and culminated before a startled world in the atomic bombs dropped on Hiroshima and Nagasaki (Fig. 21.1) and the end of World War II.

Some pessimists, for decades before there was any inkling of nuclear reactions, had predicted that when petroleum and coal were exhausted in the centuries ahead, humans would be forced to move to the warmer regions of the earth. Now, with nuclear reactions controlled by humans, habitations employing these reactions for heat and power are comfortable even in the arctic zones. Of more importance to future centuries will be the use of nuclear reactions[2] for heat and power throughout the world. The simultaneous saving of petroleum and coal will permit their continuance and enlargement as raw materials for the chemical process industries, which supply countless synthetic products, mostly organic, upon which the comforts, health, and continuance of human life depend. Synthetic fibers, rubber, plastics, medicines, fertilizers, dyes, paints and pigments, and detergents are only a few of the end uses of petroleum and coal. Obtaining energy from nuclear reactions should consequently be the principal objective of the nuclear industry, coupled with obtaining, and applying to health and to industry, radiation and radioisotopes.

[1]Harrington and Ruehle, Uranium Production Technology, p. 4, Van Nostrand, 1959.

[2]ECT (under Nuclear, an overall presentation of the fundamentals of nuclear reactions, nuclear power, radiation protection, and references).

Fig. 21.1 Uranium nuclear weapon of the Little Boy type, the kind detonated over Hiroshima, Japan, Aug. 6, 1945, in World War II. The bomb was 28 in. in diameter and 120 in. long. The first nuclear weapon ever detonated, it weighed about 9,000 lb and had a yield equivalent to approximately 20,000 tons of TNT. (*Los Alamos Scientific Laboratory.*)

There is an acute energy shortage in the world at present, and difficulty in obtaining sufficient supplies of oil, coal, and gas. Energy obtained through nuclear fission can greatly reduce the energy shortage. In the United States[3] there are 55 licensed nuclear reactors with a capacity of 36,668 MW, 76 units being built with a capacity of 76,887 MW, and 105 planned units with a capacity of 121,136 MW. In 1972, the United States used 7.2×10^{16} Btu and is projected to use 12.5×10^{16} Btu in 1980. Nuclear energy supplied 1.4% of this in 1972 and is expected to supply 15.2% in 1980.

Despite the remarkable safety record of the nuclear industry, fears of a possible catastropic accident and worries concerning the long-time storage of slow-decaying radioactive by-products have led to considerable opposition to nuclear plant construction and operation. This opposition[4] has severely hampered full exploitation of this resource.

NUCLEAR REACTIONS

RADIUM AND URANIUM The discovery of radium and its isolation by the Curies was one of the greatest events of all time. Marie Curie noticed that certain samples of uranium ore exhibited greater radioactivity than was to be expected from the uranium content. After a long series of concentrating operations based upon partial crystallization, the Curies discovered and isolated polonium first and then, in 1898, radium. The most important single application of radium has been the use of its radioactivity for curative purposes in cancer and certain other morbid conditions. However, particular isotopes made in uranium reactors are cheaper and more effective for medical radiological effects. This is shown strikingly by the fact that in obtaining uranium from pitchblende ores, the radium (1 part to each 3 million parts of uranium) is first separated from the nitric acid digestion liquors and dumped into storage for possible future use. The removal of the radioactive radium and its daughter products enables this uranium ore to be processed openly and safely, as presented below under Feed-materials Production.

FISSION REACTION[5] Matter can be transformed by a chain reaction into energy according to Einstein's law, $E = mc^2$, where E is the energy from the mass m that disappears, multiplied by the square of the velocity of light c^2. The energy released in fission is 8.20×10^{17} ergs/g of ^{235}U, or 3.527×10^{10} Btu/lb (or 3×10^6 the heating value of an equal weight of coal).

[3]As of Mar. 31, 1975, Environmental Research and Development Agency (ERDA).
[4]The Nuclear Safety Controversy, *Chem. Eng. Prog.*, **70**(11), 26 (1974).
[5]Benedict and Pigford, Nuclear Chemical Engineering, pp. 4–6, McGraw-Hill, 1957.

^{235}U or plutonium was used for atomic bombs in World War II, which exerted initially the destructive force of 20,000 tons of TNT, by converting only a small fraction of the mass of ^{235}U or plutonium into energy. Since then energy release from weapons has been greatly increased. The *fission* reaction is brought about through the reaction of neutrons and the nuclei of ^{235}U (or of plutonium) atoms. About 30 different radioactive elements of smaller atomic weight are formed, together with more neutrons, heat, and radiation. If the amount of the fissionable ^{235}U or plutonium is above a minimum critical mass, a chain reaction results from the action of the released neutrons, and exponential fission occurs. This nuclear reaction may be expressed:

$$^{235}U + \text{neutron} \longrightarrow \text{fission products} + \text{energy} + 2.43 \text{ neutrons}$$

Of the 2.43 neutrons released, 1 is needed to continue the chain reaction, and the others may bring about other nuclear reactions or be lost through leakage or through capture by elements present in the reactor. This reaction normally produces 2 fission products and energy of 3.0×10^{-4} erg. Naturally occurring uranium contains 0.71% ^{235}U, the rest being ^{238}U with a small trace of ^{234}U. To sustain the chain reaction either in a bomb or in a reactor to obtain usable energy, the ^{235}U content is increased by isotope concentration. These nuclear explosions due to fission chain reactions require slowing down or moderation of the neutrons before most reactions can occur.

The isotopes ^{238}U and ^{232}Th are called fertile materials because they can furnish useful fissionable isotopes by capture under neutron irradiation. ^{233}U and ^{239}Pu are formed in the reactors and, being fissionable, can profitably be used to extend the supply of fissionable isotopes for nuclear power. *Fertile material* is that which can be converted to useful fissionable isotopes by neutron capture; e.g., ^{232}Th gives fissile ^{233}U. *Fissile material* is that which can be caused to fission, or split, into two or more approximately equal parts.

Proper reactor design permits the use of excess neutrons beyond those necessary to maintain a steady fission rate, to convert nonfissile ^{238}U to fissile ^{239}Pu. This allows use of all the mined uranium in fission reactors. Without the conversion, only the ^{235}U isotope would be useful. When a reactor produces more fuel than it uses, it is called a *breeder.*

PLUTONIUM To supply plutonium for fission reactions, the uranium as mined must be isolated and purified as outlined under Feed Materials Production. ^{238}U undergoes thermal neutron (n) capture, with subsequent beta decay to ^{239}Np and then to ^{239}Pu:

$$^{238}_{92}U + {}^{1}_{0}n \longrightarrow {}^{239}_{92}U \xrightarrow[23 \text{ min}]{\beta,\gamma} {}^{239}_{93}Np \xrightarrow[2.3 \text{ days}]{\beta,\gamma} {}^{239}_{94}Pu$$

The plutonium isotope has a long half-life (24,000 years) and is quite important, since it has a low critical mass for fast fission, a property desirable in military applications. The neutrons needed for the production of plutonium are supplied by the fissionable ^{235}U and generate a self-sustaining reaction. Plutonium is manufactured[6] by the AEC according to this reaction from ^{238}U at its billion-dollar Hanford and Savannah River plants. Plutonium is produced as a by-product of the present generation of light-water power reactors.

FUSION REACTIONS lead to the production of energy by the conversion of hydrogen to helium. Indeed, maintenance of the heat of the sun and the stars depends on energy released during the fusion of light atoms, for example, from the matter lost when hydrogen isotopes are converted to helium by thermonuclear or fusion reactions at extremely high temperatures:[7]

$$4\,^{1}H \longrightarrow {}^{4}He + 26.7 \text{ meV or } 4.3 \times 10^{-5} \text{ erg}$$

^{2}H–^{2}H and ^{3}H–^{2}H reactions also appear promising for fusion power reactions. The large amount of energy released stems from the small fractional differences (0.7%) between the mass of the helium atom and the mass of four hydrogen atoms. The actual nuclear fusion is more complicated

[6]*Chem. Eng.* (*N.Y.*), **70**(3), 134 (1963).

[7]Fowler, Origin of the Elements, *Chem. Eng. News*, Mar. 16, 1964, pp. 91, 95.

TABLE 21.1 *U.S. U_3O_8 Mine Production, by States (In thousands of pounds)*

State	1970	1971	1972	1973	1974
New Mexico	11,574	10,567	10,808		
Wyoming	6,346	6,986	8,544		
Colorado	2,727	2,536	1,877		
Utah	1,635	1,445	1,496		
Alaska, South Dakota, Texas, and Washington	2,094	2,981	3,033		
Total	24,376	24,515	25,758	26,000*	28,500*

Source: Minerals Yearbook 1972, Dept. of the Interior, 1974. *Note:* Excludes recoverable uranium in leach, mine water, raffinate, etc. *Estimated, see *Chem. Eng. News,* May 13, 1974, p. 8.

than the simple reaction given, and the energy released in ordinary units is 100 million kWh/lb of hydrogen converted. The *nuclear* burning of hydrogen in the sun has lasted for 4.5 billion years and will probably last as long in the future. Short bursts of power have been obtained in an advanced fusion apparatus, but a reaction sustained long enough to permit significant power production has not yet been achieved.

RESERVES OF URANIUM (AND THORIUM) In the early years of the nuclear industries, the uranium ore used was pitchblende from northern Canada, and then African ores of various minerals. In later years, larger supplies of uranium ores came from the Colorado plateau of the United States and from Canada, mostly from the Blind River area north of Lake Huron. Table 21.1 summarizes the domestic uranium ore situation.

After a period in which the price of U_3O_8 fell to \$5 per pound, it started to rise slowly and in 1975 was slightly over \$10 per pound. Fuel costs were rising in 1976 and may be much higher by

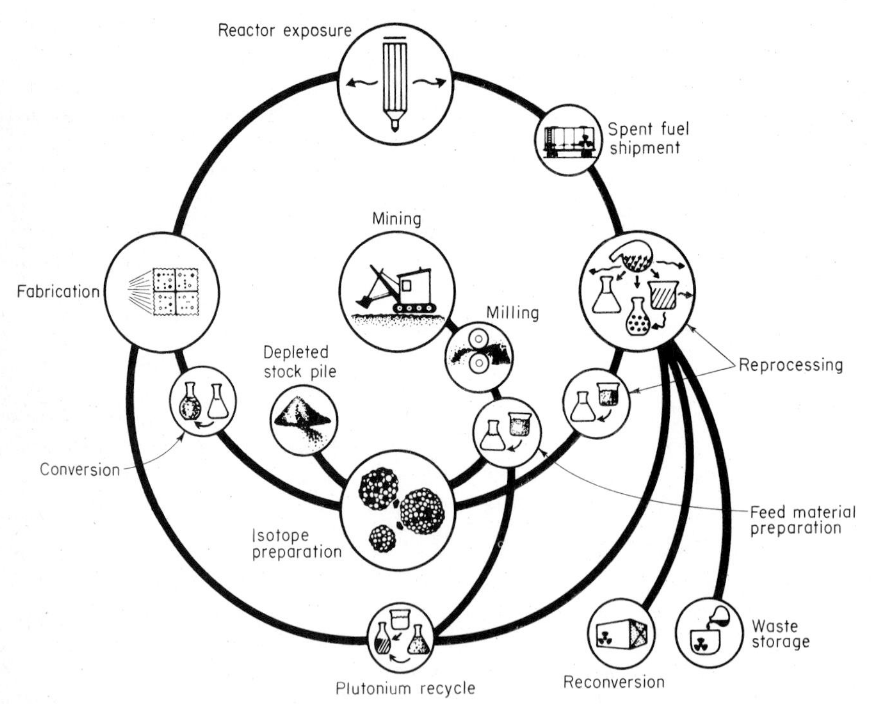

Fig. 21.2 Fuel cycle: from uranium mine through many steps, including plutonium reactor to recycle reconversion and wastes. (*R. H. Graham, World Uranium Reserves and Nuclear Power.*)

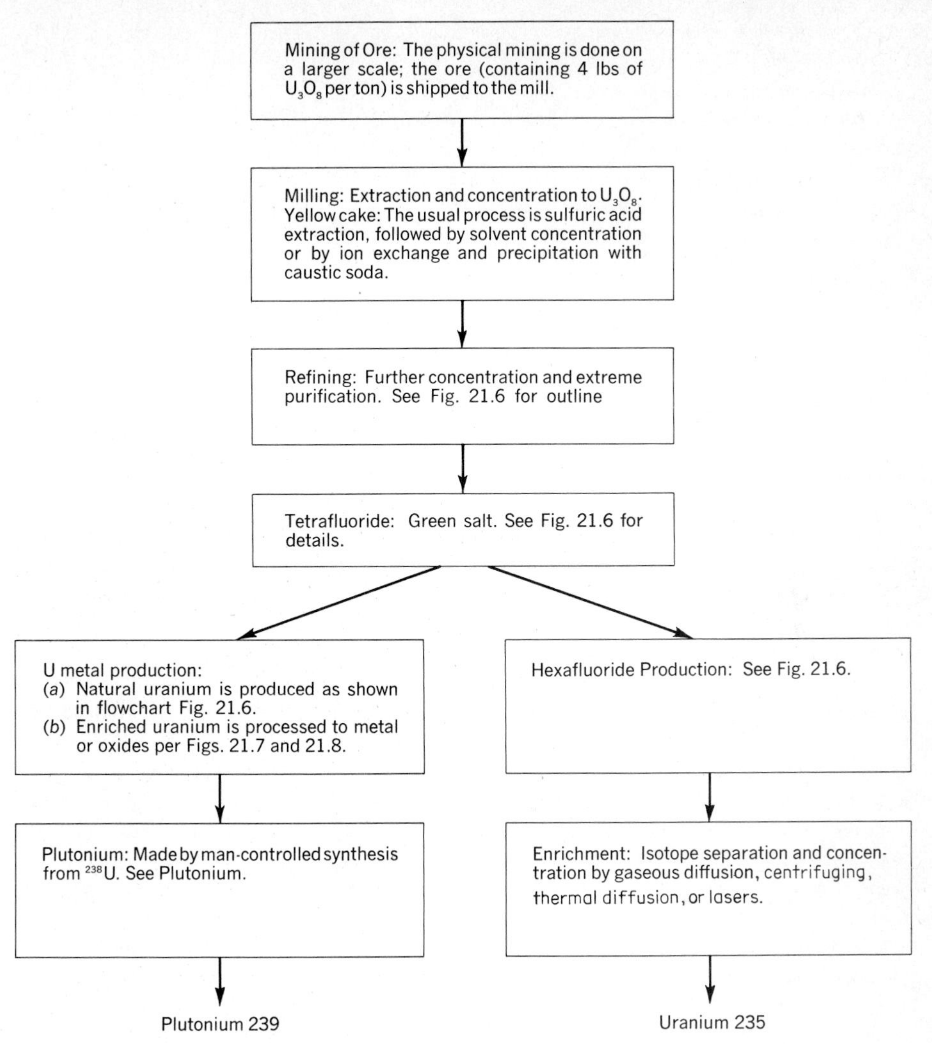

Fig. 21.3 General outline of AEC chemical flowchart from ore to usable nuclear material.

1980. In the United States, uranium is the *third* most important mineral industry, exceeded only by copper and iron. Uranium is more abundant in the earth's crust than mercury and is present in about the same amount as tin and molybdenum. The world's production (not including Communist nations) was 39,300 tons of U_3O_8 in 1973.

FEED-MATERIALS PRODUCTION OR NUCLEAR FUEL CYCLE

Figures 21.2 and 21.3 represent in outline the main steps from ore to usable nuclear materials: These may be separated into

1. *Mining* of ore.
2. *Milling*, or extraction and concentration to impure U_3O_8, or yellow cake.

3. *Refining*, or *concentration*, and *extreme purification* successively to UO_2, UF_4, and UF_6.
4. *Enrichment*[8] in ^{235}U as UF_6.
5. *Conversion* of UF_6 principally to UO_2 or uranium.
6. *Nuclear fuel* fabrication for use in reactors, usually UO_2, UC, or uranium.
7. *Mechanical shaping* into plates, rods, or tubes.
8. *Can fabrication*, cladding and systems assembly.
9. *Plutonium* by nuclear reaction from ^{238}U.

A summary of the overall reactions industrialized to obtain the first uranium used in nuclear reactions is:

$$U_3O_8 + HNO_3 \longrightarrow NO_3{-}\overset{\overset{\displaystyle O}{\|}}{\underset{\underset{\displaystyle O}{\|}}{U}}{-}NO_3 \cdot 6H_2O \longrightarrow UO_3 \longrightarrow UF_4 \longrightarrow U$$

These reactions and extraordinary care produce uranium of higher purity than that of any other substance produced on such a large scale. Mallinckrodt Chemical Works in St. Louis, Mo., first made reactor-acceptable material.

MINING OF ORE AND MILLING[9] involves extraction and concentration to impure U_3O_8, yellow cake. Mined ores carry only low percentages of uranium, in the United States an average of 4 lb of U_3O_8 per ton of ore. These are generally leached with sulfuric acid; the weak leach solutions are concentrated by solvent extraction or by ion exchange and finally precipitated by caustic soda to form yellow cake.

REFINING[10] **FOR FEED MATERIALS** involves further concentration and careful *purification*. These processes are presented in some detail in Fig. 21.4 and the accompanying text, and both lead to the production of extreme-purity uranium, UF_4, and UF_6. The yellow cake is shipped to several refineries which are purifying and concentrating plants, where the uranium metal or fluoride or oxide is prepared with the *purity* needed for nuclear reactions (before isotope enrichment). Boron is a usual contaminant which must be removed to less than 1 ppm, since it is a neutron absorber and would consequently interfere with or stop nuclear fission reactions. The process effecting this purification, a closely guarded secret until the end of World War II, involved the extraction of *pure* uranium from nitrate solutions by an organic solvent, at first ethyl ether and later tributyl phosphate, diluted with kerosine or hexane. Other organic solvents have proved to be effective in making ultrapure uranium. The organic extraction process is also useful in treating spent fuels from reactors to separate uranium and plutonium.

The supporting chemical reactions[11] are:

$$U_3O_8 + 8HNO_3(aq) \longrightarrow 3UO_2(NO_3)_2(aq) + 2NO_2 + 4H_2O \qquad \Delta H = -41.4 \text{ kcal}$$

$$UO_2(NO_3)_26H_2O \longrightarrow UO_3 + 2NO_2 + \tfrac{1}{2}O_2 + 6H_2O(g) \qquad \Delta H = +134.6 \text{ kcal}$$

$$UO_3(s) + H_2(g) \longrightarrow UO_2(s) + H_2O(g) \qquad \Delta H = -25.3 \text{ kcal}$$

$$UO_2(s) + 4HF(g) \longrightarrow UF_4(s) + 2H_2O(g) \qquad \Delta H = -43.2 \text{ kcal}$$

$$UF_4 + 2Mg \longrightarrow U + 2MgF_2 \qquad \Delta H = -83.5 \text{ kcal}$$

$$UF_4(s) + F_2(g) \longrightarrow UF_6(g) \qquad \Delta H = -60 \text{ kcal}$$

[8]Uranium Enrichment in for Decade of Expansion, *Chem. Eng.* (*N.Y.*), **82**(6), 24 (1975).

[9]Pinkney *et al.*, Chemical Processing of Uranium Ores, no. 23/24, International Atomic Energy Agency, Vienna, 1962 (costs, processes, and quantities).

[10]Harrington and Ruehle, *op. cit.*, pp. 7ff., and throughout book for production, history, details, and costs.

[11]The thermodynamics, kinetics, equilibriums, and other pertinent facts regarding these reactions are detailed by Harrington and Ruehle, *op. cit.*

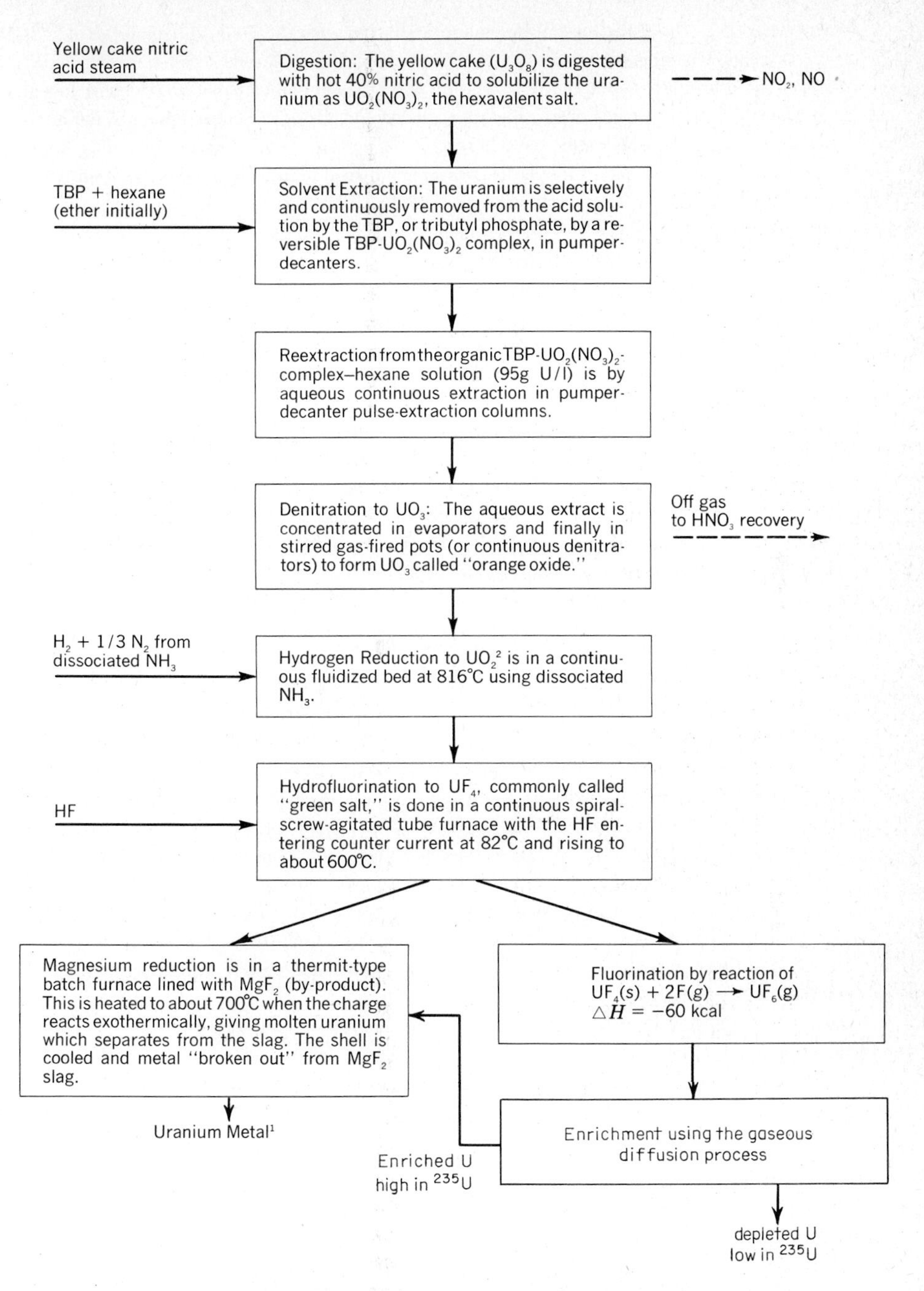

[1] At one time the uranium was remelted for purification but now (see Harrington and Ruehle *op. cit* Chapter 9, p. 271). A direct ingot, or "dingot", reduction has been practiced giving a 3,400 lb. "dingot." [2] Hedley *et al* Production of UO_2 by flame denitration. I. & E.C. Process Design Dec. 3, 1964, p. 11.

Fig. 21.4 Refining of uranium: feed-material-plant flowchart.

ENRICHMENT The only procedure used at present[12] to separate significant quantities of ^{235}U from ^{238}U is the gaseous diffusion method. UF_6 is a gas, and $^{235}UF_6$ can be separated from $^{238}UF_6$ by a method based on the difference in the diffusion rate through porous barriers. The enrichment per stage is small, so many stages in cascade are required and the power consumption of the interstage pumps is enormous. Maximum enrichment is 94% ^{235}U, with a waste containing about 0.3% ^{235}U. The United States has three large government-owned diffusion plants in operation; these use approximately 4,000 MW each and cost over $1 billion.

The industrially untested gas centrifuge process requires about one-tenth the power needed by a diffusion plant and will certainly be competitive in the immediate future. It is possible that it may replace gas diffusion entirely. Small plants are now in operation.

Other exotic separation processes continue to be suggested and/or tested. The electromagnetic separation process was used in early work and proved to be too expensive. A thermal diffusion process is possible. Laser enrichment is currently being tested and shows promise. Nozzle separation is expected to become commercial soon.

The need for isotopic separation of uranium is expected to increase 40-fold in the next 25 years, so great attention to this expensive process is essential.

The first use of purified, enriched uranium was in the atomic bomb dropped upon Hiroshima (Fig. 21.1). The first synthetic fissile element available in a usable quantity (plutonium) was employed in the atomic bomb dropped on Nagasaki. The United States has large quantities of all sizes of nuclear weapons. Stockpiling of weapons has been replaced by the peaceful uses of atom or nuclear reactions for heat, power, and multitudinous minor, yet important, applications.

NUCLEAR FUELS[13] Fissionable isotopes are used as sources of energy in nuclear reactors. Three isotopes that have a higher probability of fission than capture are ^{233}U, ^{235}U, and ^{239}Pu. These are the only common materials that can sustain the fission reaction and are therefore called *nuclear fuels*. Of these isotopes, only ^{235}U occurs in nature (1 part in 140 of natural uranium). The other two are produced artificially: ^{233}U by neutron capture by ^{232}Th, and ^{239}Pu by neutron capture by ^{238}U. ^{241}Pu and the higher odd-numbered isotopes of curium, etc., are also fissionable. Both ^{239}Pu and ^{233}U, being fissionable, can profitably be used to extend the supply of fissionable material for nuclear power.

Nuclear fuels must be specially prepared for reactor use, with consideration not only of their nuclear properties but of their physical properties as well. UO_2 enriched to varying percentages is widely preferred because of its high melting point, thermal conductivity, high density, and resistance to the effects of irradiation. The starting material is UF_6 enriched to the desired percentage. Figures 21.5 and 21.6 illustrate the conversions employed. The UO_2 is sintered into cylindrical pellets and encapsulated in zirconium or stainless-steel tubes.

THE MECHANICAL FABRICATION of uranium and plutonium fuel elements with or without enrichment involves difficult and complex procedures. These, for application to reactors, may be grouped under the headings *mechanical shaping* into plates, rods (cylinders), or tubes, generally on lathes; *core* cladding with stainless steel, zirconium, or other metals to protect from coolants; *assembly* into whatever system is being used for the type of reactor chosen.

NUCLEAR REACTORS

Nuclear reactors have been defined[14] as devices "containing fissionable material in sufficient quantity and so arranged as to be capable of maintaining a controlled, self-sustaining nuclear fission chain

[12]Uranium Enrichment in for Decade of Expansion, *Chem. Eng. (N.Y.)*, **82**(6), 24 (1975).
[13]McGraw-Hill Encyclopedia, vol. 9, p. 184, 1966.
[14]McGraw-Hill Encyclopedia, vol. 11, pp. 349–364, 1966 (see this reference for a concise but excellent description of the various types and classifications of reactors and their application); Nuclear Fission Energy, Essential, Abundant, Lower Than Ever Cost, and Safe in Capable Hands, *Chemtech.*, **4**, 531 (1974).

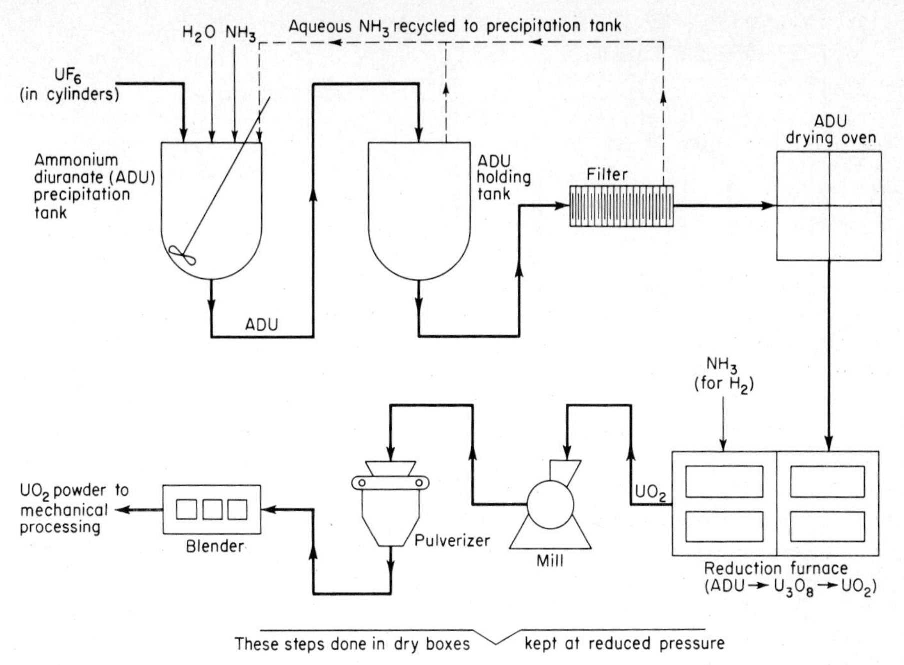

Fig. 21.5 UF_6 conversion to UO_2 by the ammonium diuranate process. (*United Nuclear Corp. and Industrial and Engineering Chemistry.*)

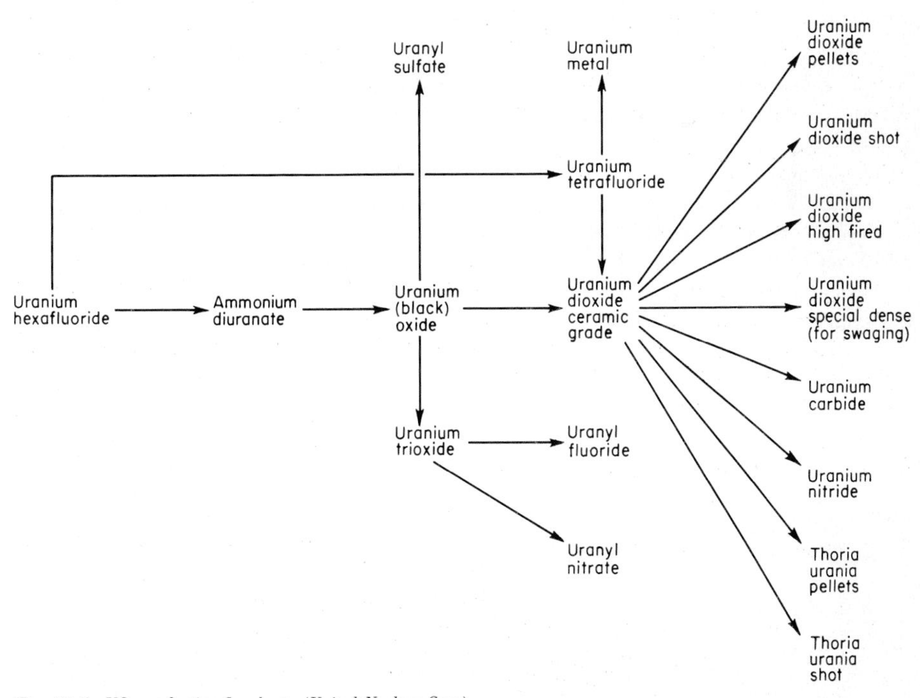

Fig. 21.6 UO_2 production flowchart. (*United Nuclear Corp.*)

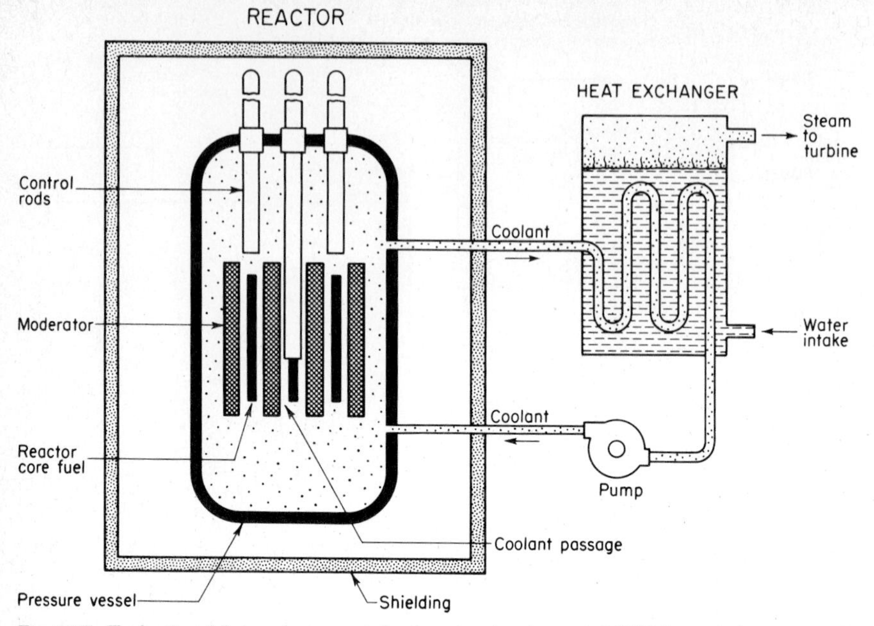

Fig. 21.7 The location of fuel, moderator, control rods, and coolant in a typical PWR (pressurized water reactor) system. These components are enclosed within a *pressure vessel* containing the various parts of the reactor. The coolant, heated to a high temperature by the nuclear fuel, flows through a heat exchanger, where it converts water, in a secondary system of pipes, into steam. The steam is then piped to a turbine, which operates an electric generator.

reaction."[15] These reactors may be classified as follows, depending on the isotopes employed as fuel and the conditions of the reaction:

1. *Burners* using uranium with enriched ^{235}U, which gradually becomes depleted (illustrated in Figs. 21.7 and 21.8). Such depleted fuel can be reworked and reused.

2. *Converters* using ^{238}U as fertile material to give ^{239}Pu, as practiced at several industrial power reactors.

3. *Breeders,*[16] which when fed a combination of fissionable and fertile fuels actually produce more fissionable fuel than they consume. The original fuel may consist of a fissionable isotope, ^{239}Pu, for example, reacting with fertile ^{238}U. In the reaction the fertile isotope is converted to plutonium to produce more fuel than the amount consumed.

Other classifications are used, as described in the literature cited in the references and footnotes. The moderator is a material such as graphite or heavy water, which slows the emitted fast neutrons so that they can react with the fuel and ^{238}U. The control rods absorb neutrons and stop the reactor.

In 1975, 30 different countries[17] had 410 power reactors built or under construction. Of these, 19 were fast reactors, 32 heavy-water reactors, 46 graphite reactors, 116 boiling-water reactors, and 197 pressurized light-water reactors. The names used generally refer to the kind of moderator used. Fuels used include natural and enriched uranium, UO_2, UC_2, UO_2-PuO_2, and UO_2-ThO_2 as rods, plates, pins, and pellets clad with zirconium, stainless steel, or other alloys. Heat is carried away by pressurized water, steam, sodium, molten salt, organic liquids, helium, and CO_2. An extensive ex-

[15]So-called critical condition.

[16]Nuclear Power, *Chem. Eng.*, **70**(9), 60 (1963); Nuclear Reactors Evolve toward Fast Breeders, *Chem. Eng.* (*N.Y.*), **76**(4), 44 (1969).

[17]*Nucl. Eng. Int.*, **20**(229), 394 (1975).

TABLE 21.2 ***Cost Comparison for Electric Energy Generation***

Fuel	*Capital cost, dollars per kWh*	*Fuel cost, cents per MBU*	*Energy cost, mills per kWh*
Coal	550	85	20
Oil	365	150	21
Nuclear	600	21	15
Nuclear breeder	650	10	15

Source: Report of the Cornell Workshops on Energy and the Environment sponsored by NSF-RANN, May 1972. *Note:* Between 1972 and 1977 capital costs and fuel costs increased rapidly for all types of electric power plants. See Chap. 4 for additional comparisons.

perimental program lasting over 30 years has permitted selection of the best combination of the above variables. The types of greatest interest are boiling water (BWR), pressurized water (PWR), gas-cooled (GCR), molten salt–cooled (MSR), and liquid metal (LMFR). Some of these reactor designs are being considered for breeder systems. Of these, development of the sodium cooled LMFBR (liquid metal fast breeder reactor) has been favored by most countries, including the United States.

The largest power-producing breeder at present is the Dounreay reactor in the United Kingdom, which produces 60 MW, thermal. Figure 21.8 presents one of the many types of reactors being built, forced recirculation BWR, which furnishes 420,000 KW, designed by General Electric Co. Atomic Products Division. This reactor was selected as a nuclear power producer[18] competitive with a conventional one fueled by oil or coal in all but the lowest-cost fossil-fuel areas. Table 21.2 shows a comparison of costs of electricity produced by a modern power reactor similar to that shown in Fig. 21.8 as contrasted with competitive units using fossil fuels.

URANIUM AS AN ENERGY SOURCE

The rising demand for energy, combined with a diminishing supply of oil and gas, have placed a sudden and relatively unexpected demand on uranium. A uranium shortage may presently develop. By 1980, the demand for U_3O_8 should be at least four times that of 1975.[19] Such swift expansion will be difficult. Table 21.3 shows that more energy is available from uranium than from any other currently known source. Costs are also very favorable. Currently, U_3O_8 sells for about $10 per pound but, using today's reactors, uranium at $50 per pound produces cheaper electricity than oil at $5 per barrel.[20] A $1 per pound increase in the price of U_3O_8 is equivalent to a 20 cents per ton

TABLE 21.3 ***U.S. Uranium Energy Available Exceeds Other Known Sources****

Resource	*Available energy × 10^{16} Btu*
Uranium	13,000
Coal and lignite	1,200
Oil shale	580
Petroleum and natural gas liquids	110
Natural gas	103

Source: ERDA. *Energy available from recoverable domestic resources.

[18]Graham, Atomic Products Division, General Electric Co., San Jose, Calif.; Nuclear kW, *Fortune*, July 1963, p. 173; McNelly *et al.*, Boiling Water Nuclear Reactors, *Chem. Eng. Prog.*, **61**(11), 72 (1965).

[19]New Vistas Stir Nuclear Industry, *Chem. Week*, July 16, 1975, p. 15.

[20]U.S. Producers Fear Uranium Imports, *Chem. Eng. News*, May 13, 1974, p. 7.

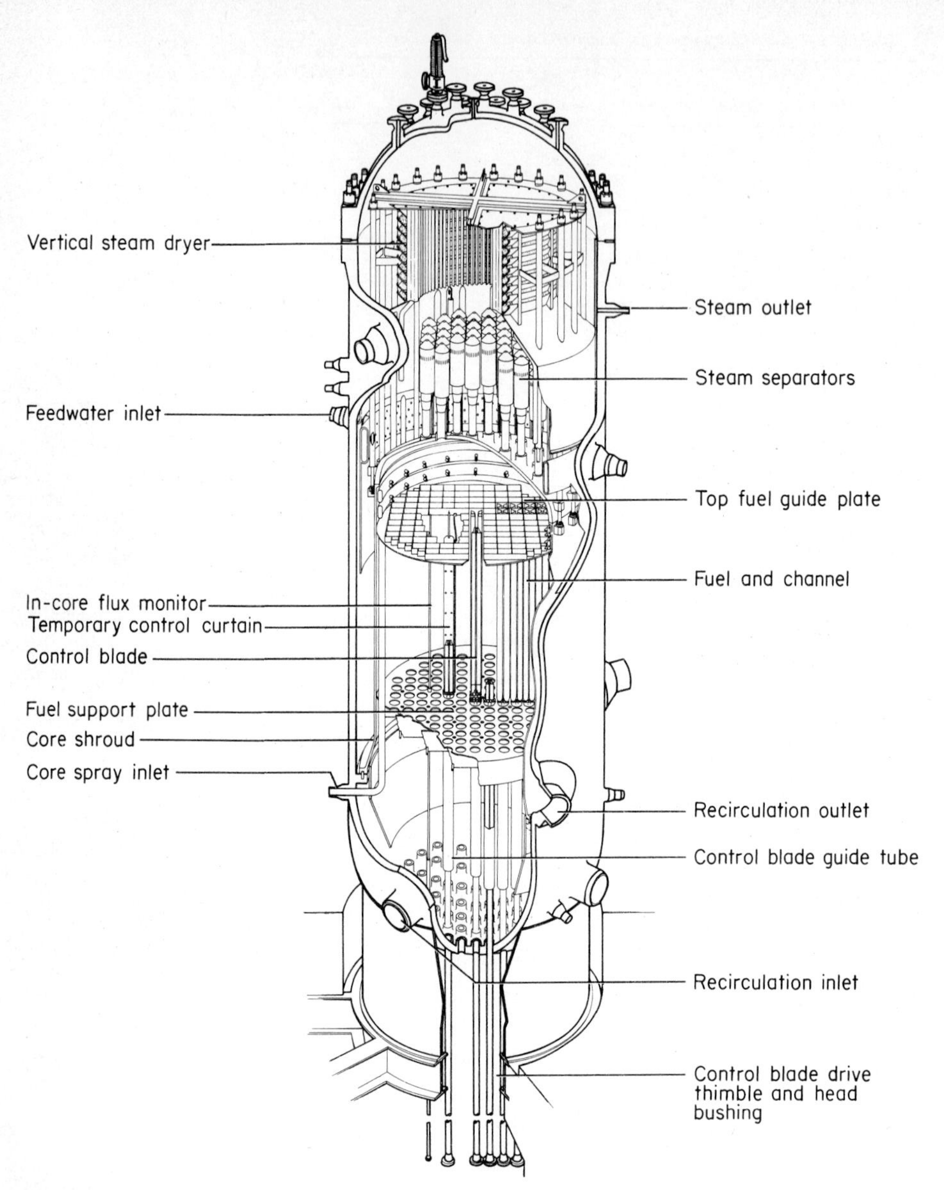

Fig. 21.8 BWR (boiling water reactor) with forced recirculation. Feedwater enters the vessel at the top of the core and, joined by recirculation coolant, flows downward around the core. The recirculation flow, which affords primary control of the load, is fed from the vessel to external pumps, which return it to the bottom of the vessel and up through the core. The generated steam passes through the internal-steam separators mounted over the core, through the vertical steam dryers, and then from the reactor vessel to the steam turbine. The fuel assemblies are located approximately midway in the reactor vessel. The mechanically driven control rods enter from the bottom of the reactor. The pressure vessel is 17 ft 9 in. in diameter and 58 ft 5 in. long and has a wall thickness of $7\frac{3}{8}$ in. (*General Electric Co.*)

increase for coal or a 5 cents per barrel increase for oil. With these figures in mind, uranium at $50 to 100 per pound would still be competitive with fossil fuels. To conserve domestic uranium reserves, the ban on foreign material has been lifted. Starting in 1977, 10% maximum imports are permitted, and all restrictions on imports expire in 1984. An official for a utility company has stated, "From both the economic and environmental standpoints, nuclear plants are clearly the best type of baseload capacity to install."

Uses for nuclear reactors, apart from weapons and as sources of energy for heat and power, involve *growing and vital medical needs for radioisotopes*, radiation processing of chemicals, instrumentation measurement, and detection of interior structure, for example, of metal welds; and new applications arise from week to week. Some examples are: (1) Several thousand licenses to use radioisotopes have been issued, of which one-third are held by industrial concerns, and the AEC is selling these isotopes in increasing amounts. (2) Such isotopes are used for radiographs of welds, to pinpoint and control the flow of gases and liquids through pipes, and for many other purposes, resulting in a total saving of more than $1 million per month in the oil industry.[21] (3) Chemical processing leading to an improved process for manufacturing 1 million lb of ethyl bromide per year has brought to the Dow Chemical Co. a merit award from *Chemical Engineering*. The Nobel prize winner, Willard E. Libby, a former member of the AEC, says, "I think there have been few occasions in the world's history when such a wealth of opportunity for technological development existed as exists now in the application of isotopes and radiation to industry, agriculture, and medicine."[22]

REPROCESSING OF NUCLEAR MATERIALS[23]

The reprocessing (or *radiochemical processing*) of used nuclear material from nuclear reactors follows somewhat the preparation of plutonium and the reworking of "burned-out" fuel from controlled fissionable materials in *nuclear reactor* power plants (Fig. 21.8). Solvent extraction is important here as elsewhere when handling fissionable material, as indicated by the flowchart in Fig. 21.9, which shows the use of tributyl phosphate. Extremely careful measures must be taken to protect personnel from radiation hazards, such as enclosing these entire operations, chemical and nuclear conversions, on all sides with several feet of concrete and with the procedures taking place under remote control. Essentially, these are large chemical plants (1) to *dissolve* with nitric acid the reaction plutonium and uranium of the irradiated reactor fuel, (2) to *separate* the radioactive waste products from the plutonium and the uranium, and (3) to purify the products and concentrate them, all with a total loss of less than 0.1%. The recovered fuel and plutonium must have a concentration of residual fission products of less than 1 part in 1 billion of the values in the irradiated fuels.

PROTECTION FROM RADIOACTIVITY

The hazards involved in nuclear industries are real and present, and the potential hazards are orders of magnitude greater than those in other industries. Radiation is greatly feared because it cannot be sensed with the ordinary human senses. Careful attention has been paid to the subject by health physicists[24] who have shown that, even with 100 power reactors operating in the United States, the extra radiation dose nearby residents are exposed to is only a small fraction of the unavoidable natural background. Reactor safety has been studied in careful detail and, as a result of closely controlled procedures, the accident rate to date has been extremely low.

[21]Radioisotopes in Chemical Processes, *Chem. Eng.* (*N.Y.*), **75**(6), 179 (1968).

[22]Radiation Processing of Chemicals, *Chem. Eng. News*, Apr. 22, 1963, p. 70; see also Chap. 38.

[23]Plutonium and Uranium via Solvent Extraction from Nuclear Fuels, *Chem. Eng.* (*N.Y.*), **71**(25), 148 (1964); HAN Gives Improved Plutonium Recovery, *Chem. Eng. News*, Dec. 14, 1970, p. 56.

[24]Population Dose from the Nuclear Industry to 2000 A.D., *Nucl. Saf.*, **15**(1), 56 (1974); Reactor Safety Study, WASH-1400, U.S. Atomic Energy Commission, Draft (August 1974).

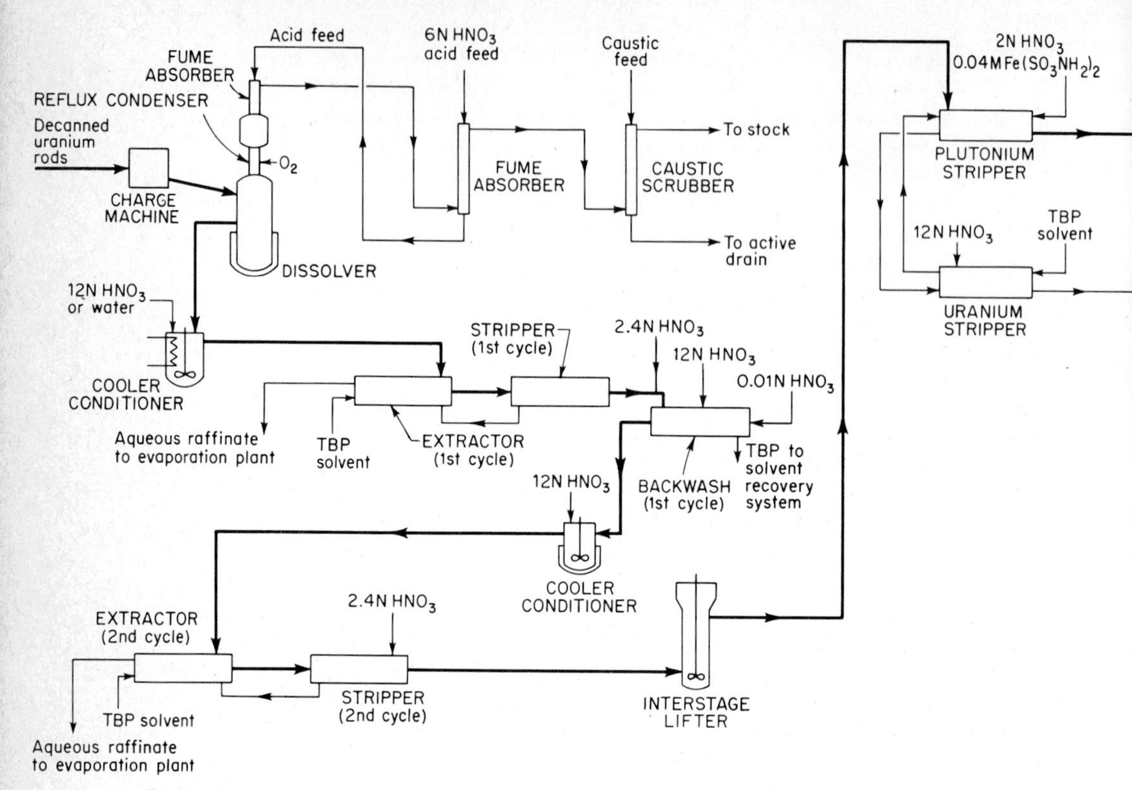

Fig. 21.9 Plutonium and uranium separation via solvent extraction from spent nuclear fuels. The flowchart describes the recovery plant of the U.K. Atomic Energy Authority at Windscale. After dissolving in aqueous HNO_3, the *fission products* are separated by extracting the uranium and plutonium nitrates in an organic solvent [tributyl phosphate (TBP) and kerosine]. This leaves the fission contamination products as "aqueous raffinate to the evaporation plant." The second solvent extraction is mainly for the separation of uranium from plutonium by adding ferrous sulfamate to reduce Pu^{4+} to Pu^{3+}, a cation unable to form strong complexes with TBP and consequently removed from the organic uranium stream. The aqueous plutonium solution is oxidized to the Pu^{4+} stage (HNO_3 and $NaNO_2$), so that plutonium can be extracted with TBP from its residual fission products

The contamination of gases, solids, and liquids by substances emitting α-, β-, and γ-radiation are all possible and must be guarded against. Shielding of only a fraction of an inch of ordinary material stops α- and β-particles, but γ-particles and neutrons are contained only by massive quantities of dense materials. ^{90}Sr and ^{137}Cs are particularly troublesome, because of their extremely long half-lives (28 and 30 years, respectively), and because of the tendency of living organisms to accumulate and concentrate them from very dilute solutions. Good analyses of environmental problems are available.[25]

WASTE DISPOSAL

The disposal of nuclear wastes is a difficult problem not yet adequately solved.[26,27] The radioactive materials must be stored for centuries until they decay to harmless materials. Disposal by burial in

[25]Environmental Survey of the Uranium Fuel Cycle, AEC, April 1974; Environmental Analysis of the Uranium Fuel Cycle (three parts), UAAEC, October 1973.

[26]Kubo and Rose, Disposal of Nuclear Wastes, *Science*, **182**, 1205 (1973).

[27]Rochlin, Nuclear Waste Disposal: Two Social Criteria, *Science*, **195**, 23 (1977).

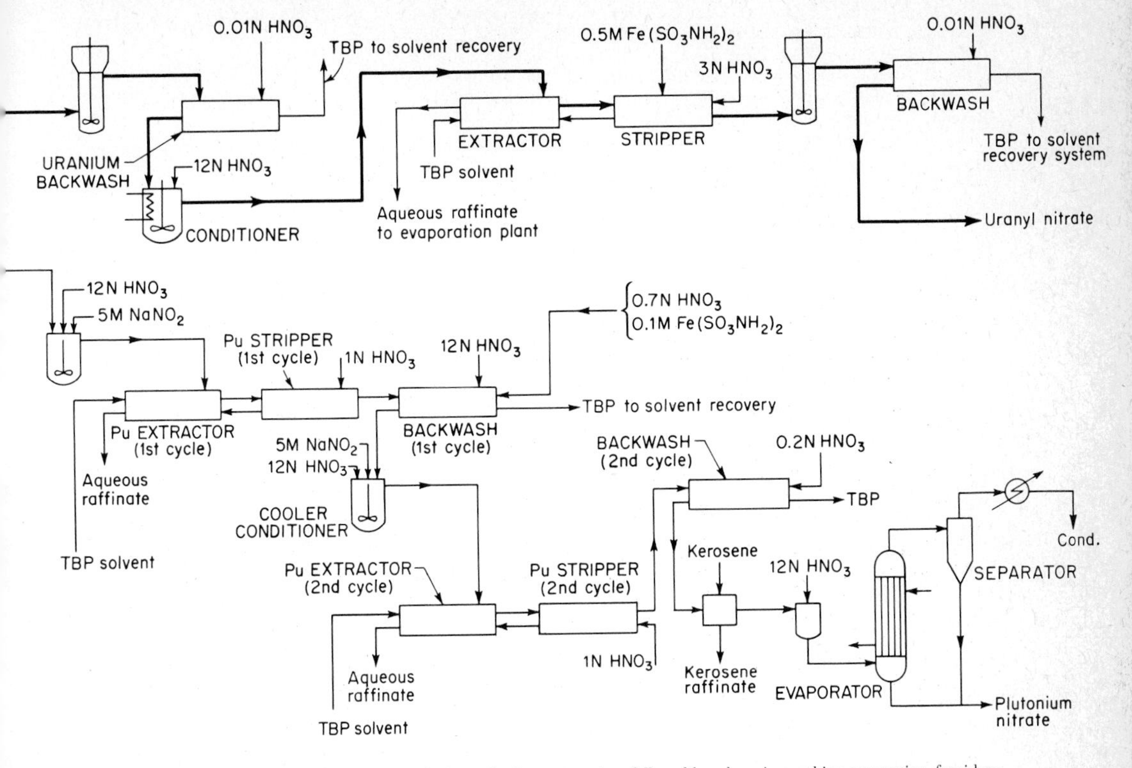

(aqueous raffinate). Successive acidification-oxidation and solvent extraction, followed by a kerosine washing evaporation, furnishes a pure solution of *plutonium nitrate. Uranium purification,* after separation from the aqueous Pu^{3+}, transfers the still impure uranium in the TBP stream by an extractive "backwash" into a dilute HNO_3 aqueous solution and the TBP recovered. Next another TBP extraction is used to remove uranium after a $3N$ HNO_3 acidification, and a second ferrous sulfamate reduction to remove traces of aqueous-soluble impurities (plutonium, etc., as "aqueous raffinate to evaporation plant"). Finally, extraction from the TBP solvent with a $0.01N$ HNO_3 solution results in a pure aqueous solution of *uranyl nitrate,* which is denitrated to UO_3. For details, see *Chem. Eng.* (*N.Y.*), **71**(25), 148 (1964).

deep wells, caves, old salt mines, the ocean, or in underground tanks has been studied. One of the difficulties is that such wastes are often liquid that could be easily, and relatively safely, temporarily stored in underground tanks; however, the radioactivity would last many years longer than the corroding tanks; hence studies are being directed toward conversion of such liquids to more safely stored solids.[28–30]

SELECTED REFERENCES

Allardice, C., and E. R. Trapnell: The Atomic Energy Commission, Praeger, 1974.
Atomic Energy in Use, U.S. Atomic Energy Commission, Supt. of Documents, Washington, D.C., 1965.
Boorse, H. A., and L. Motz: World of Atom, Basic Books, 1966.
Brownell, L. E.: Radiation Uses in Industry and Science, University of Michigan Press, 1961.

[28]High-Level Radioactive Waste Management Alternatives, U.S. Atomic Energy Commission document WASH-1297, May 1974.

[29]Schneider, Solidification and Disposal of High-Level Radioactive Waste in the United States, *Reactor Technology,* **13,** 387 (1970–1971, Winter).

[30]McLain and Blach, Disposal of Radioactive Waste in Bedded Salt Formations, *Nuclear Technology,* **24,** 398 (1974).

Chase, G. C., and J. L. Rabinowitz: Principles of Radioisotope Methodology, 2d ed., Burgess, 1962.
Cochran, T. B.: The Liquid Metal Fast Breeder Reactor, Johns Hopkins, 1974.
Coffinberry, A. S., and W. N. Miner (eds.): Metal Plutonium, University of Chicago Press, 1961.
Collins, J. C. (ed.): Radioactive Wastes: Their Treatment and Disposal, Wiley, 1961.
Cooper, A. R.: CPI in Atomic Energy Industry, Gordon and Breach, 1964.
Cronkite, E. P., and V. P. Bond: Radiation Injury in Man: Brookhaven National Laboratory, 1960.
Dukert, J. M.: Nuclear Ships of the World, Coward, McCann and Geoghegan, 1973.
Eaves, C.: Radiation Protection, Gordon and Breach, 1964.
Flagg, J. F. (ed.): Chemical Processing of Reactor Fuels, Academic, 1961.
Foell, W. K.: Small-Sample Reactivity Measurements in Nuclear Reactors, American Nuclear Society, 1972.
Fowler, E. B.: Radioactive Fallout, Soils, Plants, Foods, Man, Elsevier, 1965.
Freiling, E. C. (ed.): Radionuclides in the Environment, ACS, 1970.
Glassner, A.: Introduction to Nuclear Science, Van Nostrand, 1961 (elementary survey).
Harvey, B. G.: Nuclear Physics and Chemistry, Prentice-Hall, 1965.
Hogerton, J. F.: Atomic Energy Deskbook, Reinhold, 1963.
Jamrack, W. D.: Rare Metal Extraction by Chemical Engineering Techniques, Macmillan, 1963.
Kaufmann, A. R. (ed.): Nuclear Reactor Fuel Elements, Metallurgy and Fabrication, Wiley, 1962.
King, C. C. G.: Nuclear Power Systems, Macmillan, 1964.
Nucleonics, Handbook of Nuclear Research and Technology, McGraw-Hill, 1964.
Patton, F. E., *et. al.:* Enriched Uranium Processing, Macmillan-Pergamon, 1963.
Rand, M. H., and O. Kubaschewski: Thermochemical Properties of Uranium Compounds, Wiley, 1963.
Simnad, M. T. (ed.): Materials and Fuels for High-temperature Nuclear Energy Applications, M.I.T., 1964.
U.S. Atomic Energy Commission: Reactor Handbook, 2d ed., Wiley-Interscience, 1960–1964, 4 vols.
Weinstein, R., *et al.:* Nuclear Engineering Fundamentals, McGraw-Hill, 1964, 5 vols.
Wilkinson, W. D.: Uranium Metallurgy, Wiley, 1962, 2 vols.
Yaffe, L.: Nuclear Chemistry, Academic, 1968.

chapter 22

EXPLOSIVES, TOXIC CHEMICAL AGENTS, AND PROPELLANTS

INDUSTRIAL AND MILITARY EXPLOSIVES

Many professional chemists and chemical engineers tend to view the subject of explosives from a purely military standpoint. Although explosives have contributed much to the destruction of humans, they have also allowed many great engineering feats to be performed, which would have been physically or economically impossible without their use. Engineering projects such as Hoover Dam and the St. Bernard Tunnel would have taken hundreds of years if accomplished by hand labor alone. Explosives are among the most powerful servants of man. Mining of all kinds depends on blasting, as does such a prosaic but necessary act as the clearing of stumps and large boulders from land. Truck brakes are relined and airplanes are constructed with rivets, which have an explosive in the shank, fired by heat, resulting in consequent saving of much time and effort. Recent use of controlled underwater explosives to shape metals offers a steadily growing outlet and presents a new, versatile, and economical method for fabrication techniques.

HISTORY AND ECONOMICS An explosive mixture of sulfur, charcoal, and saltpeter, or *black powder,* was known to the Chinese centuries ago; its use in propelling missiles was demonstrated shortly after 1300. The discoveries of nitroglycerin and nitrocellulose shortly before 1850, and the invention of dynamites and the mercury fulminate blasting cap soon after, were epochal events of the *high-explosives* era. *Atomic explosives,* discussed in the preceding chapter, first detonated in 1945, mark the third broad stage in the history of explosives. Superior products, such as smokeless powder, first made in 1867, have been developed through the years; the demand for more powerful explosives for the present space program serves as a continuing challenge to chemical engineers. In peacetime large quantities of commercial explosives are consumed. In times of war tremendous quantities are required; for example, roughly 60 billion lbs were manufactured in the United States between January 1940, and V-J Day. In the decade 1960–1970 sales increased from 1,080 to 2,393 million lb/year. Comparatively few chemicals are used as explosives, because of the critical restrictions on stability, safety, and price, although a great many compounds and mixtures are explosive.

CLASSIFICATION An explosive is a material which, under the influence of thermal or mechanical shock, decomposes rapidly and spontaneously with the evolution of a great deal of heat and much gas. The hot gases cause extremely high pressure if the explosive is set off in a confined space. Explosives differ widely in their sensitivity and power. Only those of a comparatively insensitive nature, capable of being controlled and having a high energy content, are of importance commercially or in a military sense. There are three fundamental types of explosives, mechanical, atomic, and chemical; the primary concern of this chapter is chemical explosives. For purposes of classification it is convenient to place chemical explosives in two divisions in accordance with their behavior.

1. *Detonating, or high, explosives.* (*a*) Primary, or initiating, explosives (detonators): lead azide, mercury fulminate, lead styphnate (lead trinitroresorcinate). (*b*) Secondary: TNT-AN, tetryl,[1] PETN, RDX, TNT, ammonium picrate, picric acid, DNT (dinitrotoluene).
2. *Deflagrating, or low, explosives.* Smokeless powder (colloided cellulose nitrate), black powder, nitrocotton.

Cook[2] lists the major chemicals according to military or commercial use. Variations in their blasting characteristics are obtained by (1) altering physical conditions such as density and granulation and (2) combining ingredients—paraffin, aluminum, and waxes, for example. A table of nonexplosive ingredients appears in the same reference.

Initiating, or primary, high explosives are quite sensitive materials which can be made to explode by the application of fire or by means of a blow. They are very dangerous to handle and are used in comparatively small quantities to start the explosion of larger quantities of less sensitive explosives. Initiating explosives are generally used in primers, detonators, and percussion caps. They are usually inorganic salts, whereas secondary high explosives and many conventional propellants are largely organic materials.

Secondary high explosives are materials which are quite insensitive to both mechanical shock and flame but which explode with great violence when set off by an explosive shock, such as that obtained by detonating a small amount of an initiating explosive in contact with the high explosive. Decomposition proceeds by means of detonation, which is rapid chemical destruction progressing directly through the mass of the explosive. Detonation is thought to be a chain reaction and proceeds at rates frequently as high as 6,000 m/s. It is this *high rate of energy release*, rather than the total energy given off, that makes a product an explosive. Nitroglycerin has only one-eighth the energy of gasoline. On the other hand, most high explosives, when unconfined or unshocked, will merely burn if ignited.

Low explosives, or propellants, differ from high explosives in their mode of decomposition; they only burn. Burning is a phenomenon that proceeds not through the body of the material, but through layers parallel to the surface. It is quite slow in its action, comparatively speaking, rarely exceeding 0.25 m/s. The action of low explosives is therefore less shattering. Low explosives evolve large volumes of gas on combustion in a definite and controllable manner.

The industrial classification used in Table 22.1 is that of the Bureau of Mines, in which *black blasting powder* refers to all black powder having sodium or potassium nitrate as a constituent. Permissible explosives must pass certain tests[3] to ensure a minimum of flame and temperature to reduce fires in coal mines, where 99% of them are employed. The use of dynamite as the high explosive of choice in industry has greatly diminished. Most commercial explosives are based on low-priced ammonium nitrate mixtures. Such mixtures are either prepackaged or mixed on the job and offer such safety in handling that ordinary industrial machinery is used without special precautions. The explosive properties, however, remain satisfactory for blasting. Military explosives are generally quite different from industrial explosives, requiring longer storage life under poor conditions and more attention to the effects of handling by personnel.

PROPERTIES OF EXPLOSIVES In order to compare explosives for suitable use, the most important standard tests are those employed to determine sensitivity, stability, brisance, and strength.

[1]Because of higher brisance, tetryl is also used as a booster between primary and secondary high explosives.

[2]Cook, Science of High Explosives, ACS Monograph 139, p. 4, Reinhold, 1958; in Riegel, Industrial Chemistry, Reinhold, 1962, pp. 676–683, will be found an extensive list, together with characteristics and uses.

[3]Active List of Permissible Explosives and Blasting Devices, *U.S. Bur. Mines Inf. Circ.* 8493, 1973.

TABLE 22.1 *Summary of Characteristics of Explosives*

Name	*Formula*	*Products per formula weight*	Q_v, *cal/kg*	T_e, °*C*	*f, kg/cm*^{2}	*V, m/s*	*Trauzl expansion, cc/10 g*	*Potential* $\times 10^5$ *kg-m*
Gunpowder	$2KNO_3 + 3C + S$	$N_2 + 3CO_2 + K_2S$	501	2090	2,970	.....	30	2.1
Nitrocellulose	$C_{24}H_{29}O_9(NO_3)_{11}$	$20.5CO + 3.5CO_2 + 14.5H_2O + 5.5N_2$	1,250	2800	10,000	6,100	420	5.3
Nitroglycerin	$C_3H_5(NO_3)_3$	$3CO_2 + 2.5H_2O + 1.5N_2 + 0.25O_2$	1,526	3360	9,835	8,500	590	6.5
Ammonium nitrate	NH_4NO_3	$2H_2O + N_2 + 0.5O_2$	384	1100	5,100	4,100	300	1.6
TNT	$C_7H_5(NO_2)_3$	$6CO + C + 2.5H_2 + 1.5N_2$	656	2200	8,386	6,800	260	2.8
Picric acid	$C_6H_2(OH)(NO_2)_3$	$6CO + H_2O$ $0.5H_2 + 1.5N_2$	847	2717	9,960	7,000	300	3.6
Ammonium picrate	$C_6H_2(NO_2)_3ONH_4$	$6CO + H_2O + 2H_2 + 2N_2$	622	1979	8,537	6,500	230	2.6
Tetryl	$C_7H_5N_5O_8$	$7CO + H_2O + 1.5H_2 + 2.5N_2$	908	2781	10,830	7,229	320	3.9
Mercury fulminate	$Hg(ONC)_2$	$Hg + 2CO + N_2$	420	4105	5,212	3,920	213	1.8
Lead azide	PbN_6	$Pb + 3N_2$	684	3180	8,070	5,000	250	2.9

Source: After Meyer, Science of Explosives, Crowell, 1943; for more recent investigations of characteristics of explosives, cf. Cook, Science of High Explosives, p. 284, table 12.1, ACS Monograph 139, Reinhold, 1958; a more extensive list will be found in Riegel, Industrial Chemistry, 7th ed., Reinhold, pp. 570–596, 1974. *Note:* Q_v = heat of explosion at constant volume (small in comparison with fuels, but explosives exert their energy rapidly); T_e = explosion temperature; f = specific pressure, i.e., exerted by 1 kg in a volume of 1 liter at T_e heat; V = velocity of detonation wave (currently measured by high-speed photography).

Before use for commercial or military purposes can be evaluated, additional tests for volatility, solubility, density, hygroscopicity, compatibility with other ingredients, and resistance to hydrolysis are made if the explosive meets the first tests. Cost of manufacture and toxicity of materials must also be taken into consideration.

The *brisance* (shattering action) of an explosive may be measured by exploding a small quantity of it in a *sand bomb,* a heavy-walled vessel filled with standard coarse sand, which is crushed by the explosion. Screening measures the sand crushed, and from this the explosive force. Another test of a somewhat similar nature is the *Trauzl block test.* This test measures the strength of the explosive by measuring the ballooning of a soft lead cylinder in which the explosive is inserted and exploded. The standard Trauzl block is 200 mm in diameter and 200 mm high and has a central hole 25 mm in diameter and 125 mm deep. Ten grams of the explosive is used in making the test, and the results are reported in terms of cubic centimeters of increase in volume caused by detonation of the explosive. Brisance is probably a combination of strength and velocity. Cook states that there is no reason to believe that brisance is power; it seems to be directly related to detonation pressure.[4]

The *sensitivity*[5] of an explosive to *impact* is determined by finding the height from which a standard weight must be allowed to fall in order to detonate the explosive. This test is of greatest importance in the case of initiating explosives.

It is important that explosives for use in mines, particularly coal mines, be of such a type that they evolve no poisonous gases on explosion and produce a minimum of flame. This latter requirement is necessary in order that the explosive be incapable of igniting mixtures of air and coal dust, or

[4]Cook, *op. cit.,* p. 271.
[5]Macek, Sensitivity of Explosives, *Chem. Rev.,* **62,** 41 (1962) (140 refs.).

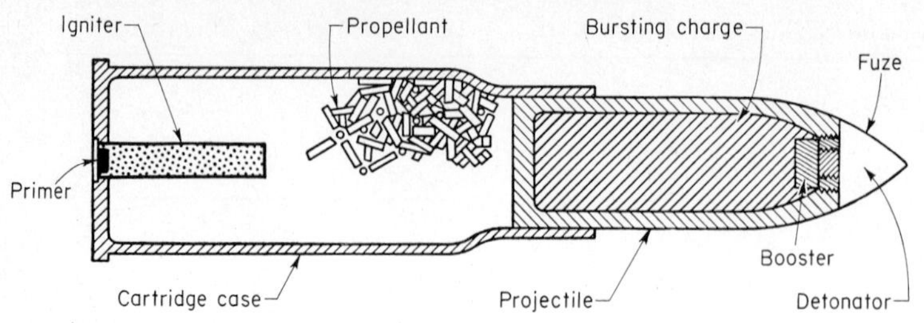

Fig. 22.1 Complete round of high-explosive ammunition.

air and methane (fire damp), which inevitably occur in coal mines. Explosives for mine use are tested and their properties specified by the Bureau of Mines; they are commonly known as *permissibles.* Permissibles differ from other explosives, particularly black powder, most markedly in the fact that they produce a flame of small size and extremely short duration. Permissibles contain coolants to regulate the temperature of their flame, hence further reduce the possibility of their igniting combustible mixtures. Permissibles are tested by explosion in a long gallery filled with coal dust, air, and methane and should not ignite this mixture.

MILITARY EXPLOSIVES

Possibly the most powerful nonatomic military explosives are cast aluminized mixtures such as Torpex and HBX (RDX, TNT, aluminum and wax). As military requirements are extremely strict, only a few explosives have survived competitive testing, and even fewer are described here. In this connection it is fundamental to understand the construction of a high-explosive artillery shell as depicted in Fig. 22.1. Such a shell consists of a thin brass or steel cartridge case holding the primer, igniter, and propellant charge. This case is designed to fit smoothly into the gun and, on explosion, to expand (obdurate), sealing the breech of the gun so that the escape of gases from the burning of the propellant charge is prevented, thus allowing the full effect of the propellant to be exerted on the projectile, or destructive half of the shell. The *primer* contains a small amount of a primary explosive or sensitive mixture [e.g., lead azide or a mixture such as $KClO_3$ + $Pb(CNS)_2$ + Sb_2S_3 + TNT + ground glass]. This mixture explodes under the impact of the firing pin and produces a flame which ignites the black-powder[6] charge in the *igniter,* in turn igniting the *propellant* charge of smokeless powder. The burning of the smokeless powder causes the rapid emission of heated gas, which ejects the *projectile* from the gun. At the target, upon impact or upon functioning of the time-fuze mechanism, a small quantity of a primary explosive (*detonator*) is set off; this causes explosion of the *booster*—an explosive of intermediate sensitivity (between that of a primary explosive and the bursting charge)—which picks up the explosive wave from the primary explosive, amplifies it, and ensures complete detonation of the bursting charge. The *bursting charge,* or high explosive, is usually TNT alone or mixed with ammonium nitrate. RDX, PETN, and ammonium picrate are also sometimes used.

NITROCELLULOSE[7] The explosive properties of nitrated cotton were recognized at an early date. The discovery of methods of colloiding the material into a dense uniform mass of resinous appearance reduced the surface and the rapidity of the explosion. With the discovery of suitable stabilization methods to prolong its storage life, nitrocellulose soon put black powder out of use as a military propellant.

[6]Black powder is easy to ignite and produces hot solids in the gases; therefore it is frequently used to ignite smokeless powder.

[7]Although the word "nitrocellulose" is commonly used for nitrated cellulose fiber, it should be noted that the material is not a nitro compound but a true nitrate ester. The name *cellulose nitrate* is therefore chemically more proper, but nitrocellulose is much more widely employed in technical spheres.

The cellulose molecule is a highly complicated one, with a molecular weight frequently as high as 300,000. Any given sample of cellulose contains a wide distribution of molecules, all having the empirical formula $[C_6H_7O_2(OH)_3]_n$. There are thus three hydroxyl groups per fundamental (glucose) unit that may be esterified with nitric acid, indicating a theoretical nitrogen content of 14%, which is higher than that of any commercially used product. The reaction may be formulated

$$C_6H_7O_2(OH)_3 + 3HONO_2 + H_2SO_4 \longrightarrow C_6H_7O_2(ONO_2)_3 + 3H_2O + H_2SO_4$$

Celluloses that have not been nitrated completely to the trinitrate are used as various industrial cellulose nitrates, as shown in Fig. 22.2. In addition to nitrate esters, some sulfate esters are formed by sulfuric acid, which is added to tie up the water resulting from the nitration reaction and to permit the reaction to progress to the right. These sulfate esters are unstable, and their decomposition would give rise to a dangerous acid condition in the stored powder if not removed; they are decomposed in the poaching process.

The finished nitrocellulose should not be allowed to become acid in use or in storage, since this catalyzes its further decomposition. A stabilizer is therefore added which reacts with any trace of nitrous, nitric, or sulfuric acid that may be released because of the decomposition of the nitrocellulose and thus stop further decomposition. For smokeless powder diphenylamine is used (diphenylurea in Great Britain), and for Celluloid, urea. The diphenylamine forms a series of innocuous compounds with the evolved gases, of which the following are examples:

O_2N–⬡–NH–⬡ ; O_2N–⬡–NH–⬡–NO_2

The commercial manufacture of cellulose nitrate is illustrated in Fig. 22.2 for the following sequences:

Cotton linters, or specially prepared wood pulp, are purified by boiling in Kiers (vats) with dilute caustic solution (Ch).

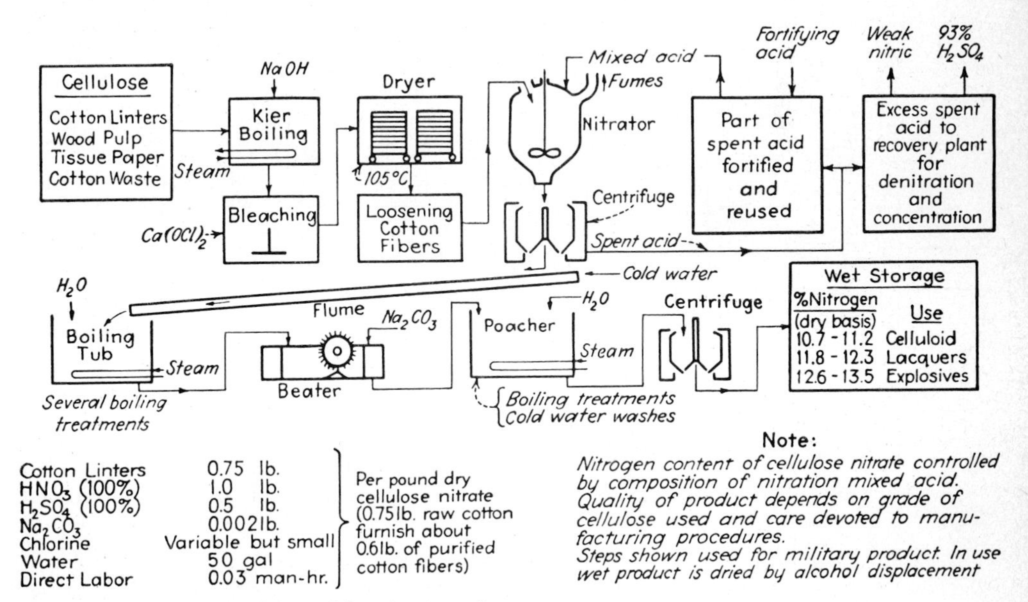

Fig. 22.2 Flowchart for nitrocellulose (cellulose nitrate) manufacture.

Bleaching is effected with CaCl·OCl, NaOCl, or $Ca(OCl)_2$ (Ch).

The cotton is dried, fluffed, and weighed (Op).

Mixed acid is made up from fortifying acid and spent acid, brought to proper temperature and run into the nitrator (Op).

Nitration (esterification) is usually conducted under carefully controlled conditions in a "mechanical-dipper" nitrator (Ch).

One nitrator charge is formed by 32 lb of purified cellulose. The cellulose is agitated with approximately 1,500 lb of mixed acid at 30°C for about 25 min. The composition of the acid used averages: HNO_3, 21%; H_2SO_4, 63%; N_2O_4, 0.5%; H_2O, 15.5%.

The entire nitrator charge is dropped into a centrifuge, where the spent acid is centrifuged from the nitrated cellulose (Op).

The spent acid is partly fortified for reuse and partly sold or otherwise disposed of, e.g., by denitration and concentration of the H_2SO_4 (Op and Ch).

The nitrated cellulose is drowned with water, washed by boiling, and again washed in a beater (Op).

In order to produce a *smokeless powder* more stable on storage, the following purification is employed in these two steps and the poaching, to destroy unstable sulfate esters and to remove free acid completely:

(1) Forty hours of boiling with at least four changes of water, and (2) pulping of fiber by means of a beater or a Jordan engine, followed by:

Poaching of the washed nitrated cotton by boiling first with a dilute Na_2CO_3 solution (5 lb of soda ash per ton of cellulose nitrate) and then many washes of boiling pure water (Ch).

The poached nitrocellulose is freed of most of its water by centrifugation. This usually results in a water content of approximately 28% (Op).

At this point the nitrocellulose is usually stored until desired, and a complete laboratory examination can be made (Op).

The water content of the nitrated cotton is reduced to a low figure by alcohol percolation under pressure dehydration (Op).

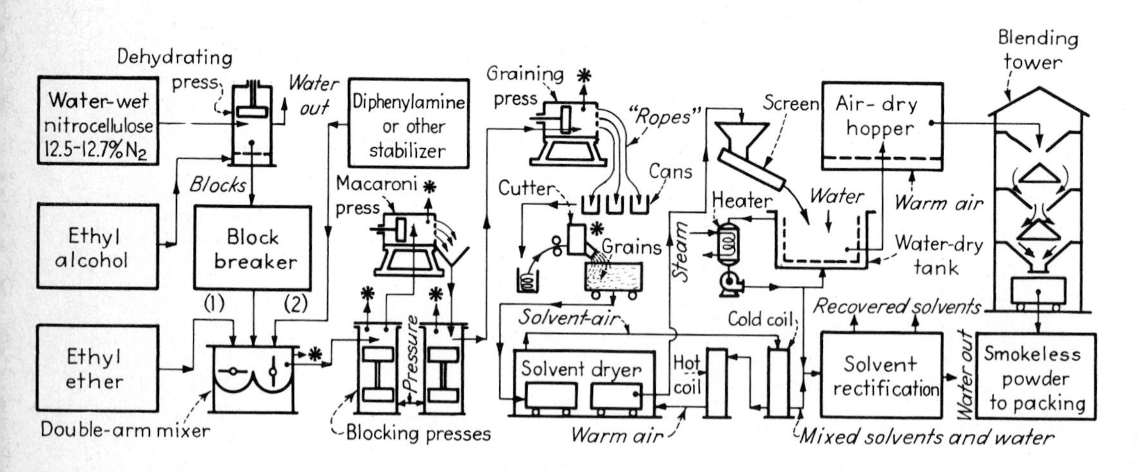

Fig. 22.3 Flowchart for smokeless powder manufacture.

The nitrated cellulose is disintegrated, then colloided by mixing with alcohol, ether, diphenylamine, and other modifying agents (Op).

Grains are formed by extrusion through dies, and these are dried and blended to form smokeless powder (Op); see Fig. 22.3.

The nitrocellulose produced in this manner contains about 12.6% nitrogen and is known as *pyrocotton.* By using a stronger acid, the nitrogen content may be made as high as 13.6%. Cotton nitrated to contain 13.2% nitrogen or greater is known as *guncotton.* Modern military smokeless powder contains about 13.15% nitrogen and is made from a blend of pyro- and guncotton.

Smokeless powder is colloided nitrocellulose containing about 1% diphenylamine, to improve its storage life, and a small amount of a plasticizer (e.g., dibutyl phthalate). The manufacturing sequences are given by Fig. 22.3. The British form a colloided double-base powder (Cordite) by treating a high, or 13% nitrogen, nitrocellulose with a mixture of acetone, nitroglycerin, and petroleum jelly. The nitroglycerin increases the heat of burning of the powder and keeps the grains from becoming brittle.

TNT (TRINITROTOLUENE) In spite of other new explosives developed, symmetric TNT remains an important military explosive, particularly in mixtures with ammonium nitrate (amatol). Its low melting point (80°C) permits loading into bombs and shells in the molten state. It does not have picric acid's tendency to form sensitive metallic salts. TNT is made by multiple-stage nitration of toluene with a mixture of sulfuric and nitric acids. Three stages nitrating to mono-, di-, and trinitrotoluene were formerly used, but continuous-flow, stirred-tank reactors and tubular units using the countercurrent flow of strong acids and toluene permit better yields and lower costs.[8]

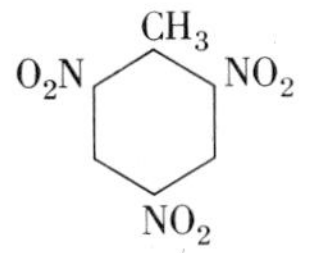

CH_3

O_2N ⬡ NO_2

NO_2

TETRYL [2,4,6-trinitrophenyl-methylnitramine, $C_6H_2(NO_2)_3NCH_3NO_2$] is chiefly used as a base charge in blasting caps, as the booster explosive in high-explosive shells, and as an ingredient of binary explosives. It is generally prepared by the action of mixed sulfuric and nitric acid on dimethylaniline in a multiple-stage nitration:

$$C_6H_5N(CH_3)_2 + 2HNO_3 \xrightarrow{H_2SO_4} C_6H_3N(CH_3)_2 \cdot NO_2 \cdot NO_2 + 2H_2O$$

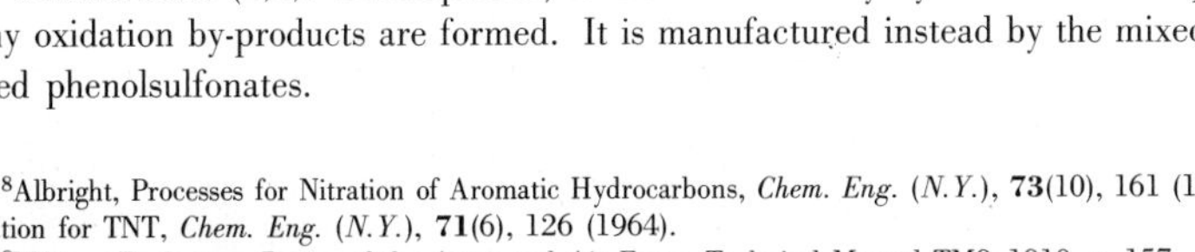

$$\underset{\text{(ring with } NO_2 \text{ at 2,4)}}{CH_3{-}N{-}CH_3} + 8HNO_3 \xrightarrow{H_2SO_4} \underset{\text{(ring with } O_2N, NO_2, NO_2 \text{ at 2,4,6)}}{O_2N{-}N{-}CH_3} + 6NO_2 + CO_2 + 6H_2O$$

It may also be made by alkylating 2,4-dinitrochlorobenzene with methylamine and then nitrating.[9] Tetryl is an extremely powerful high explosive with great shattering power.

PICRIC ACID (2,4,6-trinitrophenol) is not made directly by the nitration of phenol, because too many oxidation by-products are formed. It is manufactured instead by the mixed-acid nitration[10] of mixed phenolsulfonates.

[8]Albright, Processes for Nitration of Aromatic Hydrocarbons, *Chem. Eng.* (*N.Y.*), **73**(10), 161 (1966); Prime, Seven-stage Nitration for TNT, *Chem. Eng.* (*N.Y.*), **71**(6), 126 (1964).

[9]Military Explosives, Depts. of the Army and Air Force, Technical Manual TM9–1910, p. 157, April 1955.

[10]Formerly, strong nitric acid alone was used. Mixed acids cut down the health hazard caused by fuming, reduce the amount of acid required, and increase the yield of desired products. The use of mixed acid is now almost universal. For further details of manufacture, see Military Explosives, *op. cit.*, p. 161.

EXPLOSIVE D, OR AMMONIUM PICRATE, is made by the neutralization of a hot aqueous solution of picric acid with ammonia water. It is used in *armor-piercing shells* because of its insensitivity to shock.

PETN, pentaerythritol tetranitrate $[C(CH_2ONO_2)_4]$, is one of the most brisant and sensitive of the military high explosives. For use as a booster explosive, a bursting charge, or a plastic demolition explosive, it is desensitized by admixture with TNT (called Pentolite) or by the addition of wax. PETN may be made by the nitration of pentaerythritol with strong nitric acid (96%) at about 50°C. The reaction of decomposition is probably

$$C(CH_2ONO_2)_4 \longrightarrow 3CO_2 + 2CO + 4H_2O + 2N_2$$

RDX, cyclonite, or *sym*-trimethylene trinitramine $[(CH_2)_3N_3(NO_2)_3]$, is one of the most powerful explosives known at the present time. RDX is used in a mixture with TNT and aluminum, known as *Torpex,* for mines, depth charges, and torpedo warheads. It is also employed as an ingredient in explosives for shells and bombs and is desensitized by wax or oily materials. Pentolite, made by casting slurries of PETN with TNT, has specialized uses; for example, cast 50-50 pentolite is used as a booster for slurry blasting agents and prilled ammonium nitrate–fuel oil mixtures and as the main charge in Procore boosters. The British developed the first practical process, which involved the destructive nitration of hexamethylenetetramine with concentrated nitric acid:

$$C_6H_{12}N_4 + 3HNO_3 \longrightarrow C_3H_6O_6N_6 + 3HCHO + NH_3$$

A combination process was developed by Bachman, at the University of Michigan, who utilized the by-products to obtain a second mole of RDX. This method was developed on a *continuous scale* by Tennessee Eastman, which manufactured it the most economically (yields of 70%):

$$C_6H_{12}N_4 + 4HNO_3 + 2NH_4NO_3 + 6(CH_3CO)_2O \longrightarrow 2C_3H_6O_6N_6 + 12CH_3COOH$$

LEAD AZIDE

$$Pb\left(-N\begin{matrix} N \\ \| \\ N \end{matrix}\right)_2$$

has partially replaced mercury fulminate as an initiating, or primary, explosive for blasting caps. The fulminate had less than desirable stability, had to be manufactured in small batches, and involved the scarce raw material mercury. On the other hand, lead azide has remarkable stability, involves no strategic materials, and can be manufactured in large batches by treating sodium azide with lead acetate or nitrate. Sodium azide may be made from sodium amide and nitrous oxide:

$$NaNH_2 + N_2O \longrightarrow NaN_3 + H_2O$$

$$2NaN_2 + Pb(C_2H_3O_2)_2 \longrightarrow Pb(N_3)_2 + 2Na(C_2H_3O_2)$$

INDUSTRIAL EXPLOSIVES

Historically, black powder has been the preferred blasting agent because it is less shattering in its effects. It is an intimate mixture of KNO_3, sulfur, and charcoal in the approximate proportions 75-15-10. About 50% of the products of combustion are solids, giving its flames great igniting power, which is undesirable where combustible gases or dusts are present.

Starting in 1860, dynamite dominated the blasting industry for a century.

BLASTING AGENTS AND SLURRY EXPLOSIVES have become the principal industrial explosives because they may be handled in simple machinery almost completely without danger and because their cost is very low.

TABLE 22.2 *Classes of Well-characterized Slurry Explosives and Slurry Blasting Agents*

*Designation**	*% Sensitizer*	*Oxidizer†*	*Water, %*	
			Nominal	*Range*
SE-TNT	17–60 TNT	AN, SN, BN, SC, NaP, C	15	8–40
SE-CB	15–35 CB	AN, SN, C	15	12–16
SE-SP	20–60 SP	AN, SN, BN, SC, NaP, C	15	2–20‡
SE-HSSP	20–60 HSSP	AN, SN, C	15	2–20‡
SE-TNT/Al	5–25/0.5–40 TNT/Al	AN, SN, NaP, C	15	10–30
SE-SP/Al	10–25/1–40 SP/Al	AN, SN, C	15	12–30
SBA-Al	0.1–40 Al 0–12 fuel	AN, AN/SN, AN/NaP, NaP	15	6–30‡
SBA-fuel	4.0–15 solid fuel	AN, AN/SN, NaP, SC	15	3–16‡

Source: Ind. Eng. Chem., **60**(7), 44 (1968). Reprinted with permission of the American Chemical Society. *TNT, trinitrotoluene; CB, composition B; SP, smokeless powder; HSSP, high-strength smokeless powder; fuel, various types—sulfur, gilsonite, other solid hydrocarbons, NH_4—lignosulfonate, others. †AN, ammonium nitrate; SN, sodium nitrate; BN, barium nitrate; NaP, sodium perchlorate; SC, sodium chlorate; C, combinations (AN/SC incompatible). ‡Low percent requires water extenders, such as formamide, ethylene glycol, sugar, molasses.

The term "blasting agent" is usually applied to ammonium nitrate mixtures sensitized with nonexplosive fuels, such as oil or wax. Such mixtures are not explosive, even with conventional blasting caps, and require a powerful booster to start detonation.

"Slurry explosives" are ammonium nitrate mixtures which frequently contain another oxidizer as well and a fuel dispersed in a fluid medium which, among other functions, controls the rheology of the gel-slurry. Cook[11] lists the types and compositions of present-day industrial explosives. Explosives play such a vital part in the nation's economy that their consumption has been suggested as a reliable index of the general scientific and technical atmosphere. The current energy crisis will undoubtedly stimulate the use of explosives, particularly in mining coal and low-grade domestic ores. Competition with farmers for low-priced ammonium nitrate for ammonium nitrate–fuel oil mixtures which is the major tonnage explosive will be severe.

NITROGLYCERIN AND DYNAMITE Nitroglycerin was the first high explosive to be employed on a large scale. Nitration is effected by slowly adding glycerin of high purity (99.9+%) to a mixture having the approximate composition H_2SO_4, 59.5%; HNO_3, 40%; and H_2O, 0.5%. Nitration is accomplished in from 60 to 90 min in agitated nitrators equipped with steel cooling coils carrying brine[12] at 5°C to maintain a temperature below 10°C. After nitration, the mixture of nitroglycerin and spent acid is allowed to flow through a trough (a trough is easier to clean completely than a pipe) into separating and settling tanks some distance from the nitrator. The nitroglycerin is carefully separated from the acid and sent to the wash tank, where it is washed twice with warm water and with a 2% sodium carbonate solution to ensure complete removal of any remaining acid. Additional washes with warm water are continued until no trace of alkalinity remains. See Fig. 22.4, where the quantities of raw material are also listed. The product is really glyceryl trinitrate, and the reaction falls under the (nitrate) esterification classification.

$$\begin{array}{l} CH_2OH \\ | \\ CHOH + 3HNO_3 + (H_2SO_4) \\ | \\ CH_2OH \end{array} \longrightarrow \begin{array}{l} CH_2ONO_2 \\ | \\ CHONO_2 + 3H_2O + (H_2SO_4) \\ | \\ CH_2ONO_2 \end{array}$$

[11]Cook, Explosives, A Survey of Technical Advances, *Ind. Eng. Chem.*, **60**(7), 44 (1968).

[12]Brine can be a little cooler than 5°C, but pure nitroglycerin freezes at 12.8°C. Freezing would be hazardous, since it might interfere with the temperature control.

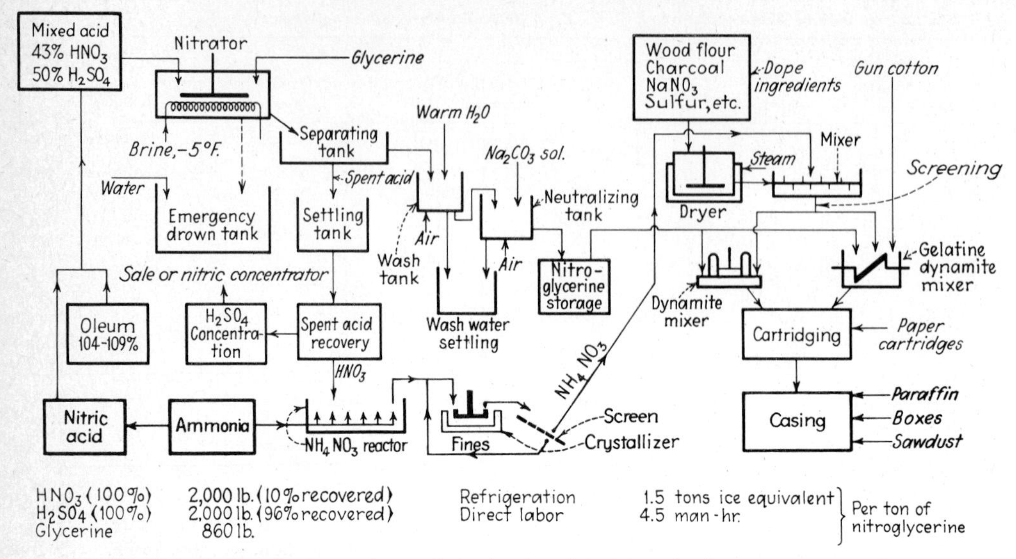

Fig. 22.4 Flowchart for the manufacture of nitroglycerin and dynamites. Usually an evaporator is needed between the ammonium nitrate reactor and the crystallizer.

Small, continuous, stirred stainless-steel nitrators (Biazzi and similar types) offer higher yields, lower operating costs, and greater safety than the batch process illustrated in Fig. 22.4.

Nitroglycerin is a liquid similar in appearance to the original glycerin. It is quite sensitive to blows and freezes at 56°F; when solid it is less sensitive. Since it has a strong tendency in this state to explode incompletely, frozen nitroglycerin must always be thawed before using. To make nitroglycerin easier and safer to handle, it is usually manufactured into dynamite. Dynamite was originally made by absorbing nitroglycerin onto kieselguhr. Modern dynamites generally use wood flour, ammonium nitrate, or sodium nitrate as the agent employed to absorb the nitroglycerin, to which an oxidizer is added. Such a mixture is easy to handle and can be made to contain as much as 75% nitroglycerin and yet retain its solid form. Because of the demand for a nonfreezing dynamite for work in cold weather, dynamites containing other material designed to lower the freezing point of the nitroglycerin are used, for example, ethylene glycol dinitrate. Such nonfreezing dynamites have potentials as great as "straight" dynamite. Nitrocellulose can be gelatinized by nitroglycerin, and the resulting firm jelly is an exceptionally powerful high explosive commonly known as *gelatin dynamite.* Almost without exception the nitro compounds and nitric acid esters used as explosives are toxic. The degree of *toxicity* varies widely with the material in question, but most are capable of causing acute distress if taken orally. Some materials used in the past have been extremely toxic; hexanitrodiphenylamine, for example, is an active vesicant. Extremely toxic properties would weigh heavily against any new explosive that may be introduced.

TOXIC CHEMICAL WEAPONS

Toxic chemical weapons, in modern parlance, include substances which, by their chemical action, produce powerful physiological effects, screening smokes, or defoliant incendiary action. Although it is currently assumed that the employment of such chemicals began in World War I, incendiary and asphyxiating substances have been used over the centuries in warfare. However, they reached a really effective stage on Apr. 22, 1915, with the release by the Germans of 600,000 lb of chlorine against an unprotected enemy along a 2-mi front. Although vast preparations for the offensive and defensive use of toxic gases were made during World War II by all countries, gas warfare was rejected by all for various reasons. Mutual distrust and fear, however, made preparedness a matter of vital importance. In World War II screening smokes and incendiaries proved to be extremely effective, as well as in the Korean conflict.

Toxic materials. Volatile substances designed to cause toxic effects when breathed commonly called *poison gases,* were tried in World War I, but it proved virtually impossible to build up and maintain casualty-producing concentrations of materials with low molecular weights, which diffuse rapidly in air. Toxic agents currently used as so-called poison gases are mostly smokes or agents in liquid form which enter the organism by adsorption through the skin. Such agents are persistent because of their high molecular weight and consequent low vapor pressure. They are difficult to protect against and would be remarkably effective defensive weapons because of their ability to deny an area to an enemy. Their use in offense is necessarily limited, because users must advance through the barrier they have prepared, decontaminating and suffering casualties from their own material in the advance.

Toxic agents designed to produce temporary incapacity are of value in controlling riots and civil disobedience. Such materials may produce sneezing, tears, vomiting, and/or nausea. Most agents suggested for chemical warfare enter the organism through respiration, ingestion, or skin penetration. The nerve agents, Tabun, Sarin, and Soman, are complex organic phosphorous-containing substances which are presumed to be effective lethal agents. From the standpoint of destroying the will to resist, casualty-causing agents should be superior to killing agents. Blistering agents such as nitrogen mustard, mustard, and lewisite are cheap, effective toxicants. The most important temporary incapacitating agent is adamsite which rapidly produces headache and nausea. Common tear gases (lachrymators) include chloroacetophenone, bromobenzylcyanide, and o-chlorobenzylmalonitrile.

Defenses. Because most toxicants do not act primarily through the lungs, masks are useless except against tear gases. Fully protective clothing, without leaks anywhere, is necessary for adequate protection. International agreements outlaw the use of toxicants, but the most formidable opponent is their lack of value as an offensive weapon except in a few exceptional circumstances and the lack of desire in most quarters to destroy people in quantity in a manner generally considered most inhumane.

DECONTAMINATION on a large scale is limited to vital installations, material, and equipment. For ships at sea, complete decontamination is required. Since air, rain, and dew aid in the natural evaporation of chemical weapons, thus reducing or eliminating hazards, only the most persistent ones require a real decontamination. Such environments restrict military operations.[13]

SCREENING SMOKES Various chemicals have been employed to produce smokes or fogs primarily designed to conceal the movements of troops or installations from enemy observations. Screening smokes are basically of two compositions: (1) dispersions of solid particles in air, which correspond to true smokes, and (2) dispersions of minute liquid droplets, which resemble natural fogs or mists. One of the important innovations of World War II was the use of smokes not dependent on water but prepared by atomization of high-boiling fractions of petroleum. Much of the hiding effect of all smokes is due to their ability to scatter light rays by reflection. This is more effective than obstructing. Smokes may be dispersed by various methods, mechanical, thermal, and chemical, even to screen a whole city. Dense clouds of smoke due to quantities of black powder used in combat as early as 1800 interfered with troop movements and obscured vision.

White phosphorus (WP) is loaded directly into shells, bombs, and grenades in the molten state. The material is dissipated by the force of the explosion and immediately burns to P_2O_5. In terms of pounds of smoke-producing agent, this is the most efficient obscuring smoke. The smoke causes

[13]Armed Forces Doctrine for Chemical and Biological Weapons, Employment and Defense, EM 101-40, NPW 36(c), AFM 355-2, LFM 03, Depts. of the Army, Navy, and Air Force, April 1964.

coughing and acid burns. These particles also have some incendiary effect. It has been improved by the development of plasticized white phosphorus (PWP).

Hexachlorethane (HC) is employed in mixtures with finely powdered aluminum and zinc oxide, which are started by a fuse.

$$2Al + 3ZnO \longrightarrow 3Zn + Al_2O_3 \qquad 3Zn + C_2Cl_6 \longrightarrow 3ZnCl_2 + 2C$$

The obscuring power here is less than that of phosphorus. The $ZnCl_2$, which is hygroscopic, attracts moisture to form a fog, the finely divided Al_2O_3 deflects the light rays, and the carbon colors the cloud gray. These mixtures are used in shells, grenades, and floating smoke pots and are known as volatile hygroscopic chlorides.

Sulfur trioxide–chlorosulfonic acid (FS) hydrolyzes in air:

$$SO_3 + H_2O \longrightarrow H_2SO_4 \qquad ClSO_3H + H_2O \longrightarrow H_2SO_4 + HCl$$

This liquid, used mainly in spray tanks by low-flying airplanes, was developed to replace the more costly and scarce titanium tetrachloride. The fumes are highly acid and cannot be used over troops, and the screening power of the smoke due to hydroscopic action is not nearly so effective as that of phosphorus. This compound is also more difficult to store and to handle.

Oil-vapor mists are high-boiling petroleum fractions which, when heated and mixed with steam, produce finely divided particles of steam and oil, resulting in a very dense white smoke. Fog oils are used in mechanical smoke generators, which are gasoline-operated portable units capable of screening huge areas for days. They are economical in cost.

Colored smokes[14] are produced by burning a pyrotechnic mixture of fuel and various colored organic dyes. Anthraquinone dyes are superior and are dispersed into the air, where they act as aerosols of brilliant hue. They are used for signaling purposes.

INCENDIARIES Flammable mixtures are probably the oldest chemical weapons known to humans; the destruction of many ancient cities was due to the use of fire. Samson tied firebrands to the tails of foxes and ran them through the Philistines' grain fields. Incendiaries were the largest single category of chemical supplies consumed during World War II and were vital factors in that victory. More casualties and property destruction were inflicted by a one-night fire-bomb raid on Tokyo than the total damage resulting from both the Hiroshima and Nagasaki nuclear-weapon attacks. Table 22.3 gives more statistics on the World War II use of flame and incendiary weapons. An *incendiary*, strictly speaking, causes ignition of combustible materials at the target, e.g., wooden buildings or stored petroleum products. Incendiaries take the form of bombs, bomblets, artillery shells, and grenades. Flame weapons, on the other hand, provide the fuel for a fire which destroys the target by heat, e.g., personnel, weapons, or electronic equipment. Flame weapons are usually massive bombs or armored, vehicular-mounted flame throwers. Flame weapons of course also ignite combustible material present at the target. Incendiaries can be divided, according to the materials used, into two classes, metallic and petroleum. Respectively, these classes also coincide with intense ignition sources and the wide scattering of flaming fuels described above. Metallic incendiaries include bombs with metallic cases (usually made of a combustible magnesium alloy although steel-cased bombs were used when magnesium became critically short) filled with a mixture of barium nitrate and aluminum, and four-fifths thermite to ignite the case. Thermite is a mixture of aluminum powder and iron oxide, which, when ignited, burns fiercely at a high temperature and *cannot be extinguished by means of water.*

$$3Fe_3O_4 + 8Al \longrightarrow 4Al_2O_3 + 9Fe \qquad \Delta H = -719 \text{ kcal}$$

Some readily ignitible material, such as black powder, is employed to ignite the thermite. Sometimes white phosphorus or a small amount of tetryl is added as a deterrent to firefighters.

[14]ECT, 2d ed., vol. 4, p. 903, 1964.

TABLE 22.3 ***Fire Bombs Used in World War II***

Type of bomb	*Filling*	*Dropped on Germany*		*Dropped on Japan*	
		Number	*Tonnage*	*Number*	*Tonnage*
M47 (s)	IM*	758,540	39,927	601,600	30,080
M76 (s)	PT†	39,370	7,874	37,755	7,151
M50 (c)	Mg	27,620,838	67,463	9,649,928	25,171
M69 (c)	Napalm	....	......	8,472,880	49,700
M74 (c)	PT	....	......	351,000	2,317

Source: Compiled from The Chemical Warfare Services in World War II, p. 74, Reinhold, 1948. *Note:* s denotes bombs dropped singly; c denotes bombs dropped in clusters. *Incendiary, metallic †Petroleum, thickened.

Petroleum incendiaries are gasoline bombs thickened with various ingredients. The first satisfactory thickener was rubber but, because of a critical rubber shortage, other thickeners were sought. One of these was isobutyl methacrylate polymer, which was dissolved in gasoline in combination with calcium soap. Perhaps the most important thickener is *napalm.* This is a granular aluminum soap prepared by precipitating aluminum sulfate in excess alkali with 2 parts of acids from coconut oil, 1 part of naphthenic acid, and 1 part of oleic acid. The soap is capable of withstanding elevated temperatures and produces a gasoline jelly at ordinary temperatures on simple mixing. Napalm surpasses rubber gels in effectiveness and is applicable in flame throwers where rubber gels would be too viscous. It was because of napalm that the flame thrower became such an important and formidable weapon.

NONMILITARY APPLICATIONS OF TOXIC CHEMICAL WEAPONS are useful in civilian life. Flame is employed for control of unwanted vegetation. Chemical smokes and fogs for crop protection against frost are increasingly used. It is also standard practice to use tear gas or sternutators to control mobs without resorting to firearms. Modern protective masks are worn in factories and mines. Self-contained breathing apparatus is more effective than masks.

PYROTECHNICS

Because of antifireworks laws, the pyrotechnics industry holds at present only a fraction of its former importance. Pyrotechnic mixtures, however, still have a number of fairly spectacular uses: parachute flares enable airplanes to land safely on strange terrain; marine signal rockets have been much improved by attachment to parachutes; the red signal flare (fusee) has become almost a requirement for modern trucks as well as trains; and many pounds of colored-light mixtures are being used for military purposes. In general, these products consist of mixtures of strong oxidizing agents, easily oxidizable materials, and various other materials which act as binders and which alter the character of the flame, together with the color-producing chemical itself. A typical composition used in the manufacture of *parachute flares* contains the following materials: barium nitrate, oxidizing agent, 34%; metallic magnesium, grained (to give heat), 36%; aluminum powder (to give strong light), 8%; sodium oxalate (to give a yellow tint), 20%; and calcium sterate, castor oil, and linseed oil (as binders), 2%. As this formula does not fire readily, an igniter consisting of a mixture of 75% black powder and 25% of the above formula is always used to start the flame. The tug that opens the parachute actuates a mechanical device which starts the combustion of the formula.

MATCHES

The manufacture of matches is an essential industry that is highly mechanized. At present, practically all matches fall within two classes: *strike-anywhere matches* and *safety matches.* The match-head composition of the former consists essentially of a material with a low kindling point, usually phos-

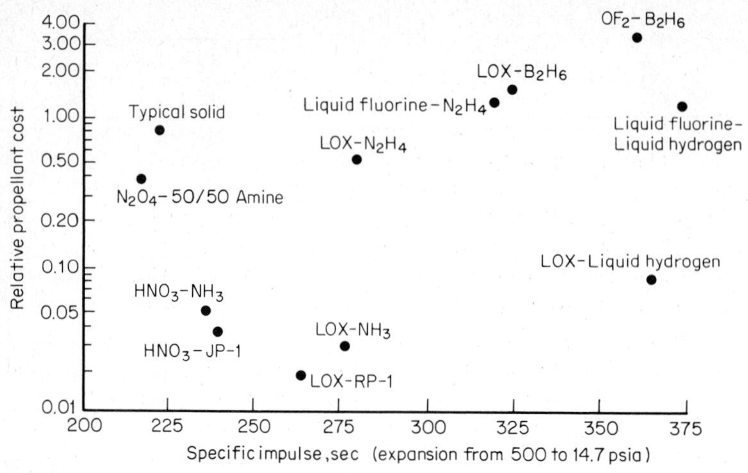

Fig. 22.5 Relative cost and specific impulse of rocket propellants. Note that the better the performance of the propellant, the higher the cost. LOX, Liquid oxygen; N_2H_4, hydrazine, B_2H_6, boroethane (diborane); OF_2, oxygen difluoride; RP-1, kerosine hydrocarbon fuel; JP-1, jet fuel. [*Chem. Eng. News, Sept. 30, 1963, p. 70 (modified, 1975).*]

phorus sesquisulfide (P_4S_3), an oxidizing agent such as potassium or barium chlorate, ground glass, and glue. *Safety matches* are ignited by the generation of heat on the striking surface of the box, the coating of which consists mainly of red phosphorus, ground glass, and glue. No phosphorus sesquisulfide is used in safety matches, but antimony sulfide is used in the heads as a flame-producing agent.

PROPELLANTS OF ROCKETS AND GUIDED MISSILES

Rocket-propulsion fuel systems derive their energy from chemical sources. Propellants[15] are low explosives and carry their own oxidant or other reactant necessary to cause the planned reaction. The thrust of the escaping hot gases pushes the device forward, according to the principle that forces act equally in opposite directions. However, these highly developed and extraordinarily engineered rockets are but an extension of the skyrocket of fireworks exhibitions. Their current importance lies in the application of their driving force to the launching of missiles and spacecraft both for exploratory and scientific missions and for military defense and offense. Careful selection of the propellant ingredients is important to give high chamber temperature and pressure, low external pressure, and low molecular weight of the gases produced. The reaction temperature may be so high that resistant structural materials are unavailable, in which case the temperature must be reduced. The fuel should be one which provides the greatest amount of heat for the smallest amount of weight. The foregoing factors are the most important ones in the equation defining *specific impulse*, or pounds of thrust per pound of weight of propellant burned per second, wherein the square root of the absolute temperature T divided by the square root of the average molecular weight M of the exhaust gases determines the value of the specific impulse, i.e., $\sqrt{T} / \sqrt{M}$.

PROPELLANT FUELS are presented under liquid and solid, or castable. Actually, there is a great variety of choices from which to select a combination of energy sources and conversion mechanisms for the design of efficient propulsion systems.

[15]Tschinkel, Propellants for Rockets, *Chem. Eng. News*, **32,** 2582 (1954) (excellent article with equations and tables); ECT, vol. 2, p. 760, 1963 (history, bibliography, equations); Penner, Combustion and Propulsion Research, *Chem Eng. News*, **40,** 74–83 (1963).

TABLE 22.4 ***Current Propellants***

Propellant	*Application*	*Specific impulse**	*Density*	*Combustion temperature, °F*
O_2/C_2H_5OH (92.5%)	Redstone	287	0.99	5,640
O_2/RP-1	Atlas, Thor, Jupiter, Titan I, Saturn, F-1 engine	301	1.02	6,150
O_2/H_2	Centaur, Saturn, J-2 engine, M-1 engine	391	0.28	4,950
WFNA†/JP-4	Small-aircraft engines	268	1.33	5,360
IRFNA‡/UDMH	Agena, Able	277	1.25	5,350
IRFNA/60% JP-4 + 40% $(CH_3)_2N_2H_2$	Bomarc A	269	1.31	5,310
N_2O_4/50% N_2H_4 + 50% $(CH_3)_2N_2H_2$	Titan II, Titan III, Trans, Apollo	288	1.21	5,590
$N_2O_4/(CH_3)N_2H_3$	Gemini	288	1.21	5,640
IRFNA/MAF§	Bullpup B	270	1.31	5,000
90% H_2O_2/JP-4	Small-aircraft engines	266	1.29	4,580
90% H_2O_2 monopropellant	Mercury, many others	151	1.39	1,400
N_2H_4 monopropellant	Mariner, Ranger	191	1.01	1,150

Source: Chem. Eng. Prog., **60**(7), 73 (1964). *Pounds of thrust per pound of weight of propellant burned per second. See text. †WFNA, White fuming nitric acid. ‡IRFNA, Inhibited red fuming nitric acid. §MAF, Mixed amine fuel.

Liquid propellants[16] are those introduced into the combustion chamber as liquids. The term includes all liquids used in a single propellant system, serving as fuel, oxidizer, and catalyst. Inert additives may be in the form of a single compound or a mixture of two or more. Liquid fuels are, for the most part, used in bipropellant systems (fuel and oxidizer stored in two fuel tanks and fed separately to the combustion chamber). Monopropellants combine the fuel and oxidizer in one mixture. A few better-known liquid propellants are listed in Table 22.4, together with some of their characteristics. Liquid propulsion systems are important means for achieving the propulsive energy required for human space travel.

Large, high-performance rockets now use LOX/LH_2 propellants, because of their high specific impulse and low cost. Greater storage efficiency has been achieved by catalytic conversion of nearly all the hydrogen to the para form. Large solid units are used where long storage times are required.

Solid propellants[17] have advantages over liquids in that they are simple in design and more easily stored, handled, and serviced. They cost less and, because the ingredients are combined in a mold which serves to confine the materials and impart the desired grain configuration, they can be launched expeditiously, and predictable burning rates can be achieved with different formulations. Originally, solids were classified in two groups: (1) *heterogeneous*, or *composite*, propellants (oxidizer and reducer present in two distinct phases), and (2) *homogeneous*, or *double-based* (oxidizer and reducer present in a single or collodial phase, e.g., nitrocellulose dissolved in nitroglycerin). Small percentages of additives are used to control the physical and chemical properties of the solid propellant. Sutton gives the properties of solid propellants:

[16]Kit, Rocket Propellant Handbook, Macmillan, 1960.

[17]Sutton, Rocket Propulsion Elements, 3d ed., Wiley, 1963; Hendel, *op. cit.;* Kit, *op. cit.;* Solid Groundwork for Solid Rockets, *Fortune,* **71,** 114–121 (March 1965) (excellent illustrations of case and nozzle fabrication for "modern and fierce compounds that strain the polymer chemist's newfound art").

TABLE 22.5 ***Solid Propellants for Rockets***

Propellant type	*Castable composite*	*Extruded double base**
Propellant system	Oxidizer-fuel, $NH_3ClO_4(C_2H_4O)_n$	NC NG Miscellaneous
Typical ingredient variation, %	$KClO_4$ (50–85), C_2H_4O (50–15)	NC (50–60) NG (30–45) Miscellaneous (1–10)
Adiabatic flame temperature, °F	2800–4500	3800–5200
Typical sea-level specific impulse	175–240	205–230
Characteristic velocity, ft/s	3,800–4,800	4,500–5,000
Specific weight, lb/in.^{3}	0.055–0.063	Average 0.058
Lower combustion limit, psi	<200	
Probable allowable operating-temperature limits, °F	−70–170	−20–140
Storage stability	Good	Fair
Smoke	Much at low oxidizer; little at high oxidizer; mist at relative humidity greater than 80%	Little
Mechanical properties	Soft and resilient to hard and tough	Hard and tough

*NC, nitrocellulose; NG, nitroglycerin.

At a 1965 conference,[18] many of the aspects of castable solid propellants were considered. One of the most interesting had the following composition: ammonium perchlorate, 70%; aluminum, 16%; elastomeric polymer binder (polybutadiene, polyurethane, etc.) 14%. Strong, fine aluminum wire was substituted for the usual aluminum powder. If properly wound throughout the mass, the aluminum wire will be oxidized uniformly and supply the same heat as previously obtained from aluminum powder. It also strengthens the grain structure, thus improving the mechanical properties of the propellant. If properly wound, the aluminum wire takes the place of part of the container vessel for the propellant and yet burns away as the reaction proceeds. The overall effect is an efficient propellant with reduced container weight. Solid propellants have a variety of growing applications: as propulsion units for missiles, target drones, and supersonic sleds. Rockets used for the separation of stages during flight and for settling ullage in the liquid-propellant tanks are usually of the solid type; for example, some details of the Minuteman ICBM solid-propellant rocket stages are available.[19] Glass-filament-wound casing for both the first and second stages of the submarine-launched Polaris carry a powerful solid propellant. The retro-rockets the Mercury capsule uses for atmosphere reentry employ solids. Flowcharts and equipment have not been released, but processing methods have been described.

Artificial satellites compose another category of spacecraft which require rocket propulsion for launching. Several of the nonmilitary type have been placed in the earth's orbit; these include vehicles for space communications (Telstar II), for meteorological observations (Zeros VII), for scientific measurements, and for human space flight (Faith VII). Many military satellites have also been sent into orbit, which include "watchdog" satellites equipped to detect nuclear explosions.

SELECTED REFERENCES

Afanas'ev, G. T., and V. K. Bobolev: Initiation of Solid Explosives by Impact, International Society Book Services, 1971.
Albright, L. F., and R. N. Shreve: Nitration, Encyclopedia of Science and Technology, vol. 9, McGraw-Hill, 1971.

[18]A. C. Eringen and H. Liebowitz (eds.), Proceeding of an International Conference on the Mechanics and Chemistry of Solid Propellants, Apr. 19, 1965, Purdue University, Lafayette, Ind., supported by the U.S. Navy, Office of Naval Research.

[19]McGraw-Hill Encyclopedia, vol. 11, p. 600, 1966.

Boyars, C. and K. Klager (eds.): Propellants Manufacture, Hazards, and Testing, ACS, 1969.
Brauer, K. O.: Handbook of Pyrotechnics, Chemical Publishing, 1974.
Carton, D. S. (ed.): Rocket Propulsion Technology, Plenum, 1962.
Casci, C. (ed.): Fuels, New Propellants, Pergamon, 1965.
Cook, M. A.: The Science of High Explosives, ACS, 1958.
Ellern, H.: Military and Civilian Pyrotechnics, Chemical Publishing, 1968.
Ellern, H.: Modern Pyrotechnics: Fundamentals of Applied Physical Pyrochemistry, Chemical Publishing, 1961.
Fedoroff, B. T., and O. E. Sheffield: Encyclopedia of Explosives and Related Items, NTIS, Dept. of Commerce, 1972.
Fry, B., and F. E. Mohrhardt: Guides to Information, vol. 1, Space Science and Technology, Wiley, 1963.
Gould, R. F. (ed.): Literature of Chemical Technology, chaps. 9, 10, and 37, ACS Monograph, 1967.
Handling and Storage of Liquid Propellants, U.S. Government Printing Office, 1963.
Heath, D. F.: Organophosphorus Poisons, Pergamon, 1961.
Hosny, A. N.: Propulsion Systems, Edwards, 1966.
James, R. W.: Propellants and Explosives, Noyes, 1974.
Kit, B., and D. S. Evered: Rocket Propellant Handbook, Macmillan, 1963–1964.
Martin, D. L. (ed.): Borax to Boranes, ACS, 1961.
Military Chemistry and Chemical Agents, Technical Manual 3-215, Air Force Manual 355-7, Dec. 6, 1963.
Morrow, C. T., *et al.* (eds.): Ballistic Missile and Aerospace Technology, Academic, 1962, 4 vols.
Penner, S. S.: Chemical Rocket Propulsion and Combustion Research, Gordon and Breach, 1962.
Ring, E. (ed.): Rocket Propellant and Pressurization Systems, Prentice-Hall, 1964.
Rothschild, J. H.: Tomorrow's Weapons: Chemical and Biological, McGraw-Hill, 1964.
Tooley, P.: Fuels, Explosives and Dyestuffs, J. Murray, London, 1971.
Urbanski, T.: Chemistry and Technology of Explosives, Pergamon, 1964–1967, 3 vols.
Vance, R. W. (ed.): Cryogenic Technology, Wiley, 1963.

chapter 23

PHOTOGRAPHIC PRODUCTS INDUSTRIES

The word photography is derived from two Greek words which mean "drawing by light." Vision is the most common method humans have at their disposal for receiving and conveying impressions of the world in which they exist—ample proof of the importance of photography to modern civilization. There is no field of human activity at the present time, whether it be industry, science, recreation, news reporting, printing, or the mere recording of family histories, which is not, in some phase or other, touched upon by the photographic process. Although it is one of the youngest of the chemical process industries, it has universal appeal.

HISTORICAL[1] The first recorded use of a lens for image formation occurred in the latter part of the sixteenth century. The effect of light on silver salts was known to the early alchemists, but it was Wedgwood, son of an English potter, who at the beginning of the nineteenth century first successfully reproduced images as negatives on paper impregnated with silver salts. In 1819, Herschel discovered the fixing properties of *sodium thiosulfate*, thus paving the way for permanent pictorial reproductions and making possible the first exhibit, in 1839, of so-called *photographs*. The same year, a Frenchman, Daguerre, released to the public his formula for the manufacture of the familiar daguerreotype, and an American, Draper, made what was probably the first photographic portrait.

The announcement of these processes created a demand for better lenses, which was soon satisfied. In 1851, Archer improved and perfected the "wet-collodion" process, but in the hands of the public all the collodion processes eventually bowed to the superior gelatin dry plate discovered by Maddox. In 1889, George Eastman introduced transparent roll film and popularized the now familiar snapshot camera. Soon after this, practical methods for producing motion pictures were invented. Color-sensitized emulsions were used as early as 1904 by the Hoechst Dye Works of Germany, and the famous Wratten panchromatic plates were introduced in 1906. Velox developing paper was announced in America at about the same time, as the result of the discoveries of Leo H. Baekeland. Portrait film was introduced about 1920, and projection printing came into general use about the same time. Since the early 1920s, developments in all phases of the art have been fundamental and rapid, culminating in the introduction of natural color film about 1928 and practical amateur color prints in 1941. It is this latter field, color photography, that now holds the most promise for the future. The present advanced position of the photographic industry was attained as a result of thorough research on the fundamentals in this field and their application.

USES AND ECONOMICS Photography finds widespread application. The amateur uses it in three major ways, for reflection prints, for home movies, and for small transparencies. Professional usage is more varied: entertainment, education, sales promotion, graphic reproduction in magazines,

[1]Neblette, Photography: Its Materials and Processes, 6th ed., Van Nostrand, 1962; Mees, Theory of the Photographic Process, 3d ed., MacMillan, 1966.

TABLE 23.1 ***Value of Photographic Products*** *(In millions of dollars)*

Class of products	Value of shipments 1971	1973	1974*
Photographic equipment	4,371	5,350	5,850
Still-picture equipment	400	550	620
Photocopying equipment	1,292	1,580	1,730
Motion-picture equipment	140	180	195
Microfilming and blueprinting	103	130	145
Photographic sensitized film	1,547	1,900	2,090
Silver halide photographic paper	310	350	375
Sensitized paper other than the silver halide type	228	250	260
Prepared photographic chemicals	201	240	255

Source: U.S. Industrial Outlook 1974, U.S. Dept. of Commerce. *Estimated by the Bureau of Competitive Assessment and Business Policy.

display advertising, industrial illustration, data recording, and medical and scientific records. Sales of photographic merchandise—equipment and supplies—reached a worldwide figure of approximately $63 billion in 1974. In the United States, 11% of the products are purchased for medical use; 13% by professional movie makers; 14% for photocopying and reproduction; 25% by business and commercial users, and 36% by amateurs. It is estimated that more than 6 billion still pictures—about 85% in color—are made by amateurs. In home movies, color film has almost completely (99%) displaced black-and-white. Table 23.1 gives the breakdown for the value of shipments of photographic equipment by type of product for the years 1971, 1973, and 1974.

PHOTOGRAPHIC PROCESS Photography is the process of producing a visible image upon a substance by the action of light or other radiant energy. Thus ultraviolet and infrared are included as initiators, and the word "light" encompasses that portion of the electromagnetic spectrum from the ultraviolet region through the visible region and into the infrared. In the case of the economically significant photographic processes, the "light"-sensitive substances employed are silver halides, diazo compounds, amorphous selenium, and zinc oxide. Silver halides are employed as the sensitive substance in the most widely used photographic materials.[2] Such materials have a natural sensitivity to ultraviolet and blue radiation. Their sensitivity to green, red, and infrared is negligible unless special sensitization to these regions is induced by adsorption of so-called sensitizing dyes to their surfaces. Sensitization to blue, green, and red is particularly important in the representation of colors as shades of gray in black-and-white photography and in their simulation in color photography. Light-sensitive silver salts are prepared in the dark in aqueous gelatin emulsions which, after proper additional treatments, are coated on a support which may be glass, paper, plastic, or some other material. Although manufacturers closely guard exact procedures employed in preparing commercial emulsions, the general principles are described in the literature.[3] The technology is complicated, and the user of photographic materials generally purchases a packaged coated product.

Exposure of the emulsion in a camera or other suitable device results in a photochemical reaction on the surface of the silver salt crystal. Photolytic silver, constituting a *latent image,* forms roughly in proportion to the amount of incident light of the color to which the crystal is sensitive. The latent image serves to catalyze further conversion of the entire crystal to a metallic silver image upon the action of a developing agent. The density of the image depends upon the number of crystals exposed and the extent to which development is permitted to occur. Crystals which did not become exposed or were insensitive to the exposing light develop slowly or not at all. Removal of these

[2]The photographic industry used 48 million troy ounces of silver in 1973; *Chem. Eng. News,* July 1, 1974, p. 7.

[3]Glafkides, Photographic Chemistry, 2d ed. (translated from the French), Macmillan, 1958–1960, 2 vols; Duffin, Photographic Emulsion Chemistry, Focal, London, 1966.

silver salts and the undeveloped portions of those which had been exposed is accomplished by the formation of soluble complexes, which are later washed from the material by water. The result of this process is an image whose density bears a direct relationship to the intensity of the initiating light energy. Thus a light object is reproduced as a dark image, and vice versa. When the sensitive material is exposed in a camera, the result of the process described above is a negative. Positives, or prints, are generally produced by using the negative to modulate the exposure of a second piece of photographic material, followed by processing as above.

Recent technological advances have brought direct positive or reversal camera films to the market. Direct-positive photographic images are obtained by first developing the negative silver image and then dissolving the silver in an oxidizing solution or "bleach." The residual complement of silver halide, which has the configuration of the positive image, is uniformly fogged either physically by light or chemically by a reducing agent and then developed to give a positive silver image.

Photographic equipment and materials. All *films* and *plates* consist essentially of an emulsion on a film support of cellulose acetate, polyester, or glass. The *emulsion* is composed of a suspension of minute silver halide crystals in gelatin, suitably sensitized by the addition of certain dyes or various classes of sulfur compounds. In addition, antifogging agents, hardening agents, and an antihalation backing are also used. Halation is fogging of the emulsion by light reflected into it from the back surface of the film. The back of the film is generally coated with a layer of hardened gelatin to prevent curling. Table 23.2 lists important photochemicals and their functions.

Processing of black-and-white materials. Modern developing solutions contain mainly four functional constituents: an organic reducing agent, a preservative, an accelerator, and a restrainer (Table 23.2). The function of the *reducing agent* is to reduce, chemically, the silver halide to metallic silver at the various points where light has produced the latent image. As this reduction is a rate process, practical *developers* are those which reduce the silver halide of the latent image to silver much faster than they reduce the silver halide unaffected by light. Chemically, they are polyhydroxide, amino hydroxide, or polyamine derivatives, and are mostly aromatic. The developing agents which are commercially important are hydroquinone, *p*-methylaminophenol (Elon, Metol), certain heterocyclic substances such as Phenidone, and *p*-phenylenediamines. The last-mentioned are particularly important in color photography, and will be discussed further. *Preservatives* guard the developer against aerial oxidation. The most common is sodium sulfite, but the bisulfite and metabisulfite are also employed. *Accelerators* increase the alkalinity of the developing solution, hence increase the activity of most of the developing or reducing agents. They include the carbonates of sodium and potassium, sodium metaborate, and borax. Antipollution laws in several states have caused the use of borates and borax to be discontinued. In order to control the speed of the developer, it is necessary to employ a *restrainer,* usually potassium bromide or a heterocyclic compound such as benzotriazole.

The theory of the development of the photographic image is connected with the properties of the emulsion. In the silver halide crystals internal dislocations act as electron traps. When a minute amount of light energy is absorbed by these supersensitive sites, a free atom of silver is liberated. This silver atom acts as a sensitive spot which absorbs more light until a positively charged stable latent image, Ag_4^+, is formed. This latent image is changed into a visible image by the developer's depositing on these nuclei sufficient metallic silver from the silver haloids of the emulsion. In the presence of sufficient sulfite a typical developer, *N*-methyl-*p*-aminophenol (Metol), probably reacts as follows:

$$CH_3NH-C_6H_4-OH + 2AgBr + Na_2SO_3 \longrightarrow CH_3NH-C_6H_3(SO_3Na)-OH + 2Ag + HBr + NaBr$$

TABLE 23.2 ***Selected Photochemicals Used in Black-and-white Photographs*** *(Chemicals represent approximately 2% of processing costs)*

Use	Common or trade name	Chemical name	Action
Developing agents			
Capable of reducing exposed crystals of silver halide in emulsion to silver without reducing unexposed crystals at the same time	Hydroquinone	*p*-Dihydroxybenzene	Slow, powerful
	Metol, Elon	*p*-Methylaminophenol	Fast, soft-working
	Phenylenediamine	*p*-Phenylenediamine	True grain (toxic)
	Phenidone	1-Phenyl-3-pyrazolidone	Forms superadditive mixtures with other developing agents
Activators			
Activate developing agent and control pH		Sodium carbonate	Forms sodium salt
		Potassium carbonate	Forms potassium salt
Preservatives			
Guard against oxidation		Sodium sulfite	Transforms oxidation products into colorless compounds
Prevent staining of emulsion		Sodium carbonate	
		Sodium citrate	
Restrainers			
Prevent chemical fog		Potassium bromide	Lowers degree of ionization of silver bromide
		6-Nitrobenzimidazole	
		Benzotriazole	
Calcium precipitants			
Prevent sludging	Calgon	Sodium hexametaphosphate	Forms complex compounds
	Quadrafos		
		Sodium tetraphosphate	Forms complex compounds
Wetting agents			
Facilitate absorption of developing solution	Denatured alcohol	Ethyl alcohol	
Neutralizer		Acetic acid	Stops development by acidification
Fixatives			
Dissolve unexposed silver halides		Ammonium thiosulfate	Converts silver halides into water-soluble compounds
		Sodium thiosulfate hypo	
Hardeners		Alum	
		Glyoxal	Hardens gelatin to insolubilize it and increase mechanical strength
Intensifiers			
Increase density	Mercuric iodide	Polymeric dialdehyde	
Reducers			
Reduce density	Potassium ferrocyanide with hypo		Varies with concentration; oxidation process
	Potassium permanganate and sulfuric acid		

Source: Augmented from Glafkides, Photographic Chemistry, Fountain, London, 1958 (modified in 1975).

Development is arrested at the desired point by immersing the photographic material in a mildly acid bath, usually dilute acetic acid, which destroys the alkaline condition necessary for development.

Modern *fixing agents* are of a nonhardening or acid-hardening type. Their main purpose is to render the silver image permanent by dissolving away the undeveloped silver halide. The type most commonly employed includes a silver halide solvent (sodium or ammonium thiosulfate), an anti-staining agent (acetic or citric acid), a preservative (sodium sulfite), and a hardening agent (potassium chrome alum or formaldehyde).

The processing of black-and-white photographic materials is concluded by a *wash* in running water to remove the silver halides solubilized by the fixing baths. Thorough washing is necessary for long-term stability of the image to avoid staining and bleaching due to decomposition of unremoved silver complexes. After washing, the materials are dried, often with heat. Where long-term stability or highest quality is not necessary, or where rapid access to the final image is desired, stabilization processing is sometimes employed. This process often consists of two baths: a conventional developer, followed by the stabilizing bath itself, which often contains sodium thiosulfate, potassium thiocyanate, or urea. The silver complexes thus formed are not washed from the material.

The photographic process may be summarized by these steps: exposure of the negative film or plate in a camera, followed by development of the latent image, fixing of the image by removal of the unaffected but still sensitive silver halides, and drying of the negative. The negative is used to make a positive, usually on paper, which is subjected to the same sequences of development, fixation, and drying.

COLOR PHOTOGRAPHY—THEORIES, MATERIALS, AND PROCESSES

Color photography is based upon the principle that the colors of nature can be adequately represented to the eye and brain by mixtures of blue, green, and red light. Such mixtures have been produced by projecting in register colored beams of light emanating from properly prepared transparent positive images or by modulating, by silver images, microscopic blue, green, and red filters which are juxtaposed on a support. In the latter method, the eye receives from any area of the picture the amounts of blue, green, and red necessary to reproduce the intended color in that area. A color television picture tube is an electronic analog of this system.

The use of blue, green, and red beams or filters is difficult in practice and is wasteful of light energy. Most methods of color photography are based on the *complements* of blue, green, and red, which are yellow, magenta, and cyan, respectively. Yellow results when blue is absent from white light; magenta, when green is absent; and cyan, when red is absent. Thus a yellow filter controls the blue component of white light while permitting green and red to pass; magenta controls green while permitting blue and red to pass; and cyan controls red while permitting blue and green to pass. Therefore combinations of various densities of yellow and magenta produce a variety of colors, including orange and red; yellow and cyan produce greens; and magenta and cyan produce blues. These combinations can be effected by superimposed dye-containing layers on a single support.

The upper cross section in Fig. 23.1 shows the arrangement of blue-, green-, and red-sensitive emulsion layers and the effect of exposure. A yellow filter layer prevents blue light from reaching the green- and red-sensitive layers, which retain sensitivity to blue. The dyes are located as shown in the other sketches, depending on the color-product objective. Dyes with properties suitable for color photography are produced when the development is accomplished by *p*-phenylenediamines in the presence of an active species called a *coupler.* The dye-forming reactions are:

$$\text{Exposed silver salt} + \text{developer} \longrightarrow \text{oxidized developer} + \text{silver}$$

$$\text{Oxidized developer} + \text{coupler} \longrightarrow \text{dye}$$

In the case of specific developer, *N,N*-diethyl-*p*-phenylenediamine, and a specific coupler, α-naphthol, the overall reaction can be written

NH_2 OH O

+ + $4Ag^+ \longrightarrow$ + $4Ag + 4H^+$

$N(C_2H_5)_2$ N

Developer Coupler

$N(C_2H_5)_5$

Cyan dye

This dye, of the indoaniline class, is cyan. From benzoylacetanilide as the coupler, a yellow dye of the azomethine type results, and from a pyrazolone, a magenta dye is formed:

O O N CH_3

C—C—C—NH N

N O N

$N(C_2H_5)_2$ N

Yellow dye C_2H_5 C_2H_5

Magenta dye

Such dyes are of the Kodachrome type and involve inclusion of the coupler in the developer solution. *Such a process requires that only one layer develop at a time and that all reactants be washed out of the photographic material before the next dye is produced.* Careful control is needed to prepare each layer for development. If, however, the bulk of the coupler is increased so that it cannot diffuse through gelatin and certain other characteristics are introduced, the coupler can be incorporated in the emulsion by the manufacturer. The exposed silver salts can then be developed in all layers simultaneously. The oxidized developer finds the proper coupler in the immediate vicinity and therefore forms the proper dye in the amount required by the subject.

There are two major types of color photographic processing, negative and reversal, as indicated in Fig. 23.1. *Negative processing* involves incorporated couplers and results in color development in the region of the exposed silver salts. A color negative results in dyes complementary to the colors of the subject. Thus a yellow dye is formed in regions exposed to blue, a magenta dye in regions exposed to green, and a cyan dye in regions exposed to red. This negative can be used to make a positive print or any number of intermediate color pictures in order to introduce special effects, as in color cinematography. In making the print, the dyes in the negative control the amounts of blue, green, and red light reaching the color-sensitive emulsion layers on the print material. The latter emulsions, like those of the negative, contain incorporated couplers.

Reversal processes involve the development of exposed silver salts to a silver negative. The remaining silver salts are exposed selectively, and color developed to yield dye layer by layer, as in Kodak Kodachrome film, or are exposed *in toto* and developed to yield dyes in all layers simultaneously, as in Kodak Ektachrome film, which contains incorporated couplers.

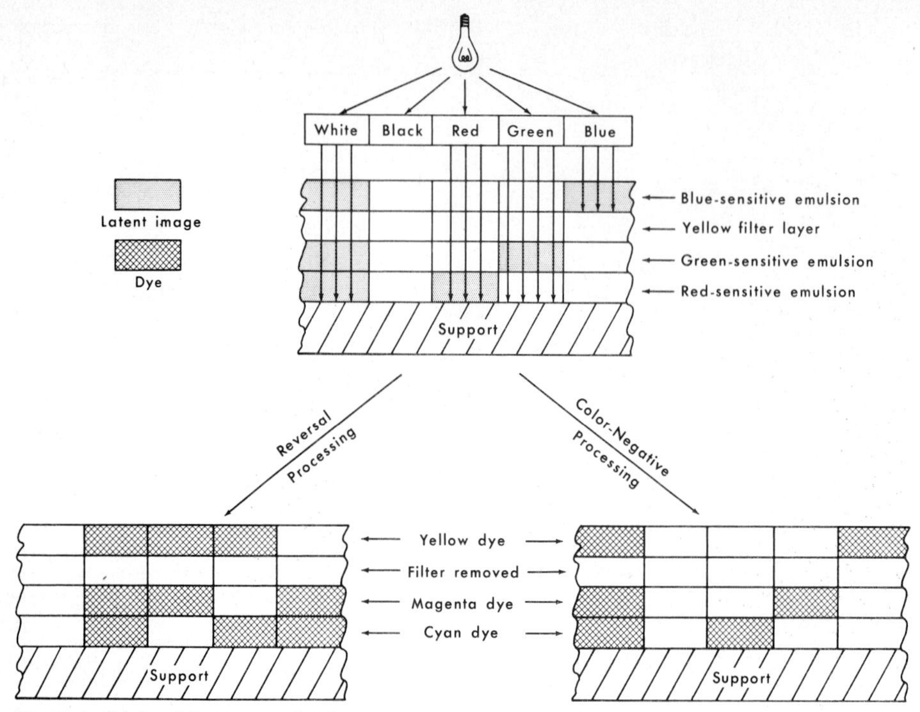

Fig. 23.1 Relation of dye images to the selective sensitivity of color photographic emulsions. (*Eastman Kodak Co.*)

New processes of color photography are being introduced. One of these, *Polacolor* (Fig. 23.2), a product of the Polaroid Corp., depends upon incorporation in the light-sensitive layer of preformed dyes to which developing agents are chemically attached. The process is accomplished as described above by sensitized emulsion layers. The dye developers are soluble in the alkaline processing solution; however, as an image is developed, the dye developer is converted to an insoluble, immobile form. Dye-developer molecules not involved with image development diffuse from the emulsions and are mordanted on a receiver, which becomes the print.

Polaroid's recently introduced SX-70 system[4] involves the exposure of the film and its almost immediate exit from the camera into many million times as much light as is used to expose it. This was accomplished by synthesizing indicator dyes that have high absorption coefficients at high pH values and that become colorless at lower pH values. The exposed film passes through the rollers of the camera which force a small quantity of alkali and titanium dioxide onto it. The alkali permeates and dissolves the layers of dye developer. The exposed silver halide causes oxidation in the developer mixture, resulting in the capture of dye developer where exposed, and then the unoxidized developer passes through to a polymeric acid layer where it becomes colorless.

MANUFACTURE OF FILMS, PLATES, AND PAPERS In the making of photographic films, plates, and papers, three distinct steps are carried out: (1) preparation of the light-sensitive emulsion, (2) manufacture of the base, or support, for the emulsion, and (3) coating of the emulsion on the base. The flowchart shown in Fig. 23.3 gives a general representation of the manufacturing steps involved.

Emulsions. The so-called photographic *emulsion* is in reality not a true emulsion, but rather a dispersion of tiny silver halide crystals in gelatin, which serves as a mechanical binder, a protective colloid, and a sensitizer for the halide grains.[5] Many different types of silver halide emulsions are

[4]Land, One-Step Photography, *Photogr. Sci. Eng.*, **16**(4), 247 (1972); U.S. Pats. 3,647,437 (1972) and 3,707,370 (1972).
[5]Wood, Role of Gelatin in Photographic Emulsions, *J. Photogr. Sci.*, **9**, 151 (1961). A review with 78 refs; Duffin *op. cit.*

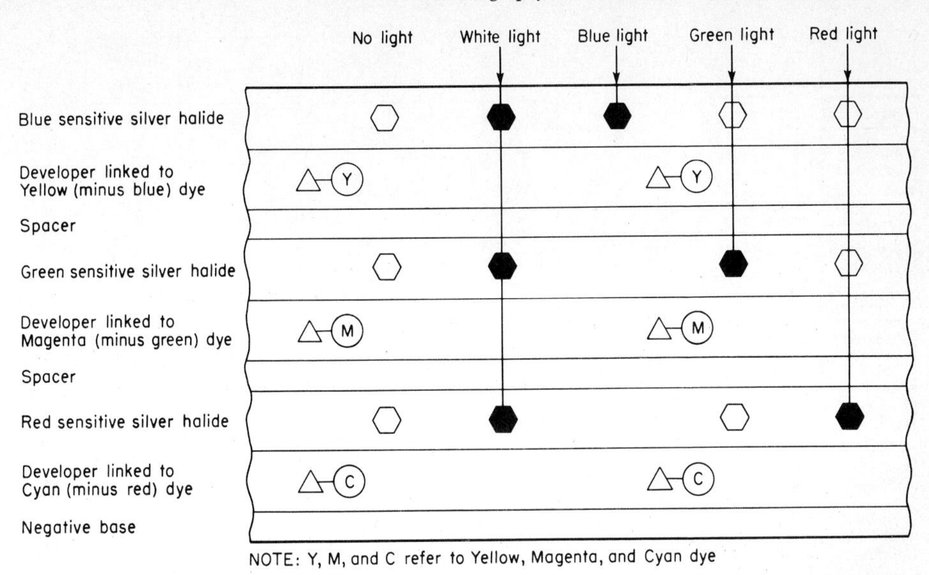

Fig. 23.2 Development process of Polaroid-Land cameras. Structure of the negative: The principal layers of the negative are shown *magnified 6,000 times.* The total thickness of all these layers (coated on a sturdy base) is less than half the thickness of a human hair.

This symbol represents an entirely new kind of molecule, which is a dye linked to a developer by an inactive atomic thread. This nonconducting leash does not allow interchange of electronic charges between the dye and developer, but it does give the developer group control over the movement of the dye. The dye part of these hitched molecules must be different in each layer.

These are unexposed grains of silver halide.

These are exposed grains of silver halide. Note, for example, how a ray of green light will pass through the blue-sensitive layer without exposing the silver halide and will expose a grain in the green layer, but not in the red layer.

In Polaroid cameras, where the exposure and development processes are carried on in the camera itself, the inventor Land uses a novel method of development for both black-and-white and color pictures, wherein the processing liquid is spread from "pods" in a thin viscous layer between the negative and the positive as the two are pressed together. [See Neblette, Photography: Its Materials and Processes, pp. 374, 378, 383, Land-developed the diffusion-transfer-reversal (DTR) process.] This process produces a finished positive black-and-white or colored picture in a remarkably short time and directly from the camera. The Polaroid color film contains an entirely new kind of molecule developed for this use, namely, a dye (yellow, magenta, or cyan) in different layers linked to a developer by an inactive thread. Upon development (pod broken), the reagent reaches all parts of the extremely thin layers and, as it reaches a linked molecule of developer and dyer, it sets the linked molecules in motion.

When a linked molecule reaches a light-exposed silver grain, it becomes involved in the development of that grain and loses its mobility and acts to form its part in the image on the positive; e.g., a ray of green light exposes a silver halide in that layer and traps a linked molecule of developer and magenta dye. Likewise, this happens for other colors of light, building up the complementary color. All these unite together in their respective layers to form the color positive. (*Polaroid Corp.*)

manufactured, the characteristics of each being dependent upon the silver halide used and the details of manufacture. In slow positive emulsions for photographic papers the bromide, chloride, and chlorobromide are chosen. All fast emulsions, usually for negatives, contain AgBr and small amounts of AgI. The iodide is essential for high-speed types, but seldom exceeds 5%. The finished emulsion generally consists of 35 to 40% silver halide and 65 to 60% gelatin. The manufacture[6] of the emulsion may be divided into four principal steps: precipitation, first ripening, washing, and second ripening, or afterripening. The following presentation deals with AgBr-AgI emulsion, and is further illustrated by Fig. 23.3.

Precipitation. The gelatin, after soaking in cold water for about 20 min, is dissolved in hot water. The requisite quantities of KBr and KI are dissolved in this solution, which is then ready

[6]Photographic Sensitizing, *Chem. Eng.* (*N.Y.*) **60**(5), 274 (1953) (pictured flowcharts).

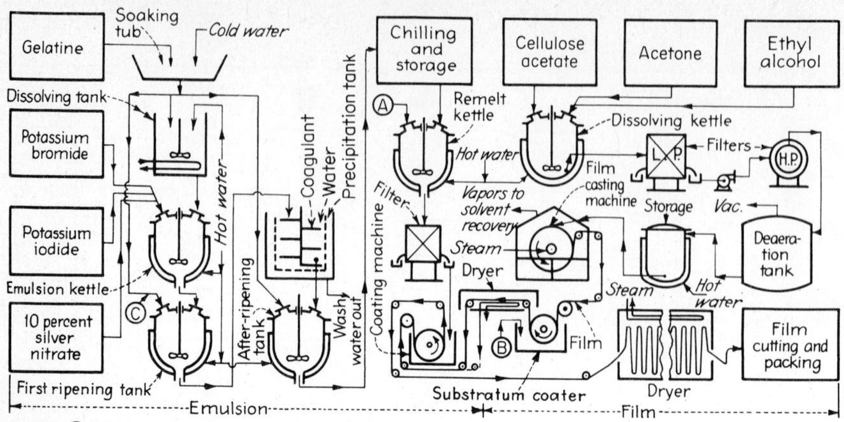

Notes: Ⓐ Chrome alum, preservative, (phenol or thymol), saponin, KBr and sensitizing dyes added here. Ⓑ Gelatine and solvent used for substratum coating.
Ⓒ Gelatine added at close of ripening period.

Potassium bromide	22.5 lb.	Silver nitrate	20 lb.	Ethyl alcohol	90 gal.	Per 10,000 sq. ft. of film; solvents mostly recovered
Potassium iodide	0.6 lb.	Cellulose acetate	450 lb.	Acetone	180 gal.	
Gelatine	80.0 lb.	Glycerine	50 lb.			

Fig. 23.3 Flowchart for the production of photographic film. Extreme cleanliness must be practiced and an appropriate safety light provided after the silver halide has been formed.

to receive the $AgNO_3$ solution. In the case of the ammonia process, a 10 to 15% excess of KBr is used, and the amount of gelatin is 20 to 60% of the total required in the finished emulsion. For neutral, or "boiled," emulsions, a 2 to 5% excess of KBr is generally used, and 10 to 20% of the total gelatin. In the ammonia process a $AgNO_3$ solution is mixed with concentrated ammonia until the initial precipitate of Ag_2O redissolves. This solution is added slowly, with thorough mixing, to the halide-gelatin solution. The mixing temperature is usually about 40°C. For neutral emulsions no ammonia is used. A 10% $AgNO_3$ solution is added to the halide-gelatin solution at 60 to 80°C. In this mixing operation an excess of KBr is necessary to produce large grain size and to prevent interaction of silver ions with the gelatin. The KBr solution is prepared with a relatively small part of the total gelatin, since too much of the latter would interfere with grain growth during the ripening step. Moreover, the principal part of the gelatin is protected from the harmful influence of heat and ammonia. Careful control of the mixing process is essential, since the concentrations of the solutions, the temperature, and the manner in which the $AgNO_3$ solution is added, all have a pronounced effect on the character of the emulsion.

First ripening. The ripening, or digesting, process is essentially a continuation of grain growth under the influence of heat, the temperature being about the same as that of the mixing step. The excess of KBr in the emulsion increases the solvent capacity for AgBr. The small crystals, being more soluble than the large ones, tend to dissolve and reprecipitate upon the large crystals. Since the sensitivity of the final product is dependent upon the grain size, the ripening step is an important one and warrants careful regulation. If this process is carried too far, the finished plate or film will be subject to fogging. After the first ripening, no more crystal growth takes place.

PRECIPITATION AND WASHING At the close of the first-ripening process the warm emulsion is coagulated with a salt solution or an organic sulfonic acid or sulfate. The density of the silver halide–gelatin mixture causes it to settle to the bottom of the vessel, and the aqueous phase is removed by decantation. The emulsion is usually reconstituted by heating the curds in water and reprecipitating. This is sufficient to remove excess halide ions. Careful control of pH, temperature, rate of addition of precipitant, and degree of stirring are all essential to produce a satisfactory material.

AFTERRIPENING (CHEMICAL SENSITIZING) The emulsion grains are still relatively insensitive to light and are treated to increase their sensitivity. Three general kinds of sensitizers are used: (1) sulfur-containing compounds (thiosulfate or thiourea), (2) reducing agents (stannous chloride, sodium sulfite), and (3) salts of precious metals (gold chloride, gold thiocyanate). The original gelatin used is inert, so that known quantities of sensitizing chemicals can be added and thus produce predictable results.

The emulsion grains, after addition of the chemical sensitizer, are digested for some time at 40 to 70°C in the afterripening or second-ripening tank as depicted in Fig. 23.3. The time may vary from a few minutes to several hours.

Before coating the emulsion on the support, it is customary to add chrome alum or formaldehyde as a hardening agent. Phenol or thymol may be introduced to prevent the growth of mold or bacterial attack. The addition of KBr at this stage aids in the prevention of fog. The introduction of amyl alcohol or saponin serves to depress the surface tension of the liquid emulsion, facilitating uniform foam-free spreading on the support. The sensitizing dyes for orthochromatic and panchromatic films are here added to the emulsion. The warm fluid emulsion is filtered and is then ready to be coated on the support.

Preparation of the support.[7] The support for the photographic emulsion can be of glass, paper, cellulose acetate, or polyester film. The use of nitrocellulose for photographic film was abandoned about 1951. Before 1888 glass was universally used. In that year Eastman designed his first roll-film camera. Glass plates are still employed for some scientific and commercial purposes, map making, and some copying and color printing jobs. Glass has the advantages of freedom from distortion and of not deteriorating with age. The glass plates are automatically cleaned by revolving brushes in a strong soda solution. They are coated with a thin substratum of gelatin containing chrome alum and dried in ovens.

The strong, chloroform-insoluble, white, flaky cellulose acetate for film has an acetyl content of 39.5 to 42.0%. To decrease the tendency of cellulose acetate to stretch in water and shrink after drying, substances such as triphenyl phosphate are generally added. This highly viscous solution is known as "dope." Filtration, aeration, and temperature adjustment are necessary before sending the dope to the film base–coating machines (Fig. 23.3). Each machine has a heated, rotating drum 20 ft in diameter and 4 to 6 ft wide. It is coated with silver. By the time the drum has made one revolution, the solvents have evaporated. The film is stripped off, wound into rolls, and transferred to the cooling rooms. The solvents may be recovered by adsorption on activated carbon. The acetate film is given a thin undercoat, or substratum ("subbing"), of gelatin and chrome alum to increase the adhesiveness of the light-sensitive emulsion.

Paper for photographic purposes, is made from rag or high-grade sulfite pulp for the specific use intended to ensure permanence. This base is coated with gelatin containing blanc fixe (precipitated barium sulfate) in order to present a smooth surface for emulsion coating or to provide the desired surface pattern. Sometimes China clay or satin white (a coprecipitate of calcium sulfate and aluminum hydroxide) is used instead of blanc fixe. Photographic paper supports having polyethylene extruded on both sides are being used for applications where rapid drying and dimensional stability are desired.

EMULSION COATING The modern method utilizes multilayer extrusion coating. The patent literature claims that up to six layers can be coextruded onto a ramp and then transferred to the film base without intermixing of the layers, which would be a catastrophe in the manufacture of color film. From an engineering and a scientific standpoint, the multilayer extrusion technique and controls necessary to obtain uniform photographic response are truly revolutionary.

Polyester film support is manufactured by the melt-casting process. The molten polymer is forced through a die to form a continuous sheet. Then this sheet is stretched severalfold in both the

[7]Stevenson, Making Photographic Films and Papers, *Automation,* **9**(6), 151–156 (1962).

machine and cross directions to give it the desired mechanical properties. A final heat treatment is required to stabilize these properties. Both the basic chemical nature and the method of manufacture of polyester support differ markedly from those of cellulose ester film supports which are cast from solvent solutions. The absence of solvent in the making of polyester film support is one of the reasons why polyester-base films show excellent dimensional stability.

SPECIAL APPLICATIONS OF PHOTOGRAPHY

PHOTOMECHANICAL REPRODUCTION FOR ILLUSTRATIONS[8] Photography has found one of its most important applications in the reproduction of photographs on the printed page by means of printing inks. These processes may be classified as (1) *relief printing,* also referred to as photoengraving, in which the raised portion of a plate receives the ink for transference to the paper, and so-called *line* plates and *halftone* plates are used; (2) *intaglio* printing, which includes photogravure, rotogravure, and metal engraving, in which the relief printing procedure is reversed and the hollow regions of the plate or metal cylinder hold the ink; (3) *planographic* printing, or lithography,[9] which makes use of the inability of a water-wet surface to take ink. The *offset printing process* utilizes lithographic plates which are particularly adaptable to illustrative work in color, and the usual separation negatives are used in their preparation.

Photopolymerization,[10] or polymerization initiated by light, has the ability (as has silver halide photography) to amplify the effect of light enormously; the multiple effect, however, occurs simultaneously with exposure, rather than in a separate processing step. This method has been used extensively in the printing field. The Kodak Photo Resist process for making photoengravings was introduced in 1954; it uses the cross-linking of a photopolymer system. Dycril photopolymer printing plates, which use photosensitive acrylic plastic bonded to metal, were marketed in 1960, and the time for making a plate has been reduced to $\frac{1}{4}$ h as compared with 4 h needed for making a conventional metal engraving. A photopolymer printing plate formed by the light-induced cross-linking of nylon has been described in the literature.[11] In 1964 a process called *panography* was introduced—the first successful use of color to create a three-dimensional effect on a flat surface. A thin film of transparent plastic material, called Epolene, is applied to produce the effect, and a special press is used for the process.[12]

PRINTING-OUT TECHNIQUES Photographic copying of documents reaches back to the earliest days of photography. Although newer methods of photocopying have largely replaced the earlier types in which visible images are produced on exposure of prepared paper, some of these *printing-out processes* are capable of producing exceptionally fine prints, and others afford a certain control over tone values unequaled by any other process. Several have been described by Neblette; among them are the silver, chromate, and iron processes.[13] Herschel developed the *blueprint process* in 1842, a method dependent on the fact that ferric ions are reduced to ferrous ions in the presence of organic matter under the influence of strong light. Paper is coated with ferric ammonium citrate and potassium ferricyanide. When a line drawing is placed over the prepared paper and then exposed to light and treated with water, an insoluble blue image is formed where the light reaches the paper; this image is in prussian blue, ferric ferrocyanide ($Fe_4[Fe(CN)_6]_3$). The cyanotype, or positive blueprint (where lines are blue and background light), uses a more radiation-sensitive ferric mixture

[8]For brief technical descriptions of these processes, see CPI 2, p. 492.

[9]Originally, lithography meant engraving on stone and printing therefrom, a process not used currently except for special work in small quantities.

[10]Revolution in Office Copying, *Chem. Eng. News,* July 13, 1964, p. 115, and July 20, 1964, pp. 91–92. This special report gives detailed information on the operation of major equipment available.

[11]*Chem. Eng. News,* July 20, 1964, pp. 91–92.

[12]Graphic Advance, *Barron's,* **47,** May 25, 1964, p. 3.

[13]Neblette, *op. cit.,* chap. 26, Printing-out Processes.

and, processed in a potassium ferricyanide solution, yields Turnbull's blue, ferrous ferricyanide ($Fe_3[Fe(CN)_6]_2$). In 1839—the same year that two other milestones in photography were reached—Breyer experimented with silver halide–coated paper to copy pages from books. However, the first successful document-copying device not requiring especially prepared originals was the Rectigraph, invented about 1900. Other projection photographic copying machines followed, including the Photostat, but all had their shortcomings. Contact-box printers were introduced in the 1930s, but they did not meet the demand for speed and easy operation. It was not until the *diffusion-transfer reversal process* was patented about 1940 that the revolution in copying methods began.[14] These copying techniques are image-forming processes which create essentially full-sized copies of the original and should not be confused with office duplicating processes such as mimeographing, multilithing, or the newer adherography. Copiers reproduce directly from an original document, while duplicators reproduce through the use of an intermediate master.

PHOTOCOPYING PROCESSES More and more business offices, governmental units, especially court record rooms, libraries, and archives find copying machines indispensable in disseminating information efficiently, rapidly, and accurately. The sales of equipment and supplies have increased from less than $50 million in 1952 to $730 million in 1974. Over 1 million photocopiers are in use.[15] Between 70 and 75% of industrial sales are for supplies, which include large amounts of chemicals such as silver halides, gelatin, developers, salts, stearates, and zinc oxide. The decline in chemicals used in *wet-copying* processes will undoubtedly be offset by increased use of premixed solutions returnable to the cartridge when not in use, as well as by increased use of chemicals in *dry-copying* processes, now on the upswing, such as dyes, resins, pigments, organic photoconductors, and waxes, both natural and synthetic. Large-volume users of equipment prefer not only permanent and legible copies, but also inexpensive ones that are almost indistinguishable from the originals, produced by compact, trouble-free machines that operate automatically. See the Xerox machine depicted in Fig. 23.4. A tabular comparison of price of equipment, time per copy, and approximate cost per copy (excluding cost of equipment, repairs, and labor, except for machine rental and depreciation on electrostatic processes) answers some questions that normally arise.[16] The early thermographic processes are described by Neblette,[17] as well as some of the other principal processes listed in Table 23.3. Several processes, such as *thin-layer silver halide film,* which has no gelatin coating, *free-radical photography,* which involves the interaction of diphenylamine and carbon tetrabromide, and a high-speed long-distance electrostatic copying system called *Videograph* have also been reviewed.[18]

Some of the newer systems are Dylux, dry silver, free-radical, and vesicular imaging.[19] An excellent discussion of the various systems using sensitized coated papers has been prepared by Diamond.[20]

The current trend is away from coated-paper copiers and toward the use of plain-paper copiers, even though the original cost of coated-paper machines is less.

Photographic miniaturization systems. Microphotography is the art of making miniature photographic facsimiles of original materials. Its current use in commercial records by in-

[14]*Ibid.,* chap. 28, Diffusion-transfer Reversal Materials (a comprehensive description of this processing technique and its variants, together with hundreds of pertinent patent numbers and 145 refs. on apparatus and applications). U.S. Pat. 2,352,014 was issued June 20, 1944, to André Rott; his first patents were assigned to Gevaert Photo-Producten, N.V., in Belgium, 1939. Weyde, a chemist in Agfa Laboratories, I. G. Farbenindustrie, applied for patents in 1941 on an improved process giving denser images. The diffusion-transfer principle is the basis of Polaroid-Land photographic film developed by Land, who was also working independently in the same period.

[15]*Chem. Eng. News,* July 20, 1964, p. 84; U.S. Industrial Outlook 1974, U.S. Dept. of Commerce.

[16]*Office,* **81**(5), 109 (1975).

[17]Neblette, *op. cit.,* chap. 9, Radiation Sensitive Systems.

[18]Revolution in Office Copying, *Chem. Eng. News,* July 20, 1964, p. 96.

[19]Weiss, Copy Papers, Noyes, 1972.

[20]Diamond, Recent Developments in Sensitized Paper Technology, *Tappi,* **50**(12), 67A (1967).

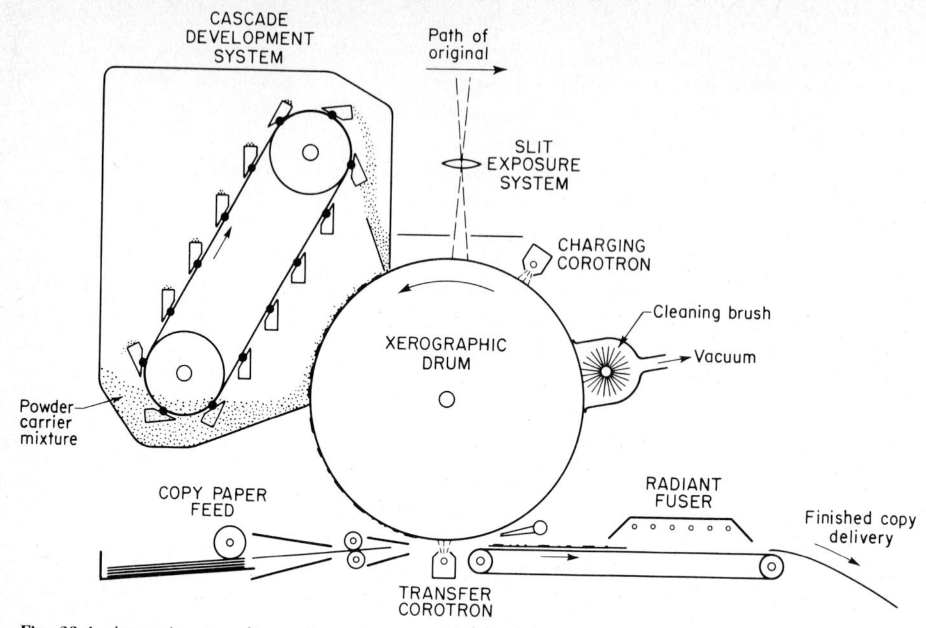

Fig. 23.4 Automatic xerographic copying equipment makes use of the photoconducting property of amorphous selenium plated on the xerographic drum. The selenium holds a positive electrostatic charge in the dark (but allows it to be dissipated when exposed to light). The revolving drum is sensitized through exposure by contact, by projection, or in a camera. The latent image is on the positively charged drum, which attracts the negatively charged black-powder resinous carrier mixture. This resinous pigment is then heated and passed from the drum to the copy paper on which the image is fused.

surance companies, banks, and engineers for active use or security purposes, as well as for historical records and newspaper files processed for permanent retention, has brought about a revolution in *microforms.* In 1934, Biblio-Film Service was organized; it operated first in the Dept. of Agriculture to make a greater volume of the world's literature available. Thirty years later, at the New York World's Fair, both the Old and New Testaments—773,746 words of the King James version—were exhibited on one sheet of transparent plastic less than 2 in. square. This miracle in miniaturization took only 4 h to reproduce in what the manufacturer called photochromic microimages.[21] Rather than silver halide, an organic dye that darkens rapidly upon exposure to ultraviolet light is the significant component of the film. Errors can be quickly erased with a flash of yellow light, and the space is reusable, which is an added feature, as is the fact that all operations can be viewed under green light. The necessity for conserving storage space and for providing physical protection for essential records, and also research in information-retrieval methods, have accelerated development of all facets of microreproduction[22] in which individual copies of any page are available by the touch of a printout button. Kalvar film,[23] introduced commercially in 1956, is used to make contact copies of microfilm and X-ray and motion-picture film. It is developed by heat rather than chemicals; if the film is coated on black paper, the image shows as a black-on-white copy. One company uses this type of film mounted on punched cards in its automatic retrieval systems. Chemical process industries therefore have a dual stake in the new graphic arts systems, as potential users of the processes and as suppliers of the chemicals.

[21]National Cash Register Company's Process Known as PCMI, U.S. Pat. 2,953,454; several other systems are mentioned in Revolution in Office Copying, *Chem. Eng. News,* July 13, 1964, p. 115, and July 20, 1964, pp. 91–92.

[22]Taylor and Kitrosser, Survey of Microfilm Processing Technology, *Proc. Nat. Microfilm Assoc.,* 1963; Path to the Moon Is Lined with Paper, *Reprographics,* **1**(1), 30–32, 35 (1963).

[23]U.S. Pat. 3,032,414 (May 1, 1962).

TABLE 23.3 ***Office Copying Processes***

Process	Paper	Cost, cents per $8\frac{1}{2} \times 11$ in. copy	Copies colors	Copies photos
Electrostatic	Plain	0.3–4	Yes	Yes
Electrostatic	Coated	1.2–4.5		
Electrofax	Coated	1.2–7	Yes	Yes
Diazo	Coated	1.0–1.5	Yes	No
Dual spectrum	Coated	2.5–8.5	Yes	Yes
Magne-dry	Coated	1.9–5.3	Yes	Yes
Thermal	Coated	0.5–4.0	No	Yes
Magne-dynamic	Coated	0.8–2	Yes	Yes
Ammonia	Coated	1.0	No	No
Transfer electrophotography	Plain	2.6	Yes	Yes
Canon N P System (liquid dry)	Plain	1.4	Yes	Yes
Photographic direct lithography	Plain	0.3–1.7	Yes	Yes
Electrographic	Plain	3.0	Yes	Yes
Xerographic	Plain	0.3–2.5	Yes	Yes

Source: Office, **81**(5), 109 (1975).

SELECTED REFERENCES

Baker, A. A.: Photography for Scientific Publications, Freeman, 1965.
Berg, W. E.: Photographic Science, Pitman, 1964.
Boni, A. (ed.): Photographic Literature, Morgan and Morgan, 1963.
Boucher, P. E.: Fundamentals of Photography, 4th ed., Van Nostrand, 1963.
Clerk, L. P.: Photography, Theory and Practice, D. A. Spencer (ed.), Focal, London, 1972.
Cox, R. J.: Photographic Sensitivity, Academic, 1973.
Cox, R. J.: Photographic Processing, Academic, 1973.
Creybecks, A. S. H.: Dictionary of Photography, Elsevier, 1965.
Croome, R. J., and F. G. Clegg: Photographic Gelatin, Focal, London, 1965.
Dentsman, H., and M. J. Schultz: Photographic Reproduction, McGraw-Hill, 1964.
Duffin, G. F.: Photographic Emulsion Chemistry, Focal, London, 1967.
Focal Encyclopedia of Photography, Focal, London, 1966, 2 vols.
Glafkides, P.: Photographic Chemistry, 2d ed., Macmillan, 1958–1960, 2 vols.
Gorokhovskii, Y. N., and T. M. Levenberg: General Sensitometry, Focal, London, 1965.
Gould, R. F. (ed.): Literature of Chemical Technology, chap. 11, Photographic Chemistry, ACS, 1967.
Hautot, A.: Photographic Theory, Pitman, 1964.
James, R. W.: Photographic Emulsions Recent Developments, Noyes, 1973.
James, T. H.: The Theory of the Photographic Process, Macmillan, New York, 1966.
John, D. H. O., and G. T. J. Field: Photographic Chemistry, Reinhold, 1963.
Kosar, J.: Light Sensitive Systems: Non-silver Halide Processes, Wiley, 1965.
Kowalski, P.: Applied Photographic Theory, Wiley, 1972.
Langford, M. J.: Basic Photography, Focal, London, 1965.
Langford, M. S.: Basic Photography, Focal, London, 1965.
Mason, L. F. A.: Photographic Processing Chemistry, Focal, London, 1967.
Morgan, W. D. (ed.): Encyclopedia of Photography, vol. 1, Greystone, 1963.
Neblette, C. B.: Photography: Its Materials and Processes, 6th ed., Van Nostrand, 1962.
Nelson, C. E.: Microfilming Technology Engineering and Related Fields, McGraw-Hill, 1965.
Saxi, R. F.: High Speed Photography, Focal, London, 1967.
Schaffert, R. M.: Electrophotography, Focal, London, 1965.
Spencer, D. A., *et al.:* Color Photography in Practice, Focal, London, 1968.
Verry, H. R.: Microcopying Methods, Pitman, 1964.
Weiss, S. G.: Copy Papers, Noyes, 1972.
Zelikman, V. L., and S. M. Levi: Making and Coating Photographic Emulsions, Focal, London, 1965.

chapter 24

SURFACE-COATING INDUSTRIES

Products of the surface-coating industries are essential for the preservation of all types of architectural structures, including factories, from ordinary attacks of weather. Uncoated wood and metal are particularly susceptible to deterioration, especially in cities where soot and sulfur dioxide accelerate such action. Aside from their purely protective action, paints, varnishes, and lacquers increase the attractiveness of manufactured goods, as well as the aesthetic appeal of a community of homes and their interiors. Here is a case where utility and art advance hand in hand. *Pigments, paint, varnish,* and *lacquers* encompass the majority of products dealt with in this chapter.

Surface-coating industries have undergone a revolution directed by *research* for the purpose of employing better quality in the constituents, whether pigments, solvents, or film formers, and to improve formulation as well as increase adaptability in application procedures. Such advances have reduced costs and both fire and health hazards, and have culminated in longer-lasting improved coatings. Principal emphasis has been placed on (1) architectural coatings, sometimes referred to as "trade sales" products, which also include maintenance paints, with some mention of (2) industrial coatings used on materials of manufacture or on those items produced in the process. Finishes in this latter classification, in contrast to the former, which are generally used on wood, are applied on a wide variety of materials, such as metal, textiles, rubber, paper, and plastics, as well as wood.

HISTORICAL The surface-coating industry is indeed an ancient one; in fact, Noah was told to use pitch within and without the Ark. The origin of paints dates back to prehistoric times, when the early inhabitants of the earth recorded their activities in colors on the walls of their caves. These crude paints probably consisted of colored earths or clays suspended in water. The Egyptians, starting very early, developed the art of painting and by 1500 B.C. had a wide number and variety of colors. About 1000 B.C. they discovered the forerunner of our present-day varnishes, using naturally occurring resins or beeswax for the film-forming ingredient. Pliny outlined the manufacture of white lead from lead and vinegar, and it is probable that this ancient procedure resembles the old Dutch process. It has been in more recent years, however, that the surface-coating industry has made its greatest strides, which can be attributed to the results of scientific research and the application of modern engineering.

USES AND ECONOMICS The manufacture of surface coatings is big business with sales near $4 billion. No one company controls over 10% of the market, but the 1,500 small companies which freely compete have over 60,000 employees. In 1971, for the first time, sales of industrial coatings exceeded those of trade coatings, indicating a trend toward more permanent, factory-applied finishes. Home application of a wide range of constantly changing types of coatings continues strong.

TABLE 24.1 ***Production of Paint and Related Products in the United States*** *(In millions of gallons)*

	1974	*1973*	*1972*	*1964*
Trade sales (consumer use)	475	424	452	395
Industrial finishes	457	473	475	330
Total	932	897	927	725

Source: Chem. Eng. News, June 2, 1975, p. 34.

PAINTS[1]

Historically, surface coatings have been divided into paint (relatively opaque solid coatings applied as thin layers, whose films are usually formed by polymerization of a polyunsaturated oil), varnishes (clear coatings), enamels (pigmented varnishes), lacquers (films formed by evaporation only), printing inks, polishes, etc. These classifications have been most useful in the past, but the introduction of plastic resins into the industry has made such classifications relatively meaningless. Table 24.2 lists the most important ingredients currently used in surface coatings.

CONSTITUENTS The constituents of paints are outlined in Tables 24.2 and 24.3. The *pigment,* although usually an inorganic substance, may also be a pure, insoluble, organic dye known as a *toner,* or an organic dye precipitated on an inorganic carrier such as aluminum hydroxide, barium sulfate, or clay, thus constituting a *lake.* Pigment *extenders,* or *fillers,* reduce the cost of paint and frequently increase its durability. The function of pigments and fillers is not to provide simply a colored surface, pleasing for its aesthetic appeal, important as that may be. The solid particles in the paint reflect many of the destructive light rays, and thus help to prolong the life of the entire paint. In general, pigments should be *opaque* to ensure good covering power, and *chemically inert* to secure stability, hence long life. Pigments should be *nontoxic,* or at least of very low toxicity, to both the painter and the inhabitants. Finally, pigments must be wet by the film-forming constituents and be of *low cost.* Different pigments possess different covering power per pound.

Without *film-forming materials,* the pigments would not be held upon the surface. Paint films are formed by the "drying" of various unsaturated oils such as those listed in Table 24.2 and further described in Chap. 28. The drying is a chemical change representing oxidation and polymerization;

TABLE 24.2 ***Paint Constituents*** *(Listed in approximate order of importance)*

Resins (film formers)
Synthetics: alkyds, acrylics, vinyls, cellulosics, rosin esters, epoxies, urea-melamines, urethanes, styrenes, phenolics, hydrocarbons, polyesters
Natural: shellac, rosin, and others
Solvents: ketones, aromatics, aliphatics, alcohols, glycol ethers, glycol ether esters, glycols, glycol esters, chlorinated products, terpenes, etc.
Drying Oils and Fatty Acids: linseed oil, soybean oil, fatty acids, tall oil, caster oil, tung oil, safflower oil, fish/marine oil, cocoanut oil, oiticica oil
Pigments and Extenders: titanium dioxide, calcium carbonate, magnesium silicate, clay, inorganic colors, barium sulfate, mica, zinc oxide, zinc dust, red lead, metallics (principally aluminum), carbon blacks, organic colors, white lead.
Driers: cobalt, manganese, lead and zinc, napthenates, resinates, linoleates, 2-ethylhexoates, tallates
Plasticizers: octyl, decyl, 2-ethyl hexyl and similar esters of phthalic, sebacic, adipic, azelaic, and similar acids.

Source: Modified from *Chem. Week,* Oct. 20, 1971, p. 37.

[1]Kiefer, The Paint Industry, *Chem. Week.* Dec. 22, 1969, p. 31; cf. Payne, Organic Coating Technology, Wiley, 1955; Mattiello (ed.), Protective and Decorative Coatings, Paints, Varnishes, Lacquers and Inks, Wiley, 1941–1946, 5 vols.; Paint Technology Manuals, Oil and Chemists Association, London, Reinhold, 1961, 6 vols.

TABLE 24.3 ***Pigments and Extenders for Surface Coatings***

Ingredients		*Function*
	1. Pigments	
White hiding pigments:	Yellow pigments:	To protect the film by reflecting the destructive ultraviolet light, to strengthen the film, and to impart an aesthetic appeal. Pigments should possess the following properties: opacity and good covering power, wettability by oil, chemical inertness, nontoxicity or low toxicity, reasonable cost
Titanium dioxide	Litharge	
Zinc oxide	Ocher	
Lithopone	Lead or zinc chromate	
Zinc sulfide	Hansa yellows	
Antimony oxide	Ferrite yellows	
Black pigments:	Cadmium lithopone	
Carbon black	Orange pigments:	
Lampblack	Basic lead chromate	
Graphite	Cadmium orange	
Iron black	Molybdenum orange	
Blue pigments:	Green pigments:	
Ultramarine	Chromium oxide	
Copper phthalocyanine	Chrome green	
Iron blues	Hydrated chromium oxide	
Red pigments:	Phthalocyanine green	
Red lead	Permansa greens	
Iron oxides	(phthalocyanine blue plus	
Cadmium reds	zinc chromate)	
Toners and lakes	Brown pigments:	
Metallics:	Burnt sienna	
Aluminum	Burnt umber	
Zinc dust	Vandyke brown	
Bronz powder	Metal protective pigments:	
	Red lead	
	Blue lead	
	Zinc, basic lead, and	
	barium potassium chromates	
	2. Extenders or inerts	
China clay	Gypsum	To reduce the pigment *cost* and in many cases to increase the covering and weathering power of pigments by complementing pigment particle size, thus improving consistency leveling and settling
Talc	Mica	
Asbestos (short fibers)	Barite	
Silica	Blanc fixe	
Whiting		
Metal stearates		

it is hastened by pretreatment of the oil and by adding driers, or catalysts, predominantly heavy-metallic soaps, which are oxygen carriers, usually soluble in oil. These driers need be used only in small amounts (1 to 2% by weight). In *emulsion-base* paints, the film-forming materials are the various lattices, with or without other additions. Film formation takes place largely through coalescence of dispersed resin particles to form a strong continuous film. Since paints are mechanical mixtures, the pigments and extenders are carried by or suspended in a *vehicle.* This vehicle is the film-forming oil, to which other liquids are added in varying amounts, not only in the factory but by the painter on the job. The *diluent,* or *thinner,* used for this purpose is a volatile naphtha. To reduce certain aspects of cracking in paints, *plasticizers* are being introduced into the formulas, largely by the proper choice of oils.

Proper paint formulation centers around the specific requirements of the particular application. These requirements may be listed as hiding, color, weather resistance, washability, gloss, metal anticorrosive properties, and consistency, as related to type of application (brushing, dipping, spraying, or roller coating). Individual requirements are met by proper choice of pigments, extenders, and

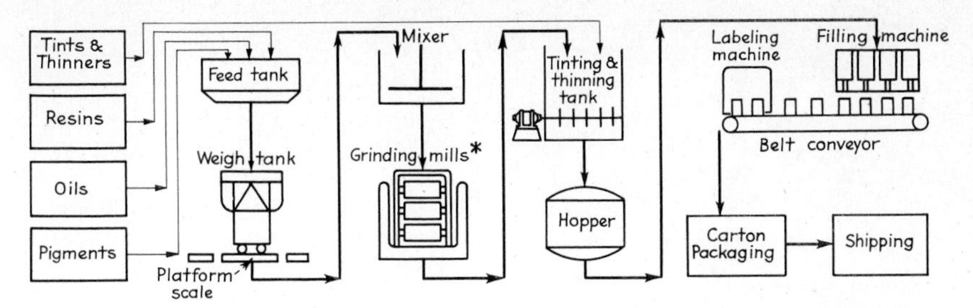

Fig. 24.1 Flowchart for the mixing of paint.

vehicles by the paint formulator. Since the techniques of paint formulation are still largely empirical, it is difficult to predict the properties of a specific formulation, and this often means that a considerable number of trials have to be run before the desired properties are obtained. However, by using simple statistical methods, it has been demonstrated that more efficient testing and development results are obtainable.[2] Almost all the major paintmakers have developed some type of automated color-control system; new methods reduce time required to match paint and, by giving closer control of pigment requirements, keep production costs down.[3] For the modern paint formulator, some authorities believe the most important concept is that of pigment volume concentration (PVC). It is defined simply as

$$\text{PVC} = \frac{\text{volume of pigment in paint}}{\text{volume of pigment in paint + volume of nonvolatile vehicle constituents in paint}}$$

The PVC largely controls such factors as gloss, reflectance rheological properties, washability, and durability. The inherent oil requirements of the pigment-extender combination being applied, however, affect the PVC used in a given formulation.

As a consequence there is usually a range of PVC for a given paint, as indicated in the following tabulation:

Flat paints	50–75%	Exterior house paints	28–36%
Semigloss paints	35–45%	Metal primers	25–40%
Gloss paints	25–35%	Wood primers	35–40%

The PVC of a given formulation serves as the guide for reformulation work, using different pigment or vehicle combinations, and as such is extremely useful to the paint formulator. However, Payne discusses the effects of variation in PVC, suggesting that there is a gradual change in coating properties with a change in pigment concentration, until a *critical concentration* (CPVC) is reached, at which the properties change markedly.

MANUFACTURING PROCEDURES The various operations needed to mix paints are wholly physical. These *unit operations* are shown in proper sequence in the flowchart in Fig. 24.1. *Chemical conversions* are involved in the manufacture of the constituents of paints as well as in the drying of the film. The manufacturing procedures illustrated in Fig. 24.1 and described here are for a mass-production paint. The weighing, assembling, and mixing of the pigments and vehicles take place on the top floor. The mixer may be similar to a large dough kneader, with sigma blades. The batch

[2]Prane *et al.*, Panel Discussion: Statistical Methods, *Off. Dig. Fed. Soc. Paint Technol.*, **31**, 303, 409 (1959).
[3]Color Technology, *Chem. Eng.* (*N.Y.*), **75**(17), 146 (1968).

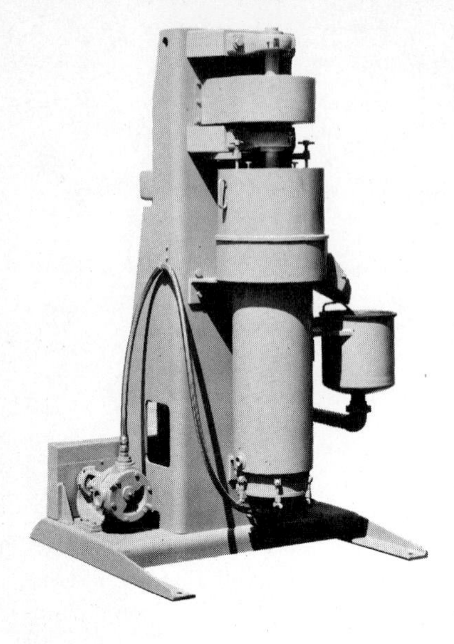

Fig. 24.2 A sand-grinding mill, illustrating the process and apparatus developed by E. I. du Pont de Nemours & Co., Wilmington, Del., covered by U.S. Pats. 2,581,414 and 2,855,156 and foreign equivalents. This mill is a combination type, with round grains of Ottawa sand which grind and disperse the pigments in the medium grinding chamber, followed by a screen which holds back the sand. See Fig. 24.3.

masses are conveyed to the floor below, where grinding and further mixing take place. A variety of grinding mills may be used.[4] One of the oldest methods is grinding, or dispersion between two buhrstones; however, ball-and-pebble mills and steel roller mills were the principal grinding mills used until recently. Sand mills, high-speed agitators, and high-speed stone mills are being used increasingly to grind paint and enamels. The types of pigments and vehicles are the dominant factors in the choice of equipment used (Figs. 24.2 to 24.4).

This mixing and grinding of pigments in oil requires much skill and experience to secure a smooth product without too high a cost. Perry[5] presents these operations in a useful manner, with pictures of various mills and tabulations of the engineering factors concerned. No one machine is universally applicable. See Figs. 24.2 and 24.3 for the sand mill. The paint is transferred to the next-lower floor, where it is thinned and tinted in agitated tanks, which may hold several-thousand-gallon batches. The liquid paint is strained into a transfer tank or directly into the hopper of the filling machine on the floor below. Centrifuges, screens, or pressure filters are used to remove non-dispersed pigments. The paint is poured into cans or drums, labeled, packed, and moved to storage, each step being completely automatic.[6]

PAINT FAILURE The failure of paints to stand up under wear may be due to several causes, and in each case there is a special term used to describe the failure. *Chalking* is a progressive powdering of the paint film from the surface inward, caused by continued and destructive oxidation of the oil after the oirginal drying of the paint. Very rapid chalking is termed *erosion. Flaking,* sometimes called *peeling,* is due to poor attachment of the paint to the surface being covered and is usually attributed to dirt or grease on the surface or to water entering from behind the paint. *Alligatoring* is a form of peeling in which the center portion of the section starting to peel remains attached to the

[4]Payne, *op. cit.*, vol. 2, pp. 991–1014 (basic principles of major types of paint-mill operations are discussed).

[5]See Perry, p. 8-42 for roller mill and pigment grinding, p. 19-24 for kneaders, p. 8-25 for ball mills, and p. 8-8 for principles of size reduction.

[6]Hanes *et al.* (eds.), *Ind. Eng. Chem.*, **46,** 2010 (1954) (describes paint manufacturing and resin processing in detail, indicating the need for inventory control; includes a flowchart for paint production, p. 2018).

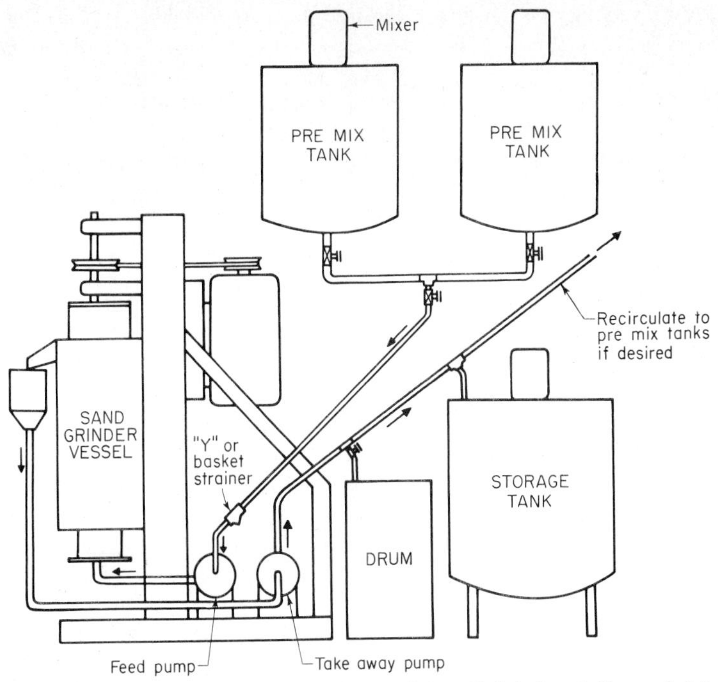

Fig. 24.3 Schematic diagram for a sand-grinding installation. (*E. I. du Pont de Nemours & Co.*)

Fig. 24.4 Three-roll mill. (*J. H. Day Co.*)

surface. *Checking* denotes a very fine type of surface cracking. As opposed to these types of paint failure, normal wear is gradual removal of paint from the surface by the elements, leaving a smooth surface.

PAINT APPLICATION[7] A surface that is efficiently painted, varnished, or lacquered often requires the application of a number of different coatings varied to suit the condition. For example, most surfaces require the use of a primer, or filler coat, to smooth over inequalities and to secure better adherence. This may then be covered with the paint proper in one or more applications. Much application of paint is still done by hand brushes and rollers, but dipping and spray-painting are gaining favor because of the ease and rapidity of spread, thus saving labor, though wasting a little material. A wide variety of types of atomizers for spraying is available, including internal-mixing and the more common external-mixing. Spray guns are operated frequently at a pressure of 40 to 60 psi. Four to eight cubic feet of free air per minute is required at 40 psi. Coating materials in aerosol containers, sold in millions, are popular because of their convenience of application. *Prefinishing* of metal and wood sheets is done by reverse or direct roll coating at the factory. This process is more economical than painting on the job, and the finishes are of better quality.

LATEX PAINTS The vehicle for this class of paints is an emulsion of binder in water; the binder may be oil, an acrylic or polyvinyl acetate resin, or another emulsifiable binder. Although water-thinned paints date back to antiquity, they were not commercially important until casein-based paints were developed about 1925. Their phenomenal growth is indicated by the large increase in U.S. production. Resin-emulsion paints have been widely used since World War I, but latex, or rubber-base paints, introduced commercially only in 1948, have had spectacular growth; more than half of interior paint sales are latex-base paints. This type of paint was developed to meet demands for greater ease of application, quick drying, low odor, easy cleaning, great durability, and impermeability to dirt. A table of water-reducible vehicles, showing trade names, producer, type, and special properties, classed as emulsions or solutions, is available.[8] The absence of volatile solvents greatly reduces air pollution.

Latex paints have as their major film-forming constituent a synthetic resin latex, with or without other film-forming constituents added, in an oil-water, emulsion-type system. The continuous phase consists of an alkali-dispersed hydrophilic colloid in water and contains two or more different types of particles in suspension.

Styrene-butadiene (SBR) copolymer was the original quality film former in latex paint. Polyvinyl acetate (PVA), acrylics, and PVA-acrylic copolymers have largely replaced SBR as film formers in today's trade (i.e., home use) paints. Quality is substantially improved.

Pigment dispersion. The dispersant and ammonia are added to water in a pony mixer, followed by pigments, premixed, and then ground in a ball mill. The pigments and extenders most used are water-dispersible grades of titanium dioxide, zinc sulfide, lithopone, and regular grades of barium sulfate, mica, diatomaceous silica, clay, and magnesium silicate. A combination of four or five inerts is generally employed. The usual colored pigments may be used for tinting, with certain exceptions such as prussian blue, chrome yellow, chrome green, and carbon black. The first three are sensitive to alkalies, and the last tends to break the emulsion. Also, sodium-free alkalies and pigments are preferred, since they minimize efflorescence caused by sodium sulfate on the paint surface. Interior pigments must, by law, be nonpoisonous if ingested.

Paint manufacture. The film formers are added to the pigment dispersion, followed by a preservative solution (usually chlorinated phenols) and an antifoam (sulfonated tallow or pine oil). The latex emulsion is stirred in slowly, followed by water. The paint is mixed, screened, and mixed again before packaging. A typical paint consists of 35% pigment and filler and about 21% film-forming ingredients.

[7]Payne, *op. cit.*, vol. 2, p. 1036.
[8]Burrell, Picture for Protective Coatings, *Ind. Eng. Chem.*, **54**, 54 (1962).

It has been shown, in the preparation of emulsion polymers, that copolymer composition, the emulsifying system, and particle size have a profound effect on such ultimate paint properties as adhesion, gloss, scrub resistance, chemical and mechanical stability, and viscosity. Formulations of PVA copolymer emulsion paints, which are used successfully on wood surfaces, are given by Payne.[9] Most SBR copolymerizations are made batchwise by charging the ingredients to a jacketed, agitated reactor, heating to reaction temperature, and cooling after the desired conversion. Acrylic ester polymerizations are carried out in the same manner, except under reflux conditions. The pigments must be pure and dispersed compatibly with the latex. The manufacture of these paints presents an excellent example of applied colloid chemistry.

PIGMENTS[10]

Pigments[11] (Table 24.3), colored, organic and inorganic insoluble substances, are used widely in surface coatings, but are also employed in the ink, plastic, rubber, ceramic, paper, and linoleum industries to impart color. A large number of pigments and dyes is consumed because different products require a particular choice of material to give maximum coverage, economy, opacity, color, durability, and desired reflectance. White lead, zinc oxide, and lithopone were once the principal white pigments; colored pigments consisted of prussian blue, lead chromates, various iron oxides, and a few lake colors. Today titanium oxide in many varieties is almost the only white pigment used. Table 24.5 presents recent figures on the production of pigments. Lead pigments, formerly of major importance, are now prohibited by law for many uses.

WHITE PIGMENTS

The oldest and formerly most important of the white pigments is white lead which is no longer permitted as a constituent of most paints. Zinc oxide, another white pigment formerly widely used, is now of only minor importance.

LITHOPONE Lithopone is a zinc sulfide pigment which had a tendency to turn temporarily gray on exposure to sunlight. However, it was discovered in 1880 that heating the product to a red heat and plunging it into water remedied this physical defect. The original light-sensitiveness has been overcome by raw-material purification and by the addition of such agents as polythionates and

Table 24.4 ***Comparison of White Pigments***

White pigment	*Refractive index*	*Gallons per 100 lb**	*Oil absorption†*	*Hiding power‡*	*Tinting strength§*	*1975 price per pound, dollars¶*
Zinc oxide	2.08	2.14	12–25	20	210	0.40
Titanium dioxide (anatase)	2.55	3.08	20–25	115	1,250	0.40
Titanium dioxide (rutile)	2.76	2.86	18–22	147	1,600	0.40

Source: Based on Payne, *op. cit.*, vol. 2, p. 753, table 1; CPI 2, p. 500. *Bulking value; gallons occupied by 100 lb of pigment. †Grams of oil required to wet 100 g of pigment. ‡In square feet per pound of pigment; relative values only, since this property is affected by the PVC. §Based on Reynolds' constant-volume method; relative values only. ¶From *Chem. Mark. Rep.*, Apr. 14, 1975.

[9]Payne, *op. cit.*, vol. 2, pp. 1113–1114.

[10]Mattiello, *op. cit.*, vol. 2, 1942, gives a very full account of pigments in all phases, with full references to the literature. See this reference for further data on any individual pigment. Also, Payne, *op. cit.*, vol. 2, p. 678; Goneck, *Off. Dig. Fed. Soc. Paint. Technol.*, **34**, 90 (1962) (discusses trends in new pigments).

[11]Specifications for pigments, ASTM Standards, vol. 8, pp. 1–903, 1958; Ammons, Dispersed Pigment Problem, *Ind. Eng. Chem.*, **55**(4), 40 (1963) (63 refs. and 4 tables, modifying and specialty pigments).

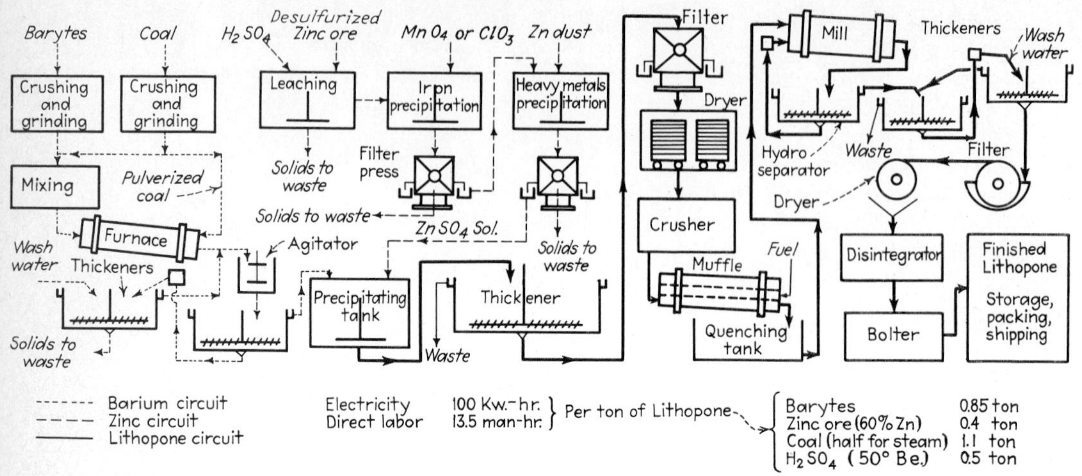

Fig. 24.5 Flowchart for lithopone manufacture.

cobalt sulfate. Lithopone is a brilliantly white, extremely fine, cheap, white pigment. It is particularly well adapted to interior coatings.

The manufacture is shown diagrammatically in Fig. 24.5, where the barium, zinc, and lithopone circuits are represented by different types of lines. The barium sulfide solution is prepared by reducing barite ore ($BaSO_4$) with carbon and leaching the resulting mass. The equation is:

$$BaSO_4 + 4C \longrightarrow BaS + 4CO$$

Scrap zinc, or concentrated zinc ores, are dissolved in sulfuric acid, and the solution purified as shown in Fig. 24.5. The two solutions are reacted, and a heavy mixed precipitate results which is 28 to 30% zinc sulfide and 72 to 70% barium sulfate.

$$ZnSO_4 + BaS \longrightarrow ZnS + BaSO_4$$

This precipitate is not suitable for a pigment until it is filtered, dried, crushed, heated to a high temperature, and quenched in cold water. The second heating in a muffle furnace at 725°C produces crystals of the right optical size.

TITANIUM DIOXIDE The best-selling white pigment is TiO_2, because of its low cost per unit of hiding power (Table 24.4). It is marketed in the rutile and anatase crystal forms and is widely employed in exterior paints and also in enamels and lacquers. A typical exterior white paint contains about 40% pigment, of which 60% is TiO_2, 5% ZnO, and 25% fillers such as mica, silica, silicates, or $CaCO_3$. Such a formulation has long life through controlled chalking (self-cleaning) and presents a good surface for subsequent repainting. Approximately 50% of the TiO_2 consumed is used in paints, varnishes, and lacquers, and the second important consumer (about 15%) is the paper industry. TiO_2 is the most widely used pigment for coloring plastics, and its use is increasing steadily. It is expected that total annual production will be more than 1 million tons by 1980.[12] The excellent hiding power of TiO_2 is ascribed to its high index of refraction (2.55 for anatase, 2.76 for rutile) with relation to that of the film former. The more stable rutile form has less chalking tendency and greater hiding power, and its yellowing tendency has been overcome. The flowchart depicted in Fig. 24.6 represents the hot *sulfuric acid leaching* process,[13] and it may be broken down into the following sequences.

[12] *Chem. Week*, Apr. 4, 1973, p. 39.
[13] Pierce *et al.*, U.S. Pat. 2,476,453 (1949); Soloducha, Can. Pat. 610,344; *Chem. Week*, July 22, 1961, p. 70.

TABLE 24.5 ***Annual Shipment and Production of Selected Pigments***
(In short tons)

Pigment	*1973*	*1972*
Titanium dioxide (composite and pure—100% TiO_2)	784,996	693,281
Chrome green	3,000*	
Chrome oxide green	7,159	6,155
Chrome yellow and orange	37,120	33,770
Molybdate chrome orange	14,057	12,410
Zinc yellow (zinc chromate)	5,307	5,657
Iron blues	5,037	5,186

Source: U.S. Dept. of Commerce. *Estimated.

The ground ilmenite ore is digested hot in large, conical, concrete or heavy-steel tanks with 66°Bé sulfuric acid. The mixture is agitated and steam-heated to 110°C. The reaction is exothermic, the heat evaporating the water (Ch and Op). The solid reaction mass is dissolved in water, giving a solution of the soluble titanium and ferrous and ferric sulfates (Op). The ferric sulfate is reduced by scrap iron (Ch). The solution is clarified in thickeners (Op). Fifty percent of the iron is removed from the solution as crystallized ferrous sulfate by cooling, crystallization, and centrifugation (Op). A second clarification removes the last traces of residues (Op). The solution is concentrated in a continuous lead-lined evaporator to the equivalent of about 200 g/l TiO_2 (Op) as soluble sulfate. This strong acid–soluble titanyl sulfate (*probably* $TiOSO_4$) is hydrolyzed. The hydrolysis reaction is dependent upon many factors, namely, quantity and quality of the seeds (*colloidal suspension* TiO_2), concentration, rate of heating, and pH. Using anatase seeds, 6 h of boiling is practiced, and with rutile seeds, 3 h is needed (Ch). The *precipitate* is vacuum-filtered, repulped, and refiltered (Op). This operation removes the rest of the iron sulfate. The filter cake is repulped, treated with a conditioning agent, and calcined to TiO_2 for 24 h. For anatase production, 0.75% K_2CO_3 (based on TiO_2 equivalent pulp content) is used as a conditioning agent (nonfritting) and to develop the greatest tinting strength and good color. In rutile production certain carbonates such as sodium,

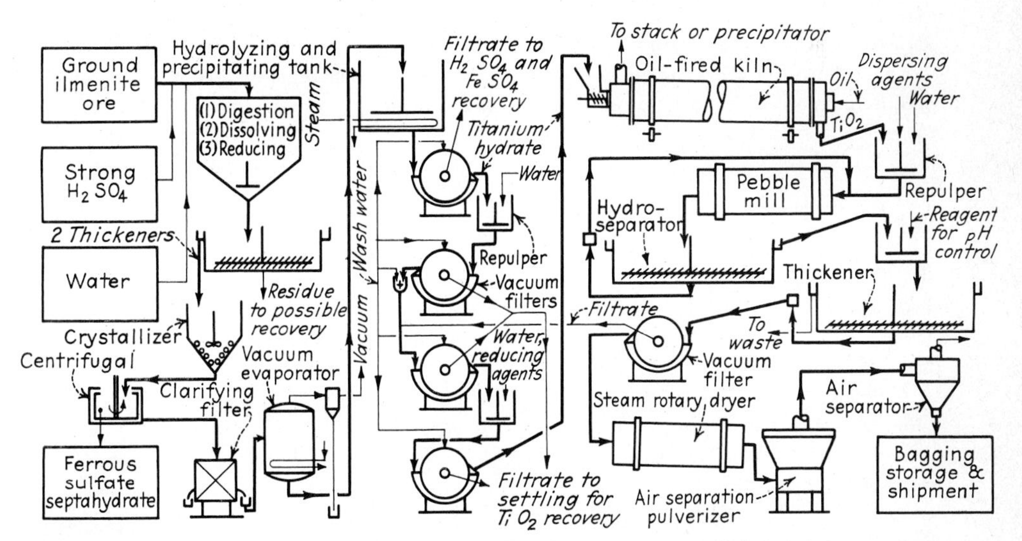

Fig. 24.6 Flowchart for the production of titanium dioxide. For reducing, scrap iron is added; for hydrolyzing, seeds of anatase or rutile and water (small amount) are added.

potassium, and lithium may be used, as well as zinc and magnesium carbonates, to promote rutilization (Ch). The hot TiO_2 is pulverized, quenched in water (dispersed), wet-ground, hydroseparated, thickened, filtered, dried, and reground (Op).

The new chloride process converts the titanium rutile ore or slag into $TiCl_4$. This process hydrolyzes the $TiCl_4$ to TiO_2 in a flame of oxygen and fuel gas (Fig. 24.7):

$$TiCl_4 + O_2 \longrightarrow TiO_2 + 2Cl_2$$

This process produces the superior flame-formed rutile crystal.[14]

The sulfate process has a difficult waste-disposal problem, because its waste contains both worthless iron sulfate and large quantities of spent sulfuric acid. Sulfate process producers have an advantage in they can start with domestically available ilmenite. Rutile is required for chloride processing and must be imported. The two processes are highly competitive at present.

BARIUM SULFATE Barium sulfate is occasionally used as a pigment because of its stability. It has very poor covering power, however. On the other hand, it is widely employed as a filler for rubber and as a coating material for certain fine papers. It is opaque to x-rays and consequently finds application in the form of the so-called *barium meal,* for x-ray visualization of the intestinal tract. For some of its cruder applications it is made by fine grinding and washing of the ore, barite. A better product, blanc fixe, results from precipitation. *Blanc fixe* is made by precipitation of a soluble barium compound, such as barium sulfide or barium chloride, by a sulfate. Barium sulfate, when intended for internal administration for x-ray diagnosis, in addition to being in the form of a very fine powder capable of giving good aqueous suspensions, must also be free of objectionable impurities such as lead, arsenic, sulfide, soluble barium salts, and the like. Since the ordinary blanc fixe of commerce is not of a purity suitable for this use, this x-ray barium sulfate must be specially manufactured. Several pigment extenders are widely used. These include silica (natural, synthetic, and fumed), various forms of whiting ($CaCO_3$), mica, kaolin, and various silicates.

BLACK PIGMENTS

The only major black pigment is carbon black which comes in several shades. Lampblack is occasionally used where its peculiar shade is wanted. Their manufacture is discussed under industrial carbon, in Chap. 8. They retard the oxidation of linseed oil and cause a slow-drying film which, under many conditions, prolongs the life of paint. Carbon pigments should not be used in direct contact with iron and steel in primer coatings, because they stimulate metal corrosion.

BLUE PIGMENTS

Ultramarine blue is a complex sodium aluminum silicate and sulfide made synthetically. Ultramarine, which has a sulfide composition, should not be used on iron or mixed with lead pigments. Ultramarine is widely used as bluing in laundering to neutralize the yellowish tone in cotton and linen fabrics. It is also applied for whitening paper and other products. Special grades, low in free sulfur, are used in inks.

Phthalocyanine blues, marketed in the United States since 1936, are particularly useful for nitrocellulose lacquers in low concentrations as a pigment highly resistant to alkalis, acids, and color change. For details of manufacture, see Chap. 39. As presently produced, it is one of the

[14]Flamed Pigment Wins Favor, *Chem. Week,* Aug. 11, 1962, p. 73; Du Pont, U.S. Pats. 2,559,638 and 2,670,275; Outlook Brightens for Titanium Dioxide Supplies, *Chem. Week,* Apr. 4, 1973, p. 39; Troubled Times for TiO_2, *Chem. Eng. (N.Y.),* **79**(9), 34 (1972).

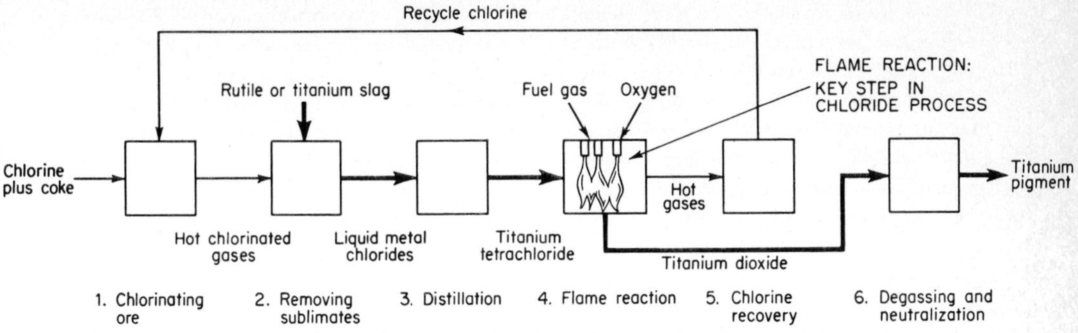

Fig. 24.7 Six basic steps in the chloride route to titanium dioxide pigments used by Du Pont and others.

most stable pigments, is resistant to crystallization in organic solvents, and is essentially free from flocculation in coatings. Both the greens and the blues have high tinting power and are used in latex paints and in printing inks. They are prepared by reacting phthalic anhydride with a copper salt in the presence of urea or by reacting phthalonitrile with a copper salt with or without ammonia.[15] Pthalocyanine blues are the most important blue pigments in use today (see Chap. 39).

The various *ferrocyanide blues* are known as prussian blue, chinese blue, milori blue, bronze blue, antwerp blue, and Turnbull's blue. As these names have lost much of their original differentiation, the more general term *iron blues* is preferred. These pigments are made in essentially the same manner by the precipitation of ferrous sulfate solutions (sometimes in the presence of ammonium sulfate) with sodium ferrocyanide, giving a white ferrous ferrocyanide, which is then oxidized to ferric ferrocyanide, $Fe_4[Fe(CN)_6]_3$, or to $Fe(NH_4)[Fe(CN)_6]$ by different reagents such as potassium chlorate, bleaching powder, and potassium dichromate. The pigment is washed and allowed to settle, since filtration is extremely difficult because of its colloidal nature. Iron blues possess very high tinting strength and good color performance; their relative transparency is an advantage in dip-coating foils and bright metal objects, and for colored granules for asphalt shingles.

RED PIGMENTS

Red lead (Pb_3O_4) has a brilliant red-orange color, is quite resistant to light, and finds extensive use as a priming coat for structural steel, particularly since it possesses corrosion-inhibiting properties. Red lead, or minium, is manufactured by the *regular process,* by oxidizing lead to litharge (PbO) in air and further oxidizing the litharge to red lead. In the *fumed* process, which produces smaller particles, molten lead is atomized by compressed air and then forced through the center of a gas flame, which in turn converts it into litharge as a fume collected in filter bags. The litharge is then oxidized to red lead.[16]

Ferric oxide (Fe_2O_3) is another red pigment employed in paints and primers, as well as in rubber formulation. A wide range of natural red oxide pigments is available and, because of durability, they are used in barn and freight-car paints. Synthetic pigment is made by heating the iron sulfate obtained from the pickling vats of steel mills. *Venetian red* is a mixture of ferric oxide with up to an equal amount of the pigment extender, calcium sulfate. This pigment is manufactured by heating ferrous sulfate with quicklime in a furnace.[17] Venetian red is a permanent and inert pigment, partic-

[15]*Am. Cyanamid Co., Tech. Bull.* 902.

[16]Payne, *op. cit.*, vol. 2, p. 835, presents physical properties of typical commercial red lead pigment produced by both processes in table 15.

[17]*Ibid.*, p. 826, states that, theoretically, two products may be obtained from calcination and that commercial production is a mixture of the two; equations are given.

ularly on wood. The calcium sulfate content, in furnishing corrosion-stimulating sulfate ions, disqualifies this pigment for use on iron. *Indian red* is a naturally occurring mineral whose ferric oxide content may vary from 80 to 95%, the remainder being clay and silica. It is made by grinding hematite and floating off the fines for use. The basic lead chromate [$PbCrO_4 \cdot Pb(OH)_2$] may also be used as an orange-red pigment; it is an excellent corrosion inhibitor. It is manufactured by boiling white lead with a solution of sodium dichromate. *Cadmium reds* are made by roasting the precipitate obtained by mixing cadmium sulfate, sodium sulfite, and sodium selenide. Cadmium colors range from light yellow to maroon and are available in pure form as well as in lithopones.[18] The larger the quantity of selenium used, the greater the shift toward red. Red pigments include a large variety of insoluble organic dyes, either in the pure state as toners or precipitated on inorganic bases as lakes. For example, *tuscan red* is a name sometimes applied to combinations of red iron oxide pigment and a light-fast organic red pigment used where bright colors are needed for exterior paints. Quinacridones are extremely durable reds, oranges, and violets comparable in servicability to pthalocyamines.

YELLOW PIGMENTS

Ocher is a naturally occurring pigment consisting of clay colored with 10 to 30% ferric hydroxide. It must be ground and levigated. At best the ochers are very weak tinting colors and are being replaced by synthetic hydrated yellow iron oxides for brighter color and better uniformity. Yellow pigments with a wide variety of shades fall in the class known as *chrome yellows;* they are the most popular yellow pigments because of exceptional brilliance, great opacity, and excellent light-fastness. They are produced by mixing a solution of lead nitrate or acetate with a solution of sodium dichromate. Extenders may be present in up to an equal weight of gypsum, clay, or barite. The pigment is of high specific gravity and settles out. The use of chrome pigments is sharply limited by their toxicity when ingested. Organic yellows are used for indoor coatings to avoid the possibility of lead or chrome poisoning of infants. *Zinc yellow,* or *chromate,* although of poor tinting power, is used because of its excellent corrosion-inhibiting effect both in mixed paints and as a priming coat for steel and aluminum. Zinc yellow is a complex of the approximate composition $4ZnO \cdot K_2O \cdot 4CrO_3 \cdot 3H_2O$. Two other yellow chromate pigments are *strontium chromate* and *barium chromate,* both used as corrosion inhibitors. *Litharge* (PbO) is made by heating melted lead in a current of air. It is used to some extent in anticorrosion paints.

GREEN PIGMENTS

The major green pigment is pthalocyanine green. One of the oldest green pigments is *chromium oxide* (Cr_2O_3). It has many disadvantages, such as high cost and lack of brilliancy and opacity. It is made by calcining either sodium or potassium dichromate with sulfur in a reverberatory furnace:

$$Na_2Cr_2O_7 + S \longrightarrow Cr_2O_3 + Na_2SO_4$$

Guignet's green (emerald green) is a hydrated chromic oxide [$Cr_2O(OH)_4$], possessing a much more brilliant green color than the oxide and yet having good permanency. It is prepared by furnacing a mixture of sodium dichromate and boric acid at a dull red heat for several hours. In addition, a green of good permanency for use in outside-trim paints may be obtained in intimate mixtures of copper phthalocyanine with zinc chromate or hansa yellow. The brightest permanent green available, chlorinated copper phthalocyanine, is expensive but durable. Water-dispersible grades and pulp colors are available for latex paints. *Chrome green,* sold under various names, is a mixture or a coprecipitation of chrome yellow and prussian blue. Inert fillers are used with this pigment in making paints. Unless carefully ground or coprecipitated, the two colors may separate when mixed in a paint.

[18] *Ibid.*, table 16, gives physical characteristics of 15 commercial cadmium colors.

BROWN PIGMENTS

The carefully controlled heating of various naturally occurring iron-containing clays furnishes the brown pigments known as burnt *sienna,* burnt *umber,* and burnt *ocher.* The iron hydroxides are more or less converted to the oxides. The umbers contain the brown manganic oxide as well as the iron oxides. These are all permanent pigments suitable for both wood and iron; however, the siennas are being replaced as tinting colors by synthetic oxides because of the latter's greater color strength and clarity.[19] *Vandyke brown* is a native earth pigment of indefinite composition, containing oxide of iron and organic matter.

TONERS AND LAKES[20]

Toners are insoluble organic dyes that may be used directly as pigments because of their durability and coloring power. Lakes result from the precipitation of organic colors, usually of synthetic origin, upon some inorganic base. They are employed in many colors. Some typical examples are given here. Para red is formed by diazotizing *p*-nitroaniline and coupling it with β-naphthol. Toluidine toner, a better and more expensive red pigment, is made by diazotizing *m*-nitro-*p*-toluidine and coupling it with β-naphthol. Hansa yellow G (lemon yellow) is manufactured by diazotizing *m*-nitro-*p*-toluidine and coupling it with acetoacetanilide. Hansa yellow 10 G (primrose yellow) is made by coupling orthochloroacetanilide with diazotized 4-chloro-2-nitroaniline. *Lakes* are really dyed inorganic pigments. The inorganic part, or base, consists of an extender such as clay, barite, or blanc fixe and aluminum hydroxide. Either the organic dye may be precipitated onto an already existing base, such as clay or barite suspended in solution, or both the dye and the base may be coprecipitated, e.g., onto blanc fixe or aluminum hydroxide. Both toners and lakes are ground in oil or applied like any other pigment.

MISCELLANEOUS PIGMENTS

In surface coatings, *metallic powders,* i.e., flaked or finely powdered metals and alloys, have been developed not only for decorative purposes but also for their durability, heat-reflective properties, and anticorrosion effects. *Aluminum flaked powder* is usually made by steel-ball-milling the granular form in a cylindrical mill to which are added a lubricant and a dispersing agent. Both this method and the millings of *aluminum granules* are described by Payne.[21] *Powdered zinc,* more often referred to as zinc dust, is used in primers, in paints for galvanized iron, and in finish coats. One method of manufacture is by distillation.[22] *Lead powders and pastes* are used in primers. Though not classed with pigments, one of the uses of *metallic stearates* is as a pigment-suspending agent. Luminescent paints have a variety of uses, particularly in advertising displays and for aircraft, because of their high visibility.

PIGMENT EXTENDERS

A number of natural and precipitated inert substances may be added to paints as fillers or extenders[23] for the purpose of giving a better paint film, improving weathering properties, furnishing a base for

[19]Payne, *op. cit.,* vol. 2, p. 821.

[20]Mattiello, *op. cit.,* vol. 2, pp. 3–286.

[21]Payne, *op. cit.,* vol. 2, chap. 22, Metallic Pigments and Metallic Stearates, p. 924; Edwards and Wray, Aluminum Paint and Powder, 3d ed., Reinhold, 1955.

[22]Newton, *Off. Dig.,* **29,** 391, 770 (1957).

[23]Harness, Natural Mineral Paint Extenders, *U.S. Bur. Mines Inf. Circ.* 7264, 1943.

the real pigment, preventing too rapid settling, or reducing the cost of the pigment (see Tables 24.3 and 24.5). These extenders should not be viewed as adulterants; they may add materially to the quality. *Flatting* is also a primary reason for adding inerts or extenders. *Barium sulfate,* in the natural crystalline form barite, but finely ground, or in the precipitated form blanc fixe, is one of the most important extenders. Finely ground calcium carbonate, sometimes called *whiting,* is another. Ordinary silicon dioxide, when finely divided, makes a good filler. Magnesium silicate is also widely used, as *asbestine* and *talc,* the latter being finely ground soapstone.

OILS

Although in the organic surface protection industries oils serve as part of the vehicle for the carrying of pigments, their chief function is to form, or to help form, the protective film and to plasticize it. For this, reactive oils[24] are used, such as those which are drying or semidrying. The drying or the hardening of these oils involves chemical reactions, which are rather complex but include oxidation as the initiating step. Some polymerization also occurs, and much cross-linkage. Oils which dry to a film possess olefinic unsaturation. For example, the acids of linseed oil contain about 9% saturated acids (palmitic and stearic), 19% oleic acid, 24% linoleic acid, and 48% linolenic acid. In drying, these oils at the first stage absorb oxygen from the air, forming peroxides or hydroperoxides at the olefinic bonds. These still-liquid products partly decompose, giving volatile oxidation products, but mainly change in the next stage of the reactions, by cross-linkages into the solid, though still elastic, films through colloidal aggregation. In the case of linseed oil, the solid, insoluble, elastic film is called *linoxyn.* Such films are not permanent, since the chemical reactions continue, though at a much slower rate, until after the course of years the film is entirely destroyed. Light, particularly ultraviolet, catalyzes these reactions, and one of the functions of the pigment in surface coatings is to reflect the light and thus help preserve the film.

Drying oils are seldom used unmodified.[25] They may be improved in a number of different ways by (1) the action of driers, (2) oil bodying, (3) fractionation and segregation, (4) isomerization or conjugation, (5) dehydration, and (6) other carbon double-bond reactions. There are also copolymer oils, poly alcohols such as polypentaerythritol, oil-modified alkyds, and synthetic oils. Bodied oils vary with the oil, but generally have better drying, wetting, and color retention. Bodying may be achieved either by heating in kettles or by blowing air in fine bubbles through the oils at 100 to 200°C for several hours. Solvent or liquid-liquid extraction separates the drying constituents of an oil from the nondrying constituents. The *isomerization* of paint oils, especially popular for linseed and soybean oils, involves partial rearrangement of the isolated double bonds (nonconjugated) into more reactive conjugated positions upon heat treatment with catalysts, such as certain metal oxides, activated nickel, or SO_2. *Dehydration,* at present, is applicable only for castor oil and is achieved by heating the oil in a vacuum in the presence of dehydrating catalysts such as alumina, fuller's earth, silica gel, H_3PO_4, or H_2SO_4. The reaction is probably as follows:

$$\underset{\text{Ricinoleic acid}}{CH_3(CH_2)_5CH(OH)CH_2CH{:}CH(CH_2)_7COOH} \longrightarrow$$

$$H_2O + \underset{\text{9,12-Linoleic acid, 59–64\%}}{CH_3(CH_2)_4CH{:}CHCH_2CH{:}CH(CH_2)_7COOH}$$

$$+ \underset{\text{9,11-Linoleic acid, 17–26\%}}{CH_3(CH_2)_5CH{:}CHCH{:}CH(CH_2)_7COOH}$$

[24] Mattiello, *op. cit.*, vol. 1, pp. 57–174; Chandler, Advances in Driers, *Paint Technol.*, **26**(9), 26 (1962); Hutchinson, Some Recent Advances in Chemistry and Technology of Drying Oils, *J. Oil Colour Chem. Assoc.*, **41**, 474 (1958).

[25] Mattil, Short Course on Drying Oils, *J. Am. Oil. Chem.' Soc.*, **36**, 477 (1959).

It was earlier thought that the 9,11-linoleic acid, with its conjugated double bonds, was formed in a larger amount, but recent investigations show the other isomer to exist in a larger quantity. The composition of the fatty acids in the thoroughly dehydrated castor oil analyzes as follows:

	Percent		*Percent*
Saturated acids	0.5–2.5	9,12-Linoleic acid	59–64
Hydroxyl acids	3–8	9,11-Linoleic acid	17–26
Oleic acid	7.5–10.5		

VARNISHES

A varnish is an unpigmented colloidal dispersion or solution of synthetic and/or natural resins in oils and/or thinners used as a protective and/or decorative coating for various surfaces and which dries by evaporation, oxidation, and polymerization of portions of its constituents. Not being pigmented, varnishes are less resistant to light than are paints, enamels, and pigmented lacquers. They furnish, however, a transparent film, which accentuates the texture of the surface coated. Varnishes are frequently oleoresinous; there are two minor classes, and spirit varnishes and japans. *Oleoresinous varnishes* are solutions of one or more natural or synthetic resins in a drying oil and a volatile solvent. The oil reduces the natural brittleness of the pure-resin film. Spirit varnishes are solutions of resins, but the solvent is completely volatile and nonfilm-forming. Oleoresinous varnishes were formerly of major importance, but alkyd and urethane varnishes have largely replaced them because of greater durability, less yellowing, ease of application, and beauty. Pressure to reduce the amount of air-polluting solvents in varnishes and paints, coupled with the desire for water cleanup of tools and spills has led to the development of water-thinned varnishes.

Spirit varnishes are solutions of resins in volatile solvents only, methanol, alcohol, hydrocarbons, ketones, and the like. Spirit varnishes dry the most rapidly but are likely to be brittle and eventually crack and peel off unless suitable plasticizers are added. The preparation of these products involves active stirring, and sometimes heating, to bring about the desired solution. An important example of a spirit varnish is *shellac* or a solution of the resin shellac in methanol or alcohol.

Japans are rarely used now. They are opaque varnishes to which asphalt or some similar material has been added for color and luster. They may be subdivided into baking, semibaking, and air-drying japans, according to their method of application.

Enamels are coatings with good leveling qualities having a high-gloss finish. They are white, tinted, or deep-colored. For exterior use alkyd resins are the more durable vehicles, whereas modified phenolic varnishes are used for interior enamels. Their composition varies widely, depending upon cost and end use. Some are applied with a brush, and others are sprayed on the surface. When hardened by baking, enamels are particularly durable. Alkyd-amine resins are used in high-quality enamels.

RESINS The original resins[26] used were *copals*, which consisted of fossil gums from various parts of the world. Another natural but present-day resin widely employed is that from the pine tree, or *rosin*. When a plant exudes these products, they are called *balsams* and, upon evaporation of the volatile constituents, they yield the resin. Thus the longleaf, yellow, and hard pines of the southern states, upon proper incision, yield a balsam which, after distillation, gives turpentine and a residue called *rosin*. The latter product is essentially abietic acid (cf. Chap. 32). Most of the natural resins employed are fossil resins that have been buried and gradually changed over the centuries. From

[26]Mattiello, *op. cit.*, vol. 1, pp. 175–498; Payne, *op. cit.*, vol. 1, p. 136.

the rosin of the present, these resins[27] reach back through dammar, copal, and kauri resins to the oldest of the fossil resins, amber. Some resins like kauri and copal are so ancient that they have changed to insoluble products and must be partly depolymerized by heat before they can be dissolved or blended with the hot oils and other constituents in the varnish kettle. An important present-day resin is *shellac,* or *lac resin.* Unlike the others, it is the product of animal life and comes from a parasitic female insect (*Coccus lacca*) which, when feeding upon certain trees in India, secretes a protective exudate that eventually coats the twigs, furnishing *stick lac.* This is collected and purified to the shellac of commerce by rolling, crushing, separating, washing, and bleaching. See Chap. 34.

Overshadowing these natural products are the more recently introduced synthetic resins (Chap. 34). The ranking of synthetic resins by use in industrial finishes is as follows: alkyds, acrylics, epoxies, nitrocellulose, phenolics, urethanes and the newer promising but expensive types now used in lesser amounts—siliconized alkyds, polyesters and acrylics, fluoropolymers and polyimides. *Phenolic resins* (phenol-formaldehyde) were the first of these to be supplied and are still in wide use since they are very resistant to water and many chemicals. In order to make these materials soluble in the oils and solvents in common use in the varnish industry, it is necessary to modify them. This may be done either by fluxing them with softer materials such as ester gum, or by controlling the reaction by choosing a para-substituted phenol, thus stopping the reaction before a final insoluble, infusible product is obtained. *Alkyd resins*[28] are formed by condensing dicarboxylic acids with polyhydric alcohols and modified with fatty acids to gain solubility. As a constituent of varnishes or enamels, they have the distinctive properties of beauty and flexibility, which show marked retention upon prolonged exposure to weather. With their introduction, the surface-coating industry became synthetic-minded. The properties of the alkyd formulation may be modified by the use of different fatty acids or oils of both drying and nondrying types, by the use of pentaerythritol for the glycerin, by the use of maleic and other dibasic anhydrides for all or part of the phthalic anhydride, and by modifying with other resins (phenolics, rosin, and the like). As such they find extremely diversified applications in various coating formulations.[29] They have largely replaced varnishes of the oleoresinous type. *Ester gum,* the product of the esterification of the abietic acid of rosin with glycerin, is another important raw material for making varnish. In addition to the coumarone-indene type, urea-formaldehyde types, and melamine-formaldehyde resins, some of the newer resins include *acrylics, silicones,* and *vinyls. Epoxide resins* are now widely used for protective and chemical-resistant coatings, but fail to meet the mass-production requirements of the furniture industry. Quick-drying *urethane coatings* have recently moved from the specialty-finish field into broader markets. Original urethane lacquers had two drawbacks, an isocyanate vapor hazard and a two-pot system; some new types may be applied from the can. The new finishes fall into two classes: (1) *Isocyanate nonreactive* oil-modified urethanes make use of toluene diisocyanate (TDI) in combination with polyols such as glycerin, methyl glucoside, linseed, and soybean oil; major uses are for exterior wood, such as for siding and shingles, boats, and also interior floors and furniture. (2) *Isocyanate reactive* types made by reacting excess TDI with a hydroxyl-bearing substance to form a prepolymer, which then reacts further with the hydroxyl groups of the polyol (first category) or with moisture (second category) to form the film. These types have been developed for heavy-duty-service applications and for various industrial finishes.[30] Other new resins are discussed in the literature.[31] Synthetic resins have numerous advantages over naturally occurring types, such as superior resistance

[27]Payne, *op. cit.,* vol. 1, p. 138. Origin and characteristics of natural resins are given in tabular form.

[28]Alkyd Resins, ECT, 2d ed. vol. 1, p. 852, 1963 (good bibliography). Because of improvement through modification with new chemicals and versatility in blending with and improving upon properties of other surface-coating materials, their importance for years to come is predicted. See Chap. 34, also, for alkyd resins.

[29]Bobolek, *Am. Paint J.,* **42,** 519 (1958).

[30]Urethane Coatings Aim High for New Jobs, *Chem. Week.,* Nov. 5, 1960, p. 27.

[31]The Paint Industry, *Chem. Eng. News,* Dec. 22, 1969, p. 32.

on exposure to weather and chemicals, the ability to be baked more rapidly at higher temperatures, and higher solid content than nitrocellulose lacquers, making a one-spray operation possible.

LACQUERS

Lacquer is a loosely used term. It refers to a coating composition based on a synthetic, thermoplastic, film-forming material dissolved in organic solvents, which dries primarily by solvent evaporation.[32] Confusion between lacquers and enamels remains in the public mind, which is not surprising in view of the standard definition of an enamel as a paint characterized by the ability to form an especially smooth film.[33] Clear lacquers, upon the addition of a pigment, become lacquer enamels, or simply pigmented lacquers.

Lacquer use is currently limited to coating furniture. When used to coat automobiles, enamels are habitually referred to as lacquers.

INDUSTRIAL COATINGS

Alkyd resins are used extensively in industrial coatings. They are widely compatible with oils and other resins, but their durability and resistance to water, sunlight, and chemicals is inferior to that of other types.

Phenolics are used to resist alcohols and food acids, particularly in cans and containers, but their use in varnish has lost out to urethanes and other film formers.

Acrylics, available as thermoplastic and thermosetting types (with mixtures compatible) represent the current optimum combination of price, durability, flexibility, and appearance. They are used in automotive topcoats.

Epoxies are used in plants where resistance to corrosion is essential. They require a curing agent and are expensive. They are used on appliances, as linings, and for prime coats.

The expensive urethanes are strongly adherent to metal and resist both chemical attack and abrasion. Their clarity and resistance to weather make them useful for severe industrial service.

Fluoropolymers represent the current maximum in weather resistance. The slick surface and good wear resistance cause them to be used as coatings for snow shovels, saws, aircraft, and chute liners; however, their cost is very high.

Polyimide resins are used to coat special pans and other material that must resist temperatures of 275°C continuously or 450°C maximum briefly.

Application Traditional application with a brush is increasingly proving too time-consuming. Roller application enjoys wide home use. Dip tanks provide cheap application, but coating is uneven and holidays in the coatings are common. Conventional spray equipment is cheap to own and maintain, but gives uniform results only in the hands of skilled operators and results in highly variable film thickness. Spraying is increasingly coming under fire because of its pollution of the air with both volatile solvents and finely divided solids. Electrostatic spraying, in which the paint droplets are made to carry a charge of one polarity while the object being painted is oppositely charged sharply reduces waste and gives an uniform coat. Its use is limited to metal objects in spray booths. Coil coating of sheets by passing them through baths, sometimes followed by leveling rollers, is efficient in the use of fluids, but edges are hard to coat properly. Electrodeposition[34] may be used with aqueous dispersions of alkyds, polyesters, epoxies, and acrylics solubilized with alkalies such as diethylamine, ammonium hydroxide, and potassium hydroxide. Great coating uniformity is

[32]ASTM Standards, pt. 29, 1974.
[33]ASTM Standards, pt. 29, 1974.
[34]Electrocoating, *Chem. Eng. News*, Aug. 26, 1974, p. 10.

achieved by anodic deposition using voltages of 100 to 450 V. The flow of current automatically stops when a coating of 1 to 1.5 mil has formed, because of the insulating properties of resins. Multiple coats are impossible by this method, but the process is most efficient in its use of material and is not air-polluting.

Fluidized-bed coating and electrostatic[35] spray coatings are used to coat metals with powders which are later fused. These techniques have been used with epoxies, vinyls, nylons 11 and 12, polyethylene, polypropylene, chlorinated polyethers, melamines and, no doubt, several other compounds. Air pollution is avoided, and coating quality is excellent, but color problems and wastage cause trouble.

Electron-beam radiation curing has been tested and shows promise. EB curing works very fast and requires no use of heat. Prepolymers are cured by using an accelerator's beam. Coatings of baked quality can be produced over heat-sensitive materials such as wood and plastic.

COATED METALS

Precoating of metals, such as steel and aluminum strips, before fabrication has now become a large market in the coating field. Roller coaters are unsurpassed in speed and surface quality for coating either one or both sides of continuous strip stock. Resins are used in the rank shown in Table 24.6.

Plasma spray coatings meet the high-temperature and wear demands of space technology. In this process extremely high-melting-point materials, including metals, are propelled to base materials such as brass, steel, graphite, and certain reinforced plastics at high velocity after being subjected to temperatures up to 30,000°F.[36] Multiple layers of glass flakes in polyester resin produce hard, thick, easily sprayed, resistant coatings for metal.[37] There is sure to be further development in the field of metal coating.

Surface coatings packed in pressurized cans (aerosols) are usually lacquers (clear, colored, or metallic), but vinyl coatings resistant to abrasion, sunlight, and moisture, epoxies, and stainless-steel flakes in vinyl or epoxy designed to prevent corrosion of machinery are available. Annual production exceeds 300 million units. Paint removers[38] sell at a rate of over $40 million per year. Two types dominate the market, those containing methylene chloride (dichloroethane) in a foam or gel base with activators and cosolvents such as methanol and monoethylamine, and the older benzol-acetone remover. These are also packaged in aerosol cans.

PRINTING INKS AND INDUSTRIAL POLISHES

Printing inks consist of a fine dispersion of pigments or dyes in a vehicle which may be a drying oil with or without natural or synthetic resins and added driers or thinners. Drying oils or petroleum oils and resins are employed, although the newer synthetic-resin systems are finding great favor because they are quick-drying and their working properties are excellent. Printing inks have a large variety of compositions and wide variation in properties.[39] This is because of the great number of different printing processes and types of papers employed. The expensive new magnetic inks, developed for use in a number of electronic machines, are the keys to a data-processing system which, if successful, may curtail the use of some of the other *specialty inks*.[40] Inks formulated with luminescent pigments achieve a superbright effect; dyes are melted into the resin and baked and, when

[35]*Hercules Chem.*, December 1972, p. 1.
[36]Plasma Arc, *Chem. Week*, Nov. 26, 1960, p. 45.
[37]Glass Flakes: Key to New, Strong Coating, *Chem. Eng.* (*N.Y.*), **67**(18), 162 (1960).
[38]Paint Remover Sales Take Off, *Chem. Week*, Oct. 20, 1971, p. 65.
[39]Haines, Printing Ink Varnishes: Old and New, *Am. Ink Maker*, **38**, 30 (1960).
[40]Magnetic Inks, *Chem. Week*, Mar, 12, 1960, p. 95.

TABLE 24.6 ***Industrial Coatings***

Industrial finishes market	*Resins in industrial finishes*
Appliances	Acrylics
Automotive, marine, and aircraft	Alkyls
Containers and closures	Epoxies
Industrial and farm equipment	Fluoropolymers
Maintenance	Nitrocellulose
Metal furniture	Phenolics
Paper and flexible packaging	Polyesters
Sheet, strip, and coil coating	Polyimides
Wood furniture	Polyurethanes
	Polyvinyls
	Siliconized polymers
	Ureas and melamines

hard, the material is easily powdered. Rhodamines, auramines, and thioflavins are the principal dyes used in heat-set inks.[41] Volatile solvents produce air pollution.[42]

Polymer polishes, which have captured 85% of the total polish market, have become increasingly popular both in industrial applications and in the $100 million household field. Present-day polishes of the self-drying, water-emulsion type had their start in the leather industry in 1926 and since then have undergone numerous improvements. During World War II, shortages sparked efforts to replace waxes with alkali-soluble resins, which in turn have been replaced by acid-sensitive polymers.

SELECTED REFERENCES

Bentley, K. W.: The Natural Pigments, vol. 4, Interscience, 1960.
Brewer, G. E. F. (ed.): Electrodeposition of Coatings, ACS, 1973.
Chapman, B. N., and J. C. Anderson (eds.): Science and Technology of Surface Coating, Academic, 1974.
Chatfield, H. W. (ed.): Science of Surface Coatings, Van Nostrand, 1962.
Drinberg, A. Y., *et al.*: Technology of Non-metallic Coating, Pergamon, 1960.
Fisk, P. M.: Physical Chemistry of Paints, Hill and Wang, 1963.
Gardner, H. A., and G. G. Sward: Paint Testing Manual, 12th ed., Gardner, 1962.
Gaylord, N. G. (ed.): Polyethers (in 3 parts), Wiley, 1962–1963.
Gould, R. F. (ed.): Literature of Chemical Technology, 16, Printing Inks, 26, Coatings, ACS, 1967.
Hicks, E.: Shellac: Its Origin and Applications, Chemical Publishing, 1961.
Huminik, J., Jr. (ed.): High-temperature Inorganic Coatings, Reinhold, 1963.
Larsen, L. M.: Industrial Printing Inks, Reinhold, 1962.
Letsky, B. M.: Practical Manual of Industrial Finishes, Reinhold, 1960.
Martens, C. R.: Emulsion and Water-soluble Paints and Coatings, Reinhold, 1964.
Martinson, C. R., and C. W. Sisler: Industrial Painting: The Engineered Approach, Reinhold, 1962.
Nylen, P., and E. Sunderland: Modern Surface Coatings, Interscience, 1965.
Paint Manual, 2d ed., Dept. of the Interior, Bureau of Reclamation, 1961.
Paints and Varnishes, OTS Selective Bibliography, SB-518, Dept. of Commerce, 1963.
Parker, D. H.: Surface Coating Technology, Wiley-Interscience, 1965.
Patton, T. C.: Resin Technology: Formulation, Calculations, Wiley-Interscience, 1962.
Patton, T. C.: Paint Flow and Pigment Dispersion, Interscience, 1964.
Patton, T. C.: Pigment Handbook, Wiley-Interscience, 1973, 3 vols.
Payne, H. F.: Organic Coating Technology, Wiley, 1961, 2 vols.
Preuss, H. P.: Pigments in Paint, Noyes, 1974.
Ranney, M. H.: Powder Coatings Technology, Noyes, 1975.
Torche, J.: Acrylic Paints, Sterling, 1966.
Turner, G. P. A.: Introduction to Paint Chemistry, Barnes and Noble, 1967.
Tysall, L. A.: Industrial Paints, Basic Principles, Macmillan-Pergamon, 1964.
Williams, A.: Paint and Varnish Removers, Noyes, 1972.

[41]Development of Improved Printing Inks, *Chem. Week*, Apr. 9, 1960, p. 92.
[42]New Inks Cure Printer's Air-Pollution Problems, *Chem. Week*, Mar. 8, 1972, p. 44.

chapter 25

FOOD AND FOOD BY-PRODUCT PROCESSING INDUSTRIES

Much of the technology used in other manufacturing industries has been applied to food. Likewise, much of the technology developed for the food industry is applicable to other industries. The demand for more preprocessing and processing of food products for home use has arisen naturally because many homemakers work away from home. The demand for uniform quality of food on a year-round basis and high quality standards, even at consumption centers remote from production, has led to improved processing methods. Increasing affluence has led to a demand for greater variety.

HISTORICAL Production and small-scale processing were formerly done on farms and in homes, but with increasing specialization and reduction in the number of people on farms, centralized processing became essential. With central processing, the establishment of grade and quality standards has become necessary. Early developments grew from cottage and community programs (e.g., canneries) into larger-scale units. Milling of grain, fluid milk processing and distribution, baking, and processing of sugar and candy products developed early. More recently, frozen processing has been applied to meat, fruits, vegetables, and manufactured food products ready to cook or serve. A whole new storage and distribution process had to be developed to permit wide-scale consumption of frozen products.

The materials produced on farms, ranches, and plantations, which were formerly consumed or retailed there, are now seldom so handled. Production enterprises are now often single-product-oriented, buying much of their basic food from manufactured sources. At each step in the movement of the product to the consumer, consideration is given to factors influencing the quality of the product and its value to the ultimate user. Costs increase and the value of the product increases, and the increase in value is related to the cost (Table 25.1).

TABLE 25.1 ***Food and Kindred Products, United States***
(In millions of dollars)

Total value of farm marketing	59,758
Value of farm marketing for food	42,385
Value added by manufacturing	34,110
Value of food products manufactured	103,631
Retail sales, grocery stores	82,800
Retail trade, eating and drinking	31,100
Expenditures for food (not including alcoholic beverages)	117,300

Source: The Statistical Abstract of the United States, 1974, which covers 1971 data. *Note:* Row 2 plus row 3 does not equal row 4 because all food manufactured in 1971 was not produced in 1971.

TABLE 25.2 *Summary of the Size of Manufacturing Industries*

	Value produced, millions of dollars	Value added by manufacturing, millions of dollars	New capital expenditures, millions of dollars	Gross book value of assets, millions of dollars		Total employees, thousands
				Total	Machinery	
All manufacturing industries	670,971	314,138	20,947	266,266	192,003	18,363
Food and kindred products	103,631	34,110	2,245	25,282	16,763	1,574
Chemicals and allied products	51,873	29,432	2,938	35,029	26,428	849
Petroleum and coal products	26,935	5,617	1,304	15,380	9,368	141
Paper and allied products	25,458	11,682	1,197	19,206	15,314	632
Textile-mill products	24,030	9,995	873	9,632	7,187	907
Rubber and plastics	17,044	9,521	742	7,787	5,885	544
Tobacco manufacturing	5,528	2,560	94	919	657	67

Source: The Statistical Abstract of the United States, 1974, which covers 1971 data.

ECONOMICS As a separate category, the food industry is twice as large as the entire chemical processing industry in terms of value of products and employment. Food and related products comprise one of the largest of the seven manufacturing industries in the United States (Table 25.2). The processing of foods is often strikingly similar to the processing of chemical substances, so chemists and chemical engineers are as important as food technologists and agricultural engineers in developing processing methods.

The food manufacturing industry transforms crude farm products into material suitable for consumption. In 1971, processing added $34,110 million in value, approximately one-third of the final value. Consumers and producers often accuse the food processing industry of excessive "middle" costs. These costs are, however, similar to the manufacturing costs of other industries (Table 25.2). Compared to the chemical process industries, the food industry invests and spends less in facilities and equipment and has more employees. This comparison, along with increasing labor costs, suggests a future increase in facilities and equipment investment and a comparative decrease in the number of employees per unit of production.

FOOD PROCESSING Processing equipment for food is similar to that for other processing. Gases, liquids, solids, and combinations must be handled without deterioration of the product or damage to the equipment. Sanitation conditions are a necessary additional consideration. Sanitary requirements are met through the correct selection of materials, proper design of equipment and

TABLE 25.3 *Size of Food Manufacturing Industries*

	Value produced, millions of dollars	Value added by manufacturing millions of dollars	New capital expenditures, millions of dollars	All employees, thousands	Payroll, millions of dollars
All food and kindred products	103,631	34,110	2,245	1,574	12,180
Meat products	26,079	4,978	288	307	2,378
Dairy products	14,813	3,909	223	193	1,517
Beverages	13,331	6,557	441	219	1,988
Canned, cured, and frozen food	11,962	4,640	315	265	1,581
Grain mills	11,210	3,426	282	108	920
Bakery products	7,357	4,141	194	237	1,932
Confectionery and related products	3,442	1,625	97	80	540
Sugar	2,857	870	66	30	255
Miscellaneous foods and related products	12,581	3,963	339	136	1,070

Source: The Statistical Abstract of the United States, 1974, which covers 1971 data.

fixtures, and appropriate use of equipment. Three-A[1] (associations) standards are the basis of design and fabrication which ensure that, through proper use, sanitary requirements will be met.

The most important developments basic to the growth of the food industry include: pasteurization, drying and dehydration, development of materials capable of withstanding food products without deleteriously affecting them, sterilization, freezing, materials handling, use of automatic controls, sanitizing and cleaning processes. Industry trends utilizing these basic principles are: cleaning in place (CIP), larger-scale operations, automation, processes to extend shelf and storage life, and reduction in environmental impact through processing—efficient use and reuse of energy, waste gases, and water.

The major material development in the last 25 years has been the general adoption of stainless steel for food processing. Without stainless steel, the development of CIP cleaning, automation, continuous operation, and aseptic processing would have been impossible or delayed. Glass and glass-coated materials have increased in use along with stainless steel. Plastics are becoming increasingly important. The use of copper, tinned copper, and zinc is almost obsolete.

Stainless steel is remarkably resistant to corrosion but, in contact with other metals, electrolytic cells can be formed which accelerate attack. Graphite packing in pumps is an often overlooked source of electrolytic corrosion. Small pockets of stagnant material, crevices in which materials can stagnate, and incomplete cleaning can yield spots where stainless may be attacked. Stainless should not be joined directly to other metals.

SANITARY EQUIPMENT DESIGN FEATURES The Three-A standards include specifications for equipment, fixtures, and fitting. Some special features for sanitary design and construction are:

1. Material, in general, should be 18-8 stainless steel, with a carbon content of not more than 0.12%, or equally corrosion-resistant material.
2. The gage of metal should be sufficient for the various applications.
3. Product surfaces fabricated from sheets should have a no. 4 finish or equivalent.[2]
4. Square corners should be avoided. Minimum radii are often specified, such as a storage tank, $\frac{1}{4}$ in. inside corners of permanent attachments; and for a tank, $\frac{3}{4}$ in.
5. No threads should be in contact with food. Acme threads should be used.
6. Surfaces should be sloped to provide drainage.
7. Designs should permit interchangeability of parts.

PRODUCT CHARACTERISTICS RELATED TO EQUIPMENT Chemical and biological properties usually play a more predominant role than mechanical properties in the design of equipment for the food industry. In addition to the usual problem of materials to resist corrosion, the character and toxicity of the products of corrosion must be considered. Good appearance must also be maintained.

The pH of fluid products is the major factor in materials selection. Products containing vinegar and salt are among the most corrosive materials. For these products, Type 316[3] stainless steel is usually required. The most acidic foods commonly processed are cranberry juice, lemon juice, sweet pickles, and cranberry sauce; all are below pH 3.0. Few foods are alkaline; generally they are below pH 8.0. Crab meat is pH 6.8, and some chicken preparations are pH 6.5. Time, temperature, pH, product roughness, velocity of flow, and surface condition (smoothness and previous usage) all affect the rate at which equipment is attacked by the material it contains.

Cleaning-in-place puts exceptional stress on materials and may require special attention. Halogens cause pitting in stainless steel, especially free chlorine dissolved in small droplets of water

[1]Three A consists of three associations: the international Association of Milk, Food and Environmental Sanitarians; the Dairy Industry Committee; and the Food and Industry Supply Association.

[2]Surface finishes are graded from no. 1 to no. 8, with 8 being most highly polished. In use, finishes above 4 generally reduce spontaneously to 4.

[3]For meaning of the AISI numbering systems see Perry, 5th ed., p. 23–15.

on the surface. Exposure to chlorine, bromine, or iodine should be limited to 20 or 30 min, and temperatures kept below 100°F.

CLEANING Stainless steel owes its corrosion resistance to a passive oxide layer on its surface. Surface contaminants can destroy this protective layer and so can cleaning agents. Exposure to air can restore the corrosion resistance under some conditions.

Material adhering to a stainless-steel surface should be removed with the least abrasive material that can do the job. Normally, a mild abrasive can be used or a stainless-steel sponge for more aggressive cleaning. Metal sponges of dissimilar metal should not be used, because particles of metal adhering to the surface may set up corrosion cells. After cleaning, the surface should be washed with hot water and left to dry. Sanitization (often incorrectly called sterilization) with 200 ppm chlorine solution should be done 30 min before use, not immediately after cleaning. This procedure is followed to avoid corrosion.

Manual cleaning involves drudgery under poor working conditions and utilizes as much as 50% of the labor used in a plant. Cleaning-in-place has eliminated this problem and is a major development. Three-A standards for cleaning-in-place are:

1. Using alkali or acid solutions appropriate for the product and equipment surface.
2. Providing a time of exposure of 10 to 60 min to remove substances without damage to the metal.
3. Utilizing a velocity of flow of 5 ft/s (3 to 10 ft/s).
4. Maintaining a slope of surface and tubing to provide for drainage ($\frac{1}{16}$ to $\frac{1}{8}$ in./ft).
5. Avoiding dead ends for flows.
6. Using connections and joints which are cleanable (welded joints, clamp-type joints, appropriate gaskets).

Stainless steels are made in a wide variety of types (2XX, 3XX, 4XX) with widely varying chemical and mechanical properties. Some are martensitic and some austinitic, and selection of the optimum material with regard to corrosion resistance and price is often complex. Complete discussions of stainless steels and their applications are available.[3]

MILK PROCESSING

FLUID MILK The major function of a milk processing plant is to pasteurize or sterilize the product. The milk is received, standardized, stored, and packaged. Most fluid milk is homogenized while being pasteurized.

The purpose of pasteurization is to kill pathogenic organisms to eliminate disease, and to inactivate enzymes to improve storage and keeping quality. The most common method of pasteurization is the high-temperature short time (HTST) method which exposes the milk to 163°F for not less than 16 s, followed by rapid cooling. Various temperatures are used for an appropriate time to kill organisms and to inactivate enzymes without excessive effects on other components of the product. Accessories in the pasteurization process include a clarifier, heater-coolers (usually plate exchangers), a homogenizer, vacuum treatment equipment, storage vats, filler, packaging equipment, controls, and cleaning equipment.

The sterilization of milk (or other fluids) consists of a more intense heat treatment than pasteurization (285°F for 6 s). The sterilized product is placed in a sterile container under aseptic conditions. Products less likely to be damaged by longer high-temperature treatment, such as evaporated or condensed milk, are usually sterilized in the container.

EVAPORATED AND CONDENSED MILK Evaporated milk is made by reducing the water content by using an evaporator. The product is reduced 3 to 1 or 2 to 1 in volume, placed in cans, and sterilized. Condensed milk is processed similarly to evaporated milk, but sugar is added to improve flavor and increase storage life. Both products are used primarily for food manufacturing or cooking. The following processing steps are usually involved:

Milk is received and cooled (Op).

Milk is sampled and evaluated for bacterial count, sediment, acidity, alcohol, appearance, flavor, and odor (Op).

Milk is clarified or filtered (Op).

Milk is stored and fat standardized to solids not fat (SNF) (Op).

Milk is preheated and held (Op).

Water is removed by evaporation (Op).

Product is homogenized (Op).

Product is cooled (Op).

(If making condensed milk, homogenization precedes evaporation and sugar is added immediately after condensation.)

Product is restandardized (Op).

Product is placed in cans which are sealed (Op).

Leaking cans are rejected (Op).

Product in cans is heated, sterilized, and cooled (Op and Ch).

Cans are cased (Op).

Cans with product are stored (Op).

Single- or multiple-effect evaporators are used. Double- and triple-effect evaporators are most common. Larger operations can justify the greater capital expense for the larger number of effects based on energy savings. The steam use and evaporating capacity of a typical double-effect milk evaporator with vapor and bleeder heaters are given in Fig. 25.1.

DRY MILK Moisture may be removed from condensed (unsweetened) milk to produce a product with less than 5% moisture. Because fat rapidly becomes rancid on storage, it is usually removed centrifugally before drying. Dry milk for human consumption should be pasteurized before drying. Spray dryers using filtered air are usually used, but products for animal consumption and some manufactured products are made on drum dryers. Treatment to speed up solubility ("instantizing") involves steaming and agglomerating to form coarse particles, followed by redrying. Nonfat dry milks (NDM) are processed at either low or high heat, resulting in differing degrees of protein denaturation.

Dry whole milk is standardized fat to solids-not-fat (SNF) in a ratio of 1:2.769. The product is homogenized before condensing. The fluid is fed to the spray dryer at 2,500 psi, and the exhaust air leaves the dryer at 180°F. The product is vacuum-packed and filled with nitrogen to maintain the storage properties of the whole milk.

ICE CREAM In 1972, 786 million gal of ice cream was manufactured. Ice cream weighs about 4 lb/gal, usually based on 100% overrun (50% air v/v). Ice cream is made of a mixture of milk products, flavors, sweeteners, stabilizers, emulsifiers, and other ingredients such as fruit and nuts. The product is frozen at 20 to 26°F, depending on the mix. A scraped surface heat exchanger, operated so that air is incorporated into the mass, is used. Quality control is extremely important.

MEAT PROCESSING Beef and pork processing are very similar. Manual handling of the various operations involved would be too expensive with present labor costs. An automated system showing the steps involved in a hog-slaughtering complex is given by Hall.[4]

POULTRY PROCESSING includes killing, bleeding, scalding, and removal of feathers. Government standards of quality have been established by the U.S. Dept. of Agriculture. Quality factors include conformation of body and limbs, flesh, fat covering, and freedom from pin feathers, cuts, bruises, and discoloration.

[4]Hall, Farrall, and Rippen, Encyclopedia of Food Engineering, p. 538, Avi Publishing, 1971.

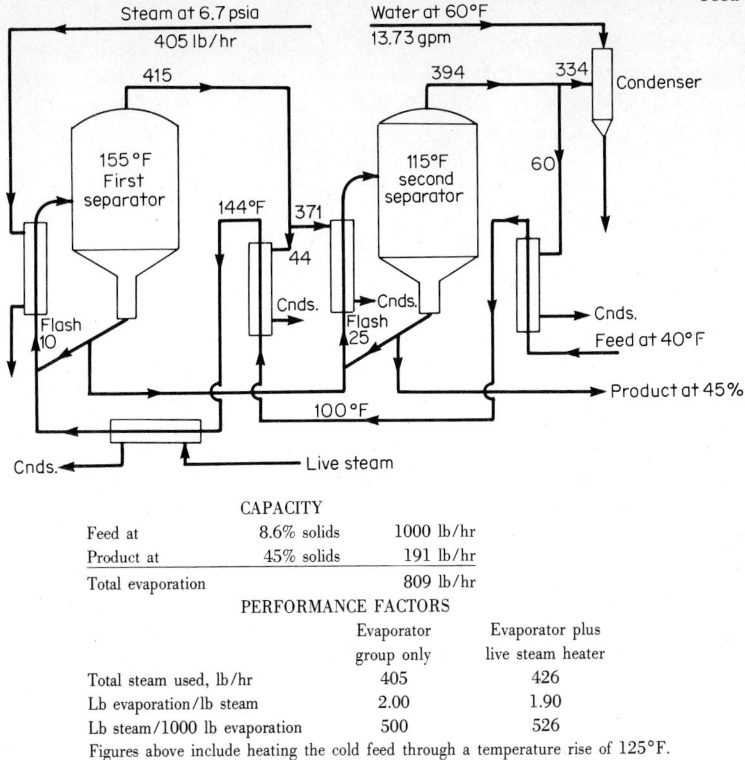

CAPACITY

Feed at	8.6% solids	1000 lb/hr
Product at	45% solids	191 lb/hr
Total evaporation		809 lb/hr

PERFORMANCE FACTORS

	Evaporator group only	Evaporator plus live steam heater
Total steam used, lb/hr	405	426
Lb evaporation/lb steam	2.00	1.90
Lb steam/1000 lb evaporation	500	526

Figures above include heating the cold feed through a temperature rise of 125°F.

Fig. 25.1 Flowchart for a double-effect evaporator with vapor and bleeder heaters.

FREEZING In 1972, 20,393 million lb of food was frozen. Juices and drinks (5,974 million lb), prepared foods (5,710 million lb), vegetables (4,981 million lb), poultry (2,236 million lb), fruits and berries (612 million lbs), and meat (510 million lb) are the principal frozen foods. The major developments in freezing have occurred since 1947. The value of products shipped is $4.5 billion. These products are normally stored below 0°F.

Rapid freezing is usually desired to avoid formation of large ice crystals in the product. Large crystals rupture the cells and decrease the freshlike quality of food as it is thawed. The following procedures are used to accomplish rapid freezing:

1. Placing packaged products over low-temperature coils; conduction cooling.
2. Fluidized-bed freezing (vegetables such as peas and beans).
3. Spraying liquid nitrogen over the product.
4. Forced convection air cooling using high-velocity air at −20 to −40°F.
5. Natural convection air cooling.

FOOD BY-PRODUCTS

The chemistry and engineering behind the leather industries are so complex that scientific control has developed comparatively slowly. The skins and bones of animals and their products—leather, gelatin, and glue—are mostly colloidal materials and have almost indefinable properties about which too little is known. The raw skin consists of a number of different complex proteins which vary in structure in different animals and even vary in different parts of the same hide. The complexity of the art of leather manufacturing is increased by the large number of active substances that affect the skins—enzymes, bacteria, alkalies, acids, tannins, tannin substitutes, oils, fats, and salts. This section is concerned with the principal phases of leather manufacture, as well as with the by-products, gelatin and glue. Discussion of the latter includes newer types of adhesives.

LEATHER

HISTORICAL Leather is one of the oldest commodities on today's world market. It has been called "the most historic of useful materials." The art of leather manufacturing from hides and skins antedates, by centuries, any scientific knowledge of chemistry. Specimens of ancient Egyptian leather have been found which are claimed to be at least 6,000 years old.[5] The color and strength of this leather are remarkably unimpaired. As judged from these well-preserved leather products, the origin of the technique must have antedated this time by many centuries. The original primitive methods for the preservation of skins probably consisted of simply drying them in air and sunlight. Later, the preservative effects of different oils were noted. Still later, probably, the accidental discovery of the tanning effects of leaves, twigs, and barks of certain trees soaked in water were observed.

ECONOMICS AND USES The sole-leather tanner spends 50% of a sales dollar for hides, and 40% for tanning materials and labor. Consequently, the leather industry is constantly searching for new chemicals with which to make better leathers at a cheaper price. Probably the greatest single advance in the leather industry was the practical application of the chrome process of tanning introduced in 1893. Because of the results of this process, more than 90% of the world's output of upper shoe leather is now chrome-tanned. It has greatly speeded up the tanning operation and gives increased strength. Vegetable tanning takes 2 to 4 months, but the chrome process has reduced this time to 1 to 3 weeks. The chief use of vegetable tanning at present is in the tanning of heavy leathers such as sole leather and leather belting; consequently, this old historical process is still widely employed. Furthermore, today most shoes are manufactured from leather that has undergone some vegetable retannage.[6]

According to the 1973 Annual Survey of Manufacturers, there were 490 leather manufacturing establishments with 23,000 employees. The value of products shipped reached $1.1 billion. About 62% of all leather tanned in this country is consumed in footwear. Of the remainder, industrial leather constitutes only 10%. It has been predicted that by 1983 the demand for leather will outstrip the world supply by 30% because the human population is rising more rapidly than the animal population from which leather comes. In 1973 the total U.S. shoe production amounted to 488 million pairs. Imports create a major problem for the industry.

SYNTHETICS Artificial leather and coated fabrics have made heavy inroads on traditional leather markets; at least 70% of all shoe sales are now synthetic. For example, Neolite soles (a combination of synthetic rubber and a special high styrene-butadiene resin) outwear leather by a wide margin and cost less. Patent leather has been extensively replaced by the superior vinyl plastic. The fact that plastics can be produced in large sheets both eliminates waste in cutting and lends itself to automation. A porous synthetic leather called Corfam was introduced in 1963. A technical but not a commercial success, it is no longer made in the United States. A variety of plastic composites continue to offer competition to leather, but fail to displace it for quality or severe applications—such as use in work shoes and boots.

[5]O'Flaherty, Roddy, and Lollar (eds.), Chemistry and Technology of Leather, vol. 2, p. 161, ACS Monograph 134, Reinhold, 1956, 4 vols.

[6]Butz, Combination Tannage, in O'Flaherty, *op. cit.*, vol. 2, p. 461.

ANIMAL SKINS

The United States is unable to produce sufficient hides and skins for its leather needs, and hides are increasingly difficult to obtain. Technically, the term *hide* is applied to the skins of larger animals, such as bulls, horses, cows, and oxen; the term *skin* refers to the skins of goats, sheep, calves, and smaller animals. The quantity of cattle hides consumed as raw material has fallen in the last decade; in 1973, 17,687,000 were consumed, or 21% less than in 1962. Greater variance has been noted for all skins in the two preceding decades; generally, the trend has been downward. In 1973, 1,262,000 calfskins and kips[7] and 3,522,000 goatskins and kidskins were used, and in 1972 (1973 figures not available), 31,397,000 sheepskins and lambskins.

Animal skins as received by the tanner[8] may be divided into three layers: the epidermis, the derma or corium, and the flesh. The epidermis, constituting about 1% of the total skin, is the outer layer and consists chiefly of the protein keratin. The hair, which grows through both the derma and epidermis, is also largely keratin. This layer is dense and chemically resistant, thus allowing the epidermis to be readily removed by chemical means. Hot water, however, causes slow solubilization by hydrolysis of the collagen to produce gelatin. The elastin is not affected by this treatment. The flesh is an extraneous layer, mostly adipose tissue, which must be removed to ensure tannin penetration on both sides of the corium.

MANUFACTURE

Figure 25.2 shows the necessary steps in the manufacture of leather.[9] These steps may be broken down into the following sequences:

The skins are opened, examined, trimmed, and graded (Op).

The skins are water-soaked for 3 to 24 h (for dried hides, longer time may be necessary), and the flesh is removed by a machine equipped with a spiral-bladed cylinder (beaming). Soaking assists are used; sodium tetrasulfide, $\frac{1}{5}$ to $\frac{2}{5}$ of 1%, hastens rehydration and produces more uniform hydration (Op and Ch).

Then the skins are treated with a saturated lime solution plus certain accelerators, such as sodium sulfide or sulfhydrate and dimethylamine, for 3 to 7 days (Op and Ch). Liming may be reduced to 24 h on light hides.

Any hair remaining is removed by machine and hand-scraped (Op).

The clean, dehaired hides are treated with certain enzymatic preparations for several hours to soften the skin and remove the lime (bating) (Ch).

After rinsing, the hides are ready for tanning (Op).

The two principal methods of tanning are shown in Fig. 25.2. The *vegetable-tanning* process consists of the following procedures:

The prepared skins are treated with organic acids to neutralize any residual lime (Ch).

The tanning is carried out in several tanning baths, starting very weak to prevent *case-hardening* and becoming progressively stronger. This requires about 1 month (Ch).

The skins are then allowed to soak in a vat containing strong tanning extracts (Ch).

The *chrome-tanning* process is somewhat different from the tanning process, as indicated in Fig. 25.2. The following pertains to one-bath processes.

[7]Kips are skins of animals smaller than a full-grown calf.

[8]McLaughlin and Theis, The Chemistry of Leather Manufacture, Reinhold, 1945.

[9]For details of manufacture, cf. O'Flaherty, *op. cit.*, vols. 1 and 2; also *Rep. Progr. Appl. Chem.* **56** (1971).

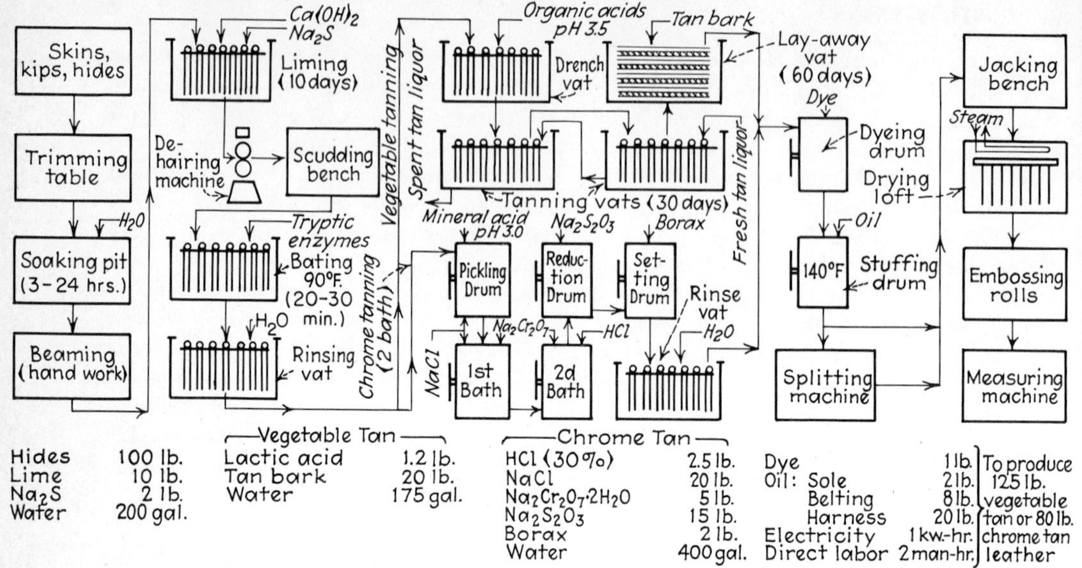

Fig. 25.2 The manufacture of leather.

The prepared skins are pickled in a bath of salt and sulfuric acid (Ch).

The pickled hides are soaked in a solution of sodium dichromate (Op and Ch).

The dichromate-saturated skins are treated with sodium thiosulfate in the reduction drum (Ch).

The chrome salt is set on the fibers by adding borax to the skins in the settling drum (Ch).

After washing, the leather is ready for the finishing operations (Op). Finishing operations are somewhat similar for both methods of tanning.

The tanned hides may be dyed (Ch).

The leather is stuffed with oil (Op).

The leather is split to produce a thinner, more supple product (Op).

Gloss is developed on the surface of the leather by ironing and glazing it with a glass cylinder (Op).

The jacked leather is dried (Op), embossed (Op), and measured (Op).

In industrial application these chemical conversions and operations are complicated and require experienced workers to carry them out. The manufacture of leather is an engineering process requiring careful chemical control in order to ensure uniform quality (Table 25.4).

PRESERVATION AND DISINFECTION OF SKINS Since raw skins decay rapidly, some method of preservation must be used to arrest bacterial action and consequent hide disintegration. Tannery experience with domestic hides has demonstrated the superiority of brine curing, in which a bactericide is employed, over curing by the conventional green-salt method. Curing is more rapid and uniform, affords greater protection prior to tanning, practically eliminates "salt stains," requires no washing of hides before "soaking back," and produces leather of plumper grain.[10] In salt curing, standard domestic practice piles the hides flesh side up in alternate layers with salt. These stacks are arranged for proper drainage of the brine. This method of curing requires a minimum of 3 or 4 weeks at about 13°C, during which the hide loses part of its moisture by dehydration and gains

[10]De Beukelaer, Preservation of Hides and Skins, in O'Flaherty, *op. cit.*, vol. 1, pp. 206–207.

TABLE 25.4 ***Comparison of Skins and Leather***

	Skins or hides	*Leather*
Pliability	Soon lose pliability and become hard and brittle	Retains pliability
Permanence	Putresce very quickly	Extremely durable, not attacked by bacteria
Water resistance	Absorb water, and are permeated easily by it	Possesses great resistance to water
Boiling water	Are converted to gelatin by hydration	Attacked with great difficulty
Mechanical strength	Good	Good

weight through salt absorption. The loss is the greater, and the overall difference is termed the "shrink." Many irregularities in leather may be traced to careless curing and storing.

PREPARATION OF HIDES FOR TANNING The first step in the tanning process is inspection of the hides for defects as they come into the tannery, and the cutting off of ears and ends.[11] The next step is the cleansing of the hides to remove dirt, manure, and salt, in order to restore them to a natural soft hydrated state, by paddle washing, drum washing, or soaking in vats.

The soaking and washing of the hides are quite important, because if moisture is not restored, the hide will not respond properly to the different tanning operations. Small amounts of sodium polysulfide and surfactants are added, which greatly accelerate the soaking. Disinfectants are used in soaks to offset the effects of bacteria present from delayed or inadequate cure. If the hide is oversoaked, it will be soft and below par. A properly soaked hide contains about 65% water. No standard methods are employed, although the objective of these procedures is the same.[12]

Liming is a means of loosening and removing the epidermis and hair from the hide and is usually done in paddle vats, drums, and pits. The hides are tied or hooked together and placed in vats containing water, 10% lime (based on the weight of the hides), and 2% sodium sulfide (based on the hide weight), which acts as an accelerating agent. Dimethylamine, sulfhydrate, and cyanide salts are also used as accelerating, or *sharpening*, agents.[13] The hides are moved ahead daily in a series consisting of three to seven vats; they remain in each for 1 day and then enter a vat with fresher lime.

The epidermis and hair are composed chiefly of keratin. Keratin is a protein containing a cystine residue which is easily attacked by alkali. Lime attacks the disulfide link of the keratin of the epidermis, the cortical layer, and the hair. This softens the hair and removes the epidermis. This action may be represented:[14]

$$RSSCH_2R + H_2O \longrightarrow RSH + RCH_2SOH$$

$$RCH_2SOH \longrightarrow RCHO + H_2S$$

After the hides have passed through this process, they are usually placed in a vat of warm water, which tends to relax them and permits easier removal of the hair in a dehairing machine. The skins are brought into contact with a roller set with dull knife blades which rub off the loose hair and epidermis, as represented in Fig. 25.2.

Bating has been one of the most mysterious processes carried out in the tannery and by far the most unpleasant. The old method consisted of placing the limed skins in a warm infusion of the dung of dogs or fowls until the plumpness of the skins disappeared. Later investigation of the important organic constituents of dung led to the use of artificial materials, and the process became

[11] O'Flaherty, *op. cit.*, vol. 1, p. 220.

[12] Goetz, Washing and Soaking, in O'Flaherty, *op. cit.*, vol. 1, p. 220.

[13] Morris, Practice of Unhairing, in O'Flaherty, *op. cit.*, vol. 1, p. 230.

[14] Lewis, Squires, and Broughton, Industrial Chemistry of Colloidal and Amorphous Materials, Macmillan, 1942.

known as *bating*.[15] Just before immersion in the tanning liquor, the hides undergo the bating, or deliming, process, which now uses ammonium sulfate or chloride as the deliming agent and proteolytic enzymes for the removal and alteration of certain proteins and for the improvement of the color of the grain. Another important reaction in bating is called *falling*—a reduction in the degree of swelling of the protein constituents of the limed skin previously swollen. The principal function of bating is to regulate the degree to which the skin structure is prepared to make it accessible to uniform absorption of the tanning agent.[16]

VEGETABLE TANNING is chiefly used for heavy leathers, for soles and belting. The wedding of the tanner's art and the chemist's knowledge has resulted in the modern technology of leathermaking.[17] In the United States, 99% of all vegetable tanning material is imported, usually in the form of solid extracts, such as quebracho wood (Argentina and Paraguay), wattle bark (South and East Africa), and eucalyptus bark (Australia). Extracts of gambier, valonia, myrobalan, and mangrove come from East India, Turkey, India, and Borneo, respectively. Chestnut tannin occurs in the wood of the American chestnut tree, but the supply is exhausted; a related European species produces a tannin similar to that of the once-plentiful domestic tree. Vegetable *tannins* may be divided into two broad groups, catechols and pyrogallols, depending on which substance is yielded on fusion with alkali.[18]

Tannins are complex mixtures of glucosides of various polyphenols. Their action is to combine with and between the collagen fibers of the skins. During the tanning process tannins liberate sugars. The acidity, or pH, is controlled by the use of sulfuric acid. These changes may be represented:

$$\text{Skin} + \text{tannin} \longrightarrow \text{leather} + \text{sugar}$$

The protein in the aggregate has the capacity to absorb large quantities of tannin. Skins may increase in weight as much as 350% during the tanning operation. This combined and absorbed material fills up the holes and stiffens the leather.

Synthetic tanning agents, or syntans, are gaining in use, as evidenced by the present production of more than 53 million lb, although vegetable tanning extracts are still widely employed. Syntans are condensation products of sulfonated phenols (or higher homologs) and formaldehyde capable of converting animal skin into leather.[19] Comparatively few syntans lack a sulfonic acid group. Syntans may be classified according to use as auxiliary, complementary, or replacement tannins. Until 1945 and the introduction of the first domestic replacement tannin in America (Orotan TV, Rohm and Haas Co.), all syntans belonged to the first two classes. They functioned as adjuncts and dye mordants and were used for bleaching chrome leather, for solubilization of vegetable tans, and for increasing the speed of penetration of natural tannins into leather. Research has been done on a potential source in the petroleum industry, and lignosulfonate by-products of the paper industry remain attractive.[20]

Vegetable tanning is usually done in 8 × 7 ft vats, 5 or 6 ft deep, which contain the tan liquors and hides. Since the hides remain in the liquor for one or more months, many vats are required, and they must therefore be cheap and durable. Hides first enter the oldest, weakest tanning liquors in order to prevent complete plugging of the surface pores and subsequent poor penetration of the tannins into the skins by a strong liquor. They pass successively through liquors of increasing strength until they leave the freshest and strongest liquor as fully tanned leather. The ratio of liquor to hide is

[15]Wilson, The Chemistry of Leather Manufacture, Reinhold, 1928.

[16]Wilson, Practice of Bating, in O'Flaherty, *op. cit.*, vol. 1, p. 344.

[17]Kay, Process of Vegetable Tannage, in O'Flaherty, *op. cit.*, vol. 2, pp. 161–197. See p. 172 for vegetable-tanning related materials, classified and tabulated as to type, together with their constituents and properties.

[18]White, Chemistry of the Vegetable Tannins, in O'Flaherty, *op. cit.*, vol. 2, pp. 100–147. (259 refs.); Haslam, Chemistry of Vegetable Tannins, Academic, 1966.

[19]Chen, Syntans and Newer Methods of Tanning, The Chemical Elements, South Lancaster, Mass., 1950; Staff Report, New Tanning Agents Based on German Technology, *Chem. Eng. News*, **26,** 1980 (1948).

[20]Turley, Replacement Synthetic Tans, in O'Flaherty, *op. cit.*, vol. 2, pp. 338–407.

kept as low as possible and is usually 6 to 4 lb of liquor to 1 lb of hide. Modern processing now speeds up the tanning time and shortens the expensive and tedious aging and mellowing in tan liquors. Much research has been devoted to expediting the process and reducing the hand labor required.

Since approximately 250 gal of *iron-free* water is required for each hide produced, water supply and disposal are important factors in locating a tannery. As in most other fields of industrial endeavor, there is no standard treatment of *waste disposal* in the tanning industry. Because of the relatively high amount of solids found in tannery wastes, some form of precipitation and sedimentation is usually included. Tannery waste disposal has proved to be expensive, but plants have been able to meet the strict antipollution standards by careful operation.

Chrome tanning. The uppers of a pair of chrome-tanned leather shoes will outwear two or three pairs made from a vegetable-tanned product. For this reason more than 90% of the world's production of light leather is chrome-tanned by the one-bath process. To condition hides and skins for subsequent mineral tannage, a process called *pickling* is utilized, wherein bated hides and skins are treated with solutions of acid, usually sulfuric, and salt. In the chrome process, combination of the chromium salts and hide fibers is much more rapid and takes place without the degree of swelling that occurs in vegetable tanning. Therefore chrome-tanned leather is more pliable and looser in structure. In sole leather, fillers are usually added to give water-resistant properties. Chrome-tanned leather is characterized by a high content of original hide proteins and mineral matter and a low content of water-soluble materials. The main mechanism of chrome tanning is the coordination of chromium to the carboxyl groups of the protein.[21] Chrome tanning is usually subdivided into two processes: the one-bath process, using basic chromic sulfate, and the two-bath process, using sodium dichromate. The chrome liquor for the one-bath process is usually made by reducing an acidified solution of sodium dichromate by slowly adding a solution of glucose to it or by bubbling sulfur dioxide through it until the reduction is complete.

$$Na_2Cr_2O_7 + 3SO_2 + H_2O \longrightarrow 2Cr(OH)SO_4 + Na_2SO_4$$

Various modifications of the two-bath process are in use, and the method shown in Fig. 25.2 serves as an illustration. After pickling, the hides are placed, successively, in two drums or paddle vats, containing about 5% sodium dichromate solution (based on the weight of the hides), with hydrochloric acid, and salt. The skins are allowed to remain in each liquor overnight and are then pulled out and allowed to drain. After an hour or two, they are placed in another paddle vat containing approximately 15% of the weight of the hides in sodium thiosulfate. The hides are left in this liquor overnight, pulled out in the morning, and washed in a drum containing water. Borax is added to reduce the acidity of the leather to the desired extent.

Finishing operations. After the tanning operation is completed, the leather, if dried rapidly, is quite stiff, usually breaks if bent sharply, and is not yet fit for use. The fibers must be lubricated to give them pliability and softness. Actually, more work is required in the operations following tannage than in all the preceding operations and processes put together. Since each of the numerous kinds of leather requires a special series of operations, details are impossible here.

If the tanned leather is not smooth, it is shaved by a machine with sharp blades. Very often it is split by means of a bandlike knife to obtain the desired thickness. (This may be done later, as shown in Fig. 25.2.) Vegetable-tanned leathers are sometimes washed, bleached, or scoured. The bleaching or washing is usually done in a revolving drum containing sodium carbonate, sodium hydroxide, or sulfuric acid and water. Other bleaching agents, such as sodium bisulfite, sulfurous acid, sulfite cellulose solutions, and various syntans, are also used.

The next operation involves the incorporation of oils and greases and is called *stuffing*, or *fat liquoring*. Sulfonated oils blended with raw oils are the most important fat liquors and have mostly

[21]Shuttleworth, Mechanism of Chrome Tannage, in O'Flaherty, *op. cit.*, vol. 2, p. 317.

replaced raw oils, natural greases, emulsions of raw oil, soaps, waxes, and resins. The leather, either dry or wet, may be stuffed by hand or in a rotating drum. A more uniform job is achieved by hand. Sole leathers are usually "loaded" with some material, such as magnesium sulfate and cellulose, to give them desired properties.

The color and shade of the finished leather are produced by a combination of dyeing and finishing operations. Dyeing may be combined with the fat-liquoring process, or it may be done before or after it. Most of the dyestuffs used now are synthetic coal-tar derivatives. The most common method of dyeing employs revolving drums or vats equipped with rotating paddles.

The finishing series of operations in the production of leather includes both mechanical treatments and applications of finishing materials to the leather. The following classes represent the more common materials used in leather finishing: (1) cellulose ethers, (2) waxes, (3) resins,[22] (4) dyes, (5) pigments, (6) lacquer materials, (7) antiseptics, and (8) miscellaneous materials such as solvents, perfumes, soaps, sulfonated oils, metal salts, plasticizers, acids, and alkalies. Mechanical operations carried out in the finishing of leathers include staking (flexing the leather over an edge to loosen fibers and soften it), glazing, buffing, graining, trimming, rolling, brushing, and plating. The finishing operations alone embrace a technique requiring great skill, wide experience, and close attention.

OIL TANNING From very early times skins have been tanned or preserved by the use of certain drying oils which have been worked into the skins and allowed to oxidize and react, thus becoming fixed. An example of this type of leather is *chamois*. Sheep- or lambskins are subjected to the usual rinsing, liming, trimming, fleshing, and bating. They are then swollen preparatory to splitting, frequently with the aid of an acid such as sulfuric at a pH of about 2.0. The pickled skins in the plumped condition can be split easily, separating the grain from the flesh layer. The former is the skiver, and the flesh layer is processed to make chamois leather. After this, depickling is carried out with an alkali such as borax, or even soda ash or ordinary chalk, with the pH at about 7.0. The damp skins are treated with about 10% of their weight of fish oil, often in a drum in which the revolving action lasts over 4 h, so that the oil can distribute itself and penetrate the skin. The skins are removed and piled in layers, during which fermentation and oxidation proceed, resulting in a reaction of collagen with liberated aldehydes formed by the oxidation of fish oil during tannage.[23] The skins are hung or hooked while the reactions proceed. The temperature rises about 20°F during this maturing process. The oiling and hooking are repeated until 40 to 50% of the oil (based on the weight of the moist skin) has been incorporated. After the oil-tanning process is completed, the chamois leather is washed with warm, alkalized water and hydraulically pressed to remove excess oil. Further washing, buffing, and drying complete the operations, except for bleaching or dyeing.

This classical method of tanning just described has been largely superseded by the "new-chamois" tanning process, wherein pretanning with formaldehyde precedes the actual fish-oil tannage; however, the latter is not superfluous and must be carefully executed. A modification which shortens the process involves the addition of water-soluble fatty agents to the fish oil. Processes in which the larger portion of the fish oil is replaced by an alkylsulfochloride are described by Kuntzel.[24]

By-products. As raw material for tanneries, skins of cattle, hogs, and sheep are themselves by-products of the packing industry; in fact, sheepskins also come from the wool industry. Since it has been estimated that, for each 100 lb of green-salted hides, only about 68 lb of leather will be produced, it is obvious that by-products, of which scrap leather is only one, must also yield revenue. For example, fleshings and trimmings are sold for glue or low-grade gelatin. From them also recent research has developed a protein supplement for use as chicken and dog feed. Beamhouse waste, which includes lime, hair, and shavings, can be converted into tiles that have many desirable features,

[22]Including synthetic monomer graft polymers of vinyl chloride, urethanes, etc.
[23]Nayudamma, Shrinkage Phenomena, in O'Flaherty, *op. cit.*, vol. 2, p. 47.
[24]Kuntzel, Oil Tanning and Sulfonylchloride Tanning, in O'Flaherty, *op. cit.*, vol. 2, pp. 432–433, 451–454.

such as tensile strength and water resistance. Spent vegetable tans, so troublesome from a standpoint of waste disposal, can be concentrated by evaporation and the concentrates sold as water-conditioning agents.[25] Hidehouse trimmings and beamhouse splits are used in the production of gelatin. Although gelatin itself is a by-product both of the packing and tanning industries, its manufacture produces several more by-products, such as calcium phosphate for the pottery industry, bone meal, and bone black.

GELATIN

Gelatin is an organic nitrogenous colloidal protein substance whose principal value depends on its coagulative, protective, and adhesive powers. Water containing only 1% high-test gelatin by weight forms a jelly when cold. There is little, if any, chemical or physical relationship between gelatin and other materials bearing this name, such as blasting gelatin formed by mixing collodion cotton with nitroglycerin (Chap. 22) or Japanese gelatin, also known as agar-agar, a mucilaginous substance extracted from seaweed. Animal *gelatin* is obtained by hydrolysis from collagen—white fibers of the connective tissues of the animal body, particularly the skin (corium), bones (ossein), and tendons. These reactions have been formulated; however, they are only indications of the complex changes that take place, proceeding from quite variable raw materials to products almost as variable.[26]

$$\underset{\text{Collagen}}{C_{102}H_{149}N_{31}O_{38}} + \underset{\text{Water}}{H_2O} \longrightarrow \underset{\text{Gelatin}}{C_{102}H_{151}N_{31}O_{39}}$$

$$\underset{\text{Gelatin}}{C_{102}H_{151}N_{31}O_{39}} + \underset{\text{Water}}{2H_2O} \longrightarrow \underset{\text{Semiglutin}}{C_{55}H_{85}N_{17}O_{22}} + \underset{\text{Hemicollin}}{C_{47}H_{70}N_{14}O_{19}}$$

USES AND ECONOMICS The industry recognizes four different kinds of gelatin, edible, technical, photographic, and pharmaceutical. Gelatin is a widely consumed food, and it is a popular dessert which is easily assimilated and even helps in the digestion of other foods by forming an emulsion with fats and proteins. Gelatin has played an important part in the rapid development of the motion-picture and photographic industries. It is coated on the film base, constituting the sensitized emulsion of the light-sensitive silver salts. Technical gelatin is quite an arbitrary name applied to small amounts used for miscellaneous purposes, such as for sizing paper, textiles, and straw hats. Gelatin is used by pharmaceutical houses for making capsules and as an emulsifier.

MANUFACTURE Fleshings, the extraneous tissue which attaches animal skin to the carcass, and cattle hides are used in large amounts for the manufacture of all four types of gelatin. Calfskin, pigskin, and animal bones are also used, and bones represent about 11% of the total amount of raw material. Although producers do not agree on which type yields the greatest quantity of gelatin, it is known that only the best-grade raw materials, such as calfskins, are used for the production of photographic and edible gelatin. Sheppard found that the presence of traces of mustard oil (allyl isothiocyanate) in gelatin, prepared from the skins of calves that had eaten wild carrots, was very effective in increasing photographic sensitivity.[27] The skins used for gelatin are usually imperfect ones, not suitable for making leather. The bones and skins frequently undergo a *pretreatment* which makes gelatin preparation easier. They are usually heated in lime and water to a temperature of about 70°C for a short time. If higher temperatures and long heating times are used, the gelatin hydrolyzes and loses some of its jellying properties. When bones are used as raw material, they should be *degreased.* This is done by heating them under steam pressure and then running off the grease layer, or by extracting with a low-boiling petroleum naphtha. Figure 25.3 shows the essential steps in the preparation of gelatin and glue from bones. The bones are then crushed, and enter a

[25]Eye, Treatment and Disposal of Tannery Wastes, in O'Flaherty, *op. cit.*, vol. 3, p. 471.
[26]Highberger, Chemical Structure . . . of Skin Proteins, in Flaherty, *op. cit.*, vol. 1, pp. 162–163.
[27]Sheppard, Photographic Gelatin, *Photogr. J.*, **65,** 380 (1925).

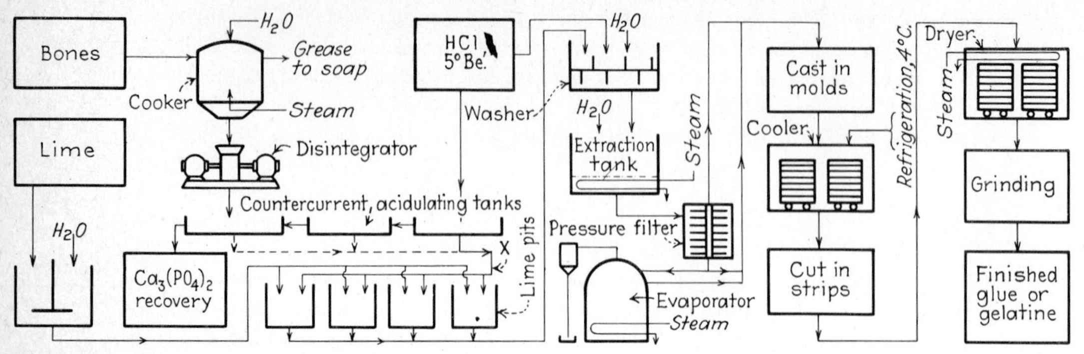

Note:- Manufacture of glue and gelatine differs largely in that poorer grades of raw materials and higher extracting temperatures are used for the former. Glue is more completely hydrolyzed than gelatine. Instead of bones, skins, hides, sinews, hide scraps, fleshings, fish stock, etc., may be used, in which case, after washing, they are introduced at X. Usually three extractions at temperatures from 60 to 75 deg.C. are used. The last extract requires concentration before molding

Bones	3.03 tons	Steam	400 lb.	To produce
Hydrochloric acid	1.14 tons	Electricity	55 kw.-hr.	Gelatine, 1 ton
Lime	0.76 tons	Direct labor	6 man-hr.	$Ca_3(PO_4)_2$ 1.67 ton
				Grease, 0.08 ton.

Fig. 25.3 Flowchart for glue and gelatin manufacture.

battery of four or more wooden tanks in series. Cold, 5°Bé (7.15%) hydrochloric acid is passed countercurrent to the bones so as to flow into the most nearly exhausted ones first. The calcium phosphate, carbonate, and other mineral matter are dissolved, leaving the organic matter, collagen, with the residue from the bones. This is now called *ossein.* The ossein is soaked in concrete vats containing milk of lime. The same treatment is also given to hides used for gelatin production. The purpose of this soaking is for *plumping* and to remove and eliminate soluble proteins (mucin, albumin). When the soaking in the lime is complete, the ossein or hides are washed in a rotary drum or conical-shaped washer, once with fresh water and then with slightly acidified water to adjust the pH for optimum hydration of the collagen. It is usually unnecessary to treat skins used in the manufacture of gelatin with hydrochloric acid, because the collagen of skins is not shielded by mineral matter such as calcium salts.

The next step is a series of *extractions* which are usually performed in tanks equipped with steam coils. The first extraction is made with water kept at 60 to 65°C by the steam coils. The pH of the hydrolyzing liquid is of importance. The optimum range is 3.0 to 4.0. The water is in contact with the ossein for approximately 8 h. An 8 to 10% solution of gelatin is obtained, which is filtered hot and cooled. Then another extraction is made at a slightly higher temperature. The filtered liquors are run into trays or on to long, stainless-steel cooling belts to jell the solutions. A third extraction of the ossein or hides is carried out at about 75°C. This is darker in color and so dilute that it will not set or jell on cooling. It is therefore necessary to evaporate this liquor under vacuum before it is cooled. Very often hydrogen peroxide or sulfurous acid is used as a bleaching agent. These extractions, usually four or five in number, are run at continually higher temperatures until 100°C is reached. The extract made at the lowest temperature is of the best quality and is chiefly used for gelatin for food or photographic purposes. The last extracts yield material suitable only for glue. The cooled, solid gelatin is cut into coarse noodles which are spread on nets and then dried by exposure to filtered air at about 40°C. On drying, the chunks shrink to a thin strip, which may be ground to a powder for further use.

The production of edible gelatin must meet the requirements of the FDA, and the manufacture of photographic gelatin must be under very rigid, specialized, technical control. The process of making photographic gelatin is much more complicated than that for producing other gelatins. It is for this reason that the price of photographic gelatin is nearly twice that of the edible variety. The reagents must be very pure. For example, lime employed in the process must be free from iron and magnesium salts, or else the gelatin cannot be used in photography.

ADHESIVES[28]

There are hundreds of adhesive preparations on the market today, supplemented by hundreds of available formulations, which are capable of holding materials together by surface attachment. Although the basic components of the products are generally obtainable, their specific combinations are not always known to the public. Many compositions are protected by patents. Modern adhesives may be classified by one of several methods, for example, their *use* in bonding various types of material, such as structural adhesives, or their *composition,* based on the principal ingredient, such as thermoplastic resin. Since the more-or-less general term adhesives embraces many types, they are discussed under their composition.

HISTORICAL Animal glue is the oldest type of adhesive, having been known for at least 3,300 years. Its manufacture began in the United States in 1808, and it has been an important article of commerce for more than 150 years, although practical application in broad industrial fields had to await progress in colloid and protein chemistry and the development of standard testing methods. Casein and starch adhesives became of commercial importance more than a generation ago, and soybean-protein adhesive in the last 30 years; synthetic-resin adhesives were developed after 1940.

MANUFACTURE

ANIMAL GLUES[29] The manufacture of glue is almost identical with the manufacture of gelatin. The procedures include grinding bones, cutting hides and scraps into small pieces, degreasing the material by percolating a grease solvent through it, liming and plumping, washing, making several extractions by hot water, filtering liquors, evaporating, chilling, and drying the jelly slabs in a tunnel. When dry, the slabs of glue are flaked or ground, blended, graded, and barreled or bagged for shipment (Fig. 25.3).

Variations of this process have been developed which do away with the tunnel drying and considerable hand labor. These consist of forcing the evaporated chilled extraction liquors containing 50% glue through a wire grill or colander instead of placing slabs in a tunnel dryer. The glue, when forced through the wire grill, is cut off into small pellets by knifelike blades. The pellets are dried in a three-stage drying system which uses bins, and rakes to stir them. In some processes the pellets are dropped from the colander or grill into a chilling bath, such as benzene, where they solidify. The adhering benzene evaporates, and the pellets are dried. This pellet form of glue is known as *pearl* glue. However, more than three-fourths of the glue produced in the United States is sold in the ground form, and the remaining fourth is sold in the flake and pellet form, mainly in the flake.

OTHER PROTEIN ADHESIVES *Fish glues,* liquid glues made from waste materials of cod, haddock, cusk, hake, and pollock, have practically the same applications as other animal glues. *Casein,* a milk-derived protein, is the basis of another large class of adhesives and can be made both water- and non-water-resistant. Casein adhesives are widely employed in the woodworking industry and in the manufacture of drinking cups, straws, and ice-cream containers. *Soya-protein adhesives,* although similar in properties to casein adhesives, are cheaper but not so good. The two are generally employed in combination, particularly in the veneer field, thus permitting a reduction in glue costs.

[28]Katz and Cagle, Adhesive Materials, Their Properties and Usage, Foster Publishing, 1971.

[29]Cf. Hull and Bangert, Animal Glue, *Ind. Eng. Chem.*, **44,** 2275 (1952) (description, accompanied by excellent pictures, tables, and a flowchart).

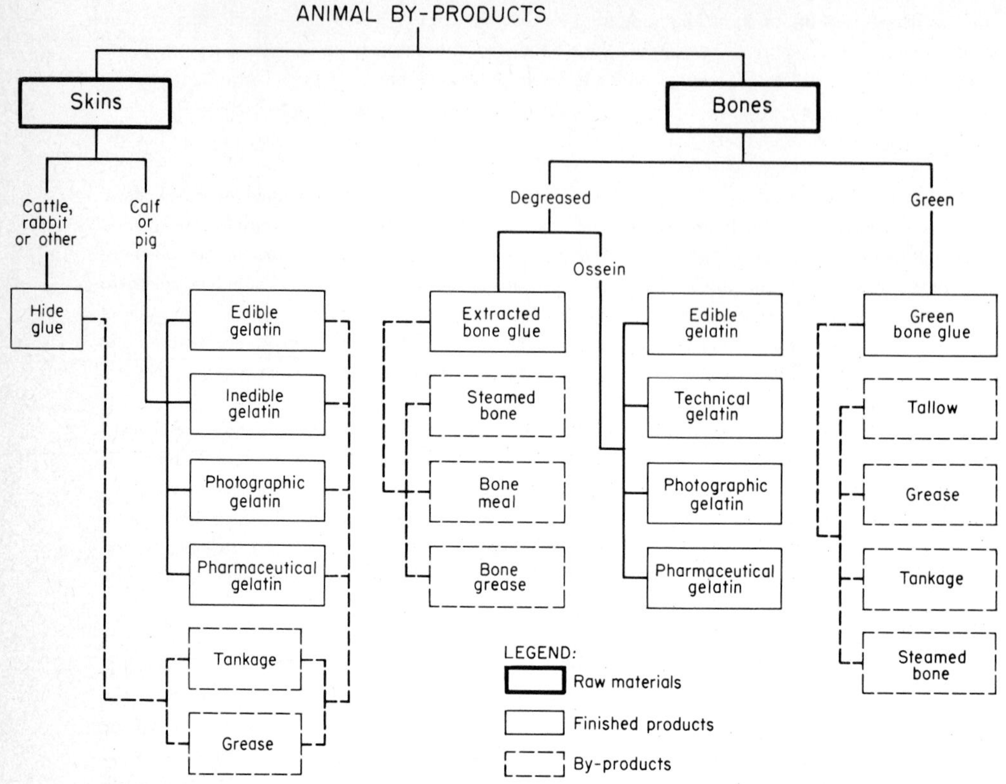

Fig. 25.4 Animal glue and gelatins. (*Adapted from U.S. Tariff Commission Report 135.*)

Albumin adhesives, from both egg and blood, find specialized uses where low film strength is not of importance yet water resistance is important. The attachment of cork pads to crown bottle caps is one of these applications. *Zein,* a corn protein, is used to a small extent as an adhesive, particularly in combination with other bases. *Peanut protein* hydrates are the newest protein-based adhesives and are suitable for making gummed tape and flexible glues for boxes and books.

STARCH ADHESIVES Starch adhesives, or glues, were first used in large-scale industrial application approximately in the 1910s. The chief kinds on the market today are made from cornstarch, tapioca flour, wheat flour, and potato starch. Starch adhesives may be applied cold and do not have the undesirable characteristic odors of some animal glues. This is one of their chief advantages over animal glues, although most of them have the disadvantage of less strength and water resistance than animal glues. Starch adhesives are less costly than synthetic-resin adhesives. *Native starch* is widely employed as an adhesive for veneer, plywood, and corrugated and laminated boards where water resistance is not important. Incorporated with 5 to 15% resins such as urea-formaldehyde, it is used for cartons where water resistance is important. Through enzyme conversions it is a base for many liquid adhesives. Native starch is the raw material for hydration to *dextrins and British gums;* these are modified starches important for envelope gums; gummed paper; labeling glues for glass, metal, and wood; cartons; laminated boards; and padding glues. Tapioca dextrin is the adhesive used for postage stamps.

FORMULATION Native starches are prepared from grains or roots (Chap. 30). Dextrins are made by heating a dry starch with dilute acid, causing partial hydration. British gums result from heating native starch with small amounts of catalysts; they are gummier and more adhesive than dextrins. In the manufacture of starch adhesives, rarely are dextrins, British gums, or starches used alone. Many chemicals may be admixed as indicated. Borax increases viscosity, gumminess, rate of tack, and speed of production. A 50% dextrin solution equals a 25% dextrin solution plus 10 to 15% borax. Sodium hydroxide accentuates borax action and also improves penetration for rosin-sized materials. Other chemicals which function in a like manner but to a lesser degree are Na_2CO_3, KOH, and K_2CO_3. Urea has an effect opposite that of borax, thus enabling manufacturers to use gummy products which would otherwise be unworkable. Other materials used in the manufacture of starch adhesives are plasticizers, defoaming agents, preservatives, "fluidifying agents" (which delay viscosity increase with aging), coloring, flavoring, and emulsifying agents. The last-mentioned are used where the surface to be glued is sized with wax or other organic-soluble agents.

SYNTHETIC-RESIN ADHESIVES[30] These adhesives are desired where water resistance is required and special conditions are to be met. *Urea-resin* adhesives consist of urea-formaldehyde alone or combined usually with starch-based adhesives where some water resistance is desirable. *Phenolic-resin* adhesives are based on phenol or its derivatives, such as resorcinol, which are condensed with aldehydes or ketones. They are thermosetting and need catalysts and curing for complete polymerization. Their chief outlet is in wood gluing, where high water resistance is required. *Epoxy resins* are synthetic resins obtained by the condensation of phenol, acetone, and epichlorohydrin. They have greater flexibility than phenolics and greater chemical resistance than alkyds. They are particularly useful in metal bonding. Alkyds, acrylates, methacrylates, allyls, and hydrocarbon polymers, such as indene, coumarone, styrene polymers, silicone resins, and natural and synthetic rubbers, all find application as additives or bases. Because of the limitless variations possible, the manufacture of synthetic-resin adhesives is not discussed here. Vinyls are a large, important class of adhesives.

Chemically, synthetic adhesives are of two types, those which are merely adhesive and those which form primary bond linkages with the materials being bonded. One classification includes rigid thermosets, rubbery thermosets, thermoplastics, copolymers and polymer mixtures, and inorganic adhesives. Application by the evaporation of volatile solvents (except water) is rapidly diminishing, because of air pollution problems. Hot-melt applications are increasing. Some compositions require the mixture of two stable solutions for activation; others are activated by oxygen or moisture in the air. Specially coated tape removal activates others. Some pressure-activated tapes are used.

Rigid thermosets include epoxies, silicones, polyesters, cynoacrylates (very fast setting), and phenolic, urea-formaldehyde, and resorcinol-formaldehyde resins. Heat and chemical resistance are usually good; some have excellent water resistance.

Rubbery thermosets. Silicone, urethane, butyl, and polysulfide rubbers are used. Many of these are used as sealants,[31] fastening materials together to exclude weather or gases, but bond strength is not highly important. Resistance to water, sunlight, and vibration for long periods is essential.

Thermoplastics. These include vinyls, polystyrenes, acrylates, polyamides, rubber-based adhesives, rosin, animal glues, and starch and dextrin glues. Most new hot-melt, nonsolvent glues belong to this group.

Copolymers and mixtures are formed to obtain better properties than those possible with single components, e.g., good adhesion and high temperature resistance. Epoxy with nylon, phenolics, polyamides, silicones, and urethanes, nitrile-phenolics, and vinyl-phenolics are all used.

[30] Adhesives, Guidebook and Directory, Noyes, 1972; Adhesive Materials, Their Properties and Usage, Foster Publishing, 1971; Adhesive Materials, Palmerton Publishing, 1972; Adhesives—The King of Fasteners, *Plast. World*, **17,** 59 (1974).

[31] Construction Sealants and Adhesives, Palmerton Publishing, 1970.

Inorganic adhesives. Glasses and ceramics are employed for very high temperatures. Sodium silicate is used for box corrugation.

Miscellaneous adhesives. Asphalt, sulfur, shellac, miscellaneous natural gums and mucilages, and a variety of cellulose esters dissolved in volatile solvents are widely used in the home and office. There is no all-purpose adhesive. Many types of industrial fasteners are being economically replaced with adhesives. Modern aircraft and automobiles (even spacecraft) use adhesives for assembly, saving money and reducing weight.

SELECTED REFERENCES

Bender, A. E.: Nutrition and Dietetic Foods, Chemical Publishing, 1973.
Bikerman, J. J.: Science of Adhesive Joints, Academic, 1961.
Blue Book 1, Adhesives, Noyes, 1972.
Bodnar, M. J. (ed.): Symposium on Adhesives for Structural Application, Wiley, 1962.
Brooker, D. B., *et al.:* Drying Cereal Grains, Avi Publishing, 1974.
Cagle, C. V. (ed.): Handbook of Adhesive Bonding, McGraw-Hill, 1973.
Daniels, R.: Breakfast Cereal Technology, Noyes, 1974.
Forest Products Laboratory: List of publications on glue, glued products, and veneer.
Gillies, M. T.: Dehydration of Natural and Simulated Dairy Products, Noyes, 1974.
Gillies, M.: Seafood Processing, Noyes, 1971.
Gould, R. F.: Contact Angle, Wettability and Adhesion, *Advan. Chem. Ser.* no. 43, ACS, 1964.
Gutcho, M.: Prepared Snack Foods, Noyes, 1973.
Gutcho, M.: Textured Foods and Allied Products, Noyes, 1973.
Gutterson, M.: Food Canning Techniques, Noyes, 1972.
Gutterson, M.: Fruit Processing, Noyes, 1971.
Gutterson, M.: Vegetable Processing, Noyes, 1971.
Guttmann, W. H.: Concise Guide to Structural Adhesives, Reinhold, 1961.
Hall, C. W., and T. I. Hendrick: Drying of Milk and Milk Products, Avi Publishing, 1971.
Hall, C. W., and G. M. Trout: Milk Pasteurization, Avi Publishing, 1968.
Hall, C. W., *et al.:* Encyclopedia of Food Engineering, Avi Publishing, 1971.
Haslam, E.: Chemistry of Vegetable Tannins, Academic, 1966.
Humphrey, G. H., *et al.:* Manufacture of Heavy Leathers, Pergamon, 1966.
Karmas, E.: Sausage Casing Technology, Noyes, 1974.
Karmas, E.: Meat Product Manufacture, Noyes, 1970.
Katz, I.: Adhesive Materials: Properties and Usage, Foster Publishing, 1964.
Martin, L. F.: Pressure Sensitive Adhesives, Noyes, 1974.
McGuire, E. P. (ed.): Adhesive Raw Materials Handbook, 1964, Padric Publishing, 1964.
O'Flaherty, F., *et al.:* Chemistry and Technology of Leather, ACS Monograph 134, Reinhold, 1956–1965, 4 vols.
Parker, R. S., and P. Taylor: Adhesion and Adhesives, Pergamon, 1967.
Partridge, J.: Chemical Treatment of Hides and Leather, Noyes, 1972.
Patrick, R. L.: Treatise on Adhesion and Adhesives, vol. 3, Dekker, 1973.
Ranney, M. W.: Epoxy and Urethane Adhesives, Noyes, 1971.
Satrianna, M. J.: Hot Melt Adhesives, Noyes, 1974.
Schwartz, M. E.: Cheese-making Technology, Noyes, 1973.
Skeist, I.: Handbook of Adhesives, Reinhold, 1962.
Torrey, M.: Dehydration of Fruits and Vegetables, Noyes, 1974.
Weiss, C. H.: Poultry Processing, Noyes, 1971.
Weiss, P.: Adhesion and Cohesion, Elsevier, 1962.
Wilcox, G.: Milk, Cream and Butter Technology, Noyes, 1971.

chapter 26

AGRICHEMICAL INDUSTRIES

This chapter is concerned with the utilization of chemicals to control either plant or animal life disadvantageous to humans and animals, to improve production of crops both in quality and quantity and, in a lesser degree, to serve as food supplements and additives. Off-the-farm production of farm chemicals such as fertilizers, pesticides, and feed supplements (Table 26.1 and Fig. 26.1), when combined with the manufacture of farm machinery and equipment and the processing and distribution of food and fiber, is larger in magnitude than the total operation of all the farms in the United States.[1] When considered in this light, agriculture in all its facets constitutes between 10 and 20% of the national economy, depending upon the yardstick employed.

The concept of agriculture as a more-or-less self-sufficient and self-contained industry was appropriate 100 years ago when the typical farm family produced its own food, fuel, farm animals, fertilizer, tools, shelter, and most of its clothing, disposing of small quantities of surplus in nonrural areas. Today, farming has changed from a subsistence to a commercial status; whereas formerly a farmer could only feed five people, the number has increased sixfold. The once 85% rural population of the United States has now become more than 90% urban, thus bringing about a significant change in employment patterns. The modern farmer confines operations, for the most part, to growing crops and raising livestock, and the other traditional functions have become highly specialized commercial operations. Farm production would be impossible, however, without farm equipment and supplies, of which agrichemicals[2] are an important part. To reach this stage in the United States, farming history may be broadly divided into three areas. The early years represent a period of grinding toil and physical growth during which farm output, as a rule, increased only in proportion to additional acres placed under cultivation. The second period overlaps the first and includes the mechanical revolution in farming which began in the nineteenth century, releasing a great percentage of acreage required for farm animal feed to food and fiber production. Machines make it possible for one farmer to plow 40 acres/day, for example, whereas three out of four farmers around the world plow less than 1 acre/day with primitive equipment. This period was also marked by the application of research to improve agricultural methods, such as crop rotation, use of cover crops, contour plowing, and modern irrigation, and by the establishment of land-grant colleges. One authority pinpointed the outstanding scientific achievement of the period as the discovery in 1888 that ticks transmit cattle fever. The third period is the current chemical age, in which the *widespread use of agrichemicals* has made an impressive beginning. For example, 54% more farm commodities were produced in 1960 on fewer acres than were in use two decades earlier.[3] The chemical industry considers the safeguarding and

[1]Davis and Goldberg, A Concept of Agribusiness, p. 2, Harvard University Graduate School of Business Administration, 1957; see Tables 25.2 and 25.3 in Chap. 25 for the value of food products vs. other industries.

[2]Term adopted by the authors to encompass the chemicals used in agriculture and the related supplying, processing, and distributing industries.

[3]Van Houweling, Chemicals in Agriculture, Eastern Plant Board Meeting, Providence, R.I., Mar. 10, 1960 (mimeographed).

TABLE 26.1 *Pesticides and Related Products, 1972*

	Production, thousands of pounds	*Unit value, per pound*
Grand total	1,157,698	1.07
Benzenoid	657,092	1.17
Nonbenzenoid	500,606	0.93
Pesticides and related products, cyclic		
Total	839,360	1.24
Fungicides	98,164	0.66
Herbicides and plant hormones	371,730	2.03
Insecticides and rodenticides	369,466	0.74
Pesticides and related products, acyclic		
Total	318,338	0.67
Fungicides	44,648	0.60
Herbicides and plant hormones	79,581	0.77
Insecticides, rodenticides, soil conditioners, and fumigants	194,109	0.65

Source: Synthetic Organic Chemicals, U.S. Tariff Commission, 1972.

improvement of our environment an absolute obligation of all segments of the responsible community, including the chemical industry itself.[4]

SAFETY REGULATIONS Chemistry pervades every area of the business of farming. Only 30 years ago, agriculture used about 50 basic farm chemicals. Today it uses hundreds in about 40,000 to 50,000 commercial formulations. Since chemicals play such a vital role in farm efficiency, chemical engineers are concerned with their effectiveness, economical production, and safety under the following specific regulatory legislation:

1. Meat Inspection Act of 1906 (as amended).
2. Food, Drug and Cosmetic Act of 1938 (as amended by the Miller Amendment of 1954 concerned with pesticide residues; Color Certification Bill of 1960; Drug Amendment of 1962 amending the so-called Delaney (or cancer) clause 409(c)(3)(A) of the Food Additives Amendment).
3. Agricultural Marketing Act of 1946 (as amended).
4. Insecticide, Fungicide and Rodenticide Act of 1947.
5. Poultry Products Inspection Act of 1957.
6. Public Law 86-139 of 1959 (which further extends item 4).
7. Federal Environmental Pesticide Control Acts of 1972 and 1975.

Increasingly stringent limitations on the use of agrichemicals are being established as the demand for them multiplies. *Potential hazards of chemicals to public health and wildlife have been considered by competent authorities and weighed against their proven value in agriculture.*[5] Laws and regulations promulgated by federal agencies, including the Dept. of Agriculture, Food and Drug Administration, Federal Trade Commission, Dept. of Interior, and Post Office Dept., and the states as well as various units within them, when coupled with standard procedures prevailing in industry, are *adequate safeguards provided label recommendations are followed intelligently and correctly by the user.*

[4]Agricultural Chemicals, Manufacturing Chemists' Association, 1963, 64 pp. (excellent summary).

[5]The possibilities of any real injury to plant and animal life as claimed by Rachel Carson in her book, Silent Spring, are minimized by the safety regulations already enacted and proposed regulations Congress and the Environmental Protection Agency (EPA) are considering; Borlang, DDT—The First Domino, *Agric. Chem.*, **27**(1), 20 (1972); Snaders, New Weapons against Insects, *Chem. Eng. News*, July 28, 1975, p. 18; Spear, Jenkins, and Mklby, Pesticide Residues and Field Workers, *Environ. Sci. Technol.*, **9**(4), 308 (1975).

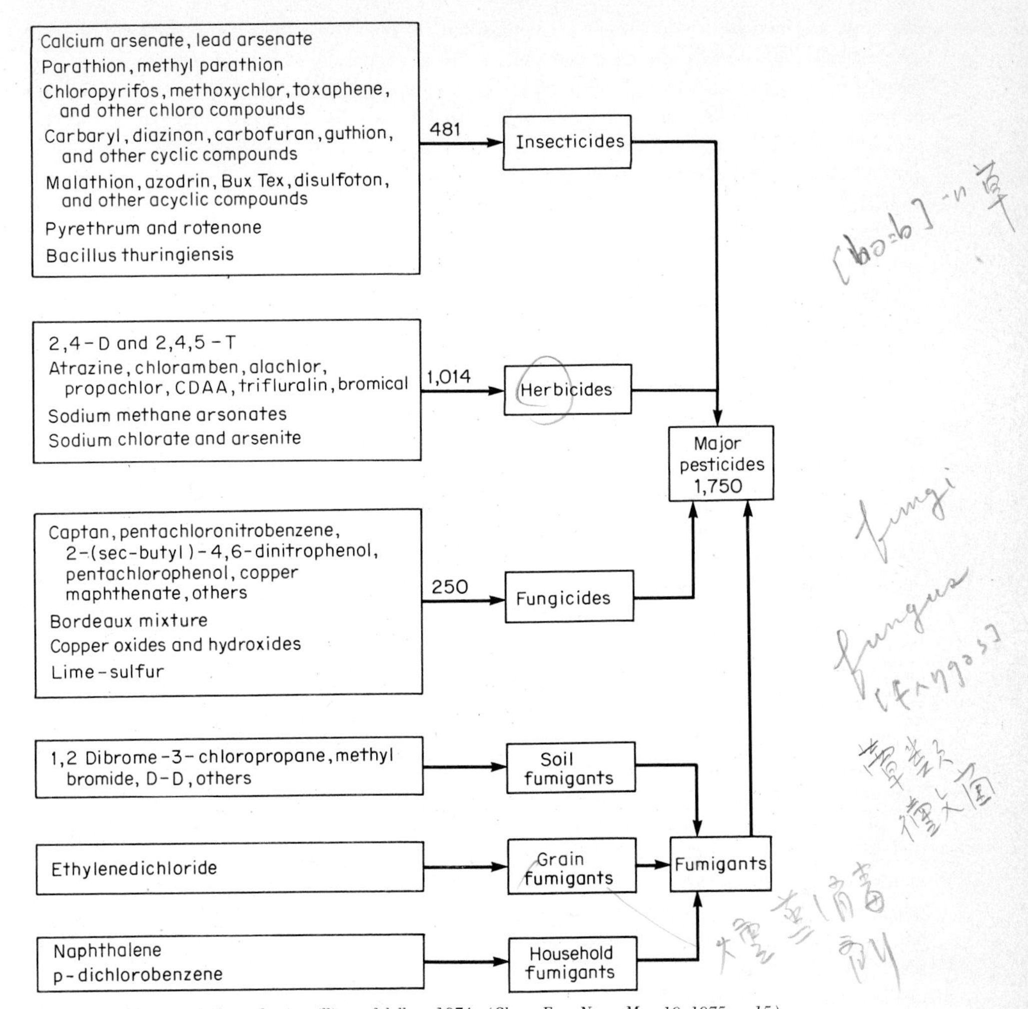

Fig. 26.1 Major pesticides, value in millions of dollars, 1974. (*Chem. Eng. News, May 19, 1975, p. 15.*)

PESTICIDES[6]

"For 300 million years, the insects have consistently been the world's foremost opportunists." They have "populated the land and the fresh waters with the greatest assemblage of species of any group of organisms." Insects are everywhere and "seem to endure with a unique kind of indestructibility."[7] This underlies not only the continued struggle between insects and humans for food and shelter, but also the human fight against insect-transmitted disease. The World Health Organization estimates that worldwide about one-third of the agricultural products grown by humans are consumed or destroyed by insects. Yellow fever, malaria, and other diseases are carried by mosquitoes. Even to control insects, insecticides *must* be used, and strong ones at that, because in the evolution of the

[6]Johnson, CW Report, Pesticides, *Chem. Week,* June 21, 1972, p. 34, and July 26, 1972, p. 8.
[7]Farb, The Insects, pp. 7–13, *Time,* 1962; Insect War Bonus, *Chem. Week,* Oct. 2, 1965, p. 69.

insect species, their bodies do not have skeletons, but are relatively stronger in relationship to their weight than other animals and are impervious to many chemicals because of their tough exterior sheath. In other words, they live within an *armor* of chitin resistant to most chemicals, which must be penetrated before they can be destroyed. It is well known, however, that some damage accompanies the use of certain chemicals which are powerful enough to be efficient insecticides. What must be done is to balance the good against the injurious and to work out new insecticides which have less deleterious effects.

It has been estimated that the annual loss attributed to agricultural pests is more than $10 billion, of which about $1 billion is caused by insects, $3 billion by disease, $1 billion by rats and rodents, $1 billion by fungi, and $4 billion by weeds. Farmers now spend about $1,750 million annually for pest-control materials, approximately 1.7% of the cash value of all farm sales. It cannot be stated too strongly that this loss would be still *larger* were not these pests kept under some measure of control by the use of chemicals, accompanied by other means, such as the proper cultivation of soil and the appropriate construction of homes and barns. Every citizen is concerned with this tremendous loss and should be interested in the chemical warfare directed against normal and constant enemies. It is a most important step in the conservation of our resources and in increasing the productivity of our soil. The chemical engineer is called upon to manufacture these agents and, under special conditions, to assist in their application.

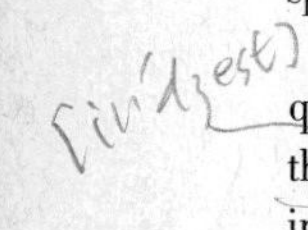

INSECTICIDES Insecticides are agents, or preparations, for destroying insects, and are frequently classified according to their method of action. *Stomach poisons* are lethal only to insects that ingest them, *contact insecticides* kill following external bodily contact, and *fumigants* act on the insect through its respiratory system. Table 26.2 lists some of the principal insecticides in each class. They may be applied as a spray if liquid, or in suspension, as a dust, or as a gas. *Systemics* are a group of insecticides which, unlike conventional insecticides, are absorbed and translocated throughout the plant. Thus they render plants toxic to aphids, red spider mites, and other sucking insects that are notoriously hard to kill. The possibility of developing, through proper selectivity, insecticides lethal to harmful insects but harmless to beneficial ones, is being pursued actively.

HISTORICAL Records show that insecticides were used as long ago as 1000 B.C. However, they were more often useless than useful since they were based on legend and superstition rather than on scientific knowledge. The essential property of early insecticides was a disagreeable odor rather than a poisonous nature. Although the toxic properties of arsenic were known as early as A.D. 40, it was not employed in the Western world until 1669. About the middle of the last century the use of paris green, lead arsenate, and other really poisonous chemicals was started as a general method of insect control.[8]

Inorganic insecticides. In recent years inorganic compounds for insecticides have been displaced by organic compounds in many applications. Arsenicals, fluorine, and phosphorus compounds are among the sufficiently toxic practical insecticides. Their major disadvantage is their comparable toxicity to man and other warm-blooded animals upon handling and as residues on food products. The *lead arsenate* commonly used as an insecticide is an acid lead arsenate ($PbHAsO_4$) and may be prepared by the following series of reactions.[9]

$$3PbO + 2CH_3CO_2H \longrightarrow (CH_3CO_2)_2Pb \cdot PbO + H_2O$$

$$(CH_3CO_2)_2Pb \cdot PbO + 2H_3AsO_4 \longrightarrow Pb_3(AsO_4)_2 + 2CH_3CO_2H + 2H_2O$$

$$Pb_3(AsO_4)_2 + 2HNO_3 \longrightarrow 2PbHAsO_4 + Pb(NO_3)_2$$

$$Pb(NO_3)_2 + H_3AsO_4 \longrightarrow PbHAsO_4 + 2HNO_3$$

[8] Sheppard, The Chemistry and Action of Insecticides, McGraw-Hill, 1951.
[9] Allen, U.S. Pat. 1,427,049 (1922).

TABLE 26.2 ***Insecticides Classified According to Method of Action***

Stomach poisons	*Contact insecticides*	*Fumigants*
BHC, DDT, methoxychlor, Meta Systox, lead arsenate, calcium arsenate, paris green, sodium fluoride, fluosilicates, compounds of phosphorus and mercury	BHC, DDT, Toxaphene, chlordane, dieldrin, aldrin, lindane, methoxychlor, nicotine preparations, lime-sulfur, oil emulsions, pyrethrins, rotenone, synthetic thiocyanates, organic phosphates such as TEPP, Malathion, parathion, and Meta Systox, Sevin	BHC, hydrogen cyanide, carbon disulfide, *p*-dichlorobenzene, nicotine, naphthalene, ethylene oxide and dichloride, methyl bromide.

The actual manufacture of $PbHAsO_4$ is much simpler than the equations indicate. Litharge is dissolved in the calculated quantities of acetic acid and nitric acid. The theoretical quantity of arsenic acid is added, the precipitated lead arsenate removed by filtration, and the mixture of acetic and nitric acids in the filtrate used over again. Three precipitations of lead arsenate can be made before the spent acid has to be strengthened or discarded. The yields range from 95 to 97%. The commercial product contains 31 to 33% arsenic trioxide and is colored pink to safeguard against confusion with foods such as flour and baking powder. Lead arsenate is widely used, chiefly for the potato beetle and for the codling moth in apple orchards. *Calcium arsenate*[10] is cheaper than lead arsenate, but does not adhere so well to leaves and thus is less effective. Its use has been drastically reduced by organic insecticides, such as methyl parathion, Sevin, and Guthion. The commercial insecticidal product is usually a mixture of tricalcium arsenate $[Ca_3(AsO_4)_2]$ and lime, called *basic* calcium arsenate. *Fluorine compounds* are important stomach-poison insecticides as substitutes for arsenicals. As they are also extremely poisonous to man, caution should be observed in their application and handling. Fluorides are too water-soluble to be used on plants, but sodium fluoride is widely employed to control roaches and poultry lice. Some fluosilicates and fluoaluminates may be applied to crops because of their decreased solubility. The principal compounds that are suitable are barium fluoaluminate (cryolite) (Chap. 20).

Sulfur and sulfur compounds are employed to some extent in the control of mites, spiders, and other insects of that type, but their chief use is as fungicides. Sulfur is one of the "workhorse" fungicides; much of it is used to check mildew on fruit trees. Suitable forms, with regard to degree of fineness, of the element are obtained by milling to 325-mesh or finer, emulsifying molten sulfur, heating mixtures of sulfur with bentonite, and using flotation sulfur obtained from the recovery of the element from hydrogen sulfide from petroleum and coal gases. Finely ground sulfur may be employed for dusting without any additives, but wetting agents are needed for the preparation of suspensions for spray purposes. *Lime sulfurs* have been widely used in the past for the control of scale insects and for control of diseases of tree fruits. These may be prepared by adding water to a dry mixture of lime and sulfur and then using the heat of reaction from slaking the lime to accelerate the reaction between calcium hydroxide and sulfur to give a self-boiled lime sulfur; by boiling a mixture of lime, sulfur, and water with external heat; or by evaporating the water from the boiled mixture to furnish a dry mix. Although the reaction is complex, it is believed that the pentasulfide is the most active fungicidally. However, the use of lime sulfur has declined rapidly in favor of organic fungicides, because it is toxic to plant foliage and causes russeting of fruits.

Hydrocyanic acid is an efficient fumigant for many pests, especially insects. Large quantities have been used in the citrus-fruit industry, and smaller amounts for greenhouse and household fumigation. When applied on a small scale, the gas is generated as needed by adding sulfuric acid to

[10]See CPI 2 for manufacture.

"eggs" of sodium cyanide, but a commercially manufactured liquid product is also available that contains up to 98% hydrocyanic acid, the rest being water. Parathion or petroleum with parathion or like products has largely replaced hydrocyanic acid, especially for citrus fruits. Care must be taken in handling this acid, since it is not only a powerful insecticide, but can also be lethal poison to humans and animals.

Plant derivatives, or natural organics, are used as insecticides and depend for toxicity upon the alkaloids they contain. These insecticides constitute a small, but important, amount of the total tonnage consumed annually.[11] As is true of the inorganic insecticide industry, many plant-derived insecticides are being supplanted by synthetic organics.

Pyrethrins. Flowers of the pyrethrum plant (a type of chrysanthemum) contain toxic, non-nitrogenous organic esters (pyrethrins). The principal sources of the flowers or of the extract are Kenya, Japan, the Congo, and Brazil. The compressed flowers are broken up, ground to a fine powder, and extracted several times with kerosine or other organic solvent. The extract is concentrated by removing the solvent in a vacuum still below 60°C, and the resulting oleoresin is employed to prepare the finished insecticide.[12] Pyrethrins are important because of their quick "knockdown" power against flies and nontoxicity to humans and warm-blooded animals. The widespread use of pyrethrins in aerosol bombs for household use has led to the preparation of highly concentrated extracts free from precipitated metals to avoid clogging the nozzle. Nitromethane is a satisfactory solvent for this use. *Synergists*, or activators (substances which increase insecticidal efficiency), are employed with pyrethrins. The most important one is piperonyl.

Allethrin,[13] the name given to the allyl homolog of cinerin I, is a component of pyrethrum; it has almost identical insecticidal properties, but lacks synergists. It was discovered in 1949 by the U.S. Dept. of Agriculture and within a year was in commercial production. It is now produced only in Japan. Other pyrethroids have been prepared experimentally that show promise of being more stable to sunlight (the major shortcoming of previous pyrethroids) and still relatively harmless to mammals and wildlife.

Nicotine is a volatile alkaloid obtained by treating by-products of the tobacco-processing industry, i.e., stems and damaged leaves, with an aqueous solution of alkali, followed by steam distillation. Because of its volatile nature, most of the nicotine is converted to the sulfate and sold as a 40% nicotine solution. Nicotine solutions are employed against aphids, leafhoppers, and thrips and are also used as fumigants.

Rotenone is the poisonous principle of the roots of several tropical and subtropical plants, chief among which is derris. It is a nonnitrogenous complex organic heterocyclic compound, $C_{22}H_{22}O_6$. Rotenone is obtained by extracting ground derris roots with chloroform or carbon tetrachloride. Other compounds related to rotenone, but not so toxic, are also present in the extract. However, no attempt is usually made to separate the rotenone from these materials. The solvent is removed, and the residue is dissolved in a water-soluble solvent such as acetone. Rotenoid compounds are effective stomach and contact poisons. They have long been used as fish poisons by natives of the East Indies and Japan, but their chief applications in the United States are as insecticides, particularly near harvest time or on milk cows, since rotenone is safe for mammals.

SYNTHETIC ORGANICS The phenomenal increase in synthetic organic compounds used for insecticides since World War II has revolutionized this industry. In 1940 the combined output of synthetic organic insecticides was but a few million pounds per year, yet in 1972 the annual production of pesticides was more than 1,150 million lb (Table 26.1).

DDT, *or dichlorodiphenyltrichlorethane.* This compound was first made by Zeidler in Germany in 1874, but its insecticidal properties were not discovered until 1937. It was extensively used during

[11]Seiferle and Frear, Insecticides Derived from Plants, *Ind. Eng. Chem.*, **40**, 683 (1948).
[12]Gnadinger and Corl, Manufacture of Concentrated Pyrethrum Extract, *Ind. Eng. Chem.*, **24**, 988 (1932)
[13]Carbide Makes Allethrin, *Chem. Eng. (N.Y.)*, **57**(6), 11 (1950).

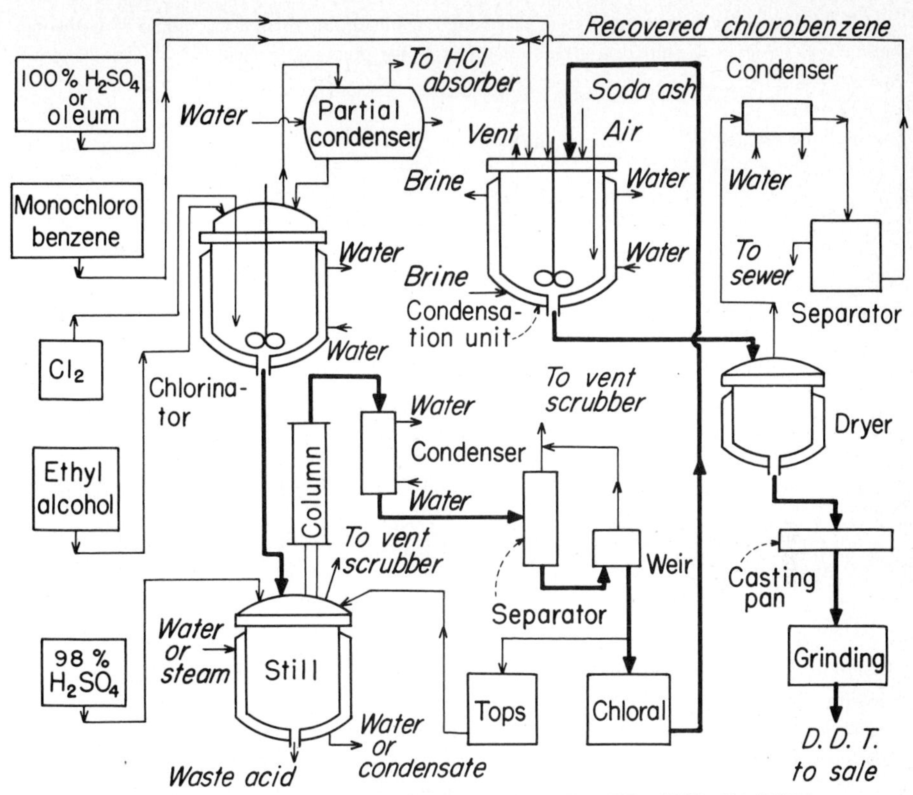

Fig. 26.2 Flowchart for the manufacture of DDT. [*Modified from Chem. Eng.* (*N.Y.*), **57**(11), 204 (1950).]

World War II to control body lice and as a mosquito larvicide. It was the first chemical to have *great enough residual contact action to be useful;* an insect can be killed by simply walking over a dried, sprayed surface. DDT has shortcomings, which naturally accelerated the development of other insecticides and led to the government ban on its use in the United States in the early 1970s. At the time it was banned it was the nation's most widely used insecticide. The ban resulted from the discovery that the slow degradation of DDT allows it to be stored in the fat of living organisms and thus to build up in the natural food chain. It can be transmitted in cow and human milk. Reports also indicated that DDT caused some birds, such as ospreys, to lay eggs with shells so thin that the percentage of eggs that hatched was sharply reduced. The final factor in the decision to ban DDT was the laboratory report that it can cause cancerlike tumors in mice and rats.[14] It is still widely used outside the United States, to control malaria-carrying mosquitoes.

Two other major difficulties are the development of DDT-resistant strains of pests and the appearance of new pests after one has been controlled. For example, in the apple industry the codling moth has been a problem for years. Now DDT gives excellent control, but the European mite and the red-banded leafroller are increasing. This is probably due to the killing of natural predators by DDT, which is ineffective on the new pests. The problems discussed here are much the same for all synthetic organics and have led to the development of many new types of compounds.

[14]Sanders, *op. cit.;* Devlin, DDT: A Renaissance, *Environ. Sci. Technol.* **8**(4), 322 (1974).

There are several commercial methods of DDT manufacture. The usual method[15] is the exothermic condensation of chloral and chlorobenzene in the presence of oleum:

$$2C_6H_5Cl + OCHCCl_3 \longrightarrow (C_6H_4Cl)_2CHCCl_3 + H_2O$$

The process as depicted in Fig. 26.2 can be broken down into the following coordinated sequences:

Alcohol is chlorinated to chloral-alcoholate in a 750-gal glass-lined chlorinator, first at below 30°C, but eventually up to 75 and 90°C (Ch). This takes place over the course of 60 to 70 h, the temperature being controlled through water in either coils or a jacket (Op).

The overhead (excess alcohol and HCl) is conducted to a partial condensor or dephlegmator, which liquefies the alcohol from the HCl that is absorbed later and the small amount of ethyl chloride that is vented (Ch).

The chloral-alcoholate is decomposed by H_2SO_4 into chloral and alcohol and purified by distillation (Op).

The chloral and chlorobenzene are condensed, using strong H_2SO_4 (100%) or oleum in a glass-lined 1,000 gal reactor (Ch). The reaction takes 5 to 6 h and is controlled at 15 to 30°C by the brine or steam coils (Ch).

The spent acid is withdrawn (Op), and the DDT is water-washed several times (Op) and neutralized with soda ash (Ch). The DDT and chlorobenzene mixture is dropped to a 500-gal dryer, where steam melts the DDT and distills any unreacted chlorobenzene overhead (Op).

The molten DDT is run to casting pans to solidify and to be ground (Op).

DDT can exist in a number of isomers. The para-para is the most potent, and manufacturing procedures are conducted to give a maximum of this isomer. However, technical grades contain considerable amounts of the ortho-para′ isomer as an impurity. For use in aerosol bombs, DDT is purified.

BHC,[16] *or benzene hexachloride.* This compound exists in a number of stereoisomers, the gamma isomer being by far the most toxic. Because of this remarkable specificity, the insecticidal value was not discovered until 1942 in England, although the compound was first prepared by Faraday in 1825. BHC, 1,2,3,4,5,6-hexachlorocyclohexane, is made by the chlorination of benzene in the presence of actinic light:

$$C_6H_6 + 3Cl_2 \longrightarrow C_6H_6Cl_6$$

Widely used to control the boll weevil in cotton, BHC, unless purified, is generally unsuitable for food crops because of its strong musty odor. *Lindane,* a refined material containing at least 99% of the gamma isomer of BHC, has largely overcome the odor problem. It may be used on food crops and is increasingly being employed to combat insects formerly controlled by DDT. However, it is also an organochlorine compound, and the EPA is currently reviewing its use to determine its safety to humans.

Methoxychlor, or bis(methoxyphenyl)trichloroethane. This compound has $—OCH_3$ groups substituted for the —Cl groups of DDT. Methoxychlor has high insecticidal efficiency, low toxicity to warm-blooded animals, and is safe for plants. It also has greater "knockdown" power than DDT. It is used safely on cattle, vegetable and forage crops, and household pests. Methoxychlor can

[15]DDT, *Chem. Eng.* (*N.Y.*), **57**(11), 204 (1950) (pictured flowchart); Cook *et al.*, Synthesis of DDT with Chlorosulfonic Acid as a Condensation Agent, *Ind. Eng. Chem.*, **39,** 868 (1947); DDT Eyes Fluosulfonic Process, *Chem. Eng.* (*N.Y.*), **59**(2), 247 (1952); Lee, Materials of Construction for Chemical Process Industries, p. 114, McGraw-Hill, 1950.

[16]Continuous Chlorination Is Key to BHC Process, *Chem. Eng.* (*N.Y.*), **61**(10), 338 (1954); Vilbrandt and Dryden, Chemical Engineering Plant Design, pp. 43–66, McGraw-Hill, 1959 (very complete flowchart and supporting data for BHC).

be made by reacting methyl chloride or dimethyl sulfate with sodium phenate to produce anisole, which is then reacted with chloral:

$$CH_3Cl + NaOC_6H_5 \longrightarrow CH_3OC_6H_5 + NaCl$$

$$2CH_3OC_6H_5 + OCHCCl_3 \longrightarrow (CH_3OC_6H_4)_2 + H_2O$$

Toxaphene, an important chlorinated camphene insecticide, kills all common cotton pests and is officially recommended by federal and many state authorities for 74 destructive insects. It is at present the nation's most widely used insecticide, but as new tests are made and new restrictions promulgated, its position is subject to change at any time. It is made by chlorinating camphene with from 67 to 69% chlorine. Camphene is produced by isomerizing α-pinene, a major constituent of turpentine. The conditions of manufacture are both corrosive and toxic. Toxaphene has the approximate empirical formula $C_{10}H_6Cl_8$. It is a yellow, waxy solid (mp 65 to 90°C) with a mild piny odor. It is sold probably to the extent of 50 million lb yearly. Other chlorinated hydrocarbons were widely sold before their use was banned by the EPA. These included aldrin, dieldrin, heptachlor, and chlordane.[17]

Malathion[18] is a popular phosphorodithionate insecticide with broad-spectrum application for nearly all fruits, vegetables and field crops, dairy livestock, and household insects. It has low mammalian toxicity. Chemically, it is the *o-o*-dimethyldithiophosphate of diethyl mercaptosuccinate.

Parathion[19] is *o-o*-diethyl-*p*-nitrophenyl thiophosphate. Its diemthyl homolog is also made. These are very powerful, broad-spectrum insecticides and acaricides which kill mites and ticks and are used largely on the South's cotton, vegetable, and citrus crops. The production of these and other organophosphorus materials amounts to more than 90 million lb annually and is growing.

ATTRACTANTS AND REPELLENTS are being developed to eliminate hazards to other forms of life endangered by potent pest killers. *Food baits* are combined with toxicants to be used against specific insects such as the fire ant. A water extract of decaying wood, made by inoculating wood with fungi, has been found to be a termite attractant. Two *sex attractants* (sex pheremones), unsaturated alcohols having a 16-carbon main chain, have been synthesized for the gypsy and silkworm moths. One for the American cockroach has been separated and identified as 2,2-dimethyl-3-isopropylidenecyclopropyl propionate. Sex pheremones have been isolated, identified, and synthesized for about 50 butterflies and moths and 15 insects of other types. The sex lure draws male insects from a distance, but a chemosterilant halts reproduction; most compounds of the latter type are derivatives of aziridine.[20]

Male insects are sometimes raised and then sterilized by exposure to gamma radiation before being released to mate with untreated females. This technique has been used to eradicate the oriental fruitfly on Guam and the screwworm in the southeastern United States. Some *repellents* against animals, as well as insect pests, are in use. *Diethyltoluamide* (USP) is an all-purpose insect repellent for all who work or play outdoors and may be applied directly to the skin. This is sold as a 15% aerosol spray under the trade name Off. An example of a microbial insecticide, really a biological poison, is Thuricide,[21] which is not toxic to animals and plants but causes disease in certain insects, for example, cabbage loopers and diamondback moth larvae on cauliflower, cabbage, and lettuce.

[17]Johnson, *op. cit.* (chemical formulas).

[18]U.S. Pat. 2,578,652.

[19]Greek and Rosenberger, Design and Construction of a Phosphate Insecticidal Plant, *Ind. Eng. Chem.*, **51**(2), 104 (1959); ECT, vol. 15, pp. 322, 328, 1968.

[20]Spotlight on Safer Insect Control, *Chem. Week*, Feb. 9, 1963, p. 55; Sanders, *op. cit.*

[21]First Bug-kill-bug Insecticide, *Chem. Eng.* (*N.Y.*), **67**(3), 42 (1960) (flowchart); Living Insecticides, *Chemtech.*, **5**, 396 (1975).

Recently interest has increased in the use of insect growth regulators, or juvenile hormone mimics. These compounds act to prevent larvae from becoming adults. One of the most successful is methoprene, which has the same carbon skeleton as natural juvenile hormone but has a methoxyl group in place of the natural hormone's epoxide ring. Methoprene is contained in 1-μm-diameter polyamide matrixes from which it diffuses over a period of 7 to 10 days.[22] It is effective against mosquitoes and flies. An interesting application is the incorporation of methoprene in animal salt licks and poultry feed. The methoprene passes through the animals and the poultry, and their manure then contains enough of it to prevent flies from developing.

FUMIGANTS are chemicals that emit poisonous vapors. Several compounds are of value in controlling soil-infesting insects: *chloropicrin*, which is highly irritating; *carbon disulfide*, now seldom used because of explosion hazards; *methyl bromide*, which is accompanied by application problems; and *ethylene dibromide*, which is effective in the control of wireworms. These halogenated alkanes comprise the bulk of soil fumigants sold. Crops in storage, particularly grains and seeds, may become infested with insects, rodents, and microorganisms.[23] Storage areas must be tight enough for fumigation. Hydrocyanic acid and carbon tetrachloride, the latter alone or in combinations, are useful fumigants. Fumigants are also employed for packaged goods.

NEMATOCIDES approach a market of $20 million/year, but lack of knowledge of these compounds, coupled with the nature of soil media, have retarded their development. In addition to the soil fumigants mentioned above, there are phosphorothioates, which act through direct control rather than by fumigation.

MITICIDES Chemicals for mite control are enjoying a major boom since insecticides have eradicated the natural enemies of mites while leaving the mites themselves unharmed. A broad range of miticides is available, including Kelthane, Tetradifon, Morocide, and Torak, which are effective in fruit mite control. Older phosphate preparations, used in oils, are still sold for dormant-season application on deciduous fruits.[24] Newer phosphate products such as ethion are gaining in use in citrus groves, and sulfur has been losing. Dinitrophenol derivatives have retained some of their markets.

RODENTICIDES are used to control certain pest animals such as mice, rats, groundhogs, squirrels, and field rodents because of their ability to do extensive property damage and to spread disease. Rodents alone cause damage in the United States of about $1 billion annually. In grain-producing states, rodenticides are widely used to control problems in food handling and distribution. *Warfarin*, 3-(α-acetonylbenzyl)-4-hydroxycoumarin, is the sales leader. Another coumarin derivative and two products based on 1,3-indandione are among those now competing with it. Compound 1080, *sodium monofluoroacetate*, is a significant product, deadly to everything, including humans. It should be handled only by experienced personnel. It is made by reacting ethyl chloroacetate and potassium fluoride in an autoclave at 200°C. The resulting fluoroacetate is saponified with a methanol solution of NaOH, and the product crystallized out. Because of the high toxicity of ethyl fluoroacetate, extreme caution must be observed in this process. *Zinc phosphide*, an older compound, is used for the same purpose. *Thallium sulfate* is effective against a variety of rodents. *Strychnine* and *red squill* are botanical rodenticides. Repeated nonlethal dosage of the former can lead to bait-shyness. Powder and extracts of the latter kill Norway rats; standardization of its content in baits has been one factor in the increased demand. *N*-3-pyridylmethyl-*N′*-*p*-nitrophenylurea, a new rodenticide comparatively harmless to humans, has been introduced under the name Vacor.

FUNGICIDES, including slimicides and wood preservatives, account for approximately 12.5% of pesticides sales. They are active against fungi, parasitic plants comprising molds, mildews, rusts, smuts, mushrooms, and allied forms capable of destroying higher plants, fabrics, and even glass, thus

[22]Allan *et al.*, Pesticides, Pollution and Polymers, *Chem. Technol.*, **3**, 171 (1973).

[23]Anderson and Alcock, Storage of Cereal Grains and Their Products, American Cereal Chemists Association, 1954; Sterilant, *Chem. Eng.* (*N.Y.*), **68**(25), 90 (1961); Borkovec, Sexual Sterilization of Insects by Chemicals, *Science*, **137**(3535), 1034 (1962).

[24]Johnson. *op. cit.*

depriving humans of valuable food and materials. They attack seeds, growing plants, plant material, and, under proper conditions, finished products such as adhesives, leather, paints, and fabrics. Fungicides for plants act by direct contact and often injure the host as well as the fungus.

The *cost of developing, testing, and preparing for market* of any of the chemicals considered in this chapter is high, partly because of the high mortality of unsuitable chemicals.

Arthur D. Little, Inc., surveyed 22 companies in 1974 and reported that typical R&D for a new pesticide required $7.4 million. This includes R&D costs for all the compounds that fail and the voluminous information and testing for effectiveness, persistence, degradation products, effect on beneficial insects, soil microorganisms, carcinogenic properties, etc., required by the EPA.

INORGANIC FUNGICIDES Elemental sulfur and compounds of heavy metals such as copper and mercury once dominated the field. Because of environmental considerations registrations for heavy metal products have either been canceled or placed under review by the EPA.

Bordeaux mixture, an important fungicide, is simple to make at home and is employed in several formulations. The 4-4-50 formula consists of 4 lb of copper sulfate, 4 lb of hydrated lime, and 50 gal of water. The copper sulfate is dissolved in one vessel, and the lime in another. Each is diluted to 25 gal and poured simultaneously through a strainer into a third container. Properly made, Bordeaux mixture consists of a light-blue, gelatinous precipitate suspended in water. It is effective for most ordinary molds and mildews. Most common fruits and vegetables can be safely treated with it.

ORGANIC FUNGICIDES vary in composition, but many new fungicides fall into these chemical classifications: dithiocarbamates, chlorinated phenols, and carboximides. Although chlorination is very important in insecticides, it is much less so in fungicides. Next to sulfur, nitrogen appears to be the most effective constituent element of an organic fungicide other than carbon and hydrogen. The first successful organic fungicide formaldehyde is sold as a 40% solution in water called *Formalin.* Because of its volatility it is still an extensively employed fumigant for seeds, soil, and greenhouses.

The dithiocarbamates are made by reacting carbon disulfide with an amine to form a dithiocarbamic acid. The acid is reacted with a metal hydroxide to give a stable salt:

$$H_2NCH_2CH_2NH_2 + 2CS_2 \longrightarrow HSCSNH(CH_2)_2NHCSSH$$

$$HSCSNH(CH_2)_2NHCSSH + 2NaOH \longrightarrow NaSCSNH(CH_2)_2NHCSSNa + 2H_2O$$

This sodium salt of ethylene bisdithiocarbamate is employed with zinc sulfate and lime and probably reacts to form the zinc salt. These salts are used to combat vegetable blights, particularly of the potato and tomato. Ferbam, or ferric dimethylthiocarbamate, is the oldest of these compounds and is particularly effective for apple rust. Ziram, or zinc dimethylthiocarbamate, and maneb, or manganese ethylenebisdithiocarbamate, are important for control of fruit rotting and leaf diseases of vegetables.

The use of dithiocarbamates has been challenged by the EPA because they tend to break down into toxic compounds such as thiourea.

Soil fungicides are currently outstanding among fungicides, as illustrated by the prediction that 2 million acres of cotton will receive such treatment for their root systems. PCNB (pentachloronitrobenzene), alone or with other fungicides such as captan, is applied at the time of planting to combat damping off or postemergence dieback; and many new such fungicides are being developed.

Industrial fungicides are illustrated by the important *coal-tar creosote,* which has long been the standard substance for wood preservation. It is effective, permanent, and inexpensive, but it is sticky and has a penetrating odor. Because of the dark stain it imparts, wood so treated cannot be painted. For many purposes, however, such as creosoted fence posts and railroad ties, these disadvantages are not of importance. Other developments in wood preservation are *chlorinated phenols:* tetra- and pentachlorophenol, chloro-*o*-phenylphenol, and β-naphthol. These are applied in 5% solutions in organic solvents. Generally, these phenols are not so powerful as creosote, but they do not have its disadvantages. Sales of pentachlorophenol have grown from 3 million lb in 1951 to 50 million lb in

1974. Several new markets have been found. 2,4,5-Trichlorophenol is another industrial fungicide whose production figures are difficult to determine, since it is also used as an intermediate for 2,4,5-T, a herbicide. Copper naphthenate is used to combat cotton fiber problems, and sodium *o*-phenylphenate is widely employed as an industrial preservative for fruits. Some compounds are taken into the plant through the roots or leaves, and the plant itself then becomes poisonous to the fungus. These are called systemic fungicides, of which benomyl, thiobendazole, and carboxin are examples.[25]

HERBICIDES Sales, which in 1961 amounted to over $100 million, were $1 billion in 1974. Whereas in the past herbicide sales were stimulated mainly by improved yields, these chemicals are being used increasingly to decrease production costs, which are otherwise mounting. In fact, many claim that only the first steps to weed control have been taken. Scarcely a month passes without the announcement of a new product; still, an average of only one compound out of every 1,800 tested meets all the required standards.

Desiccants and defoliants, included in the *Chemical Week* reports[26] listing of 106 herbicides, are discussed in this chapter under plant-growth regulators. Although cereal grains are said to represent the largest potential market, and cotton the second, practically every crop and ornamental plant has a weed problem; even the market for crabgrass killers for home lawns is large. There are two main classes of herbicidal chemicals, *contact* and *systemic*. To these a third may be added if soil sterilants are included. They may also be considered in two other categories, selective and nonselective herbicides, which can be further subdivided, usually depending upon the route of application. Nonselective herbicides have been available for many years; they have been useful in clearing weeds from railroad tracks, industrial areas, highways, and driveways. They have also been developed for *aquatic weed control* in canals, farm ponds, and irrigation ditches. Prior to the introduction of hormone-type weed killers in the early 1940s, the chemicals generally used were nonselective and included *sodium chlorate, sodium arsenite,* various borate compounds, industrial waste products, and oil sprays. Sodium chlorate (Chap. 14) still plays a major role in weed control, and another inorganic, ammonium sulfamate, whose use was begun in the early 1940s, has reached a current consumption of several million pounds per year. The latter is a foliage spray which eliminates woody growth but does not prevent the regrowth of grass. The herbicide that changed the need for hand hoeing and also proved to be the spark that ignited the present growing market was *2,4-D* (2,4-dichlorophenoxyacetic acid).[27] Historically, the results following application were phenomenal. Bindweed became limp and wilted within 24 h; within 48 h roots and rhizomes were stimulated to grow and proliferate, only to disintegrate and die, leaving uninjured apple rootstocks. Thus this herbicide proved to be selective; that is, it killed one plant, but not another.[28] Another important hormone type of weed killer is *2,4,5-T* (2,4,5-trichlorophenoxyacetic acid). All are effective at such low concentrations that the cost of treatment is small. The manufacture of 2,4-D begins by chlorinating phenol (from benzene) to 2,4-dichlorophenol. The reaction product is distilled to procure pure 2,4-dichlorophenol. After chlorinating acetic acid to monochloroacetic acid, it is converted to the sodium salt. This is reacted with the 2,4-dichlorophenol and aqueous caustic to give the sodium salt of 2,4-dichlorophenoxyacetic acid and sodium chloride. Hydrochloric acid is employed to liberate the acid from solution. 2,4,5-T is made by hydrolyzing 1,2,4,5,-tetrachlorobenzene with caustic, because phenol cannot be chlorinated to the 2-, 4-, and 5-positions. Hydrolysis yields 2,4,5-trichlorophenol directly. This is purified and converted to 2,4,5-trichlorophenoxyacetic acid by a process similar to that used for 2,4-D. Both acids are usually marketed in herbicide formulations as amine salts or as alkyl esters, since they are more

[25]Marsh, Systemic Fungicides, *Farm Chem.*, **37**(10), 29 (1974).

[26]Johnson, *op. cit.*

[27]Sanders and Prescott, 2,4-D Weed Killer and Derivatives, *Ind. Eng. Chem.*, **51,** 954 (1959) (flowchart, pictures, references); Faith, Keyes, and Clark, Industrial Chemicals, 3d ed., Wiley, 1965 (industrial data).

[28]Spotlight on Herbicides, *Farm Chem.*, **126,** 11–14, 68–69, 74, 82 (1963); Wain, Selective Herbicidal Activity, *Chemtech*, **5,** 354 (1975).

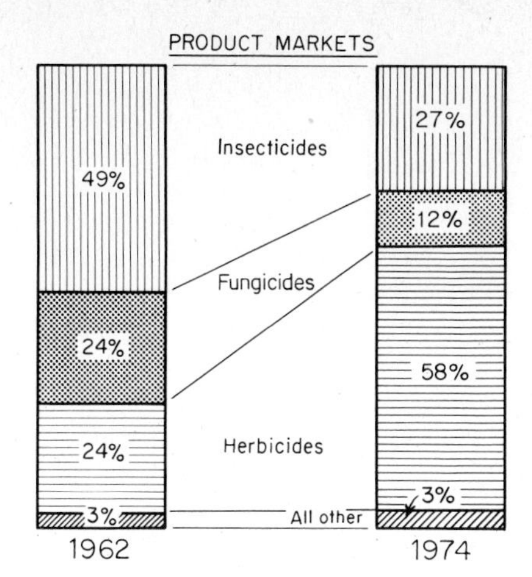

Fig. 26.3 Steady growth for herbicides. (*Chem. Eng. News, May 19, 1975, p. 15, and Chem. Week, May 25, 1963, p. 119.*)

effective as herbicides than either the straight acid or the various salts. The alkyl esters of 2,4,5-T are manufactured by reacting the appropriate alcohol with the acid, following the classical method. For the manufacture of alkanolamine salt formations, the acid is dissolved in an aqueous solution of an alkanolamine.

Other organic compounds of herbicidal interest include 2-chloro-4-ethylamino-6-isopropylamino-*s*-triazine (AAtrex, atrazine), 3-amino-2,5-dichlorobenzoic acid (Amiber, chloramben), 2-chloro-2′,6′-diethyl-*N*-(methoxymethyl)acetanilide (Lasso, alachlor), 2-chloro-*N*-isopropyl-acetanilide (Ramrod, propachlor), *N,N*-diallyl-2-chloroacetamide (Randox, DCAA), and 5-bromo-3-*sec*-butyl-6-methyluracil (Bromacil). Some of these compounds, such as 2,4-D and 2,4,5-T, are able to act as both plant-growth regulators and weed killers, depending upon the concentration used. An example of commercial importance is Treflan (trifluralin), which first had wide acceptance as a preemergent crabgrass killer used mostly in mixed fertilizers for spring application and was later cleared as a weed killer for use in cotton and soybean farming (Fig. 26.4). It is manufactured as follows:

$$\underset{\text{Parachlorobenzotrifluoride}}{CF_3\text{-}C_6H_4\text{-}Cl} \xrightarrow[\text{40°C, yield 95\%}]{HNO_3,\ H_2SO_4} \underset{\text{Mononitro intermediate}}{CF_3\text{-}C_6H_3(NO_2)\text{-}Cl} \xrightarrow[\text{20\% oleum, 120°C, yield 83\%}]{HNO_3,\ H_2SO_4} \underset{\text{Dinitro intermediate}}{CF_3\text{-}C_6H_2(NO_2)_2\text{-}Cl} \xrightarrow[Na_2CO_3,\ \text{yield 96\%}]{NH(C_3H_7)_2} \underset{\text{Trifluralin}}{CF_3\text{-}C_6H_2(NO_2)_2\text{-}N(C_3H_7)_2}$$

Charge the *nitrator* with cycle acid and fuming nitric acid (Op) and heat to 40°C (Op).

Add the *p*-chlorobenzotrifluoride for mononitration, maintaining the temperature at 40°C during the course of 4 h (Ch).

Raise the temperature and hold it at 75°C over 4 h (Ch).

Cool and settle out the spent acid from the "mono" on top (Op).

Draw off the spent acid from the "mono" on top (Op).

Draw off the spent acid to waste treatment (Op).

Charge the *dinitrator* with 100% sulfuric acid, 20% oleum, and fuming nitric acid (Op).

Raise the temperature to 120°C (Op).

Run in the "mono" at 120°C over 6 h and hold at this temperature over the course of 4 h to effect dinitration (Ch).

Cool to 20°C and settle (Op).

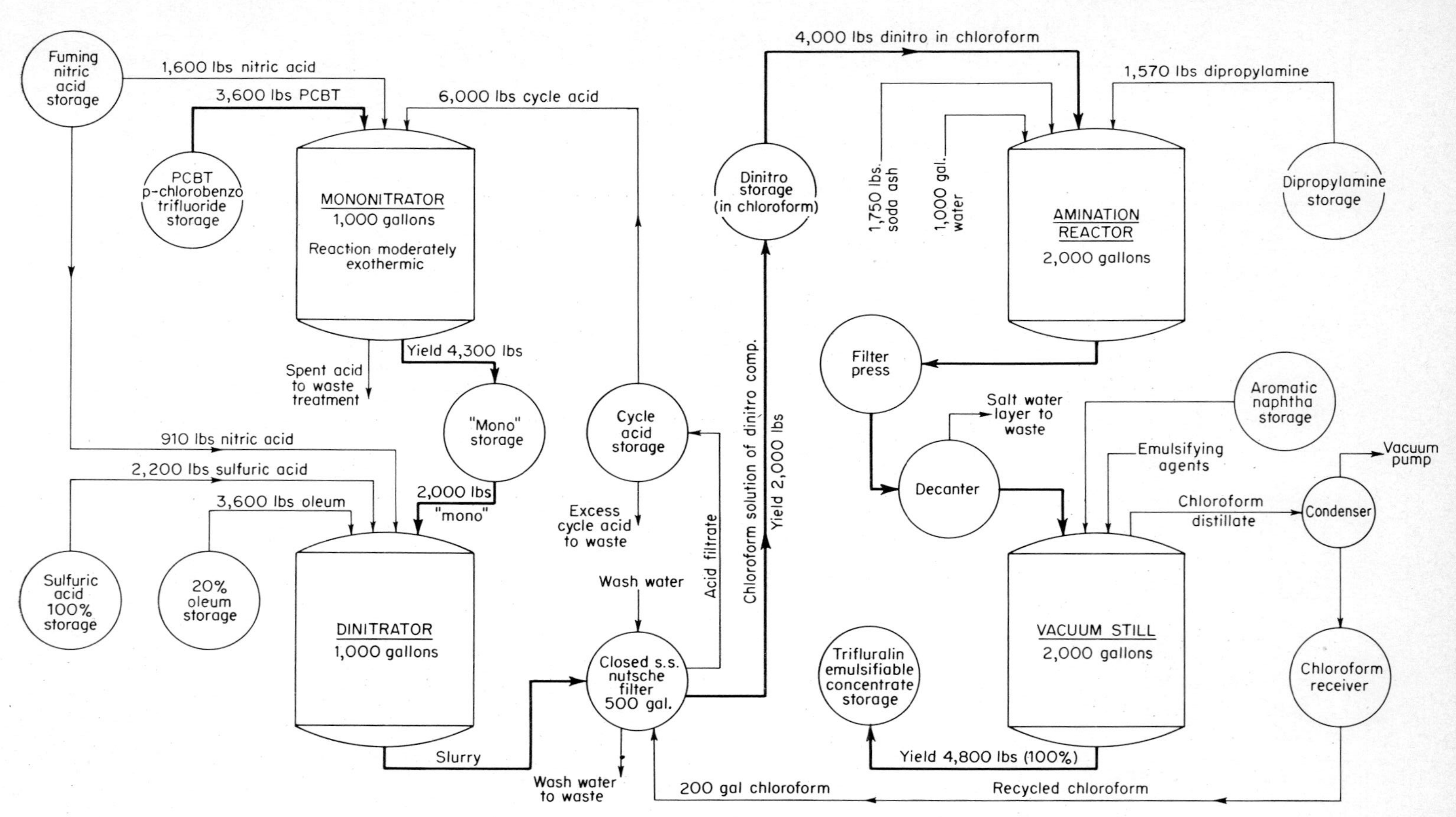

Fig. 26.4 Flowchart for the manufacture of trifluralin (Treflan). (*Eli Lilly and Co.*)

Filter off the solid "dinitro" on a closed *Nutsch filter,* running the spent acid to cycle acid storage (Op).

Wash with water, running the wash water to waste (Op).

Dissolve in chloroform (from the recovery still) (Op).

Charge into *amination reactor,* water, soda ash, and the dinitro derivative (in chloroform solution) (Op).

Adjust the temperature to 45°C (Op).

Aminate with dipropylamine added at 45 to 50°C over 4 h and heat 4 h more at 45°C (Ch).

Add the rest of the water and pass through a *filter press* and *decanter* to a *vacuum still,* the salt-water layer going to waste (Op).

In the vacuum still distill the chloroform overhead (for recycle) (Op).

Add aromatic naptha and emulsifiers to the desired formulation (Op).

PLANT NUTRIENTS AND REGULATORS

MIXED FERTILIZERS[29] Food and fiber, at the present enhanced rate of production, are largely dependent upon fertilizers containing multinutrients. If a balance is struck, showing the removal of plant nutrients from soils by crops used for food or employed as raw material in factories, when compared with the additions made by humans and nature, the paramount importance of fertilizers becomes strikingly clear. Fertilizers either place back in the soil the ingredients removed by plants, or add materials needed by native soils to make them productive or more productive.

Humans discovered early that the excreta of animal life are food for plant life. Indeed, the Chinese have maintained the fertility of their soil for 5,000 years by the application of this principle. Even in the aesthetic-oriented civilization of the United States, this need was recognized, and manure was applied on a large scale to maintain better plant yield. For example, before 1900, more than 90% of all fertilizer nitrogen was in the form of natural organics; by 1950, the proportion of the latter had decreased to 4%. The merit of wood ashes has also been recognized for centuries; they supply potassium carbonate to the soil.

In this country the fertilizer industry began in 1849, but prior to 1930 its only manufactured products were superphosphate and mixed fertilizers. The ingredients of the latter were natural minerals, by-products of other industries, or imports. Today the necessary raw materials are produced in this country. Fertilizers form an important segment of the total chemical industry, and several thousand mixing plants are located in nearly all the states. Applied scientifically and used with other fruitful farming practices, they are basic to the welfare of the nation. Fertilizers represent the *largest single item among agricultural chemicals.* Since 1940, their use has increased greatly; more than 42 million tons were marketed in 1973 at a value of approximately $2.75 billion.[30] This increase becomes even more significant when future needs are considered. The minimum arable acreage required to sustain one human being in food and fiber was established in 1945 as 2.5 acres. In the United States in 1975, 2.1 acres of arable land per person allowed surplus food production. The population is predicted to approach 400 million by the year 2000, when the arable land, including anticipated reclaimed acreage, will be 1.6 acres per person. In the world[31] as a whole, there will be less than 0.5 acre of tillable land per person.

[29]Some of the considerations to keep in mind in choosing an *ideal mixed fertilizer* are its primary high nutrient content; low cost per nutrient; chemical form of nutrients (nitrate or ammonial form of nitrogen); absence of elements toxic to plants; availability of other elements; reactions with other ingredients (hygroscopicity, low moisture content); particle size and shape (spheres, uniformity); physical properties (must flow freely in distributing equipment); alkaline influence on acid soils, and vice versa. Also, it should not dissolve too quickly when put into the soil, should not have deleterious reactions with other ingredients, and should kill weeds and insects.

[30]Commercial Fertilizers: Consumption in the U.S. and Fertilizer Trends, U.S. Industrial Outlook, Dept. of Commerce, 1975.

[31]An estimate for world population is 7,400 million for A.D. 2000, if not checked, or less than a third of an acre per person. Cf. Taylor, *Fortune,* **73,** 110 (June 1966).

TABLE 26.3 ***Fertilizer Requirements of Farm Crops***

Crop	Yield, metric tons per hectare	Nutrients taken up by crop, kg per hectare			Fertilizer required to retain fertility,* kg per hectare		
		N	P_2O_5	K_2O	*N*	P_2O_5	K_2O
Alfalfa	9.0	210	43	218	184	43	218
Apples	16.8	2	4	27	25	4	27
Corn	6.3	186	66	101	71	39	25
Cotton	1.7	91	33	53	17	18	17
Potatoes	26.9	93	36	152	67	36	152
Soybeans	2.7	165	36	49	139	36	49
Wheat	4.0	117	46	69	61	39	22

Source: Calculated from data in The Fertilizer Handbook, National Plant Food Institute, 1963.
*Assumes return of stalks, stems, leaves, etc., to the ground, and fixation of atmospheric nitrogen by bacteria at 18.4 kg/hectare. Nitrogen added by rainfall, 8.0 kg/hectare.

The three major elements needed are (1) *nitrogen,* required during the early stages of growth to promote development of stems and leaves; (2) *phosphorus,* which stimulates early growth and accelerates seeding or fruit formation in later stages of growth; (3) potash, essential to development of the starches of potatoes and grains, the sugars of fruits and vegetables, and the fibrous material of plants; an ample supply in the soil sometimes helps to prevent disease and to lessen the effects of excessive nitrogen application.

The total nutriment value of soils is surprisingly low. The plowed layer of cropland in the United States has been estimated[32] to contain 0.14% nitrogen, 0.14% P_2O_5, and 1% K_2O. Plants utilize these low concentrations for growth, and the nutriments are permanently removed when the plants are cropped. Table 26.3 indicates the amounts of nutriments removed by several common crops at usual growth levels. When crops are mature, the unused parts—straw, stems, or leaves—may be returned to the soil to help maintain fertility. Some nutriments are added by natural causes. The last three columns of Table 26.4 show the minimum amounts of fertilizer which must be added merely to maintain present soil fertility. Nitrogen added to the soil is lost due to volatilization as ammonia, conversion back to nitrogen, or leaching, so the amount recovered in crops is about 40%. P_2O_5 added is converted quickly to insoluble forms unavailable to plants except as a result of slow weathering, and the immediate-use efficiency is very low. Potash efficiency is about 50% the first year. With a 15% P_2O_5 efficiency, corn at 100 bu/acre would require 338 lb/acre of urea or 450 lb of NH_4NO_3, 1000 lb of 20% superphosphate, and 70 lb of HCl just to maintain fertility. There are very few areas of the world where fertilizer is being applied in such quantities. Chapter 18 presents the source and method of manufacture of various nitrogen compounds used both in industry and fertilizers; Chap. 16, superphosphates; Chap. 17, potassium salts. See Table 26.5 for world fertilizer statistics.

Commercial mixtures guaranteed to contain certain percentages of the three primary nutrients, nitrogen, P_2O_5, and K_2O, comprise the major portion of all types used. Fertilizers are generally spoken of as having formulas such as 3-12-12, 2-12-6, or 5-10-5, which means that they have 3, 12, and 12%, 2, 12, and 6%, or 5, 10, and 5%, respectively, total nitrogen, available phosphate (P_2O_5), and soluble potash (K_2O).[33] Note, for a memory jog, that these constituents are in alphabetical order. Another way of evaluating these fertilizer formulas is as "units." Here 1 unit is 1% of a ton, or 20 lb. Hence the first of the formulas given above contains 3 units of nitrogen, 12 units of P_2O_5,

[32]Bailey, Soil-Plant Relationships, 2d ed., Wiley, 1968.

[33]In 1960 the American Society of Agronomy voted to make all its official recommendations and decisions affecting fertilizers in terms of the elements instead of their oxides. Although this is common practice in regard to nitrogen, industry will probably continue to recommend and guarantee phosphorus and potassium in terms of their respective oxides unless compelled to do otherwise by law.

TABLE 26.4 *Important Fertilizers Consumed in the United States in 1965 and 1972 (In thousands of tons)*

Kind	*1965*	*1972*
Mixtures	18,559	21,534
N-P-K	15,536	16,996
N-P	1,517	3,007
P-K	1,181	1,151
N-K	325	380
Chemical nitrogen materials	7,695	12,801
Ammonia, anhydrous	1,563	3,637
Ammonia, aqueous	820	733
Ammonium nitrate	1,634	3,031
Ammonium nitrate, limestone	170	40
Ammonium sulfate	775	934
Nitrogen solutions	1,922	3,482
Sodium nitrate	301	97
Urea	428	785
Other	82	124
Phosphate materials	2,536	2,374
Ammonia phosphate*	720	607
Basic slag	68	112
Phosphoric acid	39	60
Phosphate rock	477	71
Superphosphate, conc.	729	1,262
Superphosphate, normal	455	218
Other	48	44
Potassium	936	2,685
Potash	814	2,462
Other	122	223
Natural organic materials	589	502
Secondary and trace nutrient material	1,521	1,310
Grand total	31,836	41,206

Source: Commercial Fertilizers: Consumption in the United States, Dept. of Agriculture, 1974. *Includes reported quantities of all grades.

TABLE 26.5 *World Fertilizer Consumption*

Millions of Metric Tons[a]	*1972*	*1980*	*Annual growth, percent*
North America	16.5	24.3	5.0%
Western Europe	17.5	22.6	3.3
Eastern Europe[b]	18.9	32.0	6.8
Asia[c]	5.4	10.8	9.1
Other Asia	4.9	9.1	8.0
Latin America	3.2	7.2	10.7
Africa[d]	1.3	2.2	6.8
Other[e]	4.3	5.5	3.1
World total	72.3	113.7	5.8%

Source: Tennessee Valley Authority. See TVA Assesses World Fertilizer Outlook, *Chem. Eng. News,* April 22, 1974, p. 13. [a]As N, P_2O_5 and K_2O [b]Includes U.S.S.R. [c]Excludes Japan and Israel [d]Excludes Republic of South Africa [e]Includes Japan, Israel, Republic of South Africa, and Oceania.

and 12 units of K_2O. These formulas are listed in the order of magnitude of their sales in 1976, the 3-12-12 being the largest sold. In addition, many commercial fertilizers contain the minor chemicals needed by plants. In a 3-12-12 mixture there is a total of only 27% active fertilizer constituents. The other 73% consists of filler, carrier, or soil-conditioning materials. Table 26.4 gives a summary of the important chemicals used commercially to supply these three fertilizing constituents.

Granulated or pelleted products. During the past two decades, fertilizer technology and production have been revolutionized. Not only has plant nutrient content increased more than 40%, but the physical form of many fertilizers has been improved, with a resulting lower cost to the farmer. Methods of application also have changed, e.g., aerial crop dusting with mixtures of fertilizer and pesticides is now used. The chemical combination of nutrients in high-analysis, water-soluble products brings about *cost savings* in both transportation and handling. Granular or pelletized products (Fig. 26.5) are less likely to cake than powdered fertilizers. Neither are they as dusty or as corrosive in the farm distributor. Uniformity of distribution increases crop yields, an important factor in their expanding use. There were no granular products in 1950. Ten years later 60% of all mixed fertilizers were marketed in granular or pellet form. This is illustrated in the flowchart in Fig. 26.6, where the

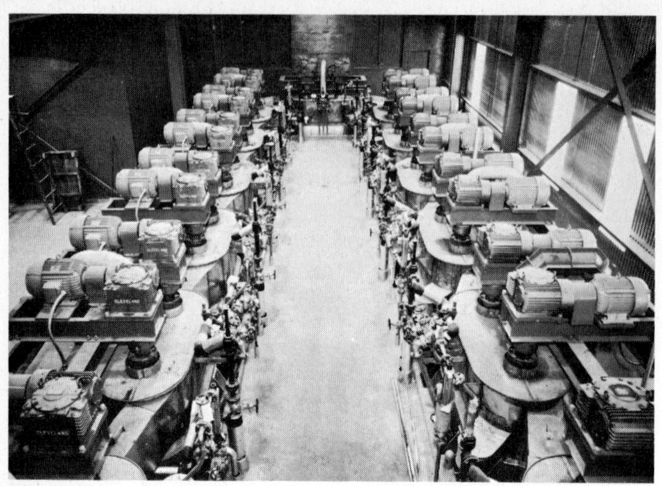

Fig. 26.5 Acidulators for complex fertilizers (see Fig. 26.6). Stainless-steel reactors homogenize raw materials at the Ortho complex pelleted fertilizer plant, Fort Madison, Iowa, to form chemically blended formulations. In these reactors, ground phosphate rock is acidulated with nitric acid to yield a slurry. Anhydrous ammonia is added to establish the final balance of nitrogen and phosphate. This mixture continues from the ammoniation stage to a mixer, where the potash content, if required, is added in the form of soluble salts, thus ensuring that every particle of the product will be uniform. The use of nitric acid as the acidulating agent for this process results in products containing two forms of both nitrogen and phosphate, quick-action and continuous-feeding components.

sperodizer used for the production of pellets is shown. The future of coated granular fertilizers to *slow down the release* of nutrients appears to be promising.

Complex fertilizers. The fertilizer industry has been supplying more concentrated fertilizers each year. A fundamental improvement along this line has been to substitute nitric[34] acid for sulfuric acid to acidulate phosphate rock. Simple acidulation of phosphate rock with nitric acid produces a hygroscopic superphosphate, since it contains calcium nitrate. TVA and others have studied and recommended commercial processes. In one, the phosphate rock is *extracted* by mixed nitric and sulfuric acids, followed by ammoniation, drying, and addition of potassium chloride (optional). Another features mixed nitric and phosphoric *acidulation,* followed by the conventional steps, and others use nitric acid alone for acidulation. These processes, as well as conditioning against moisture absorption as practiced for ammonium nitrate, have led to an extension of this acidulation with nitric acid. An example is the 600 ton/day installation[35] at the Ortho plant at Fort Madison, Iowa. To improve the position of this plant, the complex includes a 300 ton/day ammonia plant and a 250 ton/day nitric acid unit. This is the third nitrophosphate plant constructed by this company (total production 1,200 tons/day). Nitrophosphate is gaining in Europe also. Phosphate rock is decomposed with nitric acid plus a small amount of phosphoric acid. The resulting slurry is ammoniated and carbonated and, if desired, combined with potassium salts and spray-dried to yield a uniform pelletized product with both water-soluble and citrate-soluble forms of phosphate.

Figures 26.5 and 26.6 may be separated into the following coordinated sequences:

Ground phosphate rock is elevated and fed into the continuous acidulation reactors (Op).

In the acidulation reactors, the rock is decomposed by nitric acid plus a small amount of phosphoric acid (Ch).

[34]Chap. 16, under nitric and mixed acid acidulation; Yates, *et al.*, Enriched and Concentrated Superphosphates, *Ind. Eng. Chem.*, **45,** 496 (1953); Ninimaki and Orjans, Solvent Extraction Scores in Making Nitrophosphates, *Chem. Eng.* (*N.Y.*), **78**(1), 63 (1971).

[35]Weyermuller and Everingham, California Chemical Low Recycle of Complex Fertilizer, *Chem. Process.* (Chicago), Mar. 25, 1963, p. 17; Fertilizer Boom May Boost Nitrophosphate Routes, *Chem. Eng.* (*N.Y.*), **69**(25), 68 (1962).

Ammoniation takes place in the remaining reactors, the slurry flowing through these in a continuous manner (Ch).

KCl is added in the last reactor (Ch) or slurry mixing tank.

The slurry flows to a slurry surge tank, from which it is pumped and sprayed into the heated sperodizer where it is pelletized and dried (Op).

The sperodizer is a short-length, large-diameter rotary dryer whose rotation showers its contents through the hot drying gases, building up the pellets (multilayered spheres) through the accumulation of successive dry coats (simultaneous coating and drying). The in-process material at all times remains essentially dry (Op).

Not all chemicals or other constituents of fertilizers are available to all plants with equal facility, because so much depends on the nature of the chemical used, on the other materials in the soil, and in particular on the soil pH or its chemical reaction, and on the crop itself. From this it can be seen that not only the mixing, but also the application of fertilizers, must be carried out properly. Fortunately, the farmers of the nation have at their call the various county agents of the Agricultural Extension Service, and the State Agricultural Experimental Stations, that can help them select through soil tests the proper fertilizer giving the best and most economical results for each crop.

Minor constituents that research has shown to be necessary for plant life are sulfur, sodium, calcium, and magnesium compounds in small amounts. Micronutrients may also be added: boron, iron, manganese, zinc, and copper compounds. Many of these trace elements are used in special-purpose mixtures as needed to correct soil deficiencies. Some of these derivatives, such as boron, are toxic to plants if used in excess.

Fertilizer fillers, carriers, and conditioners. In a 3-12-12 mixture there is a total of only 27% active fertilizer constituents; the other 73% consists of filler, carrier, or conditioner. This resulting weight increases shipping costs of products but may serve a useful purpose such as counteracting burning. Fillers and carriers are usually sand, ground phosphate rock, or waste used to produce bulk. Conditioners are finely divided, insoluble substances such as peat or cocoa shells, usually added to counteract the tendency of fertilizer particles to cake and harden in the bag. Dolomite, a conditioner added to make a non-acid-forming fertilizer, may also be regarded as a filler, but no filler is entirely satisfactory (see spherodizer in Fig. 26.6).

Bulk blending and bulk distribution. When complete plant food is supplied by the basic manufacturer, the distributor's part in mixing raw-material ingredients is bypassed. There is also a trend toward the installation of dry-bulk blenders geared to the specific formulation required by the farmer-customer. Bulk distribution, which means loading bulk products into the farmer's truck rather than supplying fertilizer in bags, accounted for 45% of the market in 1976.

Liquid-mixed fertilizers have been rapidly and significantly developed by virtue of the convenience and timesaving in handling and application. Already such consumption is more than 27% of total combined plant nutrients. Their production involves the chemical reaction of raw materials to form a soluble solution containing plant nutrients in desired ratios. This type of fertilizer is used in row application, particularly for starter fertilization.[36] In order to apply liquid fertilizers satisfactorily, they must be free of sediment. Concentrated phosphoric acid, also known as superphosphoric acid, is employed as the basis for high-analysis liquid fertilizers. It is ammoniated to make neutral solutions of high analysis (10-34-0) which will not crystallize at low temperatures. By ammoniating under moderate pressure, ammonium polyphosphate can be made to be stored and shipped as a solid and then dissolved on the farm for liquid fertilizer.[37] A similar method of application for a single fertilizer constituent is the direct introduction into soils of the vapor from anhydrous or liquid ammonia as a side dressing for corn and other field crops.

[36]World Politics Key to Fertilizer Supply, *Chem. Eng. News*, June 17, 1974, p. 10.

[37]TVA Reveals Latest Fertilizer Technology, *Chem. Eng. News*, Aug. 20, 1962, p. 40; Pipe Reactor Improves N-P Fluid-Fertilizers, *Chem. Eng.* (*N.Y.*), **79**(18), 60 (1972).

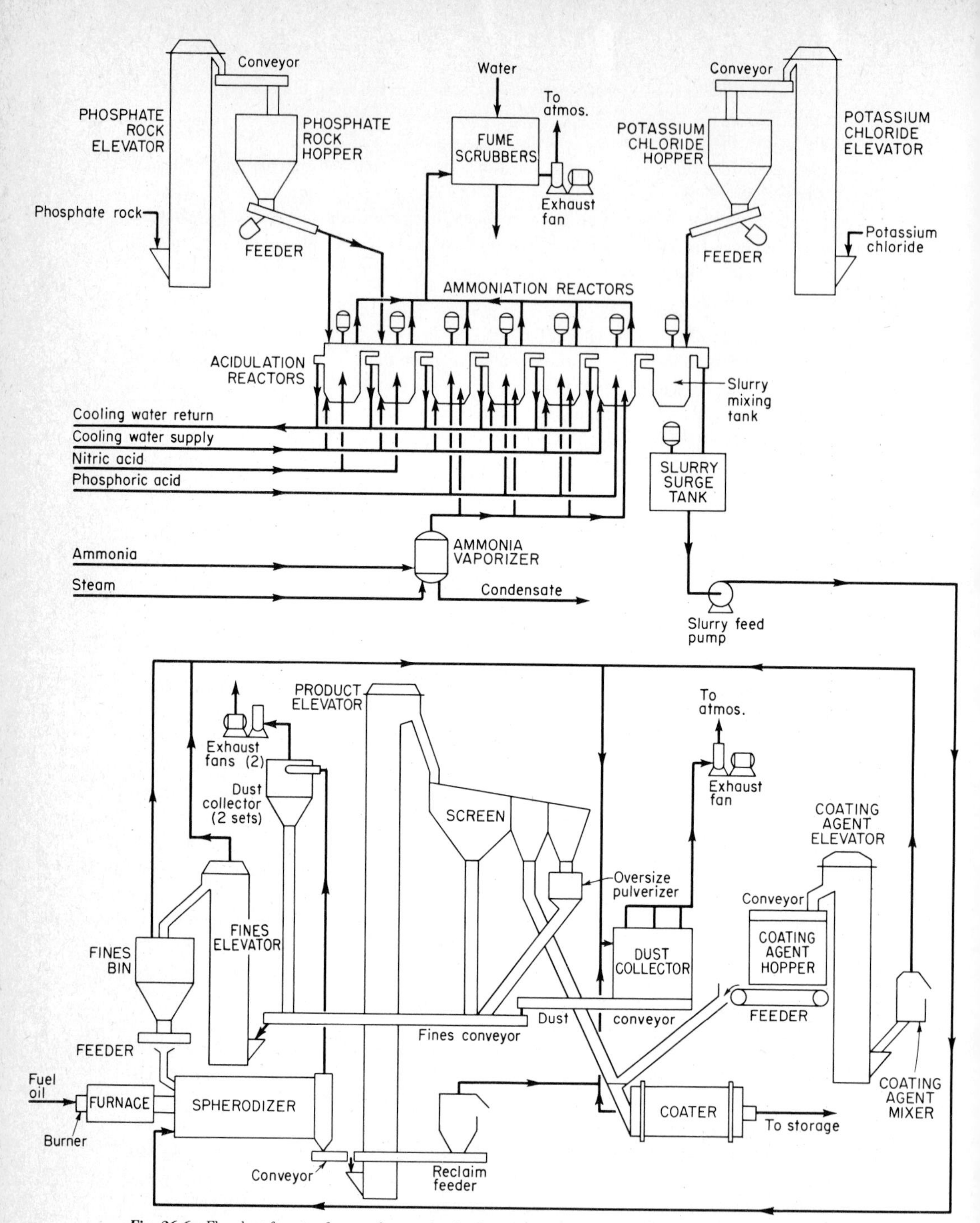

Fig. 26.6 Flowchart for manufacture of a complex fertilizer, showing a spherodizer for delivering pellets. The spherodizer is a large-diameter, short-length rotary dryer. By rotation it showers its contents through hot drying gases, which are blown through from a furnace. Simultaneously, the solution or slurry is sprayed on the showering material, forming pellets which individually increase their size through the accumulation of successive dry coats. The in-process material at all times remains essentially dry. (*Chemical and Industrial Corp.*)

TABLE 26.6 *Fertilizer 13-11-12 Grade, 200 Ton/Day Plant (Supplement to Fig. 26.6)*

Raw materials:
- 612 lb nitric acid
- 166 lb anhydrous ammonia
- 660 lb phosphate rock
- 410 lb muriate of potash
- 130 lb CO_2 at no cost, from ammonia plant
- 70 lb stabilizer

Utilities:
- Power, 40 kWh
- Water, 1,000 gal
- Fuel, natural gas, 1,500 ft^3
- Operating and maintenance supplies
- Operating labor, 7 workers per shift = 84 work-hours/ton
- Maintenance labor, 4-worker shifts per day = 0.16 work-hours/ton

Bagging and shipping:
- Operating labor, 0.6 work-hour
- Maintenance labor, 0.06 work-hour
- Operating and maintenance supplies
- Power, 9 kWh
- Bags

Base suspension fertilizer (BSF) is a high-content product developed by the TVA and fertilizer makers.[38] Since it is both a solution and a suspension, "slurry" has been suggested as a more appropriate name.[39] This product contains 13% nitrogen and 43% P_2O_5, to which potash and trace elements can be added by the farmer or custom blender. Versatility is thus added to the features of lower handling, shipping, and application costs. Storage problems due to settling and caking are avoided by maintaining the constituents in suspension by using a 1 to 3% attapulgite clay, which keeps the nutrient salt crystals small. Clogging of spray nozzles is also thus avoided. An alternative is to use the relatively impure wet-process phosphoric acid, whose impurities have a sequestering effect. The handling of fertilizers in a suspension or slurry form is a radically new method which may proceed slowly until farmers become accustomed to its requirements.

PLANT-GROWTH REGULATORS have several uses in modern agriculture and eventually may provide the industry with a billion dollar per year market. Such selective action can increase the number of seeds which germinate, cause earlier maturation of plants, stimulate root development, accelerate ripening of fruit and grain, halt undesirable flowering, produce less but larger fruit by thinning out blossoms or fruit, prevent fruit from dropping off plants or trees before ripening, produce seedless fruit, and control sprouting of potatoes. Their potential is higher yields for the farmer, better-quality foodstuffs for the consumer, superior stock for the nursery, and more attractive flowers for the home gardener.[40] An excellent table of chemicals in commercial use or under large-scale testing as plant-growth regulants is found in the literature.[41]

In an attempt to improve the isolation procedure of auxins, an entirely different product, 3-indoleacetic acid, was found; it proved to have about the same plant-growth activity as auxins. Its identification triggered the search for synthetic regulators. Two of the first found were indolebutyric acid, which stimulates root growth, and naphthaleneacetic acid, which prevents orchard fruits from

[38]TVA Ships First Tank of a "Do It Yourself" Fertilizer Concentrate, *Chem. Eng.* (*N.Y.*), **69**(19), 94 (1962).

[39]Slack and Walters, Jr., Slurry Fertilizers, *Agric. Chem.* **17**(11), 36 (1962).

[40]Open Door to Plenty, p. 20, National Agricultural Chemicals Association, 1958; ECT vol. 15, p. 675, 1968.

[41]Leopold, Auxins and Plant Growth, p. 342, University of California, 1955 (more than 800 refs.); Huffman, Chemicals to Regulate Plant Growth, *Chemtech*, **2**, 505 (1972).

dropping before ripening. This latter chemical can also be used in reverse to remove some of the fruit before it forms, allowing the remainder to attain normal size. A newer and more versatile growth regulator is isopropyl-*N*-(3-chlorophenyl) carbonate, which, at low concentration, thins excess peach fruits from trees and, at a higher concentration, acts as a preemergence weed killer. In fact, the majority of growth regulators sold at the present time are weed killers and as such are listed under herbicides.

Plant hormones are often so effective that a pinch or two alters the growth of an acre of crops and, if used expertly, can extend the growing season, ripen crops when desired, and partially offset the effects of drought, cold, and perhaps disease. They initiate the intricate chemical reactions that cause a plant to utilize its food.[42] For example, gibberellin may be used to advance the growth of a crop that must be brought to harvest in a short growing season. Although this chemical has been used successfully in a few instances, e.g., for spraying Thompson seedless grapes, it has failed to live up to its early promise and has shown curious and undesirable side effects. Other types of chemicals have a dwarfing effect and hold back stem growth or halt plants at a desired height. One of these is 2,4-dichlorotributylphosphonium chloride (phosphon). Another dwarfing compound is CCC [(2-chloroethyl) trimethylammonium chloride] which is said to affect more varieties of plants than phosphon.

SECOND-GROWTH INHIBITORS AND STERILANTS Maleic hydrazide is one of the chemicals applied as a labor saver by some tobacco growers to minimize the "sucker" problem, thus boosting yields. This practice has been opposed by tobacco buyers who prefer hand-suckered, flue-cured tobacco. Gametocides control pollinization chemically by killing plant sex cells selectively. Gamete removal is done by hand, as in detasseling corn, and it is conjectured that chemicals eventually may do the job more economically and produce hybrid seeds. A male sterilant based on sodium α, β-dichloroisobutyrate was launched in 1961 after 3 years of field testing on cotton.[43]

Defoliants are in the class of specialized chemical aids to agriculture. They cause a plant to form an abscission layer at the base of each leaf when applied to a maturing crop. The leaves later drop to the ground. This procedure is especially useful when applied to a cotton crop before mechanical harvesting, since cotton is graded according to its cleanliness.[44] Calcium cyanamid and magnesium chlorate are examples of chemicals used for this purpose. *Desiccants* play a different role in agricultural production. They are chemicals which artificially speed the drying of plant tissue. Mechanized harvesting of seed, for example, can be done earlier, with a resulting cleaner and larger yield.[45] Arsenic acid is a cotton desiccant, and DNBP (4,6-dinitro-*o-sec*-butylphenol) is a desiccant for seed crops of flax, legumes, and corn. Both of these categories of plant-growth regulators are often listed under herbicides.

FEEDSTUFF ADDITIVES AND SUPPLEMENTS

Although the field of animal nutrition is an expanding one, this section is concerned only with chemicals added to feedstuffs which are in the nonnutrient categories. For example, the chemical nature of vitamins has long been established, but they are dietary essentials, hence are not considered here. They are discussed briefly in Chaps. 31 and 40. The $110 million/year the farmer pays for feed additives helps to keep flocks and herds both healthy and profitable. This has led to the increased use of antibiotics, hormones, and other chemicals which are not nutrients but do have a marked effect on animal health and growth. Group treatment with chemicals plays an important role in health, inasmuch as it is possible to handle a specific problem in one fell swoop by adding certain

[42]Boehm, Farming with Hormones, *Fortune*, **59**, 144 (April 1959).
[43]Seeking Chemicals to Improve the Breed, *Chem. Week*, Dec. 3, 1960, p. 83.
[44]Open Door to Plenty, *op. cit.*, p. 26; Audus, Plant Growth Substances, Interscience, 1959.
[45]Johnson, *op. cit.*

chemicals to the feed or drinking water. The purpose can be therapeutic or protective, or it can involve stimulating growth or changing animal behavior in some way. Use of these new methods results in a saving of time and labor, as well as the expense of individual administration. Along with constantly increasing human food needs, a healthy and flourishing livestock population is a necessity. At least 100 feed additives have been made available to aid the farmer in reaching this goal.[46]

ANTIBIOTICS Maynard and Loosli[47] state, "There is evidence that certain antibiotics, added to good practical rations, as well as to inadequate ones, improve the early growth of chicks, turkey poults, pigs, rats, dogs, lambs, and calves under most conditions but not always." They are used as growth stimulants at low levels of feeding, on a continuous-feeding basis for egg production, and on a high-level feeding basis for disease prevention. Other uses in specific animal classes for which they are listed involve egg hatchability, stress, milk production, and stomach appetizers.[48] The measure of response of animals to antibiotics is variable, and with some products there is none. Some of the more effective include tetracyclines, procaine penicillin, oleoandomycin, tylosin, zinc bacitracin, and erythromycin. For a flowchart see Fig. 40.8.

HORMONES Stilbestrol (diethylstilbestrol), originally a hormone used therapeutically in human beings, was found to boost the growth rate of cattle. Thyroprotein (iodinated casein) is also used to promote growth in poultry and swine. The use of both of these hormones is strictly controlled by government regulations to ensure that no harmful quantities will be present in livestock and poultry meat.[49]

MISCELLANEOUS FEED ADDITIVES Nitrofurans have broad-spectrum activity against several different diseases, including coccidiosis, as well as value as growth stimulants. Arsenicals at low levels are used as growth stimulants for poultry and swine and also to aid in the prevention of dysentery in swine. On a continuing basis they are used to promote egg production in hens. Cadmium compounds, piperazine, and phenothiazine are added to feeds as worming materials. Chemicals and biologicals have played an important role in animal epidemics. The use of feed additives is comparatively new, and the manufacturer takes every precaution before marketing to see that the product will have no immediate or eventual harmful effect. At present 58 additives, including antibiotics and hormones, are allowed by the FDA.

SELECTED REFERENCES

Agricultural Chemicals, Manufacturing Chemists' Association, 1963.
Agricultural Handbook No. 313, Use of Insecticides, U.S. Govt. Printing Office, 1966.
Andus, L. J. (ed.): Physiology and Biochemistry of Herbicides, Academic, 1964.
Beroza, M. (ed.): Pest Management with Insect Sex Attractants, ACS, 1976.
Casper, M. C.: Liquid Fertilizers, Noyes, 1973.
Chemicals in Modern Food and Fiber Production, NAS, National Research Council, 1962.
Chichester, C. O. (ed.): Research in Pesticides, Academic, 1965.
Crafts, A. S.: The Chemistry and Mode of Action of Herbicides, Wiley-Interscience, 1961.
DeOng, E. R., *et al.*: Insect, Disease and Weed Control, 2d ed., Chemical Publishing, 1972.
Feedstuffs: Feed Additive Compendium, Miller Publishing, 1964.
Fertilizer Handbook, 2d ed., National Plant Food Institute, 1963.
Fletcher, W.: The Pest War, Halsted, 1974.
Frear, D. E. H. (ed.): Pesticides Index, 3d ed., College Science, 1965.
Gould, R. F. (ed.): Literature of Chemical Technology, chap. 39, Pesticides, ACS Monograph, 1967.
Gutcho, M. H.: Feeds for Livestock, Poultry and Pets, Noyes, 1973.
Hanson, L. P.: Plant Growth Regulators, Noyes, 1973.
Herber, L.: Our Synthetic Environment, Knopf, 1962.

[46]Andrews, Group Treatment with Chemicals, *Nation's Agric.*, May 1963, p. 12.
[47]Animal Nutrition, p. 265, McGraw-Hill, 1962.
[48]Feedstuffs, Feed Additive Compendium Issue, Miller Publishing, 1963.
[49]Feed Additives Compendium, Animal Health Institute, 1975.

Hesse, P. R.: Textbook of Soil Chemical Analysis, Chemical Publishing, 1972.
Irvine, D. E. G., and B. Knights: Pollution and the Use of Chemicals in Agriculture, Ann Arbor Science Publications, 1974.
Jones, H. R.: Pollution Control in the Dairy Industry, Noyes, 1974.
Klingman, G. C.: Weed Control as a Science, Wiley, 1961.
Meltzer, Y. L.: Hormonal and Attractant Pesticide Technology, Noyes, 1971.
Metcalf, R. L. (ed.): Advances in Pest Control Research, Wiley-Interscience, 1957–1966, 6 vols.
Meyer, J. H.: Aquatic Herbicides and Algaecides, Noyes, 1971.
Moritusa, E.: Organophosphorus Pesticides, CRC, 1974.
Noyes, Robert: Potash and Potassium Fertilizer, Noyes, 1966 (patients).
O'Brien, R. D.: Insecticides: Action and Metabolism, Academic, 1967.
Pratt, C. J., and R. E. Noyes: Nitrogen Fertilizers, Noyes, 1965.
Rodricks, J. V. (ed.): Mycotoxins and Other Fungal Related Food Problems, ACS, 1976.
Sauchelli, V.: Manual on Fertilizer Manufacture, 3d ed., Industry Publications, 1963.
Searle, C. E. (ed.): Chemical Carcinogens, ACS, 1976.
Sittig, M.: Agricultural Chemicals Manufacture, Noyes, 1971.
State of Food and Agriculture 1965, U.N., Columbia University Press, 1965.

chapter 27

FRAGRANCES, FLAVORS, AND FOOD ADDITIVES

THE PERFUME[1] INDUSTRY

The manufacture of perfume, cologne, and toilet water, collectively known as the fragrances, has undergone drastic changes in the past quarter century, prior to which perfumers were usually trained through apprenticeships in laboratories until, working with traditional materials in well-defined patterns, they achieved skill in mixing and blending. Only occasionally was a new and original odor developed, such as Old Spice, which immediately won spontaneous and favorable response from consumers. Not many people realize how complex the creation of an acceptable fragrance has become; it requires professional knowledge, skill, and experience, coupled with specialization in synthetic chemistry's technical problems, followed by consumer panel testing. This change has resulted from a number of factors, among which are (1) increase in the number of available raw-material ingredients, both natural and synthetic; (2) a variety of new types of products requiring fragrances; (3) innovations in packaging, especially aerosol sprays, virtually nonexistent before 1950, and other new forms of dispensing such as perfume powders, cream sachets, gels, lotions, and sticks; (4) postwar prosperity, which has widened the spending habits of the public as well as increased the consumer potential; (5) broadened channels and methods of distribution, including door-to-door selling; and (6) phenomenal growth in men's toiletries. These numerous recent developments are in contrast to the two main changes which took place before the prewar era, the introduction of synthetics and improved methods of obtaining true oils.

Perfume takes its name from the Latin word *perfumare* (to fill with smoke), since in its original form it was incense burned in Egyptian temples. Early incenses were merely mixtures of finely ground spices held together by myrrh or storax. The next advance was the discovery that, if certain spices and flowers were steeped in fat or oil, the fat or oil would retain a portion of the odoriferous principle. Thus were manufactured the ointments and fragrant unguents of Biblical fame. To Avicenna, the Arabian physician, must go the honor of discovering steam distillation of volatile oils. During his search for medical potions, he found that flowers boiled in an alembic with water gave up some of their essence to the distillate. The returning Crusaders brought to Europe all the art and skill of the Orient in perfumery, as well as information relating to sources of gums, oils, and spices. René, perfumer to Catherine de' Medici, invented many new confections to delight the queenly nose and, in his spare time, was one of the cleverest and deadliest of the famous de' Medici poisoners. Many of the finest perfumes are imported from France. The cosmetic industry dates back to 1850. Classical colognes are at least 200 years old, having originated in Cologne, Germany; they were

[1] Ullmann, Enzyklopaedie der technischen Chemie, 3d ed., vol. 14, pp. 690–776, 1963; *Cosmetics and Perfumery*, published monthly gives current research and practice.

probably the first imports into this country. It was not until the postdepression years that U.S. industry discovered that the sale of perfume odors in an alcohol-diluted form was profitable.

USES AND ECONOMICS Of the $320 million spent in 1974 for fragrances, the sale facts and figures to consumers probably represent an annual maximum of only one-third this amount; for example, for every bottle of perfume sold in this country, 40 bottles of cologne are sold. However, fragrances make a major contribution to the $4 billion per year cosmetic industry, second only to the amount used in soaps and detergents. Fragrances are used industrially in masking, neutralizing, and altering the odor of various products, as well as in creating a distinctive aroma for normally odorless objects. Cashmere shawls manufactured in Scotland are given the Hindu touch by a trace of patchouli oil applied to them. Aromatics, sometimes referred to as *reodorants*, are added to fabric sizing to disguise the glue or casein smell, leaving the product with a fine fresh odor. Leather goods and papers are scented delicately to eliminate the raw-material smell. The odor of kerosine in fly sprays is masked; artificial cedarwood is made by coating other woods with cedar oil reclaimed in pencil manufacturing. Canneries, rendering plants, municipal refuse plants, and food processing systems are other areas where reodorants are used. Paint odor during drying is masked by essential oils and fixatives introduced in small quantities in the bulk product. For their psychological effect odors are used successfully to increase customer appeal, though they are not essential to the performance of the products to which they are added.[2] A minute amount of bornyl acetate evaporated in an air-conditioning system imparts an outdoor tang to the air.

CONSTITUENTS

A perfume may be defined as any mixture of pleasantly odorous substances incorporated in a suitable vehicle. Formerly, practically all the products used in perfumery were of natural origin. Even when humans first started synthesizing materials for use in this field, they endeavored to duplicate the finest in nature. There has been a marked tendency in recent years, however, to put on the market perfumes which have no exact counterpart in the floral kingdom but which have received wide acceptance. The finest modern perfumes are neither wholly synthetic nor completely natural. The best product of the art is a judicious blend of the two in order to enhance the natural perfume, to reduce the price, and to introduce fragrances into the enchanting gamut at present available. A product made solely of synthetics tends to be coarse and unnatural because of the absence of impurities in minute amounts which finish and round out the bouquet of natural odors; however, such an eventual development is predicted. The chemist has also succeeded in creating essences of flowers which yield no natural essence or whose essence is too expensive or too fugitive to make its extraction profitable. Lily of the valley, lilac, and violet are examples. The constituents of perfumes are threefold: the vehicle or solvent, the fixative, and the odoriferous elements.

VEHICLE

The modern solvent for blending and holding perfume materials is highly refined ethyl alcohol mixed with more or less water according to the solubilities of the oils employed. This solvent, with its volatile nature, helps to project the scent it carries, is fairly inert to the solutes, and is not too irritating to the human skin. The slight natural odor of the alcohol is removed by deodorizing, or "prefixation," of the alcohol. This is accomplished by adding a small amount of gum benzoin or other resinous fixatives to the alcohol and allowing it to mature for a week or two. The result is an almost odorless alcohol, the natural rawness having been neutralized by the resins.

[2]Mitchell *et al.*, Importance of Odor as a Nonfunctional Component, Odor Symposium, New York Academy of Sciences, Nov. 7–9, 1963.

FIXATIVE

In an ordinary solution of perfume substances in alcohol, the more volatile materials evaporate first, and the odor of the perfume consists of a series of impressions rather than the desired ensemble. To obviate this difficulty, a fixative is added. Fixatives may be defined as substances of lower volatility than the perfume oils, which retard and even up the rate of evaporation of the various odorous constituents. The types of fixative considered are animal secretions, resinous products, essential oils, and synthetic chemicals. Any of these fixatives may or may not contribute to the odor of the finished product but, if they do, they must blend with and complement the main fragrance.

ANIMAL FIXATIVES Of all animal products, *castor,* or *castoreum,* a brownish-orange exudate of the perineal glands of the beaver, is employed in the greatest quantity. Among the odoriferous components of the volatile oil of castor are benzyl alcohol, acetophenone, *l*-borneol, and castorin (a volatile resinous component of unknown structure).

Civet is the soft, fatty secretion of the perineal glands of civet cats, which are indigenous to many countries, and was developed in Ethiopia. The secretions are collected about every 4 days by spooning and are packed for export in hollow horns. The crude civet is disagreeable in odor because of the skatole present. On dilution and aging, however, the skatole odor disappears, and the sweet and somewhat floral odor of civetone, a cyclic ketone, appears.

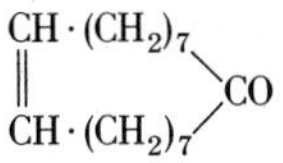

$$\begin{array}{l} CH\cdot(CH_2)_7\diagdown \\ \| \qquad\qquad\quad CO \\ CH\cdot(CH_2)_7\diagup \end{array} \qquad \begin{array}{l} \qquad\qquad CH_2 \\ CH_3—CH \qquad CO \\ \qquad\qquad (CH_2)_{12} \end{array}$$

Civetone 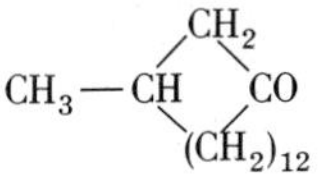 Muskone

Musk is the dried secretion of the preputial glands of the male musk deer, found in the Himalayas. The odor is due to a cyclic ketone called *muskone,* which is present to the extent of from $\frac{1}{2}$ to 2%. Musk, the most useful of the animal fixatives, imparts body and smoothness to a perfume composition even when diluted so that its own odor is completely effaced. Musk is used for its own sake in heavy Oriental perfumes.

Ambergris is the least used, but probably best known, of the animal fixatives. It is a calculus, or secretion, developed by certain whales. Ambergris is obtained by cutting open the captured whale, or it is obtained from whales stranded on a beach. It is waxy in consistency, softening at about 60°C, and may be white, yellow, brown, black, or variegated like marble. It is composed of 80 to 85% ambrein (triterpenic tricyclic alcohol), resembling cholesterol and acting merely as a binder, and 12 to 15% ambergris oil, which is the active ingredient. It is employed as a tincture, which must be matured before it is used. The odor of the tincture is decidedly musty and has great fixative powers.

Musc zibata is the newest animal fixative, derived from glands of the Louisiana muskrat. It was only during World War II that musc zibata was commercialized. About 90% of the unsaponifiable material in muskrat glands consists of large, odorless cyclic alcohols, which are converted to ketones, increasing the characteristic musk odor nearly 50 times. It is a replacement for or an addition to Asiatic musk.

RESINOUS FIXATIVES are normal or pathological exudates from certain plants, which are more important historically than commercially. These are hard resins, e.g., benzoin and gums; softer resins, e.g., myrrh and labdanum; balsams, moderately soft, e.g., Peru balsam, tolu balsam, copiaba, and storax; oleoresins, oily materials, e.g., terpenes; extracts from resins, less viscous, e.g., ambrein. All these substances, when being prepared for perfume compounding, are dissolved and aged by methods passed down by word of mouth. If solution is brought about in the cold, the mixture is called a *tincture.* If heat is required to give solution, the mixture is an *infusion.* Alcohol is the solvent, sometimes aided by benzyl benzoate or diethyl phthalate. The most important of the soft gums is *labdanum.* The leaves of a plant growing in the Mediterranean area exude this sticky substance. An extract from this gum has an odor suggestive of ambergris and is marketed as *ambrein*, having extremely good fixative value. Of the harder plant resins used in perfumes, *benzoin* is the most important. The history of chemistry was influenced by this substance. The early source of benzoin was Java, where it was called *luban jawi.* Through various contractions and linguistic modifications, it became "banjawi," "benjui," "benzoi," "benzoin," and "benjamin." In early organic chemical history an acid isolated from this gum became known as *benzoic acid,* from which compound the names of all benzo compounds of today are derived.

ESSENTIAL-OIL FIXATIVES A few essential oils are used for their fixative properties as well as their odor. The more important of these are clary sage, vetiver, patchouli, orris, and sandalwood. These oils have boiling points higher than normal (285 to 290°C).

SYNTHETIC FIXATIVES Certain high-boiling, comparatively odorless esters are used as fixatives to replace some imported animal fixatives. Among them are glyceryl diacetate (259°C), ethyl phthalate (295°C), and benzyl benzoate (323°C). Other synthetics are used as fixatives, although they have a definite odor of their own that contributes to the ensemble in which they are used. A few of these are

Amyl benzoate	Musk ketone	Heliotropin
Phenethyl phenylacetate	Musk ambrette	Hydroxycitronellal
Cinnamic alcohol esters	Benzophenone	Indole
Cinnamic acid esters	Vanillin	Skatole
Acetophenone	Coumarin	

ODOROUS SUBSTANCES

Most odorous substances used in perfumery come under three headings: (1) essential oils, (2) isolates, and (3) synthetic or semisynthetic chemicals.

ESSENTIAL OILS[3] may be defined as volatile, odoriferous oils of vegetable origin (Table 27.1). A distinction should be made, however, between natural flower oils obtained by enfleurage or solvent extraction and essential oils recovered by distillation. Distilled oils may lack some component which is not volatile enough or which is lost during distillation. Two notable examples of this are rose oil, in which phenylethyl alcohol is lost to the watery portion of the distillate, and orange flower oil, in which the distilled oil contains but a very small proportion of methyl anthranilate, whereas the extracted flower oil may contain as much as one-sixth of this constituent.

Essential oils are in the main insoluble in water and soluble in organic solvents, although enough of the oil may dissolve in water to give an intense odor to the solution, as in the case of rose water and orange flower water. These oils are volatile enough to distill unchanged in most instances, and are also volatile with steam. They vary from colorless to yellow or brown in color. An essential oil is usually a mixture of compounds, although oil of wintergreen is almost pure methyl salicylate. The refractive indexes of the oils are high, averaging about 1.5. The oils show a wide range of optical activity, rotating in both directions.

The compounds occurring in essential oils may be classified as follows:

1. Esters: mainly of benzoic, acetic, salicylic, and cinnamic acids.
2. Alcohols: linalool, geraniol, citronellol, terpinol, menthol, borneol.
3. Aldehydes: citral, citronellal, benzaldehyde, cinnamaldehyde, cuminic aldehyde, vanillin.
4. Acids: benzoic, cinnamic, myristic, isovaleric in the free state.
5. Phenols: eugenol, thymol, carvacrol.

[3]Essential Oils: Perfume and Flavoring Materials (Staff Report), *Ind. Eng. Chem.*, **53**(6), 421 (1961). International Standardization of Essential Oils, Committee ISO/TC 54 publishes standards. See *Cosmet. Perfum.*, **90**(2), 86 (1975).

TABLE 27.1 ***Important Essential Oils***

Name of oil	*Important geographical sources*	*Method of production*	*Part of plant used*	*Chief constituents*
Almond, bitter	California, Morocco	Steam	Kernels	Benzaldehyde 96–98%, HCN 2–4%
Bay	West Indies	Steam	Leaves	Eugenol 50%
Bergamot	Southern Italy	Expression	Peel	Linalyl acetate 40%, linaloöl 6%
Caraway	Northern Europe, Holland	Steam	Seed	Carvone 55%, *d*-limonene
Cassia (Chinese cinnamon)	China	Steam	Leaves and twigs	Cinnamic aldehyde 80–90%
Cedarwood	North America	Steam	Red core wood	Cedrene, cedral
Cinnamon	Ceylon	Steam	Bark	Cinnamic aldehyde 60%, eugenol 8%
Citronella, Java	Java, Ceylon	Steam	Grass	Geraniol 60–90%, citronellal
Clove	Zanzibar, Madagascar	Steam	Buds (cloves)	Eugenol 85–95%
Coriander	Central Europe, Russia	Steam	Fruit	Linaloöl, pinene
Eucalyptus	California, Australia	Steam	Leaves	Cineole (eucalyptole) 70–80%
Geranium	Mediterranean countries	Steam	Leaves	Geraniol esters 30%, citronellol
Jasmine	Grasse, France	Cold pomade	Flowers	Benzyl acetate, linaloöl, and esters
Lavender	Mediterranean area	Distillation	Flowers	Linaloöl
Lemon	California, Sicily	Expression	Peel	*d*-Limonene 90%, citral (3.5–5%)
Orange, sweet	Florida, California, Mediterranean area	Expression distillation	Peel	*d*-Limonene 90%
Peppermint	Michigan, Indiana, etc.	Steam	Leaves and tops	Menthol 45–90% and esters
Rose	Bulgaria, Turkey	Steam solvent, enfleurage	Flowers	Geraniol and citronellol 75%
Sandalwood	Mysore, East Indies	Steam	Wood	Santalol 90%, esters 3%
Spearmint	Michigan, Indiana	Steam	Leaves	Carvone 50–60%
Tuberose	France	Enfleurage, solvent extraction	Flowers	Tuberose oil
Wintergreen (gaultheria)	Eastern United States	Steam	Leaves	Methyl salicylate 99%
Ylang-ylang	Madagascar, Philippines	Steam, solvent extraction	Flowers	Esters, alcohols

6. Ketones: carvone, menthone, pulegone, irone, fenchone, thujone, camphor, methyl nonyl ketone, methyl heptenone.
7. Esters: cineole, internal ether (eucalyptole), anethole, safrole.
8. Lactones: coumarin.
9. Terpenes: camphene, pinene, limonene, phellandrene, cedrene.
10. Hydrocarbons: cymene, styrene (phenylethylene).

In living plants essential oils are probably connected with metabolism, fertilization, or protection from enemies. Any or all parts of the plant may contain oil. Essential oils are found in buds, flowers, leaves, bark, stems, fruits, seeds, wood, roots, and rhizomes and in some trees in oleoresinous exudates.

Volatile oils may be recovered from plants by a variety of methods:[4] (1) expression, (2) distillation, (3) extraction with volatile solvents, (4) enfleurage, and (5) maceration. The majority of oils are obtained by distillation, usually with steam, but certain oils are adversely affected by the temperature. Distilled citrus oils are of inferior quality; therefore they are derived by expression. For certain flowers which yield no oil upon distillation or else deteriorated oil, the last three methods are used. However, extraction with volatile solvents, a comparatively recent process, has superseded maceration (extraction with hot fat) for all practical purposes and is replacing enfleurage. Solvent extraction is the most technically advanced process and yields truly representative odors, but is more expensive than distillation.

Distillation, usually with steam. Flowers and grasses are normally charged into the still without preparation. Leaves and succulent roots and twigs are cut into small pieces. Dried materials are powdered. Woods and tough roots are sawed into small pieces or mechanically chipped. Seeds and nuts are fed through crushing rolls spaced so as to crack them. Berries are charged in the natural state, since the heat of distillation soon develops enough pressure to burst their integument. The stills employed in factories are of copper, tin-lined copper, or stainless steel and of about 600-gal capacity. They are provided with condensers of various sorts, tubular ones being the more efficient, and with a separator for dividing the oily layer from the aqueous one. Although removable baskets for holding the material to be distilled are used, the better procedure seems to be to construct the still with a perforated false plate lying just above the bottom. Underneath this false bottom are steam coils, both closed and perforated. In operating these stills, the charge is heated by steam in both the closed and the open pipes, thus effecting economical steam distillation. The aqueous layer in the condensate frequently carries, in solution, valuable constituents, as in the case of rose and orange flower oil, and is consequently pumped back into the still to supply some of the necessary water. Steam distillation usually is carried out at atmospheric pressure. If the constituents of the oil are subject to hydrolysis, the process is carried out in a vacuum. Much distillation for essential oils is done at the harvest site in extremely crude stills. These stills are converted oil drums or copper pots equipped with pipe condensers running through water tubs. The material and water are charged into the still, and a direct fire is built underneath, of dried material exhausted in previous distillations. The efficiency is low, and the oil is contaminated with pyrolysis products, such as acrolein, trimethylamine, and creosotelike substances. The crude oils obtained from the stills are sometimes further treated before use by vacuum-rectification, by fractional freezing (e.g., menthol from Japanese peppermint oil), by washing with potassium hydroxide to remove free acids and phenolic compounds, by removal of wanted or unwanted aldehydes and ketones through formation of the bisulfite addition compounds, or by formation of specific insoluble products, as in the reaction of calcium chloride with geraniol.

Expression by machine can yield an oil almost identical to the hand-pressed product. Of the hand-pressed processes, the *sponge* process is the most important, since it produces the highest-

[4]Guenther, The Essential Oils, vol. 1, chap. 3, Van Nostrand, 1948; Smith and Levi, Essential Oils, *J. Agric. Food Chem.*, **9**, 230 (1961).

quality oil. Here the fruit is halved, and the peel trimmed and soaked in water for several hours. Each peel is pressed against a sponge, and the oil is ejected into the sponge, which is periodically squeezed dry. One person can prepare only 24 oz of lemon oil a day by this method, but it is still practiced, especially in Sicily.

Enfleurage. The enfleurage process is a cold-fat extraction process used on a few types of delicate flowers (jasmine, tuberose, violet, etc.) which yield no direct oil at all on distillation. Especially in the case of jasmine and tuberose, the picked flowers continue to produce perfume as long as they are alive (about 24 h), which is utilized by this process. The fat, or base, consists of a highly purified mixture of 1 part tallow to 2 parts lard, with 0.6% benzoin added as a preservative. The process is carried out on a *chassis,* which is a rectangular wooden frame 2 in. high, 20 in. long, and 16 in. wide, supporting a glass plate coated on both sides with the fat mixture. The chassis is spread with fresh flowers about every 24 h, the old ones being removed by hand. The frames are stacked vertically, forming airtight compartments. At the end of the 8- to 10-week harvest, the fat, which is not renewed during the process, has become saturated with flower oil. This saturated fat is called a *pomade.* The fat is removed from the frames and extracted with alcohol to recover the perfume. This alcoholic solution is cooled and filtered to remove the slight amount of fat that has dissolved. The alcohol solution is the *extract,* and the residue after evaporation of the solvent is the *enfleurage absolute,* similar to the products in the old maceration process. Yields depend on both the oil and the method. Based on 5 tons of crude material, yields vary from 5 oz of violet oil to 1,800 lb of clove oil.

Extraction with volatile solvents. The most important factor in the success of this practice is the selection of the solvent. The solvent must (1) be selective, i.e., quickly and completely dissolve the odoriferous components, but only a minimum of inert matter, (2) have a low boiling point, (3) be chemically inert to the oil, (4) evaporate completely without leaving any odorous residue, and (5) be low-priced and, if possible, nonflammable. Many solvents have been used, but highly purified petroleum ether is the most successful, with benzene ranking next. The former is specially prepared by repeated rectification and has a boiling point no higher than 75°C. When benzene is employed, it is specially purified by repeated crystallization. The *extraction equipment* is complicated and relatively expensive and consists of stills for frationating the solvent, batteries for extracting the flowers, and stills for concentrating the flower-oil solutions. The two types of extractors employed are the stationary and rotary types.

In the rotary process the oil is extracted on the *countercurrent principle.* The 350-gal steam-jacketed drums revolve around a horizontal axis and are divided into compartments by perforated plates at right angles to the axis. About 300 lb of flowers are charged into the first drum, along with 150 gal of petroleum ether which has already come through the other two drums. The drum and its contents are rotated for an hour cold and for an additional half hour with steam in the jacket. The saturated solvent is pumped to the recovery still, and the flowers in the drum are treated twice more, the second time with once-used solvent and the last time with fresh solvent from the recovery still. The exhausted flowers are blown with steam to recover the adhering solvent. About 90% of the solvent is boiled off at atmospheric pressure, and the rest is removed under vacuum. After the solvent is removed in either process, the semisolid residue contains the essential oil, along with a quantity of waxes, resins, and coloring material from the blossoms. This pasty mass is known as *concrete.* In turn it is treated with cold alcohol in which most of the wax and resin are insoluble. The small amount of unwanted material that dissolves is removed by cooling the solution to −20°C and filtering it. The resulting liquid contains the essential oil and some of the ether-soluble color of the flower and is known as an *extract.* When the alcohol has been removed, an *absolute* remains.

In some oils there is a large quantity of *terpenes.* This is especially true in the case of lemon and orange oils, which have as much as 90% *d*-limonene in their normal composition. Not only are terpenes and sesquiterpenes of *exceedingly* little value to the strength and character of the oils, but they also *oxidize and polymerize rapidly* on standing to form compounds of a strong turpentine-like flavor. Furthermore, terpenes are insoluble in the lower strengths of alcohol used as a solvent and make cloudy solutions which are cleared up only with difficulty. Hence it is desirable to remove terpenes and sesquiterpenes from many oils. Such an oil, orange, for example, is 40 times as strong as the original and makes a clear solution in dilute alcohol. The oil has *now very little tendency to rancidify*, although it has not quite the freshness of the original. These treated oils are labeled "tsf" (terpene- and sesquiterpene-free). Because each oil has a different composition, *deterpenation* requires a special process. Two methods are involved, either the removal of the terpenes, sesquiterpenes, and paraffins by fractional distillation under vacuum or extraction of the more soluble oxygenated compounds (principal odor carriers) with dilute alcohol or other solvents.

Because of the complex nature and the high price commanded by so many essential oils, a great deal of *adulteration*, or sophistication, is practiced. These additions are extremely hard to detect in most cases since, whenever possible, a mixture of adulterants is used that does not change the physical constants of the oil. Common agents used are alcohol, cedar oil, turpentine, terpenes, sesquiterpenes, and low-specific-gravity liquid petroleums. The advent of so many esters of glycol and glycerol on the market has increased the difficulty of detection, since these compounds are colorless and practically odorless and in the right combination can be made to simulate almost any specific-gravity and refractive-index specifications set up for the oil they are intended to adulterate. Rose oil is sophisticated with geraniol or a mixture of geraniol and citronellol; wintergreen and sweet birch oil are mixed with large amounts of synthetic methyl salicylate; and lemon oil is often "stretched" considerably with citral from lemon grass oil.

ISOLATES are pure chemical compounds whose source is an essential oil or other natural perfume material. Notable examples are eugenol from clove oil, pinene from turpentine, anethole from anise oil, and linalool from linaloa oil (bois de rose).

SYNTHETICS AND SEMISYNTHETICS USED IN PERFUMES AND FLAVORS[5] More and more important constituents of perfumes and flavors are being made by the usual chemical synthetic procedures. *Compositions containing predominantly inexpensive synthetics now account for more than 50% of the fragrances used in perfumes.* Some constituents are chemically synthesized from an isolate or other natural starting materials and are classed as semisynthetics. Examples are vanillin, prepared from eugenol from clove oil; ionone, from citral from lemon grass oil; and terpineols, from turpentine and pine oil. Some of the significant synthetics are discussed below. The better to correlate the manufacturing methods and apparatus with other like processes, the examples here presented are grouped under the most important chemical conversion.

CONDENSATION PROCESSES[6] (Fig. 27.1)

Coumarin occurs in tonka beans and 65 other plants, but the economical source is the synthetic. It formerly was employed to amplify the flavor of vanillin, as a fixative and enhancing agent for essential oils and tobacco products, and as a masking agent for disagreeable odors in industrial products. The synthetic product may be prepared[7] in a number of different ways. One method utilizes the Perkin reaction:

$$C_6H_4(CHO)(OH) \xrightarrow[(CH_3CO)_2O]{CH_3COONa} C_6H_4(CH{=}CHCOONa)(OH) \longrightarrow C_6H_4\langle CH{=}CH{-}C({=}O){-}O \rangle$$

Salicylaldehyde, acetic anhydride, and sodium acetate are refluxed at 135 to 155°C. The reaction mixture is cooled and washed. The coumarin is recovered by solvent extraction or distillation. Other

[5]Bedoukian, Perfumery Synthetics and Isolates, Van Nostrand, 1951; Ullmann, *op. cit.;* Clark and Grande, Study of Odor Variation with Structural Change, *Cosmet. Perfum.*, **90**(6), 58 (1975); Chemicals from Trees, *Chemtech*, **75**, 236 (1975).
[6]Process-studded Plant: Key to Fragrance, *Chem. Eng. (N.Y.)*, **65**(4), 112 (1958).
[7]ECT, vol. 6, p. 425, 1965.

important methods of coumarin preparation utilize *o*-cresol as the starting material, or the Hassmann-Reimer synthesis, where coumarin-3-carboxylic is produced as an intermediate. Over 1 million lb is produced per year.

Diphenyl oxide, or *ether*, is largely used in the soap and perfume industries because of its great stability and strong geranium odor. Diphenyl oxide is obtained as a by-product in the manufacture of phenol from chlorobenzene and caustic soda.

$$2C_6H_5OH \rightleftharpoons C_6H_5OC_6H_5 + H_2O$$

Ionone and its homologs possess the so-called violet type of odor, thus constituting the base of violet perfumes. These compounds, however, are indispensable to fine perfumes, and there are but few which do not contain at least a small percentage of ionones. Annually, about 500,000 lb of ionones are produced. Because of the high price of the natural oil of violet, this was one of the first essential oils synthesized, although it has since been found in certain obscure plants. The olfactory properties of ionone are due to the presence of *dl*-α-ionone and β-ionone. Their manufacture involves two steps: First, the pseudo-ionone is prepared by the condensation of citral obtained from lemon grass oil. This is followed by an acid ring closure, and the commercial ionone is purified by distillation. Commercial ionones are generally mixtures with one form predominating, although separations are sometimes made through bisulfite compounds.

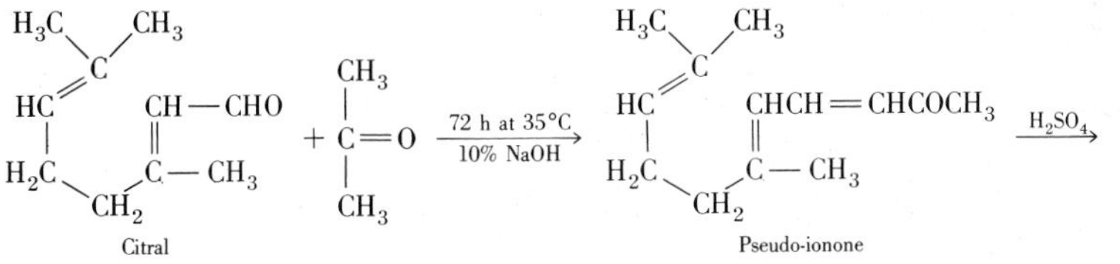

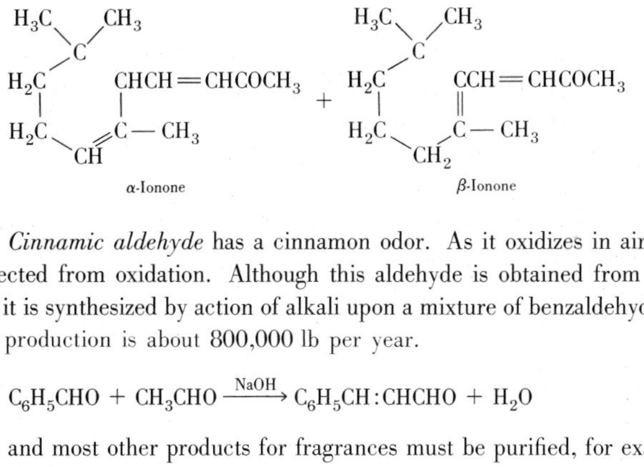

```
 H3C    CH3                                        H3C    CH3
    \  /                                              \  /
     C              CH3                                C
   //               |                                //
 HC      CH—CHO     |       72 h at 35°C           HC      CHCH=CHCOCH3        H2SO4
 |       ||     + C=O     ------------->           |       ||                 ------->
 H2C     C—CH3      |       10% NaOH               H2C     C—CH3
    \   /           CH3                               \   /
     CH2                                               CH2
   Citral                                           Pseudo-ionone

 H3C    CH3                   H3C    CH3
    \  /                         \  /
     C                            C
    /                            /
 H2C     CHCH=CHCOCH3         H2C     CCH=CHCOCH3
 |       |               +    |       ||
 H2C     C—CH3                H2C     C—CH3
    \   //                       \   /
     CH                           CH2
   α-Ionone                     β-Ionone
```

Cinnamic aldehyde has a cinnamon odor. As it oxidizes in air to cinnamic acid, it should be protected from oxidation. Although this aldehyde is obtained from Ceylon and Chinese cinnamon oils, it is synthesized by action of alkali upon a mixture of benzaldehyde and acetaldehyde (Fig. 27.2). The production is about 800,000 lb per year.

$$C_6H_5CHO + CH_3CHO \xrightarrow{NaOH} C_6H_5CH{:}CHCHO + H_2O$$

This and most other products for fragrances must be purified, for example, by vacuum fractionation (Fig. 27.3).

ESTERIFICATION PROCESSES

Benzyl benzoate has a faint aromatic odor, boils at 323 to 324°C, and is a fixative and a flavoring material. It occurs naturally in balsams (Peru, Tolú) but is prepared commercially by the esterification of benzoic acid with benzyl alcohol or by the Cannizzaro reaction with benzaldehyde.

Two esters of *salicylic acid* are very important commercially in the perfume and flavoring industries. About 350,000 lb of *amyl salicylate* is used annually in a variety of perfumes, because of its lasting quality and low price. About 4 million lb of *methyl salicylate* (synthetic wintergreen

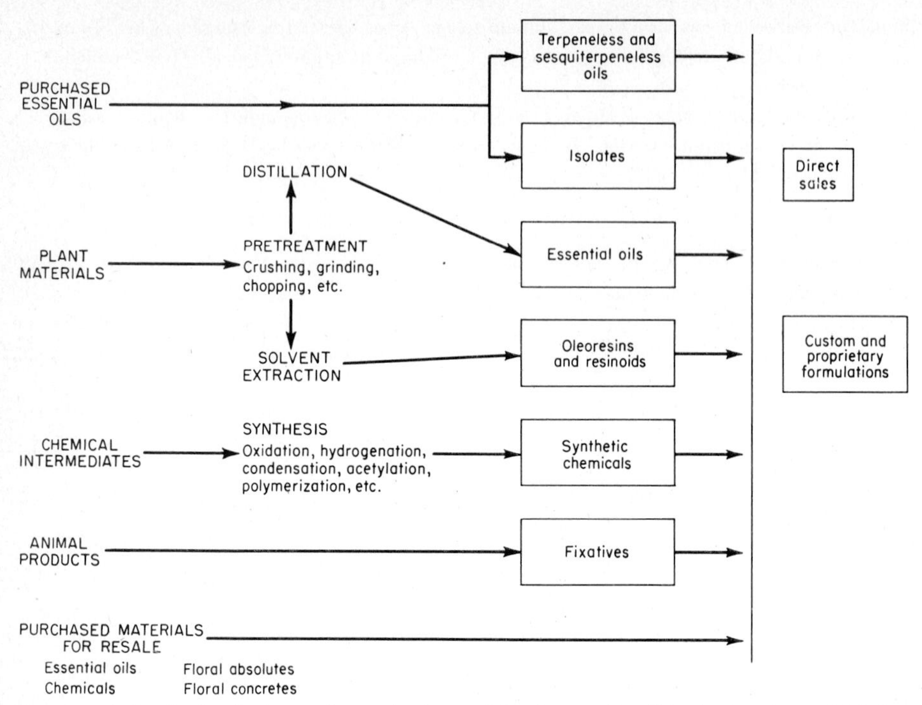

Fig. 27.1 Outline flowchart for the manufacture of perfume and flavoring materials. [*Fritzsche Bros., Inc., and Ind. Eng. Chem.*, **53**(6), 422 (1961).]

oil) is consumed annually as a flavoring ingredient. These esters are prepared as follows: Carbon dioxide and sodium phenate are reacted under pressure to obtain the salt of phenylcarbonic acid. This salt is isomerized to sodium salicylate by heating to 120 to 140°C. The esters are made from the acid and the proper alcohol (Fig. 40.3).

Benzyl acetate ($C_6H_5CH_2OCOCH_3$) is another widely used ester because of its low cost and floral odor. About 1 million lb is sold annually for soap and industrial perfumes. It is prepared by esterification of benzyl alcohol, by heating with either an excess of acetic anhydride or acetic acid with mineral acids. The product is purified by treatment with boric acid and distilled, giving a purity of over 98%. Large amounts of *benzyl alcohol* are employed in pharmaceuticals, lacquers, etc. (about 4 million lb per year). This alcohol has a much weaker odor than its esters. It is made by hydrolyzing benzyl chloride.

GRIGNARD PROCESSES

Phenylethyl alcohol has a roselike odor and occurs in the volatile oils of rose, orange flowers, and others. It is an oily liquid and is much used in perfume formulation, more than 1 million lb being sold annually. Phenylethyl alcohol can be made by a number of procedures; the Grignard reaction is used generally:

$$C_6H_5Br \xrightarrow{\text{Mg (ether)}} C_6H_5MgBr \xrightarrow{CH_2-CH_2 \text{ (O bridge)}} C_6H_5\cdot CH_2CH_2OMgBr \xrightarrow{\text{acid}} C_6H_5\cdot CH_2CH_2OH$$

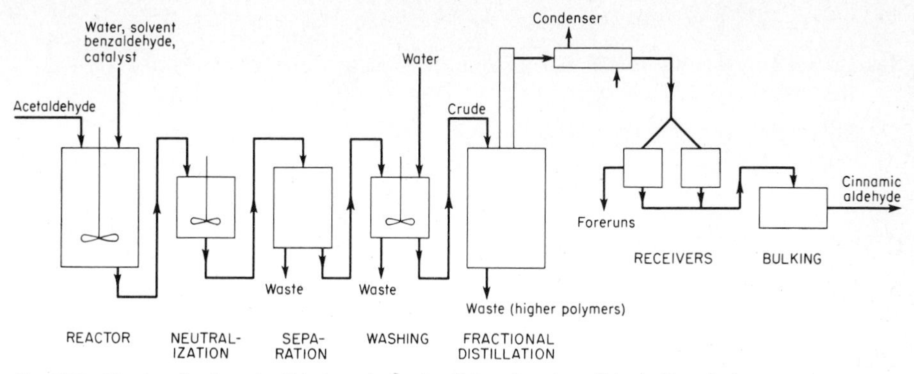

Fig. 27.2 Flowchart for cinnamic aldehyde production by aldol condensation. (*Fritzche Bros., Inc.*)

Fig. 27.3 High-temperature vacuum fractionating equipment. (*Fritzsche Bros., Inc.*)

However, the Friedel-Crafts reaction is outlined in the flowchart in Fig. 27.4 and follows the reaction

$$\underset{\text{Benzene}}{C_6H_6} + \underset{\text{Ethylene oxide}}{(CH_2)_2O} \xrightarrow{AlCl_3} C_6H_5 \cdot CH_2CH_2OH$$

HYDROGENATION

See Fig. 27.5 for citronellal from citronellol by Raney nickel hydrogenation at 200 psi.

NITRATION PROCESSES

Artificial musks comprise a number of products not identical with the natural musk, which derives its odor from macrocyclic compounds. Nitro musks are practical and economical substitutes for this expensive natural fixative, and more than 200,000 lb of musk xylene alone is manufactured annually. The reactions for the three important commercial artificial musks are

Musk ambrette:

$$\underset{m\text{-Cresol}}{\text{(OH, } CH_3 \text{ benzene ring)}} \xrightarrow[KOH]{(CH_3)_2SO_4} \text{(}OCH_3\text{, } CH_3\text{ ring)} \xrightarrow[AlCl_3]{(CH_3)_3CCl} \text{(}OCH_3\text{, } (CH_3)_3C\text{, } CH_3\text{ ring)} \xrightarrow[(CH_3CO)_2O]{HNO_3} \text{(}OCH_3\text{, } (CH_3)_3C\text{, } CH_3\text{, 2 } NO_2\text{ ring)}$$

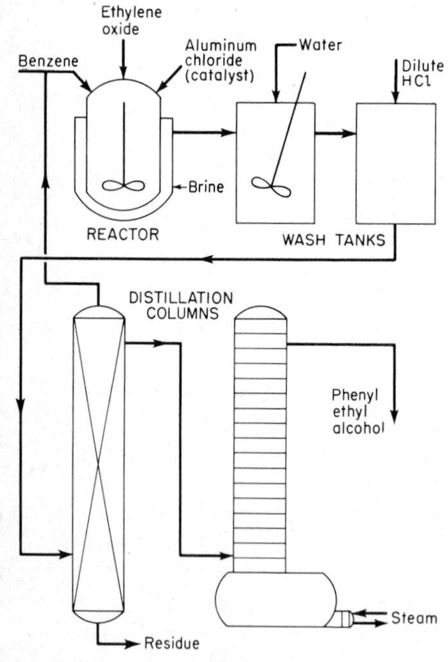

Fig. 27.4 Phenylethyl alcohol formed by the Friedel-Crafts reaction. [*Van Ameringen Haebler and Chem. Eng.* (*N.Y.*), **65**(4), 113 (1958).]

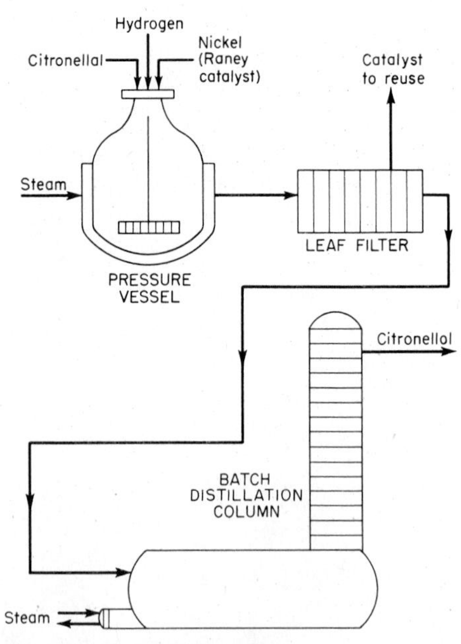

Fig. 27.5 Hydrogenation under pressure of 200 psi to convert citronellal to citronellol. [*Van Ameringen Haebler and Chem. Eng.* (*N.Y.*), **65**(4), (1958).]

Musk xylene and musk ketone:

$$C_6H_4(CH_3)_2 \xrightarrow[AlCl_3]{(CH_3)_3CCl} C_6H_3(CH_3)_2 \cdot C(CH_3)_3 \longrightarrow$$

m-Xylene

$\xrightarrow{70\%\ H_2SO_4,\ 30\%\ HNO_3,\ 35°C\ 1\frac{1}{2}\ h}$ Musk xylene (ring with CH_3, NO_2, CH_3, NO_2, $(CH_3)_3C$, O_2N)

Musk xylene

$\xrightarrow[AlCl_3]{CH_3COCl} C_6H_2(CH_3)_2 \cdot C(CH_3)_3 \cdot COCH_3 \xrightarrow{HNO_3}$ Musk ketone (ring with CH_3, $COCH_3$, CH_2, NO_2, $(CH_3)_3C$, O_2N)

Musk ketone

OXIDATION PROCESSES

Vanillin is one of the most widely used flavors, more than 1.5 million lb/year being manufactured. It is used as a flavor in perfumery and for deodorizing manufactured goods. Many processes have been employed in its manufacture, such as the following, of which the procedure is the most used:

1. From eugenol from oil of cloves, through isoeugenol, followed by oxidation to vanillin, using nitrobenzene as the oxidizing agent:

$$\text{Eugenol (OH, }OCH_3\text{, }CH_2CH{=}CH_2) \xrightarrow{KOH} \text{Isoeugenol (OH, }OCH_3\text{, }CH{=}CHCH_3) \xrightarrow{C_6H_5NO_2} \text{Vanillin (OH, }OCH_3\text{, CHO)}$$

Eugenol Isoeugenol Vanillin

2. From lignin[8] through an alkaline pressure cook at 130 to 200 lb for $\frac{1}{2}$ to 1 h of the calcium "ligninsulfonic" acid. The vanillin is purified through the sodium bisulfite compound and extraction with benzene or isopropanol (Fig. 27.6).

3. From phenol[9] or *o*-chloronitrobenzene through guaiacol, following the usual synthetic procedure.

Heliotropin, or *piperonal*, has a pleasant aromatic odor resembling heliotrope. It is produced from safrole by the following reactions:

$$\text{Safrole } (CH_2O_2C_6H_3 \cdot CH_2CH{=}CH_2) \xrightarrow{\text{heated in autoclave with 60\% KOH in alcohol, 15–20 h}} \text{Isosafrole } (CH_2O_2C_6H_3 \cdot CH{=}CHCH_3) \xrightarrow{Na_2Cr_2O_7,\ H_2SO_4,\ 14\ h,\ \text{benzene extraction}} \text{Heliotropin } (CH_2O_2C_6H_3 \cdot CHO)$$

Safrole Isosafrole, yield 90% Heliotropin, yield 85%

[8]Sandborn, U.S. Pat. 2,104,701 (1938); Devris, U.S. Pat. 2,399,607 (1946); Faith, Keyes, and Clark, Industrial Chemicals, 3d ed., pp. 796–799, Wiley, 1965 (flowchart, economics).

[9]Schwyzer, Die Fabrikation pharmazeutischer und chemischtechnischer Produkte, pp. 205–209, 279–288, Springer, Berlin, 1931; Ullmann, *op. cit.*, 3d ed., vol. 14, p. 742, 1963.

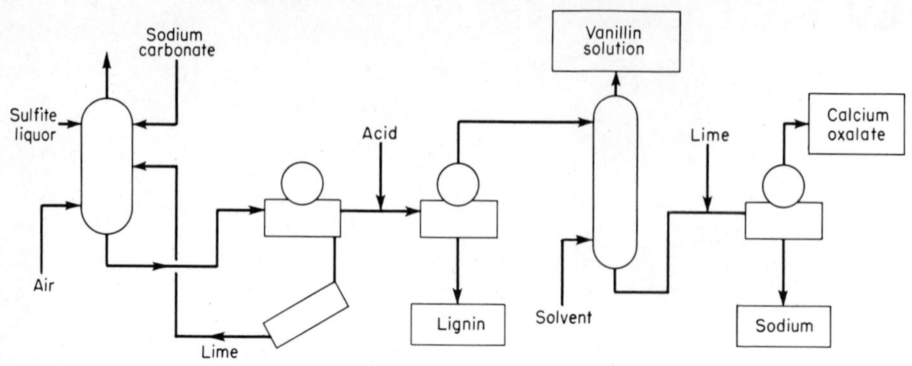

Fig. 27.6 Vanillin from lignin. (*Chem Week*, Sept. 16, 1961, p. 103.)

Anisaldehyde is a colorless oily liquid with an agreeable odor resembling coumarin, which is developed only after dilution and in mixtures. It is made by the oxidation of anethole (the chief constituent of anise, star anise, and fennel oils). Anethole has been obtained recently in this country at very low cost from higher-boiling fractions of pine oil.

$$CH_3O{-}C_6H_4{-}CH{=}CHCH_3 \xrightarrow[H_2SO_4]{Na_2Cr_2O_7} CH_3O{-}C_6H_4{-}CHO$$

Benzaldehyde is used as a flavoring agent, as an ingredient in pharmaceuticals, and as an intermediate in chemical syntheses. Commercially, it is produced by several methods and in two grades, technical and refined. The technical grade is largely used as an intermediate in the synthesis of other chemicals, such as benzyl benzoate, cinnamic aldehyde, and dyes. Most of the technical grade is made by direct vapor-phase oxidation of toluene, although some is made by chlorinating toluene to benzal chloride, followed by alkaline or acid hydrolysis. For perfume and flavoring use, the refined, chlorine-free grade is required, which is economically produced by the direct vapor-phase oxidation of toluene. This oxidation is sometimes carried out in the liquid phase.

Vapor phase: $$C_6H_5CH_3 \xrightarrow[500°C]{\text{air, catalyst}} C_6H_5CHO + H_2O \qquad \Delta H = -91.2 \text{ kcal}$$

It is claimed that a catalyst mixture of 93% uranium oxide and 7% molybdenum oxide gives relatively high yields.

Liquid phase: $$C_6H_5CH_3 \xrightarrow[40°C]{MnO_2,H_2SO_4} C_6H_5CHO + H_2O$$

Saccharin USP (o-benzosulfimide). This compound is approximately 500 times sweeter than sugar. It has been widely used by diabetics and in diet drinks and foods. However, in 1972 the FDA removed saccharin from the Generally Recognized as Safe (GRAS)[10] list and froze its use at then-current levels (4.2 million lb). A study was undertaken by the National Academy of Sciences to determine its safety for human consumption but did not provide definite conclusions.[11] In 1977 the FDA restricted the use of saccharin, a decision based on limited test data on laboratory animals. Saccharin is manufactured, employing the following reactions, with a 90% yield on the last step.

[10]Hall, GRAS—Concept and Application, *Food Technol.* **29**(1), 48 (1975).

[11]National Academy of Science Report No. PB-238137/AS, 1975 (available from National Technical Information Service, Springfield, Va.).

$$CH_3C_6H_5 + ClSO_3H \xrightarrow{0°C} o\text{-} + p\text{-}CH_3C_6H_4 \cdot SO_2Cl \quad \text{Fractionally precipitated}$$

Toluene — Chlorosulfonic acid — Mixture of toluene sulfonyl chlorides

$$o\text{-}CH_3C_6H_4SO_2Cl \xrightarrow{NH_3} o\text{-}CH_3C_6H_4SO_2NH_2 \xrightarrow[Na_2Cr_2O_7]{KMnO_4} C_6H_4(COONa)(SO_2NH_2) \xrightarrow[\text{heated}]{\text{acidified}} C_6H_4(CO)(SO_2)NH$$

o-Toluene sulfochloride, 60% — *o*-Sulfamylbenzoic acid — Saccharin

An alternate synthesis from anthranilic acid is claimed to yield a product with less bitter aftertaste.

$$o\text{-}NH_2C_6H_4COOH \xrightarrow[HOSO_2Cl]{CH_3OH} C_6H_4(COOCH_3)(SO_2Cl) \xrightarrow{NH_3} C_6H_4(CO)(SO_2)NH$$

Anthrenilic acid — Saccharin

MISCELLANEOUS PROCESSES

Sucaryl.[12] Calcium cyclamate USP $[(C_6H_{11}NHSO_3)_2Ca]$, the sodium salt, is also extremely sweet. It is prepared by sulfonating cyclohexylamine, employing chlorosulfonic acid. For purification, it is converted into the barium salt which is then treated with sulfuric acid, and the solution treated with deodorizing carbon. The acid solution is neutralized with calcium hydroxide or sodium carbonate, and the salt separated by evaporation, or salting out.

It has been banned for food uses by the FDA since 1969, because of its alleged carcinogenic properties. New laboratory studies have been conducted in an effort to have the ban lifted. The principal manufacturer, Abbott, does not feel the evidence is sufficient to warrant such a prohibition. In March 1975 the Food and Drug Administration asked the National Cancer Institute to evaluate the data.

The demand for low-calorie sweeteners by diet-conscious U.S. consumers has led to the discovery of several other sweetening compounds, some synthetic and some derived from plants. Dihydrochalcones from citrus bioflavonids are 2,000 times sweeter than sucrose, and *e*-aspartyl-*e*-phenylalanine is 200 times sweeter. Monellin, a protein from the West African serendipity berry, is 3,000 times sweeter, and a glycoprotein from the West African miracle berry is 120 times sweeter. All these substances, and others, are being studied extensively in hopes of obtaining FDA approval for their use.[13]

Terpineols are among the cheapest synthetics and are widely used in soap because of their woodsy and floral odors. Formerly, all terpineols were made from turpentine oil, which consists largely of α-pinene, but recently pine oil has become an important source. Terpineols may be manufactured directly in a one-step process from pinene by reacting with sulfuric acid and acetone for 6 h at 35 to 40°C. The product is purified by fractional distillation. The two-step method has an advantage in that the purification of the intermediate, terpin hydrate, is easier than that of terpineol. Terpin hydrate is formed by reacting pinene with dilute sulfuric acid and an emulsifying agent.

[12]For saccharin and cyclamate, cf. ECT, vol. 19, p. 593, 1969; *Chem. Eng. News,* Nov. 18, 1974, p. 7, and Nov. 25, 1974, p. 14.

[13]Sugar Substitutes Seek Sweet Smell of Success, *Chem. Eng. News,* Nov. 6, 1974, p. 37; Nov. 6, 1974, p. 37; *Food Eng.,* **46**(8), 27 (1974).

TABLE 27.2 *Composition of a Perfume*

Component	*Grams*	*Component*	*Grams*
Essential oils:		Synthetics:	
Sandalwood oil	10	Coumarin	27.5
Bergamot oil	117.5	Vanillin (from guaiacol)	20
Ylang-yland oil	40	Benzyl acetate	30
Petitgrain oil	10	Oleoresin, opopanax	2.5
Orange flower oil	10	Balsams (resinoids):	
Rose otto	15	Tolú	5
Jasmine absolute	20	Peru	7
Isolates:		Benzoin	70
Eugenol (from clove oil)	90	Animal fixative, castor tincture 1:10	12.5
Santalol (from sandalwood)	15	Synthetic fixatives:	
Semisynthetics:		Musk ketone	32.5
Isoeugenol (from eugenol)	110	Musk ambrette	12.5
Heliotropin (from safrole)	15	Vehicle, ethyl alcohol	450 kg
Methyl ionone (from citral)	237.5		

The purified hydrate is dehydrated to terpineol by oxygenated carboxylic acids. Terpineols are separated from pine oil by fractional distillation.

Menthol has long been extracted as the levo form from oil of Japanese peppermint and used in cigarettes and many other products as an antiseptic cooling flavor. The pure optically active form is made by Glidden Co. from β-pinene (from turpentine). This compound is hydrogenated, isomerized, and dehydrogenated to make β-citronellal. Catalytical conversion produces isopulegol, and hydrogenation yields *l*-menthol, after fractional distillation and crystallization.

The acetals of aldehydes have an odor only slightly modified from that of aldehydes, but have great alkali resistance. Hence these acetals are used in soaps, which are very difficult to perfume.

PERFUME FORMULATION

An actual example of a compound perfume similar to a widely sold product (Table 27.2) indicates the various components that have been discussed and shows their use in a blended product. The foundation odors are from eugenols, methyl ionone, and bergamot oil. Although the formulation given in the table shows a lower number, a single fragrance may contain 50 to 100 different compounds and subcompounds; in fact, as many as 300 ingredients may be used. Approximately 500 natural and 3,000 synthetic oils are available for perfume production.[14]

FRAGRANCE QUALITY

The majority of domestic perfume houses do not manufacture their own scents; they usually import natural floral oils and have their synthetics custom-made by aromatic firms. As the perfumer's skill and resources increase with experience and research, new equipment is developed for identifying fragrance components, even in trace quantities, such as infrared and ultraviolet spectrographs and vapor-phase chromatographs. Eventually, such instrumentation will be perfected to expedite production, but currently the perfumer's nose is regarded as the more discriminatory in arriving at a creative blend of exotic ingredients. Even he or she does not know precisely what will make one formulation successful where several hundred others may fail. Furthermore, after months of trial to attain certain objectives, a consumer panel may be unenthusiastic about the product. High quality from batch to batch is routinely ensured by standard tests, such as specific gravity, optical rotation,

[14]Aerosols; Fragrance Front-runners, *Chem. Week*, Jan. 2, 1960, p. 24.

refractive index, acid number, and ester number.[15] Provided success is met thus far, "the name, the package, the advertising must all be perfectly orchestrated" with the product.

The psychological effect of odors is successfully used primarily to increase customer appeal. Perfumed merchandise outsells its odorless counterpart by a large margin. An insurance company increased its sales of fire insurance overnight by sending out advertising blotters treated to simulate the acrid odor of a fire-gutted building. All kinds of paper are now perfumed to increase sales appeal.

THE FLAVORING INDUSTRY

There are only four basic flavors which the nerve endings in the taste buds on the tongue can detect: *sweet, sour, salty*, and *bitter*.[16] The popular conception of flavor, however, involves the combination of these four basic stimuli with concurrent odor sensations. Apple, for instance, tastes merely sour, with a trace of bitterness from the tannins present. The main concept received of an apple is due to the odor of acetaldehyde, amyl formate, amyl acetate, and other esters present in the volatile portion. The principles of perfume blending also hold good for flavor manufacturing. The best flavoring essences are natural products altered and reinforced where necessary by synthetics. In addition to alcohol as a vehicle, glycerin and isopropyl alcohol are used for liquid preparations, and emulsions of bland gums, such as tragacanth and acacia (gum arabic), for pastes. The same fixatives are employed, especially vanillin and coumarin; animal types are used sparingly. Many essential oils find application in the flavor industry, the more common being spice oils, citrus oils, peppermint, and spearmint. Almost all perfume synthetics find acceptance, plus a number made especially for flavors. The esters of ethyl, methyl, amyl, propyl, and benzyl alcohols with acetic, propionic, butyric, salicylic, caproic, formic, valeric, and anthranilic acids are widely used to characterize fruit flavors. As with flower perfumes, many chemical specialties keynote individual fruit aromas. The γ-lactone of undecylenic acid is a true representation of the fresh odor of a cut peach. A strawberry base is the ethyl ester of methylphenylglycidic acid, although it is not a true effect and partakes of a somewhat unnatural tone. Propyl succinate and ethyl butyrate are part of the makeup of many pineapple compositions.

NATURAL FRUIT CONCENTRATES

Although the essential oils used in flavoring are the same grade and source as those used for perfumes, fruit flavors are handled in a somewhat different manner. Because of the large percentage of water in most common fruits (from 75% in the banana to 90% in the strawberry) and the presence of considerable amounts of sugar and other easily fermented materials, special processes must frequently be employed, such as the following:

Distillation and extraction of the fruit. The ripe fruit is stoned and comminuted. It is then subjected to steam distillation and rectification until all the aroma is concentrated in a small portion of the aqueous distillate. This portion is then extracted with low-boiling petroleum ether, and the ether removed under vacuum to leave an *essence*, or *quintessence*, of the fruit used. Cherry, apple, strawberry, and raspberry are treated by this method.

Extraction of the juice. In this system the expressed and filtered juice is extracted directly without previous distillation. Occasionally, the juice is allowed to ferment slightly before extraction. This is supposed to result in a fuller flavor.

Concentration of the juice. The expressed and filtered juice is concentrated in vacuum evaporators with a low degree of heat until the water is largely driven off and the sugar concentration is high enough to inhibit bacterial growth (60%). This type of concentrate often has a "jam," or cooked

[15] White, Essential Oils, *Ind. Eng. Chem.*, **53**(6), 427 (1961).

[16] Hornstein and Teranishi, The Chemistry of Flavor, *Chem. Eng. News*, Apr. 3, 1967, p. 93.

flavor, especially in the case of the strawberries. An alternative method of concentration is by freezing. After reducing the temperature sufficiently, the mush of practically pure water ice is filtered off, and the partly concentrated juice is refrozen and refiltered until the requisite strength is obtained. This is the *optimum* method of producing concentrates, since there is little injury from heat, and the slight off-flavors from oxidation can be avoided by running the process in an atmosphere of carbon dioxide.

VANILLA The vanilla bean is grown principally in Madagascar, Tahiti, and Mexico. It is the immature fruit of the orchid *Vanilla planifolia* and is cultivated as a vine on trees which support it. The pods are picked when they are just starting to turn from a uniform green to yellow at the tip and have a rather disagreeable odor. The green pods undergo a curing treatment of from 3 to 5 months' duration. The cured bean is pliant, shiny, and dark-colored. The odor has become full and rich, and the sweating may have left white aromatic crystals on the outside of the bean. What has happened in substance is that the glucoside glucovanillin, present in the bean, has been acted upon by a ferment and split into glucose, vanillin, and other aromatics. Substances identified in the vanilla bean are anisic acid, alcohol, and aldehyde; vanillic acid and alcohol; cinnamic acid and its esters; vanillin, ethyl vanillin, and possibly other homologs of vanillin.

Preparation of vanilla extract. One hundred pounds of a blend of Mexican and Bourbon beans is finely cut up and macerated cold, with three successive portions of 35% ethyl alcohol of 100 lb each. These extracts are combined to make a fine vanilla extract. Other solvents may be used, and the extraction carried further, but the product is coarser and less desirable as a fine flavor (cf. Fig. 27.6).

CHOCOLATE AND COCOA The cacao bean, the seed of *Theobroma cacao* L, grows in equatorial areas on the tree in pods with from 30 to 60 beans. The pods are split open, and the watery pulp containing the seeds is allowed to ferment in boxes from 2 to 7 days, which, in addition to liquefying the pulp, kills the embryo (115°F), reduces the toughness of the bean, frees theobromine from the glucoside, and reduces the astringent tannin content. This fermentation is necessary for flavor in the final product. The fermented beans are dried and shipped to manufacturing centers. The beans are then heated in rotary roasters between 220 and 250°F, which develops the true chocolate flavor and aroma, removes unpleasant tannins and volatile matter (butyric and acetic acid, organic bases, and amines), dextrinizes the starch, and embrittles the husk. The roasted beans are quickly cooled to prevent overroasting, cracked in a conical mill, dehusked by a winnowing air stream, and degerminated. This product is known as *cacao nibs.* To work up the cacao product into chocolate, the modern method is to grind sugar in a closed-circuit disintegrator and the nibs in a separate water-cooled two-stage disk mill with closed-circuit removal of the fines. The two are then mixed. This method produces a fine, uniform product. The paste is run through a concher, which is a granite bed with reciprocally acting granite rollers. It reduces the particle size to an average of less than 1 μm. It is steam-heated to run at 135°F or can be allowed to heat itself by friction (122°F). Milk chocolates are prepared by adding condensed fresh milk or milk powder to the mill (melangeur). The finished product has a cocoa-butter content of 30 to 35% and not less than 12% milk solids. For cocoa, the roasted and ground beans are subjected to pressure in hydraulic presses to remove some of the fat content. Whereas originally roasted beans contain 55% fat, the product remaining after this treatment has the fat reduced to 20%, and is known as *cocoa.* The removal of fat makes a beverage that is not too rich and one in which the fat does not separate on top.

MONOSODIUM GLUTAMATE [MSG, $COOH(CH_2)_2CH(NH_2)COONa$] This compound is an important flavoring agent, yet has no flavor of its own. It accentuates the hidden and little-known flavors of food in which it is used. Reports that the use of MSG in foods is harmful were denied by a National Academy of Science—National Research Council report in 1971.[17] In 1974 the Food and

[17] *Chem. Eng. News,* July 12, 1971, p. 17.

Agricultural Organization of the World Health Organization approved the use of MSG as safe. Glutamic acid exists in three forms, but only the monosodium salt of *l*-glutamic acid has a flavor-accentuating capacity. Although glutamic acid is a constituent of all common proteins, the economical sources have been wheat gluten, corn gluten, and Steffens filtrate. When Steffens beet-sugar wastes are used, the principal steps involved are (1) concentration and collection of the Steffens filtrate, (2) its hydrolysis, usually with caustic soda, (3) neutralization and acidification of the hydrolysate, (4) partial removal of the inorganic salts, and (5) crystallization, separation, and purification of the glutamic acid. MSG is made from the acid as described above. The present production of MSG is largely from fermentation,[18] and a flowchart is presented in Fig. 27.7. Several U.S. companies have developed their own microbiological process or obtained this from the Japanese. The raw material is frequently dextrose or sugar. However, the Japanese are studying a synthetic process based on acrylonitrile.

FLAVOR-ESSENCE FORMULATION (Table 27.3)

A formula is given here for a type of apricot flavoring used in candy manufacture. This indicates in general how a well-balanced flavor is made up of natural products reinforced with suitable synthetics. Many formulas are to be found in print which consist mainly of esters of synthetic origin. These are harsh and unnatural, and in many cases the only resemblance to the original is in the designation of the flavor.

TABLE 27.3 *Flavor-essence Formulation*

	Grams
Apricot extract (a natural product)	1,000
γ-Lactone of undecylenic acid (persicol)	10
Nonyl aldehyde (apricolin)	1
Vanillin	0.5
Ethyl vanillin (bourbonal)	0.2
Ethyl cinnamate	0.2
Benzyl cinnamate	0.2
Tannic acid	0.1

FOOD ADDITIVES

"The food revolution in progress since the turn of the century almost defies description. The evidence is all around us—on fertile farm lands, and in lush orchards, among grazing herds and loaded refrigerator trucks and railroad cars, in magnificent food stores and, finally, in kitchens and on dining tables at home or in restaurants."[19] The food industry, estimated at over $135 billion yearly, is the country's *largest business*. The average family spends approximately 18% of its income on food. (See Chap. 25, Food and Food By-Product Processing Industries, for details on the size of this industry.) Food additives are those chemicals combined with foods by the manufacturer to effect certain modifications involving preservation, color, flavor enhancement, and stabilization, which have helped to make an astounding improvement in our food supply, as well as alleviating work in the kitchen. Additives are as old as history itself, spices, for example. *Intentional additives* are substances added in carefully controlled

[18]New Processes, *Chem. Eng.* (*N.Y.*), **69**(16), 117 (1962); *Ind. Eng. Chem.*, Annual Unit Processes Reviews: Fermentation, from 1961 on; Hoelscher, Fermentation, *Chem. Eng.* (*N.Y.*), **71**(8), 91 (1964); ECT, 2d ed., vol. 2, p. 198, 1963.

[19]Food Additives Codex, 1972, Manufacturing Chemists Association; Furia, Handbook of Food Additives, Chemical Rubber, 1968; Fernandez, Mixing Technology and Food, *New Eng.*, **3**(11), 52 (1974); Behavioral Disturbance and Food Additives, *Chemtech*, **5**, 264 (1975).

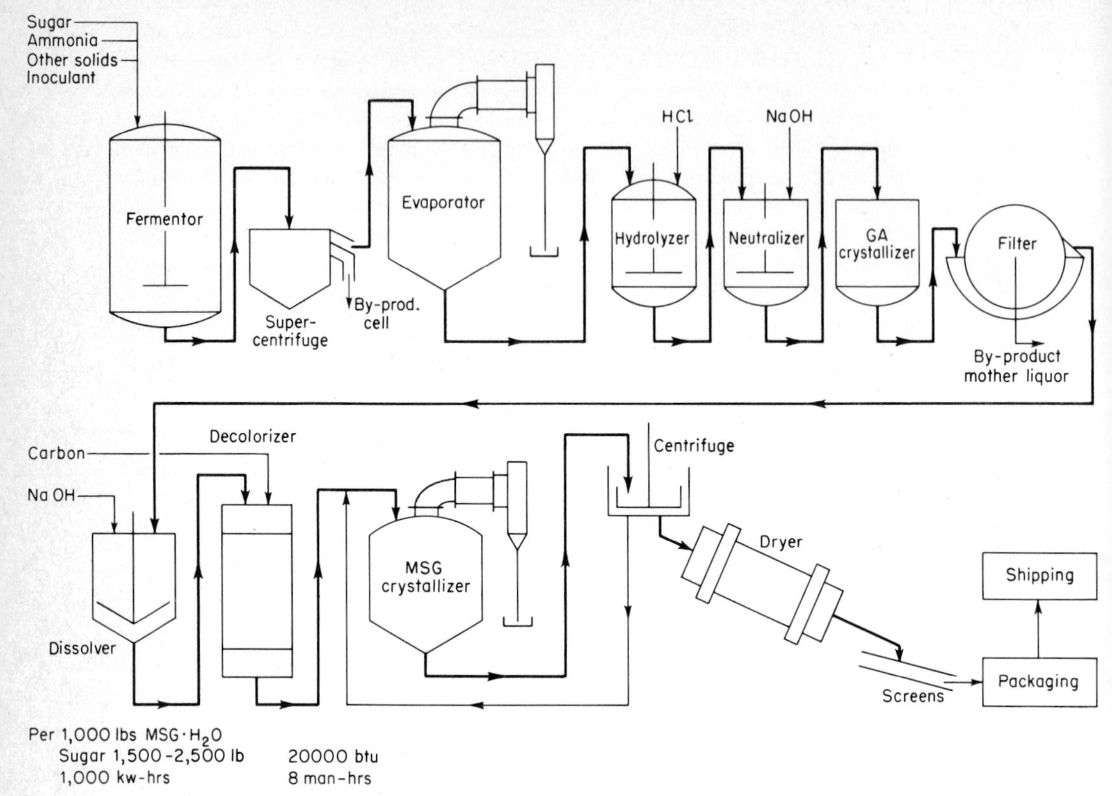

Fig. 27.7 Flowchart of monosodium glutamate production by fermentation. (*International Minerals and Chemical Corp.*)

amounts to preserve the quality of food, improve its nutritive value, or add flavor, for example, MSG (Fig. 27.7). Common kitchen staples such as vinegar, starch, and salt are in this category. *Incidental additives* are those that, although having no function in finished food, become part of it through some phase of production, processing, storage, or packaging.

Commonly used additives include chemical preservatives like propionic acid and benzoic acid; *buffers* and *neutralizing agents*, such as acetic acid and sodium citrate; *emulsifying agents* like polysorbates; *nonnutritive sweeteners*, such as saccharin; *nutrients*, among which are ascorbic acid and other vitamins; and *thickeners* like agar-agar and acacia. Here should be included *spices* like allspice and ginger and, finally, essential oils (Table 27.1) such as bitter almond and lemon. Some of these additives are permitted to be used freely, and others in limited amounts.

The FDA is conducting an on-going review of the GRAS list of substances. Contrary to popular reports, governmental regulation concerning safety in food products is not completely biased on the side of the food industry. Supplementary to the additives just listed are many other food chemicals on the market. For example, the food propellant octafluorocyclobutane does not modify the flavor of cream. There is a colorless, tasteless, and odorless carboxymethyl cellulose (Chap. 32) that adds bulk but no calories to food products; it may be thought of as a carrier rather than an additive. There are several inorganic chemicals, such as the leavening agent Py-Ran, for self-rising flour, slow-reacting sodium acid phosphate for refrigerated dough products, and fast-reacting sodium acid pyrophosphate for doughnut mixes.

NATURAL ACCESSORY CHEMICALS Foodstuffs contain small amounts of accessory chemicals, including vitamins (Chap. 40), chelating agents, and natural antioxidants. Carbohydrates consist of various sugars, starches, dextrins (Chap. 30), celluloses, and gums. Proteins are composed of amino acids. Shortenings contain esters of palmitic, oleic, linoleic, and stearic acids.[20] Natural foodstuffs contain substances of no known nutritive value, as well as substances that are harmful if taken in larger amounts than are normally used; for example, coffee and tea contain caffeine, whose pharmacological effect is well known. Even small amounts of arsenic and other toxic metals are found in food. Many laboratory-created chemicals also appear naturally in foods. Swiss cheese contains propionates, which are substances used as additives in some foods to inhibit molds. Regardless of the *source of the additive, it is a chemical.* There is a tendency on the part of people without scientific knowledge to think of the word "chemical" as something dangerous to humans, possibly even poisonous. They fail to realize that foods would be in a chaotic state if additives were suddenly to become unavailable. Experience has shown that chemical aids of the types discussed can be used safely and beneficially. Chemical processes required in the production of food additives are diversified, and many are described in other sections of this book.

SELECTED REFERENCES

Arctander, S.: Perfume and Flavor Materials of Natural Origin, Avi Publishing, 1960.
Bennett, H.: New Cosmetic Formulary, Chemical Publishing, 1970.
Daniels, R.: Sugar Substitutes and Enhancers, Noyes, 1973.
Daniels, R.: Edible Coatings and Soluble Packaging, Noyes, 1973.
De Navarre, M. C.: Chemistry and Manufacture of Cosmetics, 2d ed., Van Nostrand, 1962.
Flavor Chemistry, *Adv. Chem. Ser.*, no. 56 (1966).
Food Additives: What They Are, How They Are Used, Manufacturing Chemists' Association, 1961.
Food and Color Additives Directory, Hazelton Laboratories, 1961.
Food Chemicals Codex, National Academy of Sciences, 1966.
Fragrance Reference Guide, Fragrance Foundation, 1965.
Gillies, M. T.: Soft Drink Manufacture, Noyes, 1973.
Gould, R. F.: (ed.): Literature of Chemical Technology, chap. 40, Foods, ACS, 1967.
Hibbott, H. W.: Handbook of Cosmetic Science, Pergamon, 1963.
Howard, G. M.: Perfumes, Cosmetics and Soaps, vol. 3, Modern Cosmetics, Halsted, 1974.
James, R. W.: Fragrance Technology, Noyes, 1975.
James, R. W.: Synthetic and Natural Perfumes, Noyes, 1975.
Lawrence, A. A.: Edible Gums and Related Substances, Noyes, 1973.
Meroy, J.: Food Flavorings: Composition, Manufacture, and Use, Avi Publishing, 1960.
Parry, J. W.: Spices: Their Morphology, Histology and Chemistry, Chemical Publishing, 1962.
Pintauro, N. D.: Food Additives to Extend Shelf-Life, Noyes, 1974.
Pintauro, N. D.: Flavor Technology, Noyes, 1971.
Poucher, W. A.: Perfumes, Cosmetics and Soaps: Van Nostrand, 7th ed., 1959–1964, 3 vols.
Schultz, H. W., *et al.*: Symposium on the Chemistry and Physiology of Flavors, Avi Publishing, 1965.
Specifications for Identity and Purity of Food Additives, Columbia University Press, 1963, 2 vols.
Wieland, H.: Cocoa and Chocolate Processing, Noyes, 1972.

[20]Use of Chemicals in Food Production, Processing, Storage, and Distribution, National Academy of Sciences, National Research Council, Publ. 887, 1961. See also Chap. 28.

chapter 28

OILS, FATS, AND WAXES

At present, about 6.1 billion lb of edible vegetable and about 2.0 billion lb of edible animal fats and oils, excluding butter, are consumed in the United States each year. The total fat usage is about 12 billion lb, and only 1% is imported. Figure 28.1 depicts the relative consumption by the food, soap, and paint industries of various oils and fats. It also shows the direct competition of the chemical, soap, and paint industries with basic food production. Three types of agricultural products make up the bulk of the farm products that enter into the chemical process industries. Animal fats, both edible and inedible materials, are the largest in quantity and value. Starch (Chap. 30) runs a close second, and vegetable oils are next.[1] Fats and oils are found widely distributed in nature, in both the plant and animal kingdoms. Waxes likewise are natural products, but differ slightly from fats and oils in basic composition. Whereas fats and oils are mixtures of the glycerides of various fatty acids, waxes are mixed esters of higher polyhydric alcohols, other than glycerol, and fatty acids.

Table 28.1 indicates the characteristic composition of various important oils in regard to their fatty acid content. These acids have an even number of carbons and fall within (1) the saturated series, as exemplified by stearic acid, which is the basis for nondrying oils, (2) the monoolefinic acids, with one double bond between carbons as illustrated by oleic acid, and (3) the polyolefinic series, with more than one such double bond, as exemplified by linoleic and linolenic acids. The latter two classes of acids, being unsaturated, furnish semidrying and drying oils, according to the amounts of unsaturation present. The chief constituents of vegetable oils are 16- and 18-carbon acids, and 20-, 22-, and 24-carbon acids predominate in fish oils. Coconut oil is unique in that it consists of esters of much shorter carbon-chain acids, 12- and 14-carbon acids being present in the greatest amount.

There is a growing, recent demand for polyunsaturated oils in food products; the largest in supply, soybean oil, has 61% polyunsaturated constituents; safflower oil contains 68%; corn oil, 42%; cottonseed oil, 50%; peanut oil, 21%; and olive oil, 7%. The degree of unsaturation of the acids involved affects the melting point of the ester mixture; the more unsaturated acids give esters with lower melting points, and these are the chief constituents of oils. The more saturated esters, on the other hand, are constituents of fats. Thus we see that the factor determining whether one of these compounds is termed a *fat* or an *oil* is merely its melting point. These oils are called *fixed oils*, in distinction from the essential, or *volatile, oils* described in Chap. 27. Fixed oils cannot be distilled without some decomposition under ordinary atmospheric pressure. Differentiation should also be noted from petroleum oils, discussed in Chap. 37.

HISTORICAL Since ancient times humans have known how to remove oils and fats from their natural sources and make them fit for their own uses. Animal fats were first consumed as

[1]A Starting Point for Chemical Intermediates, *Ind. Eng. Chem.*, **54**(5), 17 (1962) (excellent tabulations); Dept. of Agriculture.

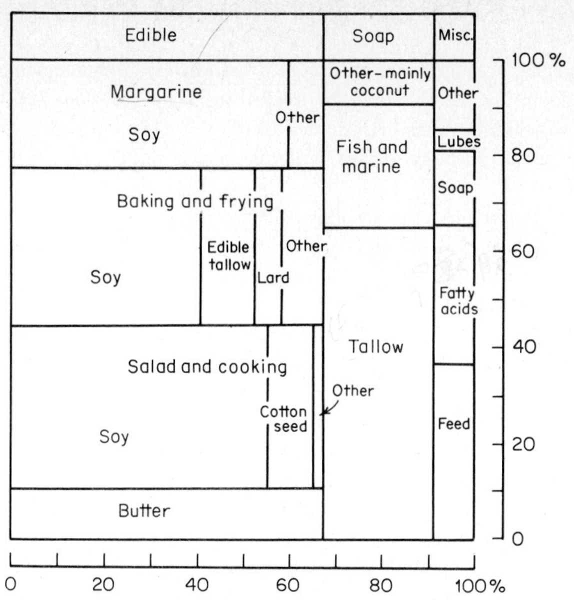

Fig. 28.1 Fats and oils, 1973.

TABLE 28.1 ***Fatty Acid Content of Various Oils and Fats*** *(In percentages)*

No. of C atoms	*Acid*	*Formula*	*Cotton-seed oil*	*Soy-bean oil*	*Corn oil*	*Men-haden oil*	*Whale oil*	*Lin-seed oil*	*Coco-nut oil*	*Mut-ton tal-low*	*Lard*
8	Caprylic	$C_7H_{15}COOH$	...	...	...	...	...	...	8.0		
10	Capric	$C_9H_{19}COOH$	...	...	...	...	...	...	7.0		
12	Lauric	$C_{11}H_{23}COOH$	...	...	...	...	...	...	48.0		
14	Myristic	$C_{13}H_{27}COOH$	0.6	...	...	7.0	8.0	...	17.5	1.0	1
14	*Myristoleic*	$C_{13}H_{25}COOH$	...	...	...	...	1.5	...	...	...	0.2
16	Palmitic	$C_{15}H_{31}COOH$	22.9	8.3	7.5	16.0	12.1	6.5	8.8	21.0	28
16	*Palmitoleic*	$C_{15}H_{29}COOH$	...	...	...	17.0	15.0	...	...	...	3
18	Stearic	$C_{17}H_{35}COOH$	2.2	5.4	3.5	1.0	2.3	4.5	2.0	30.0	13
18	*Oleic**	$C_{17}H_{33}COOH$	24.7	24.9	46.3	27.0	33.4	20.9	6.0	43.0	46
18	*Linoleic*	$C_{17}H_{31}COOH$	49.7	52.7	42.0	...	...	17.4	2.5	5.0	6
18	*Linolenic*	$C_{17}H_{29}COOH$	...	7.9	...	...	9.0	50.6	...	...	0.7
20	Arachidic	$C_{19}H_{39}COOH$	...	0.9	0.5	...	...	...	...	...	2
20	*Arachidonic*	$C_{19}H_{31}COOH$	...	...	...	20.0	8.2	0.1			
22	*Clupanadonic*	$C_{21}H_{35}COOH$	...	...	...	12.0	10.5				
24	Lignoceric	$C_{23}H_{47}COOH$	...	...	0.2						

Source: Data abbreviated and compiled from the very full Archer-Daniels-Midland fatty acid chart. Note that unsaturated acids are in italics. *Most common of all is oleic, probably 40% of total fatty acids.

food, but it was not long before the burning of the oils for light and heat was learned. Obtaining oils from vegetable sources is of ancient origin, for the natives in the tropical regions of the globe have long been removing these oils from various nuts after drying them in the sun. The utilization of marine oils began with the whaling industry, which was started by the Basques in the Bay of Biscay in the fifteenth century.

The first chemical reaction applied to fats and oils (excluding oxidation in burning) was that of saponification, to give soap. The early raw materials were mainly of animal origin, the rendering of animal flesh being an old art. The industrialization of oils and fats began with the erection of a cottonseed mill in South Carolina, about 1826. This crude industry did not expand very rapidly until after 1865. In 1850, the use of caustic soda to remove free acids from oil was introduced from France. About this time, the millers became aware of the value of the linters that clung to the hulls, and also of the hulls themselves, for cattle feed. The beginning of the oleomargarine (margarine) industry in Chicago in 1885 gave impetus to the cottonseed-oil industry. The higher quality demanded by this new market produced several processing improvements. Fuller's earth was used to decolorize the oil. In 1893, it was learned that the oil could be deodorized by blowing steam through it at high temperatures. Later it was found that deodorization under reduced pressure improved both flavor and odor. In 1900 the discovery that oils could be upgraded by *hydrogenation* to produce fats revolutionized the entire oil and fat industry and led to our modern hydrogenated shortenings. This discovery also made marketable many of the lesser-known oils. This phase of the industry has advanced until now much shortening is made by hydrogenating unsaturated oils and fats to a higher degree of saturation than originally found, but rarely to total saturation.

USES AND ECONOMICS Fats and oils have always had an essential role as food for mankind. In addition, however, our modern industrial world has found many important applications for them.

Fat or oil	*To make:*
Animal fats	Soaps, greases, paints, varnishes, syndets, fatty acids, and plasticizers
Coconut oil	Fatty alcohols, soaps, and detergents
Linseed oil	Paints, varnishes, floor coverings, lubricants, and greases
Soybean oil	Paints, varnishes, floor coverings, lubricants, and greases
Castor oil	Protective coatings, plastics, plasticizers, lubricants, hydraulic fluids
Tung oil	Paints and varnishes
Tall oil	Soaps, leather, paint, emulsifiers, adhesives, ink

Chemical uses amount to only 10 to 15% of the total cash value of nonfood consumption of farm products. See Fig. 28.1 for overall food and industrial consumption. The largest consumption of fats for chemical raw materials is in making fatty acids (Chap. 29), which have an estimated tonnage of 1,300 million lb/year. Oils are saponified, hydrogenated, epoxidized, and sulfonated to a great number of usable products, and fats are isomerized and interesterified, all to produce upgraded and more useful oils and fats.

There are two broad classifications for fats and oils: edible and inedible. The consumption of fats and oils in *edible* products represents about 73% of all uses of these materials and averages about 53 lb per capita. Various edible oils, cottonseed, olive, soybean, corn, etc., are employed for salad dressings, for other table uses, and for cooking purposes. Hydrogenated fats for cooking and baking, such as Crisco and Spry, may include a wide variety of vegetable oils, such as cottonseed, peanut, and soybean, since the hydrogenating process *improves* the color, flavor, and odor of the original crude product as well as the keeping factor. Various fish-liver oils are used in the medicinal field for their vitamin content and in the paint industry as drying oils. Castor oil is a well-known cathartic.

TABLE 28.2 ***U.S. Factory Production of Fats, Oils, and Derivatives***
(In millions of pounds) (Fat content only)

Vegetable	
Soybean	6,719
Cottonseed	1,266
Tall	131
Coconut*†	492
Linseed*	227
Corn	500
Peanut	128
Castor†	126
Tung	25
Safflower*	38
Other vegetable‡	306
Animal	
Tallow, inedible and grease*	2,543
Butter	1,431
Lard*	512
Tallow, edible*	662
Fish	
Fish oil*†	35
Sperm*†	14
Derivatives	
Cooking and salad oil	4,135
Baking or frying fats	3,486
Margarine	1,906
Soap	709
Feed	961
Paint and varnish	255
Resins	137
Lubricants and similar oils	147

Source: Data derived from Fats and Oils Production and Consumption, Dept. of Commerce, 1974. *Consumption. †Imported. ‡Principally imported.

In 1973, 27% of all oils and fats consumed in the United States was for *inedible* products, as shown in Fig. 28.1. About 20% was used in the soap industry, including tallow, coconut oil, palm oil, and certain greases. Some were slightly hydrogenated to make them suitable for soapmaking. The drying-oil industries (including paints and varnishes) consume 9% of all inedible oils, and other major consumption of linseed oil, and tung, soybean, and castor is in smaller demand. These drying oils are essentially unsaturated and produce films or coatings upon drying. They are also employed with synthetic resins and cellulose derivatives to give special types of films. Castor, linseed, soybean, rapeseed, and cottonseed oils are being used to some extent as plasticizers for lacquers and polymers. Other miscellaneous uses include the oilcloth and linoleum industry, the chemical industry for the manufacture of fat and oil derivatives, the manufacture of lubricants and greases, the printing industry, and the tin-plate industry. A wide variety of oils forms an integral part of various polishes, creams, and emulsions. Large quantities of sulfonated glycerides and sulfated fatty alcohols and derivatives serve as wetting agents and detergents. Table 28.2 details the statistics of the principal fats and oils, together with some of their important derivatives. It should be compared with Fig. 28.1. *Waxes,* such as carnauba, beeswax, and candelilla, enter into the manufacture of various polishes for floors, shoes, automobiles, and furniture. Other outlets include the making of carbon paper, candles, electrical insulation, waterproof textiles, and phonograph records.

VEGETABLE OILS

For purposes of discussion of the various technical aspects, the three classical divisions of the general subject of oils, fats, and waxes will be retained: vegetable oils, animal oils and fats, and waxes. Under each of these headings the general methods of manufacture will be discussed for the most important of the illustrative individual members. Table 28.3 indicates the yields of vegetable oils from some of the usual sources. The two general methods employed in obtaining vegetable fats and oils are expression and solvent extraction or a combination of the two. However, solvent extraction is increasing in use.

TABLE 28.3 *Approximate Oil Yields of Certain Vegetable-oil Materials*

Raw material	%	*Raw material*	%
Babassu kernels	63	Palm kernels	45
Castor beans	45	Peanuts in the shell	30–35
Corn kernels	4	Peanuts shelled	45–50
Copra	63	Perilla seed	37
Cottonseed	15	Rapeseed	35
Flaxseed	34	Sesame seed	47
Hempseed	24	Soybeans	18
Kapok seed	18	Tung nuts	50–55

Source: Mostly taken from U.S. Tariff Commission, Fats, Oils, and Oil-bearing Materials in the U.S., Dec. 15, 1941. (Modified in 1977.)

MANUFACTURING COTTONSEED OIL[2] BY EXPRESSION AND SOLVENT EXTRACTION Solvent extraction is assuming importance in virtually all vegetable-oil recovery plants, alone or in combination with prepressing. For high-oil-content seeds, such as cottonseed and safflower seed, usually both expression and extraction are utilized in the recovery systems for higher yields.[3] Obtaining crude vegetable and animal oils involves primarily *physical* changes or unit operations, but *chemical* conversions are concerned in the refining and further processing of such oils. Figure 28.2 presents a flowchart for cottonseed and similar oils. The flowchart can be broken down into the following sequences of *unit operations* (Op) (hydrogenation will be considered under its own special heading):

Cottonseeds are cleaned by screening and aspiration (Op). The lint is removed by passing the seeds through a series of linters (machines similar to cotton gins which operate on the saw-and-rib principle) (Op).[4] Each series of linters removes lint of different length, which is designated first-cut and second-cut lint. The lint cuts are aspirated and air-conveyed to separate lint beaters or cleaners which remove dirt and hull fragments from the lint before it is baled and sold.

The delinted seeds (blackseed) are cut or split in a bar-type huller, freeing the meats from the hulls, which are separated from the meats by screening and aspiration. The hulls thus removed are cleaned of attached meat particles in a beater and sent to storage for eventual consumption as roughage in animal feeds (Op).

The meats are rolled into thin flakes (about 0.010 to 0.014 in. thick) to make them easily permeable to steam in the cooking operation and are next *cooked or conditioned* in horizontal cookers at 230°F for 20 min before expression, to rupture the oil glands, to precipitate the phosphatides, to detoxify the gossypol, and to coagulate proteins (Op and Ch).[5] In direct solvent extraction, meats are conditioned before flaking.

The horizontal cookers are generally integrated with the expellers (Fig. 28.3); in prepress plants they are supplemented with a stacked cooker for additional heating capacity. The moisture is frequently raised to 12 to 14% and then gradually reduced to 5 to 7% in these units (Op).

Most of the oil from the conditioned cottonseed is prepressed in mechanical screw presses with single or double worm shafts revolving inside a heavy perforated barrel and capable of exerting

[2]For cottonseed oil, expression by the *expeller,* followed by solvent extraction, has become the most practiced. The solvent extraction (*miscella*) upgrades the oil and increases the yield, especially from damaged seed.

[3]Haines *et al.*, Filtration-extraction of Cottonseed Oil, *Ind. Eng. Chem.*, **49,** 920 (1957) (flowsheet); Decossas *et al.*, Cost Analysis (of preceding), *Ind. Eng. Chem.*, **49,** 930 (1957); Brennan, Making the Most out of Cottonseed Processing, *Chem. Eng. (N.Y.)*, **70**(1), 66 (1963); Tray and Bilbe, Solvent Extraction of Vegetable Oils, *Chem. Eng. (N.Y.)*, **54**(5), 139 (1947) (pictured flowchart); Witz and Hendrick, New Plant Puts Science into Cottonseed Processing, *Chem. Eng. (N.Y.)*, **71**(18), 48 (1964) (flowchart).

[4]Bailey, Cottonseed and Cottonseed Products, Interscience, 1948.

[5]Dunning, Unit Operations in a Mechanical Extraction Mill, *J. Am. Oil Chem. Soc.*, **33**(10), 465 (1956).

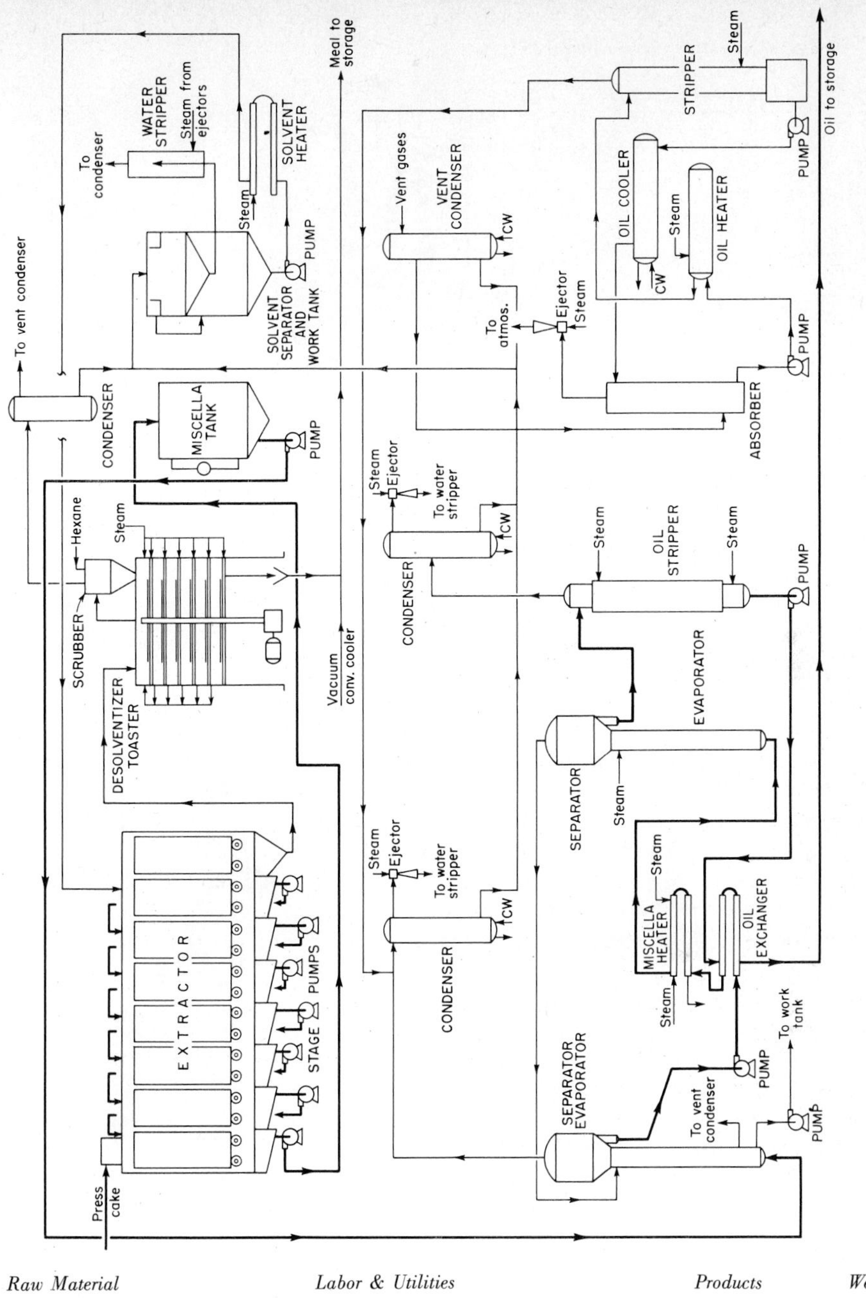

Raw Material	*Labor & Utilities*		*Products*	*Weight*
	Seed Handling	*Solvent Extraction*		
Cottonseed	2,000 lb		Oil	346 lb
Labor, man-hours	7.0	1.3	Cake	965 lb
Electricity, kw-hr	15.4	14	Hulls	370 lb
Solvent (hexane) loss, gal		0.5	Linters	207 lb
Water, gal	30†	60†	Loss & trash	112 lb
Steam, lb	260	450		
				2,000

†Recirculating cooling water

Fig. 28.2 Flowchart for cottonseed after pressing and flaking followed by solvent extraction with hexane, with a double evaporator to remove hexane from cottonseed miscella, the first effect being heated by hexane vapor and the second by steam. The cells of the extractor are mechanically moved countercurrent to the hexane. (*The Austin Co.*)

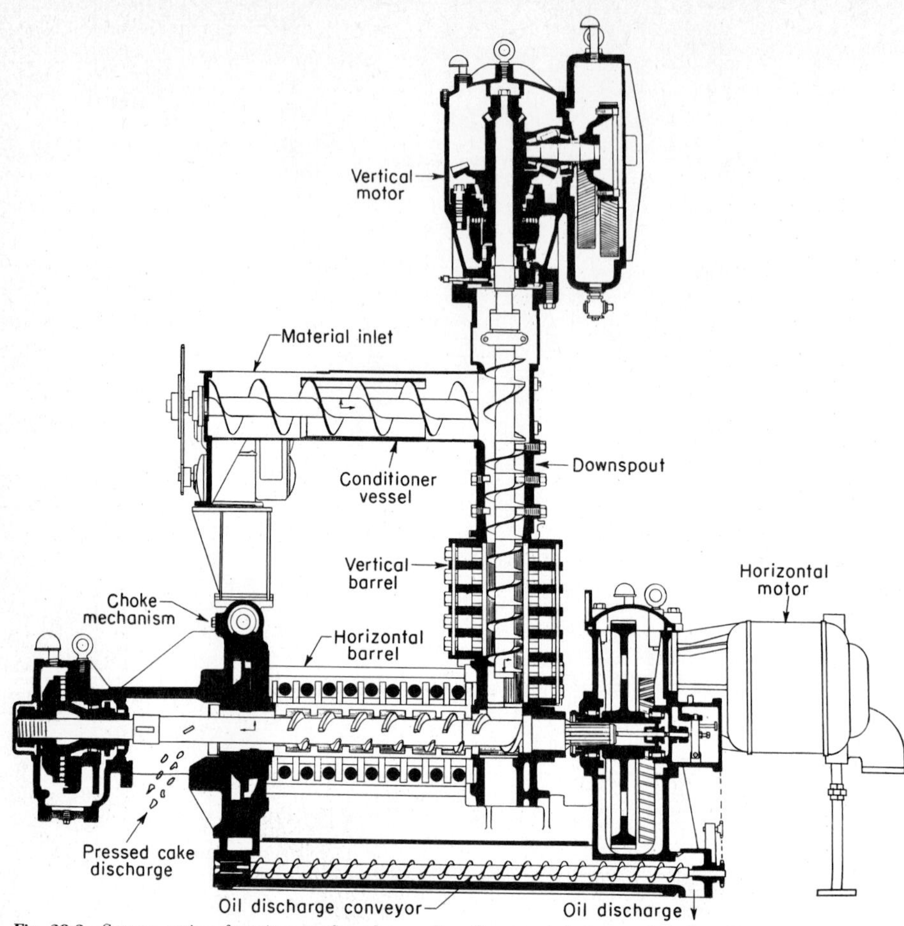

Fig. 28.3 Cutaway section of a twin-motor Superduo expeller. On cottonseed, the capacity is from 50 to 100 tons/24-h day, and for soybeans about 27 tons/24 h. To drive the propeller, 70 hp is required, and up to 5 boiler hp of steam is needed for the conditioner. (*V. D. Andersen & Co.*)

a pressure of up to 1,700 to 2,000 psi (Fig. 28.3). (These have largely displaced the hydraulic presses previously used in cottonseed plants.) The oil removed by these presses is screened, cooled, filtered, and stored for refining (Op). About 74% of all cottonseed is so processed; an additional 18% is processed by prepressing with solvent extraction and 8% by direct solvent extraction.

Solvent extraction recovers up to 98% of the cottonseed oil (Fig. 28.2), compared with 90 to 93% from screw-press expression alone. The soybean, also of low oil content, but whose physical structure is particularly suited to solvent extraction (Fig. 28.4), has been responsible for this development. But solvent extraction has assumed importance in virtually all vegetable-oil recovery processes, alone or in combination with prepressing. In Fig. 28.2 the solvent hexane is sprayed onto flakes in buckets moving horizontally in the extractor[6] countercurrent to the hexane. The hexane dissolving the oil is known as miscella and is pumped to the evaporators, the first of which is heated by hot hexane vapor and steam from the toaster and the second by steam.

[6]The V. D. Anderson Co., Cleveland, horizontal hydraulic basket extractor.

Virtually the entire cottonseed-oil production is used by edible-oil processors for shortening, margarine, and salad or cooking oils. The cake is broken or ground and used for cattle feed. The hulls provide roughage for livestock feeding. The linters which contain 70 to 85% α-cellulose are utilized as a cellulose source of high purity for rayon, plastics, and lacquer.

MANUFACTURING SOYBEAN OIL BY SOLVENT EXTRACTION

Soybean-seed preparation differs slightly from cottonseed preparation. The weighed and cleaned seeds are first cracked between corrugated rolls, then conditioned without significant change in moisture in a stacked cooker or a rotary steam-tube conditioner, and finally rolled to thin flakes (about 0.010 in. thick). Solvent extraction can recover up to 98% of the oil, compared with about

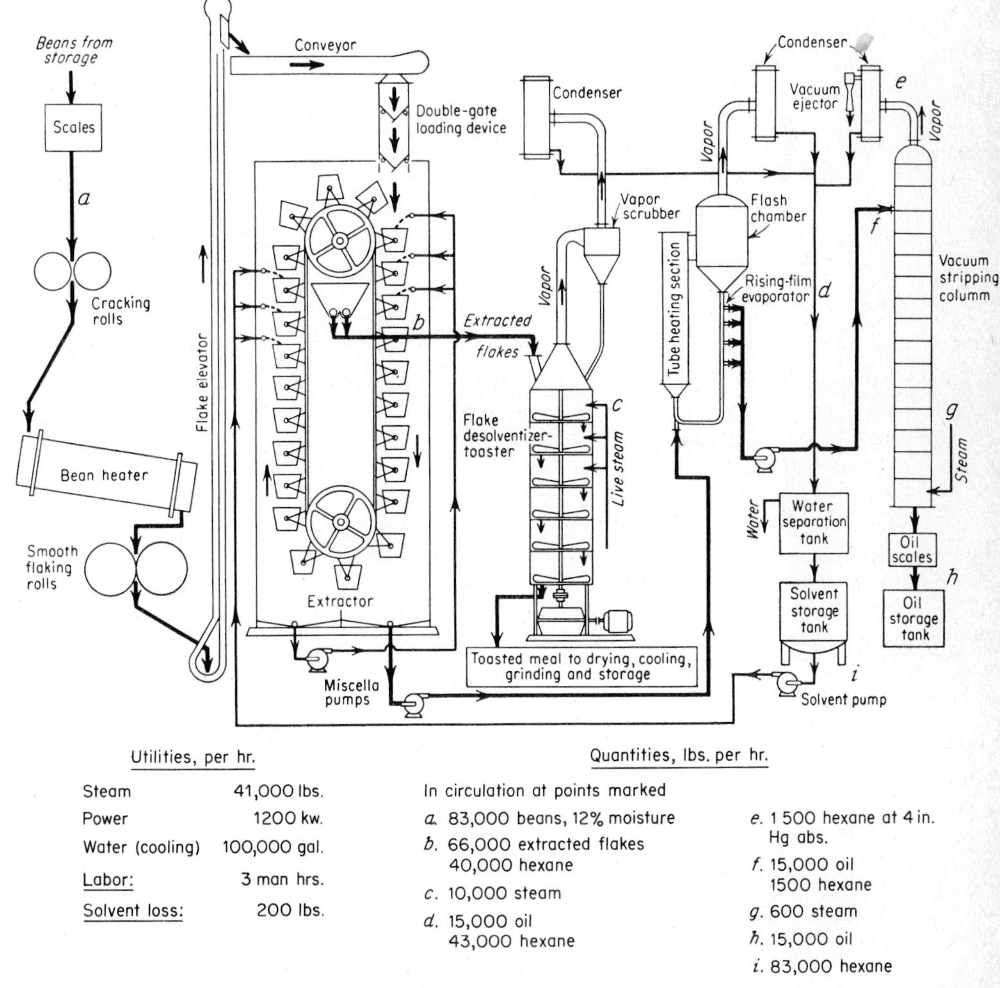

Fig. 28.4 Continuous flowchart for solvent extraction of soybeans. (*Central Soya Co., Inc.*)

80 to 90% from hydraulic or screw-press expression. The soybean, whose physical structure is particularly suited to solvent extraction, has been responsible for this development. Because of the efficiency of oil yields (hydraulic press, 8.67 lb/bu; screw presses, 9.16 lb/bu; solvent extraction, 10.94 lb/bu) virtually all new soybean installations today are solvent extractors. When solvent-extracted, soybean flakes produce meal with a protein content of 44 to 46%, which can be increased by removing the soybean *hulls* before (front-end dehulling) or after (tail-end dehulling) solvent extraction. Front-end dehulling is accomplished by screening the cracked seed and removing the hull fraction by aspiration. Small meat particles are then separated from the hull stream on specific-gravity separators. In the tail-end dehulling system the entire dried-meal stream is passed over specific-gravity separators, producing two grades of meal, one containing 41% protein and the other containing 50% protein.

Solvent extraction is carried out in a continuous countercurrent manner through a series of extraction stages. The extractors most commonly used in this country circulate solvent over the flakes, which are carried, usually in baskets, through the several extraction stages, the baskets moving in a circular, vertical, or horizontal direction. The vertical circular procedure, using hexane as solvent, is illustrated in Fig. 28.4. The basic principles of solvent extraction have been well outlined by Coefield.[7] Although milling releases some of the oil, which is immediately dissolved in the solvent, the greater portion is removed by *diffusion of the solvent* through the cell walls until equilibrium is reached. By replacing the equilibrium solution with a solvent of lower oil content, the diffusion process is again resumed. The economical limit of this procedure is about 0.5% oil remaining in the seed mass. The rate of diffusion is directly proportional to the surface area of the seed particle and in an (inverse) power function of thickness with free circulation of the solvent. After extraction, the meal is toasted to increase its nutritive value. Solvent is usually removed from the miscella phase by passing it through a rising-film evaporator followed by a steam-stripping column. Double-effect, or dual, evaporators are frequently used, with one evaporator operated under vacuum and heated by vapor from the other stage or by vapor from the desolventizing toaster.[8] Solvent losses are usually minimized by venting the process noncondensables through a refrigerated vent condenser or an oil-absorption unit. The process crude oil produced is then stored for refining or sale. Dried and toasted[9] meal from the solvent-extraction operation is ground to 10- to 12-mesh fines in a rotary pulverizer, screened, and stored for sale as feed.

THE OVERALL PROCESSING[10] of crude vegetable oils usually involves alkali refining, water washing and drying, bleaching, hydrogenation, and deodorizing. These steps, applied in a continuous manner, are illustrated in Figs. 28.5 to 28.7. In the *alkali method* free fatty acids are neutralized with caustic soda or soda ash, forming soaps, commonly called *foots*, which are removed by centrifuges, the fatty acids being recovered. The oils are bleached with adsorbent clay either batchwise or in a continuous[11] process. For salad oils, the bleached oil is subjected to a *winterizing* treatment before deodorization, which consists of cooling the oil to low temperatures (about 42°F) and filtering the solid stearin that crystalizes out. For shortening, the bleached oil is hydrogenated and bleached again before deodorization. Deodorization in both cases may be either batchwise or continuous[12] and is accomplished by blowing superheated steam through the oil (while the hydrogenated oil is still hot and in the liquid stage) under a high vacuum of about 28 in. Hg and at 400 to 500°F. Final packaging is done under a nitrogen atmosphere.

[7]Solvent Extraction of Oilseed, *Chem. Eng.* (*N.Y.*), **58**(1), 127 (1951); Perry, p. 19–42.

[8]Wingard, Theoretical Minimum Steam Consumption and How to Get There, technical paper presented at National Soybean Processors Plant Operations Symposium, May 4, 1962; Milligan and Tandy, Distillation and Solvent Recovery, *J. Am. Oil Chem. Soc.*, **51**, 347 (1974).

[9]Sipos and Witte, The Desolventizer-toaster Process for Soybean Oil Meal. *J. Am. Oil Chem. Soc.*, **38**(3), 11 (1961).

[10]Sanders, Processing of Food Fats, *Food Technol.*, **13**(1), 41 (1959); Gillies, Shortenings, Margarines and Food Oils, Noyes, 1974.

[11]ECT, p. 577, 1964.

[12]Deodorizing Edible Oils, *Chem. Eng.*, **53**(9), 134 (1946) (pictured flowcharts of both processes); Shearon *et al.*, Edible Oils, *Ind. Eng. Chem.*, **59**(6), 189 (1950) (flowcharts).

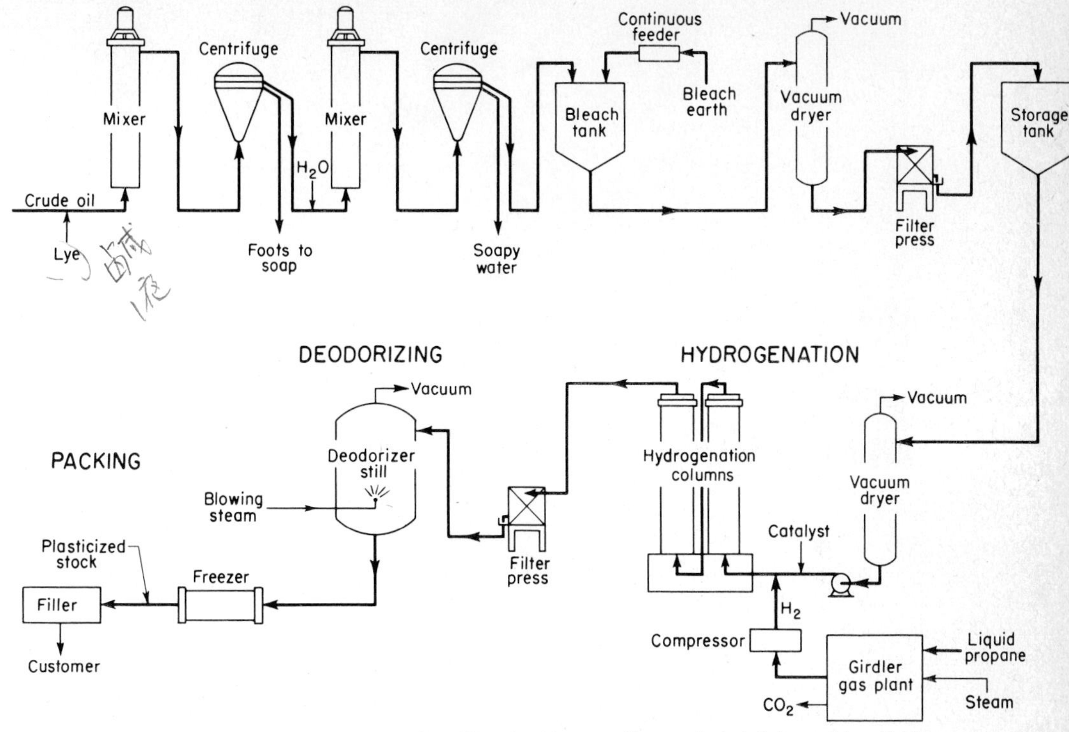

Fig. 28.5 Flowchart for continuous processing for edible oils, either vegetable or animal, including refining, bleaching, hydrogenation, and deodorizing. (*Procter & Gamble Co.*)

Linseed oil. The flaxseed produced in this country is grown largely in the Midwest, south Texas, and the Far West. Production and refining are carried out by a process similar to that used for cottonseed oil, depicted in Fig. 28.2, and Anderson expellers are first employed. The average oil content of the flaxseed is about 40%, which indicates a yield by expression of about 34% based on the weight of the seeds, leaving about 6% oil in the press cake. Newer and improved installations combine screw pressing with solvent extraction, reducing the residual oil in the cake to about 0.75%, as illustrated by Fig. 28.2.

Coconut oil. The raw material for the production of coconut oil[13] is all imported from various tropical countries, a large part coming from the Philippines. The raw material is brought in as copra, which is coconut kernel that has been shelled, cut up, and heat-dried at the point where grown. This treatment not only avoids the cost of shipping excess moisture, but also prevents deterioration of the oil. Coconuts, as they come from the tree, contain from 30 to 40% oil, and the copra contains from 65 to 70% oil. The copra is expressed in expellers or screw presses. A ton of copra yields about 1,250 lb of oil and 720 lb of cake. The oil is refined and contains from 1 to 12% free fatty acid, depending on the quality of the copra meats. Only oil of low free fatty acid content is employed for edible products, the rest (about 60% of the total receipts) being used for the production of soap and alcohols.

Corn oil. The production of corn oil differs from that of some of the other oils in certain respects. After cleaning, the corn is placed in large tanks and steeped in warm water containing SO_2, thus loosening the hulls from the kernels. The steeped corn is run through *attrition* mills, which break the germ away from the rest of the kernel. The separation of the germ and the kernel is accom-

[13]Copra Refining, *Chem. Metall. Eng.*, **52**(2), 148 (1944) (pictured flowchart).

Fig. 28.6 Continuous-hydrogenation unit. (*Procter & Gamble Co.*)

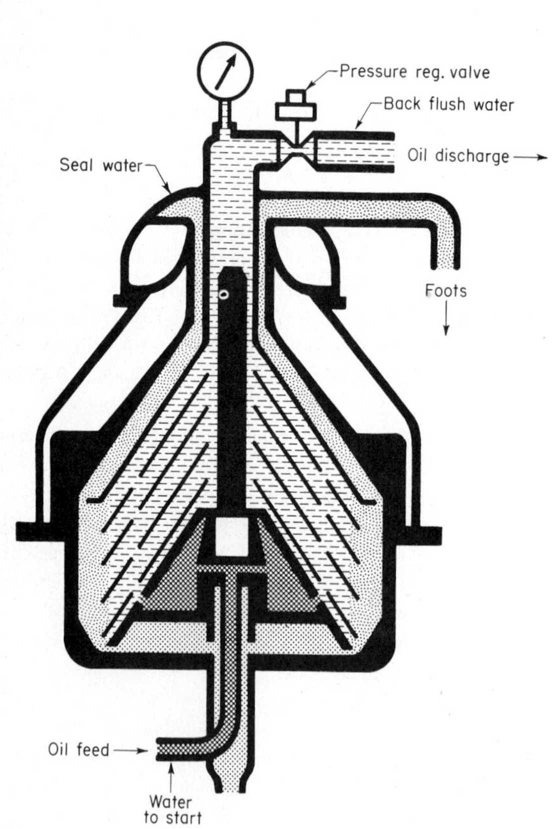

Fig. 28.7 Continuous separating centrifuge as used in the fatty-oil industry. (*Procter & Gamble Co.*)

plished by running the mixture into a tank of water, where the germ floats, because of its oil content, and is skimmed off. Before going into the ordinary grinding and expelling apparatus, the germ should be washed and thoroughly dried. The crude oil from the expellers (Fig. 28.3) is given the usual purification treatment, such as that described for cottonseed oil. The oil content of the corn kernel, exclusive of the hull, is about 4.5%. This oil is used almost exclusively as a salad oil, with lower grades going into soap manufacture.

Palm oil is prepared from the fruit of the palm tree, which has been cultivated on plantations in Indonesia, the Malay Peninsula, and elsewhere. Palms grow naturally on the west coast of Africa. The fruit is 1 to 2 in. long and oval-shaped and weighs about 6 to 8 g on the average. The oil content ranges from 40 to 50% of the kernel, or seed. The oil is obtained in two separate procedures. In the first, it is removed from the fruit, and in the second from the kernels, or seeds. The former is done at the place where the fruit is grown. The procedure consists in cooking the fruit in large steam-pressure digesters equipped with agitators. From the steaming the charge goes to basket centrifuges, where a 10-min treatment, accompanied by blowing with live steam, separates the oil. The residual fiber and kernels are dried in a continuous rotary dryer and separated by a screening operation. The nuts, or kernels, are bagged and shipped to the United States, where they are processed by the methods previously described for oil removal. The fibers are burned under the boilers of the first processing plant. In the United States most of this oil goes into soap manufacture.

Peanut oil is produced by either the hydraulic press or the Anderson expeller, from deskinned peanuts grown in the various southern states. The first expression cold (about 18%) is edible, and some is hydrogenated. The oil is hydrogenated and refined for use in the manufacture of margarine, salad and cooking oils, and some vegetable shortenings. Inedible grades of the oil result from additional hot expressions and are consumed by soapmakers.

Tung or China wood oil is obtained from the fruit of the tung tree, which grows extensively in China. Since 1923, large-scale planting has been carried out in Florida, and tung oil production has become one of the prime industries of that state. Because of the large demand for fast-drying finishes and the difficulties in obtaining this oil from China, various modifications of other oils (Chap. 24) for drying purposes are employed. This oil dries in one-third the time required by linseed oil. The oil is obtained by expression, and the cake, unfit for stock feeding, is used as fertilizer because of its high nitrogen and phosphorus content.

Castor oil. This well-known oil is obtained from the seeds, or beans, of the castor plant found in most tropical regions. The beans contain 35 to 55% oil and are expressed or solvent-extracted.[14] The finest grade of oil is reserved for medicinal purposes. The lower grades are used in the manufacture of transparent soaps, flypaper, and typewriter ink, and as a motor lubricant. Large quantities of the oil are "sulfonated" to produce the familar turkey-red oil long employed in dyeing cotton fabrics, particularly with alizarin. Dehydrated caster oil, a very important drying oil, is described in Chap. 24. This manufactured product compares quite favorably with tung oil, previously imported.

Safflower oil is the fastest growing of the edible oils, largely because of its high percentage (68%) of the polyunsaturated fatty acid, linoleic acid. The market is for special foods containing unsaturated fats, such as margarine, salad, and cooking oils, to which this oil is particularly adapted by reason of its delicate flavor. It also finds use in paints and varnishes. It grows largely in the low-rainfall states of the Great Plains, usually on land diverted from wheat. Oil-free safflower meal is used for cattle food.

ANIMAL FATS AND OILS

Much chemical processing, started early by hydrogenation and now intensified by interesterification and isomerization, has improved the quality of animal fats and oils. Chemical engineering has re-

[14]Editorial Staff Report, Now Castor Oil by Solvent Extraction, *Chem. Ind. (N.Y.)*, **64**, 926 (1949).

duced the cost of processing, particularly by changing from batch to continuous processing. Such treatment has also been applied to vegetable oils.

Neat's-foot oil. The skin, bones, and feet of cattle (exclusive of the hoofs) are cooked or rendered in water for 10 h to separate the fat. This is skimmed off the top of the water and, after filtering through cloth, heated in a kettle to 250°F for several hours. The kettle is cooled, and the contents are settled; the oil is then drawn off, filtered through flannel bags, and sent to the refinery. Here the oil is *grained,* which requires about 2 weeks at 34°F. The product is pressed once, yielding pure neat's-foot oil. The stearin from the first pressing is re-pressed to yield a second grade of oil. The pure variety is used for oiling watches and other fine machinery, and the latter in the textile and leather industries. The stearin from the second pressing is consumed in soap manufacture.

Whale oil. This oil is now obtained from modern floating factory ships which catch, butcher, and process the mammals at the scene of the catch. To prepare the oil, the blubber is stripped from the flesh and boiled in open digesters. The finest grade of oil separates first. It is practically odorless, very pale in color, and contains very little free fatty acid. Upon continued boiling, a second grade is obtained and, if the residue from this operation is cooked under pressure, a third grade is made available. All grades are centrifuged to clarify and dry them further before they are placed in storage. The oils obtained are used in the manufacture of lard substitutes and in soapmaking.[15] Both fish and whale oils contain unsaturated fatty acids of 14 to 22 carbon atoms, and as many as six double bonds.

Cod-liver oil. This oil, whose value was known long before the discovery of vitamins, was originally prepared by storing fish in barrels and allowing them to rot until the oil floated to the top. It is now rendered fresh by live steam cooking of cod and halibut livers (and other parts) until a white scum floats to the top. This usually requires about 30 min, after which the mass is settled for 5 min, decanted, and strained. The oil desired for medicinal purposes is filtered, bleached, and winterized. It is rich in vitamins A and D. Lower grades are employed in leather dressing and for poultry feed.

Shark-liver oil. Recent investigations have shown that the oil obtained from the liver of the shark *Galeorhinus zyopterus* contains more vitamins A and D than cod- or halibut-liver oil. It is rapidly becoming the most important of the fish-liver oils for use in supplying these two vitamins, although the others are still used.[16] However, in 1950, vitamin A was commercially synthesized,[17] and this process is making a sizable dent in the market formerly held by fish-liver oils. This 12-step synthesis process is based on citral, formaldehyde, and acetic acid as raw materials. See Vitamins, Chap. 40.

Fish oils. Fresh menhaden,[18] sardine, and salmon are cooked whole by steaming for a short period and pressed. The oil is settled (or centrifuged) and winterized. The remainder of the fish is dried, pulverized, and sold as meal for feed. Each fish contains, on the average, 20% oil by weight. The oils are consumed in paints, as lubricants, and in leather and soft-soap manufacture, and when sulfonated yield a variety of turkey-red oil.

The tremendous quantities of *butter* and *tallow* produced for food in the United States are diagrammed in Fig. 28.1. Tallow is rendered similarly to lard, and butter fat is obtained by centrifugal separators.

Lard and lard oil are the most important animal fats and are produced by the rendering of hog fat. Lard oil is the most important of the animal oils and is expressed from *white grease,* an inedible lard. Lard has been fundamentally upgraded by modern chemical conversions, interesterification and isomerization (q.v.).

[15]Radcliffe, The Whaling Industry, *Ind. Eng. Chem.,* **25,** 764 (1933).
[16]Grandberg, Automatic Handling Marks Vitamin Oil Extraction, *Chem. Ind.* (*N.Y.*), **65,** 41 (1950).
[17]O'Connor, Synthetic Vitamin A, *Chem. Eng.* (*N.Y.*), **57**(4), 146 (1950).
[18]Stansby, Fish Oils, Avi Publishing, 1967.

PROCESSING OF ANIMAL FATS AND OILS

There is no particular method common to the processing of all animal oils and fats, though *continuous* processing is fundamentally the most important procedure from both the economical and the chemical engineering viewpoint. New processes apply the results of recently developed chemical conversions, such as interesterification and isomerization, as well as the older hydrogenation. Both animal and vegetable oils have processing steps in common, e.g., refining, bleaching, hydrogenating, and deodorizing, as depicted in Figs. 28.5 to 28.7. Finally, both vegetable and animal oils are raw materials for further chemical conversions to produce soaps and detergents (Chap. 29), as well as improved lard and other fats and oils for food.

HYDROGENATION,[19] or hardening as applied to fats and oils, may be defined as the conversion of various unsaturated radicals of fatty glycerides into more highly or completely saturated glycerides by the addition of hydrogen in the presence of a catalyst. This constitutes one of the chief commercial examples of the chemical conversion hydrogenation.[20] Various fats and oils, such as soybean, cottonseed, fish, whale, and peanut, are converted by partial hydrogenation into fats of a composition more suitable for shortenings, margarine, and edible purposes, as well as for soapmaking and numerous other industrial uses. The object of the hydrogenation is not only to raise the melting point, but greatly to improve the keeping qualities, taste, and odor for many oils. It is frequently accompanied by isomerization (q.v.), with a significant increase in melting point, caused, for example, by oleic (cis) isomerizing to elaidic (trans) acid.

Application of the chemical conversion hydrogenation to oils has been more important and far-reaching than any other chemical improvement in this industry.

Reaction and energy changes. As the reaction itself is exothermic, the chief energy requirements are in the production of hydrogen, warming of the oil, pumping, and filtering. The reaction may be generalized:

$$(C_{17}H_{31}COO)_3C_3H_5 + 3H_2 \xrightarrow{Ni} (C_{17}H_{33}COO)_3C_3H_5 \qquad \Delta H = -180.9 \text{ Btu/lb}$$

Manufacture of hydrogenated oils requires hydrogen-generating equipment, catalyst equipment, equipment for refining the oil prior to hydrogenation, a converter for the actual hydrogenation, and equipment for posthydrogenation treatment of the fat. One type of continuous converter is pictured in Fig. 28.6. The *hydrogen* needed may be manufactured by a number of methods,[21] but the hydrocarbon-steam process has mostly replaced the others. Since traces of gaseous sulfur compounds (H_2S, SO_2, etc.) are strong catalyst poisons, as is also, to a lesser degree, carbon monoxide, it is essential that the hydrogen be completely free of these poisons, as well as of taste-producing substances. The amount of hydrogen necessary is a function of the reduction of unsaturation, as measured by the decrease in the iodine number during hydrogenation. The theoretical quantity needed to reduce the iodine number one unit is about 30 ft^3 of hydrogen per ton of oil. Weighted averages at one plant[22] show a consumption of 0.604 ft^3 of hydrogen per pound of oil hydrogenated. The *catalyst* used commercially is nickel. It is generally manufactured by the liquid, or *wet-reduced, process.*[23] One hundred parts of a highly saturated oil or the oil to be hydrogenated and one part of nickel formate are introduced into a specially designed, closed vessel equipped for accurate temperature control. The charge is heated as quickly as possible to as high as 464°F, but 375°F is a more common temperature. At about 300°F the nickel formate begins the reduction:

$$Ni(HCOO)_2 \cdot 2H_2O \longrightarrow Ni + 2CO_2 + H_2 + 2H_2O$$

[19]Albright, Theory and Chemistry for the Hydrogenation of Fatty Oils, *Chem. Eng.* (*N.Y.*), **74**(19), 197 (1967); Commercial Processes for Hydrogenating Fatty Oils, *Chem. Eng.* (*N.Y.*), **74**(21), 249 (1967).

[20]Groggins, Unit Processes in Organic Synthesis, 5th ed., pp. 555, 584, 612, 619, McGraw-Hill, 1958.

[21]See Chap. 7 for details on these various processes.

[22]Shearon *et al.*, Edible Oils, *Ind. Eng. Chem.*, **59**(6), 189 (1950) (flowcharts).

[23]Burke, Catalysts, *Chem. Week*, Nov. 8, 1972, p. 35.

The charge is held at the maximum temperature upward of 1 h and then cooled. During the reduction and cooling period, hydrogen is bubbled through the oil solely to sweep decomposition products from the oil. Upon completion of the reduction, the charge may be pumped directly to the converter or formed into blocks, flakes, or granules for later use.

Hydrogenating the oil has been carried out in batch low-pressure steel vessels but now manufacturers have changed to continuous equipment (Figs. 28.5 and 28.6) which is operated at 50 to 250 psi and 250 to 300°F, with a residence time of 1 to 5 min, using about 0.1% nickel based on the weight of the oil. The degree of hydrogenation is followed and controlled by refractometer readings to indicate the physical properties (saturation and melting point). The catalyst is filtered off and reused. As the hydrogenation is exothermic, the heat in most cases must be removed by an interchanger. Selective, or directed, hydrogenation can also be used, wherein polyunsaturated fatty acids can be largely converted to monounsaturated acids before there is significant conversion of the monounsaturated fatty acids to saturated fatty acids. Also, conditions can be changed to permit hydrogenation from both mono- and polyunsaturated fatty acids simultaneously.

INTERESTERIFICATION OF TRIGLYCERIDES OF LARD An advance in fat chemistry, which has affected the overall amount of crude fats used in shortening manufacture, has been the laboratory perfection and large-scale production of shortenings containing lard processed by random and directed interesterification.[24] For many years the blending of lard into good shortening has been blocked by certain undesirable inherent characteristics of the lard. Lard has a grainy, translucent appearance and a short plastic range, or range of temperature, over which the fat is of good workability—neither too firm nor too soft. *Graininess and short plastic range*, however, require a chemical rearrangement of the lard fat glycerides *for correction* and use in high-grade shortening. Graininess is due to the occurrence of a much greater than random proportion of disaturated triglyceride molecules containing stearic, palmitic, and unsaturated fatty acid chains. About 26% of lard triglycerides have this structure. The predominance of one particular type of triglyceride promotes the deposition of this fraction in large crystals over a short period of time. However, if lard is thoroughly dried and heated (50 to 260°C) for about 30 min with an alkaline catalyst (0.15% sodium methylate), the side chains will migrate among the glyceryl residues and, within minutes under the right conditions, will reach a random distribution. Neutralization of the catalyst stops the reaction. In such a randomizing process, the particular undesirable disaturated triglyceride structure is found to drop from 26 to about $3\frac{1}{2}$%. Thus, by *randomizing*, the graininess of the lard is counteracted. Several commercial lard-based shortenings are produced[25] from such randomized lard.

Extending the plastic range of lard requires further processing. Most of the solids in lard or randomized lard are disaturated glycerides, which give shortening little or no heat resistance, since they melt at higher atmospheric temperatures. Ideally, a process which uses disaturates to produce trisaturates and triunsaturates that do not change their physical state over the ambient temperature range but remain, respectively, solid and liquid, is desired. This has been found in what is termed *directed interesterification.*[26] If, in the presence of an active rearrangement catalyst, the randomized lard mass is chilled, fat solids in transient existence will irreversibly precipitate out. In cooling from the melt, the first solids to separate will be trisaturates. If the temperature is held slightly below the melting point of the trisaturates, they will continue to separate so long as the constantly rearranging melt continues to rearrange saturated fatty acids into trisaturates in a futile attempt to reach equilibrium. Therefore the lard composition changes with time, and trisaturates and triunsaturates increase at the expense of mono- and disaturates. When the catalyst is neutralized, the composition remains frozen in the new form.

[24]Courtesy L. H. Going, Procter & Gamble Co.; Hawley and Holman, Directed Interesterification as a New Processing Tool for Lard, *J. Am. Oil Chem. Soc.*, **33,** 29 (1956); Placek and Holman, *Ind. Eng. Chem.*, **49,** 162 (1957).

[25]Going, U.S. Pat. 2,309,949 (1943); Van der Wal *et al.*, U.S. Pat. 2,571,315 (1951); Holman *et al.*, U.S. Pat. 2,738,278 (1956).

[26]Eckey, U.S. Pats. 2,422,531 and 2,442,532 (1948); Hawley, U.S. Pat. 2,733,251 (1956); Holman *et al.*, U.S. Pats. 2,875,066 and 2,875,067 (1959).

Interesterification reactions:

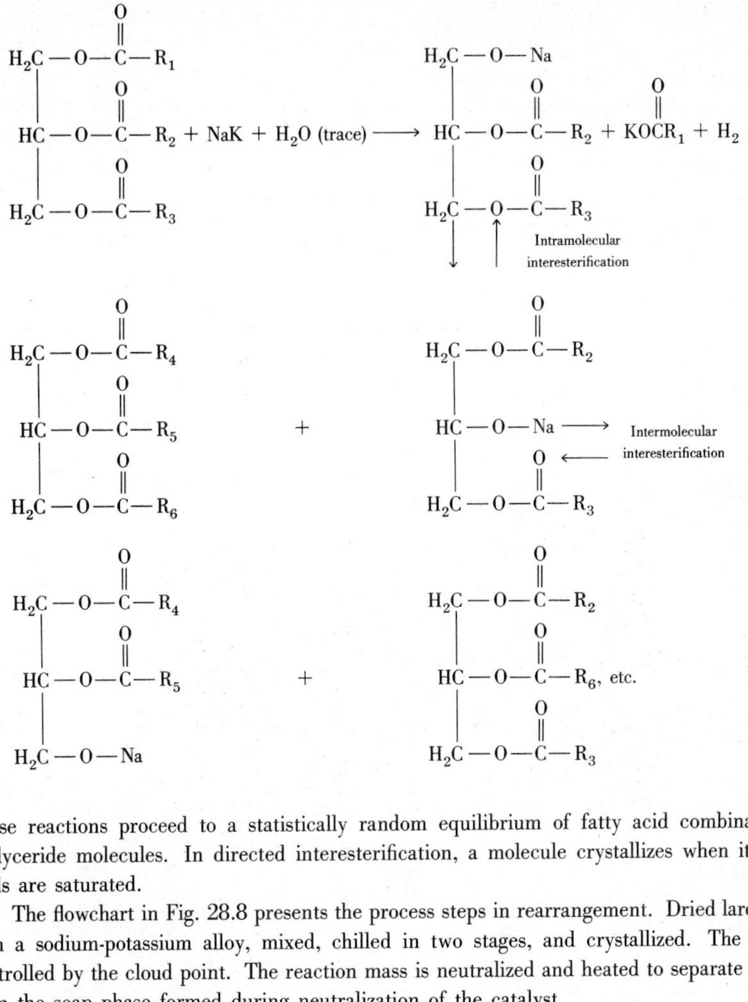

$$\begin{array}{l} H_2C-O-\overset{\overset{O}{\|}}{C}-R_1 \\ \;| \\ HC-O-\overset{\overset{O}{\|}}{C}-R_2 \\ \;| \\ H_2C-O-\overset{\overset{O}{\|}}{C}-R_3 \end{array} + NaK + H_2O\ (\text{trace}) \longrightarrow \begin{array}{l} H_2C-O-Na \\ \;| \\ HC-O-\overset{\overset{O}{\|}}{C}-R_2 \\ \;| \\ H_2C-O-\overset{\overset{O}{\|}}{C}-R_3 \end{array} + K\overset{\overset{O}{\|}}{O}CR_1 + H_2$$

$$\downarrow \uparrow \ \text{Intramolecular interesterification}$$

$$\begin{array}{l} H_2C-O-\overset{\overset{O}{\|}}{C}-R_4 \\ \;| \\ HC-O-\overset{\overset{O}{\|}}{C}-R_5 \\ \;| \\ H_2C-O-\overset{\overset{O}{\|}}{C}-R_6 \end{array} + \begin{array}{l} H_2C-O-\overset{\overset{O}{\|}}{C}-R_2 \\ \;| \\ HC-O-Na \longrightarrow \\ \;| \quad\quad\; O \longleftarrow \\ H_2C-O-\overset{\|}{C}-R_3 \end{array} \ \begin{array}{l}\text{Intermolecular} \\ \text{interesterification}\end{array}$$

$$\begin{array}{l} H_2C-O-\overset{\overset{O}{\|}}{C}-R_4 \\ \;| \\ HC-O-\overset{\overset{O}{\|}}{C}-R_5 \\ \;| \\ H_2C-O-Na \end{array} + \begin{array}{l} H_2C-O-\overset{\overset{O}{\|}}{C}-R_2 \\ \;| \\ HC-O-\overset{\overset{O}{\|}}{C}-R_6, \text{ etc.} \\ \;| \\ H_2C-O-\overset{\overset{O}{\|}}{C}-R_3 \end{array}$$

These reactions proceed to a statistically random equilibrium of fatty acid combinations in the triglyceride molecules. In directed interesterification, a molecule crystallizes when its three fatty acids are saturated.

The flowchart in Fig. 28.8 presents the process steps in rearrangement. Dried lard is catalyzed with a sodium-potassium alloy, mixed, chilled in two stages, and crystallized. The end point is controlled by the cloud point. The reaction mass is neutralized and heated to separate the fat phase from the soap phase formed during neutralization of the catalyst.

ISOMERIZATION REACTIONS In addition to the formation of more saturated compounds during hydrogenation (Figs. 28.5 and 28.6), the reaction may be accompanied by the formation of isomeric unsaturated fatty acids. These may be positional isomers resulting from the migration of double bonds, or geometric (spatial) isomers resulting from the conversion of naturally occurring cis forms to trans forms. These isomers are of considerable interest, chiefly because of their different physical properties. There is also evidence that the rate of hydrogenation of the different isomers varies, so that their presence affects the overall reaction rate and final composition.[27] One explanation of isomerization is based on the half-hydrogenation–dehydrogenation sequence. During hydrogenation a hydrogen atom can add to either end of the double bond and form a free-radical center, probably still attached to the catalyst. This free-radical center is quite unstable, and if a catalyst is only partially covered with hydrogen, a hydrogen atom may be eliminated from a carbon next to the free-radical center, reforming the double bond. In this way either a positional isomer is formed

[27]Allen and Kiess, *J. Am. Oil Chem. Soc.*, **33**, 355 (1956).

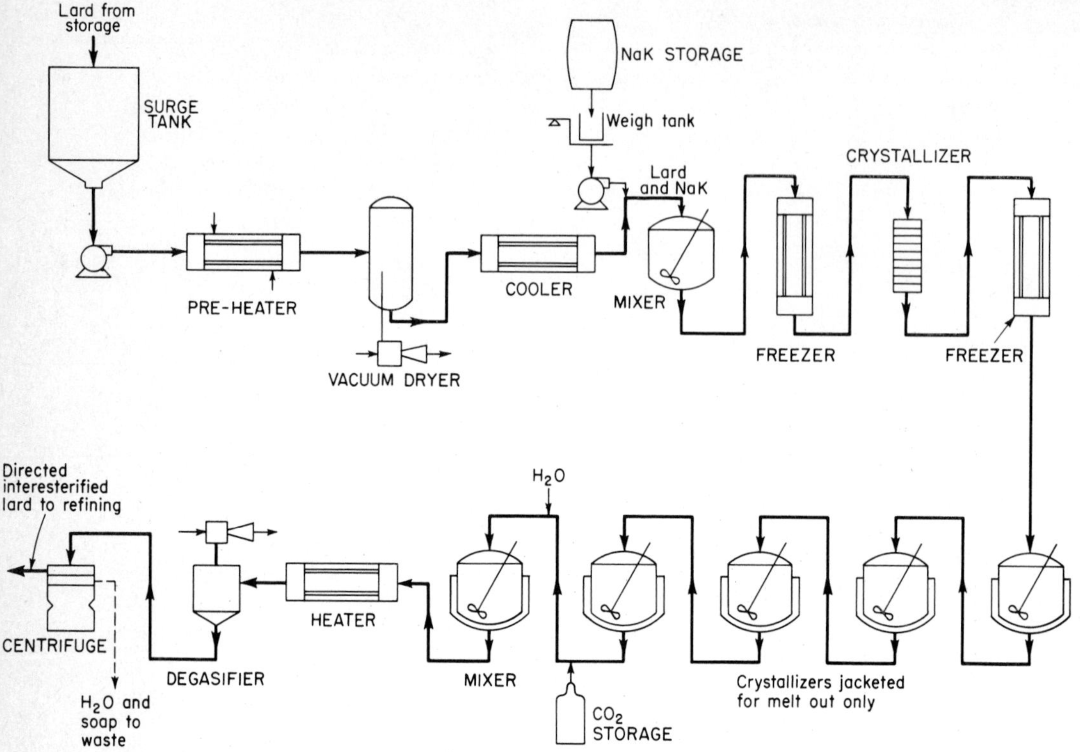

Fig. 28.8 Flowchart for continuous production of lard by directed interesterification. [*Proctor & Gamble Co. and Ind. Eng. Chem.*, **49**(2), 165 (1957).]

or the bond in the initial position is regenerated. Since the formation of a free-radical center allows free rotation, the new unsaturated compound is able to be formed in a cis or trans configuration, according to the thermodynamic stability of each form. Experimental evidence suggests a *trans*-oleic (elaidic)/*cis*-oleic equilibrium ratio of at least 2:1.[28]

To summarize: "The food fat processor can purify to a high degree the natural crude oils. He can change the character of the side chain fatty acids by hydrogenation, and change their relative positions in the triglyceride randomly or controllably by interesterification. He can create solids *in situ,* add them or remove them, and have them assume a stiffening or non-stiffening character. With such flexibility he is providing the public with a variety of palatable and nutritious foods; and if the need arises for fats with special nutritional properties, he has the means to produce them."[29]

WAXES[30]

There are animal, vegetable, mineral, and synthetic waxes, depending upon the source. Animal waxes are secreted as protective coatings by certain insects. Vegetable waxes are found as coatings on leaves, stems, flowers, and seeds. Mineral waxes are paraffin waxes obtained from petroleum, and such waxes as are yielded by coal, peat, and lignite. Mineral waxes from petroleum are not true waxes (esters) but are so classified because of their physical characteristics.

[28]For recent details see James, Homogeneous Hydrogen, Wiley-Interscience, 1975.
[29]Sanders, Processing of Food Fats, *Food Technol.*, **13**(1), 41 (1959).
[30]Tooley, Fats, Oils and Waxes, Clowes, London, 1971.

Beeswax. This is probably the best-known wax. It is made from honeycombs by solvent extraction, expression, or boiling in water. Many church candles contain more than 50% of this wax.

Carnauba wax. This wax is obtained from the carnauba palm, which grows in Brazil. The leaves are cut, dried for 3 days, and sent to the beater house. The drying loosens the wax, which can be easily beaten from the slashed leaf, and it falls to the floor, where it is gathered at the end of the day and melted. Five gallons of powdered wax yields less than 1 qt of molten wax. This is filtered hot through cheesecloth, allowed to harden, and sold. Five palm trees produce about 1 lb of wax per year. The product is used as a constituent of floor, automobile, and furniture polishes, and in carbon paper, candles, and certain molded products.

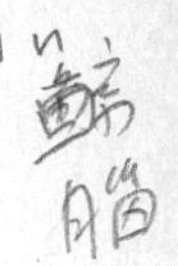

Sperm oil. The oil removed from the head cavity and parts of the blubber of the sperm whale is in reality a wax, because of its chemical composition. Sperm oil is important for lubricating. The head oil, upon chilling and setting, yields spermaceti, a solid wax. This constitutes about 11% of the original oil and is largely cetyl palmitate. It is melted, treated with a hot, dilute caustic soda solution, washed with water, and run into molds to solidify. It is translucent, odorless, and tasteless and is used chiefly as a base for ointments. By methylating sperm oil, a low-surface-tension (27 to 28 dyn) lubricant has been obtained by Archer-Daniels-Midland.

Ozocerite is the name given to certain naturally occurring mineral waxes. The ozocerite known commercially is a particular earth wax mined in eastern Europe, but important similar waxes are mined elsewhere. One variety, mined in Utah, is known as Utahwax or Utah ozocerite. It can be substituted to a great extent for the other, and is chiefly employed for electrical insulations, waterproofing, and impregnating. *Paraffin wax* is concentrated in certain lubricating-oil fractions as a result of distillation and is separated by chilling and filter-pressing (Chap. 37). Extraction of lube oil fractions with, e.g., a mixture of methyl ethyl ketone and benzene, followed by distillation into narrow-boiling fractions yields *microcrystalline waxes.* The name *montan wax* generally applies to the wax obtained from a bituminous wax solvent-extracted from bituminous lignite or shale, but a similar wax may be obtained from peat or brown coal. Its important applications include electrical insulations, polishes, and pastes. *Synthetic waxes,* developed especially by the Germans, have some of the same physical properties but are not waxlike in appearance. Many types are now on the market.[31]

Candelilla, the third most important U.S. wax in tonnage, is made by boiling the stems of a plant common in Mexico and the southwestern United States with acidulated water and skimming off the freed wax. It is an unusual wax in that a major constituent is the hydrocarbon hentriacontane $C_{31}H_{64}$.

SELECTED REFERENCES

Andersen, A. J. C.: In P. N. Williams (ed.), Refining of Oils and Fats for Edible Purposes, 2d ed., Pergamon, 1962.
Bednarcyk, N. E., and W. L. Erickson: Fatty Acids, Noyes, 1973.
Bennett, H. (ed.): Industrial Waxes, Chemical Publishing, 1974.
Devine, J., and P. N. Williams (eds.): Chemistry and Technology of Edible Oils and Fats, Pergamon, 1961.
Gillies, M. T.: Shortenings, Margarines and Food Oils, Noyes, 1974.
Gould, R. F. (ed.): Literature of Chemical Technology, chap. 16, Waxes and Polishes, ACS Monograph, 1967.
Hackett, W. J.: Maintenance Chemical Specialties, Chemical Publishing, 1972.
Hanson, L. P.: Vegetable Protein Processing, Noyes, 1974.
Holman, R. T., *et al.*: Progress in the Chemistry of Fats and Other Lipids, Macmillan, 1963.
James, B. R.: Homogeneous Hydrogenation, Wiley-Interscience, 1975.
Kirschenbauer, H. G.: Fats and Oils, 2d ed., Reinhold, 1960.
Levitt, B.: Oils, Detergents, Maintenance Specialties, Chemical Publishing, 1967, 2 vols.
Partridge, J.: Polishing Compositions, Noyes, 1972.
Piskur, M. M.: Review of Literature on Fats, Oils and Soap, *J. Am. Oil Chem. Soc.* (annually).
Swern, W. (ed.): Bailey's Industrial Oil and Fat Products, 3d ed., Wiley-Interscience, 1964.
Tooley, P.: Fats, Oils and Waxes, J. Murray, London, 1971.
Warth, A. H.: Chemistry and Technology of Waxes, 2d ed., Reinhold, 1956.

[31]Marsel, Waxes, *Chem. Ind.* (*N.Y.*), **67**, 563 (1950) (excellent table on synthetic and blended natural waxes).

chapter 29

SOAP AND DETERGENTS

The washing industry, usually known as the soap industry, has roots over 2000 years in the past, a soap factory having been found in the Pompeii excavations. However, among the many chemical process industries, none has experienced such a fundamental change in chemical raw materials as have the washing industries. Between 1940 and 1965, detergents replaced 80% of the demand for soap. By 1975, well over 80% of the market was supplied by new detergents (Fig. 29.1). These detergents utilize entirely new raw materials and chemical reactions for their manufacture.

HISTORICAL Soap itself was never actually "discovered," but instead gradually evolved from crude mixtures of alkaline and fatty materials. Pliny the Elder described the manufacture of both hard and soft soap in the first century, but it was not until the thirteenth century that soap was produced in sufficient quantities to call it an industry. Up to the early 1800s soap was believed to be a mechanical mixture of fat and alkali; then Chevreul, a French chemist, showed that soap formation was actually a chemical reaction. Domeier completed his research on the recovery of glycerin from saponification mixtures in this period. Until Leblanc's important discovery producing lower-priced sodium carbonate from sodium chloride, the alkali required was obtained by the crude leaching of wood ashes or from the evaporation of naturally occurring alkaline waters, e.g., the Nile River.

The basic process for soapmaking remained practically unchanged for 2,000 years. It involved *batchwise* saponification of oils and fats with an alkali and salting out of the soap. The major changes have been in the pretreatment of the fats and oils, in actual plant procedure, and in the processing of the finished soap, for example, spray drying. Hydrolysis hydrogenation, liquid extraction, and solvent crystallization of various fats and oils provide newer and better raw materials. *Continuous* processes date from 1937, when Procter and Gamble installed a high-pressure-hydrolysis continuous-neutralization process at Quincy, Mass. The next development was a continuous-saponification process jointly developed by Sharples and Lever Brothers and first installed at the latter's plant in Baltimore in 1945. Since then, there have been many installations of both types. These continuous soap processes, although extremely important technological developments, have been partially eclipsed by the introduction of synthetic detergents.

USES AND ECONOMICS Detergent sales in the 1970s should reach $2 billion, and soap sales should stay about constant at $400 to $500 million (Table 29.1).

DETERGENTS[1]

By reason of the much larger proportion of washing markets now supplied by detergents, these much more complex synthetic products are presented before soaps (Fig. 29.1). The comparison shown in

[1]Scientifically, the term *detergent* covers both soap and synthetic detergents, or "syndets," but widespread usage applies it to synthetic cleaning compounds as distinguished from soap. It is so used in this book. The U.S. Tariff Commission reports on detergents under the name surface-active agents, under the broader class of synthetic organic chemicals.

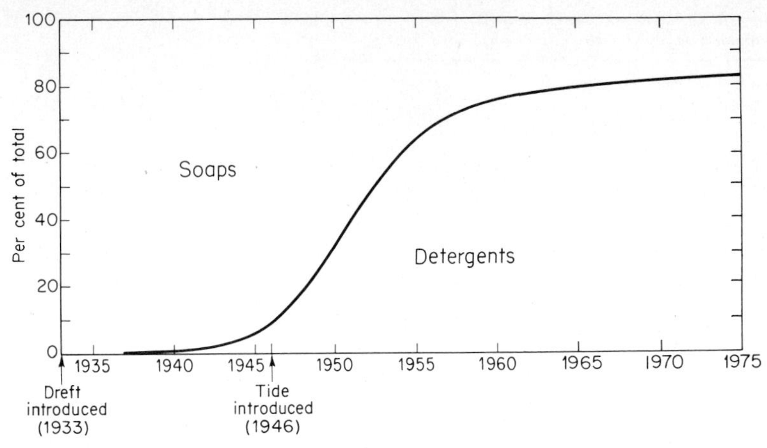

Fig. 29.1 Relative production of detergents and soap.

Table 29.2 indicates the significant *differences* in the processes used to manufacture detergents and soaps and also the *differences* in chemical composition, resulting in *differences* in action. Soaps give precipitates, hence are not efficient in hard or acid waters, whereas detergents are. Furthermore, although ordinary soaps do vary somewhat in chemical composition, they are essentially and simply the sodium or potassium salts of various fatty acids. On the other hand, detergents are very complex mixtures of many substances, all chosen to impart a *particular action in cleaning*. The modern concept of surface-active agents, or surfactants, includes soaps, detergents, emulsifiers, wetting agents, and penetrants. They all owe their activity to a modification of the properties of the surface layer between two phases in contact with another. Most surface-active agents have water-attracting, or hydrophilic, groups on one end of the molecule, water-repelling, or hydrophobic, groups on the other. Detergents excel in possessing these special properties to effect soil removal. There are light- and heavy-duty detergents, and Table 29.3 differentiates between these in composition.

TABLE 29.1 ***Production and Sales of Soaps and Detergents***

	Soap		*Detergents*		*Total*	
	Millions of dollars	*Millions of pounds*	*Millions of dollars*	*Millions of pounds*	*Millions of dollars*	*Millions of pounds*
1940	313	3,206	7	30	320	3,236
1945	527	3,782	35	150	562	3,932
1950	540	2,882	294	1,443	834	4,325
1955	379	1,590	649	2,780	1,028	4,370
1960	376	1,230	953	3,940	1,329	5,170
1965	389	1,110	1,154	4,870	1,543	5,980
1970	427	1,050	1,379	5,650	1,806	6,700
1980	500	1,100	2,000	8,102	2,500	9,202

Sources: Soap and Synthetic Detergents—Domestic Sales, Imports, Exports, and Per Capita Consumption, Chemical Economics Handbook, pp. 590.5040B-C, Stanford Research Institute, 1971–1972. Figures for 1980 are estimates. *Note:* See Fig. 29.1 for a curve of relative production of detergents and soaps. See U.S. Outlook, Dept. of Commerce (annual) for future estimates.

TABLE 29.2 ***General Differences between Soaps and Detergents in Composition and Continuous Manufacture***

Process	Product
To make synthetic detergents (continuous process):	
Tallow + methanolysis ⟶	
Tallow methyl ester + high-pressure hydrogenation ⟶	
Tallow fatty alcohol + sulfation ⟶	Detergents
Sulfated tallow fatty alcohol + NaOH ⟶	
Sodium salt of sulfated tallow fatty + alcohol + builders, etc. ⟶	
To make soap (continuous process):	
Tallow + hydrolysis (splitting fats) ⟶	
Tallow fatty acid + NaOH ⟶	Soaps
Salt of fatty acid + builders, etc. ⟶	

During the 1960s and 1970s, the composition of detergents underwent rapid changes because of environmental considerations.[2] Evidence indicated that phosphates from detergents may contribute to the eutrophication of lakes, so the use of phosphates in detergents was banned in some areas of the country. Many different substitutes were formulated into detergents, but some of these were found to be unsafe and were then banned. The position taken by the detergent industry has been that phosphates in wastewater can be removed by special treatment in sewage plants and, in view of the proved lack of toxicity of phosphates, their replacement may not be the most desirable solution. The soap and detergent industry and its suppliers face an enormous task in testing new materials for all possible effects on the environment, and extensive research will be needed before this complex problem can be solved.

DETERGENT RAW MATERIALS

A large volume of active organic compounds, or surfactants,[3] for both detergents and soap are manufactured in final form by soap and detergent companies. Examples are linear alkylbenzene sulfonate (LAS) and fatty alcohol sulfate, which these companies manufacture in hundreds of millions of pounds. The same is true for fatty acids, the basic materials for soaps. Most of the inorganic materials are purchased, such as oleum, caustic soda, various sodium phosphates, and a large number of additives, the last-mentioned amounting to 3% or less of the total product weight.

SURFACTANTS These embrace "any compound that affects (usually reduces) surface tension when dissolved in water or water solutions, or which similarly affects interfacial tension between two liquids. Soap is such a material, but the term is *most frequently* applied to organic derivatives such as sodium salts of high molecular weight alkyl sulfates or sulfonates."[4] The surfactants of both soap and synthetic detergents perform the primary cleaning and sudsing of the washing action in the same way through the reduction of surface tension. The cleaning process consists of (1) thoroughly wetting the dirt and the surface of the article being washed with the soap or detergent solution, (2) removing the dirt from the surface, and (3) maintaining the dirt in a stable solution or suspension (detergents). In wash water, soaps or detergents increase the wetting ability of the water so that it can more easily penetrate the fabrics and reach the soil. Then soil removal begins. Each molecule of the cleaning solution may be considered a long chain. One end of the chain is hydrophilic (water-loving); the

[2]Moving in Fast on a Market, *Chem. Week*, Jan. 6, 1971, p. 11; Makers Plan New Detergent Formulas, *Chem. Eng. News*, Aug. 17, 1970, p. 18; Nonionics are Set for Cleanup, *Chem. Week*, Aug. 18, 1971, p. 35; BCPAs Get Buildup as Safe Detergent Builders, *Chem. Week*, Jan. 19, 1972, p. 31; Crowding Phosphates Off the Shelf, *Chem. Week*, Oct. 25, 1972, p. 27; Enzymes Invade Laundry Products Market, *Chem. Eng. News*, Feb. 3, 1969, p. 16; Chemistry in the Economy, ACS, 1973, (see chap. 8, p. 196).

[3]Abbreviation for surface-active agents.

[4]Rose, The Condensed Chemical Dictionary, 6th ed., Reinhold, 1961.

TABLE 29.3 ***Basic Composition of Three Types of Dry Phosphate-Based Detergents*** *(Granules)*

Ingredient	Function	Ingredient on dry-solids basis, wt %		
		Light-duty high sudsers	*Heavy-duty controlled sudsers*	*Heavy-duty high sudsers*
Surfactants:				
Organic active, with suds regulators	Removal of oily soil, cleaning	25–40	8–20	20–35
Builders:				
Sodium tripolyphosphate and/or tetrasodium pyrophosphate	Removal of inorganic soil, detergent-building	2–30	30–50	30–50
Sodium sulfate	Filler with building action in soft water	30–70	0–30	10–20
Soda ash	Filler with some building action	0	0–20	0–5
Additives:				
Sodium silicate having 2.0 $\leq SiO_2/Na_2O \leq$ 3.2	Corrosion inhibitor with slight building action	0–4	6–9	4–8
Carboxymethyl cellulose	Antiredeposition of soil	0–0.5	0.5–1.3	0.5–1.3
Fluorescent dye	Optical brightening	0–0.05	0.05–0.1	*ca.* 0.1
Tarnish inhibitors	Prevention of silverware tarnish	0	0–0.02	0–0.02
Perfume and sometimes dye or pigment	Aesthetic, improved product characteristics	0.1*	0.1*	0.1*
Water	Filler and binder	1–5	2–10	3–10

Source: Van Wazer, Phosphorus and Its Compounds, vol. 2, p. 1760, Interscience, 1961. *Note:* See Ecologic Detergents, *Chem. Week*, Apr. 28, 1971, p. 10, for the composition of 16 different detergents and a comparison of phosphate and nonphosphate formulations. See also Phosphates: Use in Detergents, *Hydrocarbon Process.*, **54**(3), 85 (1975). *Approximate.

other is hydrophobic (water-hating, or soil-loving). The soil-loving ends of some of these molecules are attracted to a soil particle and surround it. At the same time the water-loving ends pull the molecules and the soil particles away from the fabric and into the wash water. This is the action which, when combined with the mechanical agitation of the washing machine, enables a soap or detergent to remove soil, suspend it, and keep it from redepositing on clothes.

CLASSIFICATION OF SURFACTANTS In most cases the *hydrophobic* portion is a hydrocarbon containing 8 to 18 carbon atoms in a straight or slightly branched chain. In certain cases, a benzene ring may replace some of the carbon atoms in the chain, for example, $C_{12}H_{25}$—, $C_9H_{19} \cdot C_6H_4$—. The *hydrophilic* functional group may vary widely and may be anionic, e.g., $—OSO_3^{\ominus}$ or $—SO_3^{\ominus}$; cationic, e.g., $—N(CH_3)_3^{\oplus}$ or $C_5H_5N^{\oplus}$—; zwitterionic, e.g., $—N^{\oplus}(CH_3)_2(CH_2)_2COO^{\ominus}$; semipolar, e.g., $—N(CH_3)_2O$; or nonionic, e.g., $—(OCH_2CH_2)_nOH$. In the *anionic* class one finds the most used compounds, namely linear alkylbenzene sulfonates from petroleum, and alkyl sulfates from animal and vegetable fats (Fig. 29.2). Other examples are alkylbenzene–ether sulfonate, fatty alcohol–ethylene oxide sulfate, alkylglycerol–ether sulfonate, alkyl esters of isothionate, and methylalkyl laurates.[5] The last-mentioned are generally employed in newer applications, such as liquids and bars, where special properties are needed and where a premium price will be paid. Soap is also anionic in character. Quaternary trimethylalkylammonium halides, of which cetyltrimethylammonium bromide is an example, are the most common *cationic* surfactants. Dialkyldimethylammonium chloride is a cationic fabric softener. Being generally weak in detergent power, although they have

[5]Kastens and Ayo, *Ind. Eng. Chem.*, **43,** 1626 (1950); U.S. Pat. 1,932,180.

good lubricating, antistatic, and germicidal properties, they are not used as household detergents. Anionics and cationics are not compatible, since they combine to form an insoluble precipitate. For the same reason cationics are not compatible with soap. Alkyl betaines are representatives of the *zwitterionics;* dimethylalkylamine oxides are *semipolar;* and ethylene oxide condensates of fatty alcohol illustrate the molecular structure of truly *nonionic* surfactants, of which there are many excellent soil-removing types and which are chiefly emulsifying agents but low sudsers, hence useful in drum-type automatic clothes and dish washers. In general, the hydrophilic nature of these functional groups decreases from ionic to nonionic. Much of the organic detergent chemical research carried out in recent years has dealt with the synthesis of new surfactant molecules as an extension of the following reactions.

Anionic, giving, in solution, surface-active ions bearing a negative charge:

$$C_{12}H_{25}OH + SO_3 \longrightarrow C_{12}H_{25}OSO_3H \xrightarrow{NaOH} C_{12}H_{25}OSO_3^{\ominus}Na^{\oplus} \qquad \text{(detergent)}$$

$$\text{Fat} + H_2O \longrightarrow C_{17}H_{35}COOH \xrightarrow{NaOH} C_{17}H_{35}COO^{\ominus}Na^{\oplus} \qquad \text{(soap)}$$

Cationic, yielding, in solution, surface-active ions positively charged:

$$C_{12}H_{25}Cl + N(CH_3)_3 \longrightarrow C_{12}H_{25}N(CH_3)_3{}^{\oplus}Cl^{\ominus}$$

Semipolar:

$$C_{12}H_{25}N(CH_3)_2 + H_2O_2 \longrightarrow C_{12}H_{25}-\underset{\underset{CH_3}{|}}{\overset{\overset{CH_3}{|}}{N}}\longrightarrow O + H_2O$$

Zwitterionic:

$$C_{12}H_{25}N(CH_3)_2 + ClCH_2\overset{\overset{O}{\|}}{C}ONa \longrightarrow C_{12}H_{25}N^{\oplus}(CH_3)_2CH_2\overset{\overset{O}{\|}}{C}O^{\ominus} + NaCl$$

$$C_{12}H_{25}N(CH_3)_2 + \underset{O\text{———}C=O}{CH_2-CH_2} \longrightarrow C_{12}H_{25}N^{\oplus}(CH_3)_2CH_2CH_2\overset{\overset{O}{\|}}{C}O^{\ominus}$$

Nonionic:

$$C_{12}H_{25}OH + n\overset{O}{\overbrace{CH_2\text{-}CH_2}} \longrightarrow C_{12}H_{25}(OCH_2CH_2)_nOH$$

SUDS REGULATORS A suds regulator—either a stabilizer or suppressor—is often used with a surfactant. These materials have no common chemical relationship and are often specific for certain surfactants. Examples of stabilizer surfactant systems are lauric ethanolamide–alkylbenzene sulfonate and lauryl alcohol–alkyl sulfate. Suds suppressors are generally hydrophobic materials, some examples of which are long-chain fatty acids, silicones, and hydrophobic nonionic surfactants.

BUILDERS Builders boost detergent power, and complex phosphates, such as sodium tripolyphosphate, have been used most extensively. These are more than water softeners which sequester water-hardening calcium and magnesium ions. They prevent redeposition on fabrics of soil from the wash water. Proper formulation with *complex phosphates* has been the key to good cleaning and a full partner with surfactants in making possible the tremendous development of detergents.

Polyphosphates (e.g., sodium tripolyphosphate and tetrasodium pyrophosphate) have a synergistic action with the surfactant and hence reduce the overall cost, together with an enhanced effectiveness. The rapid rise in the acceptance of detergents stemmed from the building action of the polyphosphates. Surfactants, suds regulators, and detergency builders make up the basic surfactant detergent formulation; however, small amounts of additives are necessary (3% or less).

ADDITIVES *Corrosion inhibitors,* such as sodium silicate, protect metal and washer parts, utensils, and dishes from the action of detergents and water. Carboxymethyl cellulose has been used as an *antiredeposition agent. Tarnish inhibitors* carry on the work of the corrosion inhibitor and extend protection to metals such as German silver. Benzotriazole has been used for this purpose. *Fabric brighteners* are fluorescent dyes which make fabrics look brighter because of their ability to convert ultraviolet light to visible light. Two materials used for such dyes are sodium 4(2*H*-naptho[1,2-*d*]triazol-2-yl)stilbene-2-sulfonate and disodium 4,4′-bis(4-anilino-6-morpholino-*S*-triazin-2-ylamino)-2,2′-stilbene disulfonate.

Bluings improve the whiteness of fabrics by counteracting the natural yellowing tendency. The ingredients used for this purpose can vary from the long-used ultramarine blue to new dye materials. *Antimicrobial* agents are ingredients not now in general use in household products; there are several, including carbanilides, salicylanilides, and cationics. Peroxygen-type *bleaches* can also be employed in laundry products, but their use in this country has been limited. Peroxygen bleach–containing detergents are common in Europe, where hypochlorite bleaches are uncommon, because of the greater effectiveness of the former under high-temperature European washing conditions. *Perfume* has become an accepted ingredient, and compositions and fragrances vary widely. However, the soap and detergent industries are the largest consumers of perfumes among the industries of the United States. *Coloring* is a means of emphasizing certain properties of a product and of obtaining distinctiveness.

MANUFACTURE OF DETERGENTS

The most widely used detergent, a heavy-duty granule, is presented in Fig. 29.2, with the quantities of materials required. The reactions are:

Linear alkylbenzene sulfonation:

1. Main reaction

$$R\text{-}C_6H_5 + H_2SO_4\cdot SO_3 \longrightarrow R\text{-}C_6H_4\text{-}SO_3H + H_2SO_4 \qquad \Delta H = -180 \text{ Btu/lb of alkylbenzene}$$

Alkylbenzene; Oleum; Alkylbenzene sulfonate; Sulfuric acid

2. Secondary reactions

$$R\text{-}C_6H_4\text{-}SO_3H + H_2SO_4\cdot SO_3 \longrightarrow R\text{-}C_6H_3(SO_3H)\text{-}SO_3H + H_2SO_4$$

Alkylbenzene sulfonate; Oleum; Disulfonate; Sulfuric acid

$$R\text{-}C_6H_4\text{-}SO_3H + R^1\text{-}C_6H_5 \longrightarrow R\text{-}C_6H_4\text{-}SO_2\text{-}C_6H_4\text{-}R^1 + H_2O$$

Alkylbenzene sulfonate; Alkylbenzene; Sulfone 1%; Water

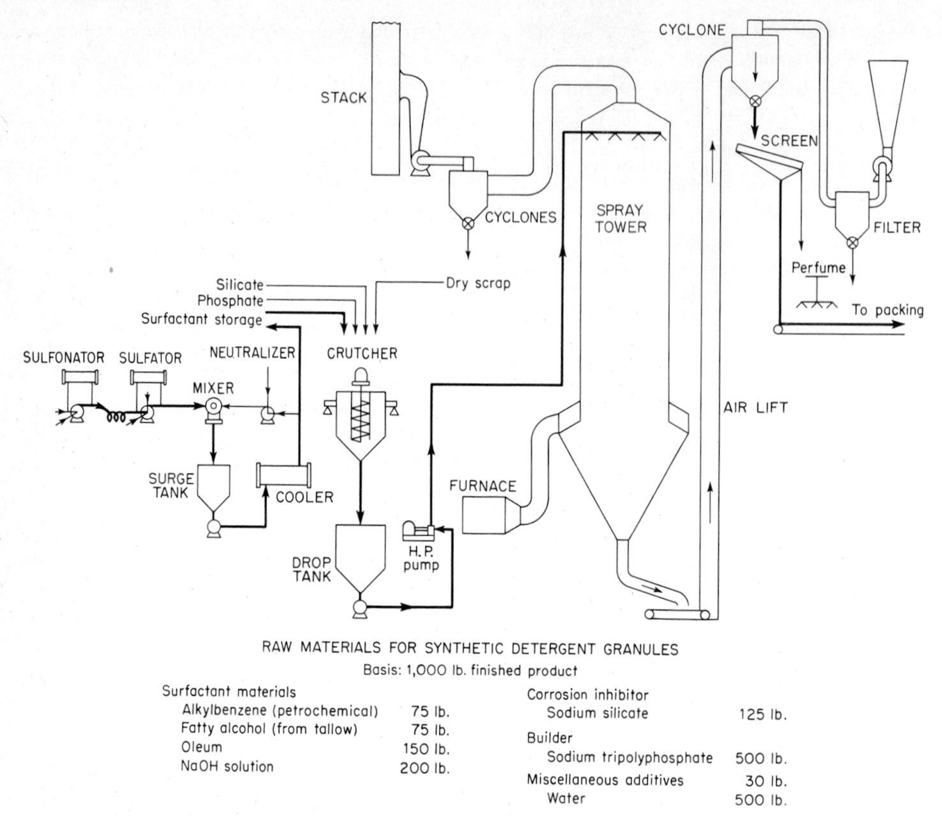

RAW MATERIALS FOR SYNTHETIC DETERGENT GRANULES
Basis: 1,000 lb. finished product

Surfactant materials		Corrosion inhibitor	
Alkylbenzene (petrochemical)	75 lb.	Sodium silicate	125 lb.
Fatty alcohol (from tallow)	75 lb.	Builder	
Oleum	150 lb.	Sodium tripolyphosphate	500 lb.
NaOH solution	200 lb.	Miscellaneous additives	30 lb.
		Water	500 lb.

Fig. 29.2 Simplified continuous flowchart for the production of heavy-duty detergent granules. (*Procter & Gamble Co.*)

Fatty alcohol sulfation:

1. Main reaction

$$R{-}CH_2OH + SO_3 \cdot H_2O \rightleftharpoons R'OSO_3H + H_2O \qquad \Delta H = -140 \text{ to } -150 \text{ Btu/lb fatty alcohol}$$

2. Secondary reactions

$$R{-}CH_2OH + R'{-}CH_2{-}OSO_3H \longrightarrow R{-}CH_2{-}O{-}CH_2{-}R' + H_2SO_4$$

$$R'{-}CH_2{-}CH_2OH + SO_3 \longrightarrow R'{-}CH{=}CH_2 + H_2SO_4$$

$$R{-}CH_2OH + SO_3 \longrightarrow RCHO + H_2O + SO_2$$

$$R{-}CH_2OH + 2SO_3 \longrightarrow RCOOH + H_2O + 2SO_2$$

This presentation is supplemented by Table 29.3, which gives the basic constituents in more detail for the three types of detergent granules. The continuous flowchart in Fig. 29.2[6] can be broken down into the following coordinated sequences:

Sulfonation-sulfation. The alkylbenzene (AB) is introduced continuously[7] into the sulfonator with the requisite amount of oleum, using the dominant bath principle shown in Fig. 29.4 to control the heat of

[6]Fedor *et al.*, Detergents Continuously, *Ind. Eng. Chem.*, **51**(1), 13 (1959) (detailed flowcharts and reference); Gushee and Scheer, Speciality Surfactants, *Ind. Eng. Chem.*, **51**(7), 798 (1959) (excellent presentation).

[7]New Process Tames Sulfur Trioxide Sulfonation, *Chem. Eng.* (*N.Y.*), **69**(5), 70 (1962) (flowchart); Gilbert, Sulfonation, Interscience, 1965.

sulfonation conversion and maintain the temperature at about 130°F (Ch and Op). Into the sulfonated mixture is fed the fatty tallow alcohol and more of the oleum. All are pumped through the *sulfater*, also operating on the dominant bath principle, to maintain the temperature at *120 to 130°F* (Ch and Op), thus manufacturing a mixture of surfactants.

Neutralization. The sulfonated-sulfated product is neutralized with NaOH solution under controlled temperature to maintain fluidity of the surfactant slurry (Ch and Op). The surfactant slurry is conducted to storage (Op).

The surfactant slurry, the sodium tripolyphosphate, and most of the miscellaneous additives are introduced into the *crutcher* (Ch and Op). A considerable amount of the water is removed, and the paste is thickened by the tripolyphosphate hydration reaction:

$$\underset{\text{Sodium tripolyphosphate}}{Na_5P_3O_{10}} + 6H_2O \longrightarrow \underset{\text{Sodium tripolyphosphate hexahydrate}}{Na_5P_3O_{10} \cdot 6H_2O}$$

This mixture is pumped to an upper story, where it is *sprayed* under high pressure into the 80-ft-high spray tower, counter to hot air from the furnace (Op). Here are formed dried granules of acceptable shape and size and suitable density. The dried granules are transferred to an upper story again by an air lift which cools from 240°F and stabilizes the granules (Op). The granules are separated in a cyclone, screened, perfumed, and packed (Op).

The sulfonation conversion is shown in Fig. 29.3 to be extremely fast. The reactions also need to have the high heats of reaction kept under control, as shown in more detail in Fig. 29.4, depicting the circulating heat exchanger, or dominant baths, for both these chemical conversions and for neutralization. The use of oleum in both cases reduces the sodium sulfate in the finished product. However, the oleum increases the importance of control to prevent oversulfonation. In particular, alkylbenzene sulfonation is irreversible and results in about 96% conversion in less than a minute when run at 130°F with 1 to 4% excess SO_3 in the oleum (Fig. 29.3). A certain minimum concentration of SO_3 in the oleum is necessary before the sulfonation reaction will start, which in this case is about 78.5% SO_3 (equivalent to 96% sulfuric acid). As both these reactions are highly exothermic and rapid, efficient heat removal is required to prevent oversulfonation and darkening. Agitation is provided by a centrifugal pump, to which the oleum is admitted. The recirculation ratio (volume of recirculating material divided by the volume of throughput) is at least 20:1 to give a favorable system. To provide the sulfonation time to reach the desired high conversion, more time is allowed by conducting the mixture through a coil (Fig. 29.4), where time is given for the sulfonation reaction to go to completion.

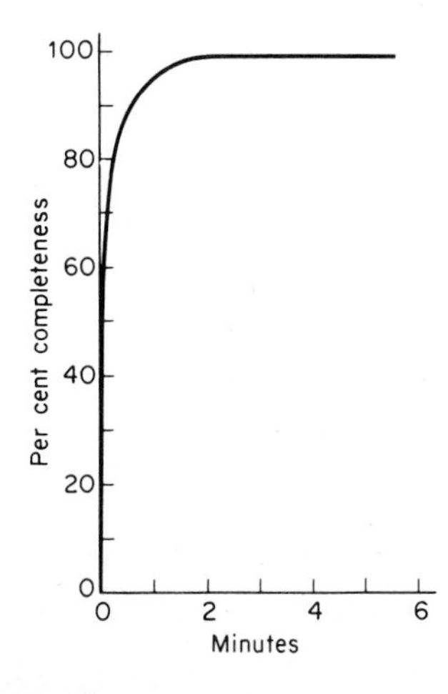

Fig. 29.3 Alkylbenzene sulfonation completeness versus time at 130°F. (*Procter & Gamble Co.*)

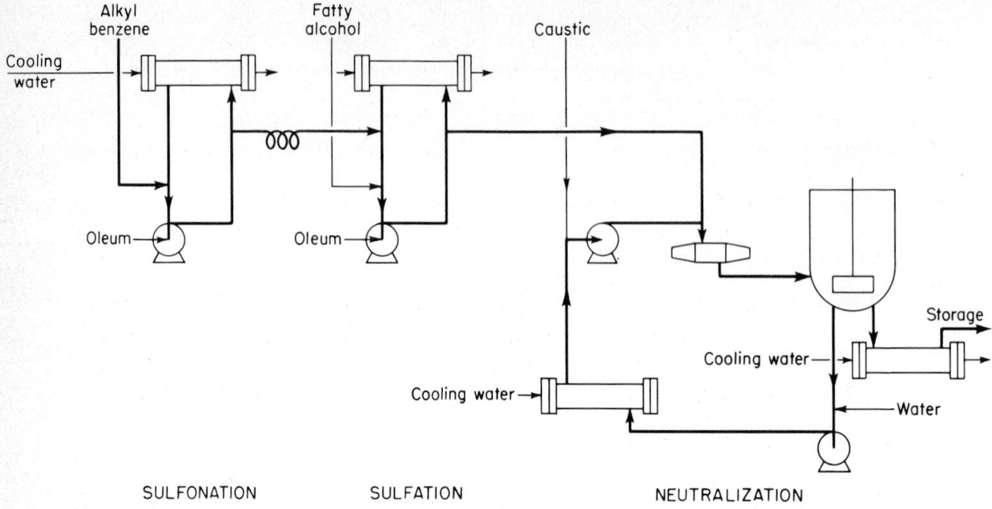

Fig. 29.4 Continuous series sulfonation-sulfation, ending with neutralization, in the circulating heat-exchanging dominant bath to control heat. (*Procter & Gamble Co.*)

Neutralization of the acid slurry releases six to eight times as much heat as the sulfonation reaction. Here a dominant bath (Fig. 29.4) is employed which *quickly* effects the neutralization, since a partly neutralized acid mix is very viscous.

Much more than half of the detergents sold are represented by the heavy-duty granules described, the aryl sulfonate having been until 1965 a branch-chain dodecylbenzene sulfonate. The next section discusses variations to improve biodegradability. Other surfactants (Table 29.4) are being used in increasing amounts, especially for detergents in the form of liquids and bars. These are more rapidly biodegradable than branched-chain benzene sulfonates.

BIODEGRADABILITY OF SURFACTANTS

In view of the attention being given to water pollution control and abatement, product-development chemists and chemical engineers in recent years have realized that surfactants being developed for use in household and industrial detergents which go down the drain to the sewer must be as readily decomposable by the microbial action (1) of sewage treatment, and (2) of surface streams, as ordinary constitutents of household wastes. This new parameter has been added to the performance, efficiency, and cost factors the detergent industry must consider in developing new products. Some surfactants, like tetrapropylene-derived alkylbenzene sulfonate, degrade *slowly*, leaving a persistent residue. Others are more readily decomposable by microorganisms and leave practically no persistent residues. The ease with which a surfactant is decomposed by microbial action has been defined as its *biodegradability*. Tests are being developed and standards are being established for biodegradability. To have broad application, such standards must recognize the breadth of variation in environmental conditions. Materials which may be only partly degraded in inefficient treatment processes can be completely decomposed by more sophisticated biological treatment systems. Tests that measure dieaway in river water, or simulate the biological processes employed in sewage treatment plants, represent only part of the extensive testing used in measuring biodegradability.

Based on research which had been under way for several years, the detergent industry set a Dec. 31, 1965, target date for more readily degradable detergents. Complete conversion was the

TABLE 29.4 ***Synthetic Surface-Active Agents, U.S. Production and Sales, 1972***

Chemical	*Production,* thousands of pounds*	*Unit sales value† per pound*
Grand total	4,038,787	0.20
Benzenoid total	1,119,229	0.20
Nonbenzenoid total	2,919,558	0.20
Amphoteric total	14,411	0.67
Anionic total	2,747,075	0.15
Carboxylic acids (and salts thereof), total	907,011	
Phosphoric and polyphosphoric acid esters (and salts therof), total	26,514	0.45
Sulfonic acids (and salts thereof), total	1,334,612	0.10
Sulfuric acid esters (and salts thereof), total		0.23
Cationic total	229,076	0.40
Amine oxide and oxygen-containing amines (except those having amine linkages), total	50,623	0.43
Amines and amine oxides having amine linkages, total	27,669	0.28
Amines not containing oxygen (and salts thereof), total	67,675	
Quaternary ammonium salts not containing oxygen, total	59,876	0.46
Nonionic total	1,048,225	0.22
Carboxylic acid amides, total	87,717	0.30
Carboxylic acid esters, total	223,597	0.29
Ethers, total	736,911	0.19

Source: Synthetic Organic Chemicals, U.S. Production and Sales of Surface-active Agents, U.S. Tariff Commission, 1974. More details in original reference. *All quantities are given in terms of 100% synthetic organic surface-active ingredient. †Calculated from rounded figures.

overall objective, and the most important step was replacing tetrapropylene benzene sulfonate (TPBS). This detergent material had been the workhorse of the detergent industry. It had earned for itself a 70% share by weight of the market for surfactants used in household dishwashing and laundry products and was consumed in amounts in the neighborhood of 500 million lb annually. TPBS is produced by alkylating benzene with a propylene tetramer, followed by sulfonation of the benzene ring. The propylene tetramer consists of a multitude of branched isomers and has few, if any, straight-chain alkyl groups. Research on a more readily degradable surfactant developed a straight-chain hydrocarbon to make the alkylbenzene. The straight-chain material gives more readily degradable detergents and has fit easily into the detergent formulations where it has replaced TPBS.

STRAIGHT-CHAIN ALKYLBENZENE

With the endeavor to improve the biodegradability[8] of detergents, chemical and engineering efforts made possible supplanting of a half billion pounds of various branched-chain alkylbenzenes by straight-chain alkyl products along the following lines, as illustrated in Fig. 29.5. These compounds are sulfonated after being made into normal, or straight-chain, alkylbenzenes.[9] The straight-chain α-olefins (preferably C_{12}) are made

1. By polymerizing ethylene to α-olefins, using Ziegler-type and other polymerizations, e.g., the Alfol process (Fig. 29.6), furnishing high-purity, but not cheap, products.

[8]Schwartz and Reid, Surface Active Agents, *Ind. Eng. Chem.*, **56**(9), 20 (1964) (theoretical and empirical); Biodegradability of Detergents, *Chem. Eng. News*, Mar. 18, 1963, p. 102; Winton, Detergent Revolution, *Chem. Week*, May 30, 1964, p. 111; Rubinfeld *et al.*, Straight-chain Alkylbenzenes, *Ind. Eng. Chem.*, **56**(11), 97 (1964).

[9]Swisher, LAS, Linear Alkylate Sulfonate: Major Development in Detergents, *Chem. Eng. Prog.*, **60**(12), 41 (1964): "LAS can be completely degraded by bacterial action"; Justice and Lamberti, Revolution in Detergents, *Chem. Eng. Prog.*, **60**(12), 35 (1964) (chemical structure, performance, and biodegradability).

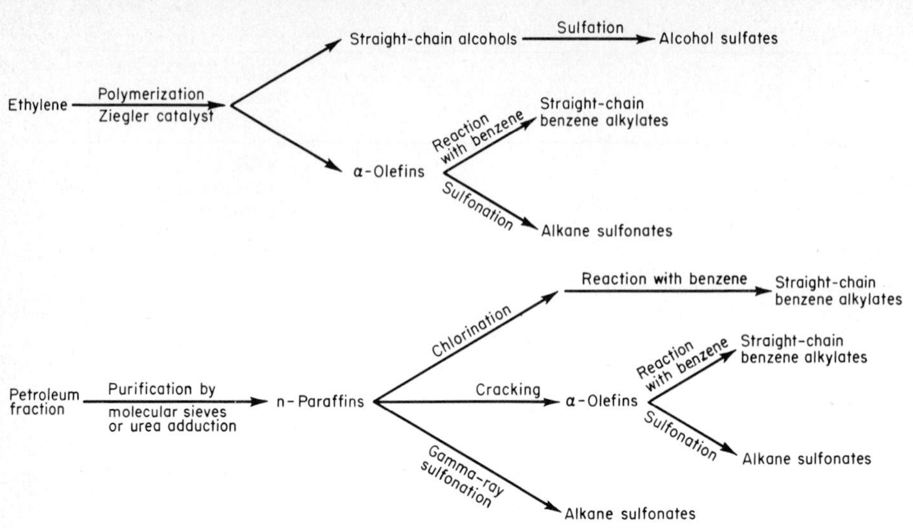

Fig. 29.5 Some possible paths to soft-detergent components. (*Chem. Eng., Aug. 5, 1963.*)

2. By molecular sieve and other means to extract straight-chain paraffins from kerosine (low-priced).[10]

3. By cracking paraffins such as wax (cheap α-olefins).

4. By obtaining fatty alcohols by improved hydrogenation of natural fats, namely, from methyl esters of fatty acids (Fig. 29.7).

α-Olefins, or paraffin halides, can be used to alkylate benzene through the Friedel-Crafts reaction, employing hydrofluoric acid or aluminum fluoride as a catalyst. Fatty alcohols can be sulfated for use in detergents as outlined in Fig. 29.2. An example of a biodegradable linear (C_{12}) alkylbenzene sulfonate is

$$CH_3{-}CH_2{-}CH_2{-}CH_2{-}CH_2{-}CH_2{-}CH_2{-}CH_2{-}CH_2{-}\underset{\displaystyle C_6H_4SO_3Na}{\underset{|}{CH_2}}{-}CH{-}CH_3$$

FATTY ACIDS AND FATTY ALCOHOLS, FOR MANUFACTURE OF DETERGENTS AND SOAPS

ECONOMICS Fatty alcohols and fatty acids are mainly consumed in the manufacture of detergents and soaps, as presented in this chapter. However, fatty acids, both saturated (e.g., stearic acid) and unsaturated (e.g., oleic) have long been employed in many industries as both free acids and, more frequently, as salts. Examples are:

Zinc or magnesium stearates in face powders.

Calcium or aluminum soaps (insoluble) employed as water repellents in waterproofing textiles and walls.

Triethanolamine oleate in dry cleaning and cosmetics.

Rosin soap consumed as a sizing for paper.

[10]Danatos, Unusual Separations, *Chem. Eng.* (*N.Y.*), **71**(25), 155 (1964).

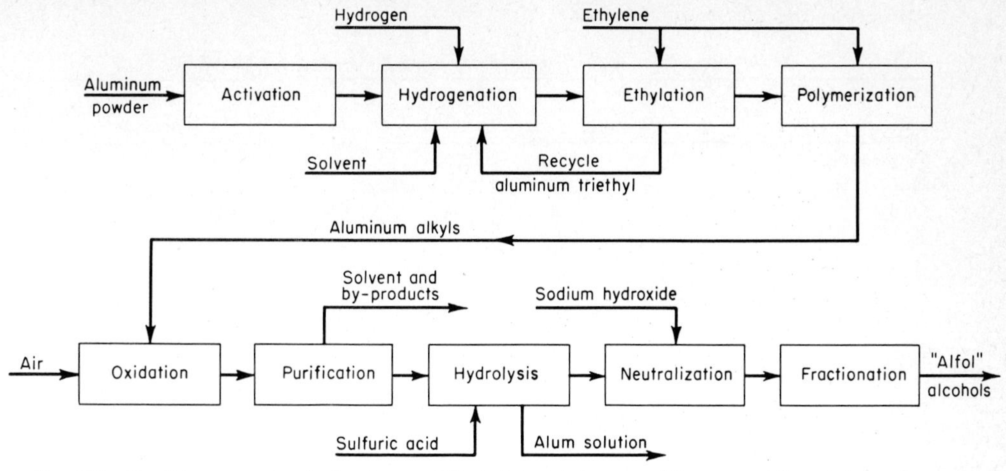

Fig. 29.6 The alfol process—schematic flow diagram. Fatty alcohols made by means of the organometallic route have carbon-chain lengths ranging from 6 to 20. The alfol process used by Conoco commences by reacting aluminum metal, hydrogen, and ethylene, all under high pressure, to produce *aluminum triethyl.* This compound is then polymerized with ethylene to form aluminum alkyls. These are oxidized with air to form aluminum alkoxides. Following purification, the alkoxides are hydrolyzed with 23 to 26% sulfuric acid to produce crude, primary, straight-chain alcohols. These are neutralized with caustic, washed with water, and separated by fractionation. Basic patents covering the process have been licensed. (*Continental Oil Co.*)

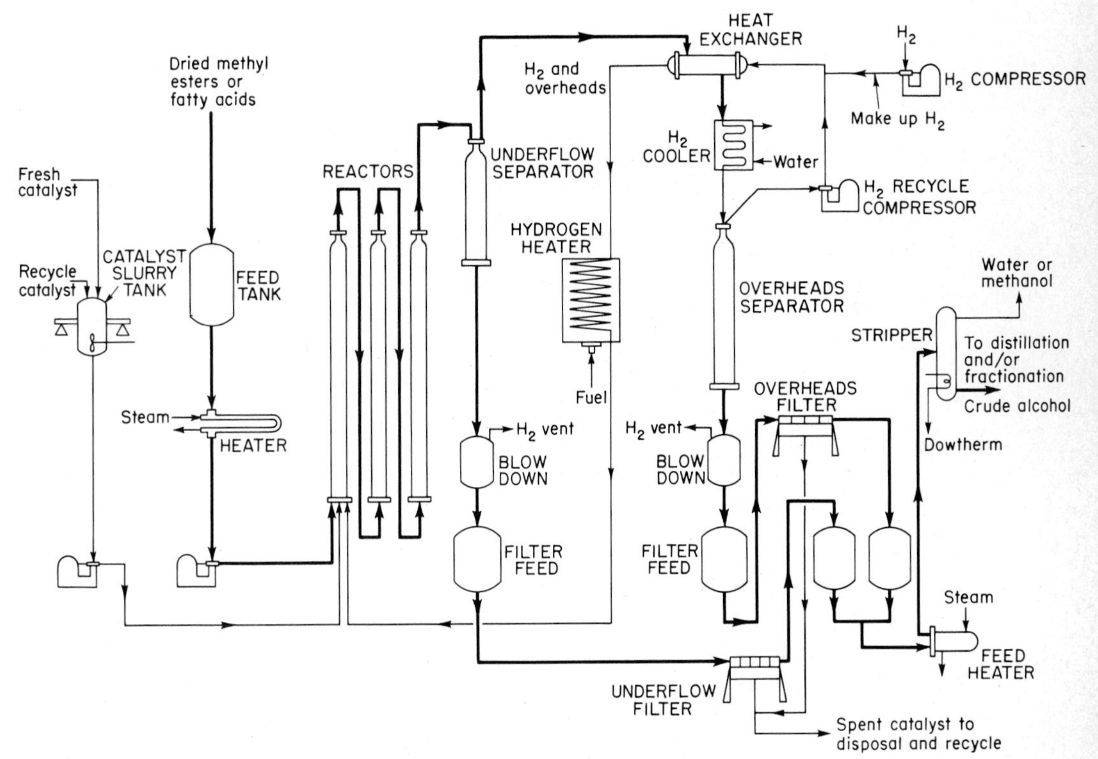

Fig. 29.7 Flowchart for the hydrogenolysis of methyl esters to obtain fatty alcohols and glycerin from natural fats. (*ECT, 2d ed., vol. 1, p. 550, 1963.*)

TABLE 29.5 *Tabular Comparison of the Various Fat-splitting Processes*

	Twitchell	*Batch autoclave*		*Continuous countercurrent**
Temperature, F	212–220	300–350	450	485 approx.
Pressure, psig	0	75–150	425–450	600–700
Catalyst	Alkyl-aryl sulfonic acids or cycloaliphatic sulfonic acids, both used with sulfuric acid 0.75–1.25% of the charge	Zinc, calcium, or magnesium oxides, 1–2%	No catalyst	Optional
Time, hr	12–48	5–10		2–3
Operation	Batch	Batch		Continuous
Equipment	Lead-lined, copper-lined, Monel-lined, or wooden tanks	Copper or stainless-steel autoclave		Type 316 stainless tower
Hydrolyzed	85–98% hydrolyzed 5–15% glycerol solution obtained, depending on number of stages and type of fat	85–98% hydrolyzed 10–15% glycerol, depending on number of stages and type of fat		97–99% 10–25% glycerol, dependent on type of fat
Advantages	Low temperature and pressure; adaptable to small scale; low first cost because of relatively simple and inexpensive equipment	Adaptable to small scale; lower first cost for small scale than continuous process; faster than Twitchell		Small floor space; uniform product quality; high yield of acids; high glycerin concentration; low labor cost; more accurate and automatic control; lower annual costs
Disadvantages	Catalyst handling; long reaction time; fat stocks of poor quality must often be acid-refined to avoid catalyst poisoning; high steam consumption; tendency to form dark-colored acids; need more than one stage for good yield and high glycerin concentration; not adaptable to automatic control; high labor cost	High first cost; catalyst handling; longer reaction time than continuous processes; not so adaptable to automatic control as continuous; high labor cost; need more than one stage for good yield and high glycerin concentration		High first cost; high temperature and pressure; greater operating skill

Source: Mostly from Marsel and Allen, Fatty Acid Processing, *CE*, **54** (6), 104 (1947). Modified in 1977. *See Fig. 29.8.

MANUFACTURE OF FATTY ACIDS[11] Long-consumed basic raw materials such as oils and fats (Chap. 28) have, since about 1955, been very extensively supplemented by improved chemical processing and by synthetic petrochemicals. A selection from these processes is given here. Table 29.5 compares three processes long used for splitting fats. Figure 29.8 illustrates the high-pressure hydrolysis mostly used in the soap industry, which is catalyzed by zinc oxide. Fatty acids are drawn off from the distillate receiver for sale or further conversion to fatty acid salts (calcium, magnesium,

[11]ECT, 2d. ed., vol. 1, p. 542, 1962.

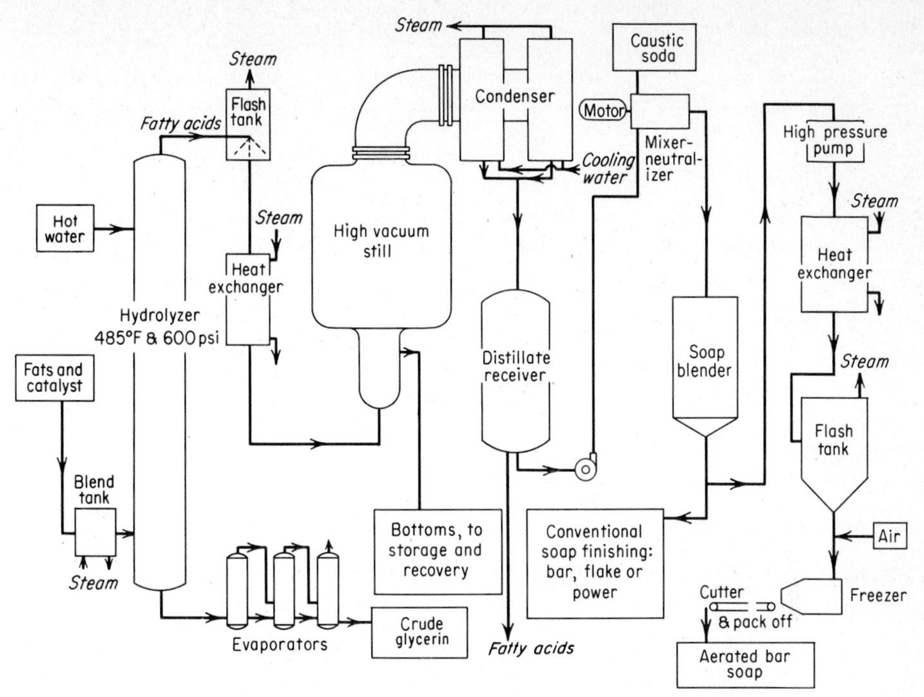

Fig. 29.8 Continuous process for the production of fatty acids and soap. (*Procter & Gamble Co.*)

zinc, etc.). Several older and less used separation methods for purifying fatty acids are panning and pressing, fractional distillation, and solvent crystallization.

MANUFACTURE OF FATTY ALCOHOLS[12] The U.S. production of higher alcohols above C_{10} is about 400 million lb annually, of which more than two-thirds is consumed in the manufacture of surfactants. The *Ziegler* catalytic procedure for converting α-olefins to fatty alcohols and the methyl ester hydrogenation process are outlined below. See also the flowchart in Fig. 29.8 and the text presented under soap for the modern continuous hydrolysis of fats to furnish fatty acids which may be hydrogenated to fatty alcohols.

The Ziegler procedure is an important one for manufacturing C_{12} to C_{18} α-olefins and fatty even-numbered straight-chain alcohols for detergents. To summarize, the Ziegler reactions are

Aluminum triethyl preparation:

$$Al + \tfrac{3}{2}H_2 + 2Al(C_2H_5)_3 \longrightarrow 3Al(C_2H_5)_2H$$

$$3Al(C_2H_5)_2H + 3C_2H_4 \longrightarrow 3Al(C_2H_5)_3$$

Build-up reaction (random or Poisson distribution of chain lengths):

$$\underset{\text{Aluminum triethyl}}{Al(C_2H_5)_3} + \underset{\text{Ethylene}}{(x + y + z)C_2H_4} \xrightarrow[250°F]{2{,}000\ psig} \underset{\text{Aluminum "trialkyl"}}{Al\begin{cases}(C_2H_4)_x + 1H \\ (C_2H_4)_y + 1H \\ (C_2H_4)_z + 1H\end{cases}}$$

ΔH is exothermic, liberating approximately 1000 Btu/lb of ethylene reacted. The "1H" is the terminal alkyl (C_2H_5) of the chain.

[12]ECT, 2d ed., vol. 1, p. 542, 1962 (excellent article with flowcharts).

Oxidation reaction:

$$\mathrm{Al}\begin{cases} R \\ R^1 \\ R^2 \end{cases} + \tfrac{3}{2}O_2 \xrightarrow[90°F]{100\ psig} \mathrm{Al}\begin{cases} OR \\ OR^1 \\ OR^2 \end{cases}$$

Aluminum "trialkyl" — Dried air — Aluminum trialkoxide

ΔH is exothermic, liberating about 1100 Btu/lb of alkyl oxidized. Its conversion is 98% to 90°F in about 2 h.

Acidolysis (hydrolysis)

$$\mathrm{Al}\begin{cases} OR \\ OR^1 \\ OR^2 \end{cases} + \tfrac{3}{2}H_2SO_4 \xrightarrow[190°F]{} \tfrac{1}{2}Al_2(SO_4)_3 + ROH + R^1OH + R^2OH$$

Aluminum trialkoxide — 23–26% sulfuric acid — Aluminum sulfate — Alkyl alcohols

Neutralization of slight excess sulfuric acid

$$H_2SO_4 + 2NaOH \longrightarrow Na_2SO_4 + 2H_2O$$

These reactions furnish a substitute for the C_{12} to C_{14} chain length and for the fatty acids in imported coconut oil or copra. Such reactions are run noncatalytically at 2,000 psig and 250°F in an inert hydrocarbon solvent such as heptane or benzene. In these solvents aluminum "trialkyl" is no longer pyrophoric at less than 40% concentration. Under the above conditions, it takes approximately 140 min to build up to a C_{12} average chain length when reacting 5 mol of ethylene for each $\frac{1}{3}$ mol of aluminum triethyl. The main side reaction is a thermal displacement reaction between aluminum alkyl and ethylene which produces α-olefins. α-Olefins can react with either aluminum triethyl or aluminum alkyls in a build-up-type reaction to form straight-chain alkyls or, to lesser extent, somewhat unstable branched alkyls.

The aluminum alkyl is *oxidized* to aluminum alkoxide by dried air at moderate temperatures of 60 to 120°F and moderate pressures in the presence of a solvent. This oxidation apparently proceeds by the addition of a molecule of oxygen to form a peroxide, which then rearranges to the alkoxide either internally or with another molecule of alkyl. The solvent and any olefins are distilled off from the aluminum alkoxides. *Acidolysis* completes the conversion of the aluminum alkoxide to fatty alcohols. Water gives a bad gel, but dilute sulfuric acid forms the soluble aluminum sulfate, which is readily marketable if kept iron-free. *Neutralization* with dilute caustic prepares the product for the final steps of dehydration, recovery of a small amount of C_2H_5OH and C_4H_9OH, and fractionation first under atomospheric conditions, and then under vacuum, to separate and obtain the fatty alcohols. The Continental Oil Co. has published[13] the *flow diagram* of its 100 million lb/year plant near Lake Charles, La., and this is reproduced in Fig. 29.6. It is based on the Ziegler reactions.

FATTY ALCOHOLS FROM METHYL ESTERS Fats have long been basic raw materials for soaps and detergents. Such fats as are available are glyceryl esters of fatty acids (C_{12} and C_{16}) and have been hydrolyzed to the acids for soaps and reduced to the alcohols by catalytic hydrogenation for detergents. The methyl esters of fatty acids[14] are also hydrogenated to fatty alcohols. These esters are prepared by reacting methanol with coconut or tallow triglyceride, catalyzed by a small amount

[13]Lobo *et al.*, The Alfol Alcohol Process, *Chem. Eng. Prog.*, **58**(5), 85 (1962); Starks *et al.*, Synthesis of Higher Aluminum Alkyls, *Ind. Eng. Chem.*, **55**(12), 89 (1963); Continental Oil Puts First "Alfol" Alcohol Process on Stream, *Chem. Eng.* (*N.Y.*), **70**(24), 14 (1963).

[14]ECT, 2d ed., vol. 1, p. 548, 1962 (details with comparisons).

of sodium. Previously, the refined oil is dried by flashing at 300°F under a vacuum of 25 in., or otherwise it will consume relatively expensive sodium and also form soap. The methyl exchange esterification takes place in about an hour, when the reaction mix is settled and separated into an upper layer rich in ester and methanol and the lower layer rich in glycerin and methanol. The ester layer is washed countercurrently and in a continuous manner to remove excess methanol, to recover glycerin, and to remove the catalyst, which would poison the hydrogenation. Yields of fatty alcohols are 90 to 95%.

Hydrogenation of methyl esters is catalyzed by an Atkins-type catalyst (made from copper nitrate, chromic oxide, and ammonia, with final roasting) and is carried out at approximately 3,000 psig and 500 to 600°F. The continuous equipment used is outlined in Fig. 29.7 and consists of three vertical reactors 40 ft high using 30 mol of heated hydrogen per mol of ester; the hydrogen serves not only for reducing but also for heating and agitation. The crude alcohols are fractionated to the specified chain length.[15]

SOAP

Soap comprises the sodium or potassium salts of various fatty acids, but chiefly of oleic, stearic, palmitic, lauric, and myristic acids. For generations its use has increased until its manufacture has become an industry essential to the comfort and health of civilized man. We may well gage the advance of modern civilization by the per capita consumption of soap and detergents. The relative and overall production of soap and detergents is shown by the curve in Fig. 29.1. *History* and *industrial statistics* are discussed in the first part of this chapter (Table 29.1).

SOAP MANUFACTURE is presented in Fig. 29.8. The long-established kettle[16] process, however, is mainly used by smaller factories or for special and limited production. As soap technology changed, *continuous* alkaline saponification was introduced. Even now computer control[17] allows an automated plant for continuous saponification by NaOH of oils and fats to produce in 2 h the same amount of soap (more than 300 tons/day) made in 2 to 5 days by traditional batch methods.

The present procedure involves *continuous splitting*, or *hydrolysis*, as outlined in Table 29.2 and detailed in Fig. 29.8. After separation of the glycerin, the fatty acids are neutralized to soap.

The basic chemical reaction in the making of soap may be expressed as *saponification;*[18]

$$\underset{\text{Caustic soda}}{3NaOH} + \underset{\text{Glyceryl stearate}}{(C_{17}H_{35}COO)_3C_3H_5} \longrightarrow \underset{\text{Sodium stearate}}{3C_{17}H_{35}COONa} + \underset{\text{Glycerin}}{C_3H_5(OH)_3}$$

The procedure is to split, or *hydrolyze*, the fat, and then, after separation from the valuable glycerin, to *neutralize* the fatty acids with a caustic soda solution:

$$\underset{\text{Glyceryl stearate}}{(C_{17}H_{35}COO)_3C_3H_5} + 3H_2O \longrightarrow \underset{\text{Stearic acid}}{3C_{17}H_{35}COOH} + \underset{\text{Glycerin}}{C_3H_5(OH)_3}$$

$$\underset{\text{Stearic acid}}{C_{17}H_{35}COOH} + \underset{\text{Caustic soda}}{NaOH} \longrightarrow \underset{\text{Sodium stearate}}{C_{17}H_{35}COONa} + H_2O$$

It should be pointed out that the usual fats and oils of commerce are not composed of the glyceride of any one fatty acid, but of a mixture. However, some individual fatty acids of 90% purity or better

[15]By use of computers, the throughput and quality have been improved. Cf. Chow and Bright, *Chem. Eng. Prog.*, **49,** 175 (1953).

[16]Full descriptions with flowcharts for the kettle process full-boiled (several days), semiboiled, and cold are available on pp. 623–625 of CPI 2 and in ECT, vol. 12, pp. 579–585, 1954.

[17]Flowchart for Crosfield Plant with Elliott 405 Computer, *Chem. Eng.* (*N.Y.*), **69**(21), 152 (1962).

[18]Although stearic acid is written in these reactions, oleic, lauric, or other constituent acids of the fats could be substituted. See Table 28.1 for fatty acid composition of various fats and oils.

are available from special processing. Since the solubility and hardness of the sodium salts (Table 29.6) of the various fatty acids differ considerably, the soapmaker chooses the raw material according to the properties desired, with due consideration of the market price.

RAW MATERIALS *Tallow* is the principal fatty material in soapmaking; the quantities used represent about three-fourths of the total oils and fats consumed by the soap industry, as shown in Fig. 28.1. It contains the mixed glycerides obtained from the solid fat of cattle by steam rendering. This solid fat is digested with steam, the tallow forming a layer above the water, so that can easily be removed. Tallow is usually mixed with coconut oil in the soap kettle or hydrolyzer in order to increase the solubility of the soap. *Greases* (about 20%) are the second most important raw material in soapmaking. They are obtained from hogs and smaller domestic animals and are an important source of glycerides of fatty acids. They are refined by steam rendering or by solvent extraction and are seldom used without being blended with other fats. In some cases, they are treated so as to free their fatty acids, which are used in soap instead of the grease itself. *Coconut oil* has long been important. The soap from coconut oil is firm and lathers well. It contains large proportions of the very desirable glycerides of lauric and myristic acids. Free *fatty acids* are utilized in soap, detergent, cosmetic, paint, textile, and many other industries. The acidification of "foots," or stock resulting from alkaline refining of oils, also produces fatty acids. The important general methods of splitting are outlined in Table 29.5. The Twitchell process is the oldest.[19] Continuous countercurrent processes are now most commonly used.

The soapmaker is also a large consumer of chemicals, especially caustic soda, salt, soda ash, and caustic potash, as well as sodium silicate, sodium bicarbonate, and trisodium phosphate. Inorganic chemicals added to the soap are the so-called *builders*. Important work by Harris of Monsanto and his coworkers[20] demonstrated conclusively that, in particular, tetrasodium pyrophosphate and sodium tripolyphosphate were unusually effective synergistic soap builders. Of considerable economic importance was the demonstration that combinations of inexpensive builders, such as soda ash, with the more effective (and expensive) tetrasodium pyrophosphate or sodium tripolyphosphate, were sometimes superior to the phosphate used alone. It was further shown that less soap could be used in these mixtures to attain the same or more effective soil removal.

In *continuous, countercurrent* splitting the fatty oil is deaerated under a vacuum to prevent darkening by oxidation during processing. It is charged at a controlled rate to the bottom of the hydrolyzing tower through a sparge ring, which breaks the fat into droplets. These towers, about 65 ft high and 2 ft in diameter, are built of Type 316 stainless steel (see Fig. 29.8). The oil in the bottom contacting section rises because of its lower density and extracts the small amount of fatty material dissolved in the aqueous glycerin phase. At the same time deaerated, demineralized water is fed to the top contacting section, where it extracts the glycerin dissolved in the fatty phase. After leaving the contacting sections, the two streams enter the reaction zone.[21] Here they are brought to reaction temperature by the direct injection of high-pressure steam, and the final phases of splitting occur. The fatty acids are discharged from the top of the splitter or hydrolyzer to a decanter, where the entrained water is separated or flashed off. The glycerin-water solution is discharged from the bottom of an automatic interface controller to a settling tank. See Fig. 29.10 for glycerin processing.

PURIFICATION OF FATTY ACIDS[22] Although the crude mixtures of fatty acids resulting from any of the above methods may be used as such, usually a separation into more useful components is

[19]This process is described in more detail in CPI 2, p. 619, and in ECT, vol. 12, pp. 579–585, 1954.

[20]*Oil Soap,* **19,** 3 (1942); Cobbs *et al., Oil Soap,* **17,** 4 (1940); Wan Wazer, Phosphorus and Its Compounds, chap. 27, Detergent Building, Interscience, 1958; cf. Sodium Phosphates, Chap. 16.

[21]Allen *et al.,* Continuous Hydrolysis of Fats, *Chem. Eng. Prog.,* **43,** 459 (1947); Fatty Acids, *Chem. Eng.* (*N.Y.*), **57**(11), 118 (1950); Ladyn, Fat Splitting, *Chem. Eng.* (*N.Y.*), **71**(17), 106 (1964) (continuous flowcharts).

[22]See this chapter under Fatty Acids and Fatty Alcohols.

TABLE 29.6 ***Solubilities of Various Pure Soaps***
(In grams per 100 g of water at 25°C)

	Stearate	*Oleate*	*Palmitate*	*Laurate*
Sodium	0.1*	18.1	0.8*	2.75
Potassium	.....	25.0	.....	70.0*
Calcium	0.004†	0.04	0.003	0.004†
Magnesium	0.004	0.024	0.008	0.007
Aluminum	*i*	*i*	*d*	

*Approximate. †Solubility given at 15°C only. *Note: i* indicates that the compound is insoluble; *d* indicates decomposition.

made. The composition of the fatty acids from the splitter depends upon the fat or oil from which they were derived. Those most commonly used for fatty acid production include beef tallow and coconut, palm, cottonseed, and soybean oil. Probably the most used of the older processes is panning and pressing. This fractional crystallization process is limited to those fatty acid mixtures which solidify readily, such as tallow fatty acid. The molten fatty acid is run into pans, chilled, wrapped in burlap bags, and pressed. This expression extracts the liquid *red oil* (mainly oleic acid), leaving the solid stearic acid. The total number of pressings indicates the purity of the product. To separate fatty acids of different chain lengths, distillation[23] is employed, vacuum distillation being the most widely used. Three fractionating towers of the conventional tray type are operated under a vacuum. Preheated, crude fatty acid stock is charged to the top of a stripping tower. While it is flowing downward, the air, moisture, and low-boiling fatty acids are swept out of the top of the tank. The condensate, with part of it redrawn as a reflux, passes into the main fractionating tower, where a high vacuum is maintained at the top. A liquid side stream, also near the top, removes the main cut (low-boiling acids), while overheads and noncondensables are withdrawn. The liquid condensate (high-boiling acids) is pumped to a final flash tower, where the overhead distillate is condensed and represents the second fatty acid fraction. The bottoms are returned to the stripping tower, reworked, and removed as pitch. The fatty acids may be sold as such or converted into many new chemicals.

The *energy* requirements that enter into the cost of producing soap are relatively unimportant in comparison with the cost of raw materials, packaging, and distribution. The energy required to transport some fats and oils to the soap factory is occasionally considerable. The reaction that goes on in the soap reactor is exothermic.

The following are the principal sequences into which the making of bar soap by water splitting and neutralization can be divided, as shown by the flowchart in Fig. 29.8:

Transportation of fats and oils (Op).

Transportation or manufacture of caustic soda (Op or Ch).

Blending of the catalyst, zinc oxide, with melted fats and heating with steam takes place in the blend tank (Op).

Hot melted fats and catalysts are introduced into the bottom of the hydrolyzer (Op).

Splitting of fats takes place countercurrently in the hydrolyzer at 485°F and 600 psi, continuously, the fat globules rising against a descending aqueous phase (Ch and Op).

The aqueous phase, having dissolved the split glycerin (about 12%), falls and is separated (Op).

The glycerin water phase is evaporated and purified (Op). See *glycerin*.

The fatty acids phase at the top of the hydrolyzer is dried by flashing off the water (Op) and further heated.

[23]Fatty Acid Distillation, *Chem. Eng.* (*N.Y.*), **55**(2), 146 (1948) (pictured flowcharts of both straight and fractional distillation of Marsel and Allen, Fatty Acid Processing, *Chem. Eng.* (*N.Y.*), **54**(6), 104 (1947); ECT, vol. 8, p. 828, 1965; Kenyon *et al.*, Chemicals from Fats, *Ind. Eng. Chem.*, **42**, 202 (1950).

In a high-vacuum still the fatty acids are distilled from the bottoms and rectified (Op).

The soap is formed by continuous neutralization with 50% caustic soda in high-speed mixer-neutralizer (Ch).

The neat soap is discharged at 200°F into a slowly agitated blending tank to even out any inequalities of neutralization (Op). At this point the neat soap analyzes: 0.002 to 0.10% NaOH, 0.3 to 0.6% NaCl, and approximately 30% H_2O. This neat soap may be extruded, milled, flaked, or spray-dried, depending upon the product desired. The flowchart in Fig. 29.8 depicts the finishing operations for floating bar soap.

These finishing operations may be detailed: the pressure on the neat soap is raised to 500 psi, and the soap is heated to about 400°F in a high-pressure steam exchanger. This heated soap is released to a flash tank at atmospheric pressure, where a partial drying (to about 20%) takes place because the soap is well above its boiling point at atmospheric pressure. This viscous, pasty soap is mixed with the desired amount of air in a mechanical scraped-wall heat exchanger, where the soap is also cooled by brine circulation in the outer shell from 220 to about 150°F. At this temperature the soap is continuously extruded in strip form and is cut into bar lengths. Further cooling, stamping, and wrapping complete the operation. This entire procedure requires only 6 h, as compared with over a week for the kettle process. The main advantages of soap manufactured by this process as compared with the kettle process are (1) improved soap color from a crude fat without extensive pretreatment, (2) improved glycerin recovery, (3) flexibility in control, and (4) less space and labor. Intimate molecular control is the key to the success of this continuous process, as, for example, in the hydrolyzer, where the desired mutual solubility of the different phases is attained by appropriate process conditions.

TYPICAL SOAPS The main classes of soap are toilet soaps and industrial soaps. These different soaps can frequently be made by one or more of the procedures described. Detergents have replaced over 80% of all soaps. Except for the purest of toilet soaps, various chemicals are added as builders to improve economically the overall cleansing quality. Practically all soap merchandised contains from 10 to about 30% water. If soap were anhydrous, it would be too hard to dissolve easily. Almost all soaps contain perfume, even though it is not apparent, serving merely to disguise the original soapy odor. Toilet soaps are made from selected materials and usually contain only 10 to 15% moisture; they have very little added material, except for perfume and perhaps a fraction of a percent of titanium dioxide as a whitening agent. *Shaving* soaps contain a considerable proportion of potassium soap and an excess of stearic acid, the combination giving a slower-drying lather. "Brushless" shaving creams contain stearic acid and fats with much less soap.

Another type of bar soap (in comparison with the floating type in Fig. 29.8) is *milled toilet soap.* The word milled refers to the fact that, during processing, the soap goes through several sets of heavy rolls, or mills, which mix and knead it. Because of the milling operation, the finished soap lathers better and has a generally improved performance, especially in cool water. The milling operation is also the way in which fragrant perfumes are incorporated into cold soap. If perfume were mixed with warm soap, many of the volatile scents would evaporate. After the milling operation, the soap is pressed into a smooth cylinder and is extruded continuously. It is then cut into bars, stamped, and wrapped as depicted in Fig. 29.9.

Crystal phases in bar soap. The physical properties of bar soap are dependent upon the crystalline soap phases present and the condition of these phases. Any of three or more phases may exist in sodium soaps, depending upon the fat used, the moisture and electrolyte composition of the system, and the *processing* conditions. Milled toilet soaps are mechanically worked to transform the omega phase, at least partially, to the translucent beta phase producing a harder, more readily soluble bar. Extruded floating soaps contain both crystals formed in the freezer and crystals that grow from the melt after it leaves the freezer. Processing conditions are adjusted for an optimum proportion of

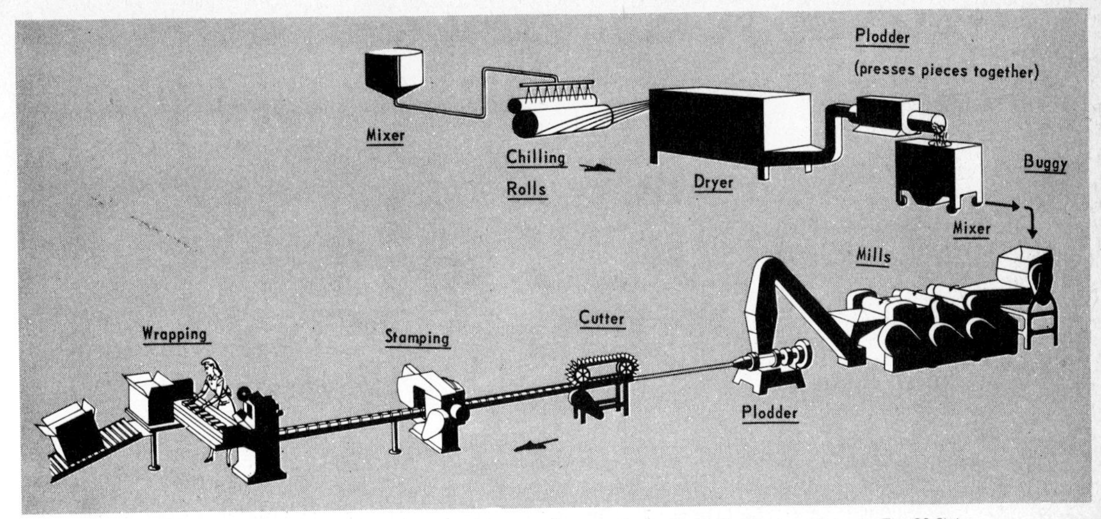

Fig. 29.9 Making soap in milled bars. Another type of bar soap (in comparison with the floating type shown in Fig. 29.8) is milled toilet soap. The word "milled" refers to the fact that, during processing, the soap goes through several sets of heavy rolls, or mills, which mix and knead it. The continuous process, as illustrated by the above isometric drawing, illustrates the application of modern chemical engineering to a process industry as old as the manufacture of soap. A much more uniform product is obtained, and much direct labor is saved. What labor is required is largely for supervision and maintenance. With proper and modern design, there are very few interruptions of continuous operation and, consequently, there is low maintenance.

crystallized matrix, which adds strength and rigidity to the bar. If necessary, the bar may be tempered by reheating to strengthen it.

GLYCERIN

HISTORICAL Glycerin[24] is a clear, nearly colorless liquid having a sweet taste but no odor. Scheele first prepared glycerin in 1779 by heating a mixture of olive oil and litharge. On washing with water, a sweet solution was obtained, giving, on evaporation of the water, a viscous heavy liquid, which the discoverer called "the sweet principle of fats." In 1846 Sobrero produced the explosive nitroglycerin for the first time, and in 1868 Nobel, by absorbing it in kieselguhr, made it safe to handle as dynamite. These discoveries increased the demand for glycerin. This was in part satisfied by the development in 1870 of a method for recovering glycerin and salt from spent soap lyes. Since about 1948, glycerol has been produced from petrochemical raw materials by synthetic processes.

USES AND ECONOMICS The production of crude glycerin is approximately 300 million lb/year. Synthetic glycerin furnishes over 50% of the market. Glycerin is supplied in several grades. USP, or CP, grade is chemically pure, contains not less than 95% glycerol, and is suitable for resins and other industrial products. Yellow distilled is used for certain processes where higher-purity types are not essential, e.g., as a lubricant in tire molds. Glycerin is employed in making, preserving, softening, and moistening a great many products, as shown in Table 29.7.

GLYCERIN MANUFACTURE

Glycerin may be produced by a number of different methods, of which the following are important: (1) the saponification of glycerides (oils and fats) to produce soap, (2) the recovery of glycerin from

[24]The term *glycerin* is chosen for the technical product containing the pure trihydroxy alcohol *glycerol*. The spelling of glycerin is that employed by the USP.

TABLE 29.7 ***Glycerin Consumption*** *(In millions of pounds)*

	1962	1964	1966	1968	1976*
Alkyds	70	75	70	62	60
Tobacco	35	40	45	45	45
Cellophane	50	50	45	38	40
Explosives	15	15	23	23	25
Drugs and toilet goods (including toothpaste)	35	46	50	52	56
Monoglycerides and goods	15	23	30	36	40
Urethane foams		10	11	13	16
Triacetin		9	10	10	10
Miscellaneous	30	21	41	26	40
Total	250	289	325	305	332

Source: Chemical Economics Handbook, Stanford Research Institute, 1972, and industry estimates. *Estimated.

the hydrolysis, or splitting, of fats and oils to produce fatty acids, and (3) the chlorination and hydrolysis of propylene and other reactions from petrochemical hydrocarbons. The new processes (2 and 3) have resulted in lowering production costs for glycerin.

In recovering glycerin from soap plants, the *energy* requirements are mostly concerned with heat consumption involved in the unit operations of evaporation and distillation, as can be seen by the steam requirements on the flowchart in Fig. 29.10. The breakdown of natural and synthetic procedures for glycerin is:

Glycerin from sweet water from hydrolyzer

Evaporation (multiple effect) for concentration (Op)
Purification with settling (Op)
Steam vacuum distillation (Op)
Partial condensation (Op)
Decoloration (bleaching) (Op)
Filtration or ion-exchange purification (Ch)

Glycerin from petroleum

Purification of propylene (Op)
Chlorination to allyl chloride (Ch)
Purification and distillation (Op)
Chlorination with HOCl (Ch)
Hydrolysis to glycerin (Ch)
Distillation (Op)

GLYCERIN RECOVERY AND REFINING Practically all natural glycerin is now produced as a coproduct of the direct *hydrolysis* of triglycerides from natural fats and oils. Hydrolysis is carried out in large continuous reactors at elevated temperatures and pressures with a catalyst. Water flows countercurrent to the fatty acid and extracts glycerol from the fatty phase. The sweet water from the hydrolyzer column contains about 12% glycerol. *Evaporation* of the sweet water from the hydrolyzer is a much easier operation compared with evaporation of spent soap lye glycerin in the kettle process. The high salt content of soap lye glycerin requires frequent shutdown of the evaporators to remove crystallized salt. Hydrolyzer glycerin contains practically no salt and is readily concentrated. The sweet water is fed to a triple-effect evaporator, as depicted by the flowchart in Fig. 29.10, where the *concentration* is increased from 12% to 75 to 80% glycerol. Usually, no additional heat (other than that present in the sweet-water effluent from the hydrolyzer) is required to accomplish the evaporation. After concentration of the sweet water to *hydrolyzer crude,* the crude is settled for 48 h at elevated temperatures to reduce fatty impurities that could interfere with subsequent processing. Settled hydrolyzer crude contains approximately 78% glycerol, 0.2% total fatty acids, and 22% water. The settled crude is *distilled* under a vacuum (60 mm Hg) at approximately 400°F. A small amount of caustic is usually added to the still feed to saponify fatty impurities and reduce the possibility of codistillation with the glycerol. The distilled glycerin is condensed in three stages at decreasing temperatures. The first stage yields the purest glycerin, usually 99% glycerol, meeting CP specifica-

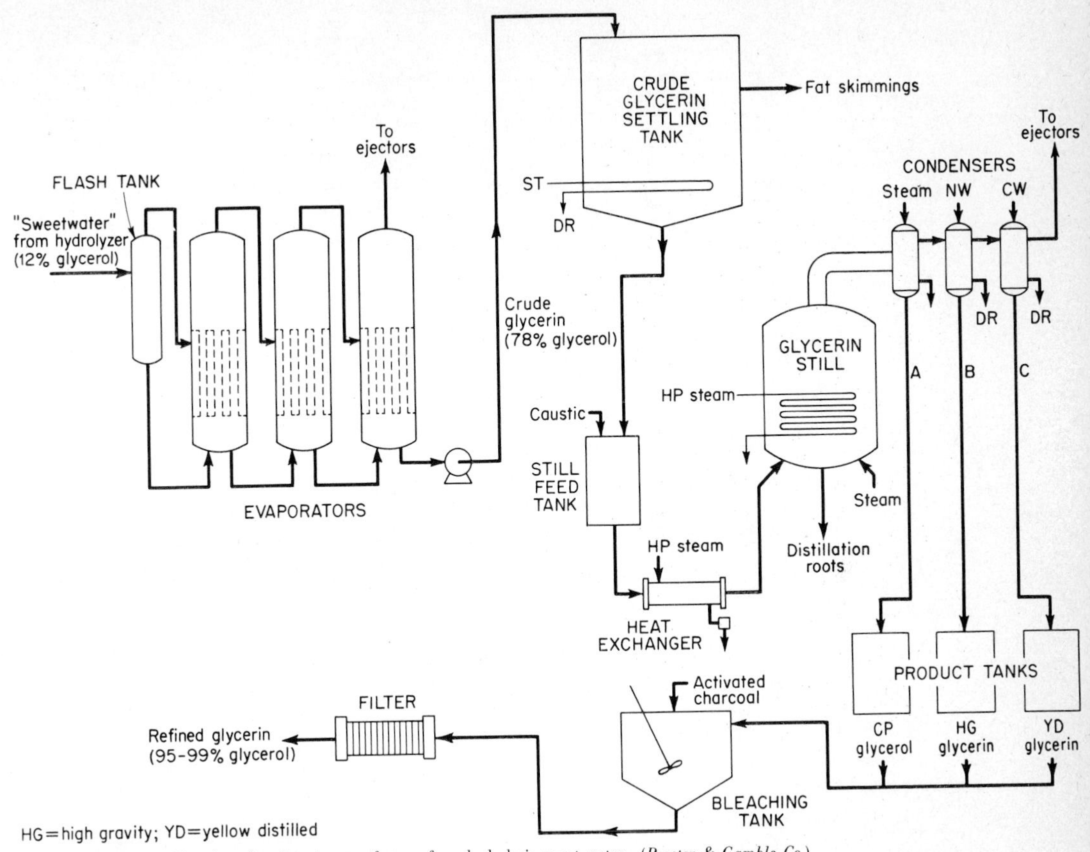

Fig. 29.10 Flowchart for glycerin manufacture from hydrolysis sweet water. (*Procter & Gamble Co.*)

tions. Lower-quality grades of glycerin are collected in the second and third condensers. Final *purification* of glycerin is accomplished by carbon bleaching, followed by filtration or ion exchange.[25]

SYNTHETIC GLYCERIN FROM PETROLEUM The growing market for glycerin, and the fact that it was a coproduct of soap and dependent upon the latter's production, had been the incentives for research into methods for producing this trihydroxy alcohol. The process of making glycerin from propylene from debutanizers of the petroleum industry procured for the Shell Development Co. the 1948 *Chemical Engineering* achievement award.[26] The propylene is chlorinated at 510°C at 1 atm to produce allyl chloride in seconds in amounts greater than 85% of theory (based on the propylene). Vinyl chloride, some disubstituted olefins, and some 1:2- and 1:3-dichloropropanes are also formed. (The reaction producing allyl chloride was new to organic synthesis, involving as it does the chlorination of an olefin by substitution instead of addition.) Treatment of the allyl chloride with hypochlorous acid at 38°C produces the glycerin dichlorohydrin ($CH_2Cl \cdot CHCl \cdot CH_2OH$), which can be hydrolyzed by caustic soda in a 6% Na_2CO_3 solution at 96°C. The glycerin dichlorohydrin can be

[25]Michalson, Ion Exchange, *Chem. Eng.* (*N.Y.*), **70**(6), 163 (1963).

[26]Hightower, Glycerin from Petroleum, *Chem. Eng.* (*N.Y.*), **55**(9), 96 (1948); Synthetic Glycerin, *Chem. Eng.* (*N.Y.*), **55**(10), 100 (1948) (pictured flowchart, pp. 134–137).

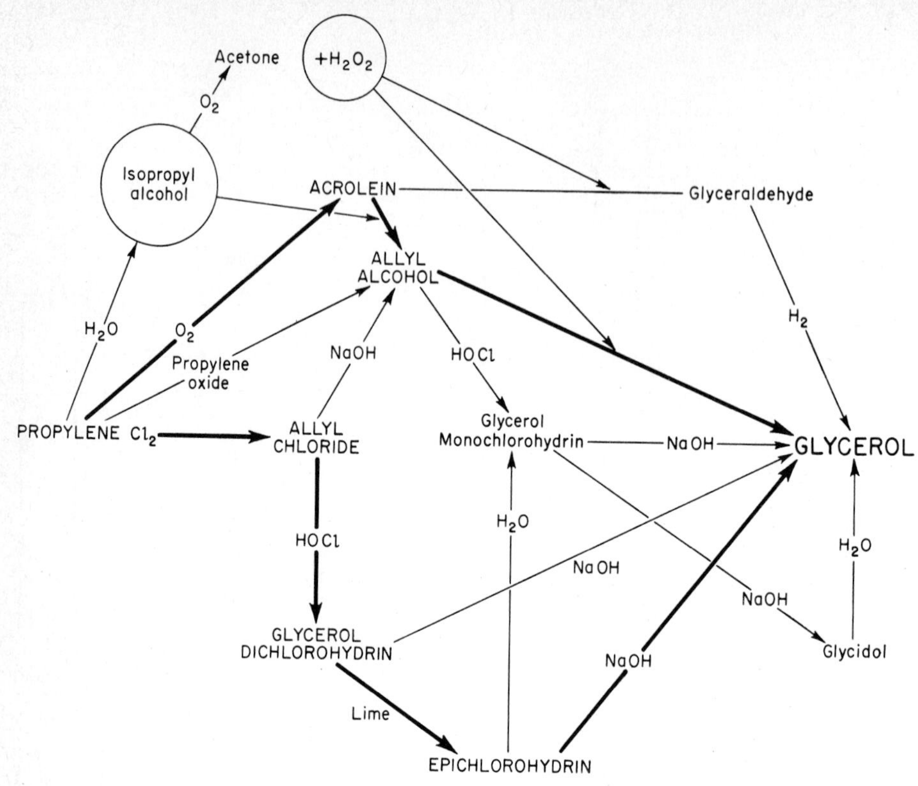

$CH_3 \cdot CH{:}CH_2$

hydrolyzed directly to glycerin, but this takes two molecules of caustic soda; hence a more economical procedure may be to react with the cheaper calcium hydroxide, taking off the epichlorohydrin as an overhead in a stripping column. The epichlorohydrin is easily hydrated to monochlorohydrin and then hydrated to glycerin[27] with caustic soda.[28] The reactions are:

$$CH_3 \cdot CH{:}CH_2 + Cl_2 \longrightarrow CH_2Cl \cdot CH{:}CH_2 + HCl \qquad (85\% \text{ yield})$$

$$CH_2Cl \cdot CH{:}CH_2 + HOCl \longrightarrow CH_2Cl \cdot CHCl \cdot CH_2OH \qquad (95\% \text{ yield})$$

$$CH_2Cl \cdot CHCl \cdot CH_2OH + 2NaOH \longrightarrow CH_2OH \cdot CHOH \cdot CH_2OH + 2NaCl$$

The overall yield of glycerin from allyl chloride is above 90%.

Shell[29] has another process for obtaining glycerin from propylene, involving the following reactions, where isopropyl alcohol and propylene furnish acetone and glycerin (through acrolein) and in good yields:

$$CH_3 \cdot CHOH \cdot CH_3 + \text{air} \xrightarrow{\text{catalyst}} CH_3 \cdot CO \cdot CH_3 + H_2O_2$$

$$CH_3 \cdot CH{:}CH_2 + \text{air} \xrightarrow{\text{catalyst}} CHO \cdot CH{:}CH_2 + H_2O$$

$$CHO \cdot CH{:}CH_2 + H_2O_2 \longrightarrow CHO \cdot CHOH \cdot CH_2OH \xrightarrow{H_2} CH_2OH \cdot CHOH \cdot CH_2OH$$

[27]Kidoo, Petrochemical Processes, *Chem. Eng.* (*N.Y.*), **59**(9), 164 (1952).
[28]Flowchart at Houston Plant of Shell Chemical Corp., *Chem. Eng. Prog.*, **44**(10), 16 (1948).
[29]*Pet. Refiner,* **36**(11), 250 (1957).

The companies that are now synthesizing glycerin are as here listed; Fig. 29.11 outlines the chemical conversions:

Producer	*Feedstock*	*1973 capacity, millions of pounds*	*1980 capacity, millions of pounds*
Shell	Propylene	175	200
Dow	Propylene	120	150
Atlas	Hexoses	25	35
FMC	Propylene oxide	40	60
Olin	Propylene	40	60

Source: Industry estimates.

SELECTED REFERENCES

Bednarcyk, N. W., and W. L. Erickon: Fatty Acids 1973, Noyes, 1973.
Burger, A. (ed.): Medicinal Chemistry: Disinfectants and Antiseptics, 2d ed., Interscience, 1960.
Durhan, K. (ed.): Surface Activity and Detergency, St. Martin's, 1961.
Garner, W.: Detergents, Elsevier, 1966.
Garrett, H. E.: Surface Active Chemicals, Pergamon, 1974.
Gould, R. F. (ed.): Literature of Chemical Technology, chap. 15, Soaps and Detergents, ACS Monograph, 1966.
Howard, G. M.: Perfumes, Cosmetics and Soaps, vol. 3, Modern Cosmetics, Halsted, 1974.
Johnson, J. C.: Antioxidants 1975, Synthesis and Applications, Noyes, 1975.
Johnson, K.: Drycleaning and Degreasing, Noyes, 1973.
Jones, H. R.: Detergents and Pollution Control, Noyes, 1972.
Levitt, B.: Oils, Detergents and Maintenance Specialties, Chemical Publishing, 1975.
Moilliet, J. L., *et al.*: Surface Activity, 2d ed., Van Nostrand, 1961.
Osipow, L. I.: Surface Chemistry: Theory and Industrial Applications, ACS Monograph 153, Reinhold, 1962.
Rubel, T.: Optical Brighteners, Noyes, 1972.
Shick, M. J.: Nonionic Surfactants, Dekker, 1967.
Shinoda, K., *et al.*: Colloidal Surfactants, Academic, 1962.
Sittig, M.: Linear Alpha Olefins, Biodegradable Detergents, Patents, Noyes, 1965.

chapter 30

SUGAR AND STARCH INDUSTRIES

Carbohydrates, sugar and starch, are synthesized by various plants with the sun's energy, using carbon dioxide and water from the atmosphere. They serve as the principal energy foods for humans and enter into countless manufacturing sequences. The present virile chemurgic movement seeks to broaden the application of these and other farm products into an ever-increasing tide of raw materials through the country's factories into manufactured consumer's goods.

SUGAR

Nature has always provided foods from her ample stores in the vegetable kingdom to supply the sweetness humans require in their diet. The average person in the United States eats his or her own weight in sugar about every year and a half. The world average is only one-third of this. This is due not solely to a liking for the sweet taste but to the demands of an active body for fuel, since sugar supplies humans with about 13% of the energy required for existence. As a result, sugar refining is an enormous industry, producing over 6,460,000 tons of sugar per year in the United States and territories and employing over 36,000 workers.

HISTORICAL[1] It is difficult to determine just when sugar first became known to humankind, but it probably traveled from New Guinea to India many centuries before Christ. Methods for extracting and purifying the sugar from the cane were very slow in being developed, but we find a record of crude methods having been brought from the East to Europe about 1400. The sugar trade between Asia and Europe was one of the most important commercial items in the early centuries. Sugar was first extracted in North America in 1689, using cane from the West Indies, and in 1751 cane was grown on the continent. From that time on, the industry increased steadily, both in size and in quality of product. Steam-driven crushing and grinding roller mills were introduced in the latter part of the eighteenth century; the vacuum pan was invented by Howard about 1824; bone-char decolorization was employed for the first time in 1812. Multieffect evaporation was proposed about 1834, and the first suspended centrifuge was developed by Weston in 1852. The use of granular activated carbon and ion-exchange processes to remove color and ash has become common. Evaporation, adsorption, centrifugation, and filtration were from the beginning the important and necessary steps in the manufacturing sequences, and much of our knowledge of these methods came from their application in the sugar industry. This study of the functioning of evaporation, adsorption, centrifugation, and filtration as exemplified in sugarmaking helped to establish the generalized concept of unit operations.

[1]Horne *et al.*, Sugar Chemistry, *Ind. Eng. Chem.*, **43,** 804 (1951); Colbert, Sugar Esters, Noyes, 1974.

In 1747 *beet sugar* was discovered, but it was not introduced into the United States until 1830, and no successful plants were operated until 1870. A great deal of time, effort, and money were expended in bringing the more complicated beet-sugar industry to the point where it could compete with the cane sugar industry. The various tariffs imposed on raw cane-sugar imports, among other things, are responsible for the continuance of both the beet-sugar and the domestic cane-sugar industries.

The first preparation of *dextrose* in 1811 led to the development of the corn-sugar industry in this country. The first manufacturing began about 1872, the product being liquid glucose. It was not until 1918, however, that appreciable quantities of pure, crystalline dextrose were produced.

At the start of the present century, the Bergius process for the production of sugar by *saccharification,* or hydrolysis, of wood received its first industrial trial. A small plant for the production of alcohol was erected in South Carolina in 1910. As far as the United States is concerned, these acid hydrolysis processes, although chemically feasible, have generally proved economically unsound, because of the abundance and low price of starch and sugar; however, they seem to be finding their place in Europe.

USES AND ECONOMICS In 1976, about 12 million tons of sugar were consumed in the United States, about 70% from cane and the remainder from beet. Exports were about 5,000 tons. About 12% of the sugar consumed annually is sold for household use in the United States; about 20% goes to institutions, and 61% to food processors. Of the latter, 23% of the 61% goes to bottlers, 13% to bakers, and 10% to confectioners, and the rest to miscellaneous users. Much sugar is sold in bulk or as sirups. *Nonfood uses of sugar* are very few and constitute only a small amount of the total output. They include the use of sugar as a sucrose octaacetate, a denaturant in ethyl alcohol; as sucrose diacetate hexaisobutyrate and octabenzoate, plasticizers; as mono- and difatty acid esters for surfactants, allyl sucrose; and as a raw material for the manufacture of glycerol and mannitol. Dextran, the polysaccharide produced from sucrose by certain bacteria, is a very effective plasma volume expander. Administered by intravenous infusion, it relieves shock and prevents loss of body fluids after extensive burns or other wounds. Other uses for sucrose derivatives are under investigation, stimulated by the International Sugar Research Foundation,[2] such as drying-oil esters for surface-coating industries and sugar-derived detergents.

No other crystalline organic product of comparable purity (99.96% anhydrous basis) has been offered on the market for so low a price as sugar and in such large volume. This, however, is merely a reflection of the progress and growth in refining methods due to applied chemical engineering within the industry, which has succeeded in reducing the price from the $4 per pound prevailing during the administration of John Adams to its present level of about 30 cents/lb.

MANUFACTURE OF SUGAR

CANE Sugarcane is a member of the grass family. It has a bamboolike stalk, grows to a height of from 8 to 15 ft, and contains 11 to 15% sucrose by weight. The sources of the raw cane refined in the United States are shown in Table 30.1. The cane is usually planted with cuttings from the mature stalk, which sprout and produce a number of new stalks. As many as seven successive crops, or *ratoons,* may be obtained from a single Cuban planting, provided conditions are favorable. The approximate period of growth of the cane in Cuba is 12 to 15 months, and nearly twice this in Hawaii and Peru. In Louisiana and Florida the growing season is 6 to 9 months. Harvesting is done by hand with machetes or by mechanical cutters following burning to remove the leaves. The workers cut off the stalks close to the ground and top the cane. The cane is loaded on tractor-pulled cane carriers and hauled to the raw-sugar plants, or *centrales.* There can be no delay in transporting the freshly

[2] Hass and Hickson, Sucrose Chemicals, International Sugar Research Foundation, 1971.

TABLE 30.1 *Raw-sugar Suppliers to United States*
(In thousands of tons)

Area	*1958*	*1966**	*1973**	*1980*†
U.S. beet	2,214	3,025	3,500	4,000
U.S. cane (mainland)	574	1,100	1,650	2,000
Hawaii	630	1,110	1,143	1,200
Puerto Rico	823	1,140	90	100
Philippines	980	1,061	1,462	1,500
Cuba	3,441			
Dominican Republic	88	354	753	800
Peru	88	283	426	500
Mexico	67	362	632	700
Brazil	....	354	634	700
Miscellaneous	6	1,011	1,510	1,500
Total	8,911	9,800	11,800	13,000

Source: International Sugar Research Foundation, 1974. *Quota. †Estimated.

cut cane to the factory, because failure to process it within less than 24 h after cutting causes loss by inversion to glucose and fructose.

The production of raw cane sugar at the factory is illustrated in Fig. 30.1 and may be divided into the following *unit operations* and *chemical conversions:*[3]

The cane is first washed to remove mud and debris (Op).

The cane is chopped and shredded by crushers in preparation for removing the juice (Op).

The juice is extracted by passing the crushed cane through a series of mills, each of which consists of three grooved rolls that exert heavy pressure. Water and weak juices may be added to help macerate the cane and aid in the extraction.[4] About 93% of the juice is extracted from the cane (Op). The spent cane (bagasse) is either burned for fuel or used to manufacture paper, hard-board, or insulating material.

The juice is screened to remove floating impurities and treated with lime to coagulate part of the colloidal matter, precipitate some of the impurities, and change the pH (Op and Ch). Phosphoric acid may be added because juices that do not contain a small amount of phosphates do not clarify well. If acid is added, an excess of lime is used. The mixture is heated with high-pressure steam and settled in large tanks called *clarifiers* or in continuous settlers or thickeners (Op).

To recover the sugar from the settled-out muds, continuous rotary-drum vacuum filters are generally used, or plate-and-frame presses (Op). The cake constitutes 1 to 4% of the weight of cane charged and is used as manure.

The filtrate, a clarified juice of high lime content, contains about 85% water. It is evaporated to approximately 40% water in triple- or quadruple-effect evaporators[5] to a thick, pale-yellow juice (Op).

The resulting thick juice goes to the first of three single-effect vacuum pans, where it is evaporated to a predetermined degree of supersaturation. Sugar-crystal nuclei are added (shock seeding) and, by the addition of thick juice and controlled evaporation, the crystals are grown to the desired size (Op) in these "strike" pans. At this optimum point, the pan is mostly filled with sugar crystals with about 10% water. The mixture of sirup and crystals (massecuite) is dumped into a crystallizer, which is a horizontal agitated tank equipped with cooling coils. Here additional sucrose deposits on the crystals already formed, and crystallization is completed (Op).

[3] See also the pictured flowsheet, Raw Sugar from Cane, *Chem. Metall. Eng.*, **48**(7), 106 (1941); Meade, Cane Sugar Handbook, 9th ed., pp. 40ff., Wiley, 1963.

[4] As with sugar beets, sugar is being extracted more economically from the chopped-up cane by *continuous countercurrent* extraction with hot water id diffusers; *Chem. Eng.* (*N.Y.*), Feb. 3, 1975, p. 32.

[5] For vacuum-evaporation equipment and accessories, heat-transfer data, and condenser water, see Perry, pp. 11-28 to 11-38.

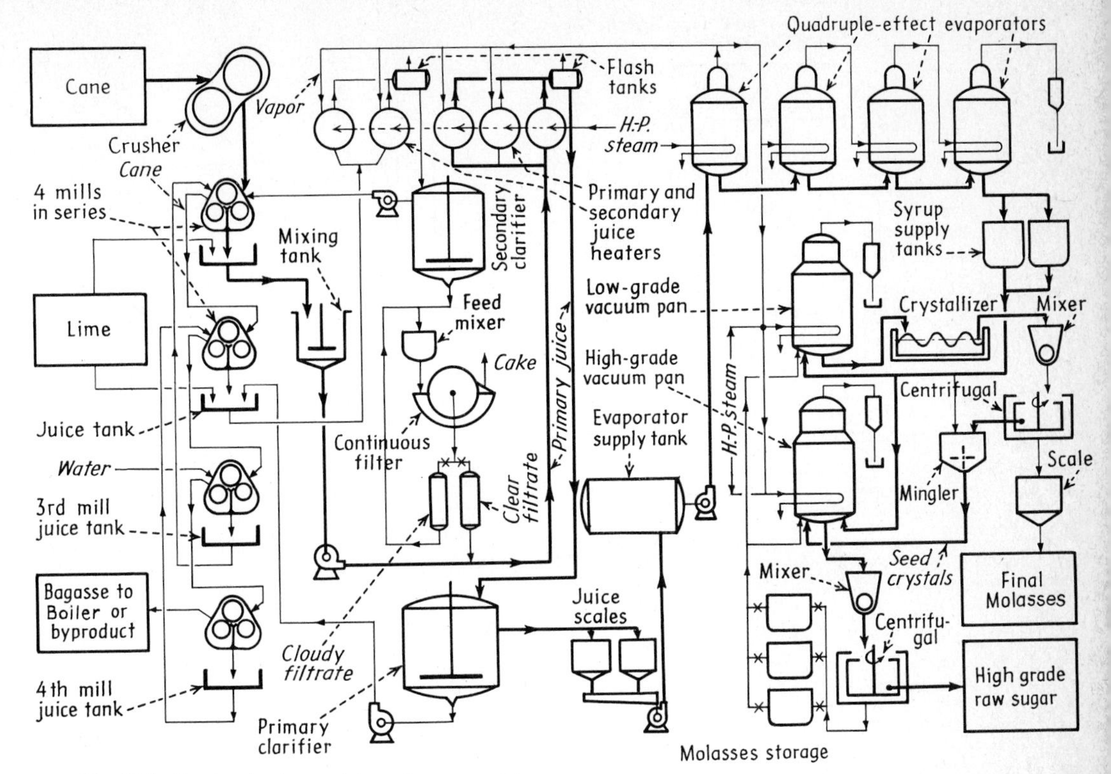

Fig. 30.1 Flowchart for the manufacture of raw cane sugar.

The massecuite is then centrifuged to remove the sirup. The crystals are quality high-grade raw sugar, and the sirup is re-treated to obtain one or two more crops of crystals. The final liquid after reworking is known as *blackstrap* molasses (Op).

The raw sugar (light brown in color), containing approximately 97.8% sucrose, is shipped in bulk to the refinery (Op).

The molasses is shipped to the United States and other countries in full-tank steamers and is used as a source of carbohydrates (a decreasing use) for cattle feed and for citric acid and other fermentations (Chap. 31).

CANE SUGAR REFINING Raw sugar is delivered to the refineries in bulk. Figure 30.2 illustrates the following sequences in the refining of cane sugar:[6]

The first step in refining is called *affination*, wherein the raw-sugar crystals are treated with a heavy sirup (60 to 80° Brix)[7] in order to remove the *film of adhering molasses*. This strong sirup dissolves little or none of the sugar but does soften or dissolve the coating of impurities. This operation is performed in *minglers*, which are heavy scroll conveyors fitted with strong mixing flights (Op).

The resulting sirup is removed by a centrifuge, and the sugar cake is sprayed with water[8] (Op).

[6]Labine, Melt House Redesigned for Automatic Flow, *Chem. Eng.* (*N.Y.*), May 2, 1960, p. 94 (new instrumented flowchart); ECT, vol. 10, p. 210, 1954 (flowchart); Shearon *et al.*, Cane Sugar Refining, *Ind. Eng. Chem.*, **43**, 552 (1951).

[7]The degree Brix is the percentage, by weight, of sucrose in a pure-sugar solution; commercially, it is the approximate percentage of solid matter dissolved in a liquid.

[8]Perry, centrifuges, pp. 19-87 to 19-100.

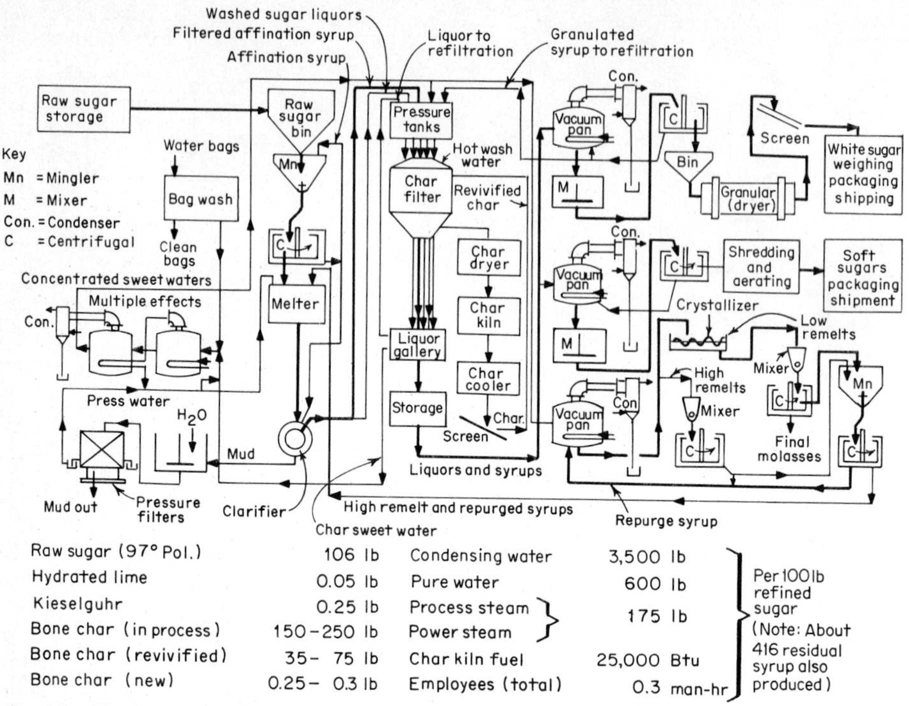

Raw sugar (97° Pol.)	106 lb	Condensing water	3,500 lb	Per 100 lb refined sugar (Note: About 416 residual syrup also produced)
Hydrated lime	0.05 lb	Pure water	600 lb	
Kieselguhr	0.25 lb	Process steam }	175 lb	
Bone char (in process)	150–250 lb	Power steam }		
Bone char (revivified)	35– 75 lb	Char kiln fuel	25,000 Btu	
Bone char (new)	0.25– 0.3 lb	Employees (total)	0.3 man-hr	

Fig. 30.2 Flowchart for refined cane sugar manufacture.

The crystals are dumped into the melter, where they are dissolved in about half their weight of hot water, part of which is sweet water from the filter presses (Op). The sirup from the centrifugals is divided, part being diluted and reused as mingler sirup and the remainder diluted to about 54° Brix and sent either to the char house for clarification and refiltration or to the pans to be boiled with remelt.

The melted and washed raw sugar (in refineries, melted means dissolved) is then treated by a process known as *clarification* or *defecation.* Either mechanical or chemical processes can be used. Mechanical clarification requires the addition of diatomaceous earth or a similar inert material; the pH is then adjusted and the mixture filtered in a press. This system gives an absolutely clear solution of unimproved color and is inherently a batch process.

The chemical system uses either a frothing clarifier or a carbonation system. Treated liquor for frothing, containing entrapped air bubbles, enters the clarifier at about 65°C. In the clarifier it is heated, causing a froth to form which rises to the surface carrying gelatinous tricalcium phosphate and entrapped impurities. The clarified liquor is filtered, if required, or subjected to additional mechanical clarification and sent to decolorization. This process reduces the coloring matter present by 25 to 45%, which greatly reduces the size of the subsequent decolorizers.

The carbonation system adds carbon dioxide from scrubbed flue gas to the melted sugar, which precipitates calcium carbonate. The precipitate carries down with it over 60% of the coloring matter present and is removed by filtration.

DECOLORIZATION-CHAR FILTRATION The clarified effluent liquor, now free of insoluble material, still retains a large amount of dissolved impurities. These impurities are removed by percolation through bone char[9] (Op). (Activated carbon is also used.) The char tanks are about 10 ft in

[9]Lewis, Squires, and Broughton, Industrial Chemistry of Colloidal and Amorphous Materials, Macmillan, 1942.

diameter and 20 ft deep. From 10 to 36 char filters are required per 1 million lb of melt. The percolation is carried out at about 180°F, and the initial product is a clear, water-white sirup. Experience teaches the operator when to shift the char-filter effluent to a lower grade of sirup.

After a certain amount of use, the char loses its decolorizing ability and must be *revivified.* This is done approximately every hour by first washing it free of sugar, removing it, and roasting it at 1000 to 2000°F in retorts (Ch). A continuous decolorizing process is also used.

The sirups from the bone-char filters are piped to the *liquor gallery,* where they are graded according to purity and strength: 99- to 99.7-deg purity, 90- to 93-deg purity, 84- to 87-deg purity, and 75- to 80-deg purity (Op).

Darker-colored liquors are treated with either bone char, synthetic bone char (Synthad), activated carbon, ion-exchange resins, or some combination to form what are known as "soft brown sugars". When the bone char loses its decolorizing power, it can be revivified by treatment in vertical pipe kilns or open-hearth furnaces. Careful heating ensures that the active surface is not oxidized away.

Activated carbon is superior to bone-char as a decolorizer because its adsorption cycle is longer, but it does not remove inorganics. Treatment with activated carbon involves stirring a small quantity of powder into the colored liquid and removing the solid in a filter press. Equipment is smaller and simpler than that needed for bone-char treatment. The carbon may be used on a single-use throwaway basis. Carbon does not have the ability of bone char to adsorb inorganics, but this can be done with ion-exchange resins which are increasingly being used. Some ion-exchange resins also remove color, and all remove both anions and cations, reducing the liquor's tendency to form scale on the equipment.

Figure 30.3 shows a cross section of the crystallizing, or "strike," vacuum pan. Here the sugar sirup is concentrated to a predetermined degree of supersaturation, when it is seeded with a measured amount of fine sugar. These small crystals are grown to a marketable size by a properly regulated rate of boiling or evaporation, as well as of agitation and sirup inflow. The rate should not be too rapid, or new crystals (false grains) will be formed and not have time to grow, with consequent loss through the screen of the centrifuge.

The purest sirups are reserved for liquid sugar (water-white), the next purest for tablets and granulated sugar, and the remainder goes to canners and bottlers, confectioners, and soft sugars (brown).

The pan is discharged into a mixer, which keeps the whole mass from sticking together, and then sent to centrifuges, where the crystals are separated from the sirup, washed, and dropped to the wet-sugar storage bin (Op).

The sirup is returned to the process for further recovery of sugar (Op).

When the purity of the sirup becomes too low, it is used for blended table sirups, the poorer lots going for animal feed. This sirup is commonly called molasses or blackstrap.

The wet sugar is dried in a granulator, which is a horizontal rotating drum about 6 ft in diameter and 25 ft long having a series of narrow shelves (flights) attached to its inner surface. These flights lift the sugar and allow it to fall through the stream of hot air flowing countercurrent to it (Op).

The dried crystals pass over a series of screens, where they are graded according to size (Op).

Various automatic packing and weighing machines put up the sugar in bags and boxes (Op), with a growing percentage in bulk.

Powdered sugars are made by grinding granulated sugar in mills (Op), but are so hygroscopic that mixing with 3% corn starch is practiced to make confectionery grades. Cube and tablet sugars are prepared by mixing certain types of granulated sugar with a heavy white sirup to form a moist mass, which is then molded and dried (Op).

Recent delivery figures are: liquid, 18.5%, bulk granulated, 17.7%, 100-lb bags, 32.2%, and household sizes, 31.6%.

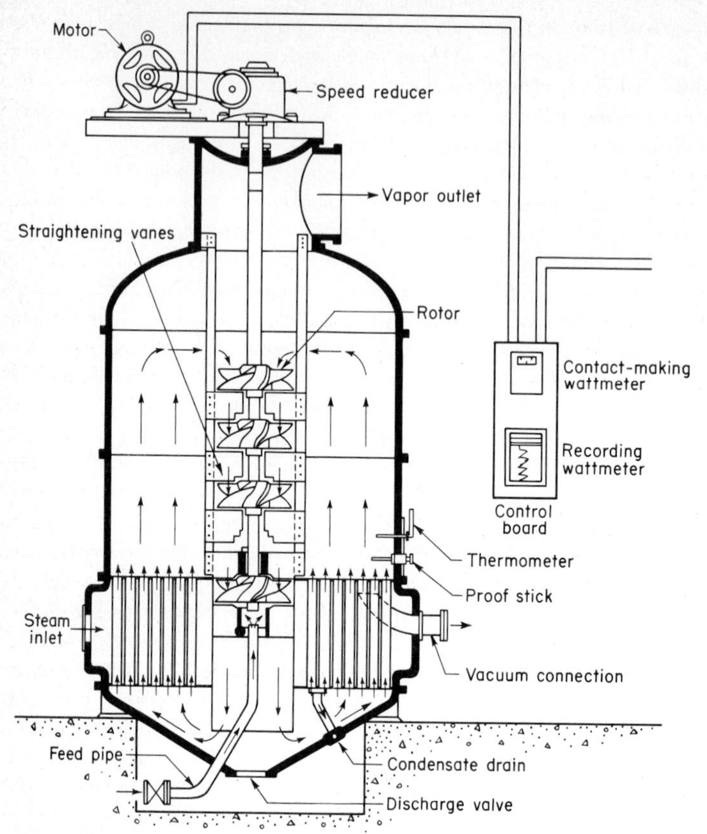

Fig. 30.3 Cross section of vacuum pan equipped with mechanical circulator. The sirup is heated by passage upward through the tubes of the steam chest, or *calandria*, near the bottom. The liquid boils at the upper surface, and vapor is drawn out through the large central pipe. A set of screw impellers forces the liquid back through the central downtake pipe. The stirrer is motor-driven and has recording and controlling wattmeters in its circuit. (*Esterline-Angus Co.*)

The yield of refined sugar obtained, based on raw sugar of 96° polarization, is usually 93 to 94%, sirup 5%, and mechanical and wash losses 0.7%.

In handling sugar, some inversion takes place according to the following reaction:

$$\underset{\text{Sucrose}}{C_{12}H_{22}O_{11}} + H_2O \longrightarrow \underset{d\text{-Glucose}}{C_6H_{12}O_6} + \underset{d\text{-Fructose}}{C_6H_{12}O_6}$$

Polarization $+66.6°$ $\quad +52.8° \quad -92.8°$

The product is called *invert sugar,* but the polarization of the pure sucrose of +66.6° (+ to the right) now reads −20.0° (− to the left) for the resulting mixture.

$$\frac{+52.8° - 92.8°}{2} = -20.0°$$

The continuous adsorption process used by The American Sugar Refining Co.[10] at its Bunker Hill refinery, using *bone char continuously,* is a fundamental improvement in the purification process

[10]U.S. Pat. 2,954,305 (Sept. 27, 1960); Marcy, American Sugar Refining Co., Adsorption by Bone Char Using Continuous Adsorption Process, 20th Annual Meeting of Sugar Industry Technicians, New York, May 1961.

tried and trusted for decades. Here, in decolorizing columns 12 ft in diameter, with 40-ft beds of bone-char adsorbent, the sugar liquor moves *upward* at a uniform rate controlled by the results desired, and the bone char moves *downward*, though at a slower rate than that of the sirup. The particles of bone char are slightly separated, or "expanded," resulting in maximum and uniform adsorption of impurities from the sugar liquor in a truly countercurrent continuous flow; the bone char of the highest adsorptive capacity is automatically in contact with the sugar liquor of the lowest level of impurities. This results in a more efficient use of the bone char, with no sugar liquid blending and no liquor gallery. Also, the color of the product liquor can be adjusted by simply adjusting the feed rate of the bone-char in relation to that of the sugar liquor. High-quality granulated sugar has been produced using "no more than 10 to 15 lbs of bone char per 100 lbs melt"—much less than normally used. The bone char is degraded in only a minor way, if at all. Finally, it is claimed that the capital required for this step has been halved and the labor reduced 90%, together with other savings. The spent char is desweetened, deashed in columns, dewatered to 20% moisture or less, and reactivated in a low-oxygen atmosphere continuously, in a multiple-hearth kiln.

BAGASSE The burning of about 70% of the bagasse produced furnishes enough steam for power heat to run the mill. Hence 30% is available to make an insulating and building board like Celotex[11] or to digest the bagasse with chemicals (NaOH, etc.) to a pulp for paper[12] manufacture on Fourdrinier machines. Furthermore, often much more than the 30% is used for such by-products when other fuel is available at a cheaper price than the $5 per ton traditionally assigned to bagasse. The amount of bagasse usually available is equal to the sugar yield.

The bagasse is conveyed by an endless belt to the rotary digesters, which are 14 ft in diameter, and is cooked under pressure in order to render the fibers pliable, loosen the encrusting material, dissolve organic material, and sterilize the fiber. The resulting pulp, in a 2 to 3% suspension, is pumped to swing-hammer shredders and washed in specially designed rotary washers in order to remove dirt, soluble compounds, and some pith. From the washers, the pulp enters half stock chests, where the sizing, usually papermakers' rosin and alum, is added and the mass stirred with powerful agitators to remove any irregularities. The fibers are refined in conical revolving Jordans, to give optimum fiber size. The refined fiber goes to stock chests, from which it is fed, as an approximately 2% suspension, to the head box of the board machine. In the head box it is diluted to about 0.5%. The board machines, although of special design, are somewhat similar to Fourdrinier machines. The stock is fed onto forming screens and led to drying felts and finally to press rolls. The sheets do not appear laminated, but are felted together to give the required thickness. The board from these machines is 13 ft wide, contains 50 to 55% water, and is produced at the rate of 200 ft/min. It is dried in a continuous sheet, at 300 to 450°F, in a gas- or steam-heated dryer 800 to 1,000 ft long. The product must be sprayed with water as it leaves the dryer in order to bring it up to its normal water content of approximately 8%. The board is then cut and fabricated.

Acoustical and structural wallboard, agricultural mulch and litter, plastic filler, furfural, and plastic reinforcing fibers have all been made from bagasse. A high-quality wax has also been extracted.

BEET SUGAR The familiar sugar sucrose can be obtained from many sources other than cane, such as maple sirup and palm sugar. However, the great commercial sources are sugarcane and sugar beets. Only a skilled chemist can tell whether a sample of refined sugar originated from tropical cane or from sugar beets grown in the temperate zone. Ordinarily speaking, there is no distinction. The sugar beet differs from the ordinary table beet in that it is much larger and is not red. Sugar beets are an important crop in many sections of the world, because of their sugar value. Also, the fertilizer used enriches the soil for other crops. Beet houses begin operation in late September and run until January or February, the farmer harvesting the crop with machines that

[11] Lathrop, The Celotex and Cane-sugar Industries, *Ind. Eng. Chem.*, **22**, 449 (1930).
[12] Cross, Making Paper from Cane Bagasse, *Chem. Eng.* (*N.Y.*), **70**(3), 74 (1962) (process flowchart).

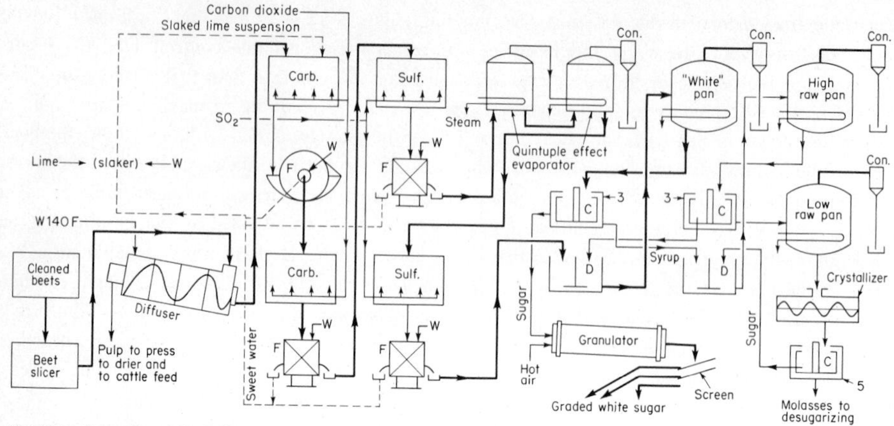

Key: Carb. = Carbonators; Sulf. = Sulfitors; F = Filters; W = Wash water; C = Centrifugals; Con. = Condensers; D = Dissolvers

Note: About half of all plants use sweet water from filter washing for slaking lime; others use fresh water

Farm land	0.4 acre	Coal (power and heating)	0.66 ton	To produce
Sugar beets	8.0 tons	Sulfur	1.6 lbs.	1 ton refined sugar
Limestone	0.35 ton	Water	9.0 tons	800 lbs. molasses (50 per cent sucrose)
Coke	0.03 ton	Direct labor	2 man-hrs.	800 lbs. dried pulp

Fig. 30.4 Flowchart for beet sugar manufacture.

uproot the beets and elevate them to trucks. The tops, good for cattle feed, may also be removed mechanically. The beets, containing from 13 to 17% sucrose and 0.8% ash, enter the factory by way of flumes, small canals filled with warm water, which not only transport them but wash them as well.

Figure 30.4 illustrates the essential steps in the making of beet sugar. This manufacture may be divided into the following sequences:[13]

The beets are rewashed, weighed, and sliced into long narrow strips called *cossettes* (Op).

The cossettes are dropped into a specially designed continuous countercurrent diffuser (Fig. 30.4). The sugar is extracted countercurrently with water at 160 to 175°F (Op). The resulting raw juice is a blue-black 10 to 12% sucrose solution with a small amount of invert sugar and 2 to 3% ash. The pulp remaining contains 0.1 to 0.3% sugar (based on the beets). This pulp is dewatered in presses, dried in a rotary dryer, and sold as cattle feed (Op). The change in the beet industry to automatic countercurrent continuous diffusers eliminates the sweet-water handling costs and reduces the labor required by the battery.[14]

The juice is given a rough screening to remove foreign materials (Op).

Milk of lime (or saccharate milk if Steffen's process is used to remove foreign materials) is added until the concentration is equivalent to about 2 to 3%. The lime aids in the precipitation of undesirable impurities. Any calcium saccharate is decomposed in carbonators by passing carbon dioxide from the lime kiln through the juice continuously. The foaming that occurs at this stage is reduced by adding a small quantity of antifoam (Ch).

The sludge produced by the lime is equal to 4 or 5% of the weight of the beets charged. This is removed by thickening and filtering on Rotary filters (Op).

Lime is added again until the concentration is equivalent to 0.5%, and the juice again carbonated, this time hot (Ch).

[13]See also pictured flowchart, Beet Sugar Production, *Chem. Metall. Eng.*, **49**(6), 110 (1942); McDill, Beet Sugar Industry: Industrial Wastes, *Ind. Eng. Chem.*, **39,** 657 (1947); McGinnis, Beet Sugar Technology, Reinhold, 1951.

[14]Havighorst, Beet Sugar: A Radical New Look, *Chem. Eng.* (*N.Y.*), **68**(20), 76 (1961), and Spreckles' New Developments in Beet Sugar, Part 1, *Chem. Eng.* (*N.Y.*), **71**(7), 72 (1964) (1961 and 1964 flowcharts with twin helical-screw continuous diffusers with adjustable slope).

[15]McGinnis *et al.*, Low-purity Beet Sugar Factory Materials, *Ind. Eng. Chem.*, **34,** 171 (1942).

It is then filtered on pressure filters (Op).

The resulting filtrate is bleached with sulfur dioxide (Ch).

The precipitate of calcium sulfite is removed by pressure or plate-and-frame filters (Op).

The purified juice is concentrated from 10 to 12% sugar to about 60% sugar in multiple-effect evaporators (Op). This increases the concentration of calcium ions again. Some calcium precipitates out.

A Spreckles' plant uses also a countercurrent decolorizing reactivated carbon adsorption system to purify and decolorize the thick juice in towers, where the juice rises against the falling carbon (Op).

The resulting thick juice is grained in vacuum pans,[15] centrifuged, washed, dried in a granulator, screened, and packed in much the same manner as described for cane sugar. (Op).

The juice from the first vacuum pan is given further treatment to recover more sugar crystals, but this is not pure enough for market and is sent back to the process for further purification (Op).

The sirup remaining after the several crystallizations, called beet *molasses,* is sold for cattle feed directly or added to the waste beet pulp. It is also an important medium for fermentation, particularly for citric acid, where the high nitrogen content of beet molasses is favorable. Processes for recovery of the remaining sugar have been worked out and are used commercially. Initial recovery was obtained by the Steffen process, which is used extensively in the United States.

Ion exchange[16] has been tried or proposed for the treatment of process liquors at several places in the sugar-manufacturing procedure. Ion exchange works well for cleaning up process liquors, but total-output treatment is frequently too expensive because of the cost of disposal of sweet waters and spent reagents (from washing). With cane sugar,[17] likewise, inversion may take place and cause losses.

ENERGY REQUIREMENTS In the direct manufacture of sugar there are few or no chemical changes that require energy. Almost all the steps in the manufacturing sequences involve physical changes or unit operations. These consume energy in the form of *power* for crushing, pumping, and centrifugation and *heat* for solution, evaporation, and drying. At the bottom of the flowcharts for refined cane sugar and for beet sugar (Figs. 30.2 and 30.4) are given some average figures for energy (steam or coal) requirements. The steady improvement by chemical engineers of the equipment necessary to make these various unit operations function efficiently has gradually reduced the energy requirements to the reasonable amounts given.

Although *bagasse* is used for fuel in many raw-sugar mills, it is becoming a more important raw material yearly for paper and hardboard (*Celotex*); hence other fuel may be employed. In all sugar-manufacturing establishments, there is an economical dual use of steam (see Table 4.11) wherein high-pressure steam from the boiler is expanded through a reciprocating engine or a turbine to furnish power and the exhaust steam is condensed to secure heat for evaporation of the juices and sirups. The condensate water resulting is pumped back to the boilers or employed in making sirups. For beet-sugar refining, Fig. 30.4 indicates 0.66 ton of coal for 1 ton of refined sugar, or 10,000 Btu/lb of sugar.[18] Some plants now produce 1 lb of refined beet sugar with only 6000 Btu.

MISCELLANEOUS SUGARS Lactose, or milk sugar, is made from waste skim milk.[19] Sorbitol is manufactured by hydrogenation of dextrose under pressure, using a nickel catalyst, or by reduction in an electrolytic cell. Most of the molasses or hydrol produced enters animal feeds, but a considerable quantity serves as the carbohydrate raw material in fermentations (Chap. 31).

[16]Michener *et al.*, Ion Exchange in Beet Sugar Factories, *Ind. Eng. Chem.*, **42,** 643 (1950); Mandru, Ion Exchange in Beet Sugar Manufacture, *Ind. Eng. Chem.*, **43,** 615 (1951); see CPI 2, pp. 652–653.

[17]ECT, Ion Exchange for Sugars, vol. 10, p. 214, 1954, and 1st suppl., pp. 400–401, 1959.

[18]Perry, p. 9-3, table 9-1, for Btu in typical coals.

[19]See CPI 2, p. 659, for flowchart; Lactose, More Uses, *Chem. Process.* (Chicago), Jan. 28, 1963, p. 26.

STARCHES AND RELATED PRODUCTS

Although starch has the formula $(C_6H_{10}O_5)_n$, it may be regarded as a chain of D-glucopyranosyl units, with n about 400 to 500, consisting of varying proportions of amylose and amylopectin:

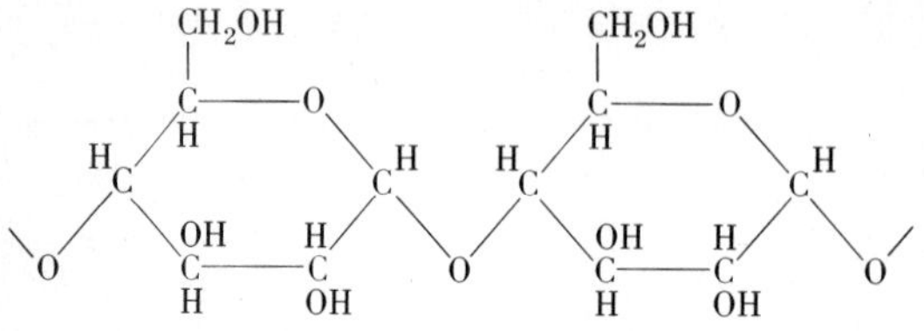

Amylose: segment of a linear chain

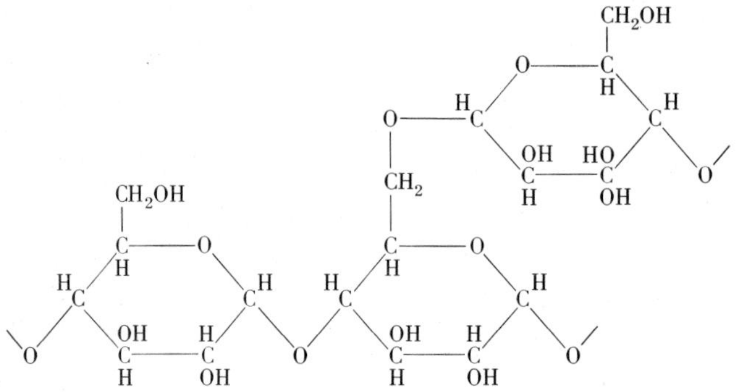

Amylopectin: segment of a branched chain

Starch is one of the most common substances existing in nature and is the major basic constituent of the average diet. Industrially, its applications[20] are numerous, and it is used in more than 300 modern industries, including the manufacture of textiles, paper, adhesives, insecticides, paints, soaps, explosives, and such derivatives as dextrins, nitrostarch, and corn sugar. Recent years have produced such derivatives as heat-resistant adhesives, esters comparable with cellulose esters, carboxylic acids from the oxidation of dextrose, and wetting agents.

HISTORICAL[21] It is a well-known fact that the ancients used starch in manufacturing paper (as an adhesive and stiffener) as early as 3500 B.C. The Egyptians of this period cemented papyrus together in this manner. Between A.D. 700 and 1300, most paper was heavily coated with starch, but the practice was abandoned toward the end of the fourteenth century and was not revived until the modern era. The use of starch in textiles began during the Middle Ages, when it was a common stiffening agent. By 1744 the English were using it in sizing and warp glazing. Textile demands soon brought about the introduction of potato starch to supplement the wheat starch solely available up to this time. In 1811 the discoveries of Kirchhoff with respect to glucose and the thinning of starches by enzymic action gave great impetus to starch manufacture through the increased fields of application created. The use of roasted starch did not begin until 1821, its usefulness being discovered as the result of a textile fire at Dublin, Ireland. It had, however, been prepared by LeGrange as early as 1804. The first starch produced in this country was white potato starch, made at Antrim, N.H., in 1831. In 1842, Kingsford began the production of corn starch, which became increasingly

[20]The Corn Industries Research Foundation, Inc., 1001 Connecticut Ave., Washington, D.C., publishes an important series of pamphlets on the corn wet-grinding industries, which are kept up to date. Titles are Corn in Industry, Corn Starch, Corn Gluten, Corn Act, Corn Syrups and Sugars.

[21]Radley, Starch and Its Derivatives, Van Nostrand, 1940.

TABLE 30.2 ***Specified Products from Corn for 1973***

Product	*1,000 units*	*Product*	*1,000 units*
Starch	2,630,000 lb	Sugar	1,144,000 lb
Sirup	3,750,000 lb	Corn oil	110,000 tons
Gluten feed and meal	1,664 tons		

Source: Estimated from Agricultural Statistics, Dept. of Agriculture, 1974.

popular until, by 1885, it had risen to the position of the leading textile starch in the field. It was in this period also that the manufacture of dextrins (roasted starches) began in the United States.

USES AND ECONOMICS In 1973 about 8.0 billion lb of starch was produced in the United States for all purposes, including conversion into sirup and sugar. Of this total about 98% was cornstarch. Some statistics on the shipment of corn products are presented in Table 30.2. Imports are principally tapioca, sago, and arrowroot starches, and exports are principally cornstarch. The largest single use for *cornstarch* is as food, about 25% being thus consumed. Industrial uses account for the remaining 75%. The paper industry utilizes cornstarch as a filler and a sizing material. Textile, laundry, foundry, air flotation, oil-well drilling, and adhesives use much starch (Chap. 25). *White potato starch* can be employed for almost every one of the uses outlined for cornstarch; it has a more desirable phosphoric acid content but is more expensive.

Wheat, rice, arrowroot, and *cassava* (*tapioca*) *starches* also have many of the same applications as cornstarch. Rice starch is particularly preferred for laundry purposes. Tapioca starch is very common as a food. In addition to the starches themselves, many further reaction products are made. These include the following products: *Dextrin,* or roasted starch, which is available in more than 100 different blends and types varying from pure white to light yellow in color; it is used to make a great number of different pastes, gums, and adhesives. *Corn sirup* is a hydrolysis product of cornstarch, containing dextrose, maltose, higher saccharides, and water. Approximately 95% is used for food, largely as sweeteners. Nonfood and industrial applications are important, as in the textile, leather-tanning, adhesive, pharmaceutical, paper, and tobacco industries. *Corn sugar,* or *dextrose,* is the sugar found in the blood and is our primary energy food. Its many food uses are dependent upon its lower rate of crystallization, lesser sweetness, and different crystal formation. Dextrose is consumed by the baking industry, and has additional uses in preserved foods, soft drinks, candy, and ice cream. It is widely employed by the medical profession for infant feeding and prescriptions in sirup form. Industrially, it is important as a constituent of the viscose-rayon spinning bath, in leather tanning, in tobacco conditioning, and in fermentation. Important by-products of the starch industry are corn gluten, feed and meal, corn-oil meal, and corn oil. Virtually all gluten and meal are used as feed, but a specially prepared gluten, very high in protein, is employed as a raw material in manufacturing plastics and lacquers. Concentrated *steep water* is consumed in the growing of penicillin and streptomycin (Chap. 40). Inositol,[22] or hexahydroxycyclohexane, a sugar substance and a member of the vitamin B complex, is made from corn steep water.

MANUFACTURE OF STARCH, DEXTRIN, AND DEXTROSE FROM CORN Corn wet refining is a large industry, consuming more than 230 million bu of corn per year. Chemically, the corn kernel consists of from 11 to 20% water, with the following average constituents, expressed in percent:[23]

Moisture	16	Oil	3.8
Starch	61	Fiber	2
Proteins	9	Sugars	1.6
Pentosans	5.3	Ash	1.3

[22]Up Output Tenfold, *Chem. Eng.* (*N.Y.*), **58**(7), 200 (1951) (flowsheet); cf. Greenfield *et al.*, Industrial Wastes Cornstarch Process, *Ind. Eng. Chem.*, **39,** 583 (1947).

[23]Corn Industries Research Foundation, *op. cit.*, 1958.

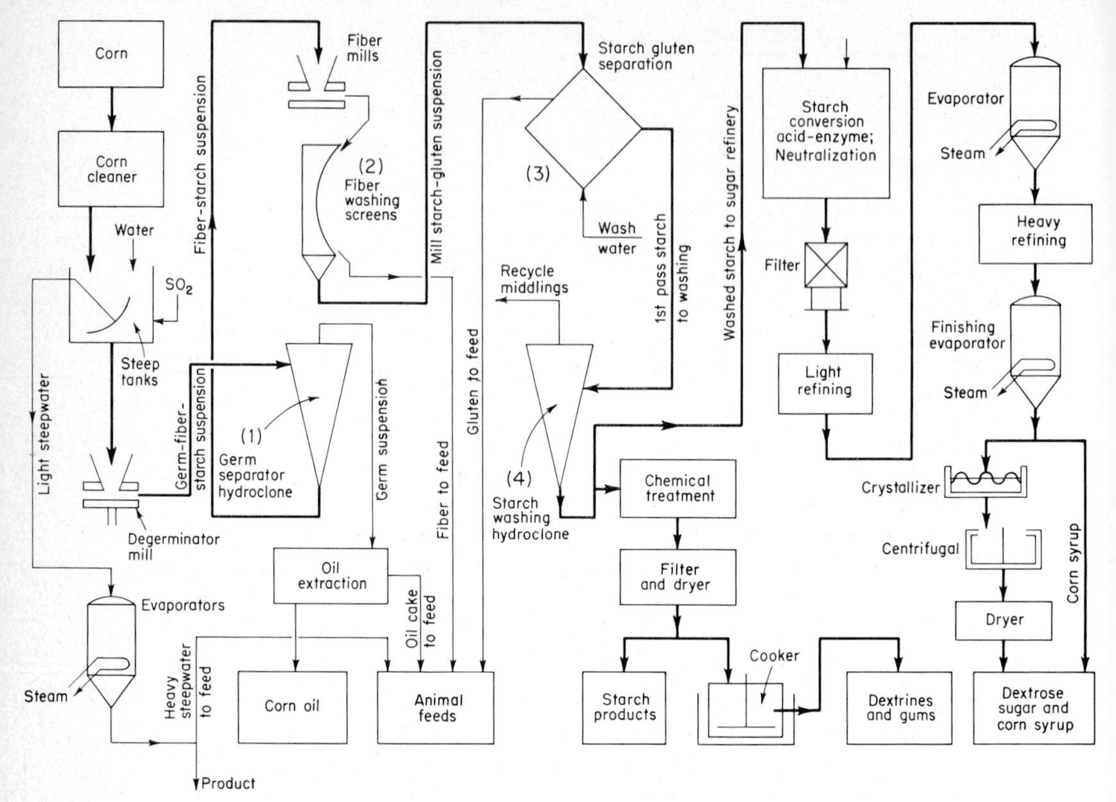

Fig. 30.5 Corn products outline flowchart. (*Corn Products Refining Corp. and Dorr-Oliver.*)

On this basis, 1 bu of corn yields about 30.8 lb of starch, 23.5 lb of by-products (gluten meal, corn bran, germ-oil meal, and steep water), and 1.7 lb of corn oil.

The sulfurous acid method of refining is employed, using shelled corn as raw material[24] and involving *physical changes* or *unit operations* (Fig. 30.5) followed by *chemical conversion* of starch to dextrins, sirups, or dextrose:

The first operation consists of cleaning the corn by means of screens, compressed air, and electromagnets. The cleaned corn is steeped for 2 days in circulating warm water (115 to 125°F) containing 0.10 to 0.30% sulfur dioxide to prevent fermentation during the soaking period. Large hopper-bottomed wood or stainless-steel steeping vats (holding 3,000 bu of corn) are employed for this operation, which softens the gluten and loosens the hulls. The steep water dissolves salts, soluble carbohydrates, and protein.

The cleaned and softened kernels are degerminated between two studded steel plates, one rotating and one stationary, which tear the kernels apart and extricate the corn germs without crushing them. The corn germs are liquid-separated from the hulls in so-called *germ separators,* formerly large, agitated tanks. Now stationary cone-shaped hydroclones[25] (Fig. 30.6) are more efficient, accentuating continuous gravity separation of the germ by the centrifugal action of a tangentially injected suspension of germ-fiber starch, with the lighter germ in the overflow (top center). This

[24]Forbath, Process Maze Yields Maize Products, *Chem. Eng.* (*N.Y.*), **68**(5), 90 (1961) (flowchart); Wet Milling of Corn, *Chem. Eng.* (*N.Y.*), **69**(16), 117 (1962).

[25]Dorr-Oliver, Hydrocyclones, *Chem. Eng.* (*N.Y.*), **71**(1), 48 (1964).

Fig. 30.6 Germ-separating 6-in.-diam hydroclones. All six shown operate in parallel with the lighter germ in the central overflows. (*Dorr-Oliver.*)

results in a cleaner and more accurate process at less capital investment. The germ is subjected to oil extraction as described in Chap. 28, using either expellers or solvent extraction.

The remainder of the corn kernel contains starch, gluten, and cellulosic fiber. It is wet-ground in impact fiber mills and passed through high-capacity stationary screens[26] called sieve bends, depicted in Fig. 30.7. In the sieve bends the starch and gluten are washed countercurrently with process water to remove them from the fiber, which is the oversized material, all of which was mechanically separated in the fiber-milling operation.

To separate the heavier starch from the gluten, the old, cumbersome, gravity starch tables have been discarded, and this separation is now done in Merco pressure-nozzle discharge centrifuges (Fig. 30.8) and purified by pumping through starch-washing hydroclones[27] (Figs. 30.9 and 30.10), from which the "middlings" are returned to the Merco centrifuge for recycling. These centrifugal devices intensify the force of gravity, require less space, and are cleaner, since they are completely enclosed. The gluten, as shown in Fig. 30.5, is sent to be mixed with oil cake and fiber for animal feed. Some gluten is partially dehydrated and sold as an adhesive or alcohol-extracted to yield zein.

The highly mechanically purified starch is dried and sold or "cooked" to heat-convert it to soluble dextrins and gums. An increasing amount of the washed starch is conveyed to the sugar

[26]Elsken and Ehenger, Stationary Screens, *Chem. Eng. Prog.*, **59**(1), 76 (1963).

[27]James, Industrial Starches, Noyes, 1974; Bradley, The Hydroclone, Pergamon, 1965.

Fig. 30.7 Fiber-washing 120° sieve bend. Stationary screens of a 120° arc. (*Dorr-Oliver.*)

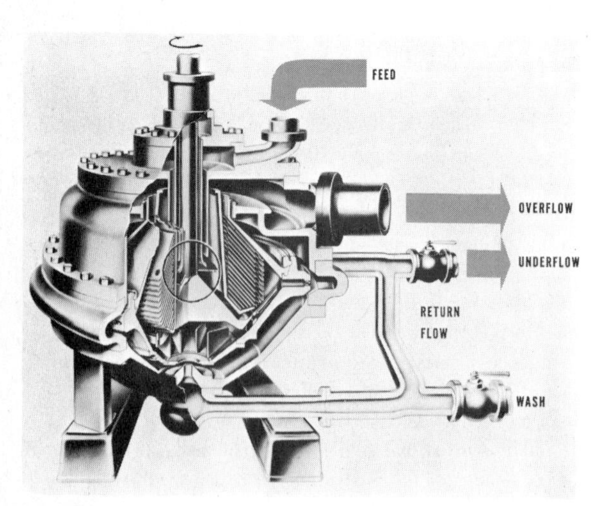

Fig. 30.8 Gluten separation (overflow) by a Merco-disc pressure-nozzle bowl centrifuge, with the suspension entering into the center feed well, at the bottom of which the impeller throws the slurry to the outer portion of the disc stack. Here centrifugal force throws the heavier starch toward the periphery rotor and, as a result, the starch particles settle on the underside of the individual discs and slide down to the outer area of the bowl, where the concentrated starch slurry is discharged through the nozzles (on the periphery) and through the underflow. A somewhat similar Merco centrifuge is used to thicken the separated gluten-water fraction. See Perry, Fig. 19-116.

Fig. 30.9 Purification of starch through gravity accentuated by centrifugal washing in multiple hydroclones. These hydroclones are much different from those in Fig. 30.6, being only 10 mm in diameter and made of nylon, with up to 480 arranged in a separate housing operating in parallel. By employing a series of nine of these multiple hydrocyclone units, both insoluble protein (gluten) and solubles are removed from the starch slurry. (*Dorr-Oliver.*)

refinery, where the starch molecule is chemically depolymerized and hydrated to dextrose or to corn sirup by the action of enzymes and acid as outlined in the flowchart. The dextrose crystallizer is depicted in Fig. 30.11.

If *commercial starch* is to be made, the starch is removed from suspension with a vacuum rotary string-discharge filter. The cake is broken and dried by flashing or in a continuous-tunnel dryer traveling countercurrent to the air. The starch enters with a moisture content of 44% and exits at 10 to 14%. This form is sold as *pearl starch.* Powdered starch is ground and screened pearl starch. *Lump,* or *gloss, starch* is made from powdered starch containing a slightly higher percent of moisture. Precooking the starch yields *gelatinized starches.* For *thick-boiling starch,* alkali conversion is used; for *thin-boiling starch,* mild acid conversion.

Another product of corn refining is *dextrin,* or roasted starch. Starch itself is not soluble in water, but its derivative, dextrin, dissolves readily to give various commercial adhesives, pastes, and gums. Conversion is carried on in round, steam-jacketed, bronze tanks equipped with a scraper and open at the top. The scraper prevents sticking during the heating period, which may vary from 2 h for some white dextrins to 15 h for certain gums. The temperature also influences the kind of dextrin being prepared.

The average yield of sugar from an acre of cane is 12,292 lb. An acre of beets usually produces about 7,500 lb of sugar. Present corn yields approach 200 bu (12,000 lb) of grain per acre.

Fig. 30.10 The multiple hydroclones are shown with one housing open, disclosing the stationary, small, relatively short nylon centrifugal devices fed through the side of the housing (not visible in the figure). Individually, the 480 hydroclones are fed tangentially, with the overflow, or lighter gluten, discharged from one housing, and the underflow, or heavier starch, discharged from the other housing. (*Dorr-Oliver.*)

The 12,000 lb of corn grain will produce 8,640 lb of starch which can be converted to 9,000 lb of glucose. Since cane requires about 18 months to produce and corn less than half that, it is apparent that, if acceptable sweeteners can be made from corn starch, yields per acre will be far greater and costs consequently less than for cane even if the value of corn oil, protein, and ensilage is ignored. This favorable economic situation has led to the conversion of nearly two-thirds of the corn starch produced to glucose sirup for food use.[28]

Starch can be hydrolyzed to glucose with acid or a combination of acid and enzymes, producing a sirup of high nutritive value but not as sweet-tasting as sucrose.

A new development commercially hydrolyzes starch slurry to D-glucose with immobilized glucoamylase and then converts the D-glucose to a mixture of approximately equal parts of D-glucose and D-fructose with glucose isomerase. The sugar solution so obtained is as sweet as sirups obtained from sucrose and considerably cheaper. This sweetener [isomerose or high-fructose corn syrup (HFCS)]

[28]Starch as a Source of Sweeteners, *Die Starke*, Nov. 7, 1973, p. 1; Wieland, Enzymes in Food Processing and Products, Noyes, 1972.

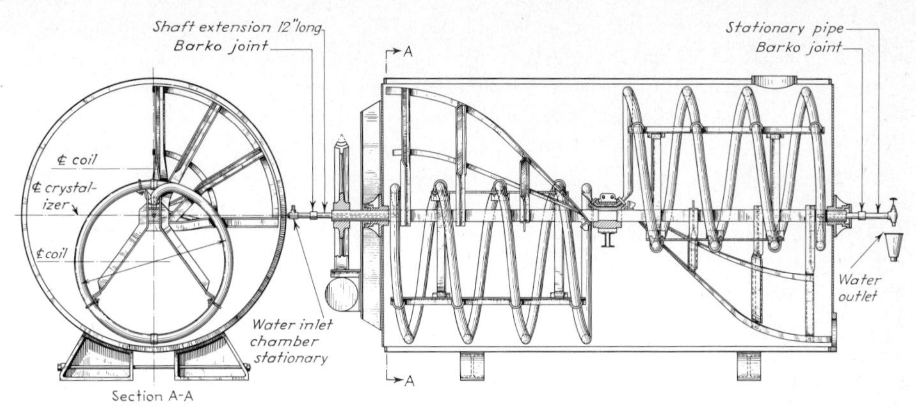

Fig. 30.11 Sugar crystallizer provided with a scraper and an eccentric coil crystallizer. Such cooling with stirring improves heat transfer while maintaining uniformity in temperature drop. Rotation $\frac{1}{8}$ rpm. Crystallization time is about 100 h for dextrose. Water is passed through the coil for cooling. (*Kilby Manufacturing Co.*)

has essentially the same functional properties with regard to sweetness, flavor enhancement, hydroscopicity, freezing-point depression, fermentability, osmotic pressure, and browning of baked goods as sucrose sirups. The largest single user of sucrose is the beverage industry, and HFCS is totally interchangeable. Over 300 million lb of such sirups (dry basis) are produced per year. Dried dextrose (D-glucose) and dextrose monohydrate are also produced by drum-drying and/or granulating.

Corn carbohydrates and their derivatives are being intensively investigated because they are completely American products of reasonable and decreasing price, now that manufacturing methods have been completely improved and modernized, as exemplified in Fig. 30.5. Such derivatives are sorbitol formed by hydrogenation and methyl glucoside which yields glucocide hydroxypropyl ethers as constituents of urethrane foam.[29] Starch phosphates and other products are appearing.

The advocates of farm chemurgy are recognizing this future enlarged outlet for the products of the American farmer from this cheap and abundant corn. There is no doubt but that the organic products of American soil, particularly corn carbohydrates and their derivatives, will be of most importance, increasingly, as raw material, in the future decades, as well as for both animal and human food. There will be a time when the wet-milling industry will be much larger, particularly because the cost of raising corn is decreasing and the cost of other raw materials is increasing.

MISCELLANEOUS STARCHES

AMYLOSE occurs to the extent of 27% in cornstarch, the other 73% being amylopectin. Amylose is a linear chain of dextrose units and resembles cellulose, which *it can supplant for many uses,* such as in films, adhesives, and paper. *Waxy* corn is almost all amylose and is being grown in increasing amounts. However, there are a number of processes[30] for separating the two starches (Staley, Corn Products, Avebe, etc.). One process heats the cornstarch slurry under pressure to prevent boiling, whereupon the mixed starch dissolves; upon slow cooling to about 120°F, the amylose crystals form and are recovered, leaving the amylopectin to be precipitated upon further cooking.

WHITE POTATO STARCH[31] White potatoes contain 10 to 30% starch. Upon being received at the factory, the potatoes are washed and disintegrated to a water pulp, using a hammer mill. The

[29]Corn Products, *Chem. Eng. News,* June 4, 1962, p. 20.

[30]Starch Competition Stiffens, *Chem. Week,* July 21, 1962, p. 135; Staley, Starch Carbohydrates, *Chem. Eng.* (*N.Y.*), **68**(5), 80 (1961).

[31]Howerton and Treadway, White Potato Starch, *Ind. Eng. Chem.,* **40,** 1402 (1948).

pulp is treated with sulfur dioxide gas, in the ratio 0.5 lb/lb of starch, and sent to a continuous horizontal centrifuge with an imperforate conical bowl and a continuous-spiral-ribbon starch remover. The protein-water mixture is separated from the starch, cellulose, and skins, and the latter three substances are resuspended in water as a milk. The suspension is sieved, and the pulp from the sieves is reground and resieved. The milk from the second sieving is again passed through a centrifuge, suspended in water, and sent to separation devices. From this point on operations are similar to those used for the manufacture of cornstarch (Fig. 30.5).

WHEAT STARCH In the manufacture of this cereal starch, the major difficulty encountered is the separation of the starch and the gluten. This may be accomplished by Martin's process, wherein wheat is ground to flour, made into a dough with about 40% of its weight of water, and stored for 1 h. The resulting mass is divided into small lumps and placed in a semicircular sieve, where a traveling roller presses out the starch, which is removed by a fine water spray. The resulting liquor is treated in much the same manner as cornstarch liquor. The gluten is also recovered and sold.

RICE STARCH This is made from "cargo rice," which still has the brown outer cuticle attached, or from broken white grains rejected as foodstuff. The rice is steeped for 24 h with caustic soda solution (1.005 specific gravity) tanks with perforated false bottoms. At the end of the period the liquor is withdrawn, the rice washed, fresh liquor added, and steeping continued for another 36 to 48 h. The resulting softened grains are ground with a caustic solution to a specific gravity of 1.24, and the mash is centrifuged. The solids obtained include all sorts of fibrous material, starch, and gluten. These are resuspended, a small amount of formaldehyde is added to inhibit fermentation, and they are recentrifuged and washed. A bleaching or bluing agent may be added at this point. The liquor is screened, adjusted to a specific gravity of 1.21, and sent to disk centrifuges. The resulting starch is dried for 2 days at 50 to 60°C.

CASSAVA (TAPIOCA) STARCH This starch is obtained from the roots and tubers of the manioc plant. Imports are about 300 million lb per year, mainly from Thailand and Brazil. The average starch content varies from 20 to 30%. In general, the roots are pulped and washed on sieves to obtain the starch. Separating and purifying operations are similar to those described for potato starch.

SAGO STARCH This is obtained from the pith of the sago palm, and also from yams in the East Indies and Borneo. Pearl sago starch is made by drying the starch so as to form a plastic dough, which is then forced through sieves and dried in the air.

SELECTED REFERENCES

Barnes, A. C.: Sugar Cane, Wiley-Interscience, 1965 (processing).
Colbert, J. C.: Sugar Esters, Noyes, 1974.
Dyke, S. F.: Chemistry of Natural Products: The Carbohydrates, vol. 5, Interscience, 1960.
Hanessian, S. (ed.): Deoxy Sugars, ACS, 1968.
Henry, J.: Technological Values of the Sugar Beet, Elsevier, 1962.
Honig, P.: Principles of Sugar Technology, Elsevier, 1953–1962, 3 vols.
Hugot, E.: Handbook of Sugar Cane Engineering, translated and revised by G. H. Jenkins, Elsevier, 1960.
Hung, M.: Patents on the Reactions of Sugars, Sugar Research Foundation, 1961.
Isbell, H. S. (ed.): Carbohydrates in Solution, ACS, 1973.
James, R. W.: Industrial Starches, Noyes, 1974.
Jenkins, G. H.: Introduction, Cane Sugar Technology, Elsevier, 1966.
Junk, R. W., and H. M. Pancoast: Handbook of Sugars, Avi Publishing, 1974.
Kerr, R. W.: Chemistry and Industry of Starch, Academic, 1950.
Pigman, W. (ed.): Carbohydrates, Chemistry, Biochemistry, Physiology, Academic, 1957.
Recent Advances in Processing Cereals, SCI Monograph 16, Gordon and Breach, 1962.
Spencer, G. L.: Cane Sugar Handbook, 9th ed., G. P. Meade (ed.), Wiley, 1963.
Technological Value of the Sugar Beet, Commission Internationale Technique de Sucrérie, Elsevier, 1962.
Whistler, R. L., *et al.* (eds.): Methods in Carbohydrate Chemistry, vol. 4, Starch, vol. 2, Reactions of Carbohydrates, vol. 5, General Polysaccharides (indexes), Academic, 1963–1964, 5 vols.

chapter 31

FERMENTATION INDUSTRIES

The employment of microorganisms to convert one substance into another is a science that is assiduously studied and energetically applied. Although the fermentation of fruits to alcohol was known to primitive humans, and although the making of various beverages out of fruits and grains has been well established for centuries, only during the past generation has wider application of this procedure been recognized. Now scientists are directing the life processes of yeasts, bacteria, and molds to the production of chemicals. Alcohol can be viewed as having been produced in this way from the earliest times, but the making of acetone and butanol, of acetic acid, lactic acid, citric acid, and many antibiotics are recent technical accomplishments.

The foundation of the scientific understanding of fermentation, indeed of the action of all microorganisms, hence of their economic control, rests firmly upon the genius of one man, Louis Pasteur.[1] He showed that fermentation is directly caused by the life processes of minute organisms. By understanding how these microorganisms function and through the recognition that varieties of yeasts, for instance, act differently and that the environment fundamentally affects even a given strain, one can control these processes of fermentation in an exact scientific manner.

Fungi are a branch of the nongreen plants and include bacteria, yeasts, and molds. These feed upon organic materials. It is this feeding that interests the manufacturer; for if certain yeasts, bacteria, or molds are supplied with the necessary fundamental energy food, together with other needed nutrients, these microvegetative organisms will not only grow and multiply but will change the food into other chemical substances.

Yeasts and bacteria are unicellular and of very small dimensions. Yeasts are irregularly oval and perhaps 0.004 to 0.010 mm in diameter. Bacteria are smaller, mostly less than 0.007 mm in the longer dimension, and more diverse in shape. Many of them, bacilli, are rod-shaped. Yeasts multiply by budding, and bacteria by binary fission. Molds are multicellular filaments and increase by vegetative growth of the filament. Sporulation provides for the next cycle, as it does also with many bacteria. The vegetative reproduction cycle of these bacteria and of yeast is short—measured in minutes. Because of this, they multiply exceedingly fast.

Many microorganisms have been brought into useful service to humans. One of the striking developments since the 1920s has been extension of the life processes of these minute vegetative organisms to the making of chemicals other than alcohol. Table 31.1 lists some of these procedures, which, on a larger industrial scale, embrace the manufacture of antibiotics, enzymes, acetone, butanol, acetic acid, lactic acid, and citric acid. This table shows the importance of fermentation in foods and feeds. Several older laboratory procedures, such as the making of citric and gluconic acids, have been developed to the industrial stage. One of the outstanding developments of World War II was

[1]Current Developments in Fermentation, *Chem. Eng.* (*N.Y.*), **81**(26), 98 (1974); ECT, vol. 6, pp. 317–375, especially table 1, and pp. 332–336 for industrial applications 1951 (62 refs.); ECT, vol. 8, pp. 871–880, 1965; Fermentation Heads for Higher Productivity, *Chem. Eng. News*, Mar. 19, 1973, p. 32; Fermentation Industry, *Chem. Week*, Dec. 16, 1967, p. 83; Fermentation Achieves New High in Flexibility, *Chem. Eng.* (*N.Y.*), **82**(10), 56 (1975).

TABLE 31.1 ***Survey of Important Fermentations***

Food and feed	*Industrial*	*Pharmaceuticals (Antibiotics)*
Beer (Y)	Acetic acid	Amphoterccin B
Bread (Y)	Acetone	Bacitracin
Cheese (M or B)	Aspartic acid	Bleomycin
Cocoa (B and Y)	2,3-Butanediol	Candicidin
Coffee (M)	*n*-Butyl alcohol	Capreamycin
Koji (M and Y)	Carbon dioxide	Cephalosporin C
MSG (B)	Citric acid	Chloramphenicol
Olives (B)	Dextran	Chlortetracycline
Pickles (B and Y)	Dihydroxyacetone	Colistin
Sauerkraut (B)	Ethyl alcohol	Cycloheximide
Single-cell protein (Y, B, or M)	Fumaric acid	Cycloserine
Tea (B)	Fusel oil	Dactinomycin
Vinegar (B and Y)	Gallic Acid	Doxorubicin
Wine (Y)	Gluconic acid	Erythromycin
Whisky (Y)	Glycerol	Gentamicin
Vitamins	Isoleucine	Griseofulvin
Ergosterol (Y and M)	Itaconic acid	Kanamycin
Riboflavin (B and Y)	2-Ketogluconic acid	Lincomycin
Vitamin A (B)	5-Ketogluconic acid	Mithramycin
Vitamin B_2 (Y)	Kojic acid	Mitomycin C
Vitamin B_{12} (B and M)	Lactic acid	Neomycin
	Lysine	Novobiocin
	Succinic acid	Nystatin
	Sulfuric acid from sulfur	Oleandromycin
	Tartaric acid	Paromomycin
	Valine	Penicillins
	Yeast	Polymycin
	Enzymes	Rifampin
	Amylase	Spectinomycin
	Cellulase	Streptomycin
	Diastase	Tetracycline
	Invertase	Vancomycin
	Maltase	Viomycin
	Zymase	

Y, Yeast; B, bacteria; M, molds.

the making of penicillin, which stimulated further important discoveries in the field of antibiotics. Many fermentation processes are frequently in direct competition with strictly chemical syntheses. Alcohol, acetone, butyl alcohol, and acetic acid produced by fermentation have largely been superseded by their synthetic counterparts. However, antibiotics have paced a recent fermentation revival and, with some exceptions, all the major antibiotics are obtained from fermentation processes. Dextran is another fermentation product. The microbiological production of vitamins has also become economically important. In Chap. 40 antibiotics, hormones, and vitamins are presented, together with several flowcharts. See Fig. 40.8 for penicillin, erythromycin, and streptomycin.

Actually, fermentation under controlled conditions involves *chemical conversions*.[2] Some of the more important processes are: *oxidation*, e.g., alcohol to acetic acid, sucrose to citric acid, and dextrose to gluconic acid; *reduction*, e.g., aldehydes to alcohols (acetaldehyde to ethyl alcohol), and sulfur to hydrogen sulfide; *hydrolysis*, e.g., starch to glucose, and sucrose to glucose and fructose and

[2]Wallen, Stodola, and Jackson, Type Reactions in Fermentation Chemistry, Dept. of Agriculture, Agricultural Research, 1959 (hundreds of reactions under 14 types).

on to alcohol; and *esterification*, e.g., hexose phosphate from hexose and phosphoric acid. Actually, certain chemical conversions can be carried out more efficiently by fermentation than by chemical synthesis.

Many chemical reactions caused by microorganisms are very complex, however, and cannot easily be classified; so the concept of fermentation itself as a chemical conversion has been developed. According to Silcox and Lee,[3] the five basic prerequisites of a good fermentation process are:

1. A microorganism that forms a desired end product. This organism must be readily propagated and be capable of maintaining biological uniformity, thereby giving predictable yields.
2. Economical raw materials for the substrate, e.g., starch or one of several sugars.
3. Acceptable yields.
4. Rapid fermentation.
5. A product that is readily recovered and purified.

According to Lee, certain factors should be stressed in relation to the fermentation chemical-conversion or unit-process concept, such as microorganism, equipment, and the fermentation itself. Certain critical factors of the fermentation are pH, temperature, aeration-agitation, pure-culture fermentation, and uniformity of yields.[4] The microorganisms should be those which flourish under comparatively simple and workable modifications of environmental conditions. See also the symposium on engineering advances in fermentation practice.[5]

In understanding, hence in correctly handling, microorganisms, a sharp differentiation should usually be made between the initial growth of a selected strain of these organisms to a sufficient quantity and the subsequent processes whereby, either through their continued *living* or as a result of *enzymes* previously secreted, the desired chemical is manufactured. To obtain a maximum chemical yield, it is frequently advisable to suppress additional increase in the quantity of the microorganism. Highly specialized microbiologists working in well-equipped laboratories are engaged in selecting and growing the particular strain of an organism that experiment has shown to produce the chemical wanted with the *greatest yields*, the *least by-product*, and at the *lowest cost*.

No longer will just any yeast do to make industrial alcohol or a fermented beverage; not only are wild yeasts excluded, but a special strain must be used.

The yeasts, bacteria, and molds employed in fermentation require specific environments and foods to ensure their activities. The concentration of the sugar or other food affects the product. The temperature most favorable varies (5 to 40°C), and the pH also has great influence. Indeed, the bacteriologist has developed acid-loving yeasts, so that *wild yeasts*, not liking acidic conditions, do not flourish. Some microorganisms require air (aerobic), and others go through their life processes without air (anaerobic). Certain anaerobes neither grow nor function in the presence of air. In directing these minute vegetative organisms, conditions can be controlled to encourage the *multiplication* of the organism first, and then its *functioning*, either directly or through the *enzymes* secreted. How important this is can be seen from the knowledge that to grow 1 g of yeast (dry basis) requires 1.5 to 2.0 g of monosaccharide per day, and 6 g to maintain it. By virtue of this growth, organic catalysts, or *enzymes*, are frequently formed that *directly cause the desired chemical change*. During the growth period, in addition to the primary, or energy food, such as monosaccharides for yeast, various *nutrients* are needed, such as small amounts of phosphates and nitrogenous compounds, as well as favorable pH and temperature. Finally, certain substances poison these useful little vegetables and their enzymes. Even the alcohol formed by the yeasts eventually reaches a concentration (varying

[3]Silcox and Lee, Fermentation, *Ind. Eng. Chem.*, **40,** 1602 (1948).

[4]Humphrey *et al.*, Fermentation, I&EC Unit Processes Review, *Ind. Eng. Chem.*, **53,** 934 (1961) (excellent); Annual Review Supplement, *Ind. Eng. Chem.*, **53,** 66 (1962); Phillips *et al.*, Oxygen Transfer in Fermentations, *Ind. Eng. Chem.*, **53,** 749–754 (1961); Deindoerfer and Humphrey, Mass Transfer from Individual Gas Bubbles, *Ind. Eng. Chem.*, **53,** 1755–1759 (1961).

[5]Articles in *Ind. Eng. Chem.*, **52**(1), 59 (1960); Gutcho, Chemicals by Fermentation, Noyes, 1973.

with the yeast from about 2 to 15%) that will suppress the activity of the organism and of the enzymes. We are, furthermore, recognizing the importance of the life processes of microorganisms in making vitamins, some of which are being recovered and sold in a concentrated form. "As with all life, the activities of microorganisms can be reduced to a consideration of enzymes acting on a substrate."[6]

INDUSTRIAL ALCOHOL

Industrial alcohol was an outgrowth of alcoholic beverages, but now it has become important by virtue of its economically useful properties as a solvent and for synthesis of other chemicals. Alcohol is sold as tax-paid[7] alcohol or, much more widely, as nontaxed denatured alcohol. The *completely denatured* formulas comprise admixtures of substances which are difficult to separate from the alcohol and which smell and taste bad, all this being designed to render the alcohol nonpotable. Such completely denatured alcohol is sold widely without bond. Factories find it an essential raw material. A typical completely denatured alcohol formula follows:

Formula No. 18. To every 100 gal of ethyl alcohol of not less than 160 proof add:
0.125 gal of Pyronate or a compound similar thereto.
0.50 gal of acetaldol (β-hydroxybutyraldehyde), 2.50 gal of methyl isobutyl ketone, and 1.00 gal of kerosine.

The federal government has recognized the needs of industry for alcohol in such form that it can center into specialized manufacturing processes where the denaturants used in completely denatured alcohols would interfere. So, since 1906, when the first U.S. denatured-alcohol law was passed, many formulas for *specially denatured* alcohol have been approved by the federal authorities. Such special formulas are limited to certain designated processes and are manufactured, stored, and used under bond, to prevent unlawful consumption. However, the 40 approved special formulas under their authorized uses enter into an exceedingly broad section of the entire industrial life of the nation.[8] Typical specially denatured formulas[9] are

To every 100 gal of ethyl alcohol, add for the designated number:
No. 1. Five gallons approved wood alcohol. Withdrawals for authorized uses: plastics, dehydrations, explosives, food products, chemicals, etc.
No. 28. One-half gallon benzene or one-half gallon rubber hydrocarbon solvent. Withdrawals for authorized uses: plastics, dehydrations, explosives, food products, chemicals, etc.
No. 29. One gal of 100% acetaldehyde or other approved denaturant. Withdrawals for authorized uses: manufacturing acetaldehyde, acetic acid, esters, ethers, etc.

In industrial nomenclature *alcohol* means ethyl alcohol, or ethanol (C_2H_5OH). It is sold by the gallon, which weighs 6.794 lb and contains 95% C_2H_5OH and 5% H_2O both by volume at 15.56 C.[10] No distinction is made as to the source of the alcohol, whether from fermentation or from synthesis.

[6] McGraw Hill Encyclopedia, vol. 7, p. 84, 1966.

[7] The federal tax is $10.50 per proof gallon, hence $19.95 on a gallon of 190-proof alcohol. The total sum collected by the Federal Alcohol Tax Unit was $5,110,001,000 in 1972, largely from beverages. A *proof gallon* (tax gallon) signifies a gallon containing 50% alcohol by volume; 100 volumes of 100-proof alcohol contain 50 volumes of absolute alcohol and 53.73 volumes of water owing to volume contraction. Ordinary alcohol of 95% strength is thus 190-proof alcohol, and pure anhydrous alcohol is 200 proof. It is interesting to observe that 100-proof alcohol is about the lower limit of burning for alcohol dilutions by direct ignition at ordinary temperatures. A *wine gallon* is a measure of volume (231 in.^{3}) of any proof.

[8] Withdrawals of specially denatured alcohol in 1965 amounted to 308.5 million wine gallons and only 2.2 million wine gallons of completely denatured alcohol.

[9] Formulas for Denatured Alcohol, U.S. Revenue Service, Part 12 of Title 26, Federal Regulations, 1961.

[10] This corresponds to 92.423% of ethyl alcohol by weight. However, when alcohol percentage strength is given, it refers to percentage by volume.

USES AND ECONOMICS Tax-paid pure alcohol is used only for medicinal, pharmaceutical, flavoring, and beverage purposes. The large consumption and the wide field of industrial application have arisen only since low-priced, tax-free denatured alcohols have become available. The trend has been toward the use of specially denatured alcohols for industrial applications. Alcohol is second only to water in solvent value and is employed in nearly all industries. In addition, it is the raw material for making hundreds of chemicals. Most important of these are acetaldehyde, ethyl acetate, acetic acid, ethylene dibromide, glycols, and ethyl chloride. During World War II, when tremendous quantities of butadiene and styrene were needed for the synthetic-rubber program, alcohol was used as a supplemental raw material for butadiene, although petroleum is the cheaper raw material. In fact, in 1945, 75% of all specially denatured alcohol entered into synthetic-rubber manufacture.

In petroleum-poor countries it is economical to mix alcohol (frequently absolute) with gasoline in order to conserve petroleum products. For such mixtures, if the alcohol is not anhydrous, a blending agent is needed. These procedures have been carried out quite extensively in Europe and in some tropical sugar-growing countries, where from 10 to 20% alcohol may be in the motor fuel, the higher amount having been employed in Sweden for many years. Similar proposals have been made for the United States, and even tried in localized areas, but were uneconomical.

The cost of alcohol is largely that of the raw material, although steam, overhead, labor, and other factors vary as to location. Raw materials from which ethyl alcohol may be made fall into five general classifications: (1) hydrocarbon gases, (2) starchy materials (cereal grains, potatoes), (3) fruit, (4) cellulosic materials (wood, agricultural residues, and waste sulfite liquors), and (5) saccharine materials (molasses from sugar beets and sugarcane). See Table 31.2. Molasses contains from 50 to 55% fermentable sugar, consisting mainly of 70% sucrose and 30% invert (a glucose-fructose mixture) sugar.

The economics of the situation have changed, and fermentation now occupies a place of minor importance as a source of *industrial* alcohol. 1950 was the first year that fermentation accounted for less than 50% of the alcohol produced, the greater part being synthetically derived. The beverage laws specify the use of grain alcohol in certain beverages (whisky), and custom demands it for some tax-paid solvent uses. Thus a demand is created for the higher-priced grain alcohol, which means that under these conditions lower-priced synthetics will not entirely displace fermentation alcohol.

TABLE 31.2 ***Ethyl Alcohol Production by Type of Raw Material, Fiscal Year Ending June 30*** *(In thousands of gallons of alcohol and spirits of 190 or more proof)*

Raw material	*1965*	*1972*	*1980**
Ethyl sulfate	447,596	124,155	100,000
Molasses	4,234	1,721	1,000
Ethylene gas	117,633	340,522	600,000
From redistillation	33,101	36,046	50,000
Grain and grain products	80,223	88,022	100,000
Sulfite liquors	7,050	6,305	6,000
Cellulose pulp, chemical and crude alcohol mixtures	2,076	n.a.	3,000
Fruit	21,728	27,120	35,000
Whey	266	423,682	600,000
Potatoes	20	n.a.	0
Marshmallows	5	n.a.	0
Total alcohol produced	713,932	1,056,573	1,495,000

Source: Alcohol and Tobacco Summary Statistics, ATF Publication 67(4-73), 1974. n.a., Not available. *Estimated.

MANUFACTURE OF INDUSTRIAL ALCOHOL

RAW MATERIALS The manufacture of alcohol from ethylene, and other synthetic manufacturing procedures, now the most important source, is discussed in Chap. 38 and illustrated by the flowchart in Fig. 38.9. Alcohol from cellulosic materials, wood, wood wastes, and sulfite liquors, is considered in Chap. 32, principally under *wood saccharification*, where yields and references are given. This procedure is not competitive except under special conditions, largely because of the cost of converting cellulosic materials to fermentable sugars. Of the 300 million gal of industrial ethanol produced in the United States in 1973, only 3% was made by fermentation. Continually changing prices of the various carbohydrate feedstocks available can cause major changes in the ratio and it is certainly possible that petroleum shortages will make this process important again. A new process[11] may prove useful.

REACTIONS The principal reaction in alcohol fermentation is

Equation of inversion:[12]

$$\underset{\text{Sucrose}}{C_{12}H_{22}O_{11}} + H_2O \xrightarrow{\text{invertase}} \underset{d\text{-Glucose}}{C_6H_{12}O_6} + \underset{d\text{-Fructose}}{C_6H_{12}O_6}$$

Equations of fermentation:[13]

$$\underset{\text{Monosaccharide}}{C_6H_{12}O_6} \xrightarrow{\text{zymase}} \underset{\text{Alcohol}}{2C_2H_5OH} + 2CO_2 \qquad \Delta H = -31.2\text{ kcal},\ \Delta F = -51.4\text{ kcal}$$

Toward the end of a fermentation, the acidity and the glycerin increase.

$$\underset{\text{Monosaccharide}}{2C_6H_{12}O_6} + H_2O \longrightarrow \underset{\text{Alcohol}}{C_2H_5OH} + \underset{\text{Acetic acid}}{CH_3COOH} + 2CO_2 + \underset{\text{Glycerin}}{2C_3H_8O_3}$$

A small amount of glycerin is always found in alcohol fermentations.

ENERGY REQUIREMENTS, UNIT OPERATIONS, CHEMICAL CONVERSIONS Plant procedures require steam heating for distillation, power for pumping, and water for condensation, and occasionally for cooling during the exothermic fermentation. The heat evolved by this fermentation reaction calculates to 88,940,000 kg-cal, or 353 million Btu/1,000 gal 95% alcohol. Its evolution usually takes about 60 h.

The manufacture of alcohol, as presented in Fig. 31.1, can be broken down into the following principal *chemical conversions* and *unit operations.* The main steps in the competitive manufacture of alcohol from petroleum cracking (cf. Fig. 38.9) are shown in parallel comparison.

[11]New Route to Alcohol Fermentation, *Chem. Eng. News,* Sept. 16, 1974, p. 20.

[12]In commercial parlance, reducing, or invert, sugars are fermented and include *d*-glucose and *d*-fructose. The product sold commercially as glucose is usually a sirup of *d*-glucose with some dextrin and maltose, whereas dextrose USP is *d*-glucose.

[13]This is the classic Gay-Lussac equation for alcohol formation. It is the principal equation in an acid or low-pH medium. When growing yeast for sale as such or for inoculation, usually with some air blown in, the yield of alcohol is lower, since it is partly changed to CO_2 and H_2O. This fermentation, however, like so many industrial reactions, is much more involved than these simple reactions indicate. Probably the first step is phosphate hexose ester formation, followed by a split in the six-carbon chain. Cf. Michaelis, Chemistry of Alcoholic Fermentation, *Ind. Eng. Chem.,* **27,** 1037 (1935). The fused oil (mixed amyl alcohols with some propyl, butyl, and hexyl alcohols and esters), amounting to 3 to 11 parts per 1,000 parts of alcohol, obtained from yeast fermentations, is held to be furnished by the protein materials in the *mash* fermented. Cf. Prescott and Dunn, Industrial Microbiology, 3d ed., chap. 4, McGraw-Hill, 1959; Underkofler (ed.), Commercial Fermentations, Chemical Publishing, 1952.

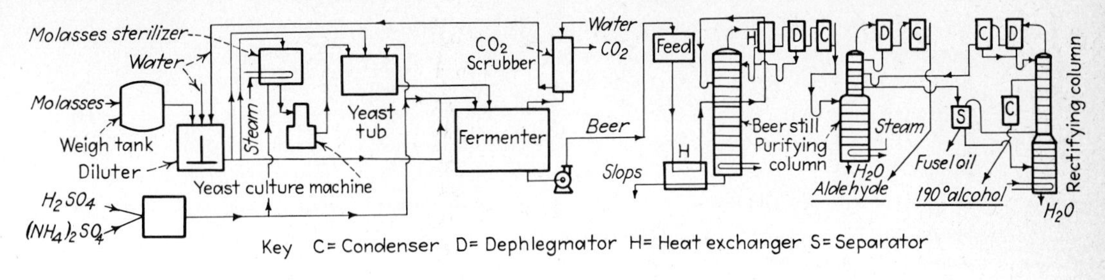

Molasses	2.5 gal.	Process water	10 gal.	Per gal. 190 proof alcohol
H_2SO_4 (60° Bé.)	0.17 lb.	Cooling water	42 gal.	
$(NH_4)_2SO_4$	0.015 lb.	Electricity	0.11 kw-hr.	
Steam	50 lb.	Direct labor	0.01 man-hr.	

Fig. 31.1 Flowchart for industrial alcohol.

Fermentation alcohol

Transportation of molasses or corn (Op)
Storage of molasses or corn (Op)
Grinding, etc., of corn (Op)
Hydrolysis by heating of cornmeal with malt or acid to make mash (Ch)
Growth of inoculating cultures (Ch)
Fermentation of diluted inverted molasses or of corn mash (Ch)
Distillation of alcohol from "beer" (Op)
Rectification and purification of alcohol (Op)
Recovery of by-products, e.g., CO_2, feed, potash salts (Op)

Alcohol from ethylene[14]

Liquefaction of petroleum gases containing ethylene (Op)
Rectification to produce pure ethylene and pure ethane (Op)
Dehydrogenation of ethane to ethylene (Ch)
Hydration over a catalyst (Ch)
Distillation of alcohol from partially converted ethylene (Op)
Rectification and purification of alcohol (Op)

MAKING OF INDUSTRIAL ALCOHOL The flowchart in Fig. 31.1 shows the various operations and conversions involved in changing molasses into saleable alcohol. Molasses, because of its strong concentration of sugar, does not support direct yeast fermentation.[15] It must first be diluted to a concentration of up to 17% sugar. This is called the *mash* and represents the carbohydrate ready for yeast inoculation. It is pumped to a large steel fermentor (60,000 to 500,000 gal), closed in modern plants, to collect the carbon dioxide evolved and to afford easier conditions for cleaning and sterilizing. An ammonium salt and sulfuric acid are added, the one to furnish a nutritive constituent deficient in molasses, and the other the right environmental pH (4.0 to 5.0) to facilitate the activity of the selected yeast and to suppress the multiplication of wild yeasts or bacteria. Magnesium sulfate is also added, when deficient, as is a small amount of phosphate (generally as superphosphate).

Meanwhile a charge of the selected yeast (about 5% of the total volume) has been growing in the yeast tub. All this is carried out under exact laboratory supervision, including the selection of the inoculating yeast strain, the sterilization of the diluted molasses, the addition of the nutrients, the pH, the temperature (76°F), and finally the cleaning and sterilizing of the yeast culture machine in readiness for the next batch. Bacteriologists have cultivated a strain of yeast that thrives under these acid conditions, whereas wild yeast and bacteria do not. As the reaction indicates, fermentation is exothermic, and so cooling is used, hence one advantage of the steel fermentor. Although the most

[14]About one-fifth of the synthetic alcohol from ethylene is made by acidic catalytic hydration at high temperature.
[15]Indeed, strong sugar sirup is a *preservative;* witness the preserving of fruits with it.

favorable temperature varies, it is usually about 70°F for starting and under 100°F at the end. By government regulations, 4 days are allowed for a fermentation cycle, though usually 36 to 50 h are used. As alcohol is formed by yeast only from monosaccharides, it is necessary to split the sucrose, $C_{12}H_{22}O_{11}$, into *d*-glucose and *d*-fructose.[16] In alcohol fermentation by yeast, this microorganism furnishes an organic catalyst, or enzyme, known as *invertase*, which effects this hydrolysis. The yeast also produces another, more important, enzyme, *zymase*, which changes monosaccharides into alcohol and carbon dioxide. The flowsheet in Fig. 7.2 outlines the procedure for purification of the carbon dioxide from fermentation. Such recovery is practiced in a number of different fermentations—industrial alcohol, whisky, and butanol-acetone.

When *starchy materials* are used for industrial alcohol, they may be converted to maltose, a disaccharide. This can be done with either malt or various mold processes. The *malt process*, except for emergencies, is too expensive for industrial alcohol. *Fungal amylase*, an enzymic material produced by the fermentation of *Aspergillus*, has been found to be an economical[17] replacement for malt for industrial alcohol. In addition, fungal amylase cannot replace barley malt for beverage production, since the flavor is not the same and because of federal regulations. During World War II, grain processing changed from batch to continuous cooking and hydrolyzing processes.[18] Two types of such cookers were used commercially: a pipeline cooker, operated at 350°F with a holding time of 1 to 2 min, and a vertical tower, utilizing a lower temperature and a longer holding time.

The liquor in the fermentors, after the action is finished, is called *beer*.[19] The alcohol is separated by *distillation*.[20] In such fermentation as pictured in Fig. 31.1, the beer, containing from 6.5 to 11% alcohol by volume, is pumped to the upper sections of the beer still, after passing several heat exchangers. As beer passes down the beer column, it gradually loses its lighter boiling constituents. The liquid discharged from the bottom of the still through a heat exchanger is known as *slop* or *stillage*.[21] It carries proteins, residual sugars and, in some instances, vitamin products, so that it is frequently evaporated and used as a constituent of animal feed. The overhead containing alcohol, water, and the aldehydes passes through a heat exchanger to the partial condenser, or dephlegmator, which condenses sufficient of the vapors to afford a reflux and also to strengthen the vapors that pass through to the condenser, where about 50% alcohol, containing the volatiles or aldehydes, is condensed. This condensate, frequently known as the *high wines*, is conducted into the aldehyde, or heads, column, from which the low-boiling impurities, or aldehydes, are separated as an overhead. The effluent liquor from part way down the aldehyde column flows into the rectifying column.

In this third column the alcohol is brought to strength and finally purified in the following manner: The overhead passing through a dephlegmator is partly condensed to keep the stronger alcohol in this column and to provide reflux for the upper plates. The more volatile products, which may still contain a trace of aldehydes and of course alcohol, are totally condensed and carried back to the upper part of the aldehyde still. Near the top of the column 95 to 95.6% alcohol is taken off through a condenser for storage and sale. Farther down the column, the higher-boiling fused oils are run off through a cooler and separator to a special still, where they are rectified from any alcohol they may carry before being sold as an impure amyl alcohol for solvent purposes. The bottom of this rectifying column discharges water.

[16]An equimolecular mixture of these is known as *invert* sugar, and results from the action of heat, acids, or enzymes.

[17]Grain Alcohol without Malt, *Chem. Week*, **68**(3), 19 (1951); Methods and Costs of Producing Alcohol from Grain by the Fungal Amylase Process on a Commercial Process, *U.S. Dep. Agric. Tech. Bull.* 1024.

[18]Stark *et al.*, Wheat as a Raw Material for Alcohol Production, *Ind. Eng. Chem.*, **35,** 133 (1943); Continuous Cooking of Cereal Grains, *Chem. Metall. Eng.* **51**(10), 142 (1944) (Pictured flowchart); Gutcho, Alcoholic Beverage Processes, Noyes, 1976.

[19]This is a general term applied to the result of any such fermentation, whether it results finally in industrial alcohol or the beverage beer, or whisky, or butyl alcohol and acetone.

[20]ECT, 2d ed., vol. 1, pp. 501–537; Bosworth, Alcohol Rectification, *Chem. Eng. Prog.*, **61**(9), 82 (1965).

[21]Reich, Molasses Stillage, *Ind. Eng. Chem.*, **37,** 534 (1945).

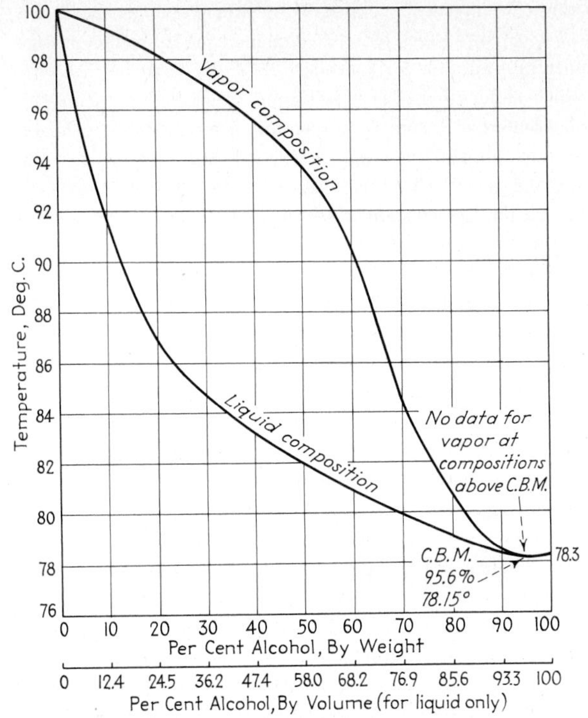

Fig. 31.2 Temperature versus composition of vapor and liquid for alcohol-water at 760 mm.

Alcohol-water mixtures are rectified to increase the strength of the alcohol component by virtue of the composition of the vapors being stronger in the more volatile constituent than the liquid from which these vapors arise. This is shown quantitatively by the curves in Fig. 31.2, where the composition of the vapor in equilibrium with the liquid is on a horizontal line. However, alcohol cannot be made stronger than 95.6% by rectification, because, as can be seen from Fig. 31.2, water forms a binary constant-boiling mixture of this composition which boils slightly lower than absolute, or anhydrous, alcohol. The principles shown here are the basis of the strengthening of the more volatile constituent of any liquid mixture by distillation.

ABSOLUTE, OR ANHYDROUS, ALCOHOL[22]

Anhydrous alcohol was made by absorbing the 4 or 5% water present in 95 to 96% industrial alcohol, using quicklime, with subsequent distillation. This process was expensive, and although it produced a very high quality of anhydrous alcohol, it has now been superseded largely by improved chemical-engineering unit operations, distillation and extraction, involving a third component. This has led to a lower cost of dehydrating operations, mostly of a continuous nature, using preferably all liquids or solutions, and has resulted in reducing the price of anhydrous alcohol to a figure only slightly in excess of the usual selling value of the alcohol contents.

[22]This is practically 100% ethyl alcohol, frequently known as *absolute* alcohol, but since the absence of water is more notable than that of other impurities, the term *anhydrous* alcohol is preferred by some.

The water in 95% alcohol is removed technically by either of two principal methods: (1) Dehydration by *distillation with a third component,*[23] which forms a minimum constant-boiling mixture in the system, boiling at a lower temperature than the 95% alcohol (78.15°C) or the water. Here (*a*) the minimum is a binary one, of which water–ethyl ether is an example,[24] or (*b*) the minimum is a ternary one, of which alcohol-water-benzene is an example. In such instances anhydrous alcohol is obtained at the bottom of the distilling column, because its vapor pressure is relatively lower than that of the constant-boiling mixture removing the water. (2) Dehydration by *countercurrent extraction,*[25] usually also in a continuous column with a third component which depresses the vapor pressure of water more than it depresses the vapor pressure of alcohol, e.g., glycerol, ethylene glycol, glycerol or glycol with dissolved salts, and a molten eutectic mixture of sodium and potassium acetates. Anhydrous alcohol comes out at the top of the extraction column.[26]

The basic principle of the process using benzene as a withdrawing agent is illustrated[27] in Fig. 31.3. There are three binary minimum constant-boiling mixtures in the system, two homogeneous ones and one heterogeneous one (between water and benzene), and a ternary minimum constant-boiling mixture which is the lowest-boiling composition in the system, boiling at 64.85°C. In Fig. 31.3 the composition of the ternary minimum constant-boiling mixture is represented by point *F*. In order that the removal of the constant-boiling mixture from the starting mixture may leave anhydrous alcohol in the still, the starting composition must lie on the straight line *CF*. If the starting mixture is to be made up by adding benzene to 95% alcohol, the starting composition must also lie on the line *EB*. Therefore the intersection *G* represents the starting composition. If enough benzene is added to 95% alcohol to bring the total composition to point *G*, continuous distillation gives the ternary constant-boiling mixture (bp 64.85°C) at the top of the column and absolute alcohol (bp 78.3°C) at the bottom of the column in a simple distillation.

An important[28] feature of the process is separation of the condensate into two liquid layers, represented in Fig. 31.3 by points *M* and *N*. The ratio of the top layer *N* to the bottom layer *M* is equal to *MF*/*FN*, or 84:16. The compositions involved are shown in Fig. 31.4, which also illustrates how this process functions. These same principles of distillation in multicomponent systems, involving constant-boiling mixtures, are used for dehydrating other organic liquids, such as propyl alcohol, and for removing the water formed in sulfonations (benzenesulfonic acid) and esterifications (ethyl acetate).[29] The fundamentals of distillation are presented here because of the extensive data on alcohol that are available.

BEERS, WINES, AND LIQUORS

The making of fermented beverages was discovered by primitive humans, and has been practiced as an art for thousands of years. Within the past century and a half it has evolved into a highly developed science. A brewer has to be an engineer, a chemist, and a bacteriologist. In common with other food industries, the factors taste, odor and, almost, individual preference exist, to force the manufacturer to exert the greatest skill and experience in producing palatable beverages of great variety. In the last analysis, the criterion of quality, with all the refinements of modern science, still lies in the human sensory organs of taste, smell, and sight.

[23]This third component is frequently called a *dehydrating*, or *withdrawing, agent*, or simply an *entrainer*. See ECT, 2d ed., vol. 2, pp. 839–859, 1963 (references), and Tassio (ed.), Extractive and Azeotropic Distillation, ACS, 1972.

[24]Othmer and Wentworth, Absolute Alcohol, *Ind. Eng. Chem.*, **32,** 1588–1593 (1940). Here the water is removed overhead by ether, the system being under 100 lb of pressure.

[25]Othmer and Trueger, Recovery of Acetone and Ethanol by Solvent Extraction, *Trans. AIChE*, **37,** 597–619 (1941).

[26]The use of a solid dehydrating agent, such as quicklime or calcium sulfate, may be looked upon as an extreme case of this method, though usually run in a discontinuous manner because of the solid involved.

[27]A somewhat similar procedure is the ether pressure system of Othmer and Wentworth, *op. cit.*

[28]Guinot and Clark, Azeotropic Distillation in Industry, *Trans. AIChE* (London), **16,** 189 (1938).

[29]Consult Perry, sec. 13, on distillation, for a rigorous and fundamental treatment of this important unit operation.

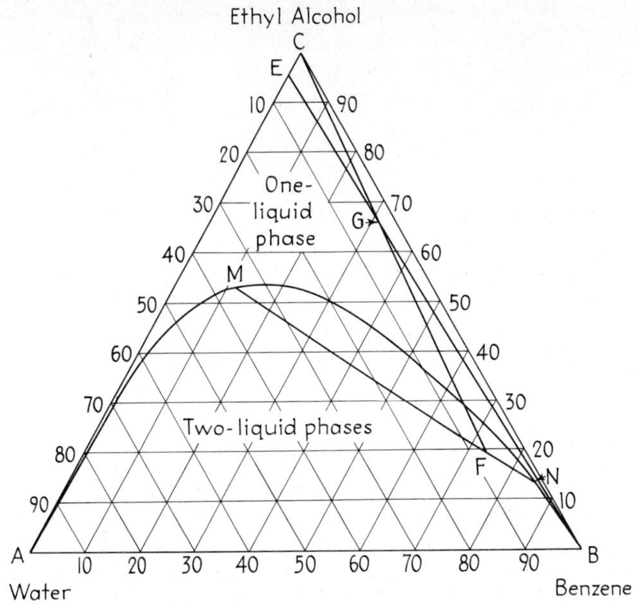

Fig. 31.3 Ternary diagram of the liquid system water-alcohol-benzene.

USES AND ECONOMICS As Table 31.3 indicates, many millions of barrels of alcoholic beverages are manufactured in the United States each year. The amount of wine production has been increasing rapidly in recent years.

RAW MATERIALS Grains and fruits supplying carbohydrates are the basic raw materials. The variety of grains and fruits employed is wide, changing from country to country or from beverage to beverage. Russia ferments potatoes and by distillation obtains vodka; similar treatment of the sap of the maguey in Mexico yields pulque; but the world's chief raw materials for fermentations are the cereals, corn, barley, and rice, and grapes.

MAKING OF BEER[30] Beer and allied products are beverages of low alcoholic content (2 to 7%) made by brewing various cereals with hops, usually added to impart a more-or-less bitter taste and to control the fermentation that follows. The cereals employed are barley, malted to develop the necessary enzymes and the desired flavor, as well as malt adjuncts: flaked rice, oats, and corn; wheat is used in Germany, and rice and millet in China. Brewing sugars and sirups (corn sugar, or glucose)

TABLE 31.3 ***U.S. Production of Alcoholic Beverages, Fiscal Years 1965–1980*** *(In thousands)*

Beverage	*1965*	*1972*	*1980**
Still, wine gal	197,257	599,000	800,000
Sparkling, wine gal	6,358	20,400	30,000
Rectified, proof gal	92,923	120,907	150,000
Beer, bbl	108,015	143,000	200,000

Source: U.S. Treasury, Internal Revenue Service. *1980 is estimated. *Note:* 1 bbl contains 31 gal.

[30]Shearon and Weissler, Brewing, *Ind. Eng. Chem.*, **43,** 1262 (1951) (many excellent pictures, tables, and diagrams); ECT, 2d ed., vol. 3, p. 297, 1964.

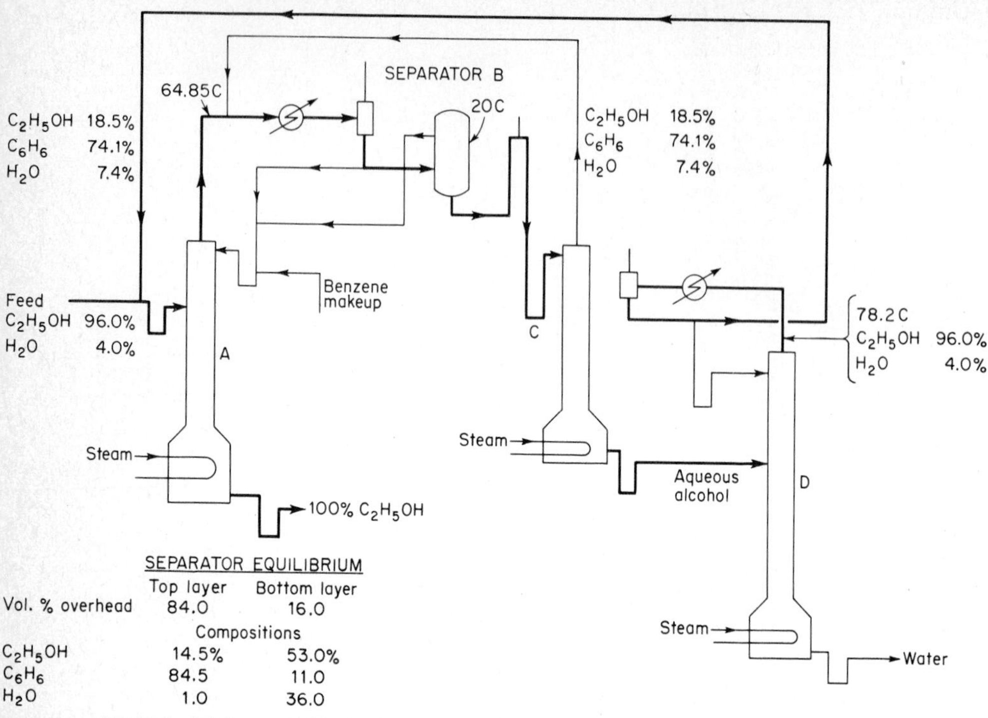

Fig. 31.4 Dehydration of 96% ethanol to absolute alcohol by azeotropic distillation with benzene at 1 atm. 96% alcohol is fed into column *A*. The ternary azeotrope is taken overhead in this column, and absolute alcohol is obtained as a bottoms product. The overhead vapors are condensed and passed to a separator (decanter) *B*, in which two liquid layers form. The upper layer, rich in benzene, is returned to column *A* as reflux, and the lower layer is fed to column *C*, which produces the ternary azeotrope as the overhead product and benzene-free aqueous alcohol as the bottoms product. This latter product is fed to column *D*, which produces by ordinary distillation an overhead product of 96% alcohol and a bottoms product of nearly pure water. The overhead from column *D* is recycled to column *A* for removal of the water. The benzene is recycled continuously in this system, and it is necessary only to make up the benzene losses from the system. This withdrawing agent is used over and over again with a loss that should not exceed 0.5% of the volume of the anhydrous alcohol produced. [*Perry, p. 13-42; Chem Eng. (N.Y.)*, **67**(10), 129 (1960).]

and yeast complete the raw materials. For beer the most important cereal is barley, which is converted into malt by partial germination.[31]

The barley is steeped in cold water and spread out on floors or in special compartments and regularly turned over for from 5 to 8 days, the layers being gradually thinned as the germination proceeds. At the proper time, when the enzymes are formed, growth is arrested by heat. During growth, oxygen is absorbed, carbon dioxide is given off, and the enzyme *diastase* is formed. The last-mentioned is the biological catalyst that changes the dissolved starch into the disaccharide maltose which, after transformation into the monosaccharide glucose by *maltase*, is directly fermentable by yeast.

The flowchart for beer manufacture in Fig. 31.5 may be divided into three groups of procedures: (1) brewing of the mash through to the cooled hopped wort, (2) fermentation, and (3) storage, finishing, and packaging for market. *Mashing* is the extraction of the valuable constituents of malt, malt adjuncts, and sugars by macerating the ground materials with 7.5 to 9 bbl of water per 1,000 lb of materials listed in Fig. 31.5 and treating with water to prevent too high a pH, which would tend to

[31]Continuous Brewery Opens in Spain, *Chem. Eng. (N.Y.)*, **74**(19), 156 (1967); cf. ECT, 2d ed., vol. 2, pp. 384–413; Chopey, What's Doing in Beer Brewing?, *Chem. Eng. (N.Y.)*, **69**(13), 94 (1962) (process flowchart with pictures).

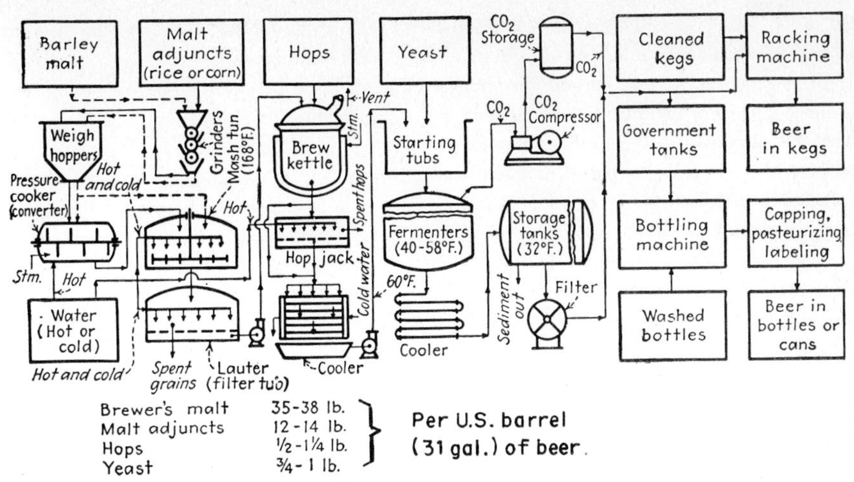

Fig. 31.5 Flowchart for the manufacture of beer.

make a dark beer. In the pressure cooker the insoluble starch is converted into liquefied starch, and the soluble malt starch into dextrin and malt sugars. The resulting boiling cooker mash, mixed with the rest of the malt in the mash tun, raising the temperature to 168°F, is used to prepare the brewers' *wort.*[32] This is carried out in the mash tun. After all the required ingredients have been dissolved from the brewing materials, the entire mash is run from the mash tun to filter presses or the lauter or straining tub, where the wort is separated from the insoluble spent grains through a slotted false bottom and run into the copper wort cooker. For complete recovery of all substances in solution, a spray of decarbonated water at 165°F is rained through the grains. This is called *sparging.*

The wort is cooked for approximately 3 h, during two of which it is in contact with hops. The purpose of boiling is to concentrate the wort to the desired strength, to sterilize it (15 min) and destroy all the enzymes, to coagulate certain proteins by heat (180°F), to modify its malty odor and to extract the hop resins' tannin and aroma from the hops, which are added during the cooking process. At the end of the 3 h the spent hops are separated from the boiling wort very quickly through a false bottom in the hop jack or strainer underneath the copper cooker. Since the spent hops retain 3 bbl of wort per 100 lb of hops, they should also be sparged. The wort is then ready to be cooled.

The cooling step is not only to reduce the temperature, but also to allow the wort to absorb enough air to facilitate the start of fermentation. In addition, the protein and hop resins are precipitated. The hot wort first may be cooled to about 150 to 160°F in a large, shallow cooler, where certain of the resins precipitate. The wort is then run over the horizontal, brine-cooled copper tubes of the open Baudelot cooler or through a shell and tube heat exchanger,[33] where aeration also takes place. Slight concentration, due to evaporation, occurs. This operation is performed under controlled conditions to prevent contamination by *wild* yeasts. Frequently, sterilized air is used.

The cooled wort is mixed with selected yeasts in the line leading to the starting tubs, between $\frac{3}{4}$ and 1 lb of yeast being used per barrel of beer. The initial *fermentation* temperature is 40 to 43°F but, as the fermentation proceeds, the temperature rises to 58°F. This is easily explained by the fact that the conversion of the sugar to carbon dioxide and ethyl alcohol by the enzymes of the yeast generates 280 Btu/lb of maltose converted. The temperature is partly controlled by attemperators

[32]The wort is the liquid resulting from the mashing process, i.e., the extracting and solubilizing of the malt and malt adjuncts. Wort composition varies from 17 to 24% solids by weight for the first wort to approximately 1% solids for the last wort removed by the sparge water.

[33]Cf. Chopey, *op. cit.*

inserted in the fermentors. The mixture is skimmed to remove the foreign substances the evolved carbon dioxide brings to the top. Thus it is quite evident that a steady evolution of gas is necessary to cleanse the beer properly. The carbon dioxide evolved is collected by using closed fermentors and stored under 250 lb of pressure for subsequent use in carbonating beer.

The yeast gradually settles to the bottom of the tub, so that at the end of 7 to 10 days the fermented beer is ready to be vatted. The liquid is very opalescent in appearance, under a cover of foam. As the beer leaves the fermenting cellar, it contains in suspension hop resins, insoluble nitrogenous substances, and a fair amount of yeast. The beer is cooled to 32°F and stored in the cellar for 3 to 6 weeks at this temperature. During this period, clarification, separation, and precipitation of hard resins and improvement in palatability (mellowing) occur. Haze on cooling may be reduced (chillproofing) by the addition of polyvinylpyrrolidone.[34] At the end of the period the beer is carbonated[35] and pumped through a pulp filter with or without such a non-taste-imparting filter aid as asbestos fiber. In the United States, public demand favors a brilliant beverage. As a result, the beer is sometimes refiltered through cotton pulp, keeping carbon dioxide on the entire system. About 97 bbl of beer is produced per 100 bbl of wort in the starting tubs. After bottling, the beer is pasteurized at 140°F.

Some beer is not pasteurized but biologically purified by membrane filtration, which removes residual yeast cells and harmful bacteria. This ultrafiltration, and several other new procedures including the addition of antimicrobials, produce so-called bottled draft beer. Beer with the carbohydrate content reduced from the usual 4% to near zero, which reduces the food content from 160 calories to 100 calories per 12-oz bottle is also available.

MAKING OF WINE Wine has been made for several thousand years by fermentation of the juice of the grape. Like other fermentations, many primitive procedures have been supplanted by improved science and engineering to reduce costs and to make more uniform products. But now, as always, the quality of the product is largely related to grape, soil, and sun, resulting in a variation in flavor, bouquet, and aroma. The color depends largely upon the nature of the grapes and whether the skins are pressed out before fermentation. Wines are classified as natural (alcohol 7 to 14%), fortified (alcohol 14 to 30%), sweet or dry, still or sparkling. Fortified wines have alcohol or brandy added. In the sweet wines some of the sugar remains.

For the manufacture of dry red wine, red or black grapes are necessary. The grapes are run through a crusher, which macerates them but does not crush the seeds, and also removes part of the stems. The resulting pulp, or *must,* is pumped into 3,000- to 10,000-gal tanks,[36] where sulfurous acid[37] is added to check the growth of wild yeast. An active culture of selected and cultivated yeast equal to 3 to 5% of the volume of juice is added. During fermentation, the temperature rises, so that cooling coils are necessary to maintain a temperature below 85°F. The carbon dioxide evolved carries the stems and seeds to the top, which is partly prevented by a grating floated in the vat. This allows extraction of the color and the tannin from the skins and seeds. When the fermentation slows up, the juice is pumped out of the bottom of the vat and back over the top. The wine is finally run into closed tanks in the storage cellar, where, during a period of 2 or 3 weeks, the yeast ferments the remainder of the sugar. The wine is given a cellar treatment to clear it, improve the taste, and decrease the time of aging. During this treatment the wine is first allowed to remain quiet for 6 weeks to remove part of the matter in suspension, and then racked for clarification.[38] Bentonite, or other

[34]*Brew. Dig.*, **47**(5), 75 (1972).

[35]The carbon dioxide should be kept free from air, which would interfere with the stability and quality of the beer. The gas is pumped in close to 32°F and amounts to between 0.36 and 0.45% of the weight of the beer.

[36]In many modern American wineries these tanks are even larger and are constructed of concrete.

[37]Potassium or sodium metabisulfite and/or sodium bisulfite may also be used.

[38]During this and the following period the new wine undergoes a complicated series of reactions, resulting in the removal of undesired constituents and development of the aroma, bouquet, and taste. Oxidation takes place, as well as precipitation of proteins and argols and esterification of the acids by alcohols. Certain modifications of this process are presented in Chemical Technology, Key to Better Wines, *Chem. Eng. News,* July 2, 1973, p. 14; Chemistry Concentrates on the Grape, *Chem. Eng. News,* June 25, 1973, p. 16.

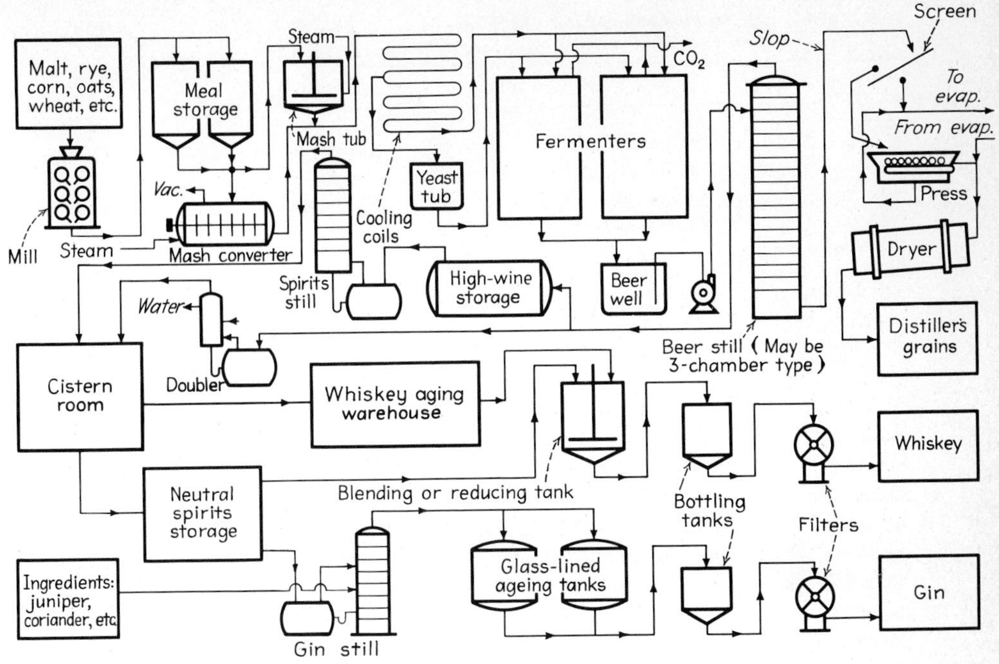

Fig. 31.6 Flowchart for the production of distilled liquors.

diatomaceous earth, may be used for clearing, 2 to 16 lb being stirred into every 1,000 gal of wine. An insoluble precipitate with the tannin is also formed. Extra tannin may also be added, and the wine racked and filtered through diatomaceous earth, asbestos, or paper pulp. The wine is corrected to commercial standards by blending it with other wines and by the addition of sugar, acids, or tannins. It is standard procedure to chill some wines for the removal of argols or crude potassium acid tartrate, which constitute the commercial source of tartaric acid and its compounds. This treatment also gives a more stable finished wine. By quick-aging methods it is possible to produce a good sweet wine in 4 months. These methods include pasteurization, refrigeration, sunlight, ultraviolet light, ozone, agitation, and aeration. The wine may be held at about freezing for 3 weeks to a month, and a small amount of oxygen gas bubbled in. Then the wine is racked, clarified, and further filtered in the usual manner. The wine trade is large and growing rapidly in the United States.

DISTILLED SPIRITS Various fermented products, upon distillation and aging, yield distilled liquors. Figure 31.6 shows the flowchart for whisky and gin, and Table 31.4 presents statistics. Brandy is distilled from wine or from the *marc*, which is the pulp left by racking or straining. Making a beer[39] from a grain mixture containing at least 51% corn and distilling and aging it yields bourbon whisky. Similarly, rye whisky must start with 51% rye in the grain to be mashed and fermented. By inspecting the flowchart in Fig. 31.6 in the light of Figs. 31.1 and 31.5, and the description accompanying them, the procedures in Fig. 31.6 for distilled liquors will be clear. In modern liquor plants the equipment,[40] up to the stills, is of steel, and the stills are of copper. By law, the aging of bourbon or rye whisky of claimed age must take place in charred new white-oak barrels of approximately 50 gal. These are kept in bonded warehouses at 65 to 85°F and at a preferred humidity of 65 to 70%,

[39]The yeast in this fermentation is grown in the presence of lactic acid to ensure proper strain and to secure the desired quality of the product (whisky).

[40]Owen, Modern Distillery Design, *Sugar*, **37**(3), 26–30 (1942); *Chem. Metall. Eng.*, **49**(11), 126 (1942) (pictured flowchart of a distillery); Stallings *et al.*, Chemical Engineering Developments in a Grain Distillery, *Trans. AIChE*, **38**, 791 (1942).

TABLE 31.4 ***U.S. Net Production of Distilled Spirits, Fiscal Year** (In tax gallons)*

	1965	*1972*	*1980**
Whisky	117,930,282	120,794,206	125,000,000
Brandy	11,521,724	14,085,875	20,000,000
Rum	2,273,811	1,908,779	2,000,000
Gin	25,560,827	28,841,563	33,000,000
Vodka	12,620,937	22,891,095	30,000,000
Alcohol and spirits:			
190 proof and over	65,864,053	67,969,431	70,000,000
Under 190 proof	36,691,486	23,496,776	30,000,000
Total	272,463,120	279,987,725	310,000,000

Source: Alcohol and Tobacco Summary Statistics, ATF Publication, **67**(4–73), 1974. *Estimated.

usually for 1 to 5 years. During this time evaporation of the contents takes place, largely through the ends of the barrel staves. By reason of more rapid capillary travel and osmosis of the smaller water molecules in comparison with the alcohol molecules, an increase in the percentage of alcohol is found in the barrel contents. The government shrinkage allowance is approximately 8% the first year, 4% the second year, 4% the third year, and 3% the fourth year. If the shrinkages are exceeded (and this is often the case), the manufacturer must pay a tax on the excess, but with the best cooperages under the best conditions these allowances can just be met. The distillate from the spirit still is under 160 proof and is subsequently diluted upon barreling to 100 to 110 proof. It is *not* pure alcohol but contains small amounts of many different constituents, generally classed together as *congenerics,* which by their reaction with each other or the alcohol, or by their absorption, both catalyzed by the char of the wood, help greatly in imparting the whisky flavor and bouquet.[41] The aging whisky also extracts color and other products from the charred white oak. Changes of a like nature occur similarly in aging brandy and rum. Here, as in other divisions of the fermentation industries, skill and scientific knowledge aid in the production of a palatable product. By law, whisky must be fermented from whole grains, so that the germs (containing the corn oil) and the husks are suspended in the liquor from the beer still in whisky manufacture. This discharge liquor is known as *slop,* or *stillage.* As shown in Fig. 31.6, it is treated to recover the values by separating the solids from the liquid slop. After vacuum evaporation of the liquid[42] portion, it is added to the solids, and the mixture is dried in rotating steam-heated dryers to produce *distillers' grains,*[43] a valuable cattle feed.

BUTYL ALCOHOL AND ACETONE

Until World War I, all the acetone produced in the United States was made by the dry distillation of calcium acetate from pyroligneous acid. Under the stimulus of the wartime demand for acetone for the manufacture of double-base smokeless powder, the important process became that developed by Weizmann[44] for the fermentation of starch-containing grains to butyl alcohol and acetone. The Commercial Solvents Corp. was organized, and it built and operated two plants in the Corn Belt to ferment corn, using *Clostridium acetobutylicum* bacteria. This fermentation, however, gave 2 parts butyl alcohol to 1 part acetone and, until the development of fast-drying nitrocellulose lacquers,

[41]Liebmann and Scherl, Changes in Whisky Maturing, *Ind. Eng. Chem.*, **41,** 534 (1949).
[42]Cf. Boruff *et al.*, Vitamin Content of Distillers' By-products, *Ind. Eng. Chem.*, **32,** 123 (1940).
[43]For a pictured flowchart, see Recovery of Grain Alcohol By-products, *Chem. Metall. Eng.*, **52**(6), 130 (1945); Boruff, Industrial Wastes: Recovery of Fermentation Residues as Feeds, *Ind. Eng. Chem.* **39,** 602 (1947).
[44]Brit. Pat 4845 (1915); U.S. Pat. 1,315,585 (1919). Weizmann used much of his large royalties from this process to help finance his interest in the Zionist movement, of which he became the head in 1920.

particularly for the automotive industry, there was virtually no market for the butyl alcohol produced. Then conditions became reversed, butyl alcohol becoming the important commodity and acetone the by-product. *This better sale of first one product and then another is characteristic of industries in the chemical field, where more than one substance results from a process and reflects the changing demand that accompanies a growing and dynamic industry.* New cultures feeding on molasses were developed, which also gave a more desirable solvent ratio (approximately 3 parts butyl alcohol to 1 part acetone). Higher costs for molasses and grain now have given the synthetic processes such a dominant position that the latter have caused cessation of the fermentation process.

RIBOFLAVIN was first produced commercially as a by-product of acetone–butyl alcohol fermentation, but the concentration was low. Several greatly improved biological processes have been dominated by the lower-cost synthetic process presently used.

VINEGAR AND ACETIC ACID

The aerobic bacterial oxidation (by the genus *Acetobacter*) of alcohol to dilute acetic[45] acid (8%) is another ancient procedure, furnishing vinegar, a flavored acetic acid solution, fermented from wine, cider, malt, or dilute alcohol. If pure dilute alcohol is fermented, pure dilute acetic acid results. The yield is 80 to 90% of theory. Air[46] must be supplied, as these formulations indicate:

$$2C_2H_5OH + O_2 \longrightarrow 2CH_3CHO + 2H_2O \qquad 2CH_3CHO + O_2 \longrightarrow 2CH_3COOH$$

Since these reactions are exothermic, either the alcohol can be slowly trickled through the apparatus, letting the heat dissipate, or it can be recirculated with special cooling. If cider, malt, or wine is fermented, the acetic acid content of the resulting vinegar rarely exceeds 5%, because of limitations of the sugar content; if dilute alcohol is the raw material, the acetic acid may rise to 12 or 14%, at which acidity the bacteria cease to thrive. If a fruit juice is turned to vinegar, certain esters are formed, varying with the raw material and thus imparting a characteristic flavor.

CITRIC ACID

Citric acid is one of our widely employed organic acids, its production amounting to about 140 million lb. Its major use is as an acidulant in carbonated beverages, jams, jellies, and other foodstuffs. Another large outlet is in the medicinal field, including the manufacture of citrates and effervescent salts. Industrial uses, relatively small, include citric acid as an iron-sequestering agent and acetyl tributyl citrate, a vinyl resin plasticizer. Citric acid is meeting competition from other organic acids, for example, fumaric, maleic, and adipic.

Except for small amounts (less than 7%) produced from citrus-fruit wastes, citric acid is manufactured[47] by aerobic fermentation of crude sugar or corn sugar by a special strain of *Aspergillus niger*, following the classical research by Currie.[48]

The overall reactions are

$$\underset{\text{Sucrose}}{C_{12}H_{22}O_{11}} + H_2O + 3O_2 \longrightarrow \underset{\text{Citric acid}}{2C_6H_8O_7} + 4H_2O$$

$$\underset{\text{Dextrose}}{C_6H_{12}O_6} + 1\tfrac{1}{2}O_2 \longrightarrow \underset{\text{Citric acid}}{C_6H_8O_7} + 2H_2O$$

[45]For a description of a large factory engaged in this process, cf. Herrick and May, *Chem. Metall. Eng.*, **42,** 142 (1935).

[46]Too much air causes losses due to further and undesired oxidation.

[47]See ECT, 1st ed., vol. 4, p. 12, and vol. 6, p. 364, for detailed presentations of the reactions and other features of citric acid made by fermentation.

[48]Currie, The Citric Acid Fermentation of *A. niger*, *J. Biol. Chem.*, **31,** 15 (1917).

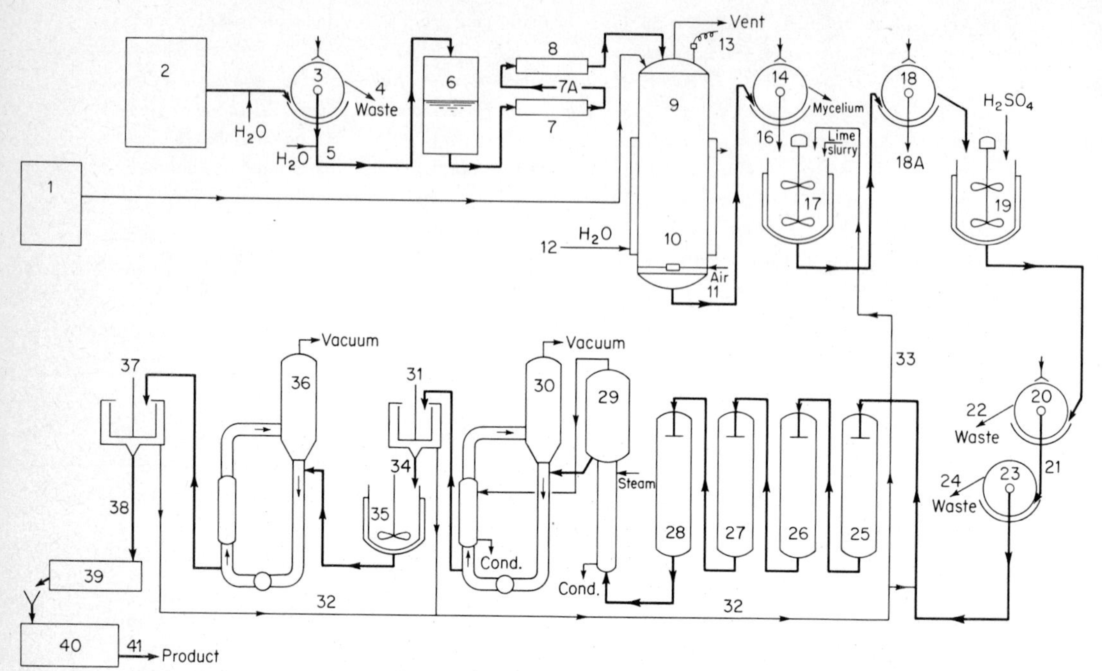

Fig. 31.7 Flowchart for commercial citric acid production. The purification of the dextrose glucose sirup is fundamental (Nos. 2 through 8), involving a rotary vacuum filter (3) to remove suspended or precipitated solids after partial dilution. This is followed by a cation-exchange cell (6) for trace-element reduction and a flash pasteurization heater (7), holding loop (7*a*), and sirup cooler (8). The sirup is pumped into the fermenter (9) plus inoculum (1). The pH adjustment is made, and nutrients added. Sterile air (11) from filters and flowmeters is sparged (10) into the fermenter (9). The fermentation process for the production of citric acid is, in its simplest terms, the conversion of a molecule of a hexose sugar to a six-carbon molecule of citric acid (9). The purification and recovery of the resultant acid are then basically an application of the lime-sulfuric scheme first used by Scheele in 1784. Today's methods involve first separating off the mycelium (14) from the broth, which contains the citric acid (16), liming it with milk-of-lime slurry (17), filtering off and washing the resultant calcium citrate (18), and finally decomposing the citrate with sulfuric acid (19). The calcium sulfate formed from this decomposition is filtered off as a waste by-product (20, 23). The more modern plants of today perform these three filtrations on some form of rotary vacuum filter. The further purification of the decomposed liquor from the sulfuric decomposition is variable from manufacturer to manufacturer. This figure indicates treatment with granular carbon in fixed beds (25, 26), followed by demineralizing beds containing cation- and anion-exchange resins (27, 28). Double- or triple-effect evaporators (29) feed a separate crystallizer (30) and centrifuge (31). [The mother liquor is recycled between feed to carbon cells (32) or to a liming tank (33).] The damp citric acid crystals are remelted (35) and vacuum-recrystallized (36). This is followed by centrifuging (37), drying (39), size classification (40), and packaging (41). The degree of purity of the initial sugar source going into the fermentation can be a factor in determining the amount of purification necessary and the need for recrystallizing the final product. (*Miles Laboratories, Inc.*)

The fermentation changes sugar and dextrose, straight-chain compounds, into branched chains. An earlier shallow-tray fermentation process was abandoned because of the expensive manual processing and the development of the submerged process.

The submerged process[49] for the manufacture of citric acid is depicted in Fig. 31.7, and this may be broken down into coordinated sequences of *biochemical conversions* (Ch) with the aid of *A. niger* (Ch) and various *unit operations* and *chemical conversions.* A selected strain of *A. niger* is grown from a test-tube slant through to a seed tank, or inoculum (Ch). This growth may take 36 to 48 h. For sequential steps see the description below Fig. 31.7.

[49]For other plants see Chopey, New Contender Enlivens Citric Acid, *Chem. Eng.* (*N.Y.*), **68**(7), 122 (1961); Faith *et al.*, p. 278, Submerged Citric Acid Fermentation of Sugar Beet Molasses, *Ind. Eng. Chem. Prod. Res. Dev.*, **1,** 59 (1962); French, Skip a Step to Citric Acid (by solvent extraction using tri-*n*-butyl phosphate), *Chem. Week*, Sept. 21, 1963, p. 59.

LACTIC ACID

Lactic acid, 2-hydroxypropionic acid, is one of the oldest known organic acids. It is the primary acid constituent of sour milk, where it derives its name, being formed by the fermentation of milk sugar (lactose) by *Streptococcus lactis.* Commercially, lactic acid is manufactured by controlled fermentation of the hexose sugars from molasses, corn, or milk. Lactates are made by synthetic methods from acetaldehyde and lactonitrile, a by-product of Monsanto's[50] acrylonitrile operation. It has been only since 1930 that lactic acid has been produced commercially from the milk by-product whey. About 2 billion lb of dry whey is produced annually from cheese or casein production, and about 1 billion lb is wasted.[51] The technical grade is employed for deliming leather in tanning. Edible grades are used primarily as acidulants for a number of foods and beverages. The small amount of lactic acid remaining is converted into plastics, solvents, and certain other chemical products. The USP grade is an old, well-established standard pharmaceutical.

DIHYDROXYACETONE ($HOCH_2COCH_2OH$) is made by the action of sorbose bacterium fermentation of glycerin.[52] This is an ingredient of an artificial-suntan lotion since it creates an artificial tan.

GLYCERIN In England, Imperial Chemical Industries makes glycerin by the fermentation process from molasses. According to a published flowchart[53] molasses is fermented anaerobically in the presence of sodium sulfite in deep tanks. After the removal of acetaldehyde and alcohol by distillation, the glycerin is concentrated, solvent-extracted, and refined.

PHARMACEUTICAL PRODUCTS The pharmaceutical industry has long employed fermentation (biosynthesis) to manufacture some of its most important medicaments. See Chap. 40, where *fermentation and life processing* are presented for antibiotics, biologicals, vitamins, and hormones. Controlled microorganisms are a most important chemical processing agent and assist in performing very complicated chemical reactions, in many cases more economically than purely chemical conversions. This is especially true for complicated structural changes to make derivatives of natural steroid hormones.

Detailed flowcharts (Fig. 40.8) are given for three antibiotics, penicillin, streptomycin, and erythromycin. Chemical synthesis is given in Chap. 40 for riboflavin, as well as fermentation processes.

ENZYMES

Tables 31.1 and 31.5 list certain enzymes, organic catalysts formed in the living cells of plants and animals, which are essential in bringing about specific biochemical reactions in living cells. Enzymes can be classified by their method of activity as follows:[53]

Catalytic function	*Name*
Oxidation-reduction	Oxidoreductases
Group transfer	Transferases
Hydrolysis	Hydrolases
Group removal	Lyases
Isomerization	Isomerases
Joining of molecules	Ligases

[50]Synthetic Lactic Acid, *Ind. Eng. Chem.*, **51**(2), 55 (1964); *Chem. Eng.* (*N.Y.*), **71**(2), 82 (1964).
[51]Fermentation Process Turns Whey into Valuable Protein, *Chem. Eng.* (*N.Y.*), **82**(6), 36 (1975).
[52]U.S. Pat. 2,948,658; *Chem. Eng. News*, Aug. 15, 1960, p. 54.
[53]*Chemtech*, **3,** 678 (1973); Wieland, Enzymes in Food Processing and Products, Noyes, 1972.

TABLE 31.5 *Typical Industrial Enzymes Sold Today*

Enzyme	*Application*
Proteases	Cheesemaking Digestive aids Meat tenderizing Beer chillproofing Cereal sirups manufacture Leather manufacture
Carbohydrases	Starch hydrolysis Sucrose inversion Fruit juice and vinegar clarification
Nucleases	Flavor control
Hydrolytic enzymes	Destruction of toxic or undesirable components in food
Oxidases	Oxidation prevention and color control in food products Clinical diagnostics

Source: Chemtech, **3,** 677 (1973), and **4,** 47, 309, 434 (1974).

Industrial enzymes originate from bacteria, fungi, and higher plants, and from animal sources. They are either extracted or prepared by fermentation (microbial enzymes).

One of the oldest and most widely sold is *rennet,*[54] which is used in the cheesemaking industry, for textile design, and for leather softening. It is prepared by extraction from washed and defatted calves' stomachs. Many enzymes are produced by the action of microorganisms and are manufactured either by submerged or surface fermentation procedures. However, continuous operations of the process are not as easily controlled as batch fermentations. The enzyme *amylase,* for converting starch into sugar, has long been made by Takamine[55] by submerged fermentation, starting with corn-steep liquor and constarch. This mixture, after proper sterilization and cooling, is inoculated with *Bacillus subtilis* and fermented. The amylase can be isolated by the precipitation of broth by isoamyl alcohol and centrifugation. This enzyme is also used for textile designing and baking and to attack food spots before dry cleaning.

SELECTED REFERENCES

Amerine, M. A., and W. V. Cruess: Technology of Wine Making, Avi Publishing, 1960.
Bergmeyer, H. U. (ed.): Methods of Enzymatic Analysis, 2d ed., Academic, 1974.
Cook, A. H. (ed.): Barley and Malt: Biology, Biochemistry, Technology, Academic, 1962.
Developments in Industrial Microbiology, vols. 1–3, Plenum, 1960–1963.
Duddington, C. L.: Micro-organisms as Allies, Macmillan, 1961.
Gutcho, S. J.: Chemicals by Fermentation, Noyes, 1973.
Gutcho, S. J.: Immobilized Enzymes, Noyes, 1974.
Kirk, R. E., and D. F. Othmer, Encyclopedia of Chemical Technology, 2d ed., vol. 12, Bibliography on Malts and Malting, pp. 861–886, Wiley-Interscience, 1967.
Lawrence, A. A.: Food Acid Manufacture Recent Developments, Noyes, 1974.
Waksman, S. A.: Actinomycetes, Williams and Wilkins, 1959–1962, 3 vols.
Webb, A. D. (ed): Chemistry of Winemaking, ACS, 1975.
Webb, F. C.: Biochemical Engineering, Van Nostrand, 1964.
Whitaker, J. R. (ed.): Food Related Enzymes, ACS, 1975.

[54]Placek *et al.*, Commercial Enzymes by Extraction (Rennet), *Ind. Eng. Chem.*, **52**(1), 2 (1960) (illustrated flowchart).
[55]Forbath, Flexible Processing Keys Enzymes' Future, *Chem. Eng.* (*N.Y.*), **64**(2), 226 (1957) (flowchart). The Takamine plant at Clifton, N.J., turns out about 30 different enzymes by batch processing.

chapter 32

WOOD CHEMICALS

Chemical wood technology is the application of chemical engineering to achieve practical results in the industries concerned with processing wood for fibers and by-products. Wood processors classify woods into two groups (1) hardwoods, which come from deciduous trees, and (2) softwoods, which come from conifers. Production of wood is high, about the same as the current use rate, although land available for forest production is steadily diminishing. As a renewable resource, wood's value is high, but waste is enormous. Inforest residues constitute up to 30% of the harvest and, while generally unsuitable for pulp and paper, could yield large amounts of chemicals. "Green junk" is the term applied to cull trees left growing in cutover land. The amount of such material has been estimated[1] at 750 million tons of hardwood in the U.S. southern states alone. Underutilization of wood products arises from the complexity of the material, the undesirably dilute forms in which by-products are frequently available, lack of economic pressure for utilization, and lack of adequate chemical knowledge. Environmental laws no longer allow dumping of unwanted residues into streams; this makes wood by-product disposal expensive. Sylvichemicals are becoming increasingly able to compete, because petroleum-based chemicals are constantly increasing in price. The current major use of wood by-products is as fuel, a very low-valued use for such a complex raw material.

Wood is a complex polymeric substance. Forty to fifty percent of wood's dry weight is cellulose, which is most valuable as fiber. The cell walls consist of polysaccharides called holocellulose and lignin. Holocellulose is a mixture of short-chain polymers of arabinose, galactose, glucose, mannose, and xylose (as anhydrides). Lignin is a complex polymer of condensed substituted phenols, which acts as a binder for the cellulose fibers. Various extractives such as turpentine, oleoresin (rosin), arabinogalactan, rubber, sugar, etc., are present in small quantities. Table 32.1 shows average values for the composition of U.S. wood.

DISTILLATION OF HARDWOOD

HISTORICAL The origin of the carbonization of wood can be traced back to antiquity. The first chemical process was probably the making of charcoal, originated by cave inhabitants to provide a smokeless fuel underground. In ancient Egypt a process for the distillation of wood was known from which not only charcoal was recovered, but also fluid wood tar and pyroligneous acid, the latter for use in embalming.

The wood distillation industry was an important source of income for the United States in earlier times. The pyroligneous acid, tar, pitch, methanol, acetic acid, and acetone produced were saleable and useful. The charcoal[2] was widely used for metallurgical purposes. Acetone was made by the dry distillation of calcium acetate made from the acetic acid. Most installations were crude

[1]Wood Draws Attention as Plastics Feedstock, *Chem. Eng. News*, Apr. 21, 1975, p. 13.
[2]Charcoal—The Model T Luxury Fuel, *Chemtech*, **1**, 383 (1971).

TABLE 32.1 ***Composition of Wood***

	Softwood %	*Hardwood %*
Holocellulose	66	76
α-Cellulose	46	49
Pentosans	8.5	19.5
Lignin	27	21

Source: Chemistry in the Utilization of Wood, Pergamon, 1967.

and inefficient. When competition from synthetic methanol and acetic acid appeared, ingenious plants for efficient by-product recovery and purification were devised but proved unable to compete. Charcoal, made from scrap lumber and frequently augmented by the addition of products from coal continues to exist as a small luxury industry providing fuel for specialty cooking. By-product methanol and acetic acid are insignificant.

NAVAL STORES INDUSTRY

The *naval stores*[3] industry, involving products from softwood, has three main divisions: gum, steam-distilled, and sulfate. See Fig. 32.3. *Gum* naval stores are produced from crude gum (oleoresin) from living longleaf and slash pine trees (5% of all rosin and 4% of all turpentine). *Steam-distilled* naval stores are derived from resin-saturated stumps of the original old-growth longleaf and slash pine forests of the Southeast (42% of all rosin and 13% of all turpentine). *Sulfate* naval stores are produced from formerly wasted sulfate, or kraft liquors (54% of all rosin as tall oil rosin and 82% of all turpentine). These products were used by the Navy in the days of wooden ships. Turpentine and rosin are complex mixtures of organic compounds.

HISTORICAL Turpentine and rosin are produced from the longleaf pine in the southeastern and southern parts of the United States; in fact, in the early days of America, the forests in this region were said to extend in a belt approximately 200 mi wide and 1,000 mi long. In 1608, what was probably the first cargo of American naval stores was shipped to England. Pine-tar manufacture became one of the first industries of the New England colonies, flourishing as early as 1650. By 1750, North Carolina had been settled and had established a growing trade in naval stores, and by 1900 Georgia was the chief producing state. In the same year a plant at Gulfport, Miss., was erected to obtain turpentine and pine oil from stumps and logging waste and also to attempt removal of the rosin present. Florida and Georgia produce a majority of today's gum turpentine and rosin.

PRODUCTS AND ECONOMICS The production of rosin and turpentine is shown in Fig. 32.1, and the consumption in Table 32.2. Although turpentine and rosin are available from different

TABLE 32.2 ***Naval Stores Consumption*** *(In millions of pounds)*

	Turpentine			*Rosin*		
	1971	*1972*	*1973*	*1971*	*1972*	*1973*
Chemicals and rubber	194	165	157	299	294	291
Ester gums and synthetic resins	0	0	0	113	127	138
Paint, varnish and lacquer	0.8	0.5	0.3	37	27	26
Paper and paper size	0	0	0.4	268	244	180
Other	0.5	1.5	1.3	26	26	27
Total	195	167	159	743	718	662

Source: U.S. Agricultural Statistics, 1974.

[3]Naval Stores Aren't Shipshape, *Chem. Week*, Dec. 2, 1970, p. 32.

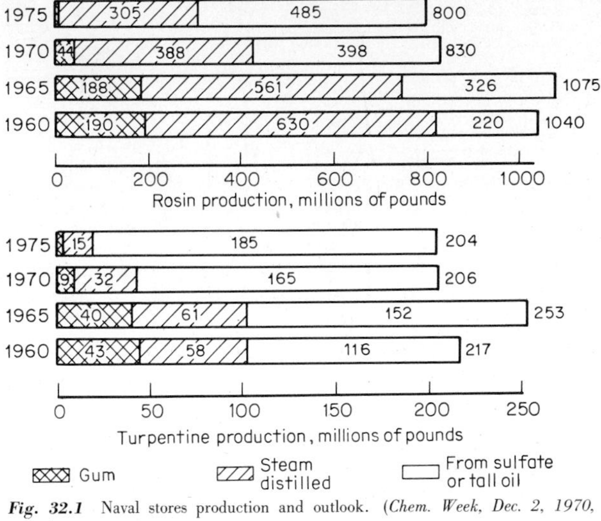

Fig. 32.1 Naval stores production and outlook. (*Chem. Week, Dec. 2, 1970, p. 32.*)

methods of production (Fig. 32.2), the uses of each product from the various sources are generally interchangeable. The major products of the naval stores industry are turpentines, rosins, pine oils, rosin oils, spirits, and pine-tar pitches and tars. The table below, Fig. 32.2, lists the products obtained from 1,000 tons of pine chips.

Turpentine is a mixture of organic compounds known as *terpenes*. A major constituent is α-pinene.[4]

```
              CH3
              |
CH ═══════════C──────────CH
|               ╲        |
|     H3C·C·CH3          |
|          |             |
CH2────────CH────────CH2
```

As a household paint and varnish thinner, turpentine has been almost totally replaced by less odoriferous and cheaper petroleum hydrocarbons. In 1974[5] the uses of turpentine were: 48% for pine oil, 16% for resins, 16% for insecticides, 9% for fragrances, and 11% for miscellaneous. Less than 1% was used as a solvent in the paint industry, the traditional use. Turpentine and rosin are both in short supply.

Pine oil is made by the fractionation of pine stump extractives, but much "synthetic" pine oil is made by acid conversion of the α-pinene in turpentine. Pine oil is used in household cleaners as a bactericide, as an odorant, for flotation, as a solvent, and in processing textiles. α-Terpineol and terpin hydrate are made by further acid degradation of turpentine.

Low-molecular-weight polyterpin resins used in pressure-sensitive adhesives, chewing gum, dry cleaning sizes, and paper sizes are made from α and β-pinenes and pyrolyzed α-pinene from turpentine.

Toxaphene insecticides are made by chlorinating camphene and camphene-α-pinene fractions.

A wide variety of flavors and fragrances is made by the isolation of specific fractions of turpentine and synthetic conversion. These include lime, lemon, peppermint, spearmint, and nutmeg

[4]Terpenes, ECT, 2d ed., vol. 14, p. 803, 1969.
[5]Chemicals from Trees, *Chemtech*, **5**, 236 (1975).

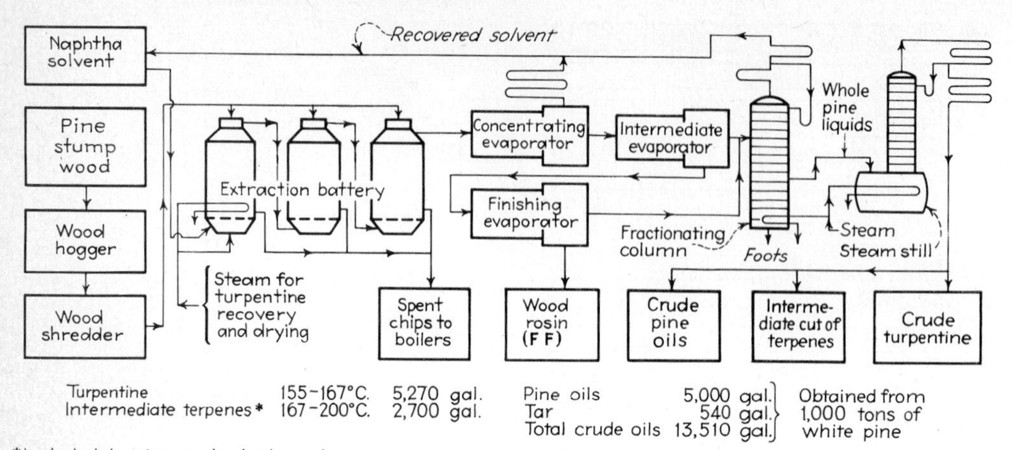

Fig. 32.2 Flowchart for wood rosin and turpentine production by steam and solvent extraction.

synthetic essential oils, and linalool, citronellol, nerol, geraniol, menthol, and ionones as semisynthetics. Some new, previously unknown, semisynthetic perfume constituents have also been prepared. Turpentine is a far more valuable and useful substance as a chemical raw material than it ever was as a solvent.

MANUFACTURE FOR NAVAL STORES

The *raw materials* for the production of naval stores include those species of wood commonly classed as coniferous. Either the gum that flows from living longleaf or slash pine or stumps from cutover pine forests, together with other resinous wood wastes, form the source of raw material for the various processes employed. Kraft, or sulfate, waste liquors, especially tall oil, are becoming increasingly important.

Processing of gum.[6] The processing of gum for naval-stores products is the earlier procedure. Originally, all gum was produced in a direct-fired copper still. The turpentine was distilled off with the water and separated by gravity; the rosin remained in the kettle and was removed in a molten, impure condition at the end of the run and strained to remove wood chips, pine needles, dirt, etc. Fire stills have largely been supplanted in this country by large *steam stills* operated either batchwise or continuously in what is known as a central plant. Gum farmers obtain the gum by chipping or wounding longleaf and slash pines and collecting the exudate by means of gutters or cups. A chemical spray on the streak increases the yield. The oleoresin so obtained is collected every 1 to 2 weeks and delivered to central gum-cleaning plants. Very little material is made from gum.

Steam and solvent process. The cutover pine forests of the South provide the raw material, mainly stumps and some other resinous waste wood, for this process. Figure 32.2 shows the essential *unit operations* in this method.[7]

The wood is first ground in a wood hog and then reduced to splinters in a shredder (Op).

The chips are loaded into a battery of extractors, where they rest upon a false bottom, below which live steam under pressure is admitted at the end of the run for solvent recovery. The ex-

[6] For an excellent article, see Turpentine, ECT, 2d ed., vol. 20, p. 748, 1969.

[7] ECT, 2d ed., vol. 17, p. 477, 1968; From Pine Stumps to Rosin and Terpene Oils, *Chem. Eng. (N.Y.)*, **54**(12), 119 (1947) (pictured flowchart, p. 150); Naval Stores: New Plant, *Chem. Eng. (N.Y.)*, **55**(8), 108 (1948); Beglinger, Distillation of Resinous Wood, no. 596, Forest Products Laboratory, 1958.

Fig. 32.3 Largest single turpentine fractionating complex in the world in 1976. This facility, located in Jacksonville, Fla., is capable of handling over 80,000,000 lb annually. The basic raw material, crude turpentine from kraft paper mills, is first split into α-pinene and β-pinene. These compounds are then made into many products, such as pine oils, rosin oils, and chemicals for perfume, flavor, and pharmaceutical uses. (*Glidden-Durkee Division of SCM Corp.*)

tractors are built of acid-resistant stainless alloys and operate at pressures of 65 to 85 psig (Op).

The solvent (so selected that it is easily separated from turpentine) countercurrently extracts the chips (Op). This solvent may be naphtha or a petroleum fraction with a boiling range of 200 to 240°F. The hot solvent is drained off, and the residual material on the chips is removed by subsequent steam distillation. The chips are used for fuel (Op).

Most of the solvent is removed from the turpentine, pine oil, and rosin in a concentrating evaporator (Op).

The residue from the first evaporator is sent to an intermediate evaporator. The vapors from this evaporator are led into the upper part of the continuous fractionating column, and the residue is sent to the finishing evaporator (Op).

Vapors from this final evaporator are combined with those of the intermediate evaporator before entering the fractionating column (Op).

Residue from the third evaporator is whole-wood rosin and may be treated by selective solvents and certain absorbents such as fuller's earth to yield light rosins[8] (Op).

The continuous fractionating column separates the pine oils and turpentine from the last of the solvent (Op). Pine liquids from the continuous column are separated into three fractions in a batch still: crude pine oil, an intermediate cut of terpenes, and crude turpentine (Op). Careful fractionation of these three cuts produces many marketable products (Op).

[8]ECT, 1st ed., vol. 14, p. 391, 1955.

Sulfate pulp turpentine. The relief gases from the digesters for sulfate (or kraft) pulp contain turpentine and pine oil. Upon condensation, a crude oil is found floating on top of the condensate. There are 2 to 10 gal of oil per ton of pulp produced. This oil contains 50 to 60% turpentine and 10 to 20% pine oil; these may easily be separated by fractional distillation. The resulting turpentine contains offensive mercaptans, which must be removed by hypochlorite solution or ethylenediamine. Another by-product of the process is tall oil, obtained upon acidification of the digester liquor.

Tall oil is the generic name for the products derived from the vile-smelling, gummy, black liquor residue from the kraft pulping process. The fats, fatty acids, rosin, and rosin acids contained in pine woods are dissolved in the cooking liquor as sodium soaps. This black liquor is concentrated in multiple-effect evaporators. The semiliquid then goes to a skim tank; the tall oil soap rises to the top, where it is removed either by skimmer rakes or by overflowing. The brown frothy curd is ready for chemical conversion (Ch) by acidulation with sulfuric acid.[9] The resultant dark-brown mixture of fatty acids, rosin, and neutral material is known as crude, or whole, tall oil; the name comes from the Swedish word *talloja,* meaning pine and oil. The fractionation process under high vacuum is now in use by the major producers of tall oil rather than the older method of acid refining. This coproduct has proved to be the cheapest organic acid available. In 1966 tall oil output was only 470,000 tons/year; by 1972 the figure had reached 750,000 tons and is still increasing. Its utilization curtails waste and stream pollution. Tall oil is an important naval-stores product, particularly as a source of turpentine (Fig. 32.2 and Table 32.2). Tall-oil fatty acids are mostly normal C_{18} acids, 75% mono- and diunsaturated, with smaller amounts of saturated and triunsaturated constituents. The fractionation of crude tall oil allows varying fatty acid content. Over 35% of all the fatty acids used in the United States are from tall oil. Protective coatings, soaps, epoxytallates, and dimer acids are made from tall oil, and from these polyamide resins for printer's ink, adhesives, coatings, and detergents.

Lignin[10] is an important industrial raw material which constitutes a large percentage of the noncellulosic material in wood. If the celluloses, hemicelluloses, sugars, starches, amino acids, inorganic salts, fats, waxes, tannins, terpenes, flavanoids, coloring matter, and other extractives are separated from wood, the 20 to 30% that still remains is *lignin.* Its principal function as part of the growing tree is that of a *cementing* substance, or a structural element contributing strength and rigidity to the wood tissue. To a kraft-mill chemical engineer, lignin is the source of an ever-growing number of chemicals.[11] To the sulfite-pulp engineer, lignin sulfonic acid represents about 65% of the solids content of spent sulfite liquor and must be washed from the pulp and withheld from stream pollution by evaporation and reuse. The bleach-pulp operator is concerned with the removal of residual lignin from the pulp to increase its visual whiteness, or *brightness.*[12] The mother liquor (red liquor) from the magnesium bisulfite pulping can be completely burned, as described in the text. Some products now derived from lignin are summarized in Fig. 32.3. The fact that new high-yield pulping processes, combined with new refining and bleaching processes, have greatly increased the use of lignin as a constituent of paper is ordinarily overlooked by those concerned with its by-product or even its waste-product use. A most profitable use would be in paper per se. The lignin recovered by the acidulation of cooking liquors differs materially from that originally present in the tree. Lignin

[9]Continuous Tall Oil Route Saves on Power and Labor, *Chem. Eng.* (*N.Y.*), **79**(2), 76 (1972); How to Increase Tall Oil Output, *Chem. Eng. Prog.*, **69**(9), 80 (1973).

[10]Glennie and McCarthy, Chemistry of Lignin, in Libby (ed.), Pulp and Paper Science and Technology, vol. 5, McGraw-Hill, 1962; Halpern, Pulp Mill Processes, Noyes, 1975.

[11]In a number of kraft mills which also make use of the neutral sulfite semichemical (NSSC) process, the liquor from the latter is processed in conjunction with black liquor.

[12]Ridding, Kraft Mill Effluent of Unwanted Color, *Chem. Eng.* (*N.Y.*), **68**(23), 110 (1961); Jones, Pollution Control and Chemical Recovery in the Pulp and Paper Industry, Noyes, 1973.

is a highly complex polymer of substituted phenylpropane units, highly aromatic in nature. It is water-insoluble and highly resistant to chemical reaction. Its molecular weight is 2,000 to 15,000, and its weight and composition vary with the species. Lignin is a potential source of many aromatic compounds but is largely burned because of lack of appropriate chemical technology to utilize it. Control of the fragmentation of large molecules has not received the study that synthesis from small ones has, so the utilization of lignin's values remains elusive. The energy requirements for severing bonds are large, which increases processing costs.

Despite the difficulties involving its chemistry, several products are currently profitably made from lignin, mostly by crude, low-yield processes. Vanillin (Fig. 27.6—7 million lb/y) is made from sulfite waste liquor, but the market is small. Dimethyl sulfide, methyl mercaptan, and dimethyl sulfoxide are made from kraft black liquor. Alkaline liquors and lignosulfonates are marketed as stabilizers for asphalt emulsions, dispersing agents, modifiers for latices, binders for printing ink, battery expander plates, oil well drilling-fluid additives, feed pellet binders, etc.

HYDROLYSIS OF WOOD

The hydrolysis of wood or its constituents, hemicellulose and cellulose, yields a considerable number and large volume of chemicals. The most important are presented in Chaps. 32 and 33. Naturally, the basic hydrolysis is followed by other appropriate chemical conversions. Such cellulosic materials as corncobs, cottonseed hulls, peanut shells, and bagasse have all been proposed as potential sources of *alcohol,*[13] following hydrolysis of cellulose to sugar—and its fermenation—however, only wood pulp and wood wastes have ever gained even a minor status, although two-thirds of their weight is cellulosic materials.

WOOD SACCHARIFICATION[14] received its first industrial trial at the beginning of this century. During World War I, two American plants manufactured alcohol from sawdust, using dilute sulfuric acid as the hydrolyzing agent. Because of the low yields, approximately 22 gal of 100% alcohol per ton of dry southern pine, the plants could not compete and were closed 2 years after the war. At the same time, Scholler was trying to produce alcohol from wood in Germany. Using almost the same methods, he obtained a 3% sugar solution, representing 60 to 70% of theory. Also, Willstätter and Zechmeister published the discovery that, at ordinary temperatures, 40% hydrocholoric acid easily puts cellulose into solution. By 1941, 30 foreign plants were in successful commercial operation. During World War II, two plants were built in this country. A plant at Bellingham, Wash., utilized waste liquors from the sulfite pulping process in which hydrolysis of cellulose to sugar had occurred. This plant continues to function as part of a highly successful complex designed to recover all possible values from cooking liquors.

In the sulfite process for the manufacture of pulp, sugars are formed by the hydrolysis of wood constituents that are dissolved out when producing usable fibers for paper. Approximately 65% of these sugars is capable of being fermented to alcohol. This is equivalent to about 1 to 2% of the sulfite waste liquor. First, the sulfite waste liquor is separated from the pulp and conditioned for fermentation.[15] After cooling to 30°C and adding lime (to adjust the pH to 4.5) and urea nutrient, the liquor is pumped into the fermentors and fermented. The alcohol is distilled out and rectified. In Sweden all the alcohol (and many derivatives therefrom) are based on waste sulfite liquors, including all beverage spirits and, during the war, alcohol for motor fuel.

[13]See Chap. 31, Industrial Alcohol.
[14]Hall *et al.* Wood Saccharification *Unasylva,* **10**(1), 7 (1956) (summary statement from Forest Products Laboratory).
[15]Forest Products and the Environment, *AIChE Symp. Ser.* no. 133, **69,** 2 (1973).

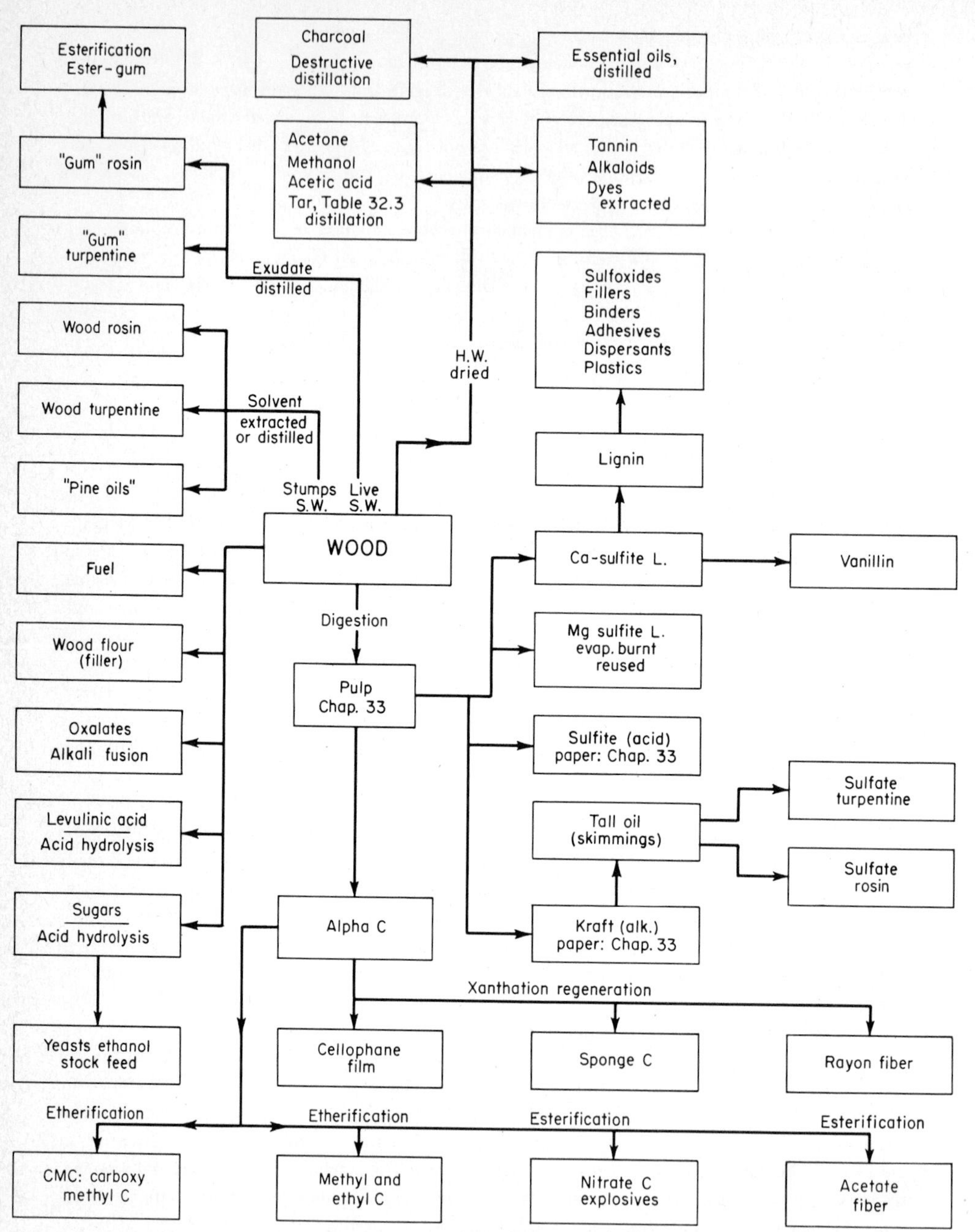

Sources: Chemicals from the Other Half of the Tree, *C&EN,* Feb. 11, 1963, p. 83; *I&EC,* **50,** 570 (1958); *ibid.,* **44,** 3808 (1952); CMC: cf. Chemical Economics Handbook, Stanford Research Institute; Hercules Powder Co., Wilmington, Del., for rosin and derivatives; Dickey, Chemistry of Cellulose, *J. Chem. Educ.,* **41** (4), 196 (1964); ECT, 2d ed., vol. 4, p. 616, 1964. Abbreviations: L-liquor, Alk-alkaline, C-cellulose, S.W.-soft wood, H.W.-hard wood.

Fig. 32.4 Wood chemicals (wood is the world's most widely used industrial raw material; see Chaps. 22, 24, and 30 through 35).

CELLULOSE DERIVATIVES

Many cellulose derivatives have attained commercial importance (Fig. 32.4) such as the ethers: ethyl cellulose,[16] methyl cellulose (Dow's Methocel), and carboxymethyl cellulose. The last-mentioned is frequently known as CMC and may be prepared by the following "alkylation" or by a variation in the proportions:

$$\underset{\text{Cellulose}}{[C_6H_7O_2(OH)_3]_x} + x\text{NaOH} \longrightarrow \underset{\text{Alkali cellulose}}{[C_6H_7O_2(OH)_2ONa]_x}$$

$$[C_6H_7O_2(OH)_2ONa]_x + \underset{\text{Sodium chloroacetate}}{x\text{ClCH}_2\text{COONa}} \longrightarrow \underset{\text{Sodium carboxymethyl cellulose}}{[C_6H_7O_2(OH)_2COONa]_x}$$

Various degrees of etherification, described in the literature, have been commercialized. Sodium CMC[17] is a white powder when dry, though it is generally produced and sold as solutions of varying concentration and viscosity. By 1970, CMC production had grown to more than 66 million lb/year. It is physiologically inert and is employed as a protective coating for textiles and paper, for sizing, for ice-cream and other emulsion stabilizers, and as an additive to impart strength to sausage casings and other films. *Hydroxyethyl cellulose*, a high-viscosity product, is useful as a thickener and a protective colloid in water-based coatings.

Among the many specialty products developed, *cellulose sponges* form an important segment. The dyed viscose cellulose sponge sirup is mixed with various-sized crystals of sodium sulfate, placed in a mold, and heated to "set" the mix around the crystals.[18] Further heating dissolves the crystals in water (that is part of their own crystallization), after which the product is removed from the mold and washed.

MISCELLANEOUS CHEMICALS FROM WOOD Much of the early industry in the United States centered around wood and wood products and often produced substances now made otherwise or forgotten. The American *chestnut* produced an extremly fine tanning bark but has been destroyed by introduced disease. Potassium salts are selectively concentrated by plants and converted to KOH and K_2CO_3 on burning, producing "pot ash" from which alkali for soapmaking was obtained for centuries. Other widely used wood extractives include cedar oil, tannin extracts, maple sugar, arabinogalactan gum, mannogalactans, quinine, sassafrass, cinnamon oil, and many others.

Bark is removed from trees at the rate of about 15 million tons/year (1974) and is generally regarded as an expensive nuisance. Some inexpensive fillers are made from bark flour, but most is burned, used as a soil improver, or wasted. A recently constructed plant[19] converts fir bark into a high-quality vegetable wax, a thermosetting-resin extender, and a phenol substitute. Cork and bast fiber can also be made. Much more extensive use of bark and wood by-products must be made in the future. Wood products should not be regarded primarily as fuels.

FIRE RETARDANTS

Chemicals are necessary in fireproofing wood and wood products, including structural boards (Chap. 33). Conventional methods include impregnating wood with various salts (which weaken the wood and also add weight) and coating with fire-retardant paint. Mono- and diammonium phosphates, ammonium sulfate and sulfamate, and mixtures of borax and boric acid are the usual impregnating salts. Better materials are needed, and their development is being speeded by new federal fire safety standards.

[16]Ethyl Cellulose, *Chem. Metall. Eng.*, **52**(9), 142 (1945) (flowchart); see Chap. 34 for cellulose derivatives used in plastics.

[17]Hader, Carboxymethylcellulose, *Ind. Eng. Chem.*, **44**, 2808 (1952) (flowchart); Synthetics as Water-soluble Resins, *Chem. Eng. News*, Apr. 25, 1960, p. 40; Turbak (ed.), Cellulose Technology Research, ACS, 1975.

[18]*Chem. Week*, Oct. 27, 1962, p. 43.

[19]Fir-Bark Conversion Route, *Chem. Eng. (N.Y.)*, **81**(11), 70 (1974).

SELECTED REFERENCES

Browning, B. L. (ed.): Chemistry of Wood, Wiley-Interscience, 1963.
Casey, J. P.: Pulp and Paper: Chemistry and Technology, Interscience, 1960–1961, 3 vols.
Duerr, W. A.: Fundamentals of Forestry Economics, McGraw-Hill, 1960.
Faurdenburg, K.: Lignin Structure and Reactions, *Advan. Chem. Ser.* no. 59 (1966).
Gould, R. F. (ed.): Literature of Chemical Technology, chap. 21, Wood, ACS Monograph, 1966.
Hajny, G. J., and E. T. Reese (eds.): Celluloses and Their Applications, ACS, 1969.
Hillis, W. E. (ed.): Wood Extractives, Academic, 1962.
Niketin, N. I.: Chemistry, Cellulose and Wood, Davey, 1967.
Oshima, M.: Wood Chemistry (Foreign Processes), Noyes, 1965.
Panshin, A. J., *et al.:* Forest Products: Their Sources, Production, and Utilization, 2d ed., McGraw-Hill, 1962.
Pearl, I. A.: Chemistry of Lignin, Dekker, 1967.
Pinder, A. R.: Chemistry of the Terpenes, Wiley, 1960.
Turbak, A. F.: Cellulose Technology Research, ACS, 1975.
Whistler, R. L. (ed.): Methods in Carbohydrate Chemistry, vol. 3, Cellulose, Academic, 1963.

chapter 33

PULP AND PAPER INDUSTRIES

Cellulose is not only the most abundant organic substance available in nature, but it is probably the most versatile, yet the simplest, replaceable organic raw material known. Its conversion to paper products is the everyday function of the pulp and paper industries, and the resulting thousands of different kinds of useful items are proof that pulp has become indispensable to modern civilization. In fact, our national per capita consumption has reached approximately 640 lb. Today more than 10,000 engineers and scientists are employed in these industries, where manufacturing processes are complicated and have many variables difficult to control. There is also a constant search for new uses and chemical modifications of paper. About 400 lb of chemicals is used to make 1 ton of paper. Manufacturers involved in the production of pulp and bleaching, papermaking, and conversion processes are responsible for the use of millions of tons of chemicals annually and thus are some of the chemical industry's most important customers.

HISTORICAL[1] The earliest attempts to record human activities were made on stone. A little later, bark, leaves, and ivory were also used. Between 2500 and 2000 B.C. writing paper was first manufactured from a tall reed growing along the Nile called *papyrus,* hence the word paper. Other early writing materials were dried calf- and goatskin parchment, the wax-covered boards of the Romans, and the clay-brick records of the Babylonians. The actual manufacture of paper was invented by the Chinese about A.D. 105, but not until the end of the fourteenth century did the process of manufacture undergo improvements and become known in southern Europe. The industry did not obtain a firm foothold in England until the seventeenth century, and in 1690 America's first paper mill was established. The Gutenberg Bible marked the beginning of book printing, hence the start of an increased demand for paper (based on rags).

About 1750 the *Hollander beater* was developed and adopted. In 1799, a Frenchman, Robert, invented a process for forming a sheet on a moving wire screen. This machine is known today as the *Fourdrinier* machine. In 1809 the *cylinder* machine was invented by Dickinson and immediately forced the Fourdrinier into the background, where it remained until about 1830, when its importance was finally realized. It was in this period that the first Fourdrinier was received in America (1827), and steam cylinders for drying were first used in 1826.[2] The expanded production made possible by the Fourdrinier and the cylinder machine so greatly increased the demand for rags that a scarcity soon developed. In 1844, however, Keller of Saxony invented a mechanical process for making pulp from wood. The soda process was developed by Watt and Burgess in 1851. In 1867 the American chemist Tilghman was granted the basic patent (U.S. 70,485) for the sulfite process. The sulfate, or kraft, process (*kraft* is German for strong) was the result of basic experiments conducted by Dahl in 1884 at Danzig. In 1909 the sulfate process was introduced into the United States. At that time,

[1]Britt (ed.), Handbook of Pulp and Paper Technology, Reinhold, 1964; Libby (ed.), Pulp and Paper Science and Technology, McGraw-Hill, 1962, 2 vols.; Stephenson (ed.), Pulp and Paper Manufacture Series, McGraw-Hill, 1950–1955, 4 vols.

[2]Hunter, Papermaking, 2d ed., Knopf, 1957.

pulp production was divided as follows: mechanical, 48%; sulfite, 40%; soda, 12%. Perhaps the greatest change in the industry in recent years has been the rise of the *sulfate* process to a position of major importance in the industry. Pulp manufacture gradually developed into an industry of its own and served industries other than the paper industry as well; for example, the manufacture of pulp for rayon manufacture has assumed major importance and become a distinctly specialized division of pulp manufacture. The control and utilization of the industry's by-products have received widespread attention (Chap. 32). The creation of useful materials from lignin and waste liquors, mostly from sulfite pulping, represents increased income for the industry, as well as a partial solution of its major problem, the avoidance of stream pollution.

USES AND ECONOMICS In 1976 the production of paper and paperboard in the United States was about 64 million tons. More than half the world's production of paper is consumed in the United States.

MANUFACTURE OF PULP FOR PAPER

There are two distinct phases in the reduction of raw wood and other materials to finished paper: the manufacture of the various pulps (Table 33.1), followed by their conversion to paper (Table 33.2). About 82% of pulping is carried out by chemical means which dissolve out the lignin from the cellulose fibers. Of all the pulping processes used in 1976, the alkaline sulfate, or kraft, process accounted for about 72%, the acid sulfite 5%, the semichemical (mainly the neutral sulfite semichemical, NSSC) 9%, groundwood 11%, and the remainder miscellaneous.

RAW MATERIALS Wood is the principal source of cellulose for papermaking. However, cotton, linen rags, and waste are also used by mills, as well as various fibers, their principal source being cordage and rough waste from the textile industry. There are two principal debarking methods. The first utilizes friction from tumbling or rotating action on a moving mass of pulpwood sticks, using either continuous rotating cylindrical drums or stationary machines with continuous agitating cam equipment to stir up the mass of logs. In the case of drums, the wood is fed into the upper end of a rotating drum immersed in a tank partly filled with water, where it is tumbled. The bark is

TABLE 33.1 ***Production and Consumption of Wood Pulp, Pulpwood, and Other Fibrous Materials*** *(In thousands of short tons, air-dry weight)*

	Production		*Consumption*	
	1972	*1971*	*1972*	*1971*
Wood pulp, total	43,621	41,120		
Special alpha and dissolving grades	1,677	1,684	1,136	1,208
Bleached sulfite	1,796	1,656	1,893	1,761
Unbleached sulfite	333	368	427	458
Bleached and semibleached sulfate	13,970	12,994	15,292	14,271
Unbleached sulfate	17,285	15,826	17,566	15,969
Soda	139	166	141	167
Groundwood (bleached and unbleached)	4,617	4,786	4,815	4,965
Semichemical	3,803	3,640	3.799	3,632
Waste paper	12,925	12,100		
	Thousands of cords of 128 ft^3			
Pulpwood, total	69,254	65,789	71,538	67,157
Softwood	. . .	. . .	53,613	50,585
Hardwood	. . .	. . .	17,925	16,572

Source: Wood Pulp Statistics, 37th ed., American Paper Institute, 1973; Bureau of the Census, Current Industrial Reports.

TABLE 33.2 ***Paper and Paperboard Products, by Grades***
(In millions of short tons)

	1973	*1970*	*1965*
Paper and paperboard, all grades	61.8	53.5	44.1
Paper	26.5	23.6	19.2
Newsprint	3.4	3.3	2.1
Coated printing and converting	3.8	3.3	2.8
Book paper, uncoated	3.1	2.6	2.1
Other printing, writing, etc.	6.4	5.1	4.2
Packaging and industrial converting	5.8	5.4	5.0
Tissue and other machine-creped	4.0	3.8	2.9
Paperboard	29.7	25.5	20.8
Unbleached kraft	13.8	11.6	7.8
Bleached fiber	4.0	3.4	2.3
Semichemical	4.3	3.4	2.7
Combination furnish	7.6	7.0	8.1
Wet machine board	0.1	0.1	0.1
Construction paper and board	5.6	4.3	3.9

Source: Statistical Abstract of U.S., 1974.

literally *rubbed off*, and the clean wood discharged at the other end. The second method for debarking uses mechanical friction, or high-pressure (about 1,400 psi) water jets applied to individual logs. Hydraulic barkers direct the water jets against the log in such a way that the bark is broken up and removed. Presses to reduce the water content of bark are widely used to facilitate its combustion.

GROUNDWOOD This process involves no chemical treatment of the pulp whatever. The chief woods employed are spruce and balsam, which are soft, coniferous species. They have the advantage that they can be floated in streams to the mill, in contrast to poplar, which sinks soon after immersion. After arrival at the mill, the wood is slashed and debarked. It is then ground in water to remove the heat of friction and to float the fibers away. The grinding is at an *acute angle* to the length of the blocks in order to furnish longer fibers by tearing rather than by right-angle cutting. For small tonnages, a "three-pocket" grinder is used, consisting of a central grindstone mounted on a steel shaft and having three chambers, or pockets, around its periphery, each chamber being surmounted by a hydraulic cylinder ram which forces the blocks against the revolving, usually artificial, stone. The pulp-and-water mixture from the grinders is dropped into a stock sewer below the grinders and passed along to the sliver screen. Here the coarser material is retained and discarded by the screen, and the fine material falls into the screened stock pit, from which it is pumped to the fine screens. The fines that pass these screens are concentrated in thickeners to give commercial mechanical pulp. The oversize from the fine screens is treated to refiners and then returned to the screens. The water overflow from the thickeners contains about 15 to 20% of the original fibers and is the so-called *white water* used in grinding and to aid flow in the stock sewer. As the process continues to operate, it is necessary to add fresh water to the system to keep down the temperature; therefore some of the white water must be removed. Before this water is sent to waste, the remaining fibers are strained from it. The fiber is returned to the thickeners. The *energy requirements* are all mechanical and consist chiefly of power for grinding. The only chemical change occurring in mechanical pulp is slight hydration of the cellulose by the long contact with water.

The *uses* of groundwood pulp are for the most part restricted to cheaper grades of paper and board, where permanency is not required. The eventual deterioration that occurs in paper made from mechanical pulp is due to chemical decomposition of the noncellulosic portions of the wood. In the manufacture of newsprint, cheap manilla, and wall, tissue, and certain wrapping papers, the mechanical pulp is usually improved by mixing with a small amount of chemical pulp.

Bleaching. More than half a million tons of groundwood pulp is bleached yearly in the United States. Many years ago its color was considerably lightened by the use of sodium or calcium bisulfite, but more recently ozone, oxygen, chlorine dioxide, hydrogen peroxide, and sodium peroxide have been used as more effective bleaching agents.[3] The bleaching of any type of pulp is an expensive process; bleaching operations must be adapted both to the results desired in terms of quality of brightness and the type of fibrous material to be bleached.[4]

KRAFT, OR SULFATE PULP is an *alkaline* process and is responsible for the major part of the pulp manufactured at the present time (Table 33.1). Almost any kind of wood may be used, hard or soft, although coniferous woods are mostly employed. The process was developed to remove the large amounts of oil and resins in these woods. For the production of dissolving pulp for rayon from hardwoods, a modified sulfate process is in use. This involves prehydrolysis to remove the pentosans and polyoses, followed by sulfate (salt-cake) treatment and multistage bleaching. With softwood being used up more extensively, more attention is being given to hardwood. The *chemical reactions* are rather indefinite, but involve hydrolysis of lignin (Table 33.3). The hydrolysis also produces mercaptans and sulfides, which are responsible for the familiar bad odor of kraft pulp mills. The *energy required* is given in Fig. 33.1, largely involving mechanical energy to chip the wood and steam to heat the chips in the digesters to the point where noncellulosic material is rapidly dissolved.

The steps in the manufacture of kraft, or sulfate, pulp are depicted in the flowchart in Fig. 33.1. These may be separated into the following sequences.[5]

The logs are slashed and debarked (1), as previously described, and conveyed to the chippers (2), which are large rotating disks holding four or more long, heavy knives, to reduce the wood to small chips (Op).

The chips are screened on either rotary or vibrating screens to separate the oversized chips, the desired product, and the sawdust. The oversized chips and slivers are put through crushers or rechippers to reduce them to the proper size (Op).

The digesters are charged with chips (3); the cooking white liquor containing essentially sodium sulfide and caustic soda is added, and live steam is turned on. Either a stationary or a rotary digester may be used, which may be constructed of stainless steel. The pressure is raised to 110 psi. The cooking period lasts about 3 h (Op and Ch).

At the end of this time the pressure is allowed to drop to 80 lb and the charge, or *brown stock*, is blown into the pit or into the hat-recovery furnace to use the steam normally wasted in a blow pit; better washing also results.

The pulp, after separation from the cooking liquor, is washed (4) (Op).

The spent cooking liquor (black liquor) is pumped to storage to await recovery of its chemicals by evaporation (5 and 6) and combustion of dissolved organic matter in the recovery furnace (8) for reuse in the process (Op and Ch).

The washed pulp is passed to the screen room, where it enters the knotters, rifflers, and screens to sieve out any small slivers of uncooked wood, and finally to filters and thickeners (Op).

The thickened pulp is next bleached, using at least one chlorine dioxide stage, followed by neutralization and calcium hypochlorite. The bleaching agent in either case oxidizes and destroys the dyes formed from the tannins in the wood and accentuated by the sulfides present in the cooking liquor (Ch).

[3]Innovations in Mechanical Pulping, *AIChE Symp. Ser.* no. 139, **70,** 40 (1974); Bleaching of Groundwood Pulp with Hydrogen Peroxide, *FMC Corp., Becco Div. Bull.* 117, 1962; Pulp Bleaching, *TAPPI*, **56,**(11), 65 (1973).

[4]O'Neil *et al.*, Bleaching, chap. 13 in Libby, *op. cit.* (tables); Parsons, Pulp Bleaching, chap. 10 in Britt, *op. cit.* (detailed classification of processes for all types of pulps, p. 274, flowchart of four-stage bleach plant, p. 279, and 63 refs); Rapid Chlorine Dioxide Bleaching, TAPPI, **57**(2), 104 (1974).

[5]Jackson, Alkaline Pulping, chap. 7 in Britt, *op. cit.*

TABLE 33.3 ***Comparison of Three Types of Chemical Pulp***

Type of process	*Kraft, or sulfate, pulp (alkaline)*	*Sulfite pulp (acid)*	*NSSC*
Cellulosic raw material	Almost any kind of wood, soft or hard	Coniferous; must be of good color and free of certain hydroxy phenolic compounds	Hardwood chiefly used, some softwood (*small chip size, fiberized*)
Principal reaction in digester	Hydrolysis of lignins to alcohols and acids; some mercaptans formed	$RC{:}CR' + Ca(HSO_3)_2 \longrightarrow RCH{-}CR'\cdot SO_3\frac{1}{2}Ca$	Lignin sulfonation and hemicellulose hydrolysis lead to formation of acetate and formate.
Composition of cooking liquor	12.5% solution of NaOH, Na_2S, and Na_2CO_3. Typical analysis of solids: 58.6% NaOH, 27.1% Na_2S, 14.3% Na_2CO_3. Dissolving action due to NaOH and Na_2S. Na_2CO_3 inactive and represents the equilibrium residue between lime and Na_2CO_3 in the formation of NaOH.	7% by weight SO_2, of which 4.5% is combined as sulfurous acid and 2.5% as calcium or $Mg(HSO_3)_2$. Cooking 1 ton of pulp requires 390 to 480 lb of SO_2 and 122 to 150 lb of MgO. Recent significant trend toward use of $Mg(OH)_2$, NH_4OH as base to speed lignin solution	Na_2S buffered with Na_2CO_3 bicarbonate, or kraft green liquor. Concentration of 90–200 g/l of Na_2S. Cooking liquor does *not* complete freeing of fibers, but mechanical treatment does
Cooking conditions	Time 2–5 h; temp. 340–350°F; pressure 100–135 psi	Time 6–12 h; temp. 257–320°F or higher; pressure 90–110 psi	Time 48–36 min; corrugating-grade pulp from mixed hardwoods 12–15 min; temp. 320–360°F, pressure 100–160 psi
Chemical recovery	Most of process is devoted to the recovery of cooking chemicals, with incidental recovery of heat through burning organic matter dissolved in liquor from wood; chemical losses from system are replenished with salt cake, Na_2SO_4	SO_2 relief gas recovered; magnesium liquor recovered and reused after wood digestion and pulp washing	Characterized by high yield—65–85%. Pulping losses 35–15% of wood components. Special recovery methods and by-product utilization
Materials of construction	Digesters, pipelines, pumps, and tanks can be made of mild steel or, preferably, of stainless	Acid liquor requires digester lining of acidproof brick; fittings of chrome-nickel steels (Type 316), lead, and bronze	Serious corrosion problems encountered in digesters and handling equipment; stainless-steel protection needed
Pulp characteristics	Brown color; difficult to bleach; strong fibers; resistant to mechanical refining	Dull white color; easily bleached; fibers weaker than kraft	Stiff, dense paper of low opacity; fibers approach chemical pulps in strength
Typical paper products	Strong brown bag and wrapping; multiwall bags, gumming paper, building paper; strong white papers from bleached kraft; paperboards such as used for cartons, containers, milk bottles, and corrugated board	White grades: book paper, bread wrap, fruit tissue, sanitary tissue	Unbleached: large percentage for corrugated board, also newsprint, specialty boards. Bleached: writing and bond papers, offset, mimeo, tissue, and toweling

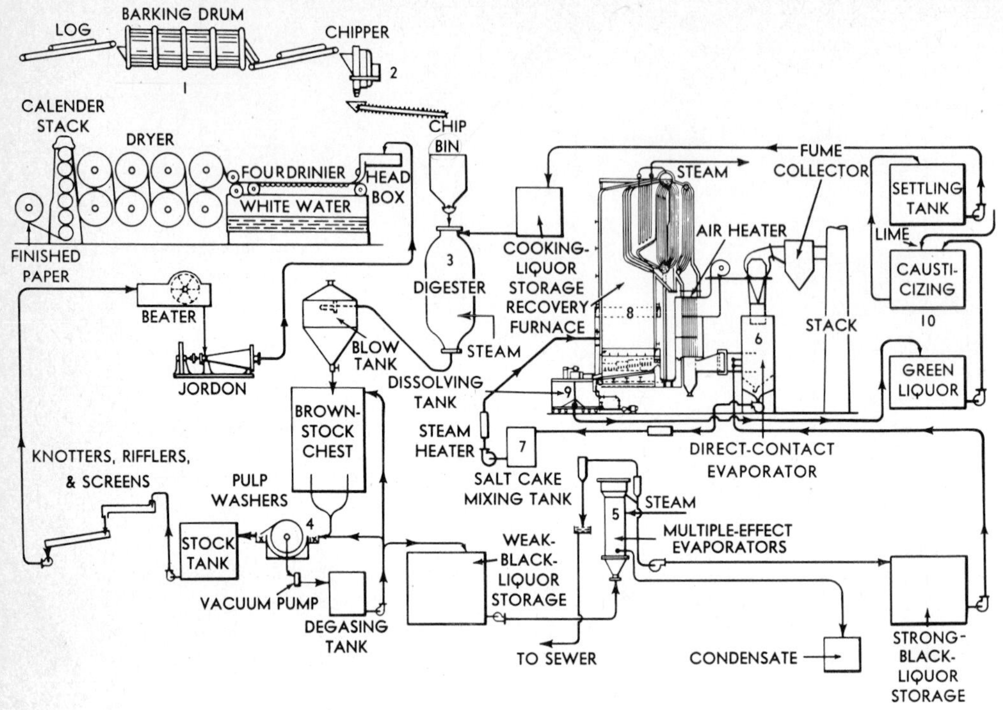

Fig. 33.1 Flowchart for the kraft, or sulfate-pulping, process with liquor recovery and reuse. Alkaline procedure. *Soda-process note:* With almost no variation, this flowchart may be taken for the soda process. The principal change is in the omission of salt cake. Make-up soda ash is added in the causticizing tank (10). (*Babcock & Wilcox Co.*)

	Per ton of dried pulp		*Per ton of dried pulp*
Wood	1.5–2 tons	Steam	13,000 lb
New lime	500 lb	Electricity	250 kWh
Soda ash	250 lb	Direct labor	5 work-hours

After bleaching, the pulp is washed and rethickened in preparation for making it into sheets dry enough to fold into a bundle called a *lap* (Op); or it may be run into the charge chest of a paper machine to give dried pulp.

Lapping is done on a wet thickener consisting essentially of a cylinder dipping into a vat filled with stock solution, an endless felt belt which carries the pulp sheet through squeeze rolls, and a series of press rolls. The resulting laps contain 35 to 45% air-dry fiber. These "wet" laps are stacked in hydraulic presses and subjected to pressures up to 3,000 psi. This product contains 50 to 60% air-dry fiber (Op).

Kraft pulp, made from coniferous woods, has the longest fibers of all the pulps. This, coupled with the fact that the chemicals used are not so harsh in their action as those employed for the other two chemical pulps, make possible the production of very strong papers. In the past, the dark color of kraft paper limited its use mainly to wrapping papers, sacks, and paperboard. Newer developments in the bleaching treatment have made possible the manufacture of light-colored and white pulps, allowing the mixing of this very high-strength pulp with other types to increase the paper strength.

Instead of large batch digesters, continuous digester systems, which offer several economic advantages, are now being used.[6]

Recovery of the black liquor.[7] An important factor in the economic balance for the kraft process has been the recovery of the spent liquor from the cooking process. The black liquor removed from the pulp in the pulp washer, or diffuser, contains 95 to 98% of the total chemicals charged to the digester. Organic sulfur compounds are present in combination with sodium sulfide. Sodium carbonate is present, as are also small amounts of sodium sulfate, salt, silica, and traces of lime, iron oxide, alumina, and potash. Total solids usually average about 20%. This black liquor is concentrated, burned, and limed as shown in Fig. 33.1. In the smelting furnace any remaining organic compounds are broken down, the carbon burned away, and the inorganic chemicals melted. At the same time, the reaction

$$Na_2SO_4 + 2C \longrightarrow Na_2S + 2CO_2$$

takes place. The molten chemical *smelt* is allowed to fall into a weak solution in tank 9 of the "dissolving liquor" coming from the causticizing plant. The chemicals dissolve immediately to give a characteristic *green liquor.* The insoluble impurities are allowed to settle out, and any carbonate is then causticized by adding slaked lime prepared from the recovered calcium carbonate. The reaction

$$Na_2CO_3(aq) + Ca(OH)_2(s) \longrightarrow 2NaOH(aq) + CaCO_3(s) \qquad \Delta H = -2.1 \text{ kcal}$$

occurs quickly. The resulting slurry is separated in settlers and rotary continuous filters, using Monel metal screens as a filtering medium. The calcium carbonate sludge, or "mud," is sent to a lime kiln to recover the calcium oxide for reuse in the process. The filtrate is the *white liquor* used in the cooking of the fibers. It contains caustic soda, sodium sulfide, and small quantities of sodium carbonate, sodium sulfate, sodium sulfite, and thiosulfate.

Among the by-products from the black-liquor recovery plant is *tall oil* (Chap. 32), a black, sticky, viscous liquid composed mainly of resin and fatty acids. The tall oil may be separated from the weak black liquor by means of centrifuges (in America) or obtained by flotation from the concentrated liquors (in Europe). It is used in the manufacture of soaps and greases and in the preparation of emulsions. The digester relief gases yield paying quantities of turpentine, from 2 to 10 gal/ton of pulp produced. This may be refined to produce *sulfate turpentine.*

SODA PULP is manufactured very similarly to sulfate pulp, both being alkaline processes. With the successful bleaching of the much stronger and cheaper kraft pulp, the old soda pulp process is of such declining importance that it does not warrant discussion here.[8]

SULFITE PULP[9] The quantity of pulp made by this process steadily diminishes despite its high quality, because of the water pollution it causes. Although spruce is the wood most commonly employed, appreciable quantities of hemlock and balsam are also used. The wood is barked, cleaned, and chipped as described for sulfate pulp, the resulting chips being about $\frac{1}{2}$ in. in length. It is then conveyed to storage bins above the digesters preparatory to being cooked. The chemistry of the sulfite digestion of cellulosic materials is extremely complicated; the thermodynamic data are incomplete and unreliable. The external *energy* required for the process includes steam for preparation of the cooking liquor and cooking the pulp, and mechanical energy for chipping the wood and pumping. The more common sulfite process consists of digestion of the wood in an aqueous solution containing

[6]Continuous Digester is Heart of Modern Pulpmill, *Chem. Eng.* (*N.Y.*), **75**(5), 94 (1968); *Chem. Week*, Feb. 10, 1968, p. 62; Halpern, Pulp Mill Processes, Noyes, 1975.

[7]Jackson, *op. cit.*, p. 171; Oxidation of Heavy Black Liquor, *AIChE Symp. Ser.* no. 139, **70,** 121 (1974).

[8]For Soda Pulp see CPI 2, p. 731; Britt, *op. cit.;* note in legend for Fig. 33.1.

[9]Magnesium-base Pulping, *Chem. Eng.* (*N.Y.*), **69**(2), 136 (1962); Ericsson *et al.*, Sulfite Pulping, chap. 10 in Libby, *op. cit.* (flowchart for calcium base process).

calcium bisulfite and an excess of sulfur dioxide. The sulfite process involves two principal types of reactions, which are probably concurrent: (1) sulfonation and solubilizing of lignin with the bisulfite, and (2) hydrolytic splitting of the cellulose-lignin complex. The hemicelluloses are also hydrolyzed to simpler compounds, and the extraneous wood components acted on. Since disposal of waste liquor (more than half of the raw material entering the process appears here as dissolved organic solids) creates a serious pollution problem, concerted attention has been turned to its removal or utilization. A slurry of magnesium oxide is substituted for lime,[10] because then chemical and heat recovery are possible, as well as a solution to the disposal problem of the waste liquor. Sodium and ammonia[11] have also been substituted for calcium as a pulping base in a limited way. The waste liquor from the calcium sulfite process cannot have its values used over again, since the calcium sulfite does not decompose to sulfur dioxide, whereas magnesium sulfite does. $CaSO_4$ is formed and lost. Hence the newer and technically more acceptable[12] sulfite process is based on magnesium bisulfite rather than the earlier used corresponding calcium compound, resulting in a greater concentration and more active combined sulfur dioxide, without danger of precipitation and of quicker separation and solution of the noncellulose wood constituents (lignin and hemicelluloses) (Fig. 33.2).

The essential reactions involved in the preparation of the cooking liquor are quite simple:

$$S + O_2 \longrightarrow SO_2$$

$$2SO_2 + H_2O + CaCO_3 \longrightarrow Ca(HSO_3)_2 + CO_2$$

$$2SO_2 + H_2O + MgCO_3 \longrightarrow Mg(HSO_3)_2 + CO_2$$

$$2SO_2 + Mg(OH)_2 \longrightarrow Mg(HSO_3)_2$$

The entire process may be divided into the following sequences, as illustrated in Fig. 33.2 for magnesium bisulfite:

Sulfur is melted in a tank heated by the rotary burner and then fed to this burner (Op) for oxidation (Ch).

Any sulfur that is vaporized in the burner enters a combustion chamber, where it is oxidized to sulfur dioxide. The amount of air in this operation is closely controlled to prevent the formation of sulfur trioxide (Ch).

The sulfur dioxide obtained is cooled quickly in a horizontal, vertical, or pond cooler consisting of a system of pipes surrounded by water (Op).

The absorption of the gas in water, in the presence of calcium or magnesium compounds, is accomplished in a series of two or more absorption towers or acid-making tanks for $Mg(HSO_3)_2$. A fine spray of water passes down through the tower system countercurrent to the sulfur dioxide gas, which is blown up through the tower (Op and Ch).

The liquor contains a certain amount of free sulfur dioxide, which is enhanced from time to time as the free sulfur dioxide vented from the digesters is bubbled through acid-making towers or tanks. The final liquor as charged to the digesters is a solution of calcium or magnesium bisulfites, analyzing about 4.5% "total" sulfur dioxide and about 3.5% "free"[13] sulfur dioxide. The digester is filled with chips, and the acid cooking liquor is pumped in at the bottom (Op). The digesters are cylindrical steel vessels with a capacity of from 1 to 23 tons of fiber and 3,000 to 51,000 gal of

[10]Hull *et al.*, Magnesia-base Sulfite Pulping, *Ind. Eng. Chem.*, **43**, 2424 (1951) (excellent article, flowcharts, pictures).

[11]Ammonia Base Sulfite Pulping at Inland Container, *Pap. Trade J.*, Nov. 20, 1972.

[12]Liquor-making Eased for Mg-base Pulp Mills, *Chem. Eng.* (*N.Y.*), **71**(15), 80 (1964); It's a Big Job to Clean up Sulfite Waste, *Chem. Week*, Apr. 26, 1972, p. 37; New Cleanup Methods Vie for Sulfite Pulping Jobs, *Chem. Week*, Nov. 22, 1972, p. 67.

[13]In the parlance of the pulp manufacturer, the "free" sulfur dioxide is the sum of the sulfurous acid and that portion which requires alkali to convert from a bisulfite to a neutral sulfite.

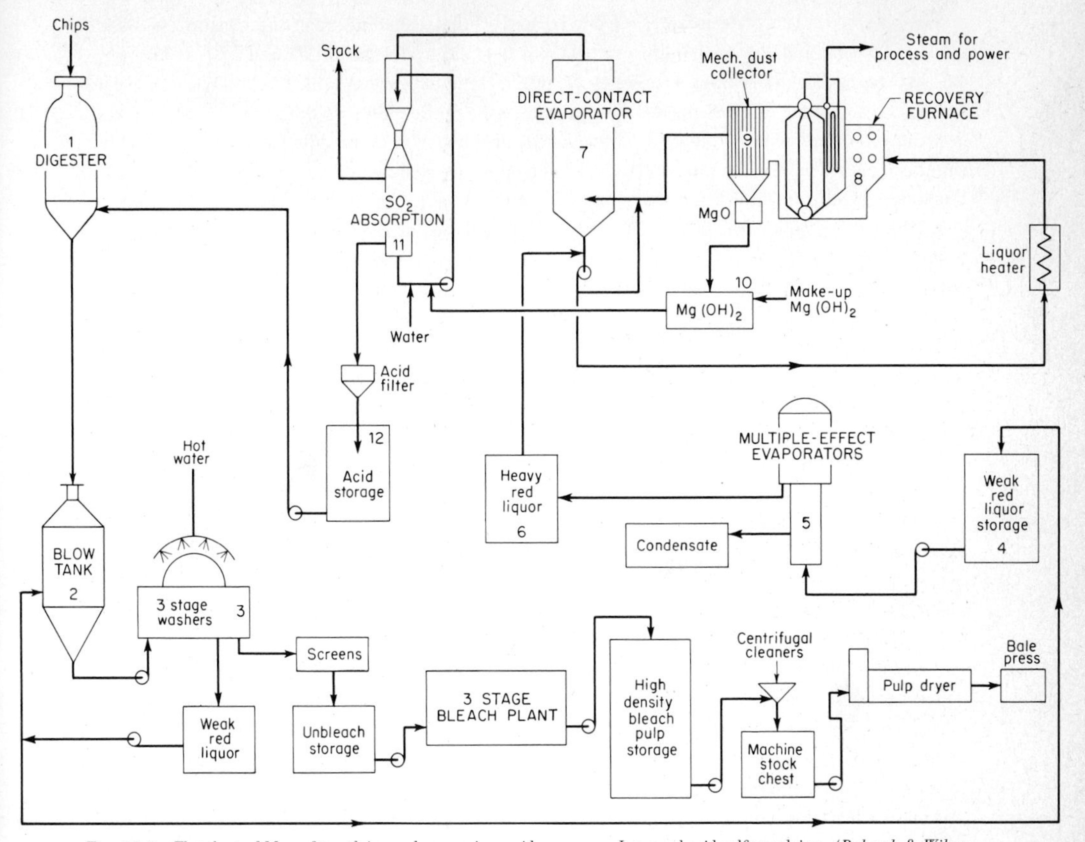

Fig. 33.2 Flowchart of Magnefite pulping and magnesium oxide recovery. Improved acid sulfite pulping. (*Babcock & Wilcox Co.*)

Pulp grades	*Dissolving*	*Newsprint*
Yield of pulp, %	35	60
Materials consumed/ton:		
Wood, lb oven-dried	5,143	3,000
Magnesium oxide, lb	122	150
Sulfur dioxide, lb	390	480
Steam, lb	2,900	2,000
Power, kW	4,035	2,785

Note: One ton of pulp contains 1,800 lb of oven- or bone-dried fiber.

"acid." A special lining of cement, crushed quartz, and acid-resisting brick is used to avoid the corrosive action of the cooking liquor.

The digester is heated with direct steam. In recent years the industry has turned to digesters, with forced outside circulation, which heat the cooking liquor in an outside stainless-steel tube heater and circulate it through the charge by means of pumps. This permits better temperature distribution through the charge and prevents dilution of the liquor with the direct steam formerly used for heating. Conditions of the cook depend on the nature of the wood, the composition of the acid, and the quality of pulp charged. The pressure varies from 70 to 160 lb, depending upon the construction of the plant. The time and the temperature range from 6 to 12 h and from 170 to 176°C (Ch).

At the end of the cooking process the digester (1) is blown to a tank (2) [a large, round tank having a false bottom and equipped with means to wash (3) the pulp with fresh water]. The cooking, or weak red liquor is evaporated (5–7) and burnt in a boiler (8,9) (to provide steam) to MgO and sulfur dioxide. The MgO is slaked (10) and pumped to the cooling and acid tower (11) down which the sulfur dioxide (regained and make-up) is passed to make fresh bisulfite liquor (12). (Op).

The pulp is pumped from the blow tank (2) to a series of screens (3), where knots and large lumps of fiber are removed (Op). The accepted stock from the screens is sent to the rifflers, or centrifuges, to remove foreign matter (Op).

The relatively pure pulp is concentrated in thickeners, which are cylindrical frames covered with 80-mesh bronze wire. The water passes through, and the pulp is retained on the screen (Op).

The pulp is sent to the bleacher, where free chlorine or chlorine dioxide is introduced. After the chlorine has been exhausted, milk of lime is added to neutralize the mass (Ch).

The stock is washed (Op), thickened (Op), and sent to the machine stock chest.

Pulp from the chest is formed into laps of about 35% dry-fiber content (Op), and the laps are dried with steam-heated rolls in a pulp dryer and baled as a product which is 80 to 90% dry fiber (Op).

The milk-of-lime system of preparing the cooking liquor consists of slaking burnt lime containing a high percentage of magnesia with warm water to produce a 1° Bé suspension. A high percentage of magnesia is desirable because the magnesium sulfites formed are more soluble than the corresponding calcium compounds. The latter tend to settle out and clog the pipes. This solution is treated with sulfur dioxide gas to produce the cooking liquor.

Sulfite pulp is a high-grade type of pulp and serves in the manufacture of some of the finest papers, including bond. It is used either alone or with some rag pulp to make writing paper and high-grade book paper. It furnishes dissolving pulps for plastics, synthetic fibers, and other products in which wood per se is unrecognizable.

Waste sulfite liquor. The disposition of the waste liquors formed in the strictly calcium bisulfite process has been the subject of much research. Until recently it was common practice to dump this liquor into a nearby stream, but legislation to prevent stream pollution has stopped this practice. Although preventing stream pollution is now the most important factor, economics emphasizes the importance of the utilization of this vast potential raw material. Magnesium bisulfite waste liquor is recoverable and reusable, as Fig. 33.2 indicates, hence reducing pollution. Among the principal products investigated, or used, are lignin (Chap. 32 and Fig. 32.3), vanillin from the lignin present (Chap. 27), tanning material, road binders, special cements, portland cement accelerator, corebinders, plastics from the lignin present, and food yeast.

SEMICHEMICAL, OR NSSC, PULPING has been developed since 1925,[14] with the production of high-quality bleached pulps. It requires smaller chips than chemical pulping, but the amount of chemicals used is less; sodium sulfate is buffered with sodium carbonate or kraft green liquor. The annual production of semichemical pulps exceeds the tonnage of sulfate pulp and is growing. Increasing improvements are being made in pulping and refining technology so that higher yields of better quality are obtained, thereby reducing waste and stream pollution. Continuous, rotary, and stationary digesters are used in the *NSSC process,* possibly the most widely used and the greatest space saver of the semichemical processes.

Chemimechanical pulping is another term for part-chemical and part-mechanical processes which have led to an increased use of hardwoods. They are two-stage processes which require both mechanical and chemical energy to separate the fibers. They call for equipment and systems of the same nature as that generally used for other categories of pulping. In the *cold-soda process,* chips

[14]Halpern, Pulp Mill Processes, Pulping, Bleaching, Recycling, Noyes, 1975.

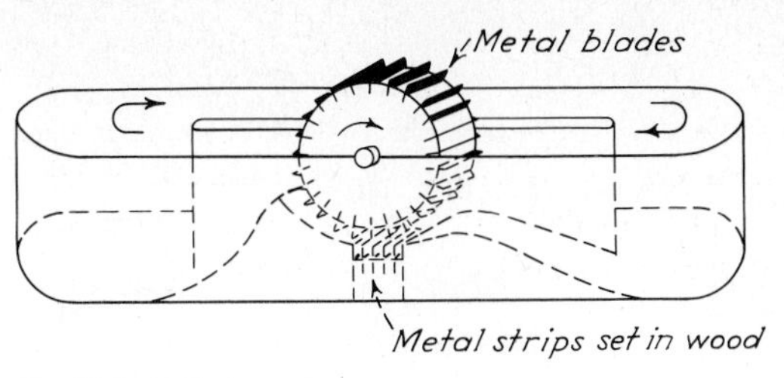

Fig. 33.3 Hollander, or beater.

are treated with sodium hydroxide at room temperature or by hydrostatic pressure, followed by fiberization in a disk refiner. The *chemigroundwood process* is an important variant in which the logs are thoroughly impregnated with a hot chemical solution before grinding. High yields are obtained in all these processes—85 to 95% in the latter two and 65 to 85% in the neutral sulfite process. These systems are described in detail with accompanying flowcharts by McGovern.[15] These semichemical processes do not employ sufficient chemicals to solubilize completely the lignin and cause fiber separation, but leave that to the accompanying mechanical treatment, which follows the partial delignification and softening effect of the chemicals used. *Wood products,* in addition to paper, are diagramed and presented in Fig. 33.3.

NEW PULPING PROCESSES Pulping with various organic solvents such as phenol and alcohols has been suggested and tried on a very small scale. Nonsulfur pulping[16] is currently being tried on a modest scale using oxygen as the primary solubilizer. The process is quite promising.

RAG PULP[17] A small but important source of fiber for the manufacture of fine paper is rags. New rags are largely scraps from textile mills and garment factories, and old rags are collected from waste-disposal domestic sources. The rags go through a thrasher, where dust is removed before they can be sorted. All are separated according to color. Colors hard to bleach are used in the manufacture of dark papers, and more easily bleached colors are employed for lighter papers. After being sorted, the rags are cut into small squares, the dust is removed, and the squares are passed over magnets to remove metallic foreign material. After this they are ready for the cook. In this operation the wax and resins in the fibers are removed, together with additional dirt and grease, and the dyes loosened from the fibers. The cooking liquors include caustic lime, caustic soda, or a mixture of caustic lime, and soda ash. The digester is a cylindrical, horizontal, rotary boiler holding about 5 tons of cloth. Cooking is continued for 10 to 12 h at 120°C. The resulting pulp is washed, bleached with chlorine and/or chloride of lime, and rewashed. Up until 1860 rag fibers (cotton and linen) were the only source of materials for papermaking. Now at least 75 grades of material are sold to mills for the manufacture of high-grade and specialty papers which command premium prices.

MANUFACTURE OF PAPER

The various pulps, even though frequently manufactured in coarse sheets, still lack those properties which are so desirable in a finished paper, such as proper surface, opacity, strength, and feel. Pulp stock is prepared for formation into paper by two general processes, beating and refining.[18] There is no sharp distinction between these two operations. Mills use either one or the other alone or both

[15]McGovern, Semichemical and Chemimechanical Pulping, chap. 8 in Britt, *op. cit.*

[16]Nonsulfur Pulping Symposium, *TAPPI*, **58**(2), 56 (1975).

[17]Atchison *et al.*, chap. 12, Rag Papermaking, in Libby, *op. cit.* (rag papermaking and other pulping processes, bibliography).

[18]Root, Stock Preparation, chap. 1 in Libby, *op. cit.*, vol. 2, pp. 1–39 (47 refs.); Halpern, Paper Manufacture, Noyes, 1975; Gould, Specialty Papers, Noyes, 1976.

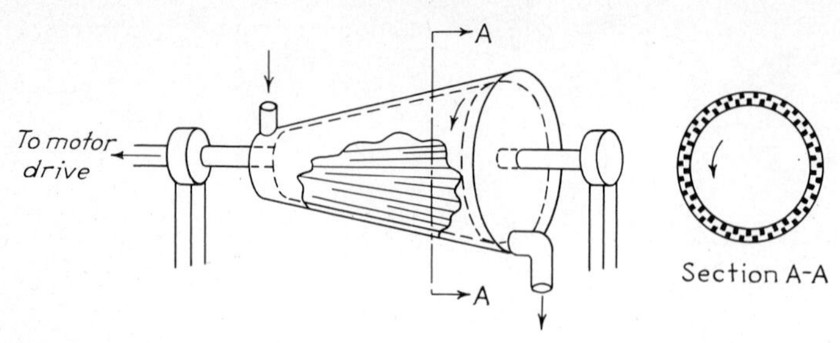

Fig. 33.4 Conical refiner, or Jordan engine.

together. However, the former operation is becoming of decreasing importance with the increasing use of continuous refining. The operation of the latter fits in well with the trend toward automatic mills.

The most generally used type of *beater* (also known as a *Hollander,* Fig. 33.3) consists of a wooden or metal tank having rounded ends and a partition part way down the middle, thus providing a channel around which the pulp circulates continuously. On one side is a roll equipped with knives or bars, and directly below this is a bedplate consisting of stationary bars. In operation, the circulating pulp is forced between the bars on the revolving roll and the stationary bars of the bedplate. The roll itself may be raised or lowered to achieve the results desired. Beating the fibers makes the paper stronger, more uniform, more dense, more opaque, and less porous.

In general shape, continuous refiners are either conical or disk. The Jordan engine (Fig. 33.4), developed 100 years ago, is still in use today—the forerunner of various modifications of conical refiners, such as the Bolton-Emerson Claflin 202 pictured in Fig. 33.5. It consists essentially of an acute conical shell and accessories to furnish maximum defibering and refining.

The *Jordan* engine (Fig. 33.4) is the standard *refiner,* consisting essentially of a conical shell on the inside of which are set stationary bars. Revolving inside the shell is a core, also set with bars. It is the action of these two sets of bars that produces the desired effect on the pulp. The latter enters the small end of the cone and, after being acted upon by the bars, passes out at the other end.

Filler, sizing, and *coloring* may be added either in the beater, which is the usual practice, in the refiner, or in both.[19] The order generally is as follows: (1) The various pulps are blended to give the desired density and uniformity; (2) the filler is added with or just after the fiber; (3) after sufficient beating the sizing is put in and mixed thoroughly; (4) the color is added and distributed well throughout the mass; (5) alum is introduced to produce coagulation and the desired coating of fibers. All papers except absorbent types, i.e., tissue or blotting paper, must have a *filler* added to them, the purpose of which is to occupy the spaces between the fibers, thus giving a smoother surface, a more brilliant whiteness, increased printability, and improved opacity. Fillers are always finely ground inorganic substances and may be either naturally occurring materials, such as talc or certain clays, or manufactured products, such as suitably precipitated calcium carbonate, blanc fixe, or titanium dioxide.

Sizing[20] is added to paper to impart resistance to penetration by liquids. Again, the only papers not so treated are blotting and other absorbent papers, where penetration is desired. The sizing may either be added during the pulping or defibering process or applied to the surface after the sheet is formed. The process of engine or stock sizing, i.e., adding size in the beater, involves addition of the sizing agent, consisting of either a soap made from the saponification of rosin with alkali or a wax emulsion, followed by precipitation of the size itself with papermaker's alum, $Al_2(SO_4)_3 \cdot 18H_2O$. This treatment produces a gelatinous film on the fiber which loses its water of hydration and produces a

[19]Halpern, *op. cit.*

[20]Davison, The Sizing of Paper, TAPPI, **58**(3), 48 (1975) (Review).

Fig. 33.5 Emerson Claflin refiner. (*A*) Plug adjustment mechanism; (*B*) bearing assemblies; (*C*) stock outlet; (*D*) packing boxes; (*E*) stock inlet; (*F*) cleanout; (*G*) oil-mist lubrication; (*H*) coupling. (*Emerson Mfg. Co.*)

hardened surface. Tub sizing, on the other hand, is carried out on the dried paper or on surfaces which may or may not have been previously and partly sized in the beating operation. The material used for this treatment must have adhesive properties, the principal substances being animal glue, modified starches, and wash sizes. The operation is carried out either on the papermaking machine itself or in a separate sizing press employing air-drying. The paper runs through a bath of the size material, then through squeezing rolls that remove excess material, and finally over drying rolls. This type of sizing operation is used to enhance the water resistance of the paper further, and especially to make it take ink evenly without blurring, even after erasures.

Wet-strength resins since 1942 have been an important additive in the manufacture of papers subjected to rain and snow, hot or cold liquids—papers that have greater than 10% wet over dry tensile strength when saturated with water. Large-scale production began with the discovery that certain synthetic resins gave papers a wet strength, which is not to be confused with water repellency. Amino-aldehyde resins have received major attention, but other wet-strength processes are used.[21]

Approximately 98% of all paper produced has a certain amount of *coloring* material added to it. Coloring, like sizing, may be added either in the beater or after the paper has been made. However, surface coloring uses less dye and requires the production of only one type of paper, which may be colored any shade later as needed. All types of dyes (acid, basic, direct, sulfur) and all types of pigments (both natural and synthetic) are used as coloring agents. Acid dyes have no affinity for the cellulose fibers and must therefore be fixed to them by means of mordants. If the paper is colored in the beater, the alum added to precipitate the size will also act as a mordant for the dye. Surface dyeing may be carried out either in the papermaking machine or in a separate piece of equipment. In either case, the process consists in passing the paper through the dye bath, removing the excess dye by means of press rolls, and drying.

Since there is a noticeable trend toward the development of specialized machines for specific operations, such as the Yankee dryer (particularly applicable in making sanitary paper products or imparting a high finish to one side of lightweight papers), there is some variation in the two types of conventional papermaking machines described here, the *Fourdrinier* and *cylinder* machines. The basic operating principles are essentially the same. The sheet is formed on a traveling wire or cylinder, dewatered under rollers, dried by heated rolls, and finished by calender rolls.

Fourdrinier machine.[22] Figure 33.6 shows the essential parts of a Fourdrinier machine. The stock from the foregoing operations, containing approximately $\frac{1}{2}$% fiber, is first sent through screens

[21]Thode *et al.*, Mechanism of Retention of Wet Strength Resins, TAPPI, **42**(3), 178 (1959), **43**(10), 861 (1960), and **44**(4), 296 (1961).

[22]Approximately 750 Computers Are Now in Use Controlling Paper Making Operations in the U.S., TAPPI, **58**(6), 71 (1975).

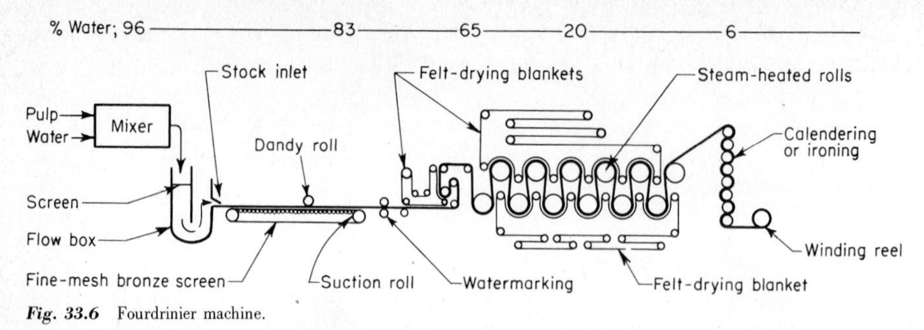

Fig. 33.6 Fourdrinier machine.

to the head box, from which it flows through the sluice onto a moving, endless, bronze-wire screen. The pulp fibers remain on the screen, while a great portion of the water drains through. As the screen moves along, it has a sidewise shaking motion which serves to orient some of the fibers and to give better felting action and more strength to the sheet. While still on the screen, the paper passes over suction boxes to remove water and under a dandy roll which smooths the top sheet. Rubber deckle straps travel along the sides of the screen at the same speed and thus serve to form the edges of the paper.

From the wire, the paper is transferred to the first felt blanket, which carries it through a series of press rolls, where more water is removed and the paper given a watermark if so desired. Leaving the first felt, the paper passes through steel smoothing rolls and is picked up by the second felt, which carries it through a series of drying rolls heated internally by steam. The paper enters the rolls with a moisture content of 60 to 70% and leaves them 90 to 94% dry. Sizing may be sprayed on the sheet at this point, in which case it must pass through another series of drying rolls before entering the calender stack, which is a series of smooth, heavy, steel rolls which impart the final surface to the paper. The resulting product, finished paper, is wound on the reel. The enormous *quantity of water used* makes it necessary to recirculate as much of it as possible for economical operation. The operation of a Fourdrinier is a complex procedure. One of the major problems is making suitable allowance in the speed of the various rolls for the shrinkage of the paper as it dries. The operating speeds of the machines vary from 200 ft/min for the finer grades of paper to 2,500 ft/min for newsprint.

Cylinder machines are employed for the manufacture of heavy paper, cardboard, and nonuniform paper. The cylinder machine has from four to seven parallel vats, into each of which similar or dissimilar dilute paper stocks are charged. This allows several similar or dissimilar layers to be united in one heavy sheet. A wire-covered rotating cylinder dips into each vat. The paper stock is deposited on the turning screen as the water inside the cylinder is removed. As the cylinder continues to revolve, the paper stock reaches the top, where the wet layer comes in contact with and adheres to a moving felt. The traveling felt, carrying the wet sheet underneath, passes under a couch roll where some of the water is pressed out. This felt and paper come in contact with the top of the next cylinder and pick up another layer of wet paper. Thus a composite wet sheet or board is built up and passed through press rolls and on to the drying and smoothing rolls. Such a composite may have outside layers of good stock, whereas the inside ones may be of groundwood pulp.

DRY PROCESSES Considerable interest in a dry process[23] for making paper and nonwoven fabrics exists because of the cost and complexity of drying equipment and the enormous process-water demand of conventional methods. Pilot-plant experiments have been made on the process.

Finishing operations. Raw paper stock delivered from the mill is converted to its end use by various means, depending on the final form required. The most extensive modification is probably

[23]Danish Firm Tests Dry Process for Paper, *Chem. Eng. News*, Feb. 6, 1967, p. 40.

the application and smoothing of coatings for specialty papers. Mineral or pigment surface coating is applied to improve the printability of paper. The basic pigment used is clay. Casein (protein adhesive) and calcium carbonate are used in large amounts, as well as small amounts of other mineral pigments. For color coatings dyes are added. Functional coatings are particularly important for packaging food products. The principal types of processes and equipment are discussed in the TAPPI *Monograph Series.*[24]

SPECIALTY PAPERS

There are hundreds of uses for paper and as many hundreds of types produced, all classified according to their broad use (Table 33.2). *Special industrial papers* are those not falling into the specific broader-use categories, and among them are the following: cigarette; filter; glassine; food containers such as paper plates, cans, cups, and wrappers coated with plastic or aluminum foil to preserve flavor and texture, prevent contamination, or inhibit moisture; Sanfordized bag material called Clupak; vegetable parchment; wallpaper and surface-waxed paper. Papermakers are currently working with textile manufacturers to develop paper suitable for disposable surgical gowns and bed sheets. A few small-volume specialties are now available in the synthetic-paper field.[25]

Nonwoven "cloth" is now of considerable importance[26] for a wide variety of uses in the hospital, industry, and home. Many such materials contain plastics in addition to wood fibers. Costs are substantially lower than for conventional woven cloth.

PAPER STOCK

The Paper Stock Institute of America defines *waste paper* as disposable waste before it is cleaned, sorted, and baled for sale to manufacturers as raw material, at which time it is designated *paper stock.* Although a large potential supply is available, which can be made into pulp at a small cost, the actual volume produced is subject to violent fluctuation. Considerable pressure from environmentalists has led to the increased recovery of used papers, but the economics of recovery and reutilization remain unclear.[27] In 1973, in the United States, salvaged wastepaper supplied was 22% of the new supply used.[28] Western Europe recycled 29%. Currently, over 25% is recovered. This represents less than 40% of the potentially recoverable paper and board. There are four broad categories of paper stock: high-grade, such as trimmings from ledger paper and envelopes; mixed paper, such as that found in wastebaskets; old corrugated-paper boxes; and used newspapers. With only mechanical methods, paper stock may be made into pulp suitable for roofing paper and boxboard. Since the product is dark in color, it is unnecessary to remove the ink and pigments; therefore shredding and beating operations suffice. For the many other uses, such as uncoated book paper, newsprint, special industrial papers, and food board, some chemical treatment is necessary, such as bleaching.

De-inking.[29] After the bales are inspected and fed to the pulper, caustic soda is the chemical ordinarily used for de-inking. Soda ash, sodium peroxide, silicate of soda, or one of the sodium phosphate compounds may be added to the formula, depending on several factors, such as type of paper stock used, local water supply, and the equipment at hand. Cooking and pulping are usually carried out rapidly in one operation at a temperature somewhat under 200°F. Foreign material such

[24]TAPPI *Monograph Ser.* (Current literature is available from TAPPI, Technical Association of the Pulp and Paper Industry.)

[25]Polymers Roll into Printing Paper Markets, *Chem. Eng.* (*N.Y.*), **78**(6), 62 (1971); Bumpy Road Ahead for Synthetic Paper, *Chem. Eng. News,* Oct. 23, 1972, p. 13.

[26]Nonwovens Symposium, TAPPI, **58**(5), 38 (1975); Gould, Specialty Papers, Noyes, 1976.

[27]Secondary Fiber Technology, 1974, *TAPPI,* **58**(4), 78 (1975).

[28]Paper Recyclers Regroup, *Chem. Eng.* (*N.Y.*), **81**(12), 44 (1974).

[29]Flowchart in CPI 2, p. 741.

as paper clips is collected as the pulp is screened prior to washing. Bleaching yields, in many cases, a snow-white product.

STRUCTURAL BOARDS

The classification of boards[30] made from wood particles is confused in the United States, and the terms *fiberboard, particleboard, flakeboard,* and *oriented board* all frequently refer to particle boards.

Fiberboards are rigid and semirigid sheets formed by the wet- or dry-felting of fibers. The low-density kinds are generally used for insulation. The new medium-density types go mainly into furniture, and the high-density types into furniture and paneling. The fiber used is usually from soft- or hardwoods, cull timber, and mill residues, and the wood is pulped by mechanical, thermo-mechanical, or explosive processes. A small amount of adhesive, generally phenol-formaldehyde resin, is used—particularly with dry-felted boards. Treatment to improve flame resistance or water resistance or to protect against insect damage and rot is common.

Particleboard manufacture uses sawdust, shavings, and wood flakes of special shapes to combine good physical properties with attractive appearance. Synthetic resin adhesives and different pressing conditions give varying density, strength, and surface embossing. A new class of high-strength boards with predetermined engineering properties is just emerging. These boards use resin adhesives and have the basic elements aligned substantially in one direction. Particleboards are used for subflooring, core stock for veneered furniture, and decorative paneling and are replacing plywood and lumber for some structural applications.

Paper-base laminates are multiple sheets of resin-treated paper bonded together under high pressure and heat; under these conditions they become rigid and have no characteristics of paper. Depending on the process used, these products are termed *resin-filled* or *resin-impregnated.* They possess several characteristics, such as stability and strength, among others, which make them useful in the building industry. Other types of polymer-modified materials are listed in Chap. 34.

SELECTED REFERENCES

Bibliography of Papermaking and U.S. Patents, Technical Association of the Pulp and Paper Industry (annual since 1936).
Britt, K. W.: Handbook of Pulp and Paper Technology, Reinhold, 1964.
Browning, B. L. (ed.): Chemistry of Wood, Interscience, 1963.
Casey, J. P. (ed.): Pulp and Paper, Chemistry and Technology, 2d ed., Interscience, 1960.
Davis, D. S.: Calculations in the Paper Industry, Franklin Publishing, 1963.
Gascoigne, J. A., and M. M. Cascoigne: Biological Degradation of Cellulose, Butterworth, London, 1960.
Gould, R. F. (ed.): Literature of Chemical Technology, chap. 20, Pulp and Paper, ACS Monograph, 1966.
Halpern, M. G.: Pulp Mill Processes, Pulping, Bleaching, Recycling, Noyes, 1975.
Higham, R. R.: A Handbook of Papermaking, Oxford, 1964.
Institute of Paper Chemistry: Various publications, including a bibliographic series.
Instrumentation in the Pulp and Paper Industries, Instrument Society of America, 1973.
Jones, H. R.: Pollution Control and Chemical Recovery in the Pulp and Paper Industry, Noyes, 1973.
Libby, C. E. (ed.): Pulp and Paper Science and Technology, McGraw-Hill, 1962, 3 vols.
McDonald M.: Paper Recycling and the Use of Chemicals, Noyes, 1971.
Paper and Pulp Mill Catalog: Engineering Handbook, Fritz Publishing, 1962 (annual).
Paper Year Book, Davidson Publishing, 1962 (annual).
Rapson, W. H. (ed.): Bleaching of Pulp, 2d ed., Technical Association of the Pulp and Paper Industry, 1963.
Rydholm, S. A.: Pulping Processes, Wiley, 1965.
Stamm, A. J.: Wood and Cellulose Science, Ronald, 1964.
TAPPI Data Sheets, Technical Association of the Pulp and Paper Industry.
Weiner, J., and R. Lillian: Energy Production and Consumption in the Paper Industry, Institute of Paper Chemistry, 1974.

[30]Symposiums on Particleboard, Washington State University, 1970–1975.

chapter 34

PLASTIC INDUSTRIES

The development of plastics from laboratory curiosities to products tailored to industry's needs has brought new and economical materials of construction to the engineer and the designer. Not only can plastics replace metals and other materials, but they can also be used with them. Today's modern alchemist is the plastics processor who, working with heat, pressure, and catalysts, mainly with petrochemicals, turns out a host of products that are taking market after market away from established, time-honored materials such as wood, metals, and natural fibers. Plastic may be defined as a material that contains as an essential ingredient a polymerized organic substance of large molecular weight, is solid in its finished state, and at some stage in its manufacture or its processing into finished articles can be shaped by flow. The common basic raw materials are coal, petrochemicals, cotton, wood, gas, air, salt, and water. Plastics lend themselves to exceedingly wide application because of their *toughness, water resistance, ease of fabrication, and remarkable color range.*

The use of a plastic material for a specific application is dependent upon its composition, its particular properties, and the design of the part. Synthetic resins are the largest source of plastics, with cellulose derivatives ranking next. The great utility of plastics may be shown by reference to a few typical applications in the various fields where these relatively new materials are being applied. In the automotive and airplane industries many different plastics find specific utilization because of beauty, strength, and oil and electrical resistance. Their resiliency and strength make them very useful in the manufacture of safety glass, laminated gears, cams, pulleys, self-lubricating bearings, door and cowl panels, carpets, seat belts, and the like. In some cases, where the plastic cannot be used by itself, it is employed in combination with metals, as in the manufacture of steering wheels. As suitable molding equipment has become available, plastics have completely replaced metals even in large parts such as automobile dashboards. Plated plastic parts are showing up increasingly in items such as small knobs and door-lock pulls. In the electrical industries, molded and laminated organic plastics are of real value as solid insulating materials because of their electrical properties and relatively high mechanical strength. The use of plastics in building construction is increasing. Flooring, paneling, and boats (glass-fiber) are commonly made from plastic, but an important future use is in large prefabricated panels of rigid corrugated and foam-cored types. Polyethylene film has even been used in California to line an artificial lake with a 2.5-mi shoreline. The packaging and fiber markets for plastics have increased. The chemical process industries employ plastics in the form of tubing and piping, tanks, absorption towers, pumps and valves, etc., because they are highly resistant to corrosive gases, liquids, and solids. In the field of the decorative arts plastics have unlimited scope. They have been used to "dress up" all types of articles by eliminating harsh contours and by adding almost any color combination conceivable. All plastic materials of construction have limitations but, when properly selected, they can be used with the same degree of assurance as metals and alloys.

CLASSIFICATION Plastics may be classified in various ways. They can best be arranged according to their broad general application into thermosetting, thermoplastic, oil-soluble, and

TABLE 34.1 *Types of Resins and Plastics, with Some Common Trade Names*

Thermosetting resins
- Phenolic resins, as Bakelite, Durez, Resinox, Catalin, Formica, Indur
- Amino resins, as Plaskon, Beetle, Cymel, Micarta, Melmac
- Alkyd resins, as Glyptal, Rezyl, Beckosol, Dulux
- Epoxy resins, as Epon, Araldite, Ren, Epocast, Marblette
- Polyester (unsaturated) and allyl resins, as Aropol, Atlac, Dapon
- Silicone resins, as Pyrotex, Dow Corning

Thermoplastic resins (including cellulose derivatives)
- Cellulose derivatives:
  - Cellulose nitrate, as Celluloid, Pyralin, Nitron
  - Cellulose acetates, as Kodapak, Tenite, Plastacele
  - Cellulose propionates, as Forticel, Reed
  - Cellulose acetate-butyrates, as Tenite II, Kodapak II
  - Ethyl cellulose, as Ethocel, Soplasco, Campco

Polymer resins:
- Acrylate or polyacrylates, as Plexiglas, Lucite, Acryloid
- Vinyls or polyvinyl resins, as Vinylite, Gelva, Butacite, Koroseal, Alvar, PVA
- Polyvinylidine resin, as Saran
- Styrenes or polystyrene resins, as Styron, Lustrex, Loalin
- Coumarone and indene resins, as Cumar, Nevindene
- Polyamides, as nylon, Zytel
- Polyethers, as Penton, Calcon, Delrin
- Polyethylene, as Polyethylene, Poly-Eth, Tygothene, Petrothene
- Polypropylene, as Poly-Pro, Pro-fax
- Fluorocarbons, as Kel-F, Teflon, Fluorosint

Oil-soluble or modified resins
- Modified alkyds, as oil-soluble Rezyl
- Modified phenolics, as Albersols, Amberol, Lewisol

Protein substances

protein products, as presented in Table 34.1 with some of the common trade names. On the basis of derivation, they may be simply grouped as natural resins, cellulose derivatives, protein products (Table 34.2), and synthetic resins. In general, except where noted, synthetic resins formed by *condensation polymerization* are of the thermosetting type (which when cured change into a substantially infusible or insoluble product), and synthetic resins formed by *addition polymerization* are of the thermoplastic type (capable of being repeatedly softened by heat and hardened by cooling). These two polymerization reactions are fundamentally different. The process of addition polymerization involves a series of conversions interrelated in a complicated manner, which produce a polymer having a recurring structural unit identical with the monomers from which it is formed. On the other hand, polymers formed by condensation reactions have recurring structural units which lack certain atoms present in the monomers from which they are formed. The reaction takes place by combining two or more units and eliminating a small molecule, usually water, methanol, or hydrogen chloride.

HISTORICAL AND GENERAL DESCRIPTION OF VARIOUS TYPES Though plastic materials had been known for about 100 years, the industry was not well established until the early part of this century, when Baekeland developed scientific control and produced commercial phenolic resins (1909). His discovery tremendously stimulated the search for new plastics and resulted in an industry which has grown into the nation's seventh largest. Even though profit margins for general-purpose plastics are low, the industry continues to invest more of its sales dollar into research and develop-

TABLE 34.2 ***Classification of Commercial Resins and Plastics by Derivation***

I. *Derivatives of natural products* (thermoplastic, except where noted)	II. *Synthetic resins formed by condensation polymerization* (thermoset, except where noted)	III. *Synthetic resins formed by addition polymerization* (thermoplastic)
A. Natural resins 1. Fossil and plant resins 2. Rosin 3. Shellac 4. Lignin (thermosetting *B.* Cellulose derivatives 1. Regenerated cellulose *a.* Viscose *b.* Cuprammonium 2. Cellulose esters *a.* Nitrate *b.* Acetate *c.* Propionate *d.* Mixed esters (nitrate-acetate, acetate-propionate, acetate-butyrate) 3. Cellulose esters *a.* Methyl *b.* Ethyl *c.* Carboxymethyl *C.* Protein derivatives *a.* Casein-formaldehyde *b.* Zein (corn protein)-formaldehyde *c.* Soybean protein-formaldehyde	*A.* Phenolic resins 1. Phenol-formaldehyde 2. Phenol-furfural 3. Resorcinol-formaldehyde *B.* Amino resins 1. Urea-formaldehyde 2. Melamine-formaldehyde *C.* Polyesters 1. Alkyd 2. Unsaturated or oil-modified alkyds 3. Polycarbonates (thermoplastic) *D.* Polyethers (thermoplastic) 1. Polyformaldehydes 2. Polyglycols *E.* Polyurethanes (thermoplastic under some conditions) *F.* Polyamides (thermoplastic) *G.* Epoxides *H.* Silicone resins (thermoplastic under some conditions) *I.* Ionomers	*A.* Polyethylene *B.* Polypropylene *C.* Polyisobutylene *D.* Fluorocarbon polymers *E.* Polyvinyl acetate and derivatives 1. Polyvinyl alcohol 2. Acetals *F.* Other vinyl polymers 1. Polyvinyl ethers 2. Divinyl polymers 3. Polyvinyl chloride *G.* Polyvinylidene chloride *H.* Polystyrene *I.* Acrylic polymers *J.* Coumarone-indene polymers

ment than most other industries. This research has resulted in better and more durable materials which have penetrated the exacting applications of the space age. Table 34.3 shows some of the general properties of the principal plastics, and Table 34.4 summarizes their properties and applications.

Cellulose nitrate, or pyroxylin, plasticized with camphor and introduced as Celluloid, was the first synthetic plastic of industrial significance.[1] Even though it is a material hazardous to handle, readily decomposed by heat, and unstable to sunlight in the unpigmented form, it has many unique properties which long made it the best thermoplastic material available for many purposes. Cellulose nitrate was discovered about the middle of the nineteenth century, and was first used as a plastic base in 1869 by Hyatt who was searching for an ivory substitute.

[1]Paist, Cellulosics, Reinhold 1958; see also Chapter 35.

TABLE 34.3 ***Properties of Plastics***

Plastic type	*Asbestos-filled phenolic*	*Asbestos-filled furane*	*Glass-fiber-filled polyester*	*Glass-fiber-filled epoxy*
Physical properties				
Specific gravity	1.7	1.7	1.5	1.5
Thermal conductivity Btu/(h)(ft^2)(°F)(ft)	0.20	0.20	0.15	0.15
Coefficient of thermal expansion, in./(in.)(°F × 10)	2.0	3.0	2.7	2.5
Specific heat, Btu/(lb)(°F)	0.3	0.3	0.3	0.3
Water absorption (24 h), %	0.2	0.2	0.2	0.1
Flammability	Self-extinguishing	Self-extinguishing	Slow-burning	Slow-burning
Mechanical properties				
Tensile strength, 1,000 psi	5–7	4–6	13–15	15–17
Compressive strength, 1,000 psi	11–13	11–13	26–28	26–28
Hardness, Rockwell	R 105–115	R 105–115	M 90–100	M 90–100
Elongation (in 2 in.), %	Low	Low	Low	Low

Source: Chem. Eng. Prog., **59**(10), 100 (1963); for other properties and other plastics see source reference and Perry.

In an effort to discover a less flammable material, research conducted by Schutzenberger, Cross, and Bevan (1894) led to the development of *cellulose acetate* and of a plastic therefrom. In 1912, cellulose acetate found an outlet in photographic film, and in World War I, it was employed in "dopes" for airplane coverings. Cellulose acetate-butyrate, cellulose propionate, and mixed esters have also come into commercial use.

Cellulose ethers were initially made and proposed as industrially useful products in 1912 in Europe. Ethyl cellulose was the first to be manufactured commercially, followed by methyl cellulose, hydroxyethyl cellulose, carboxymethyl cellulose, and sodium cellulose sulfate (Chap. 32). However, cellulosics have reached maturity, and demand may diminish as other plastics, such as acrylonitrile-butadiene-styrene polymers and polystyrene film and sheet, continue to make inroads. The first commercial synthetic resin, the *phenol-formaldehyde* condensation-polymerization product, was described and patented by Baekeland in 1909[2] in the United States. The original Bakelite Co. was organized in 1910 and is now a unit of the Carbide and Carbon Chemicals Corp. After the expiration of the original basic patents in 1926 and 1927, many new companies were formed to make phenolic resins. Furfural is used in place of formaldehyde and gives improved corrosion resistance. When resorcinol replaces phenol, a superior adhesive for shipbuilding and bonding plywood results, because the bond is fungusproof and not affected by moisture.

Furan resins are obtained from furfural and its derivatives, including furan, tetrahydrofurfuryl alcohol, and especially furfuryl alcohol. The reaction of furfural with phenols is of great interest but is included with phenolics. Furfuryl alcohol has been commercially produced since 1934 by high-pressure (1,000 to 1,500 psi) hydrogenation of furfural in the presence of a copper chromite catalyst at 175°C.

Amino plastics. Of the amino plastics, *urea-formaldehyde* resinous molding compounds were the first to appear on the U.S. market, in 1929. Their introduction extended unlimited color possibilities to the field of thermosetting plastics. The first product was a transparent sheet to replace glass but was impractical because of a tendency to crack spontaneously. Research revealed that this cracking could be controlled by the addition of a hygroscopic filler such as wood flour, bleached pulp,

[2]Baekeland, Insoluble Products of Phenol and Formaldehyde, U.S. Pats. 942,699; 942,700; 942,808; 942,809 (1909).

TABLE 34.3 ***Properties of Plastics*** *(continued)*

Acrylonitrile-butadiene-styrene	*High-density polyethylene*	*Polyvinyl chloride*	*Polytetra-fluoro-ethylene*	*Vinylidene fluoride*	*Polypropylene*
1.07	0.96	1.4	2.2	1.7	0.90
0.08	0.19	0.10	0.14	0.14	0.08
3.3	7	3.7	5.5	8.5	4.7
0.32	0.54	0.25	0.25	0.33	0.46
0.3	0.02	0.10	Nil	0.04	0.03
Burns	Burns	Self-extinguishing	Non-flammable	Non-flammable	
4–6	3–5	7–9	1–3	7	4–6
6–7	Poor	9–11	0.5–0.7	10	9–11
R 90–100	R 40–50	R 115–125	R 85–95	80 Shore D	R 85–95
26–30	100–300	4–6	150–250	300	10–20

or α-cellulose. This yielded a molding composition which could be formed into stable articles. These resins have gained wide acceptance as adhesives in the plywood and furniture industries. *Melamine-formaldehyde resins* were made in 1939 by the American Cyanamid Co. Melamine molding compounds are widely employed for tableware, which represents the largest single outlet.

Linear polyesters are basically defined as the polycondensation products of dicarboxylic acids with dihydric alcohols. The *unsaturated polyester* resins introduced in 1942 are one of the most important groups and have wide application in reinforced plastics, and are used in small electrical components as well as large boat hulls. These unsaturated polyesters are formed by a condensation reaction and then converted to the insoluble product by addition of a cross-linking agent, such as styrene. The unsaturation is introduced by the use of an unsaturated acid, such as maleic, or an unsaturated alcohol, such as allyl. The polyester resins used in organic paints are usually known as *alkyds* and are distinguished from other polyesters in that they contain fatty monobasic acids as "oil modifications." A special type of saturated polyester prepared from ethylene glycol and terephthalic acid has found application in the textile fiber field (Dacron) and as film (Mylar) (Chap. 35). Aromatic polyesters are produced from aromatic dicarboxylic acids and aromatic bisphenols (copolyesters) or from *p*-hydroxybenzoic acid (homopolyesters). Thermoplastic polyesters are the condensation products of a saturated glycol (e.g., ethylene glycol) and an aromatic dicarboxylic acid (terephthalic acid). See Chap. 35. A new high-temperature-resistant aromatic polyester (Ekonol) can be formed like metals and does not melt at 1000°F.

Ionomers contain ionized carboxyl groups which create ionic cross-links in the polymer. They have exceptional abrasion resistance and toughness.

The *polymerization of formaldehyde* was observed when an aqueous solution of formaldehyde was concentrated by evaporation and paraformaldehyde was obtained. This compound was introduced commercially by Du Pont in 1960 as Delrin, and later as Celcon by the Celanese Co. They are classed as engineering plastics, exhibiting very low frictional properties, light weight, durability, and corrosion resistance. Applications are as metal replacements in the automotive industry, in home appliances, in plumbing, and the like.[3]

[3]Engineering Resins Primer, *Plast. Eng.*, **30**(12), 18 (1974) (excellent description and chart of properties of all resin types useful as engineering resins).

Nylon, best known as a fiber (Chap. 35) and the most widely used of the polyamides, is also an important plastic because of its low density, slow burning rate, toughness, flexibility, abrasion resistance, and high tensile strength in oriented monofilaments. Nylons were developed following the classic researches of Carothers and introduced on the market in 1938 in the form of bristles and hosiery fibers. Applications are numerous, including tires, gears, bushings, wire and cable coatings, and tubing and aerosol bottles, as well as textiles.

Epoxy resins, based on ethylene oxide or its homologs or derivatives, are one of the newer thermosetting resins to be employed in the process industries (recent developments have been in thermoplastic resins). Carbowax, made by the polymerization of ethylene oxide, was the earliest commercial product. Resins made by the condensation of epichlorohydrin with bisphenol A [2,2-bis(4-hydroxyphenyl)propane] have been marketed since 1948 under various trade names. Their major applications are in the field of protective coatings and adhesives, because of their outstanding chemical resistance and high tensile strength.

TABLE 34.4 ***Summary of Resin Properties and Applications***

Resin type	*Properties*	*Applications*
Phenolics	Good strength, heat stability, and impact resistance; high resistance to chemical corrosion and moisture penetration; machinability	Impregnanting resins, brake lining, rubber resins, electrical components, structural board, laminates, glues, adhesives, binders, molds
Aminos	Good heat resistance; solvent and chemical resistance; extreme surface hardness; resistance to discoloration	Molding compounds, adhesives, laminating resins, paper coating, textile treatments, plywood dinnerware, decorating structures
Polyesters	Extreme versatility in processing; excellent heat; chemical and flame resistance; low cost; excellent mechanical and electrical properties	Construction, auto-repair putty, laminates, skis, fishing rods, boats and aircraft components, coatings, decorative fixtures
Alkyds	Excellent electrical and thermal properties; versatility in flexibility or rigidity; good chemical resistance	Electrical insulation, electronic components, putty, glass-reinforced parts, paints
Polycarbonates	High refractive index; excellent chemical and electrical properties; dimensional stability; transparent; self-extinguishing; resistant to staining; good creep resistance	Replacement for metals, safety helmets, lenses, electrical components, photographic film, die casting, insulator
Polyamides	Tough; strong and easily moldable; light; abrasion-resistant; low coefficient of friction; good chemical resistance; self-extinguishing	Unlubricated bearings, fibers, gears, appliances, sutures, fishing lines, tires, watch straps, packaging, bottles
Polyurethanes	Extreme versatility when combined with other resins; good physical, chemical, and electrical properties	Insulation, foam inner liners for clothing, rocket fuel binders, elastomers
Polyethers	Excellent corrosion resistance to common acid, alkalies, and salts; can be seam-welded and machined to fit any type, shape, or size of structure	Coatings, pump gears, water-meter parts, bearing surfaces, valves
Epoxies	Excellent chemical resistance; good adhesion properties; strong and tough with low shrinkage during cure; excellent electrical properties; good heat resistance	Laminates, adhesives, flooring, linings, propellers, surface coatings, filament-wound structures (rocket cases)
Silicones	Good thermal and oxidative stability; flexible; excellent electrical properties; general inertness; excellent repellent	Mold-release agents, rubbers, laminates, encapsulating resins, antifoaming agents, water-resistant uses
Ionomers	Excellent toughness, abrasion resistance, and transparency; outstanding low-temperature flexural properties	Skin and blister packaging, heel lifts, shoes, ski boots, automobile bumpers, and golf ball covers
Phenoxies	Ease in molding; good heat stability; low mold shrinkage; self-extinguishing; good cold flow	Surface coatings, adhesives, binders, electronic parts

TABLE 34.4 ***Summary of Resin Properties and Applications*** *(continued)*

Resin type	*Properties*	*Applications*
Polyethylene	Excellent chemical resistance; low power factor; poor mechanical strength; outstanding moisture-vapor resistance; wide degree of flexibility	Packaging films and sheets, containers, wire and cable insulation, pipe, linings, coatings, molds, toys, housewares
Polypropylene	Colorless and odorless; low density; good heat resistance; "unbreakable"; excellent surface hardness; excellent chemical resistance; good electrical properties	Housewares, medical equipment (can be sterilized), appliances, toys, electronic components, tubing and pipe (can be welded), fibers and filaments, coatings
Fluorocarbons	Low coefficient of friction; low permeability; low moisture absorption; exceptional chemical inertness; low dielectric strength	Electrical insulation, mechanical seals, gaskets, linings for chemical equipment, bearings, frying-pan coatings, cryogenic application
Polyvinyl chloride	Excellent physical properties; excellent chemical resistance; ease of processing; relatively low cost; self-extinguishing; ability to be compounded with other resins	Pipe and tubing, pipe fittings, adhesives, raincoats and baby pants, building panels, wastepaper baskets, weather stripping, shoes
Acrylics	Crystal clarity; outstanding weatherability; fair chemical resistance; good impact and tensile strength; resistant to ultraviolet exposure	Decorative and structural panels, massive glazing domes, automotive lens systems, illuminated translucent floor tiles, windows and canopies, signs, coating, adhesives, elastomers
Polystyrene	Low cost; ease of processing; excellent resistance to acids, alkalies, salts; softened by hydrocarbons; excellent clarity, versatility	Insulation, pipe, foams, cooling towers, thin-walled containers, appliances, rubbers, automotive instruments and panels
Cellulosics	Outstanding toughness; high impact strength; high dielectric strength; low thermal conductivity; high surface luster	Textile and paper finishes, thickening agents, magnetic tapes, packaging, pipe
Furanes	Excellent resistance to both acids and bases; good adhesive properties	Laminates, coatings, impregnants, linings for rocket fuels, floor tiling, abrasive wheels

Sources: Modern Plastics Encyclopedia, 1974–1975; *Modern Plastics* (monthly); Perry, table 23.9, quantitative physical and mechanical properties of plastics.

An entirely new thermoplastic resin introduced by the Union Carbide Co. in 1962 is called *phenoxy*. Its structure is similar to that of epoxy resins, except that it contains no epoxy groups, molecular weight 10 to 30 times higher, and requires no curing or chemical reaction to become useful. It is suited for extrusion, injection, or blow molding, and finds application in adhesives and coatings, behaving much like a thermosetting resin.

Silicone[4] *resins* were first made available commercially in 1942 by Dow Corning Corp. These resins represented a revolutionary advance in the electrical field because of their resistance to heat and low water absorption, combined with excellent electrical insulating properties. Chemically, silicones are semi-inorganic polymers (organopolysiloxanes) having a molecular backbone of alternate atoms of silicon and oxygen. Silicones have found wide use as defoamers, mold-release agents, dielectrics, greases or lubricants, and polishes of all types.

Polyurethanes[5] have developed rapidly in recent years since the fundamental discovery by Wurtz in 1848 that urethane can be formed by reacting isocyanate with an alcohol. The greatest market for polyurethanes is in flexible foams for use in mattresses and furniture upholstery. Rigid foams are important in insulators, particularly for refrigerators, marine equipment, picnic coolers, and the like.

Polyethylene is the simplest member of the large group of thermoplastic resins formed by the polymerization of compounds containing an unsaturated bond between two carbon atoms. After

[4]See Chap. 36 for fundamental presentation; Stone and Graham, Inorganic Polymers, Academic, 1962; Speirs, Silicones, ECT, vol. 18, p. 221, 1969.

[5]Saunders-Frisch, Polyurethanes: Chemistry and Technology, Wiley, 1963.

several years of development, quantity production was begun in 1943. Many types are available but, for convenience, they are classified according to density, low density (density range up to 0.925), medium density (0.926 to 0.940), and high density (above 0.941). More polyethylene is produced today than any other plastic material, and its rate of growth is increasing. Polyethylene as a film is used in the packaging of foods, textiles, toys, soft goods, chemicals, machinery, etc. The use of polyethylene in the molding field is increasing steadily because of its flexibility, ease of processing, ease of coloring, and low cost. Consumers are able to buy any assortment of houseware products and toys made from molded polyethylene. High electrical resistivity and good dielectric characteristics, particularly at high frequencies, have made it an important electrical insulating material.

The development of crystalling *polypropylene* is an outgrowth of the unprecedented development of catalytic techniques for the low-pressure commercial polymerization of ethylene (1957). With a specific gravity of 0.90, polypropylene is the lightest of all plastics. Its application is in many of the areas mentioned for polyethylene, as well as in textile fibers, where it is a large item.

Fluorocarbon[6] plastics are basically composed of carbon and fluorine. Several types are available: *polytetrafluoroethylene*, Teflon 1946; *polychlorotrifluoroethylene*, Kel-F 1947; and *vinylidene fluoride*. Of all the plastics, poly- and fluorocarbons possess the greatest degree of inertness. Though very expensive, they are used to solve the most difficult problems involving gaskets, seals, diaphragms, packings, valve seats, linings, and the like.

Vinyl resins have long been known; however, their commercial development has taken place in the last 30 years. The most important resins industrially are polyvinyl chloride, polyvinyl acetate, copolymers of vinyl chloride and vinyl acetate, and polyvinyl acetals. Not until 1917 was the polymerization of vinyl acetate in the presence of peroxide recorded. Ostromislensky described the preparation of vinyl chloride in 1927, the same year that polyvinyl acetate appeared commercially in Germany. Vinyl acetals, formed by alcoholizing polyvinyl acetate and condensing with aldehydes, were patented in 1934.

Polyvinylidine chloride, another resin related to this group, was marketed by the Dow Chemical Co. under the name Saran in 1939 and has found extensive uses (Chap. 35). As a group, vinyl resins rank second in production to polyethylene.

Polystyrene is one of the oldest synthetic resins and is third in production volume. Successful commercial production of this plastic material was begun by Dow Chemical Co. about 1937. Several styrene derivatives and copolymers were introduced during World War II, which had better heat resistance and impact strength than polystyrene, yet retained its excellent electrical characteristics. As an example, and excellent high-impact-strength plastic is obtained by copolymerizing a high proportion of styrene with butadiene and then blending with natural or synthetic rubber. Styrene resins are used extensively for packaging foams and for sheets.

Acrylonitrile-butadiene-styrene (ABS) resins are finding large markets in appliances and automobiles. *Acrylic-type resins* are another example of a synthetic plastic known to chemists for many years but avalable in quantity only through present-day industrial research. In 1931 the Rohm and Haas Co. began limited production of acrylates. Methyl methacrylate resin was introduced by them in 1936 in transparent sheet form, Plexiglas, and by the Du Pont Co. in 1937 as both a sheet and a molding compound under the trade name Lucite.

The shortage of natural resins during World War I led to the development of *coumarone/indene* resins as a substitute for linseed oil in Europe. Petroleum and polyterpene resins are often classed with them, because their application lies in the area of coatings such as paints, varnishes, and sealers.

USES AND ECONOMICS The uses to which plastics are put depend upon the properties of the various molding compositions. The most important of the multitudinous applications are summarized

[6]Homsy, *Ind. Eng. Chem.*, **56**(3), 21 (1964); Brady, Materials Handbook, 10th ed., McGraw-Hill, 1971.

TABLE 34.5 ***Resins Classified by Selling Price***

Type	Average selling price, dollars per pound 1937	1954	1964	1971	1980*
Polyethylenes	....	0.44	0.21	0.12	0.20
Vinyls	0.69	0.41	0.29	0.16	0.26
Styrenes	0.72	0.31	0.21	0.19	0.31
Phenolics	....	0.26	0.24	0.21	0.34
Polyesters	....	....	0.31	0.23	0.38
Urea and melamines	0.45	0.24	0.26	0.35	0.58
Coumarone-indene and petroleum resins	....	....	0.10	0.11	0.18
Alkyds	0.20	0.30	0.26	0.32	0.52
Polyamide (nylon)	....	1.60	0.93	0.55	0.90
Polyurethanes	....	....	0.58	0.46	0.75
Silicone resins	....	3.25	2.28	1.65	2.70
Polypropylene	....	....	0.24	0.13	0.21
Epoxy resins	....	....	0.59	0.47	0.77

Source: Synthetic Organic Chemicals, U.S. Tariff Commission (annual).
*Estimated.

in Table 34.4 Table 34.5 shows the relative cost of the various plastics, and Table 34.6 the U.S. production. The chemical engineer uses phenolic and furan resins reinforced with acid-washed asbestos or Fiberglas for many acid-proof vessels.[7]

RAW MATERIALS[8] The plastics industry has grown very rapidly, as Table 34.6 shows. The basic and intermediate raw materials come from all parts of the nation—mines, forests, farms, quarries, paper and textile mills, and cotton plantations, but mostly from natural gas and petroleum. Coal mines and petroleum supply the basic needs for phenol, formaldehyde, dyes, solvents, maleic and phthalic anhydrides, olefins, and many other organic chemicals. Farms are the source of lactic

TABLE 34.6 ***Plastics and Resins, U.S. Production***
(In millions of pounds)

Material	1974	1973	1971
Polyethylene			
High density	2,837	2,637	1,870
Low density	5,973	5,803	4,491
Polyvinyl alcohol	142	96	49
Polyvinyl chloride	4,850	4,562	3,437
Other vinyls	209	199	168
Styrene resins	4,152	4,146	2,278
Phenolics	1,335	1,387	1,180
Polyesters	911	1,051	707
ABS	857	877	566
Urea and melamines	1,000	1,037	623
Epoxies	249	223	167
Nylon (exclusive of fibers)	204	205	130

Source: Plast. World, **33**(5), 10 (1975); Synthetic Organic Chemicals, U.S. Tariff Commission.

[7]Barton, *Ind. Eng. Chem.*, **56**(7), 66 (1964); Ranney, Reinforced Plastics and Elastomers, Noyes, 1977; Halpern, Synthetic Paper from Fiber and Films, Noyes, 1975.

[8]Hatch, *Pet. Refiner,* **42**(4), 157 (1963), and **42**(5), 171 (1963); Kline, *Mod. Plast.* **41**(6), 131 (1964); Sherwood, *Ind. Eng. Chem.*, **54**(12), 29 (1962); Modern Plastics Encyclopedia (yearly); Raynolds, *Chem. Metall. Eng.*, **51**(3), 109 (1944); Golding, Polymers and Resins, Van Nostrand, 1959.

acid from milk or corn sugar, bean meal from soy plants, and glycerin synthesized from petrochemicals or from fatty oils. Quarries supply asbestos and other mineral fillers. The paper and textile mills provide cellulose paper products and fabrics for laminated plastics.

Plastics are made from molding compositions prepared from two or more of the following material groups:[9]

Binder. This is usually a resin or cellulose derivative.

Filler. Cellulose, wood flour, cotton fiber, asbestos, mica, glass fibers, or fabrics may be added to increase strength. These are classified in Table 34.7.

Chemical intermediates for resins are made by procedures outlined at the end of this chapter and in other chapters of this book. The following are typical important chemical intermediates in the manufacture of resins used in plastics:

Phenol, Chap. 34
Formaldehyde, Chap. 34
Hexamethylenetetramine ("hexa"), Chap. 34
Phthalic anhydride, Chap. 34
Vinyl chloride and acetate, Chap. 34
Acrylonitrile, Chap. 35
Olefins, Chaps. 37, 38
Styrene, Chap. 36
Ethylene, propylene, Chaps. 37, 38
Adipic acid, Chap. 35
Acetylene, Chap. 7
Butadiene, Chap. 36

The intermediates presented are not in any case exclusively consumed by the plastics industry. Other industries in the CPI group are heavy users, such as fibers (Chap. 35), rubber (Chap. 36), and pharmaceuticals (Chap. 40).

Plasticizers. Plasticizers are organic chemicals added to synthetic plastics and resins to (1) improve workability during fabrication, (2) extend or modify the natural properties of these resins, or (3) develop new improved properties not present in the original resins. Plasticizers reduce the viscosity of the resins and make them easier to shape and form at elevated temperatures and pressures. They also impart flexibility and other desirable properties to the finished product.

Dyes and pigments. The dyes used vary in resistance to sunlight and to the plastic binder.

Catalyst. Thermosetting resins use either an acid or basic catalyst, depending upon the desired properties, whereas resins of the vinyl group use peroxides to initiate polymerization of the monomers.

Lubricants. Lubricants, such as stearates and other metallic soaps, are used particularly in cold-molding compounds to facilitate the molding operation.

RESIN-MANUFACTURING PROCESSES

The chemical conversion for resin manufacture is *polymerization,* in which simple molecules react to form polymers. Two principal types are recognized, *condensation* and *addition.* Flory[10] has described a condensation process as occurring during a reaction between pairs of functional groups to form a group not present in the reactants, being a minor compound being split out, e.g., H_2O, and an addition process as occurring by opening of a bond in a reactant and the formation of similar bonds with other reactants without formation of a side product.

Polymerization takes place in a number of ways, namely, bulk, solution, emulsion, or suspension. *Bulk polymerization*[11] of monomers may be carried out in the liquid or gaseous state but presents some difficulties in large-scale production. Yields from polymerization of the gaseous

[9]An excellent chart of additives for plastics and their characteristics is given in *Plast. Eng.*, **31**(3), 20 (1975).

[10]Principles of Polymer Chemistry, pp. 39–40, Cornell, 1953; Arnold, Introduction to Plastics, p. 21–23, Iowa State University Press, 1968.

[11]Wohl, Bulk Polymerization, *Chem. Eng.* (*N.Y.*), **73**(16), 60, 1966.

TABLE 34.7 ***Classification of Fillers***

Organic origin	*Inorganic origin*
Cellulosic:	Mineral fillers:
Wood flour	Asbestos
Cotton	Powdered mica
α-Cellulose	Silicate clays, talc, kieselguhr
Paper pulp	Barites
Shredded textiles	Whiting
Bagasse, corn husks, seed hulls, etc.	Pumice, emery
Carbonaceous:	Zinc and lead oxides
Graphite	Cadmium and barium sulfides
Carbon black	Powdered metals:
Miscellaneous:	Iron
Powdered rubber	Lead
Plasticizers	Copper
	Aluminum
	Miscellaneous:
	Fiberglass

monomer are likely to be low unless effected by high pressure or utilization of a catalyst. Addition polymerizations are strongly exothermic and may proceed uncontrollably unless the heat evolved is removed. In some cases the polymers are soluble in their liquid monomers, as a result of which the viscosity of the solution increases rapidly, causing problems in agitation and separation. In other cases, the polymer is not soluble, and it precipitates out after a certain degree of polymerization. *Solution polymerization*[12] proceeds in a solvent suitable for both monomer and initiator. The polymer may remain in solution or may separate. Viscosity increases, and solvent removal may cause difficulties, but the heat evolved may be controlled by refluxing the solvent. In *suspension polymerization,*[13] the monomer is suspended in water by agitation, and stabilizers such as talc, fuller's earth, and Bentonite are added to stabilize the suspension and prevent polymer globules from adhering to each other. Normally, the initiator is soluble in the monomer. Each monomer globule polymerizes as a spherical pearl of high molecular weight. The heat of polymerization is removed by the water, permitting accurate temperature control. The stabilizer must be separated from the polymer, however, and sometimes, because of partial miscibility of monomer and water, subsidiary polymerization may occur in the aqueous phase, producing a low-molecular-weight polymer. *Emulsion polymerization*[14] differs from suspension polymerization. Soap is added to stabilize the monomer droplets and form aggregates called micelles. The miscelles take monomer into their interior, and the initiator dissolved in the aqueous phase diffuses into the micelle to start polymer growth. At certain degrees of polymerization, the polymer is ejected from the micelle but continues to grow. The monomer diffuses into the micelle to replenish it. Emulsion polymerizations are rapid and can be carried out at relatively low temperatures. The aqueous phase absorbs the heat evolved from the polymerization.

CONDENSATION-POLYMERIZATION PRODUCTS

Phenolic resins designate a group of synthetic resins that are probably the most varied and versatile that we know. They may be made from almost any phenolic body and an aldehyde. Phenol-formaldehyde resins constitute by far the greatest proportion, but phenol-furfural, resorcinol-formaldehyde, and

[12]Back Solution Polymerization, *Chem. Eng.* (*N.Y.*), **73**(16), 65, 1966.
[13]Church, Suspension Polymerization, *Chem. Eng.* (*N.Y.*), **73**(16), 79, 1966.
[14]Gellner, Emulsion Polymerization, *Chem. Eng.* (*N.Y.*), **73**(16), 74, 1966.

similar resins are also included in this group. The product obtained depends primarily on the concentration and chemical nature of the reactants, the nature and concentration of the catalyst used, the temperature and reaction time, and the modifying agents, fillers, and extenders. The initial reaction between the phenol and a mixture of cresols with formaldehyde, using an alkaline catalyst, produces benzyl alcohols:

$$C_6H_5OH + CH_2O \longrightarrow o\text{- or } p\text{-}C_6H_4OH \cdot CH_2OH$$

Simultaneously, additional formaldehyde may react to produce both di- and trimethylolphenols. These alcohols continue to condense and polymerize with each other rapidly and almost violently.[15]

Phenol-formaldehyde
Reactions: Condensations and Polymerizations

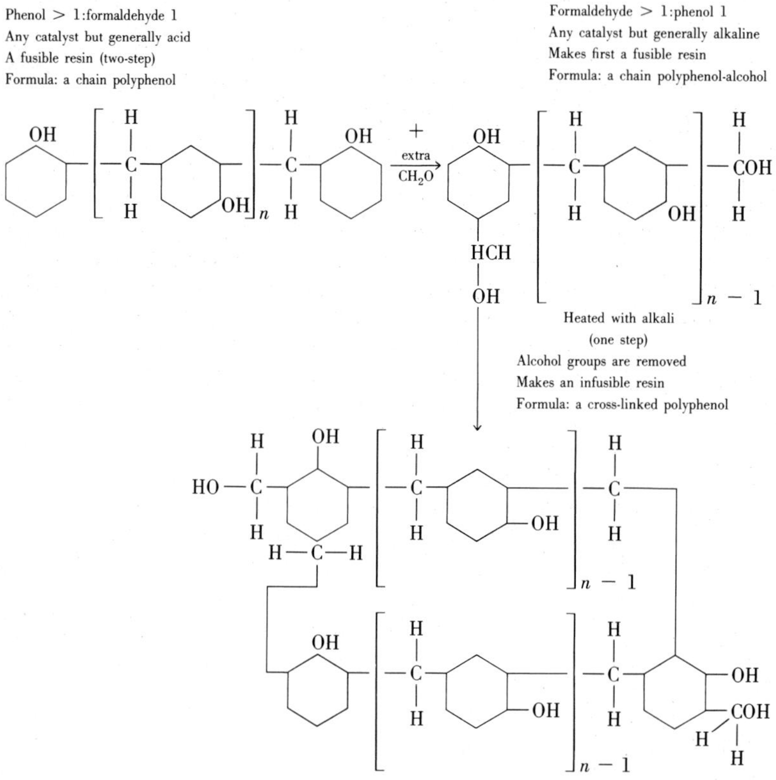

Phenol > 1:formaldehyde 1
Any catalyst but generally acid
A fusible resin (two-step)
Formula: a chain polyphenol

Formaldehyde > 1:phenol 1
Any catalyst but generally alkaline
Makes first a fusible resin
Formula: a chain polyphenol-alcohol

+ extra CH_2O →

Heated with alkali
(one step)
Alcohol groups are removed
Makes an infusible resin
Formula: a cross-linked polyphenol

These resins, when classified according to the nature of the reaction occurring during their production, are one of two fundamental types.

1. *One-step resins.* In these, all the necessary reactants (phenol, formaldehyde, catalyst) required to produce a thermosetting resin are charged into the resin kettle in the proper proportions and react together. An alkaline catalyst is used. The resin, as discharged from the kettle, is thermosetting or heat-reactive and requires only further heating to complete the reaction to an infusible, insoluble state.

[15]Butler, Phenol Resin Emulsions, *Am. Dyest. Rep.*, **32,** (128 (1943); Encyclopedia of Polymer Science and Technology, Wiley, 1964 (referred to hereafter as EPST).

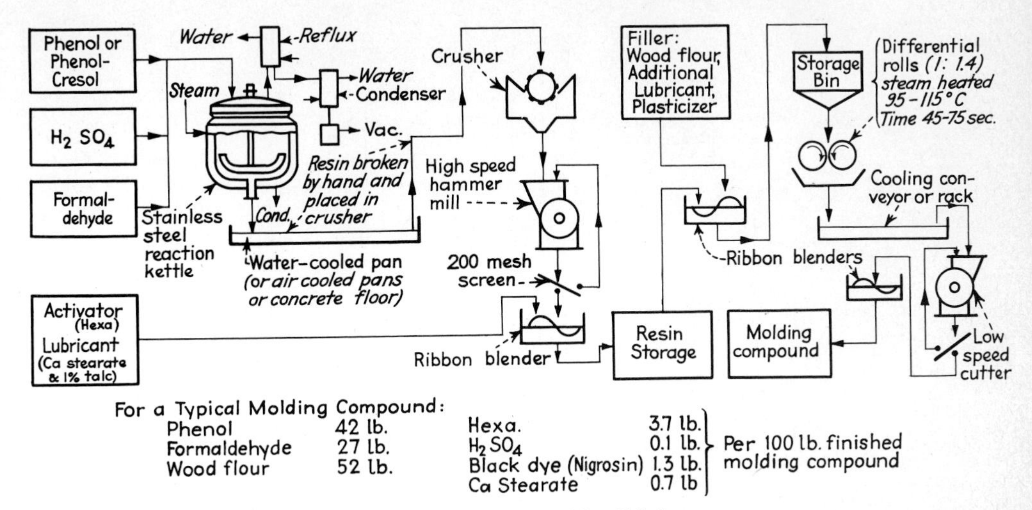

Fig. 34.1 Flowchart for production of a molding compound of the phenol-formaldehyde type.

2. *Two-step resins.* Only part of the necessary formaldehyde is added in the kettle in making these resins, and an acid catalyst is used. They are permanently fusible or thermoplastic when discharged from the kettle, but react with additional formaldehyde to produce a thermosetting resin. This additional formaldehyde is furnished by "hexa" (hexamethylenetetramine). Both one- and two-step resins are used, separately or in combination, in commercial molding materials. Both types are believed to polymerize to similar end products. The two-step scheme is represented in Fig. 34.1,[16] showing the sequences of *unit operations* and *chemical conversions*:

The phenol and formaldehyde are placed in the reaction kettle with the catalyst (sulfuric acid) and heated 3 or 4 h at a temperature of 285 to 325°F (Ch).

During *condensation,* reaction water is eliminated and forms the upper of two layers. This water of reaction is removed under vacuum without the addition of heat (Op).

The warm, dehydrated, viscous resin is run out of the kettle into shallow trays and allowed to cool and harden. The cooled, brittle resin is crushed and finely ground and becomes the resin binder for molding phenolic resins (Op).

The crushed and ground resin is blended with the activator ("hexa") (Op).

Phenolic molding compounds are molded primarily in compression and transfer molds. The powder, mixed with fillers, lubricant, and plasticizers, is further reacted on steam-heated rolls, cooled, and ground. In *compression molding* the powder is placed in hardened steel molds at a temperature of 270 to 360°F and at pressures from 2,000 to 5,000 psi (Op and Ch).

In *transfer molding* the thermosetting material is subjected to heat and pressure in an outside chamber, from which it is forced by means of a plunger into a closed mold where curing takes place (Op and Ch).

Electronic preheating (in a high-frequency electrostatic field outside the process) of the molding powders as such or in the form of pellets, prior to mold loading, helps achieve a more rapid cure with less pressure, since it promotes the flow of material in the mold cavity (Op).

The final chemical polymerization reaction, or cure, which takes place in the mold to transform the powder into a rigid, infusible shape of the finished article (CH).

[16]Phenolic Resin Process Sidesteps Kettle, *Chem. Eng.* (*N.Y.*), **72**(23), 104 (1965) (flowchart).

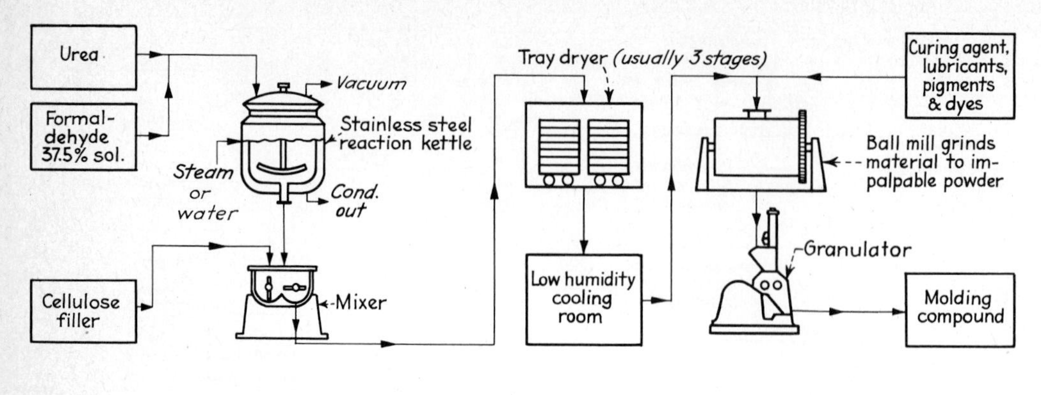

Fig. 34.2 Flowchart for manufacture of a molding compound of the urea-formaldehyde type.

Cast phenolics differ from molding compositions in that no pressure is required to make the composition flow. The phenol and formaldehyde with a basic catalyst (usually sodium or potassium hydroxide) are placed in a metal or stainless-steel kettle and heated from the boiling point to as low as 160°F for a period of 10 min to 3 h. Cooling is done by means of water in the jacket during certain phases of the exothermic reaction. At the proper time (removal of about 75% of the water formed), while the resin is still hydrophilic, an organic acid (lactic or maleic acid) is added to neutralize the resin and clarify the color. Before the final dehydration of the resin, plasticizers, pigments, and colors are added to the kettle and mixed with the resin. Dehydration is effected under vacuum at a resin temperature not in excess of 165 to 175°F, and the hot resin is withdrawn and poured into preheated lead molds. The final reaction and hardening take place by curing the resin in the molds at 185°F for periods of 3 to 10 days. The curing ovens are heated with steam under precise temperature control.

AMINO RESINS[17] *Urea-formaldehyde* and *melamine-formaldehyde* condensates are the commercially important amino resins. Other resins in this family utilize sulfanamides, aniline, and thiourea. The simplest condensates are methylolureas and methylomelamines. A typical low-stage resin is formed when urea (or melamine) is mixed with formaldehyde as shown in Fig. 34.2. The initial reaction of urea (or melamine) with formaldehyde is a simple addition to produce methylol compounds:

$$(-NH_2)_2C{=}O + CH_2O \longrightarrow O{=}C(NHCH_2OH)(NH_2)$$

Some dimethylol urea ($HOH_2CNH \cdot CO \cdot NHCH_2OH$) is also formed. The mechanism of subsequent intercondensation of the methylol compounds has not been established definitely, but there is evidence of the formation of methylene linkages, —NH, —CH_2, —NH—, or ether bridges, —$NH \cdot CH_2$—O—CH_2NH—, as condensation and curing progress. These water-soluble, water-white intermediates are employed in admixture with some form of cellulose before the final reaction and curing to form an infusible, insoluble product. A suitable catalyst and controlled temperature are also needed. Since melamine is not readily soluble in water or Formalin at room temperature, it is necessary to heat to about 80°C to obtain the methylol compounds for melamine-formaldehyde resins.

[17]Modern Plastics Encyclopedia, Cordier, p. 25, 1974–1975; ECT, 2d ed., vol. 2, p. 225, 1963; EPST, vol. 1, p. 1, 1965.

POLYESTER RESINS[18] are complex esters formed when a difunctional alcohol is reacted with an anhydride or dibasic acid. Since the reaction takes place at both ends of the chain, long molecules are possible and a plurality of ester groups is obtained. *Unsaturated polyesters* are produced when any of the reactants contain unsaturation, such as

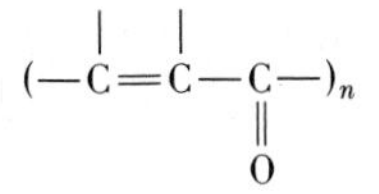

$$(-\overset{|}{C}=\overset{|}{C}-\underset{\underset{O}{\|}}{C}-)_n$$

Unsaturated polyesters can be formed in two steps, condensation of the acid and alcohol to form a soluble resin, and then addition of a cross-linking agent to form the thermosetting resin. Typical reactions are:

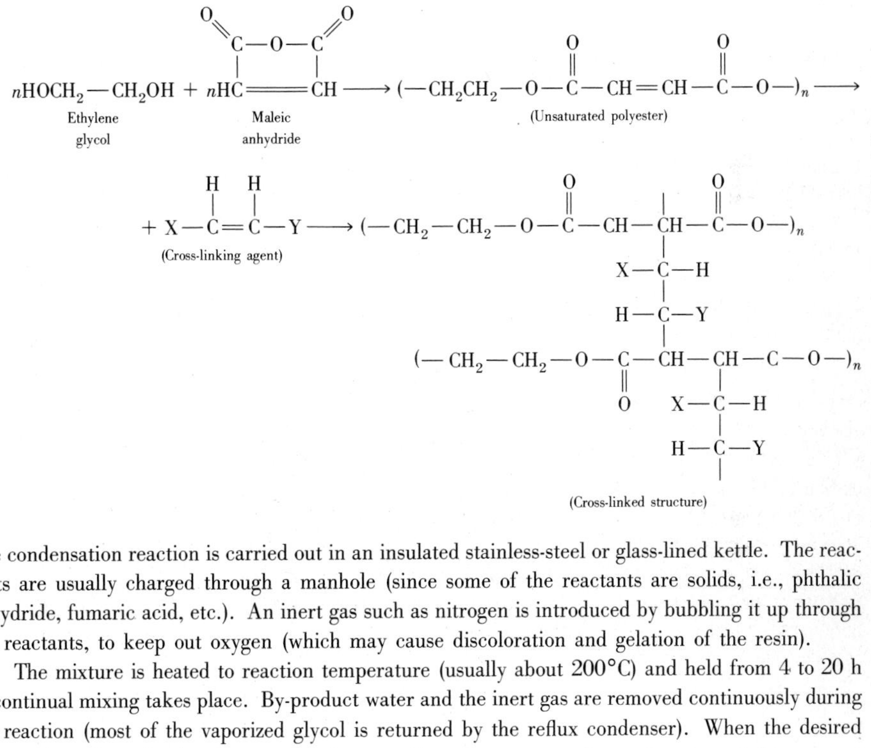

$$\underset{\text{Ethylene glycol}}{n\text{HOCH}_2-\text{CH}_2\text{OH}} + \underset{\text{Maleic anhydride}}{n\text{HC}(-\text{C}(=\text{O})-\text{O}-\text{C}(=\text{O})-)=\text{CH}} \longrightarrow \underset{\text{(Unsaturated polyester)}}{(-\text{CH}_2\text{CH}_2-\text{O}-\overset{\overset{O}{\|}}{\text{C}}-\text{CH}=\text{CH}-\overset{\overset{O}{\|}}{\text{C}}-\text{O}-)_n} \longrightarrow$$

$$+ \underset{\text{(Cross-linking agent)}}{\text{X}-\overset{\overset{H}{|}}{\text{C}}=\overset{\overset{H}{|}}{\text{C}}-\text{Y}} \longrightarrow (-\text{CH}_2-\text{CH}_2-\text{O}-\overset{\overset{O}{\|}}{\text{C}}-\text{CH}-\text{CH}(-\text{C}(\text{X})(\text{H})-\text{C}(\text{H})(\text{Y})-)-\overset{\overset{O}{\|}}{\text{C}}-\text{O}-)_n$$

$$(-\text{CH}_2-\text{CH}_2-\text{O}-\underset{\underset{O}{\|}}{\text{C}}-\text{CH}-\text{CH}(-\text{C}(\text{X})(\text{H})-\text{C}(\text{H})(\text{Y})-)-\text{C}-\text{O}-)_n$$

(Cross-linked structure)

The condensation reaction is carried out in an insulated stainless-steel or glass-lined kettle. The reactants are usually charged through a manhole (since some of the reactants are solids, i.e., phthalic anhydride, fumaric acid, etc.). An inert gas such as nitrogen is introduced by bubbling it up through the reactants, to keep out oxygen (which may cause discoloration and gelation of the resin).

The mixture is heated to reaction temperature (usually about 200°C) and held from 4 to 20 h as continual mixing takes place. By-product water and the inert gas are removed continuously during the reaction (most of the vaporized glycol is returned by the reflux condenser). When the desired degree of condensation has been attained, generally under vacuum, the product is cooled to prevent premature gelation. The polycondensation product (very viscous) is pumped to the blending tank and mixed with the cross-linking agent (usually 2 to 4 h). The resin is transferred to drums for shipment and storage (Op). For the process flowchart[19] for manufacturing polyester film (Mylar) and polyester fiber (Dacron), see Fig. 35.5.

Alkyd resins. The processing equipment (reaction kettle and blending tank) used for unsaturated polyesters can also be used for manufacturing alkyd resins. Of the four basic methods of production, namely, the fatty acid method, the fatty acid–oil method, the oil-dilution method, and the

[18]Bjorksten Research Laboratories, Polyesters and Their Applications, Reinhold, 1956; Modern Plastics Encyclopedia (yearly); Patton, Alkyd Resin Technology, Interscience, 1962; EPST, vol. 11, p. 63, 1969.

[19]*Chem. Eng.* (*N.Y.*), **70**(5), 76 (1963).

alcoholysis method, the fatty acid and alcoholysis methods are the most important and are described in the literature.[20]

Polycarbonate resins are a special variety of polyester in which a derivative of carbonic acid is substituted for adipic, phthalic, or other acid and a diphenol is substituted for the more conventional glycols. A number of methods for the preparation of polycarbonates has been described, of which the melt process and the phosgenation process are the most important.[21]

Their general formula can be represented by

$$\left[-O-C_6H_4-C(CH_3)_2-C_6H_4-O-\overset{\displaystyle O}{\overset{\|}{C}}- \right]_n$$

ADDITION-POLYMERIZATION PRODUCTS[22]

POLYOLEFINS *Polyethylene*[23] was the first, and still is the largest, in production of polyolefin plastics. It resulted from a series of accidental discoveries. Imperial Chemical Industries (ICI) of England inadvertently discovered the white, waxy solid in 1933 while attempting to react ethylene with benzaldehyde in an autoclave. The low-pressure processes [by Fischer, Ziegler, Phillips Petroleum, and Standard Oil (Indiana)], for making polyethylene were accidently discovered during an attempt to synthesize lubricants and motor oils from ethylene.

The original ICI polyethylene process operated at 500 to 3,000 atm. This was one of the *highest-pressure* processes employed in the organic chemical industry, and from 1943 to 1956 was the only type of polyethylene process in operation in the United States. Interest has naturally gravitated toward the lower-pressure processes, namely, the Ziegler[24] process operating at atmospheric pressure (Fig. 29.6 and text), the Indiana (Standard Oil) process operating between 60 and 100 atm, and the Phillips process operating at about 30 to 40 atm. The "tricks" in the manufacture of polyethylene seem to involve the choice of the catalyst used. In general, high-pressure processes use oxygen or peroxides for catalysts, whereas low-pressure processes use metal-derived catalysts (such as titanium tetrachloride in a hexane solution of aluminum triethyl, chromium oxide on a support material of a silica-alumina, etc.). A flowchart of the ICI process is shown in Fig. 34.3. Polyethylene manufacture requires high-purity ethylene, and the first step involves the demethanizer, where a mixture of methane-ethylene is removed and recycled. The feed passes to a de-ethanizer, where 99.8 to 99.9% ethylene is taken overhead and the bottoms (ethane) recycled to the ethylene plant. A free-radical-yielding catalyst, such as oxygen, is added to the high-purity ethylene, compressed to operating pressure (1,500 atm), and fed to the reactor, which is maintained at 375°F. The effluent from the reactor passes to a separator vessel, in which the unconverted ethylene is removed and recycled. The liquid from the separator is water-white polyethylene, which can be extruded, chilled and solidified, chopped, and stored.

[20]ECT, 2d ed., vol. 1, p. 851, 1964; Martins, Alkyd Resins, Reinhold, 1961; EPST, vol. 1, p. 663, 1964.

[21]ECT, 2d ed., vol. 16, p. 106, 1968; Fox and Goldberg, Aromatic Polycarbonates, Gordon Research Conference on Polymers, July 1957; Schnell, Chemistry and Physics of Polycarbonates, Interscience, 1964; Christopher and Fox, Polycarbonates, Reinhold, 1962; EPST, vol. 10, p. 710, 1969; *Mod. Plast.*, **52**(6), 16 (1975).

[22]Moore, An Introduction to Polymer Chemistry, Aldine, London; Golding, *op. cit.;* Winding and Hiatt, Polymeric Materials, McGraw-Hill, 1961.

[23]Sittig, Polyolefin Resin Processes, Gulf Publishing, 1961; Kresser, Polyethylene, Reinhold, 1957; Modern Plastics Encyclopedia, 1974–1975; ECT, 2d ed., vol. 14, p. 217, 1967; EPST, vol. 6, p. 275, 1967.

[24]The Ziegler-type polymerizations with reactions and conditions needed to build up straight-chain organic compounds are presented in Chap. 29 under detergents and alcohols.

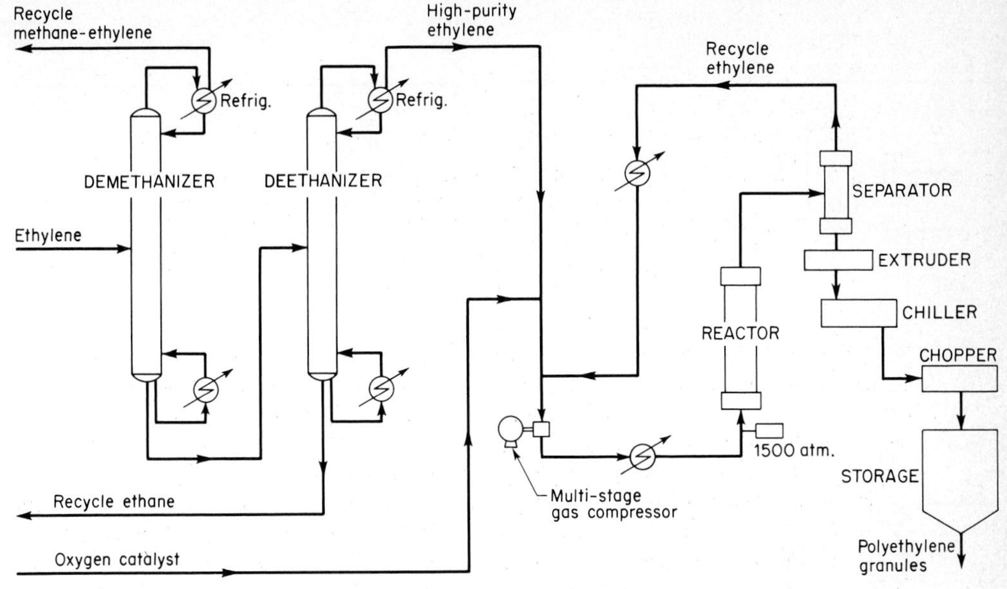

Fig. 34.3 High-pressure polyethylene resins. (*Hydrocarbon Processing and Petroleum Refiner.*)

One of the *low-pressure* processes is the Phillips process, as illustrated in Fig. 34.4. Purified ethylene and cyclohexane solvent are charged to the reactor, as well as a suitable catalyst, such as hexavalent chromium oxide (Cr_2O_3) on a silver-alumina support. Ethylene concentration is about 5% by weight, and the catalyst is about 0.5% or less. The cyclohexane serves as a medium for dissipating the heat of reaction, controls the rate of ethylene consumption to promote good polymer growth, and protects the growing polymer from chain breaks. The reactor is maintained at a temperature of 200 to 300°F and a pressure range of 100 to 500 psig. The reactor effluent is sent to a flash drum for removal of some of the excess solvent and ethylene. The polymer is then centrifuged and filtered to recover an ash-free polymer solution. The polymer is precipitated from the solution in a stripper to form a slurry. This is separated from the water by flotation, dried, and finished.

The application of some of these recently developed heterogeneous catalyst systems to *propylene polymerization* has resulted in the formation of crystalline polypropylene. Polymerization can be carried out in a batchwise or continuous process, usually in hydrocarbon diluents such as *n*-heptane at moderate temperatures (20 to 120°C) and low pressures (1 to 40 atm). The catalyst is added as a slurry, and propylene is fed at a controlled rate. The crystalline polymer is insoluble and precipitates as a finely divided granular solid. Propylene addition is continued until the slurry becomes quite thick (20 to 40% solids). The catalyst is recovered by the addition of a suitable reagent, and the polymer is freed of solvent and dried. In producing good-quality polypropylene, monomer purity is of prime importance. Catalysts are poisoned by oxygen, carbon monoxide and dioxide, sulfur compounds, and water. Polymer formation appears to require a solid catalyst phase to obtain stereospecific results, precluding the use of hydrocarbon-soluble compounds. Almost any element of the fourth, fifth, or sixth group in the periodic table is a useful metal catalyst. Titanium is the most effective metal, and titanium trichloride is the preferred halide. A typical processing scheme with a flowchart and a description is presented by Haines.[25] The polypropylene market is a growing one and amounted to about 2.2 billion lb in 1973.

[25]Haines, *Ind. Eng. Chem.*, **55**(2), 30 (1963); ECT, 2d ed., vol. 14, p. 282, 1967.

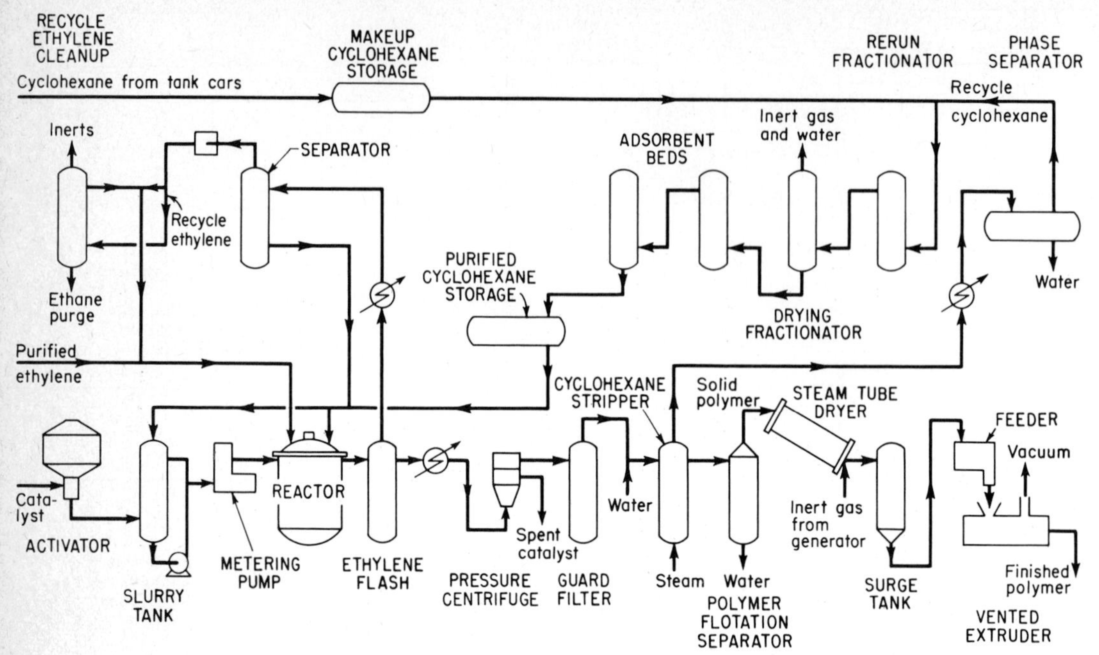

Fig. 34.4 Low-pressure polyethylene resins. (*Hydrocarbon Processing and Petroleum Refiner and Phillips Petroleum Co.*)

VINYL RESINS[26] The polyvinyl resins are synthetic materials made from compounds having a vinyl ($-CH{=}CH_2$) group. The most important members of this class are polyvinyl acetate, polyvinyl chloride, polyvinyl alcohol, and polyvinyl acetals and copolymers of vinyl chloride with vinyl acetate and vinylidene chloride. For a simplified flowchart for vinyl resins, see Fig. 34.5.

Vinyl acetate is a clear liquid, the formation of which may be expressed as

$$\begin{matrix} CH \\ ||| \\ CH \end{matrix} + HO\cdot OCCH_3 \xrightarrow[\text{salt}]{\text{Hg}} \begin{matrix} CH{=}CH_2 \\ | \\ O\cdot OCCH_3 \end{matrix}$$

See Chap. 38. A copper salt is added to the monomer to prevent premature polymerization. This salt is left behind after distillation. The monomer, a catalyst such as acetyl peroxide, and a solvent are placed in an autoclave, and heat is applied. After a short induction period polymerization begins and is allowed to proceed to a predetermined limit. Then the unreacted monomer and the solvent are removed. The polymer is a linear-chain compound which can be formulated in part as follows:

$$-\underset{\displaystyle OOCCH_3}{\underset{|}{CH}}-CH_2-\underset{\displaystyle OOCCH_3}{\underset{|}{CH}}-CH_2-\underset{\displaystyle OOCCH_3}{\underset{|}{CH}}-CH_2-$$

The dried polymer is ground and becomes the resin binder for molding compositions or a base for adhesives. See Table 34.8 for molecular weights.

Polyvinyl acetate may be prepared in this manner: A benzene solution of *vinyl acetate* containing the desired catalyst is introduced into a jacketed vessel. At a temperature of about 72°C, the mixture boils and the vapors are condensed and returned to the kettle. After about 5 h at a

[26]Winding and Hiatt, *op. cit.;* Smith, Vinyl Resins, Reinhold, 1958, Modern Plastics Encyclopedia, 1974–1975; *Mod. Plast.*, **52**(6), 16 (1975).

TABLE 34.8 ***Molecular Weights of Vinyl Copolymers***

Average molecular weight	*Use*
9,000	Surface-coating resin
10,000	Resin for calender-coating paper or injection molding
12,500	Compression molding
15,500	Stiff sheet stock
21,000	Highly plasticized sheet and extruded wire coatings
22,000	Textile fiber

gentle boil, the reaction mixture is run to a still, and the solvent and unchanged vinyl acetate are removed by steam distillation. The molten resin is then either run into drums, where it solidifies, or extruded into rods and sliced into flakes.

The fastest-growing branch of the vinyl family is *polyvinyl chloride* (PVC). PVC's popularity is due, primarily, to four factors: its excellent physical properties, its ability to be compounded for a wide range of applications, its ease of processing, and its relatively low cost.

Under normal conditions, vinyl chloride is a gas that boils at $-14°C$. It may be made by the following reactions:

$$HC{\equiv}CH + HCl \xrightarrow[100\text{–}200°C]{HgCl_2} ClCH{=}CH_2 \qquad \Delta H = -23{,}000 \text{ cal/mol} \tag{1}$$

$$CH_3{=}CH_2 + Cl_2 \longrightarrow ClCH_2CH_2Cl \xrightarrow[300\text{–}600°C]{\text{alkali}} ClCH{=}CH_2 \tag{2}$$

$$CH_2 \quad CH_2 + \tfrac{1}{2}O_2 + 2HCl \xrightarrow[500°F]{CuCl_2} ClCH{=}CH_2 + H_2O \tag{3}$$

See Fig. 34.9 for the manufacture of vinyl chloride from ethylene.

Exposure to vinyl chloride vapors, even in very small concentrations, causes some workers to develop liver cancer. In 1974 the OSHA ruled that employee exposure must be at a no-detectable level (less than 1 ppm) in all monomer, polymer, and fabricating operations. This has led to many changes in the PVC industry to comply with these strict regulations.[27]

The liquid monomer is formed into tiny globules by vigorous stirring in water containing a suspending agent. A typical recipe lists 100 parts of water, 100 parts of liquid vinyl chloride, 1 part of a persulfate catalyst, and 1.5 parts of an emulsifier such as sodium lauryl sulfate. The autoclave

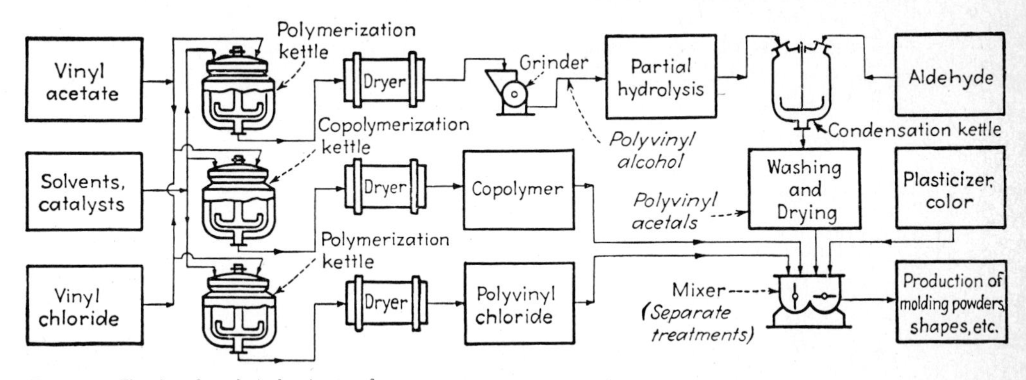

Fig. 34.5 Flowchart for polyvinyl resin manufacture.

[27]*Chem. Eng. News,* July 22, 1974, p. 9; *Chem. Week,* Sept. 18, 1974, p. 66; *Mod. Plast.* **52**(5), 49 (1975).

operates at 40 to 45°C for 72 h to give a 90% yield of polymer with a particle size of 0.1 to 1.0 μm. Recovery of these particles may be accomplished by spray-drying or by coagulation by acid addition.[28] A PVC compound can be tailor-made to achieve whatever balance of properties is desired by using plasticizers, stabilizers, lubricants, and fillers.

Copolymers of vinyl chloride and other vinyls retain the toughness and chemical resistance of PVC, but are more flexible. Studies on these copolymers were carried out because of the practical limitations (high molding temperature, poor heat stability, etc.) encountered with PVC in its early development. The monomers of both acetate and chloride are mixed with a solvent and catalyst and polymerized in an autoclave to yield a copolymer which may might be represented in part as follows:

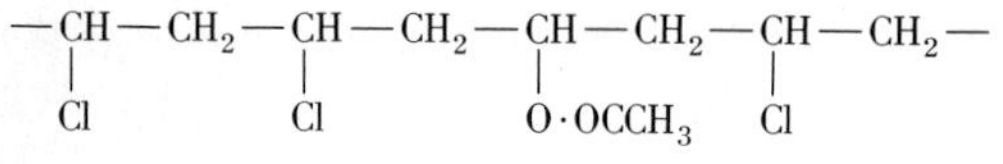

$$-\underset{\substack{|\\Cl}}{CH}-CH_2-\underset{\substack{|\\Cl}}{CH}-CH_2-\underset{\substack{|\\O\cdot OCCH_3}}{CH}-CH_2-\underset{\substack{|\\Cl}}{CH}-CH_2-$$

Although the acetate/chloride ratio can be varied over wide limits to give resinous products suitable for a great variety of applications, the more important copolymers at present contain a preponderance of chloride.

VINYL ALCOHOL RESINS These resins are made from polyvinyl acetate which is first reacted with alcohol to yield polyvinyl alcohol and then condensed with aldehydes to give a group of resins. The polyvinyl acetate is reacted with alcohol under controlled conditions with a trace of either acid or alkali, which replaces the acetyl groups with hydroxyl groups to yield polyvinyl alcohol:

$$-\underset{\substack{|\\O\cdot OCCH_3}}{CH}-CH_2-\underset{\substack{|\\O\cdot OCCH_3}}{CH}-CH_2- \xrightarrow{+ROH} -\underset{\substack{|\\OH}}{CH}-CH_2-\underset{\substack{|\\OH}}{CH}-CH_2- + CH_3CO_2R$$

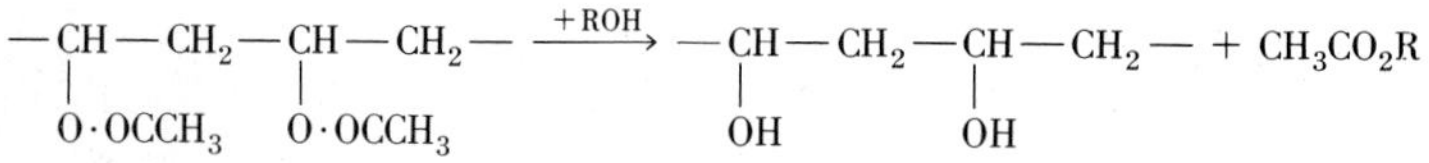

This method of preparation is used because polyvinyl alcohol cannot be prepared by direct polymerization since the monomer vinyl alcohol is an unknown compound and exists as the enol form of acetaldehyde (CH_3CHO). Polyvinyl alcohol is a unique plastic because it is plasticized by water and is completely soluble in an excess of water, making it an ideal packaging material for various soaps, detergents, and bleaches, so that the homemaker merely drops the whole package into the washing machine. Polyvinyl alcohol reacts with an aldehyde under the influence of heat and in the presence of an acid catalyst, such as sulfuric or hydrochloric acid, and the result is a typical acetal. Polyvinyl butyral is formed as follows:

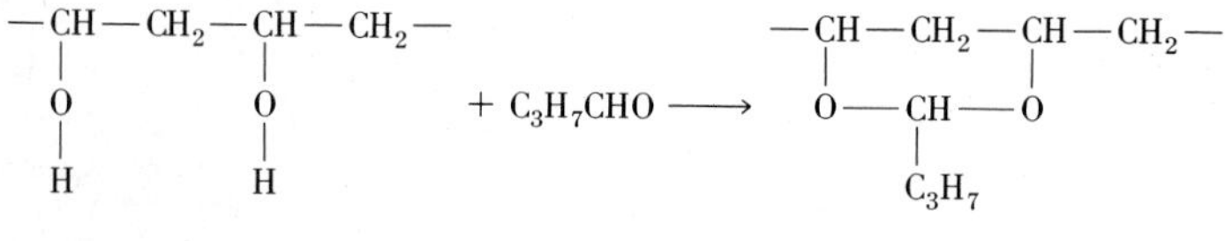

$$-\underset{\substack{|\\O\\|\\H}}{CH}-CH_2-\underset{\substack{|\\O\\|\\H}}{CH}-CH_2- + C_3H_7CHO \longrightarrow -CH-CH_2-CH-CH_2- \text{ (with } O-\underset{\substack{|\\C_3H_7}}{CH}-O \text{ bridging the two CH groups)}$$

The corresponding formal and acetal are:

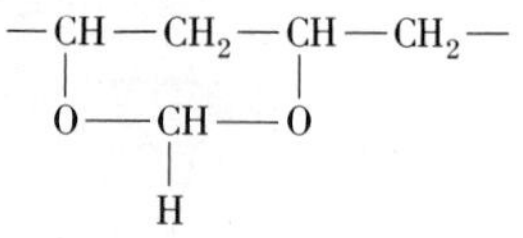

$$-CH-CH_2-CH-CH_2- \text{ (with } O-\underset{\substack{|\\H}}{CH}-O \text{ bridging)} \qquad -CH-CH_2-CH-CH_2- \text{ (with } O-\underset{\substack{|\\CH_3}}{CH}-O \text{ bridging)}$$

Formal 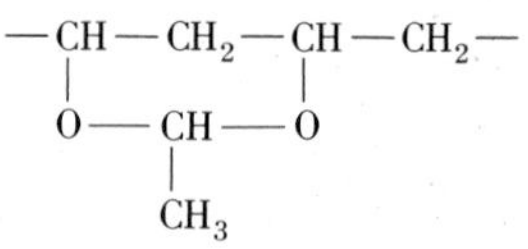 Acetal

[28]Winding and Hiatt, *op. cit.*; Krause, Mass Polymerization for PVC Resins, *Chem. Eng.* (*N.Y.*), **72**(26), 72 (1965).

In actual commercial production, the hydrolysis and condensation reactions are never carried to completion because it has been found that the presence of residual acetyl and hydroxyl groups gives the resins better properties.

Polyvinyl butyral is usually extruded as a sheet for the interlayer for safety glass. Some polyvinyl acetals can be compression-molded at temperatures from 100 to 130°C and injection-molded at temperatures from 170 to 190°C. Polyvinyl formals can be molded by either compression or injection, depending on the softening point.

VINYLIDENE CHLORIDE RESINS This class of resins was introduced in 1940 and may be formed by the polymerization of the monomeric vinylidene chloride, $CH_2{=}CCl_2$. These resins may be represented in part as

```
   H   Cl  H   Cl  H   Cl  H   Cl
   |   |   |   |   |   |   |   |
 —C—C—C—C—C—C—C—C—
   |   |   |   |   |   |   |   |
   H   Cl  H   Cl  H   Cl  H   Cl
```

or the monomer may be copolymerized with vinyl chloride to give a new product represented as follows:

```
   H   Cl  H   Cl  H   Cl  H   Cl  H   Cl
   |   |   |   |   |   |   |   |   |   |
 —C—C—C—C—C—C—C—C—C—C—
   |   |   |   |   |   |   |   |   |   |
   H   H   H   Cl  H   Cl  H   Cl  H   H
```

The resins formed by copolymerization range from a flexible material having a softening point of about 70°C to a hard, thermoplastic solid with a softening point of 180°C. Their higher softening point indicates a greater degree of crystalline character. If the fibrous crystals are not oriented, the vinylidene chloride resin has a tensile strength of 8,000 psi, which is an ordinary value. If the crystals are oriented by drawing, however, the tensile strength may be increased to about 60,000 psi. These resins may be fabricated by compression or injection molding or extrusion with the use of specially developed techniques and equipment to obtain accurate control of properties and shape.

STYRENE RESINS[29] Monomeric styrene (Fig. 36.3) is made by the pyrolysis-dehydrogenation of ethylbenzene, which is synthesized from ethylene and benzene. The factors that affect polymerization of the monomer are the temperature and purity of the styrene. Polymerization at moderate temperatures without catalysts produces resins of high average molecular weight, which impart high viscosity to their solutions. If the temperature is increased and catalysts such as benzoyl peroxide, oxygen, or stannic chloride are added, the average molecular weight and viscosity tend to decrease. Styrene of 99.5% purity, when heated at 80 to 85°C in a stainless-steel kettle, polymerizes to a 35 or 40% conversion in 40 to 60 h. The viscous solution is passed down a tower, with zones of increasing temperature up to 200°C, to strip off the unconverted monomer and to take off the polymer in a molten state. The extruded product is cooled and granulated. A *blend of acrylonitrile-butadiene rubber* with *acrylonitrile-styrene resins* has produced *ABS* plastics.[30]

ACRYLIC RESINS AND PLASTICS[31] Methyl and ethyl acrylate and methyl, ethyl, and butyl methacrylate monomers are manufactured in large tonnages. The procedures for preparing methyl

[29]Winding and Hiatt, *op. cit.;* Modern Plastics Encyclopedia, 1974–1975; *Mod. Plast.*, **52**(6), 16 (1975).

[30]*Chem. Eng.* (*N.Y.*), **69**(20), 154 (1962); *Mod. Plast.*, **52**(3), 44 (1975) (flowchart).

[31]Horn, Acrylic Resins, Reinhold, 1960; Winding and Hiatt, *op. cit.;* Golding, *op. cit.;* Modern Plastics Encyclopedia; Forbath, Simple Direct Path to Acrylates, *Chem. Eng.* (*N.Y.*), **69**(5), 96 (1962) (flowchart of Dow Badische synthesis at Freeport, Tex.); Guccione, New Developments in Acrylic Processing, *Chem. Eng.* (*N.Y.*), **73**(12), 138 (1966).

acrylate and methyl methacrylate resins, following the Dow-Badische-Reppe synthesis, are depicted in the Forbath reference and are based on the reaction

$$C_2H_2 + CO + ROH \longrightarrow CH_2{=}CH-COOR$$

These esters are polymerized under the influence of heat, light, or peroxides. The polymerization reaction is exothermic and may be carried out en masse as in castings, by suspension as in molding powders, or by emulsion, and in solution. The molecular weight decreases as the temperature and catalyst concentration are increased. Such resins are widely used because of their clarity, brilliance, ease of forming, and light weight (about half that of plastic glass). Among their applications are cockpit canopies, gun turrets, spray shields, and boats. Emulsions are widely applied as textile finishes, and paints. Plexiglas is a name applied to this plastic.

Coumarone-indene resins.[32] Coumarone-indene resins were among the first synthetic resins developed commercially. They are produced from coal-tar light oils recovered from coproduct coking operations or from various petroleum-cracking operations. The fraction distilling between 150 to 200°C containing indene, coumarone, and homologous compounds is treated with small quantities of sulfuric acid, causing polymerization.

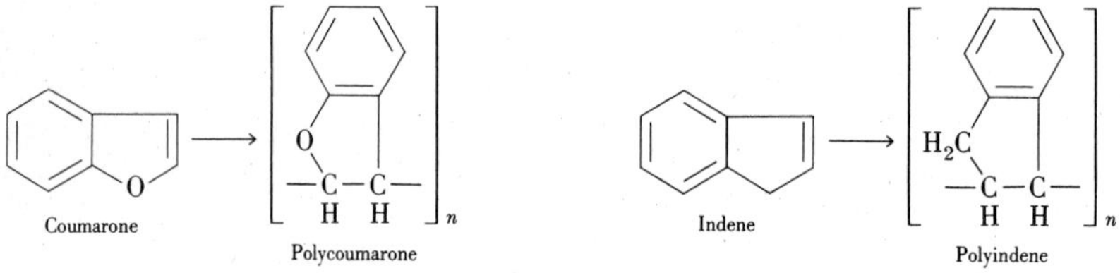

Coumarone → Polycoumarone; Indene → Polyindene

The polymerization is assumed to be a chain reaction to give polymers of the above general patterns.

NATURAL PRODUCTS AND THEIR DERIVATIVES

CELLULOSE DERIVATIVES[33] Cellulose is generally pictured as being composed of a chain of glucosidic units represented by the formula

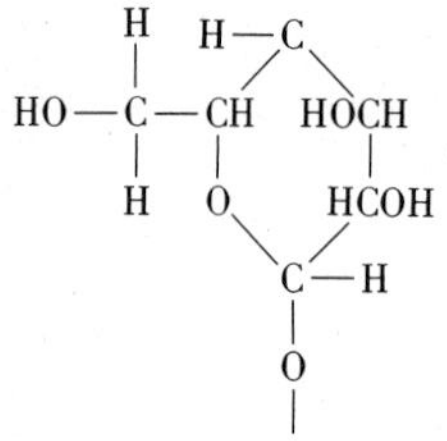

Long chains may contain 3,000 to 3,500 units. These elongated chains and the reactions of the cellulosic polyhydric alcohol are responsible for the formation of tough, flexible, cellulosic plastics. The properties of the cellulose derivatives depend upon the substituent groups, the amount of substitution, the type of pretreatment, and the degree of degradation of the long chains into shorter lengths.

CELLULOSE NITRATE was first prepared by Bracconot in 1833, and became the first synthetic "resin" for a plastic. Flowcharts for the production of cellulose nitrates are given in Fig. 22.2. Fully *nitrated cellulose* is unsuited for a plastic base because of its extremely flammable character. Hence

[32]Winding and Hiatt, *op. cit.;* Modern Plastics Encyclopedia, 1974–1975; ECT 2d., vol. 11, p. 243, 1966; EPST, vol. 4, p. 272, 1966.

[33]Golding, *op. cit.;* Winding and Hiatt, *op. cit.;* Paist, *op. cit.*

a partly nitrated product, with a nitrogen content of about 10.7 to 11.2%, is made. This cellulose nitrate is placed in large kneading mixers with solvents and plasticizers and thoroughly mixed. The standard plasticizer is *camphor,* first used by Hyatt in 1868. The compounded mixture is strained under hydraulic pressure and mixed on rolls with coloring agents. The material is pressed into blocks. Finally, the plastic is made into sheets, strips, rods, or tubes, seasoned to remove the residual solvent, and polished by pressing under low heat. It is known as Celluloid.

CELLULOSE ACETATE[34] Fully acetylated cellulose is partly hydrolyzed to give an acetone-soluble product, which is usually between the di- and triester. The esters are mixed with plasticizers, dyes, and pigments and processed in various ways, depending upon the form of plastic desired. The important properties of cellulose acetate include mechanical strength, impact resistance, transparency, colorability, fabricating versatility, moldability, and high dielectric strength. Some of the noteworthy properties of *cellulose acetate-butyrate* are low moisture absorption, high dimensional stability, excellent weathering resistance, high impact strength, availability in colors, and improved finish.

ETHYL CELLULOSE PLASTICS[35] Ethyl cellulose is a cellulose ether in which ethyl groups have replaced the hydrogen of the hydroxyl group. Commercial ethyl cellulose contains between 2.4 and 2.5 ethoxy groups per glucoside unit of the cellulose chain. The ethers are made by treating cellulose, such as wood pulp or cotton linters, with a 50% solution of sodium hydroxide. The alkali cellulose formed is alkylated with such agents as ethyl chloride or ethyl sulfate. The alkylation must be carried out under carefully controlled conditions to prevent degradation of the cellulose chain and destruction of the alkylating agent. After the reaction is complete, the excess reagents are easily washed out. The ether is purified by washing it free of soluble materials. Ethyl cellulose plastics may be molded in any manner or machined. Some of the outstanding properties are unusually good low-temperature flexibility and toughness, wide range of compatibility, stability to heat, thermoplasticity, and electrical resistance. *Methyl cellulose* is prepared by a process almost identical with that for ethyl cellulose, substituting methyl chloride or sulfate for the corresponding ethyl derivative. Water solubility, plus the ability to produce viscous solutions, have made it useful for a thickening agent in the textile, food, and adhesives industries.

SHELLAC COMPOSITIONS[36] Shellac is obtained from a resinous material secreted by an insect, *Laccifer lacca* Kerr, which is native to India. The insect produces raw lac from glands located in its skin. The resin is a solid solution of several chemical compounds having similar structures. The lac consists of two materials, a hard and a soft resin. Shellac is a unique plastic resin because it has properties of both thermoplastic and thermosetting resins. Heat curing increases its mechanical and electrical strength. The resin is compatible with phenolic resins, so that a wide variety of molding compositions can be made. Compositions are softened on steam tables at 225 to 250°F. They are molded by compression molding at temperatures from 250 to 275°F and pressures from 1,000 to 3,500 psi. The molds must be cooled to 90 to 120°F before the pressure is released. Injection molding at 230°F can be carried out in a 1- or 2-min cycle for many pieces. Some noteworthy properties of shellac compositions are low dielectric constant, high dielectric strength, arcing resistance, adhesion, hardness, high gloss, resilience, low thermal conductivity, ease of molding, ability to wet fillers, and oil resistance when cured. These compositions find use in phonograph records, protective coatings, adhesives, and electrical insulation.

LIGNIN PLASTICS[37] The lignin bond of woody material is released by the action of high-pressure steam which activates the lignin so that it can be used as a plastic binder. Wood chips are charged into suitable containers and treated with steam up to 1,200 psi for a period of seconds.

[34]Winding and Hiatt, *op. cit.;* Paist, *op. cit.;* see also Chap. 35 for manufacturing process and flowchart.
[35]Winding and Hiatt, *op. cit.;* Paist, *op. cit.*
[36]Modern Plastics Encyclopedia; Hicks, Shellac, Chemical Publishing, 1961.
[37]Golding, *op. cit.;* Winding and Hiatt, *op. cit.*

Normally, the time cycle, including the filling and emptying of these containers, is about 1 min. In this time the lignin is softened by the sudden high temperature. When the chips are suddenly released to atmospheric pressure, they are exploded by the high internal pressure into a mass of fibers and fiber bundles which still contain a natural coating of lignin. These fibers are formed into mats under pressures from 1,500 to 2,500 psi and at a temperature of 275 to 385°F. The product is usually cured at a pressure of 1,500 psi and a temperature of 175°C. The only plasticizer needed is about 4% moisture, which is present at usual humidities. The curing is very rapid.

Protein derivatives. Historically, the industrial use of proteins is quite old. Animal glues have been in use since early Egyptian civilization, yet the supply of such glues was limited until recent times because of inadequate refrigeration and transportation facilities. See Adhesives, Chap. 25.

Vegetable proteins, developed along with the soybean in the 1930s, were followed by the industrial use of zein in the 1940s. Protein plastic compositions are hygroscopic and are accordingly affected by atmospheric moisture changes. The absorption of water causes severe warping in sizes larger than 4 in.[2] The soft, uncured plastic may be molded but requires hardening in formaldehyde solution and, since thinner sections harden sooner than the thicker parts, the resulting internal stresses and strains cause severe warping. At present, protein derivatives find extensive use as glues, adhesives, and paper coatings.

MANUFACTURE OF LAMINATES AND OTHER TYPES

Laminated plastics are made largely from the thermosetting class of resins (i.e., melamine, epoxy, unsaturated polyester, silicone and, most frequently, phenolics) and have fibrous fillers such as paper, asbestos, glass, nylon, etc. The resin is usually dissolved in a suitable solvent, such as alcohol or water, with which the filler is impregnated or coated. One method of impregnating utilizes adjustable rollers or a doctor blade to control the amount of resin left on fibers after dipping in a bath. Another technique controls the concentration of the resin in the dipping tank. A third technique meters the resin applied to the fibers as they pass through a series of roll coaters. Gentle heat removes the solvent as the fibers are passed through a drying oven and partial polymerization takes place. Final cure is usually accomplished in presses or molds. This curing time varies with the thickness and type of material used, but pressures of approximately 1,000 to 2,500 psi are employed with temperatures from 130 to 175°C.

TECHNICAL EXAMPLES OF CHEMICAL INTERMEDIATES FOR RESINS[38]

PHENOL is manufactured by a number of procedures, of which the following are the most important:

1. *Sulfonation process:*

$$C_6H_6 + H_2SO_4 \xrightarrow[\text{hot}]{C_6H_6\text{ vapor}} C_6H_5SO_3H + H_2O\uparrow$$

$$C_6H_5SO_3H + 3NaOH \xrightarrow{300°C} C_6H_5ONa + Na_2SO_3 + 2H_2O$$

$$C_6H_5ONa + CO_2\ (\text{or } SO_2) + H_2O \longrightarrow C_6H_5OH + NaHCO_3\ (\text{or } NaHSO_3)$$

2. *Chlorobenzene process* (liquid phase):

$$C_6H_5Cl + 2NaOH \xrightarrow{360°C} C_6H_5ONa + NaCl + H_2O$$

$$C_6H_5ONa + HCl \longrightarrow C_6H_5OH + NaCl$$

$$2C_6H_5OH \rightleftharpoons C_6H_5OC_6H_5 + H_2O$$

[38]As in other chapters, intermediates are presented whose chief consumption is in the chemical process industry (in this case, plastics) covered by the chapter.

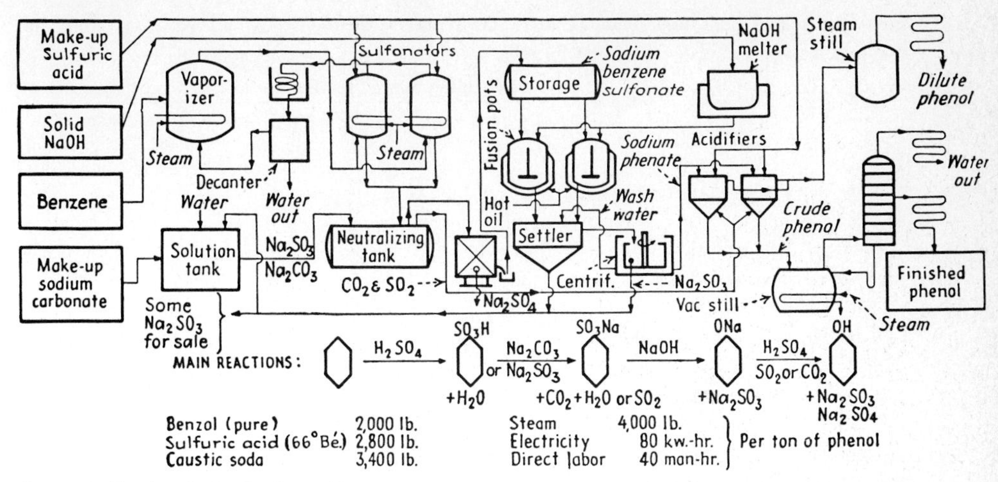

Fig. 34.6 Phenol production flowchart—sulfonation process.

3. *Catalytic vapor phase* (regenerative process, or Raschig process):

$$C_6H_6 + HCl + \tfrac{1}{2}O_2 \xrightarrow{230°C} C_6H_5Cl + H_2O \qquad \text{(exothermic)}$$

$$C_6H_5Cl + HOH \xrightarrow{425°C} C_6H_5OH + HCl \qquad \text{(endothermic)}$$

4. *Cumene hydroperoxide:*

$$\underset{\text{Cumene}}{C_6H_5CH(CH_3)_2} \xrightarrow[265°F]{\text{air}} \underset{\text{Cumene hydroperoxide}}{C_6H_5C(CH_3)_2O_2H} \xrightarrow[\text{at 130–150°F}]{\text{10–25\%, } H_2SO_4} \underset{\text{Phenol}}{C_6H_5OH} + \underset{\text{Acetone}}{CH_3COCH_3}$$

5. *Toluene oxidation* (Dow):

$$\underset{\text{Toluene}}{C_6H_5CH_3} \xrightarrow[\text{250–350°F, 30 psig}]{\text{Air, Co salts}} \underset{\text{Benzoic acid} + H_2O\text{, 90\% yield}}{C_6H_5COOH} \xrightarrow[\text{430–470°F, atm}]{\text{air, stream, Cu, Mg salts}} \underset{\text{Phenol}}{C_6H_5OH}$$

Intermediate reactions:

$$2C_6H_5COOH \xrightarrow{\text{Cu salts}} (C_6H_5COO)_2Cu \longrightarrow \underset{\text{Benzoic acid}}{C_6H_5COOH} + \underset{\text{Salicylic acid}}{C_6H_4OH \cdot COOH} \longrightarrow \underset{\text{Phenol}}{C_6H_5OH} + CO_2$$

Additional work is progressing on catalytic air-oxidation[39] products of cyclohexane and benzene: $C_6H_6 \xrightarrow{\text{air}} C_6H_5OH$ (no by-product).

The older, or *sulfonation,* process has not held its own competitively. When the concentration of the acid drops below 78% H_2SO_4 (the rest being water), the sulfonating action ceases. At one time, sufficient, strong H_2SO_4 was used to absorb the water and keep up the concentration. This, however, was a costly procedure, not only wasting acid but also requiring expensive separation of the benzenesulfonic acid. At present, water formed is removed by passing hot benzene vapor through

[39]Scientific Design, Phenol Air-Oxidation Route, *Chem. Eng. News,* Apr. 27, 1964, p. 56.

the sulfonation product, thus removing the water by *reverse steam distillation.* The ratio of benzene required may be calculated from the formula[40]

$$\text{Weight of benzene} = \text{weight of water} \times \frac{\text{molecular weight of benzene} \times \text{vapor pressure of benzene}}{\text{molecular weight of water} \times \text{vapor pressure of water}}$$

To remove the water, the vapor pressure of the benzene and the water (in the presence of H_2SO_4) must be at the temperature at which the sum of the two vapor pressures equals the pressure on the system.[41] When the water is removed from the sulfonation to keep the acid concentration above 78% H_2SO_4, the sulfonation proceeds until only a few percent of free H_2SO_4 is present, which is then directly neutralized to form the sodium salt, in conformance with the flowchart in Fig. 34.6. It should be noted that either Na_2CO_3 or Na_2SO_3 is used to make the $C_6H_5SO_3Na$ and that either the CO_2 or SO_2 is employed in the acidifiers to liberate, or "spring," the phenol. Generally, a little diluted H_2SO_4 is necessary to finish the acidification. The fusion is frequently carried out in open cast-iron fusion pots or pots with sheet-iron removable covers.

The *chlorobenzene hydrolysis*[42] *liquid phase* (procedure 2 above) involves pumping a mixture of chlorobenzene and dilute (18%) caustic soda through a steel-pipe heat exchanger, a reactor, and then through a heat exchanger for cooling. The reactor temperature is about 360°C, and the pressure 5,000 psi. The reactor itself is a length of $1\frac{1}{2}$-in. double-strength seamless steel pipe about 1 mi long arranged in a coil (in a furnace for start-up). The heat of reaction after the initial start is sufficient to maintain the desired temperature through the use of an efficient heat exchanger. This process affords excellent application of the unit operations of high pressure, pumping, and heat transfer at high pressure. In order to repress the accumulation of diphenyloxide, about 0.1 mol is added with the reactants. This diphenyloxide at first was hard to sell, but in recent years the market created by Dow has absorbed all that is made; hence this initial additional charge is now omitted.

The regenerative, or catalytic vapor-phase, process (procedure 3), is based on the Raschig patents. This procedure has been described and pictured in the literature.[43] It presents some very difficult engineering problems, particularly involving the handling of hot HCl in the presence of steam and air, requiring much expensive tantalum. The conversions are low and require recirculation. Important improvements have been made by Hooker to reduce costs and increase yields.

The most important commercial process is the one based on cumene, given in procedure 4 above. Over 90% of all U.S. production in 1974 was from cumene oxidation. Figure 34.7 is a flowchart illustrating this process. The by-product acetone furnishes the bulk of U.S. acetone, and this accounts, at least in part, for the wide use of the process in preference to others.[44]

The intermediate cumene hydroperoxide is dangerously explosive, and safety precautions must be taken in the operation of the oxidizer and the subsequent H_2SO_4 treatment.

FORMALDEHYDE[45] results from the oxidation of methanol (Fig. 38.10):

$$CH_3OH + \tfrac{1}{2}O_2 \xrightarrow{Ag} H_2CO + H_2O \qquad \Delta H = -38.0 \text{ kcal}$$

$$CH_3OH \longrightarrow H_2CO + H_2 \qquad \Delta H = +20.0 \text{ kcal}$$

[40]Derived from Avogadro's law which states that, under the same conditions of temperature and pressure, equal volumes of perfect gases contain the same number of molecules.

[41]Cf. Tyrer, U.S. Pat. 1,210,725 (1917); Guyot, *Chim. Ind. (Paris)*, **2,** 879 (1919); Zakharov, *J. Chem. Ind. USSR,* **6,** 1648 (1929).

[42]Flowchart, CPI 2, p. 882; Sittig, *op. cit.*, p. 81; Hale and Britton, Development of Synthetic Phenol from Benzene Halides, *Ind. Eng. Chem.*, **20,** 114 (1928); Groggins, Unit Processes in Organic Synthesis, 5th ed., McGraw-Hill, 1958.

[43]Hooker, New Pitch for Phenol, *Chem. Week*, Oct. 19, 1963, p. 59 (describes improvements); Raschig, U.S. Pats. 1,963,761, 2,009,023, and 2,035,917 (1935): CPI 2, p. 882, Fig. 2, shows the sequences of unit operations and chemical conversions that represent the commercialization of the basic chemical changes; Groggins, *op. cit.*, p. 793.

[44]Austin, Industrially Significant Organic Chemicals, *Chem. Eng. (N.Y.)*, **81**(15), 108 (1974); *Chem. Mark. Rep.*, Mar. 4, 1974; *Hydrocarbon Process.*, **52**(11), 158 (1973).

[45]Walker, Formaldehyde, ACS Monograph 159, Reinhold, 1964; New Expansions, *Chem. Eng. News*, Apr. 6, 1964, p. 25.

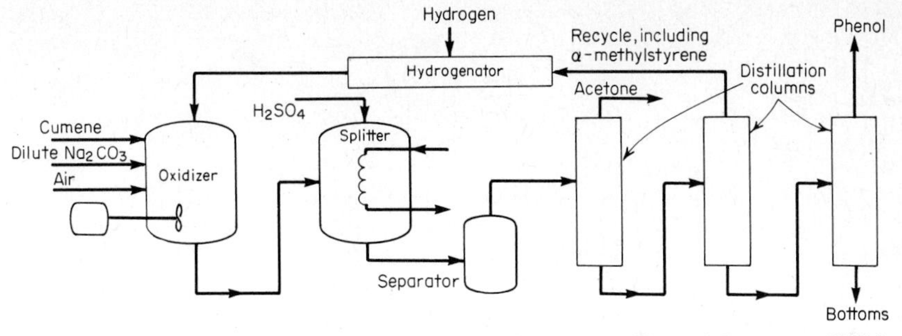

Fig. 34.7 Phenol production flowchart—Dow toluene air oxidation process [*Chemical Engineering* (*N.Y.*).]

The oxidation requires 26.7 ft^3 of air per pound of methanol reacted, a ratio that is maintained when passing separate streams of these two materials forward. Fresh and recycle methanol are vaporized, superheated, and passed into the methanol-air mixer. Atmospheric air is purified as shown in the flow diagram in Fig. 34.8, compressed, and preheated to 130°F in a finned heat exchanger. The products leave the converter at about 1150°F and at 5 to 10 psia. The converter is a small water-jacketed vessel containing several layers of silver-gauze catalyst. About 65% of the methanol is converted per pass. The reactor effluent contains about 25% formaldehyde, which is absorbed with the excess methanol and piped to the make tank. The latter feeds the methanol column for separation of recycle methanol overhead, the bottom stream containing the formaldehyde and a few percent methanol. The water intake adjusts the formaldehyde to 37% strength. The yield from the reaction is 85 to 90%, and the formaldehyde is marketed as a 37% solution sometimes called Formalin. Much of the equipment is made of stainless steel.

Another formaldehyde process involves the oxidation of methanol using an iron molybdate catalyst at essentially atmospheric pressure to produce a methanol-free product.[46]

HEXAMETHYLENETETRAMINE Evaporation of the reaction product of formaldehyde and ammonia produces hexamethylenetetramine.

$$6CH_2O + 4NH_3 \longrightarrow (CH_2)_6N_4 + 6H_2O \qquad \Delta H = -28.2 \text{ kcal}$$

(Structure printed: a cage of four N atoms, each pair bridged by CH_2 groups.)

This compound finds use as a urinary antiseptic (Urotropine), in the rubber industry, in the preparation of the explosive cyclonite (Chap. 22), and mainly in making phenolformaldehyde resins, where it is known as "hexa."

VINYL ESTERS[47] The addition of acids to acetylene furnishes esters:

$$CH{\vdots}CH + CH_3COOH \longrightarrow \underset{\text{Vinyl acetate}}{CH_3COOCH{:}CH_2}$$

$$CH{\vdots}CH + HCl \longrightarrow \underset{\text{Vinyl chloride}}{CH_2{:}CHCl}$$

[46]*Hydrocarbon Process.*, **52**(11), 135 (1973).

[47]Flowcharts and conditions for making vinyl chloride and acetate are given in Petrochemical Process Handbook, Gulf Publishing, 1963, and *Pet. Refiner*, **42**(11), 129–240 (1963).

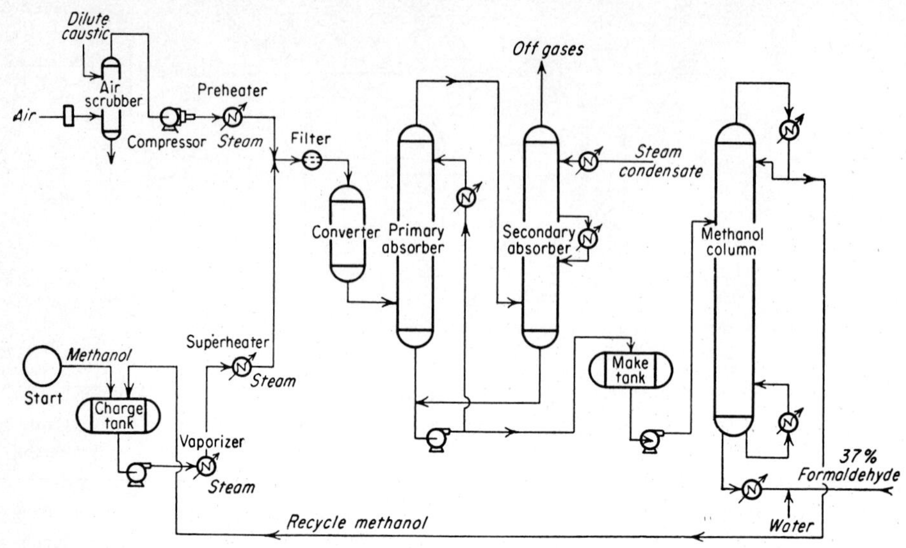

Fig. 34.8 Flowchart for the manufacture of formaldehyde.

If two molecules of acid react, a compound such as ethylidene diacetate is formed:

$$CH{-}CH + 2CH_3COOH \longrightarrow (CH_3COO)_2CHCH_3$$

The newer method, however, is to start with ethylene and by oxychlorination (dehydrochlorination) obtain apparently a cheaper procedure for *vinyl chloride.*[48] See Fig. 34.9, which presents a flowchart for the following reactions:

Chlorination

$$C_2H_4 + Cl_2 \longrightarrow \underset{\text{Dichloroethylene}}{C_2H_4Cl_2}$$

Oxychlorination

$$C_2H_4 + \tfrac{1}{2}O_2 + 2HCl \xrightarrow{CuCl_2} C_2H_4Cl_2 + H_2O$$

Dehydrochlorination

$$\underset{\text{Dichloroethylene}}{C_2H_4Cl_2} \xrightarrow[\text{moderate temp. and pressure}]{\text{cracking furnace}} \underset{\text{Vinyl chloride}}{C_2H_3Cl} + HCl$$

The $CuCl_2$ catalyst (on an inert fixed carrier) may react as follows:

$$C_2H_4 + 2CuCl_2 \longrightarrow C_2H_4Cl_2 + Cu_2Cl_2$$

$$Cu_2Cl_2 + \tfrac{1}{2}O_2 \longrightarrow CuO \cdot CuCl_2$$

$$CuO \cdot CuCl_2 + 2HCl \longrightarrow 2CuCl_2 + H_2O$$

[48]Edwards and Weaver, New Route to Vinyl Chloride, *Chem. Eng. Prog.*, **61**(1), 21 (1965) (prescribed costs relative to process starting with ethylene); Burke and Miller, Oxychlorination, *Chem. Week*, Aug. 22, 1964, p. 93 (excellent overall commercial and technical presentation and patents); Riegel *et al.*, Low Cost Hydrocarbon Chlorination, *Chem. Eng. Prog.*, **69**(10), 89 (1973).

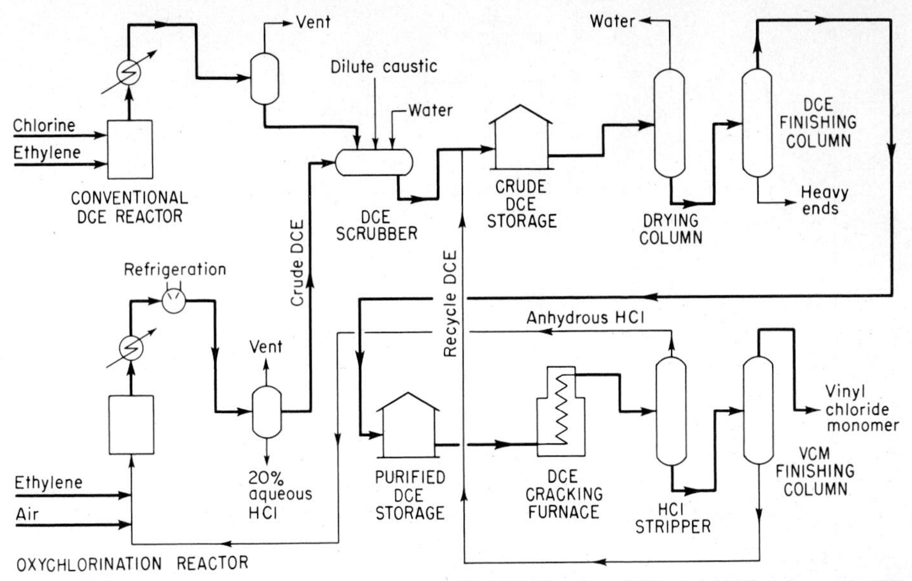

Fig. 34.9 Flowchart for the production of vinyl chloride. DCE, 1,2-dichloroethane; VCM, vinyl chloride monomer; the 20% aqueous HCl is relatively small in amount. [*Chemical Engineering Progress;* cf. Buckley, Vinyl Chloride, *Chem. Eng.* (*N.Y.*), **73**(24), 102 (1966), and Kureha *Chem. Eng.* (*N.Y.*), **74**(7), 48 (1967).]

PHTHALIC ANHYDRIDE[19] has become one of our most important intermediates. It aided in the establishment of the dye industry in America, and it has now become even more necessary for many plastics and plasticizers. It is used in making directly a number of dyes such as eosin, rhodamines, erythrosin, quinoline yellow, copper phthalocyanine, and phenolphthalein, as a step in manufacturing anthraquinone and anthraquinone derivatives by a condensation (Friedel-Crafts) procedure (Chap. 39), where it has opened up all the anthraquinone vat dyes to the American market at reasonable prices. Phthalic anhydride and phenol are two among many intermediates whose utilization has expanded far beyond dyes, and about three-quarters is consumed in the manufacture of plastics and plasticizers (Fig. 34.10).

The processes for phthalic anydride [50] are controlled catalytic air oxidation of *o*-xylene or naphthalene where $\Delta H = -5460$ Btu/lb of naphthalene oxidized. If naphthalene is burned completely to carbon dioxide and water, 18,000 Btu/lb of naphthalene is liberated. Actually, when manufacturing phthalic anydride by this process, an exothermic reaction of from 6000 to more than 10,000 Btu/lb of naphthalene is observed, owing to a certain amount of complete combustion that always occurs.

$$C_{10}H_8 + 4\tfrac{1}{2}O_2 \xrightarrow[\text{400–460°C, 0.1–0.5 s}]{V_2O_5} C_6H_4(CO)_2O + 2CO_2 + 2H_2O$$

Oxidation is one of the very useful chemical conversions in organic technology. The cheapest *agent* is air, but oxygen is sometimes employed. For liquid-phase reactions a great many oxidizing

[49]ECT, 2d ed., vol. 15, p. 444, 1968.

[50]Phthalic Anhydride, *Chem. Eng. News*, March 15, 1971, p. 18; Austin, Industrially Significant Organic Chemicals, *Chem. Eng.* (*N.Y.*), **81**(15), 109 (1974).

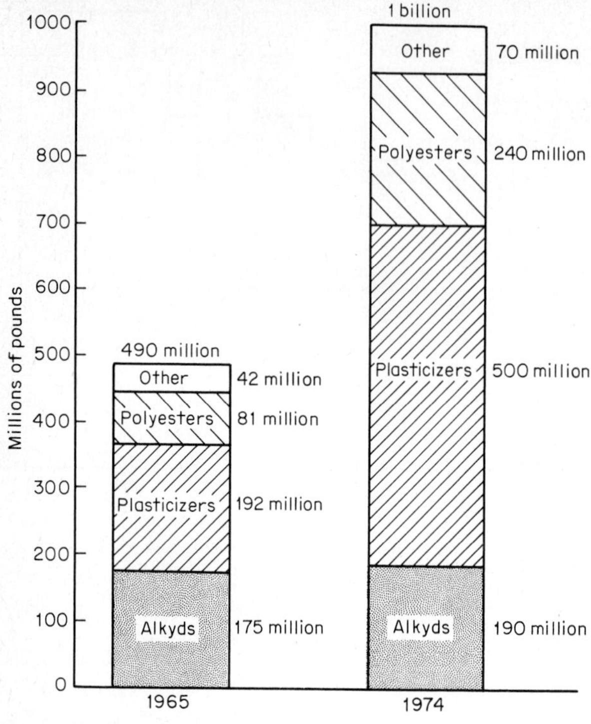

Fig. 34.10 Phthalic anhydride end-use pattern.

agents are in industrial use, such as nitric acid, permangates, pyrolusite, dichromates, chromic anhydride, hypochlorites, cholorates, lead peroxide, and hydrogen peroxide. Water and carbon dioxide and many other oxidized substances are the by-products of the main oxidation. When charcoal or carbon (amorphous) changes to carbon dioxide, $\Delta H = -96.5$ kcal/g-mol, [51] and when hydrogen burns to water (gaseous), $\Delta H = -57.8$ kcal/g-mol. These *energy changes*, although the basis of combustion, frequently accompany controlled oxidation, as in the making of phthalic anhydride and maleic acid. In most oxidations, even when the formation of carbon dioxide and water can be repressed, the energy change is exothermic and large. This requires particular care in the design and construction of the *equipment* to ensure efficient heat transfer and to prevent the controlled oxidation from becoming combustion.

The naphthalene process, with a suitable catalyst, was discovered by Gibbs and Conover.[52] It was also necessary to work out efficient equipment to remove the great amount of heat liberated and to keep the temperature within favorable narrow limits. One of the successful devices was that patented by Downs[53] and depicted in principle in Fig. 34.11. It consists of square tubes (to increase surface in relation to volume) carrying the V_2O_5 catalyst on a carrier, surrounded by mercury. Boiling removes the heat of reaction, and the hot mercury vapor preheats the entering natphthalene vapor and then passes on to the condenser. In the Downs reactor the temperature is controlled by raising or lowering the boiling point of mercury by raising and lowering the pressure of an inert gas (nitrogen)

[51]For carbon, the equivalent Btu/lb = 14,400; for hydrogen, 52,000 Btu from 1 lb.

[52]U.S. Pat. 1,285,217 (1918).

[53]U.S. Pat. 1,604,739 (1926); also Downs, U.S. Pats. 1,374,020; 1,373,021 (1921); 1,789,809 (1931); and 1,873,876 (1932).

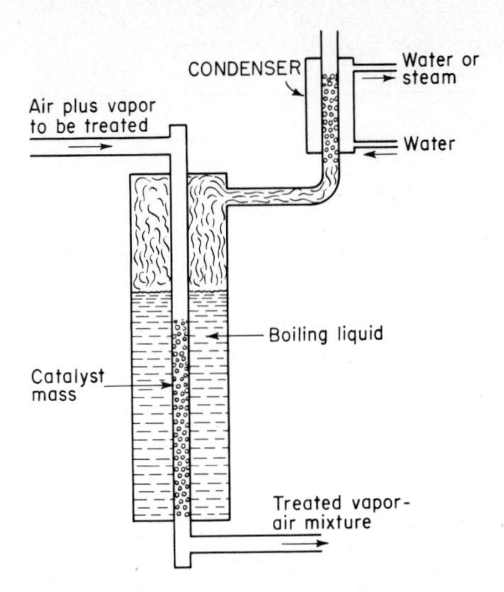

Fig. 34.11 Apparatus for the removal of reaction heat by the boiling of a liquid.

on the mercury boiling and condensing system. Other heat-transfer media have been proposed, such as water, sulfur, diphenyl, diphenyl oxide, mercury amalgams, and mixed nitrate-nitrite. The products from the reaction are rapidly cooled to about 125°C (approximately the dew point of phthalic anhydride) and then sublimed[54] or distilled.

Fundamental changes have been made in the manufacture of phthalic anhydride: The first was a change in the source of the naphthalene. The dwindling supply from coal tar was supplemented by developing a purer one from petrochemicals or by *demethylating* methylnaphthalenes (Chap. 37). The second change was to use a fluidized bed of the catalyst V_2O_5 as had long been sucessful in the fixed bed. The third change was the use of *o*-xylene in either of the above procedures. A fourth change was to develop catalysts other than V_2O_5 which would work favorably on either naphthalene or *o*-xylene. The fifth change was applied early and involved removing the large heat of reaction by a process other than one based on the latent heat of boiling mercury. The last-mentioned made it possible to use some of this heat of reaction to generate steam. Certain of these changes are illustrated in Fig. 34.12 for a fixed-bed process, and more details are available in the references.[55] Table 34.9 compares the conditions and yields of the various procedures.

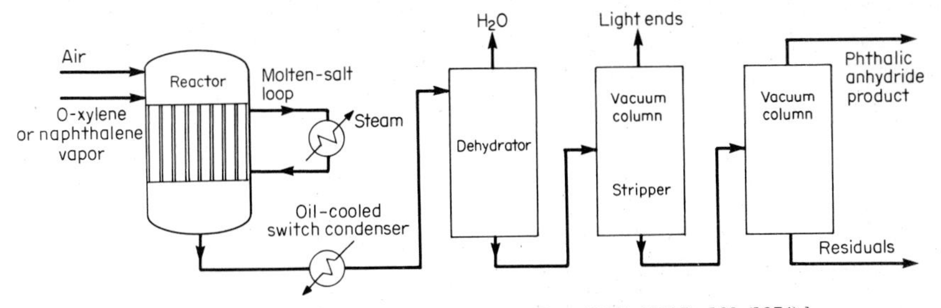

Fig. 34.12 Phthalic anhydride via fixed-bed oxidation. [*Chem. Eng.* (*N.Y.*) **81**(15), 109 (1974).]

[54]Perry, pp. 17-25 to 17-29.

[55]Chopey, Fluid-bed Phthalic Anhydride (flowchart), *Chem. Eng.* (*N.Y.*), **69**(2), 104 (1962); *Hydrocarbon Process.*, **52**(11), 159 (1973) (flowcharts); Elwood, Fixed Bed for Phthalic from *o*-Xylene, *Chem. Eng.* (*N.Y.*), **76**(12), 80 (1969) (flowchart); Zimmer, Low Air-Feedrate Cuts Phthalic Anhydride Costs, *Chem. Eng.* (*N.Y.*), **81**(5), 82 (1974).

TABLE 34.9 *Phthalic Anhydride Production and Processing* from Naphthalene or o-Xylene by Controlled Air Oxidation*

Process	*Fixed-bed*		
Raw material	Naphthalene	*o*-Xylene†	Naphthalene
Catalyst	$V_2O_5(SiO_2)$	$V_2O_5(SiO_2)$	V_2O_5, 10%
Temperature, °F	790–1020	790–1100	700
Equipment	Tubular reactor with catalyst and tubular heat exchanger		
Heat reaction:			
Theoretical, kcal	3510	2890	3510
Practical, kcal	5850–5090	6520–5260	5340–5090
Contact time, sec	0.4–0.5	0.15	19
Lb air/lb $C_{10}H_8$	25–30	20–28	15
Phthalic anhydride:			
Separation	Condenser	Condenser	Filtration
Purification	Vacuum distillation	Vacuum distillation	Vacuum distillation
Yields, weight from 100 lb	70–85	70–95	80–85
Yield, % theory	60–74	50–68	69–74

Sources: Data for production from Ullmann, Enzyklopaedie der Technischen Chemie, vol. 3, Urban & Schwartzenberg, Berlin, 1962, and Sittig, Combining Oxygen and Hydrocarbons for Profit, Gulf Publishing, 1962. *All *highly exothermic*—see reactions in text. Should be operated to oxidize all naphthalene or *o*-xylene with a minimum of complete combustion. Much heat to be recovered as steam. †See *Chem. Week*, Dec. 10, 1960, p. 100.

SELECTED REFERENCES

Akin, R. B.: Acetal Resins, Reinhold, 1962.
Albright, L. F.: Processes for Major Addition-Type Plastics and Their Monomers, McGraw-Hill, 1974.
Allen, P. E. M., and C. R. Patrick: Kinetics and Mechanisms of Polymerization Reactions, Halsted, 1974.
Baer, E.: Engineering Design for Plastics, Society of Plastics Engineers, 1964. Blue Book I: Polymer Additives, Noyes, 1972.
Boenig, H. V.: Unsaturated Polyesters: Structure and Properties, Elsevier, 1964.
Boyer, R. F. (ed.): Technological Aspects of the Mechanical Behavior of Polymers, Wiley, 1974.
Brandup, J., and E. H. Immergut (eds.): Polymer Handbook, Wiley-Interscience, 1965.
Brenner, W., *et al.*: High-temperature Plastics, Reinhold, 1962.
Bruins, P. F.: Plasticizer Technology, Reinhold, 1965.
Brydson, J. A.: Plastic Materials, Van Nostrand, 1966.
Christopher, W. F., and D. W. Fox: Polycarbonates, Reinhold, 1962.
Cowie, J. M. G.: Polymers: Chemistry and Physics of Modern Materials, Intext Educational, 1974.
Davidson, R. L., and M. Sittig: Water-soluble Resins, Reinhold, 1968.
DuBois, J. H., *et al.*: Plastics, Reinhold, 1967.
Epoxy Resin Handbook, Noyes, 1972.
Epoxy Resins, ACS, 1970.
Ferrigno, T. H.: Rigid Plastics Foams, 2d ed., Reinhold, 1967.
Fettes, E. M.: Chemical Reactions of Polymers, vol. 19, High Polymers, Wiley-Interscience, 1964.
Floyd, D. E.: Polyamide Resins, Reinhold, 1966.
Foy, G. F. (ed.): Engineering Plastics and Their Commercial Development, ACS, 1969.
Gaylord, N. G. (ed.); Polyethers, 3 pts., Wiley, 1962–1963.
Gilbert, J. *et al.*: Handbook of Technology and Engineering of Reinforced Plastics, Van Nostrand Reinhold, 1973.
Golding, B.: Polymers and Resins, Van Nostrand, 1959.
Gordon: M.: High Polymers, Addison-Wesley, 1964.
Gould, R. F. (ed): Literature of Chemical Technology, chap. 25, Resins and Plastics, ACS Monograph, 1967.
Griff, A. L.: Plastics Extrusion Technology, Reinhold, 1962.
Halpern, M. G.: Rigid Foam Laminates, Noyes, 1972.
Hicks, E.: Shellac: Its Origin and Applications, Chemical Publishing, 1961.
Jenkins, A. D.: Polymer Science, American Elsevier, 1972, 2 vols.
Johnson, J. C.: Antioxidants 1975, Synthesis and Applications, Noyes, 1975.
Kausch, H. H. (ed.): Deformation and Fracture of High Polymers, Plenum, 1973.
Mark, H. (ed.) *et al.*: Encyclopedia of Polymer Science and Technology, Wiley-Interscience, 1964, 1966.
Martens, C. R.: Alkyd Resins, Reinhold, 1961.
McKelvey, J. M.: Polymer Processing, Wiley, 1962.
Meals, R. N., and F. M. Lewis: Silicones, Reinhold, 1959.

Mellan, I.: Industrial Plasticizers, Pergamon, 1964.
Meltzer, Y. L.: Water-Soluble Polymers, Noyes, 1972.
———: Urethane Foams Technology and Applications, Noyes, 1971.
———: Plasticizers, Noyes, 1972.
Mith, M. (ed.): Manufacture of Plastics, Reinhold, 1964.
Modern Plastics Encyclopedia, McGraw-Hill (yearly).
Morgan, P. W.: Condensation Polymers, Interscience, 1965.
Nielsen, L. E. (ed.) Mechanical Properties of Polymers and Composites, Dekker, 1974, 2 vols.
Ogorkiewicz, R. M.: Thermoplastics: Properties and Design, Wiley, 1974.
Oleesky, S. S., and J. G. Mohr: Handbook of Reinforced Plastics of the SPI, Reinhold, 1964.
Platzer, N. A. (ed.): Addition and Condensation.
Platzer, N. A. (ed.): Multicomponent Polymer Systems, ACS, 1971.
Polymerization Processes, ACS, 1969.
Polymer-Plastics Technology and Engineering, Dekker, 1974.
Ranney, M. W.: Fire Resistant Flame Retardant Polymers, Noyes, 1974.
Ranney, M. W.: Reinforced Composites from Polyester Resins, Noyes, 1972.
Sandler, S. R., and W. Faro: Polymer Syntheses, vol. 1, Academic, 1974.
Saunders, J. H., and K. C. Frisch: Polyurethanes: Chemistry and Technology, Wiley, 1962–1963, 2 vols.
Schnell, H.: Chemistry and Physics of Polycarbonates, Interscience, 1964.
Schultz, J.: Polymer Materials Science, Prentice-Hall, 1974.
Simonds, H. R., and J. M. Church: Concise Guide to Plastics, 2d ed., Reinhold, 1963.
Sittig, M.: Polyacetal Resins, Gulf, Houston, Tex., 1963.
———: Acrylonitrile 1965, Noyes, 1965.
———: Acrylic Acid and Esters 1965, Noyes, 1965.
———: Vinyl Monomers and Polymers, Noyes, 1966.
———: Polyolefin Processes, Noyes, 1967.
———: Pollution Control in the Plastics and Rubber Industry, Noyes, 1975.
Skeist, I.: Reviews in Polymer Technology, Dekker, 1972.
Smith, W. M.: Manufacture of Plastics, Reinhold, 1964.
Tobolsky, A. V., and H. F. Mark: Polymer Science and Materials, Wiley, 1971.
Vollmert, B.: Polymer Chemistry, Springer-Verlag, 1973.
Williams, D. G.: Polymer Science and Engineering, Prentice-Hall, 1971.
Williams, A.: Furans, Noyes, 1973.
Yarsley, V. E. *et al.*: Cellulose Plastics, Gordon and Breach, 1964.

chapter 35

SYNTHETIC FIBER AND FILM INDUSTRIES

The ability of the chemist and the chemical engineer to create, from the test tube through the factory, products that are often superior to naturally occurring materials is one of the outstanding accomplishments of this age, and nowhere is it so graphically portrayed as by the modern synthetic fibers.[1] In the United States in 1972, \$167 million was spent for research in this field.

HISTORICAL Although the word "fiber" originally referred only to naturally occurring materials (cotton, wool, etc.), it is now used extensively for synthetic products. The latter usage includes both cellulosics and polyfibers. Cellulosics result when natural polymeric materials such as cellulose are brought into a dissolved or dispersed state and then spun into fine filaments. Viscose rayon, acrylics, and polyesters result from two methods of forming long-chain molecules: addition polymerization and condensation polymerization. The polymer is then taken from its melt, or solution, and processed into fiber form. Cellulose acetate really lies between these two classes, since in this case a natural polymer (cellulose) is converted into one of its derivatives (cellulose acetate), which is dissolved and spun into fiber form.

These synthetics, from their humble beginning in 1900, have grown to a total world production of more than 26,700 million lb in 1974 and, for the United States, to 8,100 million lb in 1974, including glass fibers. The United States started the manufacture of rayon in 1910 and produced in 1974 about 800 million lb of rayon and 380 million lb of acetate. Over 6,200 million lb of fibers was produced in 1974.[2] The list of synthetic fibers, which in 1900 included only nitrocellulose, today has many products, the *more important ones* being polyamides or nylons, fibers made from polyacrylonitrile (Orlon, Dynel, Acrilan), polyesters, (Dacron, Fortrel), olefins (Marvess, Herculon), spandex fibers (Lycra, Glospan), glass (Fiberglas, Vitron), and cellulosic fibers (viscose and acetate rayon). It is the purpose of this chapter to present a picture of the chemistry and engineering behind our moden "fashions out of test tubes." See Table 35.1.

Three of the more important general properties of fibers are length, crimp, and denier. Concerning length, there are essentially two types of fibers, continuous-filament and staple. *Continuous filaments* are individual fibers whose length is almost infinite. Silk, rayon, nylon, and most other true synthetics are manufactured in this manner. Cotton and wool are examples of natural fibers in the *staple* form, i.e., of short and more-or-less uniform length. Artificial staple fibers such as rayon, acetate, nylon, and polyester result from the cutting of tow (untwisted continuous filaments) to uniform lengths, usually between $1\frac{1}{2}$ and 6 in. *Crimp* is the curl, or waviness, placed in synthetic fibers by chemical or mechanical action, which is of great importance in the processability of staple fibers.

[1] Grove et al., Fibers, *Ind. Eng. Chem.*, Annual Review, **53**(10), 853 (1961), Annual Suppl., 1962, p. 17, and following years.
[2] *Text. Organon,* June 1975, p. 71.

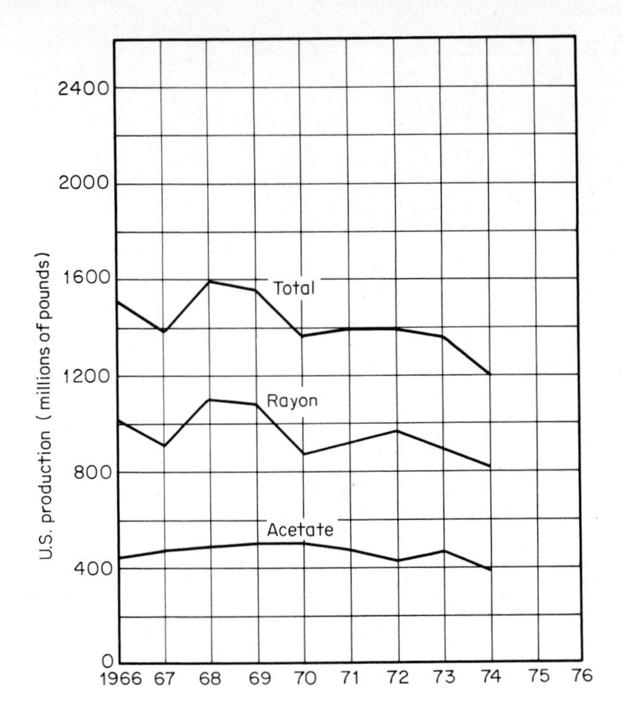

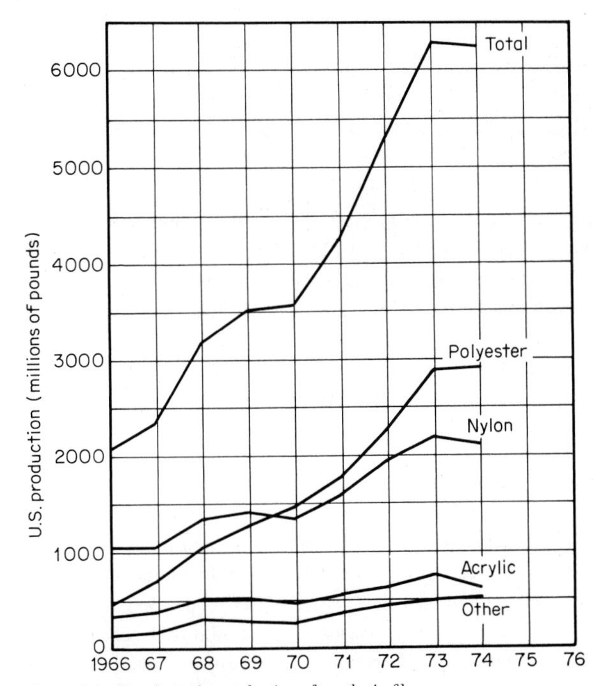

Fig. 35.1 Trends in the production of synthetic fibers.

TABLE 35.1 ***Representative Synthetic Fibers and Films****

Classification	*Spinning*†
Polyfibers and film	
Polyamides, or nylon, fibers	
Nylon 66, nylon 6, Qiana	Melt
Polyesters	
Fibers: Dacron, Trevira, Kodel, Fortrel	Melt
Films: Mylar, Cronar, Kodar, Estar	Melt
Acrylics and modacrylics	
Orlon fiber	Dry
Acrilan fiber	Wet
Creslan fiber	Wet
Dynel fiber (vinyl-acrylic)	Dry
Verel fiber	Dry
Vinyls and Vinylidines	
Saran fiber and film	Melt
Vinyon N fiber‡	Dry
Spandex	
Lycra	Dry
Numa	Wet
Glospan	Wet
Olefins	
Polyethylene films	Melt
Polypropylene fibers and films: Avisun, Herculon	Melt
Glass fibers	
Fiberglas	Melt
Cellulosic fibers and films	
Regenerated cellulose	
Fibers: Rayon (viscose), cuprammonium	Wet
Film: Cellophane	Wet
Cellulose esters	
Acetate fibers and films: Acele, Estron	Dry
Triacetate fiber: Arnel	Dry

*Completely synthetic or semisynthetic (cellulosics). †Classified as melt, dry, or wet. ‡60% vinyl and 40% acrylonitrile.

Cotton and wool possess natural crimp. Crimp is also introduced in some cases in continuous-filament yarns to alter their appearance and feel, e.g., in nylon carpet yarn. *Denier* is a measure of weight per unit length. Each fiber has a *cross section,* and special characteristics are imparted by manufacturing polyfibers with different cross sections. A fiber has a cross section corresponding to 1 denier if 9,000 m of it weighs 1 g.

Films, as well as fibers, are changing. Nitrocellulose, once largely used especially for photographic film, has given way to acetate and polyester. Many other films are suited to the demand for their special properties, e.g., cellophane, Saran, Teflon, and polyethylene.

POLYFIBERS AND FILMS

The rapidly growing and versatile noncellulosic fibers are classified chemically and by the method of spinning in Table 35.1. The manufacture of all true synthetic fibers first begins with the preparation of a polymer consisting of extremely long, chainlike molecules. The polymer is spun in one of three ways (see below), resulting in most cases in a weak, practically useless fiber until it is *stretched further to orient the molecules and set up crystalline lattices.* Although the range of any one polymer is

TABLE 35.2 *Comparison of Selected Synthetic Fibers*

Fiber	*Wool*	*Acrylic*	*Modacrylic*	*Polyesters*	*Nylon*
Tensile strength, g/denier	1.0–1.7	4.0–5.0	3.0	4.4–6.6	4.7–5.6
Elongation, %	25–35	16–21	16	18–22	25–28
Elastic recovery	0.99 at 2%	0.97 at 2%	0.80 at 2%	90–100 at 4%	100 at 8%
Strength, psi	20,000–29,000	59,000–74,000	44,000–66,000	78,000–116,000	68,000–81,000
Stiffness, g/denier*	3.9	24	30	23–63	20
Abrasion resistance†	90	330		1,570	2,520
Water absorbency‡	21.9% at 90 RH	2% at 95 RH	. . .	0.5% at 95 RH	8% at 95 RH
Effect of heat	Becomes harsh 100°C, decomposes at 130°C	Sticking point 235°C	Sticking point 235°C	Sticking point 240°C	Melts 263°C
Effect of age	Little	Little	Little	Little	Slight
Effect of sun	Weakened	Very resistant	Slight	Little	Weakened
Effect of acids (concentrated, room temp.)	Resistant	Resistant	Resistant	Resistant	Weakened
Effect of alkalies	Susceptible	Partly resistant	Resistant	Resistant	Resistant
Effect of organic solvents	Resistant	Resistant	Resistant	Resistant	Resistant
Dyeability	Good	Good	Good	Difficult	Good
Resistance to moths	None	Wholly	Wholly	Wholly	Wholly
Resistance to mildew	Good	Wholly	Wholly	Wholly	Wholly

Sources: Wool vs. Synthetics, *Chem. Week,* **69**(3), 11 (1951); Moncrief, Man-Made Fibers, 6th ed., Wiley, 1975. *Crease resistance. †Wet flex test, number of flexes. ‡RH, Relative humidity.

always limited by controlling the degree of orientation, crystallinity, and average chain length, a single polymer can be used to make a number of fibers with different mechanical properties; i.e., some can be weak and stretchy, others strong and stiff. The two elements important in determining the range of the polymer's mechanical properties are the attractive forces between the molecules and the flexibility and length of the molecular chains.

The three spinning procedures are *melt, dry,* and *wet.* Melt spinning, developed for nylon and also used for polyester, polyvinyl, polypropylene, and others, involves pumping molten polymer through spinneret jets. The polymer solidifies into filaments as it strikes the cool air. In dry spinning, as described for acetate, the polymer is dissolved in a suitable organic solvent. The solution is forced through spinnerets, and dry filaments result upon the evaporation of the solvent in warm air. Orlon and Vinyon are spun in this manner. Viscose rayon, Acrilan, and Dynel staple are examples of wet-spun fibers. Here, as the solution emerges from the spinneret, it is coagulated in a chemical bath. Several polyfibers can be spun in several ways: Table 35.1 lists the method believed to be in widest commercial use. For a comparison of selected synthetic fibers, see Table 35.2.

POLYAMIDE, OR NYLON, FIBERS

NYLON (nylon 66) This fiber was the first truly all-synthetic fiber, resulting from the brilliant research of Carothers of the Du Pont Co.[3] It opened up the entire field of true synthetic fibers, and in 1974 worldwide production was 5.7 billion lb. U.S. production was a little over 2 billion lb, and of this 70% was nylon 66 and 30% was nylon 6. In the rest of the world the split was 38% nylon 66 and 62% nylon 6. Nylon 66 is commonly produced by reacting hexamethylenediamine and adipic acid to form "nylon salt," or hexamethylene diammonium adipate. This compound by polymerization with the removal of a molecule of water, becomes polyhexamethylene adipamide, a linear polyamide. The two intermediates which form the nylon salt are currently produced from petrochemical raw materials (Fig. 35.2). The oxidation of cyclohexane is used to make adipic acid. Three processes for

[3]Bolton, Development of Nylon, *Ind. Eng. Chem.*, **34,** 53 (1942).

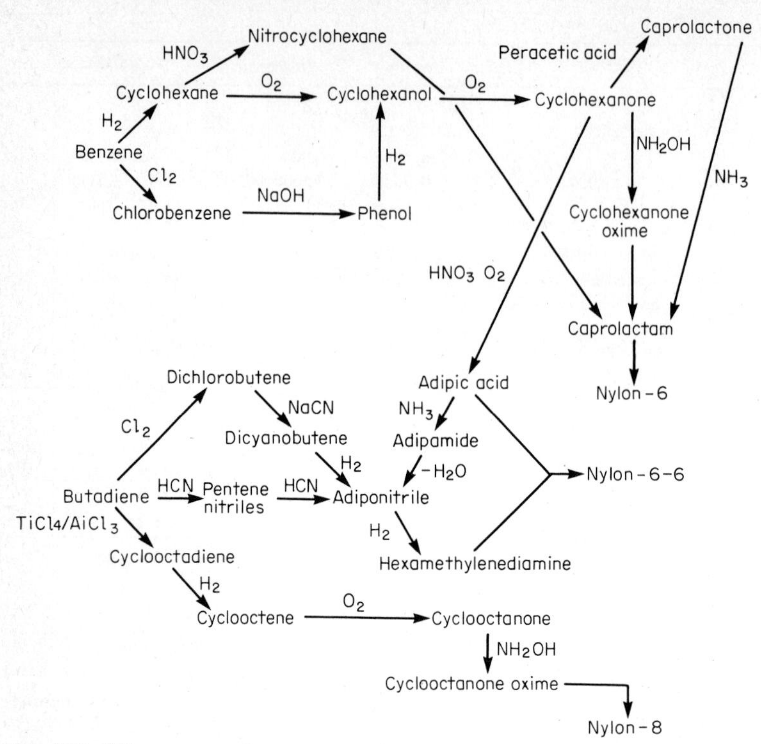

Fig. 35.2 Various routes to nylon. (*Adapted from Marshall Sittig.*)

the production of hexamethylenediamine are in current use: from butadiene, from propylene, and from adipic acid.[4]

Formulations of the reactions involved are presented in the following discussion.

NYLON INTERMEDIATES

1. *Adipic acid.* 1973 U.S. production was about 1.6 billion lb, of which 95% was used for nylon fibers and plastics.

$$\underset{\substack{\text{Cyclohexane}\\\text{from hydro-}\\\text{genation of}\\\text{benzene}}}{C_6H_{12}} \xrightarrow[\text{Mn, Co acetates}]{\text{air, }125^\circ\text{C}} \underset{\substack{\text{Cyclohexanol}\\50\%}}{C_6H_{11}OH} + \underset{\substack{\text{Cyclohexanone}\\40\%}}{C_6H_{10}O} \xrightarrow[\substack{\text{aqueous}\\HNO_3}]{80^\circ\text{C}} \underset{\text{Adipic acid}}{HOOC(CH_2)_4COOH} + xN_2O + yNO$$

2. *Hexamethylenediamine*

$$\underset{\substack{\text{Butadiene}\\\text{from dehydrogenation}\\\text{of butane}}}{CH_2{:}CH\cdot CH{:}CH_2} + Cl_2 \longrightarrow \underset{\substack{\text{1,4-Dichlorobutene}\\\text{(distill out)}}}{ClCH_2\cdot CH{:}CH\cdot CH_2Cl}$$

[4]Standish and Abrams, Adipic Acid, ECT, 2d ed., vol. 1, pp. 405–421 (119 refs.); Haines, Cyclohexane and Adipic Acid, *Ind. Eng. Chem.*, **54**(7), 23 (1962) (flowchart); Guccione, Thermal Dealkylation for Cyclohexane, *Chem. Eng.* (*N.Y.*), **70**(15), 111 (1963); Sherwood, Raw Materials for Man-made Fibers, *Ind. Eng. Chem.*, **55**(1), 37 (1963); Dufan *et al.*, High Purity Cyclohexane, *Chem. Eng. Prog.*, **60**(9), 43 (1964); Scientific Design, *Chem. Eng. News*, June 6, 1966, p. 17; U.S. Pats. 3,243,499; 3,109,864; 3,240,820; 3,239,552; Austin, *Chem. Eng.* (*N.Y.*), **81**(2), 131 (1974); **81**(11), 104 (1974).

$$\text{Latter} \xrightarrow[\text{CuCl, 80–95°C, yield 95\%}]{\text{2NaCN or 2HCN}(aq)} \underset{\text{1,4-Dicyanobutene}}{NC\cdot CH_2\cdot CH{:}CH\cdot CH_2\cdot CN} + 2NaCl$$

$$\underset{\text{1,4-Dicyanobutene}}{NC\cdot CH_2\cdot CH{:}CH\cdot CH_2\cdot CN} \xrightarrow[\text{Palladium on C, 250°C, yield 95–97\%}]{H_2} \underset{\text{Adiponitrile}}{NC(CH_2)_4CN} \xrightarrow[\text{high-pressure } NH_3]{H_2} \underset{\text{Hexamethylenediamine}}{H_2N(CH_2)_6NH_2}$$

Other routes to hexamethylenediamine:

a. $$\underset{\text{Adipic acid}}{HOOC(CH_2)_4COOH} + 2NH_3 \xrightarrow{-H_2O} \underset{\text{Adipamide}}{NH_2CO(CH_2)_4CONH_2} \xrightarrow{-2H_2O} \underset{\text{Adiponitrile}}{NC(CH_2)_4CN}$$

b. Electrolytic hydrodimerization of acrylonitrile in solubilizing concentrated solutions of tetraethylammonium *p*-toluenesulfonate:[5]

$$\underset{\text{Acrylonitrile}}{2CH_2{:}CHCN} \xrightarrow[\text{yield 90+\%}]{2H_2} \underset{\text{Adiponitrile}}{NC(CH_2)_4CN}$$

$$\underset{\text{Adiponitrile}}{NC(CH_2)_4CN} \xrightarrow[\text{100–135°C, 10,000 psi}]{4H_2Co(Cu)} \underset{\text{Hexamethylenediamine, yield 90\%}}{H_2N(CH_2)_6NH_2}$$

3. *Nylon salt and nylon*

$$\underset{\text{Adipic acid}}{xHOOC(CH_2)_4COOH} + \underset{\text{Hexamethylenediamine}}{xH_2N(CH_2)_6NH_2} \longrightarrow$$

$$\underset{\text{Hexamethylene diammonium adipate, or nylon salt}}{x(-H_3N(CH_2)_6NH_3OOC(CH_2)_4COO-)} \xrightarrow{-2xH_2O} \underset{\text{Polyhexamethylene adipamide, or nylon}}{(-HN(CH_2)_6NHOC(CH_2)_4CO-)x}$$

In the manufacture of the fiber,[6] the aqueous nylon-salt solution starts on the top level and the materials move down by gravity through the various steps, which may be divided into the following coordinated sequence of *unit operations* (Op) and *chemical conversions* (Ch) as depicted in the flowchart in Fig. 35.3:

The hexamethylene diammonium adipate solution is pumped to the top level into evaporators and concentrated (Op). Acetic acid is added to the evaporator charge to stabilize the viscosity (Ch).

After evaporation, the salt solution flows into jacketed autoclaves equipped with internal coils and heated by Dowtherm vapor (Fig. 35.4). Here the rest of the water is removed, the pigment dispersion agent (TiO_2) is added, and polymerization takes place (Op and Ch).

After polymerization is completed, the molten viscous polymer is forced out of the bottom onto a casting wheel by specially purified nitrogen at 40 to 50 psi. Each 2,000-lb batch is extruded as rapidly as possible to minimize differences due to thermal treatment of the polymer.

A ribbon of polymer about 12 in. wide and $\frac{1}{4}$ in. thick flows on the 6-ft casting drum. Water sprays on the inside cool and harden the underside of the ribbon; the outer is cooled by air and water (Op).

The ribbons are cut into small chips, or flakes, before being blended. Two or more batches are mixed to improve the uniformity of the feed to the spinning machine (Op).

The blenders empty into hoppers on a monorail which supply the spinning area (Op).

[5]New Trails to Nylon, *Chem. Week*, Oct. 12, 1963, p. 85; acrylonitrile costs half that of adipic acid; Monsanto: New Routes to Adiponitrile (from acrylonitrile), *Chem. Eng.* (*N.Y.*), **72**(20), 88 (1965), and **72**(23), 238 (1965) (*Chemical Engineering* award; excellent flowchart and photographs); New Adipo Processes, *Chem. Week*, May 12, 1971, p. 32. See Acrylonitrile, this chapter; Sitting, Polyamide Fiber Manufacture, Noyes, 1972.

[6]Lee, Nylon Production Technique, *Chem. Eng.* (*N.Y.*) **53**(3), 96 (1946) (pictured flowchart, p. 148).

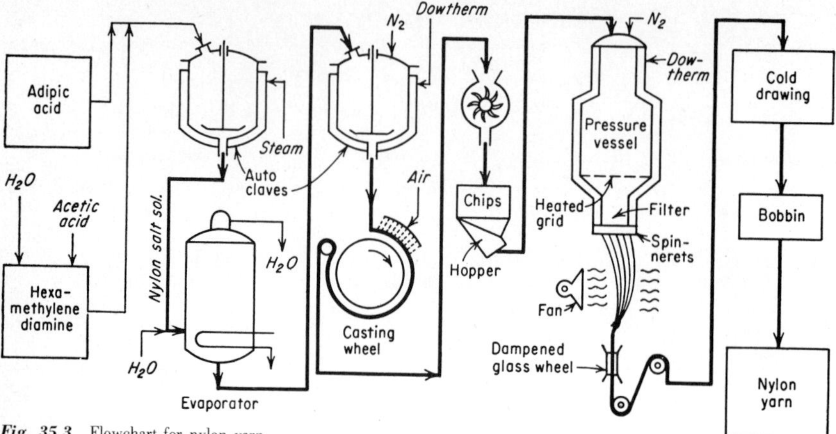

Fig. 35.3 Flowchart for nylon yarn.

A typical spinning unit is composed of a metal vessel surrounded by a Dowtherm vapor-heated jacket which keeps the temperature of the vessel above the melting point (263°C) of the nylon. Special precautions are again taken to keep the spinning oxygen-free. As the nylon flake enters the vessel it strikes a grid, where it melts and dribbles through to the melt chamber below (Op). Screw melters are also used in place of grids.

The molten polymer passes through the portholes in this chamber to the gear spinning pumps. They deliver it to a sand filter, which is followed by screens and spinnerets (Op).

The filaments are solidified by air in a cooling chimney and passed in a bundle through a steam-humidifying chamber, where the moisture content is brought to equilibrium in order to stabilize the length of the spun polymer (Op). This is not a problem after the stretching or drawing.

Fig. 35.4 Autoclaves in Du Pont nylon plants. In these receptacles the chemical process known as polymerization takes place. In other words, small molecules are joined together to make large ones, and nylon polymer is created. Plants are located at Seaford, Del., at Martinsville and Richmond, Va., at Chattanooga, Tenn., and at Camden, S.C.

After lubrication in a finish roll, the yarn is stretched or drawn to the desired degree by passing it through a roller system. Here the strength and elasticity characteristics of nylon are developed, because the molecules are oriented from their previous comparatively helter-skelter arrangement (Ch). Stretching may be from three- to sixfold, depending on the mechanical properties desired, being stronger when the orientation by stretching is greatest. The nylon filament is then shipped to various manufacturers for processing (Op).

In many plants continuous processing is more economical for larger production items. Here the pressure and temperature of the autoclave cycles are reproduced in a series of pipes and vessels. The continuous process favors uniformity of large-scale product, whereas the batch process favors flexibility among several products. Nylon is also a popular plastic molding powder and bristle material. Like all other synthetic fibers that have become competitively popular, nylon in both the filament and staple form must have certain properties that are superior to natural fibers. It is stronger than *any* natural fiber and has a wet strength of 80 to 90% of its dry strength. Its good flexing qualities make it very desirable for women's hosiery, and it has good stretch recovery. Nylon's high tenacity has made it important in parachute fabrics and related nonapparel items. Nylon can be dyed by all acid and dispersed dyes (acetate). It has a low affinity for direct cotton, sulfur, and vat dyes.

NYLON 6[7] or nylon caprolactam, is based on a polymeric fiber from only one constituent, caprolactam (Fig. 35.2), $\text{H}\underline{\text{N}\text{—}(\text{CH}_2)_5\text{C}}\text{O}$ giving the polymer $(\text{—}(\text{CH}_2)_5\text{CO—NH—})_n$. Nylon 6 had a U.S. production of about 600 million lb in 1974, or about 43% of nylon 66. Nylon 6 is widely used in Europe and Japan. It possesses an expanding market for molding resins as well as for fiber. Most processes start only with lower-cost caprolactam[8] (Fig. 35.2) from cyclohexanoneoxime from cyclohexane from benzene.

Other nylons have been prepared and have some interesting properties. Of these only nylon 610, prepared from hexamethylenediamine and sebacic acid, is produced commercially. Nylon 4 (from pyrrolidone) has received much publicity but has no commercial significance. Nylon 7 (7-aminoheptanoic acid), nylon 8 (caprylactam), and nylon 9 (9-aminopelargonic acid) are only laboratory products.

POLYESTERS

An important example of a polyester is polyethylene terephthalate,[9] the condensation product of dimethyl terephthalate or terephthalic acid and ethylene glycol (Fig. 35.5). Terephthalic acid is made by oxidation from *p*-xylene in the presence of nitric acid or air. It is then esterified with methanol. About 0.6 lb of glycol and 1 lb of ester are added to the transesterification reactor for each pound of polyester. The polymerization is carried out at 260 to 300°C, using a vacuum. The reaction releases methanol and glycol, and a polymer chain containing approximately 80 benzene rings is formed. After filtering, the material is melt-spun in a manner similar to that described for nylon. The filaments are stretched, with the application of heat, to about three to six times their original length. Table 35.2 shows some of the interesting properties of this fiber: resilience, strength, and crease recovery. It is particularly well suited for woven fabrics such as men's summer suits, men's shirts, and women's dresses and blouses, and for knitwear for sweaters and shirting. Because of its strength, this polyester is important in the tire-cord and cordage fields; it is also used for sewing thread, fire hose, and V belts. In staple form it is employed as stuffing for pillows, sleeping bags, and comforters. Its production in 1974 approached 3 billion lb.

[7]Aclion, Nylon 6 and Related Polymers, *Ind. Eng. Chem.*, **53**(10), 827 (1961); Caprolactam, *Chem. Week*, June 6, 1962, p. 74, and Mar. 14, 1964, p. 57 (outline of process).

[8]New Processes Add Flavor and Caprolactam Price, *Chem. Eng.* (*N.Y.*), **68**(6), 92 (1961). Flowchart, *Chem. Eng. News*, Aug. 28, 1967, p. 82; Caprolactam, *Chem. Eng.* (*N.Y.*), **81**(6), 54 (1974).

[9]Rossmassler, Polyester Fibers, ECT, 1st ed., vol. 13, pp. 841–847, 1954 (41 refs.); Terephthalic Route, *Chem. Week*, Mar. 9, 1963, p. 57; Brown and Reinhart, Polyester Fiber, *Science* **173**(3994), 287 (1971).

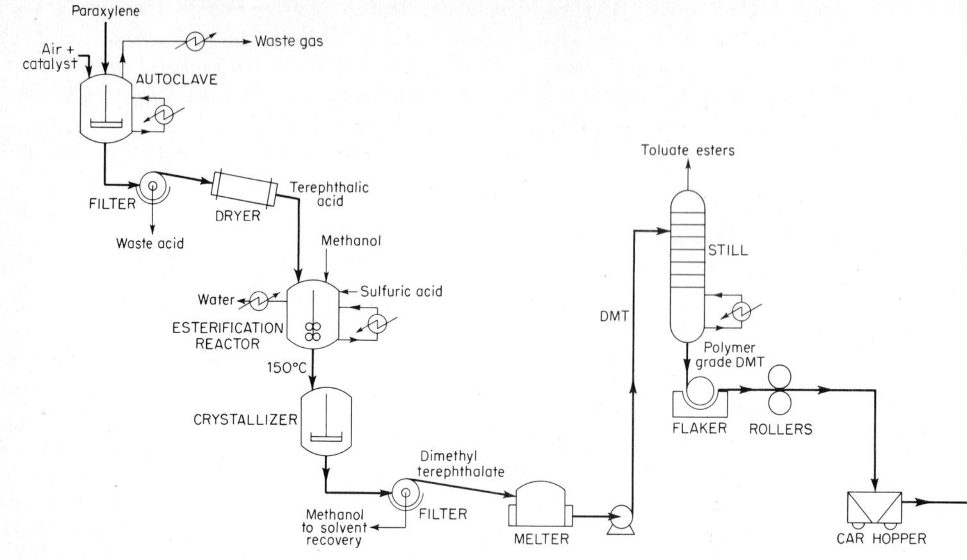

Fig. 35.5 Flowchart for the manufacture of polyester fibers. [*Chem. Eng.* (*N.Y.*), **70**(5), 76 (1963).]

It is manufactured from ethylene glycol and terephthalic acid or dimethyl terephthalate[10] and then polymerized.

Dimethyl terephthalate:

$$\underset{p\text{-xylene}}{CH_3C_6H_4CN_3} \xrightarrow[\text{or air}]{HNO_3} \underset{\text{Terephthalic acid, impure}}{p\text{-HOOC}\cdot C_6H_4\cdot COOH} \xrightarrow[\text{150°C, 0.01\% } H_2SO_4]{CH_3OH} \underset{\text{Dimethyl terephthalate, impure}}{CH_3OOC\cdot C_6H_4\cdot COOCH_3} + 2H_2O$$

The dimethyl terephthalate is purified by distillation and then subjected to alcoholysis with ethylene glycol to yield the monomer

$$CH_3OOC\cdot C_6H_4\cdot COOCH_3 + 2CH_2OH{-}CH_2OH \xrightarrow{230°C} HOC_2H_4\cdot \underset{\underset{O}{\|}}{OC}\cdot C_6H_4\cdot \underset{\underset{O}{\|}}{CO}\cdot C_2H_4OH + 2CH_3OH$$

Polymerization of the monomer:

$$\text{Monomer} \xrightarrow[\text{vacuum}]{260\text{–}300°C} \underset{\text{Polyethylene terephthalate}}{HO[C_2H_4OOC\cdot C_6H_4\cdot COO]_nC_2H_4OH} + \underset{\text{Half of glycol distilled off}}{n\text{-}\tfrac{1}{2}HOC_2H_4OH}$$

The polymer melt can be spun and stretched for polyester fiber, or cast on a wheel to form polyester film. Figure 35.5 is a flowchart of the entire process.[11]

One of the big disadvantages of polyester fiber has been its great attraction for oily soil and its moisture repellency. A new process to graft acrylic acid chemically to the polyester surface and thus provide permanent wettability and soil release has been announced. An electrical discharge in argon gas is used to effect the grafting.[12]

[10]Austin, *Chem. Eng.* (*N.Y.*), **81**(22), 114 (1974).

[11]Guccioni, Simple Chemicals Take Tortuous Route to Dacron, *Chem. Eng.* (*N.Y.*), 70(5), 76 (1963) (flowchart and description); Mitsubishi, Pure Terephthalic Acid, *Chem. Eng.* (*N.Y.*), **72**(9), 70 (1965) (toluene-benzoic-terephthalic, flowchart); Japanese Firm Banks on Toluene Route to TA, *Chem. Eng. News*, Aug. 16, 1965, p. 30.

[12]Suchecki, Route to a New Polyester, *Text. Ind.* (*Atlanta*), March 1975, p. 91.

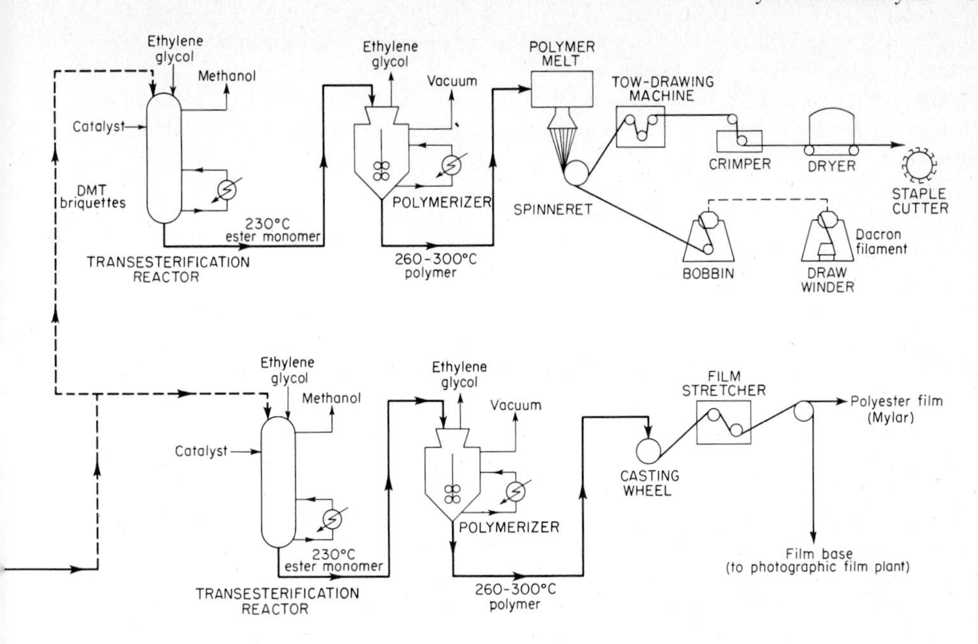

ACRYLICS AND MODACRYLICS

Polyacrylonitrile $(-CH_2CHCN-)_n$ is the major component of several industrial textile fibers, but Du Pont's Orlon was the first to attain commercial-scale operation. Dyeable acrylics are generally copolymers with modifying constituents.

Orlon is made by polymerizing acrylonitrile. The ivory-white polymer is dissolved in an organic solvent, generally dimethylformamide, although it can be dissolved in many concentrated solutions of salts, like lithium bromide or sodium sulfocyanide, or more successfully in other organic solvents, such as dimethoxyacetamide and tetramethylene cyclic sulfone. The solution is filtered and then dry-spun, utilizing the same spinning technique long used for acetate, namely, in solution through spinnerets, with the solvent evaporated to furnish the dry fiber. Unlike nylon which is drawn at room temperature, acrylics, like polyester, are drawn at elevated temperatures in a special machine. The fibers are stretched to three to eight times their original length to orient the molecules into long parallel chains for final strength. The staple fiber has esthetic properties like those of wool. The acrylics' resistance to chemical attack, and especially to weathering, makes them highly useful in several fields. Other acrylics, like Acrilan and Creslan, are spun wet into a coagulating bath. The end uses for acrylic fibers comprise sweaters, women's coats, men's winter suiting, carpets, and blankets. Blends with wool and other synthetics are common in some end uses. Acrylics are also suitable for pile fabrics and filter cloth.

MODACRYLIC FIBERS (Modified acrylics) This is a generic name for synthetic fibers in which the fiber-forming substance is any long-chain synthetic polymer composed of less than 85% but at least 35% by weight of acrylonitrile units (Federal Trade Commission).

Dynel is the name given by Carbide and Carbon to their staple copolymer modacrylic fiber made from a resin of 40% acrylonitrile and 60% vinyl chloride. The resin is converted into staple in a continuous wet-spinning process (cf. viscose rayon). The white resin powder is dissolved in acetone, filtered, and run through a spinneret, where the fibers are formed in an aqueous spinning bath. The fiber is dried, cut, and crimped. Dynel is similar to wool in many respects and has some superior characteristics. It is used for work clothing, water-softener bags, dye nets, filter cloth, blankets, draperies, sweaters, pile fabrics, etc.

ACRYLONITRILE in 1974 had a production between 1,000 and 2,000 million lb, of which about 70% went to make acrylic fibers. The —CN group sharply increases the reactivity of the unsaturated double bonds. This intermediate can be made by a number of reactions,[13] for example:

$$\text{1. } \underset{\text{Acetylene}}{CH{:}CH} + HCN \xrightarrow[\text{80°C, 15 psig}]{CuCl(aq)} \underset{\text{Acrylonitrile}}{CH_2{:}CHCN}$$

$$\text{2. } \underset{\text{Propylene}}{CH_2{:}CH\cdot CH_3} \xrightarrow{O_2} \underset{\text{Acrolein}}{CH_2{:}CH{-}CHO} \xrightarrow[\text{380 C}]{NH_3 + \frac{1}{2}O_2} CH_2{:}CHCN + \underset{\text{Yield 78\%}}{2H_2O}$$

$$\text{3. } \underset{\text{Propylene}}{CH_2{:}CH\cdot CH_3} + NH_3 + 1.5O_2 \xrightarrow[\text{380°C, yield 78\%}]{\text{Biphosphomolydate on } SiO_2} CH_2{:}CHCN + 3H_2O$$

$$\text{4. } \underset{\diagdown O \diagup}{CH_2{-}CH_2} + HCN \longrightarrow \underset{\text{Cyanohydrin}}{HO\cdot CH_2\cdot CH_2CN} \xrightarrow[\text{200°C}]{-H_2O} CH_2{:}CHCN$$

Only process 3, the Sohio process, is in commercial use in the United States (Fig. 35.6). This process accounts for about 85% of world production.

VINYLS AND VINYLIDINES

Saran is the copolymer of vinyl chloride and vinylidene chloride. It is prepared by mixing the two monomers with a catalyst and heating. Color is added by introducing the pigment into the mass. The copolymer is heated, extruded at 180°C, air-cooled, and stretched. Saran is resistant to mildew, bacterial, and insect attack, which makes it suitable for insect screens. Its chemical resistance makes it advantageous for filter-cloth applications; however, its widest use has been for automobile seat covers and home upholstery. Much film is made from Saran.

Vinyon is the trade name given to copolymers of 90% vinyl chloride and 10% vinyl acetate. The copolymer is dissolved in acetone to 22% solids and filtered, and the fibers are extruded by the dry-spinning technique. After standing, the fibers are wet-twisted and stretched. Resistance to acids and alkalies, sunlight, and aging makes Vinyon useful in heat-sealing fabrics, work clothing, filter cloths, and other related applications.

SPANDEX

Spandex, a generic name, is defined by the Federal Trade Commission as "a manufactured fiber in which the fiber-forming substance is a long chain synthetic polymer comprising at least 85% of a segmented polyurethane." Trade examples of these are Lycra, Glospan, and Numa fibers. Lycra,[14] Du Pont's spandex fiber results from the following reaction:

$$\underset{\text{Glycol}}{nHO{-}R{-}OH} + \underset{\text{Diisocyanate}}{nOCN{-}R'{-}NCO} \longrightarrow \underset{\text{Polyurethane}}{({-}OR{-}O{-}\overset{O}{\overset{\|}{C}}{-}\overset{H}{\overset{|}{N}}{-}R'{-}\overset{H}{\overset{|}{N}}{-}\overset{O}{\overset{\|}{C}}{-})_n}$$

[13]Acrylonitrile, Cost Estimate on NH_3 and HCN reaction, *Chem. Week,* July 21, 1962, p. 63; Sherwood, Raw Materials for Man-made Fibers, *Ind. Eng. Chem.,* **55**(1), 39 (1963); Sittig, Acrylic Acid and Esters, Noyes, 1965, p. 18 (flowchart); *Chem. Eng.* (*N.Y.*), **68**(2), 127 (1961); Petrochemical Complex: Goodrich New Methods Make Low-cost Vinyl, Acrylonitrile, *Chem. Week,* Aug. 29, 1964, p. 101; Guccione, Acrylonitrile's Latest Synthesis from Propylene, *Chem. Eng.* (*N.Y.*), **72**(6), 150 (1965) (SNAM process); Moncrief, *op. cit.,* chap. 28; Austin, *Chem. Eng.* (*N.Y.*), **81**(2), 131 (1974).

[14]Hicks, "Lycra" Spandex Fiber, *Am. Dyes. Rep.,* **53,** 334 (1964); Smith, Lycra Spandex Fiber, *Mod. Tex.,* **44,** 47 (1963); Hicks *et al.,* Spandex Elastic Fibers, *Science,* **147,** 373 (1965) (full account). Moncrief, *op. cit.* (1975) chap. 26.

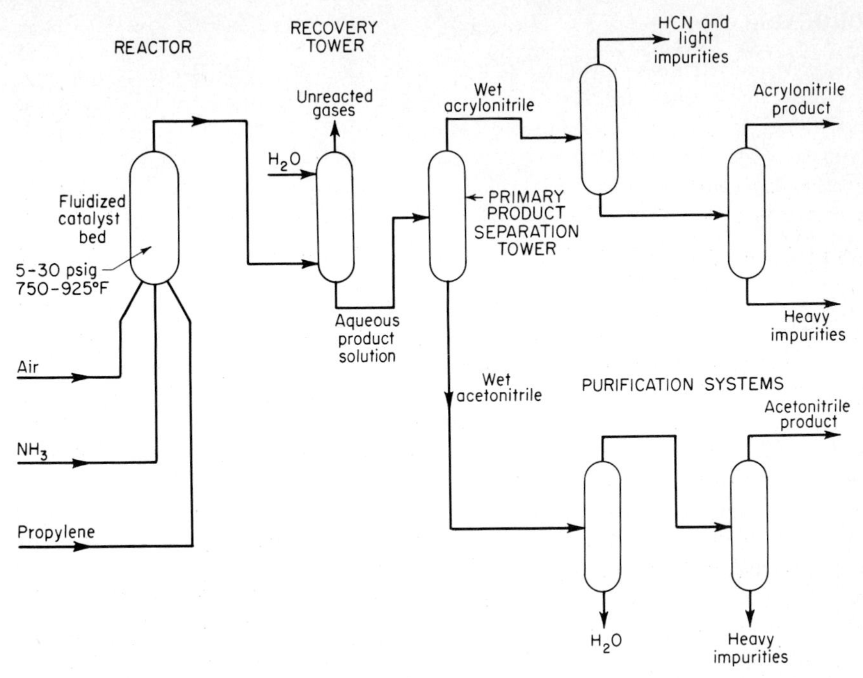

Fig. 35.6 Sohio acronitrile process. See text, reaction 3. One pound of propylene yields from 0.73 lb to an excess of 0.8 lb of acrylonitrile with no recycle, and from 0.02 to 0.11 lb of acetonitrile plus about 0.13 lb HCN. See Standard Oil Co. (Ohio) and U.S. patents, Idol, Pat. 2,904,580 (Sept. 15, 1959) and Callahan *et al.*, Pat. 3,044,966 (July 17, 1962).

The segmented polyurethane structure is achieved "by reacting diisocyanates with long chain glycols which are usually polyesters or polyethers of 1,000 to 2,000 molecular weight range." The product is then chain-extended or coupled through the use of glycol, a diamine, or perhaps water. This gives the final polymer, which is converted into fibers by dry spinning. The finished composite fiber is "one of tangled disorder," and, when stretched, the chains become oriented but exhibit a spontaneous return to the disordered state upon release of the force. This fiber is used in foundation garments, hose, swimwear, surgical hose, and other elastic products, sometimes in conjunction with other fibers, such as acrylic, polyester, acetate, rayon, or natural fibers, which are wrapped around a core of the elastic spandex fiber.

POLYOLEFINS

Polyolefin fibers and films are represented by polyethylene film, widely used as a wrap, and polypropylene fibers.[15] The latter have excelled in special cases, such as ropes, laundry nets, carpets, blankets, and backing for tufted carpets. In 1974 annual consumption in the United States was 526 million lb. Polyethylene film is extensively exploited as a heat- and water-saving mulch for cotton, strawberries, melons, and vegetables. Weed control is also claimed, and future growth appears promising.

[15]*Chem. Week*, Oct. 6, 1962, p. 89; Dyeing of Polypropylene for Textiles, *Am. Dyes. Rep.*, **54**(4), 34 (1965) (10 pp., many references); see Chap. 34.

FLUOROCARBONS

Teflon is polytetrafluorethylene [$(C_2F_4)_n$] (Chap. 20) and as fiber or film is nonflammable and highly resistant to oxidation and the action of chemicals, including strong acids, alkalies, and oxidation agents. It retains these useful properties from >450 to 550°F and is strong and tough. A very important property is its low friction, leading, with its chemical inertness, to wide use in pump packings and shaft bearings.

GLASS FIBERS

Fiberglass wearing materials were prepared as early as 1893, when a dress of fibers about five times the diameter of the present-day product was made by Owens. Since that time, however, numerous improvements have been discovered, until at present fibers as small as 0.00002 in. in diameter and of indefinite length are possible. The largest and original (1938) manufacturer of glass fibers is the Owens-Corning Fiberglas Corp., which markets its product under the trade name Fiberglas.[16] Production in 1974 was 683 million lb, with thermal and acoustical insulation accounting for about 56%, textiles for 35%, and numerous miscellaneous uses for 9%.

Glass fibers are derived by two fundamental spinning processes: blowing (glass wool) and drawing (glass textiles), as outlined in Fig. 35.7. Textile fibers[17] can be made as continuous filaments or staple fibers and are manufactured from borosilicate glasses composed of 55% silica and substantial amounts of alumina, alkaline earths for fluxing, and borates. The low strong-alkali level reduces susceptibility to corrosion. Glass for textiles must be equal in purity to the better grades of optical glass and, when molten, must be free of any seeds or bubbles that would tend to break the continuity of the fiber. In the continuous-filament process, specially prepared and inspected glass marbles are melted, or a batch of molten glass is particularly purified. Either is allowed to flow through a set of small holes (usually 102 or 204 in number) in a heated platinum bushing at the bottom of the furnace. The fibers are led through an "eye" and then gathered, lubricated, and put on high-speed winders, which rotate so much faster than the flow from the bushing that the filaments are drawn down to controlled diameters. It is used for electrical insulation in electric motors and generators, structural reinforcement for plastics, fireproof wall coverings, and tire cords.

In the production of staple fiber, glass marbles are automatically fed at regular intervals to a small electrically heated furnace. The molten glass discharges continuously through a spinneret. Directly below the orifice plate is a jet discharging high-pressure air or steam in such a manner as to seize the molten filaments and drag them downward, decreasing their diameter. The individual fibers are projected through the path of a lubricating spray and a drying torch onto a revolving drum. The resulting web of fibers on the drum is drawn off through guides and wound on tubes and, after drafting and twisting to form yarns, sent to weaving and textile fabrication. A new process treats the fabrics at high temperature to relax the fibers, thereby increasing the drapability and "hand" and removing all sizing from the yarns to make them receptive to finishing materials. Coloring can be achieved by the use of resin-bonded pigments applied during finishing. End-use applications include draperies, curtains, and bedspreads. With Owens-Corning's new Beta fiber yarns, the finest denier ever produced, other commercial uses, including wearing apparel, are being tested.

In the manufacture of glass wool for thermal and acoustical insulation, borosilicate glasses, composed of approximately 65% silica and 35% fluxing oxides and borates, are melted in large re-

[16]Games Slayter, inventor; Kozlowski, Fiber Glass Industry, *Glass Ind.* **41**(6), 1, 1960 (products, applications, companies, techniques, and properties); Koch and d'Hauteville, Glass Fibres, *Ciba Rev.*, 1963–1965; Glass Fibers Make Bid for Radial Tires, *Chem. Eng. News*, Oct. 12, 1970, p. 19.

[17]Cf. Chopey, Glass Fibers, *Chem. Eng.* (*N.Y.*), **68**(10), 136 (1961) (outline flowchart); Warlick, Carpet of Glass, *Textile Ind.* (*Atlanta*), January 1975, p. 40.

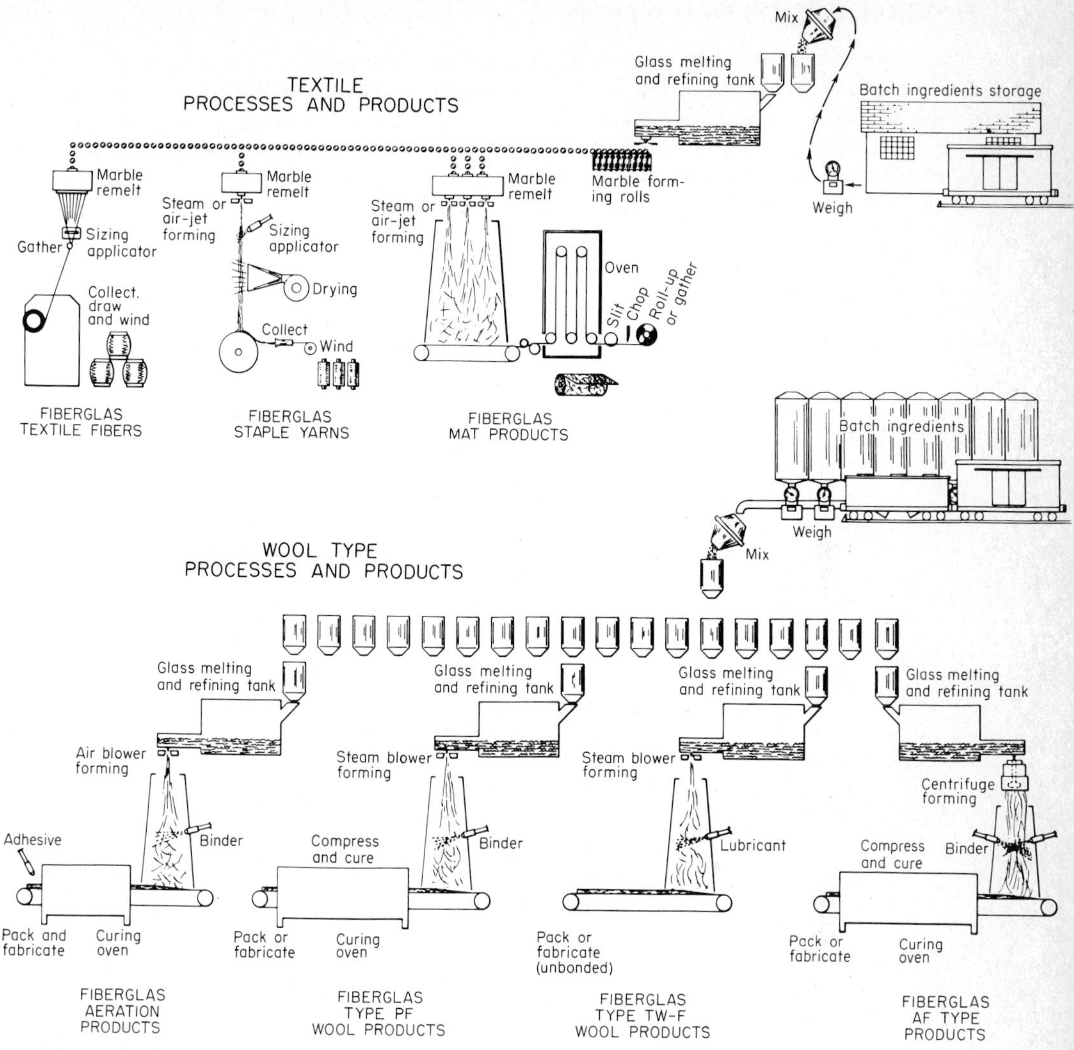

Fig. 35.7 Flowchart for Fiberglas production production. (*Owens-Corning Fiberglas Corp.*)

Basic Ingredients for Continuous-filament Fiber, %

Silicon dioxide	52–56	Boron oxide	8–13
Calcium oxide	16–25	Sodium and potassium oxides	0–1
Aluminum oxide	12–16	Magnesium oxide	0–6

generative furnaces. In blowing, molten glass at 2800°F streams through the small holes of a platinum alloy plate called bushings, or from a centrifuge, at one end of a standard glass furnace. The streams are caught by high-speed gaseous jets and yanked out into fibers that are hurled onto a moving belt. This woolly mass is impregnated with various binders and formed into countless shapes for use as insulation or, in the case of fibers as coarse as broom straw, set into frames for use as air filters. Glass-fiber mats and wool have now largely superseded rock wool for insulation.

CELLULOSIC FIBERS AND FILMS

HISTORICAL Cellulosics represent a mature industry in contrast to the rapidly developing polyfibers but are still very large in size. They were the first to be developed. In 1664, Hooke prophesied that "silk equal to, if not better than, that produced by the silkworm will be produced by mechanical means." Swan exhibited in London in 1885 nitrocellulose fabrics woven from fibers prepared by squeezing collodion through fine orifices. It remained for Chardonnet, "father of the rayon industry," to establish the first commercial unit for producing artificial silk, in 1891.

RAYON AND ACETATE

The basic process of producing cuprammonium rayon was patented in 1890, and the viscose process was discovered in 1892. The first patent for the production of cellulose acetate filaments was issued in 1894. In the United States commercial production of viscose rayon began in 1910, of nitrocellulose rayon in 1920, and of cellulose acetate in 1924. By 1926 the production of the various types of rayon and acetate exceeded the consumption of silk. Viscose, cuprammonium, and acetate yarns proved superior to denitrated nitrocellulose, which was discontinued in 1934.

USES, ECONOMICS, AND RAW MATERIALS In 1974, the U.S. production and consumption of rayon and acetate were as shown in Fig. 35.1. The 1974 world production of rayon and acetate amounted to 7,700 million lb. Percentagewise, according to process, this was viscose, 88, acetate and cuprammonium, 12. High-tenacity viscose yarn is used mainly in cords for tires, hose, and belting. The difference in strength between ordinary and high-tenacity viscose depends on the amount of orientation[18] imparted to the fiber molecules when they are made. The hydroxyl groups in the cellulose molecule allow water absorption to take place in the fiber and are the source of low wet strength. Hydroxyl groups serve as sites for hydrogen bonding and thus in the dry state serve to hold molecules together despite strong bending, resulting in fibers which tend to maintain their dry strength well even at high temperatures. The price of rayon and acetate varies according to the size of the filaments, process of manufacture, and type of finish. Textile rayon and acetate are used primarily in women's apparel, draperies, upholstery, and blends with wool in carpets and rugs.

Up to a few decades ago, the acetate process used only cotton linters as the source of cellulose. Now the viscose process is based on sulfite and a little sulfate pulp. If sheet cellulose, the form used in viscose manufacture, is desired, the sulfate pulp, after the bleaching treatment has been completed, is blended with several other batches, passed successively through a beater and a refiner (see Manufacture of Pulp for Paper in Chap. 33), and formed into sheets on a Fourdrinier.[19] Viscose rayon is a major consumer of sulfuric acid, caustic soda, and carbon disulfide. Titanium dioxide is used in delustering the yarn. Cellulose acetate employs large quantities of acetic anhydride, glacial acetic acid, sulfuric acid, and acetone. In addition to this important consumption of basic chemicals, the fiber industry needs significant quantities of dyes and other chemicals.

The viscose and the cuprammonium processes produce a filament of regenerated cellulose, and the acetate forms a thread that is a definite chemical compound of cellulose: cellulose acetate. Although each of these processes is quite different as far as details of procedure are concerned, they all follow the same general outline: solution of the cellulose through a chemical reaction, aging or ripening of the solution (peculiar to viscose), filtration and removal of air, spinning of the fiber, combining the filaments into yarn, purifying the yarn (not necessary for acetate), and finishing (bleaching, washing, oiling, and drying).

[18]Molecules lined up in one direction by stretching during spinning and drawing. See Fig. 35.8.
[19]ECT, vol. 17, p. 17, 1968.

REACTIONS

Viscose.[20]

$$\underset{\text{Cellulose}}{(C_6H_9O_4\cdot OH)_x} + xNaOH \longrightarrow \underset{\text{Alkali cellulose}}{(C_6H_9O_4\cdot ONa)_x} + xH_2O$$

$$(C_6H_9O_4\cdot ONa)_x + xCO_2 \longrightarrow \underset{\text{Cellulose xanthate}}{\left(C_6H_9O_4{-}O{-}C\begin{matrix}\nearrow S \\ \searrow SNa\end{matrix}\right)_x}$$

$$3\left(C_6H_9O_4{-}O{-}C\begin{matrix}\nearrow S \\ \searrow SNa\end{matrix}\right)_x + 2xH_2O \longrightarrow \underset{\text{Ripened xanthate}}{\left(C_{18}H_{27}O_{12}(OH)_2O{-}C\begin{matrix}\nearrow S \\ \searrow SNa\end{matrix}\right)_x} + 2xCS_2 + 2xNaOH$$

$$\left(C_{18}H_{27}O_{12}(OH)_2O{-}C\begin{matrix}\nearrow S \\ \searrow SNa\end{matrix}\right)_x + \frac{x}{2}H_2SO_4 \longrightarrow \underset{\text{Viscose rayon}}{(C_6H_9O_4\cdot OH)_{3x}} + xCS_2 + \frac{x}{2}Na_2SO_4$$

Cellulose acetate.

$$\underset{\text{Cellulose}}{[C_6H_7O_2\cdot (OH)_3]_x} + 3x(CH_3CO)_2O \longrightarrow \underset{\text{Cellulose acetate}}{[C_6H_7O_2(O_2CCH_3)_3]_x} + 3xCH_3CO_2H$$

Cuprammonium.[21]

$$\underset{\text{Cellulose}}{(C_6H_{10}O_5)_x} + \underset{\text{Schweitzer's reagent}}{Cu(OH)_2 + NH_3} + H_2O \longrightarrow \text{solution of cellulose}$$

$$\text{Cellulose solution} + NaOH\ (5\%) \longrightarrow \text{partly coagulated cellulose}$$

$$\text{Partly coagulated cellulose} + H_2SO_4(1\tfrac{1}{2}\%) \longrightarrow \underset{\text{Cuprammonium rayon}}{(C_6H_{10}O_5)_x}$$

VISCOSE MANUFACTURING PROCESS

The finished filament is pure cellulose, as shown by the equations but, because it consists of smaller molecules than the cellulose of the original wood pulp or cotton, it possesses different physical properties.

CHEMICAL CONVERSIONS AND UNIT OPERATIONS The process,[22] as shown in Fig. 35.8, can be broken down into the following sequences:

The cellulosic raw material (sheets made from sulfite or sulfate wood pulp) is charged to a steeping press containing vertical perforated steel plates and is steeped either batchwise or continuously in a caustic soda solution (17 to 20%) for about 1 h at 56 to 62°F (Ch) to dissolve α-cellulose.

[20]The cellulose molecule is composed of a large undetermined number of glucose units, here represented as $(C_6H_9O_4OH)_x$. The value of x does not remain constant throughout these reactions. Each reaction causes a reduction in the molecular weight of the cellulose molecule, so that the viscose-rayon molecule is considerably smaller than the original cellulose fed. Some CS_2 breaks away from the cellulose xanthate during the ripening process.

[21]The cuprammonium process with a flow diagram was described in CPI 2, pp. 753ff., ECT, vol. 17, p. 704, 1968. Production of this filament is minor at the present time.

[22]Burkholder and Siegman, Rayon's Latest Push into Fibers, *Chem. Eng. (N.Y.)*, **71**(14), 110 (1964); Cellophane and Viscose Rayon, *Chem. Metall. Eng.*, **46**, 25 (1939); Continuous Rayon Spinning Process, *Chem. Eng. (N.Y.)*, **54**(11), 103 (1947); Moncrief, *op. cit.*, Chap. 9.

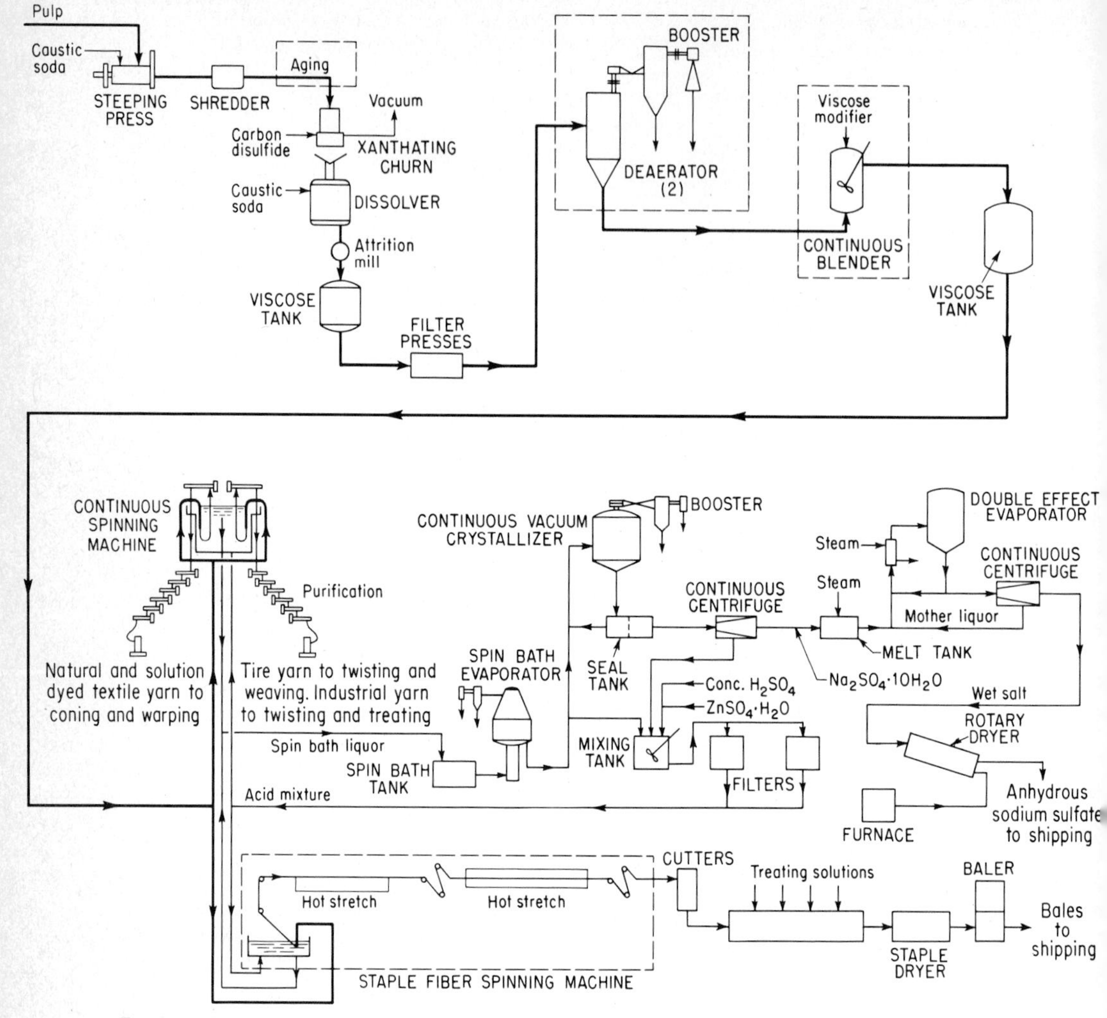

Fig. 35.8 Rayon's latest push into fibers. [*Chem. Eng.* (*N.Y.*), **71**(14), 110 (1964.)]

The excess liquor is drained off, thus removing inpurities such as cellulose degradation products.[23] The soft sheets of alkali cellulose are reduced to small crumbs in a shredder (Op). This requires 2 to 3 h, and the temperature is maintained at 65 to 68°F.

The crumbs of alkali cellulose are aged either batchwise or continuously for 24 to 48 h at 75°F in large steel cans. Some oxidation and degradation[24] occur, although the actual chemical change is unknown (Ch). Physically, correct aging produces a solution after xanthation of correct viscosity for spinning.

The aged crumbs are dropped into large, cylindrical xanthating churns. Carbon disulfide weighing between 30 to 40% of the dry recoverable cellulose is slowly added under carefully controlled

[23]See Lipsett, Industrial Wastes, chap. 4, Atlas, 1951, for chemical-waste recovery in the viscose-rayon industry. The Webcell dialyzer process is included.

[24]Lewis, Squires, and Broughton, Industrial Chemistry of Colloidal and Amorphous Materials, pp. 360ff., Macmillan, 1942.

temperature and reduced pressure over 2 h of churning, during which time the crumbs gradually turn yellow and finally deep orange, and coagulate into small balls (Ch):

$$xCS_2 + (C_6H_9O_4 \cdot ONa)_x \longrightarrow (C_6H_9O_4{-}O{-}CS{-}SNa)_x$$

Still in batch form, the cellulose xanthate balls are dropped into a jacketed dissolver (vissolver) containing dilute sodium hydroxide. The xanthate particles dissolve in the caustic, and the final product, viscose solution, contains 6 to 8% cellulose xanthate and 6 to 7% sodium hydroxide (Ch).

This reaction takes 2 to 3 h. If desired, delustering agents such as titanium dioxide or organic pigments[25] are added to the viscose solution in the mixer. The result is a sticky, golden-brown, viscous liquid. From here on this description (Fig. 35.8) through ripening and spinning is continuous (Ch).

During the ripening chemical reaction

$$3(C_6H_9O_4{-}O{-}CS{-}SNa)_x + 2xH_2O \longrightarrow [C_{18}H_{27}O_{12}(OH)_2{-}O{-}CS{-}SNa]_x$$

the proportion of combined sulfur decreases and the ease of coagulation increases (Ch). Thirty years ago, this conversion took 4 to 5 days; now improved technology has reduced it to about 24 h. In a series of tanks (only one is shown in Fig. 35.8), the reaction proceeds under deaeration and continuous blending with modifiers (viscose additives, mainly amines and ethylene oxide polymers) that control neutralization and regeneration rates (Ch).

Finally, in two continuous vacuum-flash boiling deaerators (at high vacuum and below room temperature), small air bubbles are removed that would either weaken the final yarn or cause breaks during spinning (Op).

Spinning is effected when thin streams of viscose from 750 to 2,000 holes in each thimblelike spinneret are injected under gear-pump pressure into the spin bath; they coagulate, and the cellulose is regenerated (Ch) to form fiber:

$$[C_{18}H_{27}O_{12}(OH)_2{-}O{-}CS{-}SNa]_x + \tfrac{1}{2}xH_2SO_4 \longrightarrow (C_6H_9O_4 \cdot OH)_{3x} + xCS_2 + \tfrac{1}{2}xNa_2SO_4$$

Sulfuric acid in the spinning bath neutralizes free NaOH and decomposes xanthate as shown, and various viscous by-products containing sulfur, liberating CS_2, H_2S, CO_2, and S. Salts such as $ZnSO_4$ and Na_2SO_4 coagulate the xanthate, forming relatively stable metal complexes. The sulfuric acid/salts ratio is a key control point, which, although coagulation and regeneration take place together, ensures that the xanthate gels before the acid can attack and decompose it. Four percent or more glucose prevents crystallization of salts in the filaments (Ch).

In Fig. 35.8 three separate continuous spinning and treating procedures are outlined for the coagulated filaments:

1. *Textile yarns* are twisted into continuous yarn as the filaments leave the spinneret. They are dyed (then or later) and sent to coning and warping (Op).

2. *Tire yarns* are stretched to impart strength over a series of thread-advancing reels where wash and other treatments are applied (deacidifying), desulfurizing (Na_2S), and bleaching (Op and Ch).

3. *Staple yarns* are spun on the machine in the lower part of Fig. 35.8 by combining filaments from many spinnerets without twisting and cutting them into uniform lengths. Each year more viscous filaments are made into staple fiber.

The actual process of continuous spinning has reduced the time elapsed from hours to minutes.

Three types of spinning are in general use: continuous, bucket, and bobbin. Three-quarters of United States production is from the long-used bucket type (Fig. 35.9).

[25]Fluidized Recovery Nabs CS_2, *Chem. Eng.* (*N.Y.*), **70**(8), 92 (1963).

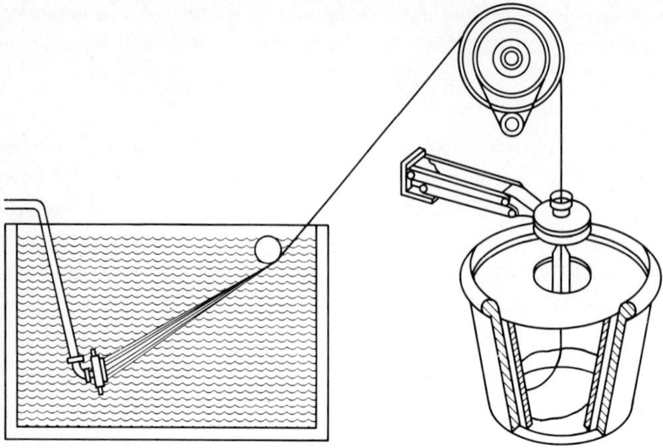

Fig. 35.9 Spinning in the viscose process.

The solution is extruded under pressure (gear pumps are used) through the spinneret into the spinning bath (Op), as described for the continuous process shown in Fig. 35.9. However, see Fig. 35.9 for general coagulation spinning.

The spinnerets are small caps of noble metal, containing minute holes through which the solution is extruded, as depicted in Fig. 35.9. Just ahead of the spinneret are the candle filters[26] to remove, in a final filtration, any foreign matter that might clog the holes.

If a bucket machine is used, the spinneret head dips horizontally into the spinning solution, and several of the filaments are gathered into a thread and fed down to a small centrifugal bucket spinning at about 7,500 rpm (Op). This is illustrated by Figs. 35.9. The bucket imparts one twist to the filaments per revolution and removes a greater portion of the occluded bath liquor through perforations in the periphery.

If a bobbin machine is used, the spinnerets point vertically upward into the spinning bath, and the filaments are wound on a revolving bobbin (Op). No twist is imparted to the thread.

Yarn from either type is washed to remove the spinning liquor (Op) and disulfurized by treating with a $\frac{3}{4}$ to 1% sodium sulfide solution (Ch).

Both types are washed (Op), bleached in hypochlorite solution (Ch), washed again, dried, and coned (Op).

PACKAGING FILMS Transparent viscose sheeting (cellophane) is manufactured from a solution similar to that used for rayon.[27] The spinning solution is extruded through a narrow slot into the coagulating bath and cast as a sheet onto a rotating drum, the lower side of which is submerged in the bath. The viscose film formed is transferred to succeeding tanks of warm water to remove the acid. It is desulfurized in a basic solution of sodium sulfide and rewashed. The characteristic yellow tinge must then be removed in a hypochlorite bleach bath, after which the sheet is rewashed. In order to impart softness and pliability, the film is passed through a glycerin bath, where it absorbs about 17% glycerin, after which it is dried. Much cellophane is subsequently made moistureproof by coating it on one side with a suitable nitrocellulose lacquer. Charch of the Du Pont Co. was awarded the 1932 Schoelkopf Medal of the Buffalo Section of the American Chemical Society for this achievement. Cellophane is widely used as a packaging and wrapping material. Viscose may be employed

[26]Candle filters are a thin roll of metal screen wrapped with filter cloth.

[27]Inskeep and Van Horn, Cellophane, *Ind. Eng. Chem.*, **44**, 2511 (1952) (many details, flowchart, pictures, and references); Hyden, Manufacture and Properties of Regenerated Cellulose Films, *Ind. Eng. Chem.*, **21**, 405 (1929).

to prepare a durable *cellulose sponge* by introducing a mixture of various sizes of Glauber's salt crystals, hemp or fibrous material, and viscose into a box which is placed in a coagulating bath to regenerate the cellulose. The blocks are then leached with warm water to remove the crystals of salt, and cut into small-sized blocks for sale.

CELLULOSE ACETATE MANUFACTURING PROCESS

Cellulose acetate (and its homologs) are esters of cellulose and not regenerated cellulose. The spinning solution may be used to produce fibers, transparent film, or photographic film, or precipitated to form molding powder for plastics and the basic constituent of cellulose lacquers.[28]

UNIT OPERATIONS AND CHEMICAL CONVERSIONS (Fig. 35.10) The raw material for the spinning solution is prepared by charging 100 lb of acetic anhydride, 100 lb of glacial acetic acid, and a small quantity of sulfuric acid as a catalyst to a jacketed, glass-lined, agitated, cast-iron acetylator (Op). The mixture is cooled to 45°F, and 35 lb of wood pulp is added slowly. The acetylation requires 5 to 8 h, and the temperature is maintained below 86°F (Ch).

The viscous fluid is diluted with equal parts of concentrated acetic acid and 10% sulfuric acid and allowed to age for 15 h at 100°F. Hydration of some of the acetate groups occurs (Ch). No method has been devised at present whereby cellulose can be converted directly to a product of the desired acetyl content. It is necessary to transform it first to the triacetate and then partly hydrolyze off the required proportion of acetate groups.[29]

The hydration is stopped by running the mixture into a large volume of water and precipitating the secondary acetate (Op). The secondary acetate is centrifugated to separate it from the still strong acetic acid which is recovered, concentrated, and used over as shown in Fig. 35.11.

The flakes are washed several times by decantation (Op) and are then ready to be used in preparing the spinning solution by dissolving the dry flakes in acetone in.a closed, agitated mixer. If desired, a delustering pigment is added (Op). Several batches are blended, filtered, and sent to the spinning machine (Op). The solution is forced through the spinnerets into a current of warm, moist air (Op). The acetone evaporates and is recovered, leaving a filament of cellulose acetate (Op).

These filaments are twisted and coned in the same manner as those of the previously described rayons (cf. Fig. 35.3) (Op). Some yarns are sold without a twist. Filament yarn is made by twisting

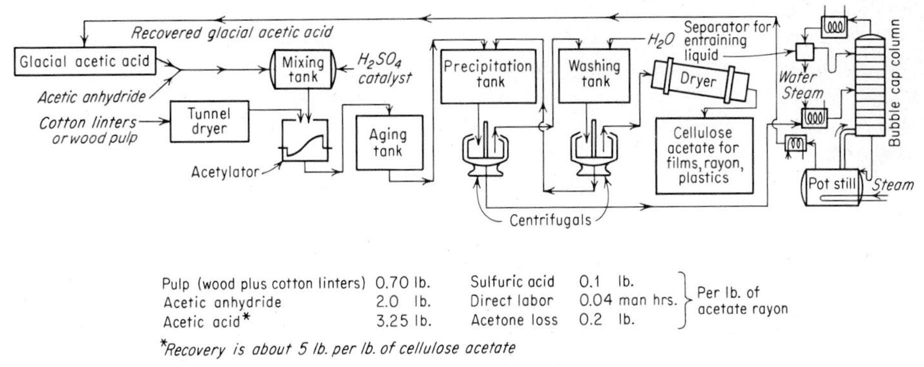

Fig. 35.10 Flowchart for cellulose acetate manufacture.

[28]Smith, Cellulose Acetate Rayons, *Ind. Eng. Chem.*, **32,** 1555 (1940); for a pictured flowchart on cellulose acetate staple yarn, see *Chem. Eng.* (*N.Y.*), **52**(1), 132 (1945).

[29]Now also the triacetate (Arnel) is dissolved and spun directly.

the threads before winding on the bobbin. Tow consists of threads gathered without twisting and is cut into short lengths for use as staple fibers.

The economical operation of the process depends on the recovery of as many of the chemicals as possible. For every pound of cellulose acetate about 4 lb of 30 to 35% aqueous acetic acid is obtained. Dilute acetic acid from various parts of the process is run through a Dorr Thickener to remove the last traces of acetate and then concentrated in a distilling unit and reconverted to acetic anhydride. The acetone-laden air from the spinning machines, for example, may be passed through activated charcoal to absorb the solvent (which is subsequently recovered by steaming and rectification) or by cooling the air in water towers and simultaneously dissolving out the acetone, the water-acetone mixture subsequently being rectified. Liquid absorption and distillation are also employed.

The manufacture of transparent cellulose acetate sheeting comparable with cellophane (viscose sheeting) is carried out by extruding an acetate "dope" of the required properties through a narrow slot onto a rotating drum, where, in the presence of warm air, the acetone solvent evaporates. The resulting sheet is pliable and moistureproof.

Cellulose triacetate fiber, Arnel, is spun and sold by Celanese Corp. It is reported to possess "resistance to glazing at high ironing temperatures, complete machine washability, low shrinkage in stretching, good crease and pleat retention, and an adaptability to a wide range of colors, designs and prints."[30] It is reported to be considerably superior to normal acetate in fiber properties and uniformity. The tricky part is to manufacture pure triacetate and then to dissolve it prior to spinning. The catalyst employed here is perchloric acid (versus sulfuric for ordinary acetate).

FINISHING AND DYEING OF TEXTILES

The textile industry, with sales of over $30 billion in 1976 is second in size to the food industry, and many of its mill operations abound in chemical-engineering problems.[31] Dyeing;[32] bleaching; printing; special finishing, such as for crease recovery, dimensional stability, resistance to microbiological attack and to ultraviolet light, and flame resistance; scouring; water treatment; and waste disposal are outstanding examples of mill treatments where such unit operations as filtering, heating, cooling, evaporation, and mixing are involved. Some mill operations are discussed in a general manner elsewhere, e.g., under water treatment and waste disposal in Chap. 3, but here brief mention is given to the use of special finishing agents. Textile processors and finishers spend over $1 billion for chemicals and dyes in each year.

The modification[33] of fibers and fabrics by special treatments to change their properties and to improve their usefulness is increasing yearly. Three important finishes consist of *flameproofing*, or fire retarding, *mildew-*, or *rotproofing*, and *water repellency*. These are by no means the only lines of treatment. Temporary flameproofing of cellulosic fibers is achieved by the application of ammonium salts or borax and boric acid. Ideal fabric flameproofing, which allows cleaning or laundering and yet maintains desirable fabric characteristics, is yet to be achieved, although much research is directed toward this aim[34] and some processes are finding commercial acceptance. Mildewproofing of cellulosic fabrics may be obtained by the use of many organic and inorganic compounds. Commonly used

[30] Grove *et al.*, Fibers, *Ind. Eng. Chem.*, **47,** 1973 (1955); Moncrief, *op. cit.*, Chap. 12.

[31] Chemical Engineering in the Textile Industry, *Chem. Eng.* (*N.Y.*), **55**(3), 127 (1948).

[32] Rose, Color Rides Synthetic Fibers, *Chem. Eng.* (*N.Y.*), **68**(6), 84 (1961); Manly, Durable Press Treatments of Fabrics, Noyes, 1976; Szilard, Bleaching Agents and Techniques Noyes, 1973.

[33] Smith, Use of Textile Auxiliaries, *Chem. Eng. News,* **29,** 548 (1951); Chemical Processing, *Chem. Eng. News,* **30,** 4153 (1952); Johnson, Antistatic Compositions for Textiles and Plastics, Noyes, 1976; Casper, Nonwoven Textiles, Noyes, 1975.

[34] Little, Flameproofing Textile Fabrics, Reinhold, 1947; Symposium, Flame Retarding of Textiles, *Ind. Eng. Chem.*, **42,** 414 (1950); Flame Resistant Textiles, *Chemist,* August 1970, p. 300; Textile Makers Spur Flammability Research, *Chem. Eng. News,* Mar. 30, 1970, p. 34.

Fig. 35.11 The manufacture of acetic anhydride and the recovery of acetic acid, important stages in the manufacture of acetate rayon, take place in these towering columns and stills at the Waynesboro, Va., plant of E. I. du Pont.

materials include acrylonitrile, chlorinated phenols, salicylanilide and organic mercurial compounds, copper ammonium fluoride, and copper ammonium carbonate. To produce water-repellent finishes durable to the usual cleaning processes, special quaternary ammonium compounds are heat-treated on the fiber. Shrinkproofing of wool employs various chlorinating processes, especially for socks, shirts, knitting yarns, and blankets. Another method for shrinkproofing woven fabrics is to coat them with a melamine formaldehyde product. Thermosetting resins are being widely used to impart crease or wrinkle resistance to cellulosic fibers. Commonly used products include urea-formaldehyde and melamine-formaldehyde resins. The fabric is treated with water-soluble precondensates, together with a condensation catalyst. The treated fabric is dried and heated at an elevated temperature to develop the resin within the fiber structure. Many other special treatments for fabrics include mothproofing, improving of resiliency, stiffening, softening, eliminating electrostatic charge during processing, sizing, lubricating, and inhibiting atmospheric gas fading. In recent years chemical finishes have actually been used to react with the fiber material, e.g., cotton, and thereby to change its properties by esterification (carboxymethylation) or amination (2-aminoethylsulfuric acid).

SELECTED REFERENCES

Alexander, A.: Man Made Fiber Processing, Noyes, 1966.
American Association of Textile Chemists and Colorists, Technical Manual, Howes, 1960 (annual).
Barden, E. L.: Flocked Materials, 1972, Noyes, 1972.
Carroll-Porczynski, C. Z.: Natural Polymer Man-made Fibers, Academic, 1961.
Casper, M. S.: Synthetic Turf and Sporting Surfaces, Noyes, 1972.
———: Nonwoven Textiles 1975, Noyes, 1975.
Continuous Glass Fibers, American Elsevier, 1973.
Cook, J. G.: Handbook of Polyolefin Fibers, Merrow Publishing, Watford, England, 1966.
Ferry, J. D.: Viscoelastic Properties of Polymers, Wiley, 1961.

Hearle, J. W. S., and R. H. Peters (eds.): Fibre Structure, Butterworth, London, 1964.
Information Services Group: Nonwoven Fabrics to 1978, Fairfield Associates, 1973.
Johnson, K.: Antistatic Agents, Noyes, 1972.
———: Dyeing of Synthetic Fibers—Recent Developments 1974, Noyes, 1974.
Jones, H. R.: Pollution Control in the Textile Industry, Noyes, 1973.
Kaswell, E. R.: Wellington-Sears Handbook of Industrial Textiles, New York, 1963.
Linton, Natural and Man-made Textile Fibers, Meredith, 1966.
Loewenstein, K. L.: The Manufacturing Technology of Continuous Glass Fibers, American Elsevier, 1973.
McDonald, M.: Nonwoven Fabric Technology 1971, Noyes, 1974.
Moncrieff, R. W.: Man-made Fibres, 6th ed., Wiley, 1975.
Morgan, P. (ed.): Glass Reinforced Plastics, 3d ed., Wiley, 1961.
Nissan, A. H.: Textile Engineering Processes, Wiley, 1960.
Ott, E.: Cellulose and Cellulose Derivatives, 2d ed., Interscience, 1954–1955, 3 vols.
Peters, R. H.: Textile Chemistry: Chemistry of Fibres, Elsevier, 1963.
Press, J. J. (ed.): Man-made Textile Encyclopedia, Interscience, 1960.
Rothenberg, G. B.: Glass Technology, Noyes, 1976.
Sitting, M.: Acrylonitrile 1965, Noyes, 1965 (patents).
———: Caprolactams, Higher Lactams, Noyes, 1965 (patents).
———: Dibasic Acids, Anhydrides, Noyes, 1966 (patents).
———: Synthetic Fibers from Petroleum, Noyes, 1967 (patents).
———: Polyamide Fiber Manufacture 1972, Noyes, 1972.
———: Acrylic and Vinyl Fibers 1972, Noyes, 1972.

chapter 36

RUBBER INDUSTRIES

Rubber has become a material of tremendous economic and strategic importance and is an excellent barometer of the industrialization of nations. In the United States, the per capita consumption of rubber is approximately 34 lb per person; in India it is scarcely $\frac{1}{2}$ lb per person. Transportation, the chemical, electrical, and electronic industries, and the space effort are all major consumers of rubber. When supplies of natural rubber were shut off because of Japan's invasion of rubber-producing areas early in World War II, the United States built up a synthetic rubber industry which has continued to expand enormously, so that at the present time 70% of the rubber consumed in the United States is of synthetic origin. The rubber industry involves the production of monomers or raw materials for natural and synthetic rubbers, the various rubbers themselves, the importation of natural rubber, the production of rubber chemicals, and finally the fabrication of rubber products.

HISTORICAL Columbus found the natives of the West Indies playing games with rubber balls. Rubber articles have been recovered from the sacred well of the Maya in Yucatan. Rubber, as we know it, is a product of the Americas but has achieved its greatest growth by transplantation to the Far East. The name "rubber" was apparently given by Priestley, the discoverer of oxygen, who first observed the ability of the material to "rub out" a pencil mark. The fact that rubberlike materials resulted from mere efforts to purify and keep such materials as styrene, butadiene, and the isoprene produced from the destructive distillation of natural rubber very early led to attempts to produce synthetic rubber. By the outbreak of World War I, inferior grades of rubber were being produced from dimethylbutadiene in Germany and in the U.S.S.R. Extensive research during the 1920–1930 period led to discovery of the emulsion copolymerization of butadiene and styrene, and of butadiene and acrylonitrile.

Goodyear is credited with the discovery of the cure, or vulcanization, of rubber with sulfur in 1839. This overcame the natural tackiness of rubber and commercialized it. Since then it has been found that many substances affect the rate of this reaction and that some subsidiary materials modify or improve the compound, making it possible to shape or fabricate an article which is then fixed into its final form by the curing reaction. The historical occurrence of most significance was the curtailment of natural rubber importation caused by Japanese invasions in 1941. This stimulated greatly both research and manufacture of various syntehtic rubbers in the years following.

STATISTICS AND ECONOMICS

The cost of natural rubber has been subject to great fluctuations over the years. The price stability of the synthetics has undoubtedly contributed to their successful invasion of the market. Table 36.1 shows the consumption of various rubbers in the United States over a 22-year period, and it is clearly seen that natural rubber is being supplemented, but not supplanted, by synthetics, and that the world would be hard-pressed to exist with natural rubber as the only source of supply.

TABLE 36.1 ***Total U.S. Consumption of Natural, Synthetic, and Reclaimed Rubber*** *(In thousands of long tons)*

Year	*Natural rubber*	*Synthetic rubber*	*Reclaimed rubber*	*Total consumption*
1950	720	538	304	1,562
1955	635	895	313	1,961
1960	479	1,079	277	2,303
1965	515	1,540	270	2,657
1970	559	1,918	200	3,001
1972	640	2,292	188	3,405

Source: Statistical Abstract of the U.S., 1974.

NATURAL RUBBER

Dandelion, guayule, goldenrod, osage orange, and numerous other plants have been proposed as sources of rubber, but none has proved to be as successful as the latex-producing tree *Hevea brasiliensis,* a native of America, which has flourished in plantings in the Far East and is now the source of most of the world's natural rubber supply, except for minor collections in Brazil and other South American and Central American countries, and the collection of balata and chicle from other species of trees. Natural rubber has become established on plantations in Malaya, Indonesia, Liberia, and their neighbors, probably because of freedom from insect and fungus diseases which beset the tree in its native areas in the Americas. Trees require approximately 7 years to reach bearing age and may then continue to yield for a number of years. Yields have been increased by the selection of improved stock to produce approximately 1,500 lb/(acre)(year). This was possible during World War II. It is believed that as much as 3,000 lb/acre may be obtained from the improved types of trees now being planted. The collection and processing of the latex require considerable labor, and the crop is obviously adaptable to the densely populated areas of the Far East and Africa but may not prove to be attractive in this hemisphere.

Latex is obtained by *tapping* the tree in such a manner as to allow the liquid to accumulate in small cups, which must be collected frequently to avoid putrefaction or contamination. The latex is carried to collection stations, where it is strained, and a preservative (NH_3) may be added. The rubber is separated by a process known as *coagulation,* which occurs when various acids or salts are added and the rubber separates from the liquid as a white, doughlike mass, which is then milled and sheeted to remove contaminants and to allow drying. The rubber of commerce is shipped in bales of convenient size, and is sufficiently stable to permit stockpiling for a period of several years.

The exact processes by which the rubber tree produces the material proved to be *cis*-polyisoprene (Table 36.2) are not known, but progress is being made in the biochemistry of latex production sufficiently to indicate that the starting material may be acetic acid. The rubber exists in minute globular particles suspended in the serum base. Latex is not the sap of the tree but serves some other function not fully understood. Considerable latex is concentrated to as high as 60 to 70% solids and shipped in that form for use in the production of foam, latex-dipped goods, adhesives, etc.

The natural-rubber molecule has now apparently been exactly duplicated by the solution polymerization of isoprene with a Ziegler-type (Chap. 29) catalyst, and with improved methods for isoprene synthesis this product is making inroads in the natural-rubber market. Natural rubber is still used in the United States for truck tires because of its outstanding heat-buildup resistance. Other reasons for its continued use are its excellent properties, such as a gumstock, and overall balance of properties. See Table 36.2 and Stereospecific Rubbers.

TABLE 36.2 ***Monomers and Polymers or Copolymers of Rubbers***

Monomer	*Polymer or copolymer unit*	*Rubber*
Isoprene, Hevea latex	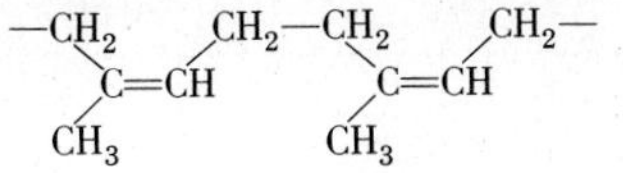 $-CH_2(CH_3)C{=}CH-CH_2-CH_2(CH_3)C{=}CH-CH_2-$ *cis*-1,4-Polyisoprene	Natural, NR 5,000 isoprene units
Isoprene 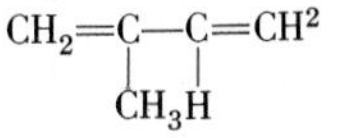 $CH_2{=}C(CH_3)-C(H){=}CH^2$		Synthetic polyisoprene, IR
Butadiene $CH_2{=}CHCH{=}CH_2$	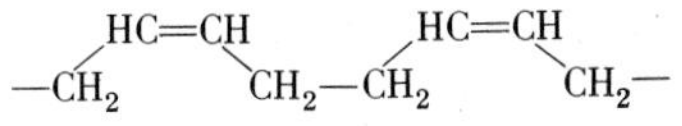 $-CH_2-HC{=}CH-CH_2-CH_2-HC{=}CH-CH_2-$ *cis*-1,4-Polybutadiene	
Styrene 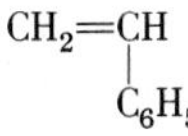 $CH_2{=}CH(C_6H_5)$		
Butadiene-styrene	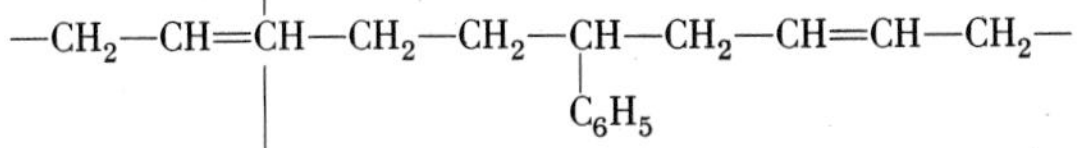 $-CH_2-CH{=}CH-CH_2-CH_2-CH(C_6H_5)-CH_2-CH{=}CH-CH_2-$	SBR (GR-S)
Acrylonitrile $CH_2{=}CHCN$		
Butadiene-acrylonitrile	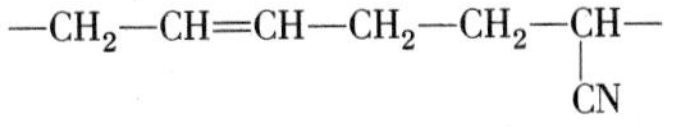 $-CH_2-CH{=}CH-CH_2-CH_2-CH(CN)-$	Nitrile, NBR
Chloroprene $CH_2{=}CCl-CH{=}CH_2$	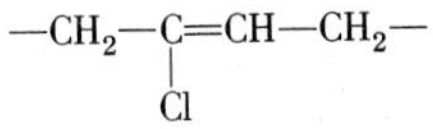 $-CH_2-C(Cl){=}CH-CH_2-$	Neoprene, CR
Isobutylene-isoprene	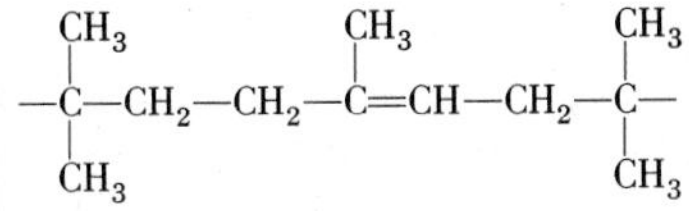 $-C(CH_3)_2-CH_2-CH_2-C(CH_3){=}CH-CH_2-C(CH_3)_2-$	Butyl, IIR
Sodium tetrasulfide-ethylene-dichloride	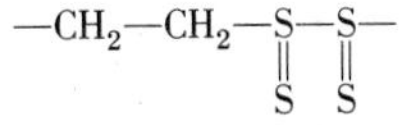 $-CH_2-CH_2-S({=}S)-S({=}S)-$	Thiokol A
Dimethylsiloxane 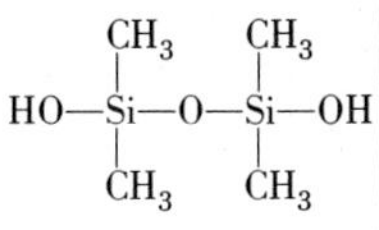 $HO-Si(CH_3)_2-O-Si(CH_3)_2-OH$	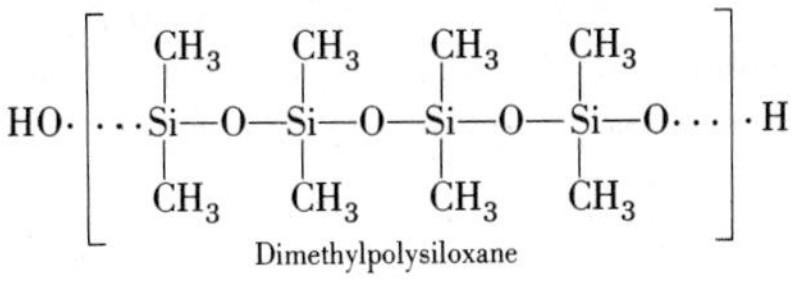 $HO\cdot[\cdots Si(CH_3)_2-O-Si(CH_3)_2-O-Si(CH_3)_2-O-Si(CH_3)_2-O\cdots]\cdot H$ Dimethylpolysiloxane	Silicone R

SYNTHETIC RUBBERS[1]

The realization that supplies of natural rubber could be shut off by enemy action led to the planning of a synthetic rubber industry in the United States prior to the outbreak of World War II, but the factories were completed only after the rubber supply had actually fallen into Japanese hands. The

[1]See Perry, Table 23-13, for physical properties of natural and synthetic rubbers; also Rubber Science and Technology in the Seventies, *Chemtech*, **1**, 401 (1971).

TABLE 36.3 ***Classification of Rubbers***

Class I. Elastomers
A. Vulcanizable
  1. Diene rubbers
  2. Nondiene rubbers
B. Nonvulcanizable and other elastomers

Class II. Hard plastics
Class III. Reinforcing resins
Class IV. Paint vehicles

Rubbers are further classified and coded from the chemical composition of the polymer chain in the following manner:

M Rubbers having a saturated chain of the polymethylene type
N Rubbers having nitrogen in the polymer chain
O Rubbers having oxygen in the polymer chain
R Rubbers having an unsaturated carbon chain; e.g., natural rubber and synthetic rubbers derived at least partly from diolefins
Q Rubbers having silicone in the polymer chain
T Rubbers having sulfur in the polymer chain
U Rubbers having carbon, oxygen, and nitrogen in the polymer chain

The R class shall be defined by inserting the name of the monomer or monomers from which it was prepared before the word "rubber" (except for natural rubber). The letter immediately preceding the letter R shall signify the diolefin from which the rubber was prepared. Any letter or letters preceding this diolefin letter signify the comonomer or comonomers.

ABR Acrylate-butadiene
BIIR Bromoisobutene-isoprene
BR Butadiene
CIIR Chloroisobutene-isoprene
CR Chloroprene
IIR Isobutene-isoprene
IR Isoprene, synthetic
NBR Nitrile-butadiene
NCR Nitrile-chloroprene
NIR Nitrile-isoprene
NR Natural rubber
PBR Pyridine-butadiene
PSBR Pyridine-styrene-butadiene
SBR Styrene-butadiene
SCR Styrene-chloroprene
SIR Styrene-isoprene

Source: ASTM Standards, Rubber and Rubber Products D1418-72a. (Reprinted by permission of the ASTM from copyrighted material.)

rubber first chosen for production was called GR-S (government rubber-styrene), admittedly derived from the German Buna-S and now falling under the category SBR. It was necessary to develop facilities for the production of butadiene and styrene as well as for the rubber product.

MONOMER PRODUCTION

The cost of the raw material monomer is usually from 60 to 80% of the cost of production of a rubber polymer. Economical methods for the production and purification of the monomer are essential. If complete conversion of the monomer to the polymer is not possible, provision for the

TABLE 36.4 ***Materials Consumed in Tires and Tire Products*** *(Estimated) (In millions of pounds)*

	1970	*1975*		*1970*	*1975*
NR	810	1,000	Plasticizers, oils	266	330
SBR	1,720	1,900	Carbon black	1,600	2,000
IIR	170	190	Inorganic pigments (mostly ZnO)	200	270
BR	560	900	Thermoplastic resins	26	32
EPDM	40	80	Thermosetting resins	3.0	3.0
IR	150	260	Rayon tire cord	100	150
Reclaim	260	260	Nylon tire cord	270	270
Masterbatch (rubber content)	175	190	Polyester tire cord	150	250
Rubber processing			Glass fiber	50	70
chemicals (except			Steel wire	5	100
sulfur and fatty acids)	190	240	Bead wire	150	190

Source: Chem. Week, May 19, 1971, p. 55.

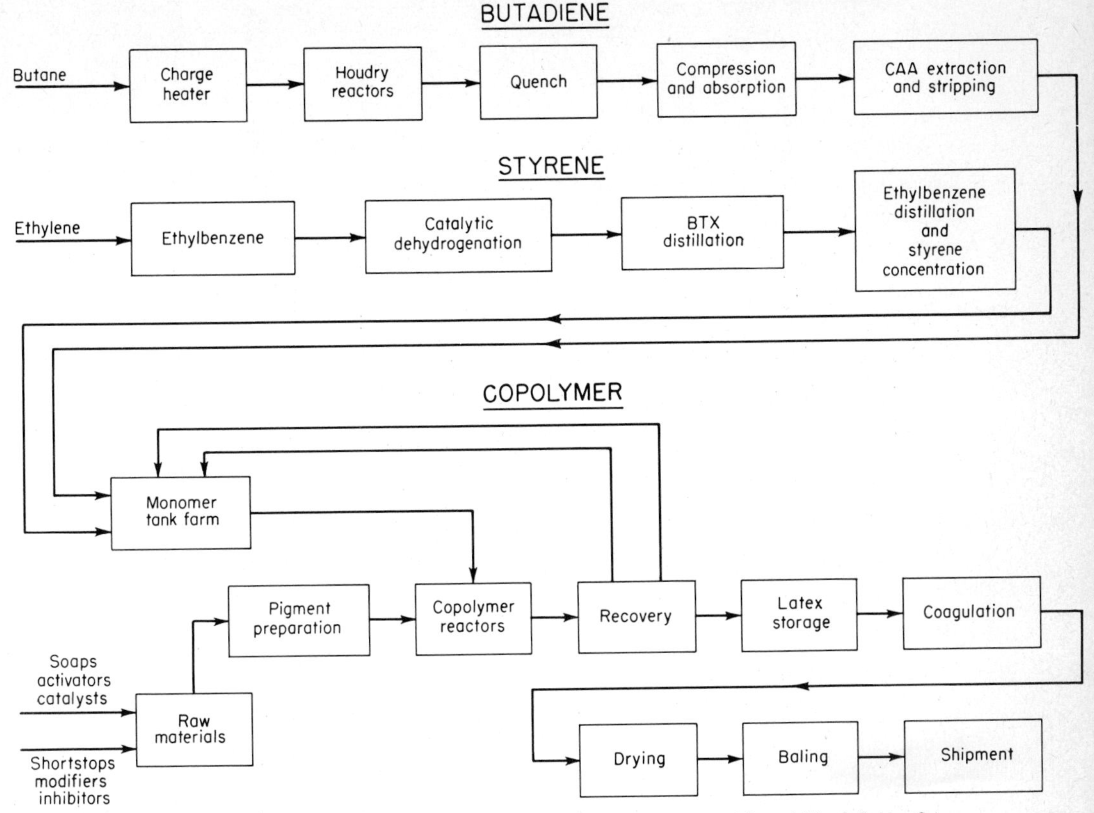

Fig. 36.1 Simplified flowchart for SBR manufacture. CAA, Cuprous ammonium acetate. (*General Tire & Rubber Co.*)

recovery and purification of unreacted monomer is usually necessary (Figs. 36.4 and 36.8). Some monomers and the related synthetic rubbers are given in Tables 36.2 and 36.4

Butadiene of rubber-grade purity has been produced by several methods,[2] including: (1) Dehydration of ethyl alcohol. (2) Hydrolysis of acetylene to acetaldehyde, followed by reaction with ethylene oxide to form acetaldol, which is then hydrogenated to 1,3-butylene glycol, and subsequently dehydrated to butadiene. (3) The reaction of acetylene with formaldehyde, followed by hydrogenation to 1,4-butanediol. The 1,4-butanediol is then hydrolyzed to butadiene. (4) From starch fermentation to 2,3-butanediol, which is then esterified to the diacetate and pyrolyzed to butadiene. (5) Catalytic dehydrogenations of normal butane or of butylenes available from refinery light-end fractions, which are the principal methods used in the United States at present. Typical flowcharts for butadiene production from *n*-butane are shown in Figs. 36.1 and 36.2.

Styrene is a monomer not only for rubber but is also used in the production of polystyrene plastics. In 1975, the total styrene capacity was almost 7.1 billion lb, and it sold it at 22 cents per pound. 3.9 billion lb was consumed in polystyrene and 1.2 billion lb in SBR copolymer. ABS resins used 16%, polyester resins 6%, styrene-acrylonitrile resins 2%, and other uses 2%. The predominant route to the production[3] of styrene is via ethylbenzene, which is alkylated from benzene

[2]ECT, 2d ed., vol. 3, p. 784,; Phillips, New Butadiene Process, *Hydrocarbon Process.*, **53**(6) 133 (1974).

[3]Styrene flowcharts; *Hydrocarbon Process.*, **52**(11), 179 (1973); Extract Styrene from Pyrolysis Gasoline, *Hydrocarbon Process.*, **52**(5), 141 (1973).

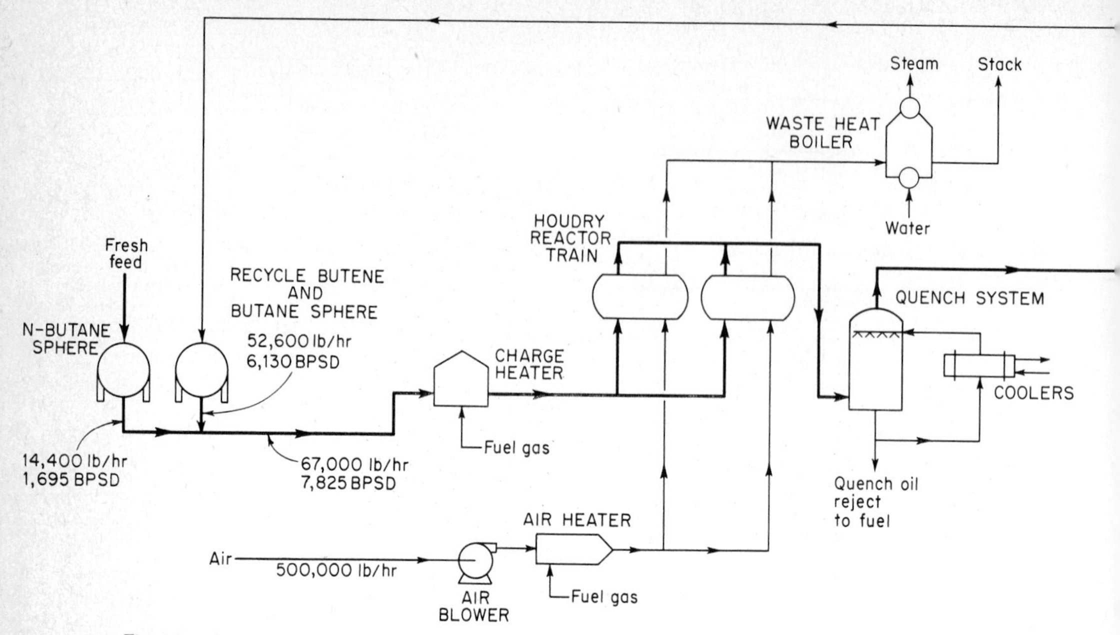

Fig. 36.2 Flowchart for the production of 29,000 metric tons of butadiene: the Houdry process for butadiene by dehydrogenation of *n*-butane. The same process can also be used for manufacturing isoprene from isopentane. (*Catalytic Construction Co.*)

with ethylene and subsequently dehydrogenated to styrene over an aluminum chloride catalyst, solid phosphoric acid, or silica-alumina catalysts (Figs. 36.1 and 36.3):

$$C_6H_6 + C_2H_4 \longrightarrow C_6H_5C_2H_5 \longrightarrow C_6H_5CH \quad CH_2 + H_2$$

Acrylonitrile has principally been made from acetylene and hydrogen cyanide, which are reacted with a liquid catalyst consisting of cuprous chloride, hydrogen chloride, and alkali chlorides at approximately atmospheric pressure:

$$CH\vdots CH + HCN \xrightarrow{CuCl_2,\ etc.} CH_2 \quad CHCN$$

A new process for manufacturing acrylonitrile from propylene, ammonia, and air has led to successful commercial operations at Standard Oil Co. (Ohio). It is understood that the catalyst is molybdenum and bismuth (Chap. 35).

Chloroprene, the monomer from which neoprene is produced, is manufactured from acetylene and hydrogen chloride. Acetylene is dimerized to monovinylacetylene, which is then reacted with hydrogen chloride to form the chloroprene monomer:

$$2CH\cdot CH \xrightarrow[65\text{–}75^\circ C,\ 2\text{–}5\ psig]{aq.\ Cu_2Cl_2, NH_4Cl} \underset{\text{Vinylacetylene}}{CH_2{:}CH\cdot C \quad CH} \xrightarrow{HCl} \underset{\text{Chloroprene}}{CH_2{:}CCl\cdot CH{:}CH_2}$$

Contact time 10–15 s, conversion 20% per pass, yield 60–65%

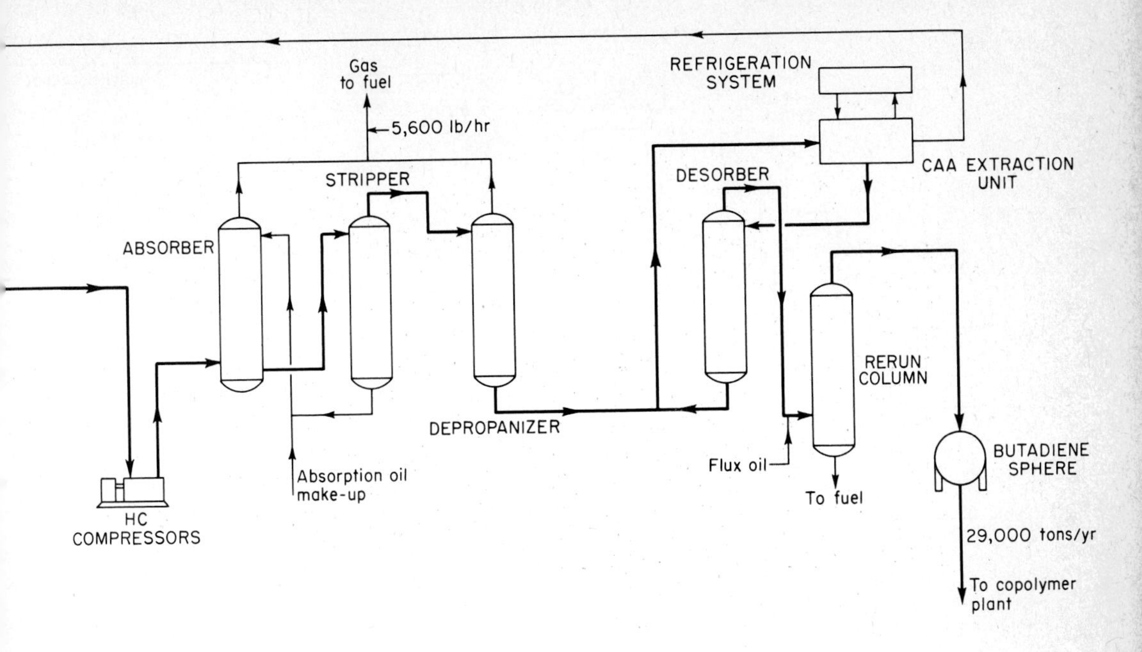

A process starting with butadiene is also commercial.

Reactions A

$CH_3—CH═CH_2$
Propylene

↓ dimerization

$CH_2═C(CH_3)—CH_2CH_2CH_3$
2-Methyl-1-pentene

↓ isomerization

$CH_3—C(CH_3)═CHCH_2CH_3$
2-Methyl-2-pentene

↓ pyrolysis

$CH_2═C(CH_3)—CH═CH_2 + CH_4$
Isoprene Methane

Reactions B

$(CH_3)_2C═CH_2 + 2HCH(═O)$
Isobutylene Formaldehyde

↓ condensation

$(CH_3)_2 > C$ ring with $CH_2—CH_2$, O, $O—CH_2$
4,4-Dimethyl-*m*-dioxane

↓ decomposition

$CH_2═C(CH_3)—CH═CH_2 + HCH(═O) + H_2O$
Isoprene

Reactions C

$HC⋮CH + CH_3—C(═O)—CH_3$
Acetylene Acetone

↓ condensation

$CH_3C(CH_3)(OH)—C≡CH$
2-Methyl-3-butyn-2-ol

↓ hydrogenation

$CH_3—C(CH_3)(OH)—CH═CH_2$
2-Methyl-3-butene-2-ol

↓ dehydration

$CH_2═C(CH_3)—CH═CH_2 + H_2O$
Isoprene

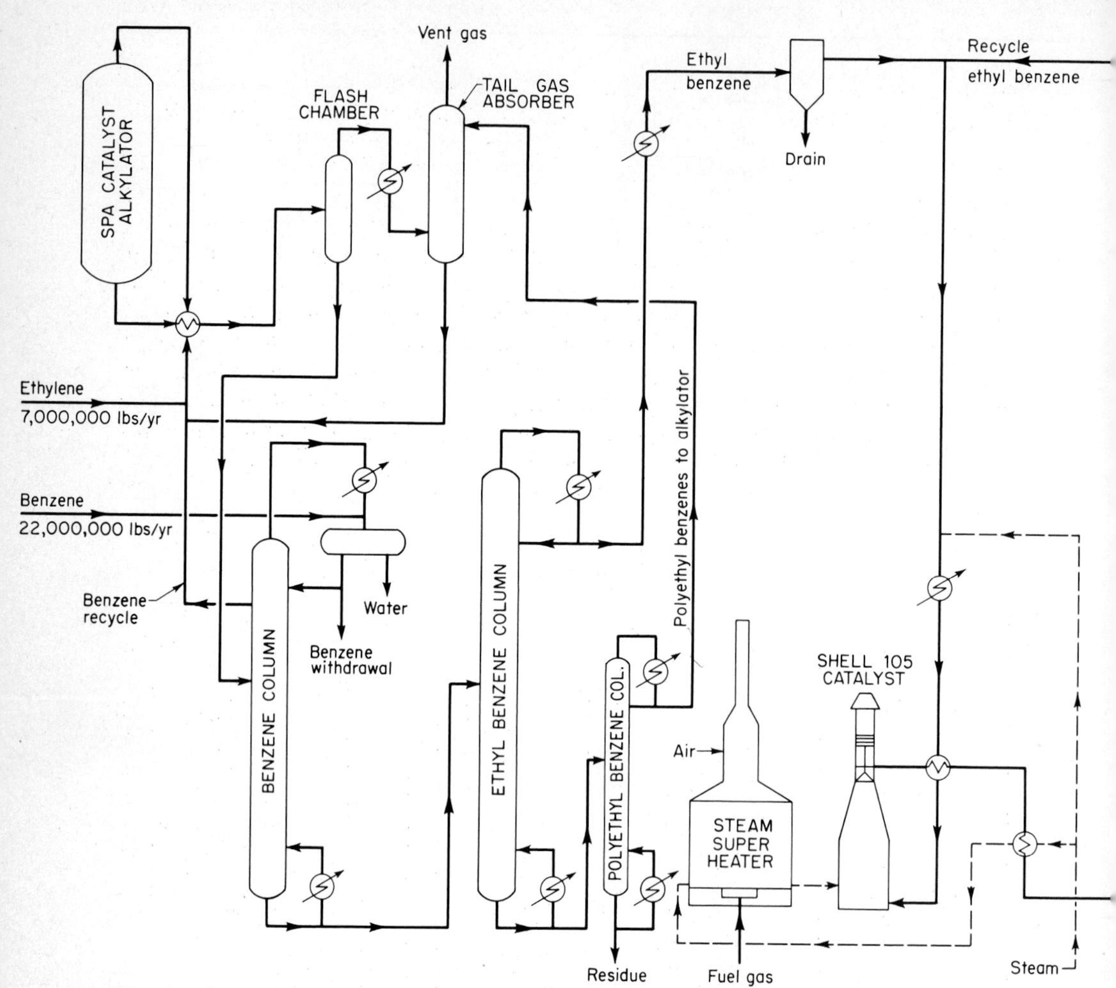

Fig. 36.3 Styrene, 9,000 metric tons from an ethylbenzene plant in West Texas. (*Petrobras.*)

Isobutylene is recovered from refinery light-end operations by conventional methods. It is the monomer for butyl rubber. See Table 36.2.

Isoprene may be produced by the dehydrogenation of isopentane in the same plant as that shown in the butadiene flowchart in Fig. 36.2. However, the presence of piperylene (for which there is no market) requires an expensive purification step. The Goodyear S-D method produces isoprene from propylene (reactions *A*). Production costs for this process are potentially quite low, and the raw material is cheap and plentiful.[4] Isoprene can be produced from isobutylene and methanol (reactions *B*). Isoprene recovered is said to be of exceptional purity. This process has been developed by the Institut Français du Pétrole. Reactions *C*[5] use 970 kg acetone, 430 kg acetylene, 400 m^3 hydrogen, and 5 kg ammonia to produce 1,000 kg of isoprene.

[4]ECT, vol. 12, p. 64, 1967; *Chem. Week*, May 6, 1961, p. 73; *Chem. Eng. News*, May 15, 1961, p. 35 (flowchart with quantities and utilities), *Pet. Refiner*, **42**(11), 187 (1963); Isoprene Technology Faces New Influences. *Chem. Eng. News*, May 17, 1965, p. 50; Better Route to Isoprene, *Hydrocarbon Process.*, **53**(7), 121 (1974).

[5]Unusual Reactants, New Isoprene Process, *Chem. Eng. (N.Y.)*, **71**(20), 78 (1964) (flowchart).

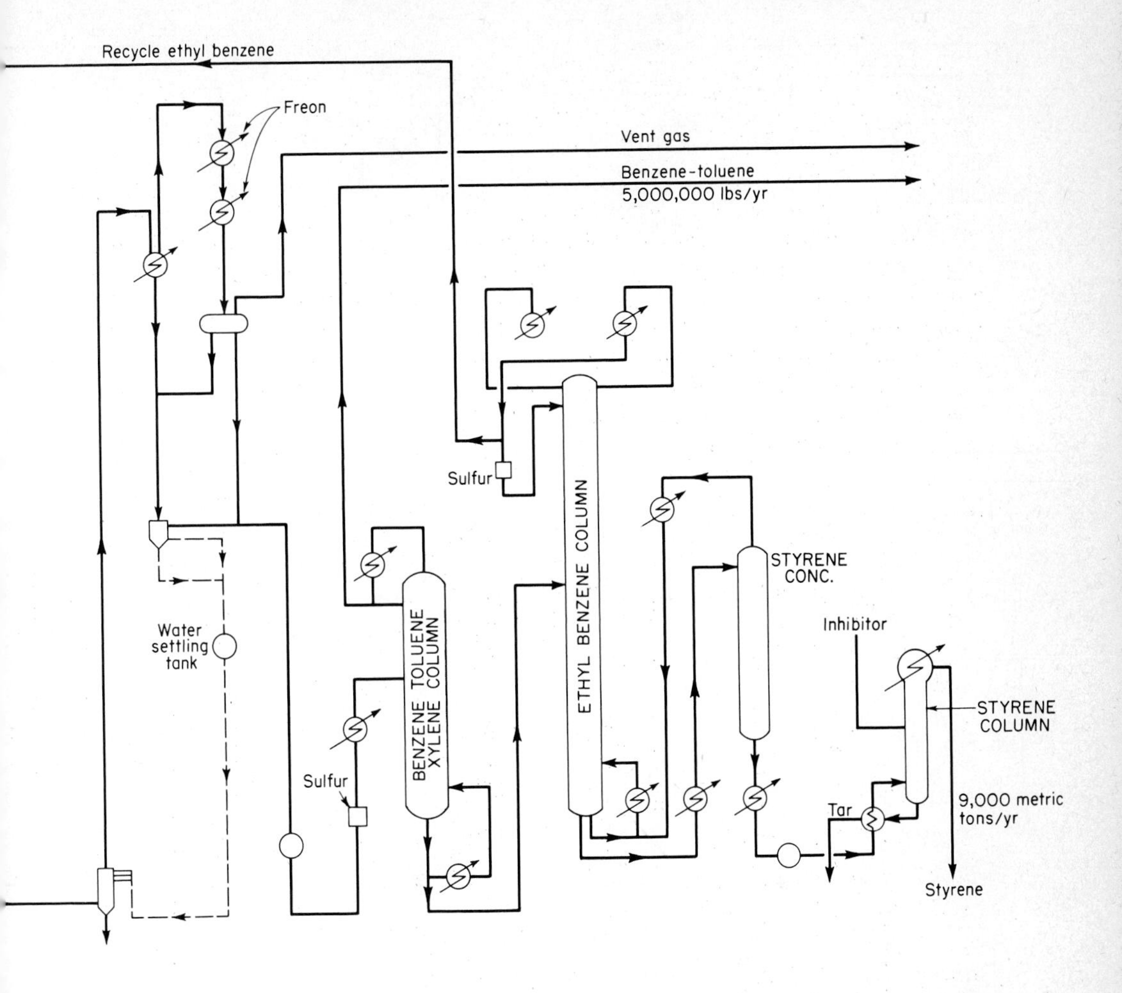

Ethylene and propylene are readily available from refinery light-end cuts or may be produced from the cracking of propane (Chap. 37). The potential low cost of these materials as monomers for rubber and the very favorable results obtained indicate a considerable future for rubbers produced from these cheap petrochemicals. It is necessary to add a third monomer to introduce unsaturation in the chain, and these include cyclopentadiene, dicyclopentadiene, and similar chemicals containing *nonconjugated* double bonds which provide sites for sulfur vulcanization or curing. Such rubbers are called EPT (ethylenepropylene tripolymers).[6] These may become competitors of SBR rubber for tires. Polyethylene itself is the raw material for Hypalon, a specialty rubber of outstanding resistance to chemical attack and fair oil resistance, made by chlorosulfonation of the polymer in a solvent reaction medium.

[6]Natta and Sartori, New Ethylene-propylene Elastomers, *Pet. Refiner,* **41,** 8, 9 (1962); Vaaler Award, *Chem. Process.* Nov. 20, 1964, p. 65; *Chem. Week,* Nov. 24, 1964, p. 67; EP Rubber, *Chem. Eng. Prog.,* **58**(7), 33 (1962); Brennan, Terpolymer Rubber, *Chem. Eng.* (*N.Y.*), **72**(14), 94 (1965) (pictured flowchart and conditions).

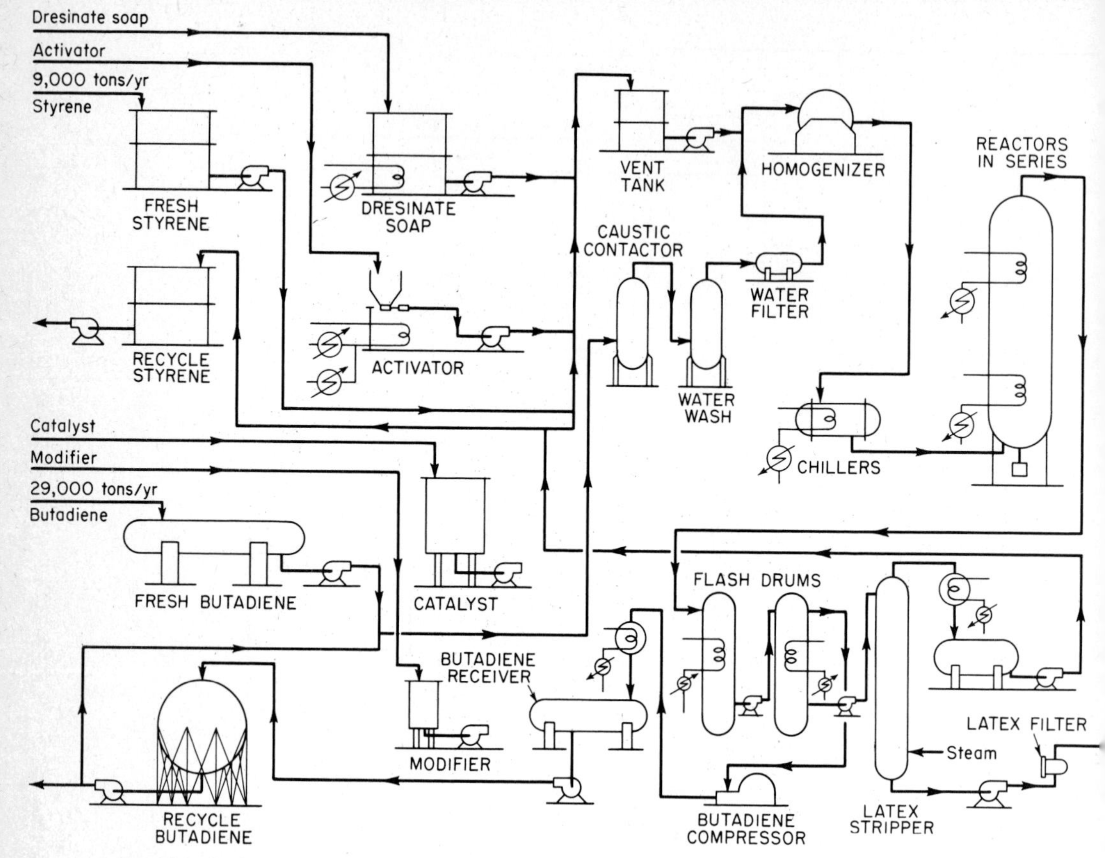

Fig. 36.4 Copolymers, etc., 40,000 metric tons. (*General Tire & Rubber Co.*)

SYNTHETIC-RUBBER POLYMERIZATION

The various methods of production of synthetic rubbers have much in common. The monomers are not particularly difficult to handle at reasonable pressures, and suitable inhibitors have been developed to impart storage stability. Flowcharts in the conventional simplified equipment form showing the material balance for the widely used SBR (Figs. 36.1, 36.4, 36.5, and 36.8) are typical, and several rubbers can be made interchangeably in such a plant. Dissipation of the heat of polymerization is frequently the controlling consideration. Adjustment of reaction rate to distribute the heat generation over a reasonable period of time, the use of refrigeration such as may be provided by ammonia coils within the reactors, and operation in dilute media such as emulsions or solutions are necessary for the adequate control of polymerization reactions (Fig. 36.5).

Control of molecular weight[7] and of molecular configuration have become the predominant quality considerations. See Stereospecific Copolymers. The ability to control molecular weight has led to the development of *oil-extended rubbers* (Fig. 36.4), which have had far-reaching effects on the industry. In late 1950, the General Tire & Rubber Co. applied for a patent and disclosed its

[7]MWD, or molecular weight distribution, is needed to control costs and quality. New means for this analysis are now available: *Chem. Eng. News*, July 16, 1965, p. 40.

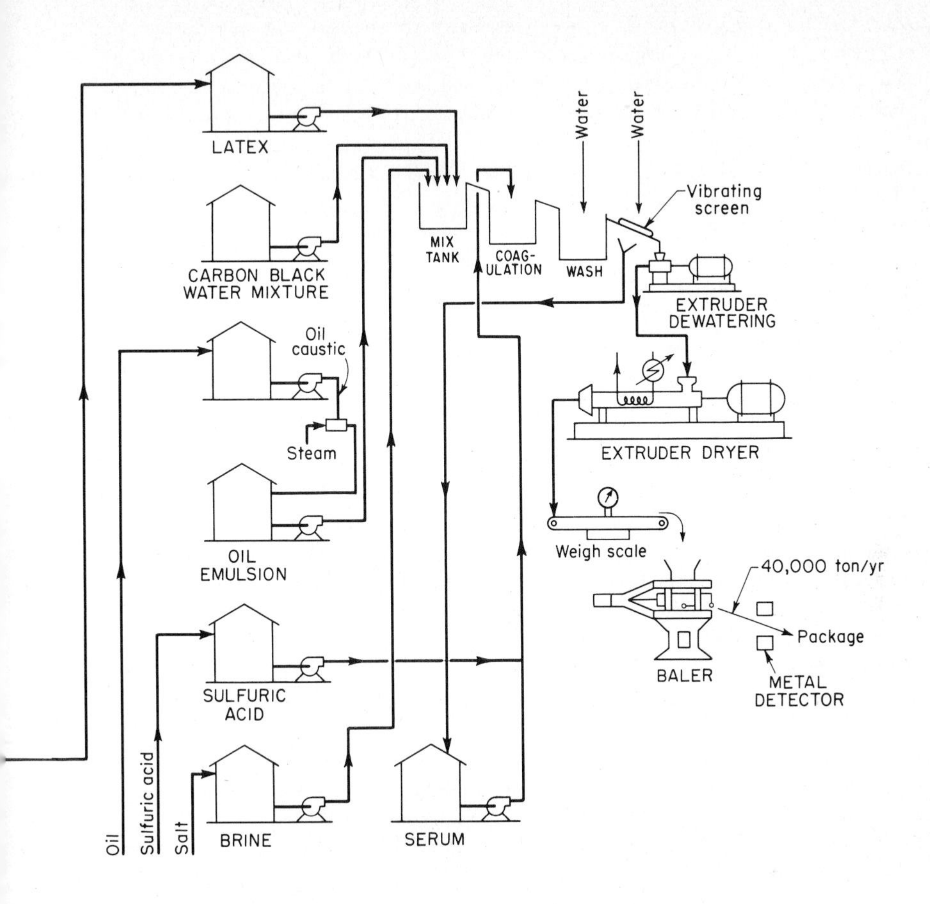

Fig. 36.5 Synthetic-rubber reactors operated in series for a continuous reaction of SBR. (*General Tire & Rubber Co.*)

process for the addition of petroleum oils to high-molecular-weight rubber. Not only does the oil render the extremely tough, high-molecular-weight rubber processible, but it allows much of the quality inherent in the tougher material to be retained and actually results in improvements in quality. Amounts of oil up to 50 parts per 100 parts rubber have become commonplace, and approximately 70% of the SBR produced in the United States is oil-extended, finding its principal use in tire-tread rubber.[8] First successfully applied to SBR, the process has assumed great importance as ways to add oils to the newer stereoregular polymers have been discovered.

MONOMER RECOVERY The rubber quality and rate of reaction both fall off as polymerization proceeds, and it is customary to stop the reaction short of complete conversion to rubber. Recovery of the unreacted monomers and purification is an essential step in economical synthetic-rubber production (Figs. 36.4, 36.8, and 36.9). Methods of recovery by steam stripping from aqueous latices or by distillation from solvent systems are employed. In some cases, recovery of the monomer can be accomplished during the drying step in a devolatilizing extruder dryer as pictured in Figs. 36.6 and 36.7. Water or solvent is removed by a combination of mechanical squeezing and by passage through a vacuum section. The space, maintenance, and labor requirements of such dryers are sufficiently less than for the conventional multipass apron dryer, so that this unit has been installed in several new rubber plants.

COAGULATION AND DRYING The finishing process usually consists of precipitating the rubber from the latex emulsion or from the solvent solution in crumb form; it is then dried and compressed into a bale (Figs. 36.6 and 36.7). Ordinary latices may be easily coagulated by the addition of sodium chloride and dilute sulfuric acid, alum, or virtually any combination of electrolyte and dilute acid. The rubber cements which result from solution polymerization can be precipitated into crumb form by adding the solution to a tank of hot water under violent agitation, with or without the addition of wetting agents, to control crumb size and prevent reagglomeration. The coagulated crumb is separated from the serum and washed on vibrating screens or rotating filters and dried at appropriate temperatures. Rubber is a difficult material to dry, and care must be taken not to overheat it or otherwise cause its deterioration. Drying times in conventional hot-air apron dryers may be as long as several hours.

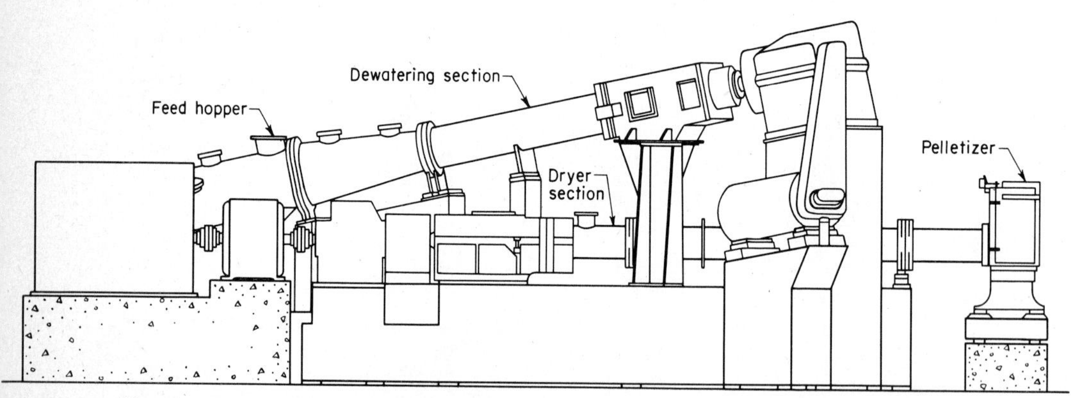

Fig. 36.6 *Welding Engineers'* 6-in. dual worm rubber dryer, or dewaterer, Model SPD-4000. This model accepts a slurry feed of 50% or more solids containing 1 to 2% hydrocarbons and devolatilizes and dries to less than 0.3% volatiles. Throughput rates are up to 5,000 lb/h, with continuous feed, using 650 hp and 300,000 Btu/h jacket heating. Typical elastomers handled by this equipment are butyl, *cis*-polybutadiene, and *cis*-polyisoprene.

[8]The Great General Tire Patent Suit, *Rubber World,* **151**(1), 33–39 (1964); U.S. Pat. 2,964,083 (Dec. 13, 1960).

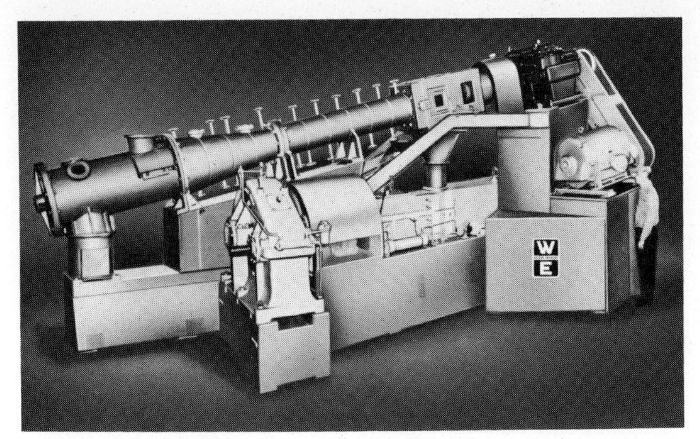

Fig. 36.7 Mechanical dewatering by an extruder-dryer capable of removing water or solvent from 6,000 to 8,000 lb/h of product. (*Welding Engineers, Inc.*)

PACKAGING Synthetic-rubber packaging is important, since problems involving adherence to the package and contamination resulting from inadequate protection are sometimes difficult to cope with because of the tendency of some of the rubber to flow and fail to retain its shape. Most rubber is now wrapped in polyethylene film as a 75-lb bale of reasonably standard dimensions. In most cases, it is possible to throw bale and polyethylene wrapping into the rubber mill or internal mixer, since the polyethylene melts at a temperature lower than the usual rubber-processing temperatures.[9]

BUTADIENE-STYRENE COPOLYMERS[10]

Copolymers of butadiene and styrene of over 50% butadiene content are known as SBR, formerly GR-S. The usual monomer ratio is 75 parts butadiene to 25 parts styrene, which results in a rubbery copolymer. As the styrene content is increased above 50%, the product becomes increasingly plastic in nature and finds use in latex paint. A typical formula for SBR is shown in the form of a material balance based on 100 lb of monomer charge. This is customarily called *one formula weight*, representing 100 parts of pure monomer and the associated materials charged with them. The formula shown in the material balance and flowchart (Fig. 36.8) is for the so-called *cold SBR* developed during World War II, and supplanted the original material, which utilized a slightly different catalyst system and polymerized at 122°F. Figure 36.8 may be used to demonstrate a useful method of handling the problem of continuous polymerization. A quantitative presentation of a steam balance for an SBR plant is shown in Fig. 36.9. Emulsion polymerization is carried out in reactors similar to those shown in Fig. 36.5, which are operated in series for reaction times of approximately 8 to 12 h. The heat of polymerization is approximately 550 Btu/lb of rubber produced and is removed by internal ammonia coils. Molecular weight is regulated by the use of a modifier such as a tertiary C_{12} mercaptan, which regulates chain growth. The reaction is ordinarily carried to only 60% conversion of the monomer to rubber, since the rate of reaction falls off beyond this point and quality begins to deteriorate. At the conclusion of the desired reaction time, a short stop is added, and the unreacted monomers are removed by steam distillation.

Latex[11] is accumulated in holding tanks and blended to the proper specifications for the grade of rubber required. It is coagulated by the addition of salt, sulfuric acid, alum, or similar materials

[9]Rubber Handling: Crumb to Bale, *Hydrocarbon Process.*, **53**(3), 79 (1974).
[10]Process Computers Crack SBR, *Chem. Eng. News*, Oct. 22, 1962, p. 62 (for analog computers).
[11]Suspension of rubber particles 400 to 2,000 Å in size.

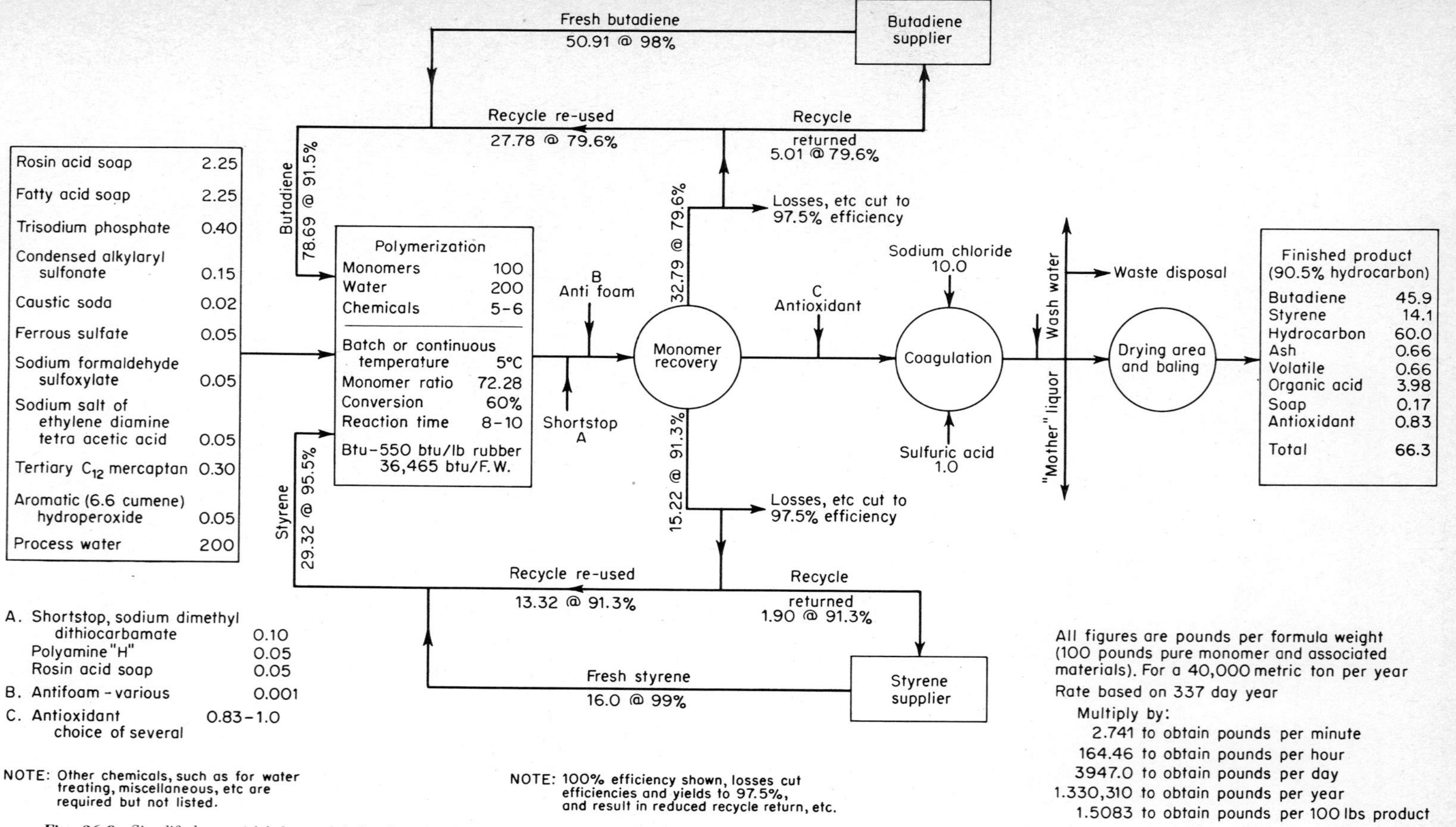

Fig. 36.8 Simplified material balance and flowchart for SBR manufacture. This can be used for computations of capacities and sizing of a typical SBR plant. (*General Tire & Rubber Co.*)

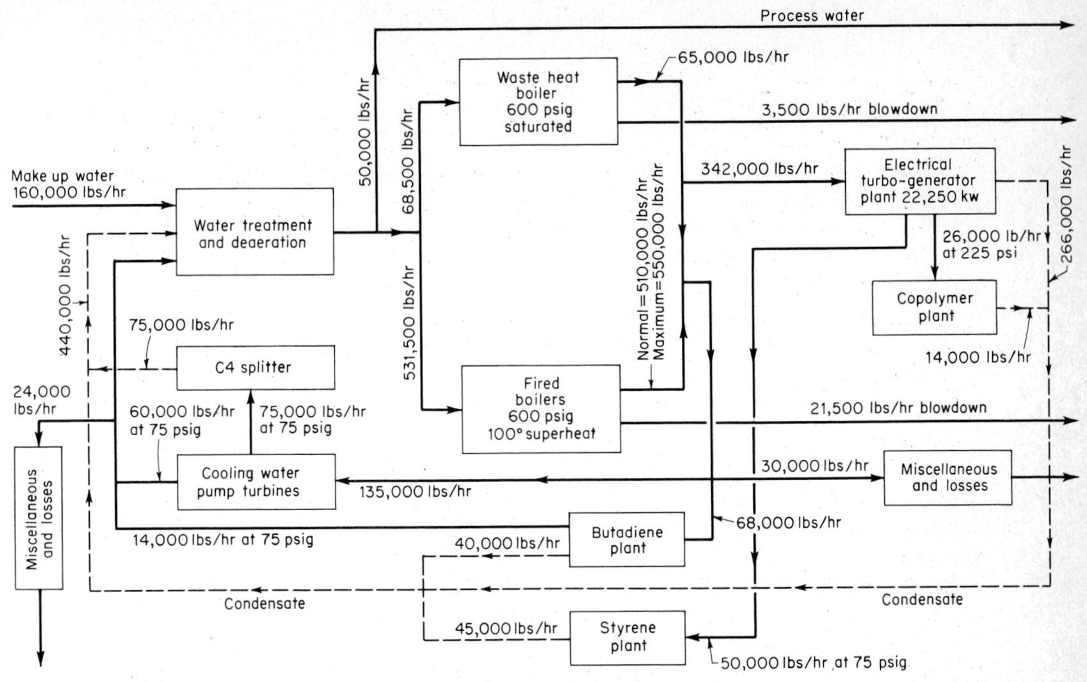

Fig. 36.9 Steam balance for SBR plant. (*General Tire & Rubber Co.*)

and extensively washed and dewatered prior to the drying operation. The development of the synthetic rubber industry during World War II was based on the original Buna-S formula, called GR-S, which was subsequently improved by the introduction of low-temperature polymerization using a redox catalyst system at 41°F. The next principal development was oil extension. Much SBR is made in the form of masterbatches.[12] Latex may be mixed with carbon black or oil before coagulation (Fig. 36.4), resulting in much more intimate mixtures than may be obtained by subsequent addition of these materials.

Oil extension was developed by the General Tire & Rubber Co. during 1950 and has rapidly assumed major importance, since oil-extended rubber is not only cheaper than rubber alone but actually provides quality advantages. Note that the greatest growth has occurred under private ownership as opposed to government control. *SBR may be said to be a general-purpose synthetic, finding use in automobile tires and mechanical goods.* Natural rubber and the various *cis*-polybutadiene rubbers have better heat-buildup properties and are used for truck tires or other thick sections. The latter competes with SBR pricewise and propertywise for use in tires. SBR *latices* can be produced in an ordinary copolymer plant, such as that shown in the flowchart in Fig. 36.4, with the addition of equipment for concentration by removal of some of the water (Figs. 36.6 and 36.7). The use of latex is an expanding development of synthetic-rubber technology, and this material is found in foam, adhesives, cements, dipped goods, carpet backing, and many similar products.

BUTADIENE-ACRYLONITRILE RUBBERS

Copolymers of butadiene and acrylonitrile made in a manner similar to that described for SBR production possess uncommonly good resistance to oil and other chemicals and are widely produced for

[12]Greek, Masterbatching, *Ind. Eng. Chem.*, **55**(7), 13 (1963).

this purpose as specialty rubbers. Terpolymers of butadiene, acrylonitrile, and styrene[13] are also produced for their outstanding impact resistance and have become known as ABS polymers.

NEOPRENE

Chloroprene rubbers were among the first synthetics produced in the United States. They were developed by Du Pont and sold commercially in 1932 under the name DuPrene[14] (Table 36.2). Following purification by fractionation, the chloroprene monomer is emulsified in water and polymerized at about 100°F under atmospheric pressure in the presence of sulfur, with potassium persulfate as a catalyst. Neoprene is a particularly versatile elastomer, because it has a combination of properties suited for many and varied applications, including those involving extreme service conditions, where the product must withstand a combination of deteriorating conditions such as oil and heat or extreme exposure to weather.

THIOKOL

Thiokol, or the polysulfide type of rubber, was first developed in this country in the early 1920s and may be said to be the first commercial synthetic rubber produced in the United States. Thiokols are prepared by the condensation polymerization of an alkaline polysulfide with a suitable organic dihalide, for example:

$$\underset{\text{Dichloroethyl formal}}{n\mathrm{ClCH_2CH_2OCH_2OCH_2CH_2Cl}} + \underset{\text{Sodium polysulfide}}{n\mathrm{Na_2S_2}} \longrightarrow \underset{\text{Thiokol (rubbery polymer)}}{(-\mathrm{CH_2CH_2OCH_2OCH_2CH_2CH_2SS}-)_n} + \underset{\text{Sodium chloride}}{2n\mathrm{NaCl}}$$

Various sodium polysulfides may be reacted with various organic dihalides to give specialty rubbers of somewhat different properties, particularly useful for linings for petroleum tanks, for building and caulking putties, cements, and sealants, and lately for rocket-fuel binders, ablation coatings, and other items requiring ease of application and good weathering resistance.

SILICONE RUBBERS

Silicone rubbers are mixed inorganic-organic polymers produced by the polymerization of various silanes and siloxanes. Although expensive, their outstanding resistance to heat makes them uniquely useful for high-temperature applications. The chain is based on alternate atoms of silicon and oxygen, with no carbon. Silicones and their derivatives are remarkable in the variety and unusualness of their properties, such as solubility in organic solvents, insolubility in water and alcohols, heat stability (−80 to 400°F), chemical inertness, high dielectric properties, relatively low flammability, solutions low in viscosity at high resin content and with low viscosity change with temperature, and nontoxicity. Because of these properties, silicones are useful in (1) fluids for hydraulics and heat transfer, (2) lubricants and greases, (3) sealing compounds for electrical applications, (4) resins for lamination and for high-temperature-resistant varnishes and enamels, (5) silicone rubber, (6) water repellents, and (7) waxes and polishes, etc.

Silicons[15] are made either by the Grignard reaction or more economically by alkylating with an organic halide, usually CH_3Cl (or C_6H_5Cl or mixtures thereof), a silicon-copper alloy, according to

[13]Chopey, Three Polymerization—Times and Yield ABS, *Chem. Eng.* (*N.Y.*), **69**(19), 154 (1962) (flowchart of graft polymerization).

[14]Carothers, *J. Am. Chem. Soc.*, **53**, 4203 (1931).

[15]ECT, vol. 18, p. 221, 1964; see Chap. 35 for plastic and insulation use; Rochow, An Introduction to the Chemistry of the Silicones, 2d ed., Wiley, 1947; Hyde, Chemical Background of Silicones, *Science*, **147**, 829 (1965).

the following reaction:

$$\text{Si (10\% Cu)} + n\text{CH}_3\text{Cl} \xrightarrow[48\ \text{h}]{200^\circ\text{C, 1–2 atm}} (\text{CH}_3)_n\text{SiCl}_{(4-n)}$$

$$\text{Principal products: } \underbrace{(\text{CH}_3)_2\text{SiCl}_2,\ (\text{CH}_3)_3\text{SiCl},\ \text{CH}_3\text{SiCl}_3}_{\text{Chlorosilanes}}$$

Chain length is varied by the percentage of R_3SiCl, which provides the end groups. Lower polymers are oils; higher polymers are solids. The silicon or copper may absorb any excess chlorine. Further reactions to silanols, siloxanes, and polymers may be expressed:

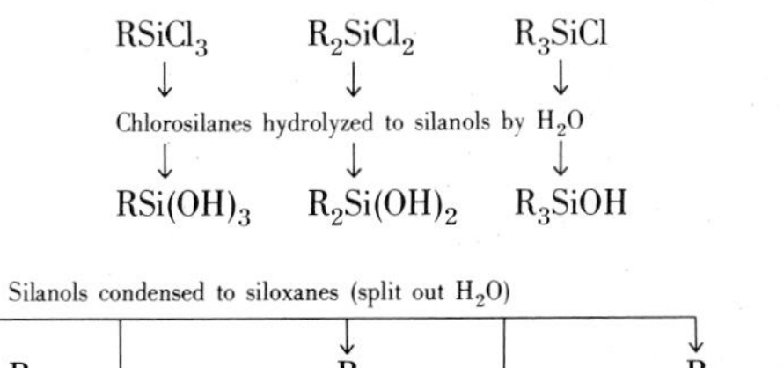

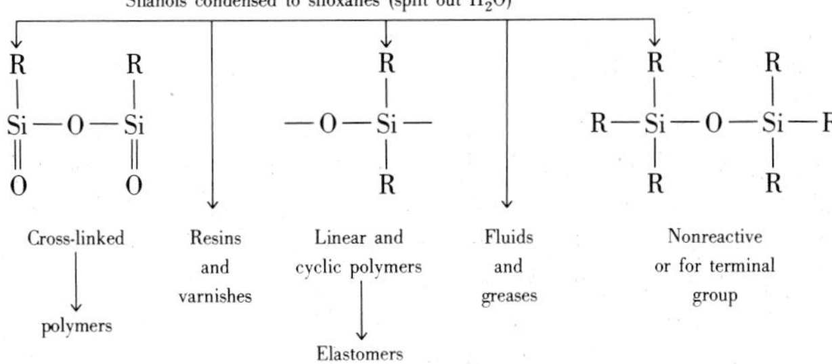

The production of silicone products for elastomers in 1971 was 95 million lb valued at about $123 million.

BUTYL RUBBER

The copolymer of isobutylene with about 2% isoprene is known as butyl rubber. The isoprene imparts sufficient unsaturation to the molecule to permit curing or vulcanization. Production is about 120,000 tons annually. Polymerization is carried out in solution at extremely low temperatures, followed by precipitation of the rubber by the addition of water, separating the rubber as crumb, which can be filtered and dried with conventional equipment. *Butyl rubber is uniquely impermeable to air and has found its major use in inner tubes and as liners of tubeless tires.* Butyl rubber is also inert to oxidation and is useful for weatherstripping and other similar exposure conditions.

URETHANE RUBBER

The reaction products of certain organic diisocyanates and polyglycols are rubbery products known as *polyurethanes.*[16] These compounds are specialty rubbers of outstanding properties, they possess high abrasion resistance and are useful at high temperatures and with high concentrations of solvents and oxygen or ozone. A major use for this type of rubber is in the production of flexible foam and elastic fibers, which result when a diisocyanate is mixed with a polyester containing both free hydroxyl and carboxyl groups. The reaction is extremely rapid, with the evolution of gas which serves to

[16]Saunders *et al.*, Polyurethanes: Chemistry and Technology, Pt. I, Chemistry, pp. 273–314, Wiley, 1962.

expand the mass, yielding foams, either hard or soft, depending on the reactants and the conditions employed. Additional gas may be supplied to regulate the density of the resultant foam. These materials have found increasing use in upholstery, mattresses, insulation, vibration damping, and other fields of application formerly restricted to foam rubber. The tensile strength and abrasion resistance of polyurethanes are extraordinary in comparison with those of hydrocarbon elastomers. Both rigid and flexible foams have encountered difficulties in some applications because of flammability. The appearance of Spandex (Chap. 35) polyurethane elastic thread has revolutionized the support-garment field because of the high holding power of fine-denier thread. The basic reactions leading to polyurethane rubbers are as follows:

$$\underset{\text{Toluene diisocyanate}}{CH_3C_6H_3(NCO)_2} + \underset{\text{Glycol}}{HO{-}R{-}OH} \longrightarrow \underset{\text{Polyurethane (tough rubbery polymer)}}{\left[-\overset{O}{\overset{\|}{C}}\cdot\overset{H}{\overset{|}{N}}{-}C_6H_3(CH_3){-}N(H)\cdot\underset{O}{\underset{\|}{C}}\cdot O\cdot R\cdot O\cdot\underset{O}{\underset{\|}{C}}\cdot\underset{H}{\underset{|}{N}}{-}C_6H_3(CH_3){-}N(H)\cdot\underset{O}{\underset{\|}{C}}\cdot O\cdot R\cdot O \right]_n}$$

HYPALON

A rubber called Hypalon results from the reaction of chlorine and sulfonyl chloride on polyethylene; a solution of polyethylene is reacted with chlorine and sulfur dioxide, transforming the thermoplastic polyethylene into a vulcanizable elastomer. Hypalon is extremely resistant to ozone attack, weathering, and heat and has excellent chemical resistance.

STEREOSPECIFIC[17] RUBBERS

POLYISOPRENE AND POLYBUTADIENE Synthesis of the natural-rubber hydrocarbon has been accomplished. The key to this development was the discovery of stereospecific catalysts of the Ziegler-Natta or alkyllithium type, which have made it possible to polymerize isoprene or butadiene in a suitable organic solvent so as to obtain the cis structure (up to 98%) characteristic of the natural-rubber molecule, and with a high degree of regularity (see Fig. 36.1, polymers).

Polyisoprene rubber has properties essentially identical to those of natural rubber and may be substituted for it in all its uses, including tires. The synthetic product has the advantage of being cleaner and more uniform in quality. Because of the abundant supply and low cost of butadiene relative to isoprene, however, *polybutadiene* is being commercialized at a much faster rate than its sister stereopolymer. Stereoregular polybutadiene (Fig. 36.10) can be produced in modified SBR plants using Ziegler-type catalysts, such as titanium halide plus aluminum alkyl, in a suitable solvent such as toluene, to yield polybutadiene of approximately 95% cis content. However, the monomer must be especially purified. Polybutadiene has proved to be difficult to use alone, but blends of polybutadiene and SBR or natural rubber have outstanding abrasion resistance, low heat buildup, and superior rebound. Polybutadiene has become one of the most rapidly increasing segments of the synthetic rubber industry, and it is estimated that yearly production is now about 500,000 long tons per year. Polybutadiene rubber is being used mainly to extend and partially replace both natural

[17]Cretides, Polyisoprene, *Ind. Eng. Chem.* **56**(5), 11 (1964); Haines, Polydienes, *Ind. Eng. Chem.*, **54**(11), 16 (1962) (flowchart); Ziegler catalysts for olefins, Chap. 29; Butyllithium, *Chem. Eng.* (*N.Y.*), **68**(1), 33 (1961); Isoprene Polymers, EPST, vol. 7, p. 782, 1967.

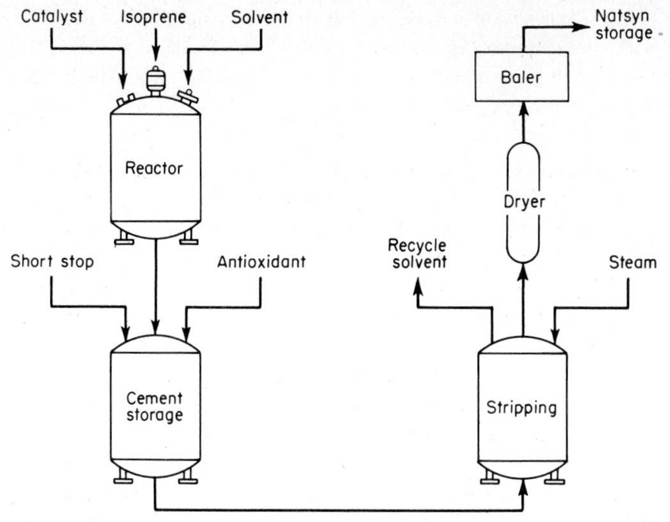

Fig. 36.10 Natsyn process for synthetic natural rubber production (synthetic polyisoprene). (*Goodyear Tire & Rubber Co.*)

rubber and SBR in tires. Early reports indicate that tread wear is improved by 35% in a 50-50 blend of polybutadiene and SBR. Higher proportions with still better mileage may be expected in the near future (Figs. 36.1 and 36.10).

$$-CH_2-C(CH_3)=CH-CH_2-CH_2-C(CH_3)=CH-CH_2-CH_2-C(CH_3)=CH-CH_2-$$

cis-1,4-Polyisoprene

The monomer must be extremely pure, and the solvent (hexane, for example) should be completely anhydrous. As noted above, the cis structure is desired, since it is usually rubbery, whereas the trans configuration is more similar to plastics or gutta-percha. Structural prerequisites of both natural and synthetic rubbers are long threadlike molecules. Their characteristic property of reversible extensibility results from the randomly coiled arrangement of the long polymer chains. When extended, the chains are distorted but, like a spring, they revert to the kinked arrangement upon removal of the stress.[18]

ETHYLENE-PROPYLENE POLYMERS[19] AND TERPOLYMERS (EPR AND EPT)

A newcomer on the rubber scene is based on the extremely cheap monomers ethylene and propylene which occur in refinery off-gases or which can be produced by cracking propane. Although not stereoregular, these polymers can be produced in solution plants and use similar catalysts. The polymer chain, based on ethylene and propylene, does not contain sufficient unsaturation for conventional vulcanization, but this property is added by the use of a third monomer in a small amount. Butadiene, cyclopentadiene, and dicyclopentadiene, among others, have proved to be suitable third

[18]McGraw-Hill Encyclopedia, vol. 11, p. 635, 1966.
[19]Natta *et al.*, Polyolefin Elastomers, *Rubber Chem. Technol.*, **36**, 1583 (1963).

monomers. The ethylenepropylene terpolymers may be expected to make inroads in the rubber industry because the monomers are potentially the cheapest available and the rubbers have given indication of sufficient quality to merit careful consideration, such as for wire and cable coatings and conveyor belts.

RUBBER-PROCESSING CHEMICALS

The rubber industry requires chemicals to modify the basic elastomers to produce desired and improved qualities in rubber goods. Such chemicals (Table 36.4) amount to more than 323 million lbs yearly, valued at $208 million. The following classification by use includes the principal chemicals used in compounding rubber; see also Table 36.4.

Vulcanizing, or curing agents. Sulfur, sulfur monochloride, selenium, tellurium, thiuram disulfides, *p*-quinone dioximes, polysulfide polymers.

Vulcanization accelerators. 2-Mercaptobenzothiazole, benzothizolyl disulfide, zinc diethyldithiocarbamate, tetramethylthiuram disulfide, tetramethylthiuram monosulfide, 1,3-diphenylguanidine.

Accelerator activators. Zinc oxide, stearic acid, litharge, magnesium oxide, amines, amine soaps.

Retarders. Salicylic acid, benzoic acid, pthalic anhydride.

Age resistors, or antioxidants. *N*-Phenyl-2-naphthylamine; alkylated diphenylamine, an acetone-diphenylamine reaction product (BLE).

Pigments. Carbon black, zinc oxide, certain clays, calcium carbonate, titanium dioxide, color pigments.

Softeners and extenders. Petroleum oils, pine tars and resins, coal-tar fractions.

Waxes. Certain petroleum waxes.

Blowing agents. Sodium or ammonium bicarbonate, diazoaminobenzene, dinitrospentamethylentetramine, fluorocarbons, azodicarbonamide, nitrosoterepthalamide.

Chemical plasticizers. 2-Napthalenethiol, *bis*(*o*-benzaminophenyl) disulfide, mixed xylene thiols, zinc salt of xylene thiols.

Peptizers. Aromatic mercaptans (thiophenols).

Accelerators. Organic accelerators reduce the time required for the vulcanization of rubber from several hours to a few minutes. In addition, less sulfur is needed and a more uniform product is obtained. The mechanism of accelerator action is not well understood, in spite of much research, but presumably involves the formation of an activated form of sulfur, which forms a "sulfur bridge" at reactive sites within the rubber molecule, linking the large molecules into a tight network structure. Most accelerators contain nitrogen and sulfur. Two thirds of all the accelerators made consist of mercaptobenzothiazole (MBT) and its derivatives. The commercial process for the manufacture of MBT comprises reacting aniline, carbon bisulfide, and sulfur at a temperature of about 250°C and a pressure of at least 450 psi. The crude product is dissolved in caustic, insoluble tars are removed, and the resulting solution is used for the manufacture of various derived accelerators, including refined MBT, benzothiazyl disulfide (MBTS), the zinc salt of MBT, and *N*-cyclohexybenzothiazole sulfanamide.

Age resistors, or antioxidants,[20] protect rubber goods from attack by oxygen and ozone in the atmosphere. They are classified as antioxidants, antiozonants, or antiflex-cracking agents. They function by combining with, and thus interrupting, free-radical chain reactions and thus prevent further chain degradation. Commercial age resistors are generally either of the amine type or the

[20]EPST, vol. 12, p. 256, 1970. Ambelang *et al.*, Antioxidants and Antiozonants for General Purpose Elastomers, *Rubber Chem. Technol.*, **36,** 1497 (1963).

phenolic type. Amines are strong protective agents and are widely used in tires and other dark-colored goods where discoloring or staining is not a problem. Examples of amine age resistors are alkylated diphenylamines, alkylated and arylated *p*-phenylenediamines, and complex reaction products of aromatic amines with aldehydes or ketones. In the case of light-colored goods, alkylated phenols and their derivatives are used to give moderate antioxidant protection with minimum discoloration. Examples are shown by the following structures:

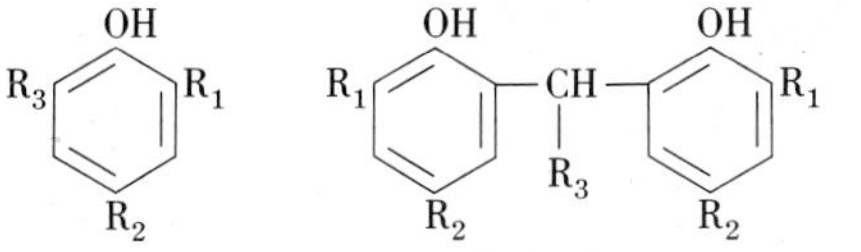

OH, R_3, R_1, R_2 — Monophenol type

OH, R_1, CH, R_3, R_2, OH, R_1, R_2 — Bisphenol type

Peptizers. Catalytic plasticizers, or peptizers, serve to reduce the viscosity of rubber to permit easier processing. When milled into the rubber, they cause chain scission, with a consequent lowering of the molecular weight. Peptizers are useful both in reclaiming vulcanized rubber and in softening high-molecular-weight crudes.

Large amounts of *inert filling materials* may be added to rubber. Some merely serve to harden or to dilute the mix. Others exert a profound influence called reinforcement, which is chemicophysical in nature. Clays, calcium carbonate, crushed coal, barytes—almost any finely divided solid—can be added to rubber, usually with a reduction in all tensile properties but with a desirable cost reduction, along with imparted hardness, retention of shape, color, and other desirable properties. However, some finely divided amorphous materials, notably *carbon black* and *silica,* unexpectedly greatly increase the strength, resilience, abrasion resistance, and other desirable properties and are properly known as reinforcing agents. The basic compounding or processing materials alone have enormously extended the uses of rubber, but many rather important properties can also be controlled during compounding. Resistance to heat, aging, and ozone attack is induced by the use of stabilizers, antioxidants, and antiozonants.

Processibility is improved by the use of reclaimed rubber, waxes, oils, factices (vulcanized vegetable oils), and mineral rubbers (asphalts, pitches, and vulcanized unsaturated hydrocarbons) or by chemical attack on the molecule, as well as by work breakdown, heat, and mastication.

Sponge structure is obtained by adding sodium bicarbonate, ammonium carbonate, and bicarbonate, urea, or organic gas-generating chemicals. *Hard rubber* may be made by greatly increasing the sulfur content and by the use of large amounts of filler.

RUBBER COMPOUNDING

Pure rubber is as useless as pure gold. The desirable properties of plasticity, elasticity, toughness, hardness, softness, abrasion resistance, impermeability, and the myriad of combinations possible are achieved by the art of the rubber compounder. A typical rubber compound may be described as follows

Ingredient	*Parts*	*Ingredient*	*Parts*
Rubber (Fig. 36.8)	100.0	Loading or filling pigment	50.0
Sulfur	2.0	Reclaim, softeners, extenders,	As required
Zinc oxide	5.0	colors, blowing agents,	
Stearic acid	3.0	antioxidants, antiozonants,	
Accelerator	1.5	odorants, etc.	

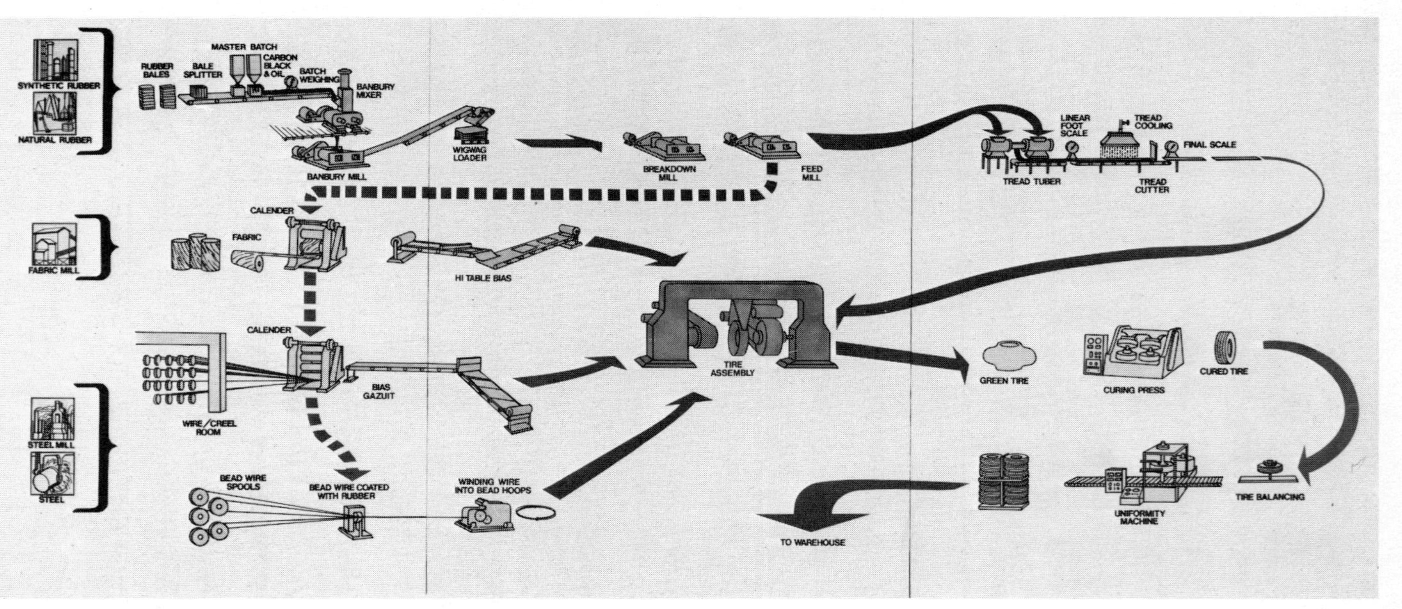

Fig. 36.11 Radial-tire manufacturing process. (*General Tire & Rubber Co.*)

RUBBER FABRICATION

As a plastic material, rubber may be spread, cemented, calendered, molded, extruded, caulked, puttied, or wrapped into virtually any shape; coated on cloth, plastic, or metal; and sandwiched or forced into cracks. Rubber is an extremely tough material. Heavy machinery is required,[21] some of which is shown in the flowchart of Fig. 36.11 and in Figs. 36.12 and 36.13. The considerable heat generated during the mixing, mastication, grinding, and extruding must be dissipated and controlled. Some typical rubber-goods manufacture is briefly described.

CALENDERING, OR COATING One of the earliest uses for rubber was for coating fabric to make it waterproof. Solutions or cements of rubber in solvents are easily spread on fabrics but, unless the ingredients necessary for cure and property control are included, the results may be quite unsatisfactory. Rubber compounds are applied to fabric by calendering, i.e., rolling the rubber compound into the fabric on multiroll calender machines. Tire cord is a special case in which cotton, rayon, nylon, or polyester cords are arranged in parallel and bound together by rubber on a calender.

MOLDING Doughlike rubber compounds may be moded into virtually any shape, which is retained by curing the compound in the mold. A tennis ball is a good example, A very high grade and resilient rubber compound is molded in "muffin tins" shaped as halves of the ball. These are cemented together (wilh a pill of a gas-releasing chemical enclosed) and vulcanized to form the core of a ball, to which the fabric cover and proper nap fiber are cemented. It is a complicated structure for very rugged service.

EXTRUDING Weatherstrip, hose, inner tubes, tire treads, gaskets, channels, and many other rubber articles are fashioned by extrusion of the plastic compound, which may be cured during this operation or later. The automobile tire serves as an example. Transportation is the major end use of rubber, and the tire its supreme product. Tire manufacture is depicted in Figs. 36.11 to 36.13. These illustrations show the many steps involved and make clear how the tire is shaped from wire,

Fig. 36.12 Radial-tire building machine. The carcass is built on a cylindrical drum and then shaped within the tread and belt. (*General Tire & Rubber Co.*)

[21]Perry, Fig. 19.27, for a Banbury mixer; also pp. 19-20 to 25.

Fig. 36.13 Twin Bag-o-matic tire curing press with segmented-mold operating cylinders. Radial tires are shaped to approximately their final appearance and vulcanized in molds such as these, which apply segmented sections.

cord, fabric, and rubber into the doughnut shape which hugs the vehicle rim, retains air for months or years, resists the heat built up by service and the attack of air, ozone, oils, and chemicals, and provides tens of thousands of miles of travel over roads and highways of all types and all of a highly abrasive nature.

The tire is *built up as a cylinder* on a collapsible, round rotating drum. Layers of cords embedded in a proper compound are applied, one layer tying the beads together in one direction and another layer in another direction. The beads—wire cables embedded in a tough, hard rubber—are "stitched" to the tire by folding the end of the cord fabric over. Last, the tread formed by extrusion is laid on, and the ends lapped together. The tire-building drum is collapsed, and the cylindrical tire removed and placed in a press. Here an inflatable rubber bag, usually made from butyl rubber and on a movable stem, is blown up inside the tire; the press mold is simultaneously closed, and the tire becomes a doughnut. Heat is applied through the mold and by steam inside the bag. Excess rubber escapes through weep holes and, after a timed cure at preselected temperature, the tire is formed (Fig. 36.12). Present-day tires may be tubeless, with the air-retentive layer built in, or an inner tube extruded from butyl rubber may be used. Butyl rubber, although a very "dead" rubber, has outstanding resistance to passage of air. The modern automobile, in addition to tires, uses rubber in window channels, weatherstrip, wiper blades, motor mounts, cushions, and sound deadening, to the extent of about 250 lb of rubber per vehicle.

Radial tires require a different method of assembly from the traditional bias cord tires as shown in Fig. 36.11. Belts of rayon, polyester, Fiberglass and steel are all being used. A tire made without cords and completely molded[22] is under development, but seems some years away. Fully automated assembly lines for tires are very near.

LATEX COMPOUNDS

Concentrated latices permit the use of rubber as a liquid which may be spread, painted, dipped, or whipped into a foam. Suspensions of compounding ingredients provide cure, but reinforcement is not nearly as effective. The use of latex is increasing, although the foam mattress, pillow, and upholstering trades have veered toward polyurethane and polyether rubbers for their supply.

RECLAIMED RUBBER

Reclaim[23] is a useful compounding ingredient produced from scrap rubber goods. Articles are ground finely, and rubber, fabric, metal, etc., separated by combinations of chemical, mechanical, and solution methods. Reclaimed rubber is partially depolymerized and contains a high percentage of carbon black (or other pigment), ash, and solvent oils. It imparts desirable processing character to rubber compounds and is not merely a cheap scrap material.[24] About 197,000 tons/year of reclaim is used in the United States, approximately 5% of the new rubber supply.

RUBBER DERIVATIVES

Rubber, particularly natural rubber, has been used as a raw material for chemical reactions to yield various derivatives of altered nature. Plastic technology is displacing these rubber-based chemicals, for rubber is obviously an expensive raw material. Halides react with rubber, adding at the double bonds to form chlorinated rubbers (or bromide and iodide may be used) which are useful as a paint ingredient and are resistant to chemical attack. The action of hydrogen chloride on a benzene solution of rubber produces rubber hydrochloride, decidedly different from rubber in many ways. A tough, transparent plastic, it forms films used for packaging. Rubber hydrochloride is resistant to chemical attack, forms excellent thin films, and is colorless, odorless, and tasteless. PVC, polyethylene, polypropylene, and newer plastics are limiting the use of expensive rubber hydrochloride. Rubber and resin blends are widely used, the rubber adding its peculiar properties of impact resistance, extensibility, and resilience.

SELECTED REFERENCES

Alliger, G., and I. J. Sjothum (eds.): Vulcanization of Elastomers, Reinhold, 1964.
Bateman, I. (ed.): Chemistry and Physics of Rubberlike Substances, Wiley, 1963.
Burlant, W. S., and A. S. Hoffman: Block and Graft Polymers, Reinhold, 1961.
Gould, R. F. (ed.): Literature of Chemical Technology, chaps. 28–35, Rubber, ACS Monograph, 1967.
Heinisheh, K. F.: Dictionary of Rubber, Halsted, 1974.
Johnson, B. L., and M. Goodman: Elastomer Stereospecific Polymerization, *Advan. Chem. Ser.* no. 52, (1966) (catalysts).
Johnson, J. C.: Antioxidants 1975, Synthesis and Applications, Noyes, 1975.
Madge, E. W.: Latex Foam Rubber, Wiley, 1962.
Maurie, M.: Rubber Technology, Van Nostrand Reinhold, 1973.
Naunton, W. J. S. (ed.): Applied Science of Rubber, St. Martin's, 1961.
Rubber Handbook, 2d ed., Rubber Manufacturer's Association, 1963.
Saunders, K. J.: Identification of Plastics and Rubbers, Barnes and Noble, 1966.
Styrene: Its Polymers, Copolymers and Derivatives, 3 pts., Hafner, 1965.
Szilard, J. A.: Reclaiming Rubber and Other Polymers, Noyes, 1973.
———: Rubber and Related Products, ASTM, 1974.

[22]Tire Rubbers: Evolution Today, Revolution Tomorrow, *Chem. Eng.* (*N.Y.*), **79**(23), 58 (1972).

[23]Nourry, Reclaimed Rubber, McLaren and Sons, London, 1962; Jackson, Recycling and Reclaiming Municipal Solid Wastes, Noyes, 1975.

[24]For flowchart see CPI 2, p. 805.

chapter 37

PETROLEUM REFINING

Except for the primary production of food and clothing from our soils, no organic industry is more important to modern technical civilization than the petroleum industry.

The United States is the largest consumer of petroleum in the world and has the greatest refining capacity, but is no longer self-sufficient.

Initially, the refining of petroleum involved simple fractionation of the constituents already present in crude oil. This was the easiest approach, and it was fortunate that the first crude oil discovered in Pennsylvania was adapted to such an operation. However, as markets were developed for more specialized products and as new fields supplying more varied crude oil were discovered, it became necessary and economical to use chemical reactions to change the molecular structure of the compounds initially present in the crude, thus opening up greatly enlarged and more adaptable markets to the petroleum industry.

For the manufacture of many consumers' chemicals, from carbon black and ammonia through ethyl alcohol and glycol to synthetic rubber, synthetic fibers, and plastics, the petroleum industry provides the cheapest raw materials. These are presented in this chapter, which also includes the conversion of the basic raw materials, paraffins, and cyclics, into *petrochemical precursors,* olefins and aromatics. Chapter 38, on Petrochemicals, and other reference chapters present the intermediate to finished products. See Table 38.2 for fundamental derivations. Although the petroleum industry is finding it economical to import more and more crude oil, it is not reasonable that the use of petroleum products as chemical raw materials will be curtailed. Rather, in the decades ahead, when the cost of petroleum products as liquid fuels for heat and power becomes comparatively greater, shale oil, oil sands, and coal can be processed.

The petroleum industry, in its design, operation, development, sales, and executive branches, has become the largest employer of chemical engineers. This has been particularly true recently because the simple distillations of the earlier years have been generally replaced by complicated refining procedures involving numerous unit physical operations and chemical conversions or unit chemical processes, frequently of great complexity and large size. All the branches of this industry are so interrelated and so technical that it requires the services of many trained engineers. Furthermore, since the petroleum industry is reaching out into many other chemical fields, supplying raw materials and using new chemicals, it is enlarging its needs. Thus chemical engineers within the petroleum industry, to be most efficient, whether in operating, constructing, or designing work, should inform themselves with regard to other fields.

HISTORICAL[1] It seems almost incredible that the petroleum industry has grown so rapidly and so extensively, from the first skimming of petroleum seepages by the Indians for use as a medicinal or rubbing oil and the first modern petroleum well drilled near Titusville, Pa., by Drake in 1859,

[1]Egloff and Alexander, Petroleum Chemistry, *Ind. Eng. Chem.* **43,** 809 (1951) (excellent historical article).

to the tremendous industry now flourishing not only in the United States but throughout the world. This development has gone through a great many market changes. At first the chief petroleum product sold was kerosine for illuminating purposes. Now kerosine is of secondary importance in comparison with the tremendous consumption of gasoline occasioned by the growth of the internal-combustion engine and its applications to transportation and to power, in the air, on our roads, in our factories, and on our farms. The synthesis of many chemicals from petroleum raw materials is the outstanding characteristic of the broad chemical and allied industries (Chap. 38).

Chemical engineering owes a great deal to the discovery and application of engineering principles on distillation, heat transfer, fluid flow, and the like, by petroleum engineers, in the designing, building, and operating of refineries. At first the almost universally used equipment was a batch still. A relatively small amount, or batch, of crude petroleum was pumped into a horizontal, direct-fired, cylindrical vessel provided with a condenser. The various constituents of the petroleum were distilled and condensed, the lightest coming out first and leaving behind in the still a residuum of tar or heavy lubricating oil. This has been supplanted almost entirely by the modern *continuous* pipe still, as exemplified by the flowcharts presented in this chapter.

ORIGIN[2] There have been many theories on the origin of petroleum, some emphasizing a vegetable or animal background and others a close relationship to coal; and certain of these theories emphasize that any organic matter of either background can be transformed into petroleum products. There seems to be general agreement that petroleum was formed from organic matter in near-shore *marine* deposits in an environment deficient in oxygen, and associated with the sediments that later solidified into rocks, limestones, dolomites, shales, and sandstones. The concentration of the organic matter may not have been high in the original deposition, but petroleum has migrated and gathered in places most favorable to its retention, such as porous standstone in domes protected by oil-impervious strata or against sealed faults in the sediments. Brooks[3] points out, referring to this organic matter, that "the proteins and soluble carbohydrates are undoubtedly quickly destroyed in the initial processes of decay or bacterial action. . . . Fatty oils are relatively resistant to bacterial action. . . . Fatty oils (or acids) are probably the chief source material from which petroleum has been derived."

EXPLORATION At one time drilling for petroleum was a hit-or-miss affair, and in such cases about 1 out of 300 wildcat[4] wells struck oil. However, by employing skilled geologists who have studied the origin and occurrence of petroleum, as well as geophysicists who are expert in the use of very delicate instruments to determine something about the geological conditions under the earth's surface, the drilling of wells has become so vastly improved that, in 1962, 1 out of 9 new wells yielded oil or gas. The geologist early recognized that petroleum occurs in oil pools caught in the anticlinal folds of sedimentary rocks. The success of the oil geologist[5] has been due to the correlation of a great deal of experience and data. This includes the study of cores from all types of wells and the accurate observation of surface indication, coupled with newer geophysical exploratory procedures, many of which have been developed and perfected by the scientific staffs of oil companies. The driller, guided by the geologist, now drills holes deeper than 4 mi and reaches gas or oil.

With the use of very sensitive instruments, geophysicists can determine the likelihood of occurrence of domes and deposits at considerable distances in the earth. The top of the arch of an anticline or a dome has, by virtue of compression, greater specific gravity than the surrounding rocks, as shown by Fig. 37.1, which depicts the various strata surrounding oil-bearing rock or sand. Oil and salt

[2]Dunstan, Nash, Brooks, and Tizard, Science of Petroleum, vol. 1, pp. 32–56, Oxford, 1938. This publication on petroleum is a monumental work and should be freely consulted. It will hereafter, in this chapter, be referred to by its title. See also Petroleum Symposium, *Ind. Eng. Chem.*, **44,** 2556–2577 (1952) (four articles on origin); Hobson and Pohl, Modern Petroleum Technology, 4th ed., Applied Science Publishers (Exxes, England), 1973.

[3]Science of Petroleum, vol. 1, p. 52; ECT, vol. 10, pp. 97–109, 1953 (52 refs. and detailed discussion).

[4]A *wildcat* is a well whose location is determined without much scientific exploration, especially in a previously untested area.

[5]See Science of Petroleum, particularly pp. 270–397, for numerous articles by experts in this field; see also Hager, Practical Oil Geology, 6th ed., McGraw-Hill, 1951; Schnacke and Drake, Oil for the World, Harper, 1950.

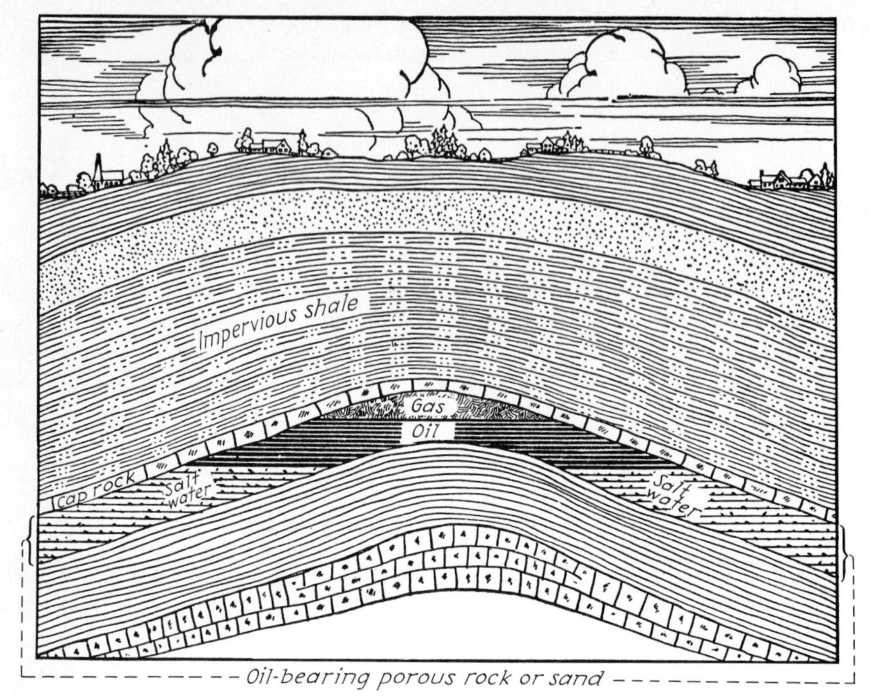

Fig. 37.1 Oil trapped in nature's reservoir. (*American Petroleum Institute.*)

deposits also have a different (lower) specific gravity than the accompanying rocks. These variations are measured by delicate gravity instruments, which record changes in the gravity constant g. The plotting of these results gives a *gravimetric survey*. Other methods of forecasting such anticlines include the exploding of charges at selected intervals, with special instruments located at various points to measure and time the reflected wave initiated by the explosion. This results in a *seismic survey*. Other methods are also used, such as determination of the electrical conductivity of the earth or the magnetic conditions. These two methods do not seem to be so successful as the gravimetric and seismic surveys. All these studies lead to the drilling of a test hole which, if successful, frequently results in the opening up of a new oil field.

PRODUCTION STATISTICS The world's production of petroleum totals over 20 billion bbl[6] per year, of which the fields in the United States produce approximately 16%, as shown in Table 37.1. No industry publishes more extensive statistical data than the petroleum industry,[7] through the American Petroleum Institute (API), the member refiners, producers, and marketers, which together provide and present such data to the public. Table 37.2 gives a comparison of the total quantities of refined products in the United States.

RESERVES AND RAW MATERIALS Estimates of oil reserves are only of an approximate nature. It should be apparent, however, that some idea, although far from exact, of the available reserves of essential minerals—petroleum included—is of vital importance in mapping their development and the future trend of related industries. The following statements apply only to proved[8]

[6]In the petroleum field a barrel holds 42 gal.

[7]Petroleum Facts and Figures, American Petroleum Institute, 1971. This is a most unusual volume on statistics, presenting the industry from the points of view of production, refining, transportation, marketing, utilization, prices, and taxation, in the form of tables. The Bureau of Mines also publishes generalized statistics on this industry in the Minerals Yearbooks.

[8]Proved reserves refers to oil that is in known and proved fields and recoverable by present production methods.

TABLE 37.1 ***World Crude Oil Production, by Regions and Countries, by Years***
(In thousands of barrels)

Country	*1962*	*1970*	*1972*	*1974**
North America				
United States	2,676,185	3,517,450	3,455,368	3,264,493
Canada	244,139	461,777	560,693	613,930
Mexico	111,830	177,599	185,011	187,428
Others	. . .	51,847	52,494	65,424
Total	3,032,154	4,208,073	4,253,666	4,131,275
South America				
Venezuela	1,167,954	1,353,420	1,178,487	1,104,125
Other	270,748	332,913	371,965	416,830
Total	1,438,702	1,686,333	1,550,452	1,520,955
Europe				
U.S.S.R.	1,357,800	2,594,550	2,895,900	3,792,350†
Others	231,043	271,146	276,679	139,138‡
Total	1,588,843	2,865,696	3,172,579	3,931,488
Africa				
Algeria	158,094	371,767	384,858	324,412
Egypt	32,321	119,165	77,592	43,180
Libya		1,209,314	819,619	620,500
Nigeria		395,836	665,282	839,500
Other	102,969	110,559	131,421	175,419
Total	293,384	2,206,614	2,078,772	2,003,011
Asia				
Iran	481,939	1,397,460	1,843,869	2,236,720
Iraq	366,832	569,726	529,419	667,694
Kuwait	669,284	1,090,039	1,201,346	949,000
Saudi Arabia	555,056	1,387,265	2,202,049	3,066,000
United Arab emirates		283,500	440,132	723,430
Qatar		132,456	176,545	199,290
Oman		121,210	103,131	108,405
People's Republic of China		146,000	216,080	
Indonesia	167,771	311,628	395,581	989,150
Others	283,994	239,183	313,852	58,706¶
Total	2,524,876	5,678,467	7,422,004	8,998,395
Oceania				
Australia and New Zealand	. . .	65,616	120,635	136,656
World Total	8,877,959	16,710,826	18,598,008	20,721,780

Source: Minerals Yearbook, 1972, Dept. of the Interior, 1974. *Estimated in *Oil Gas J.*, Dec. 30, 1974 p. 10. †Includes entire Communist group and China. ‡Non-Communist Europe. ¶Not same basis as other entries.

oil reserves and leave out of consideration the volume of alternative total oil reserves. It should be noted that up to 80% of the petroleum stored underground can be economically brought to the surface and that, although practically all natural gas can be recovered and utilized, in actual practice some oil is wasted. Proved reserves of the world in January 1974 were estimated at 716 billion bbl, of which the proved reserves in the United States amounted to 35 billion bbl.[9]

In discussing oil reserves, emphasis should be placed on improved technique in the recovery from oil wells. Ashburner, in 1887, estimated that wasteful producing operations in Pennsylvania and New York oil fields would ultimately leave eight to nine times as much of the original oil in the sands as would be recovered. Laboratory experiments indicate that, if early New York and Pennsylvania operators had been able to produce oil in an economical manner by methods now

[9]*Oil Gas J.*, Dec. 30, 1974 p. 10.

TABLE 37.2 ***U.S. Refinery Production*** *(In thousands of barrels)*

*Product**	*1968*	*1969*	*1970*	*1971*	*1972*
Liquefied refinery gas	108,641	114,362	116,527	120,952	121,182
Finished gasoline	1,933,827	2,022,407	2,099,911	2,197,550	2,315,768
Special naphthas	28,397	28,397	30,196	28,255	32,096
Jet fuel	314,651	321,700	301,892	304,665	310,029
Kerosine	100,545	101,738	94,635	86,256	79,027
Distillate fuel oil	839,373	846,863	895,656	910,727	962,405
Residual fuel oil	275,814	265,906	257,510	274,684	292,519
Petrochemical feedstocks	95,442	98,356	100,381	110,948	124,026
Lubricants	65,684	65,080	66,183	65,473	65,349
Wax	5,887	6,049	6,294	6,939	6,148
Coke	95,190	102,868	107,871	109,114	119,765
Asphalt	135,460	135,691	146,658	157,039	155,294
Still gas for fuel	149,796	160,363	163,905	156,967	170,933
All other	31,443	35,384	33,599	32,292	32,504
Processing loss	116,691	122,412	131,052	139,433	142,161

Source: Minerals Yearbook 1972, Dept. of the Interior, 1974. *Conversion factors: 1 bbl wax = 280 lb; 5 bbl coke = 1 ton; 5.5 bbl asphalt = 1 ton.

considered efficient, 2 to $2\frac{1}{2}$ times more oil than was withdrawn from the sands might have been recovered by natural flow and pumping. The key to the problem is proper conservation and use of reservoir energy.[10] Oil is underlain by water under a high head in the extraneous parts of the reservoir systems of most oil fields and overlain with gas, as diagramatically shown in Fig. 37.1. The energy available to produce oil is represented by the head of water and the pressure of the gas acting on the oil. If the withdrawal of oil from the reservoir is at a *sufficiently low rate,*[11] the entrance of water at the outcrops and expansion of large water supplies already present tend to maintain the pressure head and supply the necessary energy.

An extremely important method of conservation is the return of natural gas to the oil well. As an example, in a Sugarland, Tex., field, reintroduction of or cycling with gas at a pressure of 1,400 psi and a slower rate of production are expected to recover twice as much as would have been obtained with older methods of producing oil (a net recovery of 70% of the original store). Air mixed with the gas has certain disadvantages, because it tends to deteriorate slightly the quality of the oil produced.

Water-flooding methods are responsible for the recovery of millions of barrels of oil that former production practices left in reservoir sands. Water is usually introduced in a central well and allowed to extend gradually outward in an expanding circle to drive oil to surrounding wells.

The addition of heat and/or surface-active materials and polymers (micellar flooding) contributes to better recovery. The use of nuclear devices to fracture and heat up underground reservoirs is under study. Also, hydraulic fracturing with gels containing solid propping agents to hold fissures open is much in vogue.[12]

For years oil wells drilled into hard limestone formations have been "shot" with nitroglycerin because the resulting fissures and fractures increased the drainage of oil and gas into the well. Today, almost every well drilled into a limestone formation has the drainage channels enlarged by treatment with inhibited hydrochloric acid; sometimes in dense limestone the action of nitroglycerin is followed by acid treatment. This acid treatment has transformed many apparently nonproductive wells into rich producers. In the Illinois basin, wells drilled into oolitic limestone frequently failed to

[10]Miller and Shea, Gains in Oil and Gas Production Refining and Utilization Technology, *Natl. Resourc. Plann. Board, Tech. Pap.* 3, pp. 5ff., 1941.

[11]Schnacke and Drake, *op. cit.*, p. 50.

[12]Chemical Aid for Oil Producers: Hydrafrac Treatment, *Span*, Summer 1964, p. 16.

produce oil until an acid treatment was used. Hydrochloric acid, containing an inhibitor to prevent corrosive action on iron, is used by all service companies. A fracturing service to open up new exit passages for oil has added much recoverable oil to the reserves of the United States and Canada.

The movement of oil in underground reservoirs is closely connected to its viscosity and surface tension. These in turn are related to the amount of gas dissolved in the oil. All this is a practical application of Henry's law[13] which states that, at constant temperature, a liquid dissolves the weight of a gas proportional to its pressure.

Offshore drilling has resulted in hundreds of wells yielding gas and petroleum and represents an investment of more than $3 billion. The tar sands of Canada are now being exploited by several companies, but processing difficulties due to the arctic environment and excessive government interference have led to unprofitable operations.

OIL SHALE[14] U.S. oil shale deposits cover an area of nearly 16,500 mi^2 in Colorado, Utah, and Wyoming; they vary in thickness from 15 to over 2,000 ft and contain an overall average of 15 gal of oil per ton. However, the commercially attractive strata average 25 gal/ton and contain over 200 billion bbl of oil. Oil shale is misnamed, since it contains no oil as such, but upon pyrolysis yields shale oil containing liquid hydrocarbons with a carbon/hydrogen weight ratio of about 7:2 to 7:5 and 1.8% nitrogen and 0.8% sulfur. Shale oil is composed of 39% hydrocarbons and 61% heterohydrocarbon compounds, but furnishes only a 3% straight-run gasoline fraction upon distillation. By conventional petroleum refining techniques (e.g. thermal or hydrocracking, reforming, etc.), however, shale oil crude can be upgraded to yield adequate quantities of catalytically reformed gasoline, diesel oil, and heavy fuel stocks. There have been many attempts to develop commercially economical retorting techniques for oil shale. Much money has been spent on such processes, which are not yet profitable, and the following four merit further investigation:[15]

1. *Solids-to-solids heating.* The Oil Shale Co. and Lurgi-Ruhrgas systems.
2. *Gas-to-solids heating with internal gas combustion.* Bureau of Mines and Union Oil of California systems.
3. *Gas-to-solids heating with external heat generation.* Institute of Gas Technology, Union Oil, Petrobras and Development Engineering systems.
4. *In situ retorting.* Occidental Petroleum system.

Since the cost of producing shale oil has been constantly reduced, it is now at the point where it is believed that crude shale oil is nearly competitive with domestic petroleum.

TRANSPORTATION[16] In the United States a great pipeline transportation system moves a tremendous tonnage of petroleum and natural gas. The pipeline has reached its maximum application east of the Rocky Mountains, where trunk lines, branch lines, and pumping stations form an extremely complex system. The pipeline is the most economical method for transporting natural gas. At regular intervals pressure-booster stations are located, using a small part of the gas to drive gas engines connected to compressors.

The United States has at present over 223,000 mi petroleum pipelines and 860,000 mi of natural-gas lines. Tank ships are the "water pipelines" that connect remote oil fields with refineries and markets. As an efficient and cheap means of bulk transportation, particularly over long distances, they are less expensive than pipelines on land. The refined products of the industry are

[13]This matter is well treated, with references to the original literature, in Miller and Shea, *Oil Week.*, June 18, 1934, p. 17 (gains in oil and gas production, etc.)

[14]Cowan, A Bibliography of the U.S. Bureau of Mines, Publications Dealing with Oil Shale and Shale Oil, Bureau of Mines, Petroleum and Oil Shale Experiment Station, Laramie, Wyo., OSRD 59 (revised); Prien, Current Status of U.S. Oil Shale Technology, *Ind. Eng. Chem.*, **56**(9), 32 (1964); Perrini, Oil from Shale and Tar Sands, Noyes, 1975 (describes 213 processes relating to shale oil).

[15]Shale Oil, Process Choices, *Chem. Eng.* (*N.Y.*), **81**(10), 66 (1974).

[16]Slider, Practical Petroleum Reservoir Engineering Methods, Petroleum Publishing, 1976.

transported by pipeline and by thousands of tank cars and motor tank trucks. Huge steel or concrete reservoirs holding 100,000 bbl or more are employed for the storage of petroleum and its products. Despite a tremendous potential fire hazard, the loss through fire is among the smallest of industrial fire losses, because of care, forethought, and many automatic protective devices. Underground reservoirs are important for storing natural gas and LPG under pressure and as near as practical to large consuming areas for winter peak demands.

EARNINGS AND INVESTMENT[17] Contrary to the popular conception of petroleum as "liquid gold," the total income of the industry has been moderate. Such profits as have accrued have been more the reward of business enterprise and sound research than of good fortune. In the 5-year period, 1965–1970, the annual rate of return on net worth average 12.5%, in comparison with 13.3% from the manufacturing industries. The large investment in the petroleum field is summarized in Table 37.3. The subdivisions of this tabulation show the varied nature of this extensive business.

CONSTITUENTS OF PETROLEUM, INCLUDING PETROLEUM GASES

Crude petroleum is made up of hundreds of different individual chemicals, from methane to asphalt. Although most of the constituents are hydrocarbons (83 to 87% carbon and 11 to 15% hydrogen), ultimate analyses indicate the presence in small quantities of nitrogen (0 to 0.5%), sulfur (0 to 6%), and oxygen (0 to 3.5%)[18] Much detailed work has been carried out by research groups cooperating with the API[19] to determine the actual constituents of petroleum. The results have been published in the literature and have been excellently summarized by Sweeney,[20] Sachanen,[20] and Nelson.[20]

Hydrocarbons may be divided into two chemical classes, as described below.

OPEN-CHAIN, OR ALIPHATIC, COMPOUNDS, comprising:

n-Paraffin series, C_nH_{2n+2}. This series of hydrocarbons comprises a larger fraction of most petroleums than any of the other individual classes. Important members are, with volume-percentage occurrence, in crudes from Ponca, Okla., East Texas, and Bradford, Okla., *n*-hexane (2±%) and *n*-heptane (2.5, 1.7, 2.5%). *n*-Paraffins predominate in most straight-run gasolines.

Isoparaffin series, C_nH_{2n+2}. Branched-chain compounds are very desirable and are frequently manufactured by catalytic reforming, alkylation, and isomerization. Naturally occurring members are 2- and 3-methylpentanes (0.8, 1.5, 0.9%), 2,3-dimethylpentane, and 2-methylhexane (1.2, 1.3, 1.3%), in crudes from Ponca, Okla., East Texas, and Bradford, Okla.

TABLE 37.3 ***Estimated Capital Expenditures for U.S. Petroleum Industry*** *(In millions of dollars)*

	1962	*1961*	*1969*	*1980**
Production	4,000	3,525	4,750	10,000
Transportation	365	255	450	1,000
Refining	350	360	950	3,000
Fertilizers, chemicals	310	325	575	1,000
Marketing	600	525	1,250	2,000
Other	100	110	200	500
Total	5,725	5,100	8,175	17,500

Source: Petroleum Facts and Figures, p. 507, API, 1971. *Estimated

[17]Newendorp, Decision Analysis for Petroleum Exploration, Petroleum Publishing, 1976; Anderson, Oil Program Investments, Petroleum Publishing, 1972.

[18]Nelson, Petroleum Refinery Engineering, 4th ed., pp. 9ff., McGraw-Hill, 1958 (tables and many references).

[19]Rossini, Hydrocarbons in Petroleum, *Chem. Eng. News,* **25,** 231 (1947); Rossini and Mair, Composition of Petroleum, in Progress in Petroleum Technology, p. 334, ACS, 1951 (further references to this title are made in this chapter).

[20]Sweeney, Petroleum and Its Products, 24th Annual Priestly Lecture, Pennsylvania State College, 1950 (splendid charts and tables); Sachanen, The Chemical Constituents of Petroleum, Reinhold, 1945; ECT, vol. 10, p. 93, 1953; Nelson, *op. cit.,* pp. 9ff; McCain, Properties of Petroleum Fluids, Petroleum Publishing, 1974.

Olefin series, C_nH_{2n}. This series is either not present in crude oil or exists in very small quantities. Cracking processes produce large amounts of olefins. Olefins possess better antiknock properties than normal paraffins but have poorer properties than highly branched paraffins and aromatics. Their usefulness in mixtures is somewhat reduced by their chemical reactivity, since on storage they polymerize and/or oxidize. Olefins are the most important class of compounds *chemically derived* from petroleum in the making of other products, also by chemical processing or conversion. Examples of pure lower members are ethylene, propylene, and butylene. In cracked gasolines and in residual products are found many of the higher members of the series.

RING COMPOUNDS, comprising:

Naphthene series, C_nH_{2n}. This series, which has the same empirical formula as the olefin series, differs in that its members are completely saturated. It is the second most abundantly occurring series of compounds in most crudes. Members are methylcyclopentane (0.95, 1.3, 0.5%), cyclohexane (0.78, 0.66, 0.64%), dimethylcyclopentanes (1.75, 2.0, 1.0%), and methylcyclohexane (1.8, 2.4, 2.0%), in crudes from Ponca, Okla., East Texas, and Bradford, Okla. These naphthenes predominate in most gas oils and lubricating oils from all types of crudes. They are also present in residual products.

Aromatic, or benzene, series, C_nH_{2n-6}. Although only small amounts of aromatic compounds are present in most petroleums, Borneo and Sumatra crudes contain relatively large amounts. These compounds are produced by chemical processing and, like olefins, have high antiknock qualities. Members are benzene (0.15, 0.07, 0.06%), toluene (0.5, 0.6, 0.5%), ethylbenzene (0.18, 0.2, 0.9%), and xylenes (0.9, 1.1, 1.0%), in crudes from Ponca, Okla., East Texas, and Bradford, Okla.

Petroleum crudes are characterized by *variability* in composition and must be evaluated[21] before they can be refined. Over the years it has become usual to divide crudes into three "bases":

1. *Paraffin base.* These crudes consist primarily of open-chain compounds and furnish low-octane-number straight-run gasoline and excellent but waxy lubricating oil stocks.
2. *Intermediate base.* These crudes contain large quantities of both paraffinic and naphthenic compounds and furnish medium-grade straight-run gasolines and lubricating oils. Both wax and asphalt are found in these oils.
3. *Naphthene base.* These crudes contain a high percentage of cyclic (naphthenic) compounds and furnish relatively high-octane-number straight-run gasoline. The lubricating-oil fractions must be solvent-refined. Asphalt is present.

Petroleum products[22] have long been divided into salable cuts by fractionation in the refining operations. This is a separation by boiling ranges. Indeed, the natural separation that takes place when the petroleum leaves its underground reservoirs, into natural gas and ordinary crude, is based on the same principle. Such refinery fractions may be classified roughly as follows:

Natural (or casing-head) gasoline and natural gas
LPG

Light distillates
Motor gasolines
Solvent naphthas
Jet fuel
Kerosine
Light heating oils

Intermediate distillates
Heavy fuel oils
Diesel oils
Gas oils

Heavy distillates
Heavy mineral oils (medicinal)
Heavy flotation oils
Waxes (candles, sealing, paper treating, insulating)
Lubricating oils (large range)

Residues
Lubricating oils
Fuel oils
Petrolatum
Road oils
Asphalts
Coke

[21]Nelson and Buthod, ECT, vol. 10, p. 114, 1953; see reference to the more exact "characterization factor," Nelson, *op. cit.*, p. 81; Sitting, Aromatic Hydrocarbons, Noyes, 1976 (many manufacturing processes).

[22]Cf. table III, Summary of Petroleum Products: Their Composition, Properties and Uses, ECT, vol. 10, pp. 164–175, 1953 (excellent details with references).

Natural gas (Chap. 6) occurs as accumulations in underground, porous reservoirs, with or without petroleum oil. Though natural gas has been known for centuries, not until the early 1800s was it first discovered in the United States; since that time it has become one of this country's great industries. Although some natural gas was at first wasted, it is now being conserved, and legislation has been enacted prohibiting the stripping of "wet" gas for its small gasoline content and then wasting the gas. The total marketed production of natural gas in the United States is over 25,000 billion ft^3/year. The present annual consumption of natural gas in the United States is valued at more than $3.5 billion at the wells. Natural gas is composed chiefly of hydrocarbons of the paraffin series from methane up to pentane, carbon dioxide, nitrogen, and sometimes helium, but few if any unsaturated hydrocarbons. The most important products obtained from natural gas are fuel, natural gasoline, LPG, carbon black, helium, hydrogen, and petrochemicals.

NATURAL-GAS LIQUIDS Gasoline extracted from natural gas is differentiated from straight-run, or refinery, gasoline (which is distilled from crude oil) by the term *natural*, or *casing-head*, gasoline.[23] The recovery of natural gasoline from natural gas has a highly volatile gasoline for blending into motor fuels, particularly for easy starting in cold weather. The United States produces about 6.1 million bbl of natural gasoline per year. Natural-gasoline plants are now producing by isomerization large quantities of pure isobutane and isopentane for alkylation with light olefins such as butylene to furnish alkylated gasoline. When crude oil is forced from a well by the pressure of the natural gas, some of its lighter components are vaporized into the gas. Consequently, the composition and characteristics of the recovered natural gasoline are determined by the composition of the oil.

A large increase in the demand for light hydrocarbon liquids for petrochemical feedstocks, as well as for fuel, has been met to a large extent by greater extraction of these products from natural gas. This demand has been so great that there is hardly a single natural-gas stream which has not had these hydrocarbons extracted from it before being delivered to the ultimate consumer as fuel. Whereas only a few years ago a plant recovering approximately 70% propane was considered satisfactory, plants are now being designed to recover not only all the propane, but in many cases, from 50 to 90% of the ethane in natural gas. Two of the major processes used to achieve these high recovery rates are (1) refrigerated absorption and (2) low-temperature distillation. A typical process illustrating the refrigerated absorption procedure is presented in Fig. 37.2. The feed to some of these modern extraction plants exceeds 1 billion std ft^3/day, contrasting sharply with the 10 to 20 million std ft^3/day plants which were common before the large natural-gas transmission lines underwent rapid expansion. Also, most of these plants operate at -30 to $-50°F$, which requires a propane refrigeration system operating at a section pressure of atmospheric or slightly below. It is essential in both the absorption and low-temperature processes that this refrigeration be supplied efficiently to the process streams.

Cracked, or refinery, gases. Natural gas is devoid of unsaturated hydrocarbons and hydrogen; these compounds are, however, present in the cracked gases of refineries.[24] Because of the large quantities of unsaturated, hence chemically reactive, hydrocarbons produced by cracking processes, the *oil industry is now developing along the line of synthesis.* The olefin hydrocarbons of these cracked gases are used in the manufacture of polymerized and alkylated gasoline, antifreezes, petrochemicals, explosives, solvents, medicinals, fumigants, resins, synthetic rubber, and many other products. When olefins are not available in sufficient amounts, they are made by dehydrogenating paraffins.

Liquefied petroleum gases[25] (LPG). Light hydrocarbons such as propane and butane, which are produced as by-products from natural gasoline, are now finding wide use as "bottled," or lique-

[23]Nelson, *op. cit.*, chap. 22, text and table 22-6; Campbell, Gas Conditioning and Processing, Petroleum Publishing, 4th ed, 1976 (2 volumes).

[24]*Ibid.*

[25]Nelson, *op. cit.*, p. 860; Carney, Natural Gas Liquids, in Progress in Petroleum Technology, pp. 251–261 (references). LPG were shipped in 1972 amounting to 26,804 million gal.

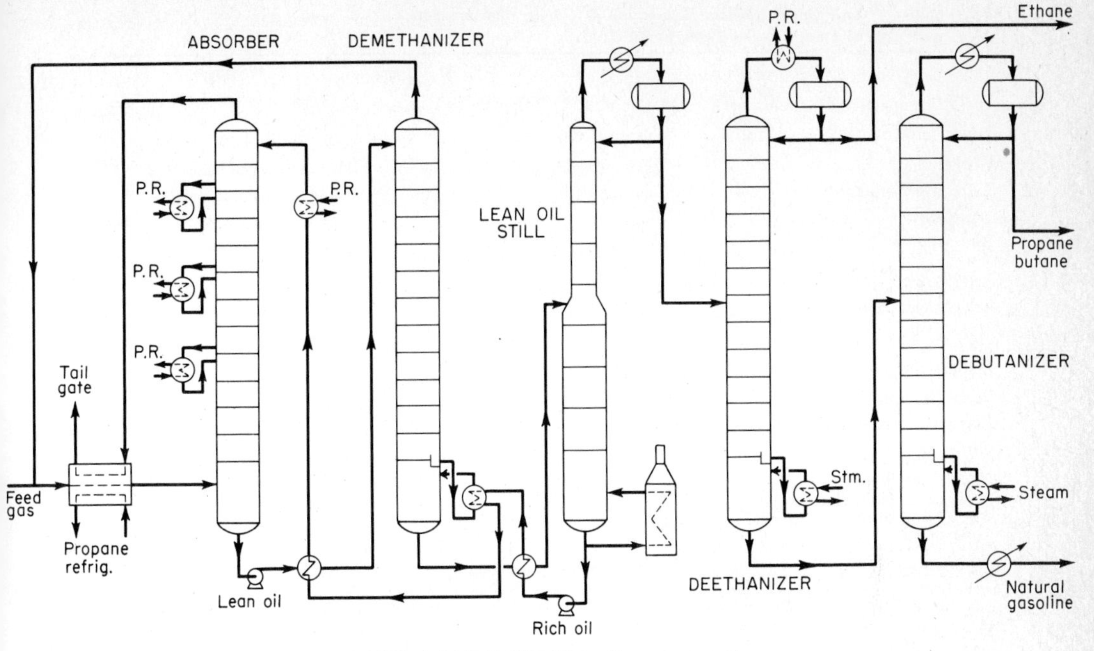

Fig. 37.2 Natural-gas plant operated by absorption and propane refrigeration. (*M. W. Kellogg Co.*)

fied, petroleum gas for domestic gas and heating and for city gas, as well as for direct motor fuel in special applications (farm and other tractors, trucks, buses). An increasing amount is being used in the manufacture of petrochemicals. Some of the gas is used as power for transportation. LPG is competitive with many types of fuel in present use. Bottled gas is used by many rural dwellers who are too far from gas mains, for cooking, lighting, heating water, and refrigeration.

PRODUCTS OF REFINING[26]

PRECURSORS OF PETROCHEMICALS are converted from gas, natural or cracked, LPG, and cyclic cuts. To distinguish and emphasize the increasing importance of acetylene, olefins, and aromatics for petrochemicals, these compounds have been designated "precursors" of intermediates leading to the finished and salable products listed in Table 38.2. These precursors are either chemically converted from natural petroleum raw materials or isolated from cracked stocks (Table 37.4). The principal precursors tabulated in Table 38.2 are

Acetylene	Propylene	Benzene	Xylenes
Ethylene	Butene	Toluene	Naphthalene

Acetylene is described in Chap. 7. *Ethylene* is presented by a generalized flowchart (Fig. 37.4), with distillate, natural gasoline, and condensate for raw materials.

[26]Cf. Petroleum Facts and Figures, API, 1971 (detailed statistics).

TABLE 37.4 *Unit Operations for Separation of Olefins and Aromatics as Precursors of Petrochemicals*

Operation	*Based on differences in:*	*Product, agent used to separate:*
Vapor-liquid		
Distillation	Vapor pressure	Ethylene from ethane
Extractive distillation	Polarizability	*n*-Butenes from butanes (aqueous acetone)
Absorption	Solubility	Ethane from methane
*Liquid-liquid**		
Solvent extraction	Solubility	Benzene from aliphatics (polyethylene glycols)
Liquid-solid		
Crystallization	Melting point	*p*-Xylenes† from other xylenes
Extractive crystallization	Clathrate‡ formation	*n*-Paraffins from other hydrocarbons (urea)
Encapsulation	Clathrate formation	*m*-Xylene
Adsorption, molecular sieves§	Surface or pore adsorption	*n*-Paraffins from isoparaffins
Vapor-solid		
Adsorption	Surface or pore adsorption	Ethylene from ethane (charcoal)

Source: Enlarged from Morrell, From Oil to Chemicals, *Esso Res. Rev. Inst. Fr. Pet.*, **14**, 80 (1959); cf. Schoen, New ChE Separation Techniques, Interscience, 1962. *Note:* In addition to the physical property differentials here tabulated, chemical differentials are employed, for example, reversible chemical reaction rates or chemical equilibriums, as exemplified by separating pure butadiene from butene, using reaction with cuprous salts. *King, Separation Processes, McGraw-Hill, 1971. †Aided by continuous pulsating separator to induce pure-crystal formation. ‡Clathrates, *Chem. Eng. News*, Jan. 13, 1958, p. 43; *Chem. Eng.* (*N.Y.*), **69**(19), 102 (1962); 1963 Petrochemical Handbook, p. 158, Gulf Publishing (flowchart for *m*-xylene). §Holm, The Petrochemical Industry, McGraw-Hill, 1970.

Ethylene is the largest volume organic chemical and can be made from feedstocks[27] varying from ethane to heavy gas oil, or even crude oil, depending on economic conditions. Conditions for its production lie somewhere between those usually associated with refining and those generally encountered in chemical production. Plants as large as 500,000 tons/year are now being built,[28] and larger ones are possible. The cracking of material diluted with an inert gas (usually steam) at a high (925°C) temperature and with a very short residence time (30 to 100 ms) yields a mixed product which must be separated to obtain useful components. A recent report shows[29] that a high-severity, short-time crack of naphtha feed yielded 1.2% H_2, 15.2% methane, 1.3% acetylene, 31.8% ethylene, 2.8% ethane, 1.2% propadiene, 11.6% propylene, 0.3% propane, 4.7% butadiene, 2.2% butylene, and 27.7% C_5 plus liquids. The gases are rapidly quenched, chilled, dehydrated, and fractionated to yield high-purity individual components, and frequently unwanted materials are recycled. Changing feedstocks and conditions can be used to manufacture propylene and butene, if these are the primary products wanted. *Ethylene* is also made according to the flowchart in Fig. 37.5, starting with refinery gas, which has a typical analysis: 25% methane, 19% hydrogen, 15% ethane, 7% ethylene, 12% propane, 6% propylene, 16% N_2, CO, CO_2, H_2S, and higher H_xC_y. This is supplemented by LPG, high-run gasoline, and ethane recycle. Ethylene had the very large consumption of 20 billion lb in 1975 and a predicted consumption of 32 billion lb for 1977.

[27]Zdonik, Bassler, and Haller, How Feedstocks Affect Ethylene, *Hydrocarbon Process.*, **53**(2), 73 (1974).
[28]Woodhouse, Samols, and Newman, The Economics and Technology of Large Ethylene Projects, *Chem. Eng.* (*N.Y.*), **81**(6), 73 (1974).
[29]Prescott, Pyrolysis Furnace Boosts the Ethylene Yield by 10–20%, *Chem. Eng.* (*N.Y.*), **82**(14) 52 (1975).

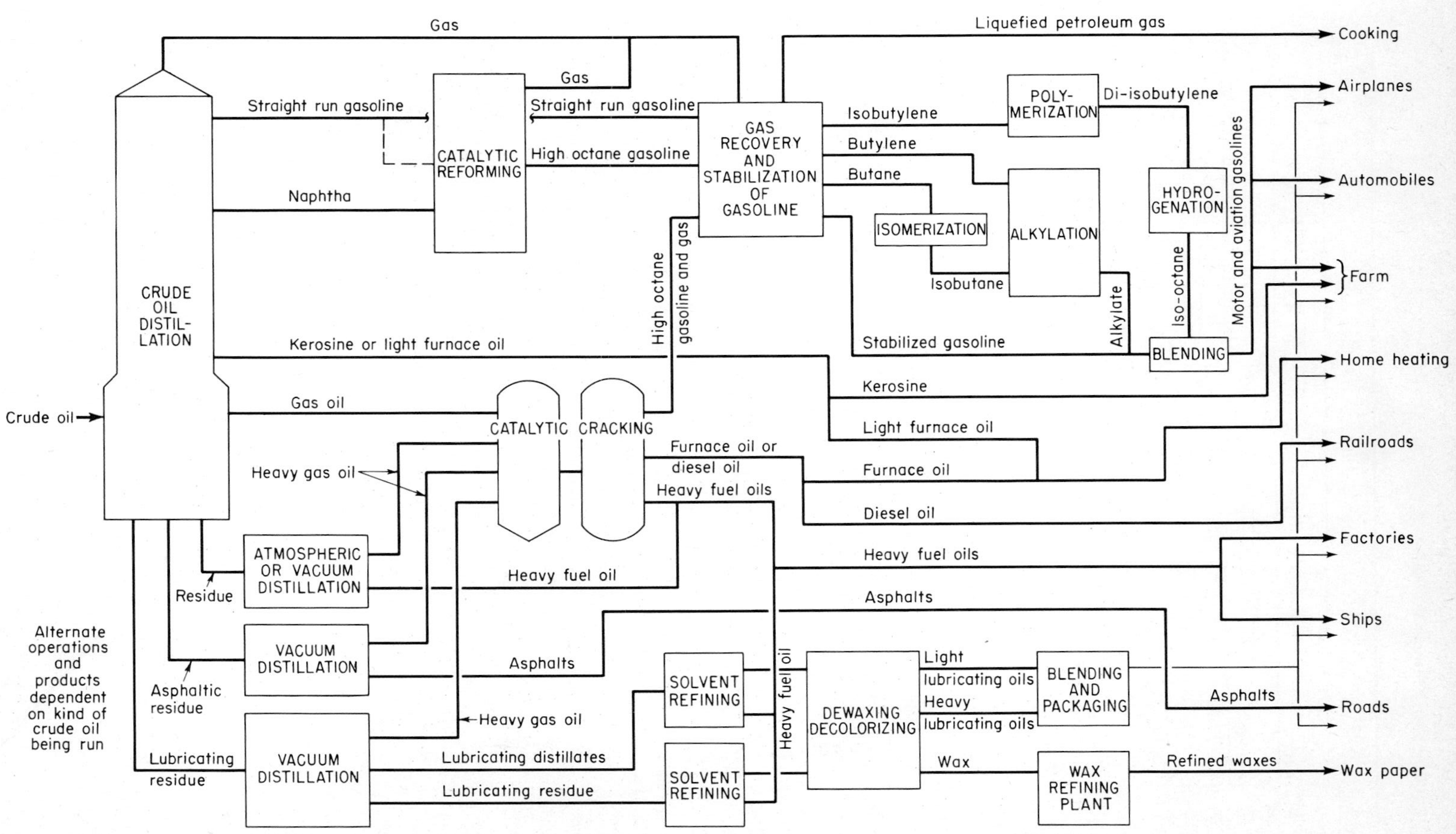

Fig. 37.3 Generalized overall refinery from crude oil to salable products. (*American Petroleum Institute.*)

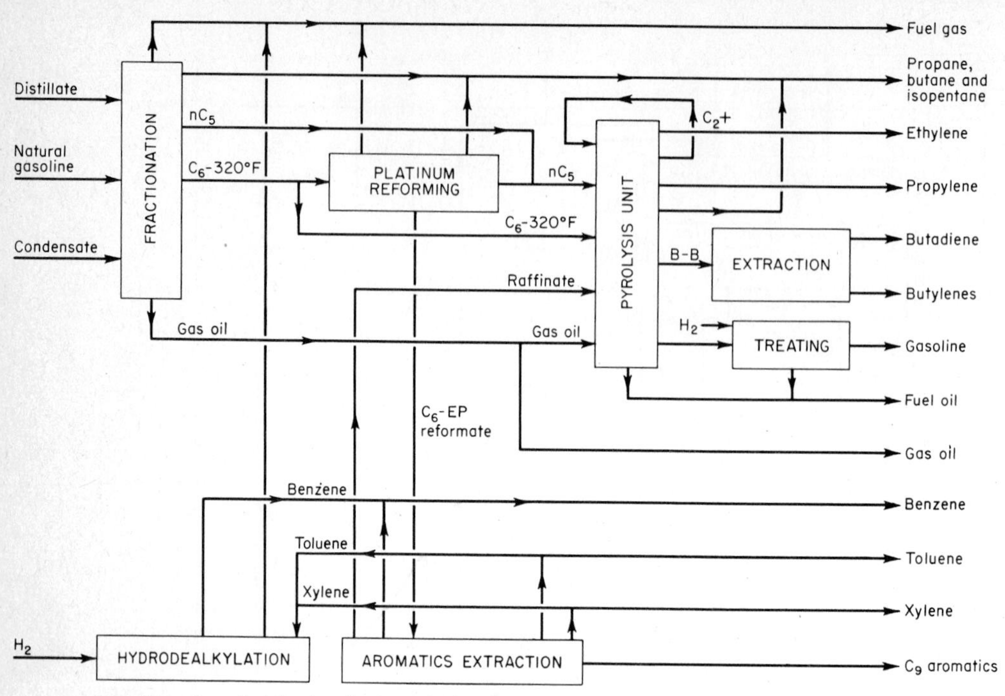

Fig. 37.4 Generalized flowchart for the production of petrochemicals. (*M. W. Kellogg Co.*)

Propylene has become a most important precursor (Table 38.5 and Fig. 38.3); 11.2 billion lb was produced in 1974, and much more is recoverable, mainly from cracking, to furnish gasoline blending stock.

The petroleum and coal sources of the *aromatic precursors* of petrochemicals, are shown in Table 37.5 together with their interrelationships (Fig. 37.6). Their production from petroleum is increasing rapidly.[30] Typical reactions for the conversion of suitable feedstocks to benzene are

Dehydrogenation:

$$\underset{\text{Substituted cyclohexane}}{C_6H_{10}R_1R_2\ (H_2C,\ CH_2,\ CHR_1,\ CHR_2\ \text{ring})} \longrightarrow \underset{\text{Benzene}}{C_6H_6} + R_1CH{=}CH_2 + R_2CH{=}CH_2 + 4H_2$$

Aromatization:

$$\underset{\text{Methylcyclopentane}}{H_2C{-}CH_2{-}CCH_3{-}CH_2{-}H_2C\ \text{(ring)}} \longrightarrow \underset{\text{Benzene}}{C_6H_6}$$

[30] Inside Hydeal Aromatics, Benzene and Toluene, *Chem. Week*, Mar. 18, 1961, p. 32 (rough flowchart); Fedor, The Major Aromatics: An Industry in Transition, *Chem. Eng. News*, Mar. 20 and 27, 1961 (details and statistical table for production and use).

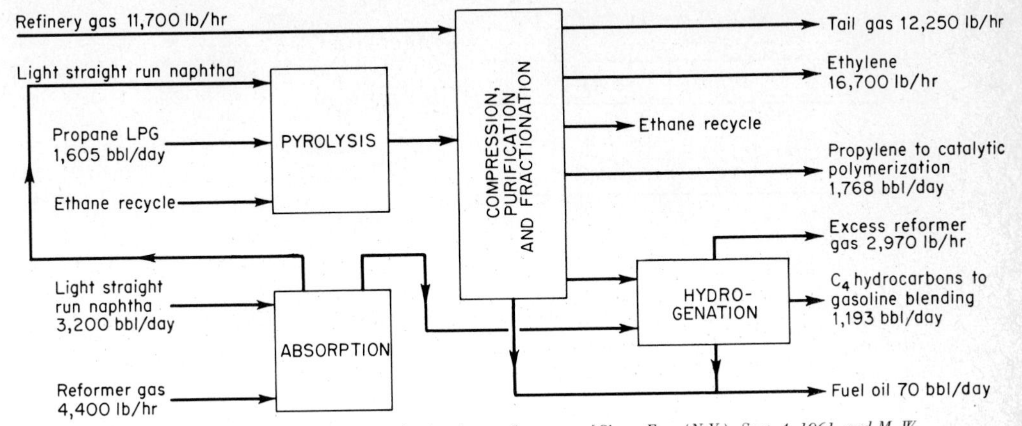

Fig. 37.5 Flowchart for ethylene and propylene production from refinery gas. [*Chem. Eng. (N.Y.), Sept. 4, 1961, and M. W. Kellogg Co;* see detailed flowcharts and data in *Kellograms* No. 1 (1962) and No. 1, (1964)]

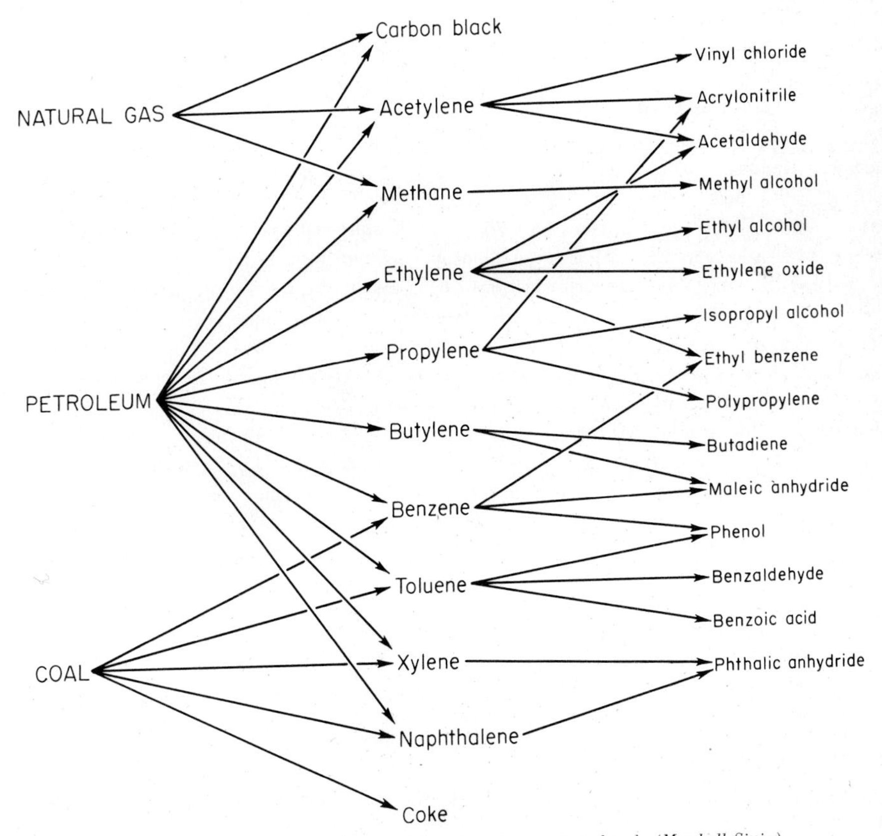

Fig. 37.6 Interrelationship of precursors from natural gas, petroleum cuts, and coal. (*Marshall Sittig.*)

TABLE 37.5 ***U.S. Aromatics Production—Over 95% Petroleum-Derived*** *(In millions of gallons per year)*

	*1980**	*1974**	*1973*	*1972*
Benzene	1,700	1,490	1,453	1,252
Toluene	1,300	1,035	958	916
Xylenes	4,000	3,625	3,386	739

Sources: Chem. Eng. News, Sept. 23, 1974, p. 17; *Chemtech*, **4**(12), 732 (1974); U.S. Tariff Commission Reports (Annual).
*Estimated.

Toluene[31] has long been made from petroleum raw materials, actually occurring in recoverable quantities from certain stocks. However, most is made by dehydrogenation of naphtha feedstock; for example, the following is a typical reaction:

Dehydrogenation:

$$\underset{\text{Methylcyclohexane}}{C_6H_{11}CH_3} \longrightarrow \underset{\text{Toluene}}{C_6H_5CH_3} + 3H_2 \qquad \text{(80\% conversion)}$$

Any excess toluene is demethylated to benzene or used to raise the octane value of gasoline.

Naphthalene with benzene has recently[32] been made by dealkylation on a heavy catalytic reformate stream at 1200°F psi, using a Cr_2O_3 and Al_2C_3 catalyst with a 10-s exposure. The naphthalene is purer than the usual product from coal tar. Many data have been published by the Ashland Oil Co.[33] about this Hydeal process, including flowcharts, yields, and costs.

LIGHT DISTILLATES These embrace naphthas and refined oils, aviation gasoline, motor gasoline, petroleum solvents, and jet-fuel kerosine. Gasoline heads the list as *the most important petroleum product.* With the advent of high-compression motors, the tendency of a fuel to knock, or detonate, violently (thought to be due to autoignition of part of the compressed charge in front of the flame) has become increasingly prevalent. Certain substances, such as tetraethyllead (TEL),[34] tetramethyllead (TML) (Chap. 38), and iron carbonyl tend to prevent knocking. *Octane number* is the percentage of isooctane (2,2,4-trimethylpentane) in a mixture with normal heptane, which as a sample fuel has the same knocking characteristics as the gasoline in question. *Aviation gasoline,* with an octane number of 100 or even higher, is composed of about one-third alkylate derived from isobutane and gaseous olefins, blended with catalytically cracked gasoline and suitable crude distillate, to which several cubic centimeters of TEL per gallon is added to reduce the knocking tendency. The amount of *sulfur* that can be safely allowed in gasoline is a controversial question. Quantities as high as 0.2% do not cause serious corrosion; the quantity of sulfur is usually limited by state regulation to 0.1%. Sulfur in a low percentage markedly reduces the effect of TEL in increasing the octane number. The color of a gasoline indicates little about its quality. Oil-soluble dyes[35] are required in leaded gasoline. Dyes also mask the off-white color of cracked gasoline. The term *naphtha* refers to any light-oil product having properties intermediate between gasoline and kerosine. Naphthas

[31]Nelson, *op. cit.*, pp. 697, 811, 815; Brownstein, U.S. Petrochemicals, Petroleum Publishing, 1972.

[32]Chopey, First Petrochemical Naphthalene, *Chem. Eng.* (*N.Y.*), **68**(9), 70 (1961) (pictured flowchart of Ashland Hydeal process); Naphthalene, Sun Oil Co. 1963 Petrochemical Handbook, p. 204, Gulf Publishing (flowchart).

[33]Hydrodealkylation Process, *Ind. Eng. Chem.*, **54**(2), 28 (1962); Dealkylation, *Chem. Eng.* (*N.Y.*), **69**(21), 80 (1963 Sun Oil); Anderson and Lamb, Naphthalene via Hydrodealkylation, *Ind. Eng. Chem. Process Des. Dev.*, **3,** 177–182 (1964) (computer and flowchart); Asselin and Erickson, Dealkylation Puts Petroleum Industry into the Benzene-naphthalene Business, *Oil Gas J.*, Mar. 19, 1962, p. 127; Paulsen, Benzene, U.S. Pat. 2,951,886 (1960); Ashland Oil & Refining Co., Ashland, Ky., special pamphlet on petronaphthalene, with flowchart and patent.

[34]Midgley, Tetraethyl Lead, *Ind. Eng. Chem.*, **17,** 827 (1925); Edgar, Tetraethyl Lead, in Progress in Petroleum Technology, pp. 221–234 (tables and properties). TEL Stand-Ins-Being Readied, *Chem. Eng.* (*N.Y.*), **81**(25), 40 (1974).

[35]Liddell, The Persistent Blue, *Color Eng.*, **1**(4), 8 (1963).

are used extensively as commercial solvents in paints, for dry cleaning, and as ethylene feedstocks. More *kerosine* is used in jet airplanes than aviation gasoline in all other airplanes. Kerosine is also used as a fuel and for illuminating purposes. Light heating oils are employed as fuel in home furnaces.

ADDITIVES TO PETROLEUM PRODUCTS The "minuscule" of additives to petroleum products are of great value in improving the performance of petroleum products in general when added to gasolines, distillates, fuels, lube oils, and brake fluids.[36]

The additives that are best known are the antiknock compounds TEL and TML. The estimated consumption of these materials was 970 million lb in 1972. Considerable environmental pressure to remove lead from gasoline exists. Unleaded gasolines require more high-octane components and more reforming to produce. Organophosphorus additives such as phosphorodithioates and tricresyl phosphate are combustion chamber deposit modifiers and are added to most premium gasolines sold today. *Antioxidants* (phenol derivatives such as 2,6-di-*tert*-butyl-4-methylphenol) stabilize the diolefins in gasoline and reduce gum formation. *Antistall* additives (carburetor de-icing) increase in volume with cold weather and frequently consist of freezing-point depressants, such as isopropyl alcohol. Among corrosion *inhibitors* are ammonium sulfonates. Additives to increase *cetane* number, such as alkyl nitrates, are often required for diesel oils. Additives for lubricating oils consist of zinc dialkyl dithiophosphates made from the reaction of phosphorus pentasulfide with alcohols or phenols, followed by neutralization with zinc oxide, for example, to inhibit corrosion and oxidation and help prevent formation of varnish or sludge in internal-combustion engines. For *detergent action* and retardation of ring sticking, alkaline-earth aromatic sulfonates, metal phosphates, or others are employed. *Pour-point depressants* and viscosity improvers are also used widely. Antifreeze and de-icing fluids should also be mentioned. Ethylene glycol is still the most used antifreeze and 1.7 billion lb. is sold yearly for this purpose.

INTERMEDIATE DISTILLATES These include gas oil, heavy furnace oil (domestic), cracking stock, diesel fuel oil, absorber oil, and distillates cracked and reformed to produce gasoline of adequate quality. Often distillates are blended with heavy tar to reduce the viscosity of the tar so that it can be marketed as fuel oil. Certain special heavy naphthas are used to reduce the viscosity of asphalt so that it can be readily applied as road oil; such material is known as *cutback asphalt.* Originally, gas oil was pyrolyzed to enrich artificial gas, but today most of it is used as a fuel or cracked into gasoline. *Diesel fuel* is a special grade of gas oil which has become an important specialty in recent years. Proper viscosity of diesel fuel is essential and should be held within rigid limits. Distillates may also be used as vehicles for insecticides.

HEAVY DISTILLATES furnish lubricating oils (also from residues), heavy oils for various purposes, and waxes. Heavy distillates are also hydrocracked to lighter distillate fuels and to gasoline. Of the 4,280 million bbl of crude oil run to stills in 1972, lubricating oils and greases accounted for 67.8 million bbl. The field of *lubricating oils*[37] is too broad to permit a complete discussion of it here. The Society of Automotive Engineers (SAE) has greatly assisted in classifying oils by introducing a number system based on viscosity and change in viscosity with temperature (viscosity index). Tests involving the flash point, viscosity, pour point, emulsibility, and resistance to sludging are useful in determining the application to which an oil may be put. Solvent extraction (Fig. 37.7 to 37.9) and chemical treatment have long been important operations in upgrading lubricants. The performance of most lubricants has been improved by the use of *additives* (0.001 to 25% or more) such as antioxidants, detergents, extreme-pressure agents, antifoam compounds, viscosity index improvers and antiscuff agents. Refined wax[38] is widely used industrially, particularly for treating

[36]Kalichevsky, Petroleum Fuel Additives, ECT, 1st Suppl., p. 648, 1959; Ranney, Corrosion Inhibitors, Noyes, 1976; Whitcomb, Non-Lead Antiknock Agents for Motor Fuels, Noyes, 1975.

[37]Nelson, *op. cit.*, (many tables and references, fine presentation, see index); Pigott and Ambrose, Petroleum Lubricants, in Progress in Petroleum Technology, pp. 235–245 (35 refs); Ranney, Synthetic Oils and Greases for Lubrication, Noyes, 1976 (over 290 processes are described).

[38]Wax, *Chem. Week*, Mar. 21, 1964, p. 47.

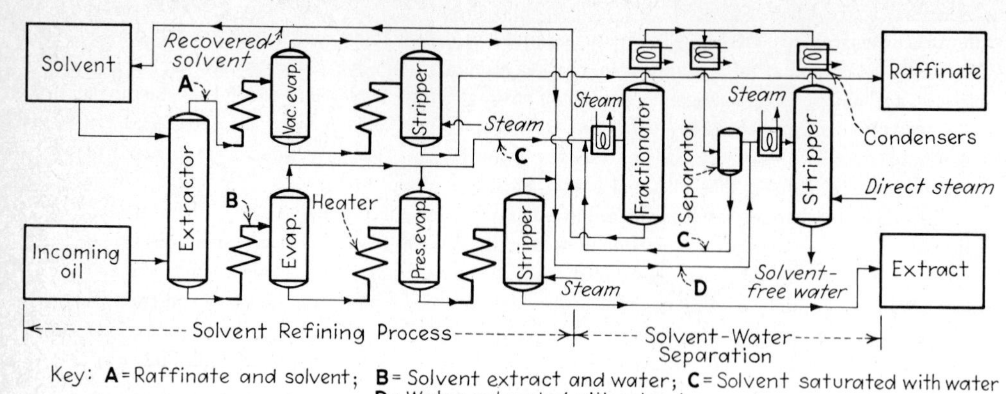

Fig. 37.7 Typical solvent-refining process employing furfural and provided with solvent-water separation and recovery.

paper. Refining may be carried out by separating into several narrow-melting fractions, by sweating, and by progressive crystallizations. Refined wax is finished by acid (sulfuric) treatment and percolation through an adsorbent clay such as the Attapulgus variety.

Residues include asphalt, residual fuel oil, coke, and petrolatum. These are by-products, or residues, from regular refining processes. Petroleum coke is used commercially for electrodes, in the manufacture of calcium carbide, in paints, and in the ceramics industry. *Asphalt,*[39] also of great importance, finds use as a paving or roofing material for waterproofing structures. The properties of asphalt may be markedly altered by heating it to a high temperature and partly oxidizing it by blowing air through it. Such material, known as *oxidized* asphalt, is more viscous and less resilient than ordinary asphalt and is widely used in the manufacture of roofing and for grouting. Very hard asphalt finds some use as a briquetting binder.

Greases, constituting a large group of different materials, fall into three classes:

1. Mixtures of mineral oil and solid lubricants.
2. Blends of waxes, fats, resin oils, and pitches.
2. Soap-thickened mineral oils.

PETROLEUM CHEMICALS, OR PETROCHEMICALS[40] are being derived from petroleum products and natural gas in increasing amounts. Examples of these are ammonia (Chap. 18), carbon black (Chap. 8), butadiene and styrene (Chap. 36), and more than several thousand inorganic and organic chemicals of increasing value (Chap. 38).

MANUFACTURE, OR REFINING

The general characteristic of petroleum refining has been an unusually economical processing of crudes into consumer products. It comes as a surprise to learn that refiners receive only 35% of the retail price of gasoline for their product. The difference between what the consumer pays and what the refiner receives is accounted for by taxation, transportation costs, and the expense of retailing.

The refining, or manufacturing, of petroleum products and of petroleum chemicals involves two major branches, physical changes, or *separation operations,* and chemical changes, or *conversion processes.*

[39] Holberg, ECT, 2d ed.; vol. 2, pp. 762–806, 1963; Hughes and Hardman, Asphalts and Waxes, in Progress in Petroleum Technology, pp. 262–277 (128 refs.); Holberg, Bituminous Materials: Asphalts, Tars, Wiley-Interscience, 1964–1966, 3 vols.
[40] ECT, vol. 10, p. 177, 1963; see table III, pp. 202–209, for more than 250 important commercial petroleum chemicals, listed with their process of manufacture and chief use.

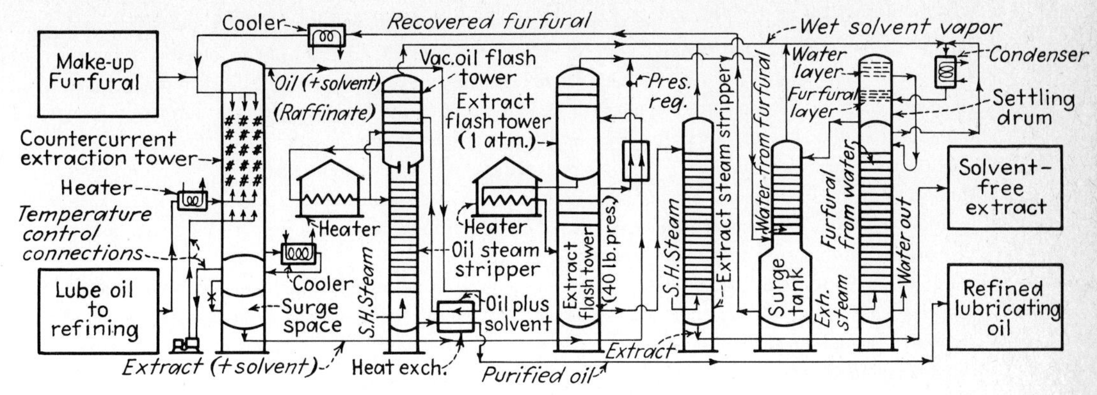

Fig. 37.8 Flowchart for lube-oil refining by furfural extraction.

Initially refining embraced separation by distillation, which involves the unit operations of fluid flow, heat transfer, and distillation. Indeed, the necessity for studying these manufacturing aspects of the petroleum field stimulated the development of this phase of chemical engineering. These purely physical separations were early supplemented by chemical conversions in the further *refining* of petroleum products. But the great stimulation to the employment of chemical change in the manufacture of petroleum products came with the growing consumption of gasoline in excess of that supplied by separation distillation. This situation, developing after 1912, forced the application of pyrolysis to petroleum products wherein, by what the industry calls *cracking*, long molecules were broken down into shorter ones suitable for gasoline. Although cracking, thermal and catalytic, is still the most important chemical change used in the petroleum industry, in recent years, to meet demands for better gasoline and for alcohol, acetone, or various chemicals from petroleum, other chemical conversions have been applied on a large scale. Among these we list alkylation, isomerization, polymerization, hydrogenation, catalytic reforming, and dehydrogenation.

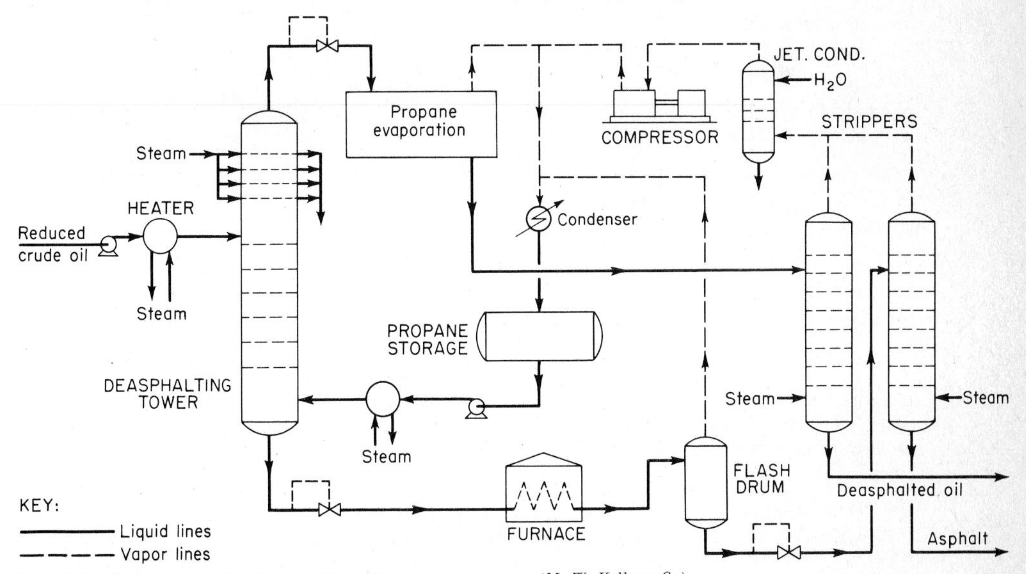

Fig. 37.9 Flowchart for lube-oil deasphalting—Kellogg tower process. (*M. W. Kellogg Co.*)

It is a long step from the initial shell-still topping[41] gasoline, and other like products from the crude, to the modern *continuous* complicated refinery wherein 200,000 bbl or more per day of crude, day in and day out, enters the manufacturing equipment by way of pipe stills and is separated in continuous fractionating towers into the various constituents from gas through petrochemicals, gasoline, kerosine, gas oil, lubricating oil, down to fuel oil, residuums, or tars. Such refineries are also equipped to carry on most of the secondary treatments in a continuous manner. Naturally, each individual treatment depends on the type of crude oil processed. Furthermore, procedures vary with market demand. With modern increased demands for gasoline, manufacturing procedures naturally are adapted to obtain a maximum economic yield of gasoline.

ENERGY CHANGES The energy changes involved in petroleum refining are both mechanical and chemical. Some of the chemical conversions are exothermic reactions, though most are endothermic. In addition to the heat[42] that must be supplied for cracking, distillation, and other processes, large amounts of mechanical energy are needed for pumping, transportation, compressing gases, and the like.

SEPARATION OPERATIONS In theory, the unit operations of oil refining seem simple; in actual practice they are quite complicated. A typical refinery consists of one or more units, called *stills*, which include: a furnace, an oil heater, a fractionating tower, steam strippers, heat-exchange equipment, coolers, and condensers; working storage tanks at the unit; batch agitators or continuous closed-tank units for treating the products to remove deleterious compounds of sulfur and produce acceptable color; filters; blending and mixing tanks for loading; a piping system for the receipt of crude oil; pumps for the transfer of oils and for loading and shipping products; storage tanks for the crude-oil supply and finished product; and a vapor-recovery system and many auxiliaries. A power plant for generation of steam, and perhaps electric power and light, should also be included. In the operation of this large entity, heat, energy, and material balances are of utmost importance in keeping rigid control of all steps in the processing of the oil.

The separation branch of refining, can be broken down into the ordinary unit operations, as follows.

Fluid flow.[43] The petroleum engineer has turned to the comprehensive literature on fluid flow of water and suitably revised this information for petroleum. Two important differences between oil and water should be recognized: oils exhibit wide change in viscosity with temperature, and they are sensitive to heat.

Heat transfer.[44] The solution of this problem is somewhat complicated because complete data are lacking. In general, the use of empirical results is desirable. Equipment should be cleaned regularly to maintain satisfactory heat transfer and to remove fouling, which frequently greatly reduces transfer rates. To reduce the amount of cooling water used, closed-loop cooling towers and water-treating equipment are essential.

Distillation.[45] This ranks among the more important unit operations. In this operation separation is based upon volatility, and the process stream may be separated by distillation into a more volatile, "lighter," component and a less volatile, "heavier," component. The older operations of batch distillation have now been almost entirely superseded by continuous ones. The presently used system involves heating the crude by pumping it through tubes placed inside a furnace (tube

[41]Topping involves the simple distillation of the more highly volatile materials, or what may be called *tops*, from the crude oil.

[42]The following references bear on this subject: Schoen, New Chemical Engineering Separation Techniques, Interscience, Wiley, 1962; Nelson, *op. cit.* (pp. 632–635 for heats of cracking; p. 657 for typical calculations of material balances); Gruse and Stevens, The Chemical Technology of Petroleum, 3d ed., McGraw-Hill, 1960, for thermodynamic data for petroleum hydrocarbons; pp. 210ff. for latent heats of evaporation and specific heat; Perry, pp. 13-2 and 13-54, for distillation; Science of Petroleum, vol. 2, p. 1256.

[43]Perry, 5th ed., sec. 5 and 6; Nelson, *op. cit.*, pp. 395–413.

[44]Perry, 5th ed., Sec. 10; Nelson, *op. cit.*, pp. 528–584, 547–555, for fouling.

[45]Perry, 5th ed. sec. 13, for distillation (should be consulted for many references to literature, presentation of fundamental principles, and applications); Nelson, *op. cit.*, pp. 226–262, 435–438.

still) and then allowing it to vaporize in a fractionating column, which is tapped at several points, allowing continuous side draw of the various boiling fractions, or products. The residuum withdrawn from the bottom of the column may be subjected to vacuum or steam distillation. Figure 37.3 is a schematic drawing of a modern refining group, showing the separation of the crude into the various fractions and the subsequent treatment of each fraction. Several products are removed from the crude tower as shown. The kerosine and naphtha fractions contain small amounts of imperfectly separated straight-run gasoline of higher volatility than the main fraction. These are removed in strippers (short columns containing only a few plates) by blowing in steam. The gasoline vaporizes out the top of the stripper and is returned to the crude tower.

In many applications of distillation to separation of petroleum products, the volatility difference is too low to be practical, and it must be enhanced by the addition of a solvent or an entrainer. The variation of distillation in which a solvent of low volatility is used to enhance separation is called *extractive distillation.*[46] An example is the separation of butenes from butanes with furfural. When a solvent of high volatility is added as an entrainer, the operation is called *azeotropic distillation.* An example of azeotropic distillation is the production of high-purity toluene using methyl ethyl ketone as the entrainer, and also distillation to produce anhydrous alcohol (Chap. 31). Sieve trays,[47] or *Linde trays,* increase distillation efficiency.

Absorption[48] is generally used to separate a higher-boiling constituent from other components of a system of vapors and gases. Usually, the absorption medium is a special-gas-oil. Absorption is widely employed in the recovery of natural gasoline from well gas and of vapors given off by storage tanks. Absorption also obtains light hydrocarbons from many refining processes (catalytic cracking, hydrocracking, coking, etc.). The solvent oil may be heavy gasoline, kerosines, or even heavier oils. The absorbed products are recovered by fractionating or steam-stripping.

Adsorption[49] is employed for about the same purpose as absorption; in the process just mentioned natural gasoline may be separated from natural gas by adsorption on charcoal. Adsorption is also used to remove undesirable colors from lubricating oil, usually employing activated clay.

Filtration[50] after chilling is the usual method for the removal of wax from wax distillates. The mixture of wax and adhering oil obtained from the press is frozen and allowed to warm slowly so that the oil drains (sweats) from the cake, thus further purifying the wax. *Contact* filtration, involving the use of clay, is the common method of purification of oils; decolorization takes place at the same time. (This is an adsorption phenomenon.) Propane and methyl ethyl ketone (MEK) solvent dewaxing are used.

Crystallization[51] ranks as one of the oldest separation operations. By means of crystallization, wax may be removed from crude oil or from lubricating oil to yield crystalline and microcrystalline waxes of low oil content. Crystallization has also been applied in the separation of *p*-xylene from other C_8 aromatics in high purity in terephthalic acid manufacture.

Extraction[52] involves the removal of a component from a liquid by means of the selective solvent action of another liquid. The procedure of selective extraction by solvents is important in the further refining of lubricating oils. Another example is the production of benzene, toluene, and xylenes by extraction from specially processed petroleum. Low-viscosity-index hydrocarbons, unstable sludges, and some colored materials may be removed from lubricating oil in this way. Usually, the extraction

[46]Perry, 5th ed., 13–43.
[47]Optimum Sieve Tray Design, *Chem. Eng. Prog.*, **71**(8), 68 (1975).
[48]Perry, 5th ed., sec. 14; Nelson, *op. cit.*, pp. 823–827, 850–861.
[49]Perry, pp. 16-2 to 16-50; Nelson, *op. cit.*, pp. 342–346; Esso Achievement, *Chem. Eng.* (*N.Y.*), **70**(23), 236 (1963); adsorption process for hydrocarbon purification wherein adsorbent is regenerated by depressurizing: Linde's Molecular Sieves, *Chem. Eng.* (*N.Y.*), **68**(23), 23 (1961).
[50]Perry, pp. 19-57 to 19-85; Nelson, *op. cit.*, pp. 308–313, 320, 323, 337–342.
[51]Perry, pp. 19-26 to 19-33.
[52]Perry, pp. 15-15ff. Solvent Extraction, International Symposium, *Chem. Eng. Prog.*, **62**(9), 49–104 (1966).

is countercurrent. Two problems are involved: obtaining solution equilibriums and separating the two immiscible phases. An example, the extractive refining of lubricating oils with furfural, is illustrated in Fig. 37.8. Furfural is particularly effective in removing color bodies, sulfur compounds, and oxygen-containing molecules from lube-oil stock. Figure 37.7 shows a typical solvent-refining process for lubricating oil. The oil is mixed with the solvent or solvents in an extractor of column. If a proper solvent[53] has been chosen, the mixture separates into two layers, one rich in solvent and containing the dissolved impurities (extract), and the other containing little solvent and most of the desirable oil (raffinate). The latter consists largely of paraffinic compounds and wax. Dewaxing with propane or MEK follows the selective solvent separation in Fig. 37.7. Sometimes deasphalting is part of the sequence (Fig. 37.9).

The procedure as shown in Fig. 37.7 and 37.8 involves the following *unit operations* when furfural is the solvent:

Continuous countercurrent extraction of the lubricating stock with furfural at temperatures between 130 and 280°F, depending on the oil used; suitable heat exchangers are provided.

Continuous separation of the raffinate fraction from the extract fraction.

Recovery of solvent (furfural) by vacuum evaporation from raffinate or refined oil.

Stripping by steam distillation of small amounts of remaining solvent from refined oil, giving wet furfural or water solution of furfural.

Recovery of solvent (furfural) from extract by atmospheric and by pressure distillations, this wet solvent (furfural) being the main recovery; fractionation leaves dry solvent behind and ready for reuse.

Stripping with steam of small amounts of solvent left in the extract, furnishing wet solvent or water solution of solvent.

Final stripping of solvent from combined aqueous solutions; overhead is chilled, and solvent is conducted to fractionator.

When furfural is used, solvent is recirculated through the system as many as 15 times each day and with a very small loss, less than 0.03%, of solvent recirculated.

Other solvents frequently used are liquid sulfur dioxide, propane and cresylic acid, dichloroethyl ether, phenol, and nitrobenzene.[54] Solvents are also widely employed for removing the wax from lubricants and thus lowering the pourpoint.[55] Figures 37.8, 37.9, and 37.10 present typical lubricating-oil purifications.

CONVERSION PROCESSES Petroleum offers such a fertile field for chemical synthesis, both for gasoline and for petrochemicals, that it is difficult to list all the conversion or unit processes that may be applied to this raw material. Even now "about 70% of U.S. crude is subjected to conversion processing."[56] Much study has gone into the fundamentals of these chemical[57] changes as applied to petroleum raw materials. Schmerling makes a division into reactions in "presence of acid type catalysts" and "those which occur thermally or are induced by peroxides." The former are explained by a carbonium-ion mechanism, and the latter by free radicals. The following are examples of a few of the more important basic reactions:[58]

[53]Nelson, *op. cit.*, pp. 347–373; Gester, Solvent Extraction, in Progress in Petroleum Technology, pp. 177–198 (excellent); Treybal, Liquid Extraction, 2d ed., McGraw-Hill, 1963.

[54]Lube Oils, *Hydrocarbon Process.*, **53**(12), 59ff. (1974).

[55]Solvent Dewaxing of Lubricants, *Chem. Metall. Eng.*, **48**(10), 106 (1941) (pictured flowchart employing methyl ethyl ketone and benzene).

[56]Sweeney, *op. cit.*, chap. 3, p. 2.

[57]Schmerling, The Mechanism of the Reactions of Aliphatic Hydrocarbons, *J. Chem., Educ.*, **28**, 562 (1951) (48 refs.); Progress in Petroleum Technology; Gorring and Weekman, Applied Kinetics and Chemical Engineering, *Ind. Eng. Chem.*, **58**(9), 18 (1966).

[58]Gruse and Stevens, *op. cit.*, chap. 3, Group Reactions in Petroleum Oils and Derived By-products; Progress in Petroleum Technology.

1. *Cracking, or pyrolysis* (Figs. 37.11 to 37.15). The breaking down of large hydrocarbon molecules into smaller molecules by heat or catalytic action. Zeolite catalysts are common.

$$\underset{\text{Heavy gas oil}}{C_7H_{15}\cdot C_{15}H_{30}\cdot C_7H_{15}} \longrightarrow \underset{\text{Gasoline}}{C_7H_{16}} + \underset{\text{Gasoline (antiknock)}}{C_6H_{12}{:}CH_2} + \underset{\text{Recycle stock}}{C_{14}H_{28}{:}CH_2}$$

2. *Polymerization.* The linking of similar molecules; the joining together of light olefins.

C   C   C   C
C—C═C + C—C═C $\xrightarrow[\text{or catalyst}]{\text{heat and pressure}}$ C—C—C—C═C + C—C—C═C—C
              C C   C C

Unsaturated short chains 82% Longer chains 18%

3. *Alkylation* (Fig. 37.18). The union of an olefin with an aromatic or paraffinic hydrocarbon. Unsaturated + isosaturated $\xrightarrow{\text{catalyst}}$ saturated branched chain, e.g., catalytic alkylation:

C═C—C—C + C—C—C (with C on middle carbon of isobutane)
1-Butylene Isobutane

C—C═C + C—C—C (with C on second carbon of each)
Isobutylene Isobutane

⟶

C—C—C—C—C (with two C on carbon 2 and one C on carbon 4)
2,2,4-Trimethylpentane, or "isooctane"

C—C—C—C—C (with one C on carbon 2 and two C on carbon 3)
2,3,3-Trimethylpentane

C—C—C—C—C (with one C on each of carbons 2, 3 and 4)
2,3,4-Trimethylpentane

C═C + C—C—C (with C on middle carbon) ⟶ C—C—C—C (with two C on carbon 2)
Ethylene Isobutane 2,2-Dimethylbutane, or neohexane

4. *Hydrogenation.* The addition of hydrogen to an olefin.

C—C—C═C—C (with two C on carbon 2 and one C on carbon 4) $\xrightarrow[\text{catalyst}]{H_2}$ C—C—C—C—C (with two C on carbon 2 and one C on carbon 4)
Diisobutylene Isooctane

5. *Hydrocracking*

$$\underset{\text{Heavy gas oil}}{C_7H_{15}\cdot C_{15}H_{30}\cdot C_7H_{15}} + H_2 \longrightarrow \underset{\text{Straight-chain}}{C_7H_{16}} + \underset{\text{Branched-chain gasoline}}{C_7H_{16}} + \underset{\text{Recycle stock}}{C_{15}H_{32}}$$

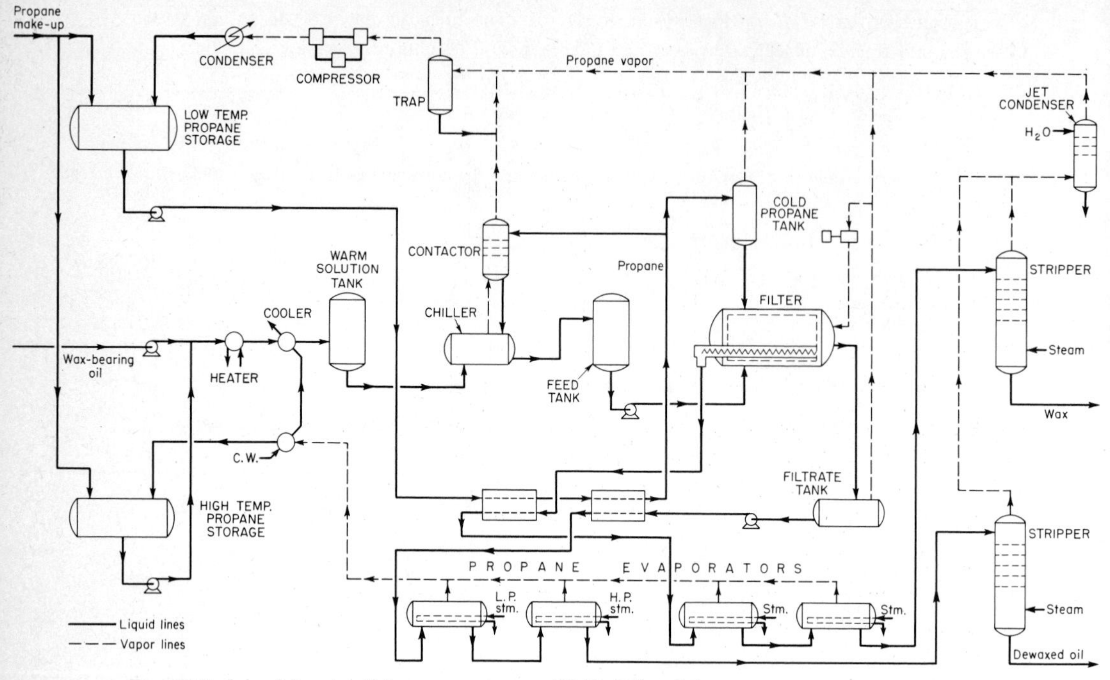

Fig. 37.10 Lube-oil dewaxing—Kellogg propane process. (*M. W. Kellogg Co.*)

6. *Isomerization.* Alteration of the arrangement of the atoms in a molecule without changing the number of atoms.

$$\underset{\text{Straight chain}}{\mathrm{C{-}C{-}C{-}C}} \xrightarrow[\mathrm{AlCl_3}]{300°\mathrm{C}} \underset{\text{Branched chain}}{\mathrm{C{-}\overset{\overset{\displaystyle C}{|}}{C}{-}C}}$$

7. *Reforming, or aromatization.* The conversion of naphthas to obtain products of higher octane number. Similar to cracking, but more volatile charge stocks are used. Catalysts usually contain rhenium, platinum or chromium.

$$\underset{\text{Methylcyclohexane}}{C_6H_{11}CH_3} \xrightarrow[\text{catalyst}]{\text{heat and}} \underset{\text{Toluene}}{C_6H_5CH_3} + 3H_2$$

$$\underset{n\text{-Heptane}}{CH_3CH_2CH_2CH_2CH_2CH_2CH_3} \xrightarrow[Cr_2O_3 \text{ on } Al_2O_3]{\text{heated with}} \underset{\text{Toluene}}{C_6H_5CH_3} + 4H_2$$

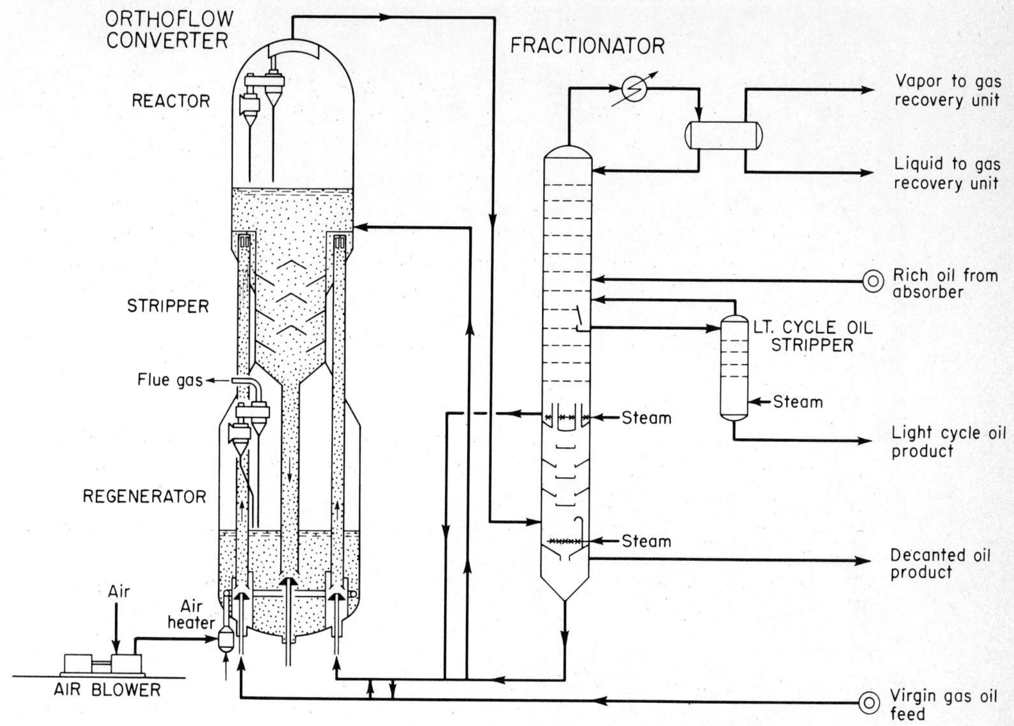

Fig. 37.11 Orthoflow catalytic-cracking converter and adjuncts. Stippled areas represent fresh or regenerated catalyst. (*M. W. Kellogg Co.*)

Operating conditions generally fall within the following ranges:

Reaction temperature, °F	885–1020
Reactor pressure, psig	10–30
Regenerator temperature, °F	1250–1400
Regenerator pressure, psig	15–35
Catalyst/oil ratio	6–20
Space velocity, W/(h)(W)	1.0–16.0

Yields from gasoline cracking can be varied within limits to suit a refiner's special local conditions. Typical yields from a Kellogg Orthoflow unit operating at a relatively high conversion are:

Charge gravity, API	27.0
Throughput ratio	1.95
Conversion, vol %	76.0
C_2 and lighter, wt %	4.0
C_3, vol %	9.7
C_4 + gasoline, vol %	67.4
Light cycle oil, vol %	21.0
Fractionator bottoms, vol %	3.0
Coke, wt %	10.0

A typical set of consistent operating conditions is:

Reactor temperature, °F	925
Reactor pressure, psig	9.0
Regenerator temperature, °F	1150
Regenerator pressure, psig	16.0
Catalyst/oil ratio	6.2
Space velocity, W/(h)(W)	2.4

Typical requirement,* unit per bbl fresh feed:	
Electricity, kWh	6†
Steam, lb	(5.2)‡
Water, cooling (40°F rise), gal	330
Water, boiler feed, gal	12
Catalyst consumption, lb	0.2–0.4

*Excludes recovery plant. †Using electric drive on the air blower. ‡There is a net production of steam. Also, 37,000 Btu heat is available for use in the recovery plant.

Fig. 37.12 Orthoflow fluid catalytic cracker. (*M. W. Kellogg Co.*)

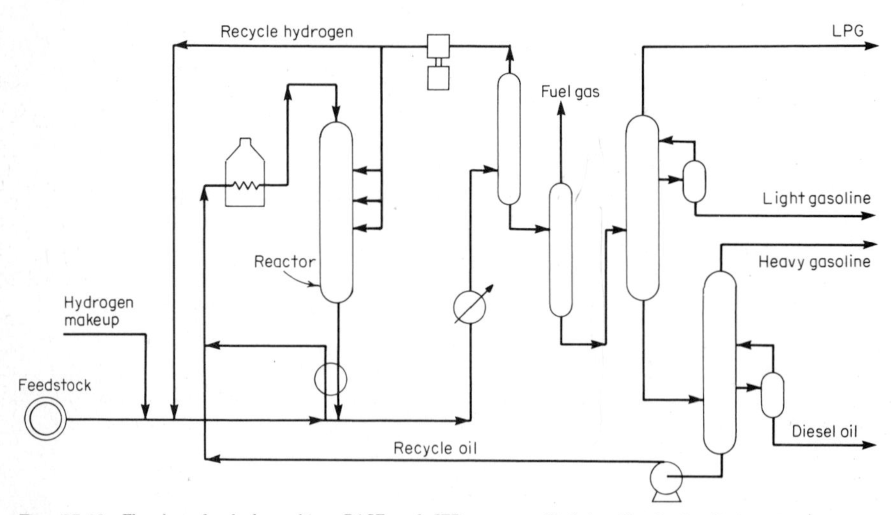

Fig. 37.13 Flowchart for hydrocracking—BASF and IFP process. [Refining Handbook, *Hydrocarbon Process.* **53**(9), 127 (1974). *Copyright © 1974 by Gulf Publishing Co., Houston, Tex. All rights reserved. Used with permission.*]

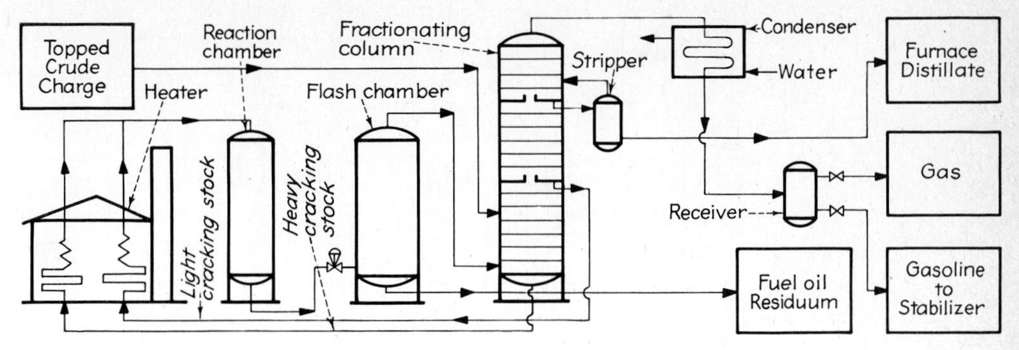

Fig. 37.14 Universal Oil Products two-coil selective cracking unit.

8. *Esterification and hydration* (Chap. 38):

$$C_2H_4 + H_2SO_4 \longrightarrow C_2H_5O\cdot HO\cdot SO_2 + (C_2H_5O)_2\cdot SO_2$$

$$C_2H_5O\cdot HO\cdot SO_2 + (C_2H_5)_2O\cdot SO_2 + H_2O \longrightarrow H_2SO_4 \text{ dil.} + C_2H_5OH + C_2H_5OC_2H_5$$

Cracking, or pyrolysis. Cracking[59] (pyrolysis) is the process of converting large molecules into smaller ones by the application of heat and/or catalysts. Olefins are always formed. A significant amount of polymerization of these smaller molecules occurs, and some carbon is formed. Nelson summarizes the reactions:

Charge stock: $C_7H_{15}\cdot C_{15}H_{30}\cdot C_7H_{15}$ (Heavy gas oil)

↓

Cracked stock: $\underset{\text{Gasoline}}{C_7H_{16}} + \underset{\text{Recycle stock}}{C_{14}H_{28}{:}CH_2} + \underset{\text{Gasoline (antiknock)}}{C_6H_{12}{:}CH_2}$

↓

More cracking: $\underset{\text{Gas}}{C_2H_6} + \underset{\text{Gasoline}}{(C_4H_8{:}CH_2 + C_8H_{18} + C_6H_{12}{:}CH_2)} + \underset{\text{Gum-forming materials}}{CH_2{:}CH\cdot CH{:}CH\cdot CH_3} + \underset{\text{Gas}}{C_2H_4}$

↓

Polymerization: $\underset{\text{Gas}}{C_2H_6} + \underset{\text{Gasoline}}{(C_4H_8{:}CH_2 + C_8H_{18})} + \underset{\text{Tar or recycle}}{C_{12}H_{22}} + \underset{\text{Gas}}{C_2H_4}$

As of Jan. 1, 1974, the U.S. cracking and reforming capacity was 14 million bbl/day. These equations are highly simplified examples of this process, for the original molecule may be broken at any one of the carbon-carbon bonds, usually with carbon formation. Thus, in cracking one pure compound, such as tetradecane, several products (heptane, heptene, hexane, octane, butane, butenes, carbon, etc.) are formed. In cracking the mixture of compounds occurring in petroleum, the number of possible products is so great that representation of the process by simple equations is impossible.

The *heat of decomposition*[60] of petroleum is difficult, if not impossible, to calculate accurately, because the necessary data are lacking. In the cracking reaction, if large amounts of gas are produced, the heat of decomposition is endothermic and relatively high. Also, the heat of reaction is often lower than theoretical, because of the exothermic reforming of some of the cracked products.

[59]ECT, vol. 10, pp. 109ff., 1953; Gruse and Stevens, *op. cit.*, chap. 10; Nelson, *op. cit.*, chap. 19, pp. 627ff., and chap. 21, pp. 759ff. (many references, tables, flowcharts); cf. Progress in Petroleum Technology; Refining Handbook Issue, *Hydrocarbon Process.*, **53**(9), 105ff. (1974).

[60]Nelson, *op. cit.*, p. 632. Discusses this subject and gives examples of typical calculations.

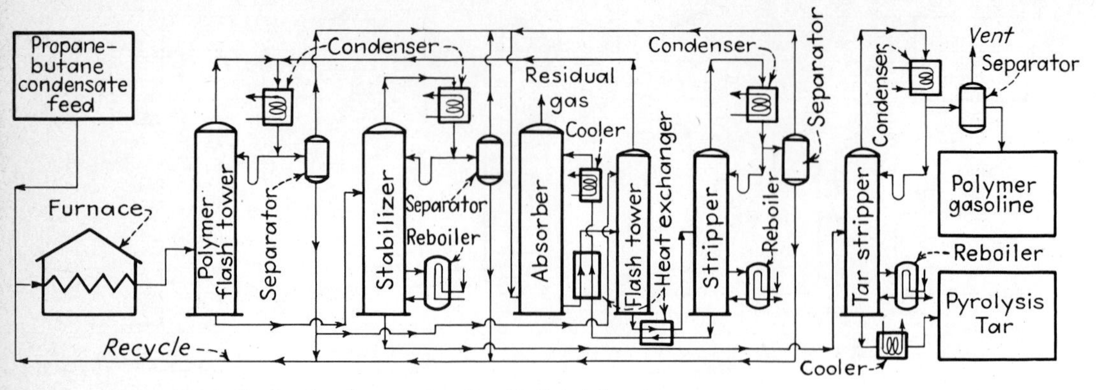

Fig. 37.15 Flowchart for polymer gasoline—Pure Oil (Union) thermal process.

"The cracking process," according to Egloff,[61] "is the greatest force for conservation that has been developed in the oil industry." Cracking is of two kinds: thermal and catalytic. In the early days of *thermal cracking*, or *reforming*, gas oil was used as the charging stock. More recent technological developments have advanced the technique of cracking, making possible its adaptation to a wide range of materials from naphthas to heavy crude residuums. In the combination distillation-cracking unit, crude oil from the storage tanks is separated into various fractions directly, and the resulting fractions are cracked in continuous operations. Selective cracking operations with a high degree of flexibility have been made possible by catalytic cracking and by thermal cracking, the latter using multiple heating coils and temperatures ranging from 900 to 1100°F and pressures from 600 to over 1,000 psi. Today, total gasoline yields of 70 to 85%[62] are obtained from crude oil, and more than one-half of the country's gasoline is produced by cracking. See Fig. 37.3 for generalized relationships of cracking in the petroleum industry.

Catalytic cracking, or reforming, has the advantage that gasoline of high quality can be produced from almost any crude oil in equipment subject to careful control and operated at a low pressure, hence at comparatively lower cost. Catalytic gasolines also have the desirable characteristics of excellent response to TEL, being low in gum formers and in corrosive sulfur compounds, and have a uniform octane rating over the boiling range of gasoline. Sittig writes,[63] "Catalytic cracking represents the largest single stride in the development of refining processes that the petroleum industry has ever made." See Table 37.6 for catalysts used in reforming. Sachanen[64] states that catalytic cracking has these advantages over thermal cracking:

1. More selective cracking and fewer end products.
2. More olefin isomerization, both of olefin bonds and carbon skeletons.
3. More controllable saturation of double bonds.
4. Greater production of aromatics.
5. Less diolefin production.
6. Relatively more economically salable coke.
7. Greater ability to tolerate high-sulfur feedstocks.

[61]Egloff, 300 Years of Oil, *Ind. Eng. Chem.*, **27,** 648 (1935).

[62]Nelson, *op. cit.*, pp. 146–152, for a tabulation of the gasoline content of many crudes; pp. 635–649, for cracking yields.

[63]Catalytic Cracking Techniques, in review reprint of *Petroleum Refiner*, Gulf Publishing 1952 (545 refs. and many diagrams and tables; excellent summary for *all* phases of catalytic cracking).

[64]Conversion of Petroleum, 2d ed., Reinhold, 1948 (as summarized by Sittig); Catalysts, CW Report, *Chem. Week*, Aug. 17, 1963, p. 51.

TABLE 37.6 *Catalysts Used in Refining*

Process	*Type*	*Millions of pounds per year*	*Millions of dollars per year*
Catalytic cracking			
Fluid bed	Zeolite-based alumina-silica	226.0	56.50
Fluid bed	Amorphous alumina-silica and clay	14.0	2.30
Moving bed	Zeolites	27.2	9.50
Moving bed	Amorphous	6.8	0.77
Total		274.0	69.07
Catalytic reforming			
Replacement	Monometallic	0.8	2.80
Replacement	Bimetallic	3.2	13.60
New	Bimetallic	1.0	4.30
New	Precious metal value	. . .	9.60
Total		5.0	30.30
Hydrotreating			
Replacement	Cobalt-molybdenum, nickel-molybdenum,	8.0	8.20
New	Nickel-tungsten and sand	2.0	2.0
Total		10.0	10.20
Hydrocracking			
New and replacement	Noble metal on amorphous support, nonnoble metal and noble metal on molecular sieves	. . .	12.80
Alkylation	Sulfuric acid	3,100.0*	39.00*
	Hydrofluoric acid	22.0	6.70
Total		3,122	45.70
Totals		3,411.0	168.07

Source: *Chem. Week*, Nov. 1, 1972, p. 26.
*Gross totals—net consumption is 12 to 14% of gross.

The *fluidized catalytic cracking* process employs a finely divided solid catalyst made by aerating the ground powder of an alumina-silica gel. This is maintained at all times as a simulated fluid by suspension in the reacting vapors or in the regenerating air. A high degree of turbulence is necessary throughout the system to ensure a uniform suspension. Under these conditions the solid catalyst similarly flows to a liquid and exerts a like pressure. Because of the even distribution of the catalyst and because of its high specific heat in relation to the vapors reacting, the entire reaction can be maintained at a remarkably exact temperature. Separation of the catalyst from gases or vapors after the reaction and regeneration processes is done largely by means of cyclone separators. Coke and tar are formed on the catalyst by the cracking. These are burned from the catalyst with air during the regeneration part of the cycle. Both cracking and regenerative cycles are continuous.

The M. W. Kellogg Co. Orthoflow fluid catalytic cracking unit[65] design carries out reaction, catalyst stripping, regeneration, and catalyst circulation in a single Orthoflow converter (Figs. 37.11 and 37.12). The reactor is mounted above the regenerator, and the catalyst stripping section is located in between, so that the stripper supports the reactor. The Orthoflow design provides straight-line flow of the catalyst between the converter compartments, virtually eliminating the erosion normally encountered in pipe bends. Within the converter, the powdered synthetic silica-alumina catalyst is maintained in fluidized beds by controlled upflowing vapors. Circulation from section to section is achieved by maintaining different concentrations in the various catalyst circulation lines: regenerated catalyst risers for fresh feed and/or recycle and spent-catalyst standpipes.

During operation the hot catalyst from the regenerator enters the feed-regenerated catalyst risers through the plug valves. Simultaneously, feed and dispersion streams are fed through the hollow stem

[65]*Oil Gas J.*, **70**(44), 49 (1972). and *Hydrocarbon Process.*, **53**(9), 119 (1974).

of the valves and mixed with the catalyst. Heat from the carefully regulated amount of hot catalyst vaporizes the oil and raises it to the required reaction temperature. As the oil and disperse-phase catalyst pass up the risers, the cracking reaction takes place. The risers terminate in the reactor, where additional cracking takes place, either in a dense bed or a disperse phase. The cracked products, steam, and inerts enter a disengaging space, from which the catalyst falls back into the bed. Reactor effluent passes upward through cyclones for separation of the catalyst from reactor overhead vapors. Spent catalyst from the cracking reaction leaves the reactor and passes through the stripper, where it is steam-stripped to remove cracked products and cycle oil trapped by the down-flowing catalyst. The stripper catalyst then falls, in a dense phase, down the spent-catalyst standpipes, through plug valves, and into the regenerator bed. Here coke laid down on the catalyst during reaction is burned off, using air. Combustion flue gas passes upward from the fluid bed through a disengaging zone and then through several stages of cyclones (usually two to three stages). The flue gas may be vented to the atmosphere or sent to CO boilers (where the CO contained in the flue gas is burned) for the generation of high pressure steam.

Since the reactor effluent is a complex mixture of inerts, gas, butanes, gasoline, and cycle-oil stocks, it must be recovered and separated into the desired product streams. Reactor overhead vapors are separated into the various gas, gasoline, cycle-oil, and bottom streams in the catalytic fractionator. The vapors leaving the catalytic-fractionator overhead reflux drum contain valuable gasoline components, and the raw unstabilized gasoline contains undesirable light hydrocarbons. The recovery, separation, and purification of these light ends and gasoline are accomplished in a gas-recovery plant which utilizes a complex arrangement of gas compression and recovery equipment in an absorption-and-fractionation system.

A valuable heating-oil product called light cycle oil is condensed in the section immediately above the heavy-cycle-oil section of the tower. The light cycle oil is generally drawn from the tower as a side stream, stripped with steam, and sent to storage. A rich oil stream is returned to the catalytic fractionation from the gas plant, where it has served as a sponge-oil medium to recover valuable hydrocarbons which otherwise would be lost to fuel gas.

This *fluid catalytic process* contributes a valuable principle[66] to the chemical industry. The use of a large amount of turbulent solid capable of absorbing much heat in the reaction mass greatly aids in the elimination of temperature variations and hot spots. If no catalytic action is desired, an inert, powdered solid may be used.

Other catalytic cracking[67] processes employ fixed beds of catalyst to treat raw or hydrogenated crude fractions to reform them to better, lighter products without the use of high temperatures and pressures. A wide variety of bed-in-place reformers and fluid-bed crackers go under such names as Isocracking, Ultracracking, Unicracking, Selectoforming, Ultracat, etc. The addition of hydrogen during the cracking process causes desulfurization and better yields of middle distillates, but hydrogen is expensive and troublesome.

Figure 37.14 shows the *Universal Oil Products two-coil selective thermal cracking process.* The charging stock is topped or reduced crude from which the lighter fractions down to gas oil have been removed by distillation (Fig. 37.3). The topped crude is charged to a fractionating tower, which is heated by vapors from the flash chamber. The lighter part of the charge is thus vaporized, but not enough heat is present to carry it out the top of the column. These vapors of light oil are condensed above the trap tray and are sent to the heating furnace. The heavy-oil fraction of the charge is removed from the bottom of the tower and also sent to the heating furnace. The furnace has two coils, one to heat the light oil and the other to heat the heavy oil. The heavy oil is easier to crack but,

[66]This principle may be applied to the control of organic oxidations that are inclined to proceed to where they become combustions. Sittig, *op. cit.*, p. 139 (30 refs); Frantz, Design for Fluidization, *Chem. Eng.* (*N.Y.*), **69**(20), 89 (1962).

[67]Flowcharts in Refining Handbook Issue of *Hydrocarbon Process.*, **53**(9), 107ff. (1974).

because it has not been vaporized, contains tars and other materials that tend to form coke and clog the furnace. Thus the heavy-oil coil of the furnace is not operated at so high a temperature as the light-oil coil and is so constructed as to permit easier cleaning.

The heated and partly cracked material from both coils is combined and sent to a downflow reaction chamber for completion of the cracking, or pyrolysis. This is simply a large tower, and most of the coke produced by the cracking operation is deposited in it. Several chambers are used in practice, though Fig. 37.14 shows only one. When a chamber becomes filled with coke, the heated material is sent from the furnace to another tower while the first one is being cleaned. The cracked material, which has lost most of its coke, is piped to the flash chamber, where the pressure is reduced. This reduction in pressure causes the lighter portions to vaporize from the residuum, which is removed at the bottom. The vapors next flow to the bottom of the fractionating tower, where their heat is used to separate the charging stock. The vapors that reach the top of the column are condensed and consist of gasoline containing dissolved gas. Part of the vapor is heavier than gasoline and condenses near the top of the column. This is called *furnace distillate.* Any material too heavy to reach either of these levels in the column is sent back to the cracking furnace. The use of thermal cracking is diminishing.

Reforming involves both cracking and isomerization. Straight-run gasoline and light naphthas usually have very low octane numbers. By sending these fractions to a reforming unit and giving them a light "crack," their octane number may be increased. This upgrading is probably due to the production of olefins and isomerization forming branched-chain compounds which have higher octane numbers.

Catalytic reforming. This involves the conversion of other hydrocarbons into aromatic compounds. Because of the high octane rating of aromatic compounds and the proved practicability of the process, catalytic reforming has now almost completely replaced thermal reforming. Catalysts such as platinum on alumina or silica-alumina and chromia on alumina are used. In 1950 the Universal Oil Products Co. showed that these expensive catalysts could be used by reducing catalyst losses. Rhenium catalysts have come into use.

Another example of catalytic reforming is the *platforming* process[68] depicted in Fig. 37.16. It was developed by the Universal Oil Products Co. as an "economical commercial method of upgrading the octane ratings of straight run, natural and thermally cracked gasolines, and for producing large quantities of benzene, toluene, xylenes and other aromatic hydrocarbons for use in chemical manufacture and in aviation fuel." The name comes from the fact that this process uses a fixed-bed Al_2O_3 catalyst containing 0.25% platinum. This catalyst has a long life without regeneration. The process is as an example of a specialized hydroforming process in which the reactions take place in the presence of recirculated hydrogen. The feed must be carefully prepared to obtain the best results and is usually a naphtha cut. The steps depicted in Fig. 37.16 may be resolved into the following *physical operations* and *chemical conversions* or *processes:*

The naphtha feed is prepared in a prefractionator (Op).

The charge is mixed with the hydrogen and introduced into the feed preheater, where the temperature is raised (Op).

The hot-feed naphtha vapors are conducted with recycle hydrogen through the four catalyst-containing reactors in series with an interheating for each reactor (Op and Ch). The temperature is 850 to 950°F and the pressure 200 to 1,000 psig.

The reactions that take place are essentially as follows (Ch):

1. Isomerization of alkylcyclopentanes to cyclohexanes.
2. Dehydrogenation of cyclohexanes to aromatics.

[68]*Hydrocarbon Process.*, **53**(9), 122 (1974), Continuous Regenerator Smooths Naphtha Reforming, *Chem. Eng.* (*N.Y.*), **79**(18), 80 (1972).

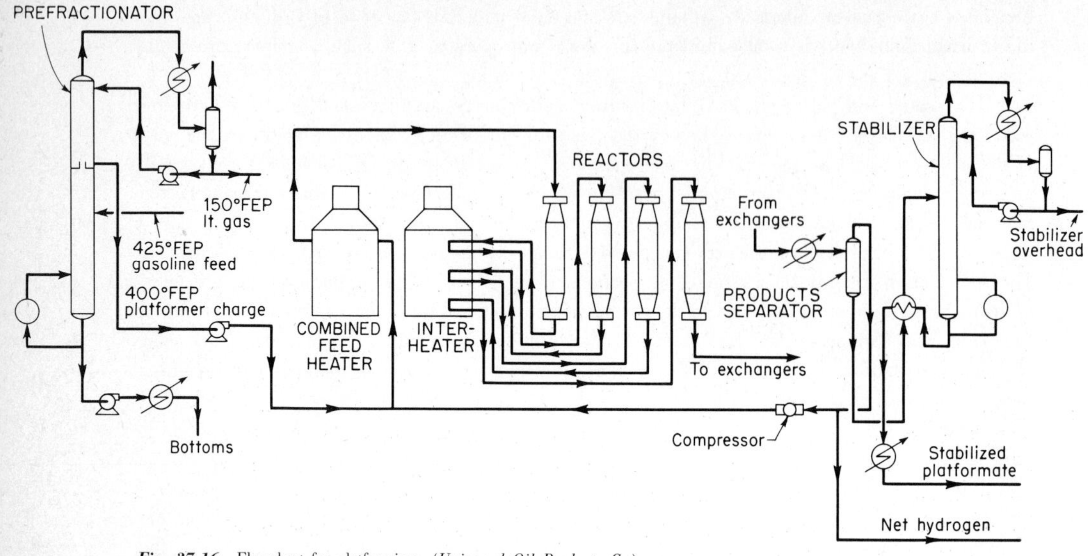

Fig. 37.16 Flowchart for platforming. (*Universal Oil Products Co.*)

3. Dehydrocyclization of paraffins to aromatics.
4. Hydrocracking of paraffins and naphthenes.
5. Hydrogenation of olefins.
6. Isomerization of paraffins.
7. Desulfurization.

From the reactor the products are cooled in heat interchangers (Op) (not shown in Fig. 37.17). From the exchanger 90% hydrogen is obtained, compressed, and recycled (Op). The main product is fractionated or stabilized after the temperature is reduced to the proper point in heat exchangers and conducted into a fractionating column (Op) or stabilizer, the overhead from which may be used as fuel.

The product is stabilized *platformate,* which may be used as a high-octane gasoline or further fractionated into its constituents, e.g., to furnish benzene, toluene, and xylenes.

Oxidation[69] -conversion as applied to petroleum gives more trouble than useful products, forming as it does gums and resins that interfere with the employment of gasolines, particularly those which contain unsaturated compounds. However, some serviceable products are obtained from petroleum by oxidation, e.g., formaldehyde by restricted oxidation of methanol and by restricted oxidation of natural gas. Much study is also being directed toward making fatty acids[70] from paraffins and even food products.

Polymerization. Rapid progress has been made in the development of polymerization processes for converting by-product hydrocarbon gases produced in cracking into liquid hydrocarbons suitable for use as high-octane motor and aviation fuels and for petrochemicals. To combine olefinic gases by polymerization to form heavier fractions, the combining fractions must be unsaturated. Hydrocarbon gases from cracking stills, particularly olefins, have been the foundation of polymerization.

[69]Gruse and Stevens, *op. cit.*, pp. 119–138, 653.

[70]Pardun and Kuchinka, Reaction Rates in the Liquid-phase Oxidation of Paraffins, *Pet. Refiner,* **22**(11), 140 (1943); Partial Oxidation for Syngas and Fuel, *Hydrocarbon Process.*, **53**(12), (1974).

The following equation is typical of polymerization reactions. Propylene, normal butylene, and isobutylene are the olefins usually polymerized in the vapor phase.

$$\underset{\text{Isobutylene}}{2CH_3-\overset{\overset{\displaystyle CH_3}{|}}{C}=CH_2} \longrightarrow \underset{\text{Diisobutylene}}{CH_3-\overset{\overset{\displaystyle CH_3}{|}}{\underset{\underset{\displaystyle CH_3}{|}}{C}}-CH_2-\overset{\overset{\displaystyle CH_3}{|}}{C}=CH_2} \xrightarrow{H_3PO_4} \underset{\substack{\text{Tetramer or}\\ \text{tetrapropylene}\\ \text{(mixed isomers)}}}{C_{12}H_{24}}$$

Vapor-phase cracking produces considerable quantities of unsaturated gases; hence polymerization units are often operated in conjunction with this type of cracking. *Thermal* polymerization is practical only for large-scale operations because of the difficulty in obtaining good heat control in small-sized units, whereas *catalytic* polymerization is practical on both a large and a small scale and is furthermore adaptable to combination with reforming to increase the quality of the gasoline (Fig. 37.16). Gasoline produced by polymerization contains a smog-producing olefinic bond. Polymer oligomers are widely used to make detergents. Since the middle and late 1950s, polymerization has been supplemented by alkylation as a method of conversion of light (C_3 and C_4) gases to gasoline fractions.

Alkylation. Alkylation[71] processes are exothermic and are fundamentally similar to polymerization; they differ in that only part of the charging stock need be unsaturated. As a result, the alkylate product contains no olefins and has a higher octane rating. These methods are based on the reactivity of the tertiary carbon of the isobutane with olefins, such as propylene, butylenes, and amylenes. The product *alkylate* is a mixture of saturated, stable isoparaffins distilling in the gasoline range, which becomes a most desirable component of many high-octane gasolines.

Alkylation is accomplished on a commercial scale with two catalysts: hydrogen fluoride and sulfuric acid. Alkylation with liquid hydrogen fluoride is illustrated by the flowchart in Fig. 37.17. Here the acid can be used repeatedly and there is no acid-disposal problem. The acid/hydrocarbon ratio in the contactor is 2:1. The temperature range, from 60 to 100°F, is cheaply maintained, since no refrigeration is necessary. The simplified reaction is as given for the conversion process of alkylation, with a two- to threefold excess of isobutane over the butylenes to minimize realkylation of the primary alkylate. The anhydrous hydrofluoric acid, when dirty, is easily regenerated by distillation from heavy alkylate. Sufficient pressure is required on the system to keep the reactants in the liquid phase. Corrosion is low, and the separated isobutane is recycled. Also, alkylation results from the combination of an olefin or alkyl halide with an aromatic hydrocarbon (Chap. 38).

Hydrogenation.[72] Mild hydrogenation of feeds ranging from light naphthas to lubricating oils is used selectively to convert sulfur, nitrogen, oxygen, halogens, diolefins, and acetylenes to less troublesome forms. Conditions vary from 400 psi or lower to 1,000 psi, with temperatures of 400 to 750°F. Nickel-containing catalysts are commonly used. Exceedingly high yields of gasoline and of lubricating oils with a high viscosity index, low carbon residue, and high resistance to oxidation can be obtained by hydrogenating heavy oils. The overall result is an upgrading of the products. The adoption of hydrogenation processes has been very general, despite the high initial cost of plant and equipment. But these methods are economical and useful in specialized procedures, and will be employed more and more in the future. Hydrogenation can be used instead of furfural solvent extraction. The diisobutylene obtained by polymerization of isobutylene is frequently hydrogenated to secure a more suitable aviation blending agent. The *Bergius process*, originated in Germany for

[71]Groggins, 5th ed., *op. cit.*, p. 804, and especially pp. 833–838 for sulfuric alkylation; Nelson, *op. cit.*, pp. 735–743 (for ethylbenzene flowchart see Chap. 36); Mrstik, Smith, and Pinkerton, Alkylation of Isobutane, in Progress in Petroleum Technology, p. 97; *Hydrocarbon Process.*, **53**(9), 206 (1974).

[72]Dehydrogenation to furnish styrene and butadiene are presented, with flowcharts, in Chap. 36.

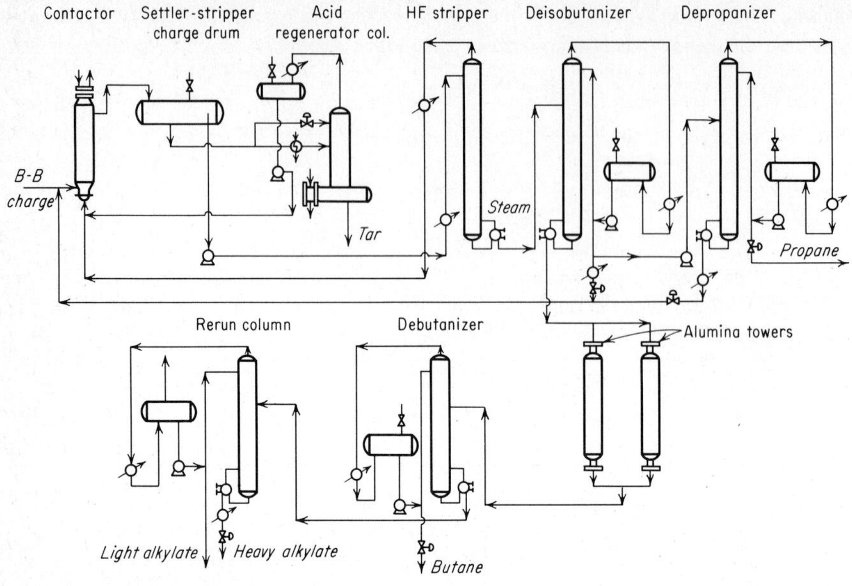

Fig. 37.17. Flowchart for HF alkylation. (*Universal Oil Products Co.*)

the production of oils by hydrogenation of coal, finds no use at present. It has been tried in this country only on a semiworks scale. The synthesis of motor fuel from carbon monoxide and hydrogen is a hydrogenation process which was industrialized in Germany as the Fischer-Tropsch process[73] to make Synthol. The largest Fischer-Tropsch plant is one in South Africa, built by the M. W. Kellogg Co.

Isomerization.[74] This conversion process has become of the utmost importance in furnishing the isobutane needed for making alkylate as a basis for aviation gasoline (see alkylation). The reaction may be formulated:

$$\underset{n\text{-Butane}}{CH_3 \cdot CH_2 \cdot CH_2 \cdot CH_2} \longrightarrow \underset{\text{Isobutane}}{CH_3 \cdot \overset{CH_3}{CH} \cdot CH_3}$$

CHEMICAL TREATMENT Some type of chemical treatment to remove or alter the impurities in petroleum products is usually necessary to produce marketable material. Depending upon the particular treatment used, one or more of the following purposes are achieved:

1. Improvement of color.
2. Improvement of odor.
3. Removal of sulfur compounds.
4. Removal of gums, resins, and asphaltic materials.
5. Improvement of stability to light and air.

Of these, removal of sulfur[75] and improvement of stability are the factors usually governing the treatment employed. With the discovery that the use of catalytic converters causes the emission of

[73]Groggins, *op. cit.*, p. 564; Kellogg, Synthol Process, *Pet. Refiner*, November 1963, p. 42 (flowchart and description).

[74]Egloff, Hulla, and Komarewsky, Isomerization of Pure Hydrocarbons, Reinhold, 1942; Liquid-Phase Isomerization, *Pet. Refiner*, **23**(2), 61 (1944); Gunners, Isomerization, Progress in Petroleum Technology, p. 109.

[75]A Sulfur in Gasoline Scare?, *Chem. Eng.* (*N.Y.*), **81**(26), 74 (1974); Ranney, Desulfurization of Petroleum, Noyes, 1975.

sulfuric acid vapors from automobile exhausts, pressure to remove or reduce sulfur in motor fuel has developed. Sulfur may be reduced by: (1) hydrogenation (which also removes metals and nitrogen), (2) treatment with caustic soda, (3) treatment with caustic soda plus a catalyst (Merox), and (4) treatment with ethanolamines. With sulfur treatment for water and the pollution treatments presently required, environmental control equipment for a refinery is approaching the size and cost of the remainder of the refinery.

Several processes are available for the alteration of objectionable sulfur and the consequent improvement in odor. The *doctor treatment* was the original method of "doctoring" the oil to reduce the odor and still remains a method of testing for the quality of treatment. This process consists in agitating the oil with an alkaline solution of sodium plumbite and a small amount of sulfur. The sulfur compounds usually found in oils are mercaptans. These give the material a disagreeable odor and cause corrosion. Doctor converts mercaptans to comparatively harmless disulfides.

$$2RSH + Na_2PbO_2 \longrightarrow (RS)_2Pb + 2NaOH$$

$$(RS)_2Pb + S \longrightarrow R_2S_2 + PbS$$

The oxidation of mercaptans to less objectionable disulfides can be accomplished with a variety of mild oxidizing agents, e.g., oxygen plus a catalyst such as copper chloride. The presence of sulfur in gasoline, even in the less objectionable disulfide form is undesirable. Not only does it increase corrosion and air pollution, but it decreases the effectiveness of lead antiknock agents. Modern processes are therefore aimed at removing sulfur-containing compounds. Hydrogen treatment reduces the sulfur in heavy fractions and resids. Hydrogen or hydrogen-donor liquids and a catalyst (or light hydrogenation) are used to treat lighter distillates, removing nitrogen, sulfur, and some metallic contaminants as well. Extraction with caustic or proprietary solutions (usually oxygen-regenerable) is frequently used on light distillates.

It is now common practice to add antioxidants to prevent the formation of gums, rather than to remove them chemically. Among the *antigumming* materials used are α-naphthol, substituted catechols, cresols, benzyl-*p*-aminophenols, and certain wood-tar and coal-tar fractions.

Research. The petroleum industry has been characterized by continuous improvement based upon research and a willingness to replace antiquated equipment or old processes by more modern ones. This development went through two phases. In the first phase, the *unit operations*, distillation, heat transfer, fluid flow, and the like, were subjected to accurate study and experimentation, resulting in the carrying out of these operations with greater efficiency and consequently less cost. Investigations of these unit operations are naturally being continued. However, the second phase[76] of research in this field applies to the study of chemical changes involving petroleum raw materials and petroleum products. The industrial application of these is recognized as *unit processes or conversion processes.* This is a currently active aspect of the development of the petroleum industry. It has yielded rich returns and is certain to yield even more products of greater usefulness in the future. Synthetic rubber is an example. No longer can petroleum companies be satisfied simply to separate into salable fractions the various products as they occur in petroleum. To maintain a competitive and well-rounded position many petroleum compounds must be chemically altered to obtain products of greater usefulness or value. Each year shows a higher percentage of petroleum sales from the action of chemical change on petroleum raw materials. Indeed, the petroleum industry has become the most important source of cheap raw materials for the entire chemical industry. Such chemical change will be more marked in the future than it has been in the past. The present sound technical position has come about through the wise dependence of the petroleum executives upon

[76]Petroleum development is often divided into three phases, the first pertaining largely to the physical operation of distillation, the second characterized by cracking, and the third distinguished by the many other chemical processes or conversions that are being applied industrially in this field.

research in the very broadest meaning of the term, *protected by patents covering the public disclosure of new discoveries.* According to Petroleum Facts and Figures, the total research and development costs for the petroleum industry were $1,067 million in 1972. In Europe, Japan, and the U.S.S.R., proteins are being grown[77] using hydrocarbons as the carbon source.

SELECTED REFERENCES

Abraham, H.: Asphalts and Allied Substances, 6th ed., Van Nostrand, 1960–1962, 5 vols.
American Society of Lubricating Engineers, *Transactions* (semiannual).
Asphalt Handbook, rev. ed., University of Maryland, Asphalt Institute, 1960.
Bland, W. F., and R. L. Davidson: Petroleum Processing, McGraw-Hill, 1967.
Braithwaite, E. R. (ed.): Lubrication and Lubricants, Elsevier, 1967.
Campbell, J. M.: Petroleum Reservoir Property Evaluation, Petroleum Publishing, 1973.
Chemistry of Petroleum Refining (a chart), American Petroleum Institute (annual).
Clark, J. A.: Chronological History of Petroleum Natural Gas Industries, Clark Book, 1964.
Frick, T. C. (ed.): Petroleum Production Handbook, McGraw-Hill, 1962, 2 vols.
Galton, L. V. G.: Technology and Economics, Noyes, 1967.
Griffin, J. M.: Capacity Measurement in Petroleum Refining, Lexington Books, 1971.
Gruse, W. A.: Motor Oils: Performance and Evaluation, Reinhold, 1967.
Gruse, W. A., and D. R. Stevens: Chemical Technology of Petroleum, 3d ed., McGraw-Hill, 1960.
Gunderson, R. C., and A. W. Hart: Synthetic Lubricants, Reinhold, 1962.
Handbook, Petroleum Chemicals Plant Refining, Gulf Publishing, 1965.
Harris, L. M., Introduction to Deepwater Floating Drilling Operations, Petroleumn Publishing, 1972.
Hengstebeck, R. J.: Distillation: Principles and Design Procedures, Reinhold, 1961.
Hobson, G. P., and W. Pohl: Modern Petroleum Technology, Wiley, 1973.
Holland, C. D.: Multicomponent Distillation, Prentice-Hall, 1963.
Johnson, J. C.: Antioxidants 1975, Synthesis and Applications, Noyes, 1975.
Jones, H. R.: Pollution Control in the Petroleum Industry, Noyes, 1973.
McDermott, J.: Drilling Mud and Fluid Additives, Noyes, 1973.
McGraw, H. G., and M. E. Charles (eds.), Origin and Refining of Petroleum, ACS, 1971.
Nelson-Smith. A.: Oil Pollution and Marine Ecology, Plenum, 1973.
Potter, J.: Disaster by Oil, Oil Spills, MacMillan, 1973.
Ranney, M. W.: Fuel Additives, Noyes, 1974.
———: Lubricant Additives, Noyes, 1973.
———: Synthetic Lubricants, Noyes, 1972.
Refining and Gas Processing: Worldwide Directory, 30th ed., Petroleum Publishing, 1972.
Rocq, M. M.: U.S. Sources of Petroleum and Natural Gas Statistics, Special Libraries Association, 1961.
Ross, W. M.: Oil Pollution as an International Problem, University of Washington Press, 1973.
Russell, C. S.: Residuals Management in Industry, Case Study of Petroleum Refining, Johns Hopkins Press, 1973.
Sittig, M.: Oil Spill Prevention and Removal Handbook, Noyes, 1974.
Spillane, L. J., and H. P. Leftin (eds.), Refining Petroleum for Chemicals, ACS, 1970.
Technical Data Book—Petroleum Refining, Pennsylvania State University Petroleum Laboratory, API, 1966 (D125).
Williams, T. J., and V. A. Lauber: Automatic Control of Petroleum Processes, Gulf Publishing, 1961.

[77]Soviets Build Large Plant for Making Protein from Oil, *Chem. Eng. News*, Aug. 19, 1974, p. 30; Single Cell Protein Moves toward Market, *Chem. Eng. News*, May 5, 1975, p. 25.

chapter 38

PETROCHEMICALS[1]

Petroleum companies with their own plants, with wholly owned subsidiaries, or in association with strictly chemical manufacturing companies, have entered into the manufacture of petrochemicals, or chemicals derived from petroleum raw materials, including natural gas. And many chemical companies, for example, Union Carbide Chemicals, Dow Chemical Co., and Monsanto Co., have been producing petrochemicals for decades.

Most of the separation processes for the original raw materials from petroleum involve highly refined physical methods.[2] However, almost all the petrochemicals actually entering into industrial usage arise from various chemical conversions,[3] as do many of the strictly petroleum products, such as high-octane gasoline and upgraded lubricating oils (Chap. 37). Petrochemicals fundamentally and economically are many basic chemical raw materials to be further chemically converted into fertilizers, such as ammonia; into plastics, which are increasing in variety and volume each year; into rubber, in all its varieties; into fibers, especially nylon, polyesters, and acrylics, and of growing importance, into paints. Petrochemicals are significant in making lubricating oil additives, pesticides, and solvents and as another and large source of aromatic chemicals. Although production plants have developed mainly in the Gulf Coast area because of economical sources there for the needed raw materials, such plants are spreading to other consuming areas. It is estimated that more than 80% of the organic chemicals of the entire chemical industry are based on petrochemicals, and 24% of the sulfur and all the carbon is derived from natural gas or petroleum products.

A definite change has taken place in the viewpoint of the managers of petroleum refiners and is a significant reason for the rapid development of petrochemicals. There has been a change in fact to the viewpoint of the chemical industry, of which petrochemicals have become the backbone. Table 38.1 reflects the contrast between the two points of view. Several decades ago, when petroleum companies first saw the possibilities of profit from petroleum chemicals, managers frequently made arrangements with chemical companies for joint ventures or at least to sell the appropriate raw materials to chemical companies. On the other hand, several of the larger petroleum companies organized their own petrochemical departments separately from their oil refineries.[4] These contributed fundamentally to the rapid development and growth of this new field. At present the seven major oil companies[5] heavily involved in petrochemicals are increasing their production capacities and expect to make more of the intermediates instead of only supplying feedstock to others.[6]

[1]The subject of petrochemicals is so extensive and involves so many hundreds of individual chemicals that this presentation is directed only toward the overall picture and toward the most important petrochemicals within the chemical process industries. Further information can be found in the Selected References; ECT, 12 vols.; Goldstein and Waddams, The Petroleum Chemicals Industry, Spon, 1967 (many references). See also Brownstein, Trends in Petrochemical Technology, Noyes, 1976.

[2]Unit Operations.

[3]Unit processes.

[4]CW Special Report, Oil Buys in 1962, *Chem. Week*, Mar. 24, 1962, p. 25 (many financial and other charts); Petrochemicals Wait for End of the Pause, *Chem. Week*, Nov. 15, 1976, p. 11.

[5]Atlantic Richfield, Amoco, Exxon, Gulf, Mobil, Shell, and Texaco.

[6]Prescott, Oil Giants Gear Up Petrochemicals Capacity, *Chem. Eng.* (*N.Y.*), **82**(7), 56 (1975).

TABLE 38.1 *Change in Economic Planning*

The oil mind	*The chemical mind*
Looks for immediate profits	Takes long-range-profit point of view
Is methodical, will not be rushed	Values timing, moves rapidly
Concentrates on single products	Believes in diversification
Is accustomed to large cash flow, big expenditures	Combines smaller cash flow with long-range investments
Produces at highest possible rate	Takes calculated risks on forecasts of greater demand
Builds the "biggest" facility	Tailors to meet present and estimated future needs
Prefers to increase amount of main products irrespective of petrochemicals	Will increase quality, number of products, to keep ahead in competition

Source: Oil Companies Reshuffle for Chemical Gain (*modified*). *CW*, Oct. 15, 1960, p. 48. To the contrast presented above it might be added for emphasis that the oil refineries generally produce mixtures of hydrocarbon products in very large tonnage, whereas the petrochemical plants produce a great many pure chemicals, mostly in smaller, but rapidly increasing, tonnage.

HISTORICAL The first petrochemical, except carbon black, manufactured on an industrial scale was isopropyl alcohol, produced by Standard Oil (N.J.) in 1920. Individual petrochemicals now number more than 3,000, many in large volume. In this chapter, about 100 are presented or references made to other chapters. The traditional division giving to coal the monopoly of supplying aromatic organic chemicals has disappeared, partly with the large production of aromatics by physical separation, but principally as a result of the chemical conversion of petroleum raw materials, such as dealkylation of methylnaphthalenes, to furnish naphthalene or re-forming for toluene. Also viewed historically, many once important processes for producing ammonia, ethyl alcohol, acetone, acetic acid, acetic anhydride, glycerin, acetylene, and other compounds are now based wholly or significantly upon petrochemicals.

RESEARCH The birth of petrochemicals has been one of the fabulous stories connected with modern chemical and chemical engineering research. Petrochemicals were neglected by the petroleum companies that years ago contracted to sell their off-gases to chemical companies. The early thinking and planning of the petroleum administrators did not proceed along chemical lines. Hence the chemical companies took the early lead in developing petrochemicals. Several notable exceptions, like Shell and Standard Oil (N.J.) (now known as Exxon), supplied the change in thinking and planning that stimulated the development of petrochemicals (Table 38.1).

ECONOMICS AND USES From the output of Standard Oil (New Jersey) of 75 tons of isopropyl alcohol in 1925, the production of petrochemicals has increased, now supplying more than 70% of all organic chemical production. *This increased rate is continuing* (Fig. 38.5). The organic chemical industries fundamentally rely on petrochemicals. The classes of end use for petrochemical products are listed in Table 38.7. In 1975 it was estimated that the oil companies alone spent nearly $740 million for the capital needed for petrochemicals. This was in addition to the large sums similarly spent by strictly chemical companies. Prescott[7] has tabulated and estimated the petrochemical growth of oil and chemical companies for 1974 and 1978. He predicts the large and profitable growth that is to be expected for petrochemicals, particularly polymers and agrichemicals (Table 38.7). Within the next 5 to 15 years, several oil companies expect to increase their share of net profits derived from chemical operations from the present 12 to 20% to over 50%. Among the many listings of petrochemicals, the tabulations[8] in the Encyclopedia of Chemical Technology and in Goldstein's book are excellent.

RAW MATERIALS, OR FEEDSTOCKS The basic raw materials (Table 38.2) supplied by petroleum refineries or natural-gas companies are LPG, natural gas, gas from cracking processes, liquid distillate (C_4 to C_9), distillates from special cracking processes, and cyclic fractions for aro-

[7]Prescott, *op. cit.*

[8]ECT, vol. 15, pp. 77–92, 1968; Goldstein, *op. cit.* See also Brownstein, U.S. Petrochemicals: Technologies, Markets, and Economics, Petroleum Publishing, 1972.

TABLE 38.2 ***Primary Precursors: Petroleum-Petrochemical Complex***

Raw materials by distillation	*Precursors (basic chemicals) by conversion*	*Intermediates by conversion*	*Finished products by conversion*
Paraffins and cyclics	Olefins, diolefins, acetylene, aromatics	Various inorganics and organics	Inorganics and organics
Natural gas			Carbon black
Sulfides	H_2S	S	H_2SO_4
Hydrogen		Synthesis gas	NH_3
Methane			Methanol
			Formaldehyde
Refinery gases	Acetylene	Acetic acid	Acetates
	Isobutene	Acetic anhydride	Fibers
Ethane*	Ethylene	Isoprene	Rubber
Propane*	Propylene	Ethylene oxide, etc.	Rubber and fiber
n-Butane*	*n*-Butenes	Butadiene	Rubber
Hexane			
Heptanes			
Refinery naphthas			
Naphthenes	Cyclopentadiene	Adipic acid	Fibers
Benzene		Ethylbenzene	Styrene
		Styrene	Rubber
		Cumene	Phenol acetone
		Alkylbenzene	
		Cyclohexane	
Toluene	Toluene	Phenol	Plastics
		Benzoic acid	
Xylenes	*o-m-,p*-xylene	Phthalic anhydride	Plastics
Methyl naphthanes	Naphthalene	Phthalic anhydride	Plastics

*From LPG and refinery cracked gas. *Note:* Aromatics are also obtained by chemical conversions (demethylation, etc.).

matics. Such mixtures are usually separated into their constituents at the petroleum refineries and chemically converted into their reactive precursors before being subjected to the chemical conversions necessary for manufacturing the various petrochemicals actively used by the manifold industries in the tonnages required.

Since the lower members of the paraffin and of the olefin series of organic raw materials are of such great importance in this area, figures and tables are presented to depict the derivatives that can be made from methane (Table 38.3 and Fig. 38.1), ethylene (Table 38.4 and Fig. 38.2), propylene, butylenes (Table 38.5 and Fig. 38.3), and cyclic petrochemicals (Table 38.6 and Fig. 38.4). These are the *precursors* of, and chief raw materials for, petrochemicals. Their consumption is increasing at a rapid rate, with consequent growth in tonnage and number of products. The manufacture of the precursors of petrochemicals, namely, olefins and aromatics, is presented in Chapt. 37, with some of the intermediates. Salable products made therefrom are depicted in the various tables and figures in this chapter, the last part describing a number of the technically important petrochemicals, all arranged under their chemical conversions and under the heading Manufacture of Petrochemicals.

The growth and the importance of petrochemicals are due[9] to

1. The abundance and relatively low cost of crude oil and natural gas.
2. R&D in petroleum technology, resulting in lower cost and sharper fractionation (separation) in distillation and various other *unit operations* (Table 37.4).

[9]McGraw-Hill Encyclopedia, vol. 10, p. 43, 1966.

TABLE 38.3 ***Petrochemicals from Methane in 1974***

Basic derivatives and sources, %	*Produced annually, millions of pounds*	*Uses, %*
Ammonia	32,000	Agricultural chemicals (such as ammonia, salts, urea), 74
Petroleum sources		Fibers, plastics, 8
Methane hydrogen, 79		Industrial explosives, 4
Refinery hydrogen, 4		Other, 14
Electrolytic hydrogen, coal, 17		
Carbon black	3,511	Rubber compounding, 96
Natural gas, ~25		Pigments, metallurgy, 4
Liquid petroleum, ~75		
Methanol	6,789	Formaldehyde (mainly for resins), 45
Petroleum sources		Solvents, 10
Methane, 63		Dimethylterephthalate, 9
Propane-butane, 7		Exports, 5
Coal, 30		Other, 31
Chloromethanes	1,048	Solvents, cleaners, 30
Methane chlorination, 50		Chlorofluorcarbons for refrigerants, aerosols, 50
Other sources, 50		
Acetylene	443	Vinyl chloride, vinyl acetate, 3
Petroleum		Acrylates, 26
		Acetylenic chemicals, 20
		Vinyl acetate, 17
		Chlorinated solvents, 4
Hydrogen cyanide	345	Methyl methacrylate, 62
Dept. of Commerce data, from methane probably, >80		Sodium nitrilotriacetate, 21
		Sodium cyanide, 10
By-product from acrylonitrile, <20		Other, 8

Sources: Chem. Mark. Rep. (1974, 1975); *Chem. Eng. News,* Apr. 28, 1975, p. 10.

3. Research and development of *conversion processes* applied to petroleum products, such as alkylation, cracking, hydrogenation, polymerization, and others, leading to purer compounds as well as to new products.

4. Supplying chemicals in larger and steadier quantities than could be obtained economically from coal, wood, or agrichemicals.

5. Concentration in supplying raw materials and products reaching into the growth area of American industry, such as antifreeze, and lube additives, as well as resins for plastics, synthetic fibers, and protective coatings.

PHYSICAL SEPARATIONS, OR UNIT OPERATIONS

Chemical engineers working in the refining of petroleum products have long been in the forefront of improvements in physical separation procedures. Their experience has been applied and extended to petrochemical raw materials and products—distillation, azeotropic and extractive distillation, superfractionation, crystallization, adsorption, and solvent extraction. Such solvents as liquid SO_2, NH_3, and many organic compounds, either pure or in more suitable mixtures, are employed. Other examples are Molex adsorption for separating straight-chain from branched-chain paraffins, Unisorb for separating aromatics from nonaromatics in treating kerosines, and light catalytic-cycle oils. Unit operations that have been used are listed in Table 37.4 and are described in the text of Chap. 37.

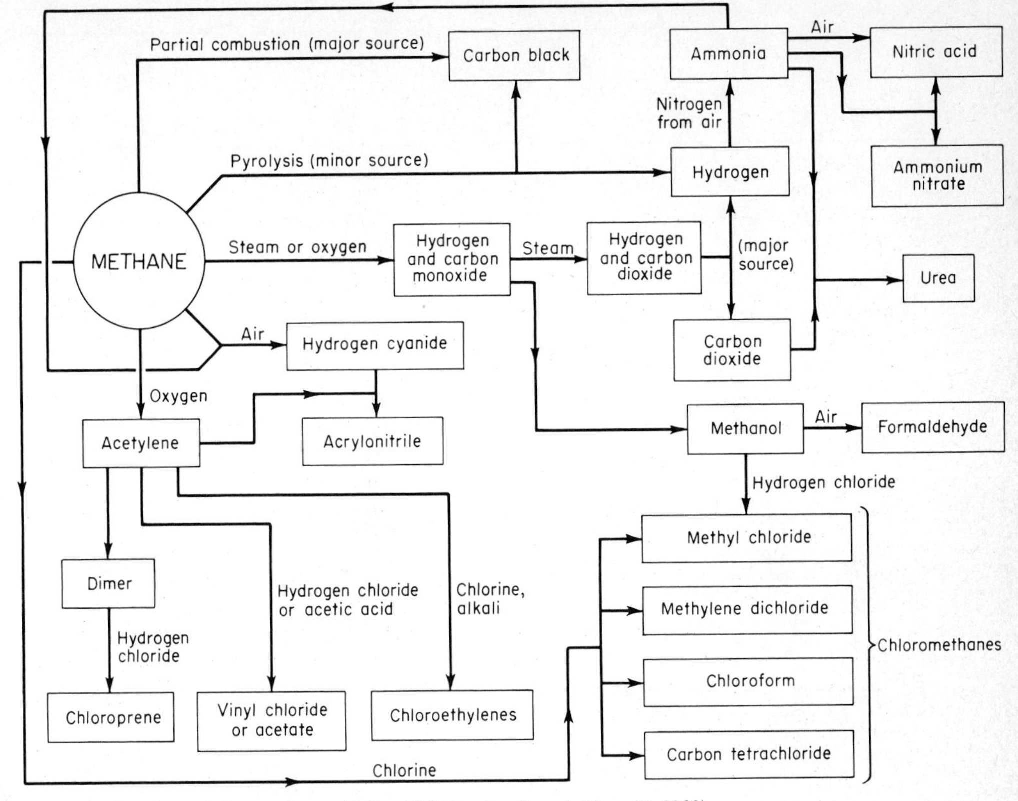

Fig. 38.1 Petrochemicals from methane. (*McGraw-Hill Encyclopedia, vol. 10, p. 45, 1966.*)

CHEMICAL CONVERSIONS[10]

Petrochemicals are largely manufactured by chemical conversions, and many of the reactions involved are complicated and novel. Most of them are characterized by *continuous processing* in large volume and under the control of modern instrumentation. Many are also system-controlled through computers (Chap. 2).

TRANSPORTATION The extension of pipelines for obtaining natural gas and LPG in quantity and at low cost, as well as the use of pipelines, barges, and ships for the delivery of products, for example, ethylene and ammonia, have been among the important and fundamental engineering bases for stimulation of the growth and geographic spread of petrochemical manufacture.

MANUFACTURE OF PETROCHEMICALS

Production of the hundreds of commercially prepared petrochemicals is based largely on efficient and, in many cases novel, chemical processing developed by the R&D departments of the various companies. For these chemical processes the basis is the chemical reaction, or the *chemical conversion.* Hence the presentation in this book is classified naturally under the principal *chemical conversion* as the foundation for the manufacture.

[10]Unit processes; cf. Table 1.2.

TABLE 38.4 *Petrochemicals from Ethylene in 1974*

Basic derivatives and sources, %	*Produced annually, millions of pounds*	*Uses, %*
Ethylene oxide	4,260	Glycols, 5
Via direct oxidation		Glycol ethers, 7
Via chlorohydrin		Ethanolamines, 6
		Nonionic detergents, 11
		Other, 11
Ethyl alcohol (industrial)	2,000	Aldehydes, 14
From ethylene, 96		Other chemicals, 26
By fermentation, 4		Solvent, 12
		Flavors, cosmetics, detergents, 30
		Other, 18
Polyethylene	8,450	Film, sheet, molding, and extrusion, 90
Styrene	6,400	Polystyrene, 51
From ethylene and benzene (or reformate)		SBR and latex, 12.5
		Miscellaneous resins, 29.5
		Other, 7
Ethyl chloride	659	TEL, ~80
Ethylene plus hydrogen chloride		Minor amounts for other ethylations
Chlorination of ethane		(such as ethyl cellulose)
Ethylene dichloride	9,300	Vinyl chloride, 78
By direct chlorination (major)		Scavenger in antiknock fluid, 3
By-product of ethylene chlorohydrin production		Other, 19
Ethylene dibromide	325	Scavenger in antiknock fluid, 50

Source: *Chem. Mark. Rep.* (1974, 1975).

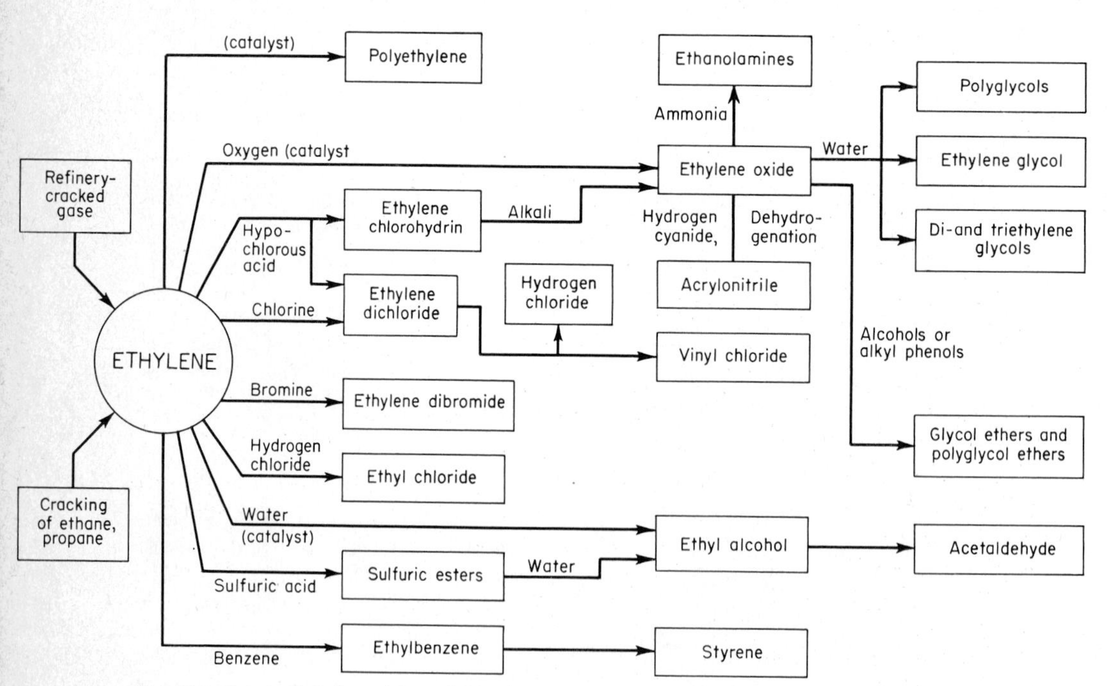

Fig. 38.2 Petrochemicals from ethylene. [*McGraw-Hill Encyclopedia, vol. 10, p. 46, 1966* (newer plants for acrylonitrile stem from propylene); see also *Chem. Eng. Deskbook*, **80**(23), 34 (1973).]

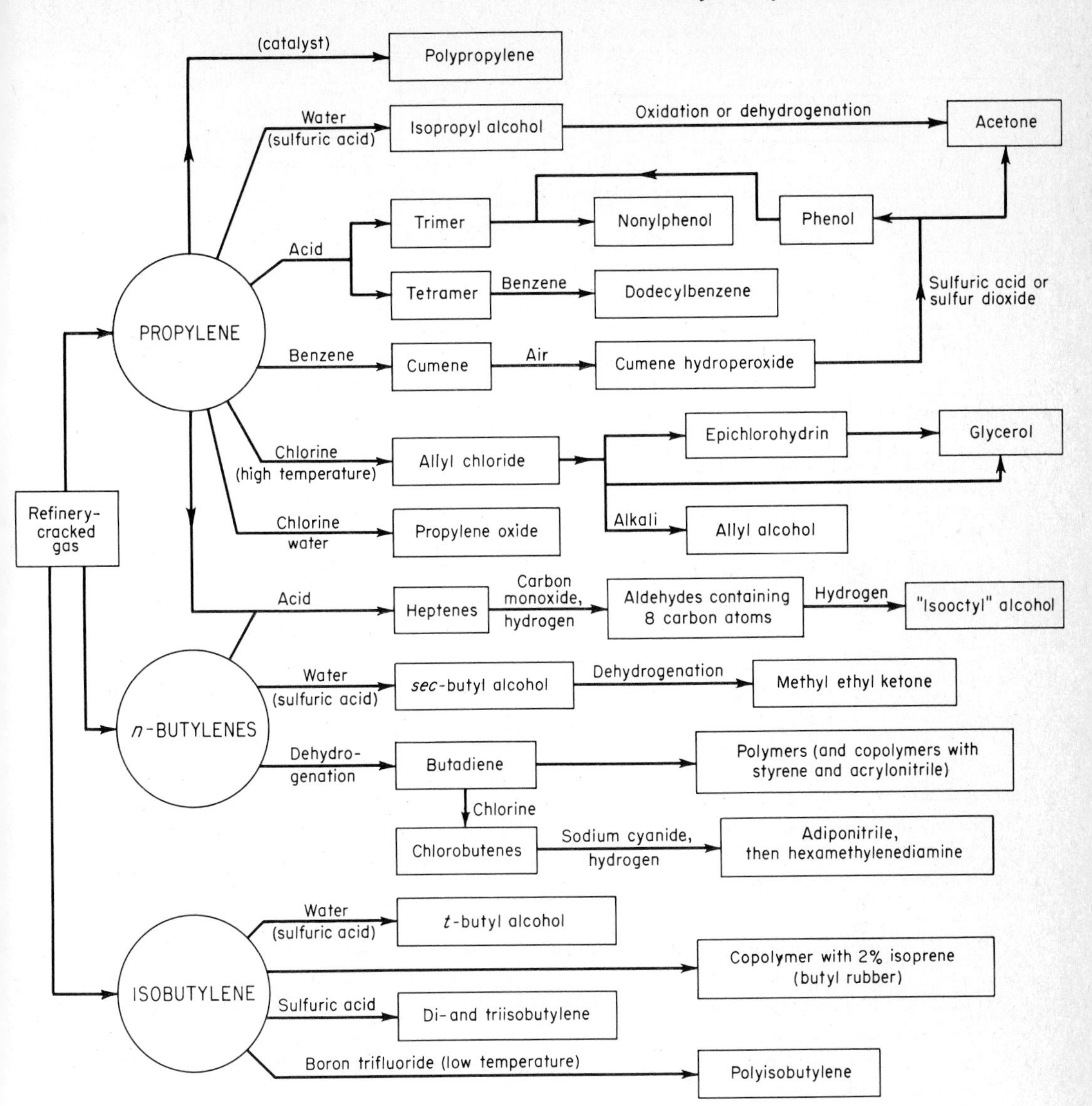

Fig. 38.3 Petrochemicals from propylene and the butylenes. [*McGraw-Hill Encyclopedia, vol. 10, p. 47, 1966;* newer plants for acrylonitrile stem from propylene—see text; butylenes, CW Report, *Chem. Week,* May 28, 1960; synthetic rubber larger consumer of butadiene; isobutylene (rubber), *Chem. Eng.* (*N.Y.*), **69**(12), 90 (1962), with flowcharts and references.]

Important as these chemical conversions are, there are many other essential factors, as outlined in the general chapters of this book, especially in Chaps. 1 and 2. See Table 2.1 and the accompanying text. Chemical engineers directing chemical processing in the present highly competitive chemical economy have called upon many branches of basic science, engineering, and management to improve their manufacturing in all phases, supplementing, but including, many former procedures. Indeed, a well-developed chemical process may be said in modern engineering conceptions to embrace a "system."[11] "Systems," according to R. R. White, "are essentially plants or factories for the

[11]White (Atlantic Refining Co.), Chemical Engineering in the U.S.A., *Proc. Jpn. Soc. Chem. Eng.*, 25th Jubilee, November 1961.

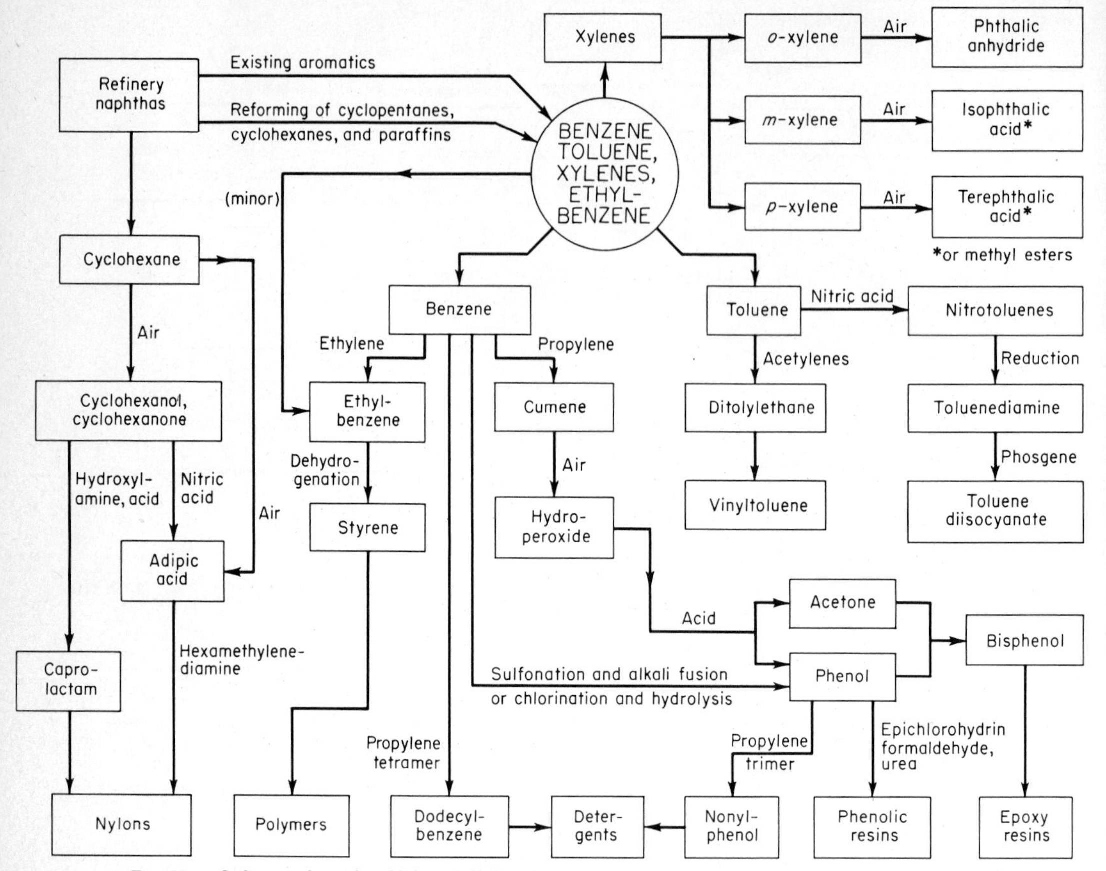

Fig. 38.4 Cyclic petrochemicals. (*McGraw-Hill Encyclopedia, vol. 10, p. 48, 1966.*)

processing of material, energy, and information. They may be thought of as doing this processing at various levels of sophistication, on the highest of which the system adapts to its environment. At the lowest level, material and energy are absorbed, products and wastes are removed, and energy is conserved and degraded. This describes a chemical plant and our present level of knowledge and practice. Information is a necessary concomitant of this process." The production of many of these petrochemicals is growing and has reached large tonnages, as illustrated by the figures, especially by Fig. 38.5, and the tables. Many chemical conversions have their rate, or the rate of a particular aspect of the reaction, accelerated by catalysts.

ALKYLATION, DEALKYLATION, AND HYDRODEALKYLATION[12] *Cumene* (isopropylbenzene): benzene is alkylated over a phosphoric acid catalyst[13] with propylene in the vapor phase at 250°C and 100 psia, frequently using a refinery propylene-propane cut. An excess of benzene minimizes polyalkylation. After flashing off the propane and unconverted propylene, the benzene and cumene are recovered and purified. Other alkylation catalysts such as aluminum chloride and sulfuric acid[14]

[12]Groggins, Unit Processes in Organic Synthesis, 5th ed., p. 804, McGraw-Hill, 1958; Sittig, Aromatic Hydrocarbons, Noyes, 1976.

[13]Matox, *Trans. AIChE,* **41,** 463 (1945); Albright, Alkylation Processes, HF as Catalyst, *Chem. Eng.* (*N.Y.*), **73**(19) 205, 1966.

[14]McAllister *et al., Chem. Eng. Prog.,* **43,** 189–196 (1947); *Pet. Process.,* **12,** 181 (1957); for flowchart, Sittig, Organic Chemical Processes, p. 190, Noyes, 1969.

TABLE 38.5 ***Petrochemicals from Propylene and the Butylenes in 1974***

Basic derivatives and sources. %	*Produced annually, millions of pounds*	*Uses.* %
Isopropyl alcohol	1,951	Acetone (by dehydrogenation or oxidation), 40
From propylene, ~100		Other chemicals, 10
Acrylonitrile	1,711	Solvent and miscellaneous, 50
		Polymer, 50
Dodecene (tetramer)	600	Detergents: dodecylbenzene sulfonate, nonylphenyl
Nonene (trimer)	509	polyglycol ethers, 80
Cumene	1,407	Phenol and acetone, 70
Oxo alcohols	1,030	Glycerol, allyl alcohol, 50
Propylene oxide	1,792	Propylene glycol, 50
Polypropylene	2,813	Film, sheet, molding, and extrusion, rubber, 90
Others	1,584	
Butadiene	4,200	Styrene-butadiene rubbers, 47
From butylenes (refinery cracked gas)		Neoprene, 8
		Acrylonitrile-butadiene rubber, 3
From butane		Polybutadiene rubber, 17
Miscellaneous hydrocarbon cracking		Adiponitrile and hexamethylenediamine for nylon, 8
		ABS resins, 6
		Other, 11
Secondary butyl alcohol	96	Methyl ethyl ketone, 60
Oxo alcohols (via C_8 aldehydes), "isooctyl" alcohol	67	Detergents, plasticizers, 90
Butyl rubber (copolymer of isobutylene and isoprene)	250	Inner tubes, mechanical rubber goods, 100
Diisobutylene	110	Detergents, plasticizers, 90
Triisobutylene	Small	
Polyisobutylene	40	Adhesives, sealants, electric insulation, lube-oil additives, 90
Tertiary butyl alcohol	30	Solvent, 90

Sources: McGraw-Hill Encyclopedia, vol. 10, Petrochemicals; Propylene, *Chem. Eng.* (*N.Y.*), **81**(19), 56 (1974); *Chem. Mark. Rep.* (1974, 1975).

may be used. In other chapters dodecylbenzene (Chap. 29) and ethylbenzene (Chap. 36) are discussed. Dealkylation is practiced in producing aromatics like benzene, toluene, xylenes, and naphthalene from petroleum (Chap. 37).

Tetraethyllead[15] (TEL) is a very important alkylated product and is the principal antiknock compound in gasolines; production is over half a billion pounds. Environmental concerns over automobile exhaust emissions have led to the widespread use of catalytic converters to clean up the atmosphere. TEL is a poison to the catalyst in the converters, and thus the use of leaded gasolines in automobiles equipped with these devices is not practical. This fact, plus the EPA's requirement that TEL be phased out of most gasolines by 1979, means that its use will probably decline drastically.[16] It is prepared commercially by the action of ethyl chloride on a lead-sodium alloy.

$$4PbNa + 4C_2H_5Cl \longrightarrow Pb(C_2H_5)_4 + 3Pb + 4NaCl$$

The autoclave for the reaction must have a heating jacket, a heavy-duty stirrer to agitate the lead alloy, and a reflux condenser. The alloy should be finely divided, and the temperature should be below

[15]Groggins, *op. cit.*, p. 852; Faith, Keyes, and Clark, Industrial Chemicals, 3d ed., p. 757, Wiley, 1965 (flowchart); Sittig, *op. cit.*, p. 636 (flowchart); CW Report on the Abundance of Metallic Organics, *Chem. Week*, Sept. 17, 1960 (excellent broad coverage).

[16]Ricci, TEL Stand-ins Being Readied, *Chem. Eng.*, **81**(25), 40 (1974).

TABLE 38.6 *Cyclic Petrochemicals in 1974*

Basic products and sources, %	*Produced annually, millions of pounds*	*Uses, %*
Benzene	11,648	Styrene, 48
From petroleum, 90		Phenol via cumene (definitely petrochemical), 19
		Cyclohexane, 13
		Miscellaneous chemicals and other uses, 30
Toluene	7,372	Motor and aviation gasoline, 60
From petroleum, 97		Explosives and other chemicals, 8
From coal, 3		Toluene diisocyanate, 4
		Benzene, 18
		Solvents for coatings, 10
Xylenes	6,400	Coatings and solvents, 11
From petroleum, 99		Aviation gasoline, 29
From coal, 1		Isomer separation (for phthalic acids), 60
Ethylbenzene	7,020	Styrene, 96
From benzene, predominant		Solvent, 2
Direct from reformate, small		Exports, 2
Cyclohexane	2,508	Polyamides* (nylons), 48
From petroleum naphthas, 70		Caprolactam, 18
From benzene by hydrogenation, 30		Exports, 22
		Miscellaneous, 12

Sources: Chem. Mark. Rep. (1974, 1975); Petroleum Facts and Figures, API 1971. *The hexamethylenediamine required comes mostly from petroleum via adipic acid or butadiene.

70°C. The reaction is heated at the start, and then must be cooled to maintain the temperature at 40 to 60°C. After 2 to 6 h, the excess ethyl chloride is distilled off, and TEL steam-distilled from the reaction mixture. Tetramethyllead (TML) is partly replacing TEL, because of higher antiknock qualities in high-compression engines with highly aromatic, high-octane gasolines. Furthermore, TML is more stable at higher temperatures. A new electrolytic alkylation process[17] involves the following conversion:

$$2RMgCl + 2RCl + Pb \xrightarrow{\text{direct current}} R_4Pb + 2MgCl_2$$

AMINATION[18] BY AMMONOLYSIS AND REDUCTION Amination by ammonolysis refers to reactions in which an amino compound is formed using ammonia or substituted ammonia as the agent; other amines are made by reduction of a nitro compound.

Ethanolamines[19] A mixture of mono-, di- and triethanolamines is obtained when ethylene oxide is bubbled through 28% aqueous ammonia at 30 to 40°C as shown by the following equations:

$$\underset{\text{O}}{CH_2\!-\!CH_2} + NH_3 \text{ (excess)} \longrightarrow \begin{array}{ll} HOCH_2CH_2NH_2 & \text{Monoethanolamine} \\ (HOCH_2CH_2)_2NH & \text{Diethanolamine} \\ (HOCH_2CH_2)_3N & \text{Triethanolamine} \end{array}$$

Yield: 95%

The rate of reaction of ethylene oxide with ammonia is about the same as with monoethanolamine or diethanolamine, hence the ratio of ethylene oxide to ammonia is the primary variable in changing the

[17]Nalco Gives TEL, TML Process Details, *Chem. Eng. News*, Dec. 7, 1964 (electrolytic alkylation, Grignard flowchart); *Chem. Eng. (N.Y.)*, **72**(13), 102 (1965), and **72**(23), 249 (1965) (reactions and flowchart); Electrolytic TEL, *Chem. Week.*, Dec. 12, 1964, p. 77.

[18]Groggins, *op. cit.*, pp. 129, 413; Amination, ECT, vol. 2, 1963; Faith, Keyes and Clark, *op. cit.*, p. 330.

[19]Ethanolamines flowchart, *Hydrocarbon Process.*, **52**(11), 120 (1973); Methylamines, *Hydrocarbon Process.*, **52**(11), 150 (1973).

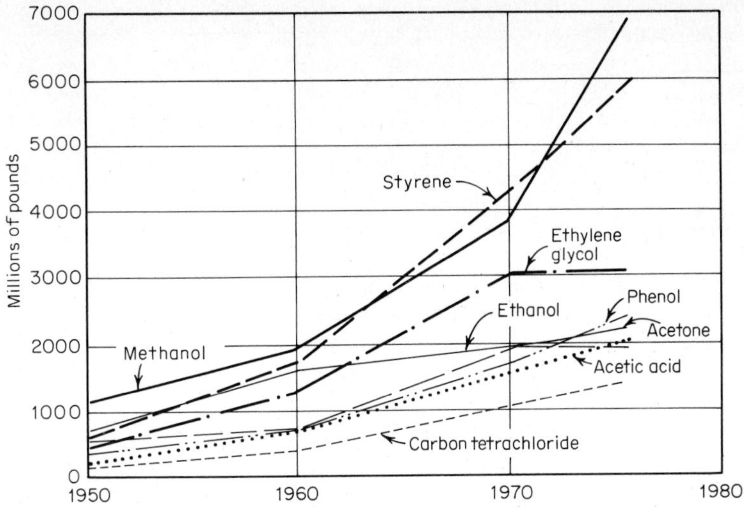

Fig. 38.5 Growth and production of important petrochemicals in the United States. (*U.S. Tariff Commission.*)

relative amounts of the ethanolamines. Since 1950 the demand for the three ethanolamines has varied, and there has been a gradual shift from tri- to di- to monoethanolamine. The process involves the exothermic, rapid, liquid-phase reaction of ethylene oxide and aqueous ammonia. A well-stirred vertical steel reactor has its contents pumped through an external cooler to maintain the reaction temperature. The reactor contents are removed at the desired rate for product recovery and purification. The unconverted ammonia is removed by stripping, and the products by fractionation under vacuum. Overall yields from ammonia and ethylene oxide are about 95%.

Methylamines production is about 160 million lb. per year. They have varied uses, mostly as intermediates in further synthesis. Their manufacture is by ammonolysis of methanol.

CRACKING, OR PYROLYSIS, THERMAL AND CATALYTIC Conversion by pyrolysis is very important in making petrochemical precursors such as olefins (Chap. 37). The chief petrochemical and product arising from this conversion is *carbon black.* This compound is so essential that a full discussion is assigned to it in Chap. 8. Isoprene (Chap. 36) can be made by cracking or synthesis.

TABLE 38.7 ***End Products of the Petrochemical Industry with Reference Chapter Number***

Product	*Chap.*	*Product*	*Chap.*
Industrial gases	7	Plastics, polymers, plasticizers	34
Industrial carbon	8	Synthetic fibers	35
Ammonia	18	Rubber, rubber chemicals	36
Nitrogen industries	18	Synthetic motor fuels	37
Sulfur and sulfuric acid	19	Lubricant and gasoline additives and synthetic lubricants	37
Aerosols	20	Flotation agents	38
Explosives	22	Antifreeze and antiknock compounds	38
Surface coatings	24	Alcohols	38
Adhesives	25	Solvents	38
Paints, varnishes, lacquers	25	Dyes, lakes, toners	39
Agrichemicals	26	Medicinal products	40
Food additives	26		
Fertilizers and pesticides	26		
Flavors and perfumes	27		
Detergents	29		

DEHYDRATION is commonly used in the production of *ethers* by the dehydration of alcohols. The more important ethers are presented in the following tabulation:

Name	*Normal boiling point, °F*
Methyl ether	−10
Ethyl ether	94
Isopropyl ether	154

Ethers can be produced by liquid-phase dehydration of the corresponding alcohol, using sulfuric acid as the catalyst, as well as by vapor-phase dehydration over acidic catalysts such as alumina. However, the chief source of ethers is as by-products of the production of the corresponding alcohol by hydration of the olefin over an acidic catalyst. Excess ether can also be recycled over an appropriate catalyst to produce the alcohol.

ESTERIFICATION[20] The making of esters is important for solvents, plasticizers, monomers, medicines, and perfumes. Also, certain nitrate esters, such as nitroglycerin and nitrocellulose, are among the most powerful explosives (Chap. 22).

ESTERIFICATION OF ORGANIC ALCOHOLS OR ACIDS *Ethyl acetate*[21] is an important solvent, particularly in lacquers. Its production in 1974 was 169.6 million lb. It is manufactured by the following reaction:

$$CH_3CH_2OH + CH_3COOH \xrightarrow{H_2SO_4} CH_3COOC_2H_5 + H_2O$$

Reactions such as this are forced to completion by removing the products as they are formed and employing an excess of one reactant. The raw materials are acetic acid, ethyl alcohol in excess, and a little sulfuric acid as catalyst. These compounds are mixed, conducted through heat exchangers, and passed through an esterification column, the overhead of which is the ternary azeotrope (bp 70.2°C) having the composition 82.16% ethyl acetate, 8.4% ethyl alcohol, and 9% water. Frequently, water is added to the condensate to remove most of the alcohol which is returned to the system. The ester, carrying about 4% water, is rectified through a column, the water going overhead as a constant-boiling mixture. It should be noted that all the acetic acid reacts by virtue of the mass action of the excess alcohol combined with removal of the products. Ethyl acetate is consumed in such large quantities that this esterification is frequently conducted in a continuous manner. The equipment is almost invariably constructed from copper.

Amyl, butyl, and isopropyl acetates[22] are all obtained from the respective alcohol and acetic acid. The total production of these esters is about 161 million lb per year, of which a large fraction was butyl acetate. All are important lacquer solvents and as such impart good "blush"-resisting qualities to drying lacquer film. Butyl acetate is the most important nitrocellulose lacquer solvent. Other esters made similarly are methyl salicylate, methyl anthranilate, diethyl phthalate, and dibutyl phthalate.

Esterification of olefins by the addition of various lower organic (or inorganic) acids is one method of producing unsaturated vinyl esters which may be used in polymerization reactions. The most important esters are vinyl esters: vinyl acetate, vinyl chloride, acrylonitrile, and vinyl fluoride. The addition reaction can be carried out in the liquid or in the vapor phase; the choice depends on the properties of the acid. In either case care must be taken to prevent polymerization of the vinyl ester product.

[20]Groggins, *op. cit.*, p. 694; Goldstein, *op. cit.*, p. 385.
[21]Groggins, *op. cit.*, p. 727; Faith, Keyes, and Clark, *op. cit.*, p. 339 (flowchart).
[22]Othmer and Leves, *Trans. AIChE*, **41**, 157 (1945); Faith, Keyes, and Clark, *op. cit.*, p. 176.

Vinyl acetate[23] (production in 1974, 1,402 million lb) is manufactured by one of the following reactions:

$$CH{\equiv}CH + CH_3COOH \longrightarrow CH_3COOCH{\equiv}CH_2$$

$$CH_2{\equiv}CH_2 + CH_3COOH + \tfrac{1}{2}O_2 \xrightarrow{Pd} CH_3COOCH{\equiv}CH_2 + H_2O$$

Vinyl acetate (bp 72°C) is produced on a commercial scale by a vapor-phase process. Acetic acid and recycle acid are mixed and heated in a stainless-steel vessel. Fresh and recycle ethylene and oxygen are mixed and bubbled through the acetic acid. By means of temperature control the desired ratio of acid to ethylene can be achieved. The vapor mixture is passed through vertical reactor tubes filled with a noble-metal catalyst. The reaction is exothermic, and the heat is used for steam generation, the steam being used in the distillation. The yield is high—about 91%. The effluent gases are cooled, and all but the unconverted ethylene and carbon dioxide are condensed. Acetic acid is recovered by absorption and scrubbing and, along with the ethylene, is recycled. Vinyl acetate is purified (99.5+%) by distillation and stabilized by the addition of an inhibitor, such as hydroquinone or diphenylamine.

The basic processing scheme as described above is widely used in the preparation of other *vinyl esters*. A different acid (HCl, HCN, HF) is used and, depending on the volatility of the raw materials and product, a liquid- or vapor-phase process may be chosen. An interesting example of addition to acetylene for the production of a monomer is illustrated by *chloroprene* (Chap. 36). Acetylene is polymerized to vinylacetylene in a horizontal agitated copper vessel, using as a catalyst a weak NH_4Cl solution of Cu_2Cl_2 and KCl. The reactor gas is passed through a condenser to remove water, and fractionated. The vinylacetylene is treated with cold, aqueous hydrochlorine and at 35 to 45°C, using aqueous Cu_2Cl_2 as a catalyst, to give chloroprene.

$$\underset{\text{Acetylene}}{2CH{\equiv}CH} \xrightarrow{55\text{–}65°C} \underset{\text{Vinylacetylene}}{CH_2{\equiv}CH{\equiv}C{\equiv}CH} \xrightarrow[Cu_2Cl_2,\ 40°C]{+\ HCl} \underset{\text{Chloroprene}}{CH_2{\equiv}CCl{\equiv}CH{\equiv}CH_2}$$

The esters of allyl alcohol, such as *diallyl phthalate*, are useful as bifunctional polymerization monomers and can be prepared by direct esterification of phthalic anhydride and allyl alcohol. Various acrylic esters, such as ethyl acrylic, are also widely employed in polymerization and can be made by direct esterification of acrylic acid with the appropriate alcohol.

HALOGENATION[24] AND HYDROHALOGENATION

CHLORINATION OF ALIPHATICS The direct chlorination of lower aliphatic hydrocarbons has been studied by Hass and coworkers[25] at Purdue University and developed by the Dow Chemical Co. The various products are obtained through both liquid- and vapor-phase reactions, using various temperatures and pressures, with or without catalysts. The mixtures of products are separated by rectification.

A new method of chlorination is known as oxychlorination and produces a whole range of C_1 and C_2 chlorinated hydrocarbons from methane and ethane by a pollution-free technique.[26] Vinyl chloride monomer can also be produced from an ethane or ethane/ethylene feed.

$$CH_4 + HCl + O_2 \longrightarrow \text{chloromethanes} + H_2O$$

The chlorine can be obtained from molecular chlorine, HCl, or waste chlorocarbons. These compounds are converted to a usable form of chlorine by a molten-salt catalyst consisting of cuprous and

[23] *Hydrocarbon Process.*, **52**(11), 189 (1973) (flowcharts); Sitting, *op. cit.* pp. 694, 695; Hahn, The Petrochemical Industry, p. 192, Petroleum Publishing, 1970; Brownstein, U.S. Petrochemicals, p. 254, Petroleum Publishing, 1972.

[24] Groggins, *op. cit.*, p. 204, Belohlar and McBee, Halogenation, Ann. Rev. Suppl., *Ind. Eng. Chem.*, p. 77, 1962.

[25] Hass, McBee *et al.*, *Ind. Eng. Chem.*, **34,** 296 (1942).

[26] Molten Salt Process Yields Chlorinated Hydrocarbons, *Chem. Eng.* (*N.Y.*), **81**(13), 114 (1974) (flowsheet).

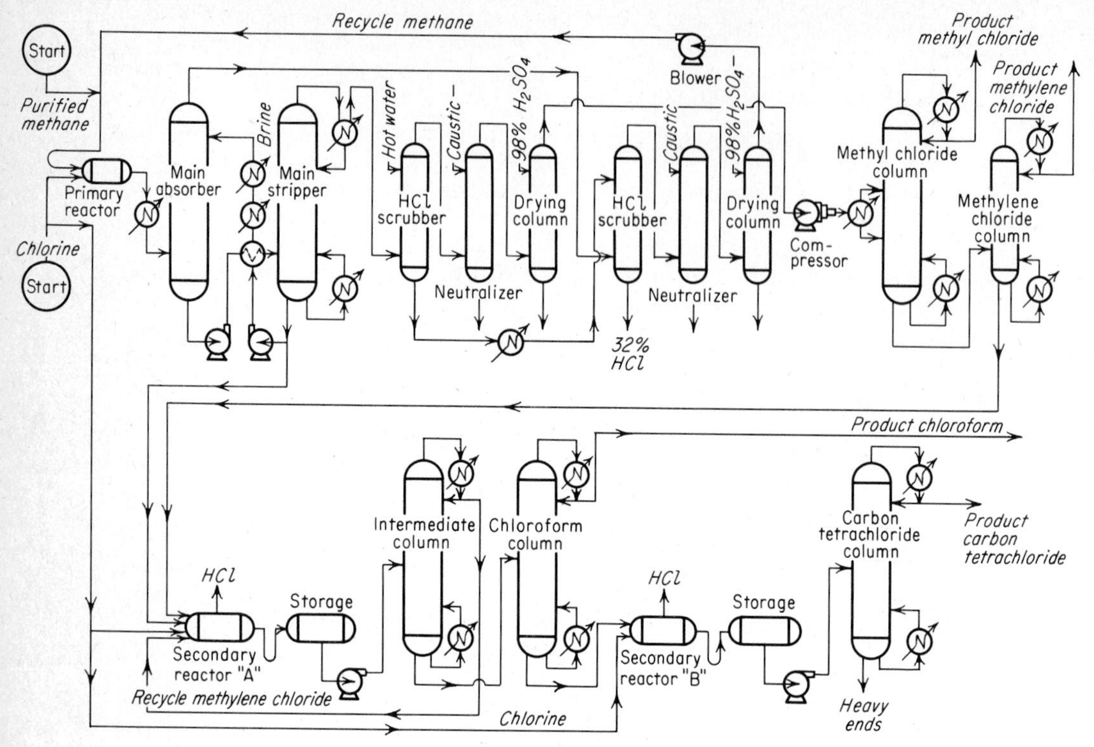

Fig. 38.6 Flowchart for chlorinating methane. The HCl scrubbers are of Karbate carbon. (*Petroleum Refiner*).

cupric chlorides and potassium chloride. The only effluents are N_2, CO_2, and H_2O, thus making this truly a pollution-free process. All waste chloro compounds are recycled and consumed. Other chlorination processes generate many pollution-causing wastes and require costly systems for their disposal.[27] These chloroaliphatics find many uses commercially, e.g., in the production of amyl alcohols from chloropentanes (see hydrolysis) and as solvents, particularly in dry cleaning under flammable conditions. The chloromethanes[28] are *methyl chloride, dichloromethane* (methylene chloride), and *chloroform;* they are produced in large quantities, the three totaling about 1,000 million lb annually. All are obtained by chlorinating methane, although methyl chloride has been made from methanol and hydrogen chloride, and chloroform by the decarboxylation of trichloroacetic acid with lime. Methyl chloride is used as an alkylating agent and refrigerant, methylene chloride as a dewaxing solvent for oils, and chloroform as a solvent, intermediate, and medicinal, but mostly to produce fluorocarbons. *Carbons tetrachloride* is also made from methane.[29] In chlorinating methane according to the flowchart in Fig. 38.6, the chlorine and methane (including recycle) are charged in the ratio of 0.6:1.0. The temperature in the primary reactor is maintained at 650 to 700°F, using the products to preheat the reactants. The chlorine conversion is essentially 100%, and the methane conversion about 65%. High velocities are maintained through narrow channels to avoid ignition. The effluent gases contain, in the indicated ratios, methyl chloride 6 parts, methylene chloride 3 parts, chloroform 1 part, and carbon tetrachloride $\frac{1}{4}$ part. Besides these are unreacted methane and

[27]Santolezi, Chlorinated Hydrocarbon Waste Disposal and Recovery Systems, *Chem. Eng. Prog.*, **69**(1), 68 (1973).
[28]Goldstein, *op. cit.*, pp. 71; Faith, Keyes, and Clark, *op. cit.*, pp. 229, 507; *Hydrocarbon Process*, **52**(11), 113 (1973).
[29]Sittig, *op. cit.*, p. 150.

the HCl, together with traces of chlorine and heavy chlorinated products. The secondary chlorination is in liquid phase at ambient temperature, and is light (mercury-arc)-catalyzed, in reactor A, converting methylene chloride to chloroform, and in reactor B, chloroform to carbon tetrachloride. Products can be varied somewhat to satisfy market demands.

Ethane[30] can be chlorinated under very similar conditions as methane to yield chlorinated ethanes, particularly *ethyl chloride*. The latter is employed extensively in making TEL (q.v.). Since polychlorinated ethanes are not desirable, a large excess of ethane is used.

$$CH_3CH_3 + Cl_2 \longrightarrow CH_3CH_2Cl + HCl$$

To utilize the by-product HCl, a mixture of ethane and ethylene may be used, and the following reaction can occur with Friedel-Crafts catalysts:

$$CH_2{=}CH_2 + HCl \xrightarrow{AlCl_3} CH_3CH_2Cl$$

The oxychlorination process is now important.

Ethylene dichloride[31] (production 7,721 million lb in 1974). The addition of halogen to unsaturates produces many derivatives, such as ethylene dichloride, ethylene dibromide, dichloroethylene, trichloroethylene, and tetrachloroethane. The preparation of ethylene dichloride with a 90% yield is a typical example.

$$CH_2{=}CH_2 + Cl_2 \xrightarrow{C_2H_4Br_2} ClCH_2CH_2Cl$$

Chlorine gas is bubbled through a tank of ethylene dibromide, and the mixed vapors sent to a chlorinating tower where they meet a stream of ethylene. Here the reaction takes place at a temperature of 40 to 50°C. The products from the tower pass through a partial condenser, followed by a separator, the crude ethylene dichloride passing off as a gas and the liquid ethylene dibromide being returned to the process. The product enters into further synthesis, as in the preparation of ethylenediamine and Thiokol rubber. It is also used as a solvent and as an additive in TEL to vaporize any lead.

Ethylene dibromide. This compound, commonly known as ethylene bromide, is prepared by the direct addition of bromine to ethylene. *Ethyl bromide* is now being manufactured by the Dow Chemical Co. to the extent of 1 million lb annually, using the following novel reaction initiated by gamma radiation from cobalt 60:

$$HBr + C_2H_4 \xrightarrow{\gamma} C_2H_5Br$$

This reaction[32] is exothermic in nature, and the radiation serves only as the initiator for the process. Since about 0.2 Mrd of gamma radiation is required for complete conversion, the radiation contributes only about 2 J/g of material and thus does not significantly alter the overall free-energy relationships of the process. The process is a two-phase, gas-liquid reaction and makes use of the recirculation system shown in Fig. 38.7; details of the reaction vessel and associated equipment are shown in Fig. 38.8. It is used in very large quantities as an additive reagent in TEL in gasoline to prevent the deposition of lead oxide on engine valves.

Perchloroethylene is used in dry cleaning and vapor degreasing to about 365 million lb/year. The reactions of the commercial process involved chlorinating ethylene and its chloro derivatives, followed by alkaline treatment. Perchloroethylene is now manufactured as a means of disposing of

[30]Shell Development, U.S. Pat. 2,246,082; Faith, Keyes, and Clark, *op. cit.*, pp. 75, 171; Sittig, *op. cit.*, pp. 304–305 (flowcharts); *Chem. Eng.* (*N.Y.*), **81**(13), 114 (1974) (flowchart).

[31]Sherwood, *Petrol. Process.*, **7**, 1809 (1952); Faith, Keyes, and Clark, *op. cit.*, p. 368, *Hydrocarbon Process.*, **52**(11), 194 (1973).

[32]Courtesy David E. Harmer of Dow Chemical Co.; Harmer *et al.*, Dow Ethyl Bromide Process, Industrial Application of Radiation Chemistry, International Conference on Large Radiation Sources in Industry, Salzburg, Austria, May 27–31, 1963; Harmer and Beale, Making Chemicals via Radiation: Dow's Ethyl Bromide Process, *Chem. Eng. Prog.*, **60**(4), 33 (1964).

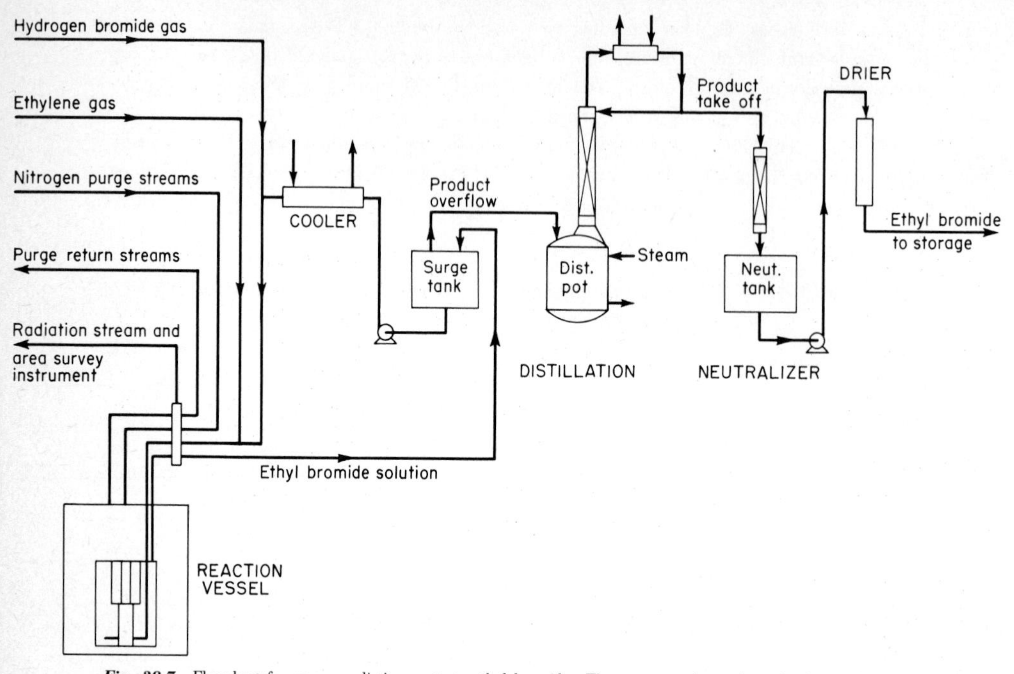

Fig. 38.7 Flowchart for gamma-radiation route to ethyl bromide. The pump continuously recirculates ethyl bromide through an external heat exchanger and then through the reaction vessel, which is located below the surface of the ground in a location external to the building which houses all other parts of the process. Feed streams of ethylene and hydrogen bromide are continuously added to the recycle ethyl bromide stream and react rapidly as they pass through the irradiated reaction vessel. Newly produced ethyl bromide accumulates in the surge tank and overflows into the distillation unit. A caustic scrubber system removes small amounts of unreacted hydrogen bromide. (*Dow Chemical Co.*)

"junk" chlorinated hydrocarbons of one to three chlorine atoms. These are reacted with excess chlorine and conducted through a reactor at 700°C, causing *chlorinolysis* to take place and yielding carbon tetrachloride (CCl_4) and perchloroethylene ($CCl_2{=}CCl_2$). Either of these can be recycled to extinction so that the supply can be controlled, depending on the markets.

$$2CCl_4 \rightleftharpoons C_2Cl_4 + 2Cl_2$$

Pentanes have long been chlorinated,[33] and the *amyl chloride* hydrolyzed to amyl alcohols.

HYDRATION AND HYDROLYSIS[34] An unusually good general presentation of this chemical conversion, or unit process, can be found in the Groggins reference. The principal examples for petrochemicals are in ethyl and other alcohols. In 1974 over 95% of all ethyl alcohol was manufactured by synthesis. Two synthetic processes are used commercially, one based on ethyl hydrogen sulfate and one on vapor-phase direct-hydration. Both start with purified ethylene, and the following are the respective reactions:

(1) $3C_2H_4 + 2H_2SO_4 \longrightarrow C_2H_5HSO_4 + (C_2H_5)_2SO_4$

$C_2H_5HSO_4 + (C_2H_5)_2SO_4 + H_2O \longrightarrow 3C_2H_5OH +$ aqueous $2H_2SO_4$

Yield 90%, with 5 to 10% ether

By-product: $2C_2H_5OH + H_2SO_4 \longrightarrow (C_2H_5)_2O + H_2SO_4 \cdot H_2O$

[33]CPI 2, p. 943, Groggins, *op. cit.*, p. 783; 1963 Petrochemical Handbook, p. 151 (flowcharts).

[34]Groggins, *op. cit.*, pp. 750, 781; Goldstein, *op. cit.*, pp. 136–153; Faith, Keyes, and Clark, *op. cit.*, p. 343.

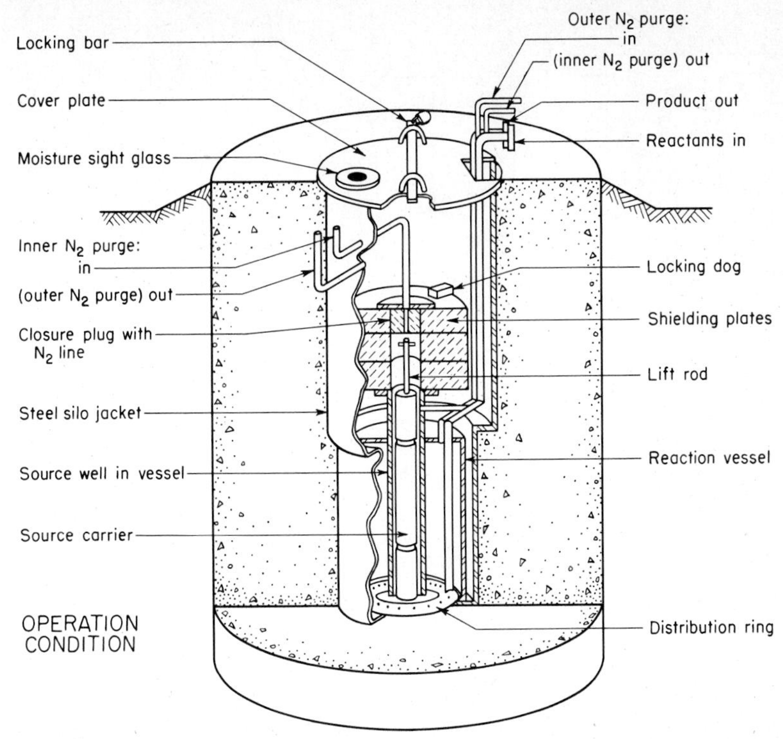

Fig. 38.8 Detailed diagram of reaction vessel for the gamma-radiation route to ethyl bromide. The cobalt-60 radiation source is contained within a source carrier which can be lowered from a lead transfer cask into the hollow center of the 40-gal reaction vessel. Because the gamma radiation from cobalt 60 functions only to bring about a chemical change, there is no induced or residual radioactivity in any of the material processed by this method. (*Dow Chemical Co.*)

$$(2)\ C_2H_4 + H_2O \xrightarrow[570°F,\ 1{,}000\ psig]{H_3PO_4} C_2H_5OH \qquad \Delta H = -9.6\ kcal$$

Yield 92% alcohol, conversion per pass 4 to 25%

The direct hydration process is depicted in Fig. 38.9. Ethylene and water are passed over a fixed-bed catalyst such as aluminum hydroxide gel, phosphoric acid on diatomaceous earth, or tungstic acid on silica gel. The conversion per pass ranges from 4 to 25%, but the overall conversion approaches 92%. This process avoids the corrosion problems of the sulfuric acid process.[35]

See Figs. 31.1 to 31.4, and the accompanying text and flowchart for the fermentation of alcohol, for diagrams and conditions for azeotropic distillation to obtain anhydrous alcohol. A parallel description of the steps is presented in Chap. 31 to compare fermentation and hydration processes for alcohol.

Isopropyl alcohol.[36] A mixture of propylene and propane (65% propylene) is absorbed in 75% sulfuric acid at 100°C and 300 to 400 psi. The alcohol is released from solution by hydrolysis and stripping and purified by a series of distillations. Propylene conversion is high, and 93 to 95% of the charged propylene yields alcohol. In 1976, about 2 billion was consumed, and about 40% converted to acetone. Under conditions only slightly less stringent than those used for propylene, 1- and

[35]Groggins, *op. cit.*, p. 788 (flowchart); Goldstein, *op. cit.*, pp. 142–143 (many equilibrium data); *Hydrocarbon Process.*, **52**(11), 121 (1973) (flowcharts).

[36]Goldstein, *op. cit.*, p. 146; Sittig, *op. cit.*, p. 393; *Hydrocarbon Process.*, **52**(11), 141 (1973).

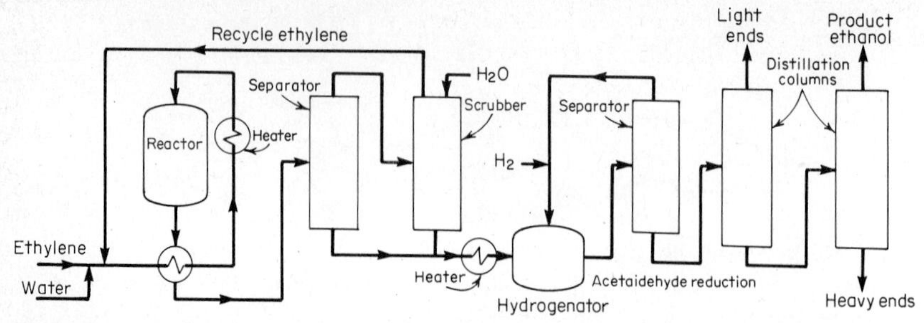

Fig. 38.9 Flowchart for synthesis of ethyl alcohol from ethylene by direct hydration. [*Chem. Eng.* (*N.Y.*), **81**(9), 144 (1974).]

2-butenes can be converted to *sec-butyl alcohol,* and isobutylene can be converted to *tert-butyl alcohol.*

Ethylene glycol[37] is the most important permanent antifreeze (in 1976 over 3,000 million lb) and is made by several processes involving hydrolysis. In the first, ethylene chlorohydrin,[38] which is produced by the reaction of chlorine water with ethylene, is treated with sodium bicarbonate solution:

$$CH_2{:}CH_2 + Cl_2 + H_2O \longrightarrow CH_2OH{-}CH_2Cl + HCl$$

$$CH_2OH{-}CH_2Cl + NaHCO_3 \longrightarrow CH_2OH{-}CH_2OH + NaCl + CO_2$$

The ethylene chlorohydrin is produced by passing ethylene and chlorine into water at 50 to 120°F. By adding caustic and rectifying the vapors, ethylene oxide can also be produced from ethylene chlorohydrin. The chlorohydrin solution is concentrated to 35 to 40% and reacted with the theoretical amount of sodium bicarbonate solution in a closed, stirred, steam-jacketed kettle. At a temperature of 70 to 80°C, the hydrolysis proceeds smoothly with the evolution of CO_2 and is complete in 4 to 6 h. The glycol solution is concentrated by distillation. As the high-boiling ethylene glycol can be separated only with difficulty from the salt, it is presumably better engineering to make ethylene oxide out of ethylene chlorohydrin by heating with caustic soda or hydrated lime. This can be hydrated to the glycol by treating with weakly acidulated water under low pressure. *Diethylene glycol* is obtained as a by-product in about 10% yield from the manufacture of ethylene glycol. Its principal use is as a humectant for tobaccos and as an emulsifying agent.

More recently the chlorohydrin[39] route to ethylene oxide and glycol has been superseded—at least, in new plant construction—by the reaction of ethylene with oxygen and water:

$$C_2H_4 + \tfrac{1}{2}O_2 \longrightarrow C_2H_4O \qquad \text{yield } 60\%$$

$$C_2H_4O + H_2O \longrightarrow CH_2OH \cdot CH_2OH \qquad \text{yield } 92\text{–}95\%$$

Ethylene, oxygen, and recycle gas are charged to a tubular reactor filled with a silver catalyst (or sulfuric acid). The reactor is operated at 200 to 300°C. Ethylene oxide is recovered from the gaseous reactor effluent by absorption in water. For ethylene glycol, the wet ethylene oxide is reacted with water in the presence of acid at 50 to 100°C. The crude glycol product is dehydrated and fractionated from di- and triglycol by-products.

Amyl alcohols[40] are produced from pentane via a two-step process involving chlorination and hydrolysis. Anhydrous pentane and chlorine vapors are mixed and charged to a furnace, where the pyrolysis of the chloropentanes is minimized by cooling the reactor effluent. The monochloropentane

[37]Matthew *et al., Chem. Eng.* (*N.Y.*), Oct. 1955, p. 280; Faith, Keyes, and Clark, *op. cit.*, p. 372; *Hydrocarbon. Process.*, **52**(11), 130 (1973) (flowcharts).

[38]Weaver, *Ind. Eng. Chem.*, **51,** 894 (1959).

[39]Sittig, *op. cit.*; *Hydrocarbon. Process.*, **52**(11), 129 (1973).

[40]Gilette, ECT, 2d ed., vol. 2, pp. 374–379, 1963 (flowchart); Kenyon *et al., Ind. Eng. Chem.*, **42,** 2388 (1950); Faith, Keyes, and Clark, *op. cit.*, 2d ed., 1961, p. 107.

(amyl chloride) is recovered from the unconverted pentane, hydrogen chloride, and polychlorinated pentanes. To form the amyl alcohol, the amyl chlorides are hydrolyzed with aqueous caustic soda with sodium oleate as a catalyst. Following the hydrolysis the brine phase is decanted, and the organic phase is run to steam-stripping columns, where pentanes and amyl chlorides are removed. The amyl alcohols are recovered by distillation.

HYDROGENATION,[41] DEHYDROGENATION Hydrogenation yields many useful products for the chemical industries. For the production of hydrogen, see Chap. 7. Besides hydrogenating double bonds, this reaction may be used to *eliminate other elements from a molecule, such as oxygen, nitrogen, sulfur, carbon, and halogens.* Hydrogenation is carried on under pressure and usually at elevated temperatures with a suitable catalyst. The temperatures are ordinarily below 400°C but may extend up to 500°C in some cases. The catalysts vary with temperature, the noble metals such as platinum and palladium being used up to 150°C, nickel and copper from 150 to 250°C, and various combinations of metals and metal oxides at higher temperatures. Pressure not only increases the rate of reaction but, in cases where there is a decrease in volume as the reaction proceeds, causes a desirable shift in the equilibrium according to Le Chatelier's principle. Dehydrogenation is also widely employed in the chemical industries. Many of the foregoing remarks on hydrogenation also apply to dehydrogenation. The main differences are in the pressure and its effect on the equilibrium and catalyst. Lower pressures are used, or a vacuum, and noble-metal (platinum) and copper and iron oxide catalysts are employed.

Methanol[42] is manufactured by one of the most important high-pressure organic syntheses used today; production in 1974 was 6,789 million lb. It is made from carbon monoxide and hydrogen (synthetic gas, Chap. 7). The essential reaction is

$$CO + 2H_2 \rightleftharpoons CH_3OH \qquad \Delta H = -24.6 \text{ kcal}$$

The catalyst most commonly employed is silver or copper, promoted with oxides of zinc, chromium, manganese, or aluminum. Contact of the gases with hot iron must be avoided, and so copper-lined reactors are used (Fig. 38.10). The theoretical proportions of the gases are mixed and compressed to 4,500 psi. This enhanced pressure (as in ammonia synthesis) results in a more favorable chemical equilibrium (Le Chatelier's principle). After oil purification, the gases go to the reactors where, at 300°C, the equilibrium yield is about 60%. The gaseous products then pass through heat exchangers to high-pressure condensers, where the methanol is condensed at 3,500 to 4,000 psi, the gases being recirculated. The conversion to alcohol is low (12 to 15%), since the presently used catalyst operates only at high temperatures where the equilibrium constant is poor, as depicted in the following tabulation:

Methanol Synthesis Data

Equilibrium constants		*Effect of pressure on equilibrium conversion*	
Temperature, °C	K_p	*Pressure, atm at 300°C*	*Percent conversion to liquid methyl alcohol in one pass*
200	1.7×10^{-2}	10	
300	1.3×10^{-4}	50	8.0
400	1.1×10^{-5}	100	24.2
		200	48.7
		300	62.3

Source: Ind. Eng. Chem., **32,** 147 (1940); Goldstean and Waddams, Petroleum Chemicals Industry, p. 55, Spon, 1967; cf. ammonia synthesis data and equipment in Chap. 18.

[41]Groggins, *op. cit.*, p. 555; Hydrogenation and Dehydrogenation, Unit Processes Reviews, *Ind. Chem. Eng.*, **53**(9), 767 (1961); Happel *et al.*, Dehydrogenation of Butanes and Butenes, Cr_2O_3-Al_2O_3 Catalyst, *Ind. Eng. Chem. Fundam.*, **5**, 289 (1966).

[42]Goldstein, *op. cit.*, p. 53, *Hydrocarbon Process.*, **52**(11), 147 (1973) (flowcharts).

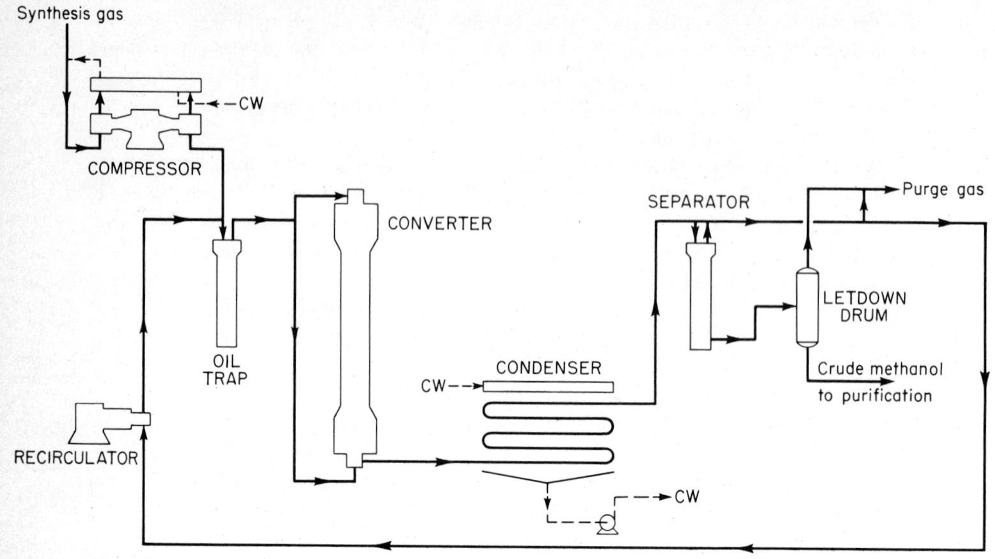

Fig. 38.10 Flowchart for the conversion of synthesis gas to methanol. (*Vulcan Manufacturing.*)

Styrene is procured from ethylbenzene by dehydrogenation (see flowchart in Fig. 36.3).

Ketones. Many lower-molecular-weight aliphatic ketones are produced by dehydrogenation of the corresponding alcohol. For example, acetone, methyl ethyl ketone, and cyclohexanone are produced in this manner. The production of cyclohexanone is discussed in Chap. 35 in connection with the caprolactam fiber process.

The most important ketone produced is *acetone*, the 1974 figure being 2,072 million lb. Acetone is used mainly for cellulose acetate and the manufacture of other chemicals, such as phorone, mesityl oxide, methyl isobutyl ketone, and methyl methacrylate, and also for solvents. Its main supply comes from catalytic dehydrogenation of isopropyl alcohol, by-products of the cumene phenol process, and several minor procedures, such as fermentation (Chap. 31) and oxidation (butane).

The conversion of isopropyl alcohol to acetone is as follows:

$$(CH_3)_2CHOH(g) \longrightarrow (CH_3)_2CO(g) + H_2(g) \qquad \Delta H_{25} = +15 \text{ kcal}$$

The reaction[43] results in an 85 to 90% conversion at temperatures of 350°C and above, over either metallic copper or zinc acetate, and is often carried out at 30 psi. The acetone is recovered by condensation from the hydrogen and is purified by simple distillation.

NITRATION[44] Nitration is one of the important means for introducing nitrogen into hydrocarbons. Except for explosives and an occasional solvent, the nitro group is usually converted by amination reduction to the desired product. Nitric acid is a powerful oxidizing agent, as well as a nitrating one. Oxidation occurs especially in difficult nitrations. *Nitrobenzene* (Chap. 39) is probably the most important technical nitration product, except for the wartime explosives (Chap. 22) such as trinitrotoluene, tetryl, picric acid, and cyclonite. The so-called nitroglycerin and nitrocellulose are nitrate esters.

[43] *Pet. Refiner*, **38,** 264 (1959); Groggins, *op. cit.*, p. 523 (flowchart).

[44] Groggins, *op. cit.*, pp. 60ff.; Goldstein, *op. cit.*, p. 85; Albright and Hanson (eds.), Industrial and Laboratory Nitrations, ACS, 1975.

Nitroparaffins.[45] A basic development in the field of nitration has been vapor-hase low-pressure nitration of aliphatic hydrocarbons, particularly propane. This compound is vaporized and mixed with vaporization nitric acid to give at least a 2:1 ratio of hydrogen to acid. The reactions, nitration and oxidation, take place from 0.1 to 5 s and at about 400°C and 100 psi. Vapors are condensed rapidly to give nitroparaffins and nitrogen oxides, with vapors of excess propane and the products of oxidation passing on. It is characteristic of these nitrations that the original carbon chain may be reacted at any point to yield a variety of products.

$$\begin{matrix}\text{Propane (in excess)} \\ + \text{ nitric acid} \longrightarrow\end{matrix} \left\{\begin{matrix} CH_3CH_2CH_2NO_2 & 25\% + CO_2 \\ CH_3CHNO_2CH_3 & 40\% + H_2O \\ CH_3CH_2NO_2 & 10\% + NO \\ CH_3NO_2 & 25\% \quad \text{etc.} \end{matrix}\right\} \text{representing 30–40\% nitric acid}$$

These reactions have been presented with a flowchart. The material of construction is essentially ceramic-lined stainless steel, arranged in a continuous series of heaters, reactors, condensers, distillation columns with the necessary pumps, and control instruments. Nitroparaffins are employed as solvents and propellants (nitroethane in the Polaris submarine) and to make derivatives, e.g., hydroxylamine. The following reactions are typical syntheses from nitroparaffins.

Reduction to amines:

$$\underset{\text{2-Nitropropane}}{CH_3CHNO_2CH_3} + \underset{\text{Hydrogen}}{3H_2} \longrightarrow \underset{\text{Isopropylamine}}{CH_3CHNH_2CH_3} + \underset{\text{Water}}{2H_2O}$$

Addition of an aliphatic aldehyde:

$$\underset{\text{Nitromethane}}{CH_3NO_2} + \underset{\text{Formaldehyde}}{3CH_2O} \xrightarrow[\text{catalyst}]{\text{basic}} \underset{\text{Tris(hydroxymethyl)nitromethane}}{CH_2OH-\overset{\overset{\displaystyle NO_2}{|}}{\underset{\underset{\displaystyle CH_2OH}{|}}{C}}-CH_2OH}$$

Addition of an aromatic aldehyde:

$$\underset{\text{2-Nitropropane}}{CH_3CHNO_2CH_3} + \underset{\text{Benzaldehyde}}{C_6H_5CHO} \xrightarrow{KOH} \underset{\text{2-Nitro-2-methyl-1-phenyl-1-propanol}}{C_6H_5CHOHC(CH_3)NO_2CH_3}$$

Formation of hydroxylamine or salts:

$$\underset{\text{1-Nitropropane}}{CH_3CH_2CH_2NO_2} + H_2SO_4 + H_2O \longrightarrow \underset{\text{Hydroxylammonium acid sulfate}}{NH_2OH \cdot H_2SO_4} + \underset{\text{Propionic acid}}{CH_3CH_2COOH}$$

OXIDATION[46] is one of the most valuable conversions for petrochemicals. The possibilities of this process are so varied that a type reaction is impossible. Generally, this involves the scission of C—C and C—H bonds with the formation of C—O bonds. The cheapest reagent, and the one often employed, is air, but more and more oxygen is being used, with the result that about 40% of the

[45]Groggins, *op. cit.*, p. 125 (thermodynamics); ECT, vol. 18, p. 864, 1967 Bachman and Pollack, *Ind. Eng. Chem.*, **46,** 715 (1954); Reidel, Propane to Nitroparaffins, *Oil Gas J.*, Jan. 9, 1956 (flowchart); Albright, Nitration of Paraffins, *Chem. Eng. (N.Y.)*, **73**(12), 149 (1966); Fear and Burnet: V. P. Nitration of Butane in Molten Salt Reactor, *Ind. Eng. Chem., Process Dev.*, **5,** 166 (1966).

[46]Groggins, *op. cit.*, p. 486; Landau, Aromatic Oxidation, *Ind. Eng. Chem.*, **53**(10), 32A (1961); Goldstein, *op. cit.*, pp. 59, 154; Toland *et al.*, Oxidation, Annual Review Suppl., *Ind. Eng. Chem.*, 1962; Sittig, Combining Oxygen and Hydrocarbons for Profit, Gulf Publishing, 1962 (essential reference source); Dumas and Bulani, Oxidation of Petrochemicals, Applied Science Publishers (Essex, England), 1974.

U.S. oxygen capacity production, or 17,000 tons/day in 1976, is consumed in the chemical industry. Sittig points out that oxygen is often more economical than air, largely because of enhanced yields, reduced recycle intervals, higher reaction rates, reduced equipment size, and reduced heat loss to nitrogen. The reactions are carried out in both the liquid and vapor phases, using a variety of catalysts, V_2O_5, for example. In the latter case, fairly high temperatures (about 400°C) are employed and, since the reactions are highly exothermic, the problem of heat removal is a large one. Oxidation gives synthesis gas (Chap. 7), but hydrogenation converts the CO to methanol, hence the latter chemical is presented under Hydrogenation (q.v.).

The Celanese Corp., at its large plants at Bishop and Pampa, Tex., oxidize propane and butane as such or as LPG and obtain a large number of oxygen-containing chemicals. Air has been used, but in recent enlarged operations, increase in production has resulted from employing oxygen. Some of the products are methanol, acetic acid, acetic anhydride, formaldehyde, acetaldehyde, acetone, butyraldehyde, ethyl acetate, butanol, and vinyl acetate. *Partial oxidation* is employed to raise the temperature and to oxidize partially a hydrocarbon, e.g., natural gas to make acetylene (Chap. 7) or under other conditions to furnish channel black (Chap. 8). It is reported that Celanese's new acetaldehyde plant at Bay City, Tex., involves the largest use of titanium yet reported.

LPG oxidation takes place with air. Greater production is obtained with oxygen for any given plant size. Products prepared without a catalyst are acetaldehyde, formaldehyde, methanol, and many oxygenated petrochemicals. The longer the normal-paraffin chain length and the higher the pressure, the easier the oxidation. Catalysts are important and selective; a cobalt catalyst with liquid-phase butane yields 76% acetic acid, and 6% formic acid, whereas a manganese catalyst furnishes 62% acetic and 23% formic acid, this being the experience of Celanese at Pampa,[47] Tex., for butane liquid-phase oxidation at 800 psig and 175°C.

Acetaldehyde[48] is produced commercially (about 1,600 million lb in 1976) by four methods:

1. By the hydration of acetylene:

$$C_2H_2 + H_2O \longrightarrow CH_3CHO$$

Yield 95%, conversion 55%

Acetylene is passed into a suspension of ferric and mercurous sulfates in sulfuric acid under a pressure of 15 psi between 70 and 100°C (maintained by the heat of reaction).

2. From ethyl alcohol, vapor-phased and oxidized to acetaldehyde by passing through a copper- or silver-gauze catalyst:

$$C_2H_5OH(g) + \tfrac{1}{2}O_2(g) \xrightarrow{540°C} CH_3CHO(g) + H_2O(g) \qquad \Delta H = -43 \text{ kcal}$$

Yield 85–95%, conversion 20–35%

The product is recovered by scrubbing with ethyl alcohol and rectifying to give 99% acetaldehyde.

3. From ethylene by the Wacker process developed by Hoechst[49] for the direct oxidation of ethylene, using either air or 99% oxygen. The catalyst is an aqueous solution of copper chloride (promoter) and palladium chloride, through which ethylene with air or oxygen is passed:

$$CH_2{:}CH_2 + PdCl_2 + H_2O \longrightarrow CH_3CHO + Pd + 2HCl$$

$$Pd + 2HCl + \tfrac{1}{2}O_2 \longrightarrow PdCl_2 + H_2O$$

[47]Sittig, Combining Oxygen and Hydrocarbons for Profit, chap. 7, pp. 101–118. Full descriptions, references, flowcharts, and tabulated variations with changes in temperature and pressure on oxygen ratio.

[48]Goldstein, *op. cit.*, p. 332; Sittig, Combining Oxygen and Hydrocarbons for Profit, pp. 232–234 (flowchart for oxidation from alcohol).

[49]*Hydrocarbon Process.* **52**(11), 91 (1973).

The acetaldehyde is vaporized from the solution and absorbed in water, separate from residual gases. The acetaldehyde is distilled. The conversion and yield are approximately 95%. A flowchart[50] is depicted in Fig. 38.11.

4. From lower paraffins. Propane and butane can be oxidized with air noncatalytically to produce, among many other compounds, acetaldehyde and acetic acid. The selectivity of the process is reasonable, only 15 to 20% of the paraffin being oxidized to CO and CO_2, and the remainder yielding condensable organic products such as acetaldehyde and acetic acid. Recovery of the products as pure components requires special distillations.

Acetic acid[51] *and anhydride* using oxygen are produced by the oxidation of acetaldehyde. Many alternative routes are available (e.g., fermentation, Chap. 31).

The acetaldehyde is oxidized by air with methyl acetate diluent and with manganese acetate catalyst present to prevent the formation of explosive amounts of peracetic acid. The pressure is about 60 psi, and the temperature at 50 to 70°C. The products, acetic anhydride and acetic acid, are separated and purified in a series of bubble-cap columns, the crude still and the anhydride still operating under vacuum. In 1974, 2,277 million lb of acetic acid were produced and, 570 million lb of acetic anhydride. The most important use of acetic anhydride is in the making of cellulose acetate. It is also employed in various other acetylations. A development[52] has been the production of acetaldehyde-monoperacetate (AMP)

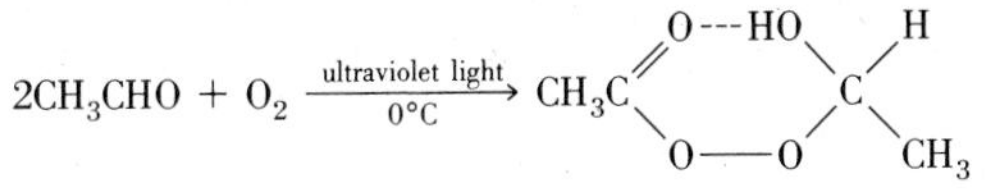

$$2CH_3CHO + O_2 \xrightarrow[0°C]{\text{ultraviolet light}} CH_3C\overset{O \cdots HO}{\underset{O - O}{\Large\langle\rangle}}C\begin{smallmatrix}H \\ CH_3\end{smallmatrix}$$

by the controlled low-temperature oxidation of acetaldehyde. The AMP can be used to produce epoxides or 2 mol of acetic acid.

Ethylene oxide[53] is an important chemical intermediate, yielding ethylene glycol (half of the ethylene oxide produced being so consumed), ethanolamines, and acrylonitrile. Until the plants wear out, about one-fifth of all ethylene oxide is still being made by the old process through chlorohydrin, but an increasing amount each year is manufactured by air or oxygen acting on pure ethylene:

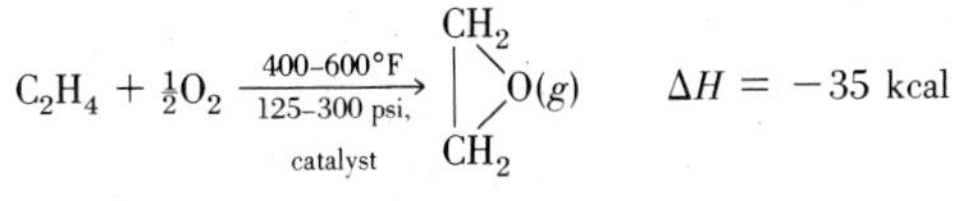

$$C_2H_4 + \tfrac{1}{2}O_2 \xrightarrow[\text{125–300 psi, catalyst}]{\text{400–600°F}} \begin{matrix}CH_2 \\ | \\ CH_2\end{matrix}\!\!>O(g) \qquad \Delta H = -35 \text{ kcal}$$

Yield 55–65%, conversion per pass 25–50%,

Production in 1974 was 4.26 billion lb.

Aromatic oxidation[54] *products* are included in various chapters of this book. As Table 37.6 indicates, there is a fundamental change in the source of aromatics from coal and coal tar to petroleum. See also Fig. 38.4 and Table 38.6.

OXO- OR HYDROFORMYLATION[55] The oxo process is a method of converting α-olefins to aldehydes and/or alcohols containing one additional carbon atom. The olefin in liquid state is reacted

[50] Hoechst Reveals Wacker Process, *Chem. Eng. News*, Apr. 17, 1961, p. 52; Petrochemical Handbook: *Pet. Refiner*, **42**(11), 1963; Guccioni, Acetaldehyde via Ethylene Oxidation. *Chem. Eng.* (*N.Y.*), **70**(25), 150 (1963).

[51] Goldstein, *op. cit.*, p. 373; Sittig, Acetic Acid and Anhydride, Noyes, 1965; Austin, Industrially Significant Organic Chemicals, *Chem. Eng.* (*N.Y.*), **81**(2), 128 (1974).

[52] Phillips *et al.*, U.S. Pat. 2,804,473 (1957).

[53] Goldstein, *op. cit.*, pp. 156–159; *Hydrocarbon Process.*, **52**(11), 129 (1973).

[54] Phthalic anhydride flowchart, Fig. 34.12; *Hydrocarbon Process.*, **52**(11), 159 (1973).

[55] Groggins, *op. cit.*, pp. 651–693 and particularly p. 683 (kinetics and flow diagram); Goldstein, *op. cit.*, pp. 204ff.; How Oxo Stays in the Race, *Chem. Week*, June 6, 1964; Hahn, *op. cit.*, pp. 104–116.

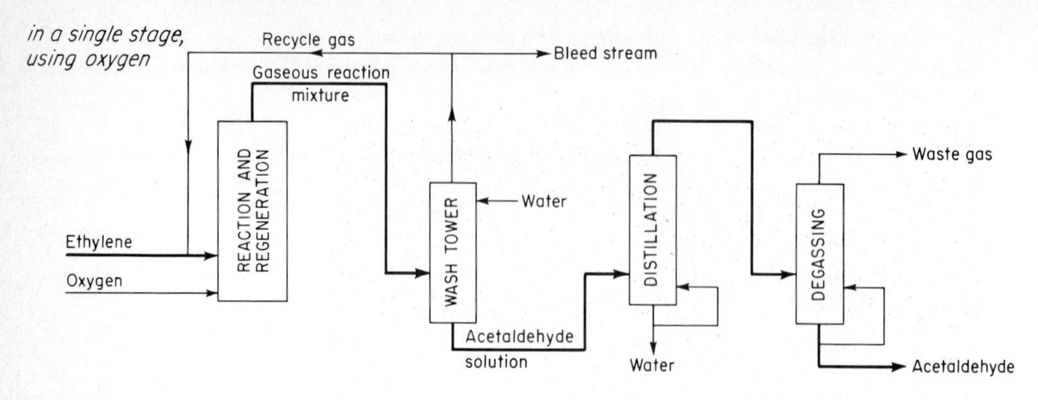

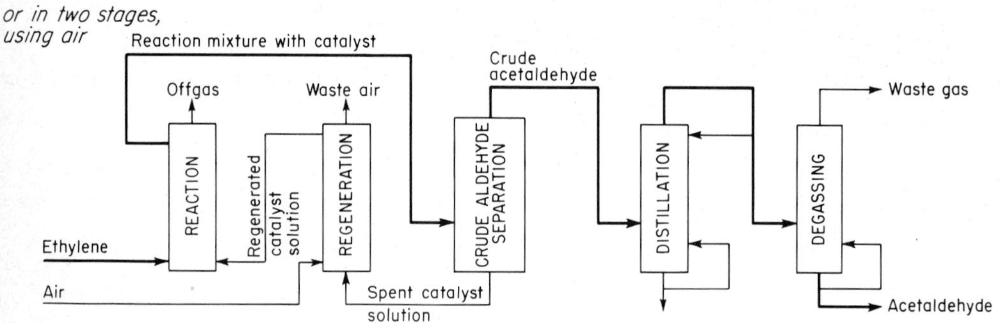

Fig. 38.11 Oxidation of ethylene to acetaldehyde by the Wacker process.

with a mixture of hydrogen and carbon monoxide in the presence of a soluble cobalt catalyst to produce the aldehyde (and some alcohol); all or part of the aldehyde is then hydrogenated over a nickel catalyst to form alcohol. The reaction is exothermic, about 35 kcal being liberated per gram-mole of olefin reacted. The usual raw materials are synthesis gas (Chap. 7) and α-olefins. The principal olefins used are propylene and the C_7 to C_{12} copolymers of propylene and isobutylene. Propylene reacts with synthesis gas to yield the following products:

$$\underset{\text{Propylene}}{CH_3CH{:}CH_2} + CO + H_2 \xrightarrow[\text{175°C, 250 atm}]{CO}$$

$$CH_3CH_2CH_2CHO + CH_3\underset{\displaystyle CH_3}{\underset{|}{C}}HCHO + CH_3CH_2CH_2CH_2OH + CH_3\underset{\displaystyle CH_3}{\underset{|}{C}}H_2CH_2OH$$

These are, respectively, *n*-butyraldehyde, isobutyraldehyde, *n*-butanol, and isobutyl alcohol.

Industrially most used is the C_7 olefin copolymer, which yields the mixture of corresponding C_8 alcohols. Addition of the CO group to the double bond is about 60% terminal and 40% secondary. A typical C_7H_{14} olefin yields the following isoctyl alcohol derivatives:

3,4-dimethyl-1-hexanol	3-methyl-1-heptanol
3,5-dimethyl-1-hexanol	5-methyl-1-heptanol
4,5-dimethyl-1-hexanol	Others

POLYMERIZATION[56] of petrochemicals falls mainly in the fields covered by Chap. 34, Plastic Industries, Chap. 35, Synthetic Fiber and Film Industries, and Chap. 36, Rubber Industries. These most important industries are based, for raw materials, upon the simple petrochemicals precursors as presented in Chap. 37.

MISCELLANEOUS A few chemicals of petrochemical origin do not fall within the preceding chemical conversions. Two important such petrochemicals are *sulfur* and *carbon disulfide.* The former is fully covered in Chap. 19, with a flowchart, and in Fig. 19.2.

Carbon disulfide is now made from methane. The long-used electrothermal process for carbon and sulfur[57] is described in CPI 2, p. 336. The reaction[58] for the methane process is

$$CH_4(g) + 4S(g) \xrightarrow[\substack{700°C, \\ 20\text{–}30 \text{ psi}}]{\substack{\text{activated} \\ Al_2O_3 + Cr_2O_3}} CS_2(g) + 2H_2S \qquad \Delta H = +19.5 \text{ kcal}$$

Yield 90% (on methane) per pass

The H_2S can be converted to sulfur and reused (Chap. 19). The main use is for viscose rayon and cellophane. Production is about 800 million lb per year.

SELECTED REFERENCES

Augustine, R. L.: Catalytic Hydrogenation, Dekker, 1965.
Barry, W. R.: Structural Analysis of Petrochemical, American Society of Mechanical Engineers, 1973.
Brownstein, A. M.: U.S. Petrochemicals, Petroleum Publishing, 1972.
Edmister, W. C.: Applied Hydrocarbon Thermodynamics, Gulf Publishing, 1961.
Gould, R. F. (ed.): Literature of Chemical Technology, chap. 12, Petroleum Chemicals, ACS Monograph, 1966.
Guthrie, V.: Petroleum Products Handbook, McGraw-Hill, 1960.
Hatch, L. F.: Isopropyl Alcohol, McGraw-Hill, 1961.
———: Higher Oxo Alcohols, Wiley, 1957.
Hatton, R. E.: Introduction to Hydraulic Fluids, Reinhold, 1962.
Hoiberg, A. J. (ed.): Asphalts, Tars and Pitches, Wiley, 1965, 2 vols.
Kobe, K. A., and J. J. McKetta, Jr. (eds.): Advances in Petroleum Chemistry and Refining, Interscience, 1958–1965, 10 vols.
Mellan, I.: Industrial Solvents Handbook, Noyes, 1970.
———: Polyhydric Alcohols, Spartan Books, 1962.
Sittig, M.: Aromatics, Noyes, 1966 (patents).
———: Chemicals from Ethylene, Noyes, 1965.
———: Chemicals from Propylene, Noyes, 1965.
———: Pollution Control in the Organic Chemical Industry, Noyes, 1974.
Smith, C. W. (ed.): Acrolein, Wiley, 1962.
Topchiev, A. V., *et al.*: Synthetic Materials from Petroleum, Pergamon, 1962.
Vervalin, C. H.: Fire Protection Manual for Hydrocarbon Processing Plants, 2d ed., Gulf Publishing, 1974.
Waddams, A. L.: Chemicals from Petroleum, Noyes, 1962.

[56]Groggins, *op. cit.*, pp. 856, 943; Golding, Polymers, chap. 34, Van Nostrand, 1959.

[57]Sittig, Combining Oxygen and Hydrocarbons for Profit, p. 37 (flowchart for thermal process) and p. 38 (flowchart for methane process); Haines, Carbon Disulfide, *Ind. Eng. Chem.*, **55**(6), 44 (1963) (flowchart).

[58]Goldstein, *op. cit.*, p. 93, reaction conditions; ECT, 2d ed., vol. 4, p. 377, 1964 (flowchart).

chapter 39

INTERMEDIATES, DYES, AND THEIR APPLICATION

Coal has been called the *black diamond.* The analogy is true not only because both consist of carbon, but because from both we obtain a rainbow of colors, one from the effect of light and the other from the action of various chemicals. Petrochemicals (Chap. 38) are now supplying, increasingly, a greater amount of the basic materials for dyes than coal. The purpose of this chapter is to follow through the manifold chemical engineering changes that transform coal, and now petrochemicals, by the action of many other chemicals into literally a thousand hues. The 1,000-odd commercial dyes we use bring variety to what would otherwise be a drab world. Furthermore, the entire dye industry throws a protective mantle over many other industries.

Because of the way in which dyes are connected with or enter into so many other phases of our industrial life, we have learned to view the industry as an *essential* one. Prior to 1914 Germany made a very large proportion of the dyes used throughout the world; indeed, propaganda insisted that only on the banks of the Rhine could dyes be made economically and of a high quality. However, since the cutting off of German dyes, beginning in 1914, and since the realization of the connection between the dye industry and other industries, even medicinal, many of the important nations have endeavored to build up within their own borders and from their own basic materials a self-contained dye industry. If we consider only a portion of our industrial life, we see how dyes enter in: in clothing, drapes, and wall coverings, and paints and varnishes, for example. In these industries, dyes make a great many products salable. We *could* go about wearing gray homespun, live in unpainted houses, and keep the sun out with unbleached or undyed curtains, but much of the attractiveness of our modern civilization would disappear. We would be living in a colorless world. One dollar's worth of dyes makes salable about $100 worth of other products. Hence the $500 million worth of dyes sold annually is essential to sales of about $50 billion from these various industries, giving employment to about 2 million men and women.

This is not the whole story. The raw materials, which in times of peace enter into the making of intermediates that are changed into dyes, in many instances are the basis of chemical explosives in time of war. This is true of benzene and toluene, which furnish dyes or explosives such as picric acid, tetryl, and TNT, and likewise the acids used in times of peace to make dyes that are essential to the making of nitroexplosives. In 1914, we imported 90% of our dyes, as shown in Table 39.1. This does not represent the real picture, for many of the necessary intermediates were imported from Germany, Switzerland, and even from England. The dye industry had been struggling for years to establish itself in the United States. Many courageous souls had tried to break in, and a few succeeded. When importations were stopped in the summer of 1914, the country became panicky over the dye shortage. The dark corners of warehouses were searched for abandoned kegs of dyes; even the floors were scraped and the dyes extracted from the scrapings.

TABLE 39.1 *Summary of Dyes, 1914*

	Employees	*No. of firms manufacturing*	*Pounds*
Dyes made	528	7	6,619,729
Dyes imported	. . .	. . .	45,950,895

Thanks to the skill of American chemists, the versatility of American engineers, and the confidence of American businesspeople, the establishment of a permanent dye industry in the United States rapidly took shape thereafter. Indeed, before the end of the war in 1918, a good start had been made. A wise temporary tariff was enacted to encourage further development, leading to the independence of the United States in dyes and intermediates. By 1920 the scales were reversed; the United States was manufacturing over 90% of the dyes needed for domestic consumption from American intermediates and American raw materials. The present situation regarding intermediate and dye production in the United States is summarized in Table 39.2, which includes medicinals, flavors, perfumes, plastics-resins, etc., because these, as well as dyes, are frequently based on the same or analogous intermediates. More details about dyes are included in Tables 39.4 and 39.5. The question may well be asked: "What do American-made dyes cost?" Ordinary competition based on improved and cheaper manufacture has brought the price down so that the consumer pays less for dyes than formerly. This is a real and lasting accomplishment. Competition in the United States has in certain instances increased the purity and the strength of commercially sold dyes. The American dye industry[1] has pioneered in bringing out new and cheaper processes for old dyes and intermediates with improved equipment. Indeed, many *new* processes for old products have been developed, together with much *new* machinery for carrying out the reactions, and also *new* intermediates and many *new* dyes with numerous improved properties. Examples are new processes for anthraquinone and new anthraquinone dyes.

This state has been achieved by research both in the fundamental chemistry involved and in the chemical engineering necessary to commercialize these reactions on an economical scale. All large American dye factories have laboratories and pilot plants either for improving old processes or for developing new ones.

RAW MATERIALS The dye industry draws upon every division of the chemical industry for the multiplicity of raw materials needed to make its finished products. However, the backbone of the raw-material sequence may be represented in one line:

Petroleum and coal ⟶ hydrocarbons ⟶ intermediates ⟶ dyes

Hydrocarbons. Chapter 38, on Petrochemicals, and Chapter 5, on Coal Chemicals, furnish discussions of the following dye hydrocarbons or precursors.

Benzene	Xylene	Anthracene
Toluene	Naphthalene	Paraffins

In developing the dye industry, the United States had to manufacture these various hydrocarbons not only of sufficient purity, but also in adequate tonnage. This enhanced demand was a tremendous incentive to increasing the installation of coproduct coke ovens and eliminating the old, wasteful beehive ovens. Obtaining coke in this manner not only provided basic raw materials for the dye and other industries, but also increased greatly the value of the products derived from coal. However, aromatic hydrocarbons from petroleum are the dominant source now. They are mostly produced by cyclization and/or dehydrogenation reactions in catalytic reforming or hydroforming plants.

[1]Lubs, The Chemistry of Synthetic Dyes and Pigments, ACS Monograph, Reinhold, 1955 (outstanding account of American dye industry by 19 Du Pont experts; excellent index, tables, and references).

TABLE 39.2 ***Synthetic Organic Chemicals and Their Raw Materials: U.S. Production and Sales***
(In millions of pounds and millions of dollars)

	Production		Sales			
			Quantity		Value	
Chemical	1971	1972	1971	1972	1971	1972
Grand total	237,961	266,419	133,666	150,818	$14,119	$16,028
Tar	6,794	7,472	3,341	3,409	36	40
Tar crudes	7,621	7,937	5,436	5,304	123	126
Crude products from petroleum and natural gas	81,043	86,792	45,752	47,900	1,078	1,177
Synthetic organic chemicals, totals	142,503	164,218	79,137	94,205	12,833	14,686
Cyclic intermediates	29,953	34,967	12,971	16,196	1,252	1,434
Dyes	244	263	230	255	423	480
Synthetic organic pigments	158	66	47	53	130	149
Medicinal chemicals	223	234	152	163	487	490
Flavor and perfume materials	96	110	85	104	84	88
Plastics and resin materials	21,071	25,921	18,473	22,946	3,507	4,258
Rubber processing materials	323	361	246	280	159	178
Elastomers (synthetic rubbers)	4,616	4,914	4,031	4,136	1,034	1,095
Plasticizers	1,494	1,708	1,404	1,637	258	291
Surface-active agents	3,828	4,039	2,186	2,258	422	451
Pesticides and other organic agricultural products	1,136	1,158	946	1,022	979	1,092
Miscellaneous chemicals	79,460	90,476	38,367	45,155	4,148	4,680

Source: U.S. Tariff Commission, Publication 681, 1975.

Inorganic chemicals. Dyes and intermediates are consumers of inorganic as well as organic chemicals. The use of these chemicals in the making of dyes and intermediates has frequently been the stimulus for improved processes by virtue of greatly increased demand, as exemplified by the development of the contact process for the manufacture of oleum. This came as a further extension of the catalytic oxidation of SO_2 to SO_3, brought about by the need for SO_3 as an oxidizing agent in the old process for converting naphthalene to phthalic anhydride, with by-product SO_2. Similarly, there is a great demand for chloro derivatives, such as chlorobenzene, requiring large tonnages of chlorine. A surprising quantity of inorganic chemicals is frequently required to make 1 lb of finished dye, it being estimated that, for some anthraquinone dyes, up to 75 lb of such chemicals is needed. Dyes and intermediates consume, in general, the following chemicals on a large scale.

Acids: Nitric, sulfuric, mixed, hydrochloric, hydrocyanic, acetic, formic, etc.

Alkalies: Caustic soda, soda ash, ammonia, slaked lime, quicklime, limestone (calcium carbonate), caustic potash, and alkylamines.

Salts: Sodium chloride, sodium sulfate (various forms), sodium nitrite, sodium sulfide, sodium cyanide, copper sulfate, potassium chloride, aluminum chloride, sodium hydrosulfite, etc.

Miscellaneous: Chlorine, bromine, iodine, hydrogen, alcohol, methanol, formaldehyde, acetylene, alkyl halides, iron, sulfur, etc.

INTERMEDIATES

The intermediates are the building stones out of which dyes, medicinals, plastics, rubbers, and synthetic fibers are directly constructed, just as our houses are made out of brick, wood, mortar, plaster, and steel, except that chemical reactions are used to connect one intermediate with another in manufacturing a dye (Table 39.3). Thus we find the producing of intermediates, economically and in a pure form, to be the *very foundation* of the organic chemical process industries, and any reduction that has come in the cost of finished products is due very largely to lowering of the cost of making pure intermediates. In the United States this has been achieved through research leading to higher yields in chemical conversions and to improved chemical engineering when commercializing these reactions. Many improved or new processes are now employed in the manufacture of intermediates, examples of which are the catalytic vapor-phase hydrogenation of nitrobenzene to aniline (Figs. 39.2 and 39.3) and the oxidation of *o*-xylene to phthalic anhydride.

The census of intermediates, taken by the U.S. Tariff Commission, shows that in 1972 the United States produced almost 35 billion lb of cyclic intermediates. Table 39.3 lists the important ones. Some intermediates, such as β-naphthol and chlorobenzene, are made in a preponderating quantity by one company, hence the quantity manufactured cannot be disclosed. However, many of the cyclic intermediates in Table 39.3 are used to only a minor degree to build dyes. For instance, part of the acetanilide, aniline, phenol, and salicyclic enter the medicinal field. But most of the phenol is consumed by plastics and resins, and practically all the dodecylbenzene becomes raw material for detergents. Much chlorobenzene and *o*- and *p*-dichlorobenzene is used for insecticides. Cyclohexane is a raw material for nylon. Phthalic anhydride (Figs. 34.10 to 34.12) is mostly consumed in

TABLE 39.3 ***Production of Important Cyclic Intermediates***
(In thousands of pounds)

	1972	*1970*
Acetanilide, technical	3,288	4,221
Aniline	409,820	398,362
7-Benz (*d, e*) anthracen-7-one (benzathrone)	1,184	1,881
Benzoic acid, technical	155,505	
Chlorobenzene	403,505	484,914
α-Chlorotoluene (benzyl chloride)	80,357	75,131
Cresols, total	106,273	91,414
Cresylic acid, refined, total	54,981	98,334
Cumene	2,292,949	1,983,349
Cyclohexane	2,298,394	1,841,052
Cyclohexanone	783,440	714,470
o-Dichlorobenzene	62,386	66,219
p-Dichlorobenzene	77,317	69,606
3,3′-Dichlorobenzidine, base and salts	4,612	3,656
4,4′-Dinitrostilbene-2,2′-disulfonic acid	9,230	10,161
Dodecylbenzenes*	524,009	553,166
Ethylbenzene	5,675,900	4,827,273
4,4′-Isopropylidenediphenol (bisphenol A)	255,189	202,171
α-Methylstyrene	37,398	19,063
Nitrobenzene	551,169	547,680
5-Nitro-*o*-toluenesulfonic acid ($SO_3H = 1$)	8,017	9,025
2-Nitro-*p*-toluidine ($NH_2 = 1$)	397	99
Nonylphenol	100,295	78,134
o-Xylene	831,782	798,706
p-Xylene	2,207,556	1,590,209
Phenol	2,096,125	1,755,278
Phthalic anhydride	952,978	734,020
Quinizarin	2,095	1,610
Salicylic acid, technical sales	47,095	
Styrene	5,940,729	4,335,313

Source: Synthetic Organic Chemicals, U.S. Tariff Commission, 1975 (annual).
*Includes branched- and straight-chain dodecylbenzene and tridecylbenzene.

the plastics and surface-coating fields. Styrene enters into plastics and rubber manufacture. The dye industry used the intermediates initially, but these are now also the building stones for many of the organic chemicals consumed by the public (summarized in Table 39.2).

UNIT, OR PHYSICAL, OPERATIONS, CHEMICAL CONVERSIONS, ENERGY CHANGES These factors[2] for the intermediate and entire organic industry are so important that, particularly for the intermediates, a presentation under each main chemical conversion is made.

NITRATION

Nitration[3] is one of the most important chemical conversions operated in the manufacture of intermediates and dyes. In the industry very few nitro groups appear in the final product, the nitro radical usually being reduced to the very reactive amine or subjected to other changes. However, nitration is frequently the initial entrance into the aromatic ring, whether benzene or more complicated rings are involved.

A typical nitration may be expressed conventionally as follows:

$$R\cdot H + HNO_3(H_2SO_4) \longrightarrow R\cdot NO_2 + H_2O + (H_2SO_4) \qquad \Delta H = -15.0 \text{ to } -35.0 \text{ kcal}$$

Although the usual nitrating acids are nitric or mixed acid, the actual nitrating *agent* is believed to be the *nitronium,* or *nitryl ion* (NO_2^+), which is formed as follows:

In mixed acid:

$$HNO_3 + 2H_2SO_4 \rightleftharpoons NO_2^+ + 2HSO_4^- + H_3O^+$$

In nitric acid:

$$2HNO_3 \rightleftharpoons H_2NO_3^+ + NO_3^-$$

$$H_2NO_3^+ \rightleftharpoons NO_2^+ + H_2O$$

Removal of the water favors the nitration reaction by shifting the equilibrium to the right. In the case of mixed acid, the sulfuric acid effects this removal by what is expressed as the dehydrating value of the sulfuric acid (DVS). Based on Groggins, the following equation is used:

$$\text{DVS} = \frac{\text{percent actual } H_2SO_4}{\text{water present in reactants + water formed by nitration}}$$

Experiments have indicated what DVS is needed for a given nitration, but it varies from 2.0 to as high as 12 for technical nitrations. Much *heat* is evolved by the nitration itself, as well as by absorption by the sulfuric acid of the water split out, but the specific heat of the sulfuric acid allows some of this heat to be absorbed. Since higher nitro compounds may be explosive, a certain amount of precaution is necessary in handling this reaction so as to keep it always under control. Nitric acid is also a powerful oxidizing agent, and this action must be minimized as much as possible by keeping temperatures as low as feasible. The *equipment* used includes cast-iron or stainless-steel nitrators employing mixed acid; but in every case the unit operations of heat transfer and of mixing should be most carefully considered. The nitration chemical conversion is applied in the making of a number of intermediates; those in the following list are important.

[2] A more detailed presentation, particularly of the chemical and thermal aspects of processes as applied to intermediates and many other similar compounds, is given in Groggins (ed.), Unit Processes in Organic Synthesis, 5th ed., McGraw-Hill, 1958. As presented in previous chapters, *chemical conversions* are used in place of unit processes, thus following the long-established usage in the important petroleum and petrochemical fields. See also Chaps. 1, 34 to 36, 38, and 40.

[3] Groggins, *op. cit.,* pp. 60ff.; Lubs, *op. cit.,* p. 12; Albright, Nitration, Chemistry and Processes, *Chem. Eng.* (*N.Y.*), **73**(10 and 11), 161 and 169 (1966).

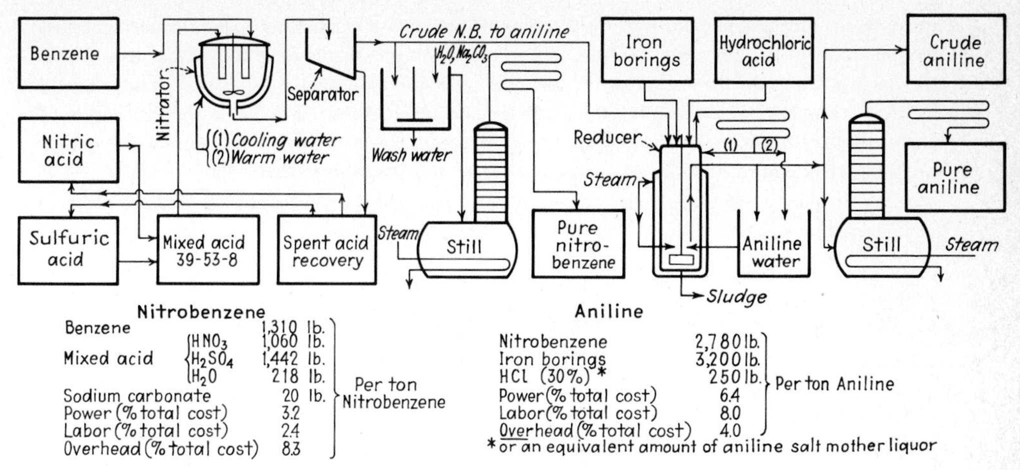

Fig. 39.1 Flowchart for nitrobenzene and aniline manufacture (see Figs. 39.2 and 39.3).

Nitration Chemical Conversion in the Manufacture of Intermediates

Chloro-2,4-dinitrobenzene, from chlorobenzene through *o*- and *p*-chloronitrobenzene by strong mixed acid.

o- and *p*-Chloronitrobenzene, from chlorobenzene by mixed acid. Separate by distillation and crystallization.

m-Dinitrobenzene, from benzene through nitrobenzene by strong mixed acid.

1,4-Dichloro-2-nitrobenzene from *p*-dichlorobenzene by mixed acid.

p-Nitroacetanilide and *p*-nitroaniline, from acetanilide by cold-mixed-acid nitration. Hydrolysis furnishes *p*-nitroaniline.

Nitrobenzene, from benzene by mixed acid.

α-Nitronaphthalene, from naphtalene by mixed acid.

o- and *p*-Nitrophenol, from phenol by nitric acid or mixed acid. Separate by steam distillation, the ortho isomer being volatile.

o- and *p*- Nitrotoluene, from toluene by mixed acid at moderate temperature to lessen oxidation. Separate by distillation and crystallization.

m-Nitro-*p*-toluidine, from acyl-*p*-toluidine by mixed acid at low temperature.

1-Nitroanthraquinone, from anthraquinone and mixed acid at 50°C.

EXAMPLE OF TECHNICAL NITRATION *Nitrobenzene* is and has been one of the most important intermediates in the dye field (655 million lb manufactured in 1974). From it are made aniline, benzidine, and metanilic acid (sulfonation and reduction), as well as dinitrobenzene and many dyes such as nigrosines and magenta. 97% is converted to aniline. As the flowchart in Fig. 39.1 indicates, nitrobenzene is made by the direct nitration of benzene, using mixed acid, according to the following reaction:

$$C_6H_6 + HNO_3(H_2SO_4) \longrightarrow C_6H_5NO_2 + H_2O(H_2SO_4) \qquad \Delta H = -27.0 \text{ kcal}^4$$

AMINATION BY REDUCTION[5]

The amino group is an exceedingly important one for dyes, since it is the one that is changed to the azo chromophore or alkylated. This amino radical is also one of the chief auxochromes. It is made by the reduction of a nitro derivative or, recently, in special cases, by ammonolysis (q.v.). The *agents* employed are iron and an acid catalyst, zinc and an alkali, sodium sulfide or polysulfide, or vapor-

[4]For mononitration of benzene, cf. Lubs, *op. cit.*, p. 12, and particularly for a continuous process, cf. Groggins, *op. cit.*, pp. 107–113; also CPI 2, pp. 870ff.

[5]Shreve, Amination by Reduction, in ECT, 2d ed., vol. 2, pp. 76–99, 1963; *FIAT Rep.* 649; U.S. Pat. 2,105,321 (1938).

phase catalytic hydrogenation, with a scattering of less important reagents. The accompanying list presents some of the aminations particularly pertaining to intermediates. With similar lists in this section it gives a bare indication of the manufacturing process. Liquid and vapor-phase catalytic hydrogenations have become quite important large-scale industrial methods for the production of naphthylamines and of aniline (Figs. 39.2 and 39.3) and of the reduction of nitrobenzene to hydrazobenzene, and of *p*-nitrophenol to *p*-aminophenol. More details on those intermediates not fully treated here can be found in the books on intermediates in the Selected References at the end of this chapter. The generalized reaction is

$$RNO_2 + 6H \longrightarrow RNH_2 + 2H_2O$$

For some technical aminations by reduction, however, a better expression is

$$4RNO_2 + 9Fe + 4H_2O \xrightarrow{\text{catalyst}} 4RNH_2 + 3Fe_3O_4$$

The *energy* of these reactions is exothermic and so large (see aniline) that much thought must be given to efficient heat transfer. The *equipment* for commercializing these chemical changes is frequently highly specialized apparatus, e.g., the reducer for aniline. In other cases, particularly when dealing with nonvolatile products like *m*-phenylenediamine, a simple well-ventilated wooden vat performs satisfactorily. This vat should have an efficient sweep stirrer located near the bottom to keep the iron in suspension.

EXAMPLE OF TECHNICAL AMINATION BY REDUCTION *Aniline.*[6] This intermediate has been of such great importance that frequently dyes are described as *aniline dyes.*[7] However, in 1966, two-thirds of its production was consumed in the rubber industries. Aniline itself is made by iron reduction of nitrobenzene, by ammonolysis of chlorobenzene, or by vapor-phase hydrogenation of nitrobenzene. A flowchart exemplifying the iron reduction is included in Fig. 39.1. The reduction can be summarized:

$$4\,C_6H_5NO_2 + 9Fe + 4H_2O \xrightarrow{\text{HCl}} 4\,C_6H_5NH_2 + 3Fe_3O_4 \qquad \Delta H = -130\ \text{kcal}$$

In the reduction, as in many other processes, the chemical reactions are much more complicated. These changes can probably be formulated with the following sequences, terminating for the iron as

$$Fe + 2H_2O \xrightarrow{FeCl_2} Fe(OH)_2 + H_2$$

$$C_6H_5NO_2 + 3Fe + 4H_2O \xrightarrow{FeCl_2} C_6H_5NH_2 + 3Fe(OH)_2$$

$$C_6H_5NH_2 + H_2O \longrightarrow C_6H_5NH_3OH \longrightarrow C_6H_5NH_3^+ + OH^-$$

$$FeCl_2 \longrightarrow Fe^{2+} + 2Cl^-$$

$$2C_6H_5NH_3^+ + 2OH^- + Fe^{2+} + 2Cl^- \longrightarrow 2C_6H_5NH_3Cl + Fe(OH)_2$$

$$C_6H_5NH_3Cl \longrightarrow C_6H_5NH_3^+ + Cl^-$$

$$2C_6H_5NH_3^+ + 2Cl^- + 2Fe(OH)_2 + 2Fe \longrightarrow 2C_6H_5NH_2 + FeCl_2 + \underset{\text{Ferroso-ferric oxide}}{Fe_3O_4} + 3H_2$$

[6]For further details on making aniline, see Groggins, *op. cit.*, pp. 33–161, 456–462; Polinski and Harvey, Aniline Production by Dual Function Catalysis, *Ind. Eng. Chem., Prod. Res. Dev.*, **10**(4), 365 (1971); Aniline, *Hydrocarbon Process.*, **52**(11), 105 (1973); Aniline from Phenol, *Chem. Eng.* (*N.Y.*), **80**(8), 42 (1973).

[7]It is essential from a manufacturing or research viewpoint to know just how many and what dyes are derived from aniline, or any other intermediate. To supply this information at the start of the American dye industry the following book was written: Shreve, Dyes Classified by Intermediates, Reinhold, 1922.

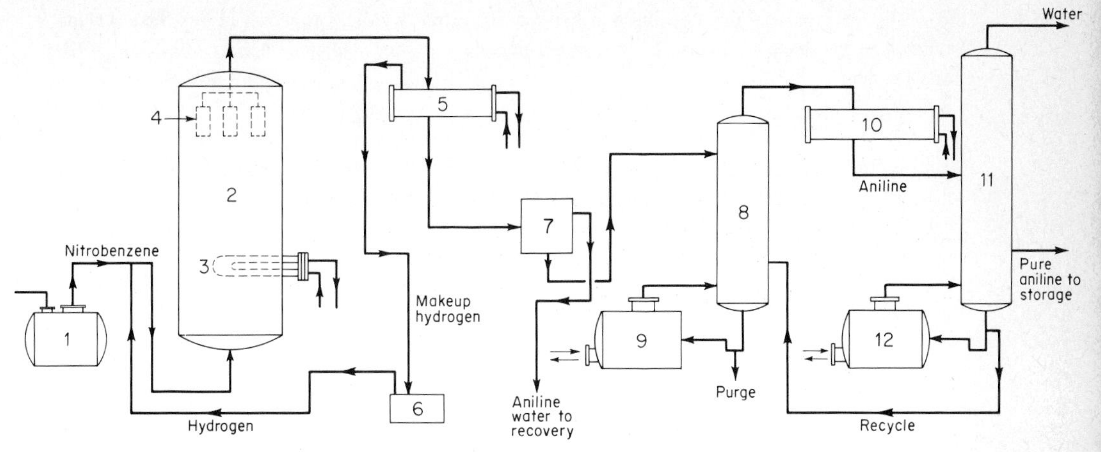

Fig. 39.2 Continuous fluid-bed vapor-phase reduction of nitrobenzene. (*American Cyanamid Co.*)

1. Nitrobenzene vaporizer
2. Reactor with fluidized catalyst bed
3. Cooling tubes
4. Catalyst filters
5. Product condenser
6. Hydrogen recycle compressor
7. Aniline-water settler and decanter
8. Crude aniline still
9. Reboiler for crude aniline still
10. Condenser
11. Aniline-finishing still
12. Reboiler for aniline-finishing still

Fig. 39.3 Overall view of the vapor-phase aniline-manufacturing facilities at Willow Island, W. Va. (*American Cyanamid Co.*)

The batch process for reducing nitrobenzene with iron to aniline as shown in Fig. 39.1 is being *superseded* by the continuous catalytic vapor-phase reduction[8] depicted in Figs. 39.2 and 39.3. The reaction is

$$C_6H_5NO_2(l) + 3H_2(g) \longrightarrow C_6H_5NH_2(l) + 2H_2O(l) \qquad \Delta H = -130 \text{ kcal}$$

The reaction is conducted at 270°C and 20 psi, with a very short contact time, using three times the theoretical hydrogen (both makeup and recirculated) introduced into the nitrobenzene vaporizer exit pipe, as illustrated in the figure. The catalyst is copper on SiO_2, on which the hydrogen is chemisorbed atomically, a molecule of nitrobenzene being absorbed on two of the hydrogenated sites. After these sites have been established, the desorption of the reaction product, water, is a rate-controlling step[9]

Amination by Reduction as Chemical Conversion in the Manufacture of Intermediates

p-Aminoacetanilide (acetyl-*p*-phenylenediamine), from *p*-nitroacetanilide by iron and acetic acid under 60°C.
α-Aminoanthraquinone, from α-nitroanthraquinone by Na_2S solution, or by ammonolysis of α-anthraquinone-sulfonic acid.
p-Aminophenol, from nitrosophenol by aqueous Na_2S, or from *p*-nitrophenol from *p*-nitrochlorobenzene.
o-Aminophenol-*p*-sulfonic acid, from *o*-nitrophenol-*p*-sulfonic acid by Na_2S, following sulfonation, nitration, and hydrolysis of chlorobenzene.
Aniline, from nitrobenzene by hydrogenation, or by iron and a little dilute HCl.
Chlorotoluidinesulfonic acids, from chlorotoluenesulfonic acids by nitration and subsequent iron reduction.[10]
Diaminostilbenedisulfonic acid, from dinitrostilbenedisulfonate by alkaline zinc reduction.
Diaminoanthraquinone, from corresponding dinitro- or aminonitroanthraquinone by reduction, or from corresponding dichloro-, disulfo-, or dihydroxyanthraquinone by ammonolysis.
2,5-Dichloroaniline and sulfonic acid, from 1,4-dichloro-2-nitrobenzene by iron and a little aqueous HCl (sulfonation).
Dianisidine by rearrangement of hydrazoanisole, from *o*-nitroanisole by alkaline zinc reduction in alcoholic solution, followed by acid rearrangement.
Benzidine by rearrangement of hydrazobenzene, from nitrobenzene by alkaline reduction by iron or zinc, followed by acid rearrangement.
Tolidine by rearrangement of hydrazotoluene, from *o*-nitrotoluene by alkaline reduction by zinc, followed by acid rearrangement.
Metanilic acid, from nitrobenzene-*m*-sulfonic acid, following oleum sulfonation of nitrobenzene.
α-Naphthylamine, from α-nitronaphthalene by iron and a little dilute HCl.
m-Nitroaniline, from dinitrobenzene by hot aqueous sodium polysulfide.
m-Phenylenediamine, from dinitrobenzene by iron and a little dilute HCl.
p-Phenylenediamine, from *p*-nitroaniline by iron and a little dilute HCl boiling at the end.
Picramic acid, from picric acid by hot sodium polysulfide solution.
Toluidines, from nitrotoluenes by iron and a little dilute HCl.
m-Tolylenediamine (or toluene-2,4-diamine), from 2,4-dinitrotoluene by iron and a little dilute HCl.
Xylidines, from nitroxylenes by iron and a little dilute HCl.

AMINATION BY AMMONOLYSIS

This chemical conversion, used for a long time to make β-naphthylamine, has been extended to aniline and *p*-nitroaniline. The conversion functions by reacting, usually at elevated temperatures and in autoclaves, with a large excess of aqueous ammonia, sometimes in the presence of sulfites (the

[8]Shreve, *op. cit.*, in ECT, 2d ed., vol. 2, pp. 80–82, 1963; Kourrs and Northcott, ECT, 2d ed., vol. 2, pp. 411–425, 1963 (63 refs.); Ribani *et al.*, Kinetics of Catalytic Vapor Phase Hydrogenation to Aniline, *Ind. Eng. Chem. Process Des. Dev.*, **4**, 403 (1965).

[9]J. W. Fulton, unpublished thesis, Oklahoma University.

[10]For chlorotoluidinesulfonic acids, cf. Venkataraman, The Chemistry of Synthetic Dyes, Academic, 1952–1975, 7 vols.

Bucherer[11] reaction). With the present low cost of ammonia, this method of amination has a much more favorable position. The *agent* used is ammonia, or a substituted ammonia, which replaces a number of different groups: —Cl, —OH, or —SO_3H. This chemical conversion is particularly useful when *labilizing* groups are present in the ortho or para positions in respect to the halogen, as in the following ammonolysis reactions:

$$\underset{\text{1,4-Dichloro-2-nitrobenzene}}{Cl_2C_6H_3NO_2} \xrightarrow{NH_3(aq)} \underset{\text{4-Chloro-2-nitroaniline}}{ClNO_2C_6H_3NH_2}$$

$$\underset{\text{1-Chloro-4-nitrobenzene}}{Cl_2C_6H_4NO_2} \xrightarrow{NH_3(aq)} \underset{p\text{-Nitroaniline}}{NH_2C_6H_4NH_2}$$

The ammonia is used as aqueous ammonia in large excess, 4 to 10 times theory. Frequently, catalysts help speed up the chemical change. Since such a large excess of ammonia is necessary for good yields, an efficient recovery system is required to keep costs low. The *equipment* required to commercialize this chemical conversion is almost always some type of agitated pressure vessel made entirely of an ammonia-resisting metal, such as iron, even to the valves. Fortunately, the all-iron valves, fittings, pumps, and other equipment previously developed for ammonia-ice machines are available at low cost for amination by ammonolysis.

TECHNICAL EXAMPLE *Aniline.* The main reactions[12] involved in the making of aniline through ammonolysis may be expressed as follows:

$$C_6H_5Cl + 2NH_3(aq) \xrightarrow[750\ \text{psi}]{180\text{–}220°C} C_6H_5NH_2 + NH_4Cl(s)$$

Amination by Ammonolysis in the Manufacture of Intermediates

β-Aminoanthraquinone, from chloranthraquinone by excess of 28% aqueous ammonia at 200°C in a stirred autoclave, or from β-anthraquinone sulfonate by excess ammonia water.

Aniline, from chlorobenzene by excess 28% aqueous ammonia at 200°C and 900 psi, in the presence of Cu_2O.

β-Naphthylamine, from β-naphthol by excess of 28% ammonia aided by ammonium sulfite at 150°C in a stirred autoclave.

2-Naphthylamine-1-sulfonic acid (Tobias acid), from 2-naphthol-1-sulfonic acid by excess ammonia water and ammonium sulfite at 150°C in an autoclave.

2-Naphthylamine-6-sulfonic acid (Broenner's acid), from the sodium salt of 2-naphthol-6-sulfonic acid (Schaeffer's acid) by 28% ammonia at 180°C in an autoclave.

2-Naphthylamine-6,8-disulfonic acid (amino-G acid), 2-naphthol-6,8-disulfonate by heating with ammonia and sodium bisulfite in an autoclave.

p-Nitroaniline, from *p*-chloronitrobenzene by excess of 28% aqueous ammonia at 170°C and 500 psi.

Phthalimide, from molten phthalic anhydride by gaseous ammonia at 240°C.

HALOGENATION

The use of halogenation[13] and chlorination in particular to effect entrance into a hydrocarbon ring, or other compound, and thus to make a more reactive derivative, is becoming of increasing importance in the field of intermediates as well as in all organic chemistry. The *agent* used is generally dried chlorine gas with or without a catalyst, though other chlorination agents such as HCl plus oxygen and (rarely) HCl alone are used. The chlorination proceeds (1) by addition to an unsaturated bond,

[11] Venkataraman, *op, cit.*

[12] More details in CPI 2, pp. 874–875, and Groggins, *op. cit.*, p. 456.

[13] Groggins, *op. cit.*, chap. 6, p. 204.

(2) by *substitution* for hydrogen, or (3) by *replacement* of another group, such as —OH or —SO_3H. Examples are:

$$CH_2{=}CH_2 + HCl \xrightarrow{\text{anhydrous}} CH_3CH_2Cl$$

$$C_6H_6 + Cl_2 \xrightarrow[\text{anhydrous}]{FeCl_3} C_6H_5Cl + HCl$$

$$C_6H_6 + HCl + \tfrac{1}{2}O_2 \longrightarrow C_6H_5Cl + H_2O$$

$$3\ C_{14}H_7O_2(SO_3Na) \text{ (anthraquinonesulfonate)} + 6HCl + NaClO_3 \longrightarrow 3\ C_{14}H_7O_2Cl \text{ (chloroanthraquinone)} + 4NaCl + 3H_2SO_4$$

Reaction (4) proceeds in hot, dilute, aqueous solution. In certain halogenation reactions much *heat* is liberated and must be removed (cf. chlorobenzene). The *equipment* necessary is much simplified if anhydrous conditions are called for, as frequently happens. Under these circumstances steel is very satisfactory unless iron and iron compounds exert a disadvantageous effect, in which case enameled steel or earthenware is used. Nickel is also employed in some special cases.

TECHNICAL EXAMPLE *Chlorobenzene* is a very important intermediate, particularly for sulfur colors. It is also employed as a solvent and to make many other products, such as aniline, phenol, chloronitrobenzene and chlorodinitrobenzene, and *o*-aminophenol-*p*-sulfonic acid. Its production in 1974 was 365 million lb. Chlorobenzene is manufactured by passing dry chlorine through benzene in iron or lead-lined iron vessels, using ferric chloride as a catalyst. The reaction is as follows:

$$C_6H_6 + Cl_2 \xrightarrow{FeCl_3} C_6H_5Cl + HCl$$

The reaction rate of monochlorination is approximately 8.5 times[14] that of dichlorination, so some dichlorobenzene is always formed. Much heat is generated and, if monochlorobenzene is the preferred product, the temperature should be kept below 60°C by circulating and cooling the benzene-chlorobenzene mixture. The hydrochloric acid escaping is washed free of chlorine by benzene and absorbed in water. The products of chlorination are distilled to separate the benzene and chlorobenzene, leaving the dichlorobenzenes (ortho and para) in the still residue.

Halogenation as a Chemical Conversion in the Manufacture of Intermediates

Chloracetic acid, from glacial acetic acid by passing in chlorine at 100°C in the presence of red phosphorus.

Chlorobenzene, from benzene by the action of chlorine in the presence of iron at 50 to 60°C.

p-Dichlorobenzene, from benzene or chlorobenzene by the action of chlorine in the presence of iron at 50 to 60°C.

2,6-Dichlorobenzaldehyde, from *o*-nitrotoluene through ring chlorination by chlorine, furnishing 2-chloro-6-nitrotoluene (40%) and 4-chloro-6-nitrotoluene (60%). These are separated, and the 2,6-isomer is subjected to reduction, diazotization, and the Sandmeyer reaction, giving 2,6-dichlorobenzal chloride, which hydrolyzes to 2,6-dichlorobenzaldehyde.

SULFONATION[15]

This is one of the most widely employed chemical conversions to effect variation in properties so as to introduce greater solubility or to make a hydrocarbon more reactive for the purpose of further synthesis. The fundamental chemistry has been carefully studied,[16] and many data are available in the literature. Frequently, isomers are formed, and the cost of a process will be high unless application can be found for the principal isomers produced. The *agents* employed are various strengths of

[14]See Groggins, *op. cit.*, for details.

[15]Groggins, *op. cit.*, chap. 7; Lubs, *op. cit.*, pp. 3, 6–13; Gilbert, Sulfonation, Interscience, 1965.

[16]See Venkataraman, Davidson, Reverdin and Fulda, Schwalbe, and Wichelhaus in Selected References at the end of this chapter; Ullmann, Enzyclopaedie der technischen Chemie, Urban & Schwarzenberg, Berlin, 1950–.

sulfur trioxide in water, from 66°Bé sulfuric acid or even weaker, to the strongest oleums. For special cases, however, sulfite and chlorosulfonic acid are very useful. As in nitrations, care should be taken to avoid oxidation by not having temperatures or concentrations too high. Excessive concentrations also form sulfones.

The reaction may be expressed as follows:

$$R\cdot H + \begin{matrix} HO \\ HO \end{matrix}\!\!>\!SO_2 \longrightarrow R\cdot SO_2\cdot OH + H_2O$$

However, recent studies indicate that an ion such as HO_3S^+ is the active sulfonating agent when sulfuric acid is used:

$$C_6H_5{-}H + {}^{+}S(\uparrow O)(\downarrow O){-}OH \longrightarrow C_6H_5(H){-}S(\nearrow O)(\downarrow O){-}OH \longrightarrow \left[C_6H_5{-}S(OH)(\downarrow O){-}OH\right]^{+} \longrightarrow C_6H_5{-}SO_3H + H^{+}$$

Depending on what product is being sulfonated, there is a critical concentration of sulfuric acid below which sulfonation ceases (phenol, Chap. 34). Therefore the removal of the water formed is important. The *energy* of the reaction is exothermic, but its removal is often not required, since many sulfonations proceed best at elevated temperatures. The *equipment*[17] for most sulfonations is fairly simple, consisting of cast-iron vessels provided with an efficient agitator, vent, condenser, and usually a jacket for heating by steam or circulating hot oil.

TECHNICAL EXAMPLES *Benzenesulfonic acid* is a step in the manufacture of phenol by the sulfonation process (phenol, Chap. 34). 1- and 2-*naphthalenesulfonic acids* are formed simultaneously by the sulfonation of naphthalene. They must be separated if they are to be used for the preparation of pure naphthols. This is a difficult procedure. However, the alpha acid can be hydrolyzed[18] to naphthalene, which is distilled out by passing dry steam into the mixed sulfonation mass at about 160°C, leaving behind pure beta acid suitable for hydrolysis or fusion to make β-naphthol.

The usual sulfonation products are more complicated to handle than those just cited, and are frequently difficult to separate from their isomers. The following list gives some of the properties employed in separations or isolations:

1. Variation in rates of hydrolysis as described for naphthalene- α- and β-sulfonic acids.
2. Variation in the solubilities of potassium salts.
3. Variation in the solubilities of calcium or barium or metal salts.
4. Differences in the solubilities of sulfonic acids in water or acids or other media.
5. Differences in solubilities of sodium sulfonic salts in the presence of a sodium salt such as sodium chloride, sulfate, nitrate, or acetate.
6. Difference in properties of derivatives such as sulfon chlorides or amides.

Sulfonation Chemical Conversions in the Manufacture of Intermediates

1-Amino-2-naphthol-4-sulfonic acid (1,2,4 acid), from 1-nitroso-2-naphthol by joint reducing and sulfonating action of sodium bisulfite.

Benzenesulfonic acid, by passing hot benzene vapor through sulfuric acid until all the sulfuric acid reacts (phenol, in Chap. 34).

[17] Shreve, Equipment for Nitration and Sulfonation, *Ind. Eng. Chem.*, **24,** 1344 (1932).

[18] Shreve, β-Naphthol, *Color Trade J.*, **14,** 42 (1924); cf. Masters, U.S. Pat. 1,922,813 (1933); Groggins, *op. cit.*, pp. 316ff., 379ff.; 1-naphthalene-sulfonic acid is the principal isomer formed when the sulfonation is carried out at about 60°C; Lubs, *op. cit.*, pp. 67ff., 77ff.

β-Naphthalenesulfonic acid, from naphthalene and 66° Bé sulfuric acid at 160° C; the α-isomer simultaneously formed is hydrolyzed to naphthalene and removed by passing in steam.

Naphthionic acid, from α-naphthylamine sulfate by baking at 170 to 180°C.

2-Naphthol-6-sulfonic acid (Shaeffer's acid), from β-naphthol by 98% sulfuric acid at 100°C and separation from the 2:8 acid (crocein acid) formed simultaneously.

2-Naphthol-3,6-disulfonic acid (R acid), from β-naphthol by excess sulfuric acid at low temperatures and separation from G acid.

2-Naphthol-6,8-disulfonic acid (G acid), from β-naphthol by excess sulfuric acid at elevated temperature and separation from R acid.

1-Naphthylamine-5-sulfonic acid (Laurent's acid) and 1-naphthylamine-8-sulfonic acid, from napthalene by low-temperature sulfonation to the alpha acid, followed by nitration to the two isomers, reduction by iron, and separation.

2-Naphthylamine-5,7-disulfonic acid, from β-naphthylamine by oleum sulfonation.

2-Naphthylamine-6,8-disulfonic and 2-naphthylamine-5,7-disulfonic acids by sulfonation of β-naphthylamine, first with the monohydrate followed by 60% oleum. These acids are separated and fused with NaOH to give, respectively, gamma acid (NOS:286) and J acid (NOS:257).

1-Naphthylamine-3,6,8-trisulfonic acid (Koch acid), from naphthalene by trisulfonation with oleum, followed by nitration and iron reduction.

p-Nitrotoluene-*o*-sulfonic acid, from *p*-nitrotoluene by oleum.

HYDROLYSIS

In the manufacture of intermediates the phase of hydrolysis[19] employed is usually alkaline fusion to replace an $-SO_3H$ group by $-OH$. However, this chemical conversion is also used to replace —Cl, particularly in making phenol. The *agent* generally used is caustic soda, though other alkalies, acids, or plain water are of industrial importance. The reaction may be formulated as follows:

$$ArSO_3Na \text{ (or ArCl)} + 2NaOH \longrightarrow ArONa + Na_2SO_3 + H_2O \text{ (or } NaCl + H_2O)$$

The *energy* involved is exothermic in many instances. In phenol from chlorobenzene, the heat evolved maintains the reaction temperature. The *equipment* required is generally either open, cast-iron fusion pots heated by gas or closed, cast-iron or welded-steel autoclaves. Especially in alkaline fusions, some loss is experienced from oxidation to tars at the temperatures employed, 300 to 325°C.

TECHNICAL EXAMPLE *β-Naphthol* is made by hydrolysis somewhat similarly to phenol (Chap. 34). The conditions for the hydrolysis (alkali fusion) of 2-naphthalenesulfonate are given in the references under sulfonation.

Hydrolysis (or in Many Instances "Alkali Fusion")
in the Manufacture of Intermediates

1-Amino-8-naphthol-3,6-disulfonic acid (H acid), from 1-naphthylamine-3,6,8-trisulfonic acid by caustic soda fusion in an autoclave.

2-Amino-5-naphthol-7-sulfonic acid (J acid), from 2-naphthylamine-5,7-disulfonic acid by caustic soda fusion in an autoclave.

2-Amino-8-napththol-6-sulfonic acid (gamma acid), from 2-napthylamine-6,8-disulfonic acid by caustic soda fusion in an autoclave.

1,5-Dihydroxyanthraquinone (anthrarufin), from anthraquinone-1,5-disulfonic acid by milk of lime under pressure.

1,8-Dihydroxynaphthalene-3,6-disulfonic acid (chromotropic acid), from 1-naphthol-3,6,8-trisulfonic acid by caustic soda fusion at 200°C.

[19] Groggins, *op. cit.*, chap. 13; Lubs, *op. cit.*, index.

2,4-Dinitrophenol, from chloro-2,4-dinitrobenzene by boiling with soda ash solution.
β-Naphthol, from purified β-naphthalenesulfonate (naphthalene and hot sulfuric acid) by caustic soda fusion.
α-Naphthol, from purified α-naphthalenesulfonate (naphthalene and cold sulfuric acid) by caustic soda fusion, or from α-naphthylamine.
1-Naphthol-4-sulfonic acid (Nevile-Winther's acid), from sodium naphthionate by heating with NH_4HSO_3 (the Bucherer reaction).
p-Nitroaniline, from *p*-nitroacetanilide, by boiling with caustic soda solution.
o- or *p*-Nitrophenol, from *o*- or *p*-chloronitrobenzene by hot, dilute NaOH solution.
Phenol from benzenesulfonate by caustic soda fusion at 320°C.
Phenol from chlorobenzene by caustic soda solution at 360°C, 5,000 psi.
Phenol from chlorobenzene by H_2O at 425°C.

OXIDATION

Oxidation[20] is controlled, or tempered, combustion and is one of the most useful chemical conversions in organic technology. The cheapest *agent* is air, but oxygen is sometimes employed. For liquid-phase reactions a great many oxidizing agents are in industrial use, such as nitric acid, permanganates, pyrolusite, dichromates, chromic anhydrides, hypochlorites, chlorates, lead peroxide, and hydrogen peroxide. H_2O, CO_2, and many other oxidized substances are the by-products of the main oxidation. When charcoal or carbon (amorphous) changes to CO_2, $\Delta H = -96.5$ kcal[21] g-mol, and when hydrogen burns to H_2O (gaseous), $\Delta H = -57.8$ dcal/g-mol. These *energy changes,* although the basis of combustion, frequently accompany controlled oxidation, as in the making of phthalic anhydride and maleic acid. In most oxidations, even when the formation of CO_2 and H_2O can be repressed, the energy is exothermic and large. This entails particular care in the design and construction of the *equipment* so as to ensure efficient heat transfer and to prevent controlled oxidation from becoming combustion.

TECHNICAL EXAMPLE *Phthalic anhydride* has become one of our most important intermediates. It is used in making directly a number of dyes such as eosin, rhodamines, erythrosin, quinoline yellow, copper phthalocyanine, and phenolphthalein, but it is of more value as a step in manufacturing anthraquinone and anthraquinone derivatives by the condensation (Friedel-Crafts) procedure, where it has opened up all the anthraquinone vat dyes to the American market at reasonable prices. These products are also consumed in industries other than dyes, as in making resins and plasticizers. Phthalic anhydride and phenol are two among many intermediates whose consumption has expanded far beyond dyes. The manufacture for both are presented in Chap. 34, on Plastics.

Oxidation Chemical Conversion in the Manufacture of Intermediates

Anthranilic acid, from phthalimide by alkaline hypochlorite.
Anthraquinone, from anthracene by chromic acid.
1,4-Dihydroxyanthraquinone (quinizarin), from anthraquinone by sulfuric acid oxidation in the presence of boric acid.
Dinitrostilbenedisulfonic acid, from *p*-nitrotoluene by sulfonation and alkaline NaOCl oxidation.
Phthalic anhydride, from naphthalene by air oxidation at 425°C in the presence of vanadium pentoxide (Chap. 34).
Dihydroxydibenzanthrone from dibenzanthrone by MnO_2 in concentrated sulfuric acid (anthraquinone vat dyes).
Phenol by air oxidation of cumene with acetone as a coproduct, or from toluene (Chap. 34).

[20]Groggings, *op. cit.*, chap. 9; Sittig, Combining Oxygen and Hydrocarbons for Profit, Gulf Publishing, 1962, especially chap. 11.
[21]For carbon, the equivalent Btu per pound = 14,400; for hydrogen, 52,000 Btu from 1 lb.

ALKYLATION

Although alkylation[22] of a hydroxyl is used occasionally in the dye field to reduce the solubility of phenol derivatives, its most extensive application in commercial dyes is in producing, with oxygen-alkylated products, dyeings that do not change their shade when exposed to dilute alkalies or acids. An outstanding example is the conversion of dihydroxydibenzanthrone to anthraquinone vat jade green (CI 59825 *vat green 1*) by alkylation with dimethyl sulfate or the methyl ester of *p*-toluenesulfonic acid. Alkylation is also used in alkylating amines. The *agents* employed are exceedingly varied, but an alcohol such as methanol, an alkyl halide, a dialkyl sulfate, or the methyl ester of *p*-toluenesulfonic acid is frequently satisfactory. The *equipment* used quite often requires heat and pressure to keep the reactants in the desired liquid state. For autoclaves of small size, jackets are frequently convenient for heating; in the larger sizes an interior coil enables more heating surface to be available than would be possible from a jacket. Furthermore, it is advantageous to have the corrosion concentrated on the more easily replaceable coil than on the more expensive shell.

TECHNICAL EXAMPLE *Dimethylaniline* is employed extensively in the manufacture of a number of triarylmethane dyes. It is prepared according to the following reaction:

$$C_2H_5NH_2 + 2CH_3OH \xrightarrow{H_2SO_4} C_6H_5N(CH_3)_2 + 2H_2O$$

Aniline, with a considerable excess of methanol and a little sulfuric acid, is heated in an autoclave at about 200°C for 5 or 6 h, the pressure rising to 525 or 550 lb. The product can be tested by noting any rise in temperature when mixed with acetic anhydride, to ascertain if any monomethylaniline is still present. If monomethyl is low or absent, the alkylation product can be discharged under its own pressure through a cooling coil, neutralized, and vacuum-distilled.

Alkylation Chemical Conversion in the Manufacture of Intermediates

Benzylethylaniline (and sulfonic acid), from ethylaniline and benzyl chrloride.
Diethylaniline, from aniline, ethyl alcohol, and a little HCl in an autoclave, or from aniline and diethyl sulfate.
Dimethylaniline, from aniline, methanol, and a little sulfuric acid at 200°C in an autoclave.
o-Nitroanisole, from *o*-chloronitrobenzene by methanol and caustic soda.

CONDENSATION AND ADDITION REACTIONS (FRIEDEL-CRAFTS)

The following list gives only a few products that are manufactured in any considerable tonnage, but these chemical conversions are used in the making of a great many different chemicals. Three of the intermediates listed, however, are the basis of some of the most useful vat dyes. The *agent* employed in this reaction is usually an acid anhydride or an acid chloride, catalyzed by aluminum chloride. Typical examples involve the making of *p*-chlorobenzoylbenzoic acid and *β*-chloroanthraquinone according to the following reactions:

Addition to furnish p-chlorobenzoylbenzoic acid:

CO, O, CO + Cl $\xrightarrow{Al_2Cl_6}$ CO, CO, OH, Cl

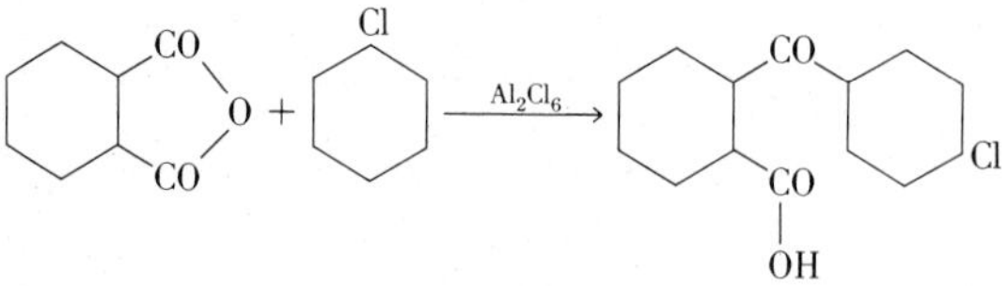

22Shreve, Alkylation, chap. 14 in Groggins, *op. cit.*; Lubs, *op. cit.*, pp. 30ff., 41ff.

Condensation, or ring closure, to furnish β-chloroanthraquinone:

$$C_6H_4(COOH)\text{-}CO\text{-}C_6H_4Cl \xrightarrow[\text{heat}]{H_2SO_4} C_6H_4(CO)_2C_6H_3Cl + H_2O$$

Typical *equipment* and conditions necessary to carry out these reactions and that of the following ring closure are sketched in Fig. 39.10. The industrial significance of these reactions, coupled with the ring closure to make reasonably priced anthraquinone or anthraquinone derivatives as the basis for fast vat dyes for cotton, cannot be overestimated. American tar distillers do not produce cheap anthracene.

Addition and Condensation Chemical Conversions in the Manufacture of Intermediates

Benzanthrone, from naphthalene, benzoyl chloride, and aluminum chloride to α-benzoylnaphthalene to benzanthrone (benzanthrone is also condensed from anthranol and glycerin with the aid of sulfuric acid—the preferred commercial procedure).
Benzoylbenzoic acid, from phthalic anhydride, benzene, and aluminum chloride.
p-Chlorobenzoylbenzoic acid, from phthalic anhydride, chlorobenzene, and aluminum chloride.
p-Methylbenzoylbenzoic acid (for 2-methylanthraquinone), from phthalic anhydride, toluene, and aluminum chloride.
3-Hydroxy-2-naphthoic anilide, from 3-hydroxy-2-naphthoic acid and aniline.
Phenylglycine, from aniline and chloroacetic acid.
Phenyl-1-naphthylamine-8-sulfonic acid, from 1-naphthylamine-8-sulfonic acid, aniline, and aniline hydrochloride by heating in an autoclave.
Tetramethyldiaminobenzophenone (Michler's ketone), from dimethylaniline (2 mols) and phosgene.
Tetramethyldiaminodiphenylmethane, from dimethylaniline (2 mols) and formaldehyde in the presence of hydrochloric acid.

Fig. 39.4 Dyestuffs intermediate still showing gauge and chart controls, at the dye works of E. I. du Pont De Nemours & Co., Deepwater Point, N.J.

MISCELLANEOUS CHEMICAL CONVERSIONS

The following list outlines a selection of the most important intermediates that have not been included in the chemical conversions considered in this section. Details of their manufacture can be readily obtained from the Selected References at the end of the chapter.

Various Miscellaneous Chemical Conversions in the Manufacture of Intermediates

Acylation:

Acetanilide, from aniline by heating with glacial acetic acid.

Acetyl-*p*-toluidine, from *p*-toluidine by heating with glacial acetic acid.

Carboxylation:

3-Hydroxy-2-naphthoic acid, from dry sodium β-naphtholate, and CO_2 under pressure at 200°C (see salicylic acid).

Salicylic acid, from dry sodium phenate and CO_2 under pressure and at 140°C (see Fig. 40.3 for flowchart).

Miscellaneous:

Aminoazotoluene (and sulfonate), from *o*-toluidine to diazoaminotoluene and molecular rearrangement.

Anthraquinone (ring closure), from *o*-benzoylbenzoic acid by sulfuric acid.

Benzoic acid, from phthalic acid by decarboxylation.

2-Chloroanthraquinone (ring closure), from *p*-chlorobenzoylbenzoic acid by sulfuric acid.

Phenylglycine from aniline, formaldehyde, and sodium cyanide.

Anthrimides from aminoanthraquinones containing halogen in the alpha position (anthraquinone vat dyes).

Dibenzanthrone from benzanthrone (anthraquinone vat dyes).

Benzoylation of aminoanthraquinones (anthraquinone vat dyes).

Arylamination of chloro-(or bromo-)anthraquinones containing alpha halogen atoms (anthraquinone vat dyes).

DYES

The complicated chemical formula representing a typical dye is often bewildering at first sight. However, an analogy can be made between houses and dyes, and dyes, as well as houses, can be viewed as being built out of separate materials put together in an orderly fashion. Such an analogy serves to simplify the concept of these complicated structures. As the architect changes the period or style of a house by varying the use of the same fundamental materials, wood, brick, stone, and steel—so the chemist makes different dyes by varying the chemical reactions to which the same intermediates are subjected. Dyes are built up out of more than 500 intermediates by a dozen or so important chemical conversions that unite one or more of these intermediates into a new chemical individual, which, if it has the right structure, becomes a dye. The "dye tree" in Fig. 39.5 gives an overall picture of dyes and some of their discovery dates.

Figures 39.6 to 39.8 present a tabular outline of the sequences of chemical conversions from basic petrochemical and coal-tar hydrocarbons, through intermediates, to a number of characteristic dyes. This chapter includes a number of typical intermediates and dyes and the chemical changes necessary for their fabrication.[23] Attention is called to the fact that the names of dyes are a combination of the chemical and the most used commercial designations. Dyes are classified both chemically (by structure) and by application (by use). The Colour Index (CI) number and name are also given.

CAUSE OF COLOR A certain amount of unsaturation in the dye molecule, with part of it at least in the form of aromatic rings, combined with the quinoid structure of minimum complexity,

[23]From the footnotes and Selected References for this chapter further details can be obtained, particularly for the intermediates, from the volumes by Groggins, Lubs, Davidson, and Cain, and for the dyes, from the books by Lubs, Venkataraman, and Georgievics; Colour Index, 3d ed., American Association of Textile Chemists and Colorists, Raleigh, N.C., 1971 (over 3000 dyes). Dyes are presented serially, chemically and by use classes. These volumes are a necessity for anyone interested in dyes. The numbers and names given in this chapter are those of the Colour Index and are labeled CI.

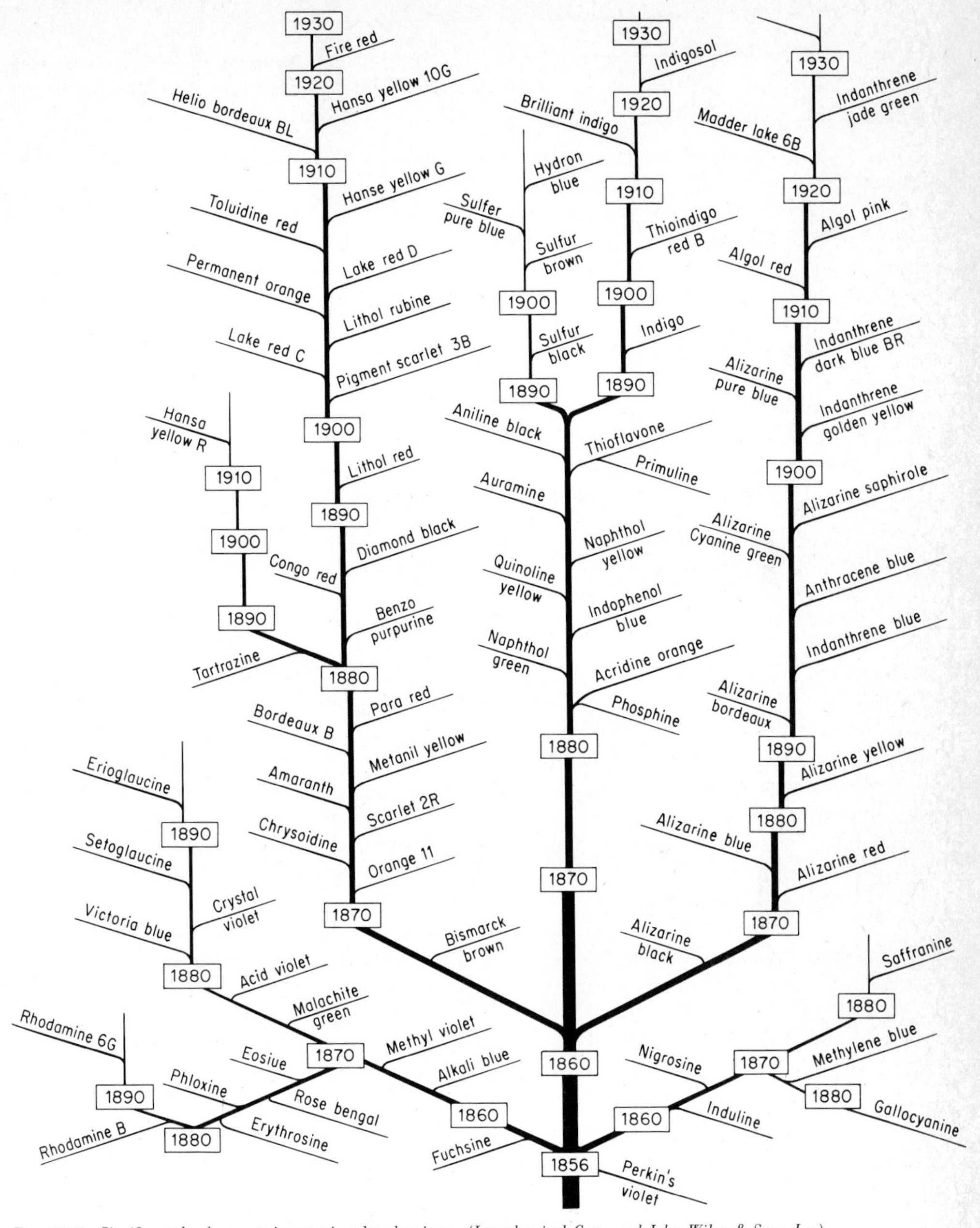

Fig. 39.5 Significant developments in organic color chemistry. (*Interchemical Corp. and John Wiley & Sons, Inc.*)

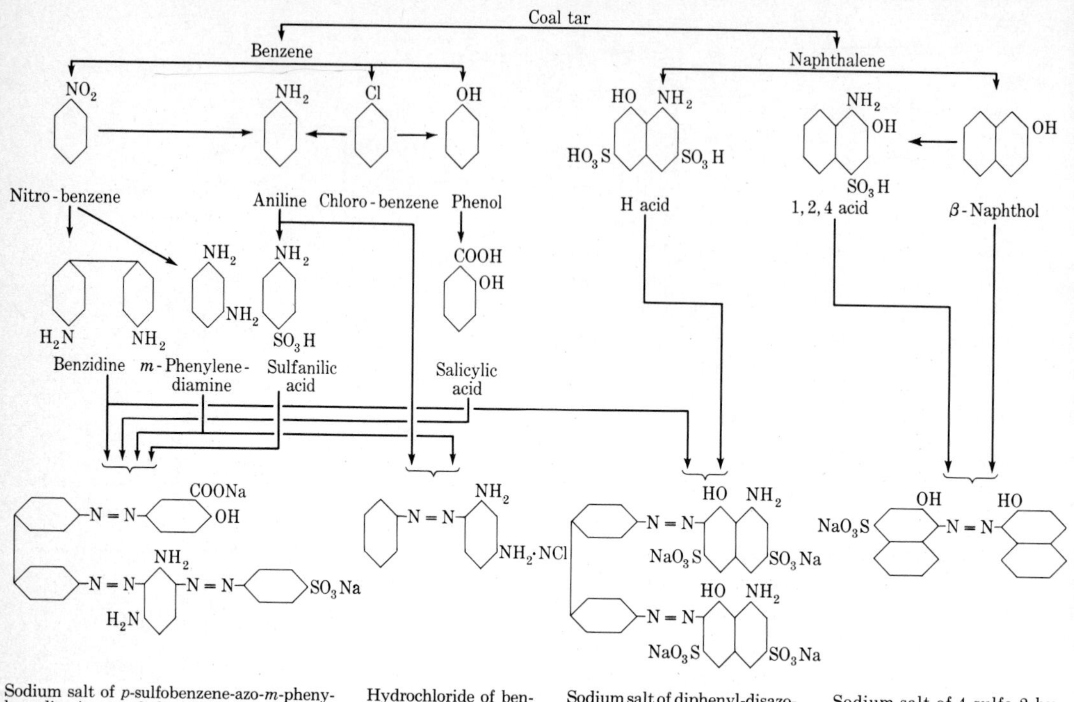

Sodium salt of *p*-sulfobenzene-azo-*m*-phenylene-diamine-azo-diphenyl-azo-salicylic acid. Sold as Direct Brown 3 GO, classified as Azo or Direct Dye, C.I. 30045, Direct Brown 1

Hydrochloride of benzene-azo-*m*-phenylenediamine. Sold as Chrysoidine Y, classified as Azo or Basic Dye, C.I. 11270, Basic Orange 2

Sodium salt of diphenyl-disazo-bis-8-amino-1-naphthol-3, 6-disulfonic Acid. Sold as Direct Blue 2 B, classified as Azo or Direct Dye, C.I. 22610, Direct Blue 6

Sodium salt of 4-sulfo-2-hydroxynaphthalene-1-azo-β-naphthol. Sold as Chrome Blue-black U, classified as Azo or Mordant and Chrome Dye, C.I. 15705, Mordant Black 17

Fig. 39.6 Relationship of coal tar, hydrocarbons, intermediates, and some azo dyes.

usually lays the foundation for dyes. Many correlations have been made between chemical structure and color,[24] and earlier conceptions promulgated by dye chemists in this field are still helpful. We may write the equation

Dye = chromogen + auxochrome

The *chromogen* is an aromatic body containing a group called a *chromophore*. By derivation, chromophore means *color giver,* and is represented by such chemical radicals as the following:

1. The nitroso group: —NO (or =N—OH)
2. The nitro group: $—NO_2$ (or =NO·OH)
3. The azo group: —N=N—
4. The ethylene group: $\rangle C{=}C\langle$
5. The carbonyl group: $\rangle C{=}O$

[24]Watson, Colour and Its Relation to Chemical Constitution, Longmans, 1918; Georgievics and Grandmougin, A Textbook of Dye Chemistry, Scott, Greenwood, London, 1920; chap. 8 of Venkataraman, *op. cit.* (many references); Davidson and Hemmendinger, Color and Color Measurement, *Color Eng.*, **1**(1), 15 (1963), and **1**(2), 19 (1963); ECT.; Lih, Color Technology, *Chem. Eng.* (*N.Y.*), **75**(17), 146 (1968).

6. The carbon-nitrogen groups: $\gt C{=}NH$ and $-CH{=}N-$

7. The sulfur groups: $\gt C{=}S$ and $\gt C-S-S-C\lt$

Such groups add color to the simpler aromatic bodies by causing displacement of, or an appearance of, absorbent bands in the visible spectrum. These chromophores are so important that we chemically classify many dyes by the chief chromophore they contain. These chromophore groups are capable of reduction and, if this is carried out, the color frequently disappears, probably because of the *removal of electron resonance.* Close packing of unsaturation, as in conjugation, also tends to produce color. Thus, even the hydrocarbon dimethylfulvene,

$$\begin{array}{l} HC{=}CH \qquad\quad CH_3 \\ |\qquad\quad \backslash\quad\ / \\ |\qquad\quad C{=}C \\ |\qquad\quad /\quad\ \backslash \\ HC{=}CH \qquad\quad CH_3 \end{array}$$

has an orange color. Although it may be colored, a chromogen may lack the chemical affinities necessary to make the color adhere to textile fibers. Thus assistant groups, or *auxochromes,* are needed, which are usually salt-forming groups such as $-NH_2$, $-OH$, their derivatives, or the solubilizing radicals $-COOH$ or $-SO_3H$. These auxochromes, chromophores, and chromogens become apparent in following the classification of dyes, but the assisting radicals such as $-OH$, $-NH_2$, $-SO_3H$, and $-COOH$ usually have more influence in the placing of a dye in a given use group directed toward dyeing a certain fiber, rather than in a chemical structural classification.

TESTING OF DYES In the use of dyes it is of the utmost importance to know such properties as fastness, solubility, and method of dyeing. There are various types of dye fastness, all relative to

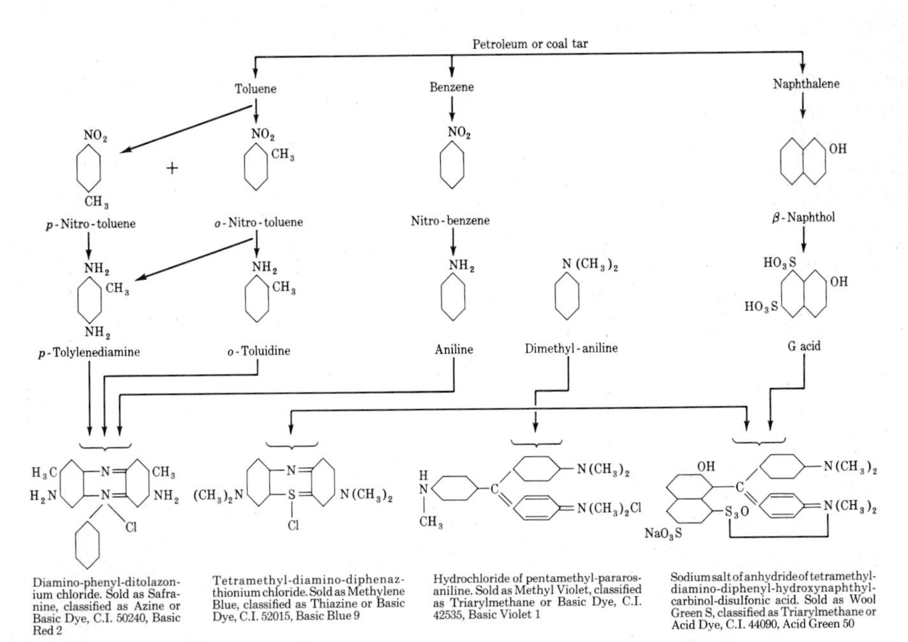

Fig. 39.7 Relationship of coal tar, hydrocarbons, and some azine, thiazine, and triarylmethane dyes.

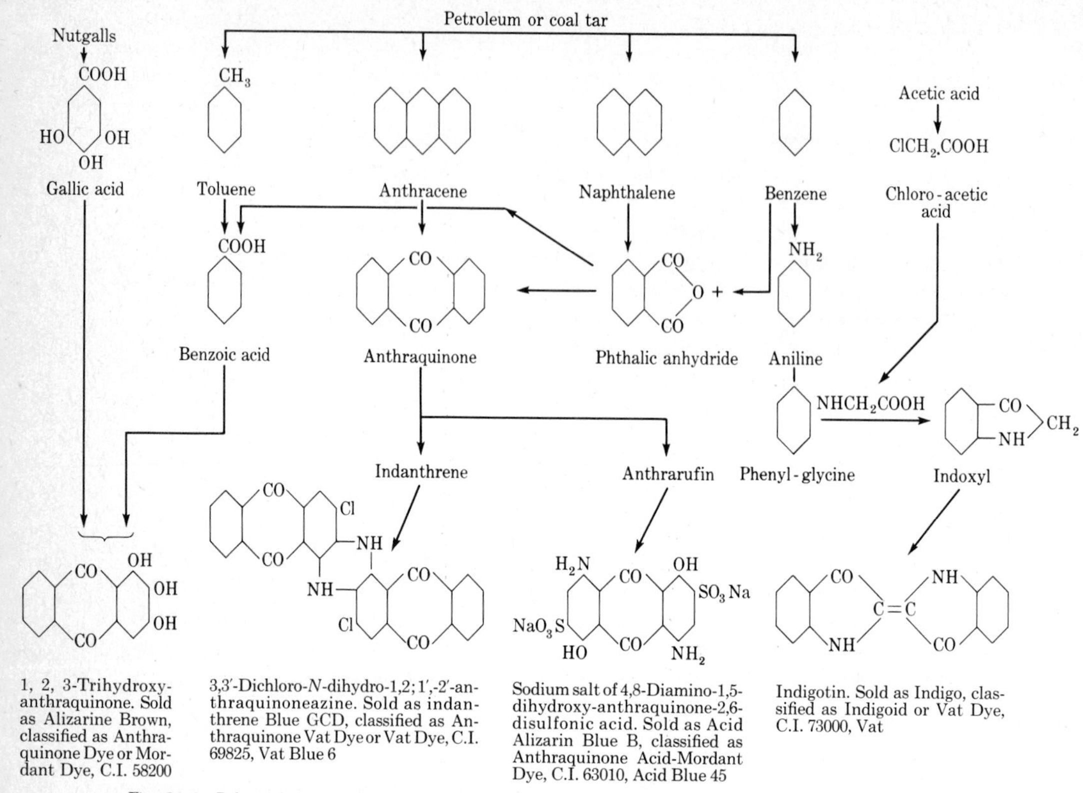

Fig. 39.8 Relationship of coal tar, hydrocarbons, intermediates, and indigo to some anthraquinone dyes.

one another. American-made dyes are as fast as any others in the world, being the same chemical compounds here as abroad.

CLASSIFICATION OF DYES Dyes are classified from both the chemical and the application, or end-use, viewpoints. Manufacturers look at dyes from the *chemical* aspect and arrange and manufacture them in groups, usually of like chemical conversions; this frequently brings similar chromophores together. Thus dyes that contain the azo chromophore are manufactured in a building that may be called the *azo building*. On the other hand, a similarity in the grouping of dyes that contain the indigoid radical is found. This very largely parallels the modern chemical-conversion method of classifying and looking at applied organic reactions. Not only do factories arrange dyes in this way, but the books presenting the properties of dyes also so classify them.

In this chapter the various dyes are discussed under this chemical classification. On the other hand, the users of dyes group them according to methods of *application*. A dyer engaged in coloring silk is particularly interested in the type of dye that gives good results on this fiber, and not specially in dyes that color only cotton. There is not always this sharp differentiation in application; indeed, certain dyes are union dyes and color more than one fiber. Three broad divisions of fibers are dyed: products of vegetable origin, those from animals, and those made synthetically. Cotton, linen, and paper are of a vegetable nature and essentially consist of cellulose. On the other hand, products of animal origin are much more reactive and consist of such substances as silk, wool, feathers, fur, and leather. Synthetic fibers consist principally of nylon,[25] Dacron, Orlon, Dynel, Acrilan, viscose rayon,

[25]See Chap. 35 for manufacture; for dyeing of such fibers, see *Chem. Eng.* (*N.Y.*), **68**(6), 84 (1961), and ECT, 1st Suppl., pp. 258–275, 1957.

the last-mentioned dyeing quite similarly to cotton, being also essentially cellulose, and acetate rayon, which is an ester and requires disperse dyes. Table 39.4 classifies the main types of dyes by *application,* and Table 39.5 by *chemical* arrangement.

Disperse[26] dyes are those which have been specially developed to dye cellulose acetate and some of the new synthetic fibers. They may be broadly divided into two general groups, embracing insoluble simple azo dyes and insoluble aminoanthraquinone colors, both of a highly dispersed type and consequently capable of penetrating, or dyeing, the fiber. Both of these general groups of dyes usually contain the ethanolamine $—NHCH_2CH_2OH$, or a similar radical, which renders them more readily dispersible in water and more easily absorbed. Early examples of dispersed dyes are the SRA, wherein the dispersing agent is sulforicinoleic acid (hence the designation SRA). Another example is CI 61100, *disperse violet 1.*

Acid dyes are used to color synthetic and natural (animal) polyamide fibers from acidified solutions. Such anionic dyes are attached to the positively charged amine group in the fiber. The acid auxochrome or solubilizing groups, $—NO_2$, $—SO_3H$, and —COOH, frequently aided by —OH, are usually present in the acid dye, whether the fundamental chemical structure is that of the azo, triarylmethane, or anthraquinone complex. These acid dyes are of importance in wool and silk dyeing. Such dyes are somewhat deficient in fastness to alkalies and soap, though usually possessed of good light resistance. Examples[27] are orange II, CI 15510, *acid orange 7;* acid black 10B, CI 20470, *acid black 1* (azos); and acid alizarine blue B, CI 63010, *acid blue 45* (anthraquinone dyes).

Azoic dyes are direct and developed dyes, applied especially on cotton. These azoic dyes are essentially *colorless azo dye intermediates,* marketed as four different groups of products: (1) Simple

TABLE 39.4 ***Comparison of U.S. Production of Dyes, by Classes of Application, 1972*** *(In thousands of pounds)*

Class of application	Production, thousands of pounds	Sales: Quantity, thousands of pounds	Sales: Value, thousands of dollars	Sales: Unit value per pound, dollars
Total	263,304	254,536	479,688	1.88
Acid	29,739	28,039	71,663	2.56
Azoic dyes and components				
Azoic compositions	2,515	2,021	3,397	1.68
Azoic diazo components, bases (fast-color bases)	1,226	743	1,352	1.82
Azoic diazo components, salts (fast-color salts)	3,569	3,178	3,811	1.20
Azoic coupling components (naphthol AS and derivatives)	2,905	2,360	5,693	2.41
Basic	17,999	17,824	48,876	2.74
Direct	37,672	34,519	59,167	1.71
Disperse	39,927	38,327	107,576	2.81
Fiber-reactive	3,699	3,562	15,582	4.37
Fluorescent brightening agents	27,321	27,442	38,269	1.39
Food, drug, and cosmetic colors	4,644	4,609	19,788	14.29
Mordant	1,465	1,711	2,567	1.50
Solvent	12,468	11,959	22,319	1.87
Vat	55,140	56,311	63,312	1.12
All other	23,015	21,931	16,316	0.74

Source: Synthetic Organic Chemicals, U.S. Tariff Commission, 1975 (annual).

[26]Venkatamaran, *op. cit.*, chap. 6; in general, the standard text in the field of dyeing or dye application is Colour Index, *op. cit.*; Knecht, Rawson, and Lowenthal, Manual of Dyeing, 8th ed., Griffin, London, 2 vols.; Lewis, Squires, and Broughton, Industrial Chemistry of Colloidal and Amorphous Materials, pp. 505–518, Macmillan, 1942 (good summary of dyeing).

[27]In these dye names, the older common name is given first, followed by the Colour Index (CI) number and the new CI ordinary name in italics.

TABLE 39.5 ***Comparison of U.S. Production and Sales of Dyes, by Chemical Classification, 1972***

Chemical class	Production, thousands of pounds	Sales: Quantity, thousands of pounds	Sales: Value, thousands of dollars	Sales: Unit value per pound, dollars
Total				
Amino ketone	63	29	172	5.93
Anthraquinone	46,589	48,105	125,399	2.61
Azo, total	92,028	87,398	191,561	2.19
Azoic	10,264	8,365	14,316	1.71
Cyanine	916	885	2,168	2.45
Ketone imine	455	611	1,318	2.16
Methine	5,576	5,171	14,752	2.85
Nitro	1,376	1,253	2,429	1.94
Oxazine	480	537	2,004	3.73
Phthalocyanine	1,381	1,474	3,365	2.28
Quinoline	2,705	2,278	7,876	3.46
Stilbene	30,898	30,538	38,122	1.25
Thiazole	352	365	1,032	2.83
Triarylmethane	8,903	8,443	20,386	2.41
Xanthene	1,167	1,052	6,262	5.95
All other	60,151	58,032	48,526	0.84

Source: Synthetic Organic Chemicals, U.S. Tariff Commission, 1975 (annual). In 1972 total dye production increased 8.0% to 263 million lb. *Includes azine, coumarin, indigoid, nitroso oxidation bases, sulfur, thiazine and others.

arylamines or their mineral acid salts (*color bases*); these need ice, sodium nitrite, and mineral acid for diazotization before coupling. (2) Stabilized diazonium compounds (*fast salts*), which are made by combining diazotized arylamines with a precipitant, such as zinc chloride. These are stable products which are soluble in water and can be coupled on the fiber with a selected component. Either the color base after diazotization or the already diazotized fast salt can be reacted by the user *in situ* with the same coupling constituent to produce the same dye. (3) The naphthol group (Du Pont's Naphthanil) dye by first impregnating the fiber with a selected coupling component, to which the diazotized arylamine is coupled *in situ*. (4) A group comprising mixtures of one selected triazene (made by combining a diazotized arylamine with a special type of secondary amine) and one selected naphthol, resulting in a *stabilized* specific insoluble azo dye produced *in situ* by special printing and acid-aging techniques. These are marketed as Rapidogens[28] (Du Pont Diagen brand) and are exclusively used for printing. Azoics are next in importance to vat dyes, which they surpass in brilliance, particularly in the bright reds.

Basic dyes are mostly amino or substituted amino derivatives, frequently from the triarylmethane or xanthene class, and are applied mostly to paper. Practically all the auramine, CI 41000, *basic yellow 2,* manufactured in America is consumed in paper dyeing. Some basic dyes, bismarck brown, CI 21010, *basic brown 4,* and chrysoidine, CI 11270, *basic orange 2,* are used for leather. Crystal violet, CI 42555, *basic violet 3,* and methyl violet, CI 42535, *basic violet 1,* and 42535 B, *solvent violet 8,* are mainly employed in typewriter ribbons, carbon paper, and duplicating inks, where some other basic dyes also find application. A significant use of solvent basic dyes is in writing and printing inks. In the sulfonated form some special types of basic dyes still find limited application in the dyeing of silk and wool, but a more important use of these sulfonated forms is their conversion to lakes for printing inks (Chap. 24).

[28]Lubs, *op. cit.*, pp. 218–220.

Direct dyes are frequently from the azo class, and are used to dye cotton and vegetable fibers. Some of them are also employed for dyeing union goods (cotton and wool or cotton and silk). As this dyeing is usually aided by the addition of common salt, or Glauber's salt, to the dye bath, such dyes have been called *salt dyes.* The salt decreases the solubility of the dye, hence causes better exhaustion of it from the dyeing solution. Like acid dyes, they are anionic; however, their differences lie in the structure of the dye molecule. To be "substantive," direct dyestuffs must be long molecules, and the aromatic rings must be capable of assuming a coplanar configuration. Examples are direct blue 2B, CI 22610, *direct blue 6;* direct black EW, CI 30235, *direct black 38;* direct brown 3GO, CI 30045, *direct brown 1;* and many others. A number of direct dyes possessing free amino groups can be *developed* on the fiber, thus increasing their insolubility, hence their fastness to washing. This developing involves *diazotizating* the free amino group and *coupling* with the developer, which may be β-naphthol.

Pigment dyes form insoluble compounds or lakes with salts of calcium, barium, chromium, and aluminum, or phosphomolybdic acid. The dye molecule frequently contains —OH or $—SO_3H$ groups. Such lakes, when ground in oil or other media, form the pigments of many paints and inks. In this use class fall such dyes as lithol red R, CI 15630, *pigment red 49;* ponceau 2R, CI 16150, *acid red 26;* Orange II, CI 15510, *acid orange 7;* and the splendid phthalocyanines (pigments). Some basic dyes are used for the tinting of paper in a water-dispersed form of phosphomolybdic (or tungstic) acid lakes. Wallpapers are frequently colored with lakes from basic dyes containing a sulfonic group. Certain insoluble dyes are widely used in a pure form, known as toners, or pigment toners, in paints, printing inks, and wallpapers, and especially for the *pigment printing* method for textiles, employing pigments of metal phthalocyanines and other types. The use of pigments in the "dope" before spinning rayon, acetate, and synthetic fibers is growing rapidly and is resulting in excellent colors of outstanding all-around fastness (Chap. 24).

Mordant dyes are applied principally to wool, wherein, by the use of a mordant, which may be chromium or less frequently aluminum or iron, fastness to light and washing is much increased. Such dyes contain —OH or —COOH radicals frequently attached to azo or anthracene (anthraquinone) complexes. The mordant dyeing is really a metallic salt or lake formed in the fiber. Examples are: Alizarin, CI 58000, *mordant red 22;* chrome blue black U, CI 15705, *mordant black 17;* Gallocyanine, CI 51030, *mordant blue 10;* and many others.

Sulfur, or *sulfide,* dyes contain a chromophore with sulfur, and are also dyed from a sodium sulfide bath, wherein the sulfur color is reduced to a colorless or light-colored leuco derivative. Sulfur dyes are usually applied on cotton and form a large, low-priced, and useful group, among which are sulfur black, CI 53185, *sulfur black 1* and analogous dyes. These dyes furnish dull shades of good fastness to light, washing, and acids. However, they are very sensitive to bleach or chlorine.

Vat dyes, including the long-used indigo, are of a chemical structure such that reduction furnishes an alkali-soluble "leuco vat," with which the fiber, generally of vegetable origin such as cotton, is impregnated. This, upon exposure to air, oxidizes back to the insoluble color. Such dyes are of complicated chemical structure, such as indanthrenes, but furnish dyes of exceptional fastness to light, alkaline washing, perspiration, and even chlorine. Vat dyes provide the fast-dyed cotton shirtings and dress goods which, although relatively expensive, are rapidly increasing in use. Vat dyes are likewise made in a paste, with an alkaline hydrosulfite-aldehyde reducing agent, printed on the cloth, and passed through an oxidizing bath of sodium bichromate or perborate. Examples are: indigo, CI 73000, *vat blue 1;* indathrene blue GCD, CI 69810, *vat blue 14;* and anthraquinone vat jade green, CI 59825, *vat green 1.* Indigo has a unique and very important use for dyeing wool in extremely dark (navy blue) shades, where it shows surprisingly good fastness properties. Quite important, also, is the use of several thioindigo dyes for the printing of rayon fabric (mainly women's dresses) in very bright shades with good fastness properties.

Solvent, or *spirit-soluble,* dyes are frequently simple azos, triarylmethane bases, or anthraquinones used to color oils, waxes, varnishes, shoe dressings, lipsticks, and gasoline.

Solvent dyeing from perchloroethylene and liquid ammonia shows much promise[29] as a method for dyeing synthetics (polyesters, polyacrylates, and triacetates) with sharply reduced problems in the disposal of spent liquors. Such processes are already commercial in Europe. *Food* dyes are of various structures, selected and tested for harmlessness and employed in coloring foods, candies, confections, and cosmetics. *Photographic, medicinal, bacteriological,* and *indicator* dyes are highly specialized products of relatively small sales volume but are of fundamental importance in the maintenance of the national economy.

Reactive dyestuffs which link chemically to cellulosic fibers were first marketed in 1956. They represented the first really new class of dyestuffs introduced since disperse dyes for acetate. The dyes are water-soluble compounds which contain reactive groups capable of combining with the hydroxyl groups of cellulose under alkaline conditions, establishing a covalent link between the dye and the fiber. The wet-fastness properties of these dyeings is high.

Fluorescent brightening agents,[30] or "optical brighteners," are used to whiten textiles, plastics, paper, soaps, and detergents (Chap. 29) and to add brightness to delicate dyeings. About $38 million worth was sold in 1976 of greatly varying chemical structure. The addition of a good white reflecting pigment like TiO_2 helps in paper or plastics. Examples of dyes are derivatives of diaminostilbene, coumarin, etc. (cf. Colour Index).

POLLUTION CONTROL in dyeworks[31] is a particularly vexing problem because of the extreme visibility of the wastewaters. Solvent dyeing offers a promise of reducing the load, but all methods of waste disposal and contaminant segregation require much expense and attention.

MANUFACTURE OF DYES

The very distinguished pioneering English chemist Sir William Henry Perkin is the father of synthetic organic chemical dyes.[32] Not only did he discover the first practical synthetic aniline dye, *mauve,* which is made of toluidine, containing aniline, by oxidation, but he organized with his father and brother a company to make synthetic dyes and went out into the dyeing establishments to show how to apply his products. In the early years, from 1856 on, Perkin *was* the dye industry. After 17 years of success, he sold out to enjoy a life of research. Meanwhile, the Germans had entered the dye industry and rapidly became the leaders in the field until 1914, when the United States, Great Britan, and other nations began to intensively cultivate this line of manufacture as an essential industry.

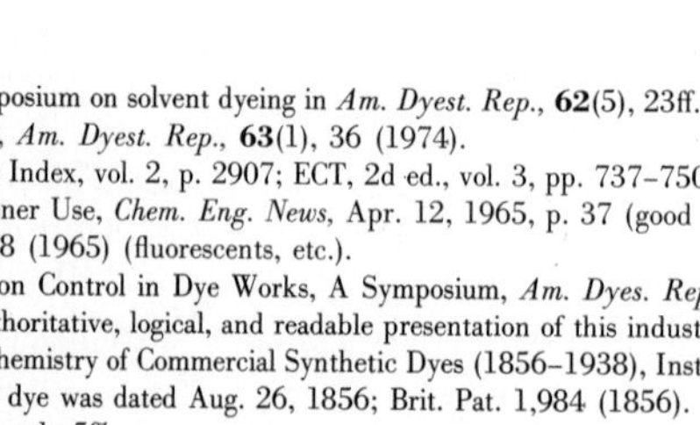

Mauve

[29]A symposium on solvent dyeing in *Am. Dyest. Rep.*, **62**(5), 23ff. (1973); Capponi, Moreau, and Somm, Continuous Dyeing with Solvents, *Am. Dyest. Rep.*, **63**(1), 36 (1974).

[30]Colour Index, vol. 2, p. 2907; ECT, 2d ed., vol. 3, pp. 737–750, 1964; *Color Eng.*, **2**(4), 12 (1964); Consumer Demand Spurs Brightener Use, *Chem. Eng. News*, Apr. 12, 1965, p. 37 (good summary); Leavitt, Advances in Organic Colorants, *Color Eng.*, **3**(4), 28 (1965) (fluorescents, etc.).

[31]Pollution Control in Dye Works, A Symposium, *Am. Dyes. Rep.*, **63**(8), 11ff. (1974).

[32]An authoritative, logical, and readable presentation of this industry is the booklet by Rowe, Two Lectures on The Development of the Chemistry of Commercial Synthetic Dyes (1856–1938), Institute of Chemistry, London, 1939. Perkin's patent for the first synthetic dye was dated Aug. 26, 1856; Brit. Pat. 1,984 (1856). Initially, his yields on mauve, or as he called it, "aniline purple," were only 5%.

Industrially, in the United States, more than 1,000 different dyes are now used, of which 300 are made in appreciable tonnage,[33] such as *direct black 38, mordant black 11, vat blue 6, vat green 1, vat green 3, vat green 8,* and *vat brown 25,* amounting in production to more than 1.5 million lb each annually, *direct black 38* heading the list (in 1974) with 6.7 million lb. Some of the outstanding dyes are arranged below by the chemical (or chromophore) classification, such as nitro, nitroso, and azo; also included is the application designation. Dyes have entered the markets of the world with a *variety of names for the same chemical individual,* as listed in the indexes and tabulations of a number of books. The *usual name,* adopted here, is the common name, together with the CI number and name which, respectively, indicate chemical class and application class. The *CI name* is that appearing in the annual reports of the U.S. Tariff Commission.

The commercial names are frequently followed by letters, some of which have special designations, among these are:

B, bluish; BB or 2B, more bluish.
G, yellowish (*gelblich*), occasionally greenish.
R, reddish.
S, bisulfite compound or sulfonic derivative, or dye for silk.
W, for wool; HW, for part wool; WS, for wool and silk.
L, easily soluble (*löslich*), lake-forming, or for linen.

From a manufacturing aspect, dyes fall naturally into their chemical classification. The following presentation is based on this long-accepted arrangement. See Table 39.5 for a production summary by chemical classification.

NITROSO DYES Chromophore: —NO (or =N—OH). Only few members of this class are produced in America, for example, naphthol green B, CI 10020, *acid green 1.* The latter is an acid dye, is used in wool and pigment dyeing, and is made by the action of nitrous acid on 2-naphthol-6-sulfonic acid and conversion to the ferric sodium salt.

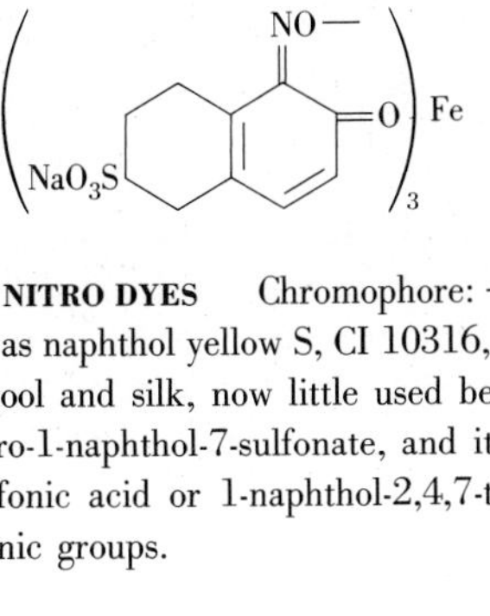

NO—
NaO_3S =O Fe
3

NITRO DYES Chromophore: $—NO_2$. A number of nitro dyes are made in the United States, such as naphthol yellow S, CI 10316, *acid yellow 1,* which is a cheap acid dye for clear yellow shades on wool and silk, now little used because of low light stability. Its chemical name is sodium 2,4-dinitro-1-naphthol-7-sulfonate, and it is manufactured by sulfonating α-naphthol to 1-naphthol-2,7-disulfonic acid or 1-naphthol-2,4,7-trisulfonic acid and then nitrating, thus replacing one or two sulfonic groups.

Naphthol yellow S, CI 10316,
acid yellow 1

ONa
NaO_3S NO_2
NO_2

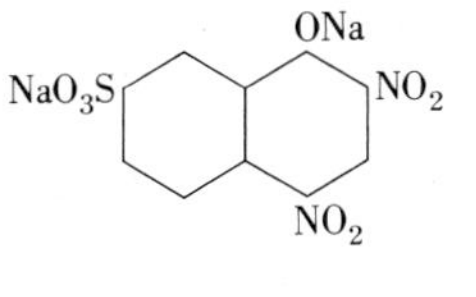

[33]Synthetic Organic Chemicals, U.S. Production and Sales, U.S. Tariff Commission (annual). This report of production and sales of dyes, intermediates, and other organic chemicals should be in the hands of everyone interested in this field, since it supplies most up-to-date statistics for the entire organic field. Not only are current tonnages available, but trends are seen by comparison of different years.

AZO DYES Chromophore:[34] —N═N—. This class exemplifies the application of the chemical conversion of diazotization and coupling. More than half of the dyes of commerce fall within this classification, and they are of various degrees of complexity, according to the number of azo groups contained or of assisting or auxochrome groups. Table 39.5 gives the production figures, indicating that of azo dyes to be 35% of the total. Based on the chemical classification, subgroups under the azo dyes are named monoazo, disazo, trisazo, or tetrakisazo dyes, depending upon whether one, two, three, or four azo groups are present. And as the chemical skeleton and the number and nature of the auxochrome groups may be varied, basic or mordant, acid or direct dyes, or indeed members of all the use classes, except vat and sulfur dyes, are obtained. After the fundamental principles of the reactions were worked out, following the basic discovery by Peter Griess[35] in 1858, literally thousands of azo dyes were made and their properties investigated. Those which survived are the azo dyes that possess suitable properties of fastness or ease of dyeing. Azo dyes can be made soluble or insoluble as the central rings, or the auxochromes, are changed. They *can* be made actually on or in the fiber,[36] or another azo group can be added to an azo dye already on the fiber (developed colors), or a stabilized diazo derivative can be coupled with a suitable intermediate, both being on the fiber (stabilized azoics). This flexibility of formation leads to useful dyes of a great variety of properties.

The fundamental reactions may be expressed as follows:

$$RNH_2 + HNO_2 + HCl(NaNO_2 + 2HCl) \longrightarrow \underset{\text{Diazonium chloride}}{R{-}N_2^+Cl^-} + 2H_2O$$

$$R{-}N_2^+Cl^- + HR'OH \text{ (or } HR'NH_2) \longrightarrow RN{=}NR'OH + HCl$$

These two reactions are known as *diazotization* and *coupling*. Specifically, the energy change in diazotizing in aqueous solution, thus forming diazobenzene hydrochloride, may be expressed:

$$C_6H_5 \cdot NH_2 \cdot HCl + HNO_2 \longrightarrow 2H_2O + C_6H_5N_2Cl \qquad \Delta H = -22.8 \text{ kcal}$$

For α-naphthylamine subjected to the same reaction, $\Delta H = -24.82$ kcal. Fortunately for their wide industrial application, these reactions can usually be carried out almost quantitatively,[37] if the temperature is kept low; this is done by the use of ice. There are specific rules for the coupling of an amine or of a phenol to a diazo body. Generally, the diazo solution is run into the arylamine, phenol, naphthol, or derivative, though there are some exceptions, such as chrome blue black U, CI 15705, *mordant black 17*, exemplified in Fig. 39.9. With phenols and naphthols, the coupling is usually *para* to the hydroxyl group or, if this position is occupied, the *ortho* place is taken. However, the very important β-naphthol couples only once and in the 1-position. Amines couple less readily than do phenols; the coupling takes place only in the para position. Of the diamines the meta derivatives react most easily, the coupling taking place para to one $—NH_2$ group and ortho to the other. This is frequently called the *chrysoidine law* and is exemplified in the making of chrysoidine, CI 11270, *basic orange 2*, wherein

$$C_6H_5N_2^+Cl^- + C_6H_4(NH_2)_2 \longrightarrow C_6H_5{-}N{=}N{-}C_6H_3(H_2N)NH_2 \cdot HCl$$

[34]Johnson et al. (Swiss Insitute of Technology), ECT, 2d ed., vol. 2, pp. 868–910, 1963.

[35]Griess, *Ann.*, **106,** 123 (1858).

[36]Ice, or ingrain, or developed colors, now called *azoics*.

[37]Boyd, Azo Dyes, pp. 67, 101 in Lubs, *op. cit.*, chap. 3 on diazotization; Cain, Chemistry and Technology of the Diazo Compounds, Arnold, London, 1920; Rys and Zollinger, Fundamentals of the Chemistry and Application of Dyes, Wiley, 1972.

Substituents and pH influence coupling; for instance, for H acid,

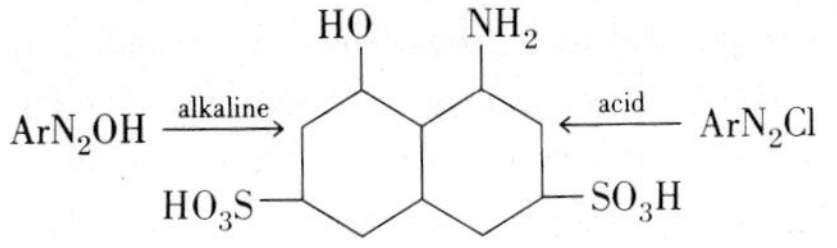

$$ArN_2OH \xrightarrow{\text{alkaline}} \underset{HO_3S \qquad\qquad SO_3H}{\overset{HO \qquad NH_2}{\text{[naphthalene ring]}}} \xleftarrow{\text{acid}} ArN_2Cl$$

the position *ortho* to the $—NH_2$ group, is activated in acid solutions, and coupling takes place there, whereas the activation in alkaline solutions is ortho to the —OH with coupling in this position; thus H acid can couple twice. The important aminonaphtholsulfonic acids, gamma acid and J acid, can couple only once, but the position also is influenced by the pH. This property is of very great technical importance in governing the formation of azo derivatives of proper shade and desired fastness. Table 39.6 lists several azo dyes under the azo subclasses and the application class, with intermediates. By following the rules of coupling just enumerated, the structure of these dyes can be determined.

The equipment in which azo dyes[38] are manufactured by diazotization and coupling is quite simple in comparison with that used for most other chemical processes. It is exemplified in Fig. 39.9, depicting the making of chrome blue black U, CI 15705, *mordant black 17*, and consists largely of wooden stirred tanks of various sizes, wooden plate-and-frame filter presses, and drying boxes wherein, on steel carts, the dye, previously placed on trays, is dried by circulating hot air. Sensitive dyes are

TABLE 39.6 ***Some Important Azo Dyes***

Colour Index No.	*Name and class of dye*	*Intermediates from which dye is made*	*Dye application class*
	Monazo Dyes: Containing R—N=N—R′		
11270	Chrysoidine Y, *basic orange 2*	Aniline *m*-Phenylenediamine	B
15510	Orange II, *acid orange 7*	Sulfanilic acid β-Naphthol	A
15705	Chrome blue black, *mordant black 17*	1-Amino-2-naphthol-4-sulfonic acid β-Naphthol	M
	Disazo Dyes: Containing R—N=N—X—N=N—R′		
20470	Acid black 10B (naphthol blue black), *acid black 1*	*p*-Nitroaniline H acid, aniline	A
22610	Direct blue 2B, *direct blue 6*	Benzidine 1-Amino-8-naphthol-3,6-disulfonic acid (alk) (2 mol)	D
24410	Direct sky blue FF, *direct blue 1*	Dianisidine 1-Amino-8-naphthol-2,4-disulfonic acid (alk) (2 mol)	D
	Trisazo Dyes: Containing R—N=N—X—N=N—Y—N=N—R′		
30295	Direct green B, *direct green 6*	Benzidine Phenol 1-Amino-8-naphthol-3,6-disulfonic acid *p*-Nitroaniline	D

Note: A, acid dye; B, basic dye; D, direct dye; M, mordant dye; S, solvent dye.

[38]For other azo flowcharts, see *Chem. Eng.* (*N.Y.*), **68**(1), 74 (1961), and **70**(17), 138 (1963).

dried in vacuum shelf dryers. The milling, mixing, and standardizing of dyes end the manufacturing sequence. The mills must be so chosen as not to ignite the sensitive dyes. Since customers require absolute uniformity in dyes, mixing and standardizing are of utmost importance. The double-cone dye mixer and blender, sometimes used, has no internal parts to hold up a dye and performs mixing and blending in an unusually satisfactory manner.

Among the monoazo dyes, chrome blue black U,[39] CI 15705, *mordant black 17,* ranks as one of the important chrome colors on wool for men's wear, particularly because its various qualities of fastness are very good (fastness to light, acids, alkalies, carbonizing, ironing, washing, perspiration, etc.). Sales were over 151,000 lb in 1976 at a low price ($1.50 per pound), because of the cheapness of the raw materials and the skill of the dyemakers. The two intermediates are 1,2,4 acid (1-amino-2-naphthol-4-sulfonic acid) and β-naphthol. The nitrous acid not only diazotizes the $—NH_2$ group but tends to oxidize the 1,2,4 acid to tars, thus throwing off the shade of the finished dye and decreasing the yield. This can be avoided by diazotizing 1-amino-naphthalene-2,4-disulfonic acid and replacing the 2-sulfonic acid by —OH through caustic soda or by diazotizing in the presence of a zinc salt. However, the oxidizing influence of the nitrous acid is diminished in the presence of a small amount of a copper salt.[40] The flowchart in Fig. 39.9 follows this procedure which involves the reaction

NH_2 OH SO_3H + $NaNO_2$ $\xrightarrow{CuSO_4}$ (N=N—OH OH SO_3H) ⟶ N N O SO_3Na + $2H_2O$

1,2,4 acid

Diazo oxide of 1,2,4 acid

N=N O SO_3Na + ONa ⟶ { Blue alkaline solution. Upon neutralization, using HCl, the dye precipitates out. ⟶

OH SO_3Na —N=N— HO + NaCl

Chrome blue black U, CI 15705, *mordant black 17*

The blue alkaline solution of the dye is precipitated the next day with diluted muriatic acid, testing, toward the end, so as to leave the mother liquor faintly acid to congo red test paper. The usual product sold, except for higher-priced concentrates, contains salt about equal to the dye content.[41]

From the flowchart and current material prices, the cost of the materials[42] for Chrome Blue Black can be calculated. Labor, sales expense, customer services, etc., frequently equal the cost of the materials.

[39]This CI 15705, like most other dyes, is sold under many names, frequently one for each manufacturer: Eriochrome Blue Black R (Gy), Calcochrome Blue Black (CCC), Pontochrome Blue Black R (DuP), Superchrome Blue B Ex (NAC). The initials in parentheses are those of the manufacturer; e.g., DuP stands for Du Pont. See also dye lists in the yearbooks of the American Association of Textile Chemists and Colorists and the Colour Index.

[40]Saunders, *op. cit.*, p. 7; Giegy, U.S. Pat. 793,743 and Brit. Pat. 10,234 (1904); O'Brien, *op. cit.*, pp. 151–154; Lubs, U.S. Pat. 2,160,882 (1939).

[41]Most dyes contain salt or sodium sulfate to standardize, since batches do not come from the plant in exactly the same strength. This is adjusted by a harmless diluent. Very pure dyes are such products as food, medicinal, and bacteriological dyes, which are costly compared with the low-priced textile dyes.

[42]This dye is sold so cheaply that even the 1,2,4 acid would have to be manufactured to compete.

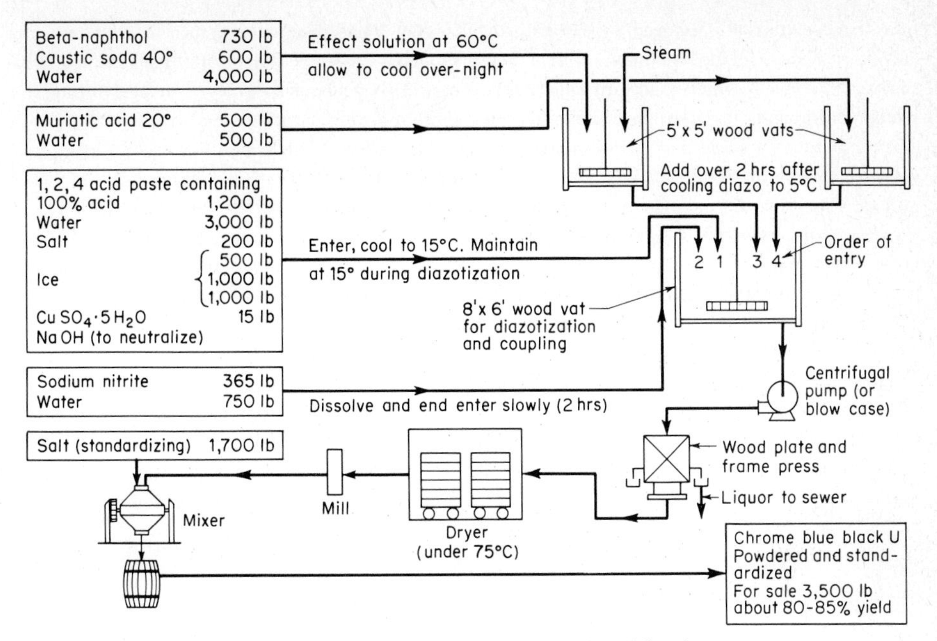

Fig. 39.9 Flowchart for chrome blue black U. Monoazo or chrome dye; batch = 5 lb-mol.

Directions for making other dyes can be found in the references.[43] The apparently complicated trisazo dye, direct green B, CI 30295, *direct green 6,* is made from four intermediates according to the scheme:

Benzidine → Phenol
Benzidine —alk.→ H acid
p-Nitroaniline —acid→ H acid

The arrows indicate coupling of the diazo body with a hydroxy or amino derivative. The resulting dye formula is

—N=N— OH
HO NH$_2$
—N=N— —N=N— NO$_2$
NaO$_3$S SO$_3$Na

Direct green B, CI 30295, *direct-green 6*

In 1880 the making of an azo dye, para red, CI 12070, *pigment red 1,* right on the fiber was undertaken. This was done by padding cotton cloth with an alkaline β-naphthol solution, followed

[43]A flowsheet for CI 24410, *direct blue 1,* is given by Woodward, p. 170, in Groggins, *op. cit.,* 3d ed.; cf. O'Brien, *op. cit.*

by an acid solution of diazotized *p*-nitroaniline. Later, the *β*-naphthol was replaced by 3-hydroxy-2-naphthoic acid or its anilide (napthol AS, CI 37505, *azoic coupling component 2*), to attain more substantivity and brighter shades of superior fastness. This generalized process enabled the more soluble constituents to penetrate into the fiber and there to form the insoluble azo dye. Such procedures are still very important and comprise a share of the made-on fiber dyes now known as azoics (see classification of dyes). Azoic prints from diazonium pastes on "naphtholated" goods are of wide utility.

For years, metallic[44] salts, particularly of chromium and of aluminum, have been used in making lakes and as mordants, having been initiated by the *neolin* of the Swiss. All dyestuff manufacturers have a line of *metalized dyes.* These dye wool from acid baths with excellent fastness to washing, milling, and light. Examples[45] are calcofast olive brown G, CI *acid brown 98,* and calcofast orange YF, CI *acid orange 69,* made by coupling nitrodiazophenols with phenylmethylpyrazolones. The metallized dye is obtained by salting out the reaction liquor, using common salt up to 25% of the aqueous solution.

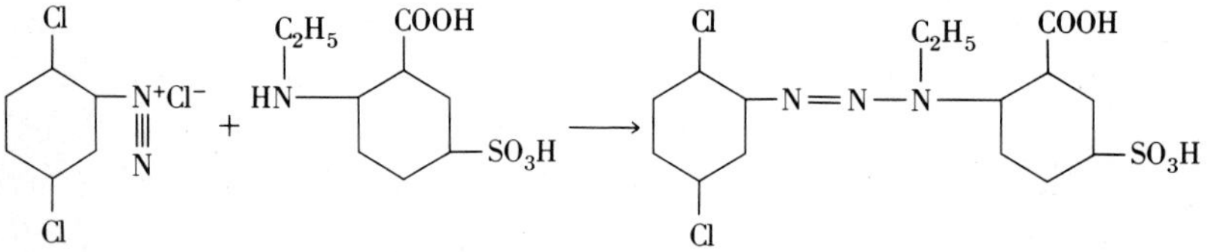

Diazotized 2,5-dichloroaniline + *N*-ethyl-5-sulfoanthranilic acid (stabilizer) → Stabilized diazo component, or triazene

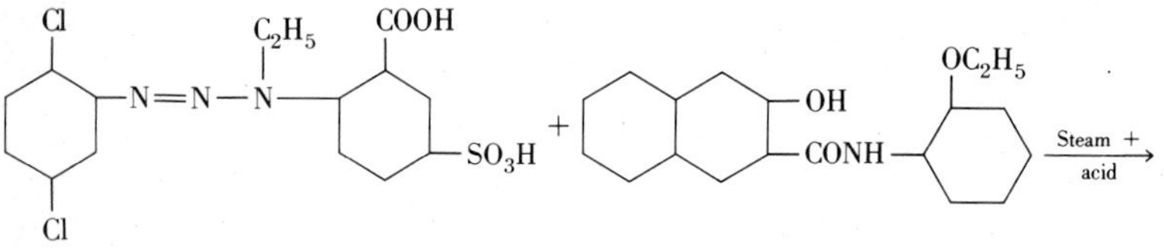

Triazene + Coupling component 14, CI 37558, 3-hydroxy-2-naphtho-*o*-phenetidide $\xrightarrow{\text{Steam + acid}}$

As applied to fabric rapidogen orange R, *azoic orange 3*

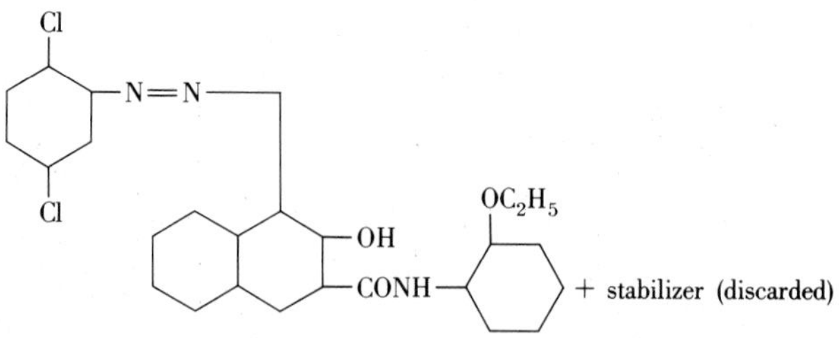

Bright orange dye on cotton + stabilizer (discarded)

STILBENE DYES Chromophores: —N=N—and=C=C=. These dyes are sometimes classed with the azo group, to which they are closely related. Although they contain an azo chromo-

[44]Lubs, *op. cit.*, p. 247; Venkataraman, *op. cit.*, metal-dye complexes, pp. 534ff., 551ff.
[45]U.S. Pat. 2,120,799 (1938) (specific examples are given).

phore, it is not made by diazotization and coupling reactions. These are direct cotton dyes, and the most used example is direct yellow R, CI 40000, *direct yellow 11,* which is made by the action of heat and caustic soda on *p*-nitrotoluene-*o*-sulfonic acid (4 mol).

PYRAZOLONE DYES Chromophores: —N═N— and ═C═C═. These are also sometimes classed with the azo group. They are acid dyes and are commonly used on silk and wool, though to some extent for lakes. The chief member is tartrazine, CI 19140, *acid yellow 23* and *food yellow 4,* which is made from dioxytartaric acid and phenylhydrazine-*p*-sulfonic acid (2 mol). Its formula is

NaO_3S—C_6H_4—N═N—C——C—COONa
HO—C═ ═N (ring closed through N)
N—C_6H_4—SO_3Na

KETONIMINE DYES Chromophore: NH═C═. This small class is so close to the next, the triarylmethanes, that the two dyes are frequently grouped as triarylmethanes. The members are auramine, CI 41000, *basic yellow 2,* and *solvent yellow 34,* and its homolog. Auramine is used primarily for paper dyeing. It is manufactured from dimethylaniline (2 mol) and formaldehyde, furnishing 4,4-bis(dimethylamino)diphenylmethane. This intermediate, upon heating with sulfur, ammonium chloride, and salt to about 200°C in a current of NH_3, yields auramine base, which is converted to the hydrochloride:

HCl·HN═C(C_6H_4—$N(CH_3)_2$)$_2$

Auramine, CI 41000, *basic yellow 2*

TRIARYLMETHANE DYES Chromophores: ═C═NH and ═C═N—. However, the cause of color in this class, with the following complex as the characteristic chemical group, is

R
|
Ar—C—Ar
|
Ar

due to more than the presence of a simple chromophore. Conjugated bonds, unsaturation, and quinoid arrangement are all important, but it is the possibility of more than one resonating form for the entire molecule (including, generally, the amino auxochromes) that is now held to be responsible for the color.[46] These dyes are basic ones, dyeing cotton with a tannin mordant or, when sulfonated, acid dyes for wool and silk. They furnish very beautiful shades but, for the most part, are not very fast to light.

[46]Lewis and Calvin, Color, *Chem. Rev.,* **25,** 273 (1939). The influence of alkylation in color change is tabulated by Shreve, p. 508 in Groggins, *op. cit.,* 3d ed.

Malachite green, CI 42000, *basic green 4,* is the chloride of *p,p'*-tetramethyldiaminotriphenylcarbinol, sold usually as a zinc chloride complex. Its formula is

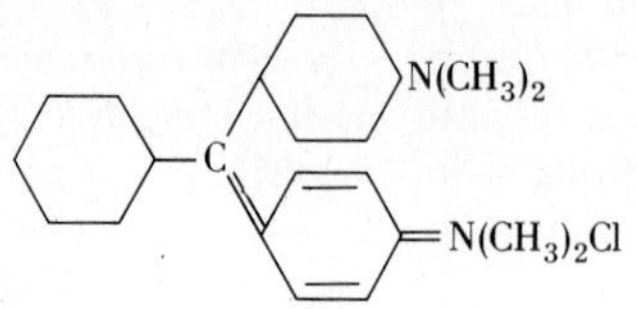

$N(CH_3)_2$

C

$=N(CH_3)_2Cl$

This dye is made by the condensation of benzaldehyde and 2 mol of dimethylaniline with the aid of heat and hydrochloric acid. This furnishes tetramethyldiaminotriphenylmethane, called the leuco base, which is then oxidized by lead peroxide to the *dye base,* the salt of which is the commercial dye (either oxalate or zinc double chloride).

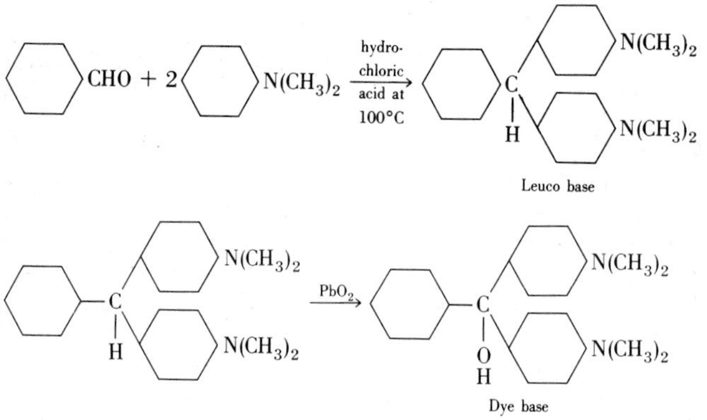

CHO + 2 $N(CH_3)_2$ —hydrochloric acid at 100°C→ C(H) $N(CH_3)_2$ $N(CH_3)_2$

Leuco base

C(H) $N(CH_3)_2$ $N(CH_3)_2$ —PbO_2→ C(OH) $N(CH_3)_2$ $N(CH_3)_2$

Dye base

Many triarylmethane dyes are carried through this same sequence.

Methyl violet, CI 42535, *basic violet 1,* is a basic dye; production is over 1,224,000 lb, for a wide variety of uses, e.g., on cotton, silk, and paper as a self-color, in calico and silk printing, in the dyeing of jute, coco fiber, wood, linen, straw, and leather, and in the manufacture of marking and stamping inks, carbon paper, copying pencils, and lacquers. This dye is the hydrochloride of pentamethyltriaminotriphenylcarbinol. It is made by the oxidation of dimethylaniline (3 mol), using cupric chloride in the presence of phenol. One methyl group goes to formaldehyde, which supplies the central carbon. The function of the phenol in this reaction is unknown. Methyl Violet has the formula

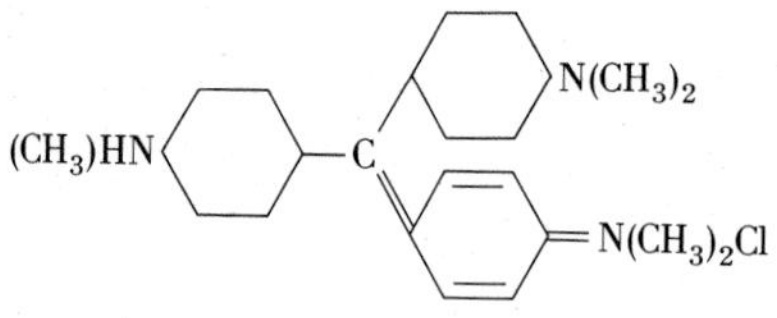

$N(CH_3)_2$

$(CH_3)HN$ C

$=N(CH_3)_2Cl$

XANTHENE DYES[47] These dyes usually are derivatives of xanthene.

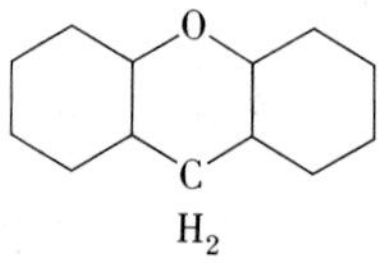

O

C

H_2

[47] Lubs, *op. cit.,* p. 291.

They are quite closely related to arylmethane dyes, and the oxygen bridge can be viewed as having been made by the elimination of water from two hydroxyl groups ortho to it. The presence of conjugation is characteristic of such chromophores as =C=O and =C=N—. One of these dyes is eosin, CI 45380, *acid red 87*, which is tetrabromofluorescein. Fluorescein is made by the condensation of resorcinol (2 mol) and phthalic anhydride in the presence of a dehydrating agent such as zinc chloride.

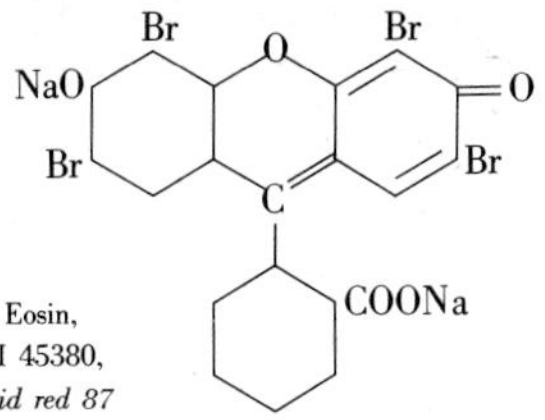

Eosin,
CI 45380,
acid red 87

Eosin is an acid dye for wool and silk and also for cotton with a tin or alum mordant. However, its chief application is in making lakes and for the preparation of ordinary red writing and stamping inks.

THIAZOLE DYES, OR PRIMULINE[48] DYES These contain sulfur in the ring:

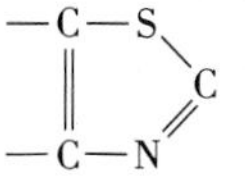

The chromophores are >C=N—, as well as—S—C≡, etc., but the arrangement in conjugated double bonds is of importance. These dyes are chiefly direct or developed dyes for cotton, though some of them are union dyes. Direct fast yellow, Cl 19555, *direct yellow 28*, is a direct color of excellent properties; it is also a union dye. It is made by sodium hypochlorite oxidation of the sodium salt of dehydrothio-*p*-toluidinesulfonic acid.

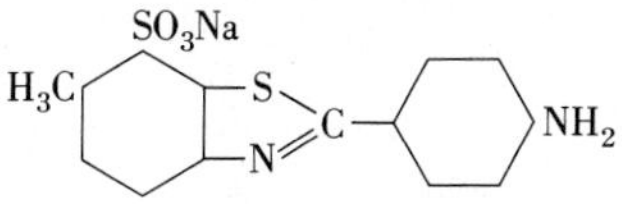

The latter is prepared by heating *p*-toluidine and sulfur, followed by sulfonation. Direct fast yellow has also an azo chromophore, as shown by the following structure.

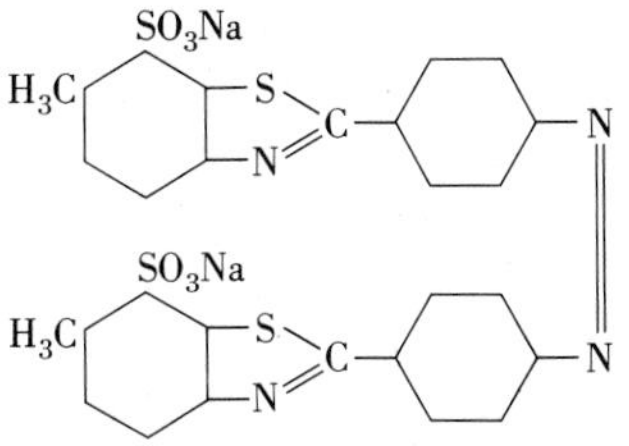

AZINE[49] DYES The azine dyes have as a mother substance phenazine:

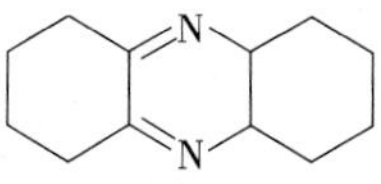

[48]Primuline is CI 49000, *direct yellow 59*.
[49]Nursten, ECT, 2d ed., vol. 2, pp. 859–868, 1963.

and we may assume as chromophores $>C{=}N{-}$; but here, as in many other of these complicated ring systems, resonance is being more and more looked upon as the basis of color. Indeed, in these simpler chromophores, like $>C{=}N{-}$, the structure contains a "fairly loosely bound chromophoric electron." Azine dyes are quite varied in their application. Nigrosines, CI50415, 50415 B, 50420, *solvent black 5, solvent black 7, acid black 2,* respectively, have large annual sales and are used after jetting with suitable yellow dyes, for blue-black polishes for leather shoes and other leather goods. Their ultimate constitution is unknown. Nigrosines are also employed for silk dyeing. They are good dyes at reasonable prices.

Although safranine, CI 50240, *basic red 2,* is not sold so largely as the nigrosines, its constitution is simple, and it may be formulated:

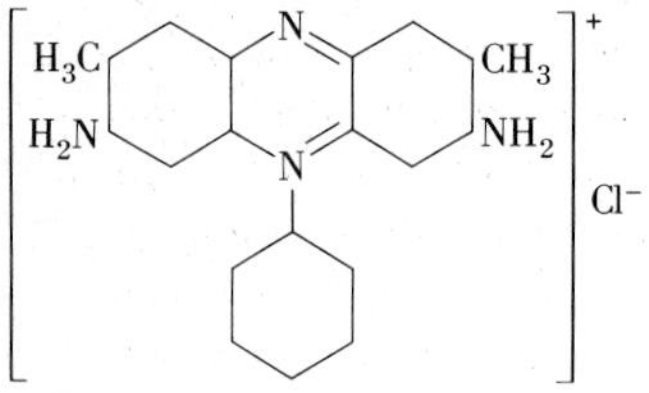

H_3C, CH_3, H_2N, NH_2, N, N, Cl^-, +

It is made by oxidation by CrO_3 in acetic acid of equimolecular proportions of *p*-tolylenediamine and *o*-toluidine to the indamine:

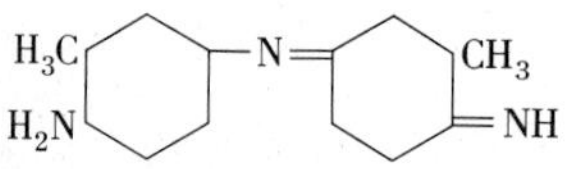

H_3C, N, CH_3, H_2N, NH

This compound is then condensed with aniline (or *o*-toluidine) in the presence of an oxidant. Safranine is a basic dye and is used largely on cotton mordanted with tannin and fixed with tartar emetic.

THIAZINE DYES These have the important member methylene blue, CI 52015, *basic blue 9,* which has the formula

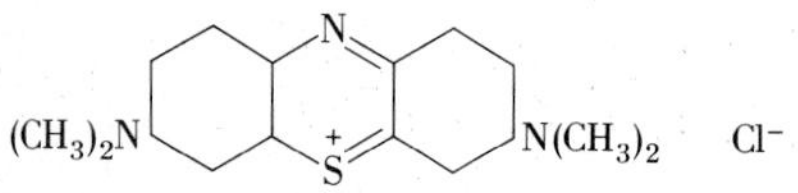

$(CH_3)_2N$, N, $N(CH_3)_2$, $\overset{+}{S}$, Cl^-

This is a basic dye and is made from dimethylaniline.

SULFUR, OR SULFIDE, DYES[50] These dyes are so complex that very little is known of their chemical structure. The chromophores present are $\equiv C{-}S{-}C\equiv$ and $\equiv C{-}S{-}S{-}C\equiv$ as bridges in complicated molecules.[51] These sulfur dyes are used to dye cotton from a sodium sulfide bath in dull but full shades of excellent fastness.

Sulfur black, CI 53185, *sulfur black 1,* etc., is widely sold, figuring on a dry basis, and is a very low-priced product. Sulfur black is made from either 2,4-dinitrochlorobenzene or 2,4-dinitrophenol by refluxing with a polysulfide solution for many hours, followed by precipitation of the dye by air oxidation. The demand for sulfur black was responsible for the first plant in America for producing chlorobenzene, installed about 1915. The time for thionation for some of these sulfur colors can be much shortened—from days to hours—by changing the solvent from alcohol or water to a monalkyl ether of diethylene glycol. Hydron blue, CI 53630, *vat blue 43,* is the product from the carbazole indophenol, using alcoholic polysulfide as the thionation agent.

[50] Lubs, *op. cit.,* pp. 302ff., 1059ff.; BIOS, 983 and 1155; *FIAT,* 1313, II, III.

[51] Empirical formulas are variously given as $C_{24}H_{16}N_6O_8S_7$ and $C_{24}H_{16}N_6O_8S_8$ for typical sulfur dyes.

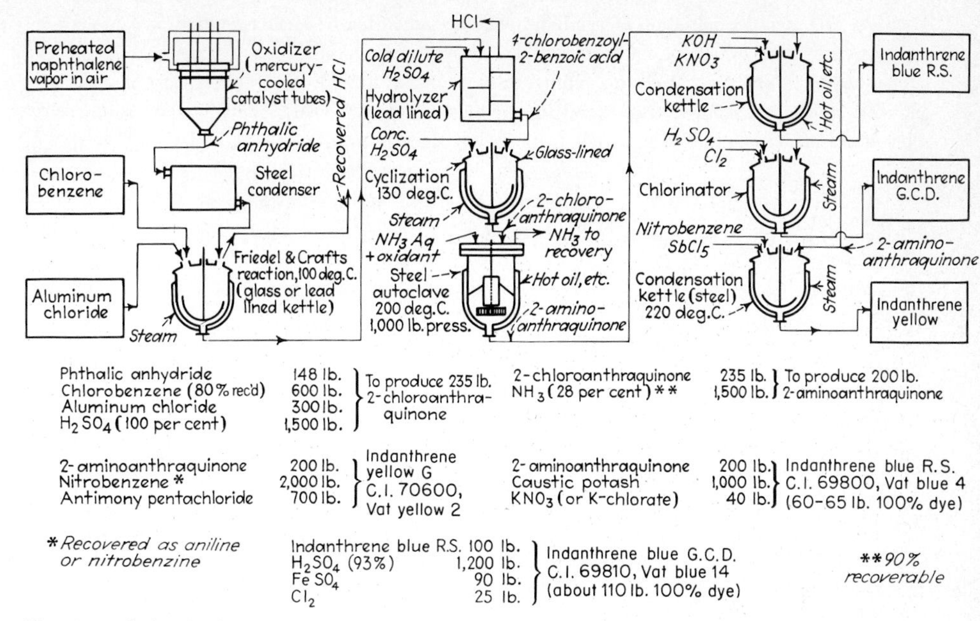

Fig. 39.10 Outline flowcharts for simple vat dyes—many steps are omitted.

ANTHRAQUINONE DYES Chromophores: =C=O and =C=C= arranged in an anthraquinone complex. The auxochromes are frequently (1) —OH, when the dyes fall into the mordant class, e.g., Alizarin, CI 58000, *Mordant Red 11*, or (2) $-SO_3H$, when we have, for example, the fast acid dye Acid Alizarin Blue B, which has the formula

OH
CO
SO_3Na
NaO_3S
CO
NH_2

and is manufactured by forming the 1,5-dihydroxyanthraquinone (anthrarufin), followed by nitration, sulfonation, and reduction.

ANTHRAQUINONE VAT DYES[52] Like the preceding class of dyes, these are also based on anthraquinone and have many of the chromophore and the fastness characteristics of that class, only accentuated in many instances. However, these dyes are capable of being reduced to an alkaline-soluble derivative called a *vat*, which oxidizes to the color (thus resembling the next class or indigo dyes). To make a commercial dye the reduced, or leuco, compound must have affinity for the fiber, and the insoluble dye must adhere firmly.[53] These conditions are met in so many of the members of this class that we have in the vat dyes a class of colors with remarkable properties of fastness and permanency. Their growth has been greatly stimulated, indeed made possible, by the American

[52]Cafrancesco, ECT, 2d ed., vol. 2, pp. 431–533, 1963; Lubs, *op. cit.*, p. 338; Venkataraman, *op. cit.*, vol. II.

[53]The basic chemical structure that gives to an anthraquinone or indigoid dye the vat-dyeing characteristics is usually the presence of two CO groups which on reduction give alkali-soluble C(OH) groups. The almost invariably used reducing agent is sodium hydrosulfite ($Na_2S_2O_4$) employed for solutions, or the formaldehyde compound formaldehyde sulfoxalate, employed for printing. The reaction may be expressed:

$$Na_2S_2O_4 + 2NaOH + H_2O = Na_2SO_4 + Na_2SO_3 + 2H_3 \quad \text{(on the dye)}$$

Lubs patented, in U.S. Pat. 2,164,930, the use of the hydrogen peroxide oxidation product of thiourea for the same purpose, in 1939.

discovery and commercialization of synthetic phthalic anhydride from air oxidation of naphthalene or of *o*-xylene, from which synthetic anthraquinone is manufactured. These vat dyes are employed largely on cotton and rayon and, to a limited extent, on silk; indeed, they furnish some of the finest and *fastest* colored fabrics. Figure 39.10 summarizes the flowcharts, depicting in outline the sequences from the main raw materials to the following three vat dyes:

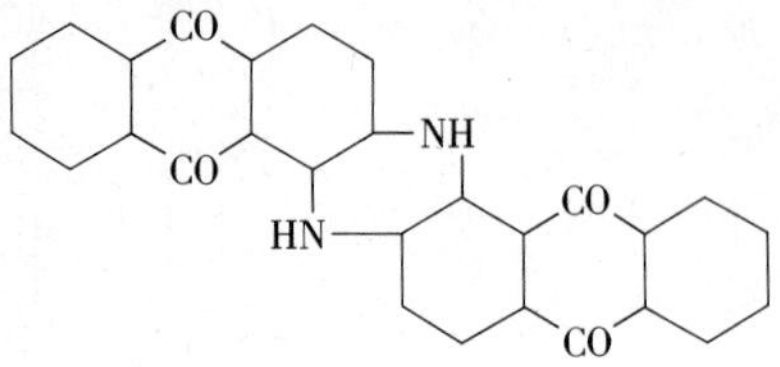

Indanthrene blue,
CI 69800, *vat blue 4*

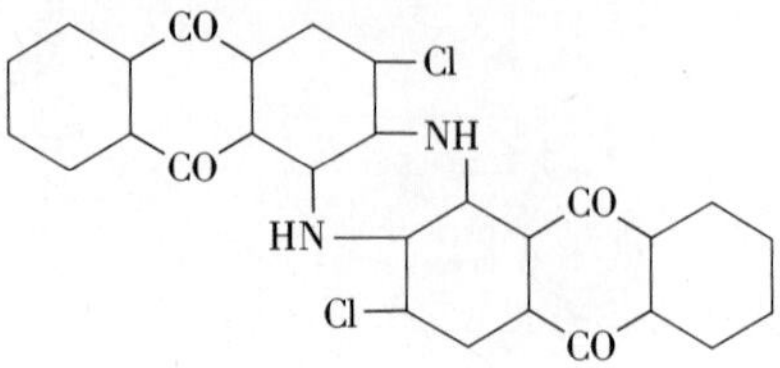

Indanthrene blue GCD,
CI 69810, *vat blue 14*

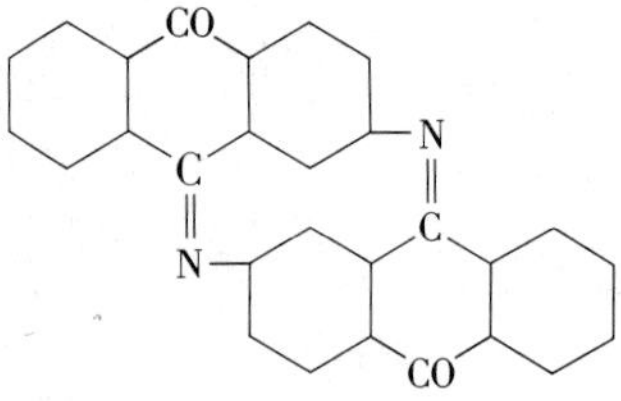

Flavanthrone,
CI 70000, *vat yellow 1*

These dyes all need β-aminoanthraquinone as an essential intermediate, proceeding by oxidizing condensation to the dyes. Indanthrene blue GCD is a chlorination product and has superior fastness properties because of this.

The anthraquinone vat dyes, because of their superior fastness, especially on cotton, are important. They are gradually supplanting dyes of less superior qualities; for example, in World War II, most of the khaki uniforms were vat-dyed and exhibited practically no fading against light and washing, in striking contrast to the bad fading of the sulfur-dyed cotton uniforms of World War I. In 1972 anthraquinone vats were produced in more than 47 million lb/year, representing about 18% of total dye production. The 18-million-lb plant of Ciba States, Ltd., at Toms River, N.J., with a normal producing capacity of 4 million lb of anthraquinone vat dyes, has been pictured and described.[54] The reference gives the fullest details ever published of any vat dye, including a diagramatic flowchart, material requirements, yields, reactions, and chemical engineering details for producing anthraquinone *vat brown 1,* CI 70800,[55] from 1,4-diaminoanthraquinone and 1-chloroanthraquinone through trianthrimide. Production is over 800,000 lb per year.

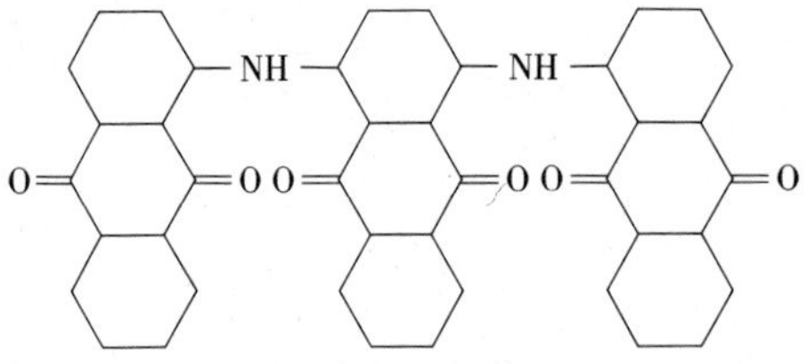

Vat Brown 1, CI 70800

[54]Bradley and Kronowett, Anthraquinone Vat Dyes, *Ind. Eng. Chem.*, **46,** 1146 (1954).
[55]Venkataraman, *op. cit.*, p. 906; see this reference for other such dyes.

Anthraquinone vat jade green, CI 59825, *vat green 1,* was produced in the largest amount in 1972 of any anthraquinone vat dye, this being 1.8 million lb of 6% paste at an average value of 79 cents per pound.

INDIGOID DYES The remarks that introduced the vat-dyes section pertain largely to this chemical class also, except that the members of this class are derivatives of indigo or thioindigo.

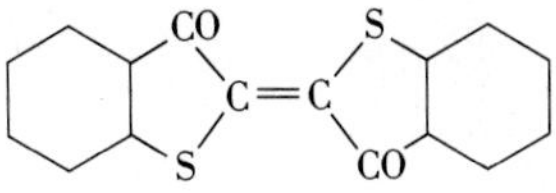

CO S C=C S CO

Thioindigo, CI 73300, *vat red 41*

These dyes are in the vat-use classification and are applied principally on cotton. A large tonnage of indigo is used each year, much of it now imported. Indigo was first a naturally cultivated dye, but German chemists worked out a profitable synthesis for it after spending over $6 million in R&D. However, this was a profitable venture for them, since their synthetic indigo is more uniform and much cheaper than that previously grown on 1.5 million acres, principally in India.

Indigo,[56] CI 73000, *vat blue 1,* itself is a water-insoluble blue dye of good properties, largely used on cotton and rayon, but sometimes on silk. Indigo produces full shades of navy blue on wool for uniform cloth of good fastness. When dyeing, it is reduced by hydrosulfite to make the alkali-soluble leuco derivative or to make indigo white.

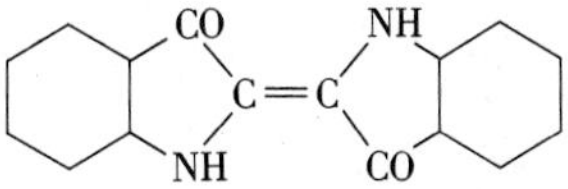

CO NH C=C NH CO $+ Na_2S_2O_4 + 6NaOH \longrightarrow$

Indigo, CI 73000, *vat blue 1*

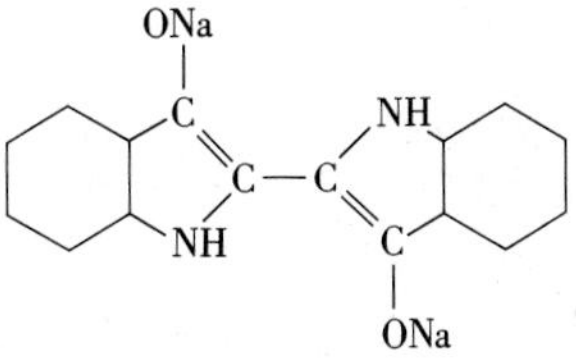

ONa C NH C—C NH C ONa $+ Na_2SO_4 + Na_2SO_3 + 3H_2O$

Indigo white

Figure 39.11 gives an outline flowchart for the manufacture of indigo by several sequences from aniline, which is the present commercial starting point for the intermediate product, phenylglycine. The following reactions summarize the chemical changes involved in this flowchart (see Fig. 39.11, also Fig. 39.8).

Indoxyl using sodium amide:

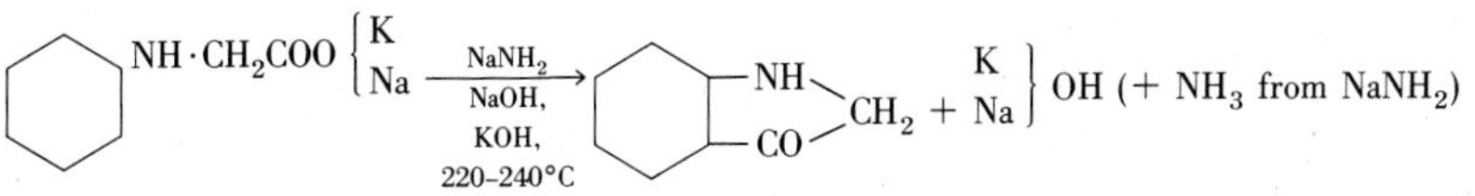

$$C_6H_5NH \cdot CH_2COO \begin{Bmatrix} K \\ Na \end{Bmatrix} \xrightarrow[\text{NaOH, KOH, 220–240°C}]{NaNH_2} C_6H_4\begin{matrix} NH \\ CO \end{matrix} CH_2 + \begin{matrix} K \\ Na \end{matrix} \Big\} OH \; (+ NH_3 \text{ from } NaNH_2)$$

Indoxyl using metallic sodium:

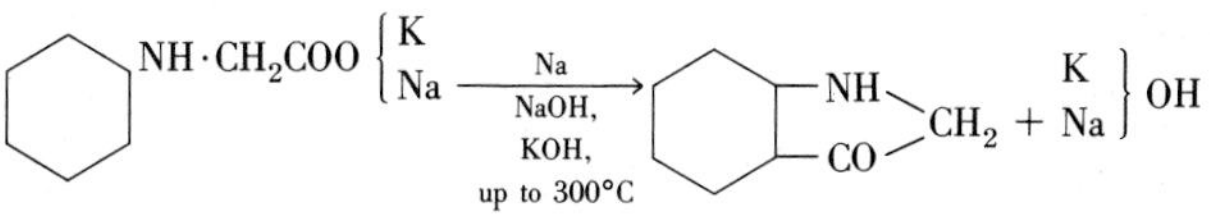

$$C_6H_5NH \cdot CH_2COO \begin{Bmatrix} K \\ Na \end{Bmatrix} \xrightarrow[\text{NaOH, KOH, up to 300°C}]{Na} C_6H_4\begin{matrix} NH \\ CO \end{matrix} CH_2 + \begin{matrix} K \\ Na \end{matrix} \Big\} OH$$

[56]Lubs, *op. cit.*, pp. 551ff.

Although apparently the sodium amide or the metallic sodium does not enter into the reaction, these two reagents and the mixed NaOH and KOH maintain the necessary fluxing and anhydrous condition for the dehydrating ring closure to proceed with good yields. The use of the *mixed* sodium-potassium salt of phenylglycine and of *mixed* NaOH and KOH gives lower fusion points, with consequent lower temperatures and enhanced yields.

Phenylglycine by chloroacetic acid:

$$C_6H_5NH_2 + ClCH_2COOH \longrightarrow C_6H_5NH\cdot CH_2\cdot COOH + HCl$$

Aniline, Chloroacetic acid, Phenylglycine

$$C_6H_5NHCH_2COOH + \left.\begin{matrix}K\\Na\end{matrix}\right\}OH \longrightarrow C_6H_5NHCH_2COO\left\{\begin{matrix}Na\\K\end{matrix}\right.$$

Phenylglycine, Mixed Na-K salt

Phenylglycine by formaldehyde and sodium cyanide:

$$H_2CO + NaHSO_3 \xrightarrow{50-80°C} H_2C(OH)SO_2\cdot ONa$$

Formaldehyde, Sodium bisulfite, Formaldehyde-bisulfite

$$C_6H_5NH_2 + HO\cdot CH_2\cdot SO_2\cdot ONa \xrightarrow{50-75°C} C_6H_5NH\cdot CH_2SO_3Na + H_2O$$

Aniline, Formaldehyde-bisulfite, Aniline-formaldehyde-bisulfite

$$C_6H_5NH\cdot CH_2SO_3Na + NaCN \xrightarrow{80°C} C_6H_5NH\cdot CH_2\cdot CN + Na_2SO_3$$

Aniline-formaldehyde-bisulfite, Sodium cyanide, Phenylgylcine-nitrile, Sodium sulfite

$$C_6H_5NH\cdot CH_2\cdot CN + 2H_2O + \left\{\begin{matrix}KOH\\NaOH\end{matrix}\right. \xrightarrow{70°C \text{ to boiling}} C_6H_5NH\cdot CH_2COO\left\{\begin{matrix}K\\Na\end{matrix}\right. + H_2O + NH_3$$

Phenylglycine-nitrile, Mixed Na-K salt of phenylgylcine

PHTHALOCYANINE[57] DYES AND PIGMENTS Figure 39.12 represents the structure of copper phthalocyanine. This group contains a chromophore of great complexity, or more likely a new arrangement of the simpler chromophores =C=N— and =C=C=. Here again, it may be that resonance will be shown to be the fundamental cause of the color. In copper phthalocyanine four isoindole units are linked by four nitrogen atoms and one copper atom to furnish a complicated ring structure, which, however, may be viewed as somewhat similar to the basic ring system of chlorophyll. This ring system and its derivatives show remarkable stability[58] to light, water, chemicals, and even heat. They sublime unchanged at 550°C. Because of this stability, their insolubility in water, and the intensity and beauty of their color, these phthalocyanines are of particular applicability in the pigment and paint fields.

[57]Lubs, *op. cit.*, chap. 9, p. 577; Moser and Thomas, Phthalocyanine Compounds, *J. Chem. Educ.*, **41**(5), 245 (1964), and Phthalocyanines, Reinhold, 1963; Dahlen, The Phthalocyanines, *Ind. Eng. Chem.*, **31,** 839 (1939) ((includes a bibliography of the work, gives patents, and describes the fundamental studies of R. P. Linstead of England, with 107 refs.); Venkataraman, *op. cit.*, vol. 5, 1971.

[58]Schwarz, The Phthalocyanines Are Faster Than Anything in Color Produced So Far, *Am. Dyes. Rep.*, **29,** 7 (1940).

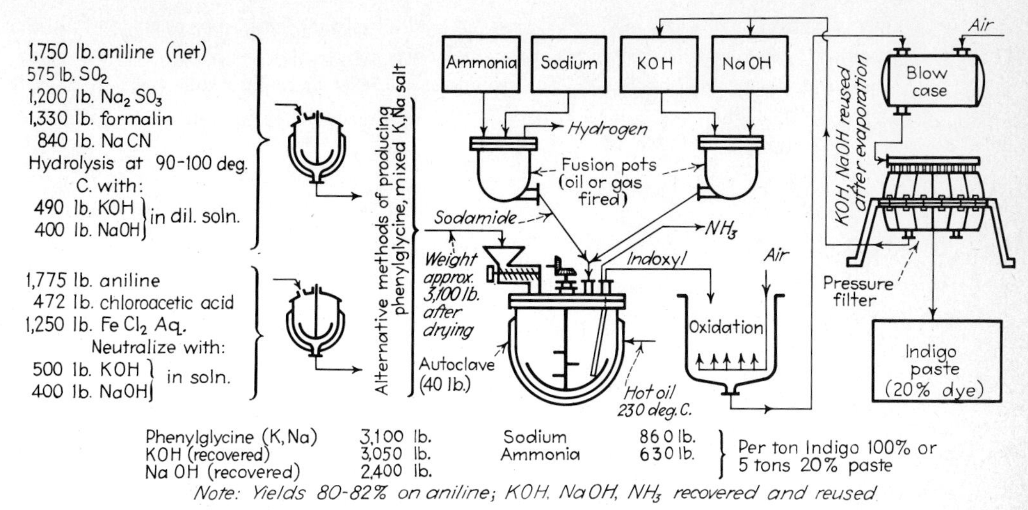

Fig. 39.11 Outline flowchart for the production of indigo.

Copper phthalocyanine is highly chlorinated and is CI 74260, *pigment green 7.* It has about 15 or 16 chlorine atoms per phthalocyanine unit. These dyes are only blue and green, but they supply a long-felt need for pigments of their excellent properties for printing inks, artists' colors, paints, lacquers, enamels, coated textiles, paper, linoleum, and rubber. Their manufacture may be summarized according to the following equation:

$$4\ C_6H_4(CN)_2 + Cu \xrightarrow{180\text{–}250°C} \text{structure as in Fig. 39.12}$$

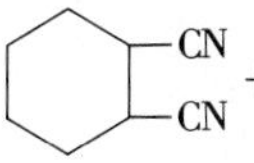

Phthalonitrile Copper Phthalocyanine (copper)

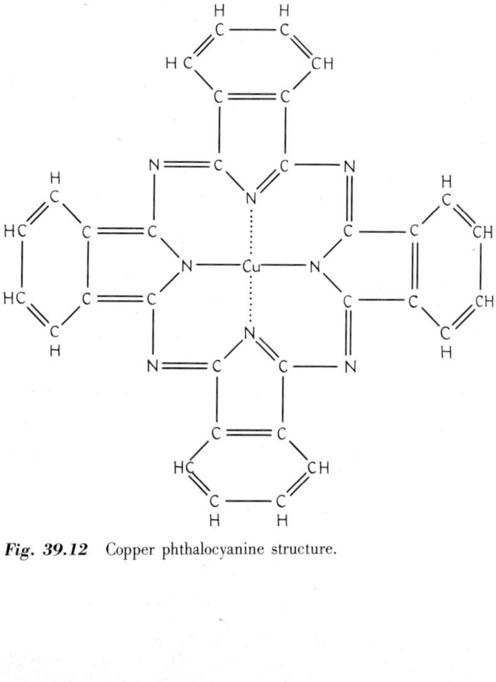

Fig. 39.12 Copper phthalocyanine structure.

CYANINE DYES are a large and varied class not used for textile dyeing. As a group, they cause sensitization of silver halide to light in regions where there is normal insensitivity. They are therefore most useful in photographic processes. The cyanine structure is typical of a large class:

```
 A⌒C—(CH=CH)n—C⌒B
 (+//          \\ )
  N             N
  |             |
  R             R'
```

NATURAL ORGANIC DYES For centuries these have been of the utmost importance. *Tyrian purple* of the ancients is well known. It was extracted from mollusks growing on the shores of the Mediterranean. It is, however, a dye, 6,6′-dibromoindigo, of the following formulation:

```
            CO        NH
     ⬡       \       /      ⬡    Br
              C  =  C
 Br          /       \
            NH        CO
```

Tyrian purple, CI 75800

It could be made synthetically, but we have other and better products to take its place. This replacement of natural, or as they have been called, *vegetable,* dyes by duplicates synthetically made in factories or by better and more cheaply made products is quite the usual thing. However, Logwood, CI 75290, *natural black 1,* in the form of chips and extract of the wood of a tree growing in Central America, has survived competition and is used extensively to give black and blue-black colors on wool and silk, mordanted with chromium or other metals. To a lesser extent fustic, cutch, and cochineal are also employed.

INORGANIC COLORING MATTERS A great many inorganic chemicals are used as pigments. These were considered under paints (Chap. 24). However, khaki shades are prepared on ducks, drills, and canvas, particularly for tents, cots, and such out-of-door use, by soaking the cotton cloth in a solution of ferrous sulfate or ferrous acetate. The iron is converted into the hydroxide, either by an alkali or by drying, and is then oxidized by a bleaching-powder solution to the darker ferric state.

SELECTED REFERENCES

American Association of Textile Chemists and Colorists, Technical Manual and Year Book, Howes Publishing Co. (annual).
Billmeyer, F. W., and M. Saltzman: Color Technology, Wiley, 1966.
Blackshaw, H., and R. Brightman: Dictionary of Dyeing and Textile Printing, Interscience, 1961.
Cockett, S. R., and K. A. Hilton: Dyeing of Cellulosic Fibres, Academic, 1961.
Colour Index, 2d ed., Society of Dyers and Colourists, and American Association of Textile Chemists and Colorists, 1956–1965, 5 vols.
Gore, T. S., *et al.* (eds.): Chemistry of Natural and Synthetic Colouring Matters, Academic, 1962.
Gould, R. F. (ed.): Literature of Chemical Technology, chap. 17, Dyes, ACS Monograph, 1967.
Gurr, E.: Synthetic Dyes in Biology, Medicine and Chemicals, Academic, 1971.
Hamer, F. M.: Cyanine Dyes and Related Compounds, Interscience, 1964.
James, R. W.: Printing and Dyeing of Fabrics and Plastics, Noyes, 1974.
Johnson, J. C.: Hair Dyes, Noyes, 1973.
Johnston, R. M., and M. Saltzman (eds.): Industrial Color Technology, ACS, 1972.
Judd, D. B., and G. Wyszecki, Color in Business Science and Industry, Wiley, 1963.
Kirk and Othmer, Encyclopedia of Chemical Technology, 2 ed., vol. 7, pp. 463–629, 1965.
Mackinney, G., and A. Little: Color of Foods, Avi Publishing, 1962.
McGregor, R.: Diffusion and Sorption in Fibers and Films, vol. 1, An Introduction with Particular Reference to Dyes, Academic, 1974.
Moser, F. H., and A. L. Thomas: Phthalocyanine Compounds, ACS Monograph 157, Reinhold, 1963.
Olin, A. D.: Sources in Searching the Literature of Synthetic Dyes, Tom's River Chemical Corp., 1963 (historical, with chart).
Rattee, I. D., and M. Breuer: The Physical Chemistry of Dye Adsorption, Academic, 1974.
Schmidlin, H. U.: Preparation and Dyeing of Synthetic Fibres, Reinhold, 1963.
Tooley, P.: Fuels, Explosives and Dyestuffs, J. Murray, London, 1971.
Venkataraman, K. (ed.): The Chemistry of Synthetic Dyes, vol. 7, Academic, 1974 (other volumes dating back to 1952).
Zollinger, H.: Azo and Diazo Chemistry: Aliphatic and Aromatic Compounds, Interscience, 1961.

chapter 40

PHARMACEUTICAL INDUSTRY

Parallel to the progress of medicine has been the development of the pharmaceutical industry. In the past two decades, emphasis has been placed on medicinal, chemical, biological, and pharmacological research to such an extent that this era is recognized, more than any other comparable period in history, for the new and effective drugs[1] that have come from scientific laboratory investigators, into the hands of physicians, who prescribe them to improve the health of the people. Their manufacture has been made possible by an industry whose highly skilled chemists, pharmacists, and chemical engineers are being challenged to discover more and better products for the benefit of humankind. The human life span has increased since 1900 from a life expectancy of 49 years to the present 70 years. In addition to prescription drugs (Table 40.1), there is an almost equal amount of nonprescription drugs, such as aspirin, antiseptics, cold remedies, and the like, that are sold by drugstores (Table 40.2), including proprietary and patent medicines.

HISTORICAL The use of drugs to relieve pain and to ward off death is interwoven with the ancient superstition that evil spirits cause disease. The healing powers of mythological personages, particularly of Aesculapius, son of Apollo, were sought in primitive cultures. The Papyrus Ebers, which takes us back to the beginning of recorded history in the Nile Valley, contains drug formulas with as many as 35 ingredients, including botanicals, minerals, and animal products. A few of the minerals, such as sulfur, magnesia, and soda, still appear in current pharmacopeias. It was the Greeks, however, who through Hippocrates and Galen, made an effort to approach therapy rationally rather than mystically. Paracelsus, born in 1493, who experimented both in the laboratory and the clinic, may be looked upon as the founder of chemotherapy.

Three centuries later, while Liebig and his students in Germany were synthesizing biologically active compounds, methods for experimental medicine were developed in France by Bernard, Magendie, and others. Although the American pharmaceutical industry had made a modest beginning in 1786, the synthetic organic chemicals ether and chloroform were not used for anesthesia until the 1840s. Three years after the end of the Civil War, the first integrated industrial synthetic organic manufacturing operation was established in the United States.[2] In the meantime, Pasteur, in studying the fermentation of beet sugar to alcohol, formulated the germ theory of disease and developed techniques for bacteriological study. The groundwork for modern pharmaceutical research was begun in 1881 with the establishment of a scientific division of Eli Lilly and Co. The shortage of important drugs, such as Veronal and novocaine, caused by the entry of the United States into World

[1]The term "drugs" as used in this chapter refers to "articles intended for use in the diagnosis, cure, mitigation, treatment, or prevention of disease in man or other animals"; Martin (ed.), Remington's Pharmaceutical Sciences, vol. 13, 1965, hereinafter cited as RPS XIII, and vol. 14, 1970, hereinafter cited as RPS XIV.

[2]Albany Aniline Chemical Co. For this and other historical facts, see Tainter and Marcelli, The Rise of Synthetic Drugs in the American Pharmaceutical Industry, *N.Y. Acad. Med.*, **35**, 387–405 (1959).

TABLE 40.1 ***Top 50 Most Prescribed Drugs in the United States in 1973***

Trademark	*Therapeutic class*	*Marketed by*	*Rank*
Valium	Ataractic	Hoffmann-La Roche	1
Librium	Ataractic	Hoffmann-La Roche	2
Darvon compound-65	Analgesic	Lilly	3
Tetracycline HCl	Antibiotic	Unspecified	4
Premarin	Hormone	Ayerst	5
Empirin compound/codeine	Analgesic, antipyretic	Burroughs Wellcome	6
Ampicillin	Antibiotic	Unspecified	7
Lasix	Diuretic	Hoechst	8
Ovral	Contraceptive	Wyeth	9
V-Cillin K	Antibiotic	Lilly	10
Indocin	Analgesic, antiinflamatory, antipyretic	Merck, Sharpe & Dohme	11
Donnatal	Antispasmodic	Robins	12
HydroDiuril	Diuretic	Merck, Sharpe & Dohme	13
Dimetapp	Antihistamine	Robins	14
Actifed	Antihistamine	Burroughs Wellcome	15
Lanoxin	Digitalizer	Burroughs Wellcome	16
Benadryl	Antihistamine	Parke-Davis	17
Phenobarbital	Barbiturate	Unspecified	18
Erythrocin	Antibiotic	Abbott	19
Ilosone	Antibiotic	Dista	20
Aldomet	Antihypertensive	Merck, Sharpe & Dohme	21
Achromycin-V	Antibiotic	Lederle	22
Diuril	Diuretic	Merck, Sharpe & Dohme	23
Sumycin	Antibiotic	Squibb	24
Lomotil	Sedative	Searle	25
Ovulen 21	Contraceptive	Searle	26
Dyazide	Diuretic, antihypertensive	Smith, Kline & French	27
Butazolidin Alka	Analgesic	Geigy	28
Thyroid	Hormone	Unspecified	29
Fiorinal	Analgesic	Sandoz	30
Librax	Ataractic	Hoffman-La Roche	31
Ornade	Antihistamine	Smith, Kline & French	32
Thorazine	Ataractic	Smith, Kline & French	33
Elavil	Antidepressant	Merck, Sharpe & Dohme	34
Dilantin sodium	Anticonvulsant	Parke-Davis	35
Mellaril	Ataractic	Sandoz	36
Orinase	Antidiabetic	Upjohn	37
Phenaphen/codeine	Analgesic	Robins	38
Triavil	Antidepressant	Merck, Sharpe & Dohme	39
Prednisone	Adrenocortical steroid	Unspecified	40
Ortho-Novum 1/50-21	Contraceptive	Ortho	41
Chlor-Trimeton	Expectorant	Schering	42
Gantrisin	Antibacterial (urinary)	Hoffmann-La Roche	43
Dalmane	Sleep inducer (hypnotic)	Hoffmann-La Roche	44
Antivert	Antihistamine	Roerig	45
Ortho-Novum 1/80-21	Contraceptive	Ortho	46
Darvon	Analgesic	Lilly	47
Digoxin	Digitalizer	Unspecified	48
Cleocin	Antibiotic	Upjohn	49
Aldoril	Antihypertensive	Merck, Sharpe & Dohme	50

Source: Pharm. Times, April 1974, p. 38.

TABLE 40.2 ***U.S. Sales by Therapeutic Class*** *(In millions of dollars)*

	1962		1970	
Pharmaceutical preparations for	*Sales*	*Percent*	*Sales*	*Percent*
Central nervous system and sense organ diseases	444.5	22.9	1,053.7	27.9
Parasitic and infectious diseases	389.7	20.1	663.4	17.5
Digestive or genitourinary system diseases	267.2	13.8	478.0	12.6
Neoplastic, endocrine, and metabolic diseases	242.2	12.5	501.5	13.3
Vitamins, nutrients, and hematinics	201.7	10.4	327.7	8.7
Respiratory system diseases	144.6	7.4	247.9	6.6
Cardiovascular system diseases	125.0	6.4	305.7	8.0
Skin diseases	55.4	2.9	116.8	3.1
Veterinary	53.4	2.7	85.7	2.3
Others	17.5	0.9	0.8	....
Total	1,941.2	100.0	3,781.2	100.0

Source: Prescription Drug Industry Fact Book, PMA, 1973.

War I, precipitated expansion of the pharmaceutical industry into a successful effort to produce the synthetic chemicals needed.

Developments of insulin, liver extract, and the short-acting barbiturates were milestones of the next decade. Sulfa drugs and vitamins were rapidly added to many product lines during the 1930s. Blood plasma, new antimalarials, and the dramatic development of penicillin resulted from the demands of war. But the spectacular surge of new products, which included steroid hormones, tranquilizers, vaccines, and broad- and medium-spectrum antibiotics, came after World War II. Research has now been stepped up to find more efficient drugs for the unconquered maladies arthritis, cancer, heart conditions, high blood pressure, hepatitis, and mental and emotional diseases, among others.

USES AND ECONOMICS Medicinal products are made available to the public in conformity with federal and state drug laws through prescription by a licensed practitioner of one of the health professions—a physician, a dentist, or a veterinarian—and dispensed by pharmacists. Some nonprescription drugs are sold directly to the consumer. Total consumer expenditures for all drugs rose each decade from $1,466 million in 1948 to $3,261 million in 1958 to $5,873 million in 1968. The $7.183 million expenditures in 1971 are expected to increase to $11,343 million in 1980. The total U.S. production of bulk synthetic organic medicinals increased from 144 million lb in 1964 to 234 million lb in 1972. Sales for 1964 were $643 million for bulk products, while in 1972 they totaled $490 million. These figures show an increase in production of 90 million lb but a net worth decrease in sales value of $153 million. These figures do not include finished preparations, such as tablets, capsules, ampules, and the like, which manufacturing pharmacists make from bulk chemicals, antibiotics, and isolates. The phenomenal increase in both sales and production is based on the fact that more than 85% of the products sold in 1965 did not even exist in the early 1930s; furthermore, of the 200 leading medicines in 1970, 159 were not on the market in 1955. The rise of the so-called wonder drugs—sulfas, antibiotics, and certain biological products such as polio vaccine—accounts for much of this meteoric increase in volume. The death rate per 100,000 population for influenza and pneumonia dropped from 102.5 in 1930 to 29.4 in 1972, and for all forms of tuberculosis from 71.1 to 2.2 during the same period. Also dramatic was the 100% decline in both poliomyelitis and whooping cough from 1958 to 1968, and in 1972 the death rate was less than one in a million population.

Economically, the largest class of therapeutic drugs is comprised of those affecting the central nervous system and the sense organs; they accounted for 27% of all U.S. sales in 1976. This group and pharmaceuticals for neoplasms captured significant sales gains between 1962 and 1976, while those for parasitic and infective diseases and vitamin preparation experienced something of a decline. In 1976, sales in the United States for veterinary preparations accounted for 2.6% of the total.

The beneficient influence of the pharmaceutical industry on animals for human food, which is often not recognized, is illustrated by this example:

Until the early 1950s, Americans considered a chicken dinner a special occasion. Total production was nearly 500 million/year—about 8.7 lb per capita. Disease was poorly controlled and losses ran up to 30% for poultry growers. Now, 3 billion birds are grown annually, with a disease loss of about 1%, and per capita consumption up to more than 37 lb. Additionally, a 3.6-lb broiler can be produced with about 8 lb of feed—half as much feed and in 40% less time than in 1950. This feed generally contains vitamins A, B_{12}, D, E, and K_1, thiamine, riboflavin, pantothenic, niacin, pyridoxine, biotin, choline, antibiotics, and medicinals. The use of medicinal products has cut down losses by protecting against blackhead, coccidiosis, infectious bronchitis, enteritis, Marek's disease, Newcastle disease, chronic respiratory diseases, and other infections.

The pharmaceutical industry employs more than 147,000 people in 1,265 firms in the United States, thus emphasizing the fact that many of these are quite small. Each of approximately 35 firms employs 1,000 or more persons globally; approximately 17 employ over 5,000; and only 6 employ over 15,000 or more. The prescription pharmaceutical industry is not dominated by any one firm. In 1976, the largest firm's share of the United States ethical market was only 7%. The 10 leading firms accounted for 51% of the total.

RESEARCH AND DEVELOPMENT Members of the Pharmaceutical Manufacturer's Association (PMA), composed of 115 firms, who account for approximately 95% of the prescription products sold in the United States, reported expenditures of over $800 million for R&D in 1973, 16 times the amount expended in 1951, up 10.8% from 1972. Over $1 billion is predicted for 1980. The PMA reported that research scientists and supporting staff employees during 1971 numbered 21,725, or 54% over 1963. About 37% of the drug industry's R&D scientists hold doctorates, and research expenditures per scientist amounted to $31,025 in 1971. Six to ten years may be required for the development and final approval of a drug for marketing. The cost of discovery and development may exceed $7 million. Of these large sums, over 98% was supplied by private industry, less than 2% coming from tax money. Research expenditures are a primary factor in determining a company's future share of industry sales and reflect the service rendered to the people. Between 1959 and 1962, a drive was made in Congress to place all aspects of the pharmaceutical industry under further control. The proposal to limit the duration of patents did not receive widespread support because of

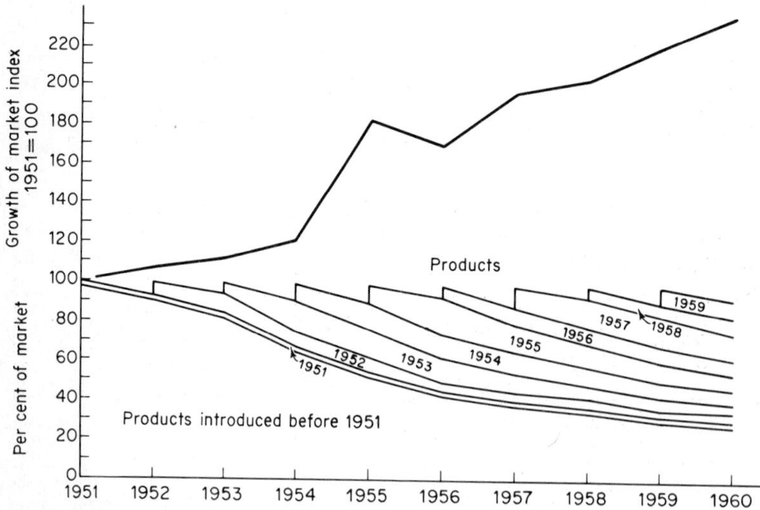

Fig. 40.1 Dynamic competition. New products of research compete with the old. This figure also indicates the rapid obsolescence of many products. (*33 PMA firms.*)

its implications for industry as a whole, as well as general public recognition that patents have been the fountainhead of progress and superiority. In the end, a bill was passed lacking the patent change which would have throttled research initiative but which provided for additional government control; this caused a marked slowdown in the introduction of new drugs, 64 products in 1972, in contrast to 199 in 1963.

BASIC RAW MATERIALS The pharmaceutical industry draws upon worldwide sources for its raw materials. Its scientists have noted and applied the healing observations of primitive people wherever found. In recent decades, large sums have been spent in worldwide exploration for new drugs, but even larger amounts for laboratory research to develop synthetics, to improve on natural drugs, and to discover new drugs with improved pharmacological properties having at the same time few or no deleterious side effects. The industry's fabrication processes involve extraction, purification, chemical synthesis, fermentation procedures, and pharmaceutical manufacturing to prepare bulk drugs for convenient use.

MANUFACTURING PHARMACY Pure bulk chemicals, or crude drugs, usually must be prepared for medical and veterinary use by the special skill of the present-day pharmaceutical manufacturing establishment into liquid preparations, extracts, tablets, capsules, pills, powders, sterile solutions for injection, and medicated aerosols. These convenient products, packages, and medicaments ensure purity, proper dosage, and adequate *sterility* until administration. Standardization of dosage by automation in comparison with hand weighing or measuring of powders and liquids saves money for users. "Identicode" (Lilly) is a safety marking device for indicating a drug with a specific formula.

CLASSIFICATION

Like most chemicals, pharmaceutical products can be arranged by use, by chemical structure, or by chemical reactions needed to manufacture the product. For the pharmaceutical industry, products are listed in the following tables:

Table 40.1 Top 50 Ethical Pharmaceutical Products,

Table 40.2 Sales of Finished Ethical Products.

Table 40.3 Selected Pharmaceuticals Dually Arranged under Therapeutic and Chemical Conversion Classifications.

These tables give an overall listing of many of the most useful pharmaceuticals. Remington (RPs XIV) is helpful in presenting important drugs arranged by therapeutic class. In presenting the *manufacture* of individual products, however, a chemical conversion classification is in accordance with the arrangement of the rest of this book and is consequently followed hereafter.

MANUFACTURE OF PHARMACEUTICAL PRODUCTS CLASSIFIED BY CHEMICAL CONVERSION PROCESSES

The pharmaceutical industry employs greater variety and more complicated steps in its manufacturing processes than almost any other of the main categories (or chapters) into which the overall chemical process industries are divided. The complexity of the chemical structure of many medicaments probably has a relationship to the even greater complexity of the ailments of human and animal bodies which the products of the pharmaceutical industry are designed to ameliorate. A foundation for this is the necessity of the long-prevailing large percentage of the profits and of sales in the pharmaceutical industry devoted to R & D (q.v.). Likewise, because of the accompanying complexity in manufacturing processes, it has been difficult to select the important chemical-conversion step under which to classify the synthesis of an individual chemical.

TABLE 40.3 ***Selected Pharmaceuticals Dually Arranged under Therapeutic and Chemical Conversion Classifications****

Therapeutic class	*Chemical conversion*	*Therapeutic class*	*Chemical conversion*
1. Analgesics and antipyritics		Erythromycin USP, Ilotycin	Fermentation
Acetanilid NF	Esterification	Oxytetracycline NF, Terramycin	Fermentation
Acetophenetidine USP, phenacetin	Esterification	Penicillin salts USP	Fermentation
Acetylsalicylic USP, aspirin	Esterification	Streptomycin salts USP	Fermentation
Aminopyrine NF, Pyramidon	Reduction, methylation	Tetracycline USP, acromycin	Dechlorination, chlortetracycline
Codeine USP	Alkylation, isolate	5. Antihistamines	
Dextro propoxyphene HCl, Darvon	Complex, esterification	Diphenylhydramine HCl, Benadryl	Alkylation
Dihydrocodeineone NF	Rearrangement (catalytic)	6. Cardiovascular drugs	
Meperidine HCl USP, Demerol	Complex, benzyl chloride	Amyl nitrate USP	Esterification
Morphine USP	Isolate	Digitalis USP	Plant: *Digitalis purpurea*
Quinine sulfate NF	Isolate	Glyceryl trinitrate USP	Esterification
Sodium salicylate USP	Carboxylation	7. Central nervous system stimulants	
2. Anesthetics		Caffeine USP	Alkylation, isolate
Chloroform USP	Halogenation	Theobromine NF	Alkylation, isolate
Cocaine HCl USP	Isolation, esterification	Theophyllin NF	Alkylation
Cyclopropane USP	Ring closure (Zn)	8. Dermatologicals	
Ether USP	Dehydration	Camphor USP	Isolate
Nitrous oxide USP	Dehydration	Menthol USP	Isolate
Procaine HCl USP, Novocaine	Alkylation	Salicylic acid USP	Carboxylation
Thiopental Na USP, Pentothal, Na	Condensation, thiourea	Sodium borate USP	Extraction
3. Antibacterials (antiseptics)		Zinc oxide USP	Oxidation
Hexylresorcinol USP	Condensation	9. Diuretics	
Hydrogen peroxide soln. USP	See Chap. 20	Chlorothiazide USP, Diuril	Complex
Isoniazid USP	Oxidation	Hydrochlorothiazide, Hydrodiuril	Hydrogenation
Methanamine NF	See Chap. 34, plastics	10. Gastrointestinal drugs	
Thimerosal NF, Merthiolate	Ethyl-mercuration thiosalicylic	Antacids	
4. Antibiotics (Antimicrobial)		Cascara sagrada USP	Dried bark
Bacitracin USP	Fermentation	Castor oil USP	See Chap. 28
Chloramphenicol USP, Chloromycetin	Complex, condensation	Phenolphthalein NF	Condensation
Chlortetracycline HCl NF, Aureomycin	Fermentation	11. Hormones	
Dihydrostreptomycin salts USP	Fermentation	Cortisone acetate USP	Complex
		Insulin USP, Iletin	Isolate
		12. Respiratory agents	
		Codeine NF	Alkylation, isolate
		13. Sedatives and hypnotics	
		Amobarbital USP, Amytal	Esterification-alkylation-condensation

TABLE 40.3 ***Selected Pharmaceuticals Dually Arranged under Therapeutic and Chemical Conversion Classifications**** *(Continued)*

Therapeutic class	*Chemical conversion*
Barbital NF	
Pentobarbital sodium USP	
Phenobarbital USP, Luminal	
14. Sulfonamides	
Sulfadiazine USP	Sulfonation-condensation
Sulfaguanidine NF	
Sulfanilamide NF	
Succinylsulfathiazole NF	
Sulfasuxidine	
Sulfathiazole NF	
15. Tranquilizers	
Chloropromazine USP, Thorazine	Condensation
Hydroxyzine NF, Aterox	Condensation
Meprobamate, Equanil, Miltown	Condensation
Chlordiazepoxide Librium	Complex
Prochlorperazine maleate USP, Compazine	Condensation
Reserpine USP	Isolation
16. Vitamins and nutrients	
Ascorbic acid salts USP	Complex: lactonization
Nicotinic acid USP, niacin	Oxidation
Riboflavin USP	Condensation, fermentation
Thiamin hydrochloride USP	Condensation
17. Pharmaceutical necessities	
Calcium cyclamate USP, Sucaryl	See Chap. 27, flavors
Dextrose USP	See Chap. 30, sugars
Phenol USP	See Chap. 34, plastics
Saccharine USP	See Chap. 27, flavors
18. All other	
Amphetamine sulf. USP, Benzedrine	Phenylacetone, formamide
Arsphenamine Salvarsan	Reduction phenylarsonic derivative
Bishydroxycoumarin USP, Dicumerol	Complex from salicylic
Ephedrine NF	Isolate ma huang
Epinephrine USP, Adrenalin	Isolate
Piperazine citrate and derivatives USP	Cyclization
Plasmochin	Condensation
Strychnine	Extraction
Tolbutamide USP, Orinase	Complex, condensation
Vaccines	Immunizing biologicals

*For further information, refer to text under the various chemical conversions, to RPS XIII; Burger, Medicinal Chemistry, Interscience, 1960; RPS XIV.

In the pharmaceutical industry greater emphasis, as a rule, is placed upon the purity of the products than in any other industry, except in certain cases in the nuclear industry (Chap. 21). Another characteristic of the pharmaceutical industry is the use of life processes as a step in the manufacture of some products which have assumed increasing importance in this area—antibiotics, biologicals, hormones, and vitamins, products of fermentation. Many of these pharmaceuticals are classified and presented under fermentation processes in this chapter, but the overall treatment of fermentation is found in Chap. 31. In the early centuries, a large proportion of the products used as medicaments either for humans or animals were natural products, mostly derived from plants. It was recognized that such products must be pure and, in the early decades of the modern pharmaceutical industry, much effort was devoted to separating and purifying the individual products extracted from plants or animals. In recent decades, extractive methods have been improved by the use of new procedures and new equipment, such as the Podbielniak contactor (Fig. 40.9), which allows countercurrent extraction to be performed efficiently and rapidly. This is essential for materials such as penicillin which are sensitive to time, temperature, and acidity. Important also is the use of specialized solvents like anhydrous ammonia, various alcohols, esters, and hydrocarbons. Ion-exchange resins facilitate the

TABLE 40.4 ***Germicidal Activity of Certain Disinfectants in the Presence of Organic Matter***

	Eberthella typhosa				*Staphylococcus aureus*			
	No organic matter added		*With 10% of horse blood*		*No organic matter added*		*With 10% of horse blood*	
Disinfectant	*Effective dilution*	*Phenol coef-ficient*	*Effective dilution*	*Phenol coef-ficient*	*Effective dilution*	*Phenol coef-ficient*	*Effective dilution*	*Phenol coef-ficient*
Phenol	1:90	1.0	1:80	0.9	1:60	1.0	1:50	0.8
Cresol compound USP	1:180	2.0	1:90	1.0	1:90	1.5	1:50	0.8
Cresylic disinfectant (stated phenol coefficient 5)[a]	1:500	5.5	1:150	1.6	1:250	4.1	1:70	1.1
Fortified cresylic disinfectant (stated phenol coefficient 5)[b]	1:500	5.5	1:150	1.6	1:250	4.1	1:70	1.1
Orthophenylphenol 15% disinfectant (Saponaceous)	1:600	6.6	1:120	1.3	1:200	3.3	1:20	0.3
Synthetic phenolic disinfectant (stated phenol coefficient 10)[c]	1:950	10.5	1:150	1.5	1:300	5.0	1:35	0.5
4-Chlorophenylphenol 8% disinfectant[d]	1:500	5.5	1:80	0.9	1:700	11.6	1:12	0.2
Pine oil disinfectant (stated phenol coefficient 4)[e]	1:225	2.5	1:250	2.7	1:20	0.3	1:20	0.3
Chlorine disinfectant I[f]	1:130	1.4	0	0	1:110	1.8	0	0
Chlorine disinfectant II[g]	1:600	6.6	1:3	0.03	1:400	6.6	1:1	0.01
Quaternary ammonium disinfectant I[h]	1:12,000	133	1:70	0.8	1:20,000	333	1:3,500	58
Quaternary ammonium disinfectant II[i]	1:27,000	300	1:3,000	33	1:10,000	166	1:4,500	75

Source: Soap Sanit. Chem., February 1944; more details and more extensive tables. Nature of tabulated disinfectant: [a]51% cresylic acid solubilized by soap; [b]26% cresylic acid, 9% *o*-phenylphenol solubilized by soap; [c]10% *o*-phenylphenol, 4% *p*-tert-amylphenol solubilized by soap; [d]8% solubilized by soap; [e]76% pine oil solubilized by soap; [f]0.48% average chlorine; [g]5.25% average chlorine; [h]hydroxy-decenyldimethylethyl ammonium bromide; [i]benzalkonium chloride, aqua.

extraction and purification of individual products, for instance, certain antibiotics and alkaloids. Sterility and specialized packaging are necessary for many pharmaceuticals.

On the other hand, as a result of the great amount of investigation carried out in the laboratories of pharmaceutical firms, many of the old drugs obtained by extraction of natural products have been supplanted by pure synthetic chemicals which may or may not be identical with the natural products, some having been modified through research and testing to give improved results. Certain individual drugs still prevail, however, such as morphine, codeine, and reserpine, extracted from plants, for which there is a continued demand on the part of physicians and for which no synthetic counterpart or substitute has been fully accepted. Darvon has now partially replaced codeine.

Typical pharmaceutical products are here presented from among the thousands manufactured by the pharmaceutical industry and used by physicians. These are selected for inclusion in this chapter because of importance, value, volume produced, or illustrative processes.[3] They are arranged under their outstanding chemical conversion process or under fermentation.

ALKYLATIONS

BARBITURIC ACID DERIVATIVES[4] (Malonylurea) The U.S. Tariff Commission reported for 1972 the total production of such sedative derivatives as 567,000 lb/year, with an average value of $5.44 per pound for that sold. Several hundred different barbituric derivatives have been prepared, with varying action. These differ in the substituents on the carbon in position 5.

PHENOBARBITAL USP, 5-ethyl-5-phenylbarbituric acid. This derivative possesses specific usefulness in epilepsy. Like the other barbituric acid derivatives, it is made from phenylethylmalonic diethyl ester, which is condensed with urea to form the product. The reactions are as follows:

$$C_6H_5CH_2Cl + NaCN + acid \xrightarrow[\text{presence of } C_2H_5OH]{\text{Hydrolysis in}} C_6H_5CH_2COOC_2H_5$$

Benzyl chloride; Ethylphenyl acetate

$$C_6H_5CH_2COOC_2H_5 \xrightarrow[\text{with ethyl oxalate in presence of } C_2H_5OH + Na]{\text{condensed}} C_2H_5O\text{-}C(=O)\text{-}C(=O)\text{-}CNa(C_6H_5)\text{-}C(=O)\text{-}OC_2H_5$$

Diethyl sodium phenyloxaloacetate

$$\xrightarrow{HCl} C_6H_5CH(COOC_2H_5)C(=O)COOC_2H_5 \xrightarrow[\downarrow CO]{\text{distilled at } 180°C} C_6H_5CH(COOC_2H_5)_2 \xrightarrow{Na + C_2H_5Cl}$$

Diethyl ester of phenylmalonic acid

$$C_6H_5(C_2H_5)C(COOC_2H_5)_2 \xrightarrow{\text{urea}} \text{Phenobarbital}$$

Diethyl ester of phenylethylmalonic acid; Phenobarbital

[3] RPS XIV gives the best overall short description of the thousands of pharmaceutical products actually employed, together with their various names. Further details of the methods of manufacture are to be found in Ullman, Enzyklopaedie der technischen Chemie, 3d ed., Urban & Schwarzenberg, Berlin, 1950–; Jenkins *et al.*, The Chemistry of Organic Medicinal Products, 4th ed., Wiley, 1957; ECT, 2d ed., Interscience-Wiley, 1963–; Burger, Medicinal Chemistry, Interscience, 1961; Colbert, Controlled Action Drug Forms, Noyes, 1974.

[4] RPS XIII, pp. 1144–1151; Schwyzer, *op. cit.*, pp. 102ff. (detailed conditions and quantities); Ross, *Ind. Eng. Chem.*, **29,** 1341 (1937); ECT, 2d ed., vol. 3, pp. 1ff. and 60ff., 1964, RPS XIV, p. 1078.

BARBITAL, diethylbarbituric acid, is also sold under the name Veronal. It is the oldest of the long-acting barbiturates and is derived through diethyl malonate by the following reactions:[5]

$$ClCH_2COOH \xrightarrow{+NaOH} ClCH_2COONa \xrightarrow[\text{boiling}]{NaCn} CH_2(CN)COONa \xrightarrow[H_2SO_4 \text{ in benzene}]{C_2H_5OH}$$

Chloroacetic acid; Sodium cyanoacetate

$$CH_2(COOC_2H_5)_2 + NH_4NaSO_4 \xrightarrow{C_2H_5ONa} (Na)(H)C(COOC_2H_5)_2 + 100\%\ C_2H_5OH \xrightarrow[100 \text{ psi}]{C_2H_5Cl} (C_2H_5)(H)C(COOC_2H_5)_2 \xrightarrow[\text{2) } C_2H_5Cl \text{, 100 psi}]{\text{1) } C_2H_5ONa}$$

Diethyl malonate; Sodio malonic ester; Diethyl ester of ethylmalonic acid

$$(C_2H_5)_2C(COOC_2H_5)_2 \xrightarrow[C_2H_5ONa]{\text{urea}} \text{Barbital}$$

Diethyl ester of diethylmalonic acid; Barbital

PROCAINE HYDROCHLORIDE USP This local anesthetic has long been dispensed under the name novocaine. It is considered to be less toxic than cocaine and furthermore does not have the danger of habituation. It is used frequently in conjunction with a vasoconstrictor like epinephrine to secure a prolonged anesthetic action. One of the principal reactions in its preparation is alkylation. It is obtained by first alkylating ethylenechlorohydrin with diethylamine, which is condensed with *p*-nitrobenzoyl chloride and reduced with tin and hydrochloric acid to obtain procaine.[6]

$$(C_2H_5)_2NH \xrightarrow{HOCH_2CH_2CL} HOCH_2CH_2N(C_2H_5)_2 \xrightarrow[\text{chloride}]{p\text{-nitrobenzoyl}}$$

Diethylamine; Ethylene chlorohydrin; Diethylamino-ethanol

$$O_2NC_6H_4COOCH_2CH_2N(C_2H_5)_2 \xrightarrow[HCl]{Sn} NH_2C_6H_4COOCH_2CH_2N(C_2H_5)_2 \cdot HCl$$

Diethylaminoethyl-*p*-nitrobenzoate; Procaine hydrochloride or the hydrochloride of diethylaminoethyl-*p*-aminobenzoate

CODEINE NF AND CODEINE PHOSPHATE USP[7] Several decades ago, sufficient codeine could be isolated from opium to supply the needs for this sedative and analgesic drug. But as the superiority of codeine for the relief of coughing and other applications, together with lesser side effects, became apparent, there being only about 10% as much codeine as morphine obtained from opium, it was necessary to methylate about 90% of the morphine produced, to supply the enhanced demand. This is a difficult alkylation, since there are three places in the morphine molecule to which a methyl group can be attached: alcoholic hydroxyl, phenolic hydroxyl, and tertiary nitrogen. To direct this alkylation to the phenolic hydroxyl and to reduce the alkylation on the tertiary nitrogen, a quaternary nitrogen alkylating agency, *phenyltrimethylammonium chloride,* is employed, with resulting yields of 91 to 93% codeine and some recovery of unalkylated morphine. The alkylation is carried out with morphine dissolved in absolute alcohol in the presence of sodium ethylate by heating with the alkyl-

[5]RPS XIII, p. 1028; Schwyzer, *op. cit.,* 102 (details, quantities); Ross, *Ind. Eng. Chem.,* **29,** 1341 (1937); ECT, 2d ed., vol. 3, pp. 1ff. and 60ff., 1964; RPS XIV, p. 1080.

[6]Details and yields are given by Schwyzer, *op. cit.*

[7]RPS XIII, pp. 1186–1187; Groggins, Unit Processes in Organic Synthesis, 5th ed., p. 847, McGraw-Hill, 1958; Schwyzer, *op. cit.,* pp. 388–390 (quantities given); RPS XIV, p. 1124.

ating agent in a copper autoclave at 60 psi to about 130°C. The dimethylaniline and solvents are recovered and reused. The reactions are as follows:

$$C_6H_5N(CH_3)_2 + CH_3Cl \longrightarrow C_6H_5N(CH_3)_3Cl$$

$$2C_2H_5OH + 2Na \longrightarrow 2C_2H_5ONa + H_2$$

$$\underset{\text{Morphine}}{HO(C_{17}H_{17}ON)OH} + \underset{\substack{\text{Phenyltrimethylammonium}\\\text{chloride}}}{C_6H_5N(CH_3)_3Cl} + \underset{\text{Sodium ethylate}}{C_2H_5ONa} \longrightarrow$$

$$\underset{\text{Codeine}}{CH_3O(C_{17}H_{17}ON)OH} + \underset{\text{Dimethylaniline}}{C_6H_5N(CH_3)_2} + \underset{\text{Ethanol}}{C_2H_5OH} + NaCl$$

Note that the basis of this reaction is morphine, which is obtained solely by isolation from opium, which is the dried exudate from the incised unripe capsule of the opium poppy *Papaver somniferum.* Naturally, all the codeine that does occur in opium is isolated, along with morphine and other useful

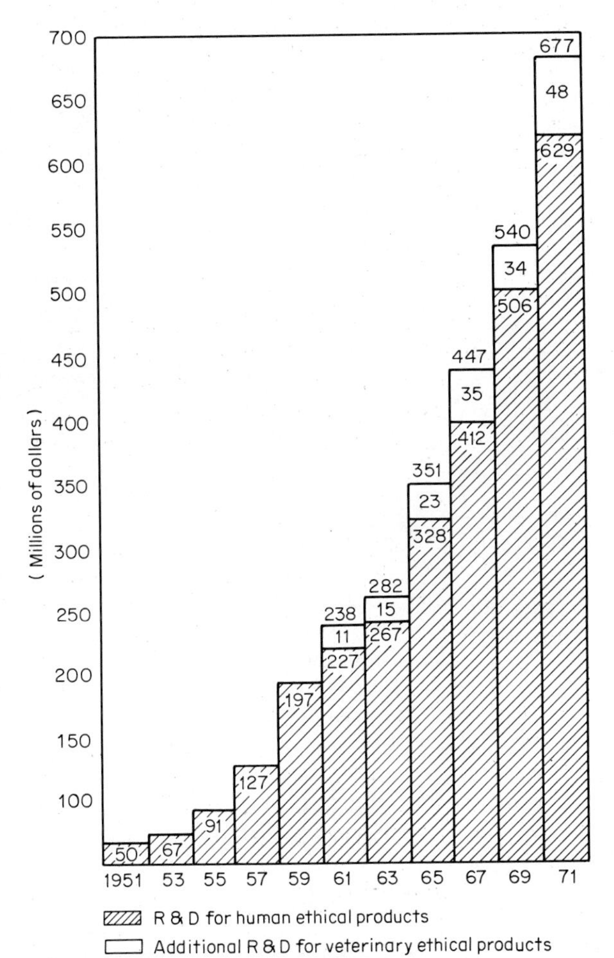

Fig. 40.2 R&D expenditures for ethical products, 1951–1971. (*Prescription Drug Industry Fact Book, PMA, 1973.*)

alkaloids (see morphine under isolates). The relationship between morphine and codeine can be seen from the following structural formulas:

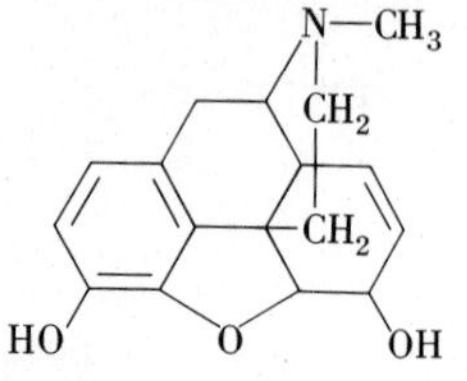

Morphine

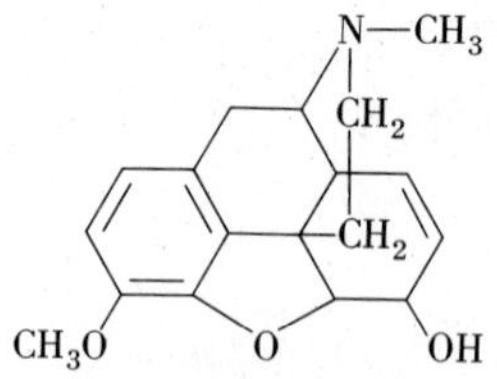

Codeine (methylmorphine)

CAFFEINE USP, THEOBROMINE NE, AND THEOPHYLLINE NF are xanthine derivatives classified as central nervous stimulants, but differing markedly in their properties. They can be extracted from a number of natural sources. Caffeine,[8] the most important, has long been obtained from waste tea and decaffeinization of coffee (see caffeine under isolates). The demand for caffeine is large in volume. Over 3 million lb of caffeine are produced each year. It can be manufactured synthetically by a number of processes. Some have been employed industrially, for instance, the methylation of theobromine and total synthesis by methylation and other reactions based upon urea. A typical sequence for the synthetic manufacture of caffeine is:

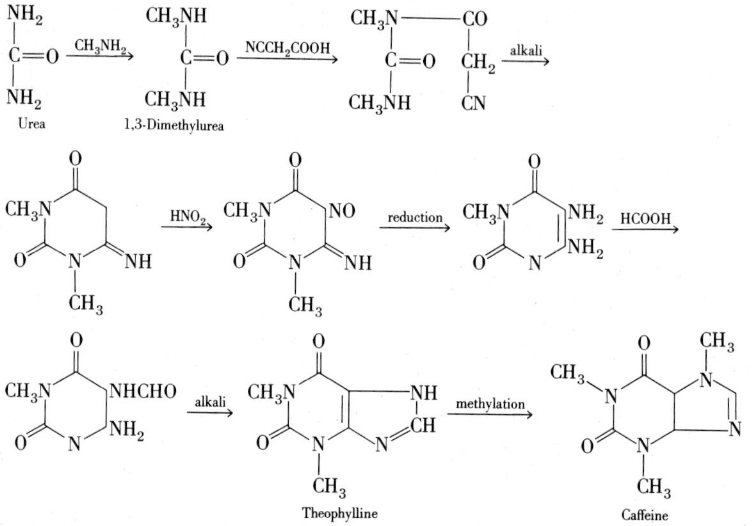

Urea $\xrightarrow{CH_3NH_2}$ 1,3-Dimethylurea $\xrightarrow{NCCH_2COOH}$ $\xrightarrow{alkali}$ $\xrightarrow{HNO_2}$ $\xrightarrow{reduction}$ $\xrightarrow{HCOOH}$ $\xrightarrow{alkali}$ Theophylline $\xrightarrow{methylation}$ Caffeine

A large demand for caffeine comes from the pharmaceutical industry, but it is also used by the soft-drink industry for mildly stimulating beverages such as Coca-Cola and Pepsi Cola. Coca-Cola contains only about one-third the amount of caffeine in a 6½-oz bottle that is present in a 5-oz cup of coffee or one-fourth the amount in a 5-oz cup of tea. Specifically, this amounts to 25 mg, or slightly less than 0.4 gr of caffeine in a 6½-oz bottle of Coca-Cola, whereas a 5-oz cup of tea contains 60.2 to 75.6 mg of caffeine. A 5-oz cup of coffee has 67.8 to 100.9 mg.

CARBOXYLATION

SALICYLIC ACID USP AND SALICYLATES USP and derivatives (aspirin, salol, methyl salicylate) have varied uses in medicinals, preservatives, rubber, perfumes, and dyes. Yearly, over 50 million lb

[8] Huber, Caffeine, ECT, 2d ed., vol. 3, pp. 911–917, 1964.

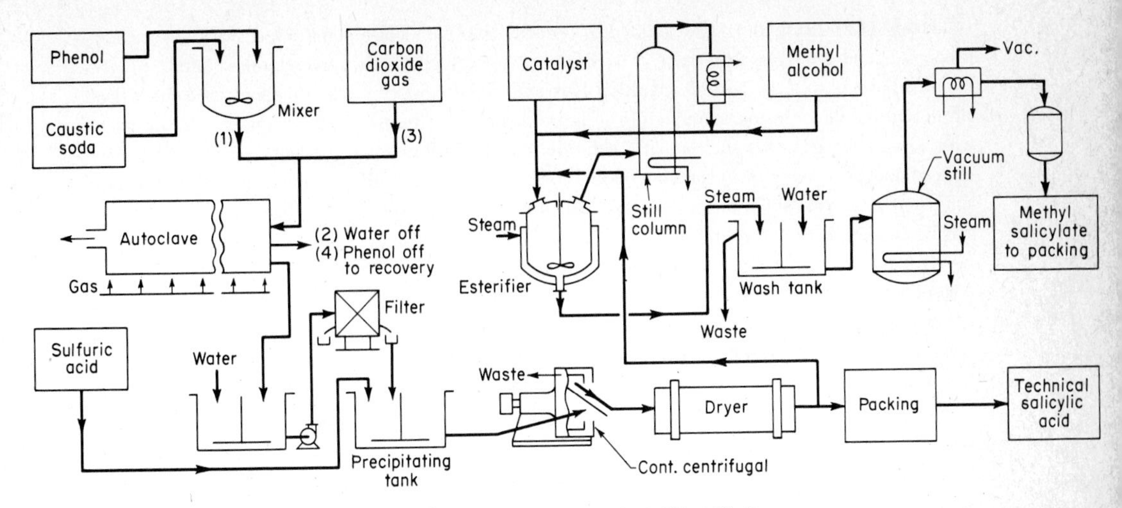

Fig. 40.3 Flowchart for salicylic acid and methyl salicylate.

Material	*Pounds/ton salicylic acid*	*Material*	*Pounds/ton salicylic acid*
Phenol	1,600	Zinc	20
Caustic soda	700	Zinc sulfate	40
Carbon dioxide	1,000	Activated carbon	40
Sulfuric acid	900		

Per 4,000 lb salicylic acid/day, Steam 10,000 lb, Power 1,000 kWh, Labor 48 h

of the acid, its sodium salt, and its derivatives are produced as such, and 40 million lb of aspirin or acetylsalicylic acid. The manufacture of salicylic acid and of salicylates follows carboxylation by the Schmitt modification of the Kolbe reaction as shown in Fig. 40.3.[9] The sodium phenolate must be finely divided and exposed to the action of the CO_2 under pressure and heat. The equipment recommended for this process is a revolving heated ball mill[10] into which the sodium phenolate solution is introduced. By revolving under vacuum and heat (130°C), the sodium phenolate is reduced to a very dry powder, after which the CO_2 is introduced under pressure (7 atm) and heat (100°C) to form, first, sodium phenyl carbonate isomerizing to sodium salicylate. This can be dissolved out of the mill, and the salicylic acid decolorized by activated carbon and precipitated by adding sulfuric acid in a wood vat to avoid iron at this stage. Sublimation has been employed to purify salicylic acid, which then can be esterified with methanol and a little sulfuric acid to form methyl salicylate.

CONDENSATION AND CYCLIZATION

The binding together of several molecules by condensation or ring closure to cyclization with or without the splitting out of a smaller molecule is a very important chemical conversion employed in the manufacture of a considerable number of pharmaceutical products.

[9] *Chem. Metall. Eng.*, **50**(8), 132–135 (1943).
[10] For ball-mill-reactor details see Groggins, *op. cit.*, p. 367.

HEXYLRESORCINOL USP, 1,3-dihydroxy-4-hexylbenzene. This chemical has marked germicidal properties, and a phenol coefficient of over 50. It is a valued odorless and stainless antiseptic commonly employed in a dilution of 1:1,000. It is one of the most efficient anthelmintics against hookworm and the like. In the manufacture of hexylresorcinol,[11] resorcinol and caproic acid are heated with a condensing agent, such as zinc chloride, and the intermediate ketone derivative is formed. This compound is purified by vacuum distillation. By reduction with zinc amalgam and hydrochloric acid, impure hexylresorcinol results which can be purified by vacuum distillation. Hexylresorcinol and its analogs were one of the series of related chemicals exhaustively studied to ascertain the *relationship between chemical constitution and bactericidal power.* It was found that this action increased up to the hexyl derivative and disappeared with the normal octyl derivative, probably because of low solubility. Figure 40.4 plots the atomic weight of the alkyl chain against the phenol coefficient [ratio of antibacterial power measured against *Eberthella typhosa* (*Salmonella typhosa*) relative to that of phenol under the same conditions]. Simultaneously with the increase in bactericidal power up to the hexyl derivative, a coincident drop in toxicity was found. Compare Table 40.4 on germicidal activity.

$$\underset{\text{Resorcinol}}{C_6H_4(OH)_2} + COOH(CH_2)_4CH_3 \xrightarrow[ZnCl_2]{\text{heat}} \underset{\text{Caproylresorcinol}}{C_6H_3(OH)_2CO(CH_2)_4CH_3} \xrightarrow{\text{Zn amalgam}} \underset{\text{Hexylresorcinol}}{C_6H_3(OH)_2CH_2(CH_2)_4CH_3}$$

PHENOLPHTHALEIN NF is a widely used cathartic, particularly in proprietary drugs. It is manufactured by adding melted phenol (10 parts) to a cooled solution of phthalic anhydride (5 parts) in concentrated sulfuric acid (4 parts) and heating the mixture 10 to 12 h at 120°C. The condensation product is poured hot into boiling water and boiled with successive changes of hot water. The condensate is dissolved in warm dilute caustic soda and precipitated with acetic acid. It may be purified by crystallization from absolute alcohol after treatment with and filtering from activated carbon.

$$C_6H_4(CO)_2O + 2C_6H_5OH \xrightarrow[120°C,\ 12\ h]{H_2SO_4} C_6H_4\langle C(C_6H_4OH)_2-O-C{=}O\rangle + H_2O(H_2SO_4)$$

PIPERAZINE CITRATE USP is used as an anthelmintic in the treatment of infections caused by pinworms and roundworms. It is also employed by veterinarians against various worms infecting domestic animals, including chickens. The U.S. Tariff Commission reported production in 1972 of piperazine and salts as 5,757,000 lb, and a sales value of about \$5.5 million. Piperazine is prepared by the cyclization of ethylene dibromide with alcohol ammonia at 100°C. The citrate is formed in aqueous solution and crystallized out.

$$2C_2H_4Br_2 + 6NH_3 \longrightarrow HN(CH_2CH_2)_2NH + 4NH_3Br$$

MEPROBAMATE,[12] 2-methyl-2-*n*-propyl-1,3-propanediol dicarbamate. Meprobamate is also sold under the trade names Miltown and Equanil. It is often dispensed as a tranquilizer. The U.S. Tariff

[11]Groggins, *op. cit.*, p. 839.
[12]Removed from USP XVII.

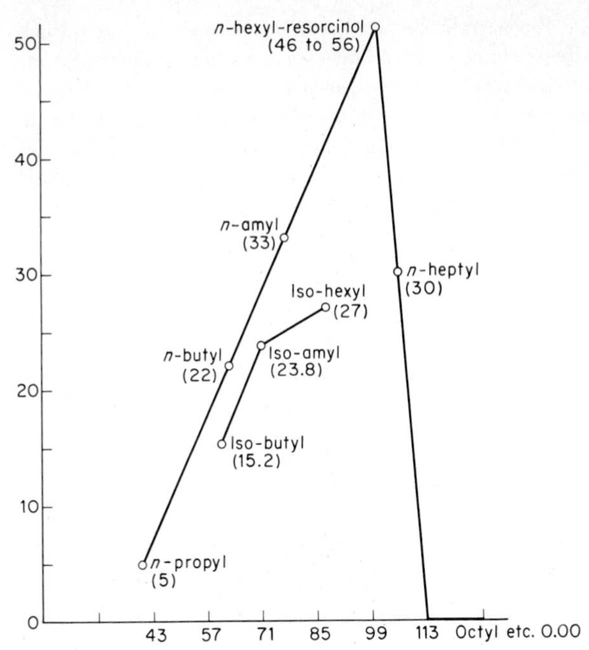

Fig. 40.4 Bacterial activity of the alkylresorcinols. The curve is obtained by plotting the phenol coefficients (U.S. Hygienic Laboratory technique) as ordinates against the sum of the atomic weights of the atoms in the alkyl chains as abscissas. [*Dohm, Cox, and Miller, J. Am. Chem. Soc.,* ***48,*** *1688 (1962) and Groggins, Unit Processes in Organic Synthesis, 5th ed., p. 389 McGraw-Hill, 1958.*]

Commission reported 1972 production as 561,000 lb, with a sales value of about $1.4 million. 2-Methyl-2-*n*-propyl-1,3-propandiol dicarbamate is prepared in toluene solution by condensation of 2-methyl-*n*-propyl-1,3-propanediol at about 0°C with phosgene in the presence of dimethylaniline to yield the dichloroformate ester, which is then subjected to ammonolysis to form the dicarbamate ester:

$$
\begin{array}{c}
CH_3 \\
| \\
-CH_2-C-CH_2-OOCNH_2 \\
| \\
CH_2CH_2CH_3
\end{array}
$$

THIAMINE HYDROCHLORIDE USP, VITAMIN B$_1$ This antineuritic vitamin is essential for bodily growth and the prevention of beriberi. Although thiamine is widely distributed in many foods, most commercially available quantities are obtained by synthesis, i.e., by the condensation of 6-amino-5-bromomethyl-2-methylpyrimidine hydrobromide with 5(β-hydroxyethyl)4-methylthiazole, which results in thiamine chloride hydrochloride. The last step is as follows:

$$
H_3C\text{-pyrimidine}(NH_2 \cdot HBr)(CH_2Br) + \text{thiazole}(CH_3)(CH_2CH_2OH) \xrightarrow[\text{then AgCl}]{\text{condense,}} \left[CH_3\text{-pyrimidine}(\overset{+}{N}H_3)-CH_2-\overset{+}{N}\text{-thiazole}(CH_3)(CH_2CH_2OH) \right] 2Cl^-
$$

Thiamine chloride hydrochloride
Vitamin B$_1$

RIBOFLAVIN USP, VITAMIN B$_2$ Riboflavin[13] is a necessary element of all living cells. It is quite stable except under excessive exposure to light. It is considered to be the growth factor of the vitamin B$_2$ complex and is added on a large scale to bread, flour, and other dietary and pharmaceutical preparations and, as such, is very important. It is also involved in the transfer of oxygen to tissues. Annual production is over 1 million lb, at about $13 per lb. This vitamin has been produced as a by-product from butyl-acetone fermentation of molasses, by direct fermentations (Chap. 31), and by the following synthesis,[14] starting with 1,3,4-xylidine, D-ribose, aniline, and alloxan:

$$(CH_3)_2C_6H_3NH_2 \xrightarrow[H_2,\ Pt\ or\ Pd]{D\text{-ribose}} (CH_3)_2C_6H_3NH\,CH_2(CHOH)_3CH_2OH \xrightarrow[aniline]{diazotized} (CH_3)_2C_6H_2(NH\,CH_2(CHOH)_3CH_2OH)(N{=}NC_6H_5) \xrightarrow[in\ boiling\ alcohol]{reduction,\ Na_2S_2O_4}$$

$$(CH_3)_2C_6H_2(NH\,CH_2(CHOH)_3CH_2OH)(NH_2) \xrightarrow[(alloxantin),\ boric\ acid]{condense,\ alloxan} \text{Riboflavin}$$

1,2-Dimethyl-4-D-ribitylamino-5-aminobenzene

Riboflavin

DEHYDRATION

ETHER USP is used for general anesthesia and as a solvent. All grades are produced in excess of 10 million lb/year. Ether and nitrous oxide, both anesthetics, are manufactured through dehydration reactions. The very simple and long-used manufacturing procedure for ether has been from alcohol (denatured with ether) by the dehydrating action of sulfuric acid. The anesthetic ether USP was especially purified. On the other hand, much ether for anesthetics, the USP grade, and commercial ether are now supplied from a by-product[15] in the manufacture of alcohol from ethylene following these reactions:

$$3C_2H_4 + \underset{95\%}{2H_2SO_4} = C_2H_5OSO_3H + (C_2H_5)_2OSO_2 \xrightarrow{H_2O} C_2H_5OH + (C_2H_5)O + 2H_2SO_4(H_2O)$$

Because of the large production of alcohol by this reaction, much more ether is produced than is industrially or pharmaceutically consumed. Therefore this excess ether is hydrated back to alcohol by passing it with water vapor over a hydrating catalyst, usually alumina.

ESTERIFICATION (ALCOHOLYSIS)

Esterification[16] as a chemical conversion process has been increasing in importance throughout all chemical process industries.

ACETYLSALICYLIC ACID USP, or aspirin, is the pharmaceutical of probably the largest poundage production, since over 40 million lb per year are made. It is widely used and taken for many purposes without scientific foundation. Aspirin, as well as the salts of salicylic acid, is employed as an antipyretic and analgesic for a variety of conditions; also for gout and acute rheumatic fever.

[13]McGraw-Hill Encyclopedia of Science and Technology, vol. 11, p. 561, 1960 (references).
[14]For other syntheses and literature references, cf. ECT, 1st ed., vol. 11, pp. 752–755, and 2d ed., vol. 17, p. 445.
[15]Groggins, *op. cit.*, p. 787 (flowchart).
[16]Groggins, *op. cit.*, p. 694.

Aspirin is more potent on a weight basis and less irritating to the gastric mucosa. The reaction follows, and a flowchart is referred to in the footnote[17] for its economical production.

$$C_6H_4(OH)(COOH) + (CH_3CO)_2O \longrightarrow C_6H_4(O \cdot COCH_3)(COOH) + CH_3COOH$$

Salicylic acid — Acetic anhydride — Acetylsalicylic acid — Acetic acid

ACETOPHENETIDINE USP This analgesic and antipyritic is employed also by veterinarians. It has been prepared from chlorobenzene[18] by Dow's process as given here. This is based on alkylating sodium phenolate with ethyl chloride and coupling the alkoxy derivative with diazotized aniline. The *p*-ethoxyazobenzene is split by reduction and, after removing the aniline to be used over again, the *p*-phenetidine is recovered and acetylated with acetic anhydride to acetophenetidine:

$$C_6H_5OC_2H_5 + C_6H_5N{=}NCl \longrightarrow C_2H_5OC_6H_4{-}N{=}N{-}C_6H_5 \xrightarrow{\text{reduction}} C_2H_5OC_6H_4NH_2 + C_6H_5NH_2$$

$$C_2H_5OC_6H_4NH_2 + \text{Acetic anhydride} \longrightarrow C_2H_5OC_6H_4NHCOCH_3$$

COCAINE This local anesthetic, an alkaloid, is the methyl ester of benzoylecgonine. Although esterification is employed in its production, the entire procedure is classified more appropriately under *isolates* (q.v.).

HALOGENATION

Chlorination, or halogenation, is used extensively as a chemical step in the manufacture of various intermediates, such as ethyl chloride or bromide, and homologs employed as intermediates in the manufacture of finished pharmaceutical products. In only a few, however, does the chlorine persist into the finished product. One such compound of long standing, used as an example here, is chloroform. Chloroform was once employed as an anesthetic by inhalation, but now only rarely. It is used in many cases as a solvent for alkaloids and other organic chemicals, in chemical analysis, and as a preservative during aqueous percolation of vegetable drugs, preventing bacterial decomposition.

CHLOROFORM USP Although alcohol and chlorinated lime once were used to make chloroform, recently acetone and calcium hypochlorite have been employed more advantageously, because the reaction is more rapid and the yield is high. The reaction may be expressed by the following simplified equations:

$$\underset{\text{Acetone}}{2CH_3COCH_3} + \underset{\text{Calcium hypochlorite}}{3Ca(OCl)_2} \longrightarrow \underset{\text{Chloroform}}{2CHCl_3} + \underset{\text{Calcium acetate}}{Ca(C_2H_3O_2)_2} + \underset{\text{Calcium hydroxide}}{2Ca(OH)_2}$$

[17] *Chem. Eng. (N.Y.)*, **60**(6), 116–120 (1953); Faith, Keyes, and Clark, p. 112.
[18] Schwyzer, *op. cit.*, pp. 211–212; RPS XIII, p. 1081; Groggins, *op. cit.*, p. 840.

Absolutely pure chloroform decomposes readily on keeping, particularly if exposed to moisture and sunlight. Therefore the USP requires the presence of a small amount of alcohol to retard this decomposition (0.5 to 1%).

OXIDATION

The chemical conversion of oxidation[19] is illustrated by two products, isoniazid and nicotinic acid.

ISONIAZID USP, isonicotinic acid hydrazide, is the most potent and selective of the tuberculostatic antibacterial agents, and it also has become known as the most effective agent in the therapy of tuberculosis. It is prepared by oxidation from γ-picoline (4-methylpyridine) as follows:

CH_3 (4-Methylpyridine) —oxidation→ $COOH$ (Isonicotinic acid) —H_2NNH_2, hydrazine→ $CONHNH_2$ (Isoniazide)

NICOTINIC ACID USP (AND NICOTINAMIDE USP) For food-labeling purposes the terms niacin and niacinamide were adopted in 1943. This antipellagra vitamin is essential to humans and other animals for growth and health. It not only aids in preventing and curing pellagra, but functions in protein and carbohydrate metabolism. In the body, niacin is converted into niacinamide, which is an essential constituent of coenzymes I and II, involved in the oxidation of carbohydrates. Niacin is the most stable of all vitamins. It is added to foods; enriched flour is one example. Over 6 million lb of niacin and niacinamide, valued at $1.81 per pound, is produced in the United States. It is estimated that the production is distributed in these proportions: animal feed, 40%; food, 25%; pharmaceuticals, 25%; and exports, 10%. These products are now prepared from pyridine derivatives, largely by oxidation of β-picoline[20] (3-methylpyridine). The reactions show air as an oxidizing agent, but potassium permanganate and electrolytic anode oxidation have also been employed. Quinoline, nicotine, and 2-methyl-5-ethylpyridine have likewise been oxidized to nicotinic acid. Nicotinamide is readily hydrolyzed to the free acid by heating in acid or alkaline solution. The methyl group in the 3-position is the most resistant to oxidation.

$COOH$ (Nicotinic acid, Niacin) ←air, V_2O_5— CH_3 (β-Picoline) —Air + NH_3, 300°C→ CN (Nitrile) —NH_4OH, 25°C→ $CONH_2$ (Nicotinamide)

SULFONATION

The outstanding drugs with sulfonation[21] as the predominant chemical conversion are sulfanilamide, sulfadiazine, sulfasuxidine, sulfaguanidine, and sulfathiazole. Sulfanilamide,[22] the original member of this class, is a very interesting chemical, long known as an intermediate in the manufacture of an

[19]Groggins, *op. cit.*, pp. 486–554.
[20]Cislak and Wheeler, U.S. Pat. 2,437,938 (Mar. 16, 1948).
[21]Groggins, *op. cit.*, pp. 303–389; Gilbert, Sulfonation and Related Reactions, Interscience, 1965.
[22]RPS XIII, pp. 1252–1263; McGraw-Hill Encyclopedia, vol. 13, p. 244.

orange-colored dye, Prontosil, before its antibactericidal properties were recognized. It was ascertained in 1935 that sulfanilamide (*p*-aminobenzenesulfonamide) was the active antimicrobial part of the dye. About 3,300 sulfonamides have been synthesized, but only a few have passed the careful testing of the pharmaceutical industry and the clinicians. The following structure characterizes all therapeutically useful sulfanilamides:

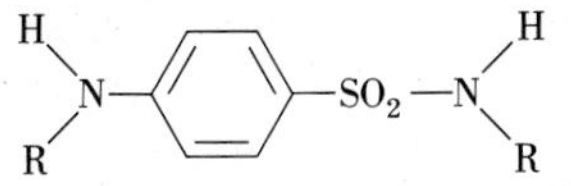

H(R)N—C₆H₄—SO_2—N(H)R

SULFADIAZINE USP, 2-sulfanilamidopyrimidine. This is the single sulfonamide most widely employed at the present time when the systemic, antibacterial actions of these agents are desired. Oral antidiabetic compounds are also sulfonamide derivatives. Manufacture by sulfonation-condensation may be represented as follows:

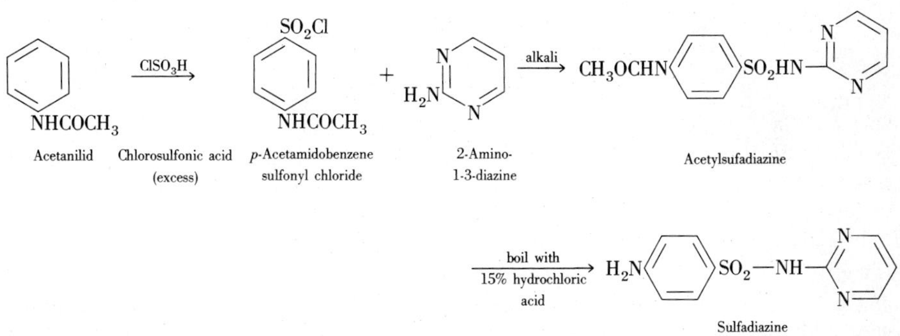

Acetanilid ($NHCOCH_3$) $\xrightarrow{ClSO_3H}$ *p*-Acetamidobenzene sulfonyl chloride (SO_2Cl, $NHCOCH_3$) + 2-Amino-1-3-diazine (H_2N) $\xrightarrow{alkali}$ Acetylsufadiazine (CH_3OCHN—C₆H₄—SO_2HN—pyrimidine)

Chlorosulfonic acid (excess)

$\xrightarrow{\text{boil with 15\% hydrochloric acid}}$ H_2N—C₆H₄—SO_2—NH—pyrimidine

Sulfadiazine

COMPLEX CHEMICAL CONVERSIONS

Many pharmaceutical chemicals are subjected to repeated, and often difficult, chemical reactions in order to obtain the desired product. Such conversions may be exemplified by ascorbic acid, chloramphenicol, and Aventyl[23] HCl. The latter, as an antidepressant, demonstrates this complex development. Twenty-six different chemicals are required to place the drug's 43 atoms in their precise molecular pattern in a six-step manufacturing process starting from phthalic anhydride. Its formula is

Aventyl hydrochloride

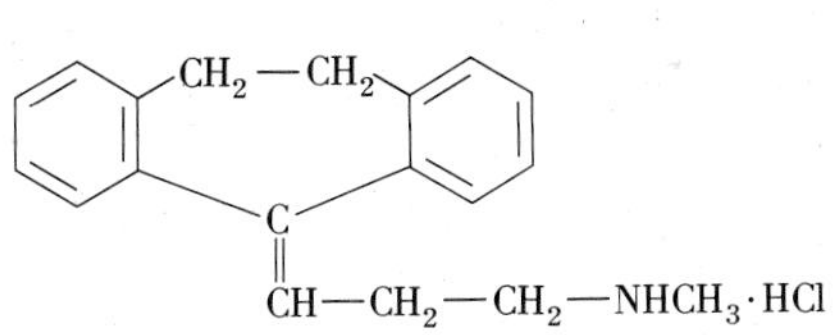

CH_2—CH_2 (bridging two benzene rings), C=CH—CH_2—CH_2—$NHCH_3\cdot HCl$

ASCORBIC ACID USP, VITAMIN C,[24] is often called the antiscorbutic vitamin. It is also needed in wound and bone healing and is a factor in resisting infection. Much of this vitamin is supplied in foods, especially fresh citrus fruits, tomatoes, and green vegetables. Ascorbic acid in solution or in foods is unstable, but the contrary is true if it is in dry form (powder or tablets). All the ascorbic acid

[23]Peters and Hennion, Synthesis of Nortriptyline, *J. Med. Chem.*, **7**, 390 (1964).
[24]RPS XIII, p. 1090; Burns, Ascorbic Acid, in ECT, 2d ed., vol. 2, pp. 747–763, 1963; RPS XIV, p. 1024.

used commercially is synthesized; 12,312,000 lb was produced in 1972 (Fig. 40.5). It is produced from glucose by these reactions:

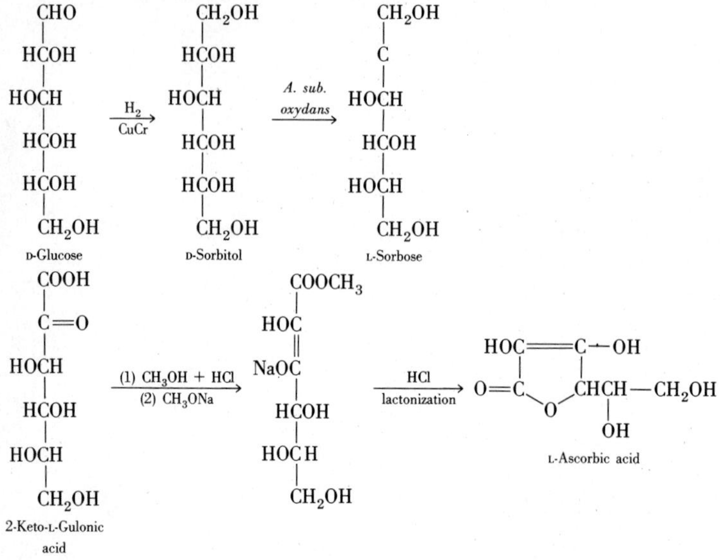

CHO–HCOH–HOCH–HCOH–HCOH–CH_2OH (D-Glucose) $\xrightarrow[\text{CuCr}]{H_2}$ CH_2OH–HCOH–HOCH–HCOH–HCOH–CH_2OH (D-Sorbitol) $\xrightarrow{\textit{A. sub. oxydans}}$ CH_2OH–C–HOCH–HCOH–HOCH–CH_2OH (L-Sorbose)

COOH–C=O–HOCH–HCOH–HOCH–CH_2OH (2-Keto-L-Gulonic acid) $\xrightarrow[\text{(2) } CH_3ONa]{\text{(1) } CH_3OH + HCl}$ $COOCH_3$–HOC=NaOC–HCOH–HOCH–CH_2OH $\xrightarrow[\text{lactonization}]{\text{HCl}}$ HOC=C–OH, O=C–O–CHCH(OH)–CH_2OH (ring) (L-Ascorbic acid)

D-Glucose

D-Sorbitol

L-Sorbose

2-Keto-L-Gulonic acid

L-Ascorbic acid

CHLORAMPHENICOL USP, Chloromycetin, has a wide spectrum of antibacterial activity against rickettsiae, certain viruses, and many bacteria. Because of serious toxic reactions, clinical use of the drug has been sharply curtailed,[25] though at one time (1951) sales by Parke, Davis & Co. were over $80 million annually. In 18 years sales dropped to about $12 million. It is believed to be the first naturally occurring pharmaceutical containing a nitro group and is a derivative of dichloroacetic acid. This compound was made initially by fermentation, which explains why it is associated with antibiotics. It is now manufactured mainly by chemical synthesis, starting with *p*-nitroacetophenone and ending with the compound

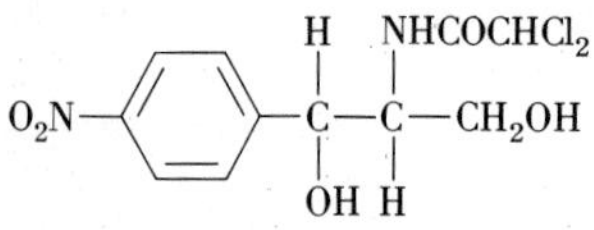

O_2N–C_6H_4–C(H)(OH)–C(H)($NHCOCHCl_2$)–CH_2OH

Chloramphenicol

DARVON (Lilly), *d*-propoxyphene HCl, is a synthetic, nonantipyritic, orally effective analgesic, similar pharmacologically to codeine. "It is approximately equal milligram for milligram to codeine in analgesic potency, produces no respiratory depressions, and has little or no antitussive activity."[26] It is unique in that it is not a narcotic yet can be substituted for codeine, and is useful in any condition associated with pain. This analgesic is not analogous *chemically* to codeine or morphine. It was discovered and synthesized commercially in the Lilly laboratories. Its synthesis starts with relatively simple chemicals, but many steps are involved in its manufacture. These are outlined in the following *chemical conversions.* The second step is detailed and illustrated in Fig. 40.6.

[25]RPS XIII, pp. 1268–1269. Properties and outline of synthesis; RPS XIV, p. 1210.
[26]RPS XIII, p. 1198.

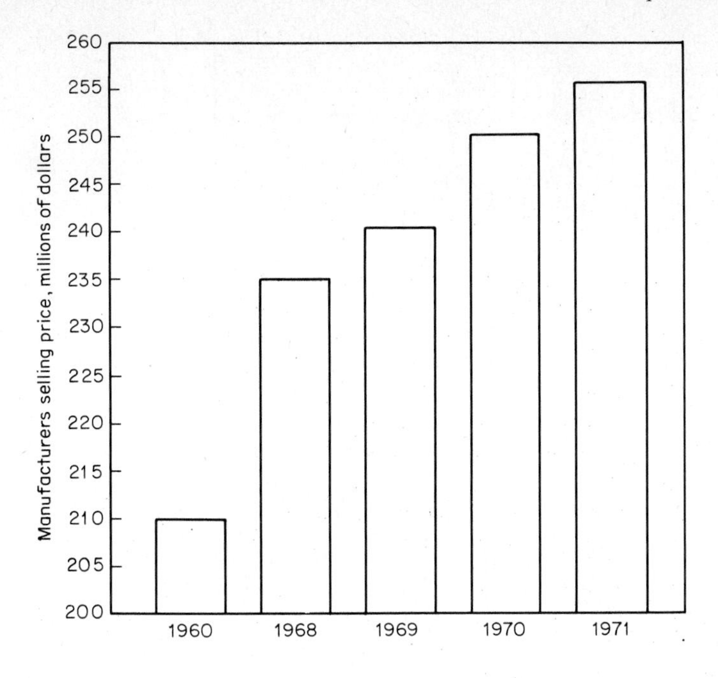

	Production 1972* thousands of pounds
Ascorbic acid and derivatives	15,588
Niacin and niacinamide	5,475
Pantothenic acid and derivatives	3,049
Vitamin B_2	1,011
Vitamin A	1,037
Vitamin B_1 (all other)	809
Vitamin E	2,905
Vitamin K	143
Vitamin D_2 & D_3	13
	30,030

Fig. 40.5 Ascorbic acid sales. Ascorbic acid leads the way in the steadily growing vitamin business. (*Chem. Eng. News, Sept. 6, 1971, p. 18.*) *Estimated.

Chemical Conversions and Reaction Summary

Step 1. Coupling to give a Mannich ketone.

$$C_6H_5\overset{\overset{\displaystyle O}{\|}}{C}—CH_2 \cdot CH_3 + (HCHO)_n + (CH_3)_2NH \cdot HCl \xrightarrow[\text{alcohol}]{\text{isopropyl}} \text{Mannich ketone}$$

Propiophenone; Paraformaldehyde; Dimethylamine HCl

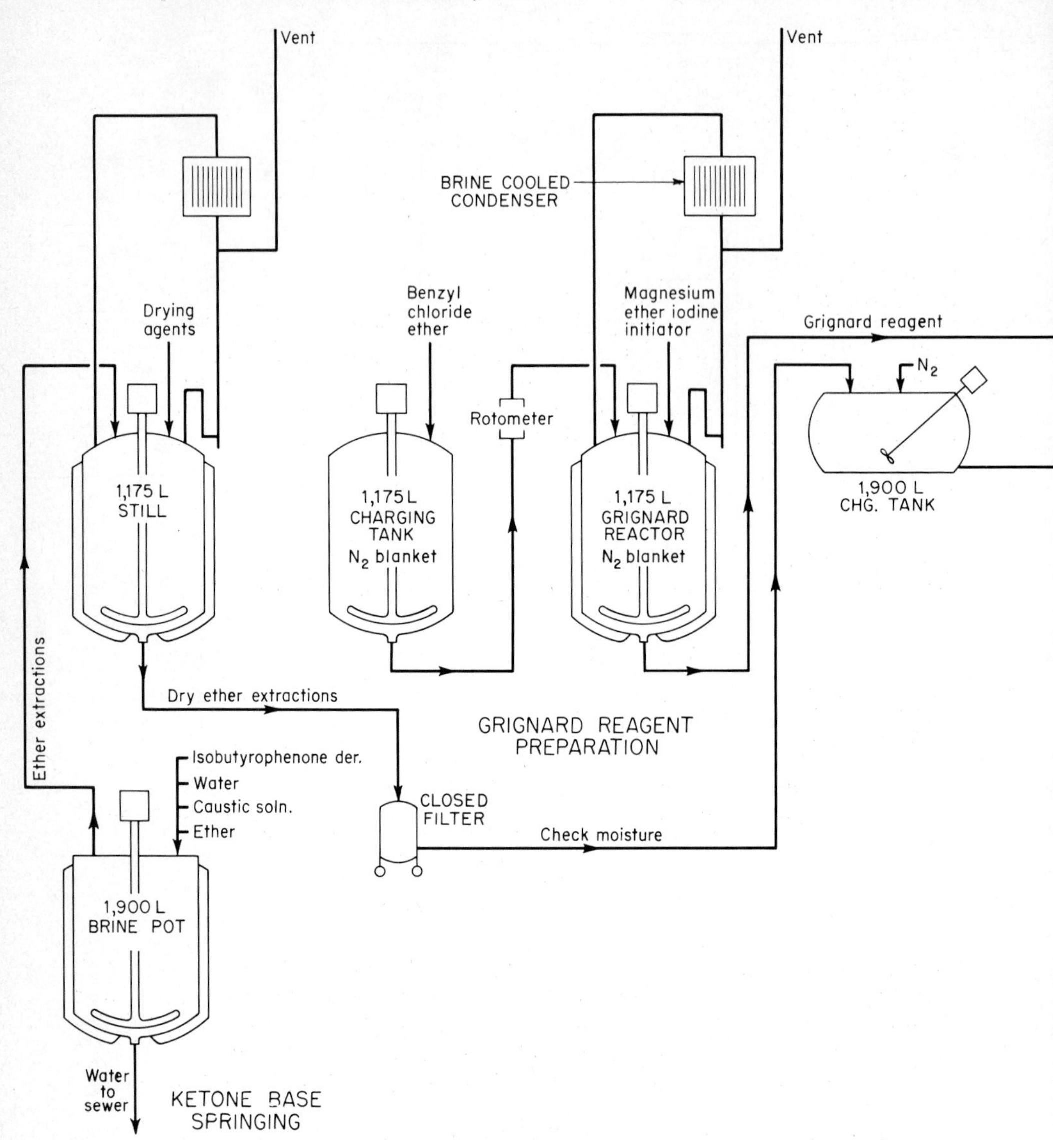

Fig. 40.6 Darvon production flowchart, showing equipment involved in step 2 described in the text. (*Eli Lilly & Co.*)

Step 2. Coupling Mannich ketone with benzyl chloride, using Grignard technique (and decomposition).

$$\underset{\text{Mannich ketone}\atop\text{Isobutyrophenone derivative}}{C_6H_5{-}\overset{O}{\overset{\|}{C}}{-}\underset{CH_3}{\underset{|}{CH}}{-}CH_2{-}N(CH_3)_2} + \underset{\text{Grignard}}{C_6H_5CH_2MgCl} \longrightarrow \underset{\text{Carbinol}}{C_6H_5{-}CH_2{-}\overset{OH}{\overset{|}{\underset{H_5C_6}{\underset{|}{C}}}}{-}\underset{CH_3}{\underset{|}{CH}}{-}CH_2N(CH_3)_2}$$

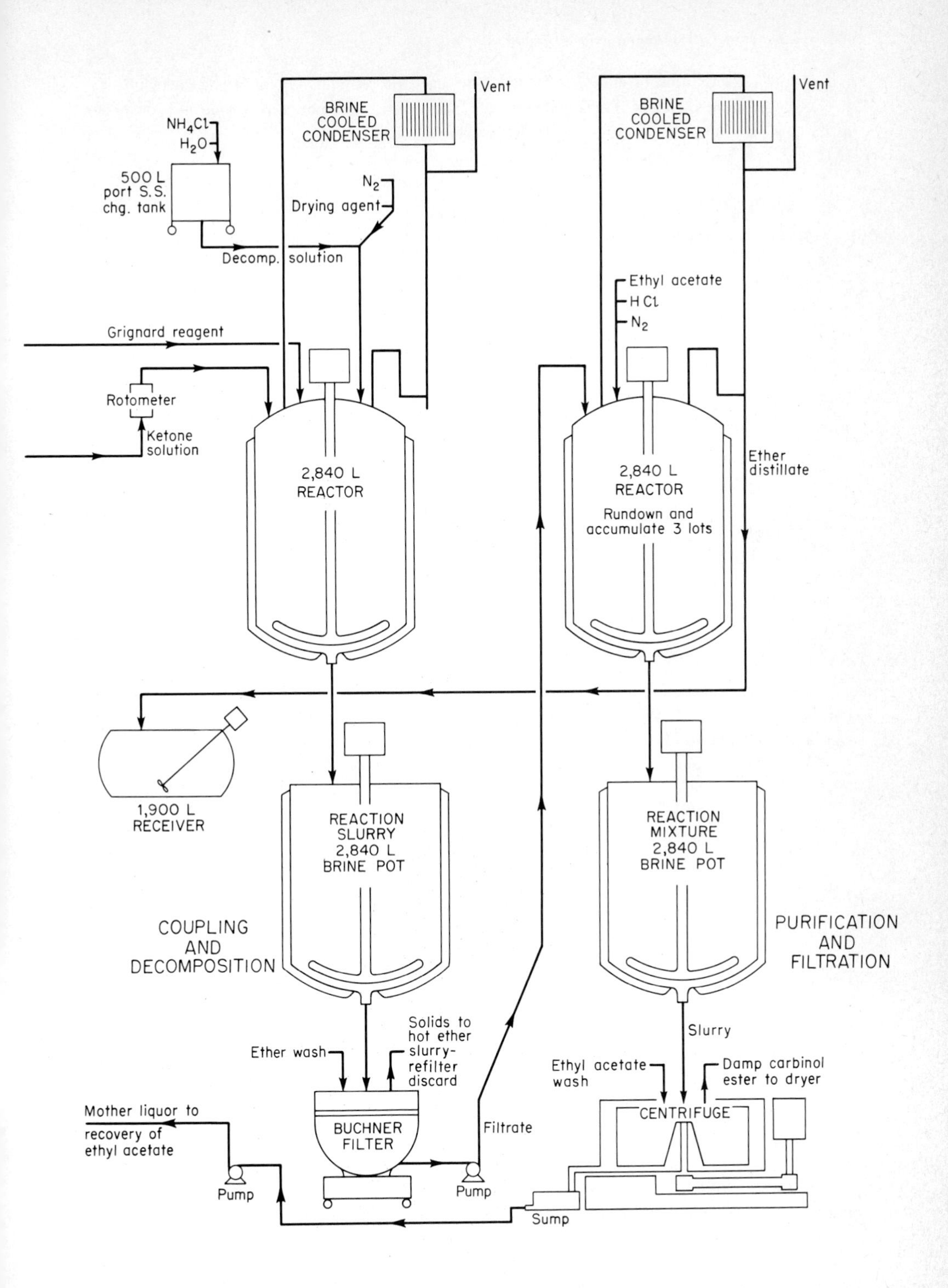

Vent
BRINE COOLED CONDENSER
NH_4Cl
H_2O
500 L port S.S. chg. tank
N_2
Drying agent
Decomp. solution
Grignard reagent
Rotometer
Ketone solution
2,840 L REACTOR
Vent
BRINE COOLED CONDENSER
Ethyl acetate
HCl
N_2
Ether distillate
2,840 L REACTOR
Rundown and accumulate 3 lots
1,900 L RECEIVER
REACTION SLURRY 2,840 L BRINE POT
REACTION MIXTURE 2,840 L BRINE POT
COUPLING AND DECOMPOSITION
PURIFICATION AND FILTRATION
Solids to hot ether slurry-refilter discard
Ether wash
Slurry
Ethyl acetate wash
Damp carbinol ester to dryer
Mother liquor to recovery of ethyl acetate
BUCHNER FILTER
Filtrate
CENTRIFUGE
Pump
Pump
Sump

Step 3. Resolution of optical isomers by the use of *d*-camphorsulfonic acid in acetone.

Step 4. Splitting off of the *d*-camphorsulfonic acid using ammonium hydroxide and conversion of the desired alpha-dextro isomer to the hydrochloride.

Step 5. Esterification of the alpha-dextro isomer with propionic anhydride.

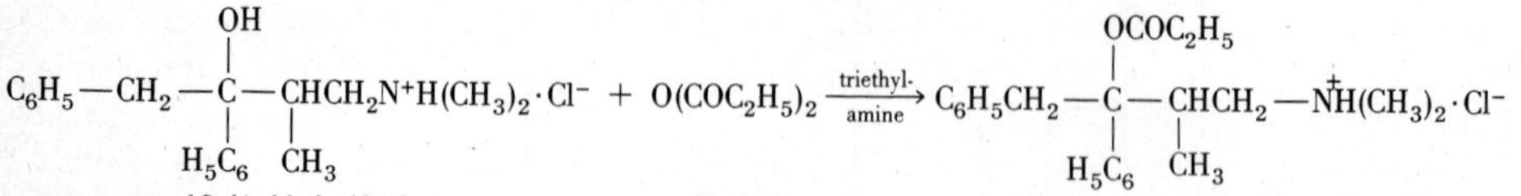

$$C_6H_5-CH_2-\underset{\underset{H_5C_6\;\;CH_3}{|\;\;\;\;|}}{\overset{\overset{OH}{|}}{C}}-CHCH_2N^+H(CH_3)_2\cdot Cl^- + O(COC_2H_5)_2 \xrightarrow[\text{amine}]{\text{triethyl-}} C_6H_5CH_2-\underset{\underset{H_5C_6\;\;CH_3}{|\;\;\;\;|}}{\overset{\overset{OCOC_2H_5}{|}}{C}}-CHCH_2-\overset{+}{N}H(CH_3)_2\cdot Cl^-$$

d-Carbinol hydrochloride — Propionic anhydride — *d*-Propoxyphene hydrochloride, or Darvon

Step 6. Isolation, filtration, drying.

LIBRIUM in 1976 represented the second largest annual sale of a prescription medicinal. See Table 40.1. Librium (chlordiazepoxide hydrochloride) is made by a complex series of chemical conversions:

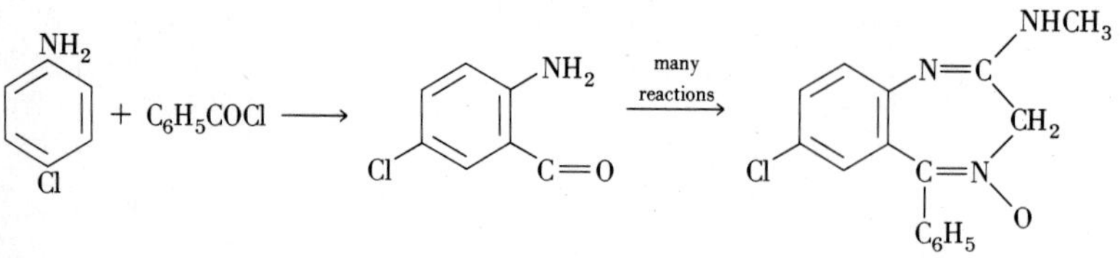

NH_2, Cl (p-chloroaniline) + C_6H_5COCl ⟶ Cl, NH_2, C=O ⟶ (many reactions) Cl, N=C, $NHCH_3$, CH_2, C=N, O, C_6H_5

Librium

CONTRACEPTIVES (oral) are taken daily by millions of women in the United States and throughout the world. Over 30 different products are marketed in the United States alone, in dosages that can be a combination of drugs, a sequence of drugs, or a single drug.

Most of the drugs in oral contraceptives are derived from the general formula:

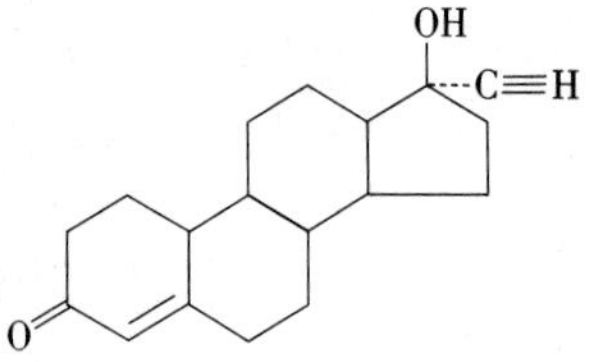

OH, C≡H, O

For instance, one of the drugs, Menstranol, has the formula:

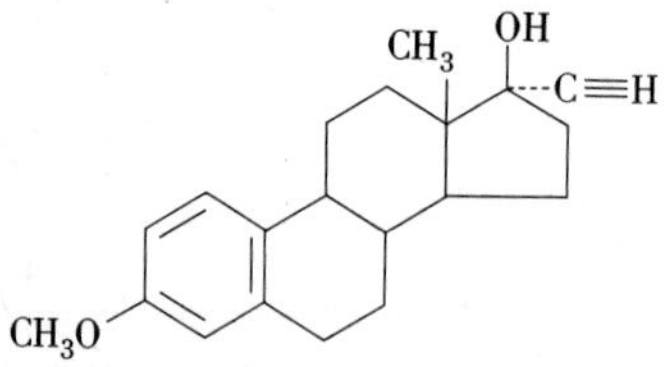

CH_3, OH, C≡H, CH_3O

RADIOISOTOPES IN MEDICINE A tabulation with 105 references, classified by "organ function" studied,[27] presents the recent and wide expansion of radioisotopes in clinical medicine. The tabulation also names the specific isotope used.

[27]Silver, Use of Radioisotopes in Medicine, *Nucleonics*, **23**(8), 106 (1965).

FERMENTATION AND LIFE PROCESSING FOR ANTIBIOTICS, BIOLOGICALS, HORMONES, AND VITAMINS

Historically the pharmaceutical industry has frequently used materials of plant or animal origin as sources of drugs. It was a logical step, although one a long time in coming, for this industry to employ the life processes of either plants or animals, and especially of microorganisms, to produce useful medicaments. This is particularly true of fermentation, in which microorganisms are permitted to grow under controlled conditions to produce valuable and often complex chemicals. However, the chemists in the laboratory have in several cases worked out synthetic or semisynthetic processes that compete with life processes. Both life processes and a chemical synthesis are competitive in the production of riboflavin and chloramphenicol. Vitamin B_{12}[28] is produced by fermentation. See Chap. 31 for general principles of fermentation and for many products made biosynthetically.

ANTIBIOTICS

The term *antibiotic* is a broad one, defined by Waksman[29] as "a substance produced by microorganisms, which has the capacity of inhibiting the growth and even of destroying other microorganisms by the action of very small amounts of the antibiotics." Approximately 150 substances, at present, come under this classification,[30] but only a few are accepted by the medical profession as being clinically useful. Penicillin, streptomycin, and its derivative dihydrostreptomycin head the list, which also includes Ilotycin (erythromycin), bacitracin, chloramphenicol, tetracycline, Aureomycin, Terramycin, Lincocin, Polycillin, and Keflin (Tables 40.2 and 40.3).

Antibiotics have profoundly affected the treatment of human disease. A "doctor with $20 worth of antibiotic drugs will often have a patient up and around after a week in bed." These unique substances are also showing great promise in nonmedical uses, for example, in animal nutrition, in the control of plant diseases, and in food preservation. Chemical engineering, with other disciplines, has contributed much to the important antibiotic industry. In a relatively few years, this technology has grown into a large enterprise backed by great capital expenditures and by extensive staffs of technically and specially trained people in laboratories, pilot plants, and factories.

Although many antibiotics are now produced commercially, processes with flowcharts for the isolation fo three representative ones are depicted in Fig. 40.8 and described in this chapter, namely, *penicillin*, Ilotycin, or *erythromycin*, and *streptomycin*. Antibiotics have presented many new problems to the chemist, the microbiologist, and the chemical engineer, particularly since they are generally unstable to heat, to wide pH ranges, and to enzymic action, and are often decomposed in solution.

PENICILLIN[31] A number of penicillins, differing only in the composition of the R group, have been isolated from natural media, and hundreds have been semisynthesized. Penicillin G USP, with

[28]Flowchart for Production of Vitamin B_{12} Feed Supplement, *Ind. Eng. Chem.*, **46,** 240 (1954).

[29]Waksman, An Institute of Microbiology: Its Aims and Purposes, *Science,* **110**(27), 839 1949; This Antibiotic Age, *Chem. Eng. News.* **29,** 1190 (1951).

[30]Federal Trade Commission, Economic Report on Antibiotics Manufacture, Washington, D.C., 1958; Baron, Handbook of Antibiotics, Reinhold, 1950; ECT, vol. 2, pp. 7–37, 1963.

[31]RPS XIII, pp. 1280–1288; ECT, vol. 9, p. 922, 1952 (many references); Penicillin, *Chem. Eng.* (*N.Y.*), **58**(4), 174 (1951) (pictured flowchart); Penicillin: Cost vs. Production, *Chem. Eng. News,* **31,** 3057 (1953); see RPS reference for other penicillins; Marshall and Elder, Chemical Engineering and Penicillin, AIChE Meeting, Atlantic City, September 1966; The History of Penicillin Production, *Chem. Eng. Prog. Symp. Ser.* no. 100, **60,** 1970.

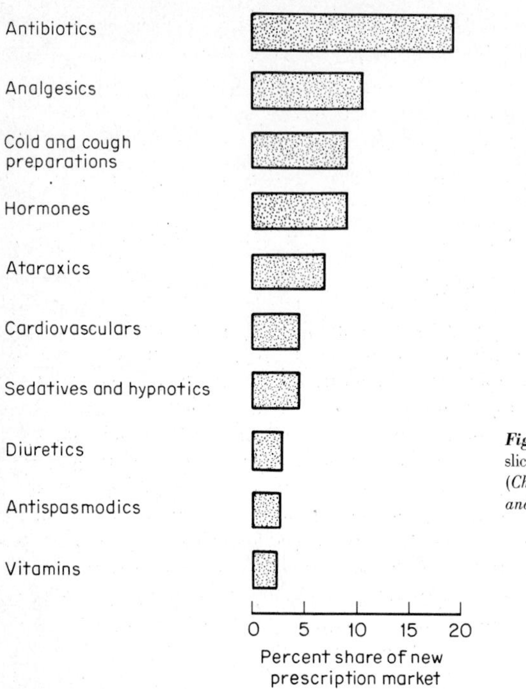

Fig. 40.7 Antibiotics have the largest slice of the total ethical drug market. (*Chem. Eng. News, Sept. 6, 1971, p. 18, and R. A. Gosselin Co.*)

benzyl for R, generally the most clinically desirable, is the type commercially available, usually combined in salt form with procaine or potassium.

```
                       S
                     /   \
      NH—CH—CH        C(CH3)2
     /       |    |          |
R—C        C——N——CHCOOH
   ||        ||
   O         O
```

Not only was penicillin the first antibiotic to be produced for widespread use, but it is important in quantities made and in general usefulness. It is practically nontoxic and is one of the most active antimicrobial agents known. Hypersensitivity reactions occur in about 10% of patients. It is also useful in animal feeds to promote growth. It is a tribute to the pharmaceutical industry that the price of penicillin has been reduced from the initial price of $25,000 per gram to $20 per pound for procaine penicillin, with a production of over 800,000 lb in 1976.

It is possible to assist the synthesis of a desired penicillin by supplying to the culture the appropriate precursor,[32] i.e., the acid of the side chain which appears in the amide linkage with 6-aminopenicillanic acid in the final product. The mold preferentially incorporates the precursor added into the corresponding penicillin to the relative exclusion of other precursors present in or formed from the nutrient raw materials in the medium. Thus, to produce benzylpenicillin, the precursor used is phenylacetic acid. Several hundred modifications[33] of the penicillin molecule not found in nature have been made by altering the side chain (R) by use of the appropriate precursor in

[32]A *precursor* is defined in the antibiotic industry as a chemical substance that can be incorporated into an antibiotic. The precursor may be used in such a way that an antibiotic modification not normally made by the organism is produced.

[33]New Research in Antibiotics, *Res. Today* (Lilly), Spring 1949.

the fermentation.[34] An important modification is penicillin V (V-cillin K, Lilly, potassium phenoxy-methyl penicillin), which is produced biosynthetically by using phenoxyacetic acid as a precursor (Fig. 40.8).

ILOTYCIN[35] (Dista, erythromycin USP[36]) is, like penicillin, also isolated by solvent-extraction methods. It is an organic base, and extractable with amyl acetate or other organic solvents under basic rather than the acidic conditions which favor penicillin extraction.

STREPTOMYCIN[37] The commercial method for producing this compound is also aerobic submerged fermentatation, as outlined in Fig. 40.8 and the appended description. Its formula is

```
                          OH
          NH               |            NH
          ||              _C_           ||
H2N—CNH—CH         CH—NHC—NH2
                 |           |
      HO—CH          CH——O——CH——CH————————O————————CH
                 \_CH_/              |     | /OH                O      CHNHCH3
                    |               O_  _C_                    |       |
                   OH                  CH     C=O              CH     CHOH
                                        |      |           CH2OH   \_ |
             Streptomycin              CH3     H                    CH
                                                                    |
                                                                    OH
```

The structure of streptomycin indicates its highly hydrophilic nature, and it cannot be extracted by normal solvent procedures. Because of the strong-base characteristics of the two substituted guanidine groups, it may be treated as a cation and removed from the filtered solution by ion-exchange techniques.

PRODUCTION AND ISOLATION OF PENCILLIN, ILOTYCIN (ERYTHROMYCIN), AND STREPTOMYCIN Figure 40.8 illustrates how these three important antibiotics of different chemical structure are formed by the life processes of three different microorganisms growing in individual pure cultures and how each one is extracted and purified. The steps in Fig. 40.8 may be separated into the following coordinated sequences of *chemical conversions* (Ch):

Isolate and purify the required pure, active culture of the respective antibiotic (Ch or Op) from the starting culture on the slant through to the inoculum tank (Fig. 40.8): for the culture of penicillin, *Penicillium chrysogesium;* of Ilotycin, *Streptomyces erythreus;* of streptomycin, *Streptomyces griseus.* Prepare and steam-sterilize at 120°C the fermentation media containing proteins, carbohydrates, lipids, and minerals (Op and Ch). Introduce the inoculum into the large, stirred, air-agitated fermenters to facilitate rapid metabolism. The air must be sterile (Op and Ch). Ferment, or "grow," the respective microorganism under conditions favorable to the planned antibiotic (Ch). Time, 100 to 150 h at 25 to 27°C. Filter off the mycellium (cells and insoluble metabolic products) of the respective microorganism and wash, using a string discharge continuous rotary filter (Op). Dry (Op) (sell for fertilizer). The filtrates from the respective microorganism contain the antibiotic which must be *initially* purified and separated, and *finally* purified according to the properties of the respective antibiotic, as in the following outline. For penicillin its purity is only 5 to 10%, impurities being medium components and metabolic products of the mold. Concentration from 5 to 10 mg/ml.

Penicillin[38]

Cool the aqueous solution and acidulate (H_2SO_4) to a pH of 2.0, resulting in an acid (and extractable) form of penicillin (Ch).

Solvent-extract in a Podbielniak countercurrent rotating contactor (Fig. 40.9) using a 1:10

[34]Behrens *et al.*, U.S. Pat. 2,562,410 (1951).
[35]Hamill *et al.*, *Antibiot. Chemother.*, **11,** 328 (1961).
[36]Bunch *et al.*, U.S. Pat. 2,653,899.
[37]RPS XIII, pp. 1289–1290.
[38]Flowcharts for Penicillin, *Chem. Eng.* (*N.Y.*), **64**(5), 247 (1957); Faith, Keyes, and Clark, p. 564.

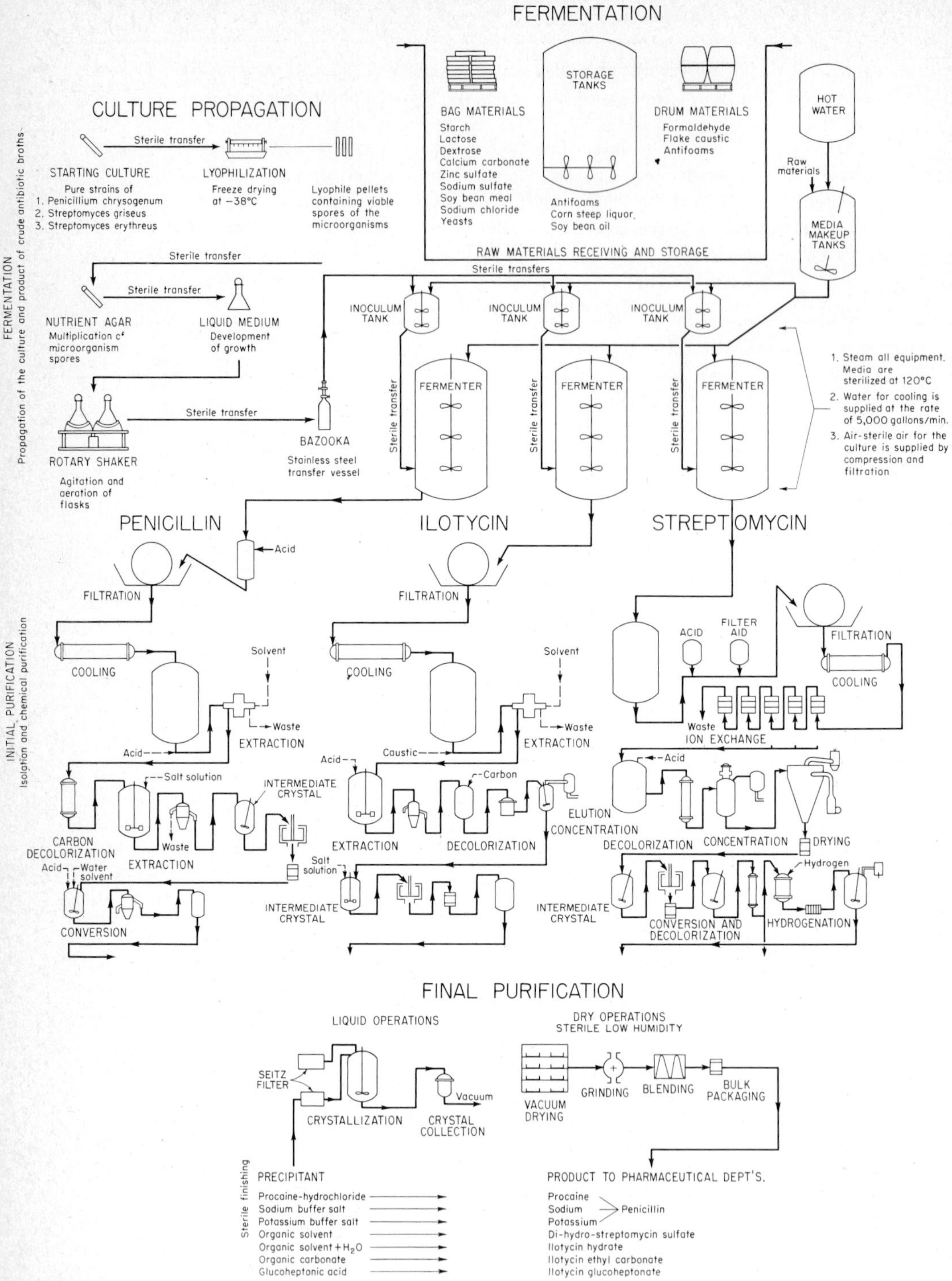

FERMENTATION

CULTURE PROPAGATION

Sterile transfer

STARTING CULTURE

Pure strains of
1. Penicillium chrysogenum
2. Streptomyces griseus
3. Streptomyces erythreus

LYOPHILIZATION

Freeze drying at −38°C

Lyophile pellets containing viable spores of the microorganisms

BAG MATERIALS

Starch
Lactose
Dextrose
Calcium carbonate
Zinc sulfate
Sodium sulfate
Soy bean meal
Sodium chloride
Yeasts

STORAGE TANKS

Antifoams
Corn steep liquor
Soy bean oil

DRUM MATERIALS

Formaldehyde
Flake caustic
Antifoams

HOT WATER

Raw materials

MEDIA MAKEUP TANKS

RAW MATERIALS RECEIVING AND STORAGE

Sterile transfers

FERMENTATION

Propagation of the culture and product of crude antibiotic broths

Sterile transfer

NUTRIENT AGAR

Multiplication of microorganism spores

LIQUID MEDIUM

Development of growth

ROTARY SHAKER

Agitation and aeration of flasks

BAZOOKA

Stainless steel transfer vessel

INOCULUM TANK

FERMENTER

1. Steam all equipment. Media are sterilized at 120°C
2. Water for cooling is supplied at the rate of 5,000 gallons/min.
3. Air-sterile air for the culture is supplied by compression and filtration

PENICILLIN

ILOTYCIN

STREPTOMYCIN

INITIAL PURIFICATION

Isolation and chemical purification

Acid

FILTRATION

COOLING

Solvent

Waste

EXTRACTION

Acid

Salt solution

INTERMEDIATE CRYSTAL

CARBON DECOLORIZATION

Waste

EXTRACTION

Acid

Water solvent

CONVERSION

Caustic

Carbon

DECOLORIZATION

Salt solution

ELUTION CONCENTRATION

ACID

FILTER AID

ION EXCHANGE

Acid

DECOLORIZATION

CONCENTRATION

DRYING

Hydrogen

CONVERSION AND DECOLORIZATION

HYDROGENATION

FINAL PURIFICATION

LIQUID OPERATIONS

SEITZ FILTER

CRYSTALLIZATION

Vacuum

CRYSTAL COLLECTION

DRY OPERATIONS STERILE LOW HUMIDITY

VACUUM DRYING

GRINDING

BLENDING

BULK PACKAGING

Sterile finishing

PRECIPITANT

Procaine-hydrochloride
Sodium buffer salt
Potassium buffer salt
Organic solvent
Organic solvent + H_2O
Organic carbonate
Glucoheptonic acid

PRODUCT TO PHARMACEUTICAL DEPT'S.

Procaine, Sodium, Potassium → Penicillin
Di-hydro-streptomycin sulfate
Ilotycin hydrate
Ilotycin ethyl carbonate
Ilotycin glucoheptonate

Fig. 40.8 Flowchart for the manufacture of antibiotics by fermentation and purification extraction and crystallization for penicillin and Ilotycin (erythromycin) with ion exchange for streptomycin. (*Eli Lilly & Co.*)

Typical penicillin fermentation, approximate batch for 12,000-gal fermenter (maintain pH preferably below 6.5 and not above 7 and temperature at 25 to 27°C):

Water	10,000 gal	Precursor	1,100 lb
Sugar	8,800 lb	Minerals	110 lb
Corn-steep liquor	1,110 lb	Time	100–150 h

Purification and isolation materials needed:

Filter aid	4,000 lb
Amyl acetate	1,000 gal
Procaine HCl	330 lb

Yield of procaine benzylpenicillin (procaine penicillin G) approximately 660 lb.

volume of amyl acetate, giving a penicillin purity 75 to 80% (Op).

Reverse the extraction at a pH of 7.5 into an aqueous solution with enhanced concentration and purity. Carbon-decolorize; add salt solution; quickly centrifuge to remove slime (Op).

Stir to form intermediate crystal (Op).

Centrifuge to recover intermediate crystal (Op).

Recover values from the mother liquor by a countercurrent acidulated solvent (amyl acetate) (Op).

In the final purification, sterilize through Seitz filters (Op).

Precipitate, e.g., with procaine HCl, to obtain procaine penicillin[39] (Ch) or as potassium or sodium salt.

Ilotycin[40] *(erythromycin)*

Alkalize (NaOH) and solvent-extract (amyl acetate) in a Podbielniak countercurrent extractor (Op).

Add acid and centrifugate the slime (Op).

Decolorize with carbon (Ch).

Salt out the intermediate crystals and centrifuge (Op).

Recrystalize or precipitate to obtain Ilotycin (Op).

Sterilize through Seitz filters (Op).

Dry under sterile conditions (Op).

Streptomycin

Extract streptomycin from the aqueous "beer" with concentrations of 10 to 15 mg/ml by ion-exchange resin (Rohm & Haas IRC-50, carboxylic acid cation exchanger) (Ch).

Elute from the resin with acidulated water (Op and Ch).

Concentrate, decolorize, crystallize, and centrifuge the intermediate crystals (Op).

Convert to desired salt (Ch) and sterilize through a Seitz filter (Op).

Dry under sterile conditions (Op).

Keflin (sodium cephalothin) is an important life-saving antibiotic, first marketed in 1964 following a long study of the first *Cephalosporium* culture, grown in 1948. Keflin is highly effective against staphylococcus bacteria resistant to other antibiotics.[41] Its complexity required the development of a special side chain essential to the final product.

[39]Rhodehamel, Water Insoluble Salt, U.S. Pat. 2,515,898 (1950).
[40]Hamel, *Antibiot. Chemother.*, **11**, 328 (1961).
[41]Keflin Fills Void in Antibiotic Therapy, *Lilly News*, Oct. 3, 1964, p. 2.

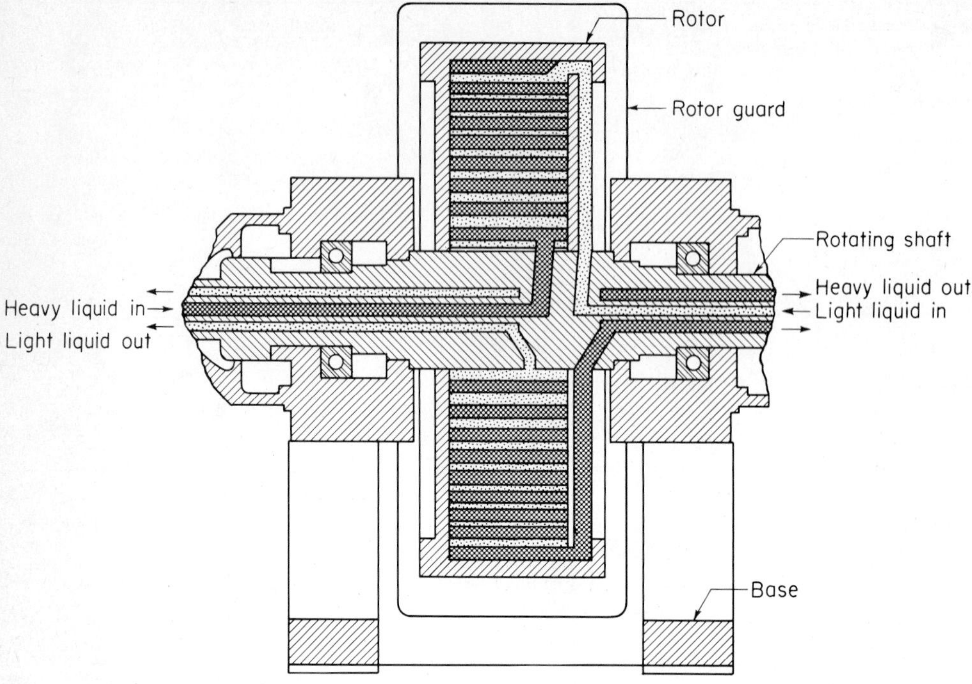

Fig. 40.9 The Podbielniak, or centrifugal, refiner, contactor, and separator. The liquids are introduced through the rotating shaft. The light liquid is led internally to the drum periphery and then out through the axis. The heavy liquid is led through the rotating shaft to the axis of the drum and out from the drum periphery. Rapid rotation causes a radial counterflow of the two liquids. [See Perry, pp. 21–29; Todd and Podbielniak, Centrifugal Extraction, *Chem. Eng. Prog.*, **61**(5), 69 (1965).]

BIOLOGICALS

"Biologic product[42] means any virus, therapeutic serum, toxin, antitoxin, or analogous product applicable to the prevention, treatment, or cure of diseases or injuries in man." They arise from the action of microorganisms, and they are used for prophylaxis, treatment, and diagnosis of infections and allergic diseases. In 1976, biologicals accounted for approximately 2.7% of sales at the manufacturer's level. Their significance greatly increased with the introduction of Salk and Asian flu vaccines, followed by the more recent oral vaccines for poliomyelitis and the various types of measles vaccines. Biological products are valuable for producing immunity to infections and preventing epidemics of contagious diseases. Between 1960 and 1976, over 100 new biologicals were marketed.

STEROID HORMONES[43]

The annual value of bulk steroids produced in the United States increases continually. Steroids[44] used in medicine may be divided into the major classes of corticoids, anabolic androgens, progestational hormones, and estrogenic hormones. Most of those sold and used in medicine today are pro-

[42]RPS XIII, pp. 1453–1492 and chaps. 79–82 for further information.
[43]Courtesy of the Upjohn Co. See also RPS XIII, pp. 901ff.
[44]Applezweig, Steroid Drugs, vol. 1, McGraw-Hill, 1962, vol. 2, Holden-Day, 1964.

duced synthetically and are not naturally occurring steroids at all, but chemical modifications of them. There are, for example, at least 16 different so-called corticoids on the market (not counting their various esters and salts), only two of which—cortisone and hydrocortisone—are identical with those found in nature.

In medicine corticoids are perhaps more widely used than steroids of other types. They are used particularly in rheumatic diseases, in inflammatory conditions of the skin, and in allergic conditions. In 1975, 50,000 lb was produced in the United States. Their use is in menstrual irregularities, menopause, and control of fertility and conception. There are many other indications for the use of steroids in medicine: in renal and cardiovascular diseases, in certain types of cancer, and in various types of stress reactions.

Outline of synthesis from soya, which contains the steroid nucleus:

$$\text{Crude soya sterols} \longrightarrow \text{stigmasterol} \xrightarrow[\text{steps}^{46}]{\text{chemical}^{45}} \text{progesterone}$$

Progesterone is converted[47] in high yield to the key intermediate for most of these sequences, 11-α-hydroxyprogesterone, by aerobic fermentation with a mold, such as *Rhizopus arrhizus*. Another fermentation step which enters into the synthesis of Medrol, Alphadrol, Oxylone, and other steroids, which have a double bond between C_1 and C_2, is dehydrogenation with a *Septomyxa* species.[48] The following chart depicts briefly the various directions in which the key intermediate may be carried.

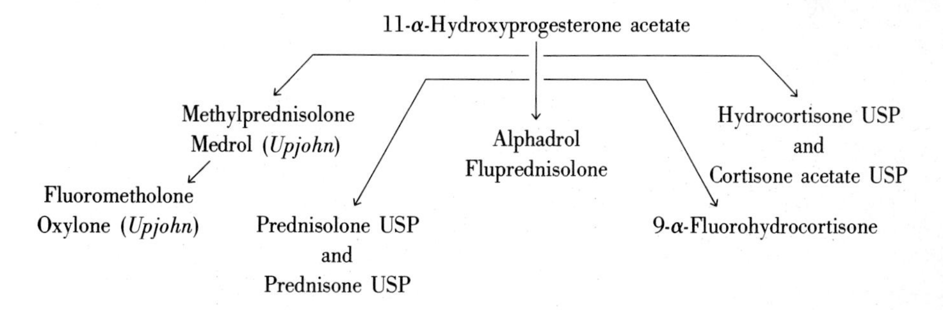

Other commerical processes for steroid hormone syntheses are based on bile acids and the plant steroid diosgenin.[49]

VITAMINS

For *vitamins*, the U.S. Tariff Commission reports a total production of 30 million lb for 1972, with an average value of $4.76 per pound. Of this quantity, ascorbic acid (vitamin C) and its derivatives total 15.6 million lb. Pantothenic acid and its derivatives have a total production of 3 million lb (Fig. 40.5). Production methods are outlined in this chapter for ascorbic acid and riboflavin by chemical synthesis, and in Chap. 31 for riboflavin as a fermentation product.. Remington[50] gives procedures for the manufacture and properties of the other vitamins.

[45]U.S. Pat. 3,005,834.
[46]U.S. Pats. 2,601,287; 2,752,337; 2,752,368.
[47]U.S. Pats. 2,602,769; 2,735,800; 2,666,070; cf. U.S. Pats. 2,790,814; 2,862,010; 2,862,011; 2,794,776; 2,752,366; 2,875,217.
[48]U.S. Pat. 2,902,410.
[49]Total Synthesis of Steroids, *Chem. Eng. News,* Aug. 26, 1963, p. 32.
[50]RPS XIII, pp. 1073–1107.

ISOLATES FROM PLANTS OR ANIMALS

Although many important pharmaceutical products result from chemical engineering–directed life processes, fermentations, and syntheses, there are some important medicaments whose sole or competitive source is through isolation from plants and animals. Quinine is made by extraction from cinchona bark. Morphine and codeine are isolated from opium, theobromine from waste cacao shells, and caffeine from the decaffeinizing of coffee or from waste tea. Insulin is isolated from pancreas.

Reserpine USP is one of the newer alkaloids widely employed for its tranquilizing effect upon the cardiovascular and central nervous systems and as an adjunct in psychotherapy. It is isolated by a nonaqueous solvent process, using, for example, boiling methanol extraction of the African root *Rauwolfia vomitoria.* Naturally, these extractions are carried out under countercurrent methods, details of which are in a flowchart published in *Chemical Engineering.*[51] The methanol extracts are concentrated and acidified with 15% acetic acid and then treated with petroleum naphtha to remove impurities. Extraction is made using ethylene dichloride. Then the solvent is neutralized with dilute sodium carbonate, evaporated to drive off the ethylene dichloride, and further evaporated to crystallize the crude reserpine crystals. These are further crystallized.

INSULIN INJECTION AND INSULIN ZINC SUSPENSION USP, ILETIN (LILLY) Insulin, a hormone, plays a key role in catalyzing the porcesses by which glucose (carbohydrates) furnishes energy or is stored in the body as glycogen or fat. The absence of insulin not only interrupts these processes, but produces depression of essential functions and, in extreme cases, even death. Its isolation and purification were started in April 1922, by Lilly, following its discovery by Banting and Best. Although approximately 750,000 patients in the United States are being treated with oral synthetic antidiabetic agents, it was estimated that about 850,000 Americans with diabetes mellitus are kept in reasonable health, and many alive, by the continued use of insulin to supply artificially a deficiency in the body processes. The usage of insulin has decreased only about 5%.

Insulin is isolated from the pancreas of beef or hogs and was one of the first proteins to be obtained in crystalline form. In different concentrations and dissociations, the molecular weight of insulin[52] varies between 36,000 and 12,000, and it has up to 51 amino acid units. Insulin protein is characterized by a high sulfur content in the form of cystine. It is unstable in alkaline solution. Insulin is isolated by extraction of minced pancreas with acidified alcohol, followed by purification, as presented in the flowchart in Fig. 40.10 and as outlined in the following description of the coordinated *unit operations* and *chemical conversions:*

Beef and pork pancreas glands are transported to the plant and refrigerated at −20°C (Op). Glands are rotoground, and the meat slurry is treated in extraction and aging tanks with ethyl alcohol after acidification (Op and Ch).

With the use of continuous countercurrent extraction in six continuous centrifuges over extraction tanks, the gland slurry is extracted with acidulated alcohol (Op).

From the sixth centrifuge, the cake is processed in a hot-fat fry tank and discharged through another centrifuge to separate waste fat from "fried residue," which is drummed and sold (Op).

The crude alcoholic extract is run from two strong extraction-centrifuge units into a collection tank from which the extract is neutralized with ammonia and filter aid added (Op and Ch).

In a continuous precoat drum filter, the cake is separated and washed, the clear liquor going to the reacidification tank (Op and Ch).

In evaporators, the first stage removes alcohol, with subsequent waste-fat separation. The extract goes to a chill tank, with filter aid added, through a filter press (Op), and into the second evaporator.

[51]*Chem. Eng.* (*N.Y.*), **64**(4), 230–233 (1957); Colbert, Prostaglandins Isolation and Synthesis, Noyes, 1973.

[52]RPS XIII, pp. 1046ff., 1028–1072, 1670; McGraw-Hill Encyclopedia, vol. 7, p. 163, 1960; Katsoyannis, *Chem. Eng. News,* July 2, 1962 (report regarding study of insulin synthesis).

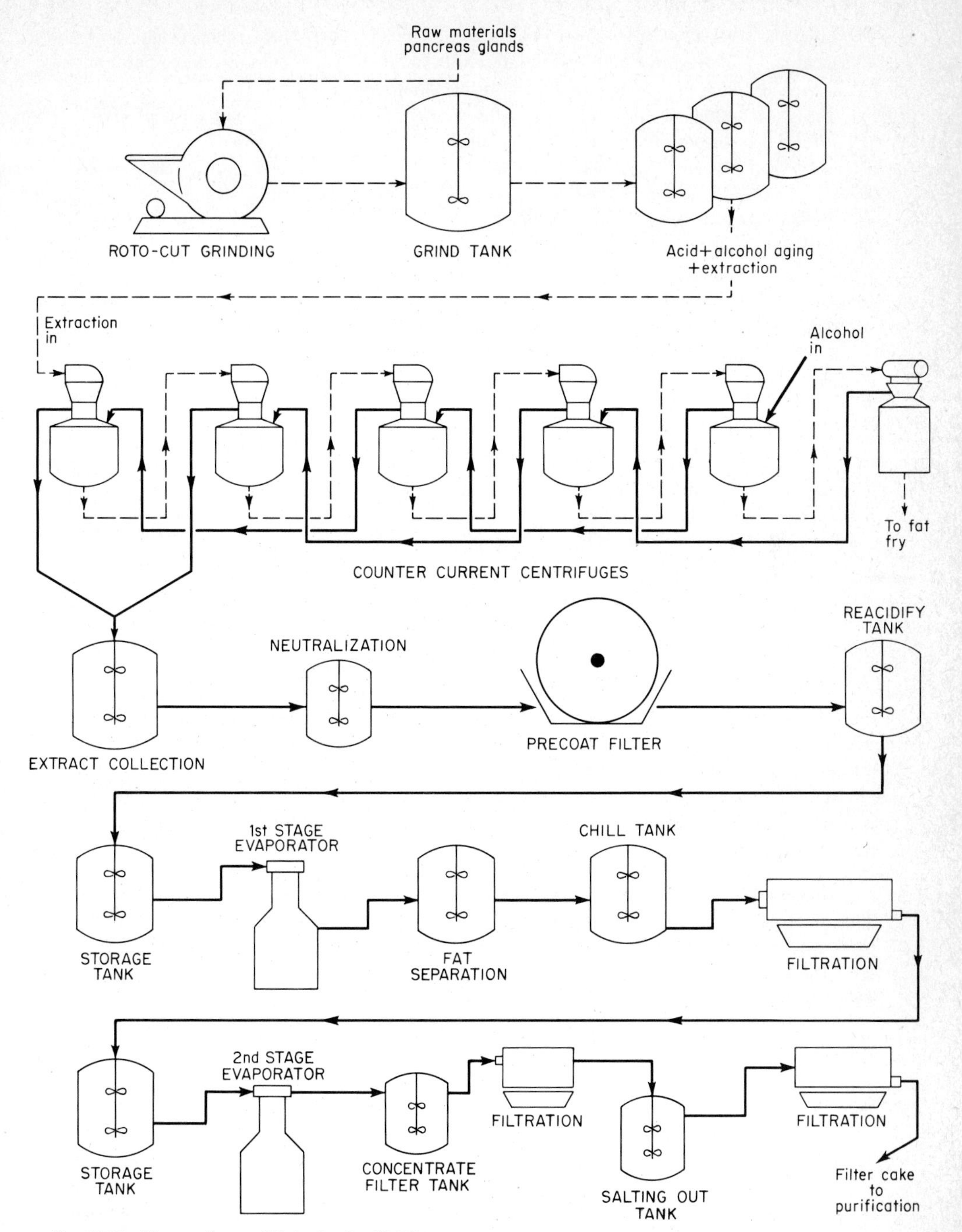

Fig. 40.10 The manufacture of Iletin (insulin, Eli Lilly).

From the second evaporator, the concentrated extract is filtered and conducted to the first salting-out tank, followed by filter-press filtration with filtrate to sewer and salt cake to purification for the second salting out. A filtration follows with filtrate to sewer (Op).

The second-salting-out product is crystallized twice to furnish (Op) Iletin (insulin) crystals.

COCAINE HYDROCHLORIDE USP This alkaloid, isolated from *Erythroxylon coca,* formerly obtained from Peruvian coca leaves containing 0.5 to 1%, is now isolated mostly from Java coca leaves containing 1.5 to 2%. In the former, cocaine was the predominating alkaloid and was prepared without great difficulty. However, the Java leaves contain a mixture of cocaine and related alkaloids as the methyl esters of cinnamyl- and benzoyl-ecgonine. The cocaine and by-alkaloids are extracted commercially by alkalizing the ground leaves with a 10% sodium carbonate solution and percolating countercurrently in large steel percolators, using kerosine or toluene. The total alkaloids are extracted from the kerosine or toluene by a process which blows them up with a 5% sulfuric acid solution in lead-lined tanks. The extracted kerosine or toluene is returned to the percolators. From the sulfuric acid solution, the mixed alkaloids are precipitated by alkalizing with sodium carbonate. The precipitated crude alkaloids are slowly boiled in an 8% sulfuric acid solution for several days to split all the alkaloids to ecgonine. During this splitting, many of the organic acids, like benzoic, are partly volatilized from the kettle with the steam. Those acids that are still suspended and those that crystallize out on cooling are filtered off. The acid aqueous solution of ecgonine is neutralized with potassium carbonate and evaporated. The low-solubility potassium sulfate is filtered hot. Upon cooling, the ecgonine crystallizes out. After drying, it is methylated, using methanol and 92% sulfuric acid, nutched, and washed with alcohol. The methylecgonine sulfate is benzoylated to cocaine in a very vigorous reaction with benzoyl chloride in the presence of anhydrous granular potassium carbonate. The cocaine is extracted with ether from the potassium salts and removed from the ether by sulfuric acid extraction, precipitated with alkali, and crystallized from alcohol. To form the hydrochloride, an alcoholic solution is neutralized with "acid alcohol" (HCl dissolved in absolute C_2H_5OH), and the cocaine HCl crystallized. The synthetic reactions are

$$
\begin{array}{l}
CH_2-CH-CHCOOH \\
\quad CH_3N \quad\quad CHOH \\
CH_2-CH-CH_2
\end{array}
\xrightarrow[\substack{(2)\ C_6H_5COCl \\ K_2CO_3}]{(1)\ CH_3HSO_4}
\begin{array}{l}
CH_2-CH-CHCOOCH_3 \\
\quad CH_3N \quad\quad CHO-\overset{}{C}(=O)-C_6H_5 \\
CH_2-CH-CH_2
\end{array}
$$

Ecgonine $C_9H_{15}NO_3$ Cocaine $C_9H_{13}NO_3(CH_3)\cdot(COC_6H_5)$

In recent years, because of decreased demand, cocaine is obtained as a by-product in the preparation of a decocained extract of *Erythroxylon coca,* this being one of the principal flavors of Coca-Cola and other cola beverages. The procedure is to alkalize the *Erythroxylon coca* and extract *all* the alkaloids with toluene. The decocainized leaf is dried and extracted with sherry wine to give the flavoring extract.

MORPHINE SULFATE USP AND CODEINE PHOSPHATE USP Remington says that "morphine is one of the most important drugs in the physician's armamentarium, and few would care to practice medicine without it, analgesia being one of the main actions. Codeine is now used to a larger extent than morphine and, while its analgesic action is only one-sixth of morphine, it is employed for its obtunding effect on the excitability of the cough reflex."[53] Morphine (about 11%) and codeine (about 1%) are extracted, along with many of the other alkaloids occurring in opium,[54] by mixing sliced opium balls or crushed dried opium with lime water and removing the alkaloidal contents by countercurrent aqueous techniques. Other solvents are also used, for instance, acetone and acetic

[53] RPS XIII, pp. 1184–1186; see also codeine under Alkylation; RPS XIV, p. 1123.

[54] Bentley, Chemistry of the Morphine Alkaloids, Oxford, 1954; Small and Lutz, Chemistry of the Opium Alkaloids, U.S. Govt. Printing Office, 1932.

acid or acidulated water. The crude morphine alkaloid is precipitated with ammonium chloride, purified by crystallization of one of its inorganic salts from water, and centrifuged; such a crystallization is repeated if necessary. The purified sulfate or hydrochloride is converted into the alkaloid by ammonia precipitation. If a still further purified alkaloid is needed, it can be prepared by crystallization from alcohol. Otherwise the morphine alkaloid is dissolved in water with sulfuric acid and crystallized in large cakes from which the mother liquor is drained and then sucked off. The sulfate is dried and cut into convenient sizes for the manufacturing pharmacist to compound it, make it into tablets, or otherwise facilitate its use by the physician.

Ths *codein* that occurs naturally in the opium is isolated from the aqueous morphine alkaloidal mother liquors by immiscible extraction with a nonaqueous solvent (Fig. 40.9). Dilute sulfuric acid is employed to extract the codeine sulfate from the nonaqueous solvent. This solution is evaporated, crystallized, and recrystallized. The alkaloid is precipitated from a sulfate solution by alkali and purified, if necessary, by alcoholic crystallization. It is converted into the phosphate by solution in phosphoric acid, evaporation, crystallization, centrifugation, and drying.

CAFFEINE USP (See caffeine under Alkylation for synthetic production.) Much caffeine has been isolated from waste tea, and in recent decades, from the decaffeinization[55] of coffee. The latter process involves extraction at 160°F of the moistened whole coffee bean with an organic solvent, frequently trichloroethylene, reducing the caffeine to about 0.03% from 1.2%. Rotating countercurrent drums are employed. The solvent is drained off, and the beans steamed to remove residual solvent. The beans are dried, roasted, packed, and sold. The extraction solvent is evaporated, and the caffeine is hot-water-extracted from the wax, decolorized with carbon, and recrystallized.

An alkaloid derived from *Vinca rosea* Linn has been the basis of two products: *Velban* (vinblastine sulfate, $C_{46}H_{58}N_4O_9 \cdot H_2SO_4$), useful in choriocarcinoma and Hodgkin's disease, and *Oncovin* (vincristine sulfate, $C_{46}H_{54}N_4O_{10} \cdot H_2SO_4$), indicated for acute leukemia in children. It takes 15 tons of periwinkle leaves and 15 weeks of precision chemical processing involving chromatography to yield a single ounce of one of these drugs. Oncovin is probably represented by the formula

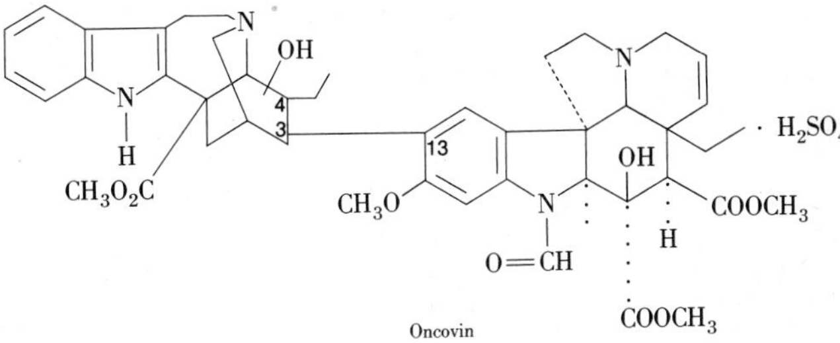

Oncovin

MEDICINALS AND HOSPITAL POPULATION are illustrated in Fig. 40.11 (p. 788) showing the decline in the number of patients in mental hospitals since tranquilizers (Tables 40.1 and 40.4) have been made available for prescriptions through the research of the pharmaceutical industry.

SELECTED REFERENCES

American Medical Association, Council on Drugs: New Drugs, 1943–1972, 39 vols.
Aiba, S., *et al.*: Biochemical Engineering, 2d ed., Academic, 1973.
Am. Ph. Ass.: Proprietary Names of Official Drugs, 1965.
Applezweig, N.: Steroid Drugs, McGraw-Hill, 1962–1964, 2 vols.
Balazs, E. A., and R. W. Jeanloz (eds.): The Amino Sugars, vol. 2A, Academic, 1965.
Bentley, K. W.: Alkaloids, Interscience, 1957–1965, 2 vols.

[55] ECT, 2d ed., vol. 3, p. 911, 1964.

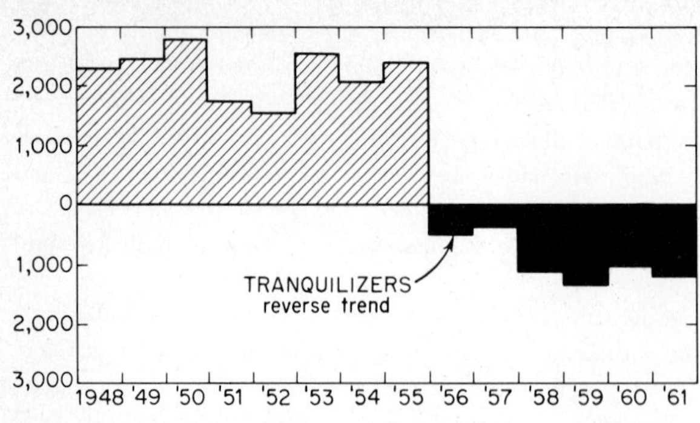

Fig. 40.11 Tranquilizing drugs have reduced the number of patients in mental hospitals. (*Key Facts, PMA.*)

Burger, A.: Medicinal Chemistry, 3d ed., Wiley, 1970.
de Haen, P.: Nonproprietary Name Index, vol. 8, 1943–1970, de Haen, 1971.
de Stevens, G.: Diuretics: Chemistry and Pharmacology, Academic, 1963.
——— (ed.): Analgesics, Academic, 1965.
Florey, H. W., *et al.:* Antibiotics: A Survey of Penicillin, Streptomycin, and other Antimicrobial Substances from Fungi, Actinomycetes, Bacteria, and Plants, Oxford, 1949, 2 vols.
Griffin, R. C., and S. Sacharow: Drug and Cosmetic Packaging, Noyes, 1975.
Harris, R. S., and I. G. Wool (eds.): Vitamins and Hormones, vol. 21, Academic, 1963.
Janssen, P. A. J.: Synthetic Analgesics, Pergamon, 1960.
Jenkins, G. S., *et al.:* Chemistry of Organic Medicinal Products, 4th ed., Wiley, 1957.
Lawrence, C. A.: Disinfection, Sterilization, and Preservation, Lea and Febiger, 1968.
Literature of Chemical Technology, chap. 10, Pharmaceutical, ACS Monograph, 1968.
Long, C. (ed.): Biochemists' Handbook, Van Nostrand, 1961.
Martin, A. N.: Physical Pharmacy, 2d ed., Lea and Febiger, 1969.
Martin, E. W. (ed.): Remington's Pharmaceutical Sciences, 14th ed., Mack Publishing, 1970 (also 13th ed., 1965).
Merck Index of Chemicals and Drugs, 8th ed., Merck, 1968.
Modell, W. (ed.): Drugs of Choice 1974–1975, Mosby, 1974.
Mule, S. J.: Cocaine, CRC Press, 1976.
Prescription Drug Industry Fact Book, Pharmaceutical Manufacturers Association, 1973.
Rinehart, K. L.: Neomycins and Related Antibiotics, Wiley, 1964.
Sebrell, W. H., and R. S. Harris (eds.): Vitamins: Chemistry, Physiology, Pathology, Academic, 1967, 7 vols.
Siegmund, O. H. (ed.): Merck Veterinary Manual, 4th ed., Merck, 1973.
U.S. Dispensatory, 27th ed., Lippincott, 1973.
Wagner, A. F., and K. Folkers: Vitamins and Coenzymes, Interscience, 1964.
Webb, F. C.: Biochemical Engineering, Van Nostrand, 1964.
Wilson, C. O., and T. E. Jones: American Drug Index, Lippincott, 1974.
Winick, C. (ed.): Sociological Aspects of Drug Dependence, CRC Press, 1974.

INDEX

Due to space limitations, no attempt has been made to index all discussions of chemical conversions, energy changes, manufacturing processes, raw materials, and product names covered in the various chapters. Inclusive paging and boldface figures are indicative of the main treatment. Footnote references are shown by *n.* following page numbers.